THE COLLECTED PAPERS OF

# M.S. Narasimhan

THE COLLECTED PAPERS OF

# M.S. Narasimhan

Volume I

1956 - 1984

**Nitin Nitsure**
Editor

Published for the

**Tata Institute of Fundamental Research**

by

HINDUSTAN
BOOK AGENCY

International Distribution by

**American Mathematical Society, USA**

HINDUSTAN BOOK AGENCY (INDIA)
P 19 Green Park Extension, New Delhi 110 016

email: hba@vsnl.com
www.hindbook.com



ISBN - 13: 978-81-85931-77-7 (2 volume set)
ISBN - 10: 81-85931-77-1 (2 volume set)

International Congress of Mathematicians, Vancouver, 1974

A. Grothendieck, K.G. Ramanathan
Yu. I. Manin, M.S. Narasimhan
TIFR, 1968

M.S. Narasimhan, G. Harder,
TIFR, 1973

1982

M. Atiyah,
B.V. Sreekantan,
M.S. Narasimhan,
TIFR, 1984

M.S. Narasimhan, A. Borel,
ICTP, 1996

ICTP, 1996

M.S. Narasimhan, C.S. Seshadri,
S. Ramanan, TIFR, 2004

ICTP, 2004

2006

Madrid, 2006

Madrid, 2006

# Contents

## Volume I

Editorial Preface xiii

Curriculum Vitae xv

*A Brief Summary of the Work of M.S. Narasimhan* xvii

*M.S. Narasimhan and his Work* by C.S. Seshadri xx

Acknowledgements xxvii

1. *The problem of limits on a Riemannian manifold*, J. Indian Math. Soc. (N.S.) **20** (1956), 291–297. 1
2. *The identity of the weak and strong extensions of a linear elliptic differential operator*, Proc. Nat. Acad. Sci. U.S.A. **43** (1957), 513–514. 8
3. *The identity of the weak and strong extensions of a linear elliptic differential operator. II*, Proc. Nat. Acad. Sci. U.S.A. **43** (1957), 620. 10
4. *The type and the Green's kernel of an open Riemann surface*, Ann. Inst. Fourier. Grenoble **10** (1960), 285–296. 11
5. *Variations of complex structures on an open Riemann surface*, Ann. Inst. Fourier Grenoble **11** (1961), 493–514, XVI–XVII. 23
6. *Existence of universal connections* (with S. Ramanan), Amer. J. Math. **83** (1961), 563–572. 45
7. *Existence of universal connections, II* (with S. Ramanan), Amer. J. Math. **85** (1963), 223–231. 55
8. *Regularity theorems for fractional powers of a linear elliptic operator* (with Takeshi Kotake), Bull. Soc. Math. France **90** (1962), 449–471. 64
9. *Stable bundles and unitary bundles on a compact Riemann surface* (with C.S. Seshadri), Proc. Nat. Acad. Sci. U.S.A. **52** (1964), 207–211. 87
10. *Holomorphic vector bundles on a compact Riemann surface* (with C.S. Seshadri), Math. Ann. **155** (1964), 69–80. 92
11. *Stable and unitary vector bundles on a compact Riemann surface* (with C.S. Seshadri), Ann. of Math. (2) **82** (1965), 540–567. 104
12. *Manifolds with ample canonical class* (with R.R. Simha), Invent. Math. **5** (1968), 120–128. 132

13. *Vector bundles on curves* (with S. Ramanan), Algebraic Geometry (Internat. Colloq., Tata Inst. Fund. Res., Bombay, 1968) pp. 335–346 Oxford Univ. Press, London, 1968. 141

14. *Moduli of vector bundles on a compact Riemann surface* (with S. Ramanan), Ann. of Math. (2) **89** (1969), 14–51. 153

15. *An analogue of the Borel-Weil-Bott theorem for hermitian symmetric pairs of non-compact type* (with K. Okamoto), Ann. of Math. (2) **91** (1970), 486–511. 191

16. *Geometry of moduli spaces of vector bundles*, Actes du Congres International des Mathematiciens (Nice, 1970), Tome 2, pp. 199–201. Gauthier-Villars, Paris, 1971. 217

17. *On the cohomology groups of moduli spaces of vector bundles on curves* (with G. Harder), Math. Ann. **212** (1975), 215–248. 220

18. *Deformations of the moduli space of vector bundles over an algebraic curve* (with S. Ramanan), Ann. of Math. (2) **101** (1975), 391–417. 254

19. *Generalised Prym varieties as fixed points* (with S. Ramanan), J. Indian Math. Soc. (N.S.) **39** (1975), 1–19. 281

20. *Geometry of Hecke cycles, I* (with S. Ramanan), C.P. Ramanujam—a tribute, pp. 291–345, Tata Inst. Fund. Res. Studies in Math., **8** Springer, Berlin-New York, 1978. 300

21. *Geometry of* SU(2) *gauge fields* (with T.R. Ramadas), Comm. Math. Phys. **67** (1979), no. 2, 121–136. 355

22. *Polarisations on an abelian variety* (with M.V. Nori), Proc. Indian Acad. Sci. Math. Sci. **90** (1981), no. 2, 125–128. 371

23. *Fibres de 't Hooft speciaux et applications* (with A. Hirschowitz), (French) [Special 't Hooft bundles and applications] Enumerative geometry and classical algebraic geometry (Nice, 1981), pp. 143–164, Progr. Math., **24** Birkhauser, Boston, Mass., 1982. 375

24. *Projective bundles on a complex torus* (with G. Elencwajg), J. Reine Angew. Math. **340** (1983), 1–5. 397

25. *Maximal subbundles of rank two vector bundles on curves* (with H. Lange), Math. Ann. **266** (1983), no. 1, 55–72. 402

## Volume II


26. *Survey of vector bundles on curves*, Singularities, representation of algebras, and vector bundles (Lambrecht, 1985), 1–8, Lecture Notes in Math., **1273** Springer, Berlin, 1987. 420

27. *2θ-linear systems on abelian varieties* (with S. Ramanan), Vector bundles on algebraic varieties (Bombay, 1984), 415–427, Tata Inst. Fund. Res. Stud. Math., **11** Tata Inst. Fund. Res., Bombay, 1987. 428

28. *Squares of ample line bundles on abelian varieties* (with H. Lange), Exposition. Math. **7** (1989), no. 3, 275–287. 441

29. *Spectral curves and the generalised theta divisor* (with Arnaud Beauville and S. Ramanan), J. Reine Angew. Math. **398** (1989), 169–179. 454

30. *Groupe de Picard des varieties de modules de fibres semi-stables sur les courbes algebriques* (with J.-M. Drezet), (French) [The Picard group of moduli varieties of semistable bundles over algebraic curves] Invent. Math. **97** (1989), no. 1, 53–94. 465

31. *Compactification of $M_{P_3}(0,2)$ and Poncelet pairs of conics* (with G. Trautmann), Pacific J. Math. **145** (1990), no. 2, 255–365. 507

32. *Rank 2 vector bundles on $P_4$ with $c_1$ odd and contact curves* (with W. Decker and F.-O. Schreyer), Math. Z. **205** (1990), no. 1, 123–136. 618

33. *The Picard group of the compactification of $M_{P_3}(0,2)$* (with G. Trautmann), J. Reine Angew. Math. **422** (1991), 21–44. 632

34. *Factorisation of generalised theta functions, I* (with T.R. Ramadas), Invent. Math. **114** (1993), no. 3, 565–623. 656

35. *Vector bundles as direct images of line bundles* (with A. Hirschowitz), K. G. Ramanathan memorial issue. Proc. Indian Acad. Sci. Math. Sci. **104** (1994), no. 1, 191–200. 715

36. *Infinite Grassmannians and moduli spaces of G-bundles* (with Shrawan Kumar and A. Ramanathan), Math. Ann. **300** (1994), no. 1, 41–75. 725

37. *Picard group of the moduli spaces of G-bundles* (with Shrawan Kumar), Math. Ann. **308** (1997), no. 1, 155–173. 760

38. *Hodge classes of moduli spaces of parabolic bundles over the general curve* (with Indranil Biswas), J. Algebraic Geom. **6** (1997), no. 4, 697–715. 779

39. *Hermitian-Einstein metrics on parabolic stable bundles* (with Jia Yu Li), Acta Math. Sin. (Engl. Ser.) **15** (1999), no. 1, 93–114. 798

40. *A note on Hermitian-Einstein metrics on parabolic stable bundles* (with Jia Yu Li), Acta Math. Sin. (Engl. Ser.) **17** (2001), no. 1, 77–80. 820

41. *A generalisation of Nagata's theorem on ruled surfaces* (with Yogish I. Holla), Compositio Math. **127** (2001), no. 3, 321–332. 824

Acknowledgements 837

# Editorial Preface

M.S. Narasimhan has been one of the major figures in Mathematics over the last five decades, having made fundamental contributions to diverse areas such as Algebraic Geometry, Differential Geometry, Representation Theory and Analysis. He is one of the principal architects of the School of Mathematics at the Tata Institute of Fundamental Research. He has also played an important role in the progress that India has made in Mathematics in these years.

Even as he approaches the age of 75 next year, Narasimhan continues to be active as a researcher, as an organiser and as a mentor for young researchers. The Tata Institute of Fundamental Research is happy to publish this 2-volume book of his collected papers. In addition to its value to the worldwide mathematical community, it is a token expression of our gratitude for all that he has done for the School of Mathematics, TIFR.

December 2006 Nitin Nitsure

# Curriculum Vitae

Born on 7 June 1932, at Thandarai in Tamil Nadu, India.

Education: Loyola College, University of Madras 1948–52. B.A. (Hons.) in Mathematics.

Joined the Tata Institute of Fundamental Research, Mumbai, as a research student in 1953.

Research Associate, CNRS Paris from 1957 to 1960. Worked with Laurent Schwartz.

Ph.D. in 1960, from University of Mumbai. Adviser: K. Chandrasekharan.

Professor at TIFR from 1965 onwards.

Professor of Eminence at TIFR 1990–92.

Director of Mathematics at ICTP, Trieste, 1993–1999.

Doctoral students: K. Gowrisankaran (1965), M.S. Raghunathan (1965), S. Ramanan (1966), M.K.V. Murthy (1970), V.K. Patodi (1971), G. Swarup (1971), R.R. Simha (1972), R. Parthasarathy (1973), S. Kumaresan (1982), T.R. Ramadas (1983), Nitin Nitsure (1987), S. Subramanian (1988), Fabrizio Coiai (2004).

*Major invitations abroad*:

1968–69: Visiting Member, Institute for Advanced Study, Princeton.

1970: Invited speaker, International Congress of Mathematicians, Nice.

1971: Visiting Professor, University of California at Los Angeles.

1973–74: Commonwealth Visiting Professor, University of Warwick.

1976: Japan Society for Promotion of Science Visiting Professorship.

1979: Visiting Professor, Université de Paris VI.

1980: Visiting Professor, Université de Nice.

1987: Visiting Professor, German Research Council.

1990: Japan Society for Promotion of Science Visiting Professorship.

1992: Visiting Professor, Université de Nice.

*Positions held in national and international mathematical organizations*:

1982 – 1986: Member, Executive Committee of International Mathematical Union.

1983 – 1987: Founder Chairman, National Board for Higher Mathematics, Government of India.

1986 – 1994: President, Commission on Development and Exchange, International Mathematical Union.

1989 – 1996: Vice-President, Centre International de Mathématiques Pures et Appliquées, Nice, France.

*Distinctions*:

1975: S.S. Bhatnagar Prize in Mathematical Sciences, Council of Scientific and Industrial Research.

1978: Meghnad Saha Award, University Grants Commission.

1987: Third World Academy of Sciences Award for Mathematics.

1988: Srinivasa Ramanujan Medal, Indian National Sciences Academy.

1989: Chevalier de l'Ordre National du Mérite, awarded by the President of France.

1990: Padma Bhushan Award by the President of India.

1994: C.V. Raman Birth Centenary Award of the Indian Science Congress.

1994: Honorary Fellow, Tata Institute of Fundamental Research.

2006: King Faisal International Prize for Science.

*Fellowships*:

Indian National Sciences Academy.

Indian Academy of Sciences.

Royal Society of London.

Third World Academy of Sciences.

# A Brief Summary of the Work of M.S. Narasimhan

This summary takes up, in turn, Narasimhan's work in algebraic geometry, differential geometry, the representation theory of semi-simple Lie groups and analysis. The numbers in square brackets refer to the order in the table of contents.

## 1 Algebraic Geometry

Here his major contributions have been to the development of the theory of moduli spaces of vector bundles on curves and higher dimensional projective varieties, beginning with the fundamental theorem regarding stable and (irreducible) unitary vector bundles on a compact Riemann surface (with C.S. Seshadri [11]). This theorem has been generalised in various significant directions by Donaldson, Uhlenbeck-Yau, Beilinson-Deligne, Hitchin and Simpson.

After this early work, there followed an extensive study, in collaboration with S. Ramanan, of the moduli spaces of vector bundles on curves. They obtained explicit descriptions in the case of low genus (of the curve) and rank (of the vector bundles). This work revealed surprising connections with classical algebraic geometry — the theory of quadratic complexes of lines, Kummer surfaces and the work of Coble on theta functions. In particular, the study of moduli spaces of rank 2 bundles on a curve of genus 3 led to an affirmative answer to a long standing question of Coble regarding the (cubic) equations defining a Kummer variety of dimension 3 [13, 14, 27].

Narasimhan and Ramanan also proved that (when the degree and rank of the vector bundles are coprime) all deformations of the moduli space of vector bundles (with fixed determinant) are obtained by deforming the curve — in particular the dimension of the deformation space is $3g - 3$ if $g > 2$ [18]. The notion of 'Hecke Correspondence', introduced and exploited in this work, has proved to be a valuable tool in subsesquent investigations by other mathematicians.

The Betti numbers of these spaces were calculated by Narasimhan and G. Harder using the Weil conjectures and Siegel's formula [17]. This work introduced and used a canonical filtration on any vector bundle (not necessarily semistable) on a curve; analogues of this filtration — now called the 'Harder-Narasimhan filtration' — have been immensely useful in numerous other contexts in Algebraic Geometry and Number Theory.

The paper [20], written jointly with S. Ramanan, contains an explicit desingularisation of the space of rank 2 bundles on a curve.

Together with H. Lange, Narasimhan investigated (in [25]) maximal subbundles of rank two bundles on curves; in the work [28] they studied squares of ample line bundles on abelian varieties.

J.M. Drezet and Narasimhan proved that the moduli spaces of vector bundles on curves are locally factorial; they also determined their Picard groups [30]. This work enables one to define the generalised theta line bundles (and generalised theta fuctions) on these spaces; these line bundles generalise the line bundle on the Jacobian of a curve determined by the Riemann theta divisor.

A "factorisation theorem", suggested by Conformal Field Theory, was proved in [34] (with T.R. Ramadas), for generalised theta functions: this relates the space of sections of powers of generalised theta bundles on the moduli space of rank two vector bundles on a curve of genus $g$ with the spaces of sections of certain generalised theta bundles on the moduli spaces of (parabolic) bundles on a curve of genus $(g-1)$.

In joint work [36] with with S. Kumar and A. Ramanathan, Narasimhan established that the space of conformal blocks (arising in Conformal Field Theory) is isomorphic to the space of generalised theta functions. This enables one to calculate the dimension of the space of generalised theta functions using the Verlinde formula.

The paper [29] (with A. Beauville and S. Ramanan) proves that a generic vector bundle of rank r on a smooth, irreducible, projective curve (over $\mathbb{C}$) is the direct image of a line bundle on a smooth $r$-sheeted covering. This result enables one to reduce certain problems on moduli spaces of vector bundles on a curve to problems on the Jacobian of a covering. A generalisation for higher dimensional varieties is contained in [35], with A. Hirshowitz.

In a paper with I. Biswas [38] he studied Hodge classes on (smooth) moduli spaces of parabolic bundles on curves, proving in particular the standard conjecture of Lefschetz type for these spaces. The paper with Y. Holla [41] generalises Nagata's theorem on the self-intersection numbers of sections of ruled surfaces to the case of principal bundles (on curves) with a reductive group as structure group and proves a boundedness result for semistable princpal bundles.

We now turn to vector bundles on varieties of higher dimension. Here Narasimhan's contributions have been to the study of the space $M(0,2)$ of rank 2 stable vector bundles with $c_1 = 0$, $c_2 = 2$ on the projective 3-space, $\mathbb{P}_3(\mathbb{C})$. He proved, with A. Hirschowitz, that $M(0,2)$ is rational [23]. The compactification of $M(0,2)$, obtained as the closure in the Maruyama scheme, was studied in detail by Narasimhan and G. Trautmann in [31]; they also determined explicitly all semi-stable torsion-free sheaves which occur as limits of vector bundles in $M(0,2)$. This work draws upon the classical theory of reguli and Poncelet pairs of conics.

The work [22], with M. Nori, proves that there are only finitely many smooth curves having a given abelian variety as the Jacobian. This fact is deduced by proving a result on the set of polarisations on an abelian variety, using reduction theory for arithmetic groups.

## 2 Differential Geometry

The basic theorem on the existence of universal connections was proved by Narasimhan and S. Ramanan; in particular, they showed that the canonical connection on the Stiefel bundle is universal for connections in principal unitary bundles [6, 7]. This result has been extensively used by physicists and mathematicians, for instance in the theory of sigma-models, in the theory of Chern-Simons invariants and in the work of Quillen on super connections.

Motivated by gauge theory, Narasimhan studied [21], with T.R. Ramadas, "moduli spaces" of connections on a principal bundle with a compact Lie group as structure group. They showed the impossibility of gauge fixing in certain cases; they also proved that the holonomy group of the Coulomb connection on the principal bundle of connections is dense in the gauge group (in the case of trivial $SU(2)$ bundle on $S^3$). This is an instance of the striking difference between abelian and non-abelian gauge theories.

A Hitchin-Kobayashi type correspondence for rank two parabolic bundles over Kahler surfaces, with parabolic struture defined along a smooth irreducible divisor, is proved with Li Jiayu in [39, 40].

## 3 Representation Theory of Lie Groups

The first breakthrough in the proof of a conjecture of Langlands on the concrete realisation of discrete series representations of Harish-Chandra was made in the joint work [15] with K. Okamoto, where they successfully treated the hermitian symmetric case.

Narasimhan formulated a conjecture, according to which the discrete series representations would be realised on the space of square integrable solutions of Dirac operators on $G/K$. This suggestion has played an important role in the further development of the theory.

## 4 Analysis

Together with T. Kotake, Narasimhan proved a theorem characterising real analytic functions via Cauchy type inequalities satisfied with respect to powers of a linear elliptic operator with analytic coefficients [8]. A special case is the well-known theorem on analyticity of solutions of a linear elliptic operator with analytic coefficients. This result has been generalised in different directions by Lions-Magenes, Baouendi-Goulaouic, Baouendi-Metivier and Bolley-Camus-Mattera. Kotake and Narasimhan also proved that fractional powers of a linear elliptic operator are (in the present-day language) pseudo-differential operators. This found application in the original proof of the Atiyah-Bott fixed point theorem.

Narasimhan played a crucial role in the formation and development of schools in algebraic geometry, differential geometry and Lie groups at the Tata Institute.

# M.S. Narasimhan and His Work[1]

C.S. Seshadri

It is indeed a great pleasure for me to talk about M.S. Narasimhan with whom my association goes back to more than 50 years. We went to the same college, Loyola College in Madras, which we joined in 1948 (one year after India obtained independence) after passing out of school. It was considered to be one of the best undergraduate institutions in India at that time and the teaching of mathematics had a special place, particularly because of a very enlightened person, a Jesuit priest Fr. C. Racine, who was a student of Elie Cartan. With Fr. Racine's encouragement, we joined in 1953 the Tata Institute of Fundamental Research which was founded a few years earlier by Homi Bhabha and whose School of mathematics was being built up with great vision by K. Chandrasekharan. It was indeed an exciting period at the Tata Institute with the presence of many talented students and a stream of outstanding visiting professors like Carl Ludwig Siegel, Laurent Schwartz, Samuel Eilenberg, Oscar Zariski, K.O. Friedrichs, Rademacher etc. We many different topics with no conscious effort at specialisation. The lectures of Laurent Schwartz on complex manifolds inspired us greatly and the lecture notes were written by Narasimhan.

Even from those days, Narasimhan began to show qualities of leadership, organising informal seminars and talks on a variety of topics. One such, which, in retrospect, turned out to be important for us, was on Riemann surfaces which we conducted with sufficient depth, studying Hermann Weyl's book. I particularly cherish the time I spent in those days, learning mathematics in his company.

Here are the titles of the first papers of Narasimhan written in 1956 and 57: (1) *Problem of limits on a Riemannian manifold* (2) *The identity of the weak and strong extensions of a linear elliptic operators I, II.* In 1957, we were sent to Paris, where we worked for three academic years. Narasimhan's supervisor was Laurent Schwartz. In the initial period of our stay in Paris, Narasimhan couldn't completely concentrate on mathematics as he was hospitalised due to a sickness. However, he told me that he used this time to read the work of Kodaira and Spencer on deformations of complex structures. During the latter part of his stay in Paris, Narasimhan did a significant piece of work in collaboration with Kotake, giving a characterisation of real analytic functions by Cauchy type of inequalities. To be more precise, let $A$ be a linear elliptic differential operator of order $m$ in a domain $\Omega \subset \mathbb{R}^n$. Let $|| \; ||$ denote the

[1]Talk given at the conference held in December 2002 in honour of M.S. Narasimhan turning 70 at ICTP, Trieste, Italy. We thank Prof. Seshadri for permission to publish the talk here.

$L^2$-norm of a function in $\Omega$. Suppose that $u \in C^\infty(\Omega)$ satisfies the inequalities:

$$||A^k u|| \leq (km)!c^{k+1}, \quad \forall k \geq 0, \text{ with a constant } c \geq 0 \text{ independent of } k.$$

Then $u$ is analytic in $\Omega$. In particular if $Au = 0$, $u$ is analytic. In this work they also study complex powers $A^s$ of $A$ ($A$ elliptic with $C^\infty$ coefficients). Their result means, in the present day language, that these operators are pseudo-differential operators. I understand that this result was used in the original proof of the Atiyah-Bott fixed point theorem and that the analyticity criterion has been generalised in several directions by many authors like Lions-Magenes, Bouendi-Goulaouic, Bouendi-Metvier and Bolly-Camus-Mattera.

In 1960, we went back to the Tata Institute. Narasimhan's presence became a source of inspiration to many young mathematicians. Ramanan became his first student. In their first joint work, they prove the existence of universal connections, namely that given a compact Lie group $G$ and an integer $n$, there exists a principal $G$-bundle $E$ and a connection $\gamma_0$ on $E$, such that any connection on a $C^\infty$ principal bundle $P$ with base of dimension $\leq n$, is the inverse image of $\gamma_0$ by a $C^\infty$ "homomorphism of $P$ onto $E$". This *generalises* the well-known results on the existence of universal bundles in topology. The Stiefel bundles carry natural connections and these connections are shown to be the universal objects (for $G = U(n)$). It is not surprising that such a result would be extensively used; it figures in the work of Chern-Simons and in a work of Quillen on super-connections.

I shall now take up the joint work of Narasimhan and myself, considered a pioneering work in the field of moduli of vector bundles, about which we started thinking around 1961–62. Let me digress a little to give an idea of the status of moduli problems around that time (i.e. early 1960's). The work of Kodaira-Spencer provided a good understanding of the infinitesimal deformation spaces, say of compact complex manifolds (e.g. smooth projective algebraic varieties) and of vector bundles on such manifolds. The big problems at that time were the global understanding and algebraisation of moduli problems, say the moduli space of curves, polarised abelian varieties, moduli spaces of vector bundles over curves (or ruled varieties). The space $H$ mod $\Gamma$, where $H$ is Siegel upper half plane and $\Gamma$ is the Siegel modular group gives the moduli space of principally polarised abelian varieties of a fixed dimension. Bailey had just proved that the Satake compactification $\overline{H \text{ mod } \Gamma}$ is algebraic. In particular this implied that $H$ mod $\Gamma$ is algebraic, but, if I am not mistaken one could not still prove that the moduli space of curves of a given genus (à priori an analytic submanifold of $H$ mod $\Gamma$) is algebraic. Regarding the moduli spaces of vector bundles, a lot of progress had been made for the case of line bundles (i.e. the study of Picard varieties) by Chow, A. Weil, Matsusaka and Chevalley (I had also made some contributions). For higher rank the only results were the following (i) Grothendieck's theorem that a vector bundle on $\mathbb{P}^1$ is a direct sum of line bundles. (ii) Atiyah's work on vector bundles on elliptic curves (curves of genus 1). (iii) A. Weil's work (1939) giving a characterisation of vector bundles on curves (or compact Riemann surfaces) which arise from representations of the fundamental group of the Riemann surface. The Grothendieck revolution ushering in the language of schemes, flat morphisms, representable functors etc. which have been fundamental for the progress of moduli problems, had just started around that time. The work of David Mumford on Geometric Invariant Theory which has turned out to be a principal tool for solving moduli problems was yet to appear.

Inspired by some remarks in the paper of A. Weil (quoted above), we first began looking at "unitary" vector bundles. If $\rho$ is a unitary representation of rank $n$ of the fundamental group $\pi$ of a compact Riemann surface $X$ of genus $g$, it defines a holomorphic vector bundle $V_\rho$ of rank $n$ and degree zero, which is referred to as a *unitary vector bundle* and called an *irreducible unitary vector bundle* if $\rho$ is irreducible unitary. It was not difficult to see that if $V_{\rho_1}$, $V_{\rho_2}$ are two unitary vector bundles, then $V_{\rho_1}$ is isomorphic to $V_{\rho_2}$ as holomorphic vector bundles if and only if $\rho_1$ and $\rho_2$ are isomorphic (i.e. equivalent) as representations; further $V_\rho$ is *simple* i.e. Aut $V_\rho \simeq \mathbb{C}^*$ if $\rho$ is irreducible. We showed that the infinitesimal deformation space of an irreducible unitary vector bundle $V_\rho$ (in the sense of holomorphic vector bundles) can be identified with the infinitesimal deformation space of $\rho$ in the sense of representations, the proof being much in the spirit of the fact that $H^1(X, \mathbb{R}) \longrightarrow H^1(X, \mathcal{O}_X)$ is an isomorphism of real vector spaces. From this it follows that on the set $S_0$ of isomorphism classes of irreducible unitary representations of $\pi$ of rank $n$, there is a natural structure of a complex manifold of the expected dimension of the moduli of vector bundles (with fixed rank and degree), namely $n^2(g-1)+1$, as well as that $S_0$ is open in $S$ — the complex manifold of simple vector bundles of rank $n$ and degree zero on $X$. Recall the classical result that when $n = 1$, $S_0$ can be identified with the Jacobian variety of line bundles of degree zero on $X$.

Soon after doing this work, we became aware of Mumford's talk at the 1962 International Congress, where he gave his definition of stable and semi-stable vector bundles on $X$ and announced the result that the isomorphism classes of stable vector bundles of fixed rank and degree, is a quasi-projective variety. Recall that a vector bundle $V$ on $X$ is said to be *semi-stable* (resp. *stable*) if for each proper sub-bundle $W$ of $V$, we have

$$\frac{\deg W}{\operatorname{rk} W} \leq \frac{\deg V}{\operatorname{rk} V} \quad \text{(resp. } <\text{)}.$$

We got also a manuscript from him where he gave the basic ingredients of his Geometric Invariant Theory (GIT), involving *stable and semi-stable points* for linear actions of reductive groups on projective varieties.

The remarkable similarities between stable bundles, irreducible unitary vector bundles (e.g. unitary bundle is semistable) and the compactification of the set of stable points by semi-stable points in Mumford's manuscript, led us to believe that stable vector bundles of rank $n$ and degree zero could coincide with the irreducible unitary vector bundles of rank $n$, when the genus of $X$ is $\geq 2$. In fact, we could initially prove this for rank 2 without much difficulty. The argument (which Narasimhan likes to call the continuity principle) runs roughly along the following lines. The set $R_0$ of isomorphism classes of stable vector bundles (of rank 2 and degree zero) has a natural structure of a *connected* complex analytic manifold. We can identify $S_0$ as an open submanifold (using our earlier work on unitary vector bundles) of $R_0$ (we have $S_0 \subset R_0 \subset S$). One can show that $S_0$ is also closed. It follows then that $R_0 = S_0$ ($S_0 \neq \emptyset$ since $g \geq 2$). A crucial point of this argument (which makes use of semi-continuity theorems) is that in an algebraic or analytic family of vector bundles, parametrised by a space $T$, the set of points of $T$ which corresponds to stable (resp. semi-stable) vector bundles is open and that the variety $J_d$ of line bundles on $X$ of degree $d$ is compact. What one needed to generalise this argument for higher rank is a "unitary characterisation" for stable vector bundles for *non-zero degrees.* A. Weil's work gave us the required clue. Let us call a *generalised unitary representation* $\rho$ the data (defining a representation of a suitable Fuchsian group)

$$A_1 B_1 A_1^{-1} B_1^{-1} \cdots A_g B_g A_g^{-1} B_g^{-1} = \lambda I$$

$A_i, B_i$ — unitary matrices of rank $n$ and $\lambda$ is an $n$th root of unity. Then we can associate a vector bundle $V_\rho$ of rank $n$ whose degree need not be zero, in fact $-n < \deg V_\rho \leq 0$. Let us call $V_\rho$ a generalised unitary vector bundle. Then Narasimhan and I showed that a generalised irreducible unitary vector bundle is stable and that any stable vector bundle of rank $n$ and degree $d$, $-n < d \leq 0$ (which we can assume by tensoring $V$ by a suitable line bundle) is given by a generalised irreducible unitary vector bundle. It followed from our result that the set of isomorphism classes of stable vector bundles on $X$ of rank $n$ and degree $d$ is a *compact complex manifold* when $n$ and $d$ are coprime. Nowadays, the fact that $S_0$ is open and closed in $R_0$, is proved in a more direct manner.

This theorem has been generalised in various directions by Donaldson, Uhlenbeck-Yau, Beilinson-Deligne, Hitchin, C. Simpson and A. Ramanathan.

After our joint work, I succeeded (using the work of Mumford, cited above) in showing that on $U(n,d)$ — the set of semi-stable vector bundles of rank $n$ and degree $d$ on $X$ ($g \geq 2$) under a certain equivalence relation, there is a natural structure of a normal projective variety. If $d = 0$, $U(n,0)$ is just the set of isomorphism classes of unitary vector bundles of rank $n$ and degree 0 and if $d \neq 0$, there is a similar description via generalised unitary representations.

After these developments, Narasimhan's work has been mainly centered around the moduli of vector bundles and particularly the study of diverse aspects of the moduli spaces $U(n,d)$. A substantial part of this work is joint work with Ramanan, of which I shall now try to give an idea.

They first showed that the smooth points of $U(n,d)$ correspond precisely to the stable vector bundles, expect for the case $n=2$, $g=2$, in which case $U(2,0)$ is smooth. They also give a concrete determination of $U(2,0)$ and $U(2,1)$ when $g=2$. For example, let $S$ be the subvariety of $U(2,0)$ consisting of vector bundles with trivial determinant. Now $X$ is canonically embedded as a divisor $\Theta$ in $J_1$. Then $S$ can be canonically identified with the complete linear system defined by $2\Theta$. In particular, it follows that $S \simeq \mathbb{P}^3$. Similarly, if $U(2,L)$ denotes the subvariety of $U(2,1)$ of vector bundles with a fixed determinant $L$ of degree one, they show that $U(2,L)$ is the intersection of two quatrics in $\mathbb{P}^5$. This was also proved by Newstead. Later Desale-Ramanan gave a similar description of $U(2,L)$ for the case of any hyper-elliptic curve. For $g=3$, $X$ non-hyperelliptic, Narasimhan and Ramanan show that $U(2,L)$ ($L$ trivial) can be identified with a *quatric* hypersurface in $\mathbb{P}^7$ and answer a question of Coble who had studied this quatric.

An interesting technical result is the non-existence of *Poincaré families* on any non-empty Zariski open subset of $U(2,0)$ for the case $g=2$, which was later proved by Ramanan for a general $U(n,d)$, $(n,d)\neq 1$. It can be shown that Poincaré families exist for $U(n,d)$ when $(n,d)=1$. Interesting relationship with classical algebraic geometry like quadratic complexes of lines figures in all this work.

A beautiful result of Narasimhan and Ramanan states that a local deformation of the moduli space $U(n,L)$ — the subvariety of $U(n,d)$ of bundles with fixed determinant $L$ for $(n,d)=1$ — identifies with a local deformation of the curve $X$. In this work, they make use of nice correspondences called "Hecke modifications" between moduli spaces e.g.

$$U(2,1) \xleftarrow{\ \phi\ } H \xrightarrow{\ \psi\ } U(2,0)$$

where $\phi$ is a $\mathbb{P}^1$ bundle and $\psi$ is a $\mathbb{P}^1$-bundle over the open subset of stable bundles.

An important contribution of Narasimhan is his joint work with Harder on the cohomology of the moduli spaces $U(n,L)$, $(n,\deg L)=1$. Newstead had obtained formulae for the Betti number of $U(2,L)$ using its unitary characterisation.

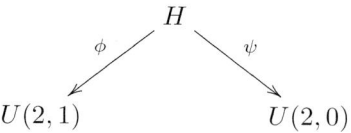

The method of Harder and Narasimhan to determine the cohomology of $U(n,d)$ is to use the Weil conjectures (which had just then been proved by Deligne) and what is known as Siegel's formula. We can suppose that $X$ is defined over a finite field $\mathbb{F}_q$ (or to say in a better way, the algebraic curve is defined over a discrete valuation ring with residue field $\mathbb{F}_q$ and we take the base change of $X$ to $\mathbb{F}_q$). We shall sketch their proof very briefly, assuming for simplicity that $n = 2$. Siegel's formula is the following

$$\sum_E \frac{1}{\#\text{Aut } E} = \frac{1}{(q-1)} q^{3g-3} \zeta(2)$$

where $\zeta$ is the zeta function of the curve and the sum runs over all vector bundles $E$ on $X$ which are defined over $\mathbb{F}_q$. We can write the LHS as $N_1 + N_2$, where $N_1$ is the sum which runs over all stable bundles $E$ and $N_2$ runs over unstable bundles. We see that $N_1 = N/(q-1)$, where $N$ denotes the cardinality of the set of $\mathbb{F}_q$-rational points of $U(2, L)$. Now by the Weil conjectures, $N$ (rather the number of all $\mathbb{F}_{q^n}$-rational points of $U(2, L)$) determines the cohomology of $U(2, L)$. Hence one has to compute $N_2$. Now if $E$ is in the sum $N_2$, $E$ is unstable and is therefore of the form:

$$0 \longrightarrow L_1 \longrightarrow E \longrightarrow L_2 \longrightarrow 0, \quad \deg L_1 > \frac{\deg E}{2}.$$

Now $L_1$ is uniquely determined and hence $L_1$ is also defined over $\mathbb{F}_q$. From these considerations one can compute $N_2$ and then one gets a formula for the Poincaré polynomial of $U(2, L)$ (see also a later paper of Desale and Ramanan). To generalise the above considerations for arbitrary rank, they defined unique filtrations for unstable bundles (now called Harder-Narasimhan filtrations). It should be mentioned that Harder had earlier employed this algebro-geometric cum number theoretic method to prove the Weil conjectures for $U(2, L)$, using the computations of Newstead.

Later Atiyah and Bott gave a different approach for the determination of the cohomology of $U(n, L)$ in their paper "Yang Mills equations over Riemann surfaces" but there are interesting formal similarities in the two approaches. The paper of Atiyah and Bott has had a great influence in fostering the connections between physics and moduli problems in Algebraic Geometry.

One has the notion of a generalised theta line bundle $\theta$ on the moduli space $U(n, d)$. Let us denote by $B(g, k)$ the space of sections of $\theta^k$ on $U(n, d)$, $g$ being the genus of $X$. Conformal Field Theory (in physics) has suggested remarkable formulae called the Verlinde formulae for $B(g, k)$, in particular they suggest a relationship between $B(g, k)$ and $B(g-1, k)$, called factorisation rules. In conformal Field Theory, $B(g, k)$ is defined as the space of invariants of sections of $L^k$, $L$ a well-defined line bundle on an infinite dimensional flag variety associated to a loop group, under the action of an infinite group $\Gamma$. With this definition, Tsuchiya-Ueno-Yamada have proved the factorisation theorem and the Verlinde formulae. The problem is to give a rigorous proof that the $B(g, k)$ of conformal Field Theory is indeed the $B(g, k)$ as has been defined above. Narasimhan, in collaboration with Shrawan Kumar and A. Ramanathan has given such a proof, even in the context of moduli spaces associated to a semi-simple algebraic group. A. Beauville and Y. Lazlo have also proved this for the case $G = SL(n)$. There is also the problem of proving the factorisation theorems by algebro-geometric methods. The relationship of $B(g, k)$ with $B(g-1, k)$ suggests that $B(g, k)$ behaves well when $X$ degenerates to a curve $X_0$ say, with one ordinary double point and then one has to relate the sections of the "generalised $\theta$-bundle" on $X_0$ with those on its normalisation. Carrying out this programme is technically complicated. Narasimhan and Ramadas have done this in rank two (the earlier work of Drezet-Narasimhan on the determination of the Picard group of $U(n, d)$ could be called the starting point for this investigation). X. Sun, a "quasi-student" of Narasimhan, has done it for arbitrary rank. Faltings has also given a proof of the Verlinde formulae.

A singular exception during all the years devoted to the study of vector bundles, is the joint work of Narasimhan with K. Okamoto in representation theory. In analogy with the Borel-Weil theorem on the realisation of finite dimensional irreducible representations of a semi-simple group, as sections of a line bundle on the flag variety associated to $G$, it was suggested by Langlands that the discrete classes of representations (in the sense of Harish Chandra) of a real semi-simple, non-compact group $G$, could be realised as square-integrable harmonic forms of type $(0, q)$ with coefficients in certain holomorphic vector bundles on $G/K$ (we suppose that $G/K$ is Hermitian symmetric). The work of Narasimhan and Okamoto is considered as the first breakthrough in the realisation of this conjecture. Narasimhan didn't pursue this subject later, but his student R. Parthasarathy has contributed significantly to this field. The starting point of Parthasarathy's work was a suggestion of Narasimhan to use Dirac operators when $G/K$ is *not* Hermitian symmetric.

I shall not go into many other interesting works of Narasimhan e.g. his work with Trautmann and Hirschowitz on the moduli spaces of rank two vector bundles on $\mathbb{P}^3$, his joint work with Ramanan on the desingularisation of $U(2, 0)$, his joint work with Beauville and Ramanan on the realisation of vector bundles on $X$ as direct images of line bundles on suitable coverings of $X$ or his work with Madhav Nori on the finiteness of curves with a given Jacobian. I hope I have given you an idea of the range and importance of his contributions. He has played a key role in the development of the subject of moduli of vector bundles on curves, which is an important part of the great developments in moduli problems in the last part of the 20th century.

With his wide mathematical interests, his capacity to make perceptive remarks and intense involvement in mathematical discussions, Narasimhan has a significant mathematical presence. This explains why he is such an excellent collaborator and has attracted such excellent students like Ramanan, Raghunathan, Patodi, Parthasarathy, Ramadas, Simha, and Nitsure and influenced many other younger mathematicians. Narasimhan is also an excellent administrator. With all these qualities he has played a great role in the development of the School of Mathematics of the Tata Institute of Fundamental Research, the National Board for Higher Mathematics in India and later in the Mathematics section at ICTP.

I wish Narasimhan very well.

# Acknowledgements

The Tata Institute of Fundamental Research gratefully acknowledges the kindness of the following institutions and individuals in granting permissions to reproduce the following.

*The problem of limits on a Riemannian manifold*
Reprinted with permission from Journal of the Indian Mathematical Society (N.S.) **20** (1956), 291–297.
Copyright © 1956 Indian Mathematical Society.

*The identity of the weak and strong extensions of a linear elliptic differential operator*
Reprinted with permission from Proceedings of the Natural Academy of Sciences U.S.A. **43** (1957), 513–514.
Copyright © M.S. Narasimhan.

*The identity of the weak and strong extensions of a linear elliptic differential operator. II*
Reprinted with permission from Proceedings of the Natural Academy of Sciences U.S.A. **43** (1957), 620.
Copyright © M.S. Narasimhan.

*The type and the Green's kernel of an open Riemann surface*
Reprinted with permission from Annales de L'Institut Fourier **10** (1960), 285–296.
Copyright © Annales de L'Institut Fourier.

*Variations of complex structures on an open Riemann surface*
Reprinted with permission from Annales de L'Institut Fourier **11** (1961), 493–514, XVI–XVII.
Copyright © Annales de L'Institut Fourier.

*Existence of universal connections* (with S. Ramanan)
American Journal of Mathematics **83** (1961), 563–572.
Reprinted with permission from Johns Hopkins University Press.
Copyright © Johns Hopkins University Press.

*Existence of universal connections, II,* (with S. Ramanan)
American Journal of Mathematics **85:2** (1963), 223–231.
Reprinted with permission from the Johns Hopkins University Press.
Copyright © Johns Hopkins University Press.

*Regularity theorems for fractional powers of a linear elliptic operator* (with Takeshi Kotake)
Reprinted with permission from Bulletin de la Societe Mathematique de France **90** (1962), 449–471.
Copyright © French Mathematical Society.

*Stable bundles and unitary bundles on a compact Riemann surface* (with C.S. Seshadri)
Reprinted with permission from Proceedings of Natural Academy of Sciences U.S.A. **52** (1964), 207–211.


*Holomorphic vector bundles on a compact Riemann surface* (with C.S. Seshadri)
Reprinted with permission from Mathematicshe Annalen **155** (1964), 69–80.


*Stable and unitary vector bundles on a compact Riemann surface* (with C.S. Seshadri)
Reprinted with permission from Annals of Mathematics (2) **82** (1965), 540–567.


*Manifolds with ample canonical class* (with R.R. Simha)
Reprinted with permission from Inventiones Mathematicae **5** (1968), 120–128.


*Vector bundles on curves* (with S. Ramanan)
Algebraic Geometry (Internat. Colloq., Tata Inst. Fund. Res., Bombay, 1968) pp. 335–346 Oxford Univ. Press, London, 1968.


*Moduli of vector bundles on a compact Riemann surface*
Reprinted with permission from Annals of Mathematics (2) **89** (1969), 14–51.


*An analogue of the Borel-Weil-Bott theorem for hermitian symmetric pairs of non-compact type*
Reprinted with permission from Annals of Mathematics (2) **91** (1970), 486–511.


*Geometry of moduli spaces of vector bundles*
Reprinted from Actes du Congres International des Mathematiciens (Nice, 1970), Tome 2, M.S. Narasimhan, pp. 199–201, Copyright © (1971) Gauthier-Villars, Paris, with permission from Elsevier.

*On the cohomology groups of moduli spaces of vector bundles on curves* (with G. Harder)
Reprinted with permission from Mathematicshe Annalen **212** (1975), 215–248.

*Deformations of the moduli space of vector bundles over an algebraic curve*
Reprinted with permission from Annals of Mathematics (2) **101** (1975), 391–417.


*Generalised Prym varieties as fixed points*
Reprinted with permission from Journal of the Indian Mathematical Society (N.S.) **39** (1975), 1–19.


*Geometry of Hecke cycles. I*
C. P. Ramanujam—A Tribute, pp. 291–345, Tata Inst. Fund. Res. Studies in Math., **8** Springer, Berlin-New York, 1978.


*Geometry of* SU(2) *gauge fields* (with T.R. Ramadas)
Reprinted with permission from Communications in Mathematical Physics **67** (1979), no. 2, 121–136.


*Polarisations on an abelian variety*
Reprinted with permission from Proc. Indian Acad. Sci. Math. Sci. **90** (1981), no. 2, 125–128.


*Fibres de 't Hooft speciaux et applications* (with A. Hirschowitz)
Enumerative geometry and classical algebraic geometry (Nice, 1981), pp. 143–164, Progr. Math., **24** Birkhauser, Boston, Mass., 1982.
Reprinted with permission from Birkhauser Publishing Ltd., Basel.


*Projective bundles on a complex torus*
Reprinted with permission from Journal für die reine und angewandte Mathematik **340** (1983), 1–5.


*Maximal subbundles of rank two vector bundles on curves* (with H. Lange)
Reprinted with permission from Mathematicshe Annalen **266** (1983), no. 1, 55–72.


*Survey of vector bundles on curves*
Singularities, representation of algebras, and vector bundles (Lambrecht, 1985), 1–8, Lecture Notes in Mathematics, **1273** Springer, Berlin, 1987.

*2θ-linear systems on abelian varieties*
Vector bundles on algebraic varieties (Bombay, 1984), 415–427, Tata Inst. Fund. Res. Stud. Math., **11** Tata Inst. Fund. Res., Bombay, 1987.
Copyright © Tata Institute of Fundamental Research.

*Squares of ample line bundles on abelian varieties*
Reprinted from Expositiones Mathematicae, **7** no. 3, H. Lange and M.S. Narasimhan, pp. 275–287, Copyright © (1989), with permission from Elsevier.

*Spectral curves and the generalised theta divisor*
Reprinted with permission from Journal für die reine und angewandte Mathematik **398** (1989), 169–179.
Copyright © 1989 Walter de Gruyter GmbH & Co. KG

*Groupe de Picard des varieties de modules de fibres semi-stables sur les courbes algebriques* (with J.-M. Drezet)
Reprinted with permission from Inventiones Mathematicae **97** (1989), no. 1, 53–94.
Copyright © 1989 Springer-Verlag.

*Compactification of* $M_{P_3}(0,2)$ *and Poncelet pairs of conics*
Reprinted with permission from Pacific Journal of Mathematics **145** (1990), no. 2, 255–365.
Copyright ©1990 Pacific Journal of Mathematics

*Rank* 2 *vector bundles on* $P_4$ *with* $c_1$ *odd and contact curves* (with W. Decker and F.-O. Schreyer)
Reprinted with permission from Mathematische Zeitschrift **205** (1990), no. 1, 123–136.
Copyright © 1990 Springer-Verlag.

*The Picard group of the compactification of* $M_{P_3}(0,2)$
Reprinted with permission from Journal für die reine und angewandte Mathematik **422** (1991), 21–44.
Copyright © 1991 Walter de Gruyter GmbH & Co. KG

*Factorisation of generalised theta functions. I* (with T.R. Ramadas)
Reprinted with permission from Inventiones Mathematicae **114** (1993), no. 3, 565–623.
Copyright © 1993 Springer-Verlag.

*Vector bundles as direct images of line bundles*
Reprinted with permission from Indian Academy of Sciences
Copyright © 1994 Proc. Indian Acad. Sci. Math. Sci.
(K. G. Ramanathan memorial issue) **104** (1994), no. 1, 191–200.

We thank the Abdus Salam International Centre for Theoretical Physics, Trieste, and Red Temática de Geometría y Física and Residencia de Estudiantes, Madrid for kindly providing some of the photographs reproduced in this collection.

# THE PROBLEM OF LIMITS ON A RIEMANNIAN MANIFOLD

*By* M. S. NARASIMHAN*

[Received March 20, 1956, and, in revised form, October 5, 1956]

1. The object of this note is to study the self-adjoint extensions of the Laplace operator, defined on the space of $C^\infty$ forms with compact support, in the Hilbert space of square summable forms on a $C^\infty$ Riemannian manifold, by using the theory of currents and the theory of linear transformations on a Hilbert space. We first obtain a necessary and sufficient condition for the closure of this operator to be self-adjoint. Associated with each self-adjoint extension of this operator and each complex number $\lambda$ we have a problem of limits, which, in special cases, reduces to the Dirichlet problem or the Neumann problem. Whether the problem admits of a unique solution or not depends on whether $\lambda$ is or is not in the resolvent of the self-adjoint extension. If $\lambda$ is in the resolvent, we associate a "Green's form" with the problem. We further determine all non-positive self-adjoint extensions whose domains consist entirely of forms with " finite energy " and for which the bilinear form defined by the self-adjoint operator coincides with the negative of the energy form. Finally we derive the Poincaré inequality.

We shall see that a theorem of Bochner-Gaffney-Yosida and a recent theorem of Lions and Schwartz fit into our general setting. We also apply the result of § 5 to prove the existence of the Green's form (for the operator $\Delta$) in a relatively compact subdomain of a Riemannian manifold, when the so-called uniqueness condition is satisfied.

2. Let $M$ be a $C^\infty$ Riemannian manifold, countable at infinity; we assume that $M$ is oriented. Let $d$, $\partial$ and $-\Delta = d\partial + \partial d$ have

* The author's thanks are due to Professor K. Chandrasekharan for his constant encouragement.

the usual meaning. Let $\mathscr{D}$ denote the space of (complex) $C^\infty$ forms with compact support, $\mathscr{E}$ the space of $C^\infty$ forms, $\mathscr{D}'$ the space of currents, $\mathscr{E}'$ the space of currents with compact support and $L_2$ the Hilbert space of square summable currents, all with the usual topology [6, 9]. Let $D : \mathscr{D} \to \mathscr{D}$ denote the operator $\Delta$ restricted to $\mathscr{D}$. $D$ is a densely defined linear operator in $L_2$. $D$ is symmetric on $\mathscr{D}$ with respect to the scalar product in $L_2$ :

$$(D\,\phi, \psi) = (\phi, D\,\psi), \quad \phi, \psi \in \mathscr{D},$$

where ( , ) is the scalar product in $L_2$. Also $D$ is non-positive on $\mathscr{D}$:

$$(D\,\phi, \phi) = (\Delta\,\phi, \phi) = -\,(d\phi,\, d\phi) - (\partial\phi,\, \partial\phi) \leqslant 0, \text{ for } \phi \in \mathscr{D}.$$

We shall consider the self-adjoint extensions of $D$ in $L_2$.

**3.** We first determine explicity the adjoint of $D$ in $L_2$. Let $L_2(\Delta)$ denote the subspace of $L_2$ consisting of those elements $T$ for which $\Delta T$ (formed in the sense of currents) are also in $L_2$. It is easy to verify that $\Delta : L_2(\Delta) \to L_2$ is the adjoint $D^*$ of $D$. The closure of $D$ in $L_2$ coincides with $D^{**}$.

By a theorem of von Neumann on abstract Hilbert spaces [10, Theorem 9.4, p. 341] we have :

$$L_2\,(\Delta) = \text{domain of } D^{**} + E(i) + E(-\,i), \quad \text{(direct sum)},$$

where $E(i)$ (resp. $E(-\,i)$) consists of all the elements $T$ in $L_2(\Delta)$ which satisfy the relation $D^*\,T = \Delta\,T = iT$ (resp. $D^*\,T = \Delta\,T = -\,iT$). By the elliptic character of $(\Delta \pm i)$ [9, Ch. V, § 6; 6, § 29] we may assume that the elements of $E(i)$ and $E(-\,i)$ are $C^\infty$ forms†. (By the elliptic character of $(\Delta + \lambda)$, where $\lambda$ is a complex number, we mean the following: if $T$ is a current and $(\Delta + \lambda)\,T$ is a $C^\infty$ form in an open set $U$, then $T$ is a $C^\infty$ form in $U$). Now, the closure of $D$ in $L_2$ is self-adjoint if and only if $E(i) = 0$. For if $E(i) = 0$, then $\Delta$ being a real operator, $E(-\,i) = 0$ so that $D^* = D^{**}$; hence the closure of $D$ is self-adjoint. Conversely, if the closure of $D$ is self-adjoint we have $D^* = D^{**}$ as $D^*$ is the adjoint of the closure of $D$;

† This shows, incidentally, that, $D^*$ is the closure in $L_2$ of the map

$$\Delta : L_2(\Delta) \cap \mathscr{E} \to L_2.$$

hence $E(i) = 0$. $E(i) = 0$ implies that $(\Delta\,\phi, \psi) = (\phi, \Delta\,\psi)$ for every $\phi, \psi \in L_2(\Delta)$, in particular for $\phi, \psi \in L_2(\Delta) \cap \mathscr{E}$. On the other hand, if $(\Delta\,\phi, \psi) = (\phi, \Delta\,\psi)$ for any two $C^\infty$ forms $\phi, \psi$ such that $\phi, \psi$, $\Delta\,\psi$ all belong to $L_2$, then $E(i) = 0$. For if $\phi \in E(i)$, $\Delta\,\phi \in L_2$ and, as stated above, $\phi \in \mathscr{E}$ so that $(\Delta\,\phi, \phi) = i(\phi, \phi) = (\phi, \Delta\,\phi) = -\,i(\phi, \phi)$ whence $(\phi, \phi) = 0$, so $\phi = 0$. Thus we have

PROPOSITION I. *A necessary and sufficient condition for the closure of $D$ in $L_2$ to be self-adjoint is that $E(i) = 0$ or, equivalently, that*

$$(\Delta\,\phi, \psi) = (\phi, \Delta\,\psi)$$

*for any two $C^\infty$ forms $\phi, \psi$ such that $\phi, \psi, \Delta\,\phi, \Delta\,\psi$ all belong to $L_2$.*

The condition

$$(\Delta\,\phi, \psi) = (\phi, \Delta\,\psi), \quad \phi, \psi \in \mathscr{E},$$

is trivially verified in the case of a compact manifold. Thus, *in the case of a compact manifold the closure of $D$ is self-adjoint*—a theorem proved differently by Bochner [1, p. 488], Gaffney [3] and Yosida [11].

4. Associated with each self-adjoint extension $\tilde{D}$ (with domain $\tilde{\mathscr{D}}$) of $D$ and a complex number $\lambda$ we have *a problem of limits* : given $f$ in $L_2$ and $h$ in $L_2(\Delta)$, find $u$ in $L_2(\Delta)$ such that

(i) $-\,\Delta\,u + \lambda\,u = f$,

(ii) $u$ satisfies the condition of limits : $h - u \in \tilde{\mathscr{D}}$.

If $\lambda$ is in the resolvent of $\tilde{D}$, the problem evidently admits of a unique solution for every given $f$ in $L_2$ and $h$ in $L_2(\Delta)$. Since $\tilde{D}$ is self-adjoint, $\lambda$ is in the resolvent of $\tilde{D}$, if $\operatorname{Im}\lambda \neq 0$. If $\lambda$ is real and the problem of limits admits of a unique solution for every $f$ in $L_2$ and $h$ in $L_2(\Delta)$, then $(-\,\tilde{D} + \lambda)^{-1}$ is defined everywhere on $L_2$, and is symmetric. By a theorem of Hellinger and Toeplitz [7, p. 295], which is an immediate consequence of Banach's closed graph theorem, $(-\,\tilde{D} + \lambda)^{-1}$ is continuous ; that is, $\lambda$ is in the resolvent of $\tilde{D}$. Hence we have

PROPOSITION II. *The problem of limits admits of a unique solution for every $f$ in $L_2$ and $h$ in $L_2$ $(\Delta)$ if and only if $\lambda$ is in the resolvent of $\tilde{D}$.*

5. Suppose that $\lambda$ is in the resolvent of $\tilde{D}$. If $\lambda$ is in the resolvent of $\tilde{D}$, so is $\bar{\lambda}$. Let $G_{\bar{\lambda}} = (-\tilde{D} + \bar{\lambda})^{-1}$. $G_{\bar{\lambda}} : L_2 \to L_2$ is continuous. Since the inclusion map $L_2 \to \mathscr{D}'$ is continuous, $G_{\bar{\lambda}} : L_2 \to \mathscr{D}'$ is continuous. $G_{\bar{\lambda}}$ maps $\mathscr{D}$ into $\mathscr{E}$; this follows from the elliptic character of $(-\Delta + \bar{\lambda})$. Since the inclusion maps $\mathscr{D} \to L_2$ and $\mathscr{E} \to \mathscr{D}'$ are continuous and $G_{\bar{\lambda}} : L_2 \to \mathscr{D}'$ is continuous, the graph of the linear map $G_{\bar{\lambda}}$; $\mathscr{D} \to \mathscr{E}$ is closed. By Banach's closed graph theorem, as extended to $(\mathscr{L}\mathscr{F})$ spaces by Dieudonné and Schwartz [2, Theorem I, p: 72], $G_{\bar{\lambda}} : \mathscr{D} \to \mathscr{E}$ is continuous. We define $G_\lambda : \mathscr{E}' \to \mathscr{D}'$ by:

$$(G_\lambda T, \phi) = (T, G_{\bar{\lambda}} \phi),\ T \in \mathscr{E}',\ \phi \in \mathscr{D}.$$

(As a matter of fact, $G_\lambda$ can be defined on the space of currents continuous in mean at infinity; for the definition of this space see [6, § 32, p. 166]; this space contains $\mathscr{E}'$ and $L_2$). If $T$ is an exterior tangent vector at a point $P$ of $M$, $T$ defines a current ("Dirac current"), also denoted by $T$, with support $P$. *$G_\lambda T$ is a fundamental solution in the large for $(-\Delta + \lambda)$ relative to the exterior tangent vector $T$; $G_\lambda T$ is a $C^\infty$ form in the complement of $P$. The double current thus obtained is the "Green's form" associated with the problem of limits corresponding to $\tilde{D}$ and $\lambda$.*

6. Let $\mathscr{E}^1_{L_2}$[5, p. 158] denote the space of elements $T$ in $L_2$ for which $dT \in L_2$ and $\partial T \in L_2$. $\mathscr{E}^1_{L^2}$ is a Hilbert space with the scalar product

$$((T_1, T_2)) = (T_1, T_2) + (dT_1, dT_2) + (\partial T_1, \partial T_2),\ T_1, T_2 \in \mathscr{E}^1_{L^2}.$$

$\mathscr{D}$ is contained in $\mathscr{E}^1_{L_2}$. We denote by $\mathscr{D}^1_{L_2}$ the closure of $\mathscr{D}$ in $\mathscr{E}^1_{L^2}$. We determine all non-positive self-adjoint extensions $\tilde{D}$ of $D$, whose domain $\tilde{\mathscr{D}}$ is contained in $E^1_{L_2}$ and for which

$$(-\tilde{D}u, v) = (du, dv) + (\partial u, \partial v)$$

for every $u, v \in \tilde{D}$. Let $V$ be a closed subspace of $\mathscr{E}^1_{L_2}$ containing $\mathscr{D}^1_{L_2}$. By applying a theorem of Friedrichs [7, p. 332] we obtain

Proposition III. *There exists one and only one self-adjoint extension* $\tilde{D}$ *of* $D$, *whose domain* $\tilde{\mathscr{D}}$ *is contained in* $V$, *and for which*

$$(-\tilde{D}u, v) = (du, dv) + (\partial u, \partial v)$$

*for every* $u \in \tilde{\mathscr{D}}$ *and* $v \in V$. *Let this extension be denoted by* $\tilde{D}(V)$ *and its domain by* $\tilde{\mathscr{D}}(V)$. *If* $\tilde{D}$ *be a non-positive self-adjoint extension whose domain* $\tilde{\mathscr{D}}$ *is contained in* $\mathscr{E}^1_{L_2}$ *and for which*

$$(-\tilde{D}u, v) = (du, dv) + (\partial u, \partial v),$$

*for* $u, v \in \tilde{\mathscr{D}}$, *and if* $V$ *is the closure of* $\tilde{\mathscr{D}}$ *in* $\mathscr{E}^1_{L_2}$, *then* $\tilde{D}(V)$ *coincides with* $\tilde{D}$.

$\tilde{D}(V)$ is a non-positive self-adjoint operator. So $\lambda$ is in the resolvent of $\tilde{D}(V)$, if it is positive [10, Theorem 5.12, p. 188]. This means that $(-\Delta + \lambda)$, $\lambda > 0$, *maps* $\tilde{\mathscr{D}}(V)$ *isomorphically onto* $L_2$. This is a theorem of Lions and Schwartz [5, p. 158].

$\tilde{D}(\mathscr{D}^1_{L_2})$ is the Friedrichs extension of $D$. The problem of limits associated with $\tilde{D}(\mathscr{D}^1_{L_2})$ and $\lambda = 0$ is the Dirichlet problem. The problem of limits associated with $\tilde{D}(\mathscr{E}^1_{L_2})$ and $\lambda = 0$ is the Neumann problem.

7. Let $\tilde{D}$ be a non-positive self-adjoint extension of $D$. By a theorem on self-adjoint transformations on a Hilbert space [10, Theorem 5.12, p. 188] we obtain

Proposition IV. *The problem of limits associated with* $\tilde{D}(\tilde{D} \leqslant 0)$ *and* $\lambda = 0$ *has a unique solution for every* $f$ *in* $L_2$ *and* $h$ *in* $L_2(\Delta)$ *if and only if*

$$(-\tilde{D}u, u) \geqslant C(u, u)$$

*for every* $u$ *in* $\tilde{\mathscr{D}}$, $C$ *being a positive constant independent of* $u$. *In the case of* $\tilde{D}(V)$, *this inequality becomes the Poincaré inequality* :

$$(du, du) + (\partial u, \partial u) \geqslant C(u, u)$$

*for every* $u$ *in* $V$, $C$ *being a positive constant independent of* $u$.

Let us suppose that zero is an eigenvalue of $\tilde{D}$, $\tilde{D}$ being again a non-positive self-adjoint extension of $D$. (This is the situation, for example, in the case of Neumann's problem for functions in a bounded domain of Euclidean $n$-space). We denote by $E_0$ the corresponding eigenspace. $E_0$ is the space of harmonic forms contained in the domain of $\tilde{D}$. The space $E_0$ and its orthogonal complement $E_0^{\perp}$ in $L_2$ reduce $\tilde{D}$. Moreover, the restriction of $\tilde{D}$ to $E_0^{\perp}$, denoted by $\tilde{D} \mid E_0^{\perp}$, is a self-adjoint transformation in $E_0^{\perp}$. We consider the problem of limits associated with $\tilde{D}$ and $\lambda = 0$, and assume, for simplicity, that $h = 0$. If the problem of limits is to have a solution, $f$ must be orthogonal to $E_0$. When the solution exists, there exists a unique solution orthogonal to $E_0$. Another application of the theorem used above [10, Th. 5.2, p. 199], this time to the operator $\tilde{D} \mid E_0^{\perp}$, yields

PROPOSITION V. *The problem of limits* ($h = 0$, $\lambda = 0$) *admits of a solution for every $f$ orthogonal to $E_0$ if and only if*

$$(-\Delta u, u) \geqslant C(u, u)$$

*for every $u$ contained in the domain of $\tilde{D}$ and orthogonal to $E_0$, $C$ being a positive constant independent of $u$.*

**8.** We now suppose that $M$ is a relatively compact subdomain of a Riemannian manifold. In this case, the spectrum of $\tilde{D}(\mathscr{D}^1_{L_2})$ consists only of eigenvalues, with finite multiplicity, which have no accumulation point. This follows immediately from the complete continuity of $(-\tilde{D}(\mathscr{D}^1_{L_2}) + \lambda)^{-1}$ for $\lambda > 0$ [8, p. 436]. We assume that the uniqueness condition holds : the only harmonic form in the domain of $\tilde{D}(\mathscr{D}^1_{L_2})$ (harmonic form which " vanishes on the boundary ") is the zero form. This implies that zero is in the resolvent of $\tilde{D}$ ($\mathscr{D}^1_{L_2}$). So, by § 5, $M$ possesses Green's form for $\Delta$; in particular, this proves the existence of a fundamental solution in the large for $\Delta$. [ 4, § 8, § 9 ].

## REFERENCES

1. S. Bochner : Completely monotone functions of the Laplace operator, *Duke Math. J.* 3 (1937), 488-502.

2. J. Dieudonné and L. Schwartz : La dualité dans les espaces $(\mathscr{F})$ et $(\mathscr{L}\mathscr{F})$, *Annals Inst. Fourier*, 1 (1949), 61-101.

3. M. P. Gaffney : Hilbert space methods in the theory of harmonic integrals, *Trans. American Math. Soc.* 78 (1955), 426-444.

4. P. R. Garabedian and D. C. Spencer : A complex tensor calculus for Kähler manifolds, *Acta Math.* 89 (1953), 279-331.

5. J. L. Lions and L. Schwartz : Problèms aux limites sur des espaces fibrés, *Acta Math.* 94 (1955), 155-159.

6. G. de Rham : *Variétés differentiables,* Paris, 1955.

7. F. Riesz and B. Sz.-Nagy : *Lecons d'analyse fonctionelle,* Budapest, 1953.

8. M. Schiffer and D. C. Spencer : *Functionals on finite Riemann surfaces*, Princeton, 1954.

9. L. Schwartz : *Théorie des distributions* I, Paris, 1950.

10. M. H. Stone : *Linear transformations in Hilbert space,* New York, 1932.

11. K. Yosida : An ergodic theorem associated with harmonic integrals, *Proc. Imp. Acad. Tokyo*, 27 (1951)), 540-543.

Tata Institute of Fundamental Research
Bombay

Reprinted from the Proceedings of the NATIONAL ACADEMY OF SCIENCES,
Vol. 43, No. 6, pp. 513–514. June, 1957.

# *THE IDENTITY OF THE WEAK AND STRONG EXTENSIONS OF A LINEAR ELLIPTIC DIFFERENTIAL OPERATOR*

BY M. S. NARASIMHAN

TATA INSTITUTE OF FUNDAMENTAL RESEARCH, BOMBAY, INDIA

*Communicated by S. Bochner, March 12, 1957*

1. Let $\Omega$ be an open subset of the Euclidean space $R^n$. Let

$$D = \sum_{j_1 + \ldots + j_n \leq m} a_{j_1 \ldots j_n}(x) \frac{\partial^{j_1 + \ldots + j}}{\partial x_1^{j_1} \ldots \partial x_n^{j_n}}$$

be a linear differential operator in $\Omega$ with indefinitely differentiable complex-valued coefficients $a_{j_1 \ldots j_n}(x)$, $x \equiv (x_1, \ldots, x_n) \in \Omega$. $D$ operates on the space of distributions in $\Omega$. Let $D_1$ denote the restriction of $D$ to the space of $C^\infty$ functions with compact supports. The *weak extension* $D_W$ of $D_1$ in $L^2(\Omega)$ is defined as follows: the domain $M$ of $D_W$ consists of all those elements $f$ in $L^2$ for which $Df$, formed in the sense of distributions, is in $L^2$, and for such $f$, $D_W f$ is defined to be $Df$. The *strong extension* $D_S$ of $D_1$ in $L^2$ is the closure in $L^2$ of the operator $D_2$, where $D_2$ is defined as follows: its domain consists of all $f$ in $L^2$ which are $C^\infty$ and for which $Df$, formed in the usual sense, belongs to $L^2$, and for such $f$, $D_2 f = Df$. It is easily seen that $D_W$ is an extension of $D_S$. It is not known whether, in general, the weak and strong extensions coincide. In this note we prove their identity for a class of differential operators.[1]

2. THEOREM 1. *Let $\Omega$ be an arbitrary open subset of $R^n$, and let $D$ be formally self-adjoint and elliptic.*[2] *The weak and strong extensions of $D_1$ in $L^2(\Omega)$ coincide.*

*Proof:* By assumption, $D_1$ is a symmetric operator in $L^2$; also, $D_W$ is the adjoint of $D_1$ in $L^2$. We have, by a theorem of von Neumann,[3]

$$\text{Domain of } D_W = \text{domain of the closure of } D_1 \oplus E(i) \oplus E(-i), \qquad (1)$$

where $E(i)$ and $E(-i)$ denote the spaces of all square-integrable solutions of the equations $DT = iT$ and $DT = -iT$, respectively. From the results of F. John and L. Schwartz,[2] the elements of $E(i)$ and $E(-i)$ are $C^\infty$ functions. It follows immediately from this and from decomposition (1) that the weak and strong extensions coincide.

3. THEOREM 2. *Assume that $\Omega$ is bounded. Let $D$ be uniformly elliptic,*[2, 4] *not necessarily formally self-adjoint. Then the weak and strong extensions of $D_1$ coincide.*

*Proof:* Since the identity of $(D + \lambda)_S$ and $(D + \lambda)_W$, for $\lambda$ real, implies that of $D_S$ and $D_W$, it is sufficient to prove that $(D + \lambda)_S \equiv (D + \lambda)_W$ for some $\lambda$. Let $D + \lambda = D'$. The domain $M$ of $D'_W$ is a Hilbert space with the scalar product

$$((f_1, f_2)) = (f_1, f_2)_{L^2} + (D'f_1, D'f_2)_{L^2},$$

and we have to show that $M \cap \mathcal{E}$ is dense in $M$ in this topology, $\mathcal{E}$ denoting the space of $C^\infty$ functions in $\Omega$. Gårding's inequality[2, 4] allows one to define the generalized Friedrich's extension[4–6] of $D_1$; we denote its domain by $N$. We consider $N$ as a Hilbert space with the scalar product induced from $M$. Then, for $\lambda$ sufficiently large, $D' | N : N \to L^2$ is a *topological isomorphism* of $N$ onto $L^2$.[5, 6] Since $C^\infty$ functions in $L^2$ are dense in $L^2$ and since $(D' | N)^{-1}: L^2 \to N$ maps $C^\infty$ functions into $C^\infty$

functions by the elliptic character of $D$, it follows therefore that $N \cap \mathcal{E}$ is dense in $N$. Now if $f \in M$, we can find a unique $f' \in N$ such that $D'f = D'f'$. It follows immediately that

$$M = N \oplus L, \tag{2}$$

where $L$ is the space of all square-integrable solutions of the equation $D'f = 0$. Again, by the elliptic character of $D$, $L \subset \mathcal{E}$. So, from decomposition (2), we see that $M \cap \mathcal{E}$ is dense in $M$.

*Remark 1:* A similar argument shows that if, in Theorem 2, the coefficients are analytic, $D_W$ is the closure of its restriction to analytic functions in the domain of $D_W$.

*Remark 2:* Actually, for the proof of Theorem 2, it is sufficient to know that $D'$ maps $M$ *onto* $L^2$. For, by Banach's open mapping theorem, the map $D'|M:M \to L^2$ will then be an open mapping, and the inverse image of the set $\mathcal{E} \cap L^2$, everywhere dense in $L^2$, will be everywhere dense in $M$.

[1] For the case of second-order elliptic operators see M. S. Birman, *Doklady Akad. Nauk S.S.S.R.*, N.S., **92,** 205–208, 1953; and for that of a class of operators with constant coefficients see L. Hörmander, *Acta Math.*, **94,** 241, 1955.

[2] L. Gårding, *Math. Scand.*, **1,** 55–72, 1953.

[3] J. von Neumann, *Math. Ann.*, **102,** 85–86, 1930.

[4] L. Nirenberg, *Communs. Pure and Appl. Math.*, **8,** 648–684, 1955, in particular pp. 657–659.

[5] J. L. Lions, *Acta Math.*, **94,** 26, 1955.

[6] P. D. Lax and A. N. Milgram, "Contributions to the Theory of Partial Differential Equations," *Ann. Math. Studies*, **33,** 167–190, 1954.

Reprinted from the Proceedings of the NATIONAL ACADEMY OF SCIENCES,
Vol. 43, No. 7, p. 620. July, 1957.


# THE IDENTITY OF THE WEAK AND STRONG EXTENSIONS OF A LINEAR ELLIPTIC DIFFERENTIAL OPERATOR: II

BY M. S. NARASIMHAN

TATA INSTITUTE OF FUNDAMENTAL RESEARCH, BOMBAY, INDIA

*Communicated by S. Bochner*

1. In a previous note under the same title[1] we proved the identity of the weak and strong extensions of certain linear elliptic operators. In this note we prove the same result for the most general linear elliptic operator with $C^\infty$ coefficients in an arbitrary open subset $\Omega$ of $R^n$. We use the same notations as in the previous paper.

We observe that, while the theorem given here includes both the theorems of the previous note, the method of proof adopted here does not yield a result which the method of proof of Theorem 2 of the previous note does (at least under certain restrictions) namely, that $D_W$ is the closure of its restriction to analytic functions in its domain if the coefficients are analytic.

2. THEOREM. *Let $\Omega$ be an arbitrary open subset of $R^n$. Let $D$ be an elliptic operator in the sense that*

$$p(x, \xi) \equiv \sum_{j_1 + \ldots + j_n = m} a_{j1 \ldots jn}(x)\, \xi_1^{j_1} \ldots \xi_n^{j_n}$$

*is different from zero for each $x \in \Omega$ and for every real nonzero vector $(\xi_1, \ldots, \xi_n)$. Then the weak and strong extensions of $D_1$ coincide.*

*Proof:*

Since $D_W$ is a closed operator in $L^2$, the domain $M$ of $D_W$ is a *Hilbert space* with the scalar product

$$((f_1, f_2)) = (f_1, f_2)_{L^2} + (Df_1, Df_2)_{L^2}, \qquad f_1, f_2 \in M.$$

We have to show that $\mathcal{E} \cap M$ is dense in $M$ in the topology of $M$, where $\mathcal{E}$ denotes the space of $C^\infty$ functions in $\Omega$. Evidently the space $\mathcal{D}$ of $C^\infty$ functions with compact support in $\Omega$ is contained in $M$. Let $H$ denote the closure of $\mathcal{D}$ in $M$, in the topology of $M$. Then if $H^\perp$ denotes the orthogonal complement of $H$ in the Hilbert space $M$, we have

$$M = H \oplus H^\perp. \tag{1}$$

We shall show that the elements of $H^\perp$ are $C^\infty$ functions. Then it would follow from decomposition (1) and the fact the $\mathcal{D}$ is dense in $H$ (in the topology induced from $M$) that $\mathcal{E} \cap M$ is dense in $M$.

Now if $f \in H^\perp$, we have, for each $\varphi \in \mathcal{D}$,

$$(f, \varphi)_{L^2} + (Df, D\varphi)_{L^2} = 0;$$

i.e., $f$ is a distribution solution of the equation $(D^*D + 1)f = 0$, where $D^*$ denotes the formal adjoint of $D$. A simple computation shows that $(D^*D + 1)$ is an elliptic operator of order $2m$ (whose characteristic form is equal to $(-1)^m|p(x, \xi)|^2$). Since $(D^*D + 1)$ is a linear elliptic operator with $C^\infty$ coefficients, every distribution solution $f$ of the equation $(D^*D + 1)f = 0$ is indefinitely differentiable.[2] Hence $H^\perp \subset \mathcal{E}$.

[1] These PROCEEDINGS, **43**, 513, 1957.
[2] L. Gårding, *Math. Scand.*, **1**, 55–72, 1953.

Ann. Inst. Fourier, Grenoble
**10** (1960), 285-296.

# THE TYPE AND THE GREEN'S KERNEL OF AN OPEN RIEMANN SURFACE

by **M. S. NARASIMHAN**

Tata Institute of Fundamental Research, Bombay
et Centre National de la Recherche Scientifique, Paris.

## 1. — Introduction.

We give in this paper a new approach to the determination of the type and the construction of the Green's function of an open Riemann surface.

We first define an open Riemann surface to be of hyperbolic type if the completion of the pre-Hilbert space of $C^\infty$ functions with compact supports endowed with the Dirichlet scalar product is a space of currents. In this case we construct in a natural way an operator $\mathcal{G}$ and call the kernel in the sense of Schwartz of $\mathcal{G}$ the Green's kernel of the open Riemann surface. We then show that an open Riemann surface is of hyperbolic type if and only if it possesses the Green's function in the classical sense and that the Green's kernel is identical, upto a scalar factor, with the Green's function in the classical sense.

The invariance of the type of an open Riemann surface under quasi-conformal maps is derived as an immediate consequence of the definition of type.

## 2. — Some spaces of currents.

Let $\Omega$ be an open Riemann surface, that is, a non-compact connected complex analytic manifold of complex dimension one. We denote by $\overset{p}{\mathcal{D}}(\Omega)$, $p = 0, 1, 2$, the space of $C^\infty$ forms of degree $p$ endowed with the topology of Schwartz [10, 11]. Let $\overset{p}{\mathcal{D}}'(\Omega)$ denote the space of currents of degree $p$ endowed with the strong topology [10, 11]. Let further $\overset{p}{\mathcal{E}}(\Omega)$ denote the space of $C^\infty$ forms of degree $p$ and $\overset{p}{\mathcal{E}}'(\Omega)$ the space of currents of degree $p$ with compact supports, each endowed with the usual topology.

The operator $*$ is defined intrinsically on 1-forms in $\Omega$ [8]. The operator $*$ is defined on the currents of degree one by the formula :

$$\langle * T, \varphi\rangle = \langle T, - * \varphi\rangle, \qquad T \in \overset{1}{\mathcal{D}}', \qquad \varphi \in \overset{1}{\mathcal{D}},$$

$\langle , \rangle$ denoting the scalar product between $\overset{1}{\mathcal{D}}'$ and $\overset{1}{\mathcal{D}}$.

We denote by $L^2(\Omega)$ the Hilbert space of measurable square integrable 1-forms with the scalar product

$$(\omega_1, \omega_2) = \int_\Omega \omega_1 \wedge * \overline{\omega}_2, \qquad \omega_1, \omega_2 \in L^2.$$

Let further $BL(\Omega)$ be the pre-Hilbert space of currents T of degree 0 for which $dT \in L^2$ endowed with the scalar product

$$(T_1, T_2)_1 = (dT_1, dT_2), \qquad T_1, T_2 \in BL.$$

If $BL^{\cdot}(\Omega)$ denotes the quotient space of BL by the subspace of constants, $BL^{\cdot}$ is a *Hilbert space* with the induced scalar product [2, p. 308].

## 3. — The Laplacian $\tilde{\Delta}$.

On a Riemann surface the Laplacian is not defined intrinsically as an operator carrying functions into functions. However we can define an operator analogous to the Laplacian carrying functions into 2 forms.

We define an operator

$$\tilde{\Delta} : \overset{0}{\mathfrak{D}}' \to \overset{2}{\mathfrak{D}}'$$

by the formula

$$\tilde{\Delta} = d * d$$

where $d$ denotes the exterior derivation. $\tilde{\Delta}$ is an elliptic operator of type $(V_1, V_2)$ where $V_1$ denotes the trivial line bundle with $\mathbf{C}$ as the fibre and $V_2$ denotes the line bundle of 2-covectors [6].

We have the following elementary formulae:

1) $\langle T, \tilde{\Delta}\varphi \rangle = \langle \tilde{\Delta} T, \varphi \rangle$ for $T \in \overset{0}{\mathfrak{D}}', \ \varphi \in \overset{0}{\mathfrak{D}}$.

2) For $T \in \overset{0}{\mathfrak{D}}', \ \varphi \in \overset{0}{\mathfrak{D}}$ define:

$$(dT, d\varphi) = \langle dT, * d\bar{\varphi} \rangle.$$

We then have

$$(dT, d\varphi) = \langle - \tilde{\Delta} T, \bar{\varphi} \rangle.$$

## 4. — The type and the Green's kernel.

Let $\mathcal{H}_0(\Omega)$ denote the vector space $\overset{0}{\mathfrak{D}}$ endowed with the Dirichlet scalar product

$$(\varphi, \psi)_1 = \int_\Omega d\varphi \wedge * d\bar{\psi}, \qquad \varphi, \psi \in \overset{0}{\mathfrak{D}}.$$

Since $\Omega$ is non-compact and connected $\mathcal{H}_0(\Omega)$ is a separated pre-Hilbert space. Let $\mathcal{H}(\Omega)$ be the completion of $\mathcal{H}_0(\Omega)$.

DÉFINITION 1. — *An open Riemann surface $\Omega$ is said to be of* hyperbolic *type if the inclusion map*

$$i : \mathcal{H}_0(\Omega) \to \overset{0}{\mathfrak{D}}'(\Omega)$$

*is continuous. Otherwise, it is said to be of* parabolic *type.*

Thus we define an open Riemann surface to be of hyperbolic type if the following condition is satisfiedd: if $\{\varphi_n\}$ is a sequence of $C^\infty$ functions with compact supports whose Dirichlet integrals tend to zero then the sequence $\{\varphi_n\}$ tends to zero in the sense of currents.

Let $\Omega$ be an open Riemann surface of hyperbolic type. Let

$$i' : \mathcal{H}(\Omega) \to \overset{0}{\mathfrak{D}}'(\Omega)$$

denote the canonical extension of the continuous inclusion

$$i : \mathcal{H}_0(\Omega) \to \overset{0}{\mathfrak{D}}'(\Omega).$$

The map $i'$ is an injection. In fact, if $T \in \mathcal{H}$ we have for every $\varphi \in \overset{0}{\mathfrak{D}}$

$$(T, \varphi)_{\mathcal{H}} = \langle -\tilde{\Delta} i'T, \bar{\varphi} \rangle.$$

Since $\overset{0}{\mathfrak{D}}$ is dense in $\mathcal{H}$ it follows from the above equality that $i'T = 0$ if and only if $T = 0$.

We identify $\mathcal{H}$ with a subspace of $\overset{0}{\mathfrak{D}}'$ by means of the injection $i'$. The completion of $\mathcal{H}_0$ is thus a space of currents. It is easily seen that $\mathcal{H}$ is contained in BL.

Since $\overset{0}{\mathfrak{D}}$ is dense in $\mathcal{H}$ the dual $\mathcal{H}'(\Omega)$ of $\mathcal{H}(\Omega)$ is canonically identified with a subspace of $\overset{2}{\mathfrak{D}}'(\Omega)$. We assert that $\tilde{\Delta}$ defines an isomorphism of $\mathcal{H}$ onto $\mathcal{H}'$. In fact let $\Lambda$ be the canonical isomorphism of $\mathcal{H}$ on the conjugate of its dual. Then for $T \in \mathcal{H}$, $\varphi \in \overset{0}{\mathfrak{D}}$ we have

$$\langle \Lambda T, \bar{\varphi} \rangle = (dT, d\varphi) = \langle -\tilde{\Delta} T, \bar{\varphi} \rangle$$

so that $\Lambda = -\tilde{\Delta}$. Hence $\tilde{\Delta} : \mathcal{H} \to \mathcal{H}'$ is an isomorphism. Let $\mathcal{G} : \mathcal{H}' \to \mathcal{H}$ be the inverse isomorphism.

Consider the spaces $\mathcal{H} \cap \overset{0}{\mathcal{E}}$ and $\mathcal{H}' \cap \overset{2}{\mathcal{E}}$; $\mathcal{H} \cap \overset{0}{\mathcal{E}}$ will be endowed with the topology upper-bound of those of $\mathcal{H}$ and $\overset{0}{\mathcal{E}}$; the same for $\mathcal{H}' \cap \overset{2}{\mathcal{E}}$. Let $\mathcal{H}' + \overset{2}{\mathcal{E}}'$ (resp. $\mathcal{H} + \overset{0}{\mathcal{E}}'$) be the strong dual of $\mathcal{H} \cap \overset{0}{\mathcal{E}}$ (resp. $\mathcal{H}' \cap \overset{2}{\mathcal{E}}$). [An element of the dual of $\mathcal{H} \cap \overset{0}{\mathcal{E}}$ can be written in the form $f + T$, $f \in \mathcal{H}'$, $T \in \overset{2}{\mathcal{E}}'$. Hence the above notation. A similar remark applies to $\mathcal{H} + \overset{0}{\mathcal{E}}'$]. Since $\tilde{\Delta}$ is an elliptic operator we see exactly as in Lions [5, p. 36] that the operator $\mathcal{G}$ can be extended to an isomorphism, still denoted by $\mathcal{G}$, of $\mathcal{H}' + \overset{2}{\mathcal{E}}'$ onto $\mathcal{H} + \overset{0}{\mathcal{E}}'$ and $\tilde{\Delta}$ is its inverse.

$\mathcal{G} : \mathcal{H}' + \overset{2}{\mathcal{E}}' \to \mathcal{H} + \overset{0}{\mathcal{E}}'$ is called the Green's operator.

DÉFINITION 2. — *Let $\Omega$ be an open Riemann surface of hyperbolic type. The kernel in the sense of Schwartz of the operator $\mathcal{G}$ is called the Green's kernel of $\Omega$.*

The Green's kernel is a bilateral elementary kernel for $\tilde{\Delta}$. The green's kernel is very regular in the sense of Schwartz [11; 4, § 12; 5].

REMARK 1. — An open Riemann surface is of hyperbolic type if and only if the inclusion map $\mathcal{H}_0 \to L^2_{loc}$ is continuous, where $L^2_{loc}$ denotes the space of locally square summable functions endowed with the topology of convergence in $L^2$ on each compact set. [See 2, p. 321, Prop. 4. 1].

REMARK 2. — Let $\mathcal{H}_1$ denote the pre-Hilbert space of $C^1$ functions with compact supports endowed with the Dirichlet scalar product. We see by regularization and Remark 1. that $\Omega$ is hyperbolic if and only if the inclusion map $\mathcal{H}_1 \to L^2_{loc}$ is continuous.

## 5. — Some properties of the Green's operator.

In this section we prove some propositions concerning the Green's operator.

PROPOSITION 1. — *$\Omega$ is of hyperbolic type if and only if $\overset{2}{\mathfrak{D}} \subset \tilde{\Delta}(BL)$.*

*Proof.* If $\Omega$ is of hyperbolic type and $\psi \in \overset{2}{\mathfrak{D}}$, then $\mathcal{G}\psi \in BL$ and $\tilde{\Delta}\mathcal{G}\psi = \psi$.

Suppose conversely that $\overset{2}{\mathfrak{D}} \subset \tilde{\Delta}(BL)$. Let $\{\varphi_n\}$, $\varphi_n \in \overset{0}{\mathfrak{D}}$ be a sequence converging to zero in $\mathcal{H}_0$. We shall show that $\langle\psi, \varphi_n\rangle \to 0$ for every $\psi \in \overset{2}{\mathfrak{D}}$. In fact let $T \in BL$ be such that $\tilde{\Delta}T = \overline{\psi}$. Then

$$\begin{aligned}\langle\psi, \varphi_n\rangle &= \langle\tilde{\Delta}\overline{T}, \varphi_n\rangle, \\ &= -(d\overline{T}, d\overline{\varphi}_n).\end{aligned}$$

Since $d\overline{T} \in L^2$ and $d\overline{\varphi}_n \to 0$ in $L^2$, we see that $\langle\psi, \varphi_n\rangle \to 0$.

PROPOSITION 2. — *Suppose that $\Omega$ is of hyperbolic type. Let $\Omega'$ be a subdomain of $\Omega$. Then $\Omega'$ is hyperbolic and there* *exists a continuous linear map* $u \to u^{\sim}$ *of* $\mathcal{H}(\Omega')$ *into* $\mathcal{H}(\Omega)$ *such that* $u^{\sim} = u$ *in* $\Omega'$ *and* $u^{\sim} = 0$ *in* $\complement\,\Omega'$. *One has*

$$(du, du)_{L^2(\Omega')} = (du^{\sim}, du^{\sim})_{L^2(\Omega)}.$$

*Proof.* For $\varphi \in \overset{0}{\mathcal{D}}(\Omega')$ let $\varphi^{\sim} \in \overset{0}{\mathcal{D}}(\Omega)$ be the function obtained by extending $\varphi$ by zero outside $\Omega'$. The map $j : \varphi \to \varphi^{\sim}$ is an isometry of $\mathcal{H}_0(\Omega')$ into $\mathcal{H}_0(\Omega)$. The inclusion map $\mathcal{H}_0(\Omega') \to \overset{0}{\mathcal{D}}{}'(\Omega')$ is the composition of the map $j$, the continuous inclusion $\mathcal{H}_0(\Omega) \to \overset{0}{\mathcal{D}}{}'(\Omega)$ and the restriction map $r$: $\overset{0}{\mathcal{D}}{}'(\Omega) \to \overset{0}{\mathcal{D}}{}'(\Omega')$ and is hence continuous. Since the map $j$ can be extended into an isometry (still denoted by $j$) the second part of the proposition follows.

We identify $\mathcal{H}(\Omega')$ with a subspace of $\mathcal{H}(\Omega)$ by means of the isometry $j$.

PROPOSITION 3. — *Let* $\Omega$ *be of hyperbolic type. Let* $\{\Omega_k\}$, $k = 1, 2, \ldots$ *be an increasing sequence of sub-domains of* $\Omega$ *such that* $\bigcup_k \Omega_k = \Omega$. *Let* $\mathcal{G}_k$ *(resp.* $\mathcal{G}$*) be the Green's operator of* $\Omega_k$ *(resp.* $\Omega$*). Let* $T \in \mathcal{H}'(\Omega)$ *and let* $T_k$ *be the restriction of* $T$ *to* $\Omega_k$. *Then* $\mathcal{G}_k T_k^{\sim} \to \mathcal{G}T$ *in* $\mathcal{H}(\Omega)$.

*Proof.* $\mathcal{H}(\Omega_k)$ is a closed subspace of $\mathcal{H}(\Omega)$ and $\mathcal{H}(\Omega)$ is the closure of $\bigcup_k \mathcal{H}(\Omega_k)$. If we verify that $\mathcal{G}_k T_k^{\sim}$ is the orthogonal projection of $\mathcal{G}T$ into the closed subspace $\mathcal{H}(\Omega_k)$ it would follow, from a known theroem on projections in a Hilbert space, that $\mathcal{G}_k T_k^{\sim} \to \mathcal{G}T$ in $\mathcal{H}(\Omega)$. Now for every $\varphi \in \overset{0}{\mathcal{D}}(\Omega_k)$ we have

$$\begin{aligned}(\mathcal{G}_k T_k, \varphi)_{\mathcal{H}(\Omega_k)} &= \langle -\tilde{\Delta}\mathcal{G}_k T_k, \bar{\varphi}\rangle \\ &= -\langle T_k, \bar{\varphi}\rangle \\ &= -\langle T, \bar{\varphi}^{\sim}\rangle.\end{aligned}$$

On the other hand if $P_k$ denotes the projection operator on $\mathcal{H}(\Omega_k)$ we have, for $\varphi \in \overset{0}{\mathcal{D}}(\Omega_k)$,

$$\begin{aligned}(P_k\mathcal{G}T, \varphi)_{\mathcal{H}(\Omega)_k} &= (d\mathcal{G}T, d\bar{\varphi}^{\sim})_{L^2(\Omega)} \\ &= \langle -\tilde{\Delta}\mathcal{G}T, \bar{\varphi}^{\sim}\rangle \\ &= -\langle T, \bar{\varphi}^{\sim}\rangle.\end{aligned}$$

Hence $\mathcal{G}_k T_k^{\sim} = P_k\mathcal{G}T$.

**Proposition 4.** — *Assume that $\Omega$ is of hyperbolic type. With the same notations as in Proposition 3, let* T *be an element of $\overset{2}{\mathcal{E}}'(\Omega)$ such that the support of* T *is contained in $\Omega_1$. Then $\mathcal{G}_k T^{\sim} \to \mathcal{G} T$ in $\overset{0}{\mathcal{D}}'(\Omega)$ and the convergence is uniform on every compact set contained in the complement of the support of* T *($\mathcal{G}_k T^{\sim}$ denotes the extension of $\mathcal{G}_k T$ to $\Omega$ by zero outside $\Omega_k$; $\mathcal{G}_k T^{\sim}$ and $\mathcal{G} T$ are functions in the complement of the support of* T*).*

*Proof.* Let $S_k = \mathcal{G}_k T^{\sim}$. To prove that $S_k \to \mathcal{G} T$ in $\overset{0}{\mathcal{D}}'(\Omega)$ it is sufficient to prove that, for every $\psi \in \overset{2}{\mathcal{D}}(\Omega)$, $\langle S_k, \psi \rangle$ tends to $\langle \mathcal{G} T, \psi \rangle$. Now $T \in \overset{2}{\mathcal{E}}'(\Omega_k)$ and $\psi \in \overset{2}{\mathcal{D}}(\Omega_k)$ for all sufficiently large $k$, say for $k \geqslant k_0$. In $\Omega_{k_0}$, we have

$$\tilde{\Delta}(\mathcal{G}\psi - \mathcal{G}_k \psi) = \psi - \psi = 0.$$

Since $\tilde{\Delta}$ is an elliptic operator, $\overset{0}{\mathcal{D}}'$ and $\overset{0}{\mathcal{E}}$ induce the same topology on the space of solutions of the equations $\tilde{\Delta} f = 0$ [6, p. 331; 11, ch. v, Th. XII]. By Proposition 3, $\mathcal{G}_k \psi \to \mathcal{G} \psi$ in $\overset{0}{\mathcal{D}}'(\Omega_{k_0})$. Hence $\mathcal{G}_k \psi \to \mathcal{G} \psi$ in $\overset{0}{\mathcal{E}}(\Omega_{k_0})$. Now

$$\langle S_k, \psi \rangle = \langle \mathcal{G}_k T, \psi \rangle = \langle T, \mathcal{G}_k \psi \rangle.$$

Since $T \in \overset{2}{\mathcal{E}}'(\Omega_{k_0})$ and $\mathcal{G}_k \psi \to \mathcal{G} \psi$ in $\overset{0}{\mathcal{E}}(\Omega_{k_0})$, we see that $\langle T, \mathcal{G}_k \psi \rangle \to \langle T, \mathcal{G} \psi \rangle$. Hence $\langle S_k, \psi \rangle$ tends to $\langle T, \mathcal{G}\psi \rangle = \langle \mathcal{G} T, \psi \rangle$.

The second part follows from the property of elliptic equations used above.

**Proposition 5.** — *Assume that $\Omega$ is hyperbolic.* Let $\Omega_0$ *be a relatively compact sub-domain of $\Omega$ bounded by a finite number of disjoint analytic Jordan curves. Let $\mathcal{G}_0$ be the Greens' operator of $\Omega_0$. Let $p \in \Omega_0$ and let $g_p$ be the Green's function in the classical sense of $\Omega_0$ with « pole »* p [7, 8]. *Then we have*

$$\mathcal{G}_0 \delta_p = -\frac{1}{2\pi} g_p$$

*where $\delta_p$ is the Dirac measure at $p$.*

*Proof.* — We first remark that $\mathcal{G}_0 \delta_p$ is the only element T of $\mathcal{H}(\Omega_0) + \overset{0}{\mathcal{E}}'(\Omega_0)$ which satisfies the equation $\tilde{\Delta} T = \delta_p$.

One knows that $\tilde{\Delta}\left(-\frac{1}{2\pi}g_p\right) = \delta_p$. The lemma will be proved if we show that $g_p \in \mathcal{H}(\Omega_0) + \overset{0}{\mathcal{E}}'(\Omega_0)$.

$g_p$ is a $C^\infty$ function in $\Omega_0$ except at $p$. Since $g_p$ attains the boundary value zero on the boundary [8, § 28. 3] the reflection principle shows that $g_p$ is $C^\infty$ in $\overline{\Omega}_0$ except at $p$. Let $\varphi \in \overset{0}{\mathcal{D}}(\Omega_0)$ equal to 1 in a neighbourhood of $p$. $\varphi g_p$ is a current of degree 0, with compact support. $(1-\varphi)\, g_p$ is $C^\infty$ in $\overline{\Omega}_0$ and hence has a finite Dirichlet integral; moreover $(1-\varphi)\, g_p$ vanishes on the boundary. It is known that such a function belongs to $\mathcal{H}_0(\Omega_0)$ [4, § 2.4; 8, § 32. 1]. Since $g_p = \varphi g_p + (1-\varphi)\, g_p$ we have $g_p \in \mathcal{H}(\Omega_0) + \overset{0}{\mathcal{E}}'(\Omega_0)$.

Remark 3. — Another method to prove Proposition 5 is to show directly, without using $g_p$, that $\mathcal{G}_0\delta_p$ attains the boundary value zero (« Regularity at the boundary »). This may be shown as in [1, ch. vii, § 4] or [4, § 12.3].

## 6. — The potential with respect to the Green's function.

The proposition proved in this section is more or less classical.

Proposition 6. — *Let $\Omega$ be an open Riemann surface which has the Green's function $g(p, q)$ in the classical sense* [7, ch. vi, § 2]. *Then for $\psi \in \overset{2}{\mathcal{D}}(\Omega)$ the function*

$$h(p) = \int_\Omega g(p, q) \wedge \psi$$

*belongs to* BL.

Before proving the proposition we prove the following

Lemma. — *Let $\Omega_0$ be a relatively compact sub-domain of $\Omega$ bounded by a finite number of disjoint analytic Jordan curves. Let $g_0(p, q)$ be the Green's function of $\Omega_0$ and $\psi \in \overset{2}{\mathcal{D}}(\Omega_0)$. Then*

$$h_0(p) = \int_{\Omega_0} g_0(p, q) \wedge \psi$$

*is $C^\infty$ in $\overline{\Omega}_0$. $h_0(p)$ is zero on the boundary and one has*

$$(dh_0,\, dh_0)_{L^2(\Omega_0)} = \langle -\tilde{\Delta} h_0,\, \overline{h_0}\rangle.$$

*Proof.* — Let K be the support of $\psi$ and $\Omega'$ a relatively compact sub-domain of $\Omega_0$ containing K. By Harnack's principle there exists, for $q_0 \in K$, a constant $k$ such *that* $g(p, q) \leqslant kg(p, q_0)$ for each $q \in K$ and $p \in \complement \Omega'$. Hence

$$|h_0(p)| \leqslant k \left(\int_{\Omega_0} |\psi|\right) g(p, q_0) \qquad \text{for} \qquad p \in \complement \Omega'.$$

Using the symmetry of the Green's function we see that

$$|h_0(p)| \leqslant k' g(q_0, p) \qquad \text{for} \qquad p \in \complement \Omega',$$

where $k'$ is a positive constant. Since $g(q_0, p)$ attains the boundary value zero we see that $h_0(p)$ is continuous in $\overline{\Omega}_0$ and is zero on the boundary. By the reflection principle $h_0$ is $C^\infty$ in $\overline{\Omega}_0$ and an application of the Green's formula yields the equality $(dh_0, dh_0)_{L^2(\Omega_0)} = \langle -\tilde{\Delta} h_0, \bar{h}_0 \rangle$.

*Proof of Proposition* 6. — Let $\{\Omega_k\}$, $k = 1, 2, \ldots$ be an exhaustion of $\Omega$ by relatively compact sub-domains $\Omega_k$ bounded by a finite number of analytic Jordan curves [8, p. 25]. We may assume that the support of $\psi$ is contained in $\Omega_1$. Let $g_k(p, q)$ be the Green's function of $\Omega_k$. Let

$$h_k(p) = \int g_k(p, q) \wedge \psi(q),$$
$$h(p) = \int g(p, q) \wedge \psi(q).$$

By the lemma,

$$\begin{aligned}(dh_k, dh_k)_{L^2(\Omega_k)} &= \langle -\tilde{\Delta} h_k, \bar{h}_k \rangle, \\ &= 2\pi \langle \psi, \bar{h}_k \rangle, \\ &= 2\pi \iint_{\Omega_k \times \Omega_k} g_k(p, q) \psi \otimes \bar{\psi}.\end{aligned}$$

Since $g_k(p, q)$ tends increasingly to $g(p, q)$ we have, for $\psi$, $\psi' \in \overset{2}{\mathfrak{D}}(\Omega)$,

$$\iint g_k \psi \otimes \psi' \to \iint g \psi \otimes \psi'.$$

Hence, if $\tilde{h}_k$ denotes the extension of $h_k$ to $\Omega$ by zero outside $\Omega_k$, $\tilde{h}_k \to h$ in $\overset{0}{\mathfrak{D}}'(\Omega)$ and $\|d\tilde{h}_k\|_{L^2(\Omega)} \leqslant C$, C being a constant independant of $k$. Since $d\tilde{h}_k$ is bounded in $L^2(\Omega)$, there exists a weakly convergent subsequence $\{dh_{k_n}\}$ converging say to

$T \in L^2(\Omega)$. Since $dh_{k_n} \to T$ weakly in $L^2$, $dh_{k_n} \to T$ in $\overset{1}{\mathcal{D}}'(\Omega)$. Since $h_k^{\sim} \to h$ in $\overset{0}{\mathcal{D}}'(\Omega)$, $dh_k^{\sim} \to dh$ in $\overset{1}{\mathcal{D}}'(\Omega)$. Hence $dh_{k_n} \to dh$ in $\overset{1}{\mathcal{D}}'(\Omega)$. Consequently $T = dh$. Since $T \in L^2(\Omega)$, $dh \in L^2(\Omega)$, that is $h \in BL$.

REMARK 4. — $h \in \mathcal{H}(\Omega)$.

## 7. — Green's kernel and the Green's function. Type

THEOREM. — *An open Riemann surface $\Omega$ is of hyperbolic type (in the sense of Definition 1) if and only if $\Omega$ possesses the Green's function in the classical sense and in this case the Green's kernel is equal to the Green's function in the classical sense multiplied by $-1/2\pi$.*

*Proof.* — Suppose that $\Omega$ is hyperbolic. Let $\{\Omega_k\}$ be an exhaustion of $\Omega$ by relatively compact subdomains $\Omega_k$ bounded by a finite number of disjoint analytic Jordan curves. Let $p \in \Omega$. We may suppose that $p \in \Omega_1$. Let $g_{k,p}$ be the Green's function of $\Omega_k$ with pole at $p$. By Proposition 5, $\mathcal{G}_k\delta_p = -\frac{1}{2\pi} g_{k,p}$ where $\mathcal{G}_k$ denotes the Green's operator of $\Omega_k$. By Proposition 4, $\mathcal{G}_k\delta_p^{\sim} \to \mathcal{G}\delta_p$ in $\overset{0}{\mathcal{D}}'(\Omega)$, the convergence being uniform on compact sets not containing $p$. Hence $-\frac{1}{2\pi} g_{k,p}^{\sim} \to \mathcal{G}\delta_p$ uniformly on compact sets not containing $p$. Hence $\Omega$ possesses a Green's function $g_p$ with pole at $p$ in the classical sense, and one has $\mathcal{G}\delta_p = -\frac{1}{2\pi} g_p$, $\mathcal{G}$ denoting the Green's operator of $\Omega$. It follows that the Green's kernel is equal to the Green's function multiplied by $-1/2\pi$.

Suppose conversely that $\Omega$ has the Green's function $g(p, q)$ in the classical sense. Let $\psi \in \overset{2}{\mathcal{D}}(\Omega)$. By Proposition 6

$$h(p) = -\frac{1}{2\pi}\int_\Omega g(p, q) \wedge \psi(q)$$

belongs to BL and one has $\ddot{\Delta} h = \psi$. By Proposition 1, $\Omega$ is of hyperbolic type.

REMARK 5. — The first part of the theorem has been proved for plane domains by Deny-Lions [2, ch. II, Th. 2.1, p. 350]. We may also refer to Weyl [12, § 7, § 8].

REMARK 6. — Another proof of the theorem may be given using the notion of the harmonic measure of the ideal boundary and Remark 4.

## 8. — The Invariance of type under quasi-conformal maps.

We shall show that the type of a Riemann surface is invariant under quasi-conformal maps. This result has been proved by Pfluger [9].

Let $\Omega_1$ and $\Omega_2$ be two open Riemann surfaces. Let $\Phi: \Omega_1 \to \Omega_2$ be a ($C^\infty$) diffeomorphism which is quasi-conformal [8, § 43.4]. Let $\varphi \in \overset{0}{\mathfrak{D}}(\Omega_2)$ and write $\varphi' = \varphi \circ \Phi$. It is easily proved [3, p. 5] that there exists a constant $k > 0$ independent of $\varphi$ such that

$$\frac{1}{k}(d\varphi,\ d\varphi)_{L^2(\Omega_2)} \leqslant (d\varphi',\ d\varphi')_{L^2(\Omega_1)} \leqslant k(d\varphi,\ d\varphi)_{L^2(\Omega_2)}.$$

That is, $\Phi$ induces an isomorphism of $\mathcal{H}_0(\Omega_2)$ onto $\mathcal{H}_0(\Omega_1)$. On the other hand $\Phi$, being a diffeomorphism, induces an isomorphism of $\overset{0}{\mathfrak{D}}{}'(\Omega_2)$ onto $\overset{0}{\mathfrak{D}}{}'(\Omega_1)$. Hence $\mathcal{H}_0(\Omega_1) \to \overset{0}{\mathfrak{D}}{}'(\Omega_1)$ is continuous if and only if $\mathcal{H}_0(\Omega_2) \to \overset{0}{\mathfrak{D}}{}'(\Omega_2)$ is continuous. Hence the type is invariant under $\Phi$.

REMARK 7. — In the above proof we assumed $\Phi$ to be $C^\infty$. If we use Remark 2, it is sufficient to assume $\Phi$ to be $C^1$.

## BIBLIOGRAPHY

[1] R. COURANT and D. HILBERT, *Methoden der Mathematischen Physik II*, Berlin, 1937.

[2] J. DENY and J. L. LIONS, Les espaces de type de Beppo-Levi, *Annales de l'Institut Fourier*, 5, (1953-1954), pp. 305-370.

[3] Mme J. LELONG-FERRAND, *Représentation conforme et transformations à intégrale de Dirichlet bornée*, Paris, 1955.

[4] J. L. LIONS, Lectures on elliptic partial differential equations, *Tata Institute of Fundamental Research*, Bombay, 1957.

[5] J. L. LIONS, Problèmes aux limites en théorie des distributions, *Acta. Math.* 94, (1955), pp. 13-153.
[6] B. MALGRANGE, Existence et approximation des solutions des équations aux dérivées partielles et des équations de convolution, *Annales de l'Institut Fourier*, VI (1955-1956), pp. 221-354.
[7] R. NEVANLINNA, *Uniformisierung*, Berlin, 1953.
[8] A. PFLUGER, *Theorie der Riemannschen Flächen*, Berlin, 1957.
[9] A. PFLUGER, Sur une propriété de l'application quasi-conforme d'une surface de Riemann ouverte, *C.R. Acad. Sci* (Paris), 227 (1948), 25-26.
[10] G. de RHAM, *Variétés différentiables*, Paris, 1955.
[11] L. SCHWARTZ, *Théorie des distributions I*, Paris, 1957.
[12] H. WEYL, The method of orthogonal projection in potential theory, *Duke Math. J.*, 7 (1940), 411-444.

Ann. Inst. Fourier, Grenoble
**11** (1961), 493-514.

# VARIATIONS OF COMPLEX STRUCTURES ON AN OPEN RIEMANN SURFACE

**by Mudumbai S. NARASIMHAN** (Bombay)

## 1. Introduction.

The purpose of this paper is to study the local properties of variations of complex structures on a relatively compact subdomain of an open Riemann surface.

Let M be an open Riemann surface and $M_1$ a relatively compact subdomain of M. Let $\mathcal{S}(t)$ be a family of complex structures on M (or on a neighbourhood of $M_1$) which depends holomorphically on $t$, $t$ being in a neighbourhood $U_1$ of $t_0$ in $\mathbf{C}^m$. We suppose that $\mathcal{S}(t_0)$ is identical with the given structure on M. Consider the family of complex structures $\mathcal{S}(t, M_1)$ induced on $M_1$ by $\mathcal{S}(t)$. The family $\mathcal{S}(t, M_1)$ defines a complex analytic structure on $M_1 \times U_1$; we denote by $\mathcal{S}(M_1 \times U_1)$ the complex analytic manifold (or structure) thus defined. The projection $\pi_1 : M_1 \times U_1 \rightarrow U_1$ defines a family of deformations of complex structures in the sense of Kodaira-Spencer.

We first prove that for every sufficiently small Stein neighbourhood U of $t_0$, $\mathcal{S}(M_1 \times U)$ is a Stein manifold (Theorem 1). We then show that the restriction of the family

$$\pi_1 : \mathcal{S}(M_1 \times U_1) \rightarrow U_1$$

to a sufficiently small neighbourhood U of $t_0$ is complex analytically homeomorphic to the family $\pi: \Omega \to \pi(\Omega) \subset C^m$, where $\Omega$ is an open Stein submanifold of the product complex manifold $M \times \mathbf{C}^m$ and $\pi: M \times \mathbf{C}^m \to \mathbf{C}^m$ is the canonical projection of $M \times \mathbf{C}^m$ onto $\mathbf{C}^m$ (Theorem 2). This result may be viewed as a sort of local triviality (« semi-triviality ») or a local imbedding theorem.

We prove also an analogue of Theorem 2 for differentiable variations of complex structures (Theorem 3).

The proofs use the theory of linear elliptic partial differential equations and some tools from functional analysis.

We now give a rough sketch of the proofs. We show that there exists a sufficiently small neighbourhood $U_2$ of $t_0$ such that any functions which is holomorphic (upto the boundary) on any fibre over a point of $U_2$ can be extended to a holomorphic function on the whole fibre system restricted to $U_2$. From this it follows easily that we can separate points on the fibre system by holomorphic functions and that there exist $(m + 1)$ holomorphic functions which form a local coordinate system at a given point. To prove the holomorph-convexity, we first prove, by considering variations of complex structures on a disc, that the fibre system, restricted to a small Stein neighbourhood of $t_0$, is « locally holomorphically convex ». Then, by solving a problem analogous to the first Cousin problem with the help of currents, the holomorph-convexity is proved.

Once Theorem 1 is proved, Theorem B on Stein manifolds assures the vanishing of certain cohomology groups; we then prove theorem 2, adopting a method of Kodaira-Spencer.

Theorem 3 (differentiable case) is proved by solving the following problem : given Cousin data on $\mathcal{G}(t, \overline{M}_1)$ which depend differentiably on the parameter, to find solutions of the (first) Cousin problem such that the solutions also depend differentiably on the parameter. The proof is inspired by a proof (unpublished) by L. Schwartz of some results concerning Cousin problems on a compact Riemann surface with varying complex structures and by some considerations in Kodaira-Spencer [2].

The author is thankful to Professor L. Schwartz for suggesting the use of Lemma 1, which simplifies the earlier demonstration of the author using power series expansions.

## 2. Statement of the theorems.

Let M be an open Riemann surface. Let $\Theta$ be the holomorphic tangent bundle of M. Let $\mathcal{E}(\Theta \otimes \overline{\Theta}^*)$ denote the space of $C^\infty(0, 1)$ forms with coefficients in $\Theta$, endowed with the natural topology [5]. If $\tilde{\mu} \in \mathcal{E}(\Theta \otimes \overline{\Theta}^*)$ and $z$ a local coordinate system then $\tilde{\mu}$ is of the form $\mu(z)\, d\bar{z} \otimes \partial/\partial z$. If we define $|\tilde{\mu}| = |\mu|$ locally, then $|\tilde{\mu}|$ is intrinsically defined as a function on M. If $\tilde{\mu} \in \mathcal{E}(\Theta \otimes \overline{\Theta}^*)$ with $|\tilde{\mu}| < 1$ then locally the forms $dz + \mu(z)\, d\bar{z}$ define a (1, 0) form for a complex structure and thus $\tilde{\mu}$ defines a complex structure on M.

Let $t_0 \in \mathbf{C}^m$ and $U_0$ be an open set in $\mathbf{C}^m$ containing $t_0$. '$t$' will denote a point in $U_0$.

For our purposes a holomorphic family $\mathcal{S}(t)$ of complex structures on M will be, by definition, a holomorphic function $\tilde{\mu}(t)$ defined in $U_0$ with values in $\mathcal{E}(\Theta \otimes \overline{\Theta}^*)$ such that $|\tilde{\mu}(t)| < 1$ and $\tilde{\mu}(t_0) = 0$. We then have on $M \times U_0$ an almost complex structure defined locally by the forms $dz + \mu(t, z)\, d\bar{z}$, $dt^1, \ldots, dt^m$ where $t^1, \ldots, t^m$ are the coordinate function in $\mathbf{C}^m$. This almost complex structure is integrable since $\tilde{\mu}(z, t)$ is holomorphic in $t$. Hence we have a complex structure on $M \times U_0$ (see also proposition 1). We denote $M \times U_0$ endowed with this complex structure by $\mathcal{S}(M \times U_0)$. The projection $\pi_1 : S(M \times U_0) \to U_0$ is holomorphic and we have a holomorphic familly of deformations of complex structures in the sense of Kodaira-Spencer [2].

If $M_1$ is a subdomain of M and V a neighbourhood of $t_0$ in $\mathbf{C}^m$ with $V \subset U_0$, we denote the manifold $M_1 \times V$ with the complex structure induced from $\mathcal{S}(M \times U_0)$ by $\mathcal{S}(M_1 \times V)$. We denote by $\mathcal{S}(t)$ the complex analytic structure on M defined by $\tilde{\mu}(t)$.

We have

THEOREM 1. — *Let $\mathcal{S}$ be a holomorphic family of complex structures on an open Riemann surface* M. *Let* $M_1$ *be a relatively compact subdomain of* M. *Then there exists a neighbourhood* V *of* $t_0$ *such that for every Stein neighbourhood* U *of* $t_0$ *contained in* V, $\mathcal{S}(M_1 \times U)$ *is a Stein manifold.*

THEOREM 2. — *Let $\mathcal{S}$ be a holomorphic family of complex structures on an open Riemann surface* M *and* $M_1$ *a relatively compact subdomain of* M. *Then there exist a neighbourhood* U *of* $t_0$, *an open Stein submanifold* $\Omega$ *of the product manifold* $M \times \mathbf{C}^m$, *a complex analytic homeomorphism* $\Phi$ *of* $\mathcal{S}(M_1 \times U)$ *onto* $\Omega$ *and a complex analytic homeomorphism* $\varphi$ *of* U *onto* $\pi(\Omega)$ ($\pi$ *denoting the projection* $M \times \mathbf{C}^m \to \mathbf{C}^m$) *such that the following diagram is commutative:*

$$\begin{array}{ccc} \mathcal{S}(M_1 \times U) & \xrightarrow{\Phi} & \Omega \\ \downarrow{\scriptstyle \pi_1} & & \downarrow{\scriptstyle \pi} \\ U & \xrightarrow{\varphi} & \pi(\Omega). \end{array}$$

REMARK. — In Theorems 1 and 2 and as well in Theorem 3, if the boundary of $M_1$ is smooth it is sufficient to assume that the variation is given only upto the boundary of $M_1$.

Let $U_0$ be an open subset in $\mathbf{R}^m$ and $t_0 \in U_0$. A differentiable family of complex structures we mean differentiable function $\tilde{\mu}(t)$ defined in $U_0$ with values in $\mathcal{E}(\Theta \otimes \overline{\Theta}^*)$ such that $|\tilde{\mu}(t)| < 1$ and $\tilde{\mu}(t_0) = 0$. (By differentiable we always mean « indefinitely differentiable ».) For a subdomain $M_1$ of M we denote by $\mathcal{S}(t, M_1)$, $t \in U_0$, the surface $M_1$ endowed with the complex structure defined by $\tilde{\mu}(t)$.

We have then

THEOREM 3. — *Let $\mathcal{S}(t)$ be a family of complex structures on* M *depending differentiably on t, t being in a neighbourhood of* $t_0$ *in* $\mathbf{R}^m$. *Let* $M_1$ *be a relatively compact subdomain of* M. *Then there exist a neighbourhood* U *of* $t_0$ *and a differentiable map* $\Phi$ *of* $M_1 \times U$ *into* M *which maps each fibre* $\mathcal{S}(t, M_1)$, $t \in U$, *biholomorphically into* M.

## 3. Some lemmas in functional analysis and potential theory.

Some of the lemmas stated in this section are more or less well-known. We state them here for convenience of reference.

We denote by $U_0$ an open set in $\mathbf{C}^m$ or $\mathbf{R}^m$ according as we consider holomorphic or differentiable variations. $t_0$ is a point of $U_0$.

Let E and F be two complete barrelled locally convex topological vector spaces. We shall say that a family of continuous linear operators $T_t$: $E \to F$, $t \in U_0$, depends holomorphically (resp. differentiably) on $t \in U_0$ if $t \to T_t$ is a holomorphic (resp. differentiable) function of $U_0$ with values in $\mathfrak{L}_s(E, F)$, where $\mathfrak{L}_s(E, F)$ denotes the space of continuous linear operators of E into F endowed with the topology simple convergence. We remark that if $T_t$ depends holomorphically (resp. differentiably, differentiably) on $t$ and $f(t)$ is a holomorphic (resp. differentiable) function with values in E then $t \to T_t f(t)$ is a holomorphic (resp. differentiable) function with values in F.

LEMMA 1. — *Let* E *and* F *be two Banach spaces and* $T_t$: $E \to F$ *depend holomorphically (resp. differentiably) on* $t$. *Assume that* $T_{t_0}$ *is an isomorphism. Then there exists a neighbourhood* $U'_0$ *of* $t_0$ *such that* $T_t$ *is an isomorphism for each* $t \in U'_0$ *and the operators* $T_t^{-1}: F \to E$ *depend holomorphically (resp. differentiably) on* $t \in U'_0$.

This lemma is a special case of implicit function theorem in Banach spaces and is proved easily.

LEMMA 2. — *Let* E *and* F *be two Banach spaces and* $T_t$: $E \to F$ *depend holomorphically (resp. differentiably) on* $t$. *Assume that* $T_{t_0}$ *admits of a right inverse. Then there exists a neighbourhood* $U'_0$ *of* $t_0$ *such that for* $t \in U'_0$, $T_t$ *admits of a right inverse depending holomorphically (resp. differentiably) on* $t$.

*Proof.* — We recall that a right inverse for $T_{t_0}$ is a continuous linear map $S_{t_0}$: $F \to E$ such that $T_{t_0} \circ S_{t_0}$ is the identity map of F. Now we apply Lemma 1 to the operators

$$T_t \mid S_{t_0}(F) : S_{t_0}(F) \to F$$

and Lemma 2 follows.

Let D be a relatively compact open subset of $\mathbf{C}$. Let $\alpha$ be a fixed real number with $0 < \alpha < 1$. Let $f$ be a complex valued function satisfying a Hölder condition of order $\alpha$ on $\overline{D}$. Put

$$\|f\|_{0,\alpha,D} = \sup_{\overline{D}} |f| + \sup_{\substack{z_1, z_2 \in \overline{D} \\ z_1 \neq z_2}} \frac{|f(z_1) - f(z_2)|}{|z_1 - z_2|^{\alpha}}.$$

We denote the space of these functions by $H_\alpha(D)$.

If $f$ is a function which is once differentiable such that its partial derivatives satisfy in $\overline{D}$ a Hölder condition of order $\alpha$ put

$$||f||_{1,\alpha,D} = \operatorname*{Sup}_{\overline{D}} |f| + \left\|\frac{\partial f}{\partial \bar{z}}\right\|_{0,\alpha,D} + \left\|\frac{\partial f}{\partial z}\right\|_{0,\alpha,D}$$

Let $H_{1,\alpha}(D)$ denote the space of such functions.

Let now D be a disc $|z| < R$, $0 < R < \infty$ in the plane. The operator $\frac{\partial}{\partial \bar{z}}$ is a continuous linear operator from the Banach space $H_{1,\alpha}(D)$ (with the norm $||f||_{1,\alpha}$) to the Banach space $H_\alpha(D)$ (with the norm $||f||_{0,\alpha}$).

LEMMA 3. — *Let* D *be a disc in the plane. The operator*

$$\frac{\partial}{\partial \bar{z}} : H_{1,\alpha}(D) \to H_\alpha(D)$$

*admits of a right inverse.*

This lemma is classical. For instance convolution with $\frac{1}{\pi z}$ yields a right inverse [1].

Let $M_0$ be a relatively compact subdomain of an open Riemann surface M such that $M_0$ is bounded by a finite number of disjoint analytic Jordan curves. We shall say, for brevity, that $M_0$ has an analytic boundary. We shall denote by $\partial M_0$ boundary of $M_0$ in M.

Let $D_1, \ldots, D_k, D_{k+1}, \ldots, D_n$ be a covering of $\overline{M}_0$ by coordinate discs $D_i$ in M with $\overline{D}_i$ compact and contained in a coordinate disc such that the following conditions are satisfied:

i) $\overline{D}_i$ is contained in $M_0$ for $i = 1, \ldots, k$

ii) if $z_j$ is the coordinate function in $D_j$ mapping $D_j$ onto $|z| < \varepsilon$, then $z_j$ maps for $j = k+1, \ldots, n$, $D_j \cap \overline{M}_0$ onto the « semi-disc » $\{|z| < \varepsilon, \operatorname{Im} z > 0\}$ and $D_j \cap \partial M_0$ onto $\{-\varepsilon < Rlz < \varepsilon\}$. Let $D'_i$ denote the covering of $\overline{M}_0$ formed by $D_1, \ldots, D_k$, $D_{k+1} \cap \overline{M}_0, \ldots, D_n \cap M_0$. Let $\{D''_i\}$ be a shrinking of the covering $\{D'_i\}$.

Let $H_{1,\alpha}(M_0)$ denote the Banach space of complex valued functions in $\overline{M}_0$ which are once differentiable in $M_0$ and whose first partial derivatives satisfy a Hölder condition of order $\alpha$

in every compact subset contained in a coordinate neighbourhood of $\overline{M}_0$ (e.g. $D_i'$) with the norm

$$\|f\|_{1,\alpha} = \operatorname*{Sup}_{i=1,\dots,n} \|f\|_{1,\alpha,D_i''}.$$

Let $\overset{0,1}{H}_\alpha(M_0)$ denote the space of (0, 1) forms whose coefficients satisfy a Hölder condition of order $\alpha$ in every compact set contained in a coordinate neighbourhood of $\overline{M}_0$. If $f \in \overset{0,1}{H}_\alpha(M_0)$ and $f = f_i\, d\bar{z}^i$ in $D'$ ($z^i$ being the coordinate function in $D_i'$) define

$$\|f\|_{0,\alpha,M_0} = \operatorname*{Sup}_{i} \|f_i\|_{0,\alpha,D_i''};$$

with this norm $\overset{0,1}{H}_\alpha(M_0)$ becomes a Banach space.

Lemma 4. — *Let* $M_0$ *be a relatively compact subdomain of* M *with analytic boundary. Then the operator*

$$d_{\bar{z}} :\ H_{1,\alpha}(M_0) \to \overset{0,1}{H}_\alpha(M_0)$$

*admits of a right inverse.*

*Proof.* — We give a sketch of the proof of this lemma. Let $M_1$ be a relatively compact subdomain of M, with analytic boundary, containing $\bigcup_{i=1,\dots,n} \overline{D}_i$. We first remark that we can find a continuous linear map $\rho : \overset{0,1}{H}_\alpha(M_0) \to \overset{0,1}{H}_\alpha(M_1)$ such that $\gamma \circ \rho =$ identity map of $\overset{0,1}{H}_\alpha(M_0)$, where $\gamma : \overset{0,1}{H}_\alpha(M_1) \to \overset{0,1}{H}_\alpha(M_0)$ denotes the restriction map. [The question being local at the boundary, locally the extension is given by reflection at the $x$-axis. For details see e.g. [4, Th. 2. 4]]. On $M_1 \times M_1$ there exists (H. Behnke-K. Stein, Math. Ann. 120, p. 436) a meromorphic differential $K(z, d\zeta)$, holomorphic for $z \neq \zeta$ such that in a coordinate disc around $z = \zeta$ we have,

$$K(z, d\zeta) = \left\{\frac{-1}{4\pi(z-\zeta)} + \text{regular function}\right\} d\zeta.$$

We may then estimate the potential

$$T_1\tilde{f} = 2i\int_{M_1} K(z, d\zeta) \wedge \tilde{f}(\zeta), \qquad \tilde{f} \in \overset{0,1}{H}_\alpha(M_1)$$

on compact subsets of $D_i$ using the estimate on a disc

for the potential with the kernel $\frac{1}{\pi(z-\zeta)}$ [1]. If $f \in \overset{0,1}{H}_\alpha(M_0)$ let $Tf$ denote the restriction of $T_1(\rho(f))$ to $M_0$. Then $d_{\bar z}Tf = f$ and

$$||Tf||_{1,\alpha,M_0} \leqslant C_1||\rho(f)||_{0,\alpha,M_1} \leqslant C_2||f||_{0,\alpha,M_0}$$

with positive constants $C_1$ and $C_2$. This proves Lemma 4.

The next lemma will be required only for the holomorphic tangent bundle of M. But we shall prove it for a general holomorphic line bundle.

Let L be a holomorphic line bundle on M. Let $H_{1,\alpha}(M_0, L)$ denote the Banach space of sections of L in $\overline{M}_0$ which are once differentiable in $M_0$ and whose first partial derivatives satisfy a Hölder condition of order $\alpha$. Let $\overset{0,1}{H}_\alpha(M_0, L)$ denote the Banach space of Hölder continuous (0, 1) forms in $M_0$ with coefficients in L (we introduce norms on $H_{1,\alpha}(M_0, L)$ and $\overset{0,1}{H}_\alpha(L)$ as on $H_{1,\alpha}(M_0)$ and $\overset{0,1}{H}_\alpha(M_0)$.

Lemma 5. — *Let* L *be a holomorphic line bundle on* M. *Let* $M_0$ *be a relatively compact subdomain with analytic boundary. Then the operator*

$$d_{\bar z} : H_{1,\alpha}(M_0, L) \to \overset{0,1}{H}_\alpha(M_0, L)$$

*admits of a right inverse.*

*Proof.* — Since M is an open Riemann surface, every holomorphic line bundle on M is holomorphically trivial. This follows for example from the exact sequence.

$$H^1(M, O) \to H^1(M, O^*) \to H^2(M, Z)$$

remarking that $H^1(M, O) = 0$, $H^2(M, Z) = 0$. (Here O denotes the sheaf of germs of holomorphic functions and $O^*$ the sheaf of germs of non-vanishing holomorphic functions.) Since L is holomorphically trivial on M there exist topological isomorphisms

$$\psi_1 : H_{1,\alpha}(M_0, L) \to H_{1,\alpha}(M_0)$$
$$\psi_2 : \overset{0,1}{H}_\alpha(M_0, L) \to \overset{0,1}{H}_\alpha(M_0)$$

such that the following diagram is commutative:

$$\begin{array}{ccc} H_{1,\alpha}(M_0,\ L) & \xrightarrow{\psi_1} & H_{1,\alpha}(M_0) \\ \downarrow d_{\bar z} & & \downarrow d_{\bar z} \\ \overset{0,1}{H}_\alpha(M_0,\ L) & \xrightarrow{\psi_2} & \overset{0,1}{H}_\alpha(M_0) \end{array}$$

Since $d_{\bar z}: H_{1,\alpha}(M_0) \to \overset{0,1}{H}_\alpha(M_0)$ admits of a right inverse it follows thats $d_{\bar z}: H_{1,\alpha}(M_0,\ L) \to \overset{0,1}{H}_\alpha(M_0,\ L)$ admits of a right inverse.

### 4. Variation of complex structures on a disc.

PROPOSITION 1. — *Let D be a disc in the plane. Let $\mu(t) = \mu(z, t)$ be a holomorphic function defined in a neighbourhood of $t_0$ in $\mathbf{C}^m$ with values in $H_\alpha(D)$ with $\mu(t_0) = 0$. Then there exist a neighbourhood $U'$ of $t_0$ and a $C^1$ function $\zeta(z, t)$ defined in $D \times U'$ such that*

i) $$\begin{cases} \dfrac{\partial\zeta(z,t)}{\partial\bar z} - \mu(z,t)\dfrac{\partial\zeta(z,t)}{\partial z} = 0, \\[2mm] \dfrac{\partial\zeta(z,t)}{\partial\bar t^i} = 0, \quad i = 1, \ldots, m. \end{cases}$$

ii) *there exist positive constants $K_1$, and $K_2$ such that one has* $K_1|z_1 - z_2| \leqslant |\zeta(z_1,\ t) - \zeta(z_2,\ t)| \leqslant K_2|z_1 - z_2|$ for $z_1,\ z_2 \in \bar D$ *and all $t \in U'$.*

iii) *If* $F(t) = F(z,\ t) = \dfrac{1}{\zeta(z,\ t) - \zeta(z_0,\ t)}$, $z_0 \in D$, *the function $t \to F(t)$ is a holomorphic function in $U'$ with values in $\mathcal{D}'(D)$, where $\mathcal{D}'(D)$ denotes the space of distributions in D; moreover for each fixed t, $F(z, t)$ is holomorphic outside $z_0$ for the complex structure defined by $dz + \mu(z, t)\,d\bar z$ ($|\mu| < 1$).*

*Proof.* — There exists a constant $C_1 > 0$ such that for $f,\ g \in H_\alpha(D)$ one has $\|fg\|_{0,\alpha} \leqslant C_1\|f\|_{0,\alpha}\|g\|_{0,\alpha}$. Hence the operator of multiplication by $\mu(z,\ t)$ is a holomorphic function of $t$ with values in $\mathcal{L}_s(H_\alpha,\ H_\alpha)$. It follows that the operators

$$T_t = \frac{\partial}{\partial\bar z} - \mu(z,\ t)\frac{\partial}{\partial z} : H_{1,\alpha}(D) \to H_\alpha(D)$$

depend holomorphically on $t$. Now $T_{t_0} = \dfrac{\partial}{\partial\bar z}\cdot$ By Lemma 3

$T_{t_0}$ admits a right inverse. Hence by lemma 2 there exists a neighbourhood $U''$ of $t_0$ and continuous linear operators

$$S_t : H_\alpha \to H_{1,\alpha}$$

depending holomorphically on $t \in U''$ such that $T_t \circ S_t =$ Identity map of $H_\alpha$. Now $\mu(t)$ is a holomorphic function with values in $H_\alpha$. Hence $f(t) = S_t(\mu(t))$ is a holomorphic function with values in $H_{1,\alpha}$. Let

$$\zeta(z, t) = z + f(z, t).$$

$\zeta(z, t)$ is of class $C^1$. Moreover

$$\begin{aligned} \frac{\partial \zeta}{\partial \bar{z}} = \frac{\partial f(z, t)}{\partial \bar{z}} &= \mu(z, t)\frac{\partial f}{\partial z} + \mu(z, t) \\ &= \mu(z, t)\left(1 + \frac{\partial f}{\partial z}\right) \\ &= \mu(z, t)\frac{\partial \zeta}{\partial z}, \end{aligned}$$

so that $\zeta(z, t)$ satisfies

$$\text{i)} \quad \begin{cases} \dfrac{\partial \zeta(z, t)}{\partial \bar{z}} - \mu(z, t)\dfrac{\partial \zeta}{\partial z} = 0, \\ \dfrac{\partial \zeta(z, t)}{\partial \bar{t}^i} = 0, \quad i = 1, \ldots, m. \end{cases}$$

To prove ii) we remark that there exists a constant $k > 0$ (depending only on D) such that for each $f \in H_{1,\alpha}$ one has

$$|f(z_1) - (z_2)| \leqslant k|z_1 - z_2| \, ||f||_{1,\alpha}, \qquad z_1, z_2 \in \bar{D}.$$

(This is proved easily applying the mean value theorem.) Since $f(t_0) = 0$ we can choose a relatively compact neighbourhood $U'$ of $t_0$ with $\overline{U'} \subset U''$ such that for $t \in U'$, $||f(t)||_{1,\alpha} \leqslant \frac{\varepsilon}{k}$, given $\varepsilon$ with $0 < \varepsilon < 1$. It is evident that there exists a constant $K_2$ such that

$$|\zeta(z_1, t) - \zeta(z_2, t)| \leqslant K_2|z_1 - z_2|, \qquad t \in U', \qquad z_1, z_2 \in \bar{D}.$$

On the other hand

$$\begin{aligned} |\zeta(z_1, t) - \zeta(z_2, t)| &= |\{z_1 + f(z_1, t)\} - \{z_2 + f(z_2, t)\}| \\ &\geqslant |z_1 - z_2| - |f(z_1, t) - f(z_2, t)| \\ &\geqslant (1 - \varepsilon)|z_1 - z_2|. \end{aligned}$$

This completes the proof of ii).

To prove iii), we note that for $t$ fixed $1/\zeta(z, t) - \zeta(z_0, t)$ is a locally summable function in D (see ii); and since

$$|F(z, t)| \leqslant K_1^{-1}/|z - z_0|,$$

$t \in U'$, we see, by Lebesgue's dominated convergence theorem, that $t \to F(t)$ is a continuous function with values in $\mathfrak{D}'(D)$. To prove that $F(t)$ is a holomorphic function with values in $\mathfrak{D}'(D)$ it is sufficient to prove that $h(t) = \langle F(t), \varphi\rangle$ is a holomorphic function of $t$ for each $\varphi \in \mathfrak{D}(D)$. [$\mathfrak{D}(D)$ denotes the space of $C^\infty$ functions with compact supports in D; $\langle F(t), \varphi\rangle$ denotes the scalar product between $F(t)$ and $\varphi$]. As was noted earlier $h(t)$ is a continuous function. Let $t_1 = (t_1^1, \ldots, t_1^m) \in U'$. We shall show that $h(t^1, t_1^2, \ldots, t_1^m)$ is differentiable at $t_1^1$ as a function of $t^1$.

Let

$$\psi(t^1) = \{h(t^1, t_1^2, \ldots, t_1^m) - h(t_1^1, t_1^2, \ldots, t_1^m)\}/(t^1 - t_1^1).$$

Then

$$\psi(t^1) = \int_K \frac{1}{t^1 - t_1^1} \times$$
$$\left\{\frac{[\zeta(z, t_1) - \zeta(z_0, t_1)] - [\zeta(z, t^1, t_1^2, \ldots, t_1^m) - \zeta(z_0, t^1, \ldots, t_1^m)]}{[\zeta(z, t^1, t_1^2, \ldots, t_1^m) - \zeta(z_0, t^1, \ldots, t_1^m)] - [\zeta(z, t_1) - \zeta(z_0, t_1)]}\right\} \varphi\, dx\, dy$$

where K is the support of $\varphi$.

We assert that there exists a constant $K_3$ such that for $t^1$ in a sufficiently small neighbourhood of $t_1^1$ we have

(A)
$$\left|\frac{[\zeta(z, t_1^1, \ldots, t_1^m) - \zeta(z_0, t_1^1, \ldots, t_1^m)] - [\zeta(z, t^1, t_1^2, \ldots, t_1^m) - \zeta(z_0, t', t_1^2, \ldots, t_1^m)]}{t^1 - t_1^1}\right| \leqslant K_3/|z - z_0|.$$

In fact, consider the function with values in $H_{1,\alpha}$ defined in a neighbourhood of $t_1^1$:

$$g(t^1) = \begin{cases} \dfrac{\zeta(t_1^1, \ldots, t_1^m) - \zeta(t^1, t_2^2, \ldots, t_1^m)}{t^1 - t_1^1} & \text{for } t^1 \neq t_1^1 \\ \left\{\dfrac{d}{dt^1}\zeta(t^1, t_1^2, \ldots, t_1^m)\right\}_{t^1 = t_1^1} & \text{for } t^1 = t_1^1 \end{cases}$$

Since $\zeta(t)$ is a holomorphic function with values in $H_{1,\alpha}$,

$g(t^1)$ is a continuous function and hence in a neighbourhood of $t_1^1$, $\|g(t^1)\|_{1,\alpha} \leqslant K_4$. Using the inequality

$$|f(z_1) - f(z_2)| \leqslant k|z_1 - z_2| \, \|f\|_{1,\alpha}$$

we obtain (A). From (A) and the first inequality in ii) we see that the integrand is majorised by $K_5/|z - z_0|$ for all $t^1$ in a sufficiently small neighbourhood of $t_1^1$. By Lebesgue's theorem we see that $\lim_{t_1 \to t_1^1} \psi(t^1)$ exists and is equal to

$$-\int_K \frac{\frac{d}{dt^1}(\zeta(t^1, \ldots, t_1^m)(z))_{t^1=t_1^1} - \left(\frac{d}{dt^1}\zeta(t^1, \ldots, t_1^m)(z_0)\right)_{t^1=t_1^1}}{\{\zeta(z, t_1^1, \ldots, t_1^m) - \zeta(z_0, t_1^1, \ldots, t_1^m)\}^2} \varphi \, dx \, dy.$$

Similary we show that the other derivatives exist. This proves that $h(t)$ is holomorphic.

From the first inequality in ii) we see that $\zeta(z, t) - \zeta(z_0, t) \neq 0$ for $z \neq z_0$, $t \in U'$. The second assertion in iii) follows immediately from this fact. This completes the proof of Proposition 1.

Remark 2. — Using i) and ii) we can show easily that if U is a polydisc contained in U′ the map $(t, z) \to (t, \zeta(t, z))$ maps $\mathcal{S}(U \times D)$ (endowed with the complex structure defined by $dz + \mu(z, t)\, d\bar{z}, dt^1, \ldots, dt^m$, $|\mu(z, t)| < 1$), biholomorphically onto a bounded domain of holomorphy in $\mathbf{C}^{m+1}$. Proposition 1 is also valid if we replace the disc by a bounded plane domain with a smooth boundary. Thus Theorems 1 and 2 are immediate consequences of Proposition 1 in the case of plane domains.

## 5. Elementary kernels for elliptic differential operators depending holomorphically on a parameter.

Let M be an open Riemann surface and let $\overset{p,q}{\mathfrak{D}}(M)$, $\overset{p,q}{\mathcal{E}}(M)$, $\overset{p,q}{\mathfrak{D}'}(M)$, $\overset{p,q}{\mathcal{E}'}(M)$ $(p = 0, 1, q = 0, 1)$ denote the space of $C^\infty$ forms of type $(p, q)$ with compact supports, $C^\infty$ forms of type $(p, q)$, currents of type $(p, q)$ and currents of type $(p, q)$ with compact supports respectively ,each endowed with the usual topology [5].

Let $\mathcal{S}(t, \mathrm{M})$ be a family of complex structures depending holomorphically on $t$. Define $\mathrm{T}_t : \overset{0,0}{\mathcal{D}}{}'(\mathrm{M}) \to \overset{0,1}{\mathcal{D}}{}'(\mathrm{M})$ by

$$\mathrm{T}_t f = d_{\bar z} f - \langle \tilde\mu(t),\ d_z f \rangle, \qquad f \in \overset{0,0}{\mathcal{D}}{}'(\mathrm{M})$$

where $\langle \tilde\mu(t),\ d_z f\rangle$ denotes the current of type $(0, 1)$ obtained by contracting $\tilde\mu(t)$ (which is of type $(-1, 1)$) and $d_z f$ (which is of type $(1, 0)$). We remark that a function $f$ is holomorphic for the structure $\mathcal{S}(t, \mathrm{M})$ if and only if $\mathrm{T}_t f = 0$. A function $f$ defined on $\mathrm{U}_0 \times \mathrm{M}$ is holomorphic for the structure $\mathcal{S}(\mathrm{U}_0 \times \mathrm{M})$ if and only if it satisfies the system of differential equations:

$$\begin{cases} \mathrm{T}_t f(z,\ t) = 0, \\ \dfrac{\partial f(z,\ t)}{\partial \bar t^i} = 0, \quad i = 1, \ldots, m. \end{cases}$$

Proposition 2. — *Let* $\mathrm{M}_0$ *be a relatively compact subdomain of* M, *with analytic boundary. Then there exists a neighbourhood* $\mathrm{U}_3$ *of* $t_0$ *and for* $t \in \mathrm{U}_3$ *continuous linear operators* $\mathrm{S}_t : \overset{0,1}{\mathcal{E}}(\mathrm{M}_0) \to \overset{0,0}{\mathcal{D}}{}'(\mathrm{M}_0)$ *depending* holomorphically *on* $t$ *such that for* $f \in \overset{0,1}{\mathcal{E}}{}'(\mathrm{M}_0)$ *one has* $\mathrm{T}_t \mathrm{S}_t f = f$ $(\mathrm{T}_t = d_{\bar z} - \langle \tilde\mu,\ d_z \rangle)$.

*Proof.* — Let $\mathrm{H}_{1,\alpha}(\mathrm{M}_0)$ and $\overset{0,1}{\mathrm{H}}_\alpha(\mathrm{M}_0)$ have the meaning given in § 3. By Lemma 4, $\mathrm{T}_{t_0} : \mathrm{H}_{1,\alpha} \to \overset{0\ 1}{\mathrm{H}}_\alpha$ has a right inverse. Let $\mathrm{S}_t : \overset{0,1}{\mathrm{H}}_\alpha \to \mathrm{H}_{1,\alpha}$ be right inverses defined in a neighbourhood $\mathrm{U}_3$ of $t_0$ depending holomorphically on $t$ (Lemma 2). $\mathrm{S}_t$ maps $\overset{0,1}{\mathcal{D}}(\mathrm{M}_0)$ into $\mathrm{H}_{1,\alpha}$. We shall show that $\mathrm{S}_t : \overset{0,1}{\mathcal{D}} \to \mathrm{H}_{1,\alpha}$ can be extended to a continuous linear map, still denoted by $\mathrm{S}_t$, of $\overset{0,1}{\mathcal{E}}{}'$ into $\overset{0,0}{\mathcal{D}}{}'$ and $\mathrm{S}_t : \overset{0,1}{\mathcal{E}}{}' \to \overset{0\,0}{\mathcal{D}}{}'$ depends holomorphically on $t$. This will prove Proposition 2, as is easy to see.

Now each $\mathrm{T}_t$ is a linear elliptic operator. By the hypo-ellipticity of $\mathrm{T}_t$, $\mathrm{S}_t$ maps $\overset{0,1}{\mathcal{D}}$ into $\overset{0,0}{\mathcal{E}}$ and $\mathrm{S}_t : \overset{0,1}{\mathcal{D}} \to \overset{0,0}{\mathcal{E}}$ is continuous by Banach's closed graph theorem. We prove that $\mathrm{S}_t : \overset{0,1}{\mathcal{D}} \to \overset{0,0}{\mathcal{E}}$ depends holomorphically on $t$. Let $\varphi \in \overset{0,1}{\mathcal{D}}$. Then the current $\mathrm{F}(z,\ t) = \mathrm{S}_t \varphi$ satisfies the system of differential equations

$$\begin{cases} \mathrm{T}_t \mathrm{F} = \varphi \\ \dfrac{\partial \mathrm{F}}{\partial \bar t^i} = 0, \qquad i = 1, \ldots, m. \end{cases}$$

Since this system is elliptic, $F(z, t)$ is a $C^\infty$ function in $M_0 \times U_3$. It follows that $t \rightarrow S_t\varphi$ is a holomorphic function with values in $\overset{0,0}{\mathcal{E}}$.

Let $T'_t : \overset{1,0}{\mathcal{D}'} \rightarrow \overset{1,1}{\mathcal{D}'}$ be the transpose of the differential operator $T_t$. Then $T_t$ is a linear elliptic differential operator with $C^\infty$ coefficients. Let $S'_t : \overset{1,1}{\mathcal{E}'} \rightarrow \overset{1,0}{\mathcal{D}'}$ be the transpose of $S_t : \overset{0,1}{\mathcal{D}} \rightarrow \overset{0,0}{\mathcal{E}}$. Then $S_t : \overset{1,1}{\mathcal{E}'} \rightarrow \overset{1,0}{\mathcal{D}'}$ depends holomorphically on $t$. By the hypo-ellipticity of $S'_t$, $S'_t$ maps $\overset{1,1}{\mathcal{D}}$ into $\overset{1,1}{\mathcal{E}}$ and $S_t : \overset{0,1}{\mathcal{D}} \rightarrow \overset{1,0}{\mathcal{E}}$ is continuous. As we proved for $S_t$, we prove, using the hypo-ellipticity of the system $\left\{S'_t, \frac{\partial}{\partial \bar{t}^1}, \dots \frac{\partial^t}{\partial \bar{t}^m}\right\}$ that $S_t : \overset{1,1}{\mathcal{D}} \rightarrow \overset{1,0}{\mathcal{E}}$ depends holomorphically on $t$. By taking transposes we obtain $S_t : \overset{0,1}{\mathcal{E}'} \rightarrow \overset{0,0}{\mathcal{D}'}$ depending holomorphically on $t$ and coinciding on $\overset{0,1}{\mathcal{D}}$ with the $S_t$ originally given.

This proves Proposition 2.

## 6. A result on the prolongation of holomorphic functions.

Proposition 3. — *Let* $M_0$ *be a relatively compact subdomain of* M *with analytic boundary. Then there exists a neighbourhood* $U_2$ *of* $t_0$ *with the following property: if* $f(z)$ *is a function which is holomorphic for the structure* $\mathcal{S}(t_1)$ *in* $\overline{M}_1$, $t_1 \in U_2$, *then there exists a function* $F(z, t)$ *in* $M_0 \times U_2$ *which is holomorphic for the structure* $\mathcal{S}(M_0 \times U_2)$ *such that* $F(z, t_1) = f(z)$.

*Proof.* — Let $S_t : \overset{0,1}{H}_\alpha(M_0) \rightarrow H_{1,\alpha}(M_0)$ be right inverses for $T_t$ depending holomorphically on $t \in U_2$. Now $f \in H_{1,\alpha}$. Define

$$F(z, t) = f - S_t T_{t_1} f, \quad (t \in U_2).$$

We then have

$$T_t(f - S_t T_t f) = T_t f - T_t S_t T_t f = T_t f - T_t f = 0;$$

since $T_{t_1}f = 0$, $F(z, t_1) = f(z)$. $F(z, t)$ satisfies the system of differential equations

$$\begin{cases} T_t F(z, t) = 0, \\ \dfrac{\partial F}{\partial \bar{t}^i} = 0, \qquad i = 1, \ldots, m. \end{cases}$$

Hence $F(z, t)$ is holomorphic for the structure $\mathcal{S}(M_0 \times U_2)$.

## 7. Proof of Theorem 1.

We now proceed to prove Theorem 1. Let $M_0$ be a relatively compact sub-domain of M with analytic boundary such that $\overline{M}_1 \subset M_0$. Let $O_1, \ldots, O_i, \ldots, O_k$ be a finite number of coordinate discs for the structure $\mathcal{S}(t_0)$ with $\overline{O}_i \subset M_0$ and $\cup_i O_i \supset \overline{M}_1$. Let $z^i$ be the coordinate function in $O_i$. Then $\tilde{\mu} = \mu_i \, d\bar{z}^i \otimes \dfrac{\partial}{\partial z^i}$. Let $V_i$, $i = 1, \ldots, k$, be neighbourhoods of $t_0$ such that functions $\zeta_i(z^i, t)$, $z^i \in O_i$, $t \in V_i$ can be defined satisfying conditions i), ii), iii) of Proposition 1. (By an obvious abuse of notation we use the letter $z^i$ to denote a point on the Riemann surface and as well its image by the coordinate function $z^i$.) Let $U_3$ and $U_2$ be neighbourhoods of $t_0$ given in Proposition 2 and 3. Let V and V′ relatively compact neghbourhoods of $t_0$ such that $\overline{V} \subset V'$ and $\overline{V}' \subset \bigcap_{i=1,\ldots,k} V_i \cap U_2 \cap U_3$. Let U be a Stein neighbourhood of $t_0$ contained in V.

We first show that holomorphic functions on $\mathcal{S}(M_1 \times U)$ separate points. Since $(t^1, \ldots, t^m)$ are holomorphic functions on $\mathcal{S}(U \times M_1)$ we have only to consider the case when the points are on the same fibre. Let then $(z_1, t_1)$, $(z_2, t_1)$ be two points $t_1 \in U$, $z_1, z_2 \in M_1$, $z_1 \neq z_2$. Now there exists a function $f(z)$ holomorphic in $\overline{M}_0$ for $\mathcal{S}(t_1)$ with $f(z_1) \neq f(z_2)$. [This is shown, for example, by taking an open set slightly larger than $M_0$ and using the fact every open Riemann surface is a Stein manifold.] By Proposition 3 there exists a function $F(z, t)$ holomorphic for the structure $\mathcal{S}(M_0 \times U)$ such that $F(z_1, t_1) = f(z)$. Hence $F(z_1, t_1) \neq F(z_2, t_1)$.

Next let $(z_1, t_1)$, $z_1 \in M_1$, $t_1 \in U$ be a point in $M_1 \times U$. We shall show that there exist $(m + 1)$ functions in $M_1 \times U$ which are holomorphic on $\mathcal{S}(M_1 \times U)$ and which form a local coordinate system at $(z_1, t_1)$. Let $f(z)$ be a function holomorphic for $\mathcal{S}(t_1)$ in $\overline{M}_0$ which forms a local coordinate system at $z_1$ in $M_0$. Let $F(z, t)$ be an extension of $f(z)$ to $M_0 \times U$ as a function holomorphic for the structure $\mathcal{S}(M_0 \times U)$. Suppose $z_1 \in O_i$. With respect to the coordinate system $(z^i, t^1, \ldots, t^m)$ the Jacobian of $(F, t^1, .., t^m)$ at $(z_1, t_1)$ is

$$\left|\left[\frac{\partial}{\partial z^i} F(z, t)\right]_{(z_1, t_1)}\right|^2 (1 - |\mu_i(z, t)|^2)$$

or

$$\left|\frac{\partial f(z)}{\partial z^i}\right|^2 (1 - |\mu_i(z, t)|^2).$$

But $\left(\frac{\partial f}{\partial z^i}\right)_{z^i = z_1} \neq 0$; for if it were zero, then $\frac{\partial f}{\partial \bar{z}^i} = \mu_i \frac{\partial f}{\partial z^i}$ would be zero so that the Jacobian of $f$ at $z_1$ would be zero. Thus the Jacobian of $(F, t^1, .., t^m)$ at $(z_1, t_1)$ is different from zero.

Finally we show that given infinite discrete set of points $\{z_n, t_n\}$, $z_n \in M_1$, $t_n \in U$ there exists a holomorphic function $\Psi$ on $\mathcal{S}(M_1 \times U)$ such that the sequence $\{\Psi(z_n, t_n)\}$ is not bounded. Now either the sequence $\{t_n\}$ contains an infinite discrete subset or the sequence $\{z_n\}$ contains an infinite discrete subset. If $\{t_n\}$ contains an infinite discrete subset we can find, since U is Stein, a holomorphic function $F(t)$ in U such that $\{F(t_n)\}$ is not bounded. Then $\Psi(z, t) = F(t)$ is holomorphic on $\mathcal{S}(M_1 \times U)$ and $\Psi(z_n, t_n)$ is not bounded. If $\{z_n\}$ contains an infinite discrete subsequence $\{z_k\}$ let $z_0 \in \overline{M}_1 \subset M_0$ be an adherent point of $\{z_k\}$, $z_0 \notin M_1$. Suppose $z_0 \in O_i$. Consider the currents of degree 0, $F(t) = 1/\zeta_i(z^i, t) - \zeta_i(z_0, t)$ in $O_i$. Since $F(t)$ is a holomorphic function with values $\overset{0}{\mathcal{D}}{}'(O_i)$ and $T_t : \overset{0}{\mathcal{D}}{}'(O_i) \to \overset{0,1}{\mathcal{D}}{}'(O_i)$ depends homorphically on $t$, $T_t F(t)$ is a holomorphic function $t$ with values in $\overset{0,1}{\mathcal{D}}{}'(O_i)$. But the supports of $T_t F(t)$ are at $z_0$ and hence $T_t F(t)$ is a holomorphic function with values in $\overset{0,1}{\mathcal{E}}{}'(M_0)$. Let $S_t$ be the operators given by Proposition 2, in $U_3$. Let $\Psi(t) = S_t(T_t F(t))$. Then

$$\begin{cases} T_t \Psi(t) = (T_t F(t)), \\ \dfrac{\partial \Psi(t)}{\partial \bar{t}^i} = 0, \quad i = 1, \ldots, m. \end{cases}$$

Since $z_0 \notin M_1$, $\Psi(t)$ defines a function in $M_1 \times U$ satisfying in $M_1 \times U$

$$\begin{cases} T_t\Psi(z, t) = 0, \\ \dfrac{\partial}{\partial \bar{t}^i}\Psi(z, t) = 0, \quad i = 1, \ldots, m; \end{cases}$$

that is $\Psi(z, t)$ is holomorphic on $\mathcal{G}(M_1 \times U)$. It remains to show that $\{\Psi(z_k, t_k)\}$ is not bounded. Let $O_i'$ be a relatively compact neighbourhood of $z_0$ such that $\overline{O_i'} \subset O_i$. We may suppose that all $z_k$ belong to $O_i'$. On $O_i \times V'$ the currents $G(z, t) = F_t(z) - \Psi_t(z)$ satisfies the system of differential equations

$$\begin{cases} T_tG(z, t) = 0 \\ \dfrac{\partial}{\partial \bar{t}^i}G(z, t) = 0, \quad i = 1, .., m. \end{cases}$$

Hence $G(z, t)$ is a $C^\infty$ function in $O_i \times V'$ and is hence bounded on $O_i' \times U$. For $z \in M_1 \times O_i$, $t \in U$

$$\Psi(z, t) = F(z, t) - G(z, t).$$

Hence

$$\Psi(z_k, t_k) = F(z_k t_k) - G(z_k, t_k) = \frac{1}{\zeta(z_k, t_k) - \zeta(z_0, t_k)} - G(z_k, t_k)$$

So

$$|\Psi(z_k, t_k) + G(z_k, t_k)| \geqslant K_2^{-1}/|z_k - z_0|.$$

by Proposition 1. Since $G(z_k, t_k)$ is bounded and $z_0$ is adherent to $z_k$, it follows that $\Psi(z_k, t_k)$ is not bounded. This completes the proof of Theorem 1.

### 8. Proof of Theorem 2.

The proof is essentially same as the proof of Theorem 5. 1 in Kodaira-Spencer [2], once we have Theorem 1. Still we give the complete proof since some changes are required in our case. It is sufficient to prove the theorem without the requirement that $\Omega$ be Stein. For, once we have a $\Phi$ with $\Omega$ an open subset of $M \times \mathbf{C}^m$ we could restrict $\Phi$ to $\mathcal{G}(M_1 \times U)$ where U is a sufficiently small Stein neighbourhood of $t_0$ and obtain Theorem 2 (since $\mathcal{G}(M_1 \times U)$ is Stein by Theorem 1).

Thus it is enough to show that there exist a neighbourhood U′ of $t_0$, an analytic homeomorphism $\Phi : \mathcal{G}(M_1 \times U') \to \Omega$ where $\Omega$ is an open submanifold of $M \times \mathbf{C}^m$, and an analytic homeomorphism $\varphi : U' \to \pi(\Omega)$ such that the following diagram is commutative:

$$\begin{array}{ccc} \mathcal{G}(U' \times M_1) & \xrightarrow{\Phi} & \Omega \\ \downarrow{\scriptstyle \pi_1} & & \downarrow \\ U' & \xrightarrow{\varphi} & \pi(\Omega) \end{array}$$

We make the following inductive assumption:

$A_{p-1}$ : If the dimension of $U_0$ is $(p - 1)$ and $M_1$, is any relatively compact subdomain of an open Riemann surface M, then there exists a neighbourhood U of $t_0$ and a holomorphic map $h$: $\mathcal{G}(M_1 \times U)$ into M which maps each fibre biholomorphically into M.

Now, assumig $A_{p-1}$ we prove $A_p$.

Let $M_0$ be a relatively compact subdomain of M such that $\overline{M}_1 \subset M_0$. Let W be a sufficiently small Stein neighbourhood of $t_0$ in $\mathbf{C}^p$, with $\overline{W}$ compact, $\overline{W} \subset U_0$. Then $\mathcal{G}(M_0 \times W)$ is a Stein manifold. If $\mathcal{F}$ denotes the holomorphic tangent bundle along the fibres, then by Theorem B on Stein manifolds $H^1(\mathcal{G}(M_0 \times W), \mathcal{F}) = 0$. From the exact sequence

$$H^0(\mathcal{G}(M_0 \times W), \Pi) \to H^0(W, T) \to H^1(\mathcal{G}(M_0 \times W), \mathcal{F})$$

($\Pi$ denotes the sheaf of germs of holomorphic vector fields which are projectable, T denotes the sheaf of germs of holomorphic vector fields on W), we see that the vector field $\frac{\partial}{\partial t^p}$ can be lifted into a holomorphic vector field X of $\mathcal{G}(M_0 \times W)$.

We may suppose $t_0 = 0$. Let $f(x) = \exp(-\pi^p(x)X)$, $x \in M_1 \times W'$ where $\pi^p(x)$ is defined as follows: if $x = (z, t^1, \ldots, t^p)$, $\pi^p(x) = t^p$. By the complex analytic analogue of Proposition 5. 1 in [2], for a neighbourhood $W' \subset W$, $f$ maps $\mathcal{G}(M_1 \times W')$ holomorphically into $\mathcal{G}(M_0 \times (t^1, \ldots, t^{p-1}, O))$ mapping the fibre at $(t^1, \ldots, t^p)$ biholomorphically into $\mathcal{G}((t^1, \ldots, t^{p-1}, O), M_0)$. $M_0$ being a relatively compact subdomain of M, there exists, by the inductive hypothesis, a holomorphic map $g$:

$$M_0 \times (t^1, \ldots, t^{p-1}, O) \to M$$

which maps each fibre biholomorphically into M. Taking the

composite $h = g \circ f$ we get a holomorphic map of $\mathcal{S}(U_4 \times M_1)$ where $U_4$ is a neighbourhood of $t_0$ in $\mathbf{C}^p$, mapping each fibre biholomorphically into M. This proves $A_p$.

Once we have proved the assertion $A_m$, consider the map

$$\Phi : \mathcal{S}(U \times M_1) \to U \times M$$

defined by $(t, z) \to (t, h(t, z))$. $\Phi$ is holomorphic and one to one. By a known theorem on holomorphic functions $\Phi$ maps $\mathcal{S}(U \times M_1)$ biholomorphically onto an open subset $\Omega$ of $U \times M$ and we have the commutative diagram

$$\begin{array}{ccc} S(U \times M_1) & \xrightarrow{\ \Phi\ } & \Omega \\ \downarrow{\scriptstyle \pi_1} & & \downarrow{\scriptstyle \pi} \\ U & \xrightarrow{\text{identity}} & U \end{array}$$

This completes the proof of Theorem 2.

## 9. Differentiable variations of complex structures. Proof of Theorem 3.

Let $\mathcal{S}(t, M)$ be a differentiable variation of complex structures on an open Riemann surface M, $t \in U_0 \subset \mathbf{R}^m$. Let $J_t$ be the almost complex structure tensor corresponding to the structure $\mathcal{S}(t, M)$. On $M \times U_0$ let J denote the tensor along the fibres composed of $\{J_t\}$. If X is a *projectable* vector field on $M \times U_0$ (with respect to the projection $M \times U_0 \to U_0$) we remark that the Lie derivative of J with respect to the vector field X, denoted by $[X, J]$, is defined as a tensor along the fibres.

Let X be a projectable vector field on $M \times U_0$ satisfying the condition $[X, J] = 0$. Let $M'$ (resp $U_0'$) be a relatively compact subdomain of M (resp. $U_0'$). If $\exp(sX)$ denotes the one parameter family of transformations associated with X, $\exp(sX)$ is a diffeomorphism of $M' \times U_0'$ into $M \times U_0$ which maps $\mathcal{S}(t, M')$, $t \in U_0'$ biholomorphically into $\mathcal{S}(\exp(s\nu)(t), M)$, where $\nu$ denotes the projection of X on $U_0$. Now referring to the proof of Theorem 2, we see that to prove Theorem 3 it is sufficient to prove

PROPOSITION 4. — *Let $\mathcal{S}(t)$ be a differentiable family of complex structures on an open Riemann surface* M. *Let* $M_1$ *be a relatively compact subdomain of* M. *Then there exists a neighbourhood* $U_1$ *of* $t_0$ *in* $R^m$ *such that every differentiable vector field (real) on* $U_1$ *can be lifted into a differentiable vector field* X *on* $M_1 \times U_1$ *satisfying the condition* $[X, J] = 0$, J *denoting the tensor along the fibres composed on the almost complex structure tensors along the fibres.*

*Proof of Proposition* 4. — Let $M_0$ be a relatively compact subdomain (of M) with analytic boundary, with $\overline{M}_1 \subset M_0$.

Let $\Theta_t$ denote the holomorphic tangent bundle of $\mathcal{S}(t, M)$. Let $\mathcal{F} = U_t\Theta_t$ be the bundle on $M \times U_0$ composed of the holomorphic tangent bundles along the fibres. If $U_2$ is a spherical neighbourhood of $t_0$ with $U_2 \subset U_0$ then $\mathcal{F}|M \times U_2$ is differentiably equivalent to the bundle $U_2 \times \Theta_{t_0}$ (Homotopy theorem). It follows that there exist isomorphisms

$$\psi_1(t): H_{1,\alpha}(M_0, \Theta_t) \to H_{1,\alpha}(M_0, \Theta_{t_0}),$$
$$\psi_2(t): \overset{0,1}{H}_\alpha(M_0, \Theta_t) \to \overset{0,1}{H}_\alpha(M_0, \Theta_{t_0}),$$

depending differentiably on $t$ such that $\psi_1(t_0) =$ identity, $\psi_2(t_0) =$ identity. Let

$$T_t = \psi_2(t)\, d_{\bar z}(t)\, \psi_1(t)^{-1} : H_{1,\alpha}(M_0, \Theta_{t_0}) \to \overset{0,1}{H}_\alpha(M_0, \Theta_{t_0})$$

where $d_{\bar z}(t)$ denotes the $d_{\bar z}$ operator with respect to the structure $\mathcal{S}(t)$. $T_t$ depends differentiably on $t$. Since $T_{t_0} = d_{\bar z}(t_0)$ admits of a right inverse by Lemma 5, there exist a neighbourhood $U_3$ of $t_0$ and operators

$$S_t: \overset{0,1}{H}_\alpha(M_0, \Theta_{t_0}) \to H_{1,\alpha}(M_0, \Theta_{t_0}), \quad t \in U_3$$

depending differentiably on $t \in U_3$ and such that $S_t$ is a right inverse of $T_t$ (Lemma 2).

Let $M_2$ be a relatively compact subdomain of $\overline{M}$ with $M_0 \subset M_2$ Let $U_4$ be a neighbourhood of $t_0$ such that there exist a finite open covering $O_1, \ldots, O_k$ of $M_2$ and diffeomorphisms $g_i$ of $O_i \times U_4$ into $\mathbf{C} \times U_4$ which maps $\mathcal{S}(t, O_i)$, $t \in U_4$ biholomorphically into in $\mathbf{C} \times (t)$. [Such a neighbourhood $U_4$ exists. This follows from the definition of differentiable variation of complex structures in the sense of Kodaira-Spencer. With our definition this follows from the differentiable analogue of Proposition 1]. We denote the coordinate function in $O_i \times U_4$ by $(z^i, t)$.

Let $U_1$ be a relatively compact neighbourhood of $t_0$ in $R^m$ such that $\overline{U}_1 \subset U_2 \cap U_3 \cap U_4$. Let $\rho = (\rho_1(t), \ldots, \rho_m(t))$ be a differentiable vector field in $U_1$. In $O_i \times U_4$ consider the vector field $\pi_i$ defined by $(O, \rho_1(t), \ldots, \rho_m(t))$ with respect to the coordinate system $(z^i, t)$. We have $[\pi_i, J] = 0$. Put $\theta_{ij} = \pi_i - \pi_j$ in $(O_i \times U_4) \cap (O_j \times U_4)$. Let $\theta'_{ij} = \theta_{ij} - iJ\theta_{ij}$. Then $\theta'_{ij}$ are sections of $\mathcal{F}$ over $(O_i \times U_4) \cap (O_j \times U_4)$ whose restriction to each fibre is holomorphic. Evidently there exist differentiable sections $f'_i(z^i, t)$ of $\mathcal{F}$ over $O_i \times U_4$ such that $f'_i - f'_j = \theta'_{ij}$, in $(O_i \times U_4) \cap (O_j \times U_4)$. If we define $\varphi(t) = d_{\bar{z}}(t) f'_i(z^i, t)$, $\varphi(t)$ is a $(0, 1)$ form on $\mathcal{S}(t, M_2)$ with values in $\Theta_t$ which depends differentiably on $t$. Let $\eta(t) = \psi_2(t)\{\varphi(t) \mid \overline{M}_0\}$ [$\psi_2(t)$ is the isomorphism defined earlier.] Then $\eta(t)$ is a differentiable function with values in $\overset{0,1}{H}_\alpha(M_0, \Theta_{t_0})$. For $t \in U_1$, let $h_1(t) = S_t\eta(t)$. Then $h_1(t)$ depends differentiably on $t$. Let $h_2(t) = \{\psi_1(t)\}^{-1}(h_1(t))$. Then $h_2(t)$ depends differentiably on $t$ and satisfies $d_{\bar{z}}(t)h_2(t) = \varphi(t)$. [It follows easily from differentiability theorem for elliptic differential equations that $h_2(z, t)$ is a differentiable vector field on $M_1 \times U_1$. See proposition 1 in [3]]. Let

$$h(z, t) = \frac{1}{2}\{h_2(z, t) + \bar{h}_2(z, t)\}$$

and

$$f_i(z, t) = \frac{1}{2}\{f'_i(z, t) + \bar{f}'_i(z, t)\}.$$

Define

$$X = \pi_i + h - f_i \qquad \text{in} \qquad (O_i \cap M_1) \times U_1.$$

Then X is globally defined on $M_1 \times U_1$, projects into $\rho$ and satisfies the equation $[X, J] = 0$. This completes the proof of Proposition 4.

## REFERENCES


[1] S. S. Chern. An elementary proof of the existence of isothermal parameters on a surface, *Proc. Amer. Math. Soc.*, 6 (1955), pp. 771-782.

[2] K. Kodaira and D. C. Spencer : On deformations of complex analytic structures I, *Annals of Math.*, 67 (1958), pp. 328-401.

[3] K. Kodaira and D. C. Spencer. On deformations of complex analytic structures, III, *Annals of Math.*, 71 (1960), pp. 43-76.
[4] J. L. Lions. Lectures on elliptic partial differential equations, *Tata Institute of Fundamental Research*, Bombay, 1957.
[5] G. de Rham. *Variétés différentiables*, Paris, 1955.

Tata Institute of Fundamental Research, Bombay
and
Centre National de la Recherche Scientifique, Paris.

[Reprinted from *American Journal of Mathematics*, Vol. LXXXIII, No. 3, July, 1961.]

# EXISTENCE OF UNIVERSAL CONNECTIONS.*

By M. S. NARASIMHAN and S. RAMANAN.

**1. Introduction.** The purpose of this paper is to prove the existence of universal connections for principal bundles with a compact Lie group as structure group. We prove (Theorem 2) that given a compact Lie group $G$ and a positive integer $n$, there exist a differentiable principal $G$-bundle $E$ *and* a connection $\gamma_0$ on $E$ such that any connection on a differentiable principal $G$-bundle $P$ with base of dimension $\leqq n$ can be obtained as the inverse image of the connection $\gamma_0$ by a differentiable bundle homomorphism of $P$ into $E$. As is well-known, the analogous problem for bundles without connections is treated in the topology of fibre bundles [1].

It is also known that the Stiefel bundles play the role of universal bundles for the unitary groups $U(k)$. One can define in a natural way a connection on every Stiefel bundle (§ 2). We prove that these connections themselves are universal for connections in $U(k)$-bundles. A precise formulation is found in Theorem 1.

In the unitary case the problem is first solved locally by explicit construction, the crucial step being the lemma in § 3. The local solutions are then pieced up with the help of a special type of covering by coordinate cells.

In the general case, the compact Lie group $G$ is identified with a closed subgroup of a unitary group. Starting from a universal connection for this unitary group, a universal connection for $G$ is constructed by generalizing the usual method of construction of an invariant connection in the principal bundle associated with a Lie group and a closed subgroup ([3], p. 45).

A theorem of A. Weil ([1], p. 57) asserts that the cohomology classes of the base of a principal $G$-bundle obtained by substitution of the curvature form of a connection on $P$ in the invariant polynomials of $G$ are independent of the connection. Our result seems to explain this invariance and in fact furnishes an alternate proof in the case of compact Lie groups.

For definitions of the notions related to connections in principal bundles we refer to [1] and [3]. We use connections and connection forms interchangeably. By "differentiable" we always mean "indefinitely differentiable."

* Received February 10, 1961.

All manifolds, bundles, bundle homomorphisms and differential forms are assumed to be differentiable. Also all manifolds that occur are paracompact.

We are thankful to Professor K. Chandrasekharan for his constant encouragement and interest.

**2. Canonical connections in Stiefel bundles.** Let $\boldsymbol{C}^N$ be the $N$-dimensional complex number space with $O$ as origin. The Stiefel manifold $V(N,k)$ (with $N \geqq k$) of all unitary $k$-frames at $O$ may then be identified with the left coset space $U(N)/I_k \times U(N-k)$ where $I_k$ is the unit $(k,k)$ matrix. To every frame $(v_1,\cdots,v_k)$ with $v_i = \sum_{j=1}^{N} b_{i,j}e_j$ where $(e_j)$ is the canonical base in $\boldsymbol{C}^N$, we associate the $(N,k)$-matrix $A = (a_{i,j})$ with $a_{i,j} = b_{j,i}$. Since $(v_1,\cdots,v_k)$ is orthonormal, we have $\sum_{\alpha=1}^{N} \bar{b}_{j,\alpha} b_{l,\alpha} = \delta_{j,l}$, i.e. $A$ satisfies the condition $A^*A = I_k$ where $A^*$ is the conjugate transpose of $A$. Thus $V(N,k)$ is identified with $(N,k)$ matrices $A$ satisfying $A^*A = I_k$. The action of $U(k)$ (resp. $U(N)$) on $V(N,k)$ to the right (resp. to the left) goes over under the above identification into multiplication of $(N,k)$ matrices by unitary $(k,k)$ matrices on the right (resp. by unitary $(N,N)$ matrices on the left). Under the action of $U(k)$, $V(N,k)$ becomes a principal $U(k)$-bundle (known as the Stiefel bundle) with the Grassman manifold $G(N,k)$ of $k$-subspaces of $\boldsymbol{C}^N$ as base. $G(N,k)$ may again be identified with the left coset space $U(N)/U(k)\times U(N-k)$.

Let $S$ be the $(N,k)$ matrix-valued function on $V(N,k)$ which associates to each frame $(v_1,\cdots,v_k)$ the matrix $A$. Consider the $(k,k)$ matrix-valued differential form $S^*dS$ on $V(N,k)$. Since $S^*S = I_k$ for every frame, on differentiation we obtain $S^*dS + (dS^*)S = 0$, or again $S^*dS + (S^*dS)^* = 0$. Hence $S^*dS$ has actually values in the Lie algebra $\mathfrak{u}(k)$ (which is the vector space of skew-Hermitian matrices) of $U(k)$.

PROPOSITION 1. *$S^*dS$ is a connection form on the Stiefel bundle $V(N,k)$ which is invariant under the action of $U(N)$.*

In fact, if $X_\xi$ is a tangent vector at $\xi \in V(N,k)$ and $s \in U(k)$, we shall denote by $X_\xi s$ the image of $X_\xi$ under the differential of the map $\xi \to \xi s$ of $V(N,k)$. Then we have

$$\begin{aligned}(S^*dS)(X_\xi s) &= S^*(\xi s)(X_\xi s)(S)\\ &= s^*S^*(\xi)(X_\xi S)s\\ &= s^{-1}(S^*dS)(X_\xi)s.\end{aligned}$$

On the other hand, if $a \in \mathfrak{u}(k)$, we identify $a$ with a tangent vector at $e$ and denote by $\xi a$ the image of $a$ under the differential of the map $s \to \xi s$ of $G$ into $V(N,k)$. Then

$$\begin{aligned}(S^*dS)(\xi a) &= S(\xi)^*(\xi a)(S)\\ &= S(\xi)^*S(\xi)a\\ &= a,\end{aligned}$$

since $S(\xi)^*S(\xi) = I_k$. Hence $S^*dS$ is a connection form on the Stiefel bundle. Moreover, if $t \in U(N)$, the left translation of the differential form by $t$ yields $(tS)^*d(tS) = S^*t^*tdS = S^*dS$.

*Remark.* This connection will hereafter be referred to as the canonical connection and will be denoted by $\gamma_0$.

The horizontal subspace for this connection at the point $\xi_0 = \begin{pmatrix} I_k \\ 0 \end{pmatrix}$ of $V(N,k)$ may be described as follows: The tangent space at $\xi_0$ can be identified with $(N,N)$ skew Hermitian matrices of the type $\begin{pmatrix} P & -Q^* \\ Q & 0 \end{pmatrix}$ where $P$ is a $(k,k)$ skew Hermitian matrix and $Q$ is a rectangular $(N,N-k)$ matrix. The horizontal vectors at $\xi_0$ for the connection $\gamma_0$ are then given by matrices of the type $\begin{pmatrix} 0 & -Q^* \\ Q & 0 \end{pmatrix}$. This description together with the invariance under the action of $U(N)$ characterises the connection $\gamma_0$ completely.

Analogous statements are true for the real Stiefel manifold $W(N,k)$ and the corresponding $O(k)$-bundle. In particular $S'dS$ (where $S'$ is the transpose of $S$) is a connection form on the Stiefel bundle, the corresponding horizontal subspace at $\xi_0$ being given by matrices of the type $\begin{pmatrix} 0 & -Q' \\ Q & 0 \end{pmatrix}$. This is easily seen to be the orthogonal complement of the vertical subspace at $\xi_0$ with respect to the killing form $\mathrm{tr}(\mathrm{ad}x\,\mathrm{ad}y)$ on $\mathfrak{o}(k)$, the Lie algebra of $O(k)$.

## 3. The local problem.

LEMMA. *Let $U$ be an open subset of $\mathbf{R}^n$ and $V$ a relatively compact open subset whose closure is contained in $U$. For every differential form $\alpha$ of degree 1 on $U$ with values in $\mathfrak{u}(k)$ (the space of skew-Hermitian matrices), there exist differentiable functions $\phi_1, \cdots, \phi_{m'}$ in $V$ with values in the space $\mathfrak{M}_k(\boldsymbol{C})$ of $(k,k)$ complex matrices such that*

i) $$\sum_{j=1}^{m'} \phi_j^* \phi_j = I_k, \quad \textit{and}$$

ii) $$\sum_{j=1}^{m'} \phi_j{}^* d\phi_j = \alpha,$$

*where* $m' = (2n+1)k^2$.

*Proof.* Let $f_1, \cdots, f_{k^2}$ be a set of positive definite matrices which form a base for the complex Hermitian matrices over the reals, such that $\| f_r \| = 1$ for every $r$ ($\| f \|$ being the norm as a linear transformation). Since $\alpha$ has values in $\mathfrak{u}(k)$, we may write $\alpha/i$ in $U$ as $\sum_{s=1}^{n}\sum_{r=1}^{k^2} \lambda_{r,s} f_r dx_s$, where $\lambda_{r,s}$ are real-valued functions and $x_s$ the coordinate functions in $\boldsymbol{R}^n$. If $a_{r,s} = \sup_V |\lambda_{r,s}|$, we have $\lambda_{r,s} = \mu_{r,s} - \nu_{r,s}$ where

$$\mu_{r,s} = \tfrac{1}{2}\{\lambda_{r,s} + a_{r,s} + 1\} \quad \text{and}$$

$$\nu_{r,s} = \tfrac{1}{2}\{a_{r,s} - \lambda_{r,s} + 1\}$$

are both strictly positive differentiable functions. Hence we may write $\mu_{r,s} = p_{r,s}{}^2$ and $\nu_{r,s} = q_{r,s}{}^2$ where $p_{r,s}$ and $q_{r,s}$ are positive differentiable functions. Clearly one may assume that $\sum_{s=1}^{n}(\mu_{r,s} + \nu_{r,s})$ is bounded by $1/2k^2$ on $V$ for every $r$, by altering the coordinate functions $x_s$ by a constant multiple, if necessary. The matrix valued function $1/k^2 I_k - \{\sum_{s=1}^{n}(\mu_{r,s} + \nu_{r,s})\} f_r$ is then positive. For,

$$\| \sum_{s=1}^{n}(\mu_{r,s} + \nu_{r,s}) f_r \| \leqq 1/2k^2 \| f_r \| < 1/k^2.$$

Let $g_r$ be the (unique) positive square-root of the positive matrix $f_r$ and $h_r$ the differentiable positive matrix-valued function satisfying

$$h_r{}^2(x) = 1/k^2 I_k - \{\sum_{s=1}^{n}(\mu_{r,s}(x) + \nu_{r,s}(x))\} f_r.$$

We now define $\mathfrak{M}_k(\boldsymbol{C})$-valued functions $\phi_j (1 \leqq j \leqq (2n+1)k^2)$ as follows.

For $1 \leqq j \leqq nk^2$, $\phi_j$ shall be the $nk^2$ functions $p_{r,s} e^{ix_s} \cdot g_r$ arranged in some order.

For $nk^2 + 1 \leqq j \leqq 2nk^2$, $\phi_j$ shall be the $nk^2$ functions $q_{r,s} e^{-ix_s} \cdot g_r$ arranged in some order.

For $2nk^2 + 1 \leqq j \leqq (2n+1)k^2$, $\phi_j$ shall be the functions $h_r$ in some order.

We have to verify that the $\phi_j$ thus defined fulfil the conditions i) and ii) of the lemma. In fact,

$$\sum_{j=1}^{m'} \phi_j{}^*\phi_j = \sum_{r,s} p_{r,s}{}^2 g_r{}^2 + \sum_{r,s} q_{r,s}{}^2 g_r{}^2 + \sum_r h_r{}^2$$
$$= \sum_{r,s} \mu_{r,s} f_r + \sum_{r,s} \nu_{r,s} f_r + I_k - \sum_{r,s} (\mu_{r,s} + \nu_{r,s}) f_r$$
$$= I_k.$$

On the other hand,

$$\sum_{j=1}^{m'} \phi_j{}^* d\phi_j = \sum_{r,s} p_{r,s} e^{-ix_s}\{p_{r,s} i e^{ix_s}(dx_s) + (dp_{r,s}) e^{ix_s}\} g_r{}^2$$
$$+ \sum_{r,s} q_{r,s} e^{ix_s}\{- q_{r,s} \cdot i e^{-ix_s}(dx_s) + (dq_{r,s}) e^{-ix_s}\} g_r{}^2$$
$$+ \sum_r h_r dhr$$
$$= \sum_{r,s} i \cdot p_{r,s}{}^2 g_r{}^2 dx_s + \sum_{r,s} p_{r,s} dp_{r,s} g_r{}^2$$
$$+ \sum_{r,s} i(- q_{r,s}{}^2) g_r{}^2 dx_s + \sum_{r,s} q_{r,s} dq_{r,s} g_r{}^2 + \sum_r h_r dh_r$$
$$= \sum_{r,s} i(\mu_{r,s} - \nu_{r,s}) f_r dx_s + \tfrac{1}{2} \sum_{r,s} d(p_{r,s}{}^2 + q_{r,s}) f_r + \sum_r h_r dh_r$$
$$= \alpha + \tfrac{1}{2} \sum_{r,s} d(\mu_{r,s} + \nu_{r,s}) f_r + \sum_r h_r dh_r.$$

But since for any $x, y \in V$, $h_r{}^2(x)$ and $h_r{}^2(y)$ commute, their positive square roots $h_r(x)$ and $h_r(y)$ also commute. It readily follows that $h_r dh_r = dh_r \cdot h_r$. Hence $\frac{1}{2} d(h_r{}^2) = h_r dh_r$. Therefore, finally we have

$$\sum_{j=1}^{m'} \phi_j{}^* d\phi_j = \alpha + \tfrac{1}{2} d\{ \sum_{r,s} (\mu_{r,s} + \nu_{r,s}) f_r + \sum_r h_r{}^2)\}$$
$$= \alpha,$$

since $\sum_{r,s} (\mu_{r,s} + \nu_{r,s}) f_r + \sum_r h_r{}^2 = I_k$, and the lemma is completely proved.

The problem is solved locally by the following

PROPOSITION 2. *Let $P$ be a principal $U(k)$-bundle over a manifold $X$ of dimension $\leqq n$ and $\gamma$ a connection form on $P$. For every relatively compact open subset $W$ of $X$ with $\bar{W}$ contained in a coordinate neighborhood $U$ of $X$ over which $P$ is trivial, there exists a differentiable bundle map $\Phi$ of $p^{-1}(W)$ into $V(m'', k)$ such that the inverse image of the canonical connection $\gamma_0$ by $\Phi$ is $\gamma$, where $m'' = (2n+1)k^3$ and $p$ is the projection $P \to X$.*

*Proof.* Let $\sigma$ be a section of $P$ over $U$ and $\alpha$ the inverse image of $\gamma$ by $\sigma$.

By the lemma, we can find differentiable $\mathfrak{M}_k(\boldsymbol{C})$-valued functions $\phi_1,\cdots,\phi_{m'}$ in $W$ such that

i) $\sum_{j=1}^{m'} \phi_j{}^*\phi_j = I_k$, and

ii) $\sum_{j=1}^{m'} \phi_j{}^* d\phi_j = \alpha$, where $m' = (2n+1)k^2$.

Define a map $\Phi$ of $P$ over $W$ into the space of $(m'', k)$-matrices by setting for $\xi \in P$, $\Phi(\xi) = \begin{pmatrix} \phi_1(p\xi) \\ \vdots \\ \phi_{m'}(p\xi) \end{pmatrix} \cdot s$ where $s \in U(k)$ is determined by $\xi = \sigma(p\xi)s$.

$\Phi$ is easily seen to be a bundle homomorphism. We then have

$$\begin{aligned} \Phi^*\Phi(\xi) &= s^*(\sum_{j=1}^{m'}(\phi_i{}^*\phi_i)(p\xi))s \\ &= s^*s, \text{ by (i)} \\ &= I_k, \text{ since } s \text{ is unitary.} \end{aligned}$$

Hence $\Phi$ maps $P|p^{-1}(W)$ actually into $V(m'', k)$. On the other hand, it is obvious that the inverse image by $\Phi$ of $\gamma_0 = S^*dS$ is given by $\Phi^*d\Phi$. But the inverse image by $\sigma$ of $\Phi^*d\Phi$ is $(\Phi\circ\sigma)^*d(\Phi\circ\sigma) = \sum_{i=1}^{m'}\phi_i{}^*d\phi_i = \alpha$ by construction. Now $\gamma$ and $\Phi^*d\Phi$ are two connections on $P|p^{-1}(W)$ such that their inverse image by the section $\sigma$ are the same. Hence $\gamma = \Phi^*d\Phi$ on $p^{-1}(W)$.

## 4. Universal connection for the unitary group.

THEOREM 1. *Let $P$ be a principal $U(k)$-bundle over a manifold $X$ of dimension $\leq n$ and $\gamma$ any connection form on $P$. Then there exists a differentiable bundle homomorphism $\Phi$ of $P$ into the Stiefel bundle $V(m,k)$ such that $\gamma$ is the inverse image by $\Phi$ of the canonical connection $\gamma_0$ on $V(m,k)$, where $m = (n+1)(2n+1)k^3$.*

*Proof.* We can find a covering of $X$ by relatively compact open sets $\{V_i\}$ such that i) each $\bar{V}_i$ is contained in a coordinate cell, and ii) the $V_i$'s can be divided into $(n+1)$ classes $\mathscr{C}_j$ in such a way that no two $V_i$'s of the same class intersect ([2], p. 61). Let $\{W_i\}$ be a shrinking of this covering, i. e., an open covering $\{W_i\}$ such that $\bar{W}_i \subset V_i$. Let $D_j$ $(j = 1,\cdots, n+1)$ be the union of the open sets $p^{-1}(W_i)$ where $\bar{W}_i \subset V_i$ with $V_i$ belonging to $\mathscr{C}_j$.

The bundle is trivial over the coordinate cells and hence, by Proposition 2, one can find differentiable bundle homomorphisms $\Phi_i$ on $p^{-1}(V_i)$ into

$V((2n+1)k^3, k)$ inducing the connection $\gamma$ on $p^{-1}(V_i)$. Corresponding to each $D_j$ there exists a $\{(2n+1)k^3, k\}$-matrix-valued differentiable function $\Psi$ on $P$ such that $\Psi$ coincides with $\Phi_i$ on $p^{-1}(W_i)$ for $V_i$ in $\mathscr{B}_j$. Let then $\Psi_1, \cdots, \Psi_{n+1}$ be the $(n+1)$ functions thus constructed. Consider a differentiable partition of unity with respect to the covering $\{D_j\}$ consisting of non-negative differentiable functions $\zeta_i$ invariant under the action of the group $U(k)$ such that the support of $\zeta_i \subset D_i$ and $\sum \zeta_i^2 = 1$.

Consider now the map $\Phi$ on $P$ defined by

$$\Phi(\xi) = \begin{pmatrix} \zeta_1(\xi)\Psi_1(\xi) \\ \vdots \\ \zeta_{n+1}(\xi)\Psi_{n+1}(\xi) \end{pmatrix} \text{for every } \xi \in P.$$

We have to prove that $\Phi$ is bundle map of $P$ into $V(m,k)$ such that $\Phi^* d\Phi = \alpha$. But

$$\begin{aligned} \Phi^*\Phi(\xi) &= \sum_{i=1}^{n+1} \zeta_i(\xi)^2 \Psi_i^*(\xi)\Psi_i(\xi) \\ &= \sum \zeta_i(\xi)^2 \Psi_i^*(\xi)\Psi_i(\xi), \end{aligned}$$

the summation being over those $i$'s for which $\xi \in D_i$. But on $D_i$, $\Psi_i^*\Psi_i = I$ and we have $(\Psi_i^*\Psi_i) = I$ for every $i$ over which the summation extends. Hence $\Phi^*\Phi(\xi) = \sum \zeta_i(\xi)^2 I = I$, since $\sum \zeta_i^2(\xi) = 1$. Moreover

$$\begin{aligned} \Phi(\xi s) &= \begin{pmatrix} \zeta_1(\xi s)\Psi_1(\xi s) \\ \vdots \\ \zeta_{n+1}(\xi s)\Psi_{n+1}(\xi s) \end{pmatrix} = \begin{pmatrix} \zeta_1(\xi)\Psi_1(\xi) s \\ \vdots \\ \zeta_{n+1}(\xi)\Psi_{n+1}(\xi) s \end{pmatrix} \\ &= \begin{pmatrix} \zeta_1(\xi)\Psi_1(\xi) \\ \vdots \\ \zeta_{n+1}(\xi)\Psi_{n+1}(\xi) \end{pmatrix} s \\ &= \Phi(\xi)\cdot s \end{aligned}$$

for every $\xi \in P$ and $s \in U(k)$.

Finally,

$$\begin{aligned} \Phi^* d\Phi &= \sum_{i=1}^{n+1} \zeta_i \Psi_i^* (d\zeta_i \Psi_i + \zeta_i d\Psi_i) \\ &= \sum_{i=1}^{n+1} \Psi_i^*\Psi_i \zeta_i d\zeta_i + \sum_{i=1}^{n+1} \zeta_i^2 \Psi_i^* d\Psi_i. \end{aligned}$$

As before, for a $\xi \in P$, the summation needs to be taken only over those $i$'s for which $\xi \in D_i$. In every such $D_i$, however, $\Psi_i^*\Psi_i = I$ and $\Psi_i^* d\Psi_i = \alpha$. Hence $\Phi^* d\Phi = \sum \zeta_i d\zeta_i I + (\sum \zeta_i^2)\alpha = \alpha$ since $\sum \zeta_i d\zeta_i = \frac{1}{2} d(\sum \zeta_i^2) = 0$.

**5. Universal connections for compact Lie groups.** Let $G_2$ be a closed subgroup of a Lie group $G_1$, and $\mathfrak{g}_2$ and $\mathfrak{g}_1$ their Lie algebras. The group $G_2$ acts on $\mathfrak{g}_1$ by the adjoint operations and $\mathfrak{g}_2$ is invariant under this representation. Suppose $\mathfrak{m}$ is a subspace of $\mathfrak{g}_1$ invariant under the action of $G_2$ which is supplementary to $\mathfrak{g}_2$. (Such a space $\mathfrak{m}$ exists if $G_2$ is compact or semi-simple.) Let $P$ be a principal bundle with group $G_1$ and $\omega_1$ a connection on $P$. $P$ is fibred by $G_2$ into a principal bundle with group $G_2$.

The direct sum decomposition $\mathfrak{g}_2 \oplus \mathfrak{m}$ of $\mathfrak{g}_1$ gives rise to a projection $\pi$ of $\mathfrak{g}_1$ onto $\mathfrak{g}_2$ which commutes with the action of $G_2$ (i. e. $\pi \circ ads = ads \circ \pi$ for every $s \in G_2$). We define a differential form $\omega_2$ on $P$ by setting $\omega_2 = \pi \circ \omega_1$. It is easy to see that $\omega_2$ is a connection on $P$ for the fibration by $G_2$. In fact, (with the notations of § 2),

$$\begin{aligned}\omega_2(X_\xi s) &= (\pi \cdot \omega_1)(X_\xi s)\\ &= \pi \cdot \omega_1(X_\xi s)\\ &= \pi \cdot \mathrm{ad}s\, \omega_1(X_\xi)\\ &= \mathrm{ad}s \cdot \pi\omega_1(X_\xi)\\ &= s^{-1}\omega_2(X_\xi)s\end{aligned}$$

for every vector $X_\xi$ at $\xi \in P$ and $s \in G_2$.

On the other hand, for every $\xi \in P$ and $a \in \mathfrak{g}_2$, we have

$$\begin{aligned}\omega_2(\xi a) &= (\pi \cdot \omega_1)(\xi a)\\ &= \pi \cdot \omega_1(\xi a)\\ &= \pi(a)\\ &= a, \text{ since } \pi \text{ is a projection.}\end{aligned}$$

THEOREM 2. *Let $G$ be a compact Lie group and $n$ a positive integer. There exist a principal $G$-bundle $B$ and a connection form $\gamma_1$ on $B$ such that for every principal $G$-bundle $P$ with base of dimension $\leq n$ and any connection form $\gamma$ on $P$, one can find a bundle homomorphism $f$ of $P$ into $B$ such that the inverse image of $\gamma_1$ by $f$ is $\gamma$.*

*Proof.* $G$ can be identified with a closed subgroup of a unitary group $U(k)$. Let $\gamma_0$ be a universal connection for $U(k)$ (for the dimension $n$) on a principal $U(k)$-bundle $B$, whose existence has been proved in Theorem 1. $G$ acts on $B$ and makes of it a principal $G$-bundle. One can define a connection $\gamma_1$ on this bundle by setting $\gamma_1 = \pi \circ \gamma_0$ where $\pi$ is a projection of $\mathfrak{u}(k)$ onto the Lie algebra $\mathfrak{g}$ of $G$, as explained above. For any principal $G$-bundle $P$, with base of dimension $\leqq n$, let $P'$ be the corresponding principal $U(k)$-bundle obtained by enlarging the group $G$. Then there is a natural inclusion $i: P \rightarrow P'$ such that $i(\xi s) = i(\xi)(s)$ for $\xi \in P$ and $s \in G$.

Moreover, if $\gamma$ is a connection form on $P$, one can define a natural connection $\gamma'$ on $P'$ such that the inverse image of $\gamma'$ by $i$ is $\gamma$ ([3], p. 35). Let $\Phi$ be a bundle map of $P$ into $B$ such that the inverse image of $\gamma_0$ by $\Phi$ is $\gamma'$. We define a bundle map $f$ of $P$ into $B$ fibred by $G$ by setting $f = \Phi \cdot i$. The inverse image of $\gamma_0$ by $f = \Phi \cdot i$ is obviously $\gamma$. But since $\gamma$ has values in $\mathfrak{g}$ and $\pi$ is identity on $\mathfrak{g}$, we have $\pi \cdot \gamma = \gamma$ and hence the inverse image of $\gamma_1$ by $f$ is $\gamma$.

## 6. Remarks.

i) We show how A. Weil's theorem on connections can be deduced from our results, at least when $G$ is compact. Let $\omega_1$, $\omega_2$ be two connections on a principal $G$-bundle $P$ with base $X$ of dimension $\leqq n-1$. Consider the bundle $P \times I' \rightarrow X \times I'$ where $I'$ is the open interval $(-\epsilon, 1+\epsilon)$, $\epsilon > 0$. Let $\alpha_1$, $\alpha_2$ be inverse images of $\omega_1$, $\omega_2$ respectively under the projection $P \times I' \rightarrow P$. The differential form $\alpha = t\alpha_1 + (1-t)\alpha_2$ where $t$ is the projection $P \times I' \rightarrow I'$, is easily seen to be a connection on $P \times I'$. Let $B$ be a principal $G$-bundle over a manifold $M$ and $\gamma_1$ a universal connection on $B$ for dimension $\leqq n$. It follows that there exists a differentiable family $F_t$ of differentiable bundle maps of $P$ into $B$ such that the inverse image of $\gamma_1$ by $F_t$ is $t\omega_1 + (1-t)\omega_2$. If $f_t$ are the corresponding maps of $X$ into $M$, then $f_0$ and $f_1$ are obviously homotopic. On the other hand, if $K_1$ and $K_2$ are the curvature forms of $\omega_1$, $\omega_2$ respectively, the 'substitution' of $K_1$, $K_2$ in each polynomial over $\mathfrak{g}$ invariant under the adjoint representation of $G$ yields closed differential forms $\beta_1$, $\beta_2$ on $X$. Then $\beta_1$ and $\beta_2$ are the inverse images under $f_0$ and $f_1$ of the form on $M$ obtained by substituting the curvature form $K$ of the universal connection in the same polynomial. Since $f_0$ and $f_1$ are homotopic, it follows that $\beta_1$ and $\beta_2$ define the same cohmology class on the base, a characteristic class of the bundle.

ii) Our method gives a universal connection for the orthogonal group $O(k)$ in particular. But the connection was defined in the complex Stiefel manifold fibred by $O(k)$ instead of the more usual real Stiefel bundle. We have already remarked (§ 2) that if the points of the real Stiefel manifold $W(N, k)$ are represented by $(N, k)$-matrices $A$ satisfying $A'A = I_k$ ($A'$ being the transpose of $A$), a connection can be defined in a canonical way on the real Stiefel bundle with the corresponding connection form $A'dA$. On the other hand, the complex Stiefel manifold $V(N,k)$ may be imbedded into the real Stiefel manifold $W(2N,k)$ by associating to each $(N,k)$ matrix $A$, the $(2N,k)$ real matrix $\tilde{A} = \begin{pmatrix} \mathrm{Rl}\ A \\ \mathrm{Im}\ A \end{pmatrix}$ where $\mathrm{Rl}\,A$, $\mathrm{Im}\,A$ are the real and imaginary parts of $A$. It is easy to see that if $A^*A = I_k$, we have $\tilde{A}'\tilde{A} = I_k$. Moreover, for every $s \in O(k) \subset U(k)$

$$\widetilde{(As)} = \begin{pmatrix} \mathrm{Rl}\ (As) \\ \mathrm{Im}\ (As) \end{pmatrix} = \begin{pmatrix} (\mathrm{Rl}\,A)s \\ (\mathrm{Im}\,A)s \end{pmatrix} = \tilde{A}\cdot s.$$

Hence the map $A \to \tilde{A}$ is a bundle map of the complex Stiefel manifold $V(N,k)$ fibred by $O(k)$, into the real Stiefel bundle. The connection form on $V(N,k)$ induced by this map is $\tilde{A}'d\tilde{A}$, but this is the same as the real part of $A^*dA$. If we take for the projection $\pi$ of § 4, the map of $\mathfrak{u}(k)$ onto the Lie algebra $\mathfrak{o}(k)$ of $O(k)$ defined as the assignment of the real part to each skew Hermitian matrix, then the corresponding connection $\gamma_1$ is the same as the real part of $A^*dA$. In other words, the canonical connection in the real Stiefel bundle is universal for $O(k)$-bundles.

TATA INSTITUTE OF FUNDAMENTAL RESEARCH, BOMBAY.

## REFERENCES.

[1] S. S. Chern, "Topics in differential geometry," *Mimeographed Notes, The Institute for Advanced Study*, Princeton, 1951.

[2] J. Nash, "The imbedding problem for Riemannian manifolds," *Annals of Mathematics*, vol. 63 (1956), pp. 20-63.

[3] K. Nomizu, "Lie groups and differential geometry," *Publications of the Mathematical Society of Japan*, 1956.

# EXISTENCE OF UNIVERSAL CONNECTIONS II.*

By M. S. NARASIMHAN and S. RAMANAN.

**1. Introduction.** In an earlier paper [3] we proved the existence of universal connections for connections in bundles with a compact Lie group as structure group. In this paper we extend this result to the case of an arbitrary connected Lie group (Theorem 1). The proof of this theorem does not depend on [3]. However, this result does not include Theorem 1 of [3] which is more precise in that it asserts that the canonical connections in the Stiefel bundles themselves are universal for connections in unitary or orthogonal bundles. The latter is useful in some applications.

Since any two connections on a principal bundle differ by a 1-form of the adjoint type, one can reduce the problem of finding a universal connection to one of finding a universal 1-form of the adjoint type (§ 3). Regarding the latter problem we prove the following more general result (Theorem 2): if $\rho$ is a finite dimensional representation of a connected Lie group $G$ and $n$ and $p$ are non-negative integers, then there exists a $n$-universal $p$-form of type $\rho$. (For the definition of forms of type $\rho$ see § 2.) This problem is essentially one for compact Lie groups (§ 6) since the structure group of a $G$-bundle $P$ can be reduced to a maximal compact subgroup $K$ of $G$, and forms of type $\rho$ on $P$ are precisely forms on $P$ obtained by extending forms of type $\rho | K$ (restriction of $\rho$ to $K$) on the reduced bundle (§ 2). It should be remarked, however, that the existence of universal connections for a connected Lie group does not follow immediately from that for compact Lie groups, since not every connection on a $G$-bundle is the extension of a connection on a reduced $K$-bundle. (For instance, the holonomy group of a connection got by extension will have to be contained in $K$). In the case of connections, Theorem 2 seems to be necessary for passing from the compact to the general case. In our procedure, however, Theorem 2 implies at once Theorem 1 without passing through the compact case.

In the case when $G$ is the orthogonal group $O(k)$ and $\rho$ is the natural representation, an $n$-universal $p$-form is constructed explicitly (§ 4). If $G$ is compact we may suppose that $\rho$ is a representation of $G$ by orthogonal matrices. This enables one to reduce the compact to the orthogonal case (§ 5).

* Received May 7, 1962.

For the notions relating to connections in principal bundles we refer to [1], [2], [5].

**2. Preliminaries.** In this section, we first fix our notation and terminology, and then give a canonical way of extending $p$-forms of a certain type to bundles obtained by extension of structure group.

By 'differentiable' we always mean 'indefinitely differentiable.' All manifolds, groups, bundles, maps and forms are assumed differentiable. We assume all our manifolds are paracompact. By a $p$-form on a manifold we mean a covariant tensor of degree $p$. If $f\colon M \to M'$ is a map and $\alpha$ a $p$-form on $M'$, $f^*(\alpha)$ will denote the inverse image of $\alpha$ by $f$.

By a $G$-bundle we always mean a *principal* bundle with structure group $G$. If $f\colon H \to G$ is a homomorphism of groups, $P_1$ a $H$-bundle, and $P_2$ a $G$-bundle, a map $h\colon P_1 \to P_2$ will be called a $f$-morphism if $h(\xi s) = h(\xi)f(s)$ for every $\xi \in P_1$, $s \in H$. When $H = G$ and $f$ is the identity map, a $f$-morphism will be called a $G$-morphism. If $\rho$ is a representation of $G$ in a finite dimensional vector space $V$, a $p$-form on a $G$-bundle with values in $V$ is said to be of type $\rho$ if i) it is equivariant for the action of $G$ and ii) it annihilates any $p$-tuple of tangent vectors one of which is vertical [2]. If $h\colon P_1 \to P_2$ is a $f$-morphism and $\alpha$ a $p$-form of type $\rho$ on $P_2$, then $h^*\alpha$ is clearly a $p$-form on $P_1$ of type $(\rho \circ f)$.

Let $G_1$, $G_2$ be two Lie groups and $f\colon G_1 \to G_2$ a homomorphism. Let $T_f$ be the functor which associates to every $G_1$-bundle $P$ the $G_2$-bundle $T_f(P)$ over the same base obtained by extension of the structure group by $f$. We recall that the total space of $T_f(P)$ is the orbit space of $P \times G_2$ under the action of $G_1$ given by $(\xi, g_2)g_1 = (\xi g_1, f(g_1)^{-1}g_2)$ for $\xi \in P$, $g_1 \in G_1$, $g_2 \in G_2$. The action of $G_2$ on $P \times G_2$ defined by $(\xi, g_2)g_2' = (\xi, g_2 g_2')$ for $\xi \in P$, $g_2, g_2' \in G_2$ commutes with the above action of $G_1$ and hence $G_2$ operates on $T_f(P)$ and makes of it a $G_2$-bundle. Moreover, if $\Phi\colon P \to P'$ is a $G_1$-morphism, the $G_2$-morphism $T_f(\Phi)\colon T_f(P) \to T_f(P')$ is induced by the map $(\xi, g_2) \to (\Phi(\xi), g_2)$ of $P \times G_2$ into $P' \times G_2$.

Let $q$ be the projection $P \times G_2 \to T_f(P)$ and $i_f$ the map $\xi \to q(\xi, e)$ of $P$ into $T_f(P)$. We then have $i_f(\xi s) = i_f(\xi)f(s)$, $\xi \in P$, $s \in G_1$; i.e., $i_f$ is a $f$-morphism.

Let $\rho$ be a finite dimensional representation of $G_2$. If $\beta$ is a $p$-form of type $\rho$ on $T_f(P)$, $i_f^*\beta$ is a $p$-form on $P$ of type $\rho \circ f$. Conversely, to every $p$-form $\alpha$ on $P$ of type $\rho \circ f$ we can associate in a natural way a $p$-form $T_f(\alpha)$ of type $\rho$ on $T_f(P)$ with $i_f^* T_f(\alpha) = \alpha$ in the following way. It is easy to check that the form $\alpha'$ on $P \times G_2$ defined by $\alpha'_{(\xi, g_2)} = \rho(g_2)^{-1}(p_1^*\alpha)_{(\xi, g_2)}$

($p_1$ being the projection $P \times G_2 \to P$) is invariant under the action of $G_1$ and annihilates any $p$-tuple of tangent vectors one of which is vertical. Hence there exists a $p$-form $T_f(\alpha)$ of type $\rho$ on $T_f(P)$ such that $q^*T_f(\alpha) = \alpha'$ and we have $i_f^*T_f(\alpha) = \alpha$. Moreover, if $\beta$ is a $p$-form of type $\rho$ on $T_f(P)$, we have $T_f(i_f^*\beta) = \beta$.

Finally, the correspondence $\alpha \to T_f(\alpha)$ is 'functorial' in the following sense: if $\Phi: P \to P'$ is a $G_1$-morphism and if $\alpha'$ is a $p$-form on $P'$ of type $\rho \circ f$, then we have

$$T_f(\Phi^*\alpha') = (T_f\Phi)^*(T_f\alpha').$$

## 3. Statement of the theorems. Proof of Theorem 1.

THEOREM 1. *Let $G$ be a connected Lie group and $n$ a positive integer. Then there exist a principal $G$-bundle $B$ and a connection form $\gamma_0$ on $B$ such that any connection form on a principal $G$-bundle $P$ with base of dimension $\leqq n$ is the inverse image of $\gamma_0$ by a $G$-morphism of $P$ in $B$.*

We deduce Theorem 1 from the following theorem which seems to be of independent interest.

THEOREM 2. *Let $G$ be a connected Lie group and $\rho$ a finite dimensional representation of $G$. Let $n$ and $p$ be two non-negative integers. Then there exist a principal $G$-bundle $E$ and a $p$-form $\alpha_0$ of type $\rho$ on $E$ such that any $p$-form of type $\rho$ on a principal $G$-bundle $P$ with base of dimension $\leqq n$ is the inverse image of $\alpha_0$ by a $G$-morphism of $P$ into $E$. Moreover, the bundle $E$ can be chosen to be classifying for dimension $\leqq n$.*

*Remarks.* 1. A $p$-form (resp. a connection) which possesses the property stated in Theorem 2 (resp. Th. 1) will be called $n$-universal.

2. Theorem 2 is also valid with "$p$-form" replaced by "exterior $p$-form." A universal exterior $p$-form is obtained by alternating a universal $p$-form.

*Proof of Theorem* 1. We now prove Theorem 1 assuming Theorem 2. Let $F$ be a differentiable $G$-bundle which is $n$-universal, and $\gamma_1$ any connection on $F$. On the other hand, let $E$ be a $G$-bundle and $\alpha_0$ a 1-form on $E$ of the adjoint type which is $n$-universal for such forms. Consider the $G$-bundle $B = F \times E$, the action of $G$ on $B$ being given by $(f, e)g = (fg, eg)$, $f \in F$, $e \in E$, $g \in G$. Let $q_1: B \to F$, $q_2: B \to E$ be the canonical projections, which are clearly $G$-morphisms. The differential form $\gamma_0 = q_1^*\gamma_1 + q_2^*\alpha_0$ is a connection form on $B$ since $q_1^*\gamma_1$ is a connection form and $q_2^*\alpha_0$ is a 1-form of the adjoint type. We assert that $\gamma_0$ is $n$-universal for connections in $G$-bundles.

In fact let $P$ be any $G$-bundle with base of dimension $\leqq n$ and $\gamma$ any connection on $P$. Since $F$ is a $n$-universal bundle, there exists a $G$-morphism $\Psi_1 : P \to F$. Then $\gamma - \Psi_1^*(\gamma_1)$ is a 1-form of the adjoint type since $\gamma$ and $\Psi_1^*(\gamma_1)$ are connection forms on $P$. Let $\Psi_2 : P \to E$ be a $G$-morphism such that $\Psi_2^*(\alpha_0) = \gamma - \Psi_1^*(\gamma_1)$. Consider the $G$-morphism $\Phi : P \to B$ defined by $q_1 \circ \Phi = \Psi_1$; $q_2 \circ \Phi = \Psi_2$. Then we have

$$\begin{aligned} \Phi^*(\gamma_0) &= \Phi^*(q_1^*\gamma_1 + q_2^*\alpha_0) \\ &= (q_1 \circ \Phi)^*\gamma_1 + (q_2 \circ \Phi)^*\alpha_0 \\ &= \Psi_1^*(\gamma_1) + \Psi_2^*(\alpha_0) \\ &= \Psi_1^*(\gamma_1) + (\gamma - \Psi_1^*(\gamma_1)) \\ &= \gamma. \end{aligned}$$

*Remark.* In the above construction, it is clear that the bundle $B$ is $n$-classifying if the $G$-bundles $E$ and $F$ are n-classifying, so that the maps induced on the bases by two $G$-morphisms $P \to B$ are homotopic. Thus the theorem of A. Weil on connections [1] is an immediate consequence of Theorem 1. However, Weil's theorem can be proved in a simpler way; for, all one requires for the proof is that any *two* given connections $\gamma_1$ and $\gamma_2$ on a bundle $P$ can be obtained as the inverse images of the *same* connection $\gamma_0$ on a bundle $B$ by morphisms whose projections on the base are homotopic. This problem is considerably simpler as can be seen by taking $B = P \times I$ and $(\gamma_0)_{(\xi,t)} = tp^*(\gamma_2) + (1-t)p^*\gamma_1$, $\xi \in P$, $t \in I$ where $p$ is the projection $P \times I \to P$ ($I$ is the open interval $[-2, 2]$). The inclusions $P \to P \times I$ given by $\xi \to (\xi, 0)$ and $\xi \to (\xi, 1)$ induce $\gamma_1$, $\gamma_2$ respectively and their projections to the base are clearly homotopic, the homotopy being induced by the identity mapping of $P \times I$ into itself [3, § 6].

**4. The orthogonal case.** In this section, we prove Theorem 2 in the case where $G$ is the real orthogonal group $O(k)$ and $\rho$ is the natural representation in $\boldsymbol{R}^k$. We identify $O(k)$ with the group of $(k, k)$ real matrices $A$ such that $A'A = I_k$ ($A'$ being the transpose of $A$) and $\boldsymbol{R}^k$ with $(k, 1)$ real matrices. $\rho$ then corresponds to left multiplication of $(k, 1)$ matrices by $(k, k)$ orthogonal matrices.

Let $W(N, k)$, $N \geqq k$, be the Stiefel bundle of $(N, k)$-real matrices $A$ such that $A'A = I_k$ ([3, § 2]). $O(k)$ acts on $W(N, k)$ by multiplication on the right. If $A \in W(N, k)$ is of the form $\begin{bmatrix} A_1 \\ \vdots \\ A_i \\ \vdots \\ A_N \end{bmatrix}$ where each $A_i$ is a $(1, k)$

matrix, the function $\sigma_i$ on $W(N,k)$ which assigns to each $A$ the $(k,1)$ matrix $A_i'$ is of type $\rho$. For

$$\sigma_i(As)=\sigma_i\begin{bmatrix}A_1s\\ \vdots\\ A_Ns\end{bmatrix}=(A_is)'=s'A_i'=s^{-1}\sigma_i(A)$$

for $A\in W(N,k)$, $s\in O(k)$.

We now construct a $n$-universal $p$-form of type $\rho$. For the rest of this section, $N$ will denote the integer $(n+1)n^p+(n+k)$. Let $V_1,\cdots,V_{n+1}$ be $(n+1)$ copies of $\boldsymbol{R}^n$ and $V_0=\boldsymbol{R}$. Consider the $O(k)$-bundle

$$E=W(N,k)\times V_0\times V_1\times\cdots\times V_{n+1}$$

where the action of $O(k)$ on $E$ is given by

$$(w,v_0,\cdots,v_{n+1})g=(wg,v_0,\cdots,v_{n+1}),\ w\in W(N,k),\ v_i\in V_i,\ g\in O(k).$$

Let $\pi$ (resp. $\pi_i$) denote the projection of $E$ onto $W(N,k)$ (resp. $V_i$). Let further $(x_i^1,\cdots,x_i^n)$ be the coordinate functions in $V_i$, $i>0$. For each multi-index $I=(i_1,\cdots,i_r,\cdots,i_p)$, $1\leqq i_r\leqq n$ and $1\leqq j\leqq n+1$, we shall denote by $\omega_j^I$ the $p$-form $\pi_j^*(dx_j^{i_1}\otimes\cdots\otimes dx_j^{i_p})$ on $E$. For convenience of notation, let us choose a bijection $\lambda$ of the set of indices $(I,j)$ with $I=(i_1,\cdots,i_p)$ $1\leqq i_r\leqq n$ and $1\leqq j\leqq n+1$, onto the set of integers $[1,(n+1)n^p]$. Obviously $\tau_j^I=\sigma_{\lambda(I,j)}\circ\pi$ is a function on $E$ with values in $(k,1)$-matrices of type $\rho$. $\pi_0$ being a real-valued function on $E$, the form $(\pi_0\cdot\tau_j^I)\omega_j^I$ is a $(k,1)$-matrix valued form which is the product of the vector valued function $\pi_0\tau_j^I$ and the real-valued form $\omega_j^I$. The form

$$\alpha_0=\sum_{I,j}(\pi_0\tau_j^I)\omega_j^I$$

is of type $\rho$. For, clearly $\alpha_0$ annihilates any $p$-tuple of vectors one of which is vertical since each $\omega_j^I$ has this property. Moreover if $X_1,\cdots,X_p$ are vectors at $\xi\in P$ and $s\in O(k)$ we have

$$\begin{aligned}\alpha_0(X_1s,\cdots,X_ps)&=\sum\pi_0(\xi s)\tau_j^I(\xi s)\omega_j^I(X_1s,\cdots,X_ps)\\&=s^{-1}\{\pi_0(\xi)\sum\tau_j^I(\xi)\omega_j^I(X_1,\cdots,X_p)\}\\&=s^{-1}\alpha_0(X_1,\cdots,X_p)\end{aligned}$$

where the $X_is$ are the vectors at $\xi s$ which are images of $X_i$ by the differential of the map $\xi\to\xi s$ of $P$ into itself.

We now proceed to prove that $\alpha_0$ is $n$-universal for $p$-forms of type $\rho$.

*Proof.* Let $P$ be a $O(k)$-bundle over a base $M$ of dimension $\leqq n$ and

$q: P \to M$ be the projection. Let $(U_i)$ be a covering of $M$ by relatively compact open sets such that

i) each $\bar{U}_i$ is contained in a coordinate cell

ii) the $U_i$'s can be divided into $(n+1)$ classes $\mathscr{C}_j$ in such a way that no two $U_i$'s of the same class intersect [4, p. 61].

Let $W_i$ be a shrinking of this covering, i.e., an open covering $W_i$ such that $\bar{W}_i \subset U_i$. Let $D_j\{j=1,\cdots,(n+1)\}$ be the union of the open sets $q^{-1}(W_i)$ for those $i$'s for which $U_i$ belongs to $\mathscr{C}_j$. Let $\zeta_j$ be a partition of unity with respect to this covering, consisting of non-negative differentiable functions $\zeta_j$ invariant under the action of $G$ with support of $\zeta_j \subset D_j$ and $\sum \zeta_j = 1$.

Let $\alpha$ be a $p$-form of type $\rho$ on $P$. It is clear that there exist functions $(f_j^1,\cdots,f_j^n)$ on $M$ whose restrictions to each $W_i$, for those $i$'s for which $U_i$ belongs to $\mathscr{C}_j$, form a coordinate system on $W_i$. Since $\alpha$ is of type $\rho$, $\alpha$ can be expressed in $D_j$ in the form $\sum \alpha_j^I q^*(df_j^I)$, where $\alpha_j^I$ are functions of type $\rho$ on $D_j$ and $df_j^I = df_j^{i_1} \otimes \cdots \otimes df_j^{i_p}$. Then it is easy to see that $\alpha = \sum \beta_j^I q^*(df_j^I)$, where $\beta_j^I = \zeta_j \alpha_j^I$ are now differentiable functions on $P$ of type $\rho$. Let $h$ be a strictly positive invariant differentiable function on $P$ such that $h(\xi)^2 > 2 \,\|\, \beta_j^I(\xi)\beta_j^I(\xi)' \,\|$ for every $\xi \in P$, where $\|\ \|$ denotes the norm as a linear operator. (The existence of such an $h$ follows for instance from the fact that $\|\, \beta_j^I(\xi)\beta_j^I(\xi)' \,\|$ is an invariant function on $P$). We have

$$\alpha = \sum h\eta_j^I q^*(df_j^I)$$

where $\eta_j^I = \frac{1}{h}\beta_j^I$. Obviously

$$\|\, \eta_j^I(\xi)\eta_j^I(\xi)' \,\| = \|\, \frac{1}{h(\xi)^2} \sum \beta_j^I(\xi)\beta_j^I(\xi)' \,\| \leqq \tfrac{1}{2}.$$

Therefore $R(\xi) = I_k - \sum \eta_j^I(\xi)\eta_j^I(\xi)'$ is a function on $P$ with values in positive definite matrices. Moreover, for $s \in O(k)$ and $\xi \in P$,

$$R(\xi s) = s^{-1}R(\xi)s.$$

For,

$$\begin{aligned} R(\xi s) &= I_k - \sum \eta_j^I(\xi s)\eta_j^I(\xi s)' \\ &= I_k - \sum s^{-1}\eta_j^I(\xi)\eta_j^I(\xi)'(s^{-1})' \\ &= s^{-1}(I_k - \sum \eta_j^I(\xi)\eta_j^I(\xi)')s \\ &= s^{-1}R(\xi)s. \end{aligned}$$

Let $S(\xi)$ be the differentiable positive matrix-valued function on $P$ such that

$S(\xi)^2 = R(\xi)$. It is clear from the uniqueness of the positive square root of a positive definite matrix that $S(\xi s) = s^{-1}S(\xi)s$ for $\xi \in P$, $s \in O(k)$.

Let $\psi: P \to W(n+k, k)$ be a $G$-morphism, the existence of which is assured by the universal bundle theorem [6, § 19]. Consider the matrix

$$\psi_1(\xi) = \begin{bmatrix} \eta_1(\xi)' \\ \vdots \\ \eta_i(\xi)' \\ \vdots \\ \eta_{(n+1)n^p}(\xi)' \\ \psi(\xi)S(\xi) \end{bmatrix}$$

where $\eta_i = \eta_j^I$ with $\lambda(I, J) = i$. Each $\eta_i'$ is a $(1, k)$ matrix and $\psi(\xi)S(\xi)$ is a $(n+k, k)$ matrix so that $\psi_1(\xi)$ is a $((n+1)n^p + n + k, k)$ matrix. The map $\xi \to \psi_1(\xi)$ is a map of $P$ into $W(N, k)$. For,

$$\begin{aligned} \psi_1'(\xi)\psi_1(\xi) &= \sum_{i=1}^{(n+1)n^p} \eta_i(\xi)\eta_i(\xi)' + S(\xi)'\psi(\xi)'\psi(\xi)S(\xi) \\ &= \sum_{(I,j)} \eta_j^I(\xi)\,(\eta_j^I(\xi))' + S(\xi)^2 \\ &= I_k - R(\xi) + S(\xi)^2. \end{aligned}$$

Hence $\psi_1'(\xi)\psi_1(\xi) = I_k$.

Moreover $\psi_1: P \to W(N, k)$ is a $G$-morphism. In fact,

$$\psi_1(\xi s) = \begin{bmatrix} \eta_1^I(\xi s)' \\ \vdots \\ \eta_i^I(\xi s)' \\ \vdots \\ \eta_{(n+1)n^p}{}^I(\xi s)' \\ \psi(\xi s)S(\xi s) \end{bmatrix} = \begin{bmatrix} (s^{-1}\eta_1(\xi))' \\ \vdots \\ (s^{-1}\eta_{(n+1)n^p}(\xi))' \\ \psi(\xi)s \cdot s^{-1}S(\xi)s \end{bmatrix} = \begin{bmatrix} \eta_1^I(\xi s)' \\ \vdots \\ \eta_{(n+1)n^p}{}^I(\xi)' \\ \psi(\xi)S(\xi) \end{bmatrix} s$$

$$= \psi_1(\xi)s.$$

Finally, we construct a $G$-morphism $\Phi$ of $P$ into $E$ such that $\Phi^*\alpha_0 = \alpha$ ($E$ and $\alpha_0$ are the bundle and $p$-form constructed in the beginning of this section). $\Phi: P \to E$ is defined by $\pi_0 \circ \Phi = h$, $\pi_j \circ \Phi = (f_j^1 \circ q, \cdots, f_j^n \circ q)'$, $(j = 1, \cdots, n+1)$ and $\pi \circ \Phi = \psi_1$. $\Phi$ is a $G$-morphism since $\psi_1$ is so and $h$, $f_j^i \circ \pi$ are invariant functions. We then have

$$\begin{aligned} \Phi^*(\alpha_0) &= \Phi^*(\sum \pi_0\tau_j^I\omega_j^I) \\ &= \sum (\pi_0 \circ \Phi)(\tau_j^I \circ \Phi)\Phi^*(\omega_j^I) \end{aligned}$$

$$= \sum h(\sigma_{\lambda(I,j)} \circ \pi \circ \Phi)\Phi^*(\omega_j{}^I)$$
$$= \sum h\eta_{\lambda(I,j)}\Phi^*\pi_j{}^*(dx_j{}^{i_1} \otimes \cdots \otimes dx_j{}^{i_p})$$

where $I = (i_1, \cdots, i_p)$ with $1 \leqq i_p \leqq n$. Hence

$$\Phi^*(\alpha_0) = \sum_{(I,j)} h\eta_j{}^I d(f_j{}^{i_1} \circ q) \otimes \cdots \otimes d(f_j{}^{i_p} \circ q) = \alpha.$$

**5. The case of a compact group.** In this section we prove Theorem 2 with $G$ compact and $\rho$ any $k$-dimensional representation of $G$. Since every representation of $G$ is equivalent to an orthogonal representation, we may assume that $\rho = j \circ f$ where $f$ is a homomorphism $G \to O(k)$ and $j$ is the natural representation of $O(k)$ in $\boldsymbol{R}^k$. Let $E_1$ be a $O(k)$-bundle together with an $n$-universal $p$-form of type $j$ (§ 4). Let $F$ be a $n$-universal $G$-bundle. We let $G$ act on $F \times E_1$ by $(v, e_1)g = (vg, e_1 f(g))$, $v \in F$, $e_1 \in E_1$, $g \in G$. This makes of $F \times E_1$ a $G$-bundle $E$. Let $q_1$ and $q_2$ be the projections of $F \times E_1$ onto $F$ and $E_1$ respectively. The $p$-form $\alpha_0 = q_2{}^*\alpha_1$ is of type $\rho$ since $q_2\colon E \to E_1$ is a $f$-morphism and $\alpha_1$ of type $j$. We now prove that $\alpha_0$ is a $n$-universal $p$-form of type $\rho$. In fact, let $P$ be a $G$-bundle over a base of dimension $\leqq n$ and $\alpha$ a $p$-form of type $\rho$ on $P$. Then there exists a $O(k)$-morphism $\Phi_2$ of $T_f(P)$ into $E_1$ such that $\Phi_2{}^*(\alpha_1) = T_f(\alpha)$. (For the definition of $T_f$ see § 2). On the other hand, $P$ admits a $G$-morphism $\Phi_1$ into $F$, since $F$ is $n$-universal. Let $\Phi\colon P \to E$ be the map defined by $q_1 \circ \Phi = \Phi_1$, $q_2 \circ \Phi = \Phi_2 \circ i_f$ ($i_f$ is the canonical map $P \to T_f(P)$, see § 2). $\Phi$ is a $G$-morphism. For, if $\xi \in P$, $s \in G$, we have

$$\begin{aligned} \Phi(\xi s) &= (\Phi_1(\xi s), \Phi_2 \circ i_f(\xi s)) \\ &= (\Phi_1(\xi)s, \Phi_2(i_f(\xi)f(s))) \\ &= \Phi(\xi)s. \end{aligned}$$

Now

$$\begin{aligned} \Phi^*(\alpha_0) = \Phi^*(q_2{}^*\alpha_1) &= (q_2 \circ \Phi)^*\alpha_1 \\ &= (\Phi_2 \circ i_f)^*\alpha_1 = i_f{}^*\Phi_2{}^*\alpha_1 = i_f{}^*T_f\alpha = \alpha. \end{aligned}$$

This completes the proof in the compact case.

**6. Proof of Theorem 2. The general case.** Let $G$ be a connected Lie group and $K$ a maximal compact subgroup of $G$. We denote by $f$ the inclusion map $K \to G$. We seek to construct a $n$-universal $p$-form of type $\rho$, where $\rho$ is any finite dimensional representation of $G$. From § 5, there exists a $n$-universal $p$-form $\alpha_1$ of type $(\rho \circ f)$ on a $K$-bundle $E_1$. Then the $p$-form $\alpha_0 = T_f(\alpha_1)$ on the $G$-bundle $E = T_f(E_1)$ is $n$-universal for $p$-forms of type $\rho$. In fact, let $P$ be a $G$-bundle over a base of dimension $\leqq n$ and $\alpha$ a $p$-form of type $\rho$ on $P$. It is well known that there exists a $K$-bundle $P_1$ such that $T_f(P_1)$ is $G$-isomorphic to $P$ (reduction of structure group to $K$, see [6. § 12]). We identify $P$ and $T_f(P_1)$ by such an isomorphism. Consider the form $i_f^*(\alpha)$ on $P_1$ which is of type $\rho \circ f$. Now let $\Phi_1 : P_1 \to E_1$ be a $K$-morphism such that $\Phi_1^*(\alpha_1) = i_f^*(\alpha)$. Consider the $G$-morphism $\Phi = T_f(\Phi_1)$ of $P$ into $E$. Then we have

$$\begin{aligned}\Phi^*(\alpha) &= (T_f(\Phi_1))^*\alpha_0 = (T_f(\Phi_1))^*T_f(\alpha_1)\\ &= T_f(\Phi_1^*\alpha_1) = T_f(i_f^*\alpha)\\ &= \alpha.\end{aligned}$$

It is clear, referring to §§ 4, 5, 6, that the bundle $E$ can be chosen to be $n$-classifying. This completes the proof of Theorem 2.

*Remark.* Theorems 1 and 2 hold even when $G$ is a Lie group with a finite number of connected components; our proofs continue to be valid in this case.

TATA INSTITUTE OF FUNDAMENTAL RESEARCH, BOMBAY.

REFERENCES.

[1] S. S. Chern, *Topics in differential geometry*, Mimeographed notes, the Institute for Advanced Study, Princeton, 1951.
[2] J. L. Koszul, *Lectures on fibre bundles and differential geometry*, The Tata Institute of Fundamental Research, Bombay, 1960.
[3] M. S. Narasimhan and S. Ramanan, "Existence of universal connections," *American Journal of Mathematics*, vol. 83 (1961), pp. 563-572.
[4] J. Nash, "The imbedding problem for Riemannian manifolds," *Annals of Mathematics*, vol. 63 (1956), pp. 20-63.
[5] K. Nomizu, *Lie groups and differential geometry*, Publications of Mathematical Society of Japan, 1956.
[6] N. Steenrod, *The Topology of fibre bundles*, Princeton, 1951.

*Bull. Soc. math. France*,
90, 1962, p. 449 à 471.

# REGULARITY THEOREMS FOR FRACTIONAL POWERS OF A LINEAR ELLIPTIC OPERATOR;

BY

TAKESHI KOTAKE AND MUDUMBAI S. NARASIMHAN.

**1. Introduction.** — Let $L$ be a linear elliptic operator with $C^\infty$ coefficients in an open subset $\Omega$ of $\mathbf{R}^n (n \geq 2)$. We suppose that $L$ admits a (strictly) positive self-adjoint realisation $\tilde{L}$ in $L^2(\Omega)$. Let $\{E_\lambda\}$ be the spectral resolution of $\tilde{L}$ so that

$$\tilde{L} = \int \lambda \, dE_\lambda.$$

We consider the family of operators $\tilde{L}^s$, depending on a complex parameter $s$, defined by

$$\tilde{L}^s = \int \lambda^s \, dE_\lambda.$$

The operators $\tilde{L}^s$ may be viewed as "fractional powers" of $L$. For $s = -1, -2, \ldots$, we obtain the Green's operator and its iterates.

We study in this paper the regularity properties of the operators $\tilde{L}^s$. For integral values of $s$, it is known that the operators $\tilde{L}^s$ define kernels which are "very regular" in the sense of SCHWARTZ ([17], chap. V, § 6) and that if further the coefficients of $L$ are analytic the kernels of $\tilde{L}^s$ are analytically very regular. For positive integral values of $s$ the results are trivial, for negative integral values of $s$ these follow from well-known regularity theorems for elliptic operators [11]. The question arises whether these results are true for all values of $s$. We prove in this paper that this is in fact the case (Theorems 2 and 3). The case of elliptic operators with constant coefficients on a torus and on $\mathbf{R}^n$ has already been dealt with respectively by S. BOCHNER [3] and L. SCHWARTZ ([16], chap. VII, § 10, ex. 7).

That the operators $\tilde{L}^s$ possess kernels follows from regularity theorems for elliptic operators. In order to prove that the kernels are very regular, we represent the kernels, for $Rl(-s)$ sufficiently large, in terms of the Green's function $G(t, x, y)$ of the associated parabolic operator. By using some results of G. BERGENDAL [1] and S. D. EIDELMAN [6] and showing that $G(t, x, y)$ and its derivatives fall off exponentially as $t \to \infty$, we then prove that the kernel $\tilde{L}^s$ is very regular.

The proof of analytic regularity, when the coefficients are analytic, is more difficult. It involves in the first instance estimates for the norms $\| A^k u \|_{L^2}$, where $u$ is a function that is to be proved to be analytic and $A$ a linear elliptic operator with analytic coefficients. Next we need to prove a general theorem (Theorem 1) to the effect that if $A$ is a linear elliptic operator of order $m$ with analytic coefficients in an open set $\Omega'$ of $\mathbf{R}^n$, and $u$ is a function satisfying the inequalities

$$\| A^k u \|_{L^2(\Omega')} \leq (km)!\, c^{k+1}$$

for every integer $k \geq 0$, with a positive constant $c$ independent of $k$, then $u$ is analytic in $\Omega'$.

This theorem is a natural one in as much as the conditions

$$\| A^k u \| \leq (km)!\, c^{k+1}$$

on every compact set are necessary for $u$ to be analytic. We notice also that this theorem contains the well-known result : if $A$ is linear elliptic operator and has analytic coefficients, and if $Au = f$ with $f$ analytic, then $u$ is analytic.

A weaker version of Theorem 1 has been proved by E. NELSON ([14], th. 7); he proves the analyticity of $u$ under the stronger assumption

$$\| A^k u \| \leq k!\, c^{k+1}.$$

Theorem 1 is proved by suitably estimating the $L^2$-norms of derivatives of order $km$ of $u$ in terms of $L^2$-norms of $u$, $Au$, ..., $A^k u$. The proof of this theorem uses some ideas of a paper of C. B. MORREY and L. NIRENBERG [13].

The use of the parabolic equation in the proofs of Theorems 2 and 3 was suggested by a paper of S. MINAKSHISUNDARAM [12].

For spaces of distributions we use the usual notation [17].

The results of this paper have been announced in [10].

**2. Statement of the theorems.** — Let $\Omega$ be an open subset of $\mathbf{R}^n$. Let $\mathcal{D}(\Omega)$ be the space of complex-valued $C^\infty$ functions with compact support in $\Omega$. $L^2(\Omega)$ is the Hilbert space of complex-valued square summable functions on $\Omega$, with scalar product $(\varphi, \psi)$ defined by

$$(\varphi, \psi) = \int_\Omega \varphi . \bar{\psi}\, dx$$

for $\varphi, \psi \in L^2(\Omega)$; $\| \varphi \|_{L^2(\Omega)}$ means $(\varphi, \varphi)^{1/2}$.

Let $A$ be a linear differential operator of order $m$,

$$A = \sum_{|\alpha| \leq m} a_\alpha(x) D^\alpha$$

with sufficiently differentiable complex-valued coefficients $a_\alpha(x)$ defined in $\Omega$, where $\alpha = (\alpha_1, \alpha_2, \ldots, \alpha_n)$, $\alpha_i$ being integer $\geq 0$ and we put :

$$|\alpha| = \alpha_1 + \alpha_2 + \ldots + \alpha_n,$$

$$D^\alpha = \left(\frac{\partial}{\partial x_1}\right)^{\alpha_1} \left(\frac{\partial}{\partial x_2}\right)^{\alpha_2} \cdots \left(\frac{\partial}{\partial x_n}\right)^{\alpha_n}.$$

We say now $A$ is an elliptic operator in $\Omega$, if the homogeneous form of order $m$

$$\sum_{|\alpha| = m} a_\alpha(x) \xi^\alpha \neq 0$$

for every $x \in \Omega$ and for every non vanishing real vector $\xi = (\xi_1, \xi_2, \ldots, \xi_n)$.

THEOREM 1. — *Let $\Omega$ be an open subset of $\mathbf{R}^n$. Let $A$ be a linear elliptic operator of order $m$ with analytic coefficients in $\Omega$. Let $A^k$ be the $k^{th}$ iterate of $A$. Suppose that a function $u$ (of class $C^\infty$) satisfies the inequality*

$$\| A^k u \|_{L^2(\Omega)} \leq (km)!\, c^{k+1}$$

*for every integer $k \geq 0$ with a positive constant $c$ independent of $k$. Then the function $u$ is analytic in $\Omega$.*

REMARK. — The above theorem is also valid for elliptic systems; the demonstration is the same as for the scalar case.

As for the following theorems, we consider a linear elliptic operator $L$ defined on $\Omega$ such that

$$(L\varphi, \psi) = (\varphi, L\psi)$$

for every $\varphi, \psi \in \mathcal{D}(\Omega)$.

Suppose further that $L$ when defined on $\mathcal{D}(\Omega)$ $(\subset L^2)$, where it is symmetric, has a strictly positive self-adjoint tion extension $\tilde{L}$.

Remark that these conditions entail that the form

$$L(x, \xi) = \sum_{|\alpha| = m} b_\alpha(x) \xi^\alpha$$

is real and definite for every $x \in \Omega$ and $\xi$ real vector, when $L = \sum_{|\alpha| \leq m} b_\alpha(x) D^\alpha$ has sufficiently smooth coefficients.

Let $\{E_\lambda\}$ be the spectral resolution of $\tilde{L}$. By the hypothesis on $\tilde{L}$, we have $\lambda > c_0 > 0$ on the spectrum.

We can now define a family of operators $\tilde{L}^s$ depending on the complex parameter $s$, by

$$\tilde{L}^s = \int \lambda^s \, dE_\lambda.$$

As we shall see in section 5, $\tilde{L}^s$ thus defined is a continuous linear map of $\mathcal{D}(\Omega)$ into the space of distributions $\mathcal{D}'(\Omega)$ for every $s$, so that $\tilde{L}^s$ defines a kernel $L^s(x, y)$ ([17], [19]); the theorems to be proved concern the regularity of the kernel $L^s(x, y)$.

THEOREM 2. — *Let $L$ be a linear elliptic differential operator with $C^\infty$ coefficients in an open set $\Omega$ of $\mathbf{R}^n$. We suppose further that $L$ admits a strictly positive self-adjoint realisation*

$$\tilde{L} = \int \lambda \, dE_\lambda.$$

*in $L^2(\Omega)$. Let $s$ be a complex number. Then the operator*

$$\tilde{L}^s = \int \lambda^s \, dE_\lambda.$$

*defines a kernel which is very regular.*

THEOREM 3. — *Let $L$ be a linear elliptic differential operator with analytic coefficients in an open set $\Omega$ of $\mathbf{R}^n$, admitting a strictly positive self-adjoint realisation $\tilde{L}$ in $L^2(\Omega)$. Then, for every complex number $s$, the kernel of the operator*

$$\tilde{L}^s = \int \lambda^s \, dE_\lambda$$

*is analytically very regular.*

For the definition of very regular kernels and analytically very regular kernels *see* ([17], chap. V, § 6).

As a consequence of the above theorems, $\tilde{L}^s(T)$ can be defined for $T$, a distribution with compact support and when $L$ has the $C^\infty$ (analytic) coefficients, $\tilde{L}^s(T)$ is an infinitely differentiable (resp. analytic) function in an open set of $\Omega$ where $T$ is an infinitely differentiable (resp. analytic) function.

**3. Preliminary lemmas.** — We consider in this section some lemmas which are required in the proof of Theorem 1.

Let $\Omega'$ be any open subset of $\Omega$. Let $u$ be of class $C^\infty$ on the closure $\overline{\Omega'}$ of $\Omega'$. Let $k$ be an integer $\geq 0$. We define the $k$-norm of $u \in C^\infty(\overline{\Omega'})$ by

$$\| u \|_{k,\Omega'} = \sum_{|\alpha| = k} \frac{k!}{\alpha!} \| D^\alpha u \|_{L^2(\Omega')},$$

where we put $\alpha! = \alpha_1! \, \alpha_2! \ldots \alpha_n!$ for $\alpha = (\alpha_1, \alpha_2, \ldots, \alpha_n)$.

LEMMA 3.1. — *Let $k$, $k'$, be given integers $\geq 0$. Then we have*

$$\| u \|_{k+k',\Omega'} = \sum_{|\alpha|=k} \frac{k!}{\alpha!} \| D^\alpha u \|_{k',\Omega'},$$

PROOF. — We have

$$\frac{(k+k')!}{\gamma!} = \sum_{\substack{|\alpha|=k \\ |\beta|=k' \\ \alpha+\beta=\gamma}} \frac{k!}{\alpha!} \frac{k'!}{\beta!}.$$

The lemma follows immediately from this equality.

The next lemma is a refined version of Friedrichs' inequality [7]. The proof is a modification of Friedrichs' proof as in [13].

We denote by $\Omega_r$ the ball $|x| < r$ of radius $r$ in $\mathbf{R}^n$.

LEMMA 3.2. — *Let $A$ be a linear elliptic operator of order $m$ with $C^\infty$ coefficients in $\Omega$. Let $r$, $\delta$ be positive numbers such that $\delta < r$ and $\Omega_{r+\delta} \subset \Omega$. Then there exists a constant $c > 0$ independent of $\delta$ such that for every $u \in C^\infty(\Omega)$ we have*

$$\| u \|_{m,\Omega_r} \leq c \{ \| A u \|_{0,\Omega_{r+\delta}} + \delta^{-m} \| u \|_{0,\Omega_{r+\delta}} \}.$$

PROOF. — Let $\zeta \in \mathcal{D}(\Omega)$ have its support in $\Omega_{r+\delta}$ and be such that $\zeta \equiv 1$ on $\Omega_r$ and satisfies

$$(3.1) \qquad \sup_{\Omega_{r+\delta}} | D^\alpha \zeta(x) | \leq c_\alpha \delta^{-|\alpha|} \qquad (\delta < r)$$

with $c_\alpha > 0$ depending only on $\alpha$.

For any $u \in C^\infty(\Omega)$, we shall consider $\zeta^m u$, which is of class $C^\infty$ having its support in $\Omega_{r+\delta}$. Since $A$ is an elliptic operator with $C^\infty$ coefficients, we have the well-known inequality [10]

$$(3.2) \qquad \| \zeta^m u \|_{m,\Omega_{r+\delta}} \leq c \{ \| A(\zeta^m u) \|_{0,\Omega_{r+\delta}} + \| \zeta^m u \|_{0,\Omega_{r+\delta}} \}$$

with a constant $c > 0$ depending only on $A$ and $\Omega_{r+\delta}$.

By using the estimate (3.1), we obtain

$$\| A(\zeta^m u) \|_{0,\Omega_{r+\delta}} \leq c' \left\{ \| \zeta^m A u \|_{0,\Omega_{r+\delta}} + \sum_{k=0}^{m-1} \delta^{-m+k} \| \zeta^k u \|_{k,\Omega_{r+\delta}} \right\},$$

$$\sum_{|\alpha|=m} \| \zeta^m D^\alpha u \|_{0,\Omega_{r+\delta}} \leq c'' \left\{ \| \zeta^m u \|_{m,\Omega_{r+\delta}} + \sum_{k=0}^{m-1} \delta^{-m+k} \| \zeta^k u \|_{k,\Omega_{r+\delta}} \right\}.$$

It follows then from (3.2),

$$(3.3)\qquad \sum_{k=0}^{m} \delta^{-m+k} \| \zeta^k u \|_{k,\Omega_{r+\delta}} \leq c \left\{ \| \zeta^m A u \|_{o,\Omega_{r+\delta}} + \sum_{k=0}^{m-1} \delta^{-m+k} \| \zeta^k u \|_{k,\Omega_{r+\delta}} \right\}$$

with $c > 0$ independent of $k$.

To complete the proof of the lemma, we need the following fact : for every $\varepsilon$, $\delta > 0$, there exists a constant $c$ independent of $\varepsilon$, $\delta$ and $u$ such that

$$(3.4)\qquad \sum_{|\alpha|=k} \| \zeta^k D^\alpha u \|_{o,\Omega} \leq \varepsilon \sum_{|\alpha|=k+1} \| \zeta^{k+1} D^\alpha u \|_{o,\Omega} + c(\varepsilon^{-1} + \delta^{-1}) \sum_{|\alpha|=k-1} \| \zeta^{k-1} D^\alpha u \|_{o,\Omega}$$

where $k \geq 1$.

In fact we have the equality

$$-(\zeta^k D^\alpha u, \zeta^k D^\alpha u) = (\zeta^{k-1} D^{\alpha'} u, \zeta^{k+1} D_1 D^\alpha u) + 2k((D_1 \zeta) \zeta^{k-1} D^{\alpha'} u, \zeta^k D^\alpha u),$$

where $\alpha' = (\alpha_1 - 1, \alpha_2, \ldots, \alpha_n)$ (we suppose $\alpha_1 \neq 0$) and $D_1 = \partial / \partial x_1$.

Now we can obtain the inequality (3.4) by Schwarz's inequality and by taking into account the estimate (3.1) for $\zeta$.

In (3.4) we take $k = m - 1$ and choose $\varepsilon$ as $\varepsilon = \delta / 2c$. Bringing the inequality thus obtained in the right side of (3.3), we have

$$(3.5)\qquad \sum_{k=0}^{m} \delta^{-m+k} \| \zeta^k u \|_{k,\Omega_{r+\delta}} \leq c \left\{ \| \zeta^m A u \|_{o,\Omega_{r+\delta}} + \sum_{k=0}^{m-2} \delta^{-m+k} \| \zeta^k u \|_{k,\Omega_{r+\delta}} \right\}$$

with $c > 0$ independent of $k$. Thus in the right side of (3.3), the terms corresponding to $k = m - 1$ can be absorbed in the left side. Repeating this procedure by using (3.4) with appropriate $\varepsilon$, we arrive finally at the desired inequality stated in the lemma.

**Lemma 3.3.** — *Let $q$ be positive integer such that $q < m$. Let $r < r_0$, $r_0$ being fixed. Then there exists a constant $c_m > 0$ depending only on $m$ and $r_0$ such that for every $\varepsilon > 0$ and $u \in C^\infty(\Omega)$ one has*

$$\| u \|_{q,\Omega_r} \leq \varepsilon \| u \|_{m,\Omega_r} + c_m \varepsilon^{-q/(m-q)} \| u \|_{o,\Omega_r}.$$

A proof of this lemma can be given by using Fourier transforms after extending the functions suitably to $\mathbf{R}^n$. Another proof can be found in [15] (Appendix).

REMARK. — Let $p$ be any integer $\geq 0$. By applying the above inequality to $D^\alpha u$ and by summing up the inequality thus obtained with respect to $\alpha$ such that $|\alpha| = mp$, we obtain from Lemma 3.1,

$$\| u \|_{pm+q,\Omega_r} \leq \varepsilon \| u \|_{(p+1)m,\Omega_r} + c_m \varepsilon^{-q/(m-q)} \| u \|_{pm,\Omega_r}$$

with the same constant $c_m$ as in the above lemma.

4. **Proof of theorem 1.** — In this section we shall prove Theorem 1. The proof is preceeded by several lemmas which permit one to estimate suitably $\| u \|_{km}$ in terms of zero-norms of $u$, $Au$, ..., $A^k u$.

We suppose throughout this section that $A$ has analytic coefficients. In this section, $c(c_1, c_2, \ldots$, etc.) will denote a positive constant, always independent of $k$, which may vary from place to place.

The first lemma gives an estimate for the commutator of the operator $D^\alpha$ and the operator of multiplication by an analytic function.

LEMMA 4.1. — *Let $a$ be an analytic function in $\overline{\Omega}'$. We define the commutator $[a, D^\alpha]$ by $[a, D^\alpha]u = a.D^\alpha u - D^\alpha(au)$, then we have for every integer $k > 0$.*

$$(4.1) \qquad \sum_{|\alpha|=k} \frac{k!}{\alpha!} \| [a, D^\alpha] u \|_{0,\Omega'} \leq k!\, c^k \sum_{p=0}^{k-1} (p!)^{-1} c^{-p} \| u \|_{p,\Omega'}$$

*with $c > 0$ independent of $k$.*

PROOF. — Since $a$ is analytic in $\overline{\Omega}'$, we have

$$(4.2) \qquad \sup_{\Omega'} | D^\alpha a | \leq \alpha!\, c^{|\alpha|+1}.$$

The Leibniz formula gives

$$D^\alpha(au) = \sum_{\beta \leq \alpha} \frac{\alpha!}{\beta!(\alpha-\beta)!} (D^\beta a)(D^{\alpha-\beta} u)$$

where $\alpha - \beta = (\alpha_1 - \beta_1, \ldots, \alpha_n - \beta_n)$ and $\beta \leq \alpha$ means $\beta_i \leq \alpha_i$ for each $i (i = 1, 2, \ldots, n)$.

From (4.2) and the definition of $[a, D^\alpha]$, it follows immediately

$$\sum_{|\alpha|=k} \frac{k!}{\alpha!} \| [a, D^\alpha] u \|_{0,\Omega'} \leq \sum_{|\alpha|=k} \sum_{\substack{\gamma \leq \alpha \\ \gamma \neq \alpha}} \frac{k!}{\gamma!} c^{k-|\gamma|} \| D^\gamma u \|_{0,\Omega_r}.$$

Now the number of $\alpha$'s such that $\alpha > \gamma$ for fixed $\gamma$ is at most of order $n^{k-|\gamma|}$, so that the right side is majorised by

$$\sum_{p=0}^{k} \frac{k!}{p!} (nc)^{k-p} \sum_{|\gamma|=p} \frac{p!}{\gamma!} \| D^\gamma u \|_{0,\Omega'};$$

this proves the lemma.

LEMMA 4.2. — *Let $r$, $\delta$ be as in the lemma 3.2. Let $\varepsilon$ be any positive number. Then there exist constants $c$, $c_1$ ($c$ depending only on $A$ and $c_1$ depending on $A$ and $\varepsilon$) such that one has for every $k$ and $u \in C^\infty(\Omega)$,*

$$(4.3)\quad \| u \|_{(k+1)m,\Omega_r} \leq c \Big\{ \| A u \|_{km,\Omega_{r+\delta}} + \delta^{-m} \| u \|_{km,\Omega_{r+\delta}} + \varepsilon \| u \|_{(k+1)m,\Omega_{r+\delta}} + ((k+1)m)!\, c_1^{k+1} \sum_{p=0}^{k} ((pm)!)^{-1} c_1^{-p} \| u \|_{pm,\Omega_{r+\delta}} \Big\}.$$

PROOF. — From Lemma 3.1 and the Friedrichs' inequality (Lemma 3.2), we have

$$(4.4)\quad \| u \|_{(k+1)m,\Omega_r} = \sum_{|\alpha|=km} \frac{(km)!}{\alpha!} \| D^\alpha u \|_{m,\Omega_r} \leq c \Big\{ \| A u \|_{km,\Omega_{r+\delta}} + \delta^{-m} \| u \|_{km,\Omega_{r+\delta}} + \sum_{|\alpha|=km} \frac{(km)!}{\alpha!} \| [A, D^\alpha] u \|_{0,\Omega_{r+\delta}} \Big\}.$$

Now, writting $A$ explicitly as $A = \sum_{|\beta| \leq m} a_\beta D^\beta$ with analytic coefficients $a_\beta$ and applying the Lemma 4.1 for $[A, D^\alpha] u = \sum_{|\beta| \leq m} [a, D^\alpha] D^\beta u$, we obtain

$$(4.5)\quad \sum_{|\alpha|=km} \frac{(km)!}{\alpha!} \| [A, D^\alpha] u \|_{0,\Omega_{r+\delta}} \leq \sum_{p=0}^{km-1} \sum_{q=0}^{m} \frac{(km)!}{p!} c_1^{km-p} \| u \|_{p+q,\Omega_{r+\delta}}.$$

Since we may suppose $c_1 > 1$ in (4.5), it follows immediately that there exists a constant $c_2 > 0$ independent of $k$ such that

$$(4.6)\quad \sum_{|\alpha|=km} \frac{(km)!}{\alpha!} \| [A, D^\alpha] u \|_{0,\Omega_{r+\delta}} \leq \sum_{s=0}^{(k+1)m-1} \frac{((k+1)m)!}{s!} c_2^{(k+1)m-s} \| u \|_{s,\Omega_{r+\delta}}.$$

We wish now to majorize the right side of (4.6), containing terms $\| u \|_s$, for $s = 0, 1, \ldots, (k+1)m - 1$, by an expression which contains only $\| u \|_{pm}$, for $p = 0, 1, \ldots, (k+1)$.

For this purpose, we write $s$ as $s = pm + q$ with $0 \leq p \leq k$, and $0 \leq q < m$. Then the remark of Lemma 3.3 gives

$$(4.7)\quad \| u \|_{pm+q,\Omega_{r+\delta}} \leq \varepsilon' \| u \|_{(p+1)m,\Omega_{r+\delta}} + c_m \varepsilon'^{-q/(m-q)} \| u \|_{pm,\Omega_{r+\delta}}$$

with $c_m$ independent of $\varepsilon'$ and $\delta$.

In (4.7), we choose $\varepsilon'$ as

$$\varepsilon' = \varepsilon \frac{(pm + q)!}{((p+1)m)!} c_2^{-(m-q)}$$

where $\varepsilon$ $(0 < \varepsilon < 1)$ is given.

Then we have

$$\varepsilon'^{-q/(m-q)} \leq \left(\frac{m}{\varepsilon}\right)^m \frac{(pm+q)!}{(pm)!} c_2^q$$

so that we obtain for $s = pm + q$,

$$(4.8) \qquad \frac{c^{-s}}{s!} \| u \|_{s,\Omega_{r+\delta}} \leq \varepsilon \frac{c_2^{-(p+1)m}}{((p+1)m)!} \| u \|_{(p+1)m,\Omega_{r+\delta}} + c_m \left(\frac{m}{\varepsilon}\right)^m \frac{c_2^{-pm}}{(pm)!} \| u \|_{pm,\Omega_{r+\delta}}.$$

Bringing this in the expression (4.6), we have

$$(4.9) \quad \sum_{|\alpha| = km} \frac{(km)!}{\alpha!} \| [A, D^\alpha] u \|_{0,\Omega_{r+\delta}} \leq m\varepsilon \| u \|_{(k+1)m,\Omega_{r+\delta}} + c'(\varepsilon) \sum_{p=0}^{k} \frac{((k+1)m)!}{(pm)!} c_2^{(k+1)m-pm} \| u \|_{pm,\Omega_{r+\delta}}$$

where we put $c'(\varepsilon) = 1 + m\varepsilon + \left(\frac{m}{\varepsilon}\right)^m c_m$. We take now in (4.9) the constant $c_2$ large enough to absorb the constant $c'(\varepsilon)$ which is independent of $k$. Then, from (4.4), the desired inequality follows.

**Definition** (*see* [13]). — Let $\lambda$ be a positive number. For each integer $k \geq 0$, we define

$$\sigma^k(u, \lambda, R) = ((km)!)^{-1} \lambda^{-k} (R - r)^{km} \sup_{R/2 \leq r < R} \| u \|_{km,\Omega_r}.$$

**Lemma 4.3.** — *Let $R < 1$. There exists a constant $\lambda$ depending only on $A$ and $R$ such that for every $k$ and $u \in C^\infty(\Omega)$ we have*

$$(4.10) \quad \sigma^{k+1}(u, \lambda, R) \leq [(km+1)\ldots((k+1)m)]\, \sigma^k(Au, \lambda, R) + \sum_{p=0}^{k} \sigma^p(u, \lambda, R).$$

**Proof.** — Multipliying by $[((k+1)m)!]^{-1} \lambda^{-(k+1)} (R - r)^{(k+1)m}$ on both sides of the inequality of Lemma 4.2 and taking the supremum for $R/2 \leq r < R$, we obtain

$$(4.11) \qquad \sigma^{k+1}(u, \lambda, R) \leq \sup_{R/2 \leq r < R} (I_1 + \varepsilon I_2 + I_3 + I_4),$$

where

$$(4.12)\qquad \begin{cases} I_1 = c[((k+1)m)!]^{-1}\lambda^{-(k+1)}(R-r)^{(k+1)m}\|Au\|_{km,\Omega_{r+\delta}}, \\ I_2 = c[((k+1)m)!]^{-1}\lambda^{-(k+1)}(R-r)^{(k+1)m}\|u\|_{(k+1)m,\Omega_{r+\delta}}, \\ I_3 = c[((k+1)m)!]^{-1}\lambda^{-(k+1)}(R-r)^{(k+1)m}\delta^{-m}\|u\|_{km,\Omega_{r+\delta}}, \\ I_4 = c\lambda^{-(k+1)}(R-r)^{(k+1)m}\sum_{p=0}^{k}\frac{c_1^{k+1-p}}{(pm)!}\|u\|_{pm,\Omega_{r+\delta}}. \end{cases}$$

We choose in what follows $\delta = \frac{R-r}{k+1}$; then we have

$$\left(\frac{R-r}{R-r-\delta}\right)^{km} = \left(1-\frac{1}{k+1}\right)^{-km} < c_2$$

with $c_2$ independent of $k$. It follows now from the definition of $\sigma^k(u, \lambda, R)$,

$$(4.13)\qquad I_1 \leq [(km+1)\dots((k+1)m)]^{-1}\left(\frac{cc_2}{\lambda}\right)\sigma^k(Au, \lambda, R).$$

Similarly

$$(4.14)\qquad I_2 \leq (cc_2)\,\sigma^{k+1}(u, \lambda, R).$$

For $I_3$, we have

$$(4.15)\qquad I_3 \leq \frac{c}{\lambda}\left(\frac{R-r}{R-r-\delta}\right)^{km}\left(\frac{R-r}{\delta}\right)^{m}\frac{(km)!}{((k+1)m)!}\sigma^k(u, \lambda, R).$$

Since we have from the definition of $\delta$.

$$\left(\frac{R-r}{\delta}\right)^{m} = (k+1)^m$$

it follows from (4.15)

$$(4.16)\qquad I_3 \leq \left(\frac{cc_2}{\lambda}\right)\sigma^k(u, \lambda, R).$$

Finally we obtain for $I_4$,

$$(4.17)\qquad I_4 \leq \left(\frac{cc_1c_2}{\lambda}\right)\sum_{p=0}^{k}\left(\frac{c_1}{\lambda}\right)^{k-p}\sigma^p(u, \lambda, R)\qquad (\lambda \geq 1).$$

It follows now for every $k \geq 0$,

$$(4.18)\qquad (1-\varepsilon c)\,\sigma^{k+1}(u, \lambda, R) \leq [(km+1)\dots((k+1)m)]^{-1}\left(\frac{c_1}{\lambda}\right)\sigma^k(Au, \lambda, R) + \left(\frac{c_1}{\lambda}\right)\sum_{p=0}^{k}\left(\frac{c_1}{\lambda}\right)^{k-p}\sigma^p(u, \lambda, R)$$

for sufficiently large constants $c$, $c_1 > 0$, $c$ being independent of $\varepsilon$, while $c_1$ depends on $\varepsilon$. After we have chosen $\varepsilon = 1/2c$ in (4.18) $c_1$ is a constant dependent only on $A$ and $R$ so that it is possible to find $\lambda$ independent of $k$ such that $\lambda > 2c_1$; thus we obtain the inequality (4.10).

**Lemma 4.4.** — *Let $\lambda$ be the same constant as in lemma 4.3; we have then*

$$(4.19)\qquad \sigma^{k+1}(u, \lambda, R) \leqq \sum_{p=0}^{k+1} 2^{k-p+1} \binom{k+1}{p} ((mp)!)^{-1} \sigma^0(A^p u, \lambda, R).$$

**Proof.** — The proof is by induction on $k$. For $k=0$, the Lemma is valid (*see* Lemma 4.3). Suppose that the lemma is valid upto $k-1$. Applying the induction hypothesis to the function $Au$, we have

$$(4.20)\qquad \sigma^k(Au, \lambda, R) \leqq \sum_{p=0}^{k} 2^{k-p} \binom{k}{p} ((pm)!)^{-1} \sigma^0(A^{p+1}u, \lambda, R).$$

Also, we have for $q \leqq k$,

$$(4.21)\qquad \sigma^q(u, \lambda, R) \leqq \sum_{p=0}^{q} 2^{q-p} \binom{q}{p} ((pm)!)^{-1} \sigma^0(A^p u, \lambda, R).$$

From Lemma 4.3, we get

$$(4.22)\qquad \sigma^{k+1}(u, \lambda, R) \leqq [(km+1)\dots((k+1)m)]^{-1} \times \sum_{p=0}^{k} 2^{k-p} \binom{k}{p} ((pm)!)^{-1} \sigma^0(A^{p+1}u, \lambda, R) + \sum_{q=0}^{k} \sum_{p=0}^{q} 2^{q-p} \binom{q}{p} ((pm)!)^{-1} \sigma^0(A^p u, \lambda, R).$$

Now, let $c_p$ be the coefficient of $\sigma^0(A^p u, \lambda, R)$. Then for $0 \leqq p \leqq k$

$$c_p = [(km+1)\dots((k+1)m)]^{-1} 2^{k-p+1} \binom{k}{p-1} [((p-1)m)!]^{-1} + \sum_{q=p}^{k} 2^{q-p} \binom{q}{p} [(mp)!]^{-1}.$$

Since

$$\sum_{q=p}^{k} 2^{q-p} \binom{q}{p} \leqq 2^{k-p+1} \binom{k}{p}$$

we get

$$c_p \leqq 2^{k-p+1} \binom{k+1}{p} ((pm)!)^{-1}.$$

On the other hand, for $p = k + 1$, we have evidently,

$$c_{k+1} = [((k+1)m)!]^{-1}.$$

Hence, it follows

$$(4.23) \qquad \sigma^{k+1}(u, \lambda, R) \leq \sum_{p=0}^{k+1} 2^{k-p+1} \binom{k+1}{p} [((pm)!)]^{-1} \sigma^0(A^p u, \lambda, R);$$

this is the inequality which we wanted to prove; thus the induction is completed.

PROOF OF THEOREM 1. — Let $u \in C^\infty(\Omega)$ such that

$$(4.24) \qquad \| A^k u \|_{L^2(\Omega')} \leq (km)!\, c^{k+1}$$

for $\Omega'$ an open set of $\Omega$ and for all $k \geq 0$ with a constant $c$ independent of $k$.

Since the analyticity is a local property, we may suppose that the origin of $\mathbf{R}^n$ belongs to $\Omega'$ and it is sufficient to prove analyticity at the origin. Take $R < 1$ with $\Omega_R \subset \Omega'$, then

$$(4.25) \qquad \sigma^0(A^k u, \lambda, R) = \| A^k u \|_{L^2(\Omega_R)} \leq km!\, c^{k+1}.$$

Now from Lemma 4.4, we have

$$(4.26) \qquad \sigma^{k+1}(u, \lambda, R) \leq \sum_{p=0}^{k+1} 2^{k-p+1} \binom{k+1}{p} [((pm)!)]^{-1} \sigma^0(A^p u, \lambda, R)$$
$$\leq \sum_{p=0}^{k+1} 2^{k-p+1} c^{p+1} \binom{k+1}{p} = c(c+2)^{k+1}.$$

From the definition of $\sigma^{k+1}(u, \lambda, R)$ we obtain

$$\| u \|_{(k+1)m, \Omega_{R/2}} \leq ((k+1)m)!\,.\, c^{k+1}$$

with a certain constant $c$ independent of $k$.

Then, Lemma 3.3 permits us to estimate $\| u \|_p$ for $p = 0, 1, \ldots$ by $\| u \|_{(k+1)m}$ for $k = 0, 1, \ldots$ and we have

$$(4.27) \qquad \| u \|_{p, \Omega_{R/2}} \leq p!\, c^{p+1}$$

for all $p\ (= 0, 1, \ldots)$, where $c$ is a constant depending only on $A$ and $\Omega_R$. Now, by Sobolev's lemma [13], we see that $u$ is analytic at the origin. Hence, the proof of Theorem 1 is completed.

5. **Regurarity of the kernel of $\tilde{L}^s$.** — We denote by $D(\tilde{L}^s)$ the domain of $\tilde{L}^s$, that is the set of elements $f \in L^2(\Omega)$ such that $\int |\lambda^s|^2 d\|E_\lambda f\| < \infty$.

Then, under our hypothesis on $\tilde{L}$, it is easy to see that

(5.1) $$D(\tilde{L}^s) \subseteq D(\tilde{L}^{s'}) \quad \text{if} \quad Rl\, s \geqq Rl\, s',$$

(5.2) for every complex number $s$,

$$\tilde{L}^s f \in \bigcap_{k=0}^{\infty} D(\tilde{L}^k) \quad \text{if} \quad f \in \bigcap_{k=0}^{\infty} D(\tilde{L}^k).$$

Let $f \in \bigcap_{k=0}^{\infty} D(\tilde{L}^k)$. It follows from (5.1), (5.2) that $f \in D(\tilde{L}^{s+k})$ and $\tilde{L}^s f \in D(\tilde{L}^k)$ for every complex number $s$ and integer $k \geqq 0$. We have then

(5.3) $$\tilde{L}^k \tilde{L}^s f = \tilde{L}^s \tilde{L}^k f = \tilde{L}^{s+k} f$$

(for these properties, *see* [16], § 228; [18], p. 222).

PROPOSITION 5.1. — *For any complex number $s$, $\tilde{L}^s$ defines a kernel $L^s(x, y)$, that is, a distribution in the product space $\Omega \times \Omega$.*

PROOF. — We first consider the case $Rl\, s < 0$. In this case, $\tilde{L}^s$ is a continuous map of $L^2(\Omega)$ into itself. For, by hypothesis on $\tilde{L} = \int \lambda\, dE_\lambda$, we have a positive constant $c_0$ such that $\lambda > c_0$ on the spectrum, hence $\lambda^{Rl\, s} \leqq c_0^{Rl\, s}$ for $\lambda \leqq 1$ and $\lambda^{Rl s} \leqq 1$ for $\lambda > 1$, since $Rl\, s < 0$.

Thus, $\lambda^s$ is bounded on the spectrum of $\tilde{L}$. Hence $\tilde{L}^s$ is a continuous linear map of $L^2(\Omega)$ into itself. *A fortiori*, $\tilde{L}^s$ is a continuous linear map of $\mathcal{D}(\Omega)$ into $\mathcal{D}'(\Omega)$. By the kernel theorem of L. SCHWARTZ [19], $\tilde{L}^s$ defines a kernel.

For general $s$, we take a positive integer $m$ such that $Rl(s - m) < 0$. Then, as seen above, $\tilde{L}^{s-m}$ is a continuous map of $\mathcal{D}(\Omega)$ into $\mathcal{D}'(\Omega)$ while $\tilde{L}^m$, $m^{\text{th}}$ iterate of $L$ with $C^\infty$ coefficients, is evidently a continuous map of $\mathcal{D}(\Omega)$ into itself.

Now, the proposition follows from (5.3), by remarking that

$$\tilde{L}^s \varphi = \tilde{L}^{s-m} \tilde{L}^m \varphi \quad \text{for} \quad \varphi \in \mathcal{D}(\Omega) \quad \text{since} \quad \mathcal{D}(\Omega) \subseteq \bigcap_{k=0}^{\infty} D(\tilde{L}^k).$$

From now on, we denote by $L^s(x, y)$ the kernel of $\tilde{L}^s$.

**Proposition 5.2.** — *For every complex number $s$, the kernel $L^s(x, y)$ is regular.*

**Proof.** — We have to prove that $\tilde{L}^s$ maps continuously $\mathcal{D}(\Omega)$ into $\mathcal{E}(\Omega)$ and can be extended to a continuous linear map of $\mathcal{E}'(\Omega)$ into $\mathcal{D}'(\Omega)$.

Suppose that for every $s$, $\tilde{L}^s$ maps continuously $\mathcal{D}(\Omega)$ into $\mathcal{E}(\Omega)$. Let $\varphi$, $\psi$ be in $\mathcal{D}(\Omega)$. We have then $(\tilde{L}^s\varphi, \psi) = (\varphi, \tilde{L}^{\bar{s}}\psi)$, $\bar{s}$ denoting the conjugate complex of $s$, this implies that $\tilde{L}^s$ can be identified on the dense subspace $\mathcal{D}(\Omega)$ of $\mathcal{E}'(\Omega)$ with the transpose of $\tilde{L}^{\bar{s}}$, while the transpose of $\tilde{L}^{\bar{s}}$ is a continuous map of $\mathcal{E}'(\Omega)$ into $\mathcal{D}'(\Omega)$ when $\tilde{L}^{\bar{s}}$ is a continuous map of $\mathcal{D}(\Omega)$ into $\mathcal{E}(\Omega)$. Hence, $\tilde{L}^s$ can be extended to a continuous map of $\mathcal{E}'(\Omega)$ into $\mathcal{D}'(\Omega)$.

It remains now to prove that $\tilde{L}^s$ maps continuously $\mathcal{D}(\Omega)$ into $\mathcal{E}(\Omega)$.

Remark first that the image of $\mathcal{D}(\Omega)$ by $\tilde{L}^s$ is contained in $\mathcal{E}(\Omega)$. For, if $\varphi \in \mathcal{D}(\Omega)$, then $\varphi \in \bigcap_{k=0}^{\infty} D(\tilde{L}^k)$, so that by (5.2) we have $\tilde{L}^s\varphi \in \bigcap_{k=0}^{\infty} D(\tilde{L}^k)$. From the regularity theorem for a linear elliptic operator with $C^\infty$ coefficients ([7], [15]), it follows that $\tilde{L}^s\varphi$ is of class $C^\infty$.

As for the continuity of the mapping $\tilde{L}^s$, it is sufficient [17] to verify that the image of every bounded set in $\mathcal{D}(\Omega)$ by $\tilde{L}^s$ is also a bounded set in $\mathcal{E}(\Omega)$.

Let $s$ be such that $Rl\, s < 0$. Let $B$ be a bounded set in $\mathcal{D}(\Omega)$. Then, by definition [17], the image $\tilde{L}^k(B)$ of $B$ by $\tilde{L}^k$ is bounded in $\mathcal{D}(\Omega)$, *a fortiori*, bounded in $L^2(\Omega)$. Now $\tilde{L}^s$ is a continuous map of $L^2(\Omega)$ into itself, so that $\tilde{L}^s\tilde{L}^k(B)$ is bounded in $L^2(\Omega)$. On the other hand, $\tilde{L}^s(B)$ is a family of $C^\infty$ functions belonging to the domain of $\tilde{L}^k$; hence it follows from (5.3) that $\tilde{L}^k\tilde{L}^s(B)$ is bounded in $L^2(\Omega)$, from this, we see, according to Lemma 3.2 and Sobolev's lemma [13], that $\tilde{L}^s(B)$ is a family of $C^\infty$ functions whose derivatives of orders $mk - \left[\frac{n}{2}\right] - 1$ are uniformly bounded on every compact of $\Omega$. Since $k$ is arbitrary, this proves that $\tilde{L}^s(B)$ is bounded in $\mathcal{E}(\Omega)$.

For general $s$, as in the proof of Proposition 5.1, choose $m$ so large that $Rl(s-m) < 0$ and remark that $\tilde{L}^s\varphi = \tilde{L}^{s-m}\tilde{L}^m\varphi$ for $\varphi \in \mathcal{D}(\Omega)$, then $\tilde{L}^m$ and $\tilde{L}^{s-m}$ map respectively $\mathcal{D}(\Omega)$ into $\mathcal{D}(\Omega)$ and $\mathcal{E}(\Omega)$ continuously. This completes the proof.

**6. Estimates for the Green's function of the associated parabolic operator.** — Consider the family of operators $G_t = \int e^{-\lambda t}\, dE_\lambda$ for $t > 0$.

$G_t$ is a bounded and Hermitian operator in $L^2(\Omega)$. Associated with these operators we have a $C^\infty$ function in $\mathbf{R}\times\Omega\times\Omega$,

$$G(t, x, y) = \int e^{-\lambda t}\, de(\lambda, x, y),$$

where $e(\lambda, x, y)$ denotes the spectral function of $\tilde{L}$ [8].

We have then

$$\left(\frac{\partial}{\partial t} + L_x\right) G(t, x, y) = 0 \qquad \text{and} \qquad \left(\frac{\partial}{\partial t} + L_y\right) \overline{G(t, x, y)} = 0$$

for $t > 0$.

The next lemma shows that the function $G(t, x, y)$ and its derivatives fall off exponentially as $t \to \infty$.

LEMMA 6.1. — *Let $H$ be a compact in $\Omega\times\Omega$. Under our assumption that $\tilde{L}$ is strictly positive operator ($\lambda > c_0 > 0$ on the spectrum), we have*

$$\left|\left(\frac{\partial}{\partial t}\right)^p D_x^\alpha D_y^\beta G(t, x, y)\right| \leq c\, e^{-c_0 t/2}$$

*for $t > 1$ and uniformly for $(x, y) \in H$, where $c$ depends on $p$, $\alpha$, $\beta$ and $H$.*

PROOF. — Denote by $\bar{L}$ the elliptic operator with conjugate complex coefficients of $L$.

Consider the operator :

$$L_x + \bar{L}_y = L\left(x, \frac{\partial}{\partial x}\right) + \bar{L}\left(y, \frac{\partial}{\partial y}\right)$$

which is evidently elliptic with $C^\infty$ coefficients in the product space $\Omega\times\Omega$.

Now, by Lemma 3.2 and Sobolev's lemma [13] applied to $(L_x + \bar{L}_y)$, it is easy to see that the desired estimate is a simple consequence of the following : let $U$ be a relatively compact open subset in $\Omega$ such that $H \subset U \times U$. Then for every positive integers $k'$, $k''$, we have

$$\left|(L_x + \bar{L}_y)^{k'} \left(\frac{\partial}{\partial t}\right)^{k''} G(t, x, y)\right| \leq c\, e^{-c_0 t/2}$$

for $t > 1$ and for $(x, y) \in U\times U$. Since

$$L_x G(t, x, y) = \bar{L}_y G(t, x, y) = -\frac{\partial}{\partial t} G(t, x, y) \qquad \text{for} \quad t > 0,$$

it is sufficient to estimate $\left(\frac{\partial}{\partial t}\right)^k G(t, x, y)$ for every positive integer $k$.

Let $m$ be a sufficiently large positive integer such that $\tilde{L}^{-m}$ has a kernel $K(x, y)$ of the Carleman type ([4], [5], [8]). For $x \in \Omega$, let $K_x \in L^2$ denote the function $K(x, \hat{})$.

Now

$$\left|\left(\frac{\partial}{\partial t}\right)^k G(t, x, y)\right| = \left|\left(\frac{\partial}{\partial t}\right)^k \int e^{-\lambda t}\, de(\lambda, x, y)\right|$$

$$= \left|\int e^{-\lambda t}(-\lambda)^k\, de(\lambda, x, y)\right|$$

$$= \left|\int e^{-\lambda t}(-\lambda)^k \lambda^{2m}\, de\,(E_\lambda K_x, K_y)\right|$$

$$\leqq e^{-c_0 t/2} \int e^{-\lambda/2} \lambda^{2m+k} \left| d(E_\lambda K_x, K_y)\right|$$

since $\lambda > c_0$ and $t \geqq 1$. Now the variation of $(E_\lambda K_x, K_y)$ in $\mathbf{R}$ is majorised by $\|K_x\|_{L^2} \|K_y\|_{L^2}$ ([16], § 126) and $\|K_x\|_{L^2} \|K_y\|_{L^2} \leqq c(U)$ for $(x, y) \in U \times U$, where $c(U)$ is a constant depending only on $U$ and $\tilde{L}$.

It follows that

$$\left|\left(\frac{\partial}{\partial t}\right)^k G(t, x, y)\right| \leqq c\, e^{-c_0 t/2}$$

for $t > 1$ and $(x, y) \in H$ with a constant $c$ depending on $k$, $H$ and $\tilde{L}$. Thus Lemma 5.1 is proved.

We next consider the behaviour of $G(t, x, y)$ and its derivatives as $t \to 0$. The required information is given by the results of G. Bergendal [1] and S. D. Eidelman [6].

Let $K$ be a relatively compact open subset of $\Omega$. Consider now the parabolic operator $\left(\frac{\partial}{\partial t} + L\right)$ on $\mathbf{R} \times K$ assiociated with $L$. According to S. D. Eidelman, we have a fundamental solution $E(t, x, y)$ of $\left(\frac{\partial}{\partial t} + L_x\right)$. It is of class $C^\infty$ in $(t, x, y)$ when $t > 0$ and satisfies near $t = 0$ the following estimate.

Lemma 6.2 (S. D. Eidelman). — *For $0 < t < 1$ and $(x, y) \in K \times K$, we have*

$$\left|\left(\frac{\partial}{\partial t}\right)^p D_x^\alpha D_y^\beta E(t, x, y)\right| \leqq c t^{-(pm+|\alpha|+|\beta|+n)/m} e^{-c_1 |x-y|^{1+\mu} t^{-\mu}},$$

*where $\mu = 1/(m-1)$ and $c_1$ depends only on $L$, $K$, while $c$ depends also on $p$, $\alpha$, $\beta$.*

As for the behaviour of $G(t, x, y)$ we have

Lemma (6.3) (G. Bergendal). — *Let $H$ be a compact subset of $\Omega \times \Omega$ such that $H \subset K \times K$. Let $E(t, x, y)$ be the same as in lemma 6.2. Then there exist positive constants $c$, $c_1$ such that*

$$\left|\left(\frac{\partial}{\partial t}\right)^p D_x^\alpha D_y^\beta [G(t, x, y) - E(t, x, y)]\right| \leqq c\, e^{-c_1 t^{-\mu}}$$

*for* $0 < t < 1$ *and for* $(x, y) \in H$, *where* $c_1$ *depends only on* $L$ *and* $H$, *while* $c$ *depends also on* $p$, $\alpha$, $\beta$.

For $p + |\alpha| + |\beta| = 0$, this is proved in [1]. The general case can be proved in a similar fashion (*see* [2], § 2.3).

## 7. A representation for the kernel $L^s(x, y)$ in terms of the Green's function $G(t, x, y)$.

PROPOSITION 7.1. — *Let* $s$ *be a complex number such that* $Rl\, s < -n/m$. *Then we have*

$$L^s(x, y) = \frac{1}{\Gamma(-s)} \int_0^\infty t^{-s-1} G(t, x, y)\, dt. \tag{7.1}$$

*The integral on the right converges uniformly on every compact subset of* $\Omega \times \Omega$ *and represents a continuous function of* $(x, y)$ *in* $\Omega \times \Omega$, *where we denote by* $\Gamma(-s)$ *the Gamma function.*

PROOF. — From Lemma 6.1 we have for $t \geq 1$ and for $(x, y) \in H$,

$$|G(t, x, y)| \leq c\, e^{-c_0 t/2} \tag{7.2}$$

while for $0 < t < 1$ and for $(x, y) \in H$, it follows from Lemma 6.2 and Lemma 6.3,

$$\begin{aligned} |G(t, x, y)| &\leq |E(t, x, y)| \\ &\quad + |(E - G)(t, x, y)| \leq c\, t^{-n/m} + c\, e^{-c_1 t^{-\mu}} \end{aligned} \tag{7.3}$$

with positive constants $c$, $c_1$ depending on $H$.

From these estimates, it is easy to see that the integral converges uniformly for $(x, y) \in H$ when $Rl\, s < -n/m$ and represents a continuous function of $(x, y)$ since $G(t, x, y)$ is of class $C^\infty$ for $t > 0$.

We shall prove now the equality stated in proposition 7.1. For $\varphi$, $\psi \in \mathcal{D}(\Omega)$, consider

$$P = \frac{1}{\Gamma(-s)} \left\langle \int_0^\infty t^{-s-1} G(t, x, y)\, dt,\ \varphi(x)\, \overline{\psi}(y) \right\rangle,$$

where $\langle\ ,\ \rangle$ denote the scalar product between $\mathcal{D}'(\Omega \times \Omega)$ and $\mathcal{D}(\Omega \times \Omega)$. By what has been seen,

$$\begin{aligned} P &= \frac{1}{\Gamma(-s)} \int_0^\infty t^{-s-1}\, dt \int_{\Omega \times \Omega} G(t, x, y)\, \varphi(x)\, \overline{\psi}(y)\, dx\, dy \\ &= \frac{1}{\Gamma(-s)} \int_0^\infty t^{-s-1}\, dt \int_{c_0}^\infty e^{-\lambda t}\, d(E_\lambda \varphi, \psi), \end{aligned}$$

where the integration $\int e^{-\lambda t}\, d(E_\lambda \varphi, \psi)$ is taken in the sense of the Radon-Stieltjes integral with respect to the complex-valued function of bounded variation $(E_\lambda \varphi, \psi)$ in $-\infty < \lambda < \infty$.

Let

$$(E_\lambda \varphi, \psi) = [\rho_1(\lambda) - \rho_2(\lambda)] + i[\rho_3(\lambda) - \rho_4(\lambda)]$$

be the canonical resolution of $(E_\lambda \varphi, \psi)$ with the real valued monotone increasing functions of bounded variation $\rho_k(\lambda)$, $k = 1, 2, 3, 4$ ([20], p. 202).

Then we have

$$\int_{c_0}^{\infty} e^{-\lambda t}\, d(E_\lambda \varphi, \psi) = \sum_{k=1}^{4} \varepsilon_k \int_{c_0}^{\infty} e^{-\lambda t}\, d\rho_k(\lambda)$$

where $\varepsilon_1 = -\varepsilon_2 = -i\varepsilon_3 = i\varepsilon_4 = 1$.

Consider now

$$\int_{c}^{\infty} t^{-s-1}\, dt \int_{c_0}^{\infty} e^{-\lambda t}\, d\rho_k(\lambda).$$

Since $t^{-s-1} e^{-\lambda t}$ is a continuous function of $(t, \lambda)$ in the integration domain: $0 < t < \infty$, $c_0 < \lambda < \infty$ and the ovbious estimate $|t^{-s-1} e^{-\lambda t}| \leqq t^{-Rl\, s-1} e^{-c_0 t}$ implies that it is integrable there with respect to the product measure $dt\, d\rho_k(\lambda)$ when $Rl\, s < 0$.

By *Fubini's theorem*, we have,

$$\int_{0}^{\infty} t^{-s-1}\, dt \int_{c_0}^{\infty} e^{-\lambda t}\, d\rho_k(\lambda) = \int_{c_0}^{\infty} d\rho_k \int_{0}^{\infty} t^{-s-1} e^{-\lambda t}\, dt.$$

Noting that $\int_{0}^{\infty} t^{-s-1} e^{-\lambda t}\, dt = \Gamma(-s)\lambda^s$ and summing up the above integral with respect to $k$, we have

$$P = \sum_{k=1}^{4} \varepsilon_k \int \lambda^s\, d\rho_k(\lambda)$$

which is equal to

$$\int \lambda^s\, d(E_\lambda \varphi, \psi) = (\tilde{L}^s \varphi, \psi).$$

This completes the proof.

8. **Proof of theorem 2.** — As in paragraph 5, we see that it is sufficient to prove Theorem 2 for $Rl\, s < -\frac{n}{m}$. Since we have already proved that $L^s(x, y)$ is regular, it is sufficient to prove that $L^s(x, y)$ is of class $C^\infty$ outside the diagonal [17].

For $Rl\,s < -\frac{n}{m}$, we have by Proposition 7.1,

$$(8.1) \qquad L^s(x, y) = \frac{1}{\Gamma(-s)} \int_0^\infty t^{-s-1} G(t, x, y)\, dt.$$

If $(x, y)$ belongs to a compact set $H$ in the complement of the diagonal we see from Lemmas 6.1, 6.2 and 6.3 that

$$(8.2) \qquad \left| \left(\frac{\partial}{\partial t}\right)^k D_x^\alpha D_y^\beta G(t, x, y) \right| \leq c\, e^{-c_1(t+t^{-\mu})} \qquad (0 < t < \infty)$$

with positive constants $c$, $c_1$, where $c_1$ is independent of $k$, $\alpha$, $\beta$.

It now follows from (8.1) and (8.2) that $L^s(x, y)$ is of class $C^\infty$ outside the diagonal, since we may differentiate under the integral sign any number of times.

## 9. Proof of theorem 3.

— In this section $c$, $c_i (i = 1, 2, \ldots)$ will denote positive constants independent of $k$. We suppose that $L$ has analytic coefficients.

To prove Theorem 3, it is sufficient to prove the following two statements:

(i) $L^s(x, y)$ is an analytic function in the complement of the diagonal in $\Omega \times \Omega$.

(ii) For each $\varphi \in \mathcal{D}(\Omega)$, $\tilde{L}^s \varphi$ is an analytic function in every open set where $\varphi$ is analytic.

Proof of (i). — $(L_x + \bar{L}_y)^k$ is a linear elliptic operator of order $m$ with analytic coefficients in $\Omega \times \Omega$. Applying Theorem 1, we see that to prove (i) it is sufficient to prove the following: for each compact set $H$ in the complet ment of the diagonal, there exists a constant $c$ independent of $k$ such tha-

$$(9.1) \qquad \sup_{(x, y) \in H} |(L_x + \bar{L}_y)^k L^s(x, y)| \leq (mk)!\, c^{k+1}.$$

It is sufficient to consider the case $Rl\,s < -\frac{n}{m}$.

As in paragraph 8, we start from the integral representation of $L^s(x, y)$:

$$(9.2) \qquad L^s(x, y) = \frac{1}{\Gamma(-s)} \int_0^\infty t^{-s-1} G(t, x, y)\, dt.$$

If $(x, y) \in H$, we have the estimate (8.2) which permits us to differentiate under the integral sign, so that we have

$$(9.3) \qquad (L_x + \bar{L}_y)^k L^s(x, y) = \frac{(-1)^k 2^k}{\Gamma(-s)} \int_0^\infty t^{-s-1} \left(\frac{\partial}{\partial t}\right)^k G(t, x, y)\, dt.$$

For we have

$$\left(\frac{\partial}{\partial t} + L_x\right) G(t, x, y) = \left(\frac{\partial}{\partial t} + \bar{L}_y\right) G(t, x, y) = 0$$

for $t > 0$. Let us first suppose that $s$ is not a negative integer. By integration by parts in (9.3) [which is permitted by (8.2)] we obtain

(9.4) $(L_x + \bar{L}_y)^k L^s(x, y)$

$$= \frac{2^k}{\Gamma(-s)}(-s-1)(-s-2)\ldots(-s-k)\int_0^\infty t^{-s-k-1} G(t, x, y)\,dt.$$

Now as a special case of (8.2) we have

$$|G(t, x, y)| \leq c\, e^{-c_1(t+t^{-\mu})}$$

uniformly for $(x, y) \in H$ with positive constants $c$, $c_1$ depending on $H$.

Remembering that $\mu = (m-1)^{-1}$, it follows from a simple calculation that

(9.5) $$\sup_{(x,y)\in H}\left|\int_0^\infty t^{-s-k-1} G(t, x, y)\,dt\right| \leq ((m-1)k)!\, c^{k+1},$$

$c$ being independent of $k$, which gives evidently, from (9.4),

$$\sup_{(x,y)\in H}|(L_x + \bar{L}_y)^k L^s(x, y)|$$

$$\leq \frac{2^k}{|\Gamma(-s)|}|(-s-1)(-s-2)\ldots(-s-k)|\,((m-1)k)!\, c^{k+1} \leq (mk)!\, c_1^k.$$

If $s$ is a negative integer, we see that the integral

$$\int_0^\infty t^{-s-1}\left(\frac{\partial}{\partial t}\right)^k G(t, x, y)\,dt \qquad (x, y) \in H$$

vanishes for all large $k$ and (9.1) is trivially valid. So (i) is proved.

PROOF OF (ii). — Let $\varphi \in \mathcal{D}(\Omega)$. We suppose $\varphi$ is analytic in an open subset $\Omega_0$ of $\Omega$. We shall show that $\tilde{L}^s\varphi$ is analytic in $\Omega_0$.

Let $\Omega_1$, $\Omega_2$ be any relatively compact open subsets of $\Omega_0$ such that

$$\bar{\Omega}_1 \subset \Omega_2 \subset \bar{\Omega}_2 \subset \Omega_0.$$

Let $\alpha \in \mathcal{D}(\Omega_0)$ and $\alpha \equiv 1$ on $\Omega_2$. One has then

$$\tilde{L}^s(\varphi) = \tilde{L}^s(\alpha\varphi) + \tilde{L}^s((1-\alpha)\varphi).$$

Now, $(1-\alpha)\varphi \in \mathcal{D}(\Omega)$ and its support does not inersect $\Omega_1$; by what has been seen in (i), $L^s(x, y)$ is an analytic function of $(x, y)$ outside the diagonal in $\Omega \times \Omega$, so that it follows immediately from the integral representation of $L^s(x, y)$ that $\tilde{L}^s((1-\alpha)\varphi)$ is analytic in $\Omega_1$.

It remains to show that $\tilde{L}^s(\alpha\varphi)$ is analytic in $\Omega_1$. It is sufficient to consider the case $Rl\,s < -\frac{n}{m}\cdot$ Then we have for each integer $k \geqq 0$

$$(9.6) \quad L^k \tilde{L}^s(\alpha\varphi)(x) = \frac{1}{\Gamma(-s)} \int_0^\infty t^{-s-1}\,dt \int_\Omega G(t, x, y)\,(\alpha L^k \varphi)_y\,dy$$

$$+ \frac{1}{\Gamma(-s)} \sum_{p=0}^{k=1} L_x^p \int_0^\infty t^{-s-1}\,dt$$

$$\times \int_\Omega G(t, x, y)\,([L, \alpha] L^{k-p-1}\varphi)_y\,dy,$$

where $[L, \alpha]$ is the commutator of $L$ and $\alpha$.

Consider the second term in the above expression, which we write as

$$(9.7) \quad \frac{1}{\Gamma(-s)} \sum_{p=0}^{k-1} F_p(x),$$

where

$$F_p(x) = L_x^p \int_0^\infty t^{-s-1}\,dt \int_\Omega G(t, x, y)([L, \alpha] L^{k-p-1}\varphi)_y\,dy.$$

Now $[L, \alpha]$ is a differential operator of order $(m-1)$ whose coefficients have their supports in $(\Omega_0 - \Omega_2)$, so that if we consider $x$ in $\Omega_1$ we may perform the differentiation $L_x^p$ under the integral sign as in paragraph 8 and we obtain,

$$(9.8) \quad F_p(x) = (-s-1)(-s-2)\ldots(-s-p) \int_0^\infty t^{-s-p-1}\,dt$$

$$\times \int_\Omega G(t, x, y)\,([L, \alpha] L^{k-p-1}\varphi)_y\,dy, \quad \text{for } s \text{ non-integral}$$

$$= 0 \text{ for all large } p \text{ if } s \text{ is a negative integer.}$$

Since the coefficients of $[L, \alpha]$ have their supports in $(\Omega_0 - \Omega_2)$ and $\varphi$ is analytic in $\Omega_0$ by hypothesis, we have

$$(9.9) \quad \sup_{x \in \Omega_0} |[L, \alpha] L^{k-p-1}\varphi| \leqq ((k-p)m)!\, c^{k-p+1}$$

with $c$ independent of $k$ and $p$. Further we have (*see* § 8)

$$(9.10) \quad \sup_{(x,y) \in \Omega_1 \times (\Omega_0 - \Omega_2)} |G(t, x, y)| \leqq c\, e^{-c_1(t + t^{-\mu})};$$

we obtain from (9.8), (9.9) and (9.10)

$$\sup_{x \in \Omega} |F_p(x)| \leqq (pm)!\,((k-p)m)!\, c^{k+1}$$

with a constant $c$ independent of $k$, $p$.

Consequently, we have

$$(9.11)\qquad \sup_{x\in\Omega_1}\left|\frac{1}{\Gamma(-s)}\sum_{p=0}^{k-1}F_p(x)\right|\leqq(km)!\,c^{k+1}.$$

On the other hand, since $\varphi$ is analytic in $\Omega_0$, we have

$$\sup_{x\in\Omega_1}|\alpha L^k\varphi|\leqq(km)!\,c^{k+1}$$

and from the results of paragraph 6 [*see* (7.2), (7.3)], we have

$$\sup_{(x,y)\in\Omega_1\times\Omega_0}|G(t,x,y)|\leqq c\,t^{-n/m}\,e^{-c_1 t}$$

so that it follows for $Rl\,s<-n/m$,

$$(9.12)\qquad \sup_{x\in\Omega_1}\left|\frac{1}{\Gamma(-s)}\int_0^\infty t^{-s-1}\,dt\int_{\Omega_0}G(t,x,y)\,(\alpha L^k\varphi)_y\,dy\right|\leqq(km)!\,c^{k+1}.$$

From (9.6), (9.11) and (9.12) we obtain finally

$$\sup_{x\in\Omega_1}|L^k\tilde{L}^s(\alpha\varphi)|\leqq(km)!\,c^{k+1}.$$

with $c$ independent of $k$; now from Theorem 1 we see that $\tilde{L}^s(\alpha\varphi)$ is analytic in $\Omega_1$, which was an arbitrary open subset of $\Omega_0$. This proves (ii) and the proof of Theorem 3 is thus completed.

## BIBLIOGRAPHY.

[1] Bergendal (G.). — On spectral functions belonging to an elliptic differential operator with variable coefficients, *Math. Scand.*, t. 5, 1957, p. 241-254.

[2] Bergendal (G.). — Convergence and summability of eigenfunction expansions connected with elliptic differential operators, *Medd. Lunds Univ. Mat. Sem.*, t. 15, 1959.

[3] Bochner (S.). — Zeta functions and Green's functions for linear partial differential operators of elliptic type with constant coefficients, *Annals of Math.*, Series 2, t. 57, 1953, p. 32-56.

[4] Browder (Felix E.). — The eigenfunction expansion theorem for the general self-adjoint singular elliptic partial differential operator, I : The analytic foundation, *Proc. Nat. Acad. Sc. U. S. A.*, t. 40, 1954, p. 454-459.

[5] Browder (Felix E.). — Eigenfunction expansions for singular elliptic operators, II : The Hilbert space argument, parabolic equations on open manifolds, *Proc. Nat. Acad. Sc. U. S. A.*, t. 40, 1954, p. 459-463.

[6] Eidelman (S. D.). — On fundamental solutions of parabolic systems [in russian], *Mat. Sbornik*, t. 38, (80), 1956, p. 51-92.

[7] Friedrichs (K. O.). — On the differentiability of the solutions of the linear elliptic differential equations, *Comm. pure and appl. Math.*, t. 6, 1953, p. 299-326.

[8] Gårding (L.). — On the asymptotic properties of the spectral function belonging to a self-adjoint semi-bounded extension of an elliptic differential operator, *Kungl. Fysiogr. Sallsk. Lund Forh.*, t. 24, 1954, n° 21, p. 1-18

[9] Gårding (L.). — Dirichlet's problem for linear elliptic partial differential equations, *Math. Scand.*, t. 1, 1953, p. 55-72.

[10] Kotake (T.) and Narasimhan (M. S.). — Sur la régularité de certains noyaux associés à un opérateur elliptique, *C. R. Acad. Sc. Paris*, t. 252, 1961, p. 1549-1550.

[11] Lions (J.-L.). — *Lectures on elliptic partial differential equations.* — Bombay, Tata Institute of fundamental Research, 1957.

[12] Minakshisundaram (S.). — A generalization of Epstein zeta functions, *Canadian J. Math.*, t. 1, 1947, p. 320-327.

[13] Morrey Jr (C. B.) and Nirenberg (L.). — On the analyticity of the solutions of linear elliptic systems of partial differential equations, *Comm. pure and appl. Math.*, t. 10, 1957, p. 271-290.

[14] Nelson (Edward). — Analytic vectors, *Annals of Math.*, t. 70, 1959, p. 572-615.

[15] Nirenberg (Louis). — Remarks on strongly elliptic partial differential equations, *Comm. pure and appl. Math.*, t. 8, 1955, p. 648-674.

[16] Riesz (F.) and Nagy (Béla, Sz). — *Leçons d'analyse fonctionnelle*, 3e éd. — Paris, Gauthier-Villars; Budapest, Akademiai Kiodo, 1955.

[17] Schwartz (Laurent). — *Théorie des distributions*, Tome 1, 2e éd. — Paris, Hermann, 1957 (*Act. scient. et ind.*, 1091-1245; *Publ. Inst. Math. Univ. Strasbourg*, 9).

[18] Schwartz (Laurent). — *Théorie des distributions*, Tome 2. — Paris, Hermann, 1951 (*Act. scient. et ind.*, 1122; *Publ. Inst. Math. Univ. Strasbourg*, 10).

[19] Schwartz (Laurent). — Théorie des noyaux, *Proc. Intern. Congr. Math.* [11, 1950. Cambridge (Mass.)]; p. 220-230. — Providence, American mathematical Society, 1952.

[20] Stone (Marshall H.). — *Linear transformations in Hilbert space and their applications to analysis.* — New York, American mathematical Society, 1932 (American mathematical Society, *Colloquim Publicalions*, 15).

(Manuscrit reçu le 30 octobre 1961.)

Takeshi Kotake,
Attaché de Recherches C. N. R. S., Paris,
Yale University,
New Haven (Conn.) U. S. A.

Mudumbai S. Narasimhan,
Tata Institute of fundamental Research,
Bombay (Inde).

Reprinted from the PROCEEDINGS OF THE NATIONAL ACADEMY OF SCIENCES
Vol. 52, No. 2, pp. 207–211. August, 1964.

# *STABLE BUNDLES AND UNITARY BUNDLES ON A COMPACT RIEMANN SURFACE*

BY M. S. NARASIMHAN AND C. S. SESHADRI

TATA INSTITUTE OF FUNDAMENTAL RESEARCH, BOMBAY, INDIA

*Communicated by D. C. Spencer, June 1, 1964*

1. D. Mumford has defined the notion of a stable vector bundle on a compact Riemann surface $X$ and announced that the set of equivalence classes of all stable vector bundles (of fixed rank and degree) has a natural structure of a nonsingular, quasi-projective, algebraic variety.[4] We prove that, if $X$ has genus $\geqslant 2$, the stable vector bundles are precisely the holomorphic vector bundles on $X$ which arise from certain irreducible unitary representations of suitably defined Fuchsian groups acting on the unit disk and having $X$ as quotient (section 3, Theorem 2).

A particular case of our result is that *a holomorphic vector bundle of degree zero on $X$ is stable if and only if it arises from an irreducible unitary representation of the undamen tal group of $X$.* It follows that a holomorphic vector bundle on $X$ arises from a *unitary* representation of the fundamental group if and only if each of its indecomposable components is of degree zero and *stable.*

2. In the following we shall mean by a vector bundle a holomorphic vector bundle.

Let $X$ be a compact Riemann surface and $W$ a vector bundle of rank $n$ on $X$. By the degree of $W$ [denoted by $d(W)$] we mean the degree of the line bundle $\overset{n}{\wedge} W$.

*Definition 1:* A vector bundle $W$ on $X$ is said to be stable [resp. semistable] if for every proper subbundle $V$ of $W$, we have (rank $W$)$\cdot d(V) <$ (rank $V$)$\cdot d(W)$ [resp. (rank $W$)$d(V) \leqslant$ (rank $V$) $d(W)$].

*Definition 2:* A vector bundle $W$ on $X$ is said to be simple if every (holomorphic) endomorphism of $W$ is a scalar multiple of the identity endomorphism.

It is easily seen that a simple vector bundle is indecomposable and that a stable bundle is simple.

We require the following theorem which is also of independent interest.

THEOREM 1. *Let $W$ be a simple bundle of rank $n$ on $X$. Then there exists a complex manifold $M$ and a holomorphic family $\{W(m)\}$, $m \in M$, of vector bundles on $X$ parametrized by $M$ such that*

(i) *the complex dimension of $M = n^2(g - 1) + 1$ ($g$ = genus of $X$);*
(ii) *there exists a point $m_0 \in M$ such that $W(m_0) = W$ (i.e., $W(m_0)$ is isomorphic to $W$);*
(iii) *if $m_1, m_2 \in M$ and $m_1 \neq m_2$, then $W(m_1)$ and $W(m_2)$ are not isomorphic;*
(iv) *given a holomorphic family $\{V(t)\}$, $t \in T$, of vector bundles parametrized by a complex manifold $T$ and a point $t_0 \in T$ with $V(t_0) = W$, there is a neighborhood $N$ of $t_0$ and a (unique) holomorphic mapping $f:N \rightarrow M$, such that $\{V(t)\}$, $t \in N$, is the inverse image of the family $\{W_m\}$, $m \in M$.*

The proof of this theorem is done in two steps. One first constructs a family having the properties (*i*), (*ii*), and (*iv*) (without uniqueness); this is done as in references 2 and 3. The property (*iii*) is then proved with the aid of the semi-continuity theorem in the form due to H. Grauert.

Let $\mathbf{S}(n, q)$ denote the isomorphism classes of simple vector bundles of rank $n$ and degree $q$ on $X$. From Theorem 1, it follows that $\mathbf{S}(n, q)$ has a natural structure of a complex manifold. (This manifold is not, in general, Hausdorff.)

3. Let $\pi$ be a discrete group acting holomorphically and properly on a simply connected Riemann surface $Y$ such that the quotient space $X = Y/\pi$ is compact. Then $X$ is a compact Riemann surface, and the canonical mapping $p:Y \rightarrow X$ is holomorphic. We assume that $p$ is ramified at most over a single point $x_0 \in X$, i.e., if $y \notin p^{-1}(x_0)$, then the isotropy group $\pi_y$ at $y$ reduces to the identity. If $y \in p^{-1}(x_0)$, the isotropy group $\pi_y$ is a finite cyclic group and its order $N$ is independent of the point $y$ in $p^{-1}(x_0)$. This integer is called the order of ramification of $p$. If $N = 1$, $p:Y \rightarrow X$ is the simply connected covering of $X$ and $\pi$ can be identified with the fundamental group of $X$.

Conversely, given a compact Riemann surface $X$ of genus $\geqslant 1$, a point $x_0 \in X$, and an integer $N \geqslant 1$, there exists a covering $p:Y \rightarrow X$, as above, which is ramified at $x_0$, with ramification order $N$.

A vector bundle $W$ on $Y$ is called a *$\pi$-bundle* if $\pi$ operates on $W$ consistent with its operation on $Y$.[1, 6] The $\pi$-bundles (of a fixed rank) form a category, with the obvious notion of $\pi$-homomorphism. If $\rho:\pi \rightarrow \text{Aut } V$ is a homomorphism, where $V$ is a finite dimensional complex vector space, then $\pi$ acts canonically on the trivial bundle $Y \times V$; this $\pi$-bundle is called the *$\pi$-bundle arising from the representation* $\rho$.

Let $W$ be a $\pi$-bundle of rank $n$ on $Y$ and $\mathbf{W}$ the sheaf of germs of holomorphic sections of $W$. Then $\pi$ operates on $\mathbf{W}$ and therefore on the direct image $p_*(\mathbf{W})$, which is a sheaf of $\mathbf{O}_X$-modules. We denote by $p_*^{\pi}(\mathbf{W})$ the subsheaf of $p_*(\mathbf{W})$ consisting of invariant elements. It is easy to see that $p_*^{\pi}(\mathbf{W})$ is a locally free sheaf of $\mathbf{O}_X$-modules of rank $n$; hence $p_*^{\pi}(\mathbf{W})$ determines a vector bundle of rank $n$ on $X$, which we denote by $p_*^{\pi}(W)$. The mapping $W \rightsquigarrow p_*^{\pi}(W)$ defines a functor from the category of $\pi$-bundles of rank $n$ on $Y$ to the category of vector bundles of rank $n$ on $X$. If $p$ is unramified, this functor is an isomorphism of categories.

Let us fix a point $y_0 \in p^{-1}(x_0)$. Let $\tau:\pi_{y_0} \rightarrow \mathbf{GL}(n;\mathbf{C})$ be a representation of the isotropy group $\pi_{y_0}$ of $\pi$ at $y_0$ such that for every $\gamma \in \pi_{y_0}$, $\tau(\gamma)$ is a *scalar multiple of the identity matrix.*

*Definition 3:* A vector bundle $W$ on $Y$ is said to be a *special $\pi$-bundle of type* $\tau$, if

(*i*) $W$ is a $\pi$-bundle arising from a representation $\rho:\pi \rightarrow \mathbf{GL}(n;\mathbf{C})$,
(*ii*) the restriction of $\rho$ to $\pi_{y_0}$ coincides with $\tau$.

Our main result is the following

THEOREM 2. *Let $X$ be a compact Riemann surface of genus $\geqslant 2$. Given integers $n$ and $q$ with $-n < q \leqslant 0$, we can choose a group $\pi$, a covering $p:Y \rightarrow X$, and a representation $\tau$, as above, with the following properties. A vector bundle $W$ on $X$ of rank $n$ and degree $q$ is stable if and only if $W = p_*^{\pi}(V)$, where $V$ is a special $\pi$-bundle of type $\tau$ such that the representation $\rho:\pi \rightarrow \mathbf{GL}(n;\mathbf{C})$, to which $V$ is associated,* is irreducible *and* unitary. *Further, two such stable bundles are isomorphic on $X$ if and only if the corresponding unitary representations of $\pi$ are equivalent.*

*Remarks 1:* The theorem characterizes, in fact, stable bundles $W$ of rank $n$ and arbitrary degree $q$, for there exists a line bundle $L$ (depending only on $n$ and $q$) such that $-n < d(W \otimes L) \leqslant 0$.

*2.* It follows from the theorem that the topology on the space of (equivalence classes of) stable bundles on $X$, of rank $n$ and degree $q$, is independent of the complex structure on $X$ and that the space is compact if $n$ and $q$ are coprime.

4. We now give a brief outline of the proof of Theorem 2. With the notation of section 3, we denote by $\mathbf{B}(\tau)$ the set of special $\pi$-bundles of type $\tau$. The following proposition plays an important role in the proof.

PROPOSITION. (i) *Let $W_1$, $W_2 \in \mathbf{B}(\tau)$. Then $p_*^{\pi}$ induces an isomorphism of the vector space of $\pi$-homomorphisms of $W_1$ into $W_2$, onto the space of homomorphisms of $p_*^{\pi}(W_1)$ and $p_*^{\pi}(W_2)$; further, $W_1$ and $W_2$ are isomorphic as $\pi$-bundles if and only if $p_*^{\pi}(W_1)$ and $p_*^{\pi}(W_2)$ are isomorphic on $X$.*

(ii) *Given integers $n$ and $q$ with $-n < q \leqslant 0$, there exists a covering $p:Y \rightarrow X$ and $\tau$, as in section 3, such that for any indecomposable vector bundle $W$ of rank $n$ and degree $q$, there exists a $V \in \mathbf{B}(\tau)$ such that $p_*^{\pi}(V)$ is isomorphic to $W$.*

The proof of (*i*) is a direct verification. For (*ii*) we require the theorem of

A. Weil which states that an indecomposable $\pi$-bundle on $Y$ is associated to a representation of $\pi$ in $\mathbf{GL}(n; \mathbf{C})$ if and only if its degree (as a $\pi$-bundle) is zero (cf. refs. 1, 6).

The proof of Theorem 2 is by induction on the rank $n$. (For $n = 1$, the theorem is classical.) Choose $\pi$, $p: Y \to X$ and $\tau$ as in (*ii*) of the above proposition. Let $R = R(n, q)$ be the set of representations $\rho: \pi \to \mathbf{GL}(n; \mathbf{C})$ such that $\rho|\pi_{y_0} = \tau$. Then $R$ has a natural structure of a complex space. By applying the functor $p_*^{\pi}$ we get a holomorphic family of vector bundles on $X$, parametrized by $R$. Let $U$ (resp. $U_0$) be the subset of $R$ corresponding to unitary (resp. irreducible unitary) representations of $\pi$. We prove the following:

(1) The bundle on $X$ corresponding to a point of $U$ is semistable;

(2) The bundle on $X$ corresponding to a point of $U_0$ is stable;

(3) If $R_s$(resp. $R_{ss}$) is the subset of $R$ consisting of elements $\rho$ such that the bundle determined by $\rho$ on $X$ is stable (resp. semistable), then $R_s$ (resp. $R_{ss}$) is open in $R$;

(4) $R_s$ consists only of simple points of $R$;

(5) The canonical map $R_s \to \mathbf{S}(n, q)$ (see sect. 2) admits of a local section at every point of the image.

Properties (2) and (3) are proved by using the induction hypothesis, and (3) is a consequence of a more general result concerning any holomorphic (or algebraic) family of vector bundles on $X$. The proof of (4) is similar to that of the results in section 2 of reference 5. The property (5) is a generalization of the theorem of A. Weil (ref. 6) to a holomorphic family of simple bundles.

Let $\Gamma_1 \subset R_{ss} \times R_{ss}$ consist of points $(\rho_1, \rho_2)$ such that there is a nontrivial homomorphism of the bundle determined by $\rho_1$ into the bundle determined by $\rho_2$. Let $\Gamma_2 \subset R_{ss} \times R_{ss}$ be the subset $(R_{ss} - R_s) \times (R_{ss} - R_s)$. Then $\Gamma = \Gamma_1 \cup \Gamma_2$ is a *closed* subset of $R_{ss} \times R_{ss}$ and defines an equivalence relation on $R_{ss}$ with the following properties: (*i*) two points $\rho_1$, $\rho_2$ in $R_s$ are equivalent if and only if the corresponding bundles on $X$ are isomorphic; (*ii*) $R_s$ is a saturated open subset of $R_{ss}$.

Let $Q$ be the quotient space $R_{ss}/\Gamma$ and $\eta: R_{ss} \to Q$, the canonical mapping. We then have: (*a*) $\eta(R_s)$ is a *connected* open subset of $Q$, (*b*) $\eta(U_0)$ is a nonempty *open* subset of $Q$. The property (*a*) is a consequence of (5) and of the existence of a "total" family of stable bundles on $X$, of rank $n$ and degree $q$, parametrized by an irreducible nonsingular algebraic variety. The property (*b*) follows from results similar to those in reference 5, and for the case $q = 0$, it is contained in reference 5.

Now $U$ is compact and hence the saturation of $U$ (for the equivalence relation $\Gamma$) is closed. It follows that $\eta(U)$ is a *closed* subset of $Q$. Then, by the connectedness of $\eta(R_s)$, it follows easily that $\eta(R_s) = \eta(U_0)$. This completes the proof of the first part of the theorem. The latter part of the theorem is quite easy to prove (cf. ref. 5).

[1] Grothendieck, A., Séminaire Bourbaki, Exposé 141 (1956–57).

[2] Kodaira, K., and D. C. Spencer, "A theorem of completeness for complex analytic fibre spaces," *Acta. Math.*, **100**, 281–294 (1958).

[3] Kodaira, K., L. Nirenberg, and D. C. Spencer, "On the existence of deformations of complex analytic structures," *Ann. Math.*, **68**, 450–459 (1958).

[4] Mumford, D., "Projective invariants of projective structures and applications," in *Proc. Intern. Cong. Math., Stockholm* (1962), pp. 526–530.

[5] Narasimhan, M. S., and C. S. Seshadri, "Holomorphic vector bundles on a compact Riemann surface," *Math. Annalen*, **155**, 69–80 (1964).
[6] Weil, A., "Généralisation des fonctions abéliennes," *J. Math. Pures et Appl.*, **17**, 47–87 (1938).

NARASIMHAN, M. S., and C. S. SESHADRI
Math. Annalen 155, 69—80 (1964)

# Holomorphic Vector Bundles on a Compact Riemann Surface

By

M. S. NARASIMHAN and C. S. SESHADRI in Bombay

## 1. Introduction

We prove in this paper that the equivalence classes of holomorphic vector bundles arising from $n$-dimensional irreducible unitary representations of the fundamental group $\pi$ of a compact Riemann surface $X$ of genus $g (\geqq 2)$ form a complex manifold $M$ of complex dimension $n^2(g-1)+1$. For $n = 1$, this complex manifold is the Picard variety of $X$. It may be remarked that the number $n^2(g-1)+1$ has been calculated heuristically by A. WEIL as the dimension of the "field of hyperabelian functions" [10].

Two holomorphic vector bundles arising from unitary representations of $\pi$ are isomorphic if and only if the representations are equivalent [Proposition 4.2]; this fact allows one to introduce a real analytic manifold structure on $M$ which is independent of the complex structure on $X$[1]). Let $\Omega = \mathrm{U}(n) \times \cdots \cdots \times \mathrm{U}(n)$ ($2g$ times) and $f: \Omega \to \mathrm{SU}(n)$ the map defined by

$$(A_1, B_1, \ldots, A_g, B_g) \to A_1 B_1 A_1^{-1} B_1^{-1} \ldots A_g B_g A_g^{-1} B_g^{-1} .$$

The set of homomorphisms of $\pi$ into the unitary group $\mathrm{U}(n)$ is identified with $f^{-1}(I)$, where $I$ is the identity matrix. We prove that the tangent vectors to $\Omega$ at $\varrho \in f^{-1}(I)$, which are mapped into zero by the differential of $f$ at $\varrho$ can be identified with the 1-cocycles of $\pi$ with respect to the representation $\mathrm{ad}\,\varrho$ of $\pi$ in $\mathfrak{u}(n)$ ($\mathrm{ad}\,\varrho$ is the composite of $\varrho$ and the adjoint representation of $\mathrm{U}(n)$ in its Lie algebra $\mathfrak{u}(n)$). As a consequence we see that $f$ is of maximal rank at a point $\varrho \in f^{-1}(I)$ if and only if $\varrho$ is an irreducible representation and that the set of irreducible unitary representations has a natural structure of a real analytic manifold. It follows that the equivalence classes of irreducible unitary representations (of a given dimension) form a real analytic manifold $M$ and that the tangent space at a point $m \in M$ can be identified with the cohomology space $H^1(\pi, \mathrm{ad}\,\varrho)$, where $\varrho$ is a unitary representation in the class $m$.

To introduce the complex structure on $M$, let $\mathrm{Ad}\,P(\varrho)$ be the vector bundle adjoint to the holomorphic $\mathbf{GL}(n; \mathbf{C})$-principal bundle $P(\varrho)$ determined by a unitary representation $\varrho$ of $\pi$ and let $H^1(X, \mathrm{Ad}\,P(\varrho))$ denote the first cohomology space of $X$ with coefficients in the sheaf of germs of holomorphic sections of

[1]) After the paper was written we came across a paper of J. IGUSA [3], in which the real analytic structure on $M$ has been considered.

$\operatorname{Ad} P(\varrho)$. We prove, by means of harmonic forms with coefficients in a local system of vector spaces, that the natural map

$$J : H^1(\pi, \operatorname{ad}\varrho) \to H^1(X, \operatorname{Ad} P(\varrho))$$

is an isomorphism of real vector spaces [Proposition 4.3]. Since $H^1(X, \operatorname{Ad} P(\varrho))$ has a natural structure of a complex vector space, $M$ thus acquires an almost complex structure. Now the map $J$ turns out, locally, to be the infinitesimal deformation map of a differentiable family of holomorphic bundles on $X$ and hence by a theorem of Kodaira-Spencer-Nakano, this almost complex structure is in fact a complex structure on $M$.

## 2. Real analytic manifold structure on the space of irreducible unitary representations

Let $X$ be a compact connected orientable surface of genus $g \geqq 2$. Let $\pi$ be the fundamental group of $X$. Let $\mathscr{R}$ denote the set of representations of $\pi$ in the unitary group $\mathrm{U}(n)$. We put on $\mathscr{R}$ the topology induced by the product topology of $\mathrm{U}(n)^\pi$. Let $M_0$ denote the subset of $\mathscr{R}$ consisting of irreducible $n$-dimensional unitary representations of $\pi$, provided with the induced topology.

As is well known, the structure of $\pi$ may be described as follows. Let $F$ be the free group on a set consisting of $2g$ elements $a_1, b_1, \ldots, a_g, b_g$; let $N$ be the normal sub-group of $F$ generated by the element $a_1 b_1 a_1^{-1} b_1^{-1} \ldots a_g b_g a_g^{-1} b_g^{-1}$. Then $\pi$ is isomorphic to the quotient group $F/N$. We identify $\pi$ with $F/N$. Let $\eta : F \to \pi$ be the canonical map.

If $A_1, B_1, \ldots, A_g, B_g$ are elements of $\mathrm{U}(n)$ (not necessarily distinct) such that $\prod_{i=1}^{g} A_i B_i A_i^{-1} B_i^{-1} = I$, there is a unique homomorphism of $\pi$ sending $\eta(a_i)$ to $A_i$ and $\eta(b_i)$ to $B_i$. Let $\Omega$ be the product of $2g$ copies $X_1, Y_1, X_2, Y_2, \ldots, X_g, Y_g$ of $\mathrm{U}(n)$. Let $p_i : \Omega \to X_i$, $q_i : \Omega \to Y_i$ denote the natural projections. The map

$$\varrho_0 \to (\varrho_0 \circ \eta(a_1), \varrho_0 \circ \eta(b_1), \ldots, \varrho_0 \circ \eta(a_g), \varrho_0 \circ \eta(b_g))$$

maps $\mathscr{R}$ bijectively, by the above remark, onto the subset $\mathscr{R}'$ of $\Omega$ consisting of $x$ in $\Omega$ for which $\prod_{i=1}^{g} p_i(x)\, q_i(x)\, p_i(x)^{-1} q_i(x)^{-1} = I$. This map is easily seen to be a homeomorphism of $\mathscr{R}$ onto $\mathscr{R}'$ (with the topology induced from the product topology on $\mathrm{U}(n) \times \cdots \times \mathrm{U}(n)$). We identify $\mathscr{R}$ and $\mathscr{R}'$. $\mathscr{R}$ is thus a real analytic subset of $\Omega$. We shall show that irreducible representations of $\pi$ are simple points of this variety.

We shall first convince ourselves that there are irreducible unitary representations of $\pi$ in each dimension.

**Proposition 2.1.** *For every $n \geqq 1$, there is an irreducible $n$-dimensional unitary representation of $\pi (g \geqq 2)$.*

*Proof:* We may suppose that $n \geqq 2$. We note that there exist two elements $A_1$ and $B_1$ in $\mathrm{U}(n)$ which form an irreducible set of matrices. For instance, let $\lambda_1, \ldots, \lambda_n$ be $n$ distinct complex numbers of modulus one. If $e_1, \ldots, e_n$ is the canonical basis in $\mathbf{C}^n$, let $A_1$ be the unitary matrix sending $e_i$ to $\lambda_i e_i$ and $B_1$ the unitary transformation defined by the cyclic permutation $e_1 \to e_2$, $e_2 \to \to e_3, \ldots, e_n \to e_1$. It is easy to check that $A_1$ and $B_1$ form an irreducible set.

Now there is a homomorphism $\varrho_0 : \pi \to \mathrm{U}(n)$ with $(\varrho_0 \circ \eta)(a_1) = A_1$, $(\varrho_0 \circ \eta)(b_1) = B_1$, $(\varrho_0 \circ \eta)(a_2) = B_1$, $(\varrho_0 \circ \eta)(b_2) = A_1$ and $\varrho_0 \circ \eta(a_i) = \varrho_0 \circ \eta(b_i) = I$ for $i > 2$, as $(A_1 B_1 A_1^{-1} B_1^{-1})(B_1 A_1 B^{-1} A^{-1}) = I$. $\varrho_0$ is clearly an $n$-dimensional irreducible unitary representation of $\pi$.

*Remark:* One can show that the set of pairs $(A_1, B_1)$ such that $(A_1, B_1)$ form an irreducible set, is dense in $\mathrm{U}(n) \times \mathrm{U}(n)$.

The next two propositions are required to prove that the irreducible representations form a manifold.

**Proposition 2.2.** *Let $\varrho$ be an $n$-dimensional unitary representation of $\pi$. Let* $\mathrm{ad}\,\varrho$ *be the representation of $\pi$ in the Lie algebra* $\mathfrak{u}(n)$ *of* $\mathrm{U}(n)$, *obtained by composing $\varrho$ and the adjoint representation of* $\mathrm{U}(n)$ *in* $\mathfrak{u}(n)$. *Let* $H^i(\pi, \mathrm{ad}\,\varrho)$ *denote the* $i$-th *cohomology space of $\pi$ with respect to the representation* $\mathrm{ad}\,\varrho$ *in* $\mathfrak{u}(n)$. *Then we have*

$$\dim_{\mathbf{R}} H^1(\pi, \mathrm{ad}\,\varrho) = 2n^2(g-1) + 2\dim_{\mathbf{R}} H^0(\pi, \mathrm{ad}\,\varrho)\,.$$

*Remark:* $\mathfrak{u}(n)$ is the space of $n \times n$ skew-hermitian matrices.

*Proof:* We first note that if $L$ is a local system of $k$-dimensional vector spaces on $X$, then $\dim_{\mathbf{R}} H^0(X,L) - \dim_{\mathbf{R}} H^1(X,L) + \dim_{\mathbf{R}} H^2(X,L) = k(2-2g)$ (The proof of this relation is similar to that of the usual Euler-Poincaré formula e.g. by using a triangulation of $X$). If we now take for $L$ the local system defined by the representation $\mathrm{ad}\,\varrho$, we have

$$(1) \qquad \dim H^1(X, L) = 2n^2(g-1) + \dim H^0(X, L) + \dim H^2(X, L)\,.$$

Since the universal covering of $X$ is a disc, we have the well-known isomorphisms ([1], Ch. XVI, § 9)

$$(2) \qquad H^i(X, L) \approx H^i(\pi, \mathrm{ad}\,\varrho)\,.$$

Moreover since the transformations $\mathrm{ad}\,\varrho(\gamma)$, $\gamma \in \pi$, leave invariant a positive definite quadratic form on $\mathfrak{u}(n)$, (e.g. the form $\mathrm{tr}(AA^*)$ on $\mathfrak{u}(n)$), $L$ is isomorphic to its dual local system. Hence by the duality theorem ([9], Th. 1.14)

$$\dim_{\mathbf{R}} H^0(X, L) = \dim_{\mathbf{R}} H^2(X, L)\,.$$

By (1) and (2) we now have

$$\dim H^1(\pi, \mathrm{ad}\,\varrho) = 2n^2(g-1) + 2\dim H^0(\pi, \mathrm{ad}\,\varrho)\,.$$

*Remark:* We shall indicate later a proof of this proposition using the Riemann-Roch theorem.

**Proposition 2.3.** *Let $\varrho : \pi \to \mathrm{U}(n)$ be a homomorphism. Then $\varrho$ is an irreducible representation if and only if*

$$\dim_{\mathbf{R}} H^1(\pi, \mathrm{ad}\,\varrho) = 2\{n^2(g-1) + 1\}\,.$$

*Proof:* $H^0(\pi, \mathrm{ad}\,\varrho)$ is the space of fixed points of the representation $\mathrm{ad}\,\varrho$ in $\mathfrak{u}(n)$ i.e., the space of skew-hermitian matrices which commute with all matrices $\varrho(\gamma)$, $\gamma \in \pi$. If $\varrho$ is irreducible these matrices are scalar, by Schur's lemma.

Therefore $\dim H^0(\pi, \operatorname{ad}\varrho) = 1$; now the formula in Proposition 2.3 follows from Proposition 2.2. If the formula in Proposition 2.3 is valid we must have $\dim H^0(\pi, \operatorname{ad}\varrho) = 1$. Since $\varrho$ is a unitary representation and hence completely reducible, it follows that $\varrho$ must be irreducible. Q. E. D.

We now proceed to prove that the $n$-dimensional irreducible unitary representations form a manifold.

Let $\Omega = \mathbf{U}(n) \times \cdots \times \mathbf{U}(n)$, $2g$ times; let $p_i$, $q_i$ denote the projections as before. Consider the map $f : \Omega \to \mathbf{SU}(n)$ defined by

$$f(x) = \prod_{i=1}^{g} p_i(x)\, q_i(x)\, p_i(x)^{-1}\, q_i(x)^{-1}\,.$$

As has been remarked earlier the space $\mathcal{R}$ of representations of $\pi$ is identified with $f^{-1}(I)$, where $I$ is the identity of $\mathbf{SU}(n)$. We shall now prove that $\varrho \in \mathcal{R}$ is irreducible if and only if $f$ is of maximal rank at $\varrho$. This will prove that $M_0$, the subset of $\mathcal{R}$ consisting of irreducible representations, is a real analytic manifold (Implicit function theorem).

Let $\varrho \in f^{-1}(I)$. Let $\Omega_\varrho$ (resp. $\mathfrak{su}(n)$) be the tangent space of $\Omega$ (resp. $\mathbf{SU}(n)$) at $\varrho$ (resp. $I$). If $D_1 : \Omega_\varrho \to \mathfrak{su}(n)$ denotes the differential of $f$ at $\varrho$, we have to calculate the dimension of the kernel of $D_1$. We shall now show that the kernel of $D_1$ can be identified with the space of 1-cocycles of $\pi$ with respect to the representation $\operatorname{ad}\varrho$ of $\pi$ in $\mathfrak{u}(n)$. To this purpose let $\omega = dA \cdot A^{-1}$ be the right-invariant Maurer-Cartan form on $\mathbf{SU}(n)$. Then $df \cdot f^{-1}$ is the inverse image of $\omega$ by $f$. Then kernel of $D_1 = \{v \in \Omega_\varrho \mid df \cdot f^{-1}(v) = 0\}$.

A tangent vector $v$ to $\Omega$ at $\varrho$ is identified (by means of right translation) with $(A_1, B_1, A_2, B_2, \ldots, A_g, B_g)$ where $A_i, B_i \in \mathfrak{u}(n)$ (remembering that $\Omega = \mathbf{U}(n) \times \cdots \times \mathbf{U}(n)$). Since $F$ is the free group on $\{a_i, b_i\}$ there is a unique 1-cocycle $\delta_v$ on $F$ with respect to the representation $\operatorname{ad}\varrho$ with $\delta_v(a_i) = A_i$, $\delta_v(b_i) = B_i$ (see e.g. ([7], § 10.9, Th. 9)). We claim that $df \cdot f^{-1}(v) = \delta_v\left(\prod_{i=1}^{g} a_i b_i a_i^{-1} b_i^{-1}\right)$. To prove this, consider the analytic manifold $\Omega \times F$, $F$ considered as a discrete manifold. Since every point of $\Omega$ defines a representation of $F$ in $\mathbf{U}(n)$, we have an analytic map $\Phi : \Omega \times F \to \mathbf{U}(n)$, $\Phi(\tilde{\varrho}, \gamma) = \tilde{\varrho}(\gamma)$, $\tilde{\varrho} \in \Omega$, $\gamma \in F$. If $v \in \Omega_\varrho$, $d\Phi \cdot \Phi^{-1}(v)$ is a function $\psi$ on $F$ with values in $\mathfrak{u}(n)$; $\psi$ is a 1-cocycle for $\operatorname{ad}\varrho$. In fact, since for $x \in \Omega$, $\gamma_1, \gamma_2 \in F$ we have $\Phi(x, \gamma_1\gamma_2) = \Phi(x, \gamma_1)\Phi(x, \gamma_2)$, we get $d\Phi(x, \gamma_1\gamma_2) = \Phi(x, \gamma_1)\, d\Phi(x, \gamma_2) + d\Phi(x_1\gamma_1)\Phi(x, \gamma_2)$ or $d\Phi(x, \gamma_1\gamma_2)\, \Phi(x, \gamma_1\gamma_2)^{-1} = d\Phi(x, \gamma_1\gamma_2)\, \Phi(x, \gamma_2)^{-1}\, \Phi(x, \gamma_1)^{-1}$

$$= \{\Phi(x, \gamma_1)\, d\Phi(x, \gamma_2) + d\Phi(x, \gamma_1)\, \Phi(x, \gamma_2)\}\, \Phi(x, \gamma_2)^{-1}\, \Phi(x, \gamma_1)^{-1}$$

$$= \Phi(x, \gamma_1)\, \{d\Phi(x, \gamma_2)\, \Phi(x, \gamma_2)^{-1}\}\, \Phi(x, \gamma_1)^{-1} + d\Phi(x, \gamma_1)\, \Phi(x, \gamma_1)^{-1}\,.$$

Hence $\psi(\gamma_1\gamma_2) = \varrho(\gamma_1)\, \psi(\gamma_2)\, \varrho(\gamma_1)^{-1} + \psi(\gamma_1)$ i.e., $\psi$ is a 1-cocycle for $\operatorname{ad}\varrho$. Moreover $d\Phi(\varrho, a_i)\, \Phi(\varrho, a_i)^{-1}(v) = dp_i p_i^{-1}(v) = A_i$, i.e., $\psi(a_i) = A_i$; similarly $\psi(b_i) = B_i$. Hence $\psi = \delta_v$. Now $\Phi(x, \prod_{i=1}^{g} a_i b_i a_i^{-1} b_i^{-1}) = f(x)$, by definition. So

$$\delta_v\left(\prod_{i=1}^{g} a_i b_i a_i^{-1} b_i^{-1}\right) = \psi\left(\prod_{i=1}^{g} a_i b_i a_i^{-1} b_i^{-1}\right) = df \cdot f^{-1}(v)\,.$$

It follows that $v$ belongs to the kernel of the differential map $D_1$ if and only if the cocycle $\delta_v$ (on $F$) is zero on $\prod_{i=1}^{g} a_i b_i a_i^{-1} b_i^{-1}$. Now a simple verification shows that the space of 1-cocycles of the representation $\operatorname{ad}\varrho$ of $\pi$ in $\mathfrak{u}(n)$ is naturally isomorphic to the space of 1-cocycles of the free group $F$, with respect to the representation $\eta \circ \operatorname{ad}\varrho$ of $F$ in $\mathfrak{u}(n)$, which vanish on the element $\prod_{i=1}^{g} a_i b_i a_i^{-1} b_i^{-1}$. Hence the kernel of $D_1$ can be identified with the space of 1-cocycles of $\pi$ with respect to the representation $\operatorname{ad}\varrho$ of $\pi$ in $\mathfrak{u}(n)$.

By Proposition 2.2,

$$\begin{aligned}\dim\ker D_1 &= \text{dimension of 1-cocycles of } \pi \text{ in } \mathfrak{u}(n) \text{ with respect to } \operatorname{ad}\varrho\\ &= \dim H^1(\pi, \operatorname{ad}\varrho) + \dim \mathfrak{u}(n) - \dim H^0(\pi, \operatorname{ad}\varrho)\\ &= 2n^2(g-1) + 2\dim H^0(\pi, \operatorname{ad}\varrho) + n^2 - \dim H^0(\pi, \operatorname{ad}\varrho)\\ &= 2n^2 g - n^2 + \dim H^0(\pi, \operatorname{ad}\varrho)\,.\end{aligned}$$

$f$ is of maximal rank at $\varrho$ if and only if $\dim\ker D_1 + \dim \mathbf{SU}(n) = \dim \Omega$ ($= 2gn^2$) i.e., if and only if $2n^2 g - n^2 + \dim H^0(\pi, \operatorname{ad}\varrho) + n^2 - 1 = 2gn^2$ i.e., if and only if $\dim H^0(\pi, \operatorname{ad}\varrho) = 1$. Since $\varrho$ is unitary, this condition is equivalent to the condition that the representation $\varrho$ of $\pi$ is irreducible. Thus we have shown that points of maximal rank for $f$ on $f^{-1}(I)$ are precisely irreducible unitary representations of $\pi$. This completes the proof that $M_0$ has a natural structure of a real analytic manifold of dimension $2\{n^2(g-1)+1\} + (n^2-1)$.

*Remark:* We have proved that the tangent space to $M_0$ at $\varrho$ can be identified with the space 1-cocycles of $\pi$ for the representation $\operatorname{ad}\varrho$.

## 3. Manifold structure on equivalence classes of irreducible unitary representations

Let $M_0$ be the real analytic manifold of all $n$-dimensional irreducible unitary representations of $\pi$. The group $\mathbf{U}(n)$ acts on $M_0$ on the right by: $(\varrho T)(\gamma) = T^{-1}\varrho(\gamma)T$, $\gamma \in \pi$, $T \in \mathbf{U}(n)$, $\varrho \in M_0$. By Schur's lemma, the compact Lie group $\mathbf{PU}(n) = \mathbf{U}(n)/(\text{centre of } \mathbf{U}(n))$ acts freely on $M_0$; the action of $\mathbf{PU}(n)$ is analytic. Since $\mathbf{PU}(n)$ is compact, the orbit space $M$ is a Hausdorff space; moreover, as is well known, $M$ has the natural structure of a real analytic manifold. The canonical map $p: M_0 \to M$ defines a principal fibre space with structure group $\mathbf{PU}(n)$.

Take $m \in M$. Let $\varrho \in M_0$ such that $p(\varrho) = m$. Let $M_{0,\varrho}$ (resp. $M_m$) be the tangent space to $M_0$ (resp. $M$) at $\varrho$ (resp. $m$). Then $M_{0,\varrho}$ is identified with the space of 1-cocycles of $\pi$ with respect to the representation $\operatorname{ad}\varrho$. If $dp: M_{0,\varrho} \to M_m$ is the differential of $p$ at $\varrho$, the kernel of $dp$ consists precisely of the space of 1-coboundaries of $\pi$ for the representation of $\operatorname{ad}\varrho$. In fact, the kernel of $dp$ is the image of $\mathfrak{u}(n)$ in $M_{0,\varrho}$ by the differential at $I$ of the map $T \to T^{-1}\varrho T$ of $\mathbf{U}(n)$ in $M_0$ and the differential at $I$ of the map $T \to (T^{-1}A_1T, \ldots, T^{-1}B_gT)$

of $\mathrm{U}(n)$ into $\Omega$ is the map $Y \to (A_1 Y A_1^{-1} - Y, \ldots, B_g Y B_g^{-1} - Y)$, $Y \in \mathfrak{u}(n)$. Since $dp$ is surjective we see that we can identify the tangent space $M_m$ with the cohomology space $H^1(\pi, \mathrm{ad}\,\varrho)$. By Proposition 4.3, $\dim M = 2\{n^2(g-1)+1\}$.

## 4. Unitary bundles on a compact complex manifold

Let $X$ be a compact connected complex manifold and $\pi$ the fundamental group of $X$. If $\varrho$ is a representation of $\pi$ in a real or complex (finite-dimensional) vector space $V$ we denote by $H^i(\pi, \varrho)$ the $i^{\text{th}}$ comhomology space of $\pi$ with respect to the representation $\varrho$ in $V$. We denote by $L(\varrho)$ the local coefficient system of vector spaces on $X$ determined by $\varrho$ and by $H^i(X, L(\varrho))$ the $i^{\text{th}}$ cohomology space of $X$ with coefficients in $L(\varrho)$.

The universal covering $\tilde{X}$ of $X$ is a holomorphic principal fibre bundle over $X$ with $\pi$ as the structure group. If $\varrho$ is a representation in a complex vector space $V$, by extension of structure group of the $\pi$-bundle $\tilde{X}$ by $\varrho$, we obtain a holomorphic principal fibre bundle $P(\varrho)$ with $\mathrm{Aut}\,V$ as the structure group. We denote by $W(\varrho)$ the associated holomorphic vector bundle. We denote by $\mathrm{Ad}\,P(\varrho)$, the holomorphic vector bundle associated to $P(\varrho)$ by the adjoint representation of $\mathrm{Aut}\,V$ in $\mathrm{End}\,V$ and $\mathrm{Ad}\,P(\varrho)$ will be referred to as the adjoint bundle. For a holomorphic vector bundle $W$ on $X$, $H^i(X, W)$ will denote the $i^{\text{th}}$ cohomology space (over $\mathbf{C}$) of $X$ with coefficients in the sheaf of germs of holomorphic sections of $W$.

**Proposition 4.1.** *Let $X$ be a compact connected complex manifold and $\varrho : \pi \to \mathrm{Aut}\,V$ a representation of $\pi = \pi_1(X)$ in a complex vector space $V$ such that $\varrho$ leaves invariant a positive definite hermitian form $H$ on $V$. Then the natural map $H^0(\pi, \varrho) \to H^0(X, W(\varrho))$ is an isomorphism.*

*Proof:* Clearly the natural map $H^0(\pi, \varrho) \to H^0(X, W(\varrho))$ is an injection. $H^0(X, W(\varrho))$ can be identified with holomorphic functions $f : \tilde{X} \to V$ satisfying $f(x\gamma) = \varrho(\gamma)^{-1} f(x)$, $\gamma \in \pi$, $x \in \tilde{X}$. If $\|f(x)\|^2 = H(f(x), f(x))$, since $\varrho$ leaves $H$ invariant, $\|f(x)\|^2$ is invariant under the action of $\pi$ on $\tilde{X}$ and hence defines a function $\varphi$ on $X$. Clearly $\varphi$ is a plurisubharmonic function on $X$ and since $X$ is compact and connected, $\varphi$ is constant. Hence $\|f(x)\|^2$ is constant on $\tilde{X}$; it follows that $f$ is constant, say $C \in V$. We have $C = \varrho(\gamma^{-1})\, C$, i.e., $C$ is fixed by all elements $\varrho(\gamma)$, $\gamma \in \pi$. This means that $C \in H^0(\pi, \varrho)$.

**Proposition 4.2.** *Let $\varrho_1$ and $\varrho_2$ be $n$-dimensional unitary representations of $\pi_1(X)$, $X$ being a compact connected complex manifold. Then the holomorphic $\mathbf{GL}(n; \mathbf{C})$-principal bundles $P(\varrho_1)$ and $P(\varrho_2)$ are isomorphic if and only if the representations $\varrho_1$ and $\varrho_2$ are equivalent.*

*Proof:* If $\varrho_1$ and $\varrho_2$ are equivalent as representations then $P(\varrho_1)$ and $P(\varrho_2)$ are clearly equivalent as holomorphic $\mathbf{GL}(n; \mathbf{C})$-bundles (see also Remark in § 5). Now suppose that $P(\varrho_1)$ and $P(\varrho_2)$ are isomorphic. This means that there exists a holomorphic function $f : \tilde{X} \to \mathbf{GL}(n; \mathbf{C})$ such that $f(x\gamma) = \varrho_1(\gamma)^{-1} f(x)\, \varrho_2(\gamma)$, $\forall\, \gamma \in \pi$.

Now $\mathbf{GL}(n;\mathbf{C})$ is a subset of $\mathfrak{gl}(n;\mathbf{C})$, the vector space of all $n \times n$ complex matrices. Consider the representation $R$ of $\pi_1$ in $\mathfrak{gl}(n;\mathbf{C})$ given by

$$R(\gamma)(A) = \varrho_1(\gamma)\, A\, \varrho_2(\gamma)^{-1},\ \gamma \in \pi_1,\ A \in \mathfrak{gl}(n;\mathbf{C})\,.$$

The elements $R(\gamma)$, $\gamma \in \pi_1$, leave invariant a positive definite hermitian form on $\mathfrak{gl}(n;\mathbf{C})$, for example the form $\operatorname{tr}(A^*A)$. By Proposition 4.1 $f$ is constant, equal to, say, $C \in \mathbf{GL}(n;\mathbf{C})$. Hence $\varrho_1(\gamma) = C\varrho_2(\gamma)\, C^{-1}$, $\forall\, \gamma \in \pi_1$. Since $\varrho_1$ and $\varrho_2$ are unitary representations, it follows, as is well known, that there exists $C_1 \in \mathbf{U}(n)$ such that $\varrho_1(\gamma) = C_1\varrho_2(\gamma)\, C_1^{-1}$, $\gamma \in \pi_1$. That is, $\varrho_1$ and $\varrho_2$ are equivalent in $\mathbf{U}(n)$.

*Remarks:* 1) Proposition 4.2 is already mentioned in [10].

2) Let $\varrho$ be an irreducible unitary representation. It follows from Proposition 4.1 and Schur's lemma that $\dim_{\mathbf{C}} H^0(X, \operatorname{Ad} P(\varrho)) = 1$. Hence, in particular, the holomorphic vector bundle $W(\varrho)$ is indecomposable.

**Proposition 4.3.** *Let $X$ be a compact, connected, Kähler manifold. Let $\varrho : \pi_1(X) \to \mathbf{U}(n)$ be a homomorphism. Then the natural map*

$$H^1(X, L(\operatorname{ad}\varrho)) \to H^1(X, \operatorname{Ad} P(\varrho))$$

*is an isomorphism of real vector spaces, where $\operatorname{ad}\varrho$ is the representation of $\pi_1(X)$ in $\mathfrak{u}(n)$ obtained by composing $\varrho$ and the adjoint representation of $\mathbf{U}(n)$ in $\mathfrak{u}(n)$.*

Since the adjoint representation of $\mathbf{U}(n)$ in $\mathfrak{gl}(n;\mathbf{C})$ is the complexification of the adjoint representation $\mathbf{U}(n)$ in $\mathfrak{u}(n)$, it is easy to see that Proposition 4.3 follows from

**Proposition 4.4.** *Let $X$ be a compact connected Kähler manifold. Let $\varrho$ be a representation of $\pi_1(X)$ in a real (finite dimensional) vector space $V$ such that $\varrho$ leaves a positive definite quadratic form $Q$ on $V$ invariant. Let $V_{\mathbf{C}}$ be the complexification of $V$ and $\tilde{\varrho}$ the natural extension of $\varrho$ to a representation in the complex vectorspace $V_{\mathbf{C}}$. Then the natural map*

$$H^1(X, L(\varrho)) \to H^1(X, W(\tilde{\varrho}))$$

*is an isomorphism of real vector spaces [$W(\tilde{\varrho})$ denotes the holomorphic vector bundle corresponding to $\tilde{\varrho}$ and $L(\varrho)$ the local system determined by $\varrho$].*

*Proof:* Let $H$ be the natural extension of $Q$ as a positive definite hermitian form on $V_{\mathbf{C}}$. The real part of $H$ defines a positive definite quadratic form $S$ on the real vector space underlying $V_{\mathbf{C}}$ and $S \mid V = Q$. The form $H$ is invariant under $\tilde{\varrho}$. Hence $H$ (resp. $S$, resp. $Q$) defines a hermitian (resp. Riemannian) metric $\tilde{H}$ (resp. $\tilde{S}$, resp. $\tilde{Q}$) along the fibres of the differentiable vector bundle $D(\tilde{\varrho})$ (resp. $D(\tilde{\varrho})$, resp. $D(\varrho)$) associated with $\tilde{\varrho}$ (resp. $\tilde{\varrho}$, resp. $\varrho$). Let $\Gamma$ be a Kähler metric on $X$ and $\Gamma_0$ the Riemannian metric on $X$ associated with $\Gamma$.

For a differentiable vector bundle $D$ on $X$, we denote by $\overset{p}{E}(D)$ the space of $C^\infty$ $p$-forms on $X$ with coefficients in $D$. We remark that since the differentiable vector bundle $D(\varrho)$ is defined by a local system, the operator of exterior differentiation $d : \overset{p}{E}(D(\varrho)) \to \overset{p+1}{E}(D(\varrho))$ is defined ([9], § 1.2). Similarly the operator $d : \overset{p}{E}(D(\tilde{\varrho})) \to \overset{p+1}{E}(D(\tilde{\varrho}))$ is defined. Let $\tilde{\Delta}_1 = d\tilde{\delta} + \tilde{\delta} d$ where $\tilde{\delta}$ is the transpose of $d$ with respect to the metric $\Gamma_0$ on $X$ and $\tilde{Q}$ on $D(\varrho)$ ([9], § 1.2). Let $\tilde{\Delta} : \overset{p}{E}(D(\tilde{\varrho})) \to \overset{p}{E}(D(\tilde{\varrho}))$ be defined similarly with respect to $\Gamma_0$ and $\tilde{S}$ and $\tilde{\Delta}_{\bar{z}} : E(D(\tilde{\varrho})) \to E(D(\tilde{\varrho}))$ be defined by means of the operator $d_{\bar{z}}$ $(= \bar{\partial})$, the Kähler metric $\Gamma$ and the hermitian metric $\tilde{H}$ on $D(\tilde{\varrho})$. ([2], § 14.4; [8]). We first prove that

i) $\tilde{\Delta}_{\bar{z}} = \tilde{\Delta}/2$

ii) If $\omega_1, \omega_2 \in \overset{p}{E}(D(\varrho))$, then

$$\tilde{\Delta}(\omega_1 + i\omega_2) = \tilde{\Delta}_1(\omega_1) + i\tilde{\Delta}_1(\omega_2), \quad (i = \sqrt{-1}).$$

To prove (i) let $\Delta$, $\Delta_{\bar{z}}$ be the usual Laplacians acting on $E(\mathbf{C})$, $\mathbf{C}$ being the trivial line bundle [8]. Choose an orthonormal basis $e_1, \ldots, e_n$ in $V_{\mathbf{C}}$ with respect to $H$. With respect to a suitable open covering $\{U_\alpha\}$ of $X$, the universal covering $\tilde{X}$ is given by transition functions $h_{\alpha\beta} \in \pi$ in $U_\alpha \cap U_\beta$ and the bundle $P(\tilde{\varrho})$ is given by constant transition functions $g_{\alpha\beta} = \tilde{\varrho}(h_{\alpha\beta})$, $g_{\alpha\beta}$ being a unitary matrix. The metric $H$ is given by the constant functions $p_\alpha = I_n$ (identity matrix) in $U_\alpha : g^*_{\alpha\beta} p_\alpha g_{\alpha\beta} = p_\beta$. It follows immediately that if $\omega = \begin{pmatrix} \varphi_1 \\ \vdots \\ \varphi_n \end{pmatrix}$ is a differentiable section of $D(\tilde{\varrho})$ in $U_\alpha$, $\varphi_i$ being ordinary (complex-valued) $p$-forms, we have $\tilde{\Delta}_{\bar{z}}\omega = \begin{pmatrix} \Delta_{\bar{z}}\varphi_1 \\ \vdots \\ \Delta_{\bar{z}}\varphi_n \end{pmatrix}$. In a similar fashion we have $\tilde{\Delta}\omega = \begin{pmatrix} \Delta\varphi_1 \\ \vdots \\ \Delta\varphi_n \end{pmatrix}$. Since $\Gamma$ is a Kähler metric, we have $\frac{\Delta}{2} = \Delta_{\bar{z}}$ ([11], p. 44, Th. 2; [8]). Hence $\frac{\tilde{\Delta}}{2} = \tilde{\Delta}_{\bar{z}}$ in each $U_\alpha$. Hence $\frac{\tilde{\Delta}}{2} = \tilde{\Delta}_{\bar{z}}$. This proves (i). To prove (ii), we have only to remark the following: $V$ and $iV$ are orthogonal with respect to $S$, $S(u, v) = S(iu, iv)$, for $u, v \in V$; $\tilde{\varrho}$ is the direct sum of $\tilde{\varrho}\,|\,V$ and $\tilde{\varrho}\,|\,iV$ (over $\mathbf{R}$) and the map $v \to iv$ of $V$ onto $iV$ is an isomorphism of $\tilde{\varrho}\,|\,V$ onto $\tilde{\varrho}\,|\,iV$.

If $\omega = \omega_1 + i\omega_2$, $\omega_j \in \overset{p}{E}(D(\varrho))$, is an element of $\overset{p}{E}(D(\tilde{\varrho}))$, let $\omega^{\#} = \omega_1 - i\omega_2$. Let $p^{0,1} : \overset{1}{E}(D(\tilde{\varrho})) \to \overset{0,1}{E}(D(\tilde{\varrho}))$ be the map which associates to each 1-form with coefficients in $D(\tilde{\varrho})$, its component of type (0, 1). Let $H'$ (resp. $H_1$) be the space of harmonic forms of degree one with respect to $\tilde{\Delta}$ (resp. $\tilde{\Delta}_1$); by (ii), $H_1 \subset H'$. Let $H^{0,1}$ be the space of $\tilde{\Delta}_{\bar{z}}$-harmonic forms of type (0, 1). If $\omega \in H'$, then, using i), $\tilde{\Delta}_{\bar{z}} p^{0,1}\omega = p^{0,1}\tilde{\Delta}_{\bar{z}}\omega = \frac{1}{2} p^{0,1}\tilde{\Delta}\omega = 0$. Thus $p^{0,1}\omega \in H^{0,1}$. We have the following commutative diagram, where all arrows indicate canonical maps:

$$\begin{array}{ccccc}
H^1(X, L(\varrho)) & \to & H^1(X, L(\tilde{\varrho})) & \to & H^1(X, W(\tilde{\varrho})) \\
\uparrow & & \uparrow & & \uparrow \\
H_1 & \xrightarrow{\;j\;} & H' & \xrightarrow{\;p^{0,1}\;} & H^{0,1}
\end{array}$$

The maps in the columns are isomorphisms ([9], § 1.2; [2], § 15.4) and $j: H_1 \to H'$ is injective. We shall prove that $p^{0,1} \circ j$ is an isomorphism over $\mathbf{R}$; this will prove Proposition 4.4. To this purpose let, for $\omega \in H^{0,1}$, $q(\omega) = \omega + \omega^{\#}$; then $q(\omega) \in \overset{1}{E}(D(\varrho))$. By (i) and (ii), $\tilde{\Delta}_1(q(\omega)) = \tilde{\Delta}(q(\omega)) = \tilde{\Delta}(\omega + \omega^{\#}) = \tilde{\Delta}\omega + (\tilde{\Delta}\omega)^{\#} = 2(\tilde{\Delta}_{\bar{z}}\omega + (\tilde{\Delta}_{\bar{z}}\omega)^{\#}) = 0$; hence $q(\omega) \in H_1$. It is simple to verify that $q: H^{0,1} \to H_1$ is the inverse of $p^{0,1} \circ j: H_1 \to H^{0,1}$. It follows that the natural map $H^1(X, L(\varrho)) \to H^1(X, W(\tilde{\varrho}))$ is an isomorphism. Q. E. D.

*Remark:* We can prove Proposition 2.2, using Propositions 4.1 and 4.4 and the Riemann-Roch theorem for holomorphic vector bundles on a compact Riemann surface ([2], [10]).

## 5. Infinitesimal deformation map of a family of bundles

Let $X$ be a compact Riemann surface of genus $g \geqq 2$. Let $M_0$ be the real analytic manifold of $n$-dimensional irreducible unitary representations of $\pi_1(X)$.

We shall now construct a differentiable (in fact, real analytic) family $\mathscr{P}_0$ of holomorphic $\mathbf{GL}(n; \mathbf{C})$ bundles on $X$ parametrised by $M_0$ such that the bundle corresponding to $\varrho \in M_0$ is the bundle $P(\varrho)$ (§ 4) and investigate the infinitesimal deformation map of this family. Let $\tilde{X}$ be the universal covering surface of $X$. Consider the action on the right of $\pi = \pi_1(X)$ on $M_0 \times \tilde{X} \times \mathbf{GL}(n; \mathbf{C})$ given by

$$(\varrho, x, g)\,\gamma = (\varrho, x\gamma, \varrho(\gamma)^{-1} g), \quad \varrho \in M_0, x \in \tilde{X}, g \in \mathbf{GL}(n; \mathbf{C}), \quad \gamma \in \pi .$$

Let $\mathscr{P}_0$ be the orbit space of $M_0 \times \tilde{X} \times \mathbf{GL}(n; \mathbf{C})$ for this action of $\pi$; we have a natural map $p_0: \mathscr{P}_0 \to M_0 \times X$. The action of $\mathbf{GL}(n; \mathbf{C})$ on $M_0 \times \tilde{X} \times \mathbf{GL}(n; \mathbf{C})$ given by

$$(\varrho, x, g)\, g_1 = (\varrho, x, g g_1), \varrho \in M_0, x \in \tilde{X}, g, g_1 \in \mathbf{GL}(n; \mathbf{C})$$

commutes with the above action of $\pi$ and hence $\mathbf{GL}(n; \mathbf{C})$ operates on $\mathscr{P}_0$ and makes of it a family of holomorphic $\mathbf{GL}(n; \mathbf{C})$ bundles on $X$ parametrised by $M_0$.

Next we consider the infinitesimal deformation map $\tau$ ([4], p. 370) of the family $\mathscr{P}_0$. Let $\varrho \in M_0$. We have seen that the tangent space $M_{0,\varrho}$ can be identified with the set of 1-cocycles $Z^1(\pi, \operatorname{ad}\varrho)$ of $\pi$ with respect to the representation ad $\varrho$ of $\pi$ in $\mathfrak{u}(n)$.

**Proposition 5.1.** *The infinitesimal deformation map $\tau_\varrho : M_{0,\varrho} \to H^1(X, \operatorname{Ad} P(\varrho))$ is the composite of the maps*

$$M_{0,\varrho} \xrightarrow{\delta} Z^1(\pi, \operatorname{ad}\varrho) \xrightarrow{\eta_0} H^1(\pi, \operatorname{ad}\varrho) \xrightarrow{\varepsilon} H^1(X, L(\operatorname{ad}\varrho)) \longrightarrow H^1(X, \operatorname{Ad} P(\varrho)),$$

*where the arrows denote the natural maps.*

*Proof:* Let $\{U_i\}$ be an open covering of $X$ such that any (non-empty) finite intersection of sets $U_i$ is contractible. With respect to this covering the $\pi$-bundle $\tilde{X}$ is represented by transition functions $\gamma_{ij} \in \pi$ satisfying $\gamma_{ij}\gamma_{jk} = \gamma_{ik}$ if $U_i \cap U_j \cap U_k \neq \emptyset$. If $t \in M_0$ ($t$ is a representation of $\pi$) put $g_{ij}(t) = t(\gamma_{ij})$. The holomorphic bundle $P(t)$ is determined by the transition functions $g_{ij}(t)$.

If $v \in M_{0,\varrho}$, $\delta(v)$ is the 1-cocycle $\delta(v)(\gamma) = d\Phi(\varrho, \gamma)\, \Phi(\varrho, \gamma)^{-1}(v)$, with the notation of § 2. In particular

$$\delta(v)(\gamma_{ij}) = (dg_{ij}(t)\, g_{ij}(t)^{-1})_{t=\varrho}(v) = (v \cdot g_{ij})_{t=\varrho}\, g_{ij}(\varrho)^{-1}.$$

On $U_i$ the local system $L(\mathrm{ad}\,\varrho)$ is trivial. Hence with respect to the trivialization in $U_i$ the element $\delta(v)(\gamma_{ij}) \in \mathfrak{u}(n)$ defines a section $f_{ij}$ of the locally simple sheaf $L(\mathrm{ad}\,\varrho)$ in $U_i \cap U_j$. The natural map $\varepsilon$ is defined by assigning to $\delta(v)$ the 1-cocycle $\{f_{ij}\}$ of the covering $\{U_i\}$ with coefficients in $L(\mathrm{ad}\,\varrho)$. On the other hand the infinitesimal deformation map $\tau_\varrho \colon M_{0,\varrho} \to H^1(X, \mathrm{Ad}\,P(\varrho))$ is defined by assigning to $v$ the 1-cocycle of $\{U_i\}$ obtained by $(v g_{ij})_{t=\varrho}\, g_{ij}(\varrho)^{-1}$, considered as a section of $\mathrm{Ad}\,P(\varrho) \,|\, (U_i \cap U_j)$ by means of the trivialisation with respect to $U_i$. Thus the Proposition is proved.

*Remark:* Consider the action of $\mathbf{U}(n)$ on $M_0 \times \tilde{X} \times \mathbf{GL}(n; \mathbf{C})$ on the left given by

$$T(\varrho, x, g) = (T \varrho T^{-1}, x, Tg),\ \varrho \in M_0,\ x \in X,\ g \in \mathbf{GL}(n; \mathbf{C}),\ T \in \mathbf{U}(n).$$

Since the actions of $\pi$ and $\mathbf{U}(n)$ on $M_0 \times \tilde{X} \times \mathbf{GL}(n; \mathbf{C})$ commute, the action of $\mathbf{U}(n)$ goes down into an (effective) action of $\mathbf{U}(n)$ on $\mathscr{P}_0$. We have for each $T \in \mathbf{U}(n)$ the commutative diagram

$$\begin{array}{ccc} \mathscr{P}_0 & \xrightarrow{\ T\ } & \mathscr{P}_0 \\ \big\downarrow{\scriptstyle p_0} & & \big\downarrow{\scriptstyle p_0} \\ M_0 \times X & \xrightarrow{\ T\ } & M_0 \times X \end{array}$$

where $T(\varrho, x) = (T \varrho T^{-1}, x)$, $\varrho \in M_0$, $x \in X$. Moreover, for each $\mathbf{T} \in U(n)$ the action of $T$ on $\mathscr{P}_0$ is a differentiable bundle automorphism which is an holomorphic isomorphism of the holomorphic bundle $p_0 : p_0^{-1}(\varrho \times X) \to X$ (the bundle $P(\varrho)$) onto the holomorphic bundle $p_0 : p_0^{-1}(T \varrho T^{-1} \times X) \to X$ (the bundle $P(T \varrho T^{-1})$).

## 6. Complex structure on the manifold $M$

Let $X$ be a compact Riemann surface ($g \geqq 2$). Let $M$ be the real analytic manifold of equivalence classes of $n$-dimensional irreducible unitary representations of $\pi_1(X)$. Let $M_0$ have the meaning of § 5 and $p : M_0 \to M$ be the natural projection. From Proposition 4.2, equivalence classes of holomorphic $\mathbf{GL}(n; \mathbf{C})$ bundles arising from irreducible representations of $\pi_1(X)$ in $\mathbf{U}(n)$ correspond to points of $M$.

We shall now define an almost complex structure on $M$. Let $m \in M$ and $\varrho_1 \in M_0$ such that $p(\varrho_1) = m$. By the remarks in § 3 we have an exact sequence

$$0 \longrightarrow B^1(\pi, \mathrm{ad}\,\varrho_1) \longrightarrow Z^1(\pi, \mathrm{ad}\,\varrho_1) \xrightarrow{\ dp\ } M_m \longrightarrow 0$$

where $Z^1(\pi, \mathrm{ad}\,\varrho_1)$ [resp. $B^1(\pi, \mathrm{ad}\,\varrho_1)$] denotes the space of 1-cocycles [resp. 1-coboundaries] of $\pi$ with respect to the representation $\mathrm{ad}\,\varrho_1$ in $\mathfrak{u}(n)$. Hence there are natural isomorphisms

$$H^1(\pi, \mathrm{ad}\,\varrho_1) \to M_m,\ I(\varrho_1) : M_m \to H^1(\pi, \mathrm{ad}\,\varrho_1).$$

By Proposition 4.3 and by the isomorphism $H^1(\pi, \mathrm{ad}\,\varrho_1) \approx H^1(X, L(\mathrm{ad}\,\varrho_1))$, the natural map $H^1(\pi, \operatorname{ad}\varrho_1) \xrightarrow{J(\varrho_1)} H^1(X, \operatorname{Ad} P(\varrho_1))$ is an isomorphism of real vector spaces. On the other hand $H^1(X, \operatorname{Ad} P(\varrho_1))$ has a natural structure of a vector space over $\mathbf{C}$. Transport this complex structure onto a complex structure on $M_m$ by means of the isomorphism $J(\varrho_1) \circ I(\varrho_1)$. We shall now prove that this complex structure on $M_m$ is independent of the point in $M_0$ lying over $m$. Namely if $\varrho_2 \in M_0$ with $p(\varrho_2) = m$, then $\varrho_2 = T\varrho_1 T^{-1}$, $T \in \mathbf{U}(n)$, and $\operatorname{ad}\varrho_2 = (\operatorname{ad} T)\,(\operatorname{ad}\varrho_1)\,(\operatorname{ad} T)^{-1}$. We easily see that we have a commutative diagram

$$\begin{array}{ccccc} & \nearrow^{I(\varrho_1)} & H^1(\pi, \operatorname{ad}\varrho_1) & \xrightarrow{J(\varrho_1)} & H^1(X, \operatorname{Ad} P(\varrho_1)) \\ M_m & & \updownarrow & & \updownarrow \\ & \searrow_{I(\varrho_2)} & H^1(\pi, \operatorname{ad}\varrho_2) & \xrightarrow{J(\varrho_2)} & H^1(X, \operatorname{Ad} P(\varrho_2)) \end{array}$$

where the maps in the columns are the isomorphisms induced by $T$ and the map $H^1(X, \operatorname{Ad} P(\varrho_1)) \to H^1(X, \operatorname{Ad} P(\varrho_2))$ is an isomorphism of *complex* vector spaces (Note that $T$ induces an isomorphism of $P(\varrho_1)$ onto $P(\varrho_2)$ (§ 5)). Hence the complex structures on $M_m$ defined by means of $J(\varrho_1) \circ I(\varrho_1)$ and $J(\varrho_2) \circ I(\varrho_2)$ are the same.

We next prove that the complex structures on $M_m$ defined above is an almost complex structure (i.e., depend differentiably on $m$) and that the almost complex structure is integrable. To this purpose let $m_1 \in M$ and $\sigma: U \to M_0$ be a section of the bundle $M_0 \to M$ in a neighbourhood $U$ of $m_1$. Let $\mathscr{P}$ be the inverse image of the family $\mathscr{P}_0 \to M_0$ by $\sigma$; $\mathscr{P}$ is a differentiable family of holomorphic bundles on $X$ parametrised by $U$. Let $\tau_m: M_m \to H^1(X, \operatorname{Ad} P_m)$ be the infinitesimal deformation map of the family $\mathscr{P}$ ([4], p. 370, (7.1)). We shall prove that

$$\tau_m = J(\sigma(m)) \circ I(\sigma(m)), \quad m \in U\,,$$

with the notation of the last paragraph. By Proposition 5.1, the following diagram is commutative:

$$\begin{array}{ccccccc} M_m & \xrightarrow{d\sigma} & M_{0,\sigma}(m) & \xrightarrow{\tau_{\sigma(m)}} & H^1(X, \operatorname{Ad} P(\sigma(m))) & \longrightarrow & H^1(X, \operatorname{Ad} P_m) \\ & & & \searrow^{\delta_1} \quad \nearrow_{J(\sigma(m))} & & & \\ & & & H^1(\pi, \operatorname{ad}\sigma(m)) & & & \end{array}$$

where $d\sigma$ denotes the differential of the map $\sigma$ and $\delta_1$ is the natural map of $M_{0,\sigma(m)}$ into $H^1(\pi, \operatorname{ad}\sigma(m))$ (§ 2); clearly $\tau_m$ is the composite of the horizontal maps. Since $I(\sigma(m)) = (\delta_1 \circ d\sigma)$, we have $\tau_m = J(\sigma(m)) \circ I(\sigma(m))$. It follows that $\tau_m$ is an isomorphism. Further we have

**Lemma.** *Let $\varrho$ be an $n$-dimensional irreducible unitary representation of $\pi_1(X)$ $(g \geq 2)$. Let $\Sigma(\varrho)$ be the holomorphic vector bundle (on $X$) of invariant tangent vector fields on $P(\varrho)$. Then*

i) *$\dim_{\mathbf{C}} H^1(X, \Sigma(\varrho))$ is independent of $\varrho$ (for fixed $n$),*

ii) *the natural map $H^1(X, \operatorname{Ad} P(\varrho)) \to H^1(X, \Sigma(\varrho))$ is injective.*

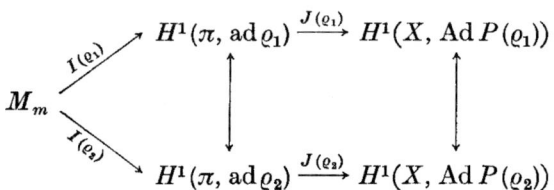

*Proof:* We have an exact sequence of vector bundles (Atiyah exact sequence)

$$(A) \quad 0 \to \operatorname{Ad} P(\varrho) \to \Sigma(\varrho) \to \Theta \to 0 ,$$

where $\Theta$ is the holomorphic tangent bundle of $X$ ([4], p. 358, $(4.5)_t$). Since $H^0(X, \Theta) = 0$ ($g \geqq 2$) and $H^2(X, \operatorname{Ad} P(\varrho)) = 0$ ($\dim_{\mathbf{C}} X = 1$), the exact cohomology sequence corresponding to $(A)$ yields the exact sequence

$$\text{(I)} \qquad 0 \to H^1(X, \operatorname{Ad} P(\varrho)) \to H^1(X, \Sigma(\varrho)) \to H^1(X, \Theta) \to 0 .$$

Since $\dim_{\mathbf{C}} H^1(X, \operatorname{Ad} P(\varrho)) = n^2(g-1) + 1$ (Prop. 2.3; Prop. 4.3), (i) and (ii) follow from (I) (In fact $\dim_{\mathbf{C}} H^1(X, \Sigma(\varrho)) = (n^2(g-1)+1) + (3g-3)$).

Q. E. D.

Now we can apply a theorem of KODAIRA-SPENCER-NAKANO ([5]; [6], Proposition 1) to conclude that the complex structures defined above on $M_m$ define a structure of a complex manifold on $M$.

The dimension of the complex manifold $M$ is clearly $n^2(g-1)+1$.

*Remark:* By using the existence of a (locally) complete family of deformations for a given vector bundle on $X$, we can introduce complex coordinates on $M$ without appealing to the theorem of KODAIRA-SPENCER-NAKANO.

## References

[1] CARTAN, H., and S. EILENBERG: Homological Algebra. Princeton: University Press 1956.

[2] HIRZEBRUCH, F.: Neue topologische Methoden in der algebraischen Geometrie. Berlin-Göttingen-Heidelberg: Springer 1956.

[3] IGUSA, J.: On a property of commutators in the unitary group. Mem. Coll. Sci. Univ. Kyoto, Ser. A Math. **26**, 45—49 (1950).

[4] KODAIRA, K., and D. C. SPENCER: On deformations of complex analytic structures. I. Ann. Math. **67**, 328—401 (1958).

[5] — — Existence of complex structure on a differentiable family of deformations of compact complex manifolds. Ann. Math. **70**, 145—166 (1959).

[6] NAKANO, S.: Parametrization of a family of bundles. Mem. Coll. Sci. Univ. Kyoto, Ser. A Math. **33**, 353—366 (1961).

[7] NORTHCOTT, D. G.: An introduction to homological algebra. Cambridge: University Press 1960.

[8] SCHWARTZ, L.: Lectures on complex analytic manifolds. Tata Institute of Fundamental Research. Bombay 1955.

[9] SESHADRI, C. S.: Generalized multiplicative meromorphic functions on a complex analytic manifold. J. Ind. Math. Soc. **21**, 149—178 (1957).

[10] WEIL, A.: Généralisation des fonctions abéliennes. J. math. pures appl. **17**, 47—87 (1938).

[11] — Introduction à l'étude des variétés kählériennes. Paris: Hermann 1958.

*(Received June 21, 1963)*

Druck: Brühlsche Universitätsdruckerei Gießen

# Stable and unitary vector bundles on a compact Riemann surface

By M. S. Narasimhan and C. S. Seshadri

## 1. Introduction

D. Mumford has defined the notion of a stable vector bundle on a compact Riemann surface $X$ and proved that the set of equivalence classes of stable bundles (of fixed rank and degree) has a natural structure of a non-singular, quasi-projective, algebraic variety [13]. We prove in this paper that, if $X$ has genus $\geqq 2$, the stable vector bundles are precisely the holomorphic vector bundles on $X$ which arise from certain *irreducible unitary* representations of suitably defined fuchsian groups acting on the unit disc and having $X$ as quotient (Theorem 2, § 12). We also prove that, if $\{W(t)\}$ is an algebraic (resp. holomorphic) family of vector bundles on $X$ parametrised by an algebraic (resp. complex) space $T$, the set of points $t \in T$ for which $W(t)$ is stable is a Zariski open subset of $T$ (resp. an open subset whose complement is an analytic subset).

It follows from our results that the space of (equivalence classes of) stable vector bundles of rank $n$ and degree $q$ on $X$ is *compact* if $n$ and $q$ are coprime.

A particular case of our result is that a holomorphic vector bundle of degree zero on $X$ is stable if and only if it arises from an irreducible unitary representation of the fundamental group of $X$. As a consequence one sees that a holomorphic vector bundle on $X$ arises from a unitary representation of the fundamental group of $X$ if and only if each of its indecomposable components is of degree zero and stable.

In this paper an essential role is played by the relationship between holomorphic vector bundles on $X$ (not necessarily of degree zero) and representations of certain fuchsian groups (which may have fixed points). This relationship (the functor $p_*^\pi$ of § 5) is already implicit in the classical paper of A. Weil [19].

For an outline of the proofs of the main results, one may refer to our note [15].

We are grateful to Professor D. Mumford for sending us a manuscript containing his results on geometric invariant theory. Although we do not use any of his results, his theory of stable and semi-stable points has been indirectly of great help to us.

## 2. Simple bundles

In the following, $X$ will denote a compact Riemann surface of genus $g$. We shall mean by a vector bundle a holomorphic vector bundle. We note that $X$ has a natural structure of an algebraic variety, and that a vector bundle on $X$ has a (uniquely determined) structure of an algebraic vector bundle on $X$[17]. If $W$ is a vector bundle of rank $n$ on $X$, the degree of the line bundle $\bigwedge^n W$ will be called the degree of $W$; we denote the degree of a vector bundle $W$ on $X$ by $d(W)$ and its rank by $r(W)$.

If $W_1$ and $W_2$ are vector bundles on $X$, $\mathrm{Hom}(W_1, W_2)$ will denote the homomorphism vector bundle; if $W_1 = W_2 = W$, we also denote $\mathrm{Hom}(W, W)$ by $\mathrm{End}\, W$. If $W$ is a vector bundle, we denote the sheaf of germs of holomorphic sections of $W$ by $\mathbf{W}$, and the $i^{\text{th}}$ cohomology space of $X$ with coefficients in $\mathbf{W}$ by $H^i(X, W)$.

If $T$ is a complex space, a holomorphic family of vector bundles on $X$ parametrised by $T$ is, by definition, a vector bundle $\mathcal{W}$ on $T \times X$; if $W(t)$ is the inverse image of $\mathcal{W}$ by the map $x \rightsquigarrow (t, x)$, $x \in X$, of $X$ into $T \times X$, we also denote the family $\mathcal{W}$ by $\{W(t)\}$. We have a similar definition of algebraic families of vector bundles on $X$.

If $T$ is a complex (or algebraic) space and $t \in T$, we denote by $T_t$ the Zariski tangent space to $T$ at $t$. If $\{W(t)\}$, $t \in T$, is a holomorphic family of vector bundles on $X$ parametrised by a complex space $T$, one can define a canonical linear mapping

$$\tau_t\colon T_t \longrightarrow H^1(X, \mathrm{End}\,(W(t)))\ ;$$

this mapping is called the infinitesimal deformation mapping of the family at $t$ [10, p. 370].

DEFINITION 2.1. A vector bundle $W$ on $X$ is said to be a *simple* bundle if $\dim_C H^0(X, \mathrm{End}\, W) = 1$.

If $W$ is a simple bundle of rank $n$ on $X$, $W$ is indecomposable and, by the Riemann-Roch theorem for vector bundles, $\dim_C H^1(X, \mathrm{End}\, W) = n^2(g-1)+1$. Using the upper semi-continuity theorem for cohomology [4], [6, § 7.3] (see also [16, Th. 5.3]), we have the following result. If $\{W(t)\}$, $t \in T$, is a holomorphic (resp. algebraic) family of vector bundles on $X$ parametrised by a complex (resp. algebraic) space $T$, then the set $S = \{t \mid t \in T,\ W(t) \text{ is simple}\}$ is an open (resp. Zariski open) subset of $T$ and $T - S$ is an analytic subset of $T$.

THEOREM 1. *Let $W$ be a simple vector bundle of rank $n$ on a compact Riemann surface of genus $g$. Then there exists a complex manifold $M$ and a holomorphic family $\{W(m)\}$, $m \in M$, of simple vector bundles on $X$ param-*

*etrised by* $M$ *such that*

(i) *there exists a point* $m_0 \in M$ *such that* $W(m_0) = W$; i.e., $W(m_0)$ *is isomorphic to* $W$;

(ii) *if* $m_1, m_2 \in M$ *and* $m_1 \neq m_2$, *then* $W(m_1)$ *and* $W(m_2)$ *are not isomorphic*;

(iii) *the family* $\{W(m)\}$ *is complete at every* $m_1 \in M$; i.e., *given a holomorphic family* $\{V(t)\}$, $t \in T$, *of vector bundles parametrised by a complex manifold* $T$ *and a point* $t_0 \in T$ *with* $V(t_0) = W(m_1)$, *there is a neighborhood* $N$ *of* $t_0$ *and a (unique) holomorphic map* $f\colon N \to M$ *such that* $\{V(t)\}$, $t \in N$, *is (isomorphic to) the inverse image by* $f$ *of the family* $\{W(m)\}$, $m \in M$;

(iv) *the complex dimension of* $M$ *is* $n^2(g-1)+1$.

For the proof we need

LEMMA 2.1. (i) *Let* $\{W(t)\}, t \in M$, *be a holomorphic family of vector bundles on* $X$ *parametrised by a complex manifold* $M$ *such that the infinitesimal deformation map* $\tau_{m_0}\colon M_{m_0} \to H^1(X, \operatorname{End} W(m_0))$ *is an isomorphism for* $m_0 \in M$. *Then the family is complete at* $m_0$.

(ii) *Let* $W$ *be a vector bundle on* $X$. *Then there exists a family of vector bundles* $\{W(m)\}$, $m \in U$, *parametrised by a complex manifold* $U$ *such that* $W(m_0) = W$ *for some* $m_0 \in U$, *and such that the infinitesimal deformation map is an isomorphism at* $m_0$.

The proof of (i) is similar to the proofs of results in [9] and [11]. The proof of (ii), while similar to [9], is far simpler in our case. In fact, let $\alpha_1, \cdots, \alpha_k$ be $C^\infty$ forms of type (0,1) on $X$ with coefficients in End $W$ such that their canonical images in $H^1(X, \operatorname{End} W)$ form a base of this vector space over $\mathbf{C}$. ($k = \dim_{\mathbf{C}} H^1(X, \operatorname{End} W)$). For $m = (m_1, \cdots, m_k) \in \mathbf{C}^k$, consider the form $\alpha(m) = m_1\alpha_1 + \cdots + m_k\alpha_k$. Since $\dim_{\mathbf{C}} X = 1$, the forms $\alpha(m)$ trivially satisfy the relation $d_{\bar z}\alpha(m) + [\alpha(m), \alpha(m)] = 0$. Hence these forms allow us to construct a family $\{W(m)\}$ of vector bundles on $X$ parametrised by a neighborhood $U$ of 0 in $\mathbf{C}^k$ such that $W(0) = W$ and whose infinitesimal deformation map at 0 is an isomorphism [9], [12, § IX. Th. 1, p. 62; § *X*, Th. 1, p. 72].

PROOF OF THEOREM 1. Construct a family $\{W(m)\}$, $m \in U$, as in (ii) of Lemma 2.1. We have $\dim_{\mathbf{C}} U = \dim_{\mathbf{C}} H^1(X, \operatorname{End} W) = n^2(g-1)+1$. Using the semi-continuity theorem, we may assume that $W(m)$ is simple for $m \in U$. Then dim $H^1(X, \operatorname{End} W(m))$ is constant for $m \in U$. Hence, applying [4, Th. 5] (see also [10]), we find that the infinitesimal deformation map is an isomorphism in a neighborhood $N'$ of $m_0$ in $U$. Hence by (i) of Lemma 2.1, the family is complete at each $m \in N'$.

We shall now prove that there exists an open set $M \subset N'$, $m_0 \in M$ for which condition (ii) of the theorem is satisfied. This will complete the proof of the theorem. Applying the semi-continuity principle for the family $\mathcal{H} = \{\mathrm{Hom}\,(W(m), W(n))\}$, $(m, n) \in N' \times N'$ parametrised by $N' \times N'$, we see that there is a neighborhood $N$ of $m_0$, $N \subset N'$ such that $\dim_C H^0(X, \mathrm{Hom}\,(W(m), W(n))) \leqq 1$ for $m, n \in N$. By the same theorem, the set

$$D = \{(m, n) \mid (m, n) \in N \times N, \dim_C H^0(X, \mathrm{Hom}\,(W(m), W(n))) = 1\}$$

is a (closed) analytic subset of $N \times N$. The set $D$ contains the diagonal $\Delta$ of $N \times N$. We shall prove that there exists a neighborhood $M$ of $m_0$, $M \subset N$, such that $D \cap (M \times M) = \Delta \cap (M \times M)$; this will prove the theorem, for, if $m, n \in M$, $m \neq n$, we would have $H^0(X, \mathrm{Hom}\,(W(m), W(n))) = 0$, so that $W(m)$ and $W(n)$ are not isomorphic.

Let $p_1$ and $p_2$ denote the projections of $N \times N$ into $N$. Let $\mathcal{W}_1$ and $\mathcal{W}_2$ denote the inverse images of the family $\{W(m)\}$ by $p_1$ and $p_2$ respectively. We assert that, for an open set $D' \subset D$, $(m_0, m_0) \in D'$, there is an isomorphism of $\mathcal{W}_1 \mid D' \times X$ onto $\mathcal{W}_2 \mid D' \times X$ such that its restriction to the bundle $W(m_0)$ is the identity. In fact, since $\dim H^0(X, \mathrm{Hom}\,(W(m), W(n))) = 1$ for $(m, n) \in D$, $\bigcup_{(m,n)\in D} H^0(X, \mathrm{Hom}\,(W(m), W(n)))$ has a natural structure of a holomorphic vector bundle on $D$ [4, Th. 5]. Hence there is a holomorphic section $\sigma$ of the vector bundle $\mathcal{H}$ in $D'' \times X$ such that $\sigma \mid \{(m_0, m_0) \times X\} =$ identity section, where $D''$ is an open subset of $D$ with $(m_0, m_0) \in D''$. There exists an open set $D' \subset D''$ with $(m_0, m_0) \in D'$ such that $\sigma$ induces an isomorphism of $\mathcal{W}_1 \mid D'$ onto $\mathcal{W}_2 \mid D'$.

Let $s_0 = (m_0, m_0)$. If $\tau^1_{s_0}$ and $\tau^2_{s_0}$ denote the infinitesimal deformation maps of $\mathcal{W}_1 \mid D'$ and $\mathcal{W}_2 \mid D'$ respectively at $s_0$ and $\tau_{m_0}$ that of $\{W(m)\}$ at $m_0$, we have the commutative diagrams

$$\begin{array}{ccc} D'_{s_0} & \xrightarrow{dp_1} & N_{m_0} \\ \downarrow{\scriptstyle \tau^1_{s_0}} & \swarrow{\scriptstyle \tau_{m_0}} & \\ H^1(X, \mathrm{End}\, W(m_0)) & & \end{array} \qquad \begin{array}{ccc} D'_{s_0} & \xrightarrow{dp_2} & N_{m_0} \\ \downarrow{\scriptstyle \tau^2_{s_0}} & \swarrow{\scriptstyle \tau_{m_0}} & \\ H^1(X, \mathrm{End}\, W(m_0)) & & \end{array}$$

where $dp_i$ $(i = 1, 2)$ denotes the differential of $p_i$ at $s_0$. Since there is an isomorphism of $\mathcal{W}_1 \mid D'$ into $\mathcal{W}_2 \mid D'$ such that its restriction to $W(m_0)$ is the identity and since $\tau_{m_0}$ is an isomorphism, it follows that, if $(v_1, v_2)$, $v_1, v_2 \in N_{m_0}$, belongs to $D'_{s_0}$, we must have $v_1 = v_2$. Thus the Zariski tangent space of $D'$ at $s_0$ coincides with that of the non-singular submanifold $\Delta$. Hence there is a neighborhood $M$ of $m_0$, $M \subset N$, such that $D' \cap (M \times M) = \Delta \cap (M \times M)$. This can be seen, for instance, as follows. We have: dimension of $D'$ at $s_0 \leqq$ dimension of the Zariski tangent space $D'_{s_0}$. On the other hand, dimension of

$\Delta$ at $s_0 = \dim \Delta_{s_0}$. Since, dimension of $D'$ at $s_0 \geqq$ dimension of $\Delta$ at $s_0$, and $\dim D'_{s_0} = \dim \Delta_{s_0}$, we must have

$$\text{dimension of } D' \text{ at } s_0 = \dim D'_{s_0} = \text{dimension of } \Delta \text{ at } s_0 .$$

So $D'$ is non-singular at $s_0$, and the germs of $D'$ and $\Delta$ at $s_0$ must coincide. This completes the proof of Theorem 1.

Let $\mathrm{S}(n, q)$ denote the set of isomorphism classes of simple vector bundles on $X$ of rank $n$ and degree $q$. Using Theorem 1, we see that $\mathrm{S}(n, q)$ has a natural structure of a complex manifold (also denoted by $\mathrm{S}(n, q)$). In fact, if $W$ is a simple bundle of rank $n$ and degree $q$ on $X$ and $\{W(m)\}$, $m \in M$, is a family of vector bundles of rank $n$ and degree $q$, constructed as in Theorem 1, the natural injective map $M \to \mathrm{S}(n, q)$ gives a local coordinate system at the isomorphism class corresponding to $W$; the condition of holomorphic compatibility on the overlaps is assured by property (iii).

Using Theorem 1, we see the following. If $\{V(t)\}$, $t \in T$, is a family of simple vector bundle of rank $n$ and degree $q$ on $X$, parametrised by a complex manifold $T$, the canonical map $T \to \mathrm{S}(n, q)$, which maps $t \in T$ to the isomorphism class of $V(t)$, is a *holomorphic* map.

### 3. Total families of vector bundles on $X$

PROPOSITION 3.1. *Given integers $n$ and $q, n \geqq 1$, there exists an algebraic family $\mathfrak{A}(n, q)$ of vector bundles of rank $n$ and degree $q$ on $X$ parametrised by an (irreducible) non-singular algebraic variety $\mathbf{A}(n, q)$, such that every indecomposable vector bundle of rank $n$ and degree $q$ is isomorphic to a vector bundle corresponding to a point of $\mathbf{A}(n, q)$.*

We first prove

LEMMA 3.1. *Let $E$ and $F$ be two vector bundles on $X$. Then there exists an algebraic family $\{W(t)\}$ of vector bundles parametrised by the affine space $H^1(X, \mathrm{Hom}(E, F))$ such that $W(t)$ is isomorphic to the extension over $E$ with kernel $F$ determined by the element $t \in H^1(X, \mathrm{Hom}(E, F))$.*

PROOF. Let $V$ denote the vector space $H^1(X, \mathrm{Hom}(E,F))$ and $p: V \times X \to X$ the canonical projection. Let $G = p^*(\mathrm{Hom}(E, F))$ be the inverse image of the vector bundle $\mathrm{Hom}(E, F)$ by $p$. Since $V$ is affine, we have, by the Künneth formula, a natural isomorphism:

$$H^0(V, \mathbf{O}_V) \otimes H^1(X, \mathrm{Hom}(E, F)) \approx H^1(V \times X, G) .$$

Let $e_1, \cdots, e_m$ be a basis of $V$ (over $\mathbf{C}$) and $x_i$ $(i = 1, \cdots, m)$ be the $i^{\text{th}}$ coordinate function on $V$ with respect to this basis. The element $\omega = \sum_{i=1}^{k} x_i \otimes e_i$ is identified, by means of the above isomorphism, with an element, still denoted by $\omega$, of $H^1(V \times X, G)$. But $G \approx \operatorname{Hom}(p^*(E), p^*(F))$. Let $W$ be the extension over $p^*(E)$ with kernel $p^*(F)$ determined by $\omega \in H^1(V \times X, G)$. The family $\mathcal{W} = \{W(t)\}$, $t \in V$, satisfies the conditions of the lemma.

**Proof of Proposition 3.1.** Let $J$ be the jacobian variety of $X$. Then there exists a family $\{\theta(j)\}$, $j \in J$, of line bundles of degree zero on $X$ parametrised by $J$ such that the class of $\theta(j)$ is $j$ [8, § 2], [3]. This proves the proposition for $n = 1$.

Let $L$ be a very ample line bundle on $X$. If $m$ is a positive integer, let $L^{-m}$ denote the $m$-fold tensor product of the dual bundle of $L$. Let $I_{n-1}(-m)$ denote the bundle $I_{n-1} \otimes L^{-m}$, where $I_{n-1}$ is the trivial vector bundle of rank $(n-1)$ on $X$.

Given integers $n$ and $q, n \geqq 2$, there exists, by a theorem of M. F. Atiyah [1, Th. 3, p. 426], a positive integer $m = m(n, q)$ such that for any indecomposable vector bundle $W$ of rank $n$ and degree $q$ we have an exact sequence of vector bundles

$$0 \longrightarrow I_{n-1}(-m) \longrightarrow W \longrightarrow L' \longrightarrow 0 ,$$

where $L'$ is a line bundle. Note that $d(L') = q + m(n-1)d(L)$ is independent of $W$. Let us denote this integer by $d$.

Let $L_1$ be a fixed line bundle of degree $d$ on $X$. Let $\{W(t)\}$, $t \in T$, be a family of vector bundles parametrised by the affine space $T = H^1(X, \operatorname{Hom}(L_1, I_{n-1}(-m)))$ such that $W(t)$ is isomorphic to the extension, over $L_1$ with kernel $I_{n-1}(-m)$, determined by the element $t \in H^1(X, \operatorname{Hom}(L_1, I_{n-1}(-m)))$ (see Lemma 3.1). Let $\mathbf{A}(n,q) = T \times J$ and $\mathcal{A}(n,q)$ the family $\{W(t) \otimes \theta(j)\}$, $t \in T, j \in J$, parametrised by $T \times J$. Given a line bundle $\theta$ of degree zero, there exists a line bundle $\varphi$ of degree zero such that (the $n$-fold tensor product) $\varphi^n = \theta$. It follows that any indecomposable vector bundle of rank $n$ and degree $q$ is of the form $W(t) \otimes \theta(j)$ for suitable $t \in T, j \in J$. Hence the family $\mathcal{A}(n, q)$ has the properties stated in the proposition.

**Remark 3.1.** Let $E$ and $F$ be vector bundles on $X$. Let $T$ be a complex space and $p: T \times X \to X$ the natural projection. Let $\mathcal{W} = \{W(t)\}$ be a vector bundle on $T \times X$ such that we have an exact sequence

$$0 \longrightarrow p^*(F) \longrightarrow \mathcal{W} \longrightarrow p^*(E) \longrightarrow 0 .$$

We then have, for each $t \in T$, an exact sequence

$$0 \longrightarrow F \longrightarrow W(t) \longrightarrow E \longrightarrow 0 ,$$

where $W(t)$ is the vector bundle on $X$ induced from $W$ by the inclusion $x \leadsto (t, x)$ of $X$ in $T \times X$. We thus obtain a map $f: T \to H^1(X, \operatorname{Hom}(E, F))$. This mapping is holomorphic and $\{W(t)\}$ is (isomorphic to) the inverse image by $f$ of the family constructed in Lemma 3.1.

REMARK 3.2. The space $\mathbf{S}(n, q)$ of simple bundles of rank $n$ and degree $q$ is *connected*. In fact let $A'$ be the subset of $\mathbf{A}(n, q)$ corresponding to simple bundles. By the semi-continuity theorem, $\mathbf{A}(n, q) - A'$ is an analytic subset; $\mathbf{A}(n, q)$ being irreducible, $A'$ is also connected. Since the canonical holomorphic map $A' \to \mathbf{S}(n, q)$ is surjective, $\mathbf{S}(n, q)$ is connected.

## 4. Stable and semi-stable vector bundles

Let $W_1$ and $W_2$ be two vector bundles of rank $n$ on the compact Riemann surface $X$. A (holomorphic) homomorphism $f: W_1 \to W_2$ is said to be of maximal rank if the canonical extension $\bigwedge^n f: \bigwedge^n W_1 \to \bigwedge^n W_2$ is a non-zero homomorphism. If $f: W_1 \to W_2$ is a homomorphism of maximal rank we have $d(W_1) \leqq d(W_2)$, and if $d(W_1) = d(W_2)$, $f$ is an isomorphism. (These statements follow from the corresponding statements for line bundles.)

Let $V$ and $W$ be two vector bundles on $X$, not necessarily of the same rank. Let $f: V \to W$ be a non-zero homomorphism. Since the structure sheaf $\mathbf{O}_X$ is a sheaf of principal ideal domains, we see that $f$ has the following canonical factorisation

$$\begin{array}{ccccccccc} 0 & \longrightarrow & V_1 & \longrightarrow & V & \xrightarrow{\eta} & V_2 & \longrightarrow & 0 \\ & & & & & & \downarrow{\scriptstyle g} & & \\ 0 & \longleftarrow & W_2 & \longleftarrow & W & \underset{i}{\longleftarrow} & W_1 & \longleftarrow & 0 \end{array}$$

where $V_1$, $V_2$, $W_1$, $W_2$ are vector bundles, each row is exact, $f = i \circ g \circ \eta$ and $g$ is of maximal rank. We call $W_1$ the subbundle of $W$ generated by the image of $f$.

DEFINITION 4.1. A vector bundle $W$ on $X$ is said to be stable [resp. semi-stable] if, for every proper sub-bundle $V$ of $W$ (i.e., $V \neq 0$, $V \neq W$), we have

$$r(W)d(V) < r(V)d(W) \qquad [\text{resp. } r(W)d(V) \leqq r(V)d(W)] \, .$$

REMARK 4.1. A semi-stable vector bundle whose degree and rank are co-prime, is stable.

PROPOSITION 4.1. *A vector bundle $W$ on $X$ is stable (resp. semi-stable) if and only if, for every proper subbundle $V$ of $W$, we have*

$$d(W^* \otimes V) < 0 \qquad (\text{resp. } d(W^* \otimes V) \leqq 0) \, ,$$

*where $W^*$ denotes the dual bundle of $W$.*

PROOF. The proposition follows from the equality

$$d(W^* \otimes V) = r(W)d(V) - r(V)d(W) .$$

As an immediate consequence we have

PROPOSITION 4.2. *Let $L$ be a line bundle on $X$. A vector bundle $W$ on $X$ is stable (resp. semi-stable) if and only if $W \otimes L$ is.*

PROPOSITION 4.3. *Let $V$ and $W$ be two semi-stable vector bundles of the same rank and degree on $X$, such that at least one of them is stable. If $f: V \to W$ is a non-zero homomorphism, then $f$ is an isomorphism.*

PROOF. If $f$ is of maximal rank, $f$ is an isomorphism as $d(V) = d(W)$. If $f$ is not of maximal rank, then $f$ has a factorization $f = i \circ g \circ \eta$

$$\begin{array}{ccccccccc} 0 & \longrightarrow & V_1 & \longrightarrow & V & \xrightarrow{\eta} & V_2 & \longrightarrow & 0 \\ & & & & & & \Big\downarrow{\scriptstyle g} & & \\ 0 & \longleftarrow & W_2 & \longleftarrow & W & \underset{i}{\longleftarrow} & W_1 & \longleftarrow & 0 , \end{array}$$

where $g$ is of maximal rank and $0 < r(V_2) = r(W_1) < r(W)$. Suppose $W$ is stable. We have

$$\begin{aligned} 0 = d(V^* \otimes V) &= d\big(V^* \otimes (V_1 + V_2)\big) \\ &= d(V^* \otimes V_1) + d(V^* \otimes V_2) . \end{aligned}$$

Since $V$ is semi-stable $d(V^* \otimes V_1) \leqq 0$ so that $d(V^* \otimes V_2) \geqq 0$. Since $d(W_1) \geqq d(V_2)$ and $r(W_1) = r(V_2)$, we see that

$$d(W^* \otimes W_1) = d(V^* \otimes W_1) \geqq d(V^* \otimes V_2) \geqq 0 .$$

Since $W$ is stable, we have $d(W^* \otimes W_1) < 0$, which is a contradiction. So $f$ must be of maximal rank, and hence an isomorphism. The case where $V$ is stable is treated similarly.

COROLLARY. *A stable vector bundle $W$ is simple (in particular, indecomposable).*

PROOF. Let $A$ be the finite dimensional algebra (over $\mathbf{C}$) of endomorphisms of $W$. By Proposition 4.3, every non-zero element of $A$ is a unit. Since $\mathbf{C}$ is algebraically closed, it follows that $A$ is of dimension one.

PROPOSITION 4.4. *Let $V$ be a semi-stable bundle on $X$. Let $W$ be a vector bundle on $X$ with $d(V)/r(V) \geqq d(W)/r(W)$ $\big($resp. $d(V)/r(V) > d(W)/r(W)\big)$, and let $f: V \to W$ be a non-zero homomorphism. Let $W_1$ be the sub-bundle of $W$ generated by the image of $f$. Then $d(W_1)/r(W_1) \geqq d(W)/r(W)$ $\big($resp. $d(W_1)/r(W_1) > d(W)/r(W)\big)$.*

PROOF. With the notation of the previous proposition, we have $d(W_1) \geqq d(V_2)$ and $d(V_2)/r(V_2) \geqq d(V)/r(V)$, as $V$ is semi-stable. If $d(V)/r(V) \geqq$

$d(W)/r(W)$, we have $d(W_1)/r(W_1) \geqq d(V_2)/r(V_2) \geqq d(V)/r(V) \geqq d(W)/r(W)$. Similarly, if $d(V)/r(V) > d(W)/r(W)$, we get $d(W_1)/r(W_1) > d(W)/r(W)$.

PROPOSITION 4.5. *If $W$ is not a stable bundle on $X$, there exists a (proper) stable sub-bundle $V$ with $d(V)/r(V) \geqq d(W)/r(W)$. If $W$ is not a semi-stable bundle, there exists a (proper) stable sub-bundle $V$ with*

$$d(V)/r(V) > d(W)/r(W) .$$

PROOF. If $W$ is not stable, there exists a proper subbundle $V'$ with $d(V')/r(V') \geqq d(W)/r(W)$. Let $V$ be a subbundle of minimal dimension among all (proper) sub-bundles $V'$ with $d(V')/r(V') \geqq d(W)/r(W)$. Then $V$ is stable. If $V$ were not stable, there would exist a proper sub-bundle $V''$ of $V$ with $d(V'')/r(V'') \geqq d(V)/r(V)$. But then $d(V'')/r(V'') \geqq d(V)/r(V) \geqq d(W)/r(W)$. This contradicts the minimal property of $V$. The second part of the proposition is proved similarly.

The last two propositions yield

PROPOSITION 4.6. *A vector bundle $W$ of rank $n$ on $X$ is not stable (resp. not semi-stable) if and only if there exists a stable vector bundle $V$ of rank $< n$ with $d(V)/r(V) \geqq d(W)/r(W)$ $\big($resp. $d(V)/r(V) > d(W)/r(W)\big)$ and a non-zero homomorphism of $V$ into $W$.*

## 5. $\pi$-bundles. The functor $p_*^\pi$

In this section we follow A. Grothendieck's exposition [5] of A. Weil's paper [19].

Let $\pi$ be a discrete group acting effectively, holomorphically and properly on a simply connected Riemmann surface $Y$ such that the quotient space $X = Y/\pi$ is compact. Then $X$ has a natural structure of a compact Riemann surface and the canonical map $p: Y \to X$ is holomorphic.

DEFINITION 5.1. A vector bundle $E$ on $Y$ is said to be a $\pi$-bundle if $\pi$ operates on $E$ (as vector bundle automorphisms) inducing the given action on $Y$.

We see that the dual of a $\pi$-bundle, the tensor product and direct sum of two $\pi$-bundles are $\pi$-bundles in a natural way. If $E_1$ and $E_2$ are $\pi$-bundles, Hom $(E_1, E_2)$ is also a $\pi$-bundle in a natural way. If $W$ is a vector bundle on $X$, the inverse image bundle $p^* W$ is canonically a $\pi$-bundle on $Y$.

DEFINITION 5.2. If $E_1$ and $E_2$ are $\pi$-bundles on $Y$, a homomorphism of $E_1$ into $E_2$ commuting with the action of $\pi$ on $E_1$ and $E_2$ is called a $\pi$-homomorphism $E_1$ into $E_2$.

We denote the vector space of $\pi$-homomorphisms of $E_1$ into $E_2$ by

$\Gamma_\pi(\text{Hom}(E_1, E_2))$.

Let $E$ be a $\pi$-bundle of rank $n$ on $Y$, and $\mathbf{E}$ the sheaf of germs of holomorphic sections of $E$. Then $\pi$ operates on $\mathbf{E}$ and therefore on the direct image sheaf $p_*(\mathbf{E})$, which is a sheaf of $\mathbf{O}_X$-modules. We denote by $p_*^\pi(\mathbf{E})$ the subsheaf of $p_*(\mathbf{E})$ consisting of elements invariant under the action of $\pi$. It is easy to see that $p_*^\pi(\mathbf{E})$ is a locally free sheaf of $\mathbf{O}_X$-modules of rank $n$. Hence $p_*^\pi(\mathbf{E})$ determines a vector bundle of rank $n$ on $X$, which we denote by $p_*^\pi(E)$. We thus obtain a functor $E \rightsquigarrow p_*^\pi(E)$ from the category of $\pi$-bundles on $Y$ to the category of vector bundles on $X$. If $p$ is unramified, this functor is fully faithful, i.e., $p_*^\pi$ induces an isomorphism of $\Gamma_\pi(\text{Hom}(E_1, E_2))$ onto $H^0(X, \text{Hom}(p_*^\pi(E_1), p_*^\pi(E_2)))$.

DEFINITION 5.2. Let $\rho$ be a representation of $\pi$ in $\mathbf{GL}(n; \mathbf{C})$. Then $\pi$ operates on the trivial bundle $Y \times \mathbf{C}^n$ by $(y, v) \rightsquigarrow (\gamma y, \rho(\gamma)v)$, $y \in Y$, $v \in \mathbf{C}^n$, $\gamma \in \pi$. We call this $\pi$-bundle, the $\pi$-bundle associated to the representation $\rho$.

We denote by $E_\pi(\rho)$ the $\pi$-bundle on $Y$ associated to $\rho$. The vector bundle $p_*^\pi(E_\pi(\rho))$ on $X$ is called the vector bundle arising from the representation $\rho$ of $\pi$.

## 6. Special $\pi$-bundles of type $\tau$

We keep the notation of the last section. In what follows *we shall assume that $p$: $Y \to X$ is ramified at most over a single point $x_0 \in X$*; i.e., if $y \notin p^{-1}(x_0)$, the isotropy group $\pi_y$ at $y$ reduces to the identity. If $y \in p^{-1}(x_0)$, the isotropy group $\pi_y$ is a finite cyclic group and its order $N$ is independent of the point $y \in p^{-1}(x_0)$. This integer is called the order of ramification of $p$ and, in what follows, $N$ will denote this integer. If $N = 1$, $p$: $Y \to X$ is the simply connected covering of $X$ and $\pi$ can be identified with the fundamental group of $X$.

Given a compact Riemann surface $X$ of genus $\geq 1$, a point $x_0 \in X$ and an integer $N \geq 1$, there exists a group $\pi$, a covering $p$: $Y \to X$ ($Y$ simply connected) as in § 5 which is ramified over $x_0$ with ramification order $N$ and unramified elsewhere; such a covering is determined uniquely up to isomorphism [5, p. 9]. For simplicity, we call such a (simply connected) covering the covering ramified over $x_0$ with ramification order $N$.

Let us fix a point $y_0 \in p^{-1}(x_0)$.

DEFINITION 6.1. Let $y_0 \in p^{-1}(x_0)$ and $\tau$ be a character of the isotropy group $\pi_{y_0}$. A homomorphism $\rho: \pi \to \mathbf{GL}(n; \mathbf{C})$ is said to be a representation of type $\tau$ if for every $\gamma \in \pi_{y_0}$ we have $\rho(\gamma) = \tau(\gamma)I$, where $I$ is the $n \times n$ identity matrix. A homomorphism $\rho: \pi \to \mathbf{U}(n)$ with $\rho(\gamma) = \tau(\gamma)I$, $\forall \gamma \in \pi_{y_0}$, is called a unitary representation of type $\tau$.

DEFINITION 6.2. Let $y_0 \in p^{-1}(x_0)$, and let $\tau$ be a character of the isotropy group $\pi_{y_0}$. A $\pi$-bundle $E$ on $Y$ is said to be a *special $\pi$-bundle of type $\tau$* if $E$ is associated to a representation $\rho: \pi \to \mathbf{GL}(n, \mathbf{C})$ of type $\tau$ (see Definitions 5.2 and 6.1).

We denote by $\mathbf{B}(\tau, n)$ the set of special $\pi$-bundles of type $\tau$ and rank $n$.

REMARK 6.1. Let $\tau$ be a character of $\pi_{y_0}$. Let $z$ be a local coordinate system at $y_0$ such that $\pi_{y_0}$ is the group of multiplication of $z$ by $\zeta^k$ where $\zeta$ is a primitive $N^{\text{th}}$ root of unity, $0 \leqq k < N$. If $\gamma_0$ is the generator $\pi_{y_0}$ corresponding to multiplication by $\zeta$, let $\tau(\gamma_0) = \zeta^s$, $0 \leqq s < N$. Then the integer $s$ depends only on $\tau$ and not on $\zeta$ and $z$. Conversely given an integer $s$ with $0 \leqq s < N$, there is a (uniquely determined) character $\tau$ of $\pi_{y_0}$ such that the associated integer is $s$. (see [5, p. 10]).

If $E$ is a special $\pi$-bundle of type $\tau$ and of rank $n$, then we have

$$d(p_*^{\tau}(E)) = -\frac{sn}{N} \qquad \text{(see [5, p. 13]).}$$

REMARK 6.2. We shall now see how the transition functions of a vector bundle on $X$ which arises from a representation of type $\tau$ look. Choose a finite open covering $\{U_i\}$ $(i = 0, 1, \cdots, m)$ of $X$ such that any non-empty intersection of the sets $U_i$ is contractible. We may assume that $x_0 \in U_0$, $x_0 \notin U_i$ for $i \neq 0$. We may also assume that the covering $\{U_i\}$ is so fine that there exist discs $D_i$, $i = 0, \cdots, m$, in $Y$ such that $y_0 \in D_0$, $U_0$ is the quotient of $D_0$ by $\pi_{y_0}$ and $p \mid D_i$ is homeomorphism of $D_i$ onto $U_i$ for $i \neq 0$. If $U_0 \cap U_i \neq \emptyset$, $i \neq 0$, $p^{-1}(U_0 \cap U_i) \cap D_0$ decomposes into $N$ connected components each of which is mapped homeomorphically by $p$ onto $U_0 \cap U_i$. Let $W_{ij,k}$ be a connected component of $p^{-1}(U_i \cap U_j) \cap D_k$. If $U_i \cap U_j \neq \emptyset$, $i \neq j$, let $\gamma_{ij}$ be the element of $\pi$ such that $\gamma_{ij} W_{ij,j} = W_{ji,i}$. (The $\gamma_{ij}$ may be looked upon as "transition functions" for the ramified covering $p: Y \to X$.)

Let $\rho$ be a representation of type $\tau$. Then a set of transition functions $g_{ij}$ of the vector bundle $p_*^{\tau}(E_\pi(\rho))$ on $X$ are given by

$$\text{(T)} \begin{cases} g_{ij} = \rho(\gamma_{ij}) & \text{in } U_i \cap U_j \text{ for } i \neq 0, j \neq 0 \\ g_{0i} = f_{0,i}\rho(\gamma_{0i}) & \text{in } U_0 \cap U_i \text{ for } i \neq 0\ , \end{cases}$$

where $f_{0,i}$ is a *scalar* function in $U_0 \cap U_i$ depending only on $\tau$ but *not on* $\rho$. To prove this, we observe that there exists an $n \times n$ matrix $F$ of meromorphic functions on $Y$ such that $\det F \not\equiv 0$, and such that $F(\gamma y) = \rho(\gamma)F(y)$, $\forall y \in Y$. Let $F_i = F \mid D_i$ (The $F_i$'s define a non-abelian divisor associated to $\rho$, in the sense of A. Weil). Then $F_0 = z^s \cdot G_0$, where $s$ is the integer associated to $\tau$ (Remark 6.1), $G_0$ is invariant under $\pi_{y_0}$, $z$ being a local coordinate system at $y_0$

valid in $D_0$. (See [5, p. 10]). We consider $G_0$ and $F_i$, $i \neq 0$, as functions on $U_0$ and $U_i$ respectively. Let $G_i = F_i$ for $i \neq 0$. Then it is easily seen that a set of transition functions $g_{ij}$ for $p_*^\pi(E_\pi(\rho))$ is given by $g_{ij} = G_i G_j^{-1}$ in $U_i \cap U_j$ [5]. Since $F(\gamma y) = \rho(\gamma)F(y)$, we obtain the relations (T) with $f_{0,i} = z^{-s} \circ (p \mid W_{0i,i})^{-1}$.

PROPOSITION 6.1. *Let $E_1 \in \mathbf{B}(\tau, m)$ and $E_2 \in \mathbf{B}(\tau, n)$. Then the natural map*

$$p_*^\pi(\mathrm{Hom}\,(E_1, E_2)) \longrightarrow \mathrm{Hom}\,(p_*^\pi(E_1), p_*^\pi(E_2))$$

*is an isomorphism. In particular, $p_*^\pi$ induces an isomorphism of $\Gamma_\pi(\mathrm{Hom}(E_1, E_2))$ onto $H^0(X, \mathrm{Hom}(p_*^\pi(E_1), p_*^\pi(E_2))$, and $E_1$ and $E_2$ are isomorphic as $\pi$-bundles if and only if the vector bundles $p_*^\pi(E_1)$ and $p_*^\pi(E_2)$ are isomorphic on $X$.*

PROOF. Let $E_1$ and $E_2$ be given by representations $\rho_1$ and $\rho_2$ of type $\tau$. Then the $\pi$-bundle $\mathrm{Hom}\,(E_1, E_2)$ is associated to the representation $\rho_1 \otimes \rho_2^*$, where $\rho_2^*$ is the representation contragredient to $\rho_2$. Now $\rho_1 \otimes \rho_2^*$ is identity on $\pi_{y_0}$, and hence the degree of $p_*^\pi(\mathrm{Hom}\,(E_1, E_2))$ is zero. On the other hand, with the notation of Remark 6.1, we have

$$d(p_*^\pi(E_1)) = -\frac{ms}{N} \quad \text{and} \quad d(p_*^\pi(E_2)) = -\frac{ns}{N}\,.$$

Hence $d(\mathrm{Hom}\,(p_*^\pi(E_1), p_*^\pi(E_2)) = d(p_*^\pi(E_1)^* \otimes p_*^\pi(E_2)) = (ms/N)\,n - (ns/N)m = 0$. The natural map $p_*^\pi(\mathrm{Hom}\,(E_1, E_2)) \rightarrow \mathrm{Hom}\,(p_*^\pi(E_1), p_*^\pi(E_2))$ is of maximal rank, as is seen by looking at a point which is not ramified. Since these two bundles have the same degree, this mapping is an isomorphism.

PROPOSITION 6.2. *Let $n$ and $q$ be integers with $-n < q \leqq 0$. Let $p\colon Y \rightarrow X$ be the simply connected covering ramified at $x_0 \in X$ with ramification order $n$, and $\pi$ the corresponding group. Let $\tau$ be a character of the isotopy group $\pi_{y_0}$, such that the integer associated to $\tau$ is $-q$* [Remark 6.1]. *Then, for any indecomposable vector bundle $W$ of rank $n$ and degree $q$ on $X$, there exists $E \in \mathbf{B}(\tau, n)$ such that $p_*^\pi(E)$ is isomorphic to $W$.*

PROOF. Let $\mathbf{L}$ be the subsheaf of the sheaf of germs meromorphic functions on $Y$ such that the stalk $\mathbf{L}_y = \mathbf{O}_{Y,y}$ if $y \notin p^{-1}(x_0)$ and $\mathbf{L}_y = z^q\mathbf{O}_{Y,y}$ if $y \in p^{-1}(x_0)$, where $z$ is a local uniformising variable at $y$ with $z(y) = 0$. Then $\mathbf{L}$ is a $\pi$-sheaf. The $\pi$-sheaf $p^*(\mathbf{W})$ (the inverse image of $\mathbf{W}$ by $p$) is a subsheaf of the $\pi$-sheaf $p^*(\mathbf{W}) \otimes_{\mathbf{O}_Y} \mathbf{L}$. Let $E$ be the $\pi$-bundle determined by $p^*(\mathbf{W}) \otimes_{\mathbf{O}_Y} \mathbf{L}$. It is seen easily that $p_*^\pi(\mathbf{L}) = p_*^\pi(\mathbf{O}_Y) = \mathbf{O}_X$. It follows that $p_*^\pi(E) = W$.

We shall now show that $E \in \mathbf{B}(\tau, n)$. Since $W$ is indecomposable, the $\pi$-bundle $E$ is also indecomposable. Further the degree $d(E, \pi)$ of the $\pi$-bundle $E$ is zero. For we have,

$$\begin{aligned} d(E, \pi) &= d(W) + q \qquad \text{(see [5, p. 13])} \\ &= 0\,. \end{aligned}$$

Hence by a theorem of A. Weil [19], [5, p. 13], $E$ is associated to a representation $\rho$ of $\pi$, which is necessarily of type $\tau$ ([5, Prop. 1]).

## 7. The space of representations of type $\tau$

Let $\tau$ be a character of $\pi_{y_0}$. Let $R(N, \tau, n)$ denote the space of $n$-dimensional representations of type $\tau$ of $\pi$. This space has a natural structure of a complex space. In fact, if $F$ is the free group on $2g + 1$ symbols ($g$ = genus of $X$) $\mathbf{a}_1, \mathbf{b}_1, \cdots, \mathbf{a}_g, \mathbf{b}_g, \mathbf{c}$, $\pi$ is isomorphic to the quotient group $F/G$, where $G$ is the normal subgroup of $F$ generated by $\mathbf{a}_1\mathbf{b}_1\mathbf{a}_1^{-1}\mathbf{b}_1^{-1} \cdots \mathbf{a}_g\mathbf{b}_g\mathbf{a}_g^{-1}\mathbf{b}_g^{-1}\mathbf{c}$ and $\mathbf{c}^N$. We denote by $a_1, b_1, \cdots, a_g, b_g, c$, respectively, the images of $\mathbf{a}_1, \mathbf{b}_1, \cdots, \mathbf{a}_g, \mathbf{b}_g, \mathbf{c}$ by this isomorphism; we may assume that $c$ is a generator of $\pi_{y_0}$. Let

$$\Omega = \mathbf{GL}(n, \mathbf{C}) \times \cdots \times \mathbf{GL}(n, \mathbf{C}) ,$$

$2g$ times and $f: \Omega \longrightarrow \mathbf{SL}(n, \mathbf{C})$ the holomorphic map

$$(A_1, B_1, \cdots, A_g, B_g) \rightsquigarrow A_1B_1A_1^{-1}B_1^{-1} \cdots A_gB_gA_g^{-1}B_g^{-1} .$$

Then $R(N, \tau, n)$ is identified with $f^{-1}(D)$, where $D$ is the matrix $\tau(c)^{-1} \times I$, $I$ denoting the $n \times n$ identity matrix. Thus $R(N, \tau, n)$ has a natural structure of a complex space. (This space may be empty. But for $n = N$, $g \geqq 2$, we shall see that it is not empty (see Prop. 9.1).)

With the above notation we have

PROPOSITION 7.1. *The map* $f: \Omega \longrightarrow \mathbf{SL}(n, \mathbf{C})$ *is of maximal rank at*

$$\rho \in f^{-1}(D) = R(N, \tau, n)$$

*if and only if* $\dim_{\mathbf{C}} H^0(\pi, \mathrm{Ad}\, \rho) = 1$, *where* $\mathrm{Ad}\, \rho$ *is the composite of the representation* $\rho$, *and the adjoint representation of* $\mathbf{GL}(n, \mathbf{C})$ *in* $\mathfrak{gl}(n, \mathbf{C})$. *A point* $\rho \in R(N, \tau, n)$ *with* $\dim_{\mathbf{C}} H^0(\pi, \mathrm{Ad}\, \rho) = 1$ *is a simple point of* $R(N, \tau, n)$.

PROOF. The second part of the proposition follows from the first.

Let $F'$ be the free group on a set consisting of $2g$ elements $\mathbf{a}_1, \mathbf{b}_1, \cdots, \mathbf{a}_g, \mathbf{b}_g$. Let $\rho = (A_1, B_1, \cdots, A_g, B_g) \in f^{-1}(D)$ and $\rho'$ the representation of $F'$ in $\mathbf{GL}(n, \mathbf{C})$ which sends $\mathbf{a}_i$ to $A_i$, and $\mathbf{b}_i$ to $B_i$, for $i = 1, \cdots, g$. The kernel of the differential of $f$ at $\rho$ is identified with the space of 1-cocycles of the representation $\mathrm{Ad}\, \rho'$ of $F'$ in $\mathfrak{gl}(n, \mathbf{C})$ (see [14, § 2]). Since $D$ is a scalar matrix, $\mathrm{Ad}\, \rho'$ is identity on $\prod_{i=1}^{g} \mathbf{a}_i\mathbf{b}_i\mathbf{a}_i^{-1}\mathbf{b}_i^{-1}$, and hence determines a representation, still denoted by $\mathrm{Ad}\, \rho'$, of the quotient group $\pi_1 = F'/N'$, where $N'$ is the normal subgroup of $F'$ generated by $\prod_{i=1}^{g} \mathbf{a}_i\mathbf{b}_i\mathbf{a}_i^{-1}\mathbf{b}_i^{-1}$. Since the adjoint representation of $\mathbf{GL}(n, \mathbf{C})$ in $\mathfrak{gl}(n, \mathbf{C})$ leaves invariant a non-degenerate bilinear form, we see, as in [14, Prop. 2.2], that

$$\dim_{\mathbf{C}} H^1(\pi_1, \mathrm{Ad}\, \rho') = 2n^2(g - 1) + 2 \dim_{\mathbf{C}} H^0(\pi, \mathrm{Ad}\, \rho') .$$

Hence

dimension of the kernel of the differential of $f$ at $\rho$

$$= 2n^2(g-1) + 2\dim_{\mathrm{C}} H^0(\pi, \mathrm{Ad}\,\rho') + \dim_{\mathrm{C}} \mathfrak{gl}(n, \mathrm{C}) - \dim_{\mathrm{C}} H^0(\pi, \mathrm{Ad}\,\rho')$$

$$= 2n^2(g-1) + n^2 + \dim_{\mathrm{C}} H^0(\pi, \mathrm{Ad}\,\rho).$$

So $f$ is of maximal rank at $\rho$ if and only if $\dim_{\mathrm{C}} H^0(\pi, \mathrm{Ad}\,\rho) = 1$.

REMARK 7.1. Let $S' = \{\rho \mid \rho \in R(N, \tau, n), \dim_{\mathrm{C}} H^0(\pi, \mathrm{Ad}\,\rho) = 1\}$. Then $S'$ is a complex manifold whose tangent space at a point $\rho \in S'$ is canonically identified with the space of 1-cocycles of the representation $\mathrm{Ad}\,\rho$ in $\mathfrak{gl}(n, \mathrm{C})$.

## 8. Family of vector bundles determined by representations of type $\tau$

We shall now construct a holomorphic family $\mathcal{F} = \mathcal{F}(N, \tau, n)$ of vector bundles on $X$ parametrised by $R = R(N, \tau, n)$ such that the bundle $F(\rho)$ corresponding to $\rho \in R$ is isomorphic to the bundle $p_*^{\pi}(E_\pi(\rho))$, where $E_\pi(\rho)$ is the $\pi$-bundle on $Y$ associated to the representation $\rho$. Consider the space $R \times Y \times \mathrm{C}^n$ and the action of $\pi$ on $R \times Y \times \mathrm{C}^n$ given by

$$\gamma(\rho, y, v) = (\rho, \gamma(y), \rho(\gamma)v)\,, \qquad \text{for } \gamma \in \pi,\ \rho \in R,\ y \in Y \text{ and } v \in \mathrm{C}^n\,.$$

Thus the trivial vector bundle on $R \times Y$ with $\mathrm{C}^n$ as fibre becomes a $\pi$-bundle $V$ on $R \times Y$. If $f\colon R \times Y \to R \times X$ is the map $(\rho, y) \leadsto (\rho, p(y))$ and $\mathbf{V}$ denotes the sheaf of germs of holomorphic sections of $V$ on $R \times Y$, then $f_*^{\pi}(\mathbf{V})$ is a locally free sheaf on $R \times X$ and determines a vector bundle $\mathcal{F}$ on $R \times X$ with the required property. (See Remark 6.2.)

Let $S = S(N, \tau, n)$ be the subset of $R(N, \tau, n)$ consisting of points $\rho$ such that $p_*^{\pi}(E_\pi(\rho))$ is a simple vector bundle on $X$, where $E_\pi(\rho)$ is the $\pi$-bundle on $Y$ associated to $\rho$. If $p_*^{\pi}(E_\pi(\rho))$ is simple, we have by Proposition 6.1,

$$\dim_{\mathrm{C}} \Gamma(\mathrm{Hom}\,(E_\pi(\rho), (E_\pi(\rho)))) = 1\,;$$

hence we must have $\dim_{\mathrm{C}} H^0(\pi, \mathrm{Ad}\,\rho) = 1$. Using Proposition 7.1 and the upper semi-continuity theorem, we conclude that $S$ is an open subset of $R$ consisting only of simple points of $R$. Thus $S$ is a complex manifold whose holomorphic tangent space $S_\rho$ at $\rho$ is canonically identified with the space of 1-cocycles of $\rho$ for the representation $\mathrm{Ad}\,\rho$ (Remark 7.1).

PROPOSITION 8.1. *Let $\mathcal{F}'$ be the restriction of the family $\mathcal{F}$ to $S$. For $\rho \in S$, let $S_\rho$ denote the tangent space to $S$ at $\rho$, and $Z^1(\pi, \mathrm{Ad}\,\rho)$ the space of 1-cocycles of $\pi$ for the representation* $\mathrm{Ad}\,\rho$. *Then*

(i) *the infinitesimal deformation map $J_\rho$ of the family $\mathcal{F}'$ at $\rho$ is the composite of the natural maps*

$$S_\rho \longrightarrow Z^1(\pi, \mathrm{Ad}\,\rho) \longrightarrow H^1(\pi, \mathrm{Ad}\,\rho)$$
$$\xrightarrow{\ \varepsilon'\ } H^1(X, p_*^{\pi}(E_\pi(\mathrm{Ad}\,\rho))) \longrightarrow H^1(X, \mathrm{End}\,(p_*^{\pi}(E_\pi(\rho))))\,;$$

(ii) *the infinitesimal deformation map*

$$J_\rho\colon S_\rho \longrightarrow H^1(X, \operatorname{End}(p^\tau_*(E_\pi(\rho))))$$

*is surjective.*

NOTE: Ad $\rho$ takes the value identity on $\pi_{y_0}$, and hence determines a local system on $X$ so that the map $\varepsilon'$ is defined (see [14, Prop. 5.1]).

PROOF. (i) The proof of (i) is similar to that of [14, Prop. 5.1] in view of Remark 6.2.

(ii) By Proposition 6.1, the natural map

$$H^1(X, p^\tau_*(E_\pi(\operatorname{Ad}\rho))) \longrightarrow H^1(X, \operatorname{End} p^\tau_*(E_\pi(\rho)))$$

is an isomorphism. So, by (i), it is sufficient to prove that the natural map

$$\varepsilon'\colon H^1(\pi, \operatorname{Ad}\rho) \longrightarrow H^1(X, p^\tau_*(E_\pi(\operatorname{Ad}\rho)))$$

is surjective. Since Ad $\rho$ is trivial on $\pi_{y_0}$, Ad $\rho$ defines a representation of the fundamental group $\pi_1(X)$ of $X$, and hence defines a local system $L(\operatorname{Ad}\rho)$ on $X$. Further, $p^\tau_*(E_\pi(\operatorname{Ad}\rho))$ is the holomorphic bundle $W(\operatorname{Ad}\rho)$ on $X$ associated to the representation Ad $\rho$ of $\pi_1(X)$ [14, § 4]. We shall prove that the natural map

$$H^1(X, L(\operatorname{Ad}\rho)) \longrightarrow H^1(X, W(\operatorname{Ad}\rho))$$

is surjective; this will complete the proof of the proposition. For this purpose, writing $W = W(\operatorname{Ad}\rho)$, we consider the exact sequence of sheaves on $X$:

$$0 \longrightarrow L(\operatorname{Ad}\rho) \longrightarrow \mathbf{W} \xrightarrow{\ d\ } \mathbf{T}^* \otimes_{\mathbf{O}_X} \mathbf{W} \longrightarrow 0$$

where $\mathbf{T}^*$ denotes the sheaf of germs of holomorphic 1-forms on $X$. The corresponding exact cohomology sequence yields, since

$$\dim_{\mathbf{C}} H^0(X, L(\operatorname{Ad}\rho)) = \dim_{\mathbf{C}} H^0(X, W) = 1\ ,$$

the exact sequence

$$0 \longrightarrow H^0(X, \mathbf{T}^* \otimes_{\mathbf{O}_X} \mathbf{W}) \longrightarrow H^1(X, L(\operatorname{Ad}\rho)) \longrightarrow H^1(X, W)\ .$$

Now, by the duality theorem,

$$\dim_{\mathbf{C}} H^0(X, \mathbf{T}^* \otimes_{\mathbf{O}_X} \mathbf{W}) = \dim_{\mathbf{C}} H^1(X, W) = n^2(g-1) + 1\ .$$

On the other hand,

$$\begin{aligned}\dim_{\mathbf{C}} H^1(X, L(\operatorname{Ad}\rho)) &= 2n^2(g-1) + 2\dim_{\mathbf{C}} H^0(\pi, \operatorname{Ad}\rho)\\ &= 2\{n^2(g-1)+1\}\end{aligned}$$

(see proof of Proposition 7.1). Hence the mapping $H^1(X, L(\operatorname{Ad}\rho)) \to H^1(X, W)$ is surjective.

COROLLARY. *The canonical map* $\tilde{f}\colon S \to \mathbf{S}(n, q)$ (§ 2, $q = -sn/N =$ *degree of any vector bundle on* $X$ *associated to* $\rho \in R$) *is of maximal rank, and hence admits of a local section at every point (S being supposed non-empty).*

PROOF. In fact, if $\rho \in S$ and $\mathfrak{W} = \{W(t)\}$ a holomorphic family constructed as in Theorem 1 in a neighborhood $M$ of $\tilde{f}(\rho)$ with $W(\tilde{f}(\rho)) = p_*^\pi(E_\pi(\rho))$, we have a commutative diagram

$$\begin{array}{ccc} S & \xrightarrow{J_\rho} & H^1(X, \mathrm{End}\, p_*^\pi(E_\pi(\rho))) \\ \Big\downarrow d\tilde{f} & & \Big\downarrow \approx \\ M_{\tilde{f}(\rho)} & \xrightarrow[J_{\tilde{f}(\rho)}]{} & H^1(X, \mathrm{End}\, W(\tilde{f}(\rho))) \end{array}$$

where $J_\rho$ (resp. $J_{\tilde{f}(\rho)}$) is the infinitesimal deformation map of the family $\mathcal{F}$ (resp. $\{W(t)\}$) at $\rho$ (resp. $\tilde{f}(\rho)$). Since $J_{\tilde{f}(\rho)}$ is an isomorphism and $J_\rho$ is surjective, it follows that $d\tilde{f}$ is surjective.

REMARK. The corollary can also be proved by introducing a holomorphic family of holomorphic connections on the $\pi$-bundles.

## 9. Unitary representations of type $\tau$

We shall assume hereafter that $g \geqq 2$ and $n = N$.

PROPOSITION 9.1. *Let $g \geqq 2$ and $\tau$ be a character of $\pi_{y_0}$. Then there exists an irreducible $N$-dimensional unitary representation of $\pi$ which is of type $\tau$.*

PROOF. To prove the proposition, it is sufficient to find $N \times N$ unitary matrices $A_1, B_1, A_2, B_2$ which form an irreducible set of matrices and which satisfy the relation

$$A_1 B_1 A_1^{-1} B_1^{-1} A_2 B_2 A_2^{-1} B_2^{-1} = \lambda I ,$$

where $\lambda$ is a given $N^{\text{th}}$ root of unity, and $I$ is the $N \times N$ identity matrix. Let $B_1$ be the $N \times N$ diagonal matrix whose diagonal elements are the distinct $N^{\text{th}}$ roots of unity. Then we can find an $N \times N$ unitary matrix $A_1$ such that $A_1 B_1 A_1^{-1} = \lambda B_1$. Let $A_2$ be the unitary matrix defined by $A_2 e_i = e_{i+1} (1 \leqq i \leqq N-1)$, $A_2 e_N = e_1$, where $e_1, \cdots, e_N$ is the canonical basis in $\mathbf{C}^N$. Let $B_2 = A_2^{-1}$. Then $A_1, B_1, A_2, B_2$ satisfy the required conditions. q.e.d.

Let $U(N, \tau, n)$ (resp. $U_0(N, \tau, n)$) be the subset of $R(N, \tau, n)$ consisting of $n$-dimensional unitary (resp. irreducible unitary) representation of type $\tau$.

One proves the following proposition as in [14, § 2 and § 3].

PROPOSITION 9.2. (i) *The space $U_0 = U_0(n, \tau, n)$ has a natural structure of a real analytic manifold of dimension $2\{n^2(g-1)+1\} + (n^2-1)$ and the tangent space at a point $\rho \in U_0$ is identified with space $Z^1(\pi, \mathrm{ad}\,\rho)$ of 1-cocycles of $\pi$, for the representation* ad $\rho$ *of $\pi$ in* $\mathfrak{u}(n)$ (ad $\rho$ *denotes the composite of $\rho$ and the adjoint representation of* $\mathbf{U}(n)$ *in its Lie algebra* $\mathfrak{u}(n)$).

(ii) *the equivalence classes of n-dimensional irreducible unitary representations of type $\tau$ form a real analytic manifold $M = M(n, \tau, n)$ of dimension $2\{n^2(g-1)+1\}$, whose tangent space at $m \in M$ is naturally isomorphic to $H^1(\pi, \mathrm{ad}\,\rho)$, where $\rho$ is a representation in the class $m$.*

PROPOSITION 9.3. *The space $M(n, \tau, n)$ is compact if $\tau$ is an isomorphism of $\pi_{y_0}$ onto the $n^{\text{th}}$ roots of unity,* i.e., *if the integer s associated with* $\tau$ (Remark 6.1) *is coprime to n.*

PROOF. This will be proved if we prove that every representation $\rho \in U(n, \tau, n)$ is irreducible (i.e., $U = U_0$). Let $A_1, B_1, \cdots, A_g, B_g$ be $n \times n$ unitary matrices such that $A_1 B_1 A_1^{-1} B_1^{-1} \cdots A_g B_g A_g^{-1} B_g^{-1} = \lambda I$, where $\lambda$ is a primitive $n^{\text{th}}$ root of unity, and $I$ is the $n \times n$ identity matrix. If we prove that $A_1, B_1, \cdots, A_g, B_g$ form an irreducible set of matrices, the proof of the proposition will be completed. If $A_1, B_1, \cdots, A_g, B_g$, were not irreducible, we could assume that $A_i = \begin{pmatrix} A_i' & 0 \\ 0 & A_i'' \end{pmatrix}$, $B_i = \begin{pmatrix} B_i' & 0 \\ 0 & B_i'' \end{pmatrix}$, where $A_i', B_i'$ are $m \times m$ matrices, $0 < m < n$. Then we have

$$\lambda^m = \det \textstyle\prod_i A_i' B_i' A_i'^{-1} B_i'^{-1} = 1 ,$$

which contradicts the fact that $\lambda$ is a primitive $n^{\text{th}}$ root of unity.

REMARK 9.1. Consider the case $n = N$. Let $q$ be an integer with $-n < q \leqq 0$. Let $\tau(q)$ be the character of $\pi_{y_0}$ such that the integer associated to $\tau(q)$ is $-q$ (see Remark 6.1). For convenience of notation, we denote $R(n, \tau(q), n)$, $U(n, \tau(q), n)$ and $\mathcal{F}(n, \tau(q), n)$ respectively by $R(n, q)$, $U(n, q)$ and $\mathcal{F}(n, q)$. When $q$ is an arbitrary integer we extend the meaning of these symbols as follows. Write $q = np + q_0$, $-n < q_0 \leqq 0$. Define $R(n, q) = R(n, q_0)$, $U(n, q) = U(n, q_0)$. We let $\mathcal{F}(n, q)$ denote the family $\{F(t) \otimes L^p\}$ parametrised by $R(n, q)$, where $\{F(t)\}$ is the family $\mathcal{F}(n, q_0)$ and $L$ is a fixed line bundle of degree 1.

## 10. Unitary bundles

Let $\rho$ be a unitary representation of type $\tau$. The vector bundle $p_*^{\tau}(E_\pi(\rho))$ is called a unitary bundle of type $\tau$. The following proposition is proved as in [14, § 4], using Proposition 6.1. (We may also deduce this proposition from [14, Prop. 4.1]).

PROPOSITION 10.1. *Let $\rho_1$ (resp. $\rho_2$) be an m-dimensional (resp. n-dimensional) unitary representation of type $\tau$. Then*

(i) *the natural map*

$$\mathrm{Hom}\,(\rho_1, \rho_2) \longrightarrow H^0(X, \mathrm{Hom}\, p_*^{\tau}(E_\pi(\rho_1), E_\pi(\rho_2)))$$

*is an isomorphism.* [Hom $(\rho_1, \rho_2)$ *denotes the space of homomorphisms*

$T: \mathbf{C}^m \to \mathbf{C}^n$ *with* $\rho_2(\gamma)T(v) = T\rho_1(\gamma)v$, $\forall \gamma \in \pi$ *and* $v \in \mathbf{C}^m$.]

(ii) *the vector bundles* $p_*^\tau(E_\pi(\rho_1))$ *and* $p_*^\tau(E_\pi(\rho_2))$ *are isomorphic if and only if the representations* $\rho_1$ *and* $\rho_2$ *are equivalent.*

COROLLARY. *If* $\rho$ *is a unitary representation of type* $\tau$, *the vector bundle* $p_*^\tau(E_\pi(\rho))$ *is a simple bundle if and only if* $\rho$ *is irreducible.*

From Proposition 6.1 and [14, Prop. 4.4] we have

PROPOSITION 10.2. *Let* $\rho \in U(n, \tau, n)$. *Then the natural map*

$$H^1(\pi, \operatorname{ad}\rho) \longrightarrow H^1(X, \operatorname{Hom}(p_*^\tau(E_\pi(\rho)), p_*^\tau E_\pi(\rho)))$$

*is an isomorphism of real vector spaces, where* ad $\rho$ *denotes the composite of* $\rho$ *and the adjoint representation of* $\mathrm{U}(n)$ *in its Lie algebra* $\mathfrak{u}(n)$.

PROPOSITION 10.3. *Let* $n = N$. *Let* $\tau$ *be a character of* $\pi_{y_0}$ *and* $q = -s$, *where* $s$ *is the integer associated with* $\tau$. *Let* $\mathfrak{U}_0$ *be the subset of the space of simple bundles* $\mathrm{S}(n, q)$ *consisting of points* $t$ *in* $\mathrm{S}(n, q)$ *such that* $t$ *is the class of a unitary vector bundle. Then* $\mathfrak{U}_0$ *is open in* $\mathrm{S}(n, q)$. *More precisely, the canonical mapping*

$$f: U_0(n, \tau, n) \longrightarrow \mathrm{S}(n, q)$$

*is a differentiable mapping of maximal rank.*

PROOF. Let $\rho \in U_0$ and $W = p_*^\tau(E_\pi(\rho))$. Let $\mathfrak{W} = \{W(t)\}$ be a family constructed as in Theorem 1 in a neighborhood $N$ of $f(\rho)$ such that $W(f(\rho)) = W$. Let $S = S(n, \tau, n)$ be the subset of $R(n, \tau, n)$ corresponding to simple bundles. Then $S$ is a complex manifold (§ 8), and let $\tilde{f}: S \to \mathrm{S}(n, q)$ be the canonical holomorphic mapping. By the corollary to Proposition 10.1, $U_0 \subset S$; we have $f = \tilde{f} \mid U_0$. Let $V$ be the real tangent space of $\mathrm{S}(n, q)$ at $f(\rho)$. Using Theorem 1, we see that we have a commutative diagram

$$\begin{array}{ccc} U_{0,\rho} = Z^1(\pi, \operatorname{ad}\rho) & \xrightarrow{J_\rho} & H^1(X, \operatorname{End} p_*^\tau(E_\pi(\rho))) \\ \Big\downarrow df & & \Big\downarrow \approx \\ V & \xrightarrow[J_{f(\rho)}]{} & H^1(X, \operatorname{End} W) \end{array}$$

where $J_\rho$ is the infinitesimal deformation map at $\rho$ of the restriction of the family $\mathcal{F}$(§ 8) to $U_0$ and $J_{f(\rho)}$ that of $W$ at $f(\rho)$. Using Proposition 8.1, we see that $J_\rho$ is the natural map

$$Z^1(\pi, \operatorname{ad}\rho) \longrightarrow H^1(X, \operatorname{End} p_*^\tau(E_\pi(\rho)))\ ;$$

in view of Proposition 10.2, this mapping is surjective. Since $J_{f(\rho)}$ is an isomorphism (of real vector spaces), it follows that $df$ is surjective. This completes the proof of the proposition.

REMARK 10.1. The mapping $f: U_0 \to S(n, q)$ induces, by Proposition 10.1 (ii) and Proposition 10.3, a diffeomorphism of $M(n, \tau, n)$ onto an open subset of $S(n, q)$. Hence the equivalence classes of unitary bundles of type $\tau$ have a natural structure of a (separated) complex manifold of complex dimension $n^2(g-1)+1$ $(g \geqq 2)$.

REMARK. That $\mathfrak{U}_0$ is open in $S(n, q)$ may also be seen as follows. The mapping $f: U_0 \to S(n, q)$ is continuous, being the restriction of the holomorphic map $\tilde{f}$. Then $f$ induces a continuous injective map $f: M(n, \tau, n) \to S(n, q)$. Since $M$ and $S(n, q)$ are manifolds of the same dimension, $f(M)$ is an open subset, by Brouwer's theorem on invariance of domain.

PROPOSITION 10.4. *Let $\rho$ be a unitary representation of type $\tau$ of $\pi$. Then $p_*^\pi(E_\pi(\rho))$ is semi-stable.*

PROOF. We shall first consider the case where $\rho$ is a unitary representation of the fundamental group $\pi_1 = \pi_1(X)$. Let $W(\rho)$ be the vector bundle on $X$ associated to $\rho$; we have $d(W(\rho)) = 0$. Let $V$ be a proper subbundle of rank $m$ of $W(\rho)$. We have to prove that $d(V) \leqq 0$. Now $\bigwedge^m V$ is a line subbundle of $\bigwedge^m W(\rho)$. But $\bigwedge^m W(\rho)$ is also a vector bundle arising from a unitary representation of $\pi_1$. So we are reduced to the case where $\rho$ is a unitary representation of $\pi_1$ and $L$ is a proper line subbundle of $W(\rho)$.

Suppose $d(L) > 0$. Let $L'$ be a line bundle of degree zero such that $L \otimes L'$ is isomorphic to the line bundle defined by the divisor $d(L)P$, where $P$ is a point of $X$. Then $H^0(X, L \otimes L') \neq 0$. Since $L'$ is associated to a character of $\pi_1$, the vector bundle $W' = W \otimes L'$ is associated to a unitary representation $\rho'$ of $\pi_1$, and $L \otimes L'$ is a subbundle of $W'$. Let $\rho'$ be the direct sum of irreducible unitary representations $\rho_1, \cdots, \rho_k$. Then $W(\rho')$ is the direct sum of the corresponding vector bundles $W(\rho_1), \cdots, W(\rho_k)$; let $p_i$ $(i = 1, \cdots, k)$ be the projection of $W(\rho')$ onto $W(\rho_i)$. Since $d(L \otimes L') > 0$, $p_i \mid L \otimes L'$ is zero if $W(\rho_i)$ is the trivial line bundle. If $\rho_j$ is not the trivial representation, then $H^0(X, W(\rho_j)) = 0$, by [14, Prop. 4.1]. If $p_j \mid L \otimes L'$ were not zero, since $H^0(X, L \otimes L') \neq 0$, we would have $H^0(X, W(\rho_j)) \neq 0$. Hence $p_i \mid L \otimes L' = 0$ for $i = 1, \cdots, k$, which is a contradiction. So $d(L) \leqq 0$.

Now consider the case of general $\pi$. Let $V$ be a subbundle of $W = p_*^\pi(E_\pi(\rho))$. Then $W^* \otimes V$ is a subbundle of $W^* \otimes W$. But by Proposition 6.1, the bundle $W^* \otimes W$ arises from the unitary representation $\rho^* \otimes \rho$ of $\pi_1$, where $\rho^*$ is contragredient to $\rho$; hence, by what has been proved above, $W^* \otimes W$ is semi-stable. So $d(W^* \otimes V) \leqq 0$. This means that $W$ is semi-stable (Proposition 4.1). This completes the proof of the proposition.

REMARK. One could give a proof of the proposition using the fact that, if $L$ is a line bundle on $X$ with $d(L) > 0$, then for some positive integer $m$, the line bundle $L^m$ is very ample. (Consider $L^m$ as a subbundle of $\bigotimes^m W(\rho)$ and decompose $\bigotimes^m W(\rho)$ into its indecomposable components.) One can also prove the proposition by a differential geometric method.

## 11. Stable and semi-stable bundles in an algebraic family

PROPOSITION 11.1. *Let $\mathcal{W} = \{W(t)\}$ be an algebraic family of vector bundles of rank $n$ on $X$ parametrised by an algebraic space $T$. Then there exists an integer $k_0$ such that, for any indecomposable vector bundle $V$ on $X$ of rank $\leqq n$ and with $d(V) > k_0$, we have*

$$H^0(X, \operatorname{Hom}(V, W(t))) = 0 , \qquad \forall t \in T .$$

PROOF. Let $L$ be a line bundle of degree 1 on $X$ corresponding to a point $P \in X$. If for a vector bundle $F$ on $X$ we have $H^0(X, F \otimes L^{-k}) = 0$, then $H^0(X, F \otimes L^{-r}) = 0$ for $r \geqq k$. Further, given a vector bundle $F$ on $X$ we can find a positive integer $k$ such that $H^0(X, F \otimes L^{-k}) = 0$. It follows from this and the upper semi-continuity theorem for cohomology that, if $\{F(t)\}$, $t \in T'$, is an algebraic family of vector bundles parametrised by an algebraic space $T'$, then there exists an integer $k$ such that $H^0(X, F(t) \otimes L^{-r}) = 0$, $\forall t \in T'$ and $\forall r \geqq k$.

Clearly it is sufficient to prove the proposition for vector bundles $V$ of fixed rank $m \leqq n$. Now, for $0 \leqq q < m$, consider the finite number of algebraic families $\mathcal{A}(m, q) = \{A(s)\}$, parametrised by $\mathbf{A}(m, q)$, constructed in Proposition 3.1. By considering the family $\{\operatorname{Hom}(A(s), W(t))\}$, $s \in \mathbf{A}$, $t \in T$, parametrised by $\mathbf{A}(m, q) \times T$, and applying the above remark, we find that there exists a positive integer $r_0$ such that $H^0(X, A^*(s) \otimes W(t) \otimes L^{-r}) = 0$ for $r \geqq r_0$, $t \in T$, $s \in \mathbf{A}(m, q)$, $0 \leqq q < m$. If $V$ is any indecomposable vector bundle of rank $m$ and $d(V) > m(1 + r_0)$, then $H^0(X, V^* \otimes W(t)) = 0$. In fact, write $d(V) = mp + q, 0 \leqq q < m$; then $L^{-p} \otimes V$ is indecomposable and hence corresponds to a point of $\mathbf{A}(m, q)$ and, since $p > r_0$, we have $H^0(X, V^* \otimes W(t)) = H^0(X, V^* \otimes L^p \otimes W(t) \otimes L^{-p}) = 0$.

PROPOSITION 11.2. *Let $\{W(t)\}$, $t \in T$, be an algebraic family of vector bundles of rank $n$ on $X$ parametrised by an algebraic space $T$. Let $T_s$ (resp. $T_{ss}$) be the subset of all points $t \in T$ such that $W(t)$ is stable (resp. semi-stable). Then $T_s$ (resp. $T_{ss}$) is a constructible subset of $T$* (i.e., $T_s$ $(T_{ss})$ *is a finite union of locally closed subsets in the Zariski topology of $T$*).

REMARK. We shall later prove that $T_s$ and $T_{ss}$ are actually Zariski open in $T$.

PROOF OF THE PROPOSITION. We shall prove that $T_s$ is constructible; the proof is similar for $T_{ss}$. The proposition is clearly true for vector bundles of rank 1. We shall assume that the proposition is true for algebraic families of vector bundles of rank $< n$.

Now every stable vector bundle of rank $m$ is indecomposable and hence occurs in some family $\mathcal{A}(m, q)$ of Proposition 3.1. Let $\mathbf{A}_s(m, q)$ denote the set of points in $\mathbf{A}(m, q)$ where the corresponding vector bundle is stable. By Proposition 11.1, there exists an integer $k_0$ such that $H^0(X, \operatorname{Hom}(V, W(t))) = 0$ for all indecomposable vector bundles $V$ with $r(V) < n$ and $d(V) \geqq k_0$. From this and Proposition 4.6 we see that there exists a *finite* number of families $\mathcal{A}(m, q)$, $(1 \leqq m < n$ and $q$ running through a finite set of integers) such that $W(t)$, $t \in T$, is not stable if and only if $H^0(X, \operatorname{Hom}(A(s), W(t))) \neq 0$ for some $s \in \mathbf{A}_s(m, q)$ for some $(m, q)$ in the above range. Consider the family of vector bundles $\{\operatorname{Hom}(A(s), W(t))\}$ parametrised by $\mathbf{A}(m, q) \times T$, $s \in \mathbf{A}(m, q)$, $t \in T$. The set of points $(s, t) \in \mathbf{A}(m, q) \times T$ such that $H^0(X, \operatorname{Hom}(A(s), W(t))) \neq 0$ is, by the upper semi-continuity theorem for cohomology, a Zariski closed subset $Z(m, q)$ of $\mathbf{A}(m, q) \times T$. Now by the induction hypothesis $\mathbf{A}_s(m, q)$ is a constructible subset of $\mathbf{A}(m, q)$ and hence $C(m, q) = \mathbf{Z}(m, q) \cap \{\mathbf{A}_s(m, q) \times T\}$ is constructible in $\mathbf{A}(m, q) \times T$. By a theorem of C. Chevalley [2, Ch. II, § 5, Prop. 5]; [7, § 1.8], the image (in $T$) of $C(m, q)$ by the canonical projection $\mathrm{pr}_T$ is constructible in $T$. Now $T - T_s$ is the union of the finite number of constructible sets $\mathrm{pr}_T(C(m, q))$ and hence is constructible. So $T_s$ is constructible.

## 12. The main theorem

Let $X$ be a compact Riemann surface of genus $g \geqq 2$. Let $n$ be a positive integer and $q$ an integer with $-n < q \leqq 0$. Let $\pi$ be a discrete group acting effectively, properly and holomorphically on the unit disc $Y$ such that $Y/\pi = X$, and such that the natural projection $p\colon Y \to X$ is ramified over a point $x_0 \in X$ with ramification order $n$ and unramified elsewhere (see § 6). Let $y_0 \in p^{-1}(x_0)$. Let $\tau = \tau(q)$ be the character of the isotropy group $\pi_{y_0}$ of $\pi$ at $y_0$ such that the integer $s$ associated to $\tau$ is $-q$ (see Remark 6.1). With this notation, our main theorem is the following:

THEOREM 2. (A) *A holomorphic vector bundle $W$ of rank $n$ and degree $q$ on a compact Riemann surface $X (g \geqq 2)$ is stable if and only if $W$ is isomorphic to $p_*^\tau(E)$, where $E$ is a special $\pi$-bundle (on $Y$) of type $\tau(q)$ such that the representation $\rho\colon \pi \to \mathbf{GL}(n, \mathbf{C})$ to which $E$ is associated is irreducible and unitary. Further, two such stable bundles are isomorphic on $X$ if and only if the corresponding unitary representations (of type $\tau$) of $\pi$ are equivalent.*

(B) *Let $T$ be an algebraic space (resp. a complex space) parametrising an algebraic (resp. holomorphic) family $\{W(t)\}$, $t \in T$, of vector bundles of rank $n$ on $X$. Let $T_s$ be the set of points $t \in T$ where $W(t)$ is stable. Then $T_s$ is Zariski open (resp. open) in $T$. Further, if $T_{ss}$ is the set of points $t \in T$ with $W(t)$ semi-stable, then $T_{ss}$ is Zariski open (resp. open in $T$).*

REMARK. One can prove that $T - T_s$ is an analytic subset (see Theorem 3 below).

PROOF OF THE THEOREM. We prove the theorem by induction on the rank $n$. For $n = 1$, property (A) is classical (for a proof see e.g. [18]) while (B) is trivial. We assume that (A) and (B) are valid for vector bundles of rank $< n$. We first prove a lemma (compare Proposition 11.1 and its proof).

LEMMA 12.1. *Let $\{W(t)\}$, $t \in T$, be a holomorphic family of vector bundles of rank $n$ parametrised by a complex space $T$, and let $K$ be a compact subset of $T$. Then there exists an integer $k_0$ such that, for any stable bundle $V$ with rank $< n$ and $d(V) > k_0$, we have*

$$H^0(X, \operatorname{Hom}(V, W(t))) = 0 , \qquad \forall t \in K .$$

PROOF. It is sufficient to prove the result for stable vector bundles $V$ of fixed rank $m < n$. Consider the finite number of families $\mathcal{F}(m, \tau(q'), m) = \{F(s)\}$ parametrised by $R(m, \tau(q'), m)$ for $-m < q' \leqq 0$ (§ 8). Then $U = (m, \tau(q'), m)$ is a compact subset of $R(m, \tau(q'), m)$. By the upper semi-continuity theorem for cohomology, there exists, since $U \times K$ is compact, a positive integer $r_0$ such that

$$H^0(X, F(s)^* \otimes W(t) \otimes L^{-r}) = 0 \qquad \text{for } r \geqq r_0,\ s \in U,\ t \in K,$$

where $L$ is a line bundle of degree 1 determined by a point $P \in X$. Now if $V$ is any stable bundle of rank $m$ and $d(V) > m(1 + r_0)$, then $H^0(X, V^* \otimes W(t)) = 0$. In fact, if $d(V) = mp + q'$, $-m < q' \leqq 0$, $L^{-p} \otimes V$ is stable and by the induction hypothesis isomorphic to some $F(s)$ with $s \in U(m, \tau(q'), m)$. Hence

$$H^0(X, V^* \otimes W(t)) = H^0(X, (V \otimes L^{-p})^* \otimes W(t) \otimes L^{-p}) = 0 ,$$

as $p \geqq r_0$. q.e.d.

We now proceed to prove (B) of the theorem. We treat the case of $T_s$; the proof in the case of $T_{ss}$ is similar. Let $\{W(t)\}$ be a holomorphic family of vector bundles of rank $n$ on $X$ parametrised by a complex space $T$. We may assume that $d(W(t))$ is constant. Let $T'$ be a relatively compact open subset of $T$ and $T'_s = T_s \cap T'$. Using the inductive hypothesis, the above lemma, Proposition 4.4, 4.5 and 4.6, and Proposition 10.4, we see the following: there exists a finite number of families $\{F(s)\} = \mathcal{F}(m, q)$ parametrised by $R(m, q)$ (with the notation of Remark 9.1) such that a holomorphic vector bundle $W(t)$, $t \in T'$, is not stable if and only if there exists $s \in U(m, q) \subset R(m, q)$ with $H^0(X, \mathrm{Hom}(F(s), W(t))) \neq 0$. Using the semi-continuity theorem, we see that the set $\Phi = \{(s, t) \mid s \in R(m, q), t \in T', H^0(X, \mathrm{Hom}(F(s), W(t))) \neq 0\}$ is closed in $R(m, q) \times T'$. Since $U(m, q)$ is compact, $\mathrm{pr}_{T'}(\Phi \cap \{U(m, q) \times T\})$ is closed. Thus $T' - T'_s$ is a finite union of closed sets and hence closed; so $T'_s$ is open. Since $T_s \cap T'$ is open for every relatively compact open subset $T'$ of $T$, $T_s$ is open.

Now if $\{W(t)\}$, $t \in T$, is an algebraic family, then by Proposition 11.2, $T - T_s$ is constructible in the Zariski topology; by what has been proved above $T - T_s$ is closed in the usual topology. That $T - T_s$ is Zariski closed (i.e., $T_s$ is Zariski open) now follows from

LEMMA 12.2. *Let $P$ be an algebraic space (defined over* C*), and $Q$ a constructible subset of $P$ which is closed in the usual topology. Then $Q$ is Zariski closed in $P$.*

PROOF. We may assume that $Q$ is non-empty. Let $Z$ be the Zariski closure of $Q$ in $P$. Let $Z_i (i = 1, \cdots, k)$ be the irreducible components of $Z$. Then $Q \cap Z_i$ is dense in $Z_i$ in the Zariski topology for each $i$ and $Q \cap Z_i$ is constructible in $Z_i$. Hence $Q \cap Z_i$ contains a non-empty Zariski open subset $U_i$ of $Z_i$. Since $Z_i$ is irreducible, $U_i$ is dense in $Z_i$ in the usual topology [20, p. 165] [17, Prop. 5]. So $Q \cap Z_i$ is dense in $Z_i$ in the usual topology; since $Q \cap Z_i$ is closed in the usual topology, we must have $Z_i = Q \cap Z_i$. Thus $Q = Z$. q.e.d.

This completes the proof of (B).

Before proceeding to prove (A), we shall prove, using (B), the following lemma which will be used in the proof of (A).

LEMMA 12.3. *Let $\mathbf{S}_s(n, q)$ denote the subset of $\mathbf{S}(n, q)$ consisting of points corresponding to stable bundles. Then $\mathbf{S}_s(n, q)$ is a connected open subset of $\mathbf{S}(n, q)$.*

PROOF. By (B), $\mathbf{S}_s(n, q)$ is open in $\mathbf{S}(n, q)$. To prove that $\mathbf{S}_s(n, q)$ is connected, let $\mathcal{A}(n, q)$ be the algebraic family of vector bundles parametrised by $\mathbf{A}(n, q)$ (Proposition 3.1). The subset $\mathbf{A}_s(n, q)$ of $\mathbf{A}(n, q)$ corresponding to stable bundles is Zariski open, by (B); since $\mathbf{A}(n, q)$ is irreducible, it follows that $\mathbf{A}_s(n, q)$ is connected. The canonical holomorphic mapping $f: \mathbf{A}_s(n, q) \rightarrow \mathbf{S}(n, q)$ maps $\mathbf{A}_s(n, q)$ onto $\mathbf{S}_s(n, q)$. Hence $\mathbf{S}_s(n, q)$ is connected. q.e.d.

We shall now prove (A) of the theorem. The latter part of (A) follows from Proposition 10.1 (ii). We have only to prove the first part. We first prove that, if $\rho$ is an irreducible $n$-dimensional unitary representation of type $\tau(q)$, then $p_*^\pi(E_\pi(\rho))$ is stable. We already know that $p_*^\pi(E_\pi(\rho))$ is semi-stable (Proposition 10.4). If $p_*^\pi(E_\pi(\rho))$ were not stable, there would exist a proper stable subbundle $V$ of rank $m < n$ such that $d(V)/m = q/n$ (Proposition 4.5).

We shall now prove that there exists an irreducible $m$-dimensional unitary representation $\chi$ of type $\tau$ of $\pi$ such that $p_*^{\pi}(E_{\pi}(\chi))$ is isomorphic to $V$. Once we have such a $\chi$, since $H^0(X, \operatorname{Hom}(V, W)) \neq 0$, we would have by Proposition 10.1, $\operatorname{Hom}(\chi, \rho) \neq 0$; since $\rho$ is irreducible, this is impossible. Hence $p_*^{\pi}(E_{\pi}(\rho))$ is stable.

We shall now prove the existence of $\chi$. Since $-m < d(V) \leqq 0$, by the induction hypothesis $V$ arises from an irreducible unitary representation $\chi'$ of type $\tau' = \tau'(d(V))$ of the group $\pi'$ acting on a simply connected Riemann surface $Y'$ with $Y'/\pi' = X$ and ramified over $x_0$ with ramification order $m$. Let $q/n = d(V)/m = q_1/n_1$, where $q_1$ and $n_1$ are coprime and $n_1$ is a positive integer. Let $N'$ be the normal subgroup of $\pi'$ generated by $\gamma_0^{n_1}$, where $\gamma_0$ is a generator of $\pi_{y_0}$, $y_0'$ being a point of $Y'$ lying over $x_0$. Then the group $\pi'' = \pi'/N'$ acts on the surface $Y'/N'$ such that the quotient is $X$ and $Y'/N'$ is ramified over $x_0$ with ramification order $n_1$. Now $\tau'$ is trivial on $N' \cap \pi'_{y_0'}$ and hence $\chi'$ is trivial on $N'$. So $\chi'$ defines a representation $\chi''$ of $\pi'/N'$. Since $n_1$ divides $n$, $Y$ covers $Y'/N'$ and $\pi'/N'$ is a quotient group of $\pi$. If $\chi$ is the representation of $\pi$ induced by $\chi''$, then clearly $p_*^{\pi}(E_{\pi}(\chi))$ is isomorphic to $p_{2*}^{\pi''}(E_{\pi''}(\chi''))$, where $p_2$: $Y'/N' \to X$ is the canonical projection. On the other hand, it is clear that $p_{2*}^{\pi''}(E_{\pi''}(\chi''))$ is isomorphic to $p_{1*}^{\pi'}(E_{\pi'}(\chi')) \approx V$, where $p_1$: $Y' \to X$ is the canonical projection. This completes the proof that $p_*^{\pi}(E_{\pi}(\rho))$ is stable if $\rho$ is an irreducible unitary representation of type $\tau$.

Let $R_s$ (resp. $R_{ss}$) be the subset of the complex space $R = R(n, q) = R(n, \tau(q), n)$ consisting of points $\rho$ such that $p_*^{\pi}(E_{\pi}(\rho))$ is stable (resp. semi-stable). Applying (B) to the family $\mathcal{F}$ parametrised by $R$ (§ 8), we see that $R_s$ and $R_{ss}$ are open in $R$. Since each stable bundle is simple (corollary to Proposition 4.1), $R_s \subset S = S(n, \tau(q), n)$, the subset of $R$ corresponding to simple bundles. Now $S$ is an open non-singular submanifold of $R$ (§ 8); hence $R_s$ is an open subset of $R$ consisting only of simple points of $R$. By what we have proved above, the set $U_0 = U_0(n, \tau(q), n)$, consisting of $n$ dimensional irreducible unitary representation of type $\tau$, is contained in $R_s$.

Let $\Gamma_1$ be the subset of $R_{ss} \times R_{ss}$ consisting of points $(\rho_1, \rho_2)$ such that there is a non-zero homomorphism of $p_*^{\pi}(E_{\pi}(\rho_1))$ into $p_*^{\pi}(E_{\pi}(\rho_2))$. By the semi-continuity theorem, $\Gamma_1$ is a closed subset of $R_{ss} \times R_{ss}$. Let $\Gamma_2 = (R_{ss} - R_s) \times (R_{ss} - R_s)$; then $\Gamma_2$ is closed is $R_{ss} \times R_{ss}$. Hence $\Gamma = \Gamma_1 \cup \Gamma_2$ is a closed subset of $R_{ss} \times R_{ss}$. Using Proposition 4.3, one checks that $\Gamma$ defines an equivalence relation on $R_{ss}$ with the following properties:

(i) two points $\rho_1$ and $\rho_2$ in $R_s$ are equivalent under $\Gamma$ if and only if

$p_*^\pi(E_\pi(\rho_1))$ and $p_*^\pi(E_\pi(\rho_2))$ are isomorphic on $X$;

(ii) $R_s$ is a saturated open subset of $R_{ss}$.

Let $Q$ be the quotient space $R_{ss}/\Gamma$ and $\eta: R_{ss} \to Q$ the canonical mapping. We claim that

(a) $\eta(R_s)$ is a connected open subset of $Q$;

(b) $\eta(U_0) = \eta(U_0(n, \tau(q), n))$ is a non-empty open subset of $\eta(R_s)$.

To prove (a), let $f: R_s \to \mathrm{S}(n, q)$ be the canonical holomorphic mapping (§ 2). By (i) and (ii), we see that there is a continuous mapping $g: \eta(R_s) \to \mathrm{S}(n, q)$ such that $f = g \circ \eta$. Since $f$ is of maximal rank at every point (corollary to Proposition 8.1), it follows that $g$ is a homeomorphism of $\eta(R_s)$ onto $\mathrm{S}_s(n, q)$, the subset of $\mathrm{S}(n, q)$ corresponding to stable bundles (that $g$ is onto $\mathrm{S}_s(n, q)$ follows from Proposition 6.2). By Lemma 12.3, $\mathrm{S}_s(n, q)$ is connected; hence $\eta(R_s)$ is connected, and by (ii), $\eta(R_s)$ is open. To prove (b), first note that $\eta(U_0)$ is non-empty by Proposition 9.1. Now $U_0 \subset R_s$, and $f(U_0)$ is open in $\mathrm{S}(n, q)$, by Proposition 10.3. Since $g$ is a homeomorphism, it follows that $\eta(U_0)$ is open in $\eta(R_s)$.

Now by Proposition 10.4, $U \subset R_{ss}$. Since $U$ is compact and $\Gamma$ is closed, the saturation of $U$ in $R_{ss}$ for the equivalence relation $\Gamma$ is closed in $R_{ss}$, i.e., $\eta(U)$ is closed in $Q$. Now $\eta(U) \cap \eta(R_s) = \eta(U_0)$ (see corollaries to Propositions 4.1 and 10.1). Thus $\eta(U_0)$ is a non-empty set which is both open and closed in $\eta(R_s)$. Since $\eta(R_s)$ is connected, it follows that $\eta(U_0) = \eta(R_s)$. This completes the proof of the theorem, in view of Proposition 6.2.

COROLLARY 1. *Let $X$ be a compact Riemann surface of genus $\geq 2$. Then a holomorphic vector bundle of degree zero on $X$ is stable if and only if it arises from an irreducible unitary representation of the fundamental group $\pi_1(X)$ of $X$.*

In fact, if $N$ is the normal subgroup of $\pi$ generated by $\pi_{y_0}$, then $\pi/N$ acts freely on the simply connected surface $Y/N$ and $\pi/N \approx \pi_1(X)$. Since $\tau(0)$ is the trivial character, any representation of type $\tau(0)$ is trivial on $N$. Hence the corollary follows from the theorem.

From Corollary 1 we have immediately

COROLLARY 2. *A holomorphic vector bundle on $X$ arises from a unitary representation of the fundamental group if and only if each of its indecomposable components is of degree zero and stable.*

REMARK 12.1. The theorem characterises, in fact, stable bundles $W$ of rank $n$ and arbitrary degree $q$, for there exists a line bundle $L$ (depending only on $n$ and $q$) such that $-n < d(W \otimes L) \leq 0$.

REMARK 12.2. We have shown in the course of the proof that the set of equivalence classes of stable bundle of rank $n$ and degree $q$ has a natural structure of a connected (separated) complex manifold. The topology on the space of stable bundles is independent of the complex structure on $X$ and the space is *compact*, if $n$ and $q$ are coprime. This follows from Theorem 2 and Proposition 9.3. We may also prove the compactness when $(n, q) = 1$ without using Proposition 9.3, as follows. With the notation used in the proof of the theorem, we see that $U \subset R_s$ since a semi-stable bundle, whose degree and rank are coprime, is stable. Since $U$ is compact, $\eta(R_s) = \eta(U_0) = \eta(U)$ is compact.

REMARK 12.3. The space of simple bundles $S(n, q)$ is not in general Hausdorff. (This fact has been observed also by D. Mumford.) Consider $S(2, 1)$. If $S_s(2, 1)$ is the subset corresponding to stable bundles, $S_s(2, 1)$ is open in $S(2, 1)$. If $S(2, 1)$ were Hausdorff, $S_s(2, 1)$, which is compact by Remark 12.2, would also be closed in $S(2, 1)$. Since $S(2, 1)$ is connected (Remark 3.2), we would have $S(2, 1) = S_s(2, 1)$. But, if $g \geqq 3$, there exist simple bundles of rank 2 and degree 1 which are not stable. In fact, let $L$ be a line bundle of degree 1 on $X$ with $H^0(X, L) = 0$, $H^1(X, L) \neq 0$. Take a non-trivial extension

$$0 \longrightarrow L \longrightarrow V \longrightarrow \mathrm{C} \longrightarrow 0\,,$$

where C denotes the trivial line bundle. Then one sees easily that $V$ is a simple bundle of degree 1 which is clearly not stable.

Finally we prove a result which supplements (B) of Theorem 2.

THEOREM 3. *Let* $\mathcal{W} = \{W(t)\}$, $t \in T$, *be a holomorphic family of vector bundles of rank* $n$ *on a compact Riemann surface* $X$, *parametrised by a complex space* $T$, *and let* $T_s$ *(resp.* $T_{ss}$*) be the set of points* $t \in T$ *such that* $W(t)$ *is stable (resp. semi-stable). Then* $T - T_s$ *and* $T - T_{ss}$ *are analytic subsets of* $T$.

PROOF. Note that the problem is local. Let $t_0 \in T$. We may assume that in a neighborhood $U_0$ of $t_0$ the family of line bundles $\{\bigwedge^n W(t)\}$ is a constant family. In fact, if $L(t) = \bigwedge^n W(t)$, we can find, in a neighborhood of $t_0$, a holomorphic family $\{L'(t)\}$ of line bundles of degree 0 such that $L'(t)^n = L(t_0)^{-1} \otimes L(t)$ and $L'(t_0) =$ trivial line bundle. Then the family $\{W'(t) = W(t) \otimes (L'(t))^{-1}\}$ has the property that $\bigwedge^n W'(t)$ is a constant family in a neighborhood of $t_0$ and $W(t)$ is stable (resp. semi-stable) if and only if $W'(t)$ is.

Let $H$ be an ample line bundle on $X$. Choose a positive integer $m$ such that $W(t_0) \otimes H^m$ is very ample so that $H^1(X, W(t_0) \otimes H^m) = 0$, and $W(t_0) \otimes H^m$ has sufficient sections. By [1, Th. 2], we have an exact sequence

$$0 \longrightarrow F \stackrel{\alpha}{\longrightarrow} W(t_0) \otimes H^m \longrightarrow E \longrightarrow 0 ,$$

where $F$ is the trivial bundle of rank $(n-1)$. Now $H^1(X, W(t) \otimes H^m) = 0$ in a neighborhood of $t_0$. Hence, by Riemann-Roch,

$$\dim H^0(X, \operatorname{Hom}(F, W(t) \otimes H^m))$$

is constant in a neighborhood of $t_0$. If $p: X \times T \to X$ is the natural projection, we see by [4, Th. 5] that $\alpha$ can be extended to a homomorphism of $p^*(F)$ into the vector bundle $\mathcal{W} \otimes p^*(H^m)$ in a neighborhood of $X \times t_0$ in $X \times T$. Hence over a neighborhood $U$ of $t_0, \mathrm{U} \subset \mathrm{U}_0$, we have an exact sequence of vector bundles

$$0 \longrightarrow p^*(F) \longrightarrow \mathcal{W} \otimes p^*(H^m) \longrightarrow \mathcal{L} \longrightarrow 0 .$$

Since $\{\Lambda^n W(t)\}$ is a constant family, we may assume that $\mathcal{L} = p^*(E)$ for a suitable line bundle $E$ on $X$. By Remark 3.1, we obtain a holomorphic map $f: \mathrm{U} \to H^1(X, \operatorname{Hom}(E, F))$ such that, if $\{G(s)\}, s \in H^1(X, \operatorname{Hom}(E, F))$, is the family parametrised by $H^1(X, \operatorname{Hom}(E, F))$ constructed in Lemma 3.1, $W(t)$ is stable (resp. semi-stable) if and only if $G(f(t))$ is so. But the family $\{G(s)\}$ is an algebraic family and by Theorem 2, $B$, the set of points for which $G(s)$ is not stable (resp. not semi-stable) is an analytic set. Hence $\mathrm{U} - \mathrm{U}_s$ and $\mathrm{U} - \mathrm{U}_{ss}$ are inverse images of analytic sets by the holomorphic map $f$ and hence are analytic subsets of U. This completes the proof of Theorem 3.

TATA INSTITUTE OF FUNDAMENTAL RESEARCH, BOMBAY

REFERENCES

1. M. F. ATIYAH, *Vector bundles over an elliptic curve*, Proc. London Math. Soc., Third series, 7 (1957), 414–452.
2. C. CHEVALLEY, Fondements de la Géométric Algébrique, Paris, 1958.
3. ———, *Sur la théorie de la variété de Picard*, Amer. J. Math., 82 (1960), 435–490.
4. H. GRAUERT, Eine Theorem der analytischen Garbentheorie und die Modulräume komplexer Strukturen, Inst. Hautes Etudes Sci., Publ. Math., 5 (1960).
5. A. GROTHENDIECK, Sur la mémoire de Weil "Généralisation des fonctions abéliennes", Séminaire Bourbaki, Exposé 141, (1956-57).
6. ———, Eléments de Géométrie Algébrique, Chapter III, Inst. Hautes Etudes Sci., Publ. Math., 17 (1963).
7. ———, Eléments de Géométrie Algébrique, Chapter IV, Inst. Hautes Etudes Sci., Publ. Math., 20 (1964).
8. K. KODAIRA, *Characteristic linear systems of complete continuous systems*, Amer. J. Math., 78 (1956), 716–744.
9. ———, L. NIRENBERG, and D. C. SPENCER, *On the existence of deformations of complex analytic structures*, Ann. of Math., 68, (1958), 450–459.
10. K. KODAIRA and D. C. SPENCER, *On deformations of complex analytic structures* I, II, Ann. of Math., 67 (1958), 328–466.
11. ———, *A theorem of completeness for complex analytic fibre spaces*, Acta. Math., 100 (1958), 281-294.

12. B. MALGRANGE, Lectures on the theory of functions of several complex variables, Tata Institute of Fundamental Research, Bombay, 1958.

13. D. MUMFORD, Projective invariants of projective structures and applications, Proc. Intern. Cong. Math., Stockholm (1962), 526-530.

14. M.S. NARASIMHAN and C.S. SESHADRI, *Holomorphic vector bundles on a compact Riemann surface*, Math. Ann., 155 (1964) 69-80.

15. ———, *Stable bundles and unitary bundles on a compact Riemann surface*, Proc. Nat. Acad. Sci. U.S.A, 52 (1964), 207-211.

16. H. ROHRL, *On holomorphic families of fibre bundles over the Riemannian sphere*, Mem. Coll. Sci. Univ. Kyoto, Ser A Math. 33 (1961), 435-477.

17. J.-P. SERRE, *Géometrie algébrique et géometrie analytique*, Ann. Inst. Fourier, 6 (1965-56), 1-42.

18. C. S. SESHADRI, *Generalized multiplicative meromorphic functions on a complex anlytic manifold*, J. Ind. Math. Soc., 21, (1957), 149-178.

19. A. WEIL, *Généralisation des fonctions abéliennes*, J. Math. Pures et Appl., 17 (1938), 47-87.

20. ———, Introduction à l'étude des variétés kähleriennes, Paris, Hermann, 1958.

(Received April 26, 1965)

Inventiones math. 5, 120—128 (1968)

# Manifolds with Ample Canonical Class

M. S. NARASIMHAN and R. R. SIMHA (Bombay, India)

## § 1. Introduction

Let $V$ be a compact connected real analytic manifold. We prove in this paper, by differential geometric methods, that the set of isomorphism classes of complex structures on $V$ with ample canonical line bundle has a natural structure of a Hausdorff complex space (Theorem in § 5).

We now briefly outline the proof. Let $V_0$ be a compact connected complex manifold with ample canonical bundle. Let $T$ be the parametrising space of the locally complete Kuranishi family $\{V_t\}_{t\in T}$ at $V_0$; we may assume that for $t\in T$ the canonical class of the manifold $V_t$ is ample. The group of automorphisms $\mathrm{Aut}(V_0)$ of $V_0$ is finite and acts on $T$. We prove that, if $T$ is restricted to be small, two points of $T$ correspond to isomorphic manifolds if and only if they are in the same orbit for the action of $\mathrm{Aut}(V_0)$ on $T$ (Lemma 4.1). The complex space which is the quotient of $T$ by $\mathrm{Aut}(V_0)$ is taken to be a local coordinate system at the isomorphism class of $V_0$; these coordinate systems patch up due to the local completeness of Kuranishi families.

The above lemma and the Hausdorff nature of the space are proved with the help of a generalisation of the Bergmann metric. Suppose that the line bundle $K^k$ is very ample on $V_0$, where $K$ is the canonical line bundle on $V_0$. We introduce a natural hermitian metric on $V_0$, depending on $k$, which reduces to the Bergmann metric if $k=1$, i.e., if $K$ is very ample; since $V_0$ gets imbedded in the projective space $P(H^0(V_0, K^k)^*)$ of 1-dimensional subspaces of the dual of $H^0(V_0, K^k)$, in order to define a metric on $V_0$ it is sufficient to introduce a hermitian form on $H^0(V_0, K^k)$. To do this, we first define a positive volume element on $V_0$. Using this volume element, we define a hermitian form on $H^0(V_0, K^k)$ by a device similar to that used in introducing the Petersson metric on modular forms. Now if $V_0$ and $V_0'$ are two manifolds with ample canonical class and $\varphi$: $V_0 \to V_0'$ is an isomorphism, then $\varphi$ is an *isometry* with respect to the metrics corresponding to large $k$. This fact, together with the continuous dependence of the metrics on parameters, yields the equicontinuity of some maps required in the proof (Lemma 3.2) and this shows in particular that limits of isomorphic manifolds are isomorphic, which proves the Hausdorff nature.

Our method gives a simple proof for the existence of a structure of a normal Hausdorff complex space on the set of isomorphism classes of compact Riemann surfaces of genus $\geqq 2$.

Professor D. MUMFORD has informed us that our theorem could also be proved by the methods of his paper with T. MATSUSAKA [8].

## § 2. Metrics on Manifolds with Ample Canonical Class

Let $V$ be a compact connected complex manifold, of complex dimension $n$, and $K$ its canonical line bundle (i.e., $K = \overset{n}{\wedge} T^*$, where $T$ is the holomorphic tangent bundle of $V$). We shall assume that the line bundle $K^k$ is very ample, for some $k \geqq 1$, and define a hermitian metric on $V$; this metric will depend on the integer $k$.

First we associate to each $s \in H^0(V, K^k)$ a continuous real $2n$-form $\theta(s)$ on $V$, which is non-negative for the natural orientation on $V$: $\theta(s) = (\kappa s \otimes \bar{s})^{1/k}$, where $\kappa = \{i^{n^2}(-1)^n\}^{-k}$. We now set

$$\|s\| = [\int_V \theta(s)]^{k/2}.$$

Clearly, $s \mapsto \|s\|$ is a continuous pseudonorm on $H^0(V, K^k)$, i.e. it is continuous and has the properties

i) $\|s\| = 0 \Leftrightarrow s = 0$

ii) $\|\lambda s\| = |\lambda| \, \|s\|$ for all $\lambda \in \mathbb{C}$.

Hence the set $\{s \in H^0(V, K^k) \mid \|s\| \leqq 1\}$ is a compact neighbourhood of 0 in $H^0(V, K^k)$. Consider the non-negative $2n$-form $\tau$ on $V$ defined by

$$\tau(x) = \sup_{\|s\|=1} \theta(s)(x), \qquad x \in V$$

where the supremum is taken in the ordered vector space of real $2n$-forms at $x \in V$. It is clear (by Lemma 1 in the Appendix) that $\tau$ is a continuous form. Since, given $x \in V$, there exists an $s \in H^0(V, K^k)$ with $s(x) \neq 0$, it follows that $\tau > 0$ everywhere on $V$, i.e. $\tau$ is a volume element. We define finally a hermitian form on $H^0(V, K^k)$ by setting

$$(s_1, s_2) = \int_V \frac{\kappa s_1 \otimes \bar{s}_2}{\tau^{k-1}}; \qquad s_1, s_2 \in H^0(V, K^k).$$

This form is clearly positive definite, and hence induces a positive definite hermitian form on the dual $H^0(V, K^k)^*$ of $H^0(V, K^k)$. Hence we can define in the standard way a hermitian metric on the projective space $P(H^0(V, K^k)^*)$ of 1-dimensional subspaces of $H^0(V, K^k)^*$. Now, since $K^k$ is very ample, $V$ gets imbedded in $P(H^0(V, K^k)^*)$, and we take on $V$ the hermitian metric induced from that on the projective space.

We shall denote this metric by $\gamma_k$. The Riemannian metric which is the real part of $\gamma_k$ defines a usual metric on $V$, which we denote by $d_k$.

Let $\varphi: V \to V'$ be an isomorphism of $V$ onto a complex manifold $V'$, and $\varphi^*: H^0(V', K^k) \to H^0(V, K^k)$ the induced isomorphism. Then it is clear that, for any $s' \in H^0(V', K^k)$, $\varphi^*(\theta(s')) = \theta(\varphi^*(s'))$, and it follows that $\varphi^*$ is unitary with respect to the hermitian forms we have introduced on $H^0(V, K^k)$ and $H^0(V', K^k)$. In particular $\varphi$ is an isometry of $(V, \gamma_k)$ onto $(V', \gamma_k)$.

**Remark.** The group $\mathrm{Aut}(V)$ of holomorphic automorphisms of $V$ is *finite* (this result is well-known, see [5]). For, on the one hand, $\mathrm{Aut}(V)$ is compact, being a closed subgroup of the group of isometries of $(V, d_k)$ which is compact. On the other hand, since every automorphism of $V$ acts naturally on the line bundle $K^k$, $\mathrm{Aut}(V)$ is an algebraic subgroup of the projective group of automorphisms of $P(H^0(V, K^k)^*)$. The finiteness of $\mathrm{Aut}(V)$ follows.

## § 3. Dependence of the Metric on a Parameter

Let $V$ be a compact connected real-analytic manifold, and $S$ a complex space. A (local) holomorphic family of complex structures on $V$ parametrised by $S$ is, by definition, a complex space structure on $S \times V$, such that the canonical projection $\pi: S \times V \to S$ is a simple morphism. (This means that every point of $S \times V$ has a neighbourhood $\mathscr{U}$ with the following property: there exists an open set $U$ in $\mathbb{C}^n$, and an isomorphism $\mathscr{U} \to \pi(\mathscr{U}) \times U$ such that the diagram

$$\begin{array}{ccc} \mathscr{U} & \longrightarrow & \pi(\mathscr{U}) \times U \\ & {\scriptstyle\pi}\searrow \quad \swarrow{\scriptstyle p} & \\ & \pi(\mathscr{U}) & \end{array}$$

commutes; here $p$ is the canonical projection of $\pi(\mathscr{U}) \times U$ onto $\pi(\mathscr{U})$.) We denote the set $S \times V$ endowed with the given complex structure by $\mathscr{V}$. For any $s \in S$, $\pi^{-1}(s)$, with the complex structure induced from $\mathscr{V}$, is a complex manifold which we denote by $V_s$. The underlying real analytic manifold of $V_s$ is just $V$. We shall also sometimes denote the family $\mathscr{V}$ by $\{V_s\}_{s \in S}$.

Now let $\mathscr{V} = \{V_s\}_{s \in S}$ be a holomorphic family of complex structures on a compact connected real analytic manifold $V$, parametrised by $S$. Then there is a line bundle $\mathscr{K}$ on $\mathscr{V}$ whose restriction to each $V_s$ is the canonical line bundle $K_s$ of $V_s$.

Suppose now that, for an $s_0 \in S$, the line bundle $K_{s_0} = K_0$ is ample. Then there exists an integer $k \geqq 1$, and a neighbourhood $S'$ of $s_0$ in $S$, such that for every $s \in S'$, $K_s^k$ is very ample on $V_s$. In fact, choose any

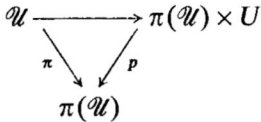

$k>1$ such that $K_0^k$ is very ample on $V_0=V_{s_0}$. Then, by the vanishing theorem of KODAIRA ([4], § 18), $H^i(V_0, K_0^k)=0$ for $i\geq 1$. Hence by GRAUERT [3] (§ 7, Satz 3, Satz 5 and Satz 6), there exists a neighbourhood $S''$ of $s_0$ in $S$ such that $\pi_0(\mathscr{K}^k)|S''$ is a vector-bundle on $S''$, whose fibres are canonically isomorphic to the $H^0(V_s, K_s^k)$. ($\mathscr{K}$ is flat over $S$, since $\mathscr{V}\to S$ is simple, and $\mathscr{K}$ is a bundle.) Hence we may assume that there exist $\sigma^{(i)}\in H^0(S'', \pi_0(K^k))$, $i=1, \ldots, N=\dim H^0(V_0, K_0^k)$, such that for each $s\in S''$, the "restrictions" $\sigma_s^{(i)}\in H^0(V_s, K_s^k)$ of the $\sigma^{(i)}$ to $V_s$ give a basis for $H^0(V_s, K_s^k)$. It is now easy to see that if $S'$ is a sufficiently small neighbourhood of $s_0$ in $S''$, then the natural "rational" mappings $V_s\to P(H^0(V, K_s^k)^*)$ are holomorphic imbeddings; in fact the $\sigma^{(i)}$ make $\mathscr{V}|S'$ isomorphic (as an $S'$-space) to a subspace of the (trivial) $P\mathbb{C}^{N-1}$-bundle over $S'$ formed by the $P(H^0(V_s, K_s^k)^*)$.

In particular, for each $s\in S'$, we can, by the procedure given in § 1, define the metric $\gamma_{k,s}=\gamma_s$ on $V_s$. We now have

**Lemma 3.1.** *The metrics $\gamma_s$ depend continuously on $s\in S'$.*

*Proof.* We keep the notation introduced above. Since $\mathscr{V}|S'$ is a subspace of the $P\mathbb{C}^{N-1}$-bundle over $S'$ formed by the $P(H^0(V_s, K_s^k)^*)$, it is sufficient to show that the hermitian forms we have defined on the $H^0(V_s, K_s^k)$ are continuous in $s$. For this purpose, let $\|\ \|_s$ denote the pseudo-norm on $H^0(V_s, K_s^k)$ defined in § 2. Then it is clear that the function $(s, \lambda_1, \ldots, \lambda_N)\mapsto\|\sum\lambda_i\sigma_s^{(i)}\|_s$ on $S'\times\mathbb{C}^N$ is continuous and satisfies the conditions of Lemma 2 of the Appendix. Hence it follows from Lemmas 1 and 2 of the Appendix that the volume elements $\tau_s$ on the $V_s$ depend continuously on $s\in S'$. It follows quickly that the hermitian forms on the $H^0(V_s, K_s^k)$ also depend continuously on $s$, Q.E.D.

In particular, using the real analytic trivialisation $\mathscr{V}=V\times S$, we can regard the Riemannian metrics associated to the $\gamma_s$ as a continuous family of Riemannian metrics on the fixed real analytic manifold $V$. If we denote the induced metrics on $V$ by $\{d_s\}_{s\in S'}$, we clearly have

**Corollary.** *For any neighbourhood $U\Subset S'$ of $s_0$, there exists a constant $C_U=C>0$ such that, for all $s\in U$ and $v_1, v_2\in V$,*

$$\frac{1}{C}\,d_s(v_1, v_2)\leq d_{s_0}(v_1, v_2)\leq C\,d_s(v_1, v_2).$$

Note that if $s_1, s_2\in S'$ and $\varphi\colon V_{s_1}\to V_{s_2}$ a holomorphic isomorphism, and if we denote by $\tilde{\varphi}$ the diffeomorphism of $V$ onto itself obtained from $\varphi$ by means of the trivialisation $\mathscr{V}=V\times S$, then $\tilde{\varphi}\colon(V, d_{s_1})\to(V, d_{s_2})$ is an isometry.

Let $\mathscr{V}=\{V_s\}_{s\in S}$, $\mathscr{V}'=\{V_t\}_{t\in T}$ be two holomorphic families of complex structures on the same (compact connected) real analytic manifold $V$.

Then with respect to fixed real analytic trivialisations $\mathscr{V}=V\times S$, $\mathscr{V}'=V\times T$, any diffeomorphism $\varphi\colon V_s\to V_t$ corresponds to a unique differentiable automorphism of $V$ which we shall also denote by $\varphi$; conversely given $s\in S$, $t\in T$, any differentiable automorphism $\varphi$ of $V$ can be regarded in a unique way as a diffeomorphism $\varphi\colon V_s\to V_t$. These identifications are to be kept in view in the following

**Lemma 3.2.** *Suppose, for an $s_0\in S$ and a $t_0\in T$, that $V_{s_0}$ and $V_{t_0}$ have ample canonical bundles. Let $\{s_n\}$, $\{t_n\}$ be sequences in $S$, $T$ converging to $s_0$, $t_0$ respectively, and $\varphi_n\colon V_{s_n}\to V_{t_n}$ holomorphic isomorphisms. Then some subsequence of $\{\varphi_n\}$ converges uniformly (with all derivatives) to a holomorphic isomorphism $\varphi\colon V_{s_0}\to V_{t_0}$.*

*Proof.* If $k$, $k'$ are integers $>1$ such that $K_{s_0}^k$ (resp. $K_{s_0}^{k'}$) is very ample on $V_{s_0}$ (resp. $V_{t_0}$), then the integer $kk'$ will serve in Lemma 3.1 for both the families $\mathscr{V}$, $\mathscr{V}'$. Thus, by suitably restricting the two families to small neighbourhoods of $s_0$, $t_0$ respectively, we will have on $V$ two families of metrics $\{d_s\}_{s\in S}$ and $\{d_t\}_{t\in T}$, and a constant $C>0$ such that, for all $s\in S$, $t\in T$ and $v_1, v_2\in V$,

$$\frac{1}{C}\,d_s(v_1,v_2)\leqq d_{s_0}(v_1,v_2)\leqq C\,d_s(v_1,v_2)$$

$$\frac{1}{C}\,d_t(v_1,v_2)\leqq d_{t_0}(v_1,v_2)\leqq C\,d_t(v_1,v_2).$$

Further if $\varphi\colon V_s\to V_t$ is a holomorphic isomorphism, $\varphi\colon (V_s,d_s)\to(V_t,d_t)$ will be an isometry. Hence it is clear that the sequence $\{\varphi_n\}$ is an equicontinuous family of mappings of $V$ into itself. Since $V$ is compact, it follows by the Ascoli-Arzela theorem that $\{\varphi_n\}$ has a subsequence $\{\varphi_{n_k}\}$ which converges uniformly to a homeomorphism $\varphi$ of $V$ onto itself. That $\varphi$ is holomorphic, and that $\varphi_{n_k}$ converges to $\varphi$ in the $C^1$-topology, now follows from the usual Montel theorem, since $\mathscr{V}$ is locally a product.

## § 4. Kuranishi Families

We state below a theorem of KURANISHI, specialised to the case when the compact complex manifold admits no continuous group of automorphisms.

**Theorem 4.1.** *Let $V_0$ be a compact complex manifold with*

$$H^0(V_0,\Theta(V_0))=0,$$

*where $\Theta(V_0)$ denotes the sheaf of germs of holomorphic vector fields on $V_0$. Then there exists a complex space $T$ and a holomorphic family $\pi\colon\mathscr{V}\to T$ of complex structures on the underlying real analytic manifold $V$ of $V_0$, and an isomorphism $i_0\colon V_0\to V_{t_0}=\pi^{-1}(t_0)$ for some $t_0\in T$, such that the following local completeness property holds:*

*Let $\mathscr{V}'=\{V_s\}_{s\in S}$ be any holomorphic family of complex structures on $V$, and suppose given an isomorphism $i\colon V_s\to V_t$, for some $s\in S$, $t\in T$. Then there exists a neighbourhood $S'$ of $s$ in $S$, and a morphism $f\colon \mathscr{V}'|S'\to\mathscr{V}$, inducing a morphism $\boldsymbol{f}\colon S'\to T$ (i.e. the diagram*

$$\begin{array}{ccc} \mathscr{V}'|S' & \xrightarrow{f} & \mathscr{V} \\ \downarrow & & \downarrow{\scriptstyle\pi} \\ S' & \xrightarrow{\boldsymbol{f}} & T \end{array}$$

*commutes, and $f|V_{s'}\colon V_{s'}\to V_{\boldsymbol{f}(s')}$ is an isomorphism for every $s'\in S'$) such that*

(i) $\boldsymbol{f}(s)=t$

(ii) $f|V_s=i$.

*Moreover $f$ is uniquely determined by* (i) *and* (ii).

*Remark 1.* In [2] and [7], the completeness property is stated only for $t=t_0$. But the proof of KURANISHI's theorem given in [2] easily gives the assertion we have made. See also [6, § 3, Theorem 3].

*Remark 2.* The uniqueness assertion in the above theorem is a consequence of the following property of the Kuranishi family $\mathscr{V}\to T$ of $V_0$, which we shall need. There exists a neighbourhood $D_0$ of $\mathrm{Id}(V)$ in the group of all differentiable automorphisms of $V$ (in the $C^1$-topology) such that, if $t_1, t_2\in T$, $\varphi\in D_0$ and $\varphi\colon V_{t_1}\to V_{t_2}$ is holomorphic, then $t_1=t_2$ and $\varphi=\mathrm{Id}(V)$. (KURANISHI [7], p. 152.) Here we have identified $V_0$ with $V_{t_0}$ by means of $i_0$.

Now suppose that $\mathrm{Aut}(V_0)$ is finite. Taking $\mathscr{V}'=\mathscr{V}$ and for $i$ an automorphism of $V_0$, we find that there is a neighbourhood $T_1$ of $t_0$ in $T$ such that $\mathrm{Aut}(V_0)$ acts on $\mathscr{V}|T_1$ and hence on $T_1$. So in this case we may assume that $\mathrm{Aut}(V_0)$ acts on $\mathscr{V}$ and $T$. Every element of $\mathrm{Aut}(V_0)$ fixes $t_0$.

We shall refer to the family given in Theorem 4.1 as the Kuranishi family at $V_0$.

**Lemma 4.1.** *Let $V_0$ be a compact connected complex manifold with ample canonical bundle. Let $\mathscr{V}\to T$ be the Kuranishi family at $V_0$, with the identification $V_0=V_{t_0}$, $t_0\in T$. We assume that* $\mathrm{Aut}(V_0)$ *acts on $\mathscr{V}$ and $T$. Then there exists a neighbourhood $T_1$ of $t_0$ in $T$ (which may be assumed* $\mathrm{Aut}(V_0)$*-stable), such that, for $t, t'\in T_1$, a diffeomorphism $\varphi\colon V_t\to V_{t'}$ is a holomorphic isomorphism if and only if $\varphi\in\mathrm{Aut}(V_0)$ and $\varphi(t)=t'$. In particular, for $t, t'\in T_1$, $V_t$ is isomorphic to $V_{t'}$ if and only if $t, t'$ are in the same orbit of* $\mathrm{Aut}(V_0)$.

*Proof.* Clearly, if $\varphi\in\mathrm{Aut}(V_0)$, then $\varphi\colon V_t\to V_{\varphi(t)}$ is an isomorphism. If the lemma were not true, we would have sequences $\{t_\nu\}$, $\{t'_\nu\}$ in $T$

converging to $t_0$, and isomorphisms $\varphi_\nu$: $V_{t_\nu} \to V_{t'_\nu}$, with $\varphi_\nu \notin \mathrm{Aut}(V_0)$. By Lemma 3.2 we may, by passing to a subsequence, assume that $\{\varphi_\nu\}$ converges in the $C^1$-topology to $\varphi_0 \in \mathrm{Aut}(V_0)$. If we put $\tau_\nu = \varphi_0^{-1}(t'_\nu)$, and $\psi_\nu = \varphi_0^{-1} \circ \varphi_\nu$, we have $\tau_\nu \to t_0$ and $\psi_\nu \to \mathrm{Id}(V_0)$. Since $\psi_\nu$: $V_{t_\nu} \to V_{\tau_\nu}$ is a holomorphic isomorphism, it follows by Remark 2 after Theorem 4.1 that $\psi_\nu = \mathrm{Id}(V_0)$ for almost all $\nu$. Hence $\varphi_\nu = \varphi_0 \in \mathrm{Aut}(V_0)$ for large $\nu$. This contradiction proves the lemma.

**Corollary.** *For* $t \in T_1$, $\mathrm{Aut}(V_t)$ *is the isotropy group at* $t$ *for the action of* $\mathrm{Aut}(V_0)$ *on* $T_1$.

## § 5. Complex Structure on the Space of Manifolds with Ample Canonical Class

**Theorem.** *Let $V$ be a compact connected real analytic manifold and $M$ the set of isomorphism classes of complex structures on $V$ with ample canonical line bundle. Then $M$ has a natural structure of a Hausdorff complex space such that if $\{V_s\}_{s\in S}$ be any family of complex structures on $V$ with ample canonical line bundle, parametrised by a complex space $S$, then the map which sends $s$ to the isomorphism class of $V_s$ is a morphism from $S$ to $M$.*

*Proof.* Let $V_0$ be a complex structure on $V$ with ample canonical line bundle. Let $\mathscr{V} \to T$ be the Kuranishi family at $V_0$. We may assume that

i) the canonical line bundle of $V_t$ is ample for $t \in T$

ii) the group of automorphisms $\mathrm{Aut}(V_0)$ of $V_0$ acts on $T$ and, for $t, t' \in T$, $V_t$ isomorphic to $V_{t'}$ if and only $t$ and $t'$ are in the same orbit of $\mathrm{Aut}(V_0)$ (Lemma 4.1).

We call such a family a *distinguished Kuranishi family at* $V_0$. Now the quotient $T/\mathrm{Aut}(V_0)$ of the complex space $T$ by the finite group $\mathrm{Aut}(V_0)$ has a natural structure of a complex space ([1]). The canonical map $T \to M$, sending $t \in T$ to the isomorphism class of $V_t$, induces an injective mapping of $T/\mathrm{Aut}(V_0)$ into $M$. This map $T/\mathrm{Aut}(V_0) \to M$ will be taken as a "local coordinate system" at the isomorphism class of $V_0$. We first verify that these local coordinate systems patch up to give the structure of a complex space on $M$. Let $V_i$, $i=1, 2$ be complex structures on $V$ with ample canonical line bundles, and let $\mathscr{V}_i \to T_i$ be a distinguished Kuranishi family at $V_i$. Let $T'_i = T_i/\mathrm{Aut}(V_i)$, and let $q_i$: $T_i \to T'_i$ be the canonical map. Let $f_i$: $T'_i \to M$ be the canonical injection. We have to prove that if $U = f_1(T'_1) \cap f_2(T'_2) \neq \emptyset$, then

(i) $f_i^{-1}(U)$ is open in $T'_i$, $i=1, 2$

(ii) $f_2^{-1} \circ f_1 : f_1^{-1}(U) \to f_2^{-1}(U)$ is an isomorphism of complex spaces.

Suppose $U \neq \emptyset$, and let $t_i \in T_i$ $i=1, 2$, be such that $V_{t_1}$ and $V_{t_2}$ are isomorphic. If $\varphi_0$: $V_{t_1} \to V_{t_2}$ is an isomorphism, we see by Theorem 4.1

that $\varphi_0$ extends to an isomorphism $\Phi$ of $\mathscr{V}_1 | W_1$ onto $\mathscr{V}_2 | W_2$, where $W_1$, $W_2$ are neighbourhoods of $t_1$, $t_2$ respectively. Since $\mathrm{Aut}(V_i)$ is finite and $\mathrm{Aut}(V_{t_i})$ is the isotropy group of $\mathrm{Aut}(V_i)$ at $t_i$, $i=1,2$, we may choose the $W_i$ such that (a) $W_i$ is $\mathrm{Aut}(V_{t_i})$-stable, (b) two points of $W_i$ are $\mathrm{Aut}(V_i)$-equivalent if and only if they are $\mathrm{Aut}(V_{t_i})$-equivalent. Now, the $q_i$ being open mappings, $q_i(W_i)$ is an open set in $T_i'$, so that (i) follows. Also, $q_i(W_i)$ is canonically isomorphic to $W_i/\mathrm{Aut}(V_{t_i})$ ([1]), so that (ii) also follows: $f_2^{-1} \circ f_1 | q_1(W_1)$ is precisely the mapping induced by the isomorphism $\Phi$: $W_1 \to W_2$ underlying $\Phi$: $\mathscr{V}_1 | W_1 \to \mathscr{V}_2 | W_2$.

Next we prove that, with the topology introduced above, $M$ is HAUSDORFF. Let $V_1$, $V_2$ be two non-isomorphic complex structures on $V$ with ample canonical line bundle. Then by Lemma 3.2, there exist distinguished Kuranishi families $\mathscr{V}_i \to T_i$ at $V_i$, $i=1,2$ such that for any $t_1 \in T_1$, $t_2 \in T_2$, $V_{t_1}$ is not isomorphic to $V_{t_2}$. Then the images of $T_1$, $T_2$ in $M$ under the canonical maps $T_i \to M$ are disjoint neighbourhoods of the isomorphism classes containing $V_1$, $V_2$ respectively.

The last part of the theorem is immediate from the completeness property of the Kuranishi family.

**Remark.** If $V_0$ is a compact Riemann surface of genus $\geqq 2$, the construction of the Kuranishi family is very easy and the parametrising space is non-singular (in fact, a neighbourhood of the origin in $H^1(V_0, \Theta)$, where $\Theta$ is the sheaf of germs of holomorphic vector fields on $V_0$). It follows that the variety of isomorphism classes of compact Riemann surfaces of genus $g \geqq 2$ has a structure of a normal complex space (see [1]).

## Appendix

We state here two lemmas which have been used above.

**Lemma 1.** *Let $f$: $X \to Y$ be a continuous, open and proper map, where $X$ and $Y$ are topological spaces. Let $\varphi$: $X \to \mathbb{R}$ be a continuous function. Then the function*

$$y \mapsto \sup_{f(x)=y} \varphi(x), \qquad y \in f(Y), \qquad x \in X,$$

*is a continuous function on $f(Y)$.*

This lemma is easy to prove and we omit the proof.

**Lemma 2.** *Let $X$ be a locally compact topological space and $(v, x) \to \| v \|_x$ be a continuous function from $\mathbb{C}^n \times X$ to $\mathbb{R}^+$ such that for any $x \in X$, $v \in \mathbb{C}^n$ and $\lambda \in \mathbb{C}$,*

$$\|v\|_x = 0 \Leftrightarrow v = 0, \qquad \|\lambda v\|_x = |\lambda| \, \|v\|_x .$$

*Let $\Sigma$ be the subset $\{\| v \|_x = 1\}$ of $\mathbb{C}^n \times X$. Then the natural projection $\pi$: $\Sigma \to X$ is proper and open.*

*Proof.* 1. To prove that $\pi$ is proper, it is enough to show that the mapping

$$(v, x) \mapsto (\|v\|_x, x)$$

of $\mathbb{C}^n \times X$ to $\mathbb{R}^+ \times X$ is proper, and this follows from the fact that, for any compact set $K \subset X$,

$$\inf_{|v|=1,\, x\in K} \|v\|_x > 0 \qquad (|\ |\text{: the usual norm on } \mathbb{C}^n).$$

2. *$\pi$ is open.* Suppose given $(v_0, x_0) \in \Sigma$, and neighbourhoods $W(v_0)$, $U(x_0)$ of $v_0, x_0$ in $\mathbb{C}^n$, $X$ respectively. Then there exists $\varepsilon > 0$ such that

$$\lambda \in \mathbb{C}, \quad |\lambda - 1| < \varepsilon \Rightarrow \lambda v_0 \in W(v_0).$$

We may assume that, for $x \in U(x_0)$

$$\big|\, \|v_0\|_x - 1 \big| < \varepsilon.$$

Then for all $x \in U(x_0)$

$$\left(\frac{v_0}{\|v_0\|_x}, x\right) \in \Sigma \cap \{W(v_0) \times U(x_0)\}$$

which proves that $\pi$ is open.

## References


1. CARTAN, H.: Quotient d'un espace analytique par un groupe d'automorphismes. Algebraic geometry and topology (A Symposium in honor of S. LEFSCHETZ), pp. 90—102. Princeton U.P. 1957.
2. DOUADY, A.: Le problème des modules pour les variétés analytiques complexes. Séminaire Bourbaki, Exposé 277, 1964—1965.
3. GRAUERT, H.: Ein Theorem der analytischen Garbentheorie und die Modulräume komplexer Strukturen. Inst. Hautes Etudes Sci., Publ. Math. **5** (1960).
4. HIRZEBRUCH, F.: Neue topologische Methoden in der algebraischen Geometrie. Berlin-Göttingen-Heidelberg: Springer 1956.
5. KOBAYASHI, S.: On the automorphism group of a certain class of algebraic manifolds. Tôhoku Math. J. **11**, 184—190 (1959).
6. KURANISHI, M.: On the locally complete families of complex analytic structures. Ann. of Math. **75**, 536—577 (1962).
7. — New proof for the existence of locally complete families of complex structures. Proceedings of the conference on complex analysis, Minneapolis 1964. Berlin-Heidelberg-New York: Springer 1965.
8. MATSUSAKA, T., and D. MUMFORD: Two fundamental theorems on deformations of polarised varieties. Amer. J. Math. **86**, 668—684 (1964).


M. S. NARASIMHAN
R. R. SIMHA
Tata Institute of Fundamental Research
Homi Bhabha Road
Bombay 5, India

*(Received October 30, 1967)*

# VECTOR BUNDLES ON CURVES

*By* M. S. NARASIMHAN *and* S. RAMANAN*

**1. Introduction.** We shall review in this paper some aspects of the theory of vector bundles on algebraic curves with particular reference to the explicit determination of the moduli varieties of vector bundles of rank 2 on a curve of genus 2 (see [3]). Later we prove, using these results, the non-existence of (algebraic) Poincaré families parametrised by non-empty Zariski open subsets of the moduli space of vector bundles of rank 2 and degree 0 on a curve of genus 2 [Theorem, §3]. This result is of interest in view of the following facts :

(i) there do exist such families when the rank and degree are coprime ;

(ii) in general (i.e. even if the degree and rank are not coprime) every stable point has a neighbourhood in the usual topology parametrising a holomorphic Poincaré family of vector bundles;

(iii) there exists always a Poincaré family of *projective* bundles parametrised by the open set of stable bundles.

The essential point in the proof of the non-existence of Poincaré families is to show that a certain projective bundle, which arises geometrically in the theory of quadratic complexes, does not come from a vector bundle. The reduction to the geometric problem is found in § 7. The geometric problem, which is independent of the theory of vector bundles, is explained in § 5 and the solution is found in § 8.

The idea of reducing this question to the geometric problem arose in our discussions with Professor D. Mumford, to whom our warmest thanks are due.

**2. The moduli variety** $U(n, d)$. Let $X$ be a compact Riemann surface or equivalently a complete non-singular irreducible algebraic

* Presented by M. S. Narasimhan.

curve defined over $\mathbf{C}$. We shall assume that the genus $g$ of $X$ is $> 2$. If $W(\neq 0)$ is a vector bundle (algebraic) on $X$ we define $\mu(W)$ to be the rational number degree $W$/rank $W$. A vector bundle $W$ will be called ***stable*** (resp. ***semi-stable***) if for every proper sub-bundle $V$ of $W$ we have $\mu(V) < \mu(W)$ (resp. $\mu(V) \leqslant \mu(W)$). D. Mumford proved that the isomorphism classes of stable bundles of rank $n$ and degree $d$ on $X$ form a non-singular quasi-projective algebraic variety (of dimension $n^2(g-1)+1$).

A characterisation of stable bundles in terms of irreducible unitary representations of certain discrete groups was given by M. S. Narasimhan and C. S. Seshadri [4]. This result implies that the space of stable bundles of rank $n$ and degree $d$ is compact if $(n, d) = 1$ and that a vector bundle of degree 0 is stable if and only if it arises from an irreducible unitary representation of the fundamental group of $X$. Moreover two such stable bundles are isomorphic if and only if the corresponding unitary representations are equivalent. These results suggest a natural compactification of the space of stable bundles, namely the space of bundles given by all unitary representations (not necessarily irreducible) of a given type.

C. S. Seshadri in [7] proved that this natural compactification is a projective variety. More precisely, Seshadri proved the following. Let $W$ be a semi-stable vector bundle on $X$. Then $W$ has a strictly decreasing filtration

$$W = W_0 \supset W_1 \supset \dots \supset W_n = (0)$$

such that, for $1 \leqslant i \leqslant n$, $W_i/W_{i-1}$ is a stable vector bundle with $\mu(W_{i-1}/W_i) = \mu(W)$. Moreover the bundle $\operatorname{Gr} W = \bigoplus_{i=1}^{n} W_{i-1}/W_i$ is determined by $W$ upto isomorphism. We say that two semi-stable bundles $W_1$ and $W_2$ are *S-equivalent* if $\operatorname{Gr} W_1 \approx \operatorname{Gr} W_2$. Obviously two stable bundles are $S$-equivalent if and only if they are isomorphic. It is proved in [7] that there is a unique structure of a normal projective variety $U(n, d)$ on the set of $S$-equivalence classes of semi-stable vector bundles of rank $n$ and degree $d$ on $X$ such that the following property holds: if $\{W_t\}_{t\in T}$ is an algebraic (resp. holomorphic) family of semi-stable vector bundles of rank $n$ and degree $d$ parametrised by an algebraic (resp. a complex) space $T$, then the mapping $T \to U(n, d)$ sending $t$ to the $S$-equivalence class of $W_t$ is a morphism.

Regarding the singularities of the varieties $U(n, d)$ we have the following result [3].

**Theorem 2.1.** *The set of non-singular points of $U(n, d)$ is precisely the set of stable points in $U(n, d)$ except when $g = 2$, $n = 2$ and $d$ even.*

It is easy to see that the above characterisation breaks down in the exceptional case. It will follow from the results quoted in § 4, that when $g = 2$, $d$ even, the variety $U(2, d)$ is actually non-singular.

Now let $L$ be a line bundle of degree $d$. Let $U_L(n, d)$ be the subspace of $U(n, d)$ corresponding to vector bundles with the determinantal bundle isomorphic to $L$. It is easy to see [4, § 3] that all stable vector bundles $V$ in $U_L(n, d)$ can be obtained as extensions

$$0 \longrightarrow E \longrightarrow V \longrightarrow (\det E)^{-1} \otimes L \longrightarrow 0,$$

where $E$ is a suitably chosen vector bundle, depending only on $U_L(n, d)$. Let $U$ be the Zariski open subset of $H^1(X, \mathrm{Hom}(L, E) \otimes \otimes \det E)$ corresponding to stable bundles. Then the natural morphism $U \to U_L(n, d)$ given by the universal property has as image the set of stable points of $U_L(n, d)$. This shows that the varieties $U_L(n, d)$ are unirational.

By a refinement of the above, it has been shown that the variety $U_L(n, d)$ is even rational if $d \equiv \pm 1 \pmod n$. The rationality of these varieties in general is not known.

**3. Poincaré families.** The next problem in the theory of vector bundles is the construction of universal (Poincaré) families of bundles on $X$ parametrised by $U(n, d)$. The existence of such a universal bundle is well-known in the case $n = 1$.

**Definition.** *Let $\Omega$ be a non-empty Zariski open subset of $U(n, d)$ or $U_L(n, d)$. A Poincaré family of vector bundles on $X$ parametrised by $\Omega$ is an algebraic vector bundle $P$ on $\Omega \times X$ such that for any $\omega \in \Omega$*

*the bundle on $X$ obtained by restricting $P$ to $\omega \times X$ is in the $S$-equivalence class $\omega$. The bundle $P$ will be called a Poincaré bundle.*

The following theorem has been proved independently by D. Mumford, S. Ramanan and C. S. Seshadri.

THEOREM. *If $n$ and $d$ are coprime, there is a Poincaré bundle on $U(n, d) \times X$.*

However we prove, in contrast, the

MAIN THEOREM. *Let $X$ be a compact Riemann surface of genus 2. Then there exists no algebraic Poincaré family parametrised by any non-empty Zariski open subset of $U(2, 0)$.*

The theorem will be proved in § 8. In the next sections we recall some results on vector bundles on a curve of genus 2 which will be used in the proof.

## 4. Vector bundles of rank 2 and degree 0 on a curve of genus 2.

THEOREM 4.1. *Let $X$ be of genus 2 and $S$ be the space of $S$-equivalence classes of semi-stable bundles of rank 2 with trivial determinant on $X$. Let $J^1$ be the variety of equivalence classes of line bundles of degree 1 on $X$ and $\Theta$ the divisor on $J^1$ defined by the natural imbedding of $X$ in $J^1$. Then $S$ is canonically isomorphic to the projective space* **P** *of positive divisors on $J^1$ linearly equivalent to $2\Theta$.*

For the proof see [3], § 6.

REMARKS. (i) The space $S$ is identified with the set of isomorphism classes of bundles of rank 2 and trivial determinant which are either stable or are of the form $j \oplus j^{-1}$, where $j$ is a line bundle of degree 0. The space of non-stable bundles in $S$, which is isomorphic to the quotient of the Jacobian $J$ of $X$ by the canonical involution of $J$, gets imbedded in **P** as a Kummer surface.

(ii) This theorem shows in particular that $S$ is non-singular. It follows easily from this that $U(2, 0)$ is non-singular if $g = 2$. In fact, $U(2, 0)$ is isomorphic to the variety of positive divisors algebraically equivalent to $2\Theta$, which is a projective bundle over the Jacobian.

(iii) This theorem suggests a close connection between $U(2, 0)$ and the variety of positive divisors on the Jacobian algebraically equivalent to $2\Theta$, when $g$ is arbitrary. This relationship has been studied when $g = 3$ and will be published elsewhere.

**5. Quadratic complexes and related projective bundles.** Before stating the next theorem it is convenient to recall certain notions connected with a quadratic complex of lines in a three dimensional projective space. For more details see [3].

Let $R$ be a four dimensional vector space over $\mathbf{C}$. Then the Grassmannian of lines $G$ in the projective space $P(R)$ is naturally embedded as a quadric in $P(\overset{2}{\wedge} R)$. Consider the tautological exact sequence

$$0 \longrightarrow L^{-1} \longrightarrow R \longrightarrow F \longrightarrow 0$$

of vector bundles on $P(R)$ where $L$ is the hyperplane bundle on $P(R)$. This leads to an exact sequence

$$0 \longrightarrow F \otimes L^{-1} \longrightarrow \overset{2}{\wedge} R \longrightarrow \overset{2}{\wedge} F \longrightarrow 0.$$

This induces an injection $P(F \otimes L^{-1}) \to P(\overset{2}{\wedge} R) \times P(R)$; the image is contained in $G \times P(R)$ and is the incidence correspondence between lines and points in $P(R)$. Consider the diagram

$$\begin{array}{ccc} & P(F \otimes L^{-1}) & \\ {\scriptstyle p_1} \swarrow & & \searrow {\scriptstyle p_2} \\ G & & P(R) \end{array}$$

The map $p_1$ is a fibration with projective lines as fibres, associated to the universal vector bundle on $G$. For $\omega \in P(R)$, $p_2^{-1}(x)$ is mapped isomorphically by $p_1$ onto a plane contained in $G$. A quadratic complex of lines is simply an element of $PH^0(G, H^2)$, where $H$ is the restriction to $G$ of the hyperplane line bundle on $P(\overset{2}{\wedge} R)$.

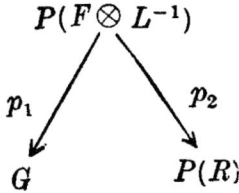

A *generic* quadratic complex in $P(R)$ is a subvariety $Q$ of $G$ defined by equations of the form

$$\begin{cases} \sum_{i=1}^{6} x_i^2 = 0, \\ \sum_{i=1}^{6} \lambda_i x_i^2 = 0, \ \lambda_i \text{ distinct}, \ \lambda_i \in \mathbf{C}, \end{cases}$$

with respect to a suitable coordinate system in $P(\overset{2}{\wedge} R)$, where $\sum_{i=1}^{6} x_i^2 = 0$ defines the Grassmannian. Let $Y = p_1^{-1}(Q)$. We then have a diagram

$$\begin{array}{ccc} & Y & \\ {}^{q_1}\swarrow & & \searrow^{q_2} \\ Q & & P(R) \end{array}$$

where $q_1$ and $q_2$ are surjective. For $\omega \in P(R)$, $q_2^{-1}(\omega)$ is imbedded in the plane $p_2^{-1}(\omega)$ as a conic. A point $\omega \in P(R)$ where $q_2^{-1}(\omega)$ is a singular conic (i.e. a pair of lines) is called a singular point of the quadratic complex $Q$. The locus $\mathscr{K}$ of singular points in $P(R)$ is a quartic surface with 16 nodes viz. a Kummer surface. Thus if $\Omega$ is a Zariski open subset of $P(R) - \mathscr{K}$, the restriction of $q_2$ to $q_2^{-1}(\Omega)$ is a projective bundle over $\Omega$. The geometric problem referred to in the introduction is whether this projective bundle is associated to an algebraic vector bundle. We shall show in §8 that this is not the case. In view of the results of §7, this will prove the main theorem.

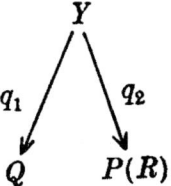

**6. Vector bundles of rank 2 and degree 1 on a curve of genus 2.** It has been shown by P. E. Newstead [6] that the space of stable bundles of rank 2 with determinant isomorphic to a fixed line bundle of degree $-1$ on a curve of genus 2, is isomorphic to the intersection of two quadrics in a 5-dimensional projective space. The following theorem, which is proved in [3], is a canonical version of this result which brings out at the same time the relationship between vector bundles (of rank 2) of degree 0 and $-1$. This relationship is of importance in the proof of non-existence of Poincaré families.

THEOREM 6.1. (i) *Let $X$ be of genus 2 and $x$ a non-Weierstrass point of $X$ (i.e. a point not fixed by the canonical rational involution on $X$). Let $S_{1,x}$ denote the variety of isomorphism classes of stable bundles of rank 2 and determinant isomorphic to $L_x^{-1}$, where $L_x$ is the line bundle determined by $x$. Let $\mathbf{P}$ be the projective space defined in Theorem 4.1 and $G$ the Grassmannian of lines in $\mathbf{P}$. Then $S_{1,x}$ is canonically isomorphic to the intersection $Q$ of $G$ and another quadric in the ambient 5-dimensional projective space.*

(ii) *The quadratic complex $Q$ is generic and the singular locus of $Q$ is the Kummer surface $\mathscr{K}$ in $\mathbf{P}$ corresponding to non-stable bundles in $S$.*

(iii) *With the identifications of $S$ with $\mathbf{P}$ and $S_{1,x}$ with $Q$, the projective bundle on $S-\mathscr{K}$ defined by the quadratic complex $Q$(see §5) is just the subvariety of $S-\mathscr{K}\times S_{1,x}$ consisting of pairs $(w, v)$ with $H^0(X, \operatorname{Hom}(V, W))\neq 0$, where $V$(resp. $W$) is in the class $v$ (resp. $w$).*

(i) and (ii) have been explicitly proved in [3], Theorem 4, §9. It has been proved there that if $v\in S_{1,x}$ and $\Lambda_v$ the line in $\mathbf{P}$ defined by $v$, then a point $w\in\mathbf{P}$ belongs to $\Lambda_v$ if and only if $H^0(X, \operatorname{Hom}(V, W))\neq 0$ where $V$(resp. $W$) is a bundle in the class $v$ (resp. $w$), (see § 9 of [3]). (iii) is only a restatement of the above.

REMARK. One can show that the space of lines on the intersection $Q$ of the two quadrics is isomorphic to the Jacobian of $X$ [3,6]. This result is to be compared with the following theorem of D. Mumford and P. E. Newstead [2]. Let $X$ be of genus $g > 2$, and $U'(2,1)$ be the subspace of $U(2, 1)$ consisting of bundles with a fixed determinant. Then the intermediary Jacobian of $U'(2, 1)$, corresponding to the third cohomology group of $U'(2, 1)$, is isomorphic to the Jacobian of $X$. The Betti numbers of $U'(2, 1)$ are determined in [5].

## 7. Reduction of the Main Theorem to a geometric problem.

LEMMA 7.1. *Let $W$ be a stable bundle of rank 2 and trivial determinant. Let $x\in X$. Let $\mathbf{O}_x=\mathbf{O}_X/\mathfrak{m}_x$ be the structure sheaf of $x$.*

(i) *If $V$ is a stable bundle of rank 2 and determinant $L_x^{-1}$ and $f\colon V \to W$ a non-zero homomorphism, then we have an exact sequence*

$$0 \longrightarrow \mathbf{V} \xrightarrow{\mathbf{f}} \mathbf{W} \longrightarrow \mathbf{O}_x \longrightarrow 0.$$

*Moreover* $\dim H^0(X, \operatorname{Hom}(V, W)) \leqslant 1$.

(ii) *If $\mathbf{W} \to \mathbf{O}_x$ is a non-zero homomorphism, then the kernel is a locally free sheaf of rank 2, whose associated vector bundle is a stable bundle with determinant $L_x^{-1}$.*

Proof. (i) It is clear that $f$ must be of maximal rank; for, otherwise the line sub-bundle of $W$ generated by the image of $f$ would have degree $\geqslant 0$, since $V$ is stable. Now the induced map $\overset{2}{\wedge} f\colon \overset{2}{\wedge} V \to \overset{2}{\wedge} W$ is non-zero and hence can vanish only at $x$ (with multiplicity 1). Hence $f$ is of maximal rank at all points except $x$ and $f$ is of rank 1 at $x$. This proves the first part of (i). Now suppose $f$ and $g$ are two linearly independent homomorphisms from $V$ to $W$; choose $y \in X$, $y \neq x$, and let $f_y$, $g_y$ be the homomorphisms $V_y \to W_y$ induced by $f$ and $g$ on the fibres of $V$ and $W$ at $y$. Then there exist $\lambda, \mu \in \mathbf{C}$, $(\lambda, \mu) \neq (0, 0)$ such that $\lambda f_y + \mu g_y$ is not an isomorphism. Then $\lambda f + \mu g$ would be a non-zero homomorphism $V \to W$ which is not of maximal rank at $y$. This is impossible by earlier remarks.

(ii) Let $V$ be the vector bundle determined by the kernel. It is clear that $\det V = L_x^{-1}$. To show that $V$ is stable we have only to show that $V$ contains no line subbundle of degree $\geqslant 0$. If $L$ were a line subbundle of $V$ of degree $\geqslant 0$, there would be a non-zero homomorphism $L \to W$, which is impossible since $W$ is stable of degree 0.

Let $p\colon P \to \Omega \times X$ be a Poincaré bundle on $\Omega \times X$, where $\Omega$ is an open subset of $S$ (see Theorem 4.1) consisting of stable points. Let $x \in X$ and let $\mathbf{O}_x = \mathbf{O}_X/\mathfrak{m}_x$ be the structure sheaf of the point $x$. Then the sheaf $\mathscr{H}om\,(\mathbf{P}, p^*{}_X\mathbf{O}_x)$ on $\Omega \times X$ is $p_\Omega$ flat. Moreover, for each $\omega \in \Omega$

$$\begin{aligned}\dim H^0(\omega \times X, \mathscr{H}om\,(\mathbf{P}, p^*{}_X\mathbf{O}_x)\,|_{\omega \times X}) \\ = \dim H^0(\omega \times X, \mathscr{H}om\,(\mathbf{P}\,|_{\omega \times X}, \mathbf{O}_x))\end{aligned}$$

$$= \dim P^*_{(\omega,x)}$$
$$= 2.$$

Hence by [1], the direct image $(p_\Omega)_* \operatorname{Hom}(\mathbf{P}, p^* \mathbf{O}_x)$ is a locally free sheaf on $\Omega$ and consequently defines a vector bundle $E$ on $\Omega$.

PROPOSITION 7.1. *There is a morphism*

$$P(E) \to \Omega \times S_{1,x}$$

*such that the diagram*

$$\begin{array}{ccc} P(E) & \longrightarrow & \Omega \times S_{1,x} \\ & \searrow \quad \swarrow & \\ & \Omega & \end{array}$$

*is commutative. Moreover this morphism is an isomorphism onto the subvariety of pairs $(W, V)$ such that $H^0(X, \operatorname{Hom}(V, W)) \neq 0$, $V \in S_{1,x}$, $W \in \Omega$.*

PROOF. Consider on $\Omega \times X$ the sheaf $\mathscr{G} = \mathscr{H}om(\mathbf{P}, p_X^*\mathbf{O}_x)$. Then we have clearly the canonical isomorphisms

$$p_*(\mathbf{T} \otimes p^* \mathscr{G}) \approx p_*(\mathbf{T}) \otimes \mathscr{G} \approx p_\Omega^*(\mathbf{E})^* \otimes \mathscr{G},$$

where $p\colon P(E) \times X \to \Omega \times X$ is the natural projection and $T$ is the tautological hyperplane bundle on $P(E) \times X$. Moreover, the direct image of $p_\Omega^*(\mathbf{E}^*) \otimes \mathscr{G}$ on $\Omega$ is isomorphic to $\mathbf{E}^* \otimes p_{\Omega_*}(\mathscr{G}) \approx \mathbf{E}^* \otimes \mathbf{E}$. Hence $H^0(P(E) \times X, (\mathbf{T} \otimes p^*\mathscr{G})) \approx H^0(\Omega, E^* \otimes \mathbf{E})$. Hence the canonical element of $H^0(\Omega, \mathbf{E}^* \otimes \mathbf{E})$ (viz. the identity endomorphism of $E$) gives rise to an element of $H^0(P(E) \times X, \mathbf{T} \otimes p^* \mathscr{G})$. In other words, we have a canonical homomorphism $p^*\mathbf{P} \to p_X^*(\mathbf{O}_x) \otimes \mathbf{T}$ of sheaves on $P(E) \times X$. Consider the commutative diagram

$$\begin{array}{ccc} P(E) \times X & \longrightarrow & P(E) \\ \Big\downarrow p & & \Big\downarrow p \\ \Omega \times X & \longrightarrow & \Omega \end{array}$$

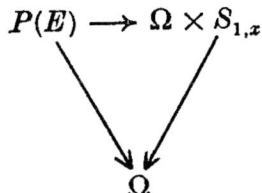

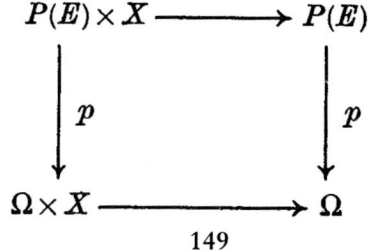

The direct image of $\mathbf{T} \otimes p^*(\mathscr{G})$ on $P(E)$ is simply $\mathbf{T} \otimes p^*(\mathbf{E})$, where $T$ also denotes the tautological bundle on $P(E)$, and the canonical element in $H^0(P(E) \times X, \mathbf{T} \otimes p^*(\mathscr{G}))$ defined above is given by the tautological element of $H^0(P(E), \mathbf{T} \otimes p^*(\mathbf{E}))$. From this we see that for $f \in P(E)$, the restriction of the homomorphism $p^*(\mathbf{P}) \to p_X^*(\mathbf{O}_x) \otimes T$ to $f \times X$ can be described as follows. The restriction of $p^*(P)$ to $f \times X$ is the restriction of $P$ to $p(f) \times X$ and hence is a stable vector bundle $W$ with trivial determinant. Moreover $f$ gives rise to a 1-dimensional subspace of $H^0(X, \mathscr{H}om(\mathbf{W}, \mathbf{O}_x))$. Any non-zero element in this 1-dimensional space gives rise to a surjective homomorphism of $p^*\mathbf{P} \mid f \times X = \mathbf{W}$ into $p_X^*\mathbf{O}_x \otimes \mathbf{T}\,|_{f \times X} \approx \mathbf{O}_x$. This homomorphism (upto a non-zero scalar) is the restriction of the canonical element. In particular it follows that the canonical homomorphism $p^*(\mathbf{P}) \to p_X^*(\mathbf{O}_x) \otimes \mathbf{T}$ is surjective. Moreover since $p_X^*(\mathbf{O}_x) \otimes \mathbf{T}$ has a locally free resolution of length 1 we see that the kernel of the homomorphism $p^*(\mathbf{P}) \to p^*(\mathbf{O}_x) \otimes \mathbf{T}$ is locally free. Let $F$ be the vector bundle on $P(E) \times X$ associated to the kernel.

LEMMA 7.2. *The restriction of the vector bundle $F$ to $f \times X$, $f \in P(E)$ is a stable vector bundle of rank 2 and determinant $L_x^{-1}$.*

In view of our earlier identification the lemma follows from Lemma 7.1.

We now complete the proof of the proposition. By Lemma 7.2 and the universal property of $S_{1,x}$ we have a morphism $q : P(E) \to S_{1,x}$. Then the morphism $(p, q): P(E) \to \Omega \times S_{1,x}$ satisfies the conditions of the proposition, in view of Lemma 7.1. The morphism is an isomorphism onto the subvariety described in Proposition 7.1, as this subvariety is non-singular by Theorem 6.1.

From Proposition 7.1 and Theorem 6.1 we have immediately the

COROLLARY. *If there is a Poincaré family on an open subset $\Omega$ of the set of stable points in $S$, then the projective bundle on $\Omega$ defined by the quadratic complex $Q = S_{1,x}$ is associated to a vector bundle.*

**8. Proof of the Main Theorem. Solution of the geometric problem.** It is easy to see that if there is a Poincaré family parametrised by a Zariski open subset of $U(2,0)$, there would exist a Poincaré family parametrised by a Zariski open subset of the space of stable points in $S$. In view of the Corollary of Proposition 7.1, the main theorem in § 3 follows from

PROPOSITION 8.1. *With the notation of § 5, let $\Omega$ be a Zariski open subset of $P(R)-\mathscr{K}$. Let $q_2: q_2^{-1}(\Omega)\to\Omega$ be the projective bundle defined in §5. Then there is no algebraic vector bundle on $\Omega$ to which this projective bundle is associated.*

PROOF. If there is such a vector bundle there would exist a Zariski open set $\Omega'$ of $\Omega$ and a section $\sigma$ over $\Omega'$ of the projective bundle $q_2^{-1}(\Omega)\to\Omega$. Let $D$ be the Zariski closure of $\sigma(\Omega')$ in $Y$. Then $D$ is a divisor of $Y$ and, since $Y$ is non-singular, $D$ defines a line bundle $L_D$ on $Y$. The restriction of the first Chern class of $L_D$ to a fibre $q_2^{-1}(\omega)$, $\omega\in\Omega'$, is the fundamental class of the fibre. On the other hand, we shall show that every element of $H^2(Y,\mathbf{Z})$ restricts to an even multiple of the fundamental class of $q_2^{-1}(\omega)$ in $H^2(q_2^{-1}(\omega), \mathbf{Z})$; this contradiction would prove the proposition. We have the commutative diagram

$$\begin{array}{ccc} H^2(P(F\otimes L^{-1}),\mathbf{Z}) & \longrightarrow & H^2(p_2^{-1}(\omega),\mathbf{Z}) \approx H^2(\mathbf{P}^2,\mathbf{Z}) \\ \downarrow & & \downarrow \\ H^2(Y,\mathbf{Z}) & \longrightarrow & H^2(q_2^{-1}(\omega),\mathbf{Z}), \end{array}$$

with the notation of § 5. We first note that the canonical mapping $H^2(G, \mathbf{Z})\to H^2(Q, \mathbf{Z})$ is an isomorphism, by Lefschetz's theorem on hypersurface sections. Moreover since $p_1: P(F\otimes L^{-1})\to G$ (resp. $q: Y\to Q$) is the projective bundle associated to a vector bundle, $H^2(P(F\otimes L^{-1}), \mathbf{Z})$ (resp. $H^2(Y, \mathbf{Z})$) is generated by the first Chern class of the tautological line bundle of the fibration $P(F\otimes L^{-1})\to G$ (resp. $Y\to Q$) and by $p_2^*(H^2(G, \mathbf{Z}))$ (resp. $q_2^* H^2(Q,\mathbf{Z})$). Since this tautological line bundle on $P(F\otimes L^{-1})$ restricts to the tautological line bundle of the fibration $Y\to Q$ and $H^2(G, \mathbf{Z})\to H^2(Q, \mathbf{Z})$ is an isomorphism, it follows that $H^2(P(F\otimes L^{-1}), \mathbf{Z})\to H^2(Y, \mathbf{Z})$ is surjective. Now from the commutativity of the diagram we see that image $H^2(Y, \mathbf{Z}) \to H^2(q_2^{-1}(\omega), \mathbf{Z})$ is contained in the image $H^2(p_2^{-1}(\omega), \mathbf{Z}) \to H^2(q_2^{-1}(\omega), \mathbf{Z})$. But $q_2^{-1}(\omega)$ is imbedded in the plane $p_2^{-1}(\omega)$ as a conic and hence the image $H^2(Y, \mathbf{Z}) \to H^2(q_2^{-1}(\omega), \mathbf{Z})$ consists of even multiples of the fundamental class of $q_2^{-1}(\omega)$. Q. E. D.

## REFERENCES


1. A. GROTHENDIECK : *Élements de Géométrie Algébrique*, Ch. III, Inst. Hautes. Etudes. Sci., Publ. Math., 17 (1963).
2. D. MUMFORD and P. E. NEWSTEAD : Periods of a moduli space of bundles on curves, *Amer. J. Math.* 90 (1968), 1201-1208.
3. M. S. NARASIMHAN and S. RAMANAN : Moduli of vector bundles on a compact Riemann surface, *Ann. of Math.* 89 (1969), 14-51.
4. M. S. NARASIMHAN and C. S. SESHADRI : Stable and unitary vector bundles on a compact Riemann surface, *Ann. of Math.* 82 (1965), 540-567.
5. P. E. NEWSTEAD : Topological properties of some spaces of stable bundles, *Topology*, 6 (1967), 241-262.
6. P. E. NEWSTEAD : Stable bundles of rank 2 and odd degree on a curve of genus 2, *Topology*, 7(1968), 205-215.
7. C. S. SESHADRI : Space of unitary vector bundles on a compact Riemann surface, *Ann. of Math.* 85 (1967), 303-336.


**Tata Institute of Fundamental Research**
**Bombay.**

# Moduli of vector bundles on a compact Riemann surface

By M. S. NARASIMHAN and S. RAMANAN

## 1. Introduction

Let $X$ be a compact Riemann surface of genus $g \geqq 2$. D. Mumford proved that the set of isomorphism classes of stable vector bundles of rank $n$ and degree $d$ on $X$ has a natural structure of a non-singular quasi-projective variety of dimension $n^2(g-1)+1$. C. S. Seshadri in [15], by introducing the notion of $S$-equivalence of semi-stable bundles, constructed a normal projective variety $U(n,d)$, which is a compactification of the space of stable bundles of rank $n$ and degree $d$. The stable points of $U(n, d)$ are simple points of this variety and, in particular, when $n$ and $d$ are coprime the variety $U(n, d)$ is non-singular. One might expect that the set of simple points of $U(n, d)$ is precisely the set of stable points in $U(n, d)$. In fact, this result is true except in the case $g = 2, n = 2, d$ even (Theorem 1, § 4). In this exceptional case, the complement of the set of stable points in $U(2, d)$ is easily seen to be of codimension 1 in the normal variety $U(2, d)$ so that the result does not hold. Theorem 1 is proved by considering the canonical mapping from the manifold of simple semi-stable bundles into $U(n, d)$.

The rest of the paper is devoted to the determination of $U(2, 0)$ and $U(2, 1)$ on a curve $X$ of genus 2. It turns out that the variety $U(2, 0)$ is actually non-singular (when $g = 2$) and is in fact a projective bundle over the Jacobian of $X$ (Theorem 3, § 7). For this one has only to consider the three dimensional sub-variety $S$ of $U(2, 0)$ consisting of bundles with trivial determinant. Let $J^1$ be the space of line bundles of degree 1 on $X$. Then $X$ is naturally imbedded in $J^1$. We denote the divisor defined by $X \subset J^1$ by $\Theta$ (and the line bundle determined by $\Theta$, by $L_\Theta$). Then we prove that $S$ is naturally isomorphic to the 3-dimensional *projective space $P$ of positive divisors on $J^1$ linearly equivalent to* $2\Theta$, i.e., to the projective space associated to $H^0(J^1, L_\Theta^2)$ (Theorem 2, § 7).

The non-stable bundles in $S$ are of the form $j \oplus j^{-1}$, where $j$ is a line bundle of degree 0, i.e., the space of non-stable bundles is isomorphic to the Kummer surface associated to the jacobian $J$ of $X$ (= quotient of $J$ by the automorphism $i: j \mapsto j^{-1}$). Now the Kummer surface gets naturally imbedded in $P$: if $j \in J$, associate to $j$ the divisor $D_j = j\Theta + j^{-1}\Theta$. {$J$ acts on $J^1$; $j\Theta$ denotes the transform of $\Theta$ by $j$; then $D_j \sim 2\Theta$.} This suggests that $P$ might be isomorphic to $S$ and also how one should define a map $S \to P$. Note that support $D_j = \{\xi \in J^1 \mid H^0(X, (j \oplus j^{-1})\xi) \neq 0\}$. Now if $W$ is a semi-stable bundle with trivial determinant let

$$C_W = \{\xi \in J^1 \mid H^0(X, W \otimes \xi) \neq 0\} .$$

Then $C_W$ is the support of a uniquely determined positive divisor $D_W$ linearly equivalent to $2\Theta$. An essential point in the proof of this result and the theorem is to identify the set of positive divisors linearly equivalent to $2\Theta$ passing through a point $\xi \in J^1$ with the projective space $P(H^1(X, \xi^{-2}))$ (Proposition 6.1). On the other hand, any element $\omega \in PH^1(X, \xi^{-2})$ gives rise to a semi-stable vector bundle $W$ with trivial determinant, obtained as an extension of $\xi$ by $\xi^{-1}$. Then we prove that $C_W$ is the support of the divisor defined by $\omega$ (see Lemma 7.1). Now this divisor depends only on $W$ and not on $\omega$, since any positive divisor linearly equivalent to $2\Theta$ is determined by its support (Proposition 6.4). The proof of the theorem is completed by showing that $W \mapsto D_W$ induces an isomorphism of $S$ onto $P$. (Our result may be expressed by saying that a semi-stable bundle of rank 2 and degree 0 is characterised up to $S$-equivalence by the set of line bundles of degree $-1$ which admit a non-zero homomorphism into it.)

Finally we consider vector bundles on $X$ of rank 2 and degree 1. It is well known that there is a close connection between line geometry in $\mathbf{P}_3$ and Kummer surfaces. In fact, it was from this view point that Kummer investigated the surfaces named after him [2, p. 212]. It is remarkable that this relationship between quadratic line complexes in $\mathbf{P}_3$ and Kummer surfaces has an exact counterpart in the determination of the variety $U(2, 1)$ in terms of the Kummer surface $J/i$ in $S = P$. There is a one-parameter family of quadratic complexes in $\mathbf{P}_3$ (which is only a family of quadratic divisors on the grassmannian $G$ of lines in $\mathbf{P}_3$) all of which have the given Kummer surface as singular locus. (See § 8 for definitions, and F. Klein [8].) This family considered as a subvariety of $\mathbf{P}_1 \times G$, will be called the Klein variety of the Kummer surface. Let $S_1'$ be the space of stable vector bundles of rank 2 on $X$, whose determinant is a positive divisor of degree 1. The canonical involution on $X$ acts on $S_1'$ as well and we prove that the quotient of $S_1'$ by this involution is just the Klein variety associated with the Kummer surface in $S$ (Theorem 4, § 9). (An outline of the proof of this result is found in § 9.) This in particular implies that the subspace of $U(2, 1)$ consisting of bundles with a fixed determinant is isomorphic to the intersection of two quadrics in $\mathbf{P}_5$.

This result was proved by different methods by P. E. Newstead [13]. We also prove that the space of lines of this intersection of quadrics is isomorphic to the jacobian of $X$. (*Added in proof*: This result is proved also in [13] and was conjectured in an earlier version of that paper.)

It is known that when $n$ and $d$ are coprime there exists a Poincaré family of vector bundles on $X$ parametrised by $U(n, d)$. In contrast, we have proved, using the results of this paper, the non-existence of algebraic Poincaré families of vector bundles parametrised by non-empty Zariski open subsets of $U(2, 0)$ in the case of a curve of genus 2. The proof will appear in the proceedings of the Bombay Colloquium on Algebraic Geometry, 1968.

*Notation.* We shall use the terms compact Riemann surface and complete non-singular irreducible algebraic curve (defined over C) interchangeably. If $D$ is a divisor on a manifold we denote by $L_D$ the line bundle defined by it; in particular if $x$ is a point of a non-singular curve we denote by $L_x$ the line bundle defined by $x$. Any locally free sheaf and its associated vector bundle will be denoted by the same symbol. If $f: X \to Y$ is a morphism and $\mathcal{F}$ is a sheaf on $X$, $f_*(\mathcal{F})$ will denote the $0^{\text{th}}$ direct image of $\mathcal{F}$ by $f$ and if $W$ is a vector bundle on $Y$, $f^*(W)$ will denote the inverse image of $W$ by $f$. We will use the following fact without explicit mention: if $f$ is proper and is a locally trivial fibration and $W$ is a vector bundle on $X$ such that $\dim H^i(f^{-1}(y), Wf^{-1}(y))$ is independent of $y$, then the $i^{\text{th}}$ direct image of $W$ by $f$ is a locally free sheaf and the fibre at $y \in Y$ of the associated vector bundle can be canonically identified with $H^i(f^{-1}(y), Wf^{-1}(y))$ [4, Th. 5]. If $E$ is a vector bundle on $X$ we shall denote by $P(E)$ the associated projective bundle of 1-dimensional subspaces of fibres of $E$. The tautological hyperplane bundle on $P(E)$ will be denoted by $\mathcal{T}(E)$ or simply by $\mathcal{T}$. The dual bundle of $E$ will be denoted by $E^*$; also, for $x \in X$, $E_x$ will denote the fibre of $E$ at $x$. For $r \geqq 1$, $S^r(E)$ will denote the $r^{\text{th}}$ symmetric power of $E$.

## 2. $S$-equivalence of semi-stable bundles

We will hereafter denote by $X$ a compact Riemann surface of genus $\geqq 2$. If $W$ is a holomorphic vector bundle $\neq 0$ on $X$, we denote by $\mu(W)$ the rational number $\deg W/\operatorname{rk} W$. A vector bundle $W$ on $X$ is said to be *semi-stable* (resp. *stable*) if for every proper sub-bundle $V$ of $W$, we have $\mu(V) \leqq \mu(W)$ (resp. $\mu(V) < \mu(W)$).

LEMMA 2.1. (i) *Let $V$ and $W$ be semi-stable bundles with $\mu(V) > \mu(W)$. Then $H^0(X, \operatorname{Hom}(V, W)) = 0$.*

(ii) *If $V$ and $W$ are stable bundles with $\mu(V) \geqq \mu(W)$, then*

$$H^0(X, \operatorname{Hom}(V, W)) = 0$$

*unless $V$ and $W$ are isomorphic. Moreover* $\dim H^0(X, \operatorname{Hom}(V, V)) = 1$.

PROOF. The lemma is an easy consequence of [12, Prop. 4.4].

Any semi-stable bundle $W$ has a (strictly decreasing) filtration

$$W = W_0 \supset W_1 \cdots \supset W_n = (0)$$

such that $W_{i-1}/W_i$ is a stable vector bundle for $1 \leqq i \leqq n$, with $\mu(W_{i-1}/W_i) = \mu(W)$. Moreover, the bundle $\operatorname{Gr} W = \bigoplus_{i=1}^n W_{i-1}/W_i$ is determined by $W$ up to isomorphism. We say that two semi-stable bundles $W_1$ and $W_2$ are *S-equivalent* if $\operatorname{Gr} W_1 \approx \operatorname{Gr} W_2$. In particular, if $W_1$ and $W_2$ are stable, they are $S$-equivalent if and only if $W_1$ and $W_2$ are isomorphic. Seshadri, in [15] introduced a structure of a normal, projective $(n^2(g-1)+1)$-dimensional variety on the set of $S$-equivalence classes of semi-stable bundles of rank $n$ and degree $d$ on $X$. We shall denote this space by $U(n, d)$. This structure is natural in the following sense [15, Th. 8.1]. (See also [14].) If $\{W_t\}_{t \in T}$ is any family of semi-stable vector bundles of rank $n$ and degree $d$ on $X$, then the map $T \to U(n, d)$ which maps $t \in T$ into the $S$-equivalence class of $W_t$ is a morphism.

Strictly speaking, the above theorem was proved in [15] only when $d = 0$, but the generalisation to arbitrary $d$ gives no trouble. Moreover, the proof in [15] can be modified to yield that operations such as direct sums, tensor products, exterior power, etc., all lead to morphisms of the corresponding spaces of $S$-equivalence classes of semi-stable bundles. In particular, the "direct sum" map $U(n_1, d_1) \times U(n_2, d_2) \to U(n_1 + n_2, d_1 + d_2)$ where $d_1/n_1 = d_2/n_2$, and the determinant map $U(n, d) \to J^d$(where $J^d$ is the space of line bundles of degree $d$) are morphisms. From these considerations, it is easy to deduce that the sub-varieties of $U(n, d)$, consisting of bundles with a fixed determinant are also normal and have universal properties similar to that of $U(n, d)$.

## 3. Generalities on extensions

An extension $0 \to E \to W \to F \to 0$, where $E, W, F$ are vector bundles on $X$, gives rise to an element $\delta(W) \in H^1(X, \operatorname{Hom}(F, E))$ which is the image of the identity homomorphism in $H^0(X, \operatorname{Hom}(F, F))$ by the connecting homomorphism $H^0(X, \operatorname{Hom}(F, F)) \to H^1(X, \operatorname{Hom}(F, E))$. This gives a one-one correspondence between the set of equivalence classes of extensions of $F$ by $E$ and $H^1(X, \operatorname{Hom}(F, E))$ [1, § 1].

LEMMA 3.1. *Let $0 \to E \to W \to F \to 0$ be an extension of $F$ by $E$. For a homomorphism $f$ of a vector bundle $V$ into $F$ to be liftable into a homo-*

*morphism $V \to W$, it is necessary and sufficient that the element*

$$\delta(W) \in H^1(X, \operatorname{Hom}(F, E))$$

*is in the kernel of the natural map*

$$H^1(X, \operatorname{Hom}(F, E)) \longrightarrow H^1(X, \operatorname{Hom}(V, E))$$

*induced by $f$.*

PROOF. Consider the commutative diagram of vector bundles

$$\begin{array}{ccccccccc} 0 & \longrightarrow & \operatorname{Hom}(F, E) & \longrightarrow & \operatorname{Hom}(F, W) & \longrightarrow & \operatorname{Hom}(F, F) & \longrightarrow & 0 \\ & & \downarrow & & \downarrow & & \downarrow & & \\ 0 & \longrightarrow & \operatorname{Hom}(V, E) & \longrightarrow & \operatorname{Hom}(V, W) & \longrightarrow & \operatorname{Hom}(V, F) & \longrightarrow & 0 \end{array}$$

where the horizontal sequences are exact and the vertical maps are induced by the homomorphism $f: V \to F$. From this we deduce the commutative diagram

$$\begin{array}{ccccc} H^0(X, \operatorname{Hom}(F, W)) & \longrightarrow & H^0(X, \operatorname{Hom}(F, F)) & \longrightarrow & H^1(X, \operatorname{Hom}(F, E)) \\ \downarrow & & \downarrow & & \downarrow \\ H^0(X, \operatorname{Hom}(V, W)) & \longrightarrow & H^0(X, \operatorname{Hom}(V, F)) & \longrightarrow & H^1(X, \operatorname{Hom}(V, E)) . \end{array}$$

The homomorphism $f$ is an element of $H^0(X, \operatorname{Hom}(V, F))$ and by the exactness, we see that $f$ can be lifted as a map $V \to W$ if and only if its image in $H^1(X, \operatorname{Hom}(V, E))$ is zero. But since $f$ is the image of the identity map $\in H^0(X, \operatorname{Hom}(F, F))$, the lemma follows from the commutativity of the above diagram.

LEMMA 3.2. *Let $0 \to E \to W \to F \to 0$ be an extension of $F$ by $E$ and $f$ a homomorphism $E \to V$. In order that $f$ be extendable to a homomorphism $g: W \to V$, it is necessary and sufficient that the element*

$$\delta(W) \in H^1(X, \operatorname{Hom}(F, E))$$

*be in the kernel of the map*

$$H^1(X, \operatorname{Hom}(F, E)) \longrightarrow H^1(X, \operatorname{Hom}(F, V))$$

*induced by $f$.*

PROOF. Consider the exact sequence

$$0 \longrightarrow F^* \longrightarrow W^* \longrightarrow E^* \longrightarrow 0 .$$

We have $\delta(W^*) = -\delta(W)$ in $H^1(X, \operatorname{Hom}(E^*, F^*)) = H^1(X, \operatorname{Hom}(F, E))$ [1, Prop. 3]. A map $g$, as in the lemma, exists if and only if the transpose of the map $f$ from $V^* \to E^*$, is liftable to a homomorphism $V^* \to W^*$. Now the lemma follows from Lemma 3.1.

LEMMA 3.3. *Let $W, W'$ be two extensions of $F$ by $E$. Then $W$ and $W'$ are isomorphic as bundles if $\delta(W) = \lambda\delta(W')$ for some $\lambda \in \mathbf{C}^*$. Moreover, if every non-zero homomorphism of $E$ into $F$ is an isomorphism and the only endomorphisms of $E$ and $F$ are scalars, then $W$ and $W'$ are isomorphic if and only if $\delta(W) = \lambda\delta(W')$ for some $\lambda \in \mathbf{C}^*$.*

PROOF. The map $\lambda I\colon F \to F$ induces from $W$ an extension $W_1$ such that we have a diagram

$$\begin{array}{ccccccccc} 0 & \longrightarrow & E & \longrightarrow & W_1 & \longrightarrow & F & \longrightarrow & 0 \\ & & {\scriptstyle I}\big\downarrow & & \big\downarrow & & \big\downarrow{\scriptstyle \lambda I} & & \\ 0 & \longrightarrow & E & \longrightarrow & W & \longrightarrow & F & \longrightarrow & 0\,. \end{array}$$

Now it is easy to see that $W_1$, which is isomorphic to $W$, is such that $\delta(W) = \lambda\delta(W_1)$. This proves the first part of the assertion. Now let $W$ and $W'$ be two extensions of $F$ by $E$ and $f\colon W' \to W$ be an isomorphism. We may clearly assume that one of $W, W'$, say $W$, is a non-trivial extension. Then $f$ maps the subbundle $E$ of $W'$ into the subbundle $E$ of $W$. Moreover, the non-zero endomorphism of $E$ induced by $f$ may be assumed to be identity. Thus we have the diagram

$$\begin{array}{ccccccccc} 0 & \longrightarrow & E & \longrightarrow & W' & \longrightarrow & F & \longrightarrow & 0 \\ & & {\scriptstyle I}\big\downarrow & & {\scriptstyle f}\big\downarrow & & \big\downarrow & & \\ 0 & \longrightarrow & E & \longrightarrow & W & \longrightarrow & F & \longrightarrow & 0\,. \end{array}$$

Our assertion follows from the fact that the map $f$ induces on $F$ an endomorphism of the type $\lambda\cdot\mathrm{Id}$, $\lambda \in \mathbf{C}^*$.

PROPOSITION 3.1. *Let $E = \{E_t\}_{t\in T}$ and $F = \{F_t\}_{t\in T}$ be two families of vector bundles on $X$ parametrised by $T$. Assume that $\dim H^1(X, \mathrm{Hom}\,(F_t, E_t))$ is independent of $t$. Let $\pi\colon V \to T$ be the bundle on $T$ associated to the first direct image of $\mathrm{Hom}\,(F, E)$. Let, moreover, $H^i(T, p_{T*}(\mathrm{Hom}\,(F, E)) \otimes V^*) = 0$, for $i = 1, 2$. Then there exists a family of extensions $\{W_v\}_{v\in V}$ on $X$ parametrised by $V$ such that $W_v$ is an extension of $F_{\pi(v)}$ by $E_{\pi(v)}$ with*

$$\delta(W) = v \in H^1\big(X, \mathrm{Hom}\,(F_{\pi(v)}, E_{\pi(v)})\big)\,.$$

*Remark.* The condition $H^i(T, p_{T*}\,\mathrm{Hom}\,(F, E) \otimes V^*) = 0$, $i = 1, 2$, is satisfied if $T$ is affine or if $H^0(X, \mathrm{Hom}\,(F_t, E_t)) = 0$ for every $t \in T$. In particular, if $\dim H^1(X, \mathrm{Hom}\,(F_t, E_t))$ is independent of $t$, every point of $T$ has neighborhood $U$ such that on $\pi^{-1}(U)$ there is a family of extensions of the above type.

PROOF OF THE PROPOSITION. We will construct an element of

$$H^1\big(V \times X, \mathrm{Hom}\,(\pi^*(F), \pi^*(E))\big)\,,$$

where $\pi: V \times X \to T \times X$ is the natural projection; the extension of $\pi^*(F)$ by $\pi^*(E)$ given by this element will have the required property. Consider the natural map

$$H^1(T \times X, \mathrm{Hom}\,(F, E) \otimes p_T^*(V^*)) \longrightarrow H^0(T, V \otimes V^*)$$

given by the Leray spectral sequence associated to $p_T$. From our assumption it follows that this map is an isomorphism. On the other hand since the fibration $\pi: V \times X \to T \times X$ has vector spaces as fibres we have an isomorphism

$$H^1(T \times X, \mathrm{Hom}\,(F, E) \otimes \pi_*(O)) \longrightarrow H^1(V \times X, \pi^* \,\mathrm{Hom}\,(F, E))$$

where $O$ denotes the structure sheaf of $V \times X$. Noting that $p_T^*(V^*) \subset \pi_*(O)$ we see that the canonical element of $H^0(T, V \otimes V^*) = H^0(T, \mathrm{Hom}\,(V, V))$ gives rise to an element of $H^1(V \times X, \mathrm{Hom}\,(\pi^*(F), \pi^*(E)))$ as required. It is easy to verify that this element has the required properties by restricting it to a fibre $\pi^{-1}(t)$ (compare [12, Lem. 3.1]).

## 4. Singularities of the space $U(n, d)$

THEOREM 1. *Let $X$ be a non-singular irreducible complete algebraic curve of genus $g \geqq 2$. Let $U(n, d)$, $n \geqq 2$, be the space of S-equivalence classes of semi-stable vector bundles of rank $n$ and degree $d$ on $X$. Then the set of singular points of the variety $U(n,d)$ is precisely the set of non-stable points, except when $g = 2$, $n = 2$ and $d$ is even.*

*Remark.* The characterisation of the singular points given above is not valid when $g = 2$, $n = 2$, and $d$ even. In fact it will be proved in § 7 (see Theorem 3) that the variety $U(2, 0)$ corresponding to a curve of genus 2 is non-singular.

LEMMA 4.1. *Let $E$ and $F$ be two stable bundles on $X$ such that $\mu(E) = \mu(F)$. If $0 \to E \to W \to F \to 0$ is any extension, then $W$ is semi-stable. If, moreover, $E$ and $F$ are not isomorphic, and the above extension non-trivial, then $W$ is simple.*

PROOF. If $W$ were not semi-stable there would exist a proper stable subbundle $V$ of $W$ such that $\mu(V) > \mu(W)$ (see [12, Prop. 4.5]). On the other hand, we have the exact sequence

$$0 \longrightarrow H^0(X, \mathrm{Hom}\,(V, E)) \longrightarrow H^0(X, \mathrm{Hom}\,(V, W))$$
$$\longrightarrow H^0(X, \mathrm{Hom}\,(V, F)) \longrightarrow H^1(X, \mathrm{Hom}\,(V, E)) \longrightarrow \cdots .$$

Since $\mu(V) > \mu(W) = \mu(E) = \mu(F)$, it follows Lemma 2.1 that

$$H^0(X, \mathrm{Hom}\,(V, E)) = 0 = H^0(X, \mathrm{Hom}\,(V, F)) .$$

This is impossible as $V \subset W$. Hence $W$ is semi-stable.

Now let $E$ and $F$ be non-isomorphic. Since $H^0(X, \text{Hom}(E, F)) = 0$, by Lemma 2.1, it follows that any endomorphism $A$ of $W$ leaves $E$ invariant and $A \mid E$ is a scalar $\lambda$. Now $A - \lambda I$ induces a map $F \to W$. But looking at the above exact sequence wherein $V$ is replaced by $F$ and noting that the 1-dimensional space $H^0(X, \text{Hom}(F, F))$ is mapped isomorphically onto the space generated by $\delta(W)$, we see that $H^0(X, \text{Hom}(F, W)) = 0$. This means that $A = \lambda I$.

*Remark.* Let $E$ and $F$ be vector bundles on $X$ such that $\mu(E) = \mu(F)$. Then by the Riemann-Roch formula

$$\dim H^1(X, \text{Hom}(F, E)) \geqq \text{rk}(F)\,\text{rk}(E)(g-1) \geqq 1 .$$

Hence there exist non-trivial extensions of $F$ by $E$.

LEMMA 4.2. *Let $W$ and $W'$ be two extensions of $F$ by $E$ where $F$ and $E$ are stable vector bundles such that $\mu(E) = \mu(F)$. Then $W$ and $W'$ are isomorphic if and only if $\delta(W) = \lambda\delta(W')$, for $\lambda \in \mathbf{C}^*$.*

PROOF. This follows from Lemma 3.3, since the conditions of Lemma 3.3 are fulfilled by Lemma 2.1 (ii).

LEMMA 4.3. *The set of stable points is open and dense in $U(n, d)$. Let $\Sigma(n, d)$ be the set of non-stable points in $U(n, d)$. Then the set of points in $\Sigma(n, d)$ which are S-equivalent to bundles of the form $E \oplus F$, where $E$ and $F$ are non-isomorphic stable bundles, is open and dense in $\Sigma(n, d)$.*

PROOF. In fact, every non-stable point in $U(n, d)$ is in the image of $U(n_1, d_1) \times U(n - n_1, d - d_1)$ under the morphism $(E, F) \mapsto E \oplus F$, where $0 < n_1 < n$ and $d_1/n_1 = d/n$. Clearly such an image is closed and its dimension is $\leqq n_1^2(g-1) + (n-n_1)^2(g-1) + 2 < n^2(g-1) + 1 = \dim U(n, d)$. Hence the set of stable points in the irreducible variety $U(n, d)$ is a non-empty Zariski open set. The second assertion follows now from the fact that elements of the form $(E, F)$, where $E$, $F$ are non-isomorphic stable bundles, is dense in $U(n_1, d_1) \times U(n - n_1, d - d_1)$.

PROOF OF THEOREM 1. Let $T$ be the set of points in $U(n, d)$ which are either stable or are $S$-equivalent to the direct sum of two non-isomorphic stable bundles. By Lemma 4.3, it is enough to prove that the set of singular points of the open set $T$ is the set of non-stable points. Now the set $S(n, d)$ of isomorphism classes of simple bundles of rank $n$ and degree $d$ has a natural structure of a complex manifold and can be covered by open sets each of which parametrises a family of bundles on $X$ [12, § 2]. Moreover the set $S_{ss}(n, d)$ of simple semi-stable bundles is an open subset of $S(n, d)$. We have then a morphism $\varphi$ of the non-singular variety $S_{ss}(n, d)$ into $U(n, d)$ given in

[15, Th. 6.1]. By Lemma 4.1 and the remark following it, the image of $\varphi$ contains $T$. The map $\varphi: \varphi^{-1}(T) \to T$ is locally injective at any stable point of $\varphi^{-1}(T)$. (In particular, this implies that stable points in $U(n, d)$ are non-singular.) Let $W$ be a simple, semi-stable bundle in $\varphi^{-1}(T)$, $S$-equivalent to $E \oplus F$, where $E$, $F$ are non-isomorphic stable bundles. Then it is clear that $W$ is a non-trivial extension of $E$ by $F$, or $F$ by $E$. Let then $W$ be an extension of $F$ by $E$ and $\delta(W)$ be the element in $PH^1(X, \mathrm{Hom}(F, E))$ corresponding to $W$. By Lemma 4.1, Proposition 3.1, and [12, Th. 1], we have a morphism $PH^1(X, \mathrm{Hom}(F, E)) \to S_{ss}(n, d)$. Moreover this morphism is injective by Proposition 4.2. Now since $H^0(X, \mathrm{Hom}(F, E)) = 0$, we have by the Riemann-Roch formula,

$$\dim PH^1(X, \mathrm{Hom}(F, E)) = \mathrm{rk}(F)\,\mathrm{rk}(E)(g-1) - 1 \geqq 1$$

if either $g > 2$ or $g = 2$ and $n \geqq 2$. Since the image of $PH^1(X, \mathrm{Hom}(F, E))$ in $S_{ss}(n, d)$ is mapped into the point $E \oplus F$ in $U(n, d)$ by $\varphi$, we see that the set of points in $\varphi^{-1}(T)$ at which $\varphi$ is locally injective is precisely the set of stable points. Now the set of non-stable points in $\varphi^{-1}(T)$ is of codimension $\geqq 2$ (under the assumption that $g > 2$, or $g = 2$, $n > 2$). For, the set of non-stable points of $\varphi^{-1}(T)$ is clearly the union of a countable family of sets $S_i$, where each $S_i$ is as follows. An element $W$ of $S_i$ is an extension of $F$ by $E$ where $E$, $F$ are non-isomorphic stable bundles belonging respectively to families parametrised by open sets $O_1$, $O_2$ in $U(n_1, d_1)$, $U(n - n_1, d - d_1)$. It is clear that $S_i$ is the image by a morphism (Proposition 3.1) of the projective bundle on $O_1 \times O_2$ which assigns to $(E, F)$ the projective space $PH^1(X, \mathrm{Hom}(F, E))$. Now the dimension of this projective bundle is

$$\begin{aligned} n_1(n - n_1)(g-1) - 1 + n_1^2(g-1) + 1 + (n - n_1)^2(g-1) + 1 \\ = n^2(g-1) - n_1(n - n_1)(g-1) + 1 \\ < n^2(g-1) = \dim U(n, d) - 1\,. \end{aligned}$$

Thus the set of points in $\varphi^{-1}(T)$ at which the map $\varphi: \varphi^{-1}(T) \to T$ is not locally injective is of codimension $\geqq 2$. That the image of the non-stable points in $\varphi^{-1}(T)$ is the singularity set in $T$ now follows from

LEMMA 4.4. *Let $f: X \to Y$ be a morphism of an $n$-dimensional complex manifold onto a normal $n$-dimensional variety $Y$. If the set $S$ of points at which $f$ is not locally injective is of* codim $\geqq 2$, *then $f(S)$ is the singularity set of $Y$.*

PROOF. It follows from the analytic analogue of $ZMT$ that $f(S)$ contains the singularity set of $Y$. If possible let $f(x)$ be a non-singular point in $Y$ for some $x \in S$. There would then exist an open neighbourhood $\mathbf{O}$ of $f(x)$ which is non-singular. Now the map $f: f^{-1}(\mathbf{O}) \to \mathbf{O}$ is not locally injective precisely at the points $x \in f^{-1}(0)$ at which $\det(df)_x = 0$. This means that the set $S \cap f^{-1}(0) = \{x \in f^{-1}(0) \colon \det(df)_x = 0\}$. So this set must be of pure codimension 1, which is a contradiction to the assumption.

*Remarks.* ( i ) At every non-stable point $E \oplus F$ of $T$, the inverse image by $\varphi$ consists of two connected components, *viz.* $PH^1(X, \operatorname{Hom}(E, F))$ and $PH^1(X, \operatorname{Hom}(F, E))$. Since on the other hand, $\varphi$ is generically an isomorphism, it follows that $\varphi^{-1}(T)$ and hence the space of simple bundles is in general non-Hausdorff.

( ii ) When $X$ is a curve of genus 2, the map $\varphi\colon \varphi^{-1}(T) \to T$ is locally injective at all points (and has actually two points over each non-stable point of $T$). This shows that $T$ is non-singular. This leaves out the subset of $U(2, 0)$ of the form $j \oplus j$, $j \in J$. By a completely different method, we prove that $U(2, 0)$ is itself non-singular in § 7.

## 5. Extensions of line bundles and semi-stable bundles of rank 2

In this section $X$ will denote a compact Riemann surface of genus $g \geq 2$ and $J^d$ the isomorphism classes of line bundles of degree $d$ on $X$. If $\xi \in J^d$ we will also denote by $\xi$ a line bundle in the class $\xi$. We shall write $J$ for $J^0$.

LEMMA 5.1. *Let $0 \to \xi^{-1} \to W \to \xi \to 0$ be any non-trivial extension of $\xi$ by $\xi^{-1}$, where $\xi$ is a line bundle of degree 1 on $X$. Then $W$ is semi-stable with trivial determinant.*

PROOF. We have only to show that there is no non-zero homomorphism of any line bundle $\zeta$ of positive degree into $W$. If $\varphi\colon \zeta \to W$ is a non-zero homomorphism then composing $\varphi$ with the surjection $W \to \xi$ we see that there is a non-zero homomorphism $\zeta \to \xi$. This is only possible if $\zeta$ is of degree 1. Moreover, if $\deg \zeta = 1$, then this map will have to be an isomorphism, which implies that the given sequence splits, contrary to assumption.

LEMMA 5.2. *Let $0 \to \xi^{-1} \to W \to \xi \to 0$ be a non-trivial extension, $\xi$ being a line bundle of degree 1 on $X$. Then $W$ is non-stable if and only if the element $\delta(W) \in H^1(X, \xi^{-2})$ corresponding to $W$ is in the kernel of the map*

$$H^1(X, \xi^{-2}) \longrightarrow H^1(X, \xi^{-2} \otimes L_x),$$

*for some $x \in X$. In this case, $W$ is S-equivalent to $\xi \otimes L_x^{-1} \oplus \xi^{-1} \otimes L_x$.*

PROOF. Let $j$ be a line bundle of degree 0 on $X$. Then it is easy to see, since $H^0(X, \operatorname{Hom}(j, \xi^{-1})) = 0$, that $H^0(X, \operatorname{Hom}(j, W)) \neq 0$ if and only if $j$ is of the form $\xi \otimes L_x^{-1}$, for some $x \in X$ and the natural map $\xi \otimes L_x^{-1} \to \xi$ can be lifted into a map $\xi \otimes L_x^{-1} \to W$. Now the lemma follows from Lemma 3.1.

*Remark.* By Lemma 5.1 every non-zero element of $H^1(X, \xi^{-2})$ gives rise to a semi-stable bundle and by Lemma 3.3 the elements of $PH^1(X, \xi^{-2})$ give rise to isomorphism classes of semi-stable bundles. Lemma 5.2 asserts that the image of $X$ in $PH^1(X, \xi^{-2})$ under the map given by the linear system $K \otimes \xi^2$ corresponds precisely to non-stable bundles.

LEMMA 5.3. *If $\varphi: \zeta \to W$ is a homomorphism of a line bundle into a vector bundle $W$, then $\varphi$ factors as a map $\zeta \to \zeta \otimes L_x \to W$, where the map $\zeta \to \zeta \otimes L_x$ is induced by the canonical section of $L_x$, if and only if $\varphi(x) = 0$.*

PROOF. Consider the exact sequence $0 \to L_x^{-1} \to O \to O/\mathfrak{M}_x \to 0$ where $O$ is the structure sheaf of $X$ and $\mathfrak{M}_x$ is the subsheaf of ideals of $O$ defined by $x$. From this, we obtain the exact sequence

$$0 \longrightarrow H^0(X, \operatorname{Hom}(\zeta \otimes L_x, W)) \longrightarrow H^0(X, \operatorname{Hom}(\zeta, W)) \longrightarrow \operatorname{Hom}(\zeta, W)_x$$

where $\operatorname{Hom}(\zeta, W)_x$ is the fibre at $x$ of the vector bundle $\operatorname{Hom}(\zeta, W)$. Lemma 5.3 is a consequence of this exact sequence.

LEMMA 5.4. *If $W$ is a vector bundle of rank 2 and*

$$\dim H^0(X, \operatorname{Hom}(\xi, W)) \geqq 2 ,$$

*then either $W \approx \xi \oplus \xi$, or $\dim H^0(X, \operatorname{Hom}(\xi \otimes L_x, W)) \neq 0$ for some $x \in X$.*

PROOF. Assume that $W \not\approx \xi \oplus \xi$. Consider the evaluation map

$$\xi \otimes H^0(X, \operatorname{Hom}(\xi, W)) \longrightarrow W .$$

Since, by assumption, this cannot be an isomorphism, there exists a non-zero homomorphism $\varphi: \xi \to W$ such that $\varphi(x) = 0$ for some $x \in X$. Applying Lemma 5.3, we get the result.

LEMMA 5.5. *For every vector bundle $W$ of rank 2 and degree 0, there exists a line bundle $\xi$ of degree $(g - 1)$ such that $H^0(X, \operatorname{Hom}(\xi^{-1}, W)) \neq 0$.*

PROOF. By the Riemann-Roch formula, $\dim H^0(X, \operatorname{Hom}(\zeta, W)) \geqq 2$ for every line bundle $\zeta$ of degree $-g$. Now the assertion follows from Lemma 5.4.

LEMMA 5.6. *Let $W$ be a semi-stable vector bundle on $X$ of rank 2 and trivial determinant. Then the set*

$$C_W = \{\xi \in J^{g-1} \mid H^0(X, \operatorname{Hom}(\xi^{-1}, W)) \neq 0\}$$

*depends only on the S-equivalence class of $W$. Moreover, if $W$ is not stable, and S-equivalent to $j \oplus j^{-1}$, $j \in J$, then $\xi \in C_W$ if and only if $\xi$ is of the form $j \otimes L_D$ or $j^{-1} \otimes L_D$ where $D$ is a positive divisor of degree $(g - 1)$.*

PROOF. If $W$ is stable, there is nothing to prove. Otherwise, we may assume $W$ to be of the form

$$0 \longrightarrow j^{-1} \longrightarrow W \longrightarrow j \longrightarrow 0 ,$$

where $j$ is a line bundle on $X$ of degree 0. We have only to show that $\xi \in C_W$

if and only if $H^0(X, \xi \otimes j) \neq 0$ or $H^0(X, \xi \otimes j^{-1}) \neq 0$. We have the exact sequence

$$0 \longrightarrow H^0(X, \xi \otimes j^{-1}) \longrightarrow H^0(X, \xi \otimes W) \longrightarrow H^0(X, \xi \otimes j) \longrightarrow H^1(X, \xi \otimes j^{-1})$$

for any $\xi \in J^{g-1}$. Obviously, if $H^0(X, \xi \otimes W) \neq 0$, then either $H^0(X, \xi \otimes j^{-1}) \neq 0$, or $H^0(X, \xi \otimes j) \neq 0$. Conversely, if $H^0(X, \xi \otimes j^{-1}) \neq 0$, then $H^0(X, \xi \otimes W) \neq 0$. Let then $H^0(X, \xi \otimes j^{-1}) = 0$ and $H^0(X, \xi \otimes j) \neq 0$. By the Riemann-Roch theorem, we see that $H^1(X, \xi \otimes j^{-1}) = 0$ and our assertion follows from the above exact sequence.

LEMMA 5.7. *Let $W$ be a vector bundle of rank 2 and trivial determinant. Then, for $\xi \in J^{g-1}$, $H^0(X, \xi \otimes W) \neq 0$ if and only if $H^0(X, K \otimes \xi^{-1} \otimes W) \neq 0$. In particular in the notation of Lemma 5.6, $C_W$ is invariant under the involutorial automorphism $\xi \mapsto K \otimes \xi^{-1}$ of $J^{g-1}$.*

PROOF. This is a trivial consequnce of the Riemann-Roch theorem, duality, and the fact that $W \approx W^*$ under our assumptions.

LEMMA 5.8. *Let $X$ be of genus 2. Let $W$ be a semi-stable vector bundle of rank 2 and trivial determinant. Let $\xi \in J^1$ such that $H^0(X, \xi \otimes W) \neq 0$. Then there exists a non-trivial extension*

$$0 \longrightarrow \xi^{-1} \longrightarrow W' \longrightarrow \xi \longrightarrow 0$$

*such that $W'$ is $S$-equivalent to $W$.*

PROOF. If $W$ is stable, the lemma follows from the fact that any non-zero homomorphism $\zeta \to W$ of a line bundle $\zeta$ of degree $-1$ is injective in each fibre, by Lemma 5.3. Now if $W$ is not stable, and $H^0(X, \xi \otimes W) \neq 0$, then $W$ is $S$-equivalent to $\xi \otimes L_x^{-1} \oplus \xi^{-1} \otimes L_x$ for some $x \in X$, by Lemma 5.6. Let $W'$ be any non-trivial extension corresponding to the kernel of the map

$$H^1(X, \xi^{-2}) \longrightarrow H^1(X, \xi^{-2} \otimes L_x) .$$

Then by Lemma 5.2, $W'$ is $S$-equivalent to $\xi \otimes L_x^{-1} \oplus \xi^{-1} \otimes L_x$, and hence to $W$.

LEMMA 5.9. *Let $X$ be of genus 2. Let $0 \to \xi^{-1} \to W \to \xi \to 0$ be a non-trivial extension, $\xi \in J^1$. Let*

$$C_W = \{\eta \in J^1 \mid H^0(X, \mathrm{Hom}\,(\eta^{-1}, W)) \neq 0\} .$$

*Let $\eta \neq \xi, \eta \neq K \otimes \xi^{-1}$ be a line bundle of degree 1. Then $\eta \in C_W$ if and only if $\delta(W)$ is in the kernel of the homomorphism*

$$H^1(X, \xi^{-2}) \longrightarrow H^1(X, \xi^{-1} \otimes \eta)$$

*induced by a non-zero section of $\xi \otimes \eta$.*

PROOF. We first note that $\dim H^0(X, \xi \otimes \eta) = 1$, under our assumptions.

The bundle $\eta \in C_W$ if and only if there exists a non-zero homomorphism of $\eta^{-1}$ into $\xi$ which can be lifted to a homomorphism of $\eta^{-1}$ into $W$. Now the lemma follows from Lemma 3.1.

*Remark*. When $X$ is a curve of genus 2 we construct in the next section a family of line bundles on $X$ parametrised by $J^1$ which assigns to $\xi \in J^1$, a line bundle isomorphic to $\xi^2$ (Lemma 6.4). Since $H^1(X, \xi^{-2})$ is of dimension 3 we would then have a vector bundle $E$ on $J^1$ such that $E_\xi \approx H^1(X, \xi^{-2})$. By Lemma 5.1 and Proposition 3.1, there exists a morphism $P(E) \to S$, where $S$ is the space of $S$-equivalence classes of semi-stable bundles with trivial determinant. The next article is devoted to the study of this map.

## 6. The divisor $2\Theta$ and its relationship with extensions

*Hereafter we assume that $X$ is a compact Riemann surface of genus* 2. Let $J$ (resp. $J^1$) be the space of equivalence classes of line bundles on $X$ of degree 0 (resp. degree 1). The spaces $J$ and $J^1$ have structures of non-singular varieties with the following characteristic property. If $\{W_t\}_{t \in T}$ is an algebraic family of line bundles on $X$ of degree 0 (resp. 1) parametrised by an algebraic space $T$, then the map which assigns to $t \in T$ the equivalence class of $W_t$ is a morphism from $T$ to $J$ (resp. $J^1$).

Let $L_\Delta$ be the line bundle on $X \times X$ associated to the diagonal divisor. Then $L_\Delta$ may be interpreted as a family of line bundles on $X$ parametrised by $X$ which assigns to $x \in X$ the line bundle $L_x$ defined by $x$. By the universal property mentioned above, we then have a morphism $t: X \to J^1$. It is easily verified that this is an imbedding of $X$ in $J^1$ and defines a non-singular divisor on $J^1$ which we denote by $\Theta$. We denote by $L_\Theta$ the line bundle on $J^1$ associated with $\Theta$.

Let $\iota$ be the involution on $J^1$ defined by $\iota(\xi) = K \otimes \xi^{-1}$, $\xi \in J^1$, where $K$ is the canonical line bundle on $X$.

LEMMA 6.1. *We have $t^* L_\Theta \approx K$, where $K$ is the canonical line bundle on $X$.*

PROOF. First we note that $t^* L_\Theta$ is simply the normal bundle of $X$ in $J^1$ for the imbedding $t$. We then have an exact sequence

$$0 \longrightarrow (t^* L_\Theta)^{-1} \longrightarrow V \longrightarrow K \longrightarrow 0 ,$$

where $V$ is the inverse image of the cotangent bundle of $J^1$ by $t$ and hence trivial. From this it follows, taking determinant of $V$, that $K \approx t^* L_\Theta$.

LEMMA 6.2. ( i ) *$\Theta$ is invariant under the involution $\iota$; the action of $\iota$ on $J^1$ lifts to $L_\Theta$. We have $\iota^*(L_\Theta) \approx L_\Theta$.*

(ii) *$L_\Theta$ is an ample line bundle; we have $H^i(J^1, L_\Theta^n) = 0$ for $i \geqq 1, n \geqq 1$ and* $\dim H^0(J^1, L_\Theta^n) = n^2$, *for $n \geqq 1$.*

(iii) *Let $\eta \in J^1$ and $Q_\eta\colon J^1 \to J^1$ be the morphism defined by*

$$Q_\eta(\xi) = \xi^2 \otimes \eta^{-1} \qquad (\xi \in J^1)\,.$$

*Then the line bundle $Q_\eta^*(L_\Theta) \otimes L_\Theta^{-2}$ on $J^1$ is algebraically equivalent to $L_\Theta^2$ and hence is ample.*

(iv) *If $\eta = L_x$, where $x$ is a Weierstrass point of $X$, then $Q_\eta^* L_\Theta \approx L_\Theta^4$.*

(v) *If $\eta = L_x$ and $\eta' = K \otimes L_x^{-1}$, $x \in X$, then $Q_\eta^* L_\Theta \otimes Q_{\eta'}^* L_\Theta \approx L_\Theta^8$. Moreover, the action of $\iota$ on $J^1$ lifts in a natural way to $L_\Theta^8$ and also to $Q_\eta^* L \otimes Q_{\eta'}^* L_\Theta$. The above isomorphism is compatible with these actions.*

*Remark.* By (i) the involution $\iota$ on $J^1$ induces an involution, also denoted by $\iota$, on $X$. The involution $\iota$ fixes six points of $X$, *viz.* the *Weierstrass points* of $X$.

PROOF OF LEMMA 6.2. (i) This follows from the fact that for $x \in X$, $\dim H^0(X, K \otimes L_x^{-1}) = 1$; in fact,

$$\begin{aligned} \dim H^0(X, K \otimes L_x^{-1}) &= \dim H^1(X, L_x) \\ &= \dim H^0(X, L_x) \qquad \text{(by the Riemann-Roch theorem)} \\ &= 1\,. \end{aligned}$$

(ii) This is well known. That $L_\Theta$ is ample is classical which gives the vanishing of higher cohomologies since the canonical class of $J^1$ is trivial. The dimension of $H^0(J^1, L_\Theta^n)$ can now be computed inductively from the exact sequence

$$0 \longrightarrow L_\Theta^{(n-1)} \longrightarrow L_\Theta^n \longrightarrow L_\Theta^n|_{\iota X} \longrightarrow 0$$

and the fact $\dim H^0(J^1, L_\Theta^n|_{\iota X}) = \dim H^0(X, K^n)$ by Lemma 6.1.

(iii) Consider the isomorphism $f_\eta\colon J^1 \to J$ defined by $\xi \mapsto \xi \otimes \eta^{-1}$. We then have the commutative diagram

$$\begin{array}{ccc} J^1 & \xrightarrow{f_\eta} & J \\ \Big\downarrow{\scriptstyle Q_\eta} & & \Big\downarrow{\scriptstyle \psi_2} \\ J^1 & \xrightarrow{f_\eta} & J \end{array}$$

where $\psi_2\colon J \to J$ is the map defined by $\psi_2(j) = j^2, j \in J$. Now if $L$ is a line bundle on $J$, $\psi_2^*(L)$ is algebraically equivalent to $L^4$ [10, Prop. 2, p. 92]. It follows that $Q_\eta^*(L_\Theta)$ is algebraically equivalent to $L_\Theta^4$ and hence $Q_\eta^*(L_\Theta) \otimes L_\Theta^{-2}$ is algebraically equivalent to $L_\Theta^2$ and is ample.

(iv) Let $\eta = L_x, x \in X$. The line bundle $Q_\eta^* L_\Theta \otimes L_\Theta^{-4}$ is algebraically trivial. The restriction of $Q_\eta^* L_\Theta$ to $X$ is only the inverse image of $L_\Theta$ by means of the map $y \mapsto L_y^2 \otimes L_x^{-1}$ of $X$ in $J^1$. It is easy to see that the inverse image of the divisor $\Theta$ consists of the six Weierstrass points of $X$ and $x$ itself. Also, it is simple to check that $x$ occurs with multiplicity $\geqq 2$. Since $Q_\eta^* L_\Theta \mid X$ is of degree 8, we have $Q_\eta^* L_\Theta |_X \approx \bigotimes_{y \text{ Weierstrass}} L_y \otimes L_x^2$. But then it is well known that the divisor consisting of all the Weierstrass points of $X$ is linearly equivalent to $K^3$. Hence $Q_\eta^* L_\Theta |_X \approx K^3 \otimes L_x^2$. In particular, if $x$ is Weierstrass, $Q_\eta^* L_\Theta |_X \approx K^4$. So $Q_\eta^* L_\Theta \otimes L_\Theta^{-4}$ is algebraically trivial and its restriction to $X$ is trivial. Hence, $Q_\eta^* L_\Theta \otimes L_\Theta^{-4}$ is trivial in this case.

(v) The first assertion is proved as in (iv). For the last part, we have only to note that the section of $K^8$ defining the divisor $\sum 2y + 2x + 2\iota(x)$, $y$ Weierstrass, is invariant under $\iota$. q.e.d.

Consider the commutative diagram

$$\begin{array}{ccc} X \times J^1 & \xrightarrow{\tau} & J \times J^1 \\ \downarrow{\scriptstyle p_X} & & \downarrow{\scriptstyle \mu} \\ X & \xrightarrow{t} & J^1 \end{array}$$

where $\tau(x, \xi) = (L_x \otimes \xi^{-1}, \xi)$, $x \in X$, $\xi \in J^1$ and $\mu(j, \xi) = j \otimes \xi$, $j \in J$, $\xi \in J^1$. Since $\mu^{-1}(tX) = \tau(X \times J^1)$ it is clear that the line bundle on $J \times J^1$ defined by the divisor $T = \tau(X \times J^1)$ is just $\mu^*(L_\Theta)$.

LEMMA 6.3. *The family of line bundles on $X$ parametrised by $J$ defined as the inverse image of $L_\Theta$ by the map $(j, x) \mapsto j \otimes L_x$ of $J \times X$ in $J^1$ assigns to each $j \in J$ a line bundle isomorphic to $K \otimes j^{-1}$, where $K$ is the canonical line bundle on $X$.*

PROOF. We have only to check that for $j \in J$ the inverse image of $L_\Theta$ by the map $l_j$: $x \mapsto j \otimes L_x$ of $X$ into $J^1$ is isomorphic to $K \otimes j^{-1}$. After Lemma 6.1 we may assume that $j \neq e$ (the identity). Now $K \otimes j^{-1}$ is a line bundle of degree 2 on $X$ which is not isomorphic to $K$. Hence dim $H^0(X, K \otimes j^{-1}) = 1$, i.e., there exists a unique pair $a, b$ of points in $X$ such that $K \otimes j^{-1} \approx L_a \otimes L_b$. If $x \in X$ such that $l_j(x) \in t(X)$, then $K \otimes j^{-1} \approx L_x \otimes L_y^{-1} \otimes K$ for some $y \in X$. Since $K \otimes L_y^{-1} \approx L_z$ for some $z \in X$ by (i), Lemma 6.2, we must have $x = a$ or $b$, i.e., the inverse image $l_j^{-1}(\Theta)$ consists only of the points $a, b$. However, degree $(l_j^* L_\Theta)$ = degree $(l_e^* L_\Theta) = 2$ by Lemma 6.1. So we must have

$$l_j^* L_\Theta \approx L_a \otimes L_b \approx K \otimes j^{-1} . \qquad \text{q.e.d.}$$

Let $\sigma$: $J \times J^1 \to J \times J^1$ be the morphism given by $(j, \xi) = (j^{-1}, \xi)$ for $j \in J$, $\xi \in J^1$. Then we have

LEMMA 6.4. *The family of line bundles on $X$ parametrised by $J^1$ defined as $\tau^* \sigma^* L_T$ assigns to each $\xi \in J^1$ a bundle isomorphic to $\xi^2$.*

PROOF. We have only to show that the inverse image of $L_\Theta$ by the map $x \mapsto \xi^2 \otimes L_x^{-1}$, where $\xi \in J^1$, is isomorphic to $\xi^2$. Since $\iota^* L_\Theta \approx L_\Theta$, where $i$ is the involution of $J^1$ defined by $\iota(\xi) = K \otimes \xi^{-1}$, it is enough to prove that the inverse image of $L_\Theta$ by the map $X \to J^1$ defined by $x \mapsto K \otimes \xi^{-2} \otimes L_x$ is isomorphic to $\xi^2$. But this is a consequence of Lemma 6.3.

LEMMA 6.5. *There exists a line bundle on $J$, denoted by $L_{2\Theta_0}$, such that*

$$p_J^* L_{2\Theta_0} \approx L_T \otimes \sigma^* L_T \otimes p_{J^1}^* L_\Theta^{-2} .$$

*Moreover for $\eta \in J^1$, $L_{2\Theta_0}$ is isomorphic (non-canonically) to the line bundle $L_{\eta^*\Theta} \otimes L_{(K\otimes\eta^{-1})^*\Theta}$, where $\eta^*\Theta$ (resp. $(K \otimes \eta^{-1})^*\Theta$) is the divisor on $J$ which is the inverse image of $\Theta$ by the map $j \mapsto j \otimes \eta$ (resp. $j \mapsto K \otimes \eta^{-1} \otimes j$).*

PROOF. For $j \in J$, the line bundle $L_T \otimes \sigma^* L_T\,|_{j\times J^1}$ is isomorphic to $L_\Theta^2$. This follows from the fact that the divisor $j^*\Theta + (j^{-1})^*\Theta$ is linearly equivalent to the divisor $2\Theta$, by the theorem of the square [10, Cor. 2, Th. 4, p. 73]. (Here $j^*\Theta$ denotes the divisor consisting of elements $j^{-1} \otimes L_x$, $x \in X$). Hence $L_T \otimes \sigma^* L_T \otimes p_{J^1}^* L_\Theta^{-2}$ is isomorphic to $p_J^* L_{2\Theta_0}$ for a suitable line bundle $L_{2\Theta_0}$ on $J$, by the see-saw principle. For the second part we simply observe that $L_T\,|_{J\times\{\eta\}} \approx L_{\eta^*\Theta}$ and that

$$\sigma^* L_T\,|_{J\times\{\eta\}} \approx L_{(K\otimes\eta^{-1})^*\Theta}$$

using $\iota^* L_\Theta \approx L_\Theta$ (Lemma 6.1). Compare [11, § 3, Prop. 1].

*Remark.* Although $L_{\Theta_0}$ is not defined canonically on $J$, $L_{2\Theta_0}$ is.

Consider now the exact sequence

$$0 \longrightarrow L_T^{-1} \longrightarrow O \longrightarrow O \mid T \longrightarrow 0$$

of sheaves on $J \times J^1$, where $O$ is a structure sheaf of $J \times J^1$ and $O \mid T$ that of $\tau(X\times J^1)$. Tensoring this with $L_T \otimes \sigma^* L_T \otimes p_{J^1}^*(L_\Theta^{-2}) \approx p_J^*(L_{2\Theta_0})$, we obtain the exact sequence

$$\text{(A)} \qquad 0 \longrightarrow \sigma^* L_T \otimes p_{J^1}^*(L_\Theta^{-2}) \longrightarrow p_J^*(L_{2\Theta_0}) \longrightarrow p_J^*(L_{2\Theta_0})\,|_T \longrightarrow 0 ,$$

where the sheaf $p_J^*(L_{2\Theta_0}) \mid T$ has to be interpreted as $\tau_*(\tau^* p_J^*(L_{2\Theta_0}))$.

We wish to study the $0^{\text{th}}$ direct image $(p_{J^1})_*$ of the exact sequence (A) by means of the projection $p_{J^1}\colon J \times J^1 \to J^1$. For $\xi \in J^1$, we have $\sigma^* L_T \mid J \times \{\xi\}$ is ample, being the inverse image of $L_\Theta$ by the isomorphism $j \mapsto j^{-1} \otimes \xi$ of $J$ onto $J^1$. Hence $H^1(J \times \{\xi\}, \sigma^* L_T) = 0$ for every $\xi \in J^1$. This implies that the first direct image of $\sigma^* L_T \otimes p_{J^1} L_\Theta^{-2}$ by $p_{J^1}$ is 0. Therefore taking direct images of the exact sequence (A) by $p_{J^1}$ we obtain the exact sequence of vector bundles on $J^1$,

$$0 \longrightarrow (p_{J^1})_*(\sigma^* L_T) \otimes L_\Theta^{-2} \longrightarrow J^1 \times H^0(J, L_{2\Theta_0}) \longrightarrow (p_{J^1})_*(\tau^* p_J^* L_{2\Theta_0}) \longrightarrow 0 .$$

Moreover, it is obvious that $\dim H^0(J \times \{\xi\}, \sigma^* L_T) = \dim H^0(J^1, L_\Theta)$, which is equal to 1, by (ii), Lemma 6.2. Hence $(p_{J^1})_*(\sigma^* L_T)$ is a line bundle and moreover $\sigma^* L_T$ has a section whose restriction to each fibre $J \times \{\xi\}$, $\xi \in J^1$, is non-zero. In other words, $(p_{J^1})_*(\sigma^* L_T)$ admits a section which is nowhere zero. This implies that $(p_{J^1})_*(\sigma^* L_T)$ is trivial. Hence the above exact sequence reduces to the exact sequence

$$\text{(B)} \qquad 0 \longrightarrow L_\Theta^{-2} \longrightarrow J^1 \times H^0(J, L_{2\Theta_0}) \longrightarrow (p_{J^1})_*(\tau^* p_J^* L_{2\Theta_0}) \longrightarrow 0 .$$

Now we shall interpret the last term $F$ in (B), as the $0^{\text{th}}$ direct image of a family of line bundles on $X$ parametrised by $J^1$, which assigns to $\xi \in J^1$ a bundle isomorphic to $K \otimes \xi^2$. In particular, it would follow that rank $F =$ dim $H^0(X, K \otimes \xi^2) = 3$ by the Riemann-Roch theorem. Now

$$F = (p_{J^1})_*(\tau^* p_J^* L_{2\Theta_0}) = (p_{J^1})_*(\tau^* L_T \otimes \tau^* \sigma^* L_T \otimes p_{J^1}^* L_\Theta^{-2}) .$$

First we see that $\tau^* L_T \approx p_X^* K$, for

$$\tau^* L_T = \tau^* \mu^* L_\Theta = (\mu \circ \tau)^* L_\Theta = (t \circ p_X)^* L_\Theta = p_X^* t^* L_\Theta \approx p_X^* K$$

by Lemma 6.1. On the other hand, by Lemma 6.4, $\tau^* \sigma^* L_T$ may be considered a family of line bundles on $X$ which assigns to $\xi \in J^1$ a bundle isomorphic to $\xi^2$. This gives the interpretation of $F$ as asserted above. Now we have

LEMMA 6.6. *For $\eta \in J^1$, the image of the fibre $(L_\Theta^{-2})_\eta$ at $\eta$ of $L_\Theta^{-2}$ in $H^0(J, L_{2\Theta_0})$ by the inclusion in* (B) *is the image of*

$$H^0(J, L_{\eta * \Theta}) \otimes H^0(J, L_{(K \otimes \eta^{-1}) * \Theta}) \longrightarrow H^0(J, L_{2\Theta_0})$$

*given by the isomorphism*

$$L_{\eta * \Theta} \otimes L_{(K \otimes \eta^{-1}) * \Theta} \approx L_{2\Theta_0} .$$

PROOF. We have

$$\begin{aligned} (L_\Theta^{-2})_\eta &= H^0(J \times \{\eta\}, L_T^{-1} \otimes p_J^* L_{2\Theta_0}) \\ &= H^0(J, L_{\eta * \Theta}^{-1} \otimes L_{2\Theta_0}) \\ &= H^0(J, L_{(K \otimes \xi^{-1}) * \Theta}) \qquad \text{by Lemma 6.5.} \end{aligned}$$

Now the map $(L_\Theta^{-2})_\eta \to H^0(J, L_{2\Theta_0})$ is induced by the non-zero section of $L_{T\,|J \times \{\eta\}}$, which can be identified with a non-zero section of $H^0(J, L_{\eta * \Theta})$. This proves Lemma 6.6.

Now by dualising the exact sequence (B), we obtain the exact sequence

$$\text{(C)} \qquad 0 \longrightarrow E = F^* \longrightarrow J^1 \times H^0(J, L_{2\Theta_0})^* \longrightarrow L_\Theta^2 \longrightarrow 0 ,$$

where the 3-dimensional vector bundle $E$ may be considered the first direct image bundle on $J^1$ of a family of line bundles on $J^1$, assigning to $\xi \in J^1$, a bundle isomorphic to $\xi^{-2}$.

Now we shall show that the natural map $H^0(J, L_{2\Theta_0})^* \to H^0(J^1, L_\Theta^2)$ given by (C) is an isomorphism. Since $\dim H^0(J, L_{2\Theta_0})^* = 4$ and $\dim H^0(J^1, L_\Theta^2) = 4$ by (ii), Lemma 6.2, we have only to show that $H^0(J^1, E) = 0$. We shall prove this by checking that $H^2(J^1, F) = 0$. From our interpretation of

$$M = \tau^* L_T \otimes \tau^* \sigma^* L_T \otimes p_{J^1}^* L_\Theta^{-2} ,$$

it follows that its first and second direct images on $J^1$ are 0. Hence we have the isomorphism $H^2(J^1, F) \approx H^2(X \times J^1, M)$. Now for each $x \in X$, $M \mid x \times J^1$ is isomorphic to the tensor product of $L_\Theta^{-2}$ and the inverse image of $L_\Theta$ by the map $\xi \mapsto \xi^2 \otimes L_x^{-1}$ of $J^1$ into $J^1$. It follows by (iii), Lemma 6.2 that the first and second direct images of $M$ by $p_X$ are 0. Thus we have an isomorphism

$$H^2(X \times J^1, M) \approx H^2(X, (p_X)_* M) = 0 .$$

We have therefore proved

PROPOSITION 6.1. *There is an exact sequence*

$$0 \longrightarrow E \longrightarrow J^1 \times H^0(J^1, L_\Theta^2) \longrightarrow L_\Theta^2 \longrightarrow 0$$

*where the surjection is the evalution map and E can be identified with the first direct image on $J^1$ of a family of line bundles on X, parametrised by $J^1$, which assigns to $\xi \in J^1$ a line bundle isomorphic to $\xi^{-2}$.*

Compare the exact sequence (A) with the exact sequence (A′) by means of the commutative diagram (D):

$$\text{(A}'\text{)} \qquad \begin{array}{ccccccccc} 0 & \longrightarrow & p_J^* L_{\eta^*\Theta} \otimes L_T^{-1} & \longrightarrow & p_J^* L_{\eta^*\Theta} & \longrightarrow & p_J^* L_{\eta^*\Theta}|_T & \longrightarrow & 0 \\ & & \downarrow & & \downarrow & & \downarrow & & \\ 0 & \longrightarrow & p_J^* L_{2\Theta_0} \otimes L_T^{-1} & \longrightarrow & p_J^* L_{2\Theta_0} & \longrightarrow & p_J^* L_{2\Theta_0}|_T & \longrightarrow & 0 \end{array}$$

the vertical maps being induced by the non-zero homomorphism $L_{\eta^*\Theta} \rightarrow L_{2\Theta_0}$, determined up to a scalar multiple by the isomorphism of $L_{\eta^*\Theta} \otimes L_{(K\otimes\eta^{-1})^*\Theta}$ with $L_{2\Theta_0}$ given by Lemma 6.5. (Note that

$$\dim H^0(J, L_{(K\otimes\eta^{-1})^*\Theta}) = \dim H^0(J^1, L_\Theta) = 1$$

by (ii) Lemma 6.2).)

Now $\tau^* p_J^* L_{\eta^*\Theta}$ is a family of line bundles on $X$, parametrised by $J^1$, which assigns to $\xi \in J^1$, a bundle isomorphic to $K \otimes \xi \otimes \eta^{-1}$. This is because for each $\xi \in J^1$, the bundle $\tau^* p_J^* L_{\eta^* \otimes \Theta}|_{X \times \{\xi\}}$ is the inverse image of $L_\Theta$ by the map $X \rightarrow J^1$ given by $x \mapsto \eta \otimes \xi^{-1} \otimes L_x$. Now our assertion follows from Lemma 6.3. Moreover, the map $p_J^* L_{\eta^*\Theta}|_T \rightarrow p_J^* L_{2\Theta_0}|_T$ in (A′) is induced by a non-zero section of $p_J^* L_{(K\otimes\eta^{-1})^*\Theta}|_T$. Hence we see that the induced map

$$(p_{J^1})_*(p_J^* L_{\eta^*\Theta}|_T) \longrightarrow (p_{J^1})_*(p_J^* L_{2\Theta_0}|_T)$$

can be interpreted, for each $\xi \in J^1$ such that $\xi \neq \eta,\ K \otimes \eta^{-1}$, as the map

$$H^0(X, K \otimes \xi \otimes \eta^{-1}) \longrightarrow H^0(X, K \otimes \xi^2)$$

given by a non-zero section of $\xi \otimes \eta$. This follows from the fact that if $\xi \neq \eta$, $K \otimes \eta^{-1}$, the dimensions of $H^0(X, K \otimes \xi \otimes \eta^{-1})$ and $H^0(X, \xi \otimes \eta)$ remain constant (and equal to 1). On the other hand, the map

$$H^0(J, L_{\eta^*\Theta}) \longrightarrow H^0(X, K \otimes \xi \otimes \eta^{-1})$$

given by the surjection in (A′) is obtained by taking the inverse image of sections of $L_{\eta^*\Theta}$ by means of the map $x \mapsto L_x \otimes \xi^{-1}$. Since $\xi \neq \eta, K \otimes \eta^{-1}$, this map is an isomorphism. Thus we get from (D) the commutative diagram

$$\begin{array}{ccc} H^0(J, L_{\eta^*\Theta}) & \longrightarrow & H^0(X, K \otimes \xi \otimes \eta^{-1}) \\ \downarrow & & \downarrow \\ H^0(J, L_{2\Theta_0}) & \longrightarrow & H^0(X, K \otimes \xi^2) = F_\xi \,. \end{array}$$

By Lemma 6.6, we may replace $H^0(J, L_{\eta^*\Theta})$ by $(L_\Theta^{-2})_\eta$ and the map $H^0(J, L_{\eta^*\Theta}) \to H^0(J, L_{2\Theta_0})$ by the inclusion of $(L_\Theta^{-2})_\eta$ in $H^0(J, L_{2\Theta_0})$ given by (B).

On dualising, we obtain, from the commutative diagram above,

PROPOSITION 6.2. *The following diagram is commutative* ($\eta \in J^1$)

$$\begin{array}{ccc} (L_\Theta^2)_\eta & \longleftarrow & H^1(X, \xi^{-1} \otimes \eta) \\ \uparrow & & \uparrow \\ H^0(J^1, L_\Theta^2) & \longleftarrow & H^1(X, \xi^{-2}) \end{array}$$

*for* $\xi \neq \eta, K \otimes \eta^{-1}, \xi \in J^1$. *Here, the map* $H^0(J^1, L_\Theta^2) \to (L_\Theta^2)_\eta$ *is the evaluation map at* $\eta$, *the map* $H^1(X, \xi^{-2}) \to H^1(X, \xi^{-1} \otimes \eta)$ *is induced by a non-zero section of* $\xi \otimes \eta$, *the map* $H^1(X, \xi^{-2}) \to H^0(J^1, L_\Theta^2)$ *is the inclusion of the fibre* $E_\xi$ *in Proposition* 6.1, *and* $H^1(X, \xi^{-1} \otimes \eta) \to (L_\Theta^2)_\eta$ *is an isomorphism.*

PROPOSITION 6.3. *The line bundle* $L_\Theta^2$ *on* $J^1$ *induces a morphism of* $J^1$ *into the projective space* $PH^0(J^1, L_\Theta^2)^*$, *associated to the dual of* $H^0(J^1, L_\Theta^2)$ *and this imbeds, as a quartic surface, the quotient of* $J^1$ *by the involution* $\iota$ *on* $J^1$ *defined by* $\iota(\xi) = K \otimes \xi^{-1}$, *for* $\xi \in J^1$. *Similarly, the space* $J/i$, *where* $i$ *is the involution* $i(j) = j^{-1}$, *for* $j \in J$, *gets imbedded in* $PH^0(J^1, L_\Theta^2)$ *as a quartic surface.*

PROOF. We shall only prove the Proposition in the case of $J^1$. The proof in the case of $J$ is similar. On identifying $H^0(J^1, L_\Theta^2)^*$ with $H^0(J, L_{2\Theta_0})$, the map $J^1 \to PH^0(J, L_{2\Theta_0})$ is given by Proposition 6.1 as assigning to $\xi \in J^1$, the subspace $(L_\Theta^{-2})_\xi$ of $H^0(J, L_{2\Theta_0})$. This however, has been identified by Lemma 6.6 as the image of $H^0(J, L_{\xi^*\Theta}) \otimes H^0(J, L_{(K\otimes\xi^{-1})^*\Theta})$ in $H^0(J, L_{2\Theta_0})$. This means that we associate to $\xi$ the divisor $\xi^*\Theta + (K \otimes \xi^{-1})^*\Theta$ in $J$. From this, it follows easily that the map $J^1/\iota \to PH^0(J^1, L_\Theta^2)^*$ is injective. We shall prove that this is an imbedding at all points $\xi \in J^1$ such that $\xi^2 \neq K$. The two-fold symmetric product $S^2(X)$ of $X$ may be considered as the variety obtained by blowing up $\xi$ in $J^1$, by means of the map $p: S^2(X) \to J^1$ defined as $(x, y) \mapsto L_x \otimes L_y \otimes K^{-1} \otimes \xi$. It is easily checked that $p^*(L_\Theta^2) \otimes L_D^{-2}$, where $D$ is the blown down divisor, is the line bundle associated to the sum of the divisors $\{a\} \times X$ and $\{b\} \times X$ where $a, b$ are points of $X$ with $L_a \otimes L_b = (K \otimes \xi^{-1})^2$. Hence $H^1(S^2(X), p^*(L_\Theta^2) \otimes L_D^{-2})$ is isomorphic to the space of invariants in

$$H^1(X, L_a \otimes L_b) \otimes H^0(X, L_a \otimes L_b) \oplus H^0(X, L_a \otimes L_b) \otimes H^1(X, L_a \otimes L_b)$$

for the natural action of $\mathbf{Z}_2$. (This is seen by taking the inverse image on $X \times X$ and applying [6, Cor. on p. 202].) But by our assumption $L_a \otimes L_b \not\approx K$ and hence $H^1(X, L_a \otimes L_b) = 0$. This proves that $H^1(S^2(X), p^* L_\Theta^2 \otimes L_D^{-2}) = 0$ and therefore implies that the map $J^1 \to PH^0(J^1, L_\Theta^2)^*$ is of maximal rank at $\xi$ [9, § 3]. It follows that the image which is a hypersurface in $PH^0(J^1, L_\Theta^2)^*$, has only finitely many singularities and hence is normal. Now our assertion follows by Zariski's main theorem. Finally, it is easy to see that the image is a quartic surface.

PROPOSITION 6.4. *If $\sigma$ is a section of $L_\Theta^2$, we denote by $D_\sigma$ the set $\{\eta \in J^1 \mid \sigma(\eta) = 0\}$. Then $D_{\sigma_1} = D_{\sigma_2}$ if and only if $\sigma_1 = \lambda\sigma_2$ for a non-zero scalar $\lambda$.*

*Remark.* In other words, any positive divisor in $J^1$ linearly equivalent to $2\Theta$ is characterised by its support.

PROOF. We need to prove the statement only in the case $D = D_{\sigma_1} = D_{\sigma_2}$ is reducible. If the space of sections of $L_\Theta^2$ vanishing on $D$ is of dim $\geqq 2$, it is clear that under the map of $J^1$ into the projective space of $H^0(J^1, L_\Theta^2)^*$ given in Proposition 6.3, the set $D$ gets mapped into a projective line. Hence $D$ is invariant under $\iota$ and the quotient of $D$ by $\iota$ is rational (by Proposition 6.3). This implies that $D$ has at most two components which are interchanged by $\iota$ and each component is rational. This however is impossible since there is no non-constant morphism of the projective line into $J^1$.

## 7. Determination of $U(2, 0)$ on a curve of genus 2

THEOREM 2. *Let $X$ be a complete, non-singular irreducible algebraic curve of genus 2. Let $S$ be the space of S-equivalence classes of semi-stable bundles of rank 2 with trivial determinant on $X$. Let $J^1$ be the space of equivalence classes of line bundles of degree 1 on $X$. Let $\Theta$ be the divisor on $J^1$, defined by the natural imbedding of $X$ in $J^1$ and $L_\Theta$ the associated line bundle on $J^1$. For a semi-stable bundle $W$ of rank 2 and trivial determinant, the set $C_W = \{\xi \in J^1 \mid H^0(X, \xi \otimes W) \neq 0\}$ is the support of a uniquely determined positive divisor $D_W$ on $J^1$ linearly equivalent to $2\Theta$. The divisor $D_W$ depends* *only on the S-equivalence class of W and the induced map* $D\colon S \to PH^0(J^1, L_\Theta^2)$ *is an isomorphism.*

For the proof of the theorem, we need the following lemma which is a consequence of the results of § 5 and § 6.

LEMMA 7.1. *Let* $0 \to \xi^{-1} \to W \to \xi \to 0$ *be a non-trivial extension,* $\xi \in J^1$ *and* $\delta(W)$ *the corresponding element in* $H^1(X, \xi^{-2})$. *Let* $E_\xi = H^1(X, \xi^{-2})$ *be the fibre at* $\xi$ *of the vector bundle E of Proposition* 6.1. *Let the image of* $\delta(W)$ *in* $J^1 \times H^0(J^1, L_\Theta^2)$ *be* $(\xi, \sigma)$, *where* $\sigma$ *is a section of* $L_\Theta^2$. *Then we have*

$$C_W = \{\eta \in J^1 \colon \sigma(\eta) = 0\} .$$

PROOF. Since by assumption $\xi \in C_W$, it follows by Lemma 5.7 that $K \otimes \xi^{-1} \in C_W$. On the other hand, by the exact sequence in Proposition 6.1, we have $\sigma(\xi) = 0$, and from Proposition 6.3, it follows that $\sigma(K \otimes \xi^{-1}) = 0$. Hence we may assume that $\eta \neq \xi, K \otimes \xi^{-1}$ and prove that $\eta \in C_W \Leftrightarrow \sigma(\eta) = 0$. By Lemma 5.9, we see that $\eta \in C_W$ if and only if $\delta(W)$ belongs to the kernel of the map $H^1(X, \xi^{-2}) \to H^1(X, \xi^{-1} \otimes \eta)$ induced by a non-zero section of $\xi \otimes \eta$. Using the commutative diagram (Proposition 6.2)

$$\begin{array}{ccc} H^1(X, \xi^{-2}) & \longrightarrow & H^1(X, \xi^{-1} \otimes \eta) \\ \downarrow & & \downarrow \\ H^0(J^1, L_\Theta^2) & \longrightarrow & (L_\Theta^2)_\eta \end{array}$$

and noting that the image of $\delta(W)$ in $H^0(J^1, L_\Theta^2)$ is $\sigma$, we see that $\eta \in C_W$ if and only if the image of $\sigma$ in $(L_\Theta^2)_\eta$ is zero.

PROOF OF THE THEOREM. By Lemmas 5.5 and 5.8, for every $\omega \in S$, there exists a representative $W$, which is an extension of $\xi$ by $\xi^{-1}$ for some $\xi \in J^1$. Now the fact that $C_W$ is the support of a positive divisor $D_W$ linearly equivalent to $2\Theta$ follows from Lemma 7.1 above. That $D_W$ is uniquely determined by $C_W$ is a consequence of Proposition 6.4. But $C_W$ itself depends only on $\omega$ (Lemma 5.6). Now consider the diagram

$$\begin{array}{ccc} P(E) & \xrightarrow{\ f\ } & PH^0(J^1, L_\Theta^2) = P \\ & {}_s\searrow & \uparrow D \\ & & S \end{array}$$

where $P(E)$ is the projective bundle associated to $E$ and $P$ the projective space associated to $H^0(J^1, L_\Theta^2)$, $s$ associates to each element $\neq 0$ of $H^1(X, \xi^{-2})$ the $S$-equivalence class of the corresponding extension of $\xi$ by $\xi^{-1}$, and $f$ is the map induced by the injection of the exact sequence in Proposition 6.1. From

Lemma 7.1, we conclude that this diagram commutes. Moreover, since $P(E) \to P$ is obviously surjective, it follows that $D$ is surjective. We shall now show that $D$ is injective. Let $\omega_1, \omega_2 \in S$ such that $D(\omega_1) = D(\omega_2)$. Let $\xi \in \operatorname{Supp} D(\omega_1) = \operatorname{Supp} D(\omega_2)$. There exist then (by Lemma 5.8) $v_1, v_2 \in P(E)_\xi$ such that $s(v_1) = \omega_1$ and $s(v_2) = \omega_2$. Hence $f(v_1) = f(v_2)$. But then the map $f: P(E)_\xi \to P$ is an inclusion. This implies that $v_1 = v_2$ and hence that $\omega_1 = \omega_2$. Thus we have shown that $D$ is a bijective map.

Finally, by Proposition 3.1, and the universal property of $S$ (see § 2), it follows that the map $s: P(E) \to S$ is a morphism. Since $S$ is normal (§ 2), the map $D: S \to P$ is also a morphism and since $P$ is non-singular, an isomorphism, by Zariski's main theorem.

As a corollary, we obtain

THEOREM 3. *Let $X$ be a complete, non-singular algebraic curve of genus 2. Let $U(2, 0)$ be the space of S-equivalence classes of semi-stable vector bundles of rank 2 and degree 0. The group $\Gamma$ of elements of order 2 in $J$ acts on $PH^0(J^1, L_\Theta^2)$ in a natural way; let $A$ be the associated projective bundle on $J/\Gamma \approx J$. Then $U(2, 0)$ is canonically isomorphic to $A$. In other words, $U(2, 0)$ is canonically isomorphic to the space of positive divisors on $J^1$ algebraically equivalent to $2\Theta$.*

PROOF. We have a natural morphism $S \times J \to U(2, 0)$ given by tensor product. It is clear that $U(2, 0)$ is the quotient of $S \times J$, for the diagonal action of $\Gamma$. On the other hand, since the projective bundle described above, is the quotient of $P \times J$ under the diagonal action of $\Gamma$, we have only to check that the isomorphism $D: S \to P$ preserves the action of $\Gamma$. But this is obvious since $C_{W \otimes j} = jC_W$, for $j \in \Gamma$. That the space of positive divisors algebraically equivalent to $2\Theta$ is isomorphic to this projective bundle is classical and may be verified, for instance, by restricting the algebraic family to $X$.

*Remarks* 1. Considerations similar to the above show that the space of $S$-equivalence classes of rank 2 and determinant $K$ is isomorphic to $PH^0(J, L_{2\Theta_0}) = P^*$.

2. It follows from Theorem 2, that the involution $\iota$ acts trivially on $S$. In particular, if $W$ is a *stable* bundle of rank 2 and trivial determinant, we have $\iota^* W \approx W$. However, it is not true that $\iota^* W \approx W$ for any semi-stable vector bundle of rank 2 and trivial determinant. In fact, if $E$ is a non-trivial extension of $j$ by $j^{-1}$, $j \in J$, then $\iota^* E \approx E$ if and only if $j$ is of order 2.

## 8. Quadratic complexes; Kummer surfaces and Klein variety

In this section we review the classical theory of quadratic line complexes in $\mathbf{P}_3$ and its relationship with the geometry of Kummer surfaces in $\mathbf{P}_3$ with particular reference to the determination by Klein of all quadratic complexes which have a given Kummer surface in $\mathbf{P}_3$ as "singularity locus". We shall later tie up these results with the theory of stable bundles of rank 2 on a curve of genus 2.

Let $R$ be a four dimensional vector space over $C$. Then the grassmannian $G$ of lines in $P(R)$ is naturally imbedded as a quadric in $P(\bigwedge^2 R)$. Let $H$ denote the hyperplane bundle on $P(\bigwedge^2 R)$ and also its restriction to $G$. It is then easy to see that the restriction map gives an isomorphism

$$H^0(P(\bigwedge^2 R), H) \longrightarrow H^0(G, H)$$

and a surjection $H^0(P(\bigwedge^2 R), H^2) \to H^0(G, H^2)$ with one-dimensional kernel. An element of $PH^0(G, H) \approx P(\bigwedge^2 R^*)$ is called a *linear complex* of lines in $P(R)$, and an element of $PH^0(G, H^2)$ is called a *quadratic complex* of lines in $P(R)$. In other words, a quadratic complex is a divisor on $G$, obtained as the intersection with $G$ of a quadric in $P(\bigwedge^2 R)$.

Let $\{l_\alpha\}_{\alpha \in A}$ be a family of linear complexes indexed by $A$. We say $\{l_\alpha\}_{\alpha \in A}$ is *mutually apolar* with respect to $G$ if there exists a basis $(x_\alpha)_{\alpha \in A}$ of $\bigwedge^2 R^*$ such that $x_\alpha$ represents $l_\alpha$ for every $\alpha \in A$ and $G$ is defined by the equation $\sum_{\alpha \in A} x_\alpha^2 = 0$. For later use, we make the following observation.

LEMMA 8.1. *Let $(l_\alpha)$, $1 \leqq \alpha \leqq 6$, be a family of linear complexes. We shall assume that any five of the $l_\alpha$'s are in general position in $P(\bigwedge^2 R))^*$.* (i.e. *any five of the $l_\alpha$'s generate a four dimensional linear subspace*). *Also let the image of the elements $l_\alpha^2 \in PH^0(P(\bigwedge^2 R), H^2)$ be contained in a five dimensional subspace of $PH^0(G, H^2)$. Then $(l_\alpha)_{1 \leqq \alpha \leqq 6}$ is a set of mutually apolar complexes for $G$.*

PROOF. If $(x_\alpha)_{1 \leqq \alpha \leqq 6} \in (\bigwedge^2 R)^*$ represent $l_\alpha$, then it easily follows from our assumptions that the quadratic from $\sum \lambda_\alpha x_\alpha^2$ is non-zero if $\lambda_\alpha \neq 0$ for some $\alpha$. This implies that $G$ is defined by an equation of the form $\sum_{\alpha=1}^{6} \lambda_\alpha x_\alpha^2 = 0$. Now since $G$ is non-singular, $(x_\alpha)_{1 \leqq \alpha \leqq 6}$ must form a basis for $\bigwedge^2 (R)^*$ and $\lambda_\alpha \neq 0$ for every $\alpha$. This proves the lemma.

In what follows, a quadratic complex will be called *generic* if the characteristic roots of the defining bilinear form (when $G$ is taken to be defined by the identity form) are distinct. This is equivalent to requiring that there exists a (unique) system of linear complexes, mutually apolar with respect to $G$, such that the divisor on $G$ is given by the equation $\sum \lambda_\alpha x_\alpha^2 = 0$, with $\lambda_\alpha$ distinct, in the associated coordinate system. It is easy to see that a generic quadratic complex in $G$ is irreducible, non-singular, and does not contain any plane.

The lines in $P(R)$ through any point which belong to a generic quadratic complex form a quadratic cone. A point in $P(R)$ at which this quadratic cone breaks up into a pair of planes is called a *singular point* of the quadratic complex. Since the quadratic complex is generic, we see easily that not every point of $P(R)$ is singular. The locus of singular points (which we shall define more precisely below, making it obvious that it is a quartic surface) is called the *Kummer surface* associated to the quadratic complex.

Consider the tautological exact sequence

$$0 \longrightarrow L^{-1} \longrightarrow R \longrightarrow F \longrightarrow 0$$

of vector bundles of $P(R)$, where $L$ is the hyperplane bundle on $P(R)$. This leads to the exact sequence [7, Satz 4.1.3]

$$0 \longrightarrow F \otimes L^{-1} \longrightarrow \bigwedge^2 (R) \longrightarrow \bigwedge^2 (F) \longrightarrow 0 .$$

This induces a map $P(F \otimes L^{-1}) \rightarrow G \subset P(\bigwedge^2 (R))$. The inverse image of the hyperplane bundle $H$ on $G$ is the tautological hyperplane bundle $\mathcal{T}$ on $P(F \otimes L^{-1})$. We have a map, for every $r$,

$$H^0(G, H^r) \longrightarrow H^0(P(F \otimes L^{-1}), \mathcal{T}^r) = H^0(P(R), S^r(F \otimes L^{-1})^*) .$$

LEMMA 8.2. *The map*

$$H^0(G, H^r) \longrightarrow H^0(P(F \otimes L^{-1}), \mathcal{T}^r) = H^0(P(R), S^r(F \otimes L^{-1})^*)$$

*is an isomorphism.*

PROOF. This follows from the fact that the map $P(F \otimes L^{-1}) \rightarrow G$ has projective lines as fibres.

*Remark.* The image of a section of $H^2$ by the map

$$H^0(G, H^2) \longrightarrow H^0(P(F \otimes L^{-1}), \mathcal{T}^2) = H^0(P(R), S^2(F \otimes L^{-1})^*)$$

can be looked upon as the precise version of the assignment to each point of $P(R)$, of the quadratic cone at that point defined by the quadratic complex.

Now, there is associated to the vector bundle $W = F \otimes L^{-1}$, a canonical element of $H^0(S^2(W^*), \pi^*(\det W)^{-2})$, where $\pi$ is the projection onto $P(R)$, *viz.* the "discriminant function". The set of zeros of this section is precisely the bundle of degenerate bilinear forms on the fibres of $W$. We may hence associate to each section $s$ of $H^2$, the inverse image of this discriminant by the section of $S^2(F \otimes L^{-1})^*$ determined by $s$ as above. Thus we have obtained a map

$$k\colon H^0(G, H^2) \longrightarrow H^0(P(R), (\det W^{-2})) .$$

(Note, however, that $k$ is not linear, for $k(\lambda s) = \lambda^3 k(s)$ for $\lambda \in C$.) Finally we note that

$$(\det W)^{-2} = \det(F \otimes L^{-1})^{-2} = (\det F \otimes L^{-3})^{-2} \approx L^4$$

and hence we get a map

$$k\colon H^0(G, H^2) \longrightarrow H^0(P(R), L^4)\ .$$

If $s \in H^0(G, H^2)$ represents a generic quadratic complex, then $k(s)$ is non-zero. The quartic surface in $P(R)$ defined by $k(s) \in H^0(P(R), L^4)$ is called the *Kummer surface* associated to the quadratic complex.

Let now $s$ be any generic quadratic complex. F. Klein [8] determined all the quadratic complexes which have the Kummer surface of $s$ as their singular locus. In fact, let $(l_\alpha)_{\alpha \in A}$ be the family of mutually apolar linear complexes determined by $s$. Then $l_\alpha^2 \in PH^0(G, H^2)$ span a 4-dimensional projective subspace which contains $s$. It is easy to see that the elements $l_\alpha^2$ and $s$ are in general position and hence there exists a unique rational (normal) quartic curve (i.e., parametrised by $\mathbf{P}_1$) which passes through the seven points and lies in the space spanned by them. Then Klein proved

*A generic quadratic complex has the Kummer surface of* $s$ *as singular locus if and only if it lies on this quartic curve.*

DEFINITION. Let a Kummer surface in $P(R)$ be given. The family of quadratic complexes defined above, considered as a subvariety of $\mathbf{P}_1 \times G$, is called the *Klein variety* associated to the Kummer surface.

*Remark.* We shall not need the above results of Klein. In our case we will explicitly construct a rational quartic curve of quadratic complexes and prove that all generic quadratic complexes in this family have the same singular locus. As a matter of fact, the existence of such a curve corresponding to any Kummer surface can be deduced from our results. We shall not make any use of the fact that these are *all* the quadratic complexes which have the given Kummer surface as singular locus.

LEMMA 8.3. *Let* $q\colon \mathbf{P}_1 \to PH^0(G, H^2)$ *be a normally imbedded quartic curve through six points of the form* $l_\alpha^2$, *where* $\{l_\alpha\}$ *is a mutually apolar system of linear complexes and contained in the projective subspace generated by them. Then every point in* $q(P^1)$ *other than* $l_\alpha^2$ *represents a generic quadratic complex. The subvariety of* $G \times P^1$ *that the family* $q$ *induces is irreducible, normal.*

PROOF. If we choose $x_\alpha \in (\bigwedge^2 R)^*$ representing $l_\alpha$, so that $\sum x_\alpha^2 = 0$ on $G$, then $q(x)$, $x \in \mathbf{P}_1$ is given by a linear combination $\sum \lambda_\alpha x_\alpha^2$. If two of the $\lambda_\alpha$'s are equal, say $\lambda_1 = \lambda_2 = \lambda$, then $q(x)$ is also represented by $\sum_{\alpha=3}^{6} (\lambda_\alpha - \lambda) x_\alpha^2$. This however implies that $q(x)$ lies in the hyperplane spanned by $l_\alpha^2$, $\alpha \geqq 3$. Since $q(\mathbf{P}_1)$ is a quartic curve, this is impossible, if $q(x)$ is not one of the $l_\alpha^2$.

Moreover, the family of quadratic complexes gives rise to a subvariety of codimension 1 in $\mathbf{P}_1 \times G$. It is easily checked that the singularity set of this subvariety is of codimension 2 and is in fact contained in the union of the divisors on $G$ defined by $l_\alpha^2$. This proves that the variety is normal. Since on the other hand, each of the quadratic complexes $q(x)$ is connected, it follows that the above variety is also connected and being normal, is irreducible.

## 9. Statement of Theorem 4 and outline of proof

THEOREM 4. *Let $X$ be a complete non-singular irreducible algebraic curve of genus 2. Let $S_1'$ (resp. $S_{1,a}$) be the space of isomorphism classes of stable bundles over $X$ of rank 2, whose determinant bundles are of the form $L_x^{-1}$, for some $x \in X$ (resp. whose determinant is isomorphic to $L_a^{-1}$). Then the quotient of the jacobian $J$ by the canonical involution has a natural imbedding in $PH^0(J^1, L_\Theta^2) = P$ as a Kummer surface, and the quotient of $S_1'$ by the canonical involution is canonically isomorphic to the Klein variety associated to the Kummer surface. In particular, if $a$ is a non-Weierstrass point of $X$, then the space $S_{1,a}$ is canonically isomorphic to the intersection of the grassmannian $G$ of lines in $P$ (which is imbedded as a non-singular quadric in $P(\bigwedge^2 H^0(J^1, L_\Theta^2))$ with another generic quadric.*

The rest of this paper will, in the main, be devoted to the proof of Theorem 4. We identify the space $S = P$ with isomorphism classes of vector bundles of rank 2 with trivial determinant which are either stable or direct sum of line bundles of degree 0. We first prove that for $V \in S_1'$, the set of elements $W$ in $P$ such that $H^0(X, \mathrm{Hom}(V, W)) \neq 0$ is a projective line $\Lambda_V$ in $P$. (See Proposition 10.1.) Then the map $S_1' \to G \times X$ given $V \mapsto (\Lambda_V, (\det V)^{-1})$ is seen (Proposition 10.2) to induce an injective map $\Lambda\colon S_1'/\iota \to G \times X/\iota$. The image of $S_1'/\iota$ is a divisor in $G \times X/\iota$. That this divisor is a family of quadratic complexes on $G$ is proved as follows.

The map $f\colon P(E) \to P$ of Theorem 2 induces a map $P(\bigwedge^2 E) \to G$. We will construct in § 11, a quotient bundle $E_1'$ of $p_{J^1}^*(E)$ on $X \times J^1$ which associates to each $(x, \xi) \in X \times J^1$, the vector space $H^1(X, \xi^{-2} \otimes L_x)$. Each non-zero element of $H^1(X, \xi^{-2} \otimes L_x)$ gives rise to a stable bundle of rank 2 and determinant $L_x^{-1}$ which is an extension of $\xi \otimes L_x^{-1}$ by $\xi^{-1}$ (Lemma 10.1). This leads to a commutative diagram (Proposition 11.1)

$$\begin{array}{ccc} P(E_1') & \longrightarrow & X \times P(\bigwedge^2 E) \\ \downarrow & & \downarrow \\ S_1/\iota & \xrightarrow{\Lambda} & X/\iota \times G . \end{array}$$

Now $P(E_1')$ may be considered as a divisor on $X \times P(\bigwedge^2 E)$, and the sum of this divisor and its transform by the involution $\iota$ on $X$ is seen to be given by a section $\sigma$ of the line bundle $p^*_{X/\iota}(H^4) \otimes \mathcal{T}^2$ on $X/\iota \times P(\bigwedge^2 E)$, where $H$ is the hyperplane bundle on the projective line $X/\iota$ and $\mathcal{T}$ the tautological hyperplane bundle on $P(\bigwedge^2 E)$(Proposition 11.2). This line bundle is the inverse image of $p^*_{X/\iota}(H^4) \otimes p^*_G(H^2)$ where $H$ also denotes the hyperplane bundle of $G$. Finally we show that the section $\sigma$ goes down into a section of $p^*_{X/\iota}(H^4) \otimes p^*_G(H^2)$ on $X_{/\iota} \times G$. (Proposition 12.1.) Clearly $\Lambda(S_1')$ is contained in the support of the divisor $\mathcal{K}$ defined by this section.

The proof of Theorem 4 is completed by checking that all these quadratic complexes (barring those corresponding to the Weierstrass points of $X$) are generic and have $J/\iota$ as their singular locus (Lemma 13.2). It follows that $\mathcal{K}$ is normal and irreducible (Lemma 8.3) and that $\Lambda\colon S_1'/\iota \to \mathcal{K}$ is an isomorphism.

## 10. Injectivity of the map $S_1'/\iota \to G \times X/\iota$

LEMMA 10.1. (i) *Let $0 \to \xi^{-1} \to V \to \xi \otimes L_x^{-1} \to 0$ be a non-trivial extension for $\xi \in J^1$. Then $V$ is a stable bundle with determinant $L_x^{-1}$.*

(ii) *Moreover, two extensions $V, W$ of $\xi \otimes L_x^{-1}$ by $\xi^{-1}$ are isomorphic if and only if $\delta(V) = \lambda\delta(W)$ for some $\lambda \in \mathbf{C}^*$.*

PROOF. (i) Similar to Lemma 5.1. (ii) follows from the fact that $\dim H^0(X, \xi \otimes V) \leq 1$, by Lemma 5.4.

LEMMA 10.2. *Let $V$ be any stable vector bundle of rank 2 and degree $-1$. Then there exists $\xi \in J^1$ such that $H^0(X, \operatorname{Hom}(\xi^{-1}, V)) \neq 0$.*

PROOF. By the Riemann-Roch theorem, we see that $\dim H^0(X, L \otimes V) \geq 3$, for any line bundle $L$ of degree 3. Let then $\pi$ be a subspace of $H^0(X, L \otimes V)$ of dimension 3. Associate to each $x \in X$, the subspace $H_x$ of sections of $L \otimes V$ in $\pi$ which vanish at $x$. Then the subspace $H_x$ is contained in the image under the injective map

$$H^0(X, L \otimes V \otimes L_x^{-1}) \longrightarrow H^0(X, L \otimes V) \qquad \text{by Lemma 5.3.}$$

If $H_x$ is of dimension $\geq 2$, this implies that $\dim H^0(X, L \otimes V \otimes L_x^{-1}) \geq 2$ and our assertion follows from Lemma 5.4. We shall therefore assume that $H_x$ is of dimension 1 for all $x$. This means that we have a non-constant morphism $X \to P\pi$. Now, for example by Plücker's formula, it is easily seen that such a map cannot be an imbedding, since $X$ is of genus 2. Let then $\sigma$ be a singular point of the image $X \to P$. It is easy to see that $\sigma$ is the image of a section of $L \otimes L_y^{-1} \otimes L_z^{-1} \otimes V$ by the natural map $L \otimes L_y^{-1} \otimes L_z^{-1} \otimes V \to L \otimes V$, for some $y, z \in X$. This completes the proof of Lemma 10.2.

We follow the notation in Theorem 2, § 7. The space $S$ is identified with isomorphism classes of vector bundles of rank 2 with trivial determinant which are either stable or direct sum of line bundles of degree 0. If $W \in S$, we denote sometimes by $W$ also a vector bundle of the above type in the class $W$.

PROPOSITION 10.1. *Let $V$ be a stable bundle of rank 2 and determinant $L_x^{-1}$ for some $x \in X$. Let $l_1(V)$ be the subspace of $S$ defined by*

$$\{W \in S \mid H^0(X, \operatorname{Hom}(V, W)) \neq 0\}$$

*($W$ is either stable or is of the form $j \oplus j^{-1}, j \in J$). Let $\Lambda_V$ be the image of $l_1(V)$ in $P$ by the canonical isomorphism $D$. Then $\Lambda_V$ is a projective sub-line of $P$.*

PROOF. In view of Lemma 10.2, Proposition 10.1 follows from

LEMMA 10.3. *Let $\delta \in PH^1(X, \xi^{-2} \otimes L_x)$, $\xi \in J^1$, $x \in X$ and let $V$ be the associated vector bundle obtained as an extension of $\xi \otimes L_x^{-1}$ by $\xi^{-1}$. Then $\delta$ defines a projective line $\Lambda_\delta$ in $PH^1(X, \xi^{-2})$ as the inverse image of $\delta$ by the canonical surjection*

$$H^1(X, \xi^{-2}) \longrightarrow H^1(X, \xi^{-2} \otimes L_x) .$$

*The image of $\Lambda_\delta$ in $PH^0(J^1, L_\Theta^2)$ by the injection $f \mid P(E_\xi)$ (Theorem 2, § 7) is then $\Lambda_V$.*

PROOF. We first prove that $f(\Lambda_\delta) \subset \Lambda_V$. Since by definition $\delta$ is the kernel of the map

$$H^1(X, \xi^{-2} \otimes L_x) \longrightarrow H^1(X, \xi^{-1} \otimes V^*)$$

induced by the surjection $V \to \xi \otimes L_x^{-1}$, we have $\Lambda_\delta =$ kernel of the map

$$H^1(X, \xi^{-2}) \longrightarrow H^1(X, \xi^{-1} \otimes V^*)$$

induced by the composite of the maps $V \to \xi \otimes L_x^{-1}$ and $\xi \otimes L_x^{-1} \to \xi$. Let

$$0 \longrightarrow \xi^{-1} \longrightarrow W \longrightarrow \xi \longrightarrow 0$$

be any non-trivial extension, with $\delta(W) \in \Lambda_\delta$. By Lemma 3.1, the non-zero map $V \to \xi$ defined above can be lifted into a homomorphism $V \to W$. This implies that $f(\delta(W)) \in \Lambda_V$. In fact, if $W$ is stable, $f(\delta(W))$ is the isomorphism class of $W$ and there is nothing to prove. If on the other hand, $W$ is of the form

$$0 \longrightarrow j^{-1} \longrightarrow W \longrightarrow j \longrightarrow 0$$

with $j \in J$, the fact that $H^0(X, \operatorname{Hom}(V, W)) \neq 0$ implies that

$$H^0(X, \operatorname{Hom}(V, j)) \neq 0$$

or $H^0(X, \operatorname{Hom}(V, j^{-1})) \neq 0$. This means that $f(\delta(W)) = j \oplus j^{-1} \in \Lambda_V$.

Conversely, let $W$ be a stable bundle such that $W \in \Lambda_V$. The restriction to $\xi^{-1}$ of a non-zero homomorphism of $V$ into $W$ is non-zero, since otherwise it would lead to a non-zero homomorphism $\xi \otimes L_x^{-1} \to W$. Now the map $\xi^{-1} \to W$ is an injection since $W$ is stable, by Lemma 5.3. Thus we get a diagram

$$\begin{array}{ccccccccc} 0 & \longrightarrow & \xi^{-1} & \longrightarrow & V & \longrightarrow & \xi \otimes L_x^{-1} & \longrightarrow & 0 \\ & & \downarrow & & \downarrow & & \downarrow & & \\ 0 & \longrightarrow & \xi^{-1} & \longrightarrow & W & \longrightarrow & \xi & \longrightarrow & 0 . \end{array}$$

Clearly we may assume that $\xi^{-1} \to \xi^{-1}$ is the identity map. Moreover, the map $\xi \otimes L_x^{-1} \to \xi$ is non-zero since otherwise it would lead to a non-zero homomorphism $V \to \xi^{-1}$ which is impossible since $V$ is stable. Finally, by changing the extension $W$ by a constant multiple if necessary, we may assume that $\xi \otimes L_x^{-1} \to \xi$ is the canonical map. It follows then from Lemma 3.1 that $\delta(W) \in \Lambda_\delta$. Hence $W = f(\delta(W)) \in f(\Lambda_\delta)$.

Suppose now that $W = j \oplus j^{-1} \in \Lambda_V$. Then there is a non-zero homomorphism of $V$ into $j$ or $j^{-1}$, say $j$. Since $V$ is stable, this leads to an exact sequence

$$0 \longrightarrow j^{-1} \otimes L_x^{-1} \longrightarrow V \longrightarrow j \longrightarrow 0 .$$

If $j = \xi \otimes L_x^{-1}$, then by (ii), Lemma 10.1, this extension corresponds to the element $\delta$ in $PH^1(X, \xi^{-2} \otimes L_x)$. Now the kernel of the map $H^1(X, \xi^{-2}) \to H^1(X, \xi^{-2} \otimes L_x)$ gives an extension $W'$ of $\xi$ by $\xi^{-1}$, which is $S$-equivalent to $W$ by Lemma 5.2. By definition, $\delta(W') \in \Lambda_\delta$ and hence $W = f(\delta(W')) \in f(\Lambda_\delta)$. If on the other hand, $j \neq \xi \otimes L_x^{-1}$, then the restriction to $\xi^{-1}$ of the map $V \to j$ is non-zero and hence $j = \xi^{-1} \otimes L_y$ for some $y \in X$. Moreover, by Lemma 3.2, $\delta$ is in the kernel of

$$H^1(X, \xi^{-2} \otimes L_x) \longrightarrow H^1(X, \xi^{-2} \otimes L_x \otimes L_y) ,$$

and since $\xi^{-2} \otimes L_x \otimes L_y = j \otimes \xi^{-1} \otimes L_x$ is non-trivial by assumption, $\delta$ is actually the kernel of this map. Hence $\Lambda_\delta$ is the kernel of

$$H^1(X, \xi^{-2}) \longrightarrow H^1(X, \xi^{-2} \otimes L_x \otimes L_y)$$

and this implies that the kernel of

$$H^1(X, \xi^{-2}) \longrightarrow H^1(X, \xi^{-2} \otimes L_y)$$

is contained in $\Lambda_\delta$. But then this kernel gives rise to an extension $W'$ of $\xi$ by $\xi^{-1}$, which is $S$-equivalent to $j \oplus j^{-1}$, by Lemma 5.2. As before, $W = f(\delta(W')) = f(\Lambda_\delta)$.

*Remark.* If $0 \to \xi^{-1} \to V \to \xi \otimes L_x^{-1} \to 0$ is a non-trivial extension, it follows from the proof of Lemma 10.3 that there are only finitely many line bundles of degree 0 (in general four) into which there is a non-zero homomorphism from $V$. We have a projective line bundle $R$ on $J^1$ whose fibre at $\xi \in J^1$ is isomorphic to $P(H^1(X, \xi^{-2} \otimes L_x))$ (see § 11). We have a morphism from $R \to S_{1,x}$ whose fibres are finite. Since $\dim R = \dim S_{1,x} = 3$, $R$ compact, $S_{1,x}$ non-singular and connected, it follows that this morphism is surjective. This gives another proof of Lemma 10.2.

We have a natural morphism (determinant) from $U(2, -1) \to J^{-1}$ ($J^{-1}$ is the space of line bundles of degree $-1$.) Let $S_1'$ be the inverse image of the set of line bundles of the form $L_x^{-1}$, $x \in X$. Let $G$ be the grassmannian of lines in $P$. By Proposition 10.1 we have a natural map $\Lambda: S_1' \to G \times X$, defined by $V \mapsto (\Lambda_V, (\det V)^{-1})$. The involution $\iota$ on $X$ defines an action, still denoted by $\iota$, on $S_1'$ and $J^{-1}$ compatible with the map $U(2, -1) \to J^{-1}$. Noting that $\iota$ acts trivially on $P$ and hence trivially on $G$, we obtain a map

$$\Lambda: S_1'/\iota \longrightarrow G \times (X/\iota)$$

($S_1'/\iota$ is the quotient of $S_1'$ by $\iota$). We then have

PROPOSITION 10.2. *The map* $\Lambda: S_1'/\iota \to G \times X/\iota$ *is injective.*

PROOF. Let $x$ be a non-Weierstrass point of $X$. We shall prove that the natural map $V \mapsto \Lambda_V$ of $S_{1,x}$ to $G$ is injective. Let $V$ and $V'$ be two stable bundles of rank 2 with determinant $L_x^{-1}$ such that $\Lambda_V = \Lambda_{V'}$. Suppose there exists $\xi \in J^1$ with $\xi^{-1} \subset V$ and $\xi^{-1} \subset V'$. Let $\delta, \delta'$ be the corresponding elements in $PH^1(X, \xi^{-2} \otimes L_x)$. By Lemma 10.3 we have $\Lambda_\delta = \Lambda_{\delta'}$. This proves that $V \approx V'$, in this case.

Now by Lemma 10.2 there exists an exact sequence

$$0 \longrightarrow \xi^{-1} \longrightarrow V \longrightarrow \xi \otimes L_x^{-1} \longrightarrow 0$$

for $\xi \in J^1$. This implies that $\xi \otimes L_x^{-1} \oplus \xi^{-1} \otimes L_x \in \Lambda_V = \Lambda_{V'}$ and hence there is a non-zero homomorphism of $V'$ into $\xi \otimes L_x^{-1}$ or $\xi^{-1} \otimes L_x$. In the former case, the map $V' \to \xi \otimes L_x^{-1}$ is a surjection since $V'$ is stable and this means that $\xi^{-1} \subset V'$. Hence, in that case, we are through. So, let $V' \to \xi^{-1} \otimes L_x$ be a non-zero homomorphism. As before, this leads to an exact sequence

$$0 \longrightarrow \xi \otimes L_x^{-2} \longrightarrow V' \longrightarrow \xi^{-1} \otimes L_x \longrightarrow 0 .$$

If now $\xi^2 = L_x^2$, we have $\xi \otimes L_x^{-2} = \xi^{-1} \subset V'$ and again, our assertion is proved. We may therefore suppose that $\xi^2 \neq L_x^2$. Since, on the other hand, $x$ is non-Weierstrass, we have $L_x^2 \neq K$ and thus $\xi^{-1} \otimes L_x^2 \neq K \otimes \xi^{-1}$. Hence the planes $f(P(E)_\xi)$, $f(P(E)_{\xi^{-1} \otimes L_x^2})$ in $P$ are distinct by Proposition 6.3. These planes therefore intersect along a line. In other words, there exists exactly one line each in $P(E)_\xi$ and $P(E)_{\xi^{-1} \otimes L_x^2}$ which have the same image in $P$ by $f$. This line in $P(E)_\xi$ is easy to identify. This is in fact the kernel of

$$H^1(X, \xi^{-2}) \longrightarrow H^1(X, \xi^{-2} \otimes L_x^2)$$

given by the canonical section of $L_x^2$. For, if $p \in PH^1(X, \xi^{-2} \otimes L_x)$ is given by the kernel of $H^1(X, \xi^{-2} \otimes L_x) \to H^1(X, \xi^{-2} \otimes L_x^2)$, then $\Lambda_p$ is the line defined above. Moreover, the vector bundle $V_0$ given by $p$ can also be defined as an extension $q$ of $\xi^{-1} \otimes L_x$ by $\xi \otimes L_x^{-2}$. To see this, we have only to prove the existence of a non-zero homomorphism $V_0 \to \xi^{-1} \otimes L_x$. Now the non-zero map $\xi^{-1} \to \xi^{-1} \otimes L_x$ can be lifted to a map $V_0 \to \xi^{-1} \otimes L_x$ by Lemma 3.1 and our assertion is proved. Thus $f(\Lambda_p) = \Lambda_{V_0} = f(\Lambda_q)$. Since by assumption

$$\Lambda_V = \Lambda_{V'} = f(P(E_\xi)) \cap f(P(E_{\xi^{-1} \otimes L_x^2})) ,$$

it follows that $\Lambda_\delta = \Lambda_p$ and $\Lambda_{\delta'} = \Lambda_q$. Hence, by what we have proved above, $V \approx V_0 \approx V'$.

Now let $V, V'$ be two stable bundles of rank 2 and determinant $L_x^{-1}$, $x$ Weierstrass, such that $\Lambda_V = \Lambda_{V'}$. Then we have to show that $V$ is isomorphic to $V'$ or to $\iota^*(V')$. Let $\delta \in PH^1(X, \xi^{-2} \otimes L_x)$, $\xi \in J^1$, define $V$ (Lemma 10.2). Again, as before, there exists a non-zero homomorphism of $V'$ into $\xi \otimes L_x^{-1}$ or $\xi^{-1} \otimes L_x$. In the first case, we see that $V \approx V'$, as in the earlier case. In the second case, we have an exact sequence

$$0 \longrightarrow \xi \otimes L_x^{-2} \longrightarrow V' \longrightarrow \xi^{-1} \otimes L_x \longrightarrow 0 .$$

But since by assumption $L_x^2 = K$, we have $(K \otimes \xi^{-1})^{-1} \subset V'$ and so, $\xi^{-1} \subset \iota^*(V')$. Since $\Lambda_{\iota^*(V')} = \iota^* \Lambda_{V'} = \Lambda_{V'}$, this shows that $\Lambda_V = \Lambda_{\iota^*(V')}$. Moreover $\xi^{-1} \subset V$ and $\xi^{-1} \subset \iota^*(V')$ and it follows that $V \approx \iota^*(V')$.

## 11. A family of divisors on $P(\bigwedge^2 E)$

We shall construct a bundle $E_1'$ on $X \times J^1$ such that the fibre of $E_1'$ at $(x, \xi)$, $x \in X$, $\xi \in J^1$ is identified with $H^1(X, \xi^{-2} \otimes L_x)$. The dual $F_1'$ of $E_1'$ is defined as the $0^{\text{th}}$ direct image on $X \times J^1$ of a family of line bundles on $X$ parametrised by $X \times J^1$ constructed as follows. This family is obtained as the tensor product of the family $(L_x^{-1})_{x \in X}$ given by the diagonal divisor $\Delta$ on $X \times X$ and the family $(K \otimes \xi^2)_{\xi \in J^1}$ given by the line bundle $M$ of § 6. Consider the exact sequence on $X \times J^1$,

$$0 \longrightarrow M \otimes L_\Delta^{-1} \longrightarrow M \longrightarrow M|_{\Delta \times J^1} \longrightarrow 0 ,$$

where $M$ and $L_\Delta$ have been used to denote also the inverse images of $M$ and $L_\Delta$ in $X \times X \times J^1$. Noting that the first direct image of $M \otimes L_\Delta^{-1}$ on $X \times J^1$ (where $X$ here denotes the $1^{\text{st}}$ component in $X \times X \times J^1$) is zero since $H^1(X, K \otimes \xi^2 \otimes L_x^{-1}) = 0$, for all $x \in X$, $\xi \in J^1$, we obtain the following exact sequence of vector bundles on $X \times J^1$

$$0 \longrightarrow F_1' \longrightarrow p_{J1}^*F \longrightarrow M \longrightarrow 0 \,.$$

On dualising, this leads to the exact sequence

$$0 \longrightarrow M^{-1} \longrightarrow p_{J1}^*E \longrightarrow E_1' \longrightarrow 0 \,.$$

This is turn gives rise to the exact sequence [7, Satz 4.1.3]

$$\text{(D)} \qquad 0 \longrightarrow E_1' \otimes M^{-1} \longrightarrow X \times \bigwedge^2 E \longrightarrow \bigwedge^2(E_1') \longrightarrow 0 \,.$$

Thus $P(E_1' \otimes M^{-1})$ gives rise to a divisor on $X \times P(\bigwedge^2 E)$.

*Remark.* The fibre $P(\bigwedge^2 E)_\xi = P(\bigwedge^2 H^1(X, \xi^{-2}))$ can be considered as the space of lines in $PH^1(X, \xi^{-2})$ or, what is the same, as the set of 2-dimensional subspaces of $H^1(X, \xi^{-2})$. For $x \in X$, we have associated above a divisor in $P(\bigwedge^2 E)$. The restriction of this divisor to $P(\bigwedge^2 E)_\xi$ is simply the set of 2-dimensional subspaces of $H^1(X, \xi^{-2})$ which contain the kernel of $H^1(X, \xi^{-2}) \to H^1(X, \xi^{-2} \otimes L_x)$. In other words, this restriction is just the set of lines in $P(H^1(X, \xi^{-2}))$ which pass through the image of $x$ under the mapping $X \to PH^1(X, \xi^{-2})$ given by the linear system $K \otimes \xi^2$.

LEMMA 11.1. *The involution $\iota \times \iota$ on $X \times J^1$ has a natural lift to the line bundle $M$. This induces an action on $\mathcal{T} \otimes \pi^*(M \otimes p_{J1}^*(L_\Theta^{-2}))$, where $\mathcal{T}$ is the tautological hyperplane bundle on $X \times P(\bigwedge^2 E)$ and $\pi$ is the projection $X \times P(\bigwedge^2 E) \to X \times J^1$. Moreover, the divisor $P(E_1' \otimes M^{-1})$ in $X \times P(\bigwedge^2 E)$ is given by a section of $\mathcal{T} \otimes \pi^*(M \otimes p_{J1}^*(L_\Theta^{-2}))$ invariant under the above action.*

PROOF. It is clear that the action $\iota \times \iota$ on $X \times J^1$ lifts naturally to the exact sequence (D). Hence the map $X \times \bigwedge^2 E \to \bigwedge^2(E_1')$ gives rise to an invariant element of

$$H^0(X \times J^1, p_{J1}^*(\bigwedge^2 E)^* \otimes \bigwedge^2 E_1') = H^0(X \times P(\bigwedge^2 E), \mathcal{T} \otimes \pi^*(\bigwedge^2 E_1')) \,.$$

But

$$\bigwedge^2(E_1') = \det(E_1') = M \otimes p_{J1}^*(\det E) = M \otimes p_{J1}^*(L_\Theta^{-2}) \,, \qquad \text{q.e.d.}$$

The fibre at $(x, \xi)$ of $P(E_1' \otimes M^{-1}) = P(E_1')$ consists of the 1-dimensional subspaces of $H^1(X, \xi^{-2} \otimes L_x)$ and hence each point of this fibre can be interpreted as an extension

$$0 \longrightarrow \xi^{-1} \longrightarrow V \longrightarrow \xi \otimes L_x^{-1} \longrightarrow 0 \,,$$

where $V$ is stable by Lemma 10.1. Moreover, by the universal property of $S_1'$ and Proposition 3.1, we have a morphism $P(E_1' \otimes M^{-1}) \to S_1'$ which is surjective by Lemma 10.2.

PROPOSITION 11.1. *The following diagram is commutative*

$$\begin{array}{ccc} P(E_1' \otimes M^{-1}) & \longrightarrow & X \times P(\bigwedge^2 E) \\ \big\downarrow & & \big\downarrow I \times \varphi \\ S_1' & \xrightarrow{\Lambda} & X \times G \end{array}$$

*where the map* $P(E_1' \otimes M^{-1}) \to X \times P(\bigwedge^2 E)$ *is given by the exact sequence* (D), $\varphi$ *is the canonical map* $P(\bigwedge^2 E) \to G$ *induced by the inclusion* $E \subset H^0(J^1, L_\Theta^2)$ *and* $\Lambda$ *is defined by* $\Lambda(V) = ((\det V)^{-1}, \Lambda_V))$. *Moreover,* $\Lambda$ *is a morphism.*

PROOF. The map $P(E' \otimes M^{-1})_{(x,\xi)} \to P(\bigwedge^2 E)$ is obtained by taking the inverse images of 1-dimensional subspaces of $H^1(X, \xi^{-2} \otimes L_x)$ by means of the natural map $H^1(X, \xi^{-2}) \to H^1(X, \xi^{-2} \otimes L_x)$. In other words, the composite of this map with $I \times \varphi$ is simply the map $\delta \mapsto (x, \Lambda_\delta)$, for $\delta \in PH^1(X, \xi^{-2} \otimes L_x)$. Hence the commutativity of the diagram follows from Lemma 10.3. That $\Lambda$ is a morphism now follows from Zariski's main theorem.

PROPOSITION 11.2. *The divisor on* $X \times P(\bigwedge^2 E)$ *which is the sum of* $P(E_1' \otimes M^{-1})$ *and its transform by the automorphism* $\iota \times \mathrm{Id}$ *of* $X \times P(\bigwedge^2 E)$ *is given by an element of* $H^0(X \times P(\bigwedge^2 E), p_X^* K^4 \otimes \mathcal{T}^2)$. *Actually this section is invariant under the natural actions of* $\iota$ *on the bundle* $K^4$ *over* $X$ *and on* $\mathcal{T}^2$ *over* $P(\bigwedge^2 E)$.

PROOF. This follows from Lemma 11.1 and the following

LEMMA 11.2. *Let* $M$ *be the line bundle* $\tau^*(L_T) \otimes \tau^*\sigma^*(L_T) \otimes p_{J^1}^*(L_\Theta^{-2})$ *on* $X \times J^1$ *and* $I$ *the involution on* $X \times J^1$ *defined by*

$$I(x, \xi) = \big(\iota(x), \xi\big)\,, \qquad x \in X,\ \xi \in J^1\,,$$

$\iota$ *being the canonical involution on* $X$. *Then we have*

$$M \otimes I^*(M) \approx p_X^*(K^4) \otimes p_{J^1}^*(L_\Theta^4)\,.$$

*Moreover, the actions of* $\iota$ *on* $X$ *and* $J^1$ *lift naturally to both* $M \otimes I^*M$ *and* $p_X^* K^4 \otimes p_{J^1}^*(L_\Theta^4)$. *The above isomorphism is compatible with these actions.*

PROOF. For $x \in X$, the restriction of $M$ (resp. $I^*(M)$) to $\{x\} \times J^1$ is isomorphic to $Q_x^*(L_\Theta) \otimes L_\Theta^{-2}$ (resp. $Q_{\iota(x)}^*(L_\Theta) \otimes L_\Theta^{-2}$), where $Q_x \colon J^1 \to J^1$ is the map $\xi \mapsto \xi^2 \otimes L_x^{-1}$. Hence the restriction of $M \otimes I^*(M)$ to $\{x\} \times J^1$ is isomorphic to $Q_x^*(L_\Theta) \otimes Q_{\iota(x)}^*(L_\Theta) \otimes L_\Theta^{-4}$, which, by (v), Lemma 6.2 is isomorphic to $L_\Theta^4$. Thus the restriction $M \otimes I^*(M) \otimes p_{J^1}^*(L_\Theta^{-4})$ to each $\{x\} \times J^1$ is trivial. Hence this bundle is the inverse image of a line bundle $L'$ on $X$. To calculate $L'$, note that, for $\xi \in J^1$, the ristriction of $\tau^*(L_T) \otimes \tau^*(\sigma^* L_T)$ to $\{\xi\} \times X$ is isomorphic to $K \otimes \xi^2$ while the restriction of $I^*(\tau^*(L_T) \otimes \tau^*\sigma^*(L_T))$ is isomorphic to $K \otimes (K \otimes \xi^{-1})^2$. Hence $L' \approx K^4$. This completes the proof of the lemma, noting that the above isomorphisms are compatible with the actions of $\iota$ on $X$ and $J^1$ and their lifts.

## 12. A family of quadratic complexes

PROPOSITION 12.1. *Consider the map* $\eta \times \varphi: X \times P(\bigwedge^2 E) \to X/\iota \times G$ *where* $\eta: X \to X/\iota$ *is the natural map and* $\varphi: P(\bigwedge^2 E) \to G$ *the map defined in Proposition* 11.2. *Then the divisor on* $X \times P(\bigwedge^2 E)$ *given in Proposition* 11.2 *is the inverse image of a divisor* $\mathcal{K}$ *on* $X/\iota \times G$. *Moreover,* $\mathcal{K}$ *is given by a section of* $p^*_{X/\iota}(H^4) \otimes p^*_G(H^2)$ *where* $H$ *denotes both the hyperplane bundle on* $X/\iota$ *and the restriction of the hyperplane bundle on* $P(\bigwedge^2 H^0(J^1, L^2_\Theta))$ *to* $G$.

*Remark.* The divisor $\mathcal{K}$ on $X/\iota \times G$ would be the Klein variety we are looking for. This divisor gives a family of quadratic complexes parametrised by $X/\iota$.

PROOF OF PROPOSITION 12.1. We note that, $\varphi^* H = \mathcal{T}$, where $\mathcal{T}$ is the tautological hyperplane bundle on $P(\bigwedge^2 E)$. Hence we have only to show that the above divisor on $X \times P(\bigwedge^2 E)$ is given by an element in the image of the map

$$(\eta \times \varphi)^*: H^0(X/\iota, H^4) \otimes H^0(G, H^2) \longrightarrow H^0(X, K^4) \otimes H^0(P \bigwedge^2 E, \mathcal{T}^2) .$$

Since $(\eta \times \varphi)^* = \eta^* \otimes \varphi^*$, in view of Proposition 11.2, Proposition 12.1 follows from

PROPOSITION 12.2. (i) *The map* $\eta^*: H^0(X/\iota, H^4) \to H^0(X, K^4)$ *is an isomorphism onto the space of invariants of* $H^0(X, K^4)$ *for the action* $\iota$.

(ii) *For* $r = 1,2$, *the map* $\varphi^*: H^0(G, H^r) \to H^0(P(\bigwedge^2 E), \mathcal{T}^r)$ *is an isomorphism onto the space of invariants for the action* $\iota$ *on* $H^0(P(\bigwedge^2 E), \mathcal{T}^r)$.

PROOF. (i) is clear.

(ii) By Proposition 6.3, $J^1/\iota$ is imbedded in $P^* = PH^0(J^1, L^2_\Theta)^*$ as a quartic surface and the exact sequence

$$0 \longrightarrow \bigwedge^2 E \longrightarrow \bigwedge^2 H^0(J^1, L^2_\Theta) \longrightarrow E \otimes L^2_\Theta \longrightarrow 0$$

on $J^1$ is the inverse image of the exact sequence on $P^*$

$$0 \longrightarrow \bigwedge^2 \tilde{E} \longrightarrow \bigwedge^2 H^0(J^1, L^2_\Theta) \longrightarrow \tilde{E} \otimes H \longrightarrow 0 ,$$

where $H$ is the hyperplane bundle on $P^*$ and $\tilde{E}$ is defined by

$$0 \longrightarrow \tilde{E} \longrightarrow H^0(J^1, L^2_\Theta) \longrightarrow H \longrightarrow 0 .$$

These exact sequences give rise to a commutative diagram

$$\begin{array}{ccc} P(\bigwedge^2 E) & \longrightarrow & G \\ \downarrow & \nearrow & \\ P(\bigwedge^2 \tilde{E}) . & & \end{array}$$

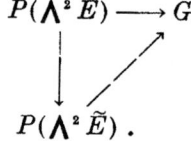

The map $P(\bigwedge^2\tilde{E})\to G$ has projective lines as fibres and hence induces an isomorphism $H^0(G, H^r)\to H^0(P(\bigwedge^2\tilde{E}), \mathcal{T}^r)$. Now using the canonical isomorphisms

$$H^0(P(\bigwedge^2 E), \mathcal{T}^r) \approx H^0(J^1, S^r(\bigwedge^2 E)^*)$$

and

$$H^0(P(\bigwedge^2\tilde{E}), \mathcal{T}^r) \approx H^0(P^*, S^r(\bigwedge^2 \tilde{E})^*) ,$$

we see that $\varphi^*$ can be replaced by the map

$$H^0(P^*, S^r(\bigwedge^2 \tilde{E})^*) \longrightarrow H^0(J^1, S^r(\bigwedge^2 E)^*)$$

induced by the imbedding of $J^1/\iota$ in $P^*$. To prove (ii), Proposition 12.2, we have only to show that the image consists precisely of the invariants of $\iota$ in $H^0(J^1, S^r(\bigwedge^2 E)^*)$. Now this space of invariants is simply $H^0(J^1/\iota, S^r(\bigwedge^2 E)^*)$. Since $J^1/\iota$ is a quartic surface, we have an exact sequence

$$0 \longrightarrow H^{-4}\otimes S^r(\bigwedge^2 \tilde{E})^* \longrightarrow S^r(\bigwedge^2 \tilde{E})^* \longrightarrow S^r(\bigwedge^2 E^*) \mid J^1/\iota \longrightarrow 0 .$$

From this we deduce the exact cohomology sequence

$$\begin{aligned} 0 &\longrightarrow H^0(P^*, H^{-4}\otimes S^r(\bigwedge^2 \tilde{E})^*) \longrightarrow H^0(P^*, S^r(\bigwedge^2 \tilde{E})^*) \\ &\longrightarrow H^0(J^1/\iota, S^r(\bigwedge^2 E^*)) \longrightarrow H^1(P^*, H^{-4}\otimes S^r(\bigwedge^2 \tilde{E})^*) \longrightarrow \cdots . \end{aligned}$$

Now Proposition 12.2 follows from

LEMMA 12.1. *For* $r = 1,2$, $H^i(P^*, H^{-4}\otimes S^r(\bigwedge^2 \tilde{E})^*) = 0$ *for all* $i \geqq 0$.

PROOF. This can be checked, for instance, using the theorem of Bott [3]. If $P^*$ is identified with the quotient $U(4)/U(3)\times U(1)$, we see that the bundle $H$ is induced by the character on $U(3)\times U(1)$ defined by $(A, \lambda)\mapsto \lambda^{-1}$. Also, the bundle $\tilde{E}$ is defined by the representation $(A,\lambda)\mapsto {}^tA^{-1}$. Hence the highest weight (with respect to the natural lexicographic order) of the representation yielding $H^{-4}\otimes S^r(\bigwedge^2 \tilde{E})^*$ is

$$\begin{pmatrix} \lambda_1 & & & \\ & \lambda_2 & & 0 \\ & & \lambda_3 & \\ 0 & & & \lambda_4 \end{pmatrix} \longmapsto r(\lambda_1+\lambda_2)+4\lambda_4 .$$

Hence the highest weight plus the half sum of positive roots of $U(4)$ is

$$\left(r+\frac{3}{2}\right)\lambda_1 + \left(r+\frac{1}{2}\right)\lambda_2 - \frac{1}{2}\lambda_3 + \frac{5}{2}\lambda_4 .$$

Since this is not a regular element for $r = 1,2$, the lemma follows from Bott's theorem [3].

## 13. Completion of the proof of Theorem 4

The family $\mathcal{K}$ of quadratic complexes defined in § 12 leads to the commutative diagram

$$\begin{array}{ccc} & PH^0(X/\iota, H^4)^* & \\ {}^{q}\nearrow & & \searrow \\ X/\iota & \xrightarrow{\quad k \quad} & PH^0(G, H^2) \end{array}$$

where $q$ is the normal quartic imbedding of the projective line $X/\iota$ and the map $PH^0(X/\iota, H^4)^* \to PH^0(G, H^2)$ is the rational map induced by the element in $H^0(X/\iota, H^4) \otimes H^0(G, H^2)$ given in Proposition 12.1. If $x \in X/\iota$ is weierstrassian, then $k(x)$ is actually represented by the square of a section of $H$ over $G$. For, from Lemma 11.1, we see that the divisor in $P(\bigwedge^2 E)$ corresponding to $x$ is given by a section of $\mathcal{T} \otimes \pi^*(M|_{\{x\}\times J^1} \otimes L_\Theta^{-2})$. But $M|_{\{x\}\times J^1} \otimes L_\Theta^{-2}$ is trivial by (iv), Lemma 6.2. Since $H^0(P(\bigwedge^2 E), \mathcal{T}) \approx H^0(G, H)$ by Proposition 8.2, our assertion is proved. Thus corresponding to each Weierstrass point of $X$, we have defined an element $l_\alpha$ of

$$PH^0(G, H) = P \bigwedge^2 H^0(J^1, L_\Theta^2)^* .$$

First, we shall note that any five of the $l_\alpha$'s, say $\{l_\alpha\}_{1 \leq \alpha \leq 5}$ are in general position. If they were not, one easily sees that there exists $\gamma \in G$ such that the hyperplanes determined by $\{l_\alpha\}_{1\leq\alpha\leq 5}$ pass through $\gamma$. Let $d \in P(\bigwedge^2 E)$ such that $\varphi(d) = \gamma$. This means that considered as a line in $P(E)_\xi$, $d$ contains the points in $PH^1(X, \xi^{-2}) = P(E)_\xi$ represented by the kernels of the maps

$$H^1(X, \xi^{-2}) \longrightarrow H^1(X, \xi^{-2} \otimes L_{x_\alpha}) , \qquad 1 \leqq \alpha \leqq 5 .$$

In other words, the inverse image of the line $d$ by the map $X \to PH^1(X, \xi^{-2})$ given by the linear system $K \otimes \xi^2$, contains $(x_\alpha)_{1\leq\alpha\leq 5}$ which is impossible, since $K \otimes \xi^2$ is of degree 4.

LEMMA 13.1. *The map $PH^0(X/\iota, H^4)^* \to PH^0(G, H^2)$ defined above is injective. The set of elements $\{l_\alpha\}$ forms a system of mutually apolar complexes for $G$.*

PROOF. That $\{l_\alpha\}$ forms a system of mutually apolar complexes for $G$ follows from the above remark and Lemma 8.1. Since, in particular, the image of $PH^0(X/\iota, H^4)^*$ contains the squares $l_\alpha^2$, which span a four dimensional projective subspace of $PH^0(G, H^2)$, the injectivity of the map $PH^0(X/\iota, H^4)^* \to PH^0(G, H^2)$ is proved.

Thus we have a rational normal quartic curve of quadratic complexes passing through $l_\alpha^2$, parametrised by $X/\iota$. The associated divisor on $X/\iota \times G$ is just the divisor $\mathcal{K}$ defined in Proposition 12.1. Then it is clear from the commutative diagram (see Proposition 11.1)

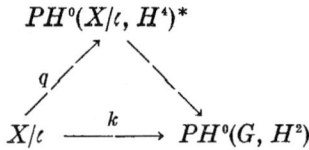

$$\begin{array}{ccc} P(E_1' \otimes M^{-1}) & \longrightarrow & X \times P(\bigwedge^2 E) \\ \downarrow & & \downarrow \\ S_1'/\iota & \xrightarrow{\Lambda} & X/\iota \times G \end{array}$$

that $\Lambda(S_1'/i) \subset \mathcal{K}$. Since by Lemma 8.3, $\mathcal{K}$ is normal and irreducible and $\Lambda$ is injective by Proposition 10.1, and $\dim \mathcal{K} = \dim S_1'/\iota$, it follows that $\Lambda$ is an isomorphism of $S_1'/\iota$ onto $\mathcal{K}$.

Finally the proof of the Theorem 4 will be completed by

LEMMA 13.2. *Let $x$ be non-Weierstrass point in $X$. Then the quadratic complex $k(\eta x)$ is generic and the singular locus of $k(\eta x)$ in $PH^0(J^1, L_\Theta^2)$ is $J/\iota$ (with the natural imbedding). In particular, $J/\iota$ is a Kummer surface. If $j \in J$, then the two planes into which the quadratic cone at $j \oplus j^{-1} \in P$ breaks are the images of $P(E)_{j\otimes L_x}$ and $P(E)_{j^{-1}\otimes L_x}$ under the map $f: P(E) \to P$.*

PROOF. That $k(\eta x)$ is generic follows from Lemma 8.3. Since the singular locus of $k(\eta x)$ is a quartic surface and $J/\iota$ is also a quartic surface, it is enough to prove that $J/\iota$ is contained in the singular locus of $k(\eta x)$, i.e., the cone of lines belonging to $k(\eta x)$ through a point of $J/\iota$ contains a plane. Let then $j \oplus j^{-1}$ be a point of $P, j \in J$. We claim that the image of the plane $P(E)_{j\otimes L_x}$ in $P$ under the map $P(E) \to P$ is contained in the cone of lines through $j \oplus j^{-1}$. The image of $x$ in $P(E)_{j\otimes L_x}$ under the map

$$X \longrightarrow PH^1(X, j^{-2} \otimes L_x^{-2})$$

given by the linear system $K \otimes j^2 \otimes L_x^2$ represents the bundle $j \oplus j^{-1}$ (Lemma 5.2). Thus the image of $P(E)_{j\otimes L_x}$ is a plane through $j \oplus j^{-1}$ and by Remark preceding Lemma 11.1, the lines in this plane through $j \oplus j^{-1}$ belong to the quadratic complex. This completes the proof of Lemma 13.2.

## 14. Jacobian and lines on intersection of two quadrics

THEOREM 5. *Let $X$ be a non-singular complete irreducible curve genus 2. Let $x \in X$ be a non-Weierstrass point of $X$. Let $k(x)$ be the quadratic complex associated to $x$. Then the space of lines in $P(\bigwedge^2 H^0(J^1, L_\Theta^2))$ contained in $k(x)$ is canonically isomorphic to the jacobian $J$ of $X$.*

PROOF. Consider the subset of $P \times P^*$ consisting of points $(p, \pi)$ where $p$ is a point of $P$ belonging to the hyperplane $\pi$ in $P$. To each such $(p, \pi)$, we could associate the line in $P(\bigwedge^2 H^0(J^1, L_\Theta^2))$ consisting of all lines in $P$ lying in $\pi$ and passing through $p$. This is clearly contained in $G$ and the above assignment identifies the space of lines in $G$ with this divisor. With this identification, the space of lines in $k(x)$, is simply the set $(p, \pi) \in P \times P^*$ such that $p$ is a point of the singular locus of $k(x)$ and $\pi$ is any one of the planes into which the quadric cone through $p$ breaks. Now consider the map $J \to P \times P^*$ given by $j \mapsto (j \oplus j^{-1}, f(P(E)_{j\otimes L_x}))$. From Proposition 6.3 it is easy to see that this is an imbedding, since $x$ is not a Weierstrass point. By Lemma 13.2, it follows that the image is the space of lines in $k(x)$.

*Remark.* Through every point $V$ of $k(x)$ there pass four lines lying in $k(x)$ and hence $V$ determines four points in $J$. From our identification of $S_{1,x}$ with $k(x)$, it is easy to see that these are the line bundles $j$ such that $H^0(X, \text{Hom}(V, j)) \neq 0$ (see Remark following Lemma 10.3). It follows that a vector bundle $V$ of rank 2 and determinant $L_x^{-1}$ is characterised by the line bundles $j$ of degree zero such that there exists a non-zero homomorphism of $V$ into $j$.

TATA INSTITUTE OF FUNDAMENTAL RESEARCH

REFERENCES

[1] M. F. ATIYAH, *Complex analytic connections in fibre bundles*, Trans. Amer. Math. Soc. **75** (1953), 428-443.
[2] H. F. BAKER, Principles of Geometry, Vol. 4, Cambridge, 1940.
[3] R. BOTT, *Homogeneous vector bundles*, Ann. of Math. **66** (1957), 203-248.
[4] H. GRAUERT, *Eine Theorem der analytischen Garbentheorie und die Modulräume Komplexer Strukturen*, I.H.E.S. Publ. Math. 5 (1960).
[5] A. GROTHENDIECK, Eléments de Géométrie Algébrique, Ch. III, I.H.E.S. Publ. Math., 17 (1963).
[6] ———, *Sur quelques points d'algèbre homologique*, Tohoku Math. J. **9** (1957), 119-221.
[7] F. HIRZEBRUCH, Neue topologische Methoden in der algebraichen Geometrie, Springer-Verlag, 1962.
[8] F. KLEIN, *Zur Theorie der Linencomplexe des ersten und zweiten Grades*, Math. Ann. **2** (1870), 198-226.
[9] K. KODAIRA, *On Kähler varieties of restricted type*, Ann. of Math. **60** (1954), 28-48.
[10] S. LANG, Abelian Varieties, Interscience (1959).
[11] D. MUMFORD, *On the equations defining abelian varieties*, Invent. Math. **1** (1966), 287-354.
[12] M. S. NARASIMHAN, and C. S. SESHADRI, *Stable and unitary vector bundles on a compact Riemann surface*, Ann. of Math. **82** (1965), 540-567.
[13] P. E. NEWSTEAD, *Stable bundles of rank 2 and odd degree over a curve of genus 2*, Topology **7** (1968), 205-215.
[14] M. RAYNAUD, Familles de fibrés vectoriels sur une surface de Riemann, Séminaire Bourbaki, Exposé 316 (1966).
[15] C. S. SESHADRI, *Space of unitary vector bundles on a compact Riemann surface*, Ann. of Math. **85** (1967), 303-336.

(Received December 5, 1967)

# An analogue of the Borel-Weil-Bott theorem for hermitian symmetric pairs of non-compact type*

By M. S. NARASIMHAN and K. OKAMOTO**

### Introduction

Let $G$ be a connected non-compact semi-simple Lie group admitting a finite dimensional faithful representation. Let $K$ be a maximal compact subgroup of $G$. We suppose that $G/K$ is hermitian symmetric.

When $G = SL(2, \mathbf{R})$, V. Bargmann [3] constructed the discrete series for $G$ on spaces of square-integrable holomorphic functions on $G/K$ (which is isomorphic to the unit disc) and in the general case Harish-Chandra (cf. [5]) constructed part of the discrete series for $G$ in a similar way. (See also [6].) However, in the general case, this method does not yield all the discrete series. In analogy with the Borel-Weil-Bott theorem [4, 8(a)], it was suggested in [9] that all the discrete classes (which were obtained by Harish-Chandra in [7(f)]) might be realized on spaces of square-integrable harmonic forms of type $(0, q)$ ("square-integrable $\bar{\partial}$-cohomology spaces") with coefficients in holomorphic vector bundles on $G/K$ arising from finite dimensional irreducible unitary representations of $K$. (When $q = 0$, we have the case which was considered in [5].) We prove in this paper that most of the discrete classes are obtained in this way.

We proceed to describe the results of this paper in more detail. Let $\mathfrak{g}$ and $\mathfrak{k}$ be the Lie algebras of $G$ and $K$ respectively. Let $\mathfrak{h}$ be a Cartan subalgebra of $\mathfrak{k}$. Then $\mathfrak{h}$ is also a Cartan subalgebra of $\mathfrak{g}$. Choose an ordering on the roots of $(\mathfrak{g}^C, \mathfrak{h}^C)$ compatible with the complex structure on $G/K$. For an irreducible unitary representation $\tau_\Lambda$ of $K$ with the highest weight $\Lambda$, we denote by $E_\Lambda$ the holomorphic vector bundle on $G/K$ associated to the contragredient representation. Let $H_2^{0,q}(E_\Lambda)$ denote the Hilbert space of square-integrable harmonic forms of type $(0, q)$ with coefficients in $E_\Lambda$ (these are the "square-integrable $\bar{\partial}$-cohomology spaces attached to $E_\Lambda$" defined in [11]). The unitary representation $\pi_\Lambda^q$ of $G$ on $H_2^{0,q}(E_\Lambda)$ decomposes into a finite number of irreducible representations each of which belongs to the discrete series for $G$ (Proposition 3.1). Let $\rho$ denote the half sum of positive roots of $(\mathfrak{g}^C, \mathfrak{h}^C)$ and let $q_\Lambda$

* Supported in part by National Science Foundation Grant GP-7952X.
** The latter author was partially supported also by the Matsunaga Science Foundation and by the Sakkokai Foundation.

denote the number of non-compact positive roots $\alpha$ such that $\langle \Lambda + \rho, \alpha \rangle > 0$. We prove in Theorem 1, § 6, that the alternating sum of the characters of $\pi^q_\Lambda$ is equal to $(-1)^{q_\Lambda}\Theta_{\omega(\Lambda+\rho)^*}$, where $\omega(\Lambda + \rho)^*$ denotes the class of the discrete series contragredient to the one associated with the form $\Lambda + \rho$ and $\Theta_{\omega(\Lambda+\rho)^*}$ its character (see Lemma 2.4). This shows that, in any case, all the discrete classes are realized on a subspace of $H^{0,q}_2(E_\Lambda)$ for some $\Lambda$ and some $q$.

Define the constant $c$ as in Corollary 1 to Theorem 2, § 7. We prove that

$$H^{0,q}_2(E_\Lambda) = 0 \qquad \text{if } q \neq q_\Lambda$$

provided that $|\langle \Lambda+\rho, \alpha\rangle| > c$ for any non-compact positive root $\alpha$. (A slightly more general result is proved in Theorem 2, § 7.) The alternating sum formula, together with the vanishing theorem, shows that, if $\Lambda$ satisfies the above condition, the class of the representation $\pi^{q_\Lambda}_\Lambda$ on $H^{0,q_\Lambda}_2(E_\Lambda)$ is precisely $\omega(\Lambda + \rho)^*$ (Theorem 3, § 8).

We now briefly outline the proof of the alternating sum formula. Let $L^{0,q}_2(E_\Lambda)$ be the space of square-integrable forms of type $(0, q)$ with coefficients in $E_\Lambda$. The group $G$ acts, *via* the induced representation, on this space and let $L^{0,q}_2(E_\Lambda)_d$ be the discrete part of this space. The representation of $G$ on $L^{0,q}_2(E_\Lambda)_d$ decomposes into a finite sum of discrete classes (Lemma 1.2). The laplacian $\square$ has only finitely many eigenvalues on $L^{0,q}_2(E_\Lambda)$ and the sum of the eigenspaces coincides with $L^{0,q}_2(E_\Lambda)_d$ (Proposition 5.1). We then have a complex

$$\cdots \longrightarrow L^{0,q}_2(E_\Lambda)_d \xrightarrow{\bar\partial} L^{0,q+1}_2(E_\Lambda)_d \longrightarrow \cdots$$

and the cohomology of this complex at the $q^{\text{th}}$ stage is isomorphic to $H^{0,q}_2(E_\Lambda)$ (Proposition 5.2). If $\varphi$ is a $K$-finite $C^\infty$-function with compact support on $G$, $\varphi$ defines an operator $T^q_\varphi$ on $L^{0,q}_2(E_\Lambda)_d$ and $\{T^q_\varphi\}$ form an endomorphism of finite rank of the above complex. So it suffices to calculate the alternating sum of the traces of the operators $T^q_\varphi$ on $L^{0,q}_2(E_\Lambda)_d$. The alternating sum of the traces of $T^q_\varphi$ is given by an integral over the union of conjugacy classes of a compact Cartan subgroup. The integral involves the character of the representation of $K$ and the integral is evaluated by using Weyl's character formula for irreducible representations of $K$ and coincides with Harish-Chandra's character formula for the discrete series for $G$. This gives the required formula (Theorem 1).

We are grateful to Professors A. Borel and Harish-Chandra for helpful discussions.

## 1. Definitions and notation

Let $G$ be a connected non-compact semi-simple Lie group which has a finite dimensional faithful representation and $K$ a maximal compact subgroup of $G$.

We denote by $\mathfrak{g}$ the Lie algebra of left invariant vector fields on $G$ and by $\mathfrak{k}$ the subalgebra of $\mathfrak{g}$ corresponding to the subgroup $K$. Let $\mathfrak{g}^C$ be the complexification of $\mathfrak{g}$. We put

$$\mathfrak{p} = \{Y \in \mathfrak{g};\ B(X, Y) = 0 \text{ for all } X \in \mathfrak{k}\}$$

where $B$ denotes the Killing form of the Lie algebra $\mathfrak{g}^C$. Then we have

$$\mathfrak{g} = \mathfrak{k} + \mathfrak{p}\,, \quad \mathfrak{k} \cap \mathfrak{p} = 0\,, \quad [\mathfrak{p}, \mathfrak{p}] \subset \mathfrak{k}\,, \quad [\mathfrak{k}, \mathfrak{p}] \subset \mathfrak{p}\,.$$

For any subset $\mathfrak{m}$ of $\mathfrak{g}^C$ we denote by $\mathfrak{m}^C$ the complex subspace of $\mathfrak{g}^C$ spanned by $\mathfrak{m}$.

Now we assume that the homogeneous space $G/K$ has a $G$-invariant complex structure. Then there exist abelian subalgebras $\mathfrak{p}_+$ and $\mathfrak{p}_-$ of $\mathfrak{g}^C$ such that

$$\mathfrak{p}^C = \mathfrak{p}_+ \oplus \mathfrak{p}_-\,, \quad \mathfrak{p}_+ \cap \mathfrak{p}_- = 0\,, \quad \mathfrak{p}_+ = \bar{\mathfrak{p}}_- \text{ and } [\mathfrak{k}^C, \mathfrak{p}_+] \subset \mathfrak{p}_+\,.$$

Moreover, the space of all anti-holomorphic tangent vectors at the origin $K$ can be canonically identified with $\mathfrak{p}_+$ (cf. [11]). Since we assumed that $G$ has a finite dimensional faithful representation, we can consider a complex form $G^C$ of $G$. We denote by $K^C$ (resp. $P_+$, $P_-$) the complex analytic subgroup of $G^C$ corresponding to $\mathfrak{k}^C$ (resp. $\mathfrak{p}_+$, $\mathfrak{p}_-$). Put $U = K^C P_+$. Then $U$ is a complex analytic subgroup of $G^C$ and $P_+$ is a normal subgroup of $U$. Consider the complex homogeneous space $G^C/U$. Then $G/K$ can be canonically identified with the open submanifold of $G^C/U$ which is the $G$-orbit of $U$ in $G^C/U$. We put $\mathfrak{g}_u = \mathfrak{k} + \sqrt{-1}\mathfrak{p}$. Then $\mathfrak{g}_u$ is a compact real form of $\mathfrak{g}^C$. If $\theta$ denotes the conjugation of $\mathfrak{g}^C$ with respect to $\mathfrak{g}_u$, we write $X^* = -\theta X$ for $X \in \mathfrak{g}^C$. We define the inner product $(\,,)$ in $\mathfrak{g}^C$ by

$$(X, Y) = B(X, Y^*) \qquad (X, Y \in \mathfrak{g}^C)\,.$$

We notice that $\mathfrak{k}$ contains a Cartan subalgebra $\mathfrak{h}$ of $\mathfrak{g}$. Let $\Sigma$ be the set of all non-zero roots of $\mathfrak{g}^C$ with respect to $\mathfrak{h}^C$. Then for each $\alpha \in \Sigma$ we can choose an eigenvector $X_\alpha$ corresponding to the root $\alpha$ such that $(X_\alpha, X_\alpha) = 1$. A root $\alpha \in \Sigma$ is called compact or non-compact according as $X_\alpha \in \mathfrak{k}^C$ or $X_\alpha \in \mathfrak{p}^C$. As is easily seen, there exists a unique subset $P_n \subset \Sigma$ such that $\mathfrak{p}_+ = \sum_{\alpha \in P_n} CX_\alpha$. Moreover, it is easy to see that we can introduce a linear order on $\Sigma$ such that $P_n$ is contained in the set $P$ of all positive roots. We denote by $P_k$ the set of all compact positive roots. Then we have $P = P_k \cup P_n$. For any linear form $\lambda$ on $\mathfrak{h}^C$, we shall denote by $H_\lambda$ the element of $\mathfrak{h}^C$ such that $B(H_\lambda, H) = \lambda(H)$ for all $H \in \mathfrak{h}^C$. For any pair $(\lambda, \mu)$ of linear forms on $\mathfrak{h}^C$, we put $\langle\lambda, \mu\rangle = \lambda(H_\mu)$. Let $\mathcal{F}$ be the set of all integral forms on $\mathfrak{h}^C$, i.e., the set of linear forms $\lambda$ such that $2\langle\lambda, \alpha\rangle/\langle\alpha, \alpha\rangle$ is an integer for every root $\alpha$. We put

$$\mathcal{F}' = \{\lambda \in \mathcal{F};\ \langle\lambda + \rho, \alpha\rangle \neq 0 \quad (\alpha \in \Sigma)\}\ ,$$
$$\mathcal{F}'_0 = \{\lambda \in \mathcal{F}';\ \langle\lambda + \rho, \alpha\rangle > 0 \quad (\alpha \in P_k)\}$$

where $\rho = \frac{1}{2}\sum_{\alpha \in P} \alpha$. Then it is easy to see that $\lambda \in \mathcal{F}'$ is in $\mathcal{F}'_0$ if and only if $\lambda$ is dominant with respect to $P_k$ (i.e., $\langle\lambda, \alpha\rangle \geqq 0$ for all $\alpha \in P_k$).

Henceforward we assume, for convenience, that $G^C$ is simply connected. For any $\Lambda \in \mathcal{F}'_0$ we denote by $\tau_\Lambda$ the irreducible unitary representation of $K$ with highest weight $\Lambda$ on the representation space $V_\Lambda$. Then $\tau_\Lambda$ is uniquely extended to a holomorphic representation of $U$ which is trivial on $P_+$. Let $\tau^*_\Lambda$ be the representation contragredient to $\tau_\Lambda$ on the dual space $V^*_\Lambda$ of $V_\Lambda$. We regard $G^C$ as a principal fibre bundle over the base space $G^C/U$ with structure group $U$. Let $\tilde{E}_\Lambda$ denote the holomorphic vector bundle on $G^C/U$ associated with the holomorphic representation $\tau^*_\Lambda$ of $U$. We denote by $E_\Lambda$ the restriction of $\tilde{E}_\Lambda$ to the open submanifold $G/K$ of $G^C/U$. Since the structure group of $E_\Lambda$ is reduced to $\tau^*_\Lambda(K)$ which is a subgroup of the unitary group, we have a canonical hermitian metric on $E_\Lambda$. We notice that the restriction of $(\ ,\ )$ on $\mathfrak{p}$ is a $K$-invariant inner product. Using the canonical identification of $\mathfrak{p}$ with tangent space of $G/K$ at $K$, we can introduce a $G$-invariant hermitian metric on $G/K$. Thus we have constructed a hermitian vector bundle $E_\Lambda$ on $G/K$ for each $\Lambda \in \mathcal{F}'_0$.

Fix an orthonormal base $(v_1, \cdots, v_r)$ of $V_\Lambda$. Let $(v^*_1, \cdots, v^*_r)$ be the dual base of $V^*_\Lambda$. Select an orthonormal base $(H_1, \cdots, H_l)$ of $\mathfrak{h}^C$ where $l = \operatorname{rank} G$. Let $P_n = \{\alpha_1, \cdots, \alpha_m\}$ and $P_k = \{\alpha_{m+1}, \cdots, \alpha_{m+k}\}$. To simplify the notation, we put

$$z_i = \begin{cases} X_{\alpha_i} & (i = 1, \cdots, m + k) \\ H_{i-m-k} & (i = m + k + 1, \cdots, m + k + l)\ , \\ X^*_{\alpha_{n-i+1}} & (i = m + k + l + 1, \cdots, n) \end{cases}$$

where $n = \dim G$. Then we have

$$z_i \in \begin{cases} \mathfrak{p}_+ & (i = 1, \cdots, m)\ , \\ \mathfrak{k}^C & (i = m + 1, \cdots, n - m)\ , \\ \mathfrak{h}^C & (i = m + k + 1, \cdots, m + k + l)\ , \\ \mathfrak{p}_- & (i = n - m + 1, \cdots, n)\ . \end{cases}$$

We put

$$C^q(G, V^*_\Lambda) = \bigwedge\nolimits^q \mathfrak{p}_- \otimes C^\infty(G) \otimes V^*_\Lambda\ ,$$
$$L^q_2(G; V^*_\Lambda) = \bigwedge\nolimits^q \mathfrak{p}_- \otimes L_2(G) \otimes V^*_\Lambda\ .$$

where $C^\infty(G)$ (resp. $L_2(G)$) denotes the space of $C^\infty$ (resp. square-integrable) complex valued functions on $G$. Let $C^{0,q}(E_\Lambda)$ (resp. $L^{0,q}_2(E_\Lambda)$) denote the space of all $C^\infty$ (resp. square-integrable) differential forms of type $(0, q)$ with coefficients in $E_\Lambda$. We define an injective mapping

$$\eta\colon C^{0,q}(E_\Lambda) \longrightarrow C^q(G,\ V_\Lambda^*)$$

by putting, for $\varphi \in C^{0,q}(E_\Lambda)$,

$$\eta(\varphi) = \sum_{1\leq i_1<\cdots<i_q\leq m} \sum_{j=1}^r z_{i_1}^* \wedge \cdots \wedge z_{i_q}^* \varphi(z_{i_1}, \cdots, z_{i_q})(v_j) \otimes v_j^* \ .$$

For any $u, u' \in L_2^q(G,\ V_\Lambda^*)$ we set

$$(u,\ u') = \sum_{1\leq i_1<\cdots<i_q\leq m} \sum_{j=1}^r \int_G f_{i_1\cdots i_q, j}(g)\overline{f'_{i_1\cdots i_q, j}(g)}dg$$

where

$$u = \sum_{1\leq i_1<\cdots<i_q\leq m} \sum_{j=1}^r z_{i_1}^* \wedge \cdots \wedge z_{i_q}^* \otimes f_{i_1\cdots i_q, j} \otimes v_j^* \ ,$$
$$u' = \sum_{1\leq i_1<\cdots<i_q\leq m} \sum_{j=1}^r z_{i_1}^* \wedge \cdots \wedge z_{iq}^* \otimes f_{i_1\cdots i_q, j} \otimes v_j^* \ .$$

Then $(\,,)$ defines an inner product in $L_2^q(G,\ V_\Lambda^*)$. Let $r$ (resp. $l$) be the right (resp. left) regular representation of $G$ on $L_2(G)$. We denote by $\mathrm{Ad}_+^q$ (resp. $\mathrm{Ad}_-^q$) the representation of $K$ on $\bigwedge^q \mathfrak{p}_+$ (resp $\bigwedge^q \mathfrak{p}_-$) induced by the adjoint action of $K$ on $\mathfrak{p}_+$ (resp. $\mathfrak{p}_-$). Put

$$C^q(G,\ V_\Lambda^*)^0 = \{u \in C^q(G,\ V_\Lambda^*);\ (\mathrm{Ad}_-^q \otimes r \otimes \tau_\Lambda^*)(k)u = u\ (k \in K)\} \ ,$$
$$L_2^q(G,\ V_\Lambda^*)^0 = \{u \in L_2^q(G,\ V_\Lambda^*);\ (\mathrm{Ad}_-^q \otimes r \otimes \tau_\Lambda^*)(k)u = u\ \ (k \in K)\} \ .$$

Then $\eta$ maps $C^{0,q}(E_\Lambda)$ isomorphically onto $C^q(G,\ V_\Lambda^*)^0$. Moreover, it is easy to see that the mapping $\eta$ induces an isometry of $L_2^{0,q}(E_\Lambda)$ onto $L_2^q(G,\ V_\Lambda^*)^0$. We define a representation $\nu$ of $\mathfrak{g}^C$ on $C^\infty(G)$ by

$$\nu(X)f = Xf \qquad (X \in \mathfrak{g}^C,\ f \in C^\infty(G)) \ .$$

Let $\mathfrak{G}$ be the univeral enveloping algebra of $\mathfrak{g}^C$. Then $\nu$ defines a representation of $\mathfrak{G}$ on $C^\infty(G)$. We denote by $\bar\partial\colon C^{0,q}(E_\Lambda) \to C^{0,q+1}(E_\Lambda)$ the usual operator of exterior derivation with respect to conjugate complex variables. Let $\vartheta$ be the formal adjoint operator of $\bar\partial$. We put

$$\square = (\bar\partial\vartheta + \vartheta\bar\partial) \ .$$

LEMMA 1.1. *Let $\Omega$ be the Casimir operator of $G$. Then we have*

$$\square u = \frac{1}{2}\{\langle\Lambda + 2\rho,\ \Lambda\rangle - 1 \otimes \nu(\Omega) \otimes 1\}u \ ,$$

*if $u \in C^q(G,\ V_\Lambda^*)^0$.*

For a proof, see [11, Th. 4.1].

Let $\mathcal{E}_K$ be the set of all equivalence classes of irreducible representations of $K$. If $\pi$ is a unitary representation of $G$ or $K$ we denote by $[\pi]$ the equivalence class which contains $\pi$. For any unitary representation $\sigma$ of $K$ and $d \in \mathcal{E}_K$ we denote by $(\sigma : \mathfrak{d})$ the multiplicity with which $\mathfrak{d}$ occurs in $\sigma$. Wewr ite $([\sigma] : \mathfrak{d}) =$

$(\sigma: \mathfrak{d})$. For any unitary representation $\pi$ of $G$ let $\pi \mid K$ denote the restriction of $\pi$ to $K$. Then $(\pi \mid K: \mathfrak{d})(\mathfrak{d} \in \mathcal{E}_K)$ depends only on the equivalence class $[\pi]$. We also write $([\pi] \mid K: \mathfrak{d})$ instead of $(\pi \mid K: \mathfrak{d})$. For any unitary representation $\pi$ of $G$ on the representation space $\mathfrak{H}$, let $\mathfrak{H}_d$ be the smallest closed invariant subspace of $\mathfrak{H}$ which contains every irreducible closed invariant subspace of $\mathfrak{H}$. We denote by $\pi_d$ the restriction of $\pi$ to $\mathfrak{H}_d$. We call $\pi_d$ (resp. $\mathfrak{H}_d$) the discrete part of $\pi$ (resp. $\mathfrak{H}$). For any unitary representation $\sigma$ and a finite dimensional representation $\tau$ of $K$, we put

$$([\sigma]: [\tau]) = \sum_{\mathfrak{d} \in \mathcal{E}_K} ([\sigma]: \mathfrak{d})([\tau]: \mathfrak{d}) .$$

For any unitary representation $\pi$ of $G$ or $K$, let $\pi^*$ be the representation contragredient to $\pi$. It is obvious that $[\pi^*]$ depends only on $[\pi]$. In this case we write $[\pi]^*$ instead of $[\pi^*]$. For any $g \in G$ we denote by $\tilde{\pi}^q_\Lambda(g)$ the canonical action of $g$ on $L_2^{0,q}(E_\Lambda)$. Then $\tilde{\pi}^q_\Lambda$ is the unitary representation of $G$ which is induced by the representation $\mathrm{Ad}^q_- \otimes \tau^*_\Lambda$ of $K$. Let $\mathcal{E}$ be the set of equivalence classes of irreducible unitary representations of $G$. We call $\omega \in \mathcal{E}$ a discrete class if $\omega$ contains a representation equivalent to the right (or left) regular representation restricted to a closed invariant subspace of $L_2(G)$. We denote by $\mathcal{E}_d$ the set of discrete classes in $\mathcal{E}$ (cf. [7 (c), § 37]). $\mathcal{E}_d$ is called the discrete series for $G$.

LEMMA 1.2. *Put* $\tau'^q_\Lambda = \mathrm{Ad}^q_+ \otimes \tau_\Lambda$. *Then we have*

$$[(\tilde{\pi}^q_\Lambda)_d] = \bigoplus_{\omega \in \mathcal{E}_d} (\omega \mid K: [\tau'^q_\Lambda])\omega^* .$$

(The sum is actually finite).

PROOF. The lemma is an immediate consequence of a modified version of [10, Lem. 1], in view of the definition of the discrete part of $\pi^q_\Lambda$. This modification is necessary as "the induced representation" was defined in [10] using the action of $K$ on the left, while in this paper we have used the action on the right. The finiteness follows from Lemma 2.1 below.

## 2. Some results of Harish-Chandra

We denote by $\mathcal{C}(G)$ the Schwartz space of $G$ (for definition, see [7 (f), § 9]). Define the convolution $f*g$ $(f, g \in \mathcal{C}(G))$ as usual so that

$$(f*g)(x) = \int_G f(y)g(y^{-1}x)dy \qquad (x \in G) .$$

Then, every $\varphi \in \mathcal{C}(G)$ defines a continuous linear mapping $\mathcal{C}(G) \to \mathcal{C}(G)$ by $f \mapsto \varphi*g$. For any $\alpha \in C^\infty(K)$ and $f \in C^\infty(G)$ we define

$$\alpha*f = \int_K \alpha(k)l(k)fdk ,$$

and

$$f * \alpha = \int_K \alpha(k^{-1}) r(k) f dk .$$

For any $\mathfrak{d} \in \mathcal{E}_K$ we put $\alpha_{\mathfrak{d}} = d(\mathfrak{d})\bar{\chi}_{\mathfrak{d}}$ where $d(\mathfrak{d})$ is the degree of $\mathfrak{d}$ and $\chi_{\mathfrak{d}}$ is the character of $\mathfrak{d}$. For any finite subset $F$ of $\mathcal{E}_K$, define

$$\alpha_F = \sum_{\mathfrak{d} \in F} \alpha_{\mathfrak{d}} .$$

Then if $f \in C^\infty(G)$ is $K$-finite there exists a finite $F$ of $\mathcal{E}_K$ such that $\alpha_F * f * \alpha_F = f$.

LEMMA 2.1. *For any finite dimensional representation $\sigma$ of $K$, define*

$$\mathcal{E}_d(\sigma) = \{\omega \in \mathcal{E}_d ; (\omega \mid K : [\sigma]) \neq 0\} .$$

*Then $\mathcal{E}_d(\sigma)$ is a finite set.*

This follows from [7 (f), Lem. 70]. (See also the remark before Lemma 3.1 below.)

Let $\mathfrak{Z}$ be the center of the universal enveloping algebra $\mathfrak{G}$ of $\mathfrak{g}$ and let $\Theta$ be a distribution on $G$. Then $\Theta$ is said to be $\mathfrak{Z}$-finite, if the space of all distributions of the form $z\Theta (z \in \mathfrak{Z})$ has finite dimension. Then an invariant and $\mathfrak{Z}$-finite distribution is actually a locally summable function which is analytic on the open set $G'$ of all regular elements of $G$ (see [7 (d), Th. 2]).

LEMMA 2.2. *Let $f \in L_2(G)$ be both $K$-finite and $\mathfrak{Z}$-finite. Then $f \in \mathcal{C}(G)$.*

For a proof, see [7 (f), Cor. 1 to Lem. 65].

Let $E$ denote the orthogonal projection of $L_2(G)$ onto $L_2(G)_d$. For any $\omega \in \mathcal{E}_d$ we denote by $d(\omega)$ the formal degree of $\omega$ (for definition, see [7 (c), § 3]) and by $\Theta_\omega$ the character of $\omega$. Then $\Theta_\omega$ is an invariant eigen-distribution which is tempered. Hence $\Theta_\omega$ is a locally summable function on $G$ which is analytic on $G'$.

LEMMA 2.3. *For any $f \in \mathcal{C}(G)$ put*

$$^\circ f(x) = \sum_{\omega \in \mathcal{E}_d} d(\omega)\Theta_{\omega^*}(r(x)f) \qquad (x \in G) .$$

*Then $^\circ f$ is a continuous function on $G$ and $^\circ f = Ef$.*

For a proof, see [7 (f), Cor. 3 to Lem. 69 and a remark on p. 100].

Let $W_G$ be the subgroup of the Weyl group with respect to $(\mathfrak{g}^C, \mathfrak{h}^C)$, which is generated by reflections with respect to the compact roots. Put

$$\mathfrak{h}' = \{H \in \mathfrak{h}; \alpha(H) \neq 0 \quad (\alpha \in \Sigma)\} .$$

For any $\lambda \in \mathcal{F}'$ we denote by $\Theta_{\lambda+\rho}$ the unique tempered invariant eigen-distribution corresponding to $\lambda + \rho$ which was defined in [7 (c), Th. 3], (see also its introduction and [7 (f), Th. 7 and Lem. 73]). Then we have

$$\Delta(\exp H)\Theta_{\lambda+\rho}(\exp H) = \sum_{s \in W_G} \varepsilon(s) e^{s(\lambda+\rho)(H)} \qquad (H \in \mathfrak{h})$$

where

$$\Delta(\exp H) = \prod_{\alpha \in P} (e^{\alpha(H)/2} - e^{-\alpha(H)/2}).$$

For any $\lambda \in \mathcal{F}'$ we put

$$\varepsilon(\lambda + \rho) = \operatorname{sign}\left(\prod_{\alpha \in P} \langle \lambda + \rho, \alpha \rangle\right).$$

Then $\mathcal{E}_d$ is canonically parametrized by $\mathcal{F}'_0$ as follows.

LEMMA 2.4. *For any $\lambda \in \mathcal{F}'_0$, there exists a unique element $\omega(\lambda + \rho) \in \mathcal{E}_d$ such that $\Theta_{\omega(\lambda+\rho)} = (-1)^m \varepsilon(\lambda + \rho)\Theta_{\lambda+\rho}$ where $m = -\frac{1}{2}\dim G/K$. Moreover, the mapping $\mathcal{F}'_0 \to \mathcal{E}_d$ given by $\lambda \mapsto \omega(\lambda + \rho)$ is bijective.*

This follows from [7 (f), Th. 16], noting that any $\lambda \in \mathcal{F}'_0$ is dominant with respect to $P_k$.

For any $\omega \in \mathcal{E}_d$ and $f \in \mathcal{C}(G)$, put

$$\Phi_\omega(h) = \Delta(h)\Theta_\omega(h)$$

and

$$F_f(h) = \Delta(h)\int_G f(xhx^{-1})dx$$

for $\mathfrak{h} \in T' = \exp \mathfrak{h}'$.

Then $\Phi_\omega$ can be extended to an analytic function on $T$ (cf. [7 (d), Lem. 31]) and $F_f$ is a $C^\infty$-function on $T'$ and the restriction of $F_f$ to each connected component of $T'$ can be extended as a continuous function to its closure. (See [7 (f), Lem. 27 and its proof].)

LEMMA 2.5. *Let $f$ be a $\mathfrak{Z}$-finite function in $\mathcal{C}(G)$. Then*

$$\Theta_\omega(f) = (-1)^m [W_G]^{-1} \int_T F_f(h)\Phi_\omega(h)dh \qquad (\omega \in \mathcal{E}_d),$$

*where $[W_G]$ is the order of $W_G$ and $m = \frac{1}{2}\dim G/K$.*

For a proof, see [7 (f), Lem. 79].

LEMMA 2.6. *Let $\omega$, $\omega'$ be two elements in $\mathcal{E}_d$. Then*

$$\Theta_\omega(f) = \begin{cases} d(\omega)^{-1} f(1) & \text{if } \omega' = \omega^*, \\ 0 & \text{otherwise} \end{cases}$$

*for $f \in \mathcal{C}_{\omega'}(G)$. (For the definition of $\mathcal{C}_{\omega'}(G)$, see the next section.)*

For a proof, see [7 (f), Lem. 81].

### 3. Some lemmas

In this section we derive some immediate consequences of the results stated in the previous section.

For any $\omega \in \mathcal{E}_d$, we define a closed subspace $L_2(G)_\omega$ of $L_2(G)$ as follows. Fix $\pi \in \omega$ and let $\mathfrak{H}$ be the representation space of $\pi$. We denote by $L_2(G)_\omega$ the smallest closed subspace of $L_2(G)$ containing all functions $f$ of the form

$$f(x) = (\varphi, \pi(x)\psi) \qquad (x \in G)$$

where $\varphi, \psi \in \mathfrak{H}$. Then it is clear that this definition is independent of the particular choice of $\pi$ and that $L_2(G)_\omega$ is stable under both left and right regular representations. Put

$$\mathcal{C}_\omega(G) = L_2(G)_\omega \cap \mathcal{C}(G) .$$

Then $\mathcal{C}_\omega(G)$ is dense in $L_2(G)_\omega$ (see [7 (f), Th. 14]). We denote by $\mathfrak{H}_\mathfrak{d} (\mathfrak{d} \in \mathcal{E}_K)$ the subspace of $\mathfrak{H}$ consisting of all vectors which transform according to $\mathfrak{d}$ under the restriction of $\pi$ to $K$. Choose an orthonormal base $\psi_i (i \in J)$ for $\mathfrak{H}$ such that every $\psi_i$ lies in $\mathfrak{H}_\mathfrak{d}$ for some $\mathfrak{d} \in \mathcal{E}_K$. Let $J_\mathfrak{d}$ be the set of all $i$ such that $\psi_i \in \mathfrak{H}_\mathfrak{d}$. Then $J_\mathfrak{d}$ is a finite set (cf. [7 (c), Th. 4]). It follows from the definition of $\Theta_\omega$ that

$$\Theta_\omega(f) = \sum_{i \in J} \int_G f(x)(\psi_i, \pi(x)\psi_i)dx \qquad (f \in C_c^\infty(G)).$$

where $C_c^\infty(G)$ denotes the space of complex valued $C^\infty$-functions on $G$ with compact support. For any $i \in J$ we put

$$f_i(x) = (\psi_i, \pi(x)\psi_i) \qquad (x \in G) .$$

Then in view of Lemma 2.2, $f_i \in \mathcal{C}(G)$. It follows that $f_i \in \mathcal{C}_\omega(G)$. For any $\omega \in \mathcal{E}_d$ and $\mathfrak{d} \in \mathcal{E}_K$, we define

$$\Theta_{\omega,\mathfrak{d}}(f) = \Theta_\omega(f * \alpha_\mathfrak{d}) \qquad (f \in \mathcal{C}(G)) .$$

For any finite subset $F$ of $\mathcal{E}_K$ we put

$$\Theta_{\omega,F} = \sum_{\mathfrak{d} \in F} \Theta_{\omega,\mathfrak{d}} .$$

Then it is easy to see that

$$\Theta_{\omega,F} = \sum_{i \in J_F} f_i .$$

where

$$J_F = \sum_{\mathfrak{d} \in F} J_\mathfrak{d} .$$

It follows that $\Theta_{\omega,F} \in \mathcal{C}_\omega(G)$. For any $\varphi \in \mathcal{C}(G)$ put

$$\check{\varphi}(g) = \varphi(g^{-1}) \qquad (g \in G) .$$

Then it is clear that $\check{\varphi} \in \mathcal{C}(G)$. Hence, we have $\check{\varphi} * f \in \mathcal{C}(G)$. Therefore, for any $\varphi \in \mathcal{C}(G)$ we can define

$$(\Theta_\omega * \varphi)(f) = \Theta_\omega(f * \check{\varphi}) \qquad (f \in \mathcal{C}(G)) .$$

Then it is easy to see that

$$\Theta_\omega * \varphi = \Theta_{\omega,F} * \varphi \qquad (\omega \in \mathcal{E}_d)$$

if $\varphi$ is a $K$-finite function in $\mathcal{C}(G)$ such that $\varphi = \alpha_{F^*} * \varphi$ where $F$ is a finite subset of $\mathcal{E}_K$.

*Remark.* Let $\mathfrak{d} \in \mathcal{E}_K$. Then $\Theta_{\omega,\mathfrak{d}} \neq 0$ if and only if $(\omega \mid K: \mathfrak{d}) \neq 0$.

LEMMA 3.1. *Let $\varphi$ be a $K$-finite function in $\mathcal{C}(G)$. Then*

$$\Theta_\omega * \varphi \in \mathcal{C}_\omega(G) .$$

*Moreover, we have*

$$\Theta_\omega * \varphi = 0$$

*except for finitely many $\omega \in \mathcal{E}_d$.*

PROOF. Choose a finite subset $F$ of $\mathcal{E}_K$ such that $\varphi = \alpha_{F^*} * \varphi$. Then we know that $\Theta_{\omega,F} * \varphi = \Theta_\omega * \varphi$. Notice that $\Theta_{\omega,F} \in \mathcal{C}_\omega(G)$. It follows that $\Theta_{\omega,F} * \varphi \in L_2(G)_\omega$. Hence

$$\Theta_\omega * \varphi = \Theta_{\omega,F} * \varphi \in L_2(G)_\omega \cap \mathcal{C}(G) = \mathcal{C}_\omega(G) .$$

The second assertion follows from Lemma 2.1 and the above remark.

LEMMA 3.2. *Let $\varphi$ be a $K$-finite function in $\mathcal{C}(G)$. Then ${}^\circ\varphi$ is a $\mathfrak{Z}$-finite function which belongs to $\mathcal{C}(G)$. Moreover, we have*

$$\Theta_\omega({}^\circ\varphi) = \Theta_\omega(\varphi) \qquad (\omega \in \mathcal{E}_d) .$$

PROOF. First, we notice that

$$\Theta_\omega(x^{-1}) = \overline{\Theta_\omega(x)} = \Theta_{\omega^*}(x) \qquad (x \in G')$$

for any $\omega \in \mathcal{E}_d$. This implies that $\Theta_\omega(\check{f}) = \Theta_{\omega^*}(f)$ for all $f \in \mathcal{C}(G)$. Next, remark that $\Theta_{\omega^*}(r(x)\varphi) = (\Theta_\omega * \varphi)(x)$ $(x \in G)$. Then, from the definition of ${}^\circ\varphi$ (see Lemma 2.3), we have

$${}^\circ\varphi = \sum\nolimits_{\omega \in \mathcal{E}_d} d(\omega)\Theta_\omega * \varphi .$$

Since every element of $\mathcal{C}_\omega(G)$ is an eigenfunction of $\mathfrak{Z}$, it follows from Lemma 3.1 that ${}^\circ\varphi$ is a $\mathfrak{Z}$-finite function and that ${}^\circ\varphi \in \mathcal{C}(G)$. Moreover, from Lemma 2.6 we have

$$\begin{aligned}\Theta_\omega({}^\circ\varphi) &= \sum\nolimits_{\omega' \in \mathcal{E}_d} d(\omega')\Theta_\omega(\Theta_{\omega'} * \varphi) \\ &= (\Theta_{\omega^*} * \varphi)(1) = \Theta_{\omega^*}(\check{\varphi}) = \Theta_\omega(\varphi) .\end{aligned}$$

This completes the proof of the lemma.

LEMMA 3.3. *For any $f \in C_c^\infty(G)$, we have*

$$\int_G \varphi(g)\big(l(g)Ef\big)(x)dg = \int_G {}^\circ\varphi(xy^{-1})f(y)dy .$$

(For definition of ${}^\circ\varphi$ see Lemma 2.3.)

PROOF. For any $f \in \mathcal{C}(G)$ and $x \in G$ we have

$$\begin{aligned}(r(x)\check{f})^{\vee}(g) &= (r(x)\check{f})(g^{-1}) \\ &= f(x^{-1}g) = r(x^{-1})f(x^{-1}gx) \qquad (g \in G)\,.\end{aligned}$$

Since $\Theta_\omega$ is an invariant distribution we see that

$$\Theta_{\omega^*}(r(x)\check{f}) = \Theta_\omega((r(x)\check{f})^{\vee}) = \Theta_\omega(r(x^{-1})f)\,.$$

Noticing that $\omega \mapsto \omega^*$ is a bijection of $\mathcal{E}_d$ onto itself it follows from Lemma 2.3 that $Ef = (E\check{f})^{\vee}$. It is easy to see that

$$\int_G \varphi(g)l(g)f(x)dg = (r(x)\bar{f}, \check{\varphi})$$

where we put $\bar{f}(g) = \overline{f(g)}$ $(g \in G)$. We remark that $\overline{Ef} = E\bar{f}$ (if $f \in C_c^\infty(G)$). Since $E$ and $r(x)$ $(x \in G)$ commute, we have

$$\begin{aligned}&\int_G \varphi(g)(l(g)Ef)(x)dg \\ &= (r(x)\overline{Ef}, \check{\varphi}) = (Er(x)\bar{f}, \check{\varphi}) = (r(x)\bar{f}, E\check{\varphi}) = (r(x)\bar{f}, (E\varphi)^{\vee}) \\ &= \int_G {}^0\varphi(g)l(g)f(x)sg = \int_G {}^0\varphi(xy^{-1})f(y)dy\,.\end{aligned}$$

This completes the proof of the lemma.

## 4. The space of square-integrable harmonic forms

Let $H_2^{0,q}(E_\Lambda)$ denote the subspace of $L_2^{0,q}(E_\Lambda)$ consisting of elements $\varphi \in L_2^{0,q}(E_\Lambda)$ such that $\Box\varphi = 0$, $\Box\varphi$ being formed in the sense of distributions. Then $H_2^{0,q}(E_\Lambda)$ is a closed subspace of $L_2^{0,q}(E_\Lambda)$. We see that $\varphi$ in $L_2^{0,q}(E_\Lambda)$ belongs to $H_2^{0,q}(E_\Lambda)$ if and only if $\bar\partial\varphi = 0$ and $\vartheta\varphi = 0$ (in the sense of distributions). This is a consequence of the completeness of the metric on $G/K$ (see [1], [11]). For the interpretation of $H_2^{0,q}(E_\Lambda)$ as square-integrable $\bar\partial$-cohomology spaces see [11].) We also denote by $H_2^{0,q}(E_\Lambda)$ the subspace of $L_2^q(G, V_\Lambda^*)^0$ corresponding to the space of harmonic forms under the identification mapping $\eta$.

We denote by $\pi_\Lambda^q$ the unitary representation of $G$ on $H_2^{0,q}(E_\Lambda)$.

PROPOSITION 4.1. *Put*

$$\mathcal{E}_d(\Lambda) = \{\omega \in \mathcal{E}_d,\ \chi_\omega(\Omega) = \langle\Lambda + 2\rho, \Lambda\rangle\}$$

*where $\chi_\omega$ denotes the infinitesimal character of $\omega$, and $\Omega$ the Casimir operator. Then we have*

$$\pi_\Lambda^q = \bigoplus_{\omega \in \mathcal{E}_d(\Lambda)} (\omega \mid K: [\tau_\Lambda^q])\omega^* \qquad (\textit{finite sum})\,.$$

PROOF. Let $\mathcal{E}$ be the set of all equivalence classes of irreducible unitary representations of $G$. For any real number $c$, define $\mathcal{E}_c = \{\omega \in \mathcal{E};\ \chi_\omega(\Omega) = c\}$.

Then it is known (cf. [10]) that $\mathcal{E}_c - \mathcal{E}_d$ is of measure zero with respect to the Plancherel measure for $G$ if rank $G/K = 1$. In [10] it was conjectured that this would hold in general. Recently, it was proved by Harish-Chandra that this is true in general. We use this fact in the following. First, we notice that [10, Th. 1] holds for any finite dimensional representation $\sigma$ of $K$ if we define $(\omega \mid K: [\sigma])$ as in § 1. Next we remark that the canonical action of $G$ on $L_2^q(G, V_\Lambda^*)^0$ is nothing but the induced representation of $G$ generated by $(\tau_\Lambda^q)^*$. Consider $\mathfrak{A} = \mathbf{C}[\Omega]$ (the algebra of all polynomials in $\Omega$) and define $\chi \in \mathrm{Hom}(\mathfrak{A}, \mathbf{C})$ by

$$\chi(\Omega) = \langle \Lambda + 2\rho, \Lambda \rangle .$$

Then the assertion of the proposition is an immediate consequence of [10, Th. 1] and Lemma 1.1.

*Remark* 1. In view of Proposition 4.1, $\pi_\Lambda^q$ is a finite sum of irreducible representations and hence the character of $\pi_\Lambda^q$, denoted by Trace $\pi_\Lambda^q$, is defined as a distribution on $G$.

*Remark* 2. Proposition 4.1 implies in particular that $\pi_\Lambda^q = (\pi_\Lambda^q)_d$. Namely, the reduction of $\pi_\Lambda^q$ into irreducible components has no "continuous spectrum".

## 5. The discrete part of the induced representation and eigenforms of the laplacian

Let $\tilde{\Box}^q$ denote the maximal (or weak) realization of $\Box$ on $L_2^q(G, V_\Lambda^*)^0$: the domain $D(\tilde{\Box}^q)$ of $\tilde{\Box}^q$ consists of all $\varphi \in L_2^q(G, V_\Lambda^*)^0$ for which $\Box\varphi$, formed in the sense of distributions, belongs to $L_2^q(G, V_\Lambda^*)^0$ and $\tilde{\Box}^q\varphi = \Box\varphi$ for $\varphi \in D(\tilde{\Box}^q)$. Since the metric on $G/K$ is complete we may apply the results of [1, § 6]. These results show that if $\varphi \in D(\tilde{\Box}^q)$ then $\bar{\partial}\varphi \in L_2^{q+1}(G, V_\Lambda^*)^0$, and $\vartheta\varphi \in L_2^{q-1}(G, V_\Lambda^*)^0$, and that $\tilde{\Box}^q$ is symmetric. It follows easily that $\tilde{\Box}^q$ is in fact self-adjoint.

PROPOSITION 5.1. *The operator $\tilde{\Box}^q$ has only finitely many eigenvalues in $L_2^q(G, V_\Lambda^*)^0$ and the discrete part $L_2^q(G, V_\Lambda^*)_d^0$ coincides with the sum of all eigenspaces of $\tilde{\Box}^q$.*

PROOF. An argument similar to the proof of Proposition 4.1 shows that each eigenspace is a finite sum of discrete classes. Let $\mathfrak{H}$ be an irreducible component of $L_2^q(G, V_\Lambda^*)_d^0$ and $\pi$ denote the restriction of $\tilde{\pi}_\Lambda^q$ to $\mathfrak{H}$. We shall show that $\mathfrak{H}$ is contained in an eigenspace. Suppose $\varphi \in \mathfrak{H}$ is a $C^\infty$-vector for $\pi$. Then $\varphi \in C^q(G, V_\Lambda^*)^0$ and $\pi(\Omega)\varphi = (1 \otimes \nu(\Omega) \otimes 1)\varphi$, where $\Omega$ is the Casimir operator. Since $\pi$ is irreducible, $\pi(\Omega)\varphi = \chi_\pi(\Omega)\varphi$ where $\chi_\pi$ is the infinitesimal character of $\pi$. Using Lemma 1.1 we have $\Box\varphi = \lambda\varphi$ where $\lambda$ is a scalar depending on $\pi$ but not on $\varphi$. This implies that $\varphi$ belongs to $D(\tilde{\Box}^q)$ and $\tilde{\Box}^q\varphi =$

$\lambda\varphi$. Now let $\varphi_0$ be an arbitrary element in $\mathfrak{H}$. Then, as is well known, there exists a sequence $\{\varphi_n\}$ of $C^\infty$-vectors for $\pi$ such that $\varphi_n \to \varphi_0$ in $\mathfrak{H}$ (as $n \to \infty$). By what has been proved above, $\varphi_n \in D(\widetilde{\Box}^q)$ and $\widetilde{\Box}^q\varphi_n = \lambda\varphi_n$. So, $\widetilde{\Box}^q\varphi_n \to \lambda\varphi_0$. Since $\widetilde{\Box}^q$ is a closed operator it follows that $\varphi_0 \in D(\widetilde{\Box}^q)$ and $\widetilde{\Box}^q\varphi_0 = \lambda\varphi_0$. Hence $\mathfrak{H}$ is contained in the eigenspace corresponding to $\lambda$. It follows from Lemma 1.2 that $\widetilde{\Box}^q$ has only finitely many eigenvalues and $L_2^q(G, V_\Lambda^*)_d^0$ coincides with the sum of all eigenspaces of $\widetilde{\Box}^q$.

Put $V^q = L_2^q(G, V_\Lambda^*)_d^0$. From Proposition 5.1 and a remark made earlier, we see that $\bar\partial$ maps $V^q$ into $V^{q+1}$.

PROPOSITION 5.2. *The cohomology space $\mathcal{H}^q$ of the complex*

$$\cdots \longrightarrow V^q \xrightarrow{\bar\partial} V^{q+1} \longrightarrow \cdots$$

*at the $q^{\text{th}}$ stage is naturally isomorphic to the space $H_2^{0,q}(E_\Lambda)$ of square integrable harmonic forms of type* $(0, q)$.

PROOF. We use Proposition 5.1 in the proof. Let $\varphi \in H_2^{0,q}(E_\Lambda)$. Then $\bar\partial\varphi = 0$ and $\vartheta\varphi = 0$. So we have a natural map $H_2^{0,q}(E_\Lambda) \to \mathcal{H}^q$. This map is injective. For if $\varphi = \bar\partial\varphi'$, $\varphi' \in V^{q-1}$, then $(\varphi, \varphi) = (\varphi, \bar\partial\varphi') = (\vartheta\varphi, \varphi') = 0$ (using [1, § 6]), so that $\varphi = 0$. On the other hand, let $\alpha \in V^q$ with $\bar\partial\alpha = 0$. Let $H\alpha$ be the projection of $\alpha$ onto $H_2^{0,q}(E_\Lambda)$ with respect to the decomposition

$$V^q = \bigoplus_i H_{\lambda_i}^q ,$$

where $H_{\lambda_i}^q$ denotes the eigenspace corresponding to the eigenvalue $\lambda_i$, $\lambda_i$ being distinct. Then $\alpha - H\alpha = \sum \alpha_i$, $\alpha_i \in H_{\lambda_i}^q$, $\lambda_i \neq 0$. Since $\bar\partial(\alpha - H\alpha) = 0$, we have $\sum \bar\partial\alpha_i = 0$. Since $\bar\partial\alpha_i \in H_{\lambda_i}^{q+1}$, it follows that $\partial\alpha_i = 0$ for $i$. Since $\Box\alpha_i = \lambda_i\alpha_i$ and $\bar\partial\alpha_i = 0$, we have $\alpha_i = \bar\partial(\lambda_i^{-1}\vartheta\alpha_i)$. Since $\Box\vartheta\alpha_i = \vartheta\Box\alpha_i = \lambda_i\vartheta\alpha_i$, we have $\lambda_i^{-1}\vartheta\alpha_i \in V^{q-1}$, so that the $\alpha - H\alpha$ is a coboundary in this complex. Hence, the natural map $H_2^{0,q}(E_\Lambda) \to \mathcal{H}^q$ is also surjective.

## 6. The alternating sum formula

In this section we fix $\Lambda \in \mathcal{F}_0'$ once for all and use the notation in §1 ~ §3.

Consider the representation $T^q$ of $G$ on $L_2^q(G, V_\Lambda^*)$ defined by $T^q = 1 \otimes l \otimes 1$, where $l$ denotes the left regular representation of $G$. Then it is easy to see that the canonical action $\tilde\pi_\Lambda^q(g)$ of $g \in G$ on $L_2^{0,q}(E_\Lambda)$ goes over, by means of the identification map $\eta$, into $T_g^q$ on $L_2^q(G, V_\Lambda^*)^0$. Fix any $K$-finite function $\varphi \in C_c^\infty(G)$. Define

$$T_\varphi^q = \int_G \varphi(g) T_g^q dg .$$

We remark that the orthogonal projection $L_2^q(G, V_\Lambda^*) \to L_2^q(G, V_\Lambda^*)_d$ is given by $1 \otimes E \otimes 1$ where $E$ is the orthogonal projection of $L_2(G)$ onto $L_2(G)_d$. Put

$$E_0 = \int_K \mathrm{Ad}^q_-(k) \otimes r(k) \otimes \tau^*_\Lambda(k) dk \,.$$

Then $E_0$ gives the orthogonal projection $L^q_2(G, V^*_\Lambda) \to L^q_2(G, V^*_\Lambda)^0$. It is clear that $L^q_2(G, V^*_\Lambda)_d$ is stable under $E_0$ and that the restriction of $E_0$ to $L^q_2(G, V^*_\Lambda)_d$ gives the orthogonal projection onto $L^q_2(G, V^*_\Lambda)^0_d$. It follows that $E_0 \circ (1 \otimes E \otimes 1)$ is the orthogonal projection of $L^q_2(G, V^*_\Lambda)$ onto $L^q_2(G, V^*_\Lambda)^0_d$.

PROPOSITION 6.1. *The composition* $\widetilde{K}^q_\varphi = T^q_\varphi \circ E_0 \circ (1 \otimes E \otimes 1)$ *is an operator of finite rank from* $L^q_2(G, V^*_\Lambda)$ *into itself and coincides with an integral operator with an* End $(\bigwedge^q \mathfrak{p}_- \otimes V^*_\Lambda)$*-valued* $C^\infty$ *kernel function* $K^q_\varphi$ *which is given by*

$$K^q_\varphi(x, y) = \int_K {}^\circ\varphi(xky^{-1})\, \mathrm{Ad}^q_-(k) \otimes \tau^*_\Lambda(k) dk$$

*for* $(x, y) \in G \times G$. *Moreover,* $\int_G \mathrm{Trace}\, K^q_\varphi(x, x) dx$ *exists and coincides with the trace of the operator* $\widetilde{K}^q_\varphi$.

PROOF. Put

$$C^q_c(G, V^*_\Lambda) = \bigwedge^q \mathfrak{p}_- \otimes C^\infty_c(G) \otimes V^*_\Lambda \,.$$

Any $u \in C^q_c(G, V^*_\Lambda)$ can be identified with a $(\bigwedge^q \mathfrak{p}_- \otimes V^*_\Lambda)$ valued $C^\infty$-function on $G$ with compact support. Then, from Lemma 3.3, we have

$$\begin{aligned}
&\int_G K^q_\varphi(x, y) w(y) dy \\
&= \int_G dg \int_K \{\mathrm{Ad}^q_-(k) \otimes r(k)\varphi(g) l(g) E \otimes \tau^*_\Lambda(k)\} w(x) dk \\
&= \int_G dg \int_K \{\mathrm{Ad}^q_-(k) \otimes \varphi(g) l(g) r(k) E \otimes \tau^*_\Lambda(k)\} w(x) dk \\
&= \{T^q \circ E_0 \circ (1 \otimes E \otimes 1)\} w(x)
\end{aligned}$$

for any $x \in G$ and $w \in C^q_c(G, V^*_\Lambda)$. Since we assumed that $\varphi$ is $K$-finite, there exists a finite subset $F$ of $\mathcal{E}_K$ such that $\varphi = \alpha_F * \varphi * \alpha_F$. Put $T^q_{\alpha_F} = \int_K \alpha_F(k) T^q_k dk$. Then it is easy to see that $T^q_{\alpha_F} T^q_\varphi T^q_{\alpha_F} = T^q_{(\alpha_F * \varphi * \alpha_F)} = T^q_\varphi$. We notice that the canonical action of $g \in G$ on $L^{0,q}_2(E_\Lambda)_d$ goes over, by means of the identification mapping $\eta$, into the restriction of $T^q_g$ to $L^q_2(G, V^*_\Lambda)^0_d$. Moreover, it is obvious that $T^q_{\alpha_F}$ defines a projection operator of $L^q_2(G, V^*_\Lambda)$ onto the subspace of $L^q_2(G, V^*_\Lambda)^0_d$, spanned by all vectors which transform according to some $\mathfrak{d} \in F$. Let $P$ be the operator $T^q_{\alpha_F} \circ E_0 \circ (1 \otimes E \otimes 1)$ and let $R(P)$ denote its range. We then have a diagram

$$L^q_2(G, V^*_\Lambda) \xrightarrow{P} R(P) \xrightarrow{T^q_\varphi} L^q_2(G, V^*_\Lambda)^0_d \xrightarrow{T^q_{\alpha_F}} R(P) \,.$$

By Lemma 1.2 and Lemma 2.1, $R(P)$ is finite dimensional. It follows that $\widetilde{K}^q_\varphi$

is an operator of finite rank.

The operator $P$ is the orthogonal projection of $L_2^q(G, V_\Lambda^*)$ onto $R(P)$. So the operator $\tilde{K}_\varphi^q$ is zero on the orthogonal complement of $R(P)$. Hence there exisits a finite number of elements $\{u_i\}_{1\leq i\leq r_1}$, $\{v_j\}_{1\leq j\leq r_2}$ in $R(P)$ such that

$$\tilde{K}_\varphi^q w = \sum_{i,j}(v_j, w)u_i\,, \qquad \text{for } w\in L_2^q(G, V_\Lambda^*)\,.$$

Since $R(P)\subset L_2^q(G, V_\Lambda^*)_d^0$ it follows from Proposition 5.1 that $u_i, v_j \in C^q(G, V_\Lambda^*)\cap L_2^q(G, V_\Lambda^*)$. So there exists an End $(\bigwedge^q\mathfrak{p}_-\otimes V_\Lambda^*)$-valued $C^\infty$-function $B$ on $G\times G$ such that

$$\tilde{K}_\varphi^q w(x) = \int_G B(x, y)w(y)dy\,, \qquad \text{for } w\in L_2^q(G, V_\Lambda^*)\,.$$

We have

$$\tilde{K}_\varphi^q w(x) = \int_G K_\varphi(x, y)w(y)dy\,, \qquad \text{for } w\in C_c^\infty(G, V_\Lambda^*)\,,$$

as was proved at the beginning of the proof of this proposition. Hence $K_\varphi^q(x,y) = B(x, y)$ $(x, y\in G)$.

To prove the last part we note that

$$\text{Trace } K_\varphi^q(x, x) = \text{Trace } B(x, x) = \sum_{i,j}(v_j(x), u_i(x))\,.$$

So $\int_G \text{Trace } K_\varphi^q(x, x)dx$ exists and

$$\text{Trace } \tilde{K}_\varphi^q = \sum_{i,j}(v_j, u_i) = \sum_{i,j}\int_G (v_j(x), u_i(x))dx$$
$$= \int_G \text{Trace } K_\varphi^q(x, x)dx\,.$$

This completes the proof of Proposition 6.1.

PROPOSITION 6.2. *Let $q_\Lambda$ be the number of the non-compact positive roots $\alpha$ such that $\langle\Lambda+\rho, \alpha\rangle > 0$. Then we have*

$$\sum_{q=0}^m(-1)^q \text{Trace } \tilde{K}_\varphi^q = (-1)^{q_\Lambda}\Theta_{\omega(\Lambda+\rho)^*}(\varphi)\,.$$

(For the definition of the right hand side, see Lemma 2.4).

PROOF. By Proposition 6.1, we have

$$\text{Trace } \tilde{K}_\varphi^q = \int_G \text{Trace } K_\varphi^q(x, x)dx\,.$$

We put

$$A = \sum_{q=0}^m(-1)^q \text{Trace } \tilde{K}_\varphi^q\,.$$

Then, by Proposition 6.1 we have

$$A = \sum_{q=0}^m(-1)^q\int_G dx\int_K \text{Trace Ad}_-^q(k)^\circ\varphi(xkx^{-1})\overline{\chi_\Lambda(k)}dk$$

where $\chi_\Lambda$ is the character of the irreducible representation with highest weight $\Lambda$. On the other hand one has

$$\sum_{q=0}^{m} (-1)^q \text{ Trace } \mathrm{Ad}_-^q (k) = \det (1 - \mathrm{Ad}_-^1 (k))$$

so that

$$A = \int_G dx \int_K \det (1 - \mathrm{Ad}_-^1 (k))^\circ \varphi(xkx^{-1}) \overline{\chi_\Lambda(k)} dk .$$

We now use Weyl's integral formula: for $f \in C^\infty(K)$,

$$\int_K f(k)dk = \frac{1}{[W_G]} \int_T |\Delta_k(h)|^2 dh \int_K f(khk^{-1})dk$$

where

$$\Delta_k(h) = \prod_{\alpha \in P_k} (e^{\alpha(H)/2} - e^{-\alpha(H)/2})$$

($|\Delta_k(h)|^2$ is well defined on $T$). We then obtain

$$A = \frac{1}{[W_G]} \int_G dx \int_{T \times K} \det (1 - \mathrm{Ad}_-^1 (h))^\circ \varphi(xkhk^{-1}x^{-1}) \overline{\chi_\Lambda(h)} \, |\Delta_k(h)|^2 dk .$$

We have

$$\chi_\Lambda(h) = \sum_{s \in W_G} \frac{\varepsilon(s) e^{s(\Lambda + \rho_k)(H)}}{\Delta_k(h)} \qquad \text{(Weyl's character formula)}$$

where $\rho_k$ is the half sum of compact positive roots. Using this formula and that $s\rho_n = \rho_n$ ($s \in W_G$) where $\rho_n$ is the half sum of non-compact positive roots, we obtain

$$\chi_\Lambda(h) = \prod_{\alpha \in P_n} (1 - e^{-\alpha(H)}) \sum_{s \in W_G} \frac{\varepsilon(s) e^{s(\Lambda + \rho)(H)}}{\Delta(h)} .$$

By Lemma 2.4, and observing that $(-1)^m \varepsilon(\Lambda + \rho) = (-1)^{q_\Lambda}$ we have

$$\Theta_{\omega(\Lambda+\rho)}(h) = (-1)^{q_\Lambda} \sum_{s \in W_G} \frac{\varepsilon(s) e^{s(\Lambda+\rho)(H)}}{\Delta(h)} \qquad (h \in T') .$$

This gives

$$\prod_{\alpha \in P_n} (1 - e^{-\alpha(H)}) \overline{\chi_\Lambda(h)} = (-1)^{q_\Lambda} |\Delta_n(h)|^2 \Theta_{\omega(\Lambda+\rho)^*}(h) \qquad (h \in T')$$

where $\Delta_n(h) = \prod_{\alpha \in P_n} (e^{\alpha(H)/2} - e^{-\alpha(H)/2})$($|\Delta_n(h)|^2$ is well defined). It is clear that

$$\det (1 - \mathrm{Ad}_-^1 (h)) = \prod_{\alpha \in P_n} (1 - e^{-\alpha(H)})$$

for all $h = \exp H \in T$. It follows that

$$A = \frac{(-1)^{q_\Lambda}}{[W_G]} \int_G dx \int_{T \times K} (-1)^m \Phi_{\omega(\Lambda+\rho)^*}(h) \Delta(h)^\circ \varphi(xkhk^{-1}x^{-1}) dh dk$$

where

$$\Phi_{\omega(\Lambda+\rho)^*}(h) = \Delta(h) \Theta_{\omega(\Lambda+\rho)^*}(h) \qquad (h \in T') .$$

(Remark that $\Phi_{\omega(\Lambda+\rho)^*}$ can be extended to a $C^\infty$-function on $T$.) Put

$$F(h) = |\Delta(h)| \int_G |{}^o\varphi(xkhk^{-1}x^{-1})|\, dx \qquad (h \in T',\ k \in K)\ .$$

Since, ${}^o\varphi \in \mathcal{C}(G)$ by Lemma 3.2, we see that $F$ is bounded in $T'$, using the definition of $\mathcal{C}(G)$ and [7 (f), Th. 5]. We may therefore apply Fubini's theorem and we have

$$\begin{aligned} A &= \frac{(-1)^{m+q}}{[W_G]} \int_{T'\times K} \Phi_{\omega(\Lambda+\rho)^*}(h)\Delta(h)dhdk \int_G {}^o\varphi(xkhkx^{-1})dx \\ &= \frac{(-1)^{m+q_\Lambda}}{[W_G]} \int_T \Phi_{\omega(\Lambda+\rho)^*}(h)F_{{}^o\varphi}(h)dh \\ &= (-1)^{q_\Lambda}\Theta_{\omega(\Lambda+\rho)^*}({}^o\varphi)\ . \end{aligned}$$

The last equality follows from Lemma 2.5 and Lemma 3.2. But we have from Lemma 3.2

$$\Theta_{\omega(\Lambda+\rho)^*}({}^o\varphi) = \Theta_{\omega(\Lambda+\rho)^*}(\varphi)\ .$$

This completes the proof of the proposition.

LEMMA 6.1. *Let $\{T^q\}$, $T^q\colon V^q \to V^q$, be an endomorphism of finite rank of the complex $\{V^q\}$ $(q = 0, 1, \cdots, m)$. Then*

$$\sum_{q=0}^m (-1)^q \operatorname{Trace} T^q = \sum_{q=0}^m (-1)^q \operatorname{Trace} \mathcal{H}^q(T^q)$$

*where $\mathcal{H}^q(T^q)$ is the linear map induced by $T^q$ on the cohomology space $\mathcal{H}^q$.*

The proof of this lemma is elementary (see [2, Prop. 2.1]).

THEOREM 1. *Let $\pi_\Lambda^q$ denote the unitary representation of $G$ on the space $H_2^{0,q}(E_\Lambda)$ of square-integrable harmonic forms of type $(0, q)$ with coefficients in the vector bundle $E_\Lambda$. Let $q_\Lambda$ be the number of the non-compact positive roots $\alpha$ such that $\langle\Lambda + \rho, \alpha\rangle > 0$. Then we have*

$$\sum_{q=0}^m (-1)^q \operatorname{Trace} \pi_\Lambda^q = (-1)^{q_\Lambda}\Theta_{\omega(\Lambda+\rho)^*}\ ,$$

*where* Trace $\pi_\Lambda^q$ *denotes the character of $\pi_\Lambda^q$ and $\Theta_{\omega(\Lambda+\rho)^*}$ denotes the character of the discrete class $\omega(\Lambda+\rho)^*$ contragredient to the discrete class determined by the form $\Lambda+\rho$* (see Lemma 2.4).

PROOF. Let $\varphi$ be a $K$-finite function in $C_c^\infty(G)$. Let $\tilde{T}_\varphi^q$ be the operator on $V^q = L_2^q(G, V_\Lambda^*)_d^0$ defined by

$$\tilde{T}_\varphi^q = \int_G \varphi(g)(\tilde{\pi}_\Lambda^q)_d(g)dg$$

where $(\tilde{\pi}_\Lambda^q)_d$ denotes the restriction of the representation $\tilde{\pi}_\Lambda^q$ to $V^q$. Then $\{\tilde{T}_\varphi^q\}$ form an endomorphism of finite rank of the complex defined in Proposition 5.2 (see the proof of Proposition 6.1). By Lemma 6.1 we have

$$\sum_{q=0}^{m}(-1)^q \text{ Trace } \tilde{T}_\varphi^q = \sum_{q=0}^{m}(-1)^q \text{ Trace } \mathcal{H}^q(\tilde{T}_\varphi^q) ,$$

where $\mathcal{H}^q(T_\varphi^q)$ denotes the linear map induced by $\tilde{T}_\varphi^q$ on the cohomology space $\mathcal{H}^q$. The operator $\tilde{T}_\varphi^q$ leaves the space $H_2^{0,q}(E_\Lambda)$ invariant. Using Proposition 5.2 it is immediate that

$$\text{Trace } \mathcal{H}^q(\tilde{T}_\varphi^q) = \text{Trace } (\tilde{T}_\varphi^q \mid H_2^{0,q}(E_\Lambda)) ,$$

where $\tilde{T}_\varphi^q \mid H_2^{0,q}(E_\Lambda)$ denotes the restriction of $\tilde{T}_\varphi^q$ to $H_2^{0,q}(E_\Lambda)$. Remark that

$$\tilde{T}_\varphi^q \mid H_2^{0,q}(E_\Lambda) = \int_G \varphi(g)\pi_\Lambda^q(g)dg = \pi_\Lambda^q(\varphi)$$

and that

$$\text{Trace } \tilde{T}_\varphi^q = \text{Trace } \tilde{K}_\varphi^q$$

(see the proof of Proposition 6.1).

Hence, we have, using Proposition 6.2,

$$\sum_{q=0}^{m}(-1)^{q_\Lambda} \text{ Trace } \pi_\Lambda^q(\varphi) = \sum_{q=0}^{m}(-1)^q \text{ Trace } \tilde{K}_\varphi^q$$
$$= (-1)^{q_\Lambda}\Theta_{\omega(\Lambda+\rho)^*}(\varphi) .$$

Since $K$-finite functions are dense in $C_c^\infty(G)$, the theorem follows.

*Remark.* Since the mapping $\lambda \mapsto \omega(\lambda + \rho)$ in Lemma 2.4 is a bijection of $\mathcal{F}_0'$ onto $\mathcal{E}_d$, Theorem 1 shows that all the discrete classes are realized as subrepresentations of $\pi_\Lambda^q$ for some $\Lambda$ and some $q$.

## 7. Vanishing theorem

Fix any $\Lambda \in \mathcal{F}_0'$ and $q$ $(0 \leqq q \leqq m)$. For any subset $Q$ of $P_n$, we define

$$\langle Q \rangle = \sum_{\alpha \in Q} \alpha .$$

(If $Q$ is empty, $\langle Q \rangle = 0$.)

Put

$$Q_\Lambda = \{\alpha \in P_n; \langle \Lambda + \rho, \alpha \rangle > 0\} .$$

Then, by definition, $[Q_\Lambda] = q_\Lambda$ where $[Q_\Lambda]$ denotes the number of elements in $Q_\Lambda$. We define

$$\Gamma_q = \{\langle Q \rangle : Q \subset P_n, [Q] = q\} .$$

Put

$$c_\Lambda^q = \frac{1}{2} \text{Max}_{s \in W_G, \gamma \in \Gamma_q} \{\langle 2s\gamma_\Lambda - \rho - \gamma, \rho - \gamma \rangle + \langle \rho, \rho \rangle\}$$

where $\gamma_\Lambda = \langle Q_\Lambda \rangle$. We recall that $\tau_\Lambda^q = \text{Ad}_+^q \otimes \tau_\Lambda$.

LEMMA 7.1. *Assume that $(\tau_\Lambda^q : \mathfrak{d}) \neq 0$ $(\mathfrak{d} \in \mathcal{E}_K)$. Then there exist an $s \in W_G$ and a $\lambda \in \Gamma_q$ such that the highest weight of $\mathfrak{d}$ is given by $s(\Lambda + \rho_k) - \rho_k + \gamma$*

*where* $\rho_k = \frac{1}{2}\sum_{\alpha \in P_k} \alpha$.

PROOF. Weyl's character formula says that

$$\text{Trace}\, \tau_\Lambda (\exp H) = \frac{\sum_{s \in W_G} \varepsilon(s) e^{s(\Lambda+\rho_k)(H)}}{\Delta_k (\exp H)} \qquad (H \in \mathfrak{h}') .$$

On the other hand, it is clear that

$$\text{Trace Ad}_+^q (\exp H) = \sum_{\gamma \in \Gamma_q} e^{\gamma(H)} .$$

It follows that if $(\tau_\Lambda^q : \mathfrak{d}) \neq 0$ then the highest of $\mathfrak{d}$ is of the form $s(\Lambda + \rho_k) - \rho_k + \gamma$ for some $s \in W_G$ and $\gamma \in \Gamma_q$.

THEOREM 2. *Let $H_2^{0,q}(E_\Lambda)$ denote the space of square-integrable harmonic forms of type $(0, q)$ with coefficients in the vector bundle $E_\Lambda$. Assume that $\Lambda$ and $q$ satisfy the condition;*

$$|\langle \Lambda + \rho, \alpha \rangle| > c_\Lambda^q \qquad \textit{for all } \alpha \in P_n .$$

*Then*

$$H_2^{0,q}(E_\Lambda) = 0 \qquad \textit{if } q \neq q_\Lambda .$$

PROOF. Assume that $H_2^{0,q}(E_\Lambda) \neq 0$. Then it follows from Proposition 4.1 that there exists an $\omega \in \mathcal{E}_d$ which has the following properties;

(1) $\chi_\omega(\Omega) = \langle \Lambda + 2\rho, \Lambda \rangle$ ,

(2) $(\omega \mid K : \tau_\Lambda^q) \neq 0$.

The second condition implies that $\omega \mid K$ and $\tau_\Lambda^q$ contain a certain $\mathfrak{d} \in \mathcal{E}_K$ in common. We fix one such $\mathfrak{d} \in \mathcal{E}_K$ and let $\mu$ be the highest weight of $\mathfrak{d}$. Let $\pi \in \omega$. We denote by $\mathfrak{H}$ the representation space of $\pi$ and by $\mathfrak{H}_\mathfrak{d}$ the subspace spanned by the elements which transform under $\pi \mid K$ according to $\mathfrak{d}$. Fix a unit weight vector $\psi_\mu \in \mathfrak{H}_\mathfrak{d}$ belonging to the highest weight $\mu$. Then, since $\psi_\mu$ is infinitely differentiable under $\pi$ (see [7 (a), Th. 6]), we have

$$\pi(\Omega)\psi_\mu = \chi_\omega(\Omega)\psi_\mu .$$

Since $(\tau_\Lambda^q : \mathfrak{d}) \neq 0$ it follows from Lemma 7.1 that there exist an $s \in W_G$ and a $\gamma \in \Gamma_q$ such that $\mu = s(\Lambda + \rho_k) - \rho_k + \gamma$.

We put

$$Q'_\Lambda = P_n - Q_\Lambda$$

and

$$P'_n = (-sQ_\Lambda) \cup sQ'_\Lambda .$$

Then it is obvious that

$$P'_n \cup (-P'_n) = P_n \cup (-P_n) \quad \text{and} \quad P'_n \cap (-P'_n) = \varnothing ,$$

where $\varnothing$ denotes the empty set.

We choose a Weyl basis $\{E_\alpha\}_{\alpha\in\Sigma}$ of $\mathfrak{g}^C$ (mod $\mathfrak{h}^C$) with respect to the compact real form $\mathfrak{g}_u = \mathfrak{k} + \sqrt{-1}\mathfrak{p}$ so that we have

$$[E_\alpha, E_{-\alpha}] = H_\alpha ,$$

and

$$E_\alpha - E_{-\alpha},\ \sqrt{-1}(E_\alpha + E_{-\alpha}) \in \mathfrak{g}_u$$

for all $\alpha \in \Sigma$. Then it is clear that $E_\alpha^* = E_{-\alpha}$ $(\alpha \in \Sigma)$. Moreover, since $\pi$ is a unitary representation of $G$, we have

$$\pi(E_\alpha^*) = \begin{cases} \pi(E_\alpha)^* & (\alpha \in P_k) , \\ -\pi(E_\alpha)^* & (\alpha \in P_n) \end{cases}$$

where $\pi(E_\alpha)^*$ denotes the adjoint operator of $\pi(E_\alpha)$.

We notice that

$$\begin{aligned} \Omega - \Omega_K &= \textstyle\sum_{\alpha\in P_n} (E_\alpha E_{-\alpha} + E_{-\alpha}E_\alpha) = \sum_{\alpha\in P'_n} (E_\alpha E_{-\alpha} + E_{-\alpha}E_\alpha) \\ &= \textstyle\sum_{\alpha\in P'_n} [E_\alpha, E_{-\alpha}] + 2\sum_{\alpha\in P'_n} E_{-\alpha}E_\alpha \\ &= H_{2\rho'_n} + 2\textstyle\sum_{\alpha\in P'_n} E_\alpha^* E_\alpha \end{aligned}$$

where $\rho'_n = \frac{1}{2}\sum_{\alpha\in P'_n} \alpha$ and $\Omega_K$ denotes the Casimir operator of $K$. Since $\Omega_K \psi_\mu = \langle \mu + 2\rho_k, \mu\rangle \psi_\mu$, it follows that

$$\begin{aligned} \pi(\Omega)\psi_\mu &= \langle \mu + 2\rho_k, \mu\rangle\psi_\mu + \langle 2\rho'_n, \mu\rangle + 2\textstyle\sum_{\alpha\in P'_n} \pi(E_\alpha^*)\pi(E_\alpha)\psi_\mu \\ &= \langle \mu + 2\rho'_n + 2\rho_k, \mu\rangle\psi_\mu - 2\textstyle\sum_{\alpha\in P'_n} \pi(E_\alpha)^*\pi(E_\alpha)\psi_\mu . \end{aligned}$$

Since $sP_n = P_n (s \in W_G)$ we have $s\rho_n = \rho_n$. Hence, we have

$$\begin{aligned} &-2\textstyle\sum_{\alpha\in P'_n} \|\pi(E_\alpha)\psi_\mu\|^2 \\ &= \langle \Lambda + 2\rho, \Lambda\rangle - \langle \mu + 2\rho'_n + 2\rho_k, \mu\rangle \\ &= \langle \Lambda + 2\rho, \Lambda\rangle - \langle s(\Lambda + \rho_k) + \gamma + 2\rho'_n + \rho_k, s(\Lambda + \rho) + \gamma - \rho_k\rangle \\ &= 2\langle s(\Lambda + \rho), \rho_n - \rho'_n - \gamma\rangle - \langle \rho_n - \gamma - 2\rho'_n - \rho_k, \rho - \gamma\rangle - \langle \rho, \rho\rangle . \end{aligned}$$

By the definition of $\Gamma_q$ there exists a subset $Q$ of $P_n$ such that $[Q] = q$ and $\langle Q\rangle = \gamma$. Since $[\mathfrak{k}, \mathfrak{p}_+] \subset \mathfrak{p}_+$, we have $s^{-1}Q \subset P_n$. Put

$$\begin{aligned} Q_0 &= s^{-1}Q \cap Q_\Lambda , \\ Q'_0 &= s^{-1}Q \cap Q'_\Lambda . \end{aligned}$$

Since $\rho_n - s^{-1}\rho'_n = \langle Q_\Lambda\rangle$ we have

$$\rho_n - s^{-1}\rho'_n - s^{-1}\gamma = \langle Q_\Lambda - Q_0\rangle - \langle Q'_0\rangle .$$

On the other hand, we see that

$$\begin{aligned} &\langle \rho_n - \gamma - 2\rho'_n - \rho_k, \rho - \gamma\rangle + \langle \rho, \rho\rangle \\ &\qquad = \langle 2s\gamma_\Lambda - \rho - \gamma, \rho - \gamma\rangle + \langle \rho, \rho\rangle \leqq 2c_\Lambda^q . \end{aligned}$$

It follows that

$$\langle \Lambda + \rho, \langle Q_\Lambda - Q_0 \rangle - \langle Q_0' \rangle \rangle - c_\Lambda^q \leqq 0 .$$

If $Q \neq Q_0$ or $Q_0' \neq \emptyset$ there exists an $\alpha \in P_n$ such that

$$\langle \Lambda + \rho, \langle Q_\Lambda - Q_0 \rangle - \langle Q_0' \rangle \rangle \geqq |\langle \Lambda + \rho, \alpha \rangle| .$$

Hence, we have $|\langle \Lambda + \rho, \alpha \rangle| \leqq c_\Lambda^q$. This contradicts the assumption of the theorem. Thus we have

$$Q_\Lambda = Q_0 \quad \text{and} \quad Q_0' = \emptyset .$$

This implies that

$$q = [Q] = [sQ_0] = [Q_0] = [Q_-] = q_\Lambda .$$

This completes the proof of the theorem.

Put

$$c_\Lambda = \mathrm{Max}_{0 \leqq q \leqq m}\, c_\Lambda^q \quad \text{and} \quad \Gamma = \bigcup\nolimits_{0 \leqq q \leqq m} \Gamma_q .$$

Define

$$c = \frac{1}{2} \mathrm{Max}_{\gamma_1, \gamma_2, \in \Gamma} \{\langle 2\gamma_1 - \rho - \gamma_2, \rho - \gamma_2 \rangle + \langle \rho, \rho \rangle\} .$$

Then we have

$$c_\Lambda^q \leqq c_\Lambda \leqq c .$$

The second inequality follows from the fact that $s\Gamma \subset \Gamma$ for any $s \in W_G$.

The following corollary is an immediate consequence of Theorem 2.

COROLLARY 1. *Let $c$ be the constant defined above. Assume that $|\langle \Lambda + \rho, \alpha \rangle| > c$ for all $\alpha \in P_n$. Then we have*

$$H_\Lambda^{0,q}(E_\Lambda) = 0 \qquad \textit{if } q \neq q_\Lambda .$$

COROLLARY 2. *Suppose that $\Lambda$ and $q$ satisfy the same condition as in Theorem 2. Assume that an $\omega \in \mathcal{E}_d$ has the following properties:*

(1) $\chi_\omega(\Omega) = \langle \Lambda + 2\rho, \Lambda \rangle$,

(2) *$\omega \mid K$ and $\tau_\Lambda^q$ contain an irreducible representation $\mathfrak{d}$ of $K$, with highest weight $\mu$, in common.*

*Then we have*

(1) $q = q_\Lambda$,

(2) $\mu = \Lambda + \gamma_\Lambda$

*where* $\gamma_\Lambda = \sum_{\alpha \in Q_\Lambda} \alpha$.

PROOF. We keep the notation in the proof of Theorem 2. First we prove that $\gamma_\Lambda$ is dominant with respect to $P_k$. Put

$$\Phi_\Lambda = \bigwedge\nolimits_{\alpha \in Q_\Lambda} X_\alpha .$$

Then it is clear that $\Phi_\Lambda$ is a weight vector in $\bigwedge_{\mathfrak{p}_+}^{q_\Lambda}$ which belongs to the weight

$\gamma_\Lambda$. Since $[\mathfrak{k}^C, \mathfrak{p}_+] \subset \mathfrak{p}_+$, for any $\alpha \in P_n$ and $\beta \in P_k$ we have $\alpha + \beta \in P_n$ if $\alpha + \beta$ is a root. By the definition of $Q_\Lambda$ we have

$$\langle \Lambda + \rho, \alpha \rangle > 0 \qquad \text{for all } \alpha \in Q_\Lambda .$$

Since we assumed that $\Lambda \in F_0'$, $\langle \Lambda + \rho, \beta \rangle > 0$ for all $\beta \in P_k$. It follows that

$$\{\alpha + \beta \in P_n; \alpha \in Q_\Lambda, \beta \in P_k\} \subset Q_\Lambda .$$

This shows that for any $\beta \in P_k$ there exists at least one $\alpha \in Q_\Lambda$ such that $\alpha + \beta$ is not a root. Hence, we have

$$\mathrm{Ad}_+^{q_\Lambda}(X_\beta)\Phi_\Lambda = 0$$

for all $\beta \in P_k$. This implies that $\gamma_\Lambda$ is the highest weight of some irreducible component of $\mathrm{Ad}_+^{q_\Lambda}$. Therefore, $\gamma_\Lambda$ is dominant with respect to $P_k$. Under our assumption we have seen in the proof of Theorem 2 that $q = q_\Lambda$, $Q_\Lambda = Q_0 = s^{-1}Q$ and $\mu + \rho_k = s(\Lambda + \rho_k) + \langle Q \rangle$ where $s \in W_G$ and $Q$ is a subset of $P_n$. It follows that $\mu + \rho_k = s(\Lambda + \gamma_\Lambda + \rho_k)$. By what has been proved above, $\Lambda + \gamma_\Lambda$ is dominant with respect to $P_k$ as well as $\mu$ so that we have $s = 1$. This completes the proof of the corollary.

## 8. The representation of $G$ on the space of square-integrable harmonic forms

Fix $\Lambda \in \mathcal{F}_0'$ such that $|\langle \Lambda + \rho, \alpha \rangle| > c_\Lambda$ for all $\alpha \in P_n$ where we put $c_\Lambda = \mathrm{Max}_{0 \leq q \leq m}\, c_\Lambda^q$ (for the definition of $c_\Lambda^q$, see § 7). Then from Theorem 2 we have

$$H_2^{0,q}(E_\Lambda) = 0 \qquad \text{if } q \neq q_\Lambda .$$

It follows from Theorem 1 that

$$\mathrm{Trace}\, \pi_\Lambda^{q_\Lambda} = \Theta_{\omega(\Lambda+\rho)^*} .$$

On the other hand, by Proposition 4.1, $\pi_\Lambda^{q_\Lambda}$ is a finite sum of irreducible unitary representations of $G$. Using [7 (c), Th. 6 and its corollary] we obtain the following theorem.

THEOREM 3. *Let $q_\Lambda$ be the number of non-compact positive roots $\alpha$ such that $\langle \Lambda + \rho, \alpha \rangle > 0$. We denote by $[\pi_\Lambda^{q_\Lambda}]$ the equivalence class of the representation of $G$ on $H_2^{0,q_\Lambda}(E_\Lambda)$. Assume that $\Lambda$ satisfies the condition*

$$|\langle \Lambda + \rho, \alpha \rangle| > c_\Lambda \quad \textit{for all non-compact positive roots } \alpha .$$

*Then we have*

$$[\pi_\Lambda^{q_\Lambda}] = \omega(\Lambda + \rho)^*$$

*where $\omega(\Lambda + \rho)^*$ denotes the discrete class defined in Theorem 1.*

COROLLARY. *Under the same condition of Corollary 1 to Theorem 2, we have*

$$[\pi_{\Lambda}^{q_\Lambda}] = \omega(\Lambda + \rho)^* .$$

PROPOSITION 8.1. *Suppose that $\Lambda \in \mathcal{F}_0'$ satisfies the condition of Theorem 3. Then an $\omega \in \mathcal{E}_d$ is equal to $\omega(\Lambda + \rho)$ if and only if the following two conditions are satisfied:*

(1) $\chi_\omega(\Omega) = \langle \Lambda + 2\rho, \Lambda \rangle$,
(2) $(\omega \mid K: [\tau_{\Lambda}^{q_\Lambda}]) \neq 0$.

PROOF. Put

$$\mathcal{E}_d(\Lambda)^{q_\Lambda} = \{\omega \in \mathcal{E}_d;\ \chi_\omega(\Omega) = \langle \Lambda + 2\rho, \Lambda \rangle, (\omega \mid K: [\tau_{\Lambda}^{q_\Lambda}]) \neq 0\} .$$

We have to prove that $\mathcal{E}_d(\Lambda)^{q_\Lambda} = \{\omega(\Lambda + \rho)\}$. Proposition 4.1 shows that

$$\pi_{\Lambda}^{q_\Lambda} = \bigoplus_{\omega \in \mathcal{E}_d(\Lambda)^{q_\Lambda}} (\omega \mid K: [\tau_{\Lambda}^{q_\Lambda}])\omega^* .$$

It follows from Theorem 3 that

$$(\omega \mid K: [\tau_{\Lambda}^{q_\Lambda}]) = 0 \quad (\omega \in \mathcal{E}_d(\Lambda)^{q_\Lambda}) \qquad \text{unless } \omega = \omega(\Lambda + \rho) .$$

and that

$$(\omega(\Lambda + \rho) \mid K: [\tau_{\Lambda}^{q_\Lambda}]) = 1 .$$

Hence,

$$\mathcal{E}_d(\Lambda)^{q_\Lambda} = \{\omega(\Lambda + \rho)\} .$$

This completes the proof of the proposition.

Put $\gamma_\Lambda = \sum_{\alpha \in Q_\Lambda} \alpha$ where $Q_\Lambda = \{\alpha \in P_n;\ \langle \Lambda + \rho, \alpha \rangle > 0\}$. Then we know that $\gamma_\Lambda$ is dominant with respect to $P_k$ (see the proof of Corollary 2 to Theorem 2). Since we assumed that $\Lambda \in \mathcal{F}_0'$, it follows that $\Lambda + \gamma_\Lambda$ is dominant with respect to $P_k$.

COROLLARY. *Under the notation and the condition of Theorem 3 we have*

$$(\omega(\Lambda + \rho) \mid K: \tau_{\Lambda+\gamma_\Lambda}) = 1$$

*where $\tau_{\Lambda+\gamma_\Lambda}$ is the irreducible representation of $K$ with the highest weight $\Lambda + \gamma_\Lambda$.*

PROOF. We use the notation of the proof of Proposition 8.1. In the proof of Proposition 8.1, we have seen that

$$(\omega(\Lambda + \rho) \mid K: [\tau_{\Lambda}^{q_\Lambda}]) = 1 .$$

On the other hand from Corollary 2 to Theorem 2 we have

$$(\omega(\Lambda + \rho)\,|_K: [\tau_{\Lambda}^{q_\Lambda}]) = (\omega(\Lambda + \rho)\,|_K: [\tau_\mu])([\tau_{\Lambda}^{q_\Lambda}]: [\tau_\mu])$$

where $\mu = \Lambda + \gamma_\Lambda$. Hence, we have $(\omega(\Lambda + \rho)\,|_K: \tau_{\Lambda+\gamma_\Lambda}) = 1$, which proves the corollary.

*Remark* 1. In [11] it was conjectured that $(\omega(\Lambda + \rho)\,|_K: \tau_{\Lambda+\gamma_\Lambda}) \neq 0$ (see [11, remark on Th. 4.3]). The above corollary shows that this is true if $\Lambda$

satisfies the condition in Theorem 3.

*Remark* 2. Fix any $\Lambda \in \mathcal{F}_0'$ and put $\lambda = \Lambda + \rho$. Then, by the definition of $\mathcal{F}_0'$, $\lambda$ is regular. Hence, there exists the unique linear order on the set of all non-zero roots of $(\mathfrak{g}^C, \mathfrak{h}^C)$ under which $\lambda$ is dominant. We denote by $\rho_n^\lambda$ (resp. $\rho_k^\lambda$) the half sum of non-compact (resp. compact) positive roots under this linear order. Then, as is easily seen, we have $\gamma_\Lambda = \rho_n + \rho_n^\lambda$ and $\rho_k = \rho_k^\lambda$ so that $\Lambda + \gamma_\Lambda = \lambda + \rho_n^\lambda - \rho_k^\lambda$. Therefore, the above corollary can be restated as follows: $\omega(\lambda) \mid K$ contains the irreducible representations of $K$ with the highest weight $\lambda + \rho_n^\lambda - \rho_k^\lambda$ with multiplicity one. Harish-Chandra has recently proved that this is true if $\lambda$ satisfies a certain condition, without assuming the existence of a $G$-invariant complex structure on $G/K$.[1]

## 9. Final remarks

We keep the notation of the previous sections. Let $\sigma$ be an irreducible (unitary) representation of $K$ on a representation space $V$ and let $\mathfrak{A}$ be a subalgebra of the center of the universal enveloping algebra $\mathfrak{G}$ of $\mathfrak{g}$ such that $\Omega \in \mathfrak{A}$, where $\Omega$ is the Casimir operator of $G$. For any $\chi \in \mathrm{Hom}(\mathfrak{A}, \mathbf{C})$, we denote by $\mathfrak{H}(\sigma, \chi)$ the space of all $V$-valued $C^\infty$-functions $f$ on $G$ which satisfy the following conditions:

$$f(kx) = \sigma(k)f(x) \qquad (k \in K,\ x \in G)\ ,$$

$$\| f \|^2 = \int_G \| f(x) \|_V^2\, dx < \infty$$

and

$$zf = \chi(z)f \qquad (z \in \mathfrak{A})$$

where $\| \ \|_V$ denotes the norm in $V$. For any $g \in G$, we define

$$\big(U_g(\sigma, \chi)f\big)(x) = f(xg) \qquad \big(f \in \mathfrak{H}(\sigma, \chi),\ x \in G\big)\ .$$

Then we can prove that $\mathfrak{H}(\sigma, \chi)$ is a Hilbert space (see [10, Prop. 1]) and that $U(\sigma, \chi)$ is a unitary representation of $G$ (see [10]). Put

$$\mathcal{E}(\sigma) = \{\omega \in \mathcal{E};\ (\omega \mid K: [\sigma]) \neq 0\}$$

and

$$\mathcal{E}(\chi) = \{\omega \in \mathcal{E};\ \chi_\omega|_{\mathfrak{A}} = \chi\}\ .$$

In [10], the question of non-triviality and irreducibility of $U(\sigma, \chi)$ was discussed. The following proposition answers partially, but to some extent satisfactorily, this question.

PROPOSITION 9.1. *Assume that* $\Lambda \in \mathcal{F}_0'$ *satisfies the condition:*

[1] Our proof of the vanishing theorem was inspired by Professor Harish-Chandra's proof of this fact, which proof was kindly communicated by him to us.

$|\langle\Lambda+\rho,\alpha\rangle|>c_\Lambda$ *for all* $\alpha\in P_n$. *Let* $\omega(\Lambda+\rho)$ *be the discrete class defined in Lemma* 2.4, *and let* $\mathfrak{A}=\mathbf{C}[\Omega]$ (*the algebra of all polynomials in* $\Omega$). *Then we have*

$$[U(\tau_{\Lambda+\gamma_\Lambda},\chi_\Lambda)]=\omega(\Lambda+\rho)$$

*where* $\tau_{\Lambda+\gamma_\Lambda}$ *is the irreducible representation of* $K$ *with highest weight* $\Lambda+\gamma_{\Lambda+\gamma_\Lambda}(\gamma_\Lambda=\sum_{\alpha\in Q_\Lambda}\alpha)$ *and* $\chi_\Lambda$ *is the unique element of* Hom $(\mathfrak{A},\mathbf{C})$ *such that*

$$\chi_\Lambda(\Omega)=\langle\Lambda+2\rho,\Lambda\rangle\ .$$

PROOF. From Corollary 2 to Theorem 2 and Proposition 8.1 we have

$$\mathfrak{S}(\tau_{\Lambda+\gamma_\Lambda})\cap\mathfrak{S}(\chi_\Lambda)\cap\mathfrak{S}_d=\{\omega(\Lambda+\rho)\}\ .$$

On the other hand, from the corollary to Proposition 8.1 we have

$$(\omega(\Lambda+\rho)\,|\,K\colon\tau_{\Lambda+\gamma_\Lambda})=1\ .$$

Since $\Omega\in\mathfrak{A}$, $\mathfrak{S}(\tau_{\Lambda+\gamma_\Lambda})\cap\mathfrak{S}(\chi_\Lambda)-\mathfrak{S}_d$ is of measure zero with respect to the Plancherel measure for $G$, as we mentioned in the proof of Proposition 4.1. Therefore, the conditions (A. 1), (A. 2), and (A. 3) in the corollary to Theorem 1 in [10] are satisfied if we put $\sigma=\tau_{\Lambda+\gamma_\Lambda}$, $\chi=\chi_\Lambda$ and $\omega=\omega(\Lambda+\rho)$. Hence we have

$$[U(\tau_{\Lambda+\gamma_\Lambda},\chi_\Lambda)]=\omega(\Lambda+\rho)\ .$$

*Remark* 1. In the case of the universal covering group of De Sitter group R. Takahashi [12] constructed all the discrete series for $G$ by a similar method (See also [10, Prop. 4].)

*Remark* 2. There is a misprint in [11]; p. 96, line, 9, read "$(\bar\partial)^*$" instead of "$(\bar\partial_0)^*$".

Theorem 1.2 in [11] is not true in general unless the metric on $M$ is complete. However, the theorem is true if one defines

$$H_2^{n-p,n-q}(E^*)=\{\varphi\in L_2^{n-p,n-q}(E^*);\,\vartheta_1\varphi=0,\,\bar\partial_1\varphi=0\}$$

where $\vartheta_1=(\bar\partial_0)^*$ and $\bar\partial_1=\vartheta_1^*$. This is because $*\sharp$ maps the domain of $\bar\partial$ (resp. $\vartheta$) onto the domain of $\vartheta_1$ (resp. $\bar\partial_1$).

*Remark* 3. The analogy of Theorem 2 and Theorem 3 with [8 (a), Th. 6.4] is obvious.

If we put $U=K^CP_-$ (instead of putting $U=K^CP_+$) and if we consider the holomorphic vector bundle $E_\Lambda$ associated with $\tau_\Lambda$, then (without taking the contragredient representation $\tau_\Lambda^*$) the statement of Theorem 2 is still correct but in Theorem 3 one must replace $\omega(\Lambda+\rho)^*$ by $\omega(\Lambda+\rho)$.

TATA INSTITUTE OF FUNDAMENTAL RESEARCH AND INSTITUTE FOR ADVANCED STUDY
OSAKA UNIVERSITY AND INSTITUTE FOR ADVANCED STUDY

REFERENCES

[1] A. ANDREOTTI and E. VESENTINI, *Carleman estimates for the Laplace-Beltrami equations on complex manifolds*, Inst. Hautes Etudes Sci. Publ. Math. **25**, (1965), 313-362.

[2] M. F. ATIYAH and R. BOTT, *A Lefschetz fixed point formula for elliptic complexes*, Ann. of Math. **86** (1967), 374-407.

[3] V. BARGMANN, *Irreducible unitary representations of the Lorentz group*, Ann. of Math. **48** (1947), 568-640.

[4] R. BOTT, *Homogeneous vector bundles*, Ann. of Math. **66** (1957), 203-248.

[5] F. BRUHAT, Travaux de Harish-Chandra, Seminaire Bourbaki, exposé 143 (1957), 1-9.

[6] M. I. GRAEV, *Unitary representations of real semisimple Lie groups*, Trudy, **7** (1958), 335-389.

[7] HARISH-CHANDRA,

(a) *Representations of a semi-simple Lie group on a Banach space*: I, Trans. Amer. Math. Soc. **75** (1953), 185-243.

(b) *Representations of semisimple Lie groups*: III, Trans. Amer. Math. Soc. **76** (1954), 243-253.

(c) *Representation of semisimple Lie groups*; V, Amer. J. Math. **78** (156), 1-41.

(d) *Invariant eigendistribution on a semisimple Lie group*, Trans. Amer. Math. Soc. **119** (1965), 457-508.

(e) *Discrete series for semisimple Lie groups*: I, Acta Math. **113** (1965), 241-318.

(f) *Discrete series for semisimple Lie groups*: II, Acta Math. **116** (1966), 1-111.

[8] B. KOSTANT,

(a) *Lie algebra cohomology and the generalized Borel-Weil theorem*, Ann. of Math. **74** (1961), 329-387.

(b) "Orbits, symplectic structures, and representation theory", in Proc. U.S.-Japan Sem. on Diff. Geom., Kyoto, 1965, p. 71.

[9] R. P. LANGLANDS, "Dimension of spaces of automorphic forms", in Proc. of Symposia in Pure Math. IX, Algebraic Groups and Discontinuous Subgroups, Providence (1966), 253-257.

[10] K. OKAMOTO, *On induced representations*, Osaka J. Math. **4** (1967), 85-94.

[11] ——— and H. OZEKI, *On square-integrable $\bar{\partial}$-cohomology spaces attached to Hermitian symmetric space*, Osaka J. Math. **4** (1967), 95-110.

[12] R. TAKAHASHI, *Sur représentations unitaires des groupes de Lorentz généralises*, Bull. Soc. Math. France **91** (1963), 289-433.

(Received March 18, 1969)

*Actes, Congrès intern. Math.*, 1970. Tome 2, p. 199 à 201.

# GEOMETRY OF MODULI SPACES OF VECTOR BUNDLES

by M. S. NARASIMHAN

**Moduli spaces of vector bundles.**

I shall speak about certain varieties which arise in the moduli problem for holomorphic vector bundles on a compact Riemann surface.

Let $X$ be a compact Riemann surface of genus $g \geqslant 2$. By a vector bundle we shall always mean a holomorphic vector bundle. If $W$ is a vector bundle on $X$ we shall denote by $d(W)$ its degree and by $n(W)$ its rank.

As is well known, the classification of line bundles of degree zero on $X$ is achieved by the Jacobian $J$ of $X$. The underlying differentiable manifold for $J$ is the space of characters of the first homology group $H_1(X, \mathbf{Z})$. Moreover, the holomorphic tangent space to $J$ at any point is identified with the cohomology space $H^1(X, \mathcal{O})$, where $\mathcal{O}$ is the sheaf of germs of holomorphic functions on $X$. We remark that there is a natural positive definite hermitian form on $H^1(X, \mathcal{O})$ : the space $H^1(X, \mathcal{O})$ is identified with the space $H$ of closed $(0, 1)$ forms on $X$ and if $\omega, \eta \in H$ we set

$$(\omega, \eta) = \frac{1}{i} \int_X \omega \wedge \overline{\eta} \, .$$

Passing on to vector bundles of higher rank, to obtain good moduli varieties one has to restrict the class of vector bundles. A vector bundle $W$ on $X$ is said to be stable if for every proper subbundle $V$ of $W$ one has $d(V)/n(V) < d(W)/n(W)$. D. Mumford proved that the isomorphism of classes of stable bundles of rank $n$ and degree $d$ form a non-singular quasi-projective variety.

Let $\pi_1$ be the fundamental group of $X$. If $\rho$ is a representation of $\pi_1$ in the unitary group $\mathrm{U}(n)$, $\rho$ defines a holomorphic vector bundle $W(\rho)$ of rank $n$ and degree 0 on $X$. Moreover if $\rho_1$ and $\rho_2$ are unitary representations of $\pi_1$, then the holomorphic vector bundles $W(\rho_1)$ and $W(\rho_2)$ are isomorphic if and only if $\rho_1$ and $\rho_2$ are equivalent representations. It was proved in [2] that a vector bundle of degree 0 on $X$ is stable if and only if it arises from an *irreducible* unitary representation of $\pi_1$. This result suggests a natural compactification for the space of stable bundles of degree 0. In fact C.S. Seshadri proved [4] that there is a natural structure of a *projective* variety on the space equivalence classes of all $n$-dimensional unitary representations of $\pi_1$. This structure depends in general on the complex structure on $X$. In general this space has singularities and the singular points have been determined in [3]. The singular points correspond precisely to reducible representations except when $g = 2$, $n = 2$, in which case the variety is non-singular. Other moduli spaces (corresponding to vector bundles of degree $\neq 0$) are obtained by considering unitary representations of suitabley defined Fuchsian groups [2].

**Non-singular moduli spaces.**

The moduli spaces are non-singular for vector bundles whose degree and rank are coprime. These varieties are constructed in the following way. Let $\pi$ be a discrete group acting effectively, properly and holomorphically on the unit disc $Y$ such that $Y/\pi = X$ and such that the natural projection $p : Y \to X$ is ramified over a single point $x_0 \in X$ with ramification order $n$. Let $y_0 \in p^{-1}(x_0)$ and $\pi_{y_0}$ be the isotropy group at $y_0$. Let $\tau$ be a character of $\pi_{y_0}$ such that $\tau$ is an isomorphism of $\pi_{y_0}$ onto the $n^{\text{th}}$ roots of unity. A representation $\rho$ of $\pi$ into unitary group $U(n)$ is said to be of type $\tau$ if $\rho \mid \pi_{y_0} = \tau . I_n$, where $I_n$ is the $n \times n$ identity matrix. Such a representation is irreducible. To each $\rho$ of type $\tau$ we can associate a holomorphic vector bundle $W(\rho)$ of rank $n$ and degree $d$ on $X$ (Here $d$ is an integer coprime to $n$, associated with $\tau$). There is a natural structure of a compact complex manifold ([1], [2, Remark 10.1]) on the set $M$ of equivalence classes of $n$-dimensional unitary representations of type $\tau$ of $\pi$. (In fact $M$ is a projective variety).

Let $m \in M$ and $\rho$ a representation of type $\tau$ in the class $m$. Let Ad$\rho$ denote the representation of $\pi$ in $\mathfrak{gl}(n, C)$ obtained by composing $\rho$ and the adjoint representation of $U(n)$ in $\mathfrak{gl}(n, C)$. Ad$\rho$ is a representation of $\pi_1$. Let $W(\text{Ad}\rho)$ be the holomorphic vector bundle on $X$ associated with Ad$\rho$. Then the holomorphic tangent space to $M$ at $m$ is naturally identified with the cohomology space $H^1(X, W(\text{Ad}\rho))$ [1].

**Canonical hermitian metrics**

We now introduce a hermitian metric on M. To do this it is sufficient to introduce a positive definite hermitian form on $H^1(X, W(\text{Ad}\rho))$. Since $W(\text{Ad}\rho)$ is given by a local system, the operator of exterior differentiation, d, is well defined on $C^\infty$ differential forms with values in $W(\text{Ad}\rho)$. Let $T(\rho)$ denote the space of $d$-closed $C^\infty$ forms of type $(0, 1)$ with coefficient in $W(\text{Ad}\rho)$. Then $T(\rho)$ is canonically isomorphic to $H^1(X, W(\text{Ad}\rho))$. So it suffices to introduce a positive definite hermitian form on $T(\rho)$. If $\omega \in T(\rho)$, let $\omega^\#$ denote the $(1, 0)$ form with coefficients in $W(\text{Ad}\rho)$ obtained by using the conjugation $A \mapsto A^*$ in $\mathfrak{gl}(n, C)$. ($A^*$ denotes the conjugate transpose of $A$. Locally, if $\omega = A(z)\,d\bar{z}$, $\omega^\# = A^*(z)\,dz$). Define the hermitian scalar product in $T(\rho)$ by

$$(\omega_1, \omega_2) = \frac{1}{i}\int_X \text{Trace}\,(\omega_1, \omega_2^\#),\ \omega_1, \omega_2 \in T(\rho),$$

where the exterior product is taken with respect to the multiplication

$$\mathfrak{gl}(n, C) \times \mathfrak{gl}(n, C) \to \mathfrak{gl}(n, C).$$

This defines a hermitian metric on $M$. This metric is Kählerian. In fact, using the forms in $T(\rho)$ we can construct at each point $P$ of $M$ a geodesic holomorphic coordinate system (i.e., one in which all first derivatives of the components of the metric tensor are zero at $P$).

## REFERENCES

[1] NARASIMHAN M.S. and SESHADRI C.S. — Holomorphic vector bundles on a compact Riemann surface, *Math. Ann*, 155, 1964, p. 69-80.

[2] NARASIMHAN M.S. and SESHADRI C.S. — Stable and unitary vector bundles on a compact Riemann surface, *Ann. of Math.*, 82, 1965, p. 540-567.

[3] NARASIMHAN M.S. and RAMANAN S. — Moduli of vector of bundles on a compact Riemann surface, *Ann. of Math.*, 89, 1969, p. 19-51.

[4] SESHADRI C.S. — Space of unitary vector bundles on a compact Riemann surface. *Ann. of Math.*, 85, 1967, p. 303-336.

School of Mathematics
Tata Institute of Fundamental Research
Bombay 5
Inde

Math. Ann. 212, 215—248 (1975)


# On the Cohomology Groups of Moduli Spaces of Vector Bundles on Curves

G. Harder and M. S. Narasimhan

## Introduction

In this paper we take up the idea in [9] which established a connection between the cohomology groups of certain moduli spaces of vector bundles on projective non-singular curves and the Tamagawa number of $SL_2$. This method gained strength through the recent results of Deligne on Weil conjectures.

We now state the main results of the paper. Let $k$ be an algebraically closed field and $Y/k$ a non-singular projective algebraic curve. Let $L_0/Y$ be a line bundle with $\deg(L_0)=r$ and $n \geqq 2$ an integer with $(n, r)=1$. We consider the moduli scheme $M/k = M(n, L_0)/k$ of stable vector bundles on $Y$ of rank $n$ and determinant isomorphic to $L_0$. It is known that $M$ is a non-singular projective variety. We suppose that $(p, n)=1$, where $p$ is the characteristic of the field $k$. The group $T_n$ of $n$-division points of the jacobian $J/k$ of $Y/k$ acts on $M$ by tensorisation.

**Theorem 1.** *If $l$ is a prime number with $(l, p)=(l, n)=1$, then $T_n$ acts trivially on the $l$-adic cohomology groups $H^i(M, \mathbb{Q}_l)$.*

If $k=\mathbb{C}$ then an analogous result holds for the ordinary cohomology with complex coefficients; this follows immediately from Artin's comparison theorem. This result, for $k=\mathbb{C}$, was conjectured by Narasimhan and Ramanan in [12].

Let us now assume that $Y$ is defined over a finite field $\mathbb{F}_q$. If we choose $L_0$ to be in $Pic(Y)/\mathbb{F}_q$, we may assume, by going over to a finite extension of $\mathbb{F}_q$ if necessary, that $M$ is defined over $\mathbb{F}_q$. The group scheme $T_n/\mathbb{F}_q$ acts on $M/\mathbb{F}_q$ and the quotient $N/\mathbb{F}_q = (M/T_n)/\mathbb{F}_q$ exists. Regarding the number of rational points on these schemes one has

**Theorem 2.** *We have*

$$|M_{\mathbb{F}_q}| = |N_{\mathbb{F}_q}|$$

*where $|M_{\mathbb{F}_q}|$ and $|N_{\mathbb{F}_q}|$ denote the number of rational points over $\mathbb{F}_q$ of $M$ and $N$ respectively.*

Theorem 2 is proved by an extension of the methods of [9]. Theorem 2, combined with the results of Deligne on the Weil conjectures and a specialisation argument, implies Theorem 1.

We shall also give some results on the Betti numbers of $\bar{M} = M \times_{\mathbf{F}_q} \bar{\mathbb{F}}_q$. We mention only

**Theorem 3.** *The third Betti number,* $\dim H^3(\bar{M}, \mathbb{Q}_l)$, *is* $2g$, *where* $g$ *is the genus of* $Y/\mathbb{F}_q$.

This result has been proved in [11] except when $g=2$, $n=3$.

The second named author would like to thank the „Sonderforschungsbereich für Theoretische Mathematik an der Universität Bonn" and the Mathematics Institute, University of Warwick, for their hospitality at the time of the preparation of this paper.

## I. Vector Bundles and Projective Bundles over Curves

### *1.1. The Moduli Spaces of Stable Vector and Projective Bundles*

Let $k$ be a field and $Y/k$ be a projective non-singular absolutely irreducible curve over $k$. Let $K/k$ be the function field of $k$. If $\bar{k}$ is an algebraic closure of $k$, we set $\bar{Y} = Y \times_k Spec\,\bar{k}$.

We shall consider locally free $\mathcal{O}_Y$ modules of finite rank and by abuse of language we shall call them vector bundles. If $E/Y$ is a vector bundle then $rkE$ will denote its rank. If $rkE = n$ we put $\det E = \bigwedge^n E$ and call this line bundle the determinant of $E$. We define $\deg E = \deg(\det E)$. For $E \neq 0$ we introduce the rational number

$$\mu(E) = \deg E / rkE\,.$$

We shall also consider projective bundles $X \to Y$ i.e., locally trivial bundles with respect to the Zariski topology whose fibre is $\mathbb{P}^{n-1}/k$. More precisely, there exists a covering $Y = \cup_i U_i$ by Zariski open subsets of $Y$ such that for every $i$ we have a commutative diagram

$$\begin{array}{ccc} X \times_Y U_i & \xrightarrow{\sim} & \mathbb{P}^{n-1} \times_k U_i \\ & \searrow \quad \swarrow & \\ & U_i & \end{array}$$

Each vector bundle $E/Y$ gives rise to a projective bundle which is the scheme of lines in $E/Y$. On the other hand, it is easy to see that a projective bundle $X/Y$ comes from a vector bundle $E/Y$ and that this vector bundle is unique up to a tensorisation with a line bundle. Since for a line bundle $L/Y$ and a vector bundle $E/Y$ we have $\deg(E \otimes L) = (rkE) \cdot (\deg L) + \deg E$ we see that $\underline{\deg}\, X \equiv \deg E \pmod{n}$ is well defined. We call $\underline{\deg}\, X \in \mathbb{Z}/(n)$ the degree of the projective bundle $X/Y$.

Let us assume for the moment that $k$ is algebraically closed. A vector bundle $E/Y$ is called stable (resp. semi-stable) if for all proper subbundles

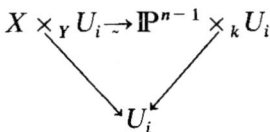

$F\,(\neq 0, E)$ we have $\mu(F)<\mu(E/F)$ (resp. $\mu(F)\leqq\mu(E/F)$). One observes that these inequalities are equivalent to

$$\mu(F)<\mu(E)\,(\text{resp. } \mu(F)\leqq\mu(E))\,.$$

A projective bundle is called stable (resp. semi-stable) if the corresponding vector bundle is stable (resp. semi-stable). This notion is well defined since a vector bundle $E$ is stable (resp. semi-stable) if and only if $E\otimes L$ is stable (resp. semi-stable) where $L$ is a line bundle. It is easy to see that if $E$ is semi-stable and $(\deg E, rk E)=1$, then $E$ is stable.

Let $n$ and $r$ be integers $n\geqq 1$ with $(n,r)=1$ and $L_0$ be a fixed line bundle of degree $r$ on $Y$. It is known that the set of isomorphism classes of vector bundles $E/Y$ of rank $n$ with $\det E\approx L_0$ is parametrised by a non-singular projective variety $M=M(n,L_0)$ over $k$, [17, 18]. Let us assume that the characteristic of $k$ is prime to $n$. If $L$ is a line bundle on $Y$ such that its $n^{\text{th}}$ power $L^{\otimes n}$ is trivial; then $\det(E\otimes L)=\det E\otimes L^{\otimes n}=\det E$, so that the group $T_n$ of $n$-division points of the jacobian $J/k$ of $Y$ acts on $M/k$. The quotient of $M$ for this action of $T_n$ is again a projective scheme $N/k$ and $N/k$ provides the solution for the moduli problem for projective bundles of degree $r$ mod $n$.

We now drop the assumption that $k$ is algebraically closed. A vector bundle $E/Y$ is called stable (resp. semi-stable) if $\bar{E}/\bar{Y}$ is stable (resp. semi-stable) where $\bar{E}/\bar{Y}$ is the extension $E$ to $\bar{Y}/\bar{k}$, $\bar{k}$ denoting the algebraic closure of $k$. If the line bundle $L_0$ is given over $Y/k$ we may assume, by passing to a finite extension if necessary, that the moduli spaces $M$ and $N$ are defined over $k$ i.e., we have

$$\begin{array}{ccc} M & \longrightarrow & N \\ & \searrow \quad \swarrow & \\ & k & \end{array}$$

However $M/k$ and $N/k$ are only coarse moduli schemes in the sense of Mumford [10] i.e., for any field $L$, $k\subset L\subset\bar{k}$, if $M_L$ and $N_L$ denotes the set of $L$-rational points of $M$ and $N$ respectively, we have maps

$$\varphi_L:\left\{\begin{array}{l}\text{Set of isomorphism classes of stable}\\ \text{bundles } E\to Y\times_k L \text{ with } rk E=n \text{ and}\\ \det E\approx L_0\end{array}\right\}\to M_L$$

$$\psi_L:\left\{\begin{array}{l}\text{Set of isomorphism classes of projective}\\ \text{bundles } X\to Y\times_k L \text{ with } rk X=n-1\\ \text{and } \underline{\deg} X\equiv r \bmod n\end{array}\right\}\to N_L$$

which are functorial in $L$ and which become bijections if $L=\bar{k}$; but for $L=k$ these maps are neither injective nor surjective in general.

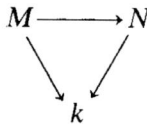

### *1.2. Moduli Schemes over a Finite Field*

We now assume that $k$ is a finite field $\mathbb{F}_q$ and investigate the maps $\varphi_k$ and $\psi_k$ defined above.

**Proposition 1.2.1.** *The map $\varphi_{\mathbb{F}_q}$ is bijective and the map $\psi_{\mathbb{F}_q}$ is surjective.*

*Proof.* Let $m$ be a point of $M_{\mathbb{F}_q}$ and let $\bar{E}\to\bar{Y}$ be a stable vector bundle in the class $m$. Since $m$ is a rational point over $\mathbb{F}_q$, we have for all $\sigma\in\operatorname{Gal}(\bar{\mathbb{F}}_q/\mathbb{F}_q)$ an isomorphism

$$\lambda_\sigma : \bar{E}\to\bar{E}^\sigma ,$$

where $\operatorname{Gal}(\bar{\mathbb{F}}_q/\mathbb{F}_q)$ denotes the Galois group of the extension $\bar{\mathbb{F}}_q/\mathbb{F}_q$. Since the automorphisms of a stable bundle are given by multiplication by scalars [13, Corollary to Proposition 4.3], we see that

$$\lambda_\tau^\sigma \circ \lambda_\sigma = \lambda_{\tau\sigma}\cdot a_{\sigma\tau}$$

where $a_\sigma^\tau\in\bar{\mathbb{F}}_q^*$ and $\lambda_\tau^\sigma$ denotes the transform of $\lambda_\tau$ by $\sigma$. It is clear that $a_{\sigma\tau}$ is a 2-cocycle. Since the Brauer group $H^2(\mathbb{F}_q, G_m)$ is trivial [15, p. 170] we can modify the $\lambda_\sigma$ in such a way that $a_{\sigma,\tau}=1$. But then it follows from the theory of descent that there exists a vector bundle $E$ on $Y$ such that $\bar{E}$ is isomorphic to $E\times_Y\bar{Y}$. Using the fact that $H^1(\mathbb{F}_q, G_m)=0$ we see that $E$ is unique and $\det E\approx L_0$. This proves that $\varphi_{\mathbb{F}_q}$ is bijective.

To prove that $\psi_{\mathbb{F}_q}$ is surjective we apply similar arguments for projective bundles. If $\bar{X}\to\bar{Y}$ is a projective stable bundle, then the group of automorphisms $\operatorname{Aut}(\bar{X})$ of $\bar{X}/\bar{Y}$ is a subgroup of the group $T_n$ of $n$-division points of $J_{\mathbb{F}_q}$. In fact, let $\bar{E}$ be a vector bundle on $\bar{Y}$ giving rise to $\bar{X}$. Since $\bar{E}$ is stable, the group of automorphisms of $\bar{E}$ is $\bar{k}^*$ and it follows that $\operatorname{Aut}(\bar{X})$ is isomorphic to the group of isomorphism classes of line bundles $L/\bar{Y}$ with $L\otimes\bar{E}\approx\bar{E}$, $L$ necessarily satisfying the condition that $L^{\otimes n}$ is trivial (see [6], Corollary to Proposition 2). Now if $\bar{X}\simeq\bar{X}^\sigma$ for all $\sigma\in\operatorname{Gal}(\bar{\mathbb{F}}_q/\mathbb{F}_q)$ then $\operatorname{Aut}(\bar{X})$ is defined over $\mathbb{F}_q$ and the obstruction to descent is contained in $H^2(\mathbb{F}_q, \operatorname{Aut}(\bar{X}))$. But $H^2(\mathbb{F}_q, \operatorname{Aut}\bar{X})=0$ [16, Chapter II, 3.3] and hence $\psi_{\mathbb{F}_q}$ is surjective. This proves Proposition 1.2.1.

Next we consider how many points are mapped by $\psi_{\mathbb{F}_q}$ into the same point in $N_{\mathbb{F}_q}$, i.e., we consider how often does it happen that $X\not\approx X'$ but $\bar{X}\approx\bar{X}'$ where $X$ and $X'$ are projective bundles on $Y$. Given $X\to Y$ it is well known that the number of $\mathbb{F}_q$ forms of $X$ is equal to the order of $H^1(\mathbb{F}_q, \operatorname{Aut}(X))$. On the other hand, it is known (and this fact is crucial for us) that

$$|H^1(\mathbb{F}_q, \operatorname{Aut} X)| = |\operatorname{Aut}(X)_{\mathbb{F}_q}|$$

[15, Chapter VIII, Proposition 8]. [To apply this proposition we have to take for $G$ a sufficiently large quotient $G = \mathrm{Gal}(\mathbb{F}_q r/\mathbb{F}_q)$ of the Galois group. Actually we require that $\mathrm{Aut}(X)_{\mathbb{F}_{q^r}} = \mathrm{Aut}(X)_{\bar{\mathbb{F}}_q}$ and that the norm mapping $\mathrm{Aut}(X)_{\mathbb{F}_{q^r}} \to \mathrm{Aut}(X)_{\mathbb{F}_q}$ is zero.] Thus we have

**Proposition 1.2.2.** *Let $m \in N_{\mathbb{F}_q}$. Then the number of points in $\psi_{\mathbb{F}_q}^{-1}(m)$ is equal to the order of* $\mathrm{Aut}(X)_{\mathbb{F}_q}$, *where $X \to Y$ is any projective bundle whose isomorphism class is mapped into $m$ by $\psi_{\mathbb{F}_q}$.*

### *1.3. Canonical Filtrations on Non-Semistable Bundles*

We first assume that $k$ is algebraically closed and collect some results on the structure of non-semistable bundles. It will turn out at the end that the assumption that $k$ is algebraically closed is superfluous.

*Definition 1.3.1.* Let $E$ be a vector bundle on $Y$ which is not semi-stable. A subbundle $F$ of $E(F \neq 0, E)$ is said to be SCSS in $E$ ("strongly contradicting semi-stability") if the following two conditions are fulfilled:

a) $F$ is semi-stable.

b) For every subbundle $F'$ of $E$ with $F \subsetneqq F' \subset E$ we have $\mu(F) > \mu(F')$.

*Remark 1.3.2.* Condition b) is equivalent to b') for any subbundle $Q$ of $E/F$ with $0 \neq Q \subset E/F$ we have $\mu(Q) < \mu(F)$ and b') is equivalent to b") for any *stable* subbundle $Q$ of $E/F$ with $0 \neq Q \subset E/F$ we have $\mu(Q) < \mu(F)$.

Clearly b) and b') are equivalent and b') implies b"). If b") is fulfilled, let $0 \neq Q \subset E/F$ be subbundle of $E/F$. Then there exists a *stable* subbundle $Q' \neq 0$ of $Q$ with $\mu(Q') \geqq \mu(Q)$ [13, Proposition 4.5]. Hence $\mu(Q) \leqq \mu(Q') < \mu(F)$.

*Remark 1.3.3.* If $E$ is not semi-stable and $F \neq 0$ a subbundle of $F$ satisfying i) $F$ and $E/F$ are semi-stable and ii) $\mu(F) > \mu(E/F)$ then $F$ is SCSS (in $E$). In fact for any bundle $0 \neq Q \subset E/F$ we have $\mu(Q) \leqq \mu(E/F) < \mu(F)$ so that the Condition b') in Remark 1.3.2 is fulfilled.

**Proposition 1.3.4.** *If $E/Y$ is not semi-stable then it contains a unique SCSS subbundle.*

For the proof we need two lemmas

**Lemma 1.3.5.** *Let $F_1$ and $F_2$ be subbundles of $E$ such that $F_1$ is semi-stable and $F_2$ satisfies Condition* b) *of Definition 1.3.1. If $F_1$ is not contained in $F_2$ then we have $\mu(F_2) > \mu(F_1)$.*

*Proof.* Consider the canonical map $F_1$ to $E/F_2$ which is non-zero by assumption. Since $Y$ is a non-singular curve we have a factorisation

$$\begin{array}{ccc} F_1 \to & F_1' & \to 0 \\ & \downarrow & \\ E/F_2 \leftarrow & F_1'' & \leftarrow 0 . \end{array}$$

where $F_1' \to F_1''$ is an isomorphism on a non-empty open set [13, §4]. Since $F_1$ is semi-stable, $\mu(F_1) \leqq \mu(F_1')$ and since $F_2$ satisfies Condition b) we have $\mu(F_1'') < \mu(F_2)$. On the other hand $\mu(F_1') \leqq \mu(F_1'')$ as $\deg F_1' \leqq \deg F_1''$ and $rk F_1' = rk F_1''$. It follows that $\mu(F_1) < \mu(F_2)$.

**Lemma 1.3.6.** *If $F_1$ and $F_2$ are subbundles of $E$ which are SCSS then $F_1 = F_2$.*

*Proof.* If $F_1$ is not contained in $F_2$ we have, by Lemma 1.3.5, $\mu(F_2) > \mu(F_1)$. Applying again the lemma, we must then have $F_2 \subsetneqq F_1$. But since $F_1$ is semi-stable we have $\mu(F_2) \leqq \mu(F_1)$, which is a contradiction. Thus $F_1 \subsetneqq F_2$ and similarly $F_2 \subsetneqq F_1$.

*Proof of Proposition 1.3.4.* Since the uniqueness of a SCSS bundle follows from Lemma 1.3.6 we need only to prove the existence of a SCSS bundle. Let $m = \sup_{F \subset E, F \neq 0} \mu(F)$. Since $E$ is not semi-stable, we have $m > \mu(E)$. Among all subbundles $F$ for which $\mu(F) = m$ (the set of such $F$ is non-empty since the values of $\mu$ are discrete and bounded from above) we choose one, say $F_0$, which has maximal rank. If $0 \neq F' \subset F_0$ is subbundle we have $\mu(F') \leqq m = \mu(F_0)$ so that $F_0$ is semi-stable. If on the other hand we have a subbundle $F'$ with $F_0 \subsetneqq F' \subset E$ then $rk F' > rk F_0$ and by the choice of $F_0$ we have $\mu(F') < \mu(F_0)$. Thus $F_0$ also satisfies Condition b) i.e., $F_0$ is SCSS. This proves Proposition 1.3.4.

**Lemma 1.3.7.** *If a vector bundle $E$ is not semi-stable we have a flag*

$$0 = F_0 \subsetneqq F_1 \subsetneqq \cdots \subsetneqq F_i \subsetneqq \cdots \subsetneqq F_k = E$$

*satisfying the conditions*

$$\text{(A)} \begin{cases} \text{i) } F_i/F_{i-1} \text{ is semi-stable for } i = 1, \ldots, k, \\ \text{ii) } F_i/F_{i-1} \text{ is SCSS in } E/F_{i-1} \text{ for } i = 1, \ldots, k-1. \end{cases}$$

*Moreover such a flag is uniquely determined.*

*Proof.* The existence follows from Proposition 1.3.4. In fact, let $F_1$ be a subbundle of $E$ which is SCSS. If $E/F_1$ is semi-stable we are through. Otherwise we find $F_2' \subset E/F_1$ which is SCSS in $E/F_1$ and define $F_2$ to be the inverse of $F_2'$ by the map $E \to E/F_1$. By repeating this construction we find a flag satisfying A. The uniqueness is proved by induction on $\dim E$ applying Proposition 1.3.4 and noting that $\{F_i/F_{i-1}\}$, $i \geqq 2$, form a filtration of $E/F_1$ satisfying A.

**Lemma 1.3.8.** *Let*

$$0 = F_0 \subsetneqq F_1 \subsetneqq \cdots \subsetneqq F_k = E$$

*be a flag. Then Conditions* (A) *in Lemma 1.3.7 are equivalent to the conditions:*

$$(\mathrm{B}) \begin{cases} \text{i}') \ F_i/F_{i-1} \text{ is semi-stable for } i=1,\dots k, \\ \text{ii}') \ \mu(F_i/F_{i-1}) > \mu(F_{i+1}/F_i) \text{ for } i=1,\dots k-l. \end{cases}$$

*Proof*. Suppose that Conditions (A) are satisfied. We have the exact sequence

$$0 \to F_i/F_{i-1} \to E/F_{i-1} \to E/F_i \to 0\,.$$

Since $F_{i+1}/F_i \subset E/F_i$ and $F_i/F_{i-1}$ is SCSS in $E/F_{i-1}$ we must have $\mu(F_i/F_{i-1}) > \mu(F_{i+1}/F_i)$ [see Remark 1.3.2, Condition b′)].

Now suppose that Conditions (B) are satisfied. We first show that $F_{k-1}/F_{k-2}$ is SCSS in $E/F_{k-2}$. Consider the exact sequence

$$0 \to F_{k-1}/F_{k-2} \to E/F_{k-2} \to E/F_{k-1} \to 0\,.$$

We have by Condition ii′) in (B)

$$\mu(F_{k-1}/F_{k-2}) > \mu(E/F_{k-1})\,;$$

since $E/F_{k-1}$ and $F_{k-1}/F_{k-2}$ are semi-stable, we see, by Remark 1.3.3, that $F_{k-1}/F_{k-2}$ is SCSS in $E/F_{k-2}$. We proceed to prove that Condition ii) in (A) is satisfied, by downward induction on i. Consider the exact sequences

$$0 \to F_i/F_{i-1} \to E/F_{i-1} \to E/F_i \to 0$$

and

$$0 \to F_{i+1}/F_i \to E/F_i \to E/F_{i+1} \to 0\,.$$

To prove that $F_i/F_{i-1}$ is SCSS in $E/F_{i-1}$ it is sufficient to prove, by Remark 1.3.2, b″), that for any stable subbundle $Q \neq 0$ of $E/F_i$ we have $\mu(F_i/F_{i-1}) > \mu(Q)$.

Now if $Q \subset F_{i+1}/F_i$ we will have $\mu(Q) \leqq \mu(F_{i+1}/F_i)$ as $F_{i+1}/F_i$ is semi-stable and by hypothesis we have $\mu(F_{i+1}/F_i) < \mu(F_i/F_{i-1})$ so that $\mu(Q) < \mu(F_i/F_{i-1})$.

Suppose $Q$ is not contained $F_{i+1}/F_i$. By induction hypothesis we may assume that $F_{i+1}/F_i$ is SCSS in $E/F_i$. Since $Q$ is semi-stable and $Q \subset F_{i+1}/F_i$, we have, by Lemma 1.3.5, $\mu(Q) < \mu(F_{i+1}/F_i)$. Since by hypothesis $\mu(F_{i+1}/F_i) < \mu(F_i/F_{i+1})$ it follows that $\mu(Q) < \mu(F_i/F_{i-1})$.

Combining Lemmas 1.3.7 and 1.3.8 we have

**Proposition 1.3.9.** *Let E be a vector bundle which is not semi-stable. Then E contains a uniquely determined flag*

$$0 = F_0 \subsetneqq F_1 \subsetneqq \cdots \subsetneqq F_k = E$$

*satisfying*

a) $F_i/F_{i-1}$ *is semi-stable for* $i=1,\dots k$

*and*

b) $\mu(F_i/F_{i-1}) > \mu(F_{i+1}/F_i)$ *for* $i=1,\dots,k-1$.

Now it is clear that the assumption that $k$ is algebraically closed can be dropped. In fact, if $E/Y$ is not semi-stable then by definition $\bar{E}/\bar{Y}$, where $\bar{Y} = Y \times_k \bar{k}$, is not semi-stable. Hence $\bar{E}/\bar{Y}$ has a unique flag satisfying the conditions of Proposition 1.3.9. But then it is clear, at least when $k$ is perfect, that this flag is already defined over $k$ i.e., it is induced by a flag in $E$.

*Definition 1.3.10.* A flag

$$0 = F_0 \subsetneqq F_1 \cdots \subsetneqq F_i \subsetneqq \cdots \subsetneqq F_k = E$$

is said to satisfy Condition (N) ("Numerical Condition") if

$$\mu(F_i/F_{i-1}) > \mu(F_{i+1}/F_i) \quad \text{for} \quad i = 1, \dots k-1$$

i.e., $\mu(F_1) > \mu(F_2/F_1) > \cdots > \mu(E/F_{k-1})$.

**Lemma 1.3.11.** *If a flag satisfies Condition* (N) *of Definition 1.3.10, then it also satisfies the condition*

$$\mu(F_1) > \mu(F_2) > \cdots > \mu(E).$$

We first note the following lemma whose proof is trivial.

**Lemma 1.3.12.** *Let* $0 \to E_1 \to E_2 \to E_3 \to 0$ *be an exact sequence of vector bundles* $E_i \neq 0$. *Then the following conditions are equivalent:*

i) $\mu(E_1) > \mu(E_3)$.
ii) $\mu(E_1) > \mu(E_2)$.
iii) $\mu(E_2) > \mu(E_3)$.

*Proof of Lemma 1.3.11.* First consider the exact sequence

$$0 \to F_1 \to F_2 \to F_2/F_1 \to 0.$$

Since by hypothesis $\mu(F_1) > \mu(F_2/F_1)$ it follows from Lemma 1.3.12 that $\mu(F_1) > \mu(F_2)$. Let us assume by induction that $\mu(F_{i-1}) > \mu(F_i)$ and consider the exact sequences

$$0 \to F_{i-1} \to F_i \to F_i/F_{i-1} \to 0$$

$$0 \to F_i/F_{i-1} \to F_{i+1}/F_{i-1} \to F_{i+1}/F_i \to 0.$$

Since by hypothesis $\mu(F_i/F_{i-1}) > \mu(F_{i+1}/F_i)$, we have by Lemma 1.3.12

$$\mu(F_i/F_{i-1}) > \mu(F_{i+1}/F_{i-1}) > \mu(F_{i+1}/F_i)$$

and the assumption $\mu(F_{i-1}) > \mu(F_i)$ implies $\mu(F_i) > \mu(F_i/F_{i-1})$. Hence we have $\mu(F_i) > \mu(F_{i+1}/F_i)$; this in turn implies that $\mu(F_i) > \mu(F_{i+1})$ on applying Lemma 1.3.12 to the exact sequence

$$0 \to F_i \to F_{i+1} \to F_{i+1}/F_i \to 0.$$

Thus Lemma 1.3.11 is proved.

The next two definitions are motivated by considerations in § 3.

*Definition 1.3.13.* Let $\mathscr{F}_1 = \{0 = F_0 \subset F_1 \cdots \subset F_k = E\}$ be a flag on $E$ and let $\mathscr{F}_2$ be a flag on $E$ which is a refinement of $\mathscr{F}_1$. We say that the pair $(\mathscr{F}_1, \mathscr{F}_2)$ satisfies the Condition $(\mathscr{C})$ if for each $i$, $1 \leqq i \leqq k$, the flag on $F_i/F_{i-1}$ induced by $\mathscr{F}_2$ satisfies Condition $(N)$ of Definition 1.3.10.

*Definition 1.3.14.* Let $C = \{\mathscr{F}_1, \ldots, \mathscr{F}_l\}$ be a chain of flags on $E$, i.e., $\mathscr{F}_{j+1}$ is a refinement of $\mathscr{F}_j$ for $j = 1, \ldots\ l-1$. We say that the chain $C$ satisfies the Condition (Num) if

a) the flag $\mathscr{F}_1$, satisfies Condition (N) of Definition 1.3.10 and

b) for each $j$, $1 \leqq j \leqq l-1$, the pair $(\mathscr{F}_j, \mathscr{F}_{j+1})$ satisfies Condition $(\mathscr{C})$ of Definition 1.3.13.

**Proposition 1.3.15.** *Let $F$ be a subbundle of $E$ occuring in a chain of flags satisfying the Condition* (Num) *of Definition 1.3.14. Then if $F \neq E$ we have $\mu(F) > \mu(E)$.*

For the proof we need

**Lemma 1.3.16.** *Let $0 \to E_1 \to E \to E/E_1 \to 0$ be an exact sequence of vector bundles with $\mu(E_1) > \mu(E)$. Let $F$ be a subbundle of $E$ containing $E_1$ and satisfying $\mu(F/E_1) > \mu(E/E_1)$. Then we have $\mu(F) > \mu(E)$.*

*Proof.* We use Lemma 1.3.12 several times in the proof. From the exact sequence

$$0 \to F/E_1 \to E/E_1 \to E/F \to 0$$

we see that the condition $\mu(F/E_1) > \mu(E/E_1)$ implies that $\mu(E/E_1) > \mu(E/F)$. Since $\mu(E_1) > \mu(E)$ we have $\mu(E) > \mu(E/E_1)$ so that $\mu(E) > \mu(E/F)$ which in turn implies that $\mu(F) > \mu(E)$.

*Proof of Proposition 1.3.15.* We prove the proposition by induction on $rk E$, the proposition being clear for $rk E = 2$. Consider the flag

$$\mathscr{F}_1 = \{0 = F_0 \subset F_1 \subset \cdots \subset F_k = E\}.$$

Since by definition, $\mathscr{F}_1$ satisfies Condition $(N)$ of Definition 1.3.10, we have, by Lemma 1.3.11

$$\mu(F_1) > \cdots > \mu(F_i) > \cdots > \mu(F_k) = \mu(E).$$

In particular the proposition is proved for $F = F_i$ and we are through if the chain consists only of $\mathscr{F}_1$. Now suppose $F \neq F_i$ for any $i$. If $F \subset F_{k-1}$, we have, by the induction hypothesis, $\mu(F) > \mu(F_{k-1})$; since $\mu(F_{k-1}) > \mu(E)$ it follows that $\mu(F) > \mu(E)$. If $F \supset F_{k-1}$ we have, by the induction hypothesis, $\mu(F/F_{k-1}) > \mu(F_k/F_{k-1})$. Since we also have $\mu(F_{k-1}) > \mu(F_k)$ it follows from Lemma 1.3.16 that we must have $\mu(F) > \mu(F_k) = \mu(E)$.

## II. Tamagawa Numbers and Siegel's Formula

### *2.1. Tamagawa Numbers of* $\mathrm{SL}_n$ *and* $\mathrm{PL}_n$

We first recall some results on Tamagawa numbers. We assume that the field of constants is a finite field $\mathbb{F}_q$ and denote the field of functions on the curve $Y/\mathbb{F}_q$ by $K/\mathbb{F}_q$. If $G/K$ is a connected affine algebraic group, we denote the group of adeles of $G/K$ by $G_A$. The group of $K$-rational points of $G_A$ is a discrete subgroup of the locally compact group $G_A$. If $\omega_A$ is a right invariant measure on $G_A$ then it induces a measure on $G_A/G_K$ and we may consider the volume

$$\mathrm{Vol}_{\omega_A}(G_A/G_K) = \int_{G_A/G_K} \omega_A$$

which may be infinite.

There exists a procedure to construct a right invariant measure $\omega_A$ starting from a right invariant non-zero differential form of highest degree on $G/K$, defined over $K$ [19, § 2.3]. This gives the so-called Tamagawa measure on $G_A$, which in some cases is uniquely determined (e.g. if $G$ is semi-simple or unipotent) and which in some cases depends on the choice of a system of convergence factors. The volume

$$\tau(G) = \int_{G_A/G_K} \omega_A^\tau$$

of $G_A/G_K$ with respect to this measure is called the Tamagawa number of $G/K$.

Let $\mathrm{GL}_n$ (resp. $\mathrm{SL}_n$) denote the full (resp. special) linear group over $K$. Let $\mathrm{PL}_n$ denote the projective group, namely the quotient of $\mathrm{GL}_n$ by its centre which is the multiplicative group $G_m/K$. It has been proved in [19, Theorem 3.3.1] that

$$\begin{aligned} \tau(\mathrm{SL}_n) &= 1 \\ \tau(\mathrm{PL}_n) &= n\,. \end{aligned} \tag{2.1.1}$$

One checks easily that the canonical mapping $\pi: \mathrm{GL}_n(A) \to \mathrm{PL}_n(A)$ is surjective. If $\underline{x} \in \mathrm{PL}_n(A)$ and $\pi(\underline{y}) = \underline{x}$, then $\det(\underline{y}) \in G_m(A) = I_K$, where $I_K$ is the idele group of $K$. The idele norm of $\underline{t} = \det \underline{y}$ is

$$|\det \underline{y}| = |\underline{t}| = q^{-\deg \underline{t}},$$

where $\deg \underline{t} = \Sigma\, n_{\mathfrak{p}} \operatorname{ord}_{\mathfrak{p}}(t_{\mathfrak{p}})$, where $n_{\mathfrak{p}}$ denotes the degree over $\mathbb{F}_q$ of the of the residue field at a closed point $\mathfrak{p}$ of $Y$. Since for an element

$$\underline{u} = \begin{pmatrix} \underline{a} & & 0 \\ & \ddots & \\ 0 & & \underline{a} \end{pmatrix} \in \mathrm{GL}_n(A)$$

of the centre we have $\det \underline{u} = \underline{a}^n$, we see that $\deg(\det \underline{y}) \bmod n$ does not depend on the choice of $\underline{y}$ and we obtain a homomorphism

$$\underline{\deg} : \mathrm{PL}_n(A) \to \mathbb{Z}/n\mathbb{Z} .$$

For $v \in \mathbb{Z}/n\mathbb{Z}$ we put

$$\mathrm{PL}_n^v(A) = \{\underline{x} \in \mathrm{PL}_n(A) \mid \underline{\deg}(x) = v\} ;$$

this set is invariant under the right action of $\mathrm{PL}_n(K)$

**Lemma 2.1.2.** *For $v \in \mathbb{Z}/n\mathbb{Z}$, we have*

$$\int_{\mathrm{PL}_n^v(A)/\mathrm{PL}_n(K)} \omega_A^\tau = 1 .$$

*Proof.* The Lemma is obvious since in this case the Tamagawa measure is left invariant and since the map deg is surjective.

## 2.2. *Truncated Tamagawa Numbers of Parabolic Subgroups of* $\mathrm{SL}_n$ *and* $\mathrm{PL}_n$

Let us now consider a parabolic subgroup $P/K$ of $\mathrm{SL}_n/K$. Without loss of generality we may assume that $P$ consists of the matrices $p$ in $\mathrm{SL}_n$ of the form

$$p = \begin{pmatrix} a_{11} & & & * \\ & a_{22} & & \\ & & \ddots & \\ 0 & & & a_{kk} \end{pmatrix}$$

where $a_{ii} \in \mathrm{GL}_{m_i}$ and $\Sigma\, m_i = n$. We introduce the characters

$$\gamma_v = \gamma_v^P : P \to G_m$$

$$\gamma_v : p \mapsto \prod_{i=1}^{v} \det a_{ii}$$

and $\lambda_v : p \mapsto \det a_{vv}$. Set $\chi_{ij} = \lambda_i^{m_j} \lambda_j^{-m_i}$ and put $\alpha_i = \chi_{i,i+1}$.

If $\underline{p} \in P_A$ then $\gamma_v(\underline{p}) \in I_K$ we define $\delta_v(\underline{p})$ by

$$|\gamma_v(\underline{p})| = q^{-\delta_v(\underline{p})} .$$

We then obtain a surjective homomorphism $\delta : P_A \to \mathbb{Z}^{k-1}$ by setting

$$\delta(\underline{p}) = (\delta_1(\underline{p}) \ldots, \delta_{k-1}(\underline{p})) .$$

If $\underline{n} \in \mathbb{Z}^{k-1}$ we define

$$P_A(\underline{n}) = \{\underline{p} \in P_A \mid \delta(\underline{p}) = \underline{n}\} .$$

Then $P_A(0)$ is a subgroup of $P_A$. We also define a map $\delta': P_A \to \mathbb{Q}^{k-1}$ by setting

$$\delta'(\underline{p}) = (\delta'_1(\underline{p}), \ldots, \delta'_{k-1}(\underline{p})),$$

where $|\alpha_i(\underline{p})| = q^{-\delta'_i(\underline{p})}$.

We now construct a Tamagawa measure on $P_A$; in this case we have to introduce convergence factors. The algebraic group $P/K$ is obtained by base extension from an algebraic group $P_0/\mathbb{F}_q$, $P_0 \subset \mathrm{SL}_n/\mathbb{F}_q$. Let us choose a right invariant differential form $\omega_0 \neq 0$ of highest degree on $P_0/\mathbb{F}_q$; this gives also such a form $\omega$ on $P/K$. For each closed point $\mathfrak{p}$ in $Y$, the form $\omega$ defines a measure on $P_{\hat{K}_\mathfrak{p}}$, where $\hat{K}_\mathfrak{p}$ denotes the completion of $K$ with respect to the valuation defined by $\mathfrak{p}$ [19, § 2.2]. If $\hat{\mathcal{O}}_\mathfrak{p}$ denotes the ring of integers in $K_\mathfrak{p}$ and $P_{\hat{\mathcal{O}}_\mathfrak{p}} = P_{\hat{K}_\mathfrak{p}} \cap \mathrm{SL}_n(\hat{\mathcal{O}}_\mathfrak{p})$ is the group of integral points in $P_{K_\mathfrak{p}}$ we then have

$$\mathrm{vol}_{\omega_\mathfrak{p}}(P_{\hat{\mathcal{O}}_\mathfrak{p}}) = |P_0(k(\mathfrak{p}))| \, (N\mathfrak{p})^{-\dim P}$$

where $k(\mathfrak{p})$ is the residue field at $\mathfrak{p}$ and $N\mathfrak{p} = |k(\mathfrak{p})|$. (Compare [19], Theorem 2.2.5 and its proof.)

Now we have obviously

$$|P_0(k(\mathfrak{p})| = (N\mathfrak{p})^{\dim U_0} \, |M_0(k(\mathfrak{p}))| \, (N\mathfrak{p} - 1)^{k-1}$$

where $U_0$ is the unipotent radical of $P$ and $M_0$ is the semi-simple group

$$M_0 = \left\{ \begin{pmatrix} a_{11} & & & 0 \\ & a_{22} & & \\ & & \ddots & \\ 0 & & & a_{kk} \end{pmatrix} \middle| \, a_{ii} \in \mathrm{SL}_{m_i} \right\}.$$

It follows that $\lambda_\mathfrak{p} = (1 - 1/N_\mathfrak{p})^{-k+1}$ is a system of convergence factors since $\prod_\mathfrak{p} (|M_0(k(\mathfrak{p})|/(N_\mathfrak{p})^{\dim M_0})$ is well known to be convergent [19]. We define the Tamagawa measure $\omega_A^\tau$ on $P_A$ by taking $\lambda_\mathfrak{p}$ as a system of convergence factors.

**Proposition 2.2.1.** *We have*

a) $$\int_{P_A(0)/P_K} \omega_A^\tau = \left( \frac{|J_{\mathbb{F}_q}|}{(q-1)q^{g-1}} \right)^{k-1}$$

*where* $|J_{\mathbb{F}_q}|$ *denotes the number of* $\mathbb{F}_q$*-rational points of the Jacobian* $J/\mathbb{F}_q$ *of* $Y$.

b) *For* $\underline{p} \in P_A$, *we have*

$$\int_{\underline{p}P_A(0)/P_K} \omega_A^\tau = q^{-\sum_{i=1}^{k-1} f_i \delta_i(\underline{p})} \int_{P_A(0)/P_K} \omega_A^\tau = q^{-\sum_{i=1}^{k-1} a_i \delta_i'(\underline{p})} \int_{P_A(0)/P_K} \omega_A^\tau$$

*where* $f_i = m_i + m_{i+1}$ *and* $a_i = s_{i+1}(n - s_{i+1})/m_i m_{i+1}$ *with* $s_i = \sum_{i \leqq l \leqq k} m_l$.

*Proof*. Let $U \subset P$ be the unipotent radical of $P$; then $H = P/U$ is a reductive group over $K$. We apply Theorem 2.4.4 in [19] and obtain

$$\int_{P_A(0)/P_K} \omega_A^\tau = \int_{U_A/U_K} \omega_{U,A}^\tau \int_{H_A(0)/H_K} \omega_{H,A}^\tau = \tau(H_A(0)/H_K)$$

since normalisation in Tamagawa measure gives measure 1 for the unipotent group $U_A$. To compute $\tau(H_A(0)/H_K)$ consider the exact sequence

$$1 \to M \to H \xrightarrow{\gamma} G_m^{k-1} \to 1$$

where $\gamma = (\gamma_1 \ldots \gamma_{k-1})$. We apply again Theorem 2.4.4 in [19] to this exact sequence. If $S/K \approx G_m^{k-1}/K$ then the maps $H_A \to S_A$, $H_K \to S_K$ are surjective since $H^1(K, \mathrm{SL}_n) = 0$ ([15], Chapter X, Proposition 3, Corollary). Now we take for $f$ in Theorem 2.4.4 of [19], the characteristic function of $S_A(0)/S_K$ and get

$$\int_{H_A(0)/H_K} \omega_{H,A}^\tau = \tau(\Pi\, \mathrm{SL}_{m_i}) \int_{S_A(0)/S_K} \omega_{S,A}^\tau = \int_{S_A(0)/S_K} \omega_{S,A}^\tau$$

since $\tau(\mathrm{SL}_{m_i}) = 1$ by 2.1.1. To evaluate $\int \omega_{S,A}^\tau$, let $\mathfrak{A}$ denote the canonical maximal compact subgroup of $S_A$. Then the number of double cosets $\mathfrak{A} \backslash S_A(0)/S_K$ is $|J_{\mathbb{F}_q}|^{k-1}$ and, by the choice of our convergence factors, $\operatorname{vol} \omega_{S,A}^\tau(\mathfrak{A}) = q^{(1-g)(k-1)}$. Now the exact sequence

$$1 \to S_A(0)/\mathfrak{A} S_K \to S_A(0)/S_K \to \mathfrak{A} S_K/S_K \to 1$$

shows that

$$\int_{S_A(0)/S_K} \omega_{S,A}^\tau = (|J_{\mathbb{F}_q}|/(q-1))^{k-1} \cdot q^{(1-g)(k-1)}$$

on remarking that

$$\mathfrak{A} S_K/S_K = \mathfrak{A}/S_K \cap \mathfrak{A} = \mathfrak{A}/(\mathbb{F}_q^*)^{k-1}.$$

This proves a).

To prove b) we note that the measure $\omega_A^\tau$ is not left invariant and that the modulus of the left translation by $\underline{p}$ is precisely $q^{-\Sigma f_i \delta_i(p)} = q^{-\Sigma a_i \delta_i'(p)}$. (See e.g. [3], Chapter VII, § 3, No. 3.)

Next we consider the corresponding parabolic subgroups $\underline{P}$ of $\mathrm{PL}_n/K$ and $\tilde{P}$ in $\mathrm{GL}_n/K$. The characters $\lambda_i = \det a_{ii}$ are defined on $\tilde{P}_A$ and the roots $\chi_{ij} = \lambda_i^{m_j} \lambda_j^{-m_i}$ are defined on $\underline{P}_A$. We set $\alpha_i = \chi_{i,i+1}$. We define

$$\underline{P}_A^0(0) = \left\{ \underline{p} \in \underline{P}_A \;\middle|\; \begin{array}{ll} \underline{\deg}(\underline{p}) = 0 & \text{and} \\ |\chi_{ij}(\underline{p})| = 1 & \text{for all } i, j \end{array} \right\}.$$

We see that

$$\underline{P}_A^0(0)=\{\underline{p}\in\underline{P}_A \mid \underline{\deg}(\underline{p})=0 \quad\text{and}\quad |\alpha_i(\underline{p})|=1 \quad\text{for}\quad 1\leqq i\leqq k-1\}.$$

We also note that the character

$$R=\prod_{i<j}\chi_{ij}=\prod\gamma_i^{f_i}=\prod\alpha_i^{a_i}$$

is defined on $\underline{P}_A$.

**Proposition 2.2.2.** *We have*

a) $$\int_{\underline{P}_A^0(0)/\underline{P}_K}\omega_{\underline{P},A}^{\tau}=\left(\frac{|J_{\mathbb{F}_q}|}{(q-1)q^{g-1}}\right)^{(k-1)}$$

b) *For* $\underline{p}\in\underline{P}_A^0$ *one has*

$$\int_{\underline{p}\underline{P}_A^0(0)/\underline{P}_K}\omega_{\underline{P},A}^{\tau}=q^{-\Sigma a_i\delta_i'(\underline{p})}\int_{\underline{P}_A^0(0)/\underline{P}_K}\omega_{\underline{P},A}^{\tau}$$

where the $a_i$ have the same meaning as in Proposition *2.2.1*.

*Remark.* The main assertion of the proposition is that these volumes are the same as those corresponding to the parabolic subgroup $P$.

*Proof.* Referring to the proof of Proposition 2.2.1 we see easily that we have to prove that

$$\int_{H_A(0)/H_K}\omega_{H,A}^{\tau}=\int_{\underline{H}_A^0(0)/\underline{H}_K}\omega_{\underline{H},A}^{\tau}$$

where $\underline{H}\subset\underline{P}$ is the reductive part in $\underline{H}$ corresponding to $H$. We consider the diagram of algebraic groups

$$\begin{array}{c} 1\\ \downarrow\\ G_m\\ \downarrow\\ 1\to \mathrm{SL}_n\to\mathrm{GL}_n\to G_m\to 1\\ \downarrow\pi\\ \mathrm{PL}_n\\ \downarrow\\ 1\end{array}$$

Let $\tilde{H}$ be the reductive subgroup of $\mathrm{GL}_n$ corresponding to $H$ and $\underline{H}$ i.e. $\tilde{H}$ is the inverse image of $\underline{H}$. Let

$$\tilde{H}_A(0)=\{\underline{h}\in\tilde{H}_A \mid |\det\underline{h}|=1, \pi_A(\underline{h})\in\underline{H}_A(0)\}.$$

We then have a diagram of exact sequences

$$\begin{array}{ccccccccc} & & & & 1 & & & & \\ & & & & \downarrow & & & & \\ & & & & I_k^0 & & & & \\ & & & & \downarrow & & & & \\ 1 & \to & H_A(0) & \to & \tilde{H}_A^0(0) & \to & I_k^0 & \to & 1 \\ & & & & \downarrow & & & & \\ & & & & \underline{H}_A^0(0) & & & & \\ & & & & \downarrow & & & & \\ & & & & 1 & & & & \end{array}$$

where $I_K^0 = G_{m,A}(0) = \{x \in I_K \mid |x| = 1\}$. It now follows by [19, Theorem 2.4.4] that

$$\int_{H_A(0)/H_K} \omega_{H,A}^{\tau} = \int_{\underline{H}_A^0(0)/\underline{H}_K} \omega_{\underline{H},A}^{\tau} .$$

## 2.3. *Siegel's Formula*

Let $E_0/Y$ be a vector bundle of rank $n$ over the curve $Y/\mathbb{F}_q$. Put $L_0 = \det E_0$. Let $V/K$ be the generic fibre of $E_0$; $V/K$ is the $n$-dimensional vector space of all meromorphic sections of $E_0/Y$. For a closed point $\mathfrak{p}$ we denote by $\mathcal{O}_{\mathfrak{p}}$ the ring of integers at $\mathfrak{p}$ and by $\hat{\mathcal{O}}_{\mathfrak{p}}$ and $\hat{K}_{\mathfrak{p}}$ the completions of $\mathcal{O}_{\mathfrak{p}}$ and $K$ respectively with respect to the valuation defined by $\mathfrak{p}$. If $\hat{E}_{0,\mathfrak{p}}$ denotes the completion (with respect to the valuation defined by $\mathfrak{p}$) of the stalk $E_{0,\mathfrak{p}} = \varinjlim_{U \ni \mathfrak{p}} \Gamma(U, E_0)$, then $\hat{E}_{0,\mathfrak{p}}$ is an $\hat{\mathcal{O}}_{\mathfrak{p}}$ lattice in $V \otimes_K \hat{K}_{\mathfrak{p}}$. We can reconstruct $E$ from the family of lattices $\{\hat{E}_{0,\mathfrak{p}}\}$ (see [9, § 2]).

The bundle $E_0$ defines a maximal compact subgroup $\mathfrak{K}$ of $\mathrm{SL}(V)_A$; namely, $\mathfrak{K} = \Pi\, \mathfrak{K}_{\mathfrak{p}}$ where $\mathfrak{K}_{\mathfrak{p}} = \mathrm{SL}(\hat{E}_{0,\mathfrak{p}}) \subset \mathrm{SL}(V \otimes \hat{K}_{\mathfrak{p}})$.

For any $\underline{x} \in \mathrm{SL}(V)_A$ we consider the family of lattices

$$\{\hat{E}_{0,\mathfrak{p}}^{\underline{x}}\} = \{x_{\mathfrak{p}}^{-1} \hat{E}_{0,\mathfrak{p}}\} .$$

Then this family defines a locally free sheaf $E_0^{\underline{x}}$ by

$$\Gamma(U, E_0^{\underline{x}}) = \{v \in V \mid v \in \underline{x}_{\mathfrak{p}}^{-1} \hat{E}_{0,\mathfrak{p}}\}$$

for all $\mathfrak{p} \in U$. It is easy to see that the isomorphism class of the vector bundle $\hat{E}_0^{\underline{x}}$ depends only on the double coset $\mathfrak{K}\underline{x}\mathrm{SL}(V)_K$ and that the mapping

$$\mathfrak{K}\backslash \mathrm{SL}(V)_A/\mathrm{SL}(V)_K \to \left\{ \begin{array}{l} \text{Set of isomorphism classes of bundles} \\ E/Y \text{ of rank } n \text{ and } \det E \approx L_0 \end{array} \right\} \quad (2.3.1)$$

is surjective [9, § 2]. Let $\tilde{\mathfrak{K}}$ be the maximal compact subgroup of $\mathrm{GL}(V)_A$ defined by $E_0$. Then the following facts are obvious

1) The group of automorphisms of $E_0^{\underline{x}}$ is $\underline{x}^{-1}\tilde{\mathfrak{K}}\underline{x}\cap\mathrm{GL}(V)_K$.

2) $E_0^{\underline{x}}\approx E_0^{\underline{y}}$ if and only if $\underline{y}\in\tilde{\mathfrak{K}}\underline{x}\mathrm{GL}(V)_K$.

These facts give information about the number of points in the inverse image of an isomorphism class under the mapping in 2.3.1. Let $D=\mathfrak{K}\underline{x}\mathrm{SL}(V)_K$ be a double coset and $D'=\mathfrak{K}\underline{y}\mathrm{SL}(V)_K$ a double coset defining the same isomorphism class. If $\underline{y}=\underline{k}\underline{x}a$ with $\underline{k}\in\tilde{\mathfrak{K}}$, $a\in\mathrm{GL}(V)_K$ then $\underline{x}^{-1}\underline{y}=\underline{x}^{-1}\underline{k}\underline{x}a$. Hence $\det(\underline{x}^{-1}\underline{k}\underline{x})$. $\det a=1$ and $\det(a)^{-1}$ $=\det(\underline{x}^{-1}\underline{k}\underline{x})\in\mathbb{F}_q^*$. If $\underline{y}=\underline{k}_1\underline{x}a_1$, $\underline{k}_1\in\tilde{\mathfrak{K}}$, $a_1\in\mathrm{GL}(V)_K$ then $\underline{k}_1\underline{x}a_1=\underline{k}\underline{x}a$ so that $\underline{x}^{-1}\underline{k}^{-1}\underline{k}_1\underline{x}=aa_1^{-1}$; this means that $aa_1^{-1}\in\underline{x}^{-1}\tilde{\mathfrak{K}}\underline{x}\cap\mathrm{GL}(V)_K$ $=\mathrm{Aut}\,E_0^{\underline{x}}$. Hence, if we consider the map

$$\det\colon \mathrm{Aut}\,E_0^{\underline{x}}=\underline{x}^{-1}\tilde{\mathfrak{K}}\underline{x}\cap\mathrm{GL}(V)_K\to\mathbb{F}_q^*$$

we see that the pair $(D,D')$ defines a class in $\mathbb{F}_q^*/\det\mathrm{Aut}(E_0^{\underline{x}})$. Moreover $\underline{y}\in\mathfrak{K}\underline{x}\mathrm{SL}(V)_K$ if and only if this class is trivial; in fact if $\underline{y}=\underline{k}\underline{x}a$ and $\det a=\det b$ for $b=\underline{x}^{-1}\underline{k}_1\underline{x}\in\mathrm{Aut}\,E_0^{\underline{x}}$, $\underline{k}_1\in\tilde{\mathfrak{K}}$, then $\underline{y}=\underline{k}\underline{k}_1\underline{x}b^{-1}a$; since $\det b^{-1}a=\det\underline{x}=\det\underline{y}=1$ we have $\det(\underline{k}\underline{k}_1)=1$ so that $\underline{y}\in\mathfrak{K}\underline{x}\mathrm{SL}(V)_K$. On the other hand it is clear that given $\underline{x}\in\mathrm{SL}(V)_A$ we can find $\underline{y}\in\mathrm{SL}(V)_K$ such that $\underline{y}=\underline{k}\underline{x}a$, $\underline{k}\in\tilde{\mathfrak{K}}$, $a\in\mathrm{GL}(V)_K$ and $\det(a)$ is a given element in $\mathbb{F}_q^*$. Therefore we see that for a given $\underline{x}\in\mathrm{SL}(V)_A$ there are exactly $|\mathbb{F}_q^*/\mathrm{Aut}\,\underline{E}_0^{\underline{x}}|$ double cosets in $\mathfrak{K}\backslash\mathrm{SL}(V)_A/\mathrm{SL}(V)_K$ which map into the same isomorphism class of vector bundles. Since the automorphisms of a stable bundle consists only of scalars, we have, from the above considerations, the following

**Lemma 2.3.2.** *The number of double cosets in* $\mathfrak{K}\backslash\mathrm{SL}(V)_A/\mathrm{SL}(V)_K$ *which are mapped into the same isomorphism class of stable bundles is* $|\mathbb{F}_q^*/\mathbb{F}_q^{*n}|$.

We now exploit the fact that the Tamagawa number of $\mathrm{SL}(V)_K$ is 1:

$$\int_{\mathrm{SL}(V)_A/\mathrm{SL}(V)_K}\omega_A^\tau=1\,.$$

Decomposing $\mathrm{SL}(V)_A$ into double cosets $\mathfrak{K}\underline{x}\mathrm{SL}(V)_A$ we have

$$1=\int_{\mathrm{SL}(V)_A/\mathrm{SL}(V)_K}\omega_A^\tau=\sum_{\underline{x}}\int_{\mathfrak{K}\cdot\underline{x}\mathrm{SL}(V)_K/\mathrm{SL}(V)_K}\omega_A^\tau$$

where $\underline{x}$ runs through representatives of double cosets. Since

$$(\mathrm{vol}(\mathfrak{K}\underline{x}\mathrm{SL}(V)_K/\mathrm{SL}(V)_K))=(\mathrm{vol}_{\omega_A^\tau}(\mathfrak{K}))\cdot|\underline{x}^{-1}\mathfrak{K}\underline{x}\cap\mathrm{SL}(V)_K|^{-1}$$

we obtain

$$(2.3.3)\quad 1/\mathrm{vol}_{\omega_A^\tau}(\mathfrak{K})=\sum_{\underline{x}}\left(\frac{1}{|\underline{x}^{-1}\mathfrak{K}\underline{x}\cap\mathrm{SL}(V)_K|}\right)$$

where $\underline{x}$ runs through representatives of double cosets.

We now assume that $\deg(L_0)=r$ is coprime to $n$. In this case all semi-stable bundles are stable. Moreover, if $\underline{x}\in \mathrm{SL}(V)_A$ is such that $E_0^{\underline{x}}$ is stable, we have $\mathrm{Aut}\, E_0^{\underline{x}}=\mathbb{F}_q^*$ and hence

$$\begin{aligned}|\underline{x}^{-1}\mathfrak{K}\underline{x}\cap \mathrm{SL}(V)_K| &= \text{number of } n\text{th roots of unity in } \mathbb{F}_q\\ &= |\mathbb{F}_q^*/\mathbb{F}_q^{*n}|\,.\end{aligned}$$

It then follows from Lemma 2.3.2, that the contribution to the sum in (2.3.3) from the double cosets for which $E_0^{\underline{x}}$ is stable is exactly the number of isomorphism classes of stable bundles $E/Y$ with $\det E\approx L_0$. But by Proposition 1.2.1 this number is the same as the number $|M_{\mathbb{F}_q}|$ of $\mathbb{F}_q$ rational points of the corresponding moduli scheme $M/\mathbb{F}_q$. On the other hand

$$\begin{aligned}\mathrm{vol}_{\omega_A^\tau}(\mathfrak{K}) &= q^{-(n^2-1)(g-1)}\Big(\prod_{\mathfrak{p}} \mathrm{vol}(\mathfrak{K}_{\mathfrak{p}})\Big)\\ &\equiv q^{-(n^2-1)(g-1)}\prod_{\mathfrak{p}}\left(1-\frac{1}{(N\mathfrak{p})^2}\right)\cdots\left(1-\frac{1}{(N\mathfrak{p})^n}\right)\\ &\qquad \text{(see [19], pp. 22, 33)}\\ &= q^{-(n^2-1)(g-1)}\zeta(2)^{-1}\ldots\zeta(n)^{-1}\end{aligned}$$

where $\zeta$ denotes the Zeta function of $K$. Thus we obtain

**Proposition 2.3.4.** *We have*

$$|M_{\mathbb{F}_q}| = q^{(n^2-1)(g-1)}\zeta(2)\ldots\zeta(n)-\sum_{\underline{x}}\frac{1}{|\underline{x}^{-1}\mathfrak{K}\underline{x}\cap \mathrm{SL}(V)_K|}$$

*where $\underline{x}$ runs through representatives of double cosets (in $\mathfrak{K}\backslash \mathrm{SL}(V)_A/\mathrm{SL}(V)_K$) such that $E_0^{\underline{x}}$ is not stable.*

Next we consider the projective bundle $X_0/Y$ where $X_0=\mathbb{P}(E_0)$. Then the scheme of automorphisms of $X_0/Y$ is locally isomorphic to $\mathrm{PL}_n/Y$ and the generic fibre is $\mathrm{PL}(V)$.

Let

$$\underline{\mathfrak{K}} = \prod_{\mathfrak{p}} (\mathrm{Aut}\, X_0)_{\hat{\mathcal{O}}_{\mathfrak{p}}} \subset \mathrm{PL}(V)_A\,.$$

We easily see that

$$\underline{\mathfrak{K}}\backslash \mathrm{PL}^0(V)_A/\mathrm{PL}(V)_K \simeq \left\{\begin{array}{l}\text{Isomorphism classes of projective}\\ \text{bundles } X/Y \text{ with fibre dimension}\\ (n-1) \text{ and } \underline{\deg}(X)=(\deg L_0) \bmod n\,.\end{array}\right\}$$

(This is proved by the same kind of arguments as used for $\mathrm{SL}(V)_A$. In contrast to that case we have bijectivity here since $\mathrm{PL}_n=\mathrm{Aut}\,\mathbb{P}(V)$ while in the case SL, $\mathrm{Aut}\, V=\mathrm{GL}$ but we take double cosets in SL.)

We get, as above,

$$\frac{1}{\mathrm{vol}_{\omega\xi_A}(\mathfrak{K})} = \sum_{\underline{x} \in \mathfrak{K}\backslash \mathrm{PL}^0(V)_A/\mathrm{PL}(V)_K} \frac{1}{|\underline{x}^{-1}\mathfrak{K}\underline{x} \cap \mathrm{PL}(V)_K|}.$$

Now $\underline{x}^{-1}\underline{\mathfrak{K}}\underline{x} \cap \mathrm{PL}(V)_K = \mathrm{Aut}(X/Y)_{\mathbb{F}_q}$, where $X = X_{\mathfrak{d}}^{\underline{x}}$.

It may happen that a stable $X/Y$ has non-trivial automorphisms; but in that case there are exactly $|\mathrm{Aut}(X)_{\mathbb{F}_q}|$ isomorphism classes of projective bundles which are mapped into the same point as $X$ on the moduli scheme $N/\mathbb{F}_q$ (Proposition 1.2.2).

Hence

$$\sum_{\underline{x} \in \underline{\mathfrak{K}}\backslash \mathrm{PL}^0(V)_A/\mathrm{PL}(V)_K} 1/|\underline{x}^{-1}\underline{\mathfrak{K}}\underline{x} \cap \mathrm{PL}(V)_K| = |N_{\mathbb{F}_q}|$$

$$+ \sum_{\substack{\underline{x} \\ \underline{x} \in \underline{\mathfrak{K}}\backslash \mathrm{PL}^0(V)_A/\mathrm{PL}(V)_K \\ X_{\mathfrak{d}}^{\underline{x}} \text{ not stable}}} 1/|\underline{x}^{-1}\underline{\mathfrak{K}}\underline{x} \cap \mathrm{PL}(V)_K|$$

On the other hand

$$1/\mathrm{vol}_{\underline{\omega}\xi_A}(\underline{\mathfrak{K}}) = q^{(n^2-1)(g-1)}\zeta(2)\dots\zeta(n).$$

Thus we obtain

**Proposition 2.3.5.**

$$|N_{\mathbb{F}_q}| = q^{(n^2-1)(g-1)}\zeta(2)\dots\zeta(n) - \sum_{\underline{x}} \frac{1}{|\underline{x}^{-1}\underline{\mathfrak{K}}\underline{x} \cap \mathrm{PL}(V)_K|}$$

*where $\underline{x}$ runs through representatives of double cosets in $\underline{\mathfrak{K}}\backslash \mathrm{PL}^0(V)_A/\mathrm{PL}(V)_K$ such that $X_{\mathfrak{d}}^{\underline{x}}$ is not stable.*

## III. Proof of the Theorems

### *3.1. The Summation over the Unstable Part in Siegel's Formula. Proof of Theorem 2*

We prove Theorem 2 by showing that the summation over unstable bundles in Proposition 2.3.4 and 2.3.5 are the same.

To prove this let us assume that

$$E_0 = L_0 \oplus \underbrace{\mathcal{O}_Y \oplus \cdots \oplus \mathcal{O}_Y}_{(n-1)\text{ summands}}.$$

We define a (complete) flag $0 \subset F_{0,1} \cdots \subset F_{0,n-1} \subset E_0$ on $E_0$ by setting $F_{0,i} = L_0 \oplus \underbrace{\mathcal{O}_Y \oplus \cdots \oplus \mathcal{O}_Y}_{(i-1)\text{ summands}}$ and call it the standard flag on $E_0$. If $m_i \geqq 1$

are integers with $\sum_{i=1}^{k} m_i = n$, we call the stabilizer of the flag $0 \subset F_{0,m_1} \subset F_{0,m_1+m_2} \subset \cdots \subset E_0$ a standard parabolic subgroup (of type $(m_1, \ldots, m_k)$).

Suppose that $\underline{x} \in \mathrm{SL}(V)_A$ is such that $E_0^{\underline{x}}$ is not stable. Then by Proposition 1.2.9 we have a uniquely determined flag

$$0 \subset F_1 \subset F_2 \cdots \subset F_{k-1} \subset F_k = E_0^{\underline{x}}$$

such that $F_i/F_{i-1}$ is semi-stable and $\mu(F_1) > \mu(F_2/F_1) \cdots > \mu(E_0^{\underline{x}}/F_{k-1})$. If $\dim(F_1) = m_1$, $\dim F_2 = m_1 + m_2, \ldots$, then consider the flag on $E_0$ defined by $0 \subset F_{0,m_1} \subset F_{0,m_1+m_2} \cdots \subset E_0$ and the stabilizer of this flag in $\mathrm{SL}(V)_A$, which is a (standard) parabolic subgroup. Thus to each $\underline{x}$ with $E_0^{\underline{x}}$ not stable there corresponds a standard parabolic subgroup and this parabolic group depends only the double coset containing $\underline{x}$. Thus if $\mathrm{Inst} \subset \mathfrak{K} \backslash \mathrm{SL}(V)_A / \mathrm{SL}(V)_K$ is the set of double cosets which give rise to vector bundles which are not stable, then we have a decomposition

$$\mathrm{Inst} = \bigcup_P \mathrm{Inst}^P$$

where $P$ runs through the different parabolic subgroups fixing subflags of the standard flag. A similar decomposition holds for the set of double cosets $\underline{\mathrm{Inst}} \subset \underline{\mathfrak{K}} \backslash \mathrm{PL}^0(V)_A / \mathrm{PL}(V)_K$ giving rise to projective bundles which are not stable.

If $P$ is a standard parabolic subgroup of type $(m_1, \ldots, m_k)$ and $\underline{p} \in P_A$, then $E_0^{\underline{p}}$ has a canonical flag

$$0 \subset F_{0,m}^{\underline{p}} \subset F_{0,m_1+m_2}^{\underline{p}} \cdots \subset E_0^{\underline{p}}$$

and an analogous assertion is clear for $\underline{P} \subset \mathrm{PL}(V)/K$. Now we see easily that, writing $v_i = m_1 + \cdots + m_i$,

$$\begin{aligned} \deg(F_{0,v_i}^{\underline{p}}/F_{0,v_{i-1}}^{\underline{p}}) &= \deg(F_{0,v_i}/F_{0,v_{i+1}}) + (n_i - n_{i-1}) \\ &= (n_i - n_{i-1}) + d_i \end{aligned}$$

where $d_1 = r$ and $d_i = 0$ for $i > 1$ and $n_i = \delta_i(\underline{p})$ with the notation of § 2.2.

We then define a subset $P_{A,\mathrm{num}}$ of $P_A$ by

$$P_{A,\mathrm{num}} = \left\{ \underline{p} \in P_A \,\middle|\, \frac{l_1(\underline{p})}{m_1} > \cdots > \frac{l_k(\underline{p})}{m_k} \right\}$$

where $l_i(\underline{p}) = d_i + n_i - n_{i-1}$. We easily verify that

$$P_{A,\mathrm{num}} = \{\underline{p} \in P_A \mid |\alpha_i(\underline{p})| < C_i \quad \text{for} \quad 1 \leqq i \leqq k-1\}$$

where $\alpha_i = \lambda_i^{m_{i+1}} \lambda_{i+1}^{-m_i}$ and $C_i = q^{(d_i m_{i+1} - d_{i+1} m_i)}$.

We also define $\underline{P}^0_{A,\mathrm{num}}$ similarly:

$$\underline{P}^0_{A,\mathrm{num}} = \{\underline{p} \in \underline{P}^0_A \mid |\alpha_i(\underline{p})| < C_i \quad \text{for} \quad 1 \leqq i \leqq k-1\}$$

noting that $\alpha_i$ are defined on $\underline{P}$. Moreover we define

$$P^{sst}_{A,\mathrm{num}} = \{\underline{p} \mid \underline{p} \in P_{A,\mathrm{num}} \quad \text{and} \quad F^{\underline{p}}_{\mathfrak{O},\nu_i}/F^{\underline{p}}_{\mathfrak{O},\nu_{i-1}} \quad \text{is semi-stable}\}$$

and

$$\underline{P}^{0,sst}_{A,\mathrm{num}} = \{\underline{p} \mid \underline{p} \in \underline{P}^0_{A,\mathrm{num}} \quad \text{and} \quad P(F^{\underline{p}}_{\mathfrak{O},\nu_i}/F^{\underline{p}}_{\mathfrak{O},\nu_{i-1}}) \quad \text{is semi-stable}\}.$$

It is now clear that we have mappings

$$\mathfrak{K} \cap P_A \backslash P^{sst}_{A,\mathrm{num}}/P_K \to \mathrm{Inst}^P$$

and

$$\underline{\mathfrak{K}} \cap \underline{P}_A \backslash \underline{P}^{0,sst}_{A,\mathrm{num}}/\underline{P}_K \to \mathrm{Inst}^P$$

and these mappings are bijective by Proposition 1.3.9. Moreover we have

$$\underline{p}^{-1}\mathfrak{K}\underline{p} \cap \mathrm{SL}(V)_K = \underline{p}^{-1}\mathfrak{K}\underline{p} \cap P_K \quad \text{for} \quad \underline{p} \in P^{sst}_{A,\mathrm{num}}$$

and

$$\underline{p}^{-1}\underline{\mathfrak{K}}\underline{p} \cap \mathrm{PL}(V)_K = \underline{p}^{-1}\underline{\mathfrak{K}}\underline{p} \cap P_K \quad \text{for} \quad \underline{p} \in P^{0,sst}_{A,\mathrm{num}},$$

since the flag structure is unique. Thus we get

$$\sum_{\underline{x} \in \mathrm{Inst}^P} \frac{1}{|\underline{x}^{-1}\mathfrak{K}\underline{x} \cap \mathrm{SL}(V)_K|} = \sum_{P_A \cap \mathfrak{K} \backslash P^{sst}_{A,\mathrm{num}}/P_K} \frac{1}{|\underline{p}^{-1}\mathfrak{K}\underline{p} \cap P_K|}$$

and an analogous formula for $\mathrm{PL}(V)/K$.

If we introduce right invariant measures $\omega_{P,A}$ and $\omega_{\underline{P},A}$ on $P$ and $\underline{P}$ such that the maximal compact subgroups $\mathfrak{K} \cap P_A$ and $\underline{\mathfrak{K}} \cap \underline{P}_A$ have volume 1 we see that the values of the summation over the unstable part are

$$\sum_P \int_{P^{sst}_{A,\mathrm{num}}/P_K} \omega_{P,A} \quad \text{for} \quad \mathrm{SL}(V)_K$$

and

$$\sum_{\underline{P}} \int_{P^{0,sst}_{A,\mathrm{num}}/P_K} \omega_{\underline{P},A} \quad \text{for} \quad \mathrm{PL}(V)_K$$

where $P$ (resp. $\underline{P}$) runs through standard parabolic subgroups. But still it is not clear that these two expressions are equal; the difficulty is due to the fact that the condition for $\underline{p}$ to be in $P^{sst}_A$ is difficult to handle. Therefore we write

$$\int_{P^{sst}_{A,\mathrm{num}}/P_K} \omega_{P,A} = \int_{P_{A,\mathrm{num}}/P_K} \omega_{P,A} - \int_{(P_{A,\mathrm{num}}/P_K) - (P^{sst}_{A,\mathrm{num}}/P_K)} \omega_{P,A} \, .$$

(It will be seen later that the integrals are convergent.) To understand the second ("error") term on the right, we observe that an element $\underline{p} \in P_{A,\mathrm{num}} - P^{sst}_{A,\mathrm{num}}$ gives rise to a bundle $E^{\underline{p}}_0$ with a flag

$$0 \subset F^{\underline{p}}_{0,\nu_1} \subset F^{\underline{p}}_{1,\nu_2} \subset \cdots \subset F^{\underline{p}}_{0,\nu_k} = E^{\underline{p}}_0$$

for which at least one of the quotients $F^{\underline{p}}_{0,\nu_i}/F^{\underline{p}}_{0,\nu_{i-1}}$ is not semi-stable. Therefore we can find, by Proposition 1.3.9 unique flags in the non-semi-stable quotients satisfying the conditions of that proposition. Thus we see that

$$\int_{P_{A,\mathrm{num}}/P_K} \omega_{P,A} - \int_{P^{sst}_{A,\mathrm{num}}/P_K} \omega_{P,A} = \sum_{Q \subsetneqq P} \int_{Q^*_A/Q_K} \omega_{Q,A}$$

where $Q$ runs through proper standard parabolic subgroups of $P$ and $Q^*_A$ is the subset of $Q_A$ consisting of elements $\underline{q} \in Q_A$ satisfying the following two conditions:

i) $\underline{q} \in P_{A,\mathrm{num}}$,

ii) if $0 \subset V_l \cdots \subset V_k = V$ is the flag defined by $P$ then the image $\bar{Q}^{(i)}$ of $Q$ in $\mathrm{GL}(V_i/V_{i-1})$ by the canonical homomorphism $P \to \mathrm{GL}(V_i/V_{i-1})$ is a parabolic subgroup of $\mathrm{GL}(V_i/V_{i-1})$ and the second condition is $\bar{q}_i \in \bar{Q}^{(i)\,sst}_{A,\mathrm{num}}$ where $\bar{q}_i$ denotes the image of $q$.

These considerations lead to the following definitions. (Compare Definitions 1.3.12 and 1.3.13.)

*Definition 3.1.1.* Let $Q$ and $P$ be standard parabolic subgroups with $Q \subset P$. We say that $\underline{q} \in Q_A$ satisfies Condition $\mathscr{C}$ with respect to $\underline{P}_A$ if the following holds: If $0 \subset V_1 \subsetneqq \cdots \subsetneqq V_k = V$ is the flag defined by $P$, $\bar{Q}_i$ is the image $Q$ in $\mathrm{GL}(V_i/V_{i-1})$ and $\bar{q}_i \in \bar{Q}^{(i)}_A$ is the image of $\underline{q} \in \bar{Q}_{i,A}$ then we have $\underline{\bar{q}}_i \in Q^{(i)}_{A,\mathrm{num}}$ for $i = 1, \ldots, k$.

*Definition 3.1.2.* Let $C$ be a chain of parabolic subgroups: $Q = Q^1 \subsetneqq \cdots \cdots \subsetneqq Q^\lambda \subsetneqq Q^{\lambda+1} \cdots \subsetneqq Q^l$. Then we denote by $Q^C_{A,\mathrm{num}}$ the subset of $Q_A$ defined by

$$Q^C_{A,\mathrm{num}} = \left\{ \underline{q} \in Q_A \;\middle|\; \begin{array}{l} \underline{q} \in Q^l_{A,\mathrm{num}} \text{ and } \underline{q} \in Q^\lambda_A \text{ satisfies} \\ \text{Condition } \mathscr{C} \text{ (of Definition 3.1.1) with} \\ \text{respect to } Q^{\lambda+1}_A \text{ for } \lambda = 1, \ldots\ l-1\ . \end{array} \right\}$$

We then have

**Proposition 3.1.3.** *Let $C$ be a chain of parabolic subgroups of $\mathrm{SL}(V)_A$ and let us denote also by $C$ the corresponding chain of parabolic subgroups in $\mathrm{PL}^0_A$. We then have*

1) $$\int_{Q^C_{A,\mathrm{num}}/Q_K} \omega_{Q,A} = \int_{\underline{Q}^{0,C}_{A,\mathrm{num}}/\underline{Q}_K} \omega_{\underline{Q},A} < \infty\,.$$

2) $$\sum_{\underline{x} \in \mathrm{Inst}} \frac{1}{|\underline{x}^{-1} \Re \underline{x} \cap \mathrm{SL}(V)_K|} = \sum_C (-1)^{|C|+1} \int_{Q^C_{A,\mathrm{num}}/Q_K} \omega_{Q,A}$$

$$= \sum_{\underline{x} \in \underline{\mathrm{Inst}}} \frac{1}{|\underline{x}^{-1} \underline{\Re} \underline{x} \cap \mathrm{PL}(V)_K|}$$

*where $C$ runs through chains of (standard) parabolic subgroups and $|C|$ denotes the length of the chain.*

*Proof*. From the considerations above we see that

$$\sum_{\underline{x}\in \mathrm{Inst}} \frac{1}{|\underline{x}^{-1}\mathfrak{K}\underline{x}\cap \mathrm{SL}(V)_K|} = \sum_C (-1)^{|C|+1} \int_{Q^C_{A,\mathrm{num}}/Q_K} \omega_{Q,A}$$

and

$$\sum_{\underline{x}\in \underline{\mathrm{Inst}}} \frac{1}{|\underline{x}^{-1}\underline{\mathfrak{K}}\underline{x}\cap \mathrm{PL}(V)_K|} = \sum_C (-1)^{|C|+1} \int_{\underline{Q}^{0,C}_{A,\mathrm{num}}/\underline{Q}_K} \omega_{\underline{Q},A}$$

and (2) follows from (1).

We first show that $\int_{Q^C_{A,\mathrm{num}}/Q_K} \omega_{Q,A}$ is finite. We decompose $Q^C_{A,\mathrm{num}}$ into the fibres of the map $\underline{q}\mapsto(\delta_1(\underline{q}), \dots \delta_{k-1}(\underline{q}))\in \mathbb{Z}^{k-1}$. We have, by Proposition 2.2.1 b

$$\int_{\underline{q}Q_A(0)/Q_K} \omega_{Q,A} = q^{-\Sigma f_i n_i} \int_{Q_A(0)/Q_K} \omega_{Q,A}$$

where $n_i=\delta_i(\underline{q})$ and $f_i=m_i+m_{i+1}$. Each $n_i$ is, up to an additive constant, equal to the degree of the bundle $F^{\underline{q}}_{0,i}\supset E^{\underline{q}}_0$ and as $\underline{q}\in Q^C_{A,\mathrm{num}}$ these degrees are bounded below by Proposition 1.3.15. It follows that $\int_{Q^C_{A,\mathrm{num}}/Q_K} \omega_{Q,A}$ is finite.

Now we have $Q^C_{A,\mathrm{num}}/Q_A(0)\approx Q^{0,C}_{A,\mathrm{num}}/Q^0_A(0)$. In fact if we consider the map $Q_A\to \mathbb{Q}^{k-1}$ given by $\underline{q}\mapsto(\delta'_1(\underline{q}), \dots \delta'_{k-1}(\underline{q}))$ where $|\alpha_i(\underline{q})|=q_i^{-\delta'(\underline{q})}$, then $Q^C_{A,\mathrm{num}}$ is the inverse image of a certain subset $X^C_Q$ of $\mathbb{Q}^{k-1}$ (defined by certain inequalities) and we see that $Q^{0,C}_{A,\mathrm{num}}$ is also the inverse image of $X^C_Q$ by the corresponding map $Q^0_A\to\mathbb{Q}^{k-1}$ given by the roots $\alpha_i$ of $Q^0_A$. Moreover we have, for $\underline{q}\in Q^0_{A,\mathrm{num}}$

$$\int_{\underline{q}Q_A(0)/Q_K} \omega_{Q,A} = q^{-\Sigma a_i\delta_i(\underline{q})} \int_{Q_A(0)/Q_K} \omega_{Q,A} \text{ by Proposition 2.2.1},$$

and for $\underline{q}\in Q^{0,C}_{A,\mathrm{num}}$ we have

$$\int_{\underline{q}Q^0_A(0)/Q_K} \omega_{Q,A} = q^{-\Sigma a_i\delta'_i(\underline{q})} \int_{\underline{Q}^0_A(0)/\underline{Q}_K} \omega_{Q,A} \text{ by Proposition 2.2.2}.$$

Thus to complete the proof of the proposition it is enough to show that

$$\int_{Q_A(0)/Q_K} \omega_{Q,A} = \int_{\underline{Q}^0_A(0)/\underline{Q}_K} \omega_{\underline{Q},A}\,.$$

By Propositions 2.2.1 and 2.2.2 we have equality if we take the Tamagawa measures instead of the measures $\omega_{Q,A}$ and $\omega_{\underline{Q},A}$.

So we have to prove that

$$\int_{\mathfrak{R}\cap Q_A} \omega^{\tau}_{Q,A} = \int_{\mathfrak{R}\cap \underline{Q}_A} \omega^{\tau}_{\underline{Q},A}.$$

To prove this we observe that

$$\int_{\mathfrak{R}\cap U_A} \omega^{\tau}_{U,A} = \int_{\mathfrak{R}\cap \underline{U}_A} \omega^{\tau}_{\underline{U},A} = q^{(1-g)\dim U + p(Q)}$$

where $U$ denotes the unipotent radical and $p(Q)$ is the numerical invariant of the parabolic group scheme $Q/Y$ defined by the flag in $E_0$ (see [8, § 1, 3]). Thus it is enough to show that

$$\int_H \omega^{\tau}_{H,A} = \int_{\underline{H}} \omega^{\tau}_{\underline{H},A}$$

where $\mathfrak{R}_H$ and $\mathfrak{R}_{\underline{H}}$ are the obvious maximal compact subgroups in $H_A = Q_A/U_A$ and $\underline{H}_A = \underline{Q}_A/\underline{U}_A$ defined by $E_0$ and the flag. But this equality is clear in view of [19, Theorem 2.2.5]. This completes the proof of Proposition 3.1.3.

Propositions 2.3.4, 2.3.5, and 3.1.3 together yield Theorem 2.

### *3.2. The Action of $T_n$ on the Etale Cohomology of M*

We first recall briefly some results on the $l$-adic cohomology and zeta functions of algebraic varieties. Let $X/\mathbb{F}_q$ be a projective variety over the finite field $\mathbb{F}_q$. Let $\bar{X} = X_{\mathbb{F}_q} \times \bar{\mathbb{F}}_q$. If $l$ is a prime coprime to $q$ and $\mathbb{Q}_l$ is the field of $l$-adic numbers then the $l$-adic cohomology groups

$$H^i(\bar{X}, \mathbb{Q}_l) = \left(\varprojlim_{\nu} H^i(\bar{X}; \mathbb{Z}/l^{\nu}\mathbb{Z})\right) \otimes \mathbb{Q}_l$$

of $\bar{X}$ are defined by means of etale cohomology [1, 7] [5, Exposé III]. The Frobenius map $\varphi: X \to X$ defines an endomorphism

$$\varphi_i^*: H^i(\bar{X}, \mathbb{Q}_l) \to H^i(\bar{X}, \mathbb{Q}_l).$$

The spaces $H^i(\bar{X}, \mathbb{Q}_l)$ are finite dimensional and vanish for $i > 2\dim X$. We define

$$Z_X(t) = \prod_{i=0}^{2N} \det(\mathrm{Id} - \varphi_i^* t)^{(-1)^{i+1}} = \frac{P_1(t) \dots P_{2N-1}(t)}{P_0(t) \dots P_{2N}(t)}$$

where $N = \dim X$. By a theorem of Grothendieck we have

$$\frac{Z'_X(t)}{Z_X(t)} = \sum_{n=1}^{\infty} |X_{\mathbb{F}_{q^n}}| \, t^{n-1}$$

where $|X_{\mathbb{F}_{q^n}}|$ denotes the numbers of $\mathbb{F}_{q^n}$ rational points [9].

Recently Deligne [4] has proved the *Weil Conjecture: If $X/\mathbb{F}_q$ is projective and smooth then the eigenvalues of*

$$\varphi_i^* : H^i(\bar{X}, \mathbb{Q}_l) \to H^i(\bar{X}, \mathbb{Q}_l)$$

*are algebraic integers of absolute value $q^{i/2}$.* (This means each conjugate has absolute value $q^{i/2}$.)

This has the following consequence. If $X/\mathbb{F}_q$ is smooth, then the polynomials $P_i(t) = \det(\mathrm{Id} - \varphi_i^* t)$ are pairwise coprime and this implies that the nuberator $\prod_{i \text{ odd}} P_i(t)$ and the denominator $\prod_{i \text{ even}} P_i(t)$ are determined by $Z_X(t)$ i.e., by the number of rational points $|X_{\mathbb{F}_{q^n}}|$ for all $n$.

We need the following proposition for which we could not find a reference.

**Proposition 3.2.1.** *Let $X/\bar{\mathbb{F}}_q$ be a projective variety and let $G$ be a finite group of automorphisms of $\bar{X}$. Then $\bar{X}/G$ exists as a projective variety. If* $(|G|, l) = 1$, *the mapping*

$$H^i(\bar{X}/G, \mathbb{Q}_l) \to H^i(\bar{X}, \mathbb{Q}_l)^G$$

*is an isomorphism, where $H^i(\bar{X}, \mathbb{Q}_l)^G$ denotes the invariants of $G$ in $H^i(\bar{X}, \mathbb{Q}_l)$.*

*Proof.* The existence of $\bar{X}/G$ is well known [14, Chapter III, No. 12]. To prove the second part, it is sufficient to prove that

$$H^i(\bar{X}/G, \mathbb{Z}/l^\nu\mathbb{Z}) \xrightarrow{\sim} H^i(X, \mathbb{Z}/l^\nu\mathbb{Z})^G .$$

Let $f : \bar{X} \to \bar{X}/G$ denote the projection. It is known [1, Exposé VIII, Proposition 5.5] that

$$R^q f_*(\mathbb{Z}/l^\nu\mathbb{Z}) = 0 \quad \text{for} \quad q > 0 .$$

The sheaf $F = R^0 f_*(\mathbb{Z}/l^\nu\mathbb{Z})$ is constructible and the group $G$ acts on $F$. Moreover it is clear that the constant sheaf $\mathbb{Z}/l^\nu\mathbb{Z}$ on $\bar{X}/G$ injects into $F$. We claim that $\mathbb{Z}/l^\nu\mathbb{Z} \xrightarrow{\sim} F^G$. To see this we consider the fibre at any geometric point $\bar{y} \in \bar{X}/G$ (Compare [1], Exposé VIII). Then we have to show that $(\mathbb{Z}/l^\nu\mathbb{Z})_{\bar{y}} = \mathbb{Z}/l^\nu\mathbb{Z} = F_{\bar{y}}^G$.

Let $\bar{X}_y = f^{-1}(\bar{y})$, then we have, for $f_{\bar{y}} : \bar{X}_{\bar{y}} \to \bar{y}$,

$$F_{\bar{y}} = R^0 f_{\bar{y}}(\mathbb{Z}/l^\nu\mathbb{Z}) = \mathfrak{C}(\bar{X}_y, \mathbb{Z}/l^\nu\mathbb{Z})$$

where the last term denotes the set of mappings from the underlying set $X_{\bar{y}}$ into $\mathbb{Z}/l^\nu\mathbb{Z}$. The group $G$ acts on $X_{\bar{y}}$ and $\mathfrak{C}$; the only invariants in $\mathfrak{C}$ are the constants.

Since $(|G|, l) = 1$, we see that we get a decomposition

$$F = F^G \oplus R = \mathbb{Z}/l^\nu\mathbb{Z} \oplus R$$

where $R$ is the sheaf of elements of trace zero. By the Leray spectral sequence [1, Exposé VII, 1.5] we have

$$H^i(\bar{X}, \mathbb{Z}/l^\nu\mathbb{Z}) = H^i(\bar{X}/G, R^0 f_*(\mathbb{Z}/l^\nu\mathbb{Z})):$$

since

$$H^i(\bar{X}/G, R^0 f_*(\mathbb{Z}/l^\nu\mathbb{Z})) = H^i(\bar{X}/G, \mathbb{Z}/l^\nu\mathbb{Z}) \oplus H^i(\bar{X}/G, R)$$

the proposition is proved.

We now proceed to prove Theorem 1 in the case $Y$ is defined over a finite field $\mathbb{F}_q$. We apply Proposition 3.2.1 to the projection $M/\mathbb{F}_q \to (M/T_n)/\mathbb{F}_q = N/\mathbb{F}_q$ and obtain

$$H^i(\bar{M}, \mathbb{Q}_l)^{T_n} = H^i(\bar{N}, \mathbb{Q}_l).$$

Then, if

$$\psi_i^* = \mathrm{res}_{H^i(\bar{N}, \mathbb{Q}_l)}(\varphi_i : H^i(\bar{M}, \mathbb{Q}_l) \to H^i(\bar{M}, \mathbb{Q}_l)),$$

we see that $\det(\mathrm{Id} - \psi_i^* t)$ divides $\det(\mathrm{Id} - \varphi_i^* t)$. On the other hand, we have, by Theorem 2, $Z_M(t) = Z_N(t)$ and we know from Weil conjectures that there are no cancellations in the expression for $Z_M(t)$. It follows that $\det(\mathrm{Id} - \psi_i^* t) = \det(\mathrm{Id} - \varphi_i^* t)$.

This proves Theorem 1 when $Y$ is defined over a finite field.

We now remove the restriction of the field of constants to be a finite field. Let $k$ by any field with $\mathrm{char}(k) \nmid n$ and let $Y/k$ be a smooth projective curve. We shall derive the validity of Theorem 1 for $\bar{Y} = Y \times_k \bar{k}$ — where $\bar{k}$ is an algebraic closure of $k$ — by means of a specialisation argument. One tool will be the proper base change theorem of Artin ([2], Exposé XVI, Corollary 2.2). The other tool needed is the theory of Mumford and Seshadri on the action of reductive groups on projective schemes, which is essential for the construction of our moduli schemes. At this place it would be very convenient if we knew that Seshadri's theory [18] works also for families of curves, especially for curves over valuation rings with unequal characteristics. Unfortunately we do not know whether this is actually true; in any case it does not seem to be obvious. To avoid this difficulty we shall use a rather crude argument which shows that over a Dedekind ring the construction of the moduli scheme is almost everywhere compatible with specialisation.

**Lemma 3.2.2**[1]. *Let $A$ be a Dedekindring with quotient field $k$. Let $Y \to Spec(A)$ be a smooth projective curve. Let $L'_0$ be a line bundle on $Y \times_{Spec(A)} k$ of degree $r$ prime to $n$. This line bundle has a unique extension to a line bundle $L_0$ over $Y/Spec(A)$. Then there exists a non empty open*

[1] We are thankful to C. S. Seshadri for pointing out an inaccuracy in an earlier version of the proof of this lemma.

*subset $U \supset Spec(A)$ and a projective smooth scheme $M \to U$, such that for all points $\mathfrak{p} \in U$ with residue field $k(\mathfrak{p})$ the scheme $M \underset{U}{\times} Spec(k(\mathfrak{p})) = M_{\mathfrak{p}}$ is isomorphic to the moduli scheme $M(n, L_0)/Spec(k(\mathfrak{p}))$* (Compare 1.1).

*If moreover the residue characteristics for all $\mathfrak{p} \in U$ are prime to $n$, we have an action of $T_n$ on $M \to U$, which induces on all fibres the standard action.*

*Proof.* We have to analyse Seshadri's construction of the moduli scheme $M(n, L_0)$ over an algebraically closed field (Compare [17, 18]). We choose first of all an ample line bundle $L$ on $Y/A$. Moreover we choose $N$ points $P_1, \ldots, P_N$ on $Y_{\bar{k}}$ such that the divisor $\sum_{i=1}^{N} P_i$ is rational over $k$ and such that these points are pairwise distinct and remain pairwise distinct after reduction mod a prime $\mathfrak{p} \in U$, where $U \subset Spec(A)$ is suitably chosen. We also choose an integer $m > 0$ such that for this choice of $m$ and $P_1 \ldots P_N$, the Corollary 7.1 in [17] will be true for any specialisation of our curve induced by a homomorphism $A \to \bar{k}$ or $A \to \bar{k}(\mathfrak{p})$ for $G \in U$. Then we get a functorial mapping $\tau$

$$\left\{\begin{array}{l}\text{isomorphism classes of}\\ \text{stable vector bundles}\\ \text{with determinant } L_0 +\\ \text{basis on } H^0(\;\;, \otimes L^m)\end{array}\right\} \xrightarrow{\tau} \mathcal{G}$$

where $\mathcal{G} \to U$ is a twisted form of the projective scheme $\mathcal{G}^N_{d,n} \to U$ and where the twisting comes in since the $P_i$ are not necessarily defined over $k$. Here $d = \dim H^0(Y, \otimes L^m)$.

The group $\mathrm{GL}_d/U$ acts on $\mathcal{G}/U$ in the usual way (Compare [17, § 4]) and we choose on $\mathcal{G}/U$ the standard ample line bundle which has a $\mathrm{GL}_d$-linearisation [10].

We claim that there exists a closed subscheme $\mathcal{G}_{ns} \subset \mathcal{G}/U$ whose geometric points are exactly the non-stable geometrie points of $\mathcal{G}/U$. To see this we refer to the proof of Theorem 3.1 in [18]. We first choose a split maximal torus $T \subset \mathrm{GL}_d/U$ (for example the standard diagonal torus) and a Borel subgroup $B \subset T$. Then we know that there exists a finite set of one parameter subgroups in $T$ such that a geometric point which is stable with respect to this finite set of one parameter groups is also stable with respect to all other one parameter subgroups of $B$. Therefore these finitely many one parameter subgroups define a closed subscheme $\mathcal{W} \subset \mathcal{G}/U$ of non-stable points. Then we proceed as in the proof of Theorem 3.1 in [18] by using the action of $\mathrm{GL}_d$ and the completeness of $\mathrm{GL}_d/B$.

We set $\mathcal{G}_s/U = (\mathcal{G} - \mathcal{G}_{ns})/U$. This is a quasi-projective smooth scheme over $U$ which is $\mathrm{GL}_d$ invariant. Then we know that $\mathcal{G}_s \times_U k$ and $\mathcal{G}_s \times_U k(\mathfrak{p})$ are the schemes of stable points respectively. It follows from [17], that the image of $\tau$ is a closed smooth subscheme

$$\begin{array}{ccc} X & \longrightarrow & \mathcal{G}_s \\ & \searrow \quad \swarrow & \\ & U & \end{array}$$

Moreover it is clear that the quotient $X \times_U k/\mathrm{GL}_d \times_U k = M'$ exists. (A priori $M'$ is defined only over $\bar{k}$; but by a standard argument of descent we see that it is defined over $k$.) We know that in this special case $M'/k$ is a smooth projective scheme and that

$$p' : X \times_U k \to M'$$

is a principal fibre space with structure group $\mathrm{PGL}_d$. Making $U$ smaller if necessary we can extend $M'$ to a smooth projective scheme $M/U$ and we can extend

$$X \xrightarrow{p} M$$

such that $X$ is still a $\mathrm{PGL}_d$ principal fibre space over $M$. This follows by standard arguments from the fact that $p'$ is locally trivial with respect to the etale topology. But then it is clear that $M \times_U k(\mathfrak{p})$ is the quotient of $X \times_U k(\mathfrak{p})$ by the action of $\mathrm{GL}_d \times_U k(\mathfrak{p})$ for $\mathfrak{p} \in U$ because

$$X \times_U k(\mathfrak{p}) \to M \times_U k(\mathfrak{p})$$

is still a $\mathrm{PGL}_d \times k(\mathfrak{p})$ principal fibre space. Therefore

$$M \times k(\mathfrak{p}) \xrightarrow{\sim} M(n, L_{0,\mathfrak{p}})/Spec(k(\mathfrak{p})) .$$

The action of $T_n$ on $M \times_U k = M'$ extends to an action on $M$ if we make $U$ smaller if necessary. Then it has to induce on $M \times_U k(\mathfrak{p})$ the action we want, since we know what happens on the bundles, which are the points on $M$. This completes the proof of Lemma 3.2.2.

Now it is more or less clear how we can deduce Theorem 1 for an arbitrary algebraically closed ground field. If $\bar{Y}/\bar{k}$ is any projective non-singular curve we find a field $k \subset \bar{k}$ which is finitely generated over the prime field in $\bar{k}$ and such that $\bar{Y} = Y' \times_k \bar{k}$. We may find a Dedekind ring

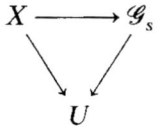

$A \subset k$ with quotient field $k$ and infinitely many closed prime ideals and such that $Y'/k$ extends to a smooth curve $Y/Spec(A)$. We apply Lemma 3.2.2 and get from Artin's proper base change theorem ([2], Exposé XVI, Corollary 2.2) that Theorem 1 is true for $\overline{Y/k}$ if it is true for $Y \underset{Spec(A)}{\times} \overline{k(\mathfrak{p})}$ where $\mathfrak{p}$ runs over the closed prime ideals in $A$ and $\overline{k(\mathfrak{p})}$ is an algebraic closure of $A/\mathfrak{p} = k(\mathfrak{p})$. But the transcendence degree of $k(\mathfrak{p})$ over the prime field is less than the trancendence degree of $k$ over the prime field if the characteristic does not change. Therefore we can reduce Theorem 1 by induction to the case that $k = \mathbb{Q}$ or $k = \mathbb{F}_q$. Only the first case is still of interest for us. Now we apply again the method above but this time we will have a change of the characteristic; our Dedekind ring will be the ring of integers in an algebraic number field and the residue field is finite. Therefore we can reduce the first case to the second one.

### 3.3. *The Computation of Some Betti Numbers*

We shall now derive some explicit formulas for Betti numbers of $\bar{M}/\bar{\mathbb{F}}_q$ in low and high dimensions. This will be done by estimating the sum over the instable part in our formula for $|M_{\mathbb{F}_q}|$ (Proposition 2.3.3). It turns out that this part has a lower order of magnitude than the term $q^{(n^2-1)(g-1)}\zeta(2)\ldots\zeta(n)$ and therefore we can read off the Betti numbers in a certain range from the expansion of that term. To obtain these estimates we have to consider the expressions (Proposition 3.1.3)

$$\int_{Q^C_{A,\mathrm{num}}/Q_K} \omega_{Q,A}\,.$$

Let us assume that $Q$ is of type $(m_1, \ldots, m_k)$.

We consider the map

$$\delta : Q_A \to \mathbb{Z}^{k-1}\,,$$

$$\delta : \underline{q} \mapsto (\delta_1(\underline{q}) \ldots \delta_{k-1}(\underline{q})) = (n_1, \ldots, n_{k-1}) = \underline{n}\,.$$

We know that $Q^C_{A,\mathrm{num}}$ is the inverse image of a certain subset $Y^C_Q \subset \mathbb{Z}^{k-1}$ which is described by inequalities derived from the Definition 3.1.2; the actual shape of $Y^C_Q$ is not of interest for us at the moment. It follows from the considerations in the proof of Proposition 3.1.3 that

$$\int_{Q^C_{A,\mathrm{num}}/Q_K} \omega_{Q,A} = \sum_{\underline{n} \in Y^C_Q} q^{-f_1 n_1 - \cdots - f_{k-1} n_{k-1}} \int_{Q_A(0)/Q_K} \omega_{Q,A}$$

where $f_i = m_i + m_{i+1}$. If $H = Q/U$ where $U$ is the unipotent radical of $Q$ then we see as in the proof of Proposition 3.1.3 that

$$\int_{Q_A(0)/Q_K} \omega_{Q,A} = \frac{1}{\mathrm{vol}_{\omega^\tau_{Q,A}}(\mathfrak{R}_Q)} \int_{Q_A(0)/Q_K} \omega^\tau_{Q,A}$$

where $\omega^{\tau}_{Q,A}$ is the Tamagawa measure (Compare 2.2) and $\mathfrak{K}_Q = \mathfrak{K} \cap Q_A$ is the compact subgroup defined by means of $E_0$ (Compare 2.3). We use the formula a) in Proposition 2.2.1 and get

$$\int_{Q_A(0)/Q_K} \omega_{Q,A} = \frac{1}{\mathrm{vol}_{\omega^{\tau}_{Q,A}}(\mathfrak{K}_Q)} \left( \frac{|J_{\mathbb{F}_q}|}{(q-1)\,q^{(g-1)}} \right)^{(k-1)}$$

If $U$ is the unipotent radical of $Q$ we find

$$\mathrm{vol}_{\omega^{\tau}_{Q,A}}(\mathfrak{K}_Q) = q^{(1-g)\dim U + p(Q)}\, \mathrm{vol}_{\omega^{\tau}_{H,A}}(\mathfrak{K}_H)$$

(Compare proof of Proposition 3.1.3) and for the last factor we obtain

$$\mathrm{vol}_{\omega^{\tau}_{H,A}}(\mathfrak{K}_H) = q^{(1-g)\dim H} \times \begin{array}{l}\text{product of values of } \zeta_K \text{ at some}\\ \text{of the arguments } 2, 3, \ldots, n\,,\end{array}$$

since we have only to take into account the semi simple part of $H$ which is isogeneous to a product $\Pi\, \mathrm{SL}_{m_i}$. The central part of $H$ gives contribution 1 because of the choice of the convergence factors. This shows that the order of magnitude is

$$\int_{Q_A(0)/Q_K} \omega_{Q,A} = q^{(\dim Q)(g-1) - p(Q)} \left(1 + O(q^{-\frac{1}{2}})\right).$$

We shall abbreviate this and just write

$$\int_{Q_A(0)/Q_K} \omega_{Q,A} \sim q^{(\dim Q)(g-1) - p(Q)}$$

and we shall say that the integral on the left hand side has the order of magnitude $q^{(\dim Q)(g-1) - p(Q)}$.

Therefore we get in total

$$\int_{Q^G_{A,\mathrm{num}}/Q_K} \omega_{Q,A} = \left( \sum_{\underline{n} \in Y^C_Q} q^{-f_1 n_1 - \cdots - f_{k-1} n_{k-1}} \right) q^{\dim Q(g-1) - p(Q)}.$$

To estimate the infinite sum in the bracket we use a very crude majorisation of $Y^C_Q$. Let us put $d_i = m_1 + m_2 + \cdots + m_i$; then $f_i = d_{i+1} - d_{i-1}$. We started from a flag

$$0 \subset F_{0,1} \subset F_{0,2} \subset \cdots \subset F_{0,k} = E_0$$

where

$$F_{0,i} = L_0 \oplus \mathcal{O}_Y \cdots \oplus \mathcal{O}_Y \quad \text{(Compare 3.1)}.$$

First of all we observe that for $\underline{q} \in Q_A$ we have

$$\deg F^{\underline{q}}_{0,i} = \deg L_{0,i} + n_i = r + n_i$$

if $n_i = \delta_i(\underline{q})$. Then it follows from Proposition 1.3.15 that $\underline{q} \in Q^C_{A,\mathrm{num}}$ implies

$$\mu(F^{\underline{q}}_{0,i}) = \frac{r+n_i}{d_i} > \mu(E_0) = \frac{r}{n}$$

and this is equivalent to

$$n_i > d_i \frac{r}{n} - r .$$

This is a necessary condition for a point $\underline{n}$ to be in $Y^C_Q$. We remark that for a maximal parabolic subgroup this condition is also sufficient (Lemma 1.3.12). Therefore we see that the order of magnitude of

$$\sum_{\underline{n} \in Y^C_Q} q^{-n_i f_i - \cdots - n_{k-1} f_{k-1}}$$

is less than or equal to

$$q^{-\sum_{i=1}^{k-1} f_i \left( \left[ \frac{d_i r}{n} \right] - r \right)}$$

where $\left[\frac{d_i r}{n}\right]$ is the smallest integer which is greater than or equal to $\frac{d_i r}{n}$. On the other hand one checks easily from the definition that

$$p(Q) = \sum_{i=1}^{k-1} f_i \deg F_{0,i} - d_{k-1} \deg E_0 = \left( \sum_{i=1}^{k-1} f_i \right) r - d_{k-1} r .$$

This altogether gives that the order of magnitude of

$$\int_{Q^C_{A,\mathrm{num}}/Q_K} \omega_{Q,A}$$

is less than or equal to

$$q^{\dim P(g-1) - \sum_{i-1}^{k-1} f_i \left[ \frac{d_i r}{n} \right] + d_{k-1} r} .$$

This gives the exact order of magnitude if $Q$ is maximal parabolic as we see from the above remark.
We have

$$|M_{\mathbb{F}_q}| = q^{(n^2-1)(g-1)} \zeta(2) \ldots \zeta(n) - \sum_{\underline{x} \in \mathrm{Inst}} \frac{1}{|\underline{x}^{-1} \mathfrak{R} \underline{x} \cap \mathrm{SL}(n,k)|}$$

and we obtained above an estimate for the second term on the right hand side. The difference in the exponents in $q^{\cdots}$ for both terms is

$$\geqq (\operatorname{codim}_G Q)(g-1) + \sum_{i=1}^{k-1} f_i \left[ \frac{d_i r}{n} \right] - r d_{k-1}$$

and therefore we have to look for the minimum of this expression if $Q$ runs over all parabolic subgroups.

We have

**Proposition 3.3.1.** *We assume $g \geqq 2$ and $0<r<n$. Then the minimum of the expression*

$$(\operatorname{codim}_G Q)(g-1)+\sum_{i=1}^{k-1} f_i\left[\frac{d_i r}{n}\right]-r d_{k-1},$$

*where $Q$ runs over the parabolic subgroups, is obtained only once. The parabolic subgroup for which the minimum is obtained is maximal and defined by a line bundle or a hyperplane bundle. The value of that minimum is*

$$\max(r, n-r)+(n-1)(g-1).$$

*Proof.* Without loss of generality we may assume that $r<\frac{n}{2}$, otherwise we pass to the dual situation. We perform some easy calculations:

$$\begin{aligned}
&\sum_{i=1}^{k-1} f_i\left[\frac{d_i r}{n}\right]-d_{k-1} r \\
&=\sum_{i=1}^{k-1} d_{i+1}\left[\frac{d_i r}{n}\right]-\sum_{i=1}^{k-1} d_{i-1}\left[\frac{d_i r}{n}\right]-d_{k-1} r \\
&=\sum_{i=1}^{k-1} d_{i+1}\left[\frac{d_i r}{n}\right]-\sum_{i=1}^{k-2} d_i\left[\frac{d_{i+1} r}{n}\right]-d_{k-1} r \\
&=\sum_{i=1}^{k-2}\left(d_{i+1}\left[\frac{d_i r}{n}\right]-d_i\left[\frac{d_{i+1} r}{n}\right]\right)+d_k\left[\frac{d_{k-1} r}{n}\right]-d_{k-1} r \\
&=\sum_{i=1}^{k-1}\left(d_{i+1}\left\{\frac{d_i r}{n}\right\}-d_i\left\{\frac{d_{i+1} r}{n}\right\}\right)
\end{aligned}$$

where $\{x\}=[x]-x$. Then our expression above becomes

$$d_2\left\{\frac{d_1 r}{n}\right\}+\sum_{i=2}^{k-1} f_i\left\{\frac{d_i r}{n}\right\}.$$

This last expression is strictly positive and since it is an integer it is $\geqq 1$.

We check first what happens if $Q$ is maximal and defined by a line bundle or a hyperplane bundle. In that case we have $d_2=n$ and

$$\begin{aligned}
&(\operatorname{codim}_G Q)(g-1)+n\left\{\frac{d_i r}{n}\right\} \\
&=\begin{cases}(\operatorname{codim}_G Q)(g-1)+r & \text{if } d_1=1 \\ (\operatorname{codim}_G Q)(g-1)+n-r & \text{if } d_1=n-1 .\end{cases}
\end{aligned}$$

Therefore our assumption $r < \frac{n}{2}$ implies that the minimum is obtained in the case $d_1 = 1$. It is now sufficient to show that for the parabolic subgroups that are different from these two maximal parabolic subgroups we have

$$(\operatorname{codim}_G Q)(g-1)+1 > (n-1)(g-1)+r.$$

*Case 1.* $Q$ itself is maximal and therefore of type $(m)$ where $1 < m < n-1$. Then we have to check that

$$(n-m)\,m(g-1)+1 > (n-1)(g-1)+r.$$

It is clear that the left hand side takes its minimum value at $m=2$, therefore we must check that

$$2(n-2)(g-1)+1 > (n-1)(g-1)+r$$

or

$$(n-3)(g-1) > r-1.$$

This is certainly allright if $n \geqq 6$. For $n=5$ and $n=4$ it is easily checked and for $n=3$, 2 this first case cannot occur.

*Case 2.* $Q$ is the intersection of the two maximal parabolic groups defined by $d_1 = 1$ and $d_1 = n-1$. In this case we have to prove that

$$(2n-3)(g-1)+1 > (n-1)(g-1)+r.$$

But it is clear that this is covered by our considerations in the first case since $2n-4 < 2n-3$.

*Case 3.* $Q$ is arbitrary but different from the two maximal ones defined by $d_1 = 1$, $d_1 = n-1$. In this case it is clear that we can find a parabolic subgroup $Q' \supset Q$, which is covered by one of the first two cases; since the codimension decreases we are through.

Now Proposition 3.3.1 yields that for $r < \frac{n}{2}$

$$|M_{\mathbb{F}_q}| = q^{(n^2-1)(g-1)}\,\zeta(2)\ldots\zeta(n) - q^{(n^2-n)(g-1)-r} + O(q^{(n^2-n)(g-1)-r-\frac{1}{2}}).$$

We express the $\zeta$-function in the usual form

$$\zeta(s) = \frac{\Pi(1-\omega_i q^{-s})}{(1-q^{-s})(1-q\cdot q^{-s})}.$$

In the expression for $\zeta(m)$, $m=2,\ldots n$, we substitute $(-T)$ for each of the $\omega_i$ and $T^2$ for $q$. Then we get rational functions

$$Z_m(T) = \frac{(1+T^{1-2m})^{2g}}{(1-T^{-2m})(1-T^{2(1-m)})}.$$

Let us define

$$P(T) = T^{2(n^2-1)(g-1)} Z_2(T) \ldots Z_n(T).$$

This is a Laurent series in the variable $T^{-1}$ and define $b_\nu (\nu = 0, 1, \ldots)$ by

$$P(T) = T^{2(n^2-1)(g-1)} \left( \sum_{\nu=0}^{\infty} b_\nu \cdot T^{-\nu} \right).$$

It is now obvious that for $0 \leqq \nu < 2(n-1)(g-1) + r$ the number $b_\nu$ is equal to the number of terms in our formula for $|M_{\mathbb{F}_q}|$ which have absolute value $q^{(n^2-1)(g-1)-\frac{\nu}{2}}$.

For $\nu = 2((n-1)(g-1)+r)$ this number of terms is equal to $b_\nu - 1$. Using the Weil conjectures and Poincaré duality this gives us

**Theorem 3.3.2.** $\dim H^\nu(\bar{M}\ \mathbb{Q}_l) = b_\nu$ for $0 \leqq \nu < 2((n-1)(g-1)+r)$.

$$\dim H^\nu(\bar{M}, \mathbb{Q}_l) = b_\nu - 1 \quad \text{for} \quad \nu = 2((n-1)(g-1)+r).$$

**Corollary 3.3.3.** *The third Betti number of $\bar{M}/\bar{F}_q$ is always equal to $2g$.*

**Corollary 3.3.4.** *If $r \not\equiv \pm r' \bmod n$ then the corresponding moduli spaces (of vector bundles of rank $n$) are not topologically equivalent.*

## References


1. Artin, M., Grothendieck, A., Verdier, J. L.: Théorie des topos et cohomologie étale des schémas, (SGA4), Tom 2. Lecture Notes in Mathematics, No. 270, Berlin-Heidelberg-New York: Springer 1972
2. Artin, M., Grothendieck, A., Verdier, J. L.: Theorie des topos et cohomologie étale des schémas, (SGA4), Tom 3. Lecture Notes in Mathematics, No. 305, Berlin-Heidelberg-New York: Springer 1973
3. Bourbaki, N.: Eléments de mathématiques, Integration, Livre **VI**. Paris: Hermann 1969
4. Deligne, P.: La conjecture de Weil I. Publ. Math. IHES, No. 43, Paris: Presses Universitaires de France 1974
5. Giraud, J., *et al.*: Dix exposés sur la cohomologie des schémas. Amsterdam: North-Holland Publ. Co. 1968
6. Grothendieck, A.: Géométrie formelle et géométrie algébrique. Seminaire Bourbaki, t11, (1958/59), No. 182
7. Grothendieck, A.: Formule de Lefschetz et rationalité de fonctions *L*. Séminaire Bourbaki (1964—65), No. 279
8. Harder, G.: Minkowskische Reduktionstheorie über Funktionenkörpern. Invent. math. **7**, 33—54 (1969)
9. Harder, G.: Eine Bemerkung zu einer Arbeit von P. E. Newstead. Jour. für Math. **242**, 16—25 (1970)

10. Mumford, D.: Geometric invariant theory. Ergebnisse der Mathematik. Berlin-Heidelberg-New York: Springer 1965
11. Narasimhan, M. S., Ramanan, S.: Deformations of moduli spaces of Vector bundles on curves (to appear in Ann. of Math.)
12. Narasimhan, M. S., Ramanan, S.: Generalized Prym varieties as fixed points (to appear in Jour. Ind. Math. Society)
13. Narasimhan, M. S., Seshadri, C. S.: Stable and unitary vector bundles on a compact Riemann surface. Ann. of Math. **82**, 540—567 (1965)
14. Serre, J. P.: Groups algébriques et Corps de classes. Paris: Hermann 1959
15. Serre, J. P.: Corps locaux. Paris: Hermann 1968
16. Serre, J. P.: Cohomologie galoisiénne. Springer Lecture Notes 5, Berlin-Göttingen-Heidelberg: Springer 1964
17. Seshadri, C. S.: Space of unitary vector bundles on a compact Riemann surface. Ann. of Math. **85**, 303—336 (1967)
18. Seshadri, C. S.: Quotient spaces modulo reductive algebraic groups. Ann. of Math. **95**, 511—556 (1972)
19. Weil, A.: Adeles and algebraic groups. Lecture Notes. Princeton: 1961

G. Harder
Sonderforschungsbereich 40
Mathem. Institut
D-5300 Bonn
Wegelerstr. 10
Federal Republic of Germany

M. S. Narasimhan
Tata Institute for Fundamental Research
Bombay 5, India

(Received August 3, 1974)

# Deformations of the moduli space of vector bundles over an algebraic curve

By M. S. NARASIMHAN and S. RAMANAN

## 1. Introduction and statement of results

Let $X$ be a non-singular projective curve of genus $g \geqq 2$ defined over the field C of complex numbers. Let $\xi$ be a line bundle of degree $d$ on $X$ and $n \geqq 2$ an integer coprime to $d$. We shall denote by $U(n, \xi)$ the projective non-singular variety of isomorphism classes of stable bundles on $X$, of rank $n$ and determinant isomorphic to $\xi$([18]). Our main result is that the group of automorphisms of $U(n, \xi)$ is finite and that the number of moduli of $U(n, \xi)$ is the same as that of the curve $X$.

THEOREM 1. a) *The group of automorphisms of* $U(n, \xi)$ *is finite and*

$$H^i(U(n, \xi), \Theta) = 0 \qquad \text{for } i \neq 1 ,$$

*where* $\Theta$ *is the tangent sheaf.*

b) $$\dim H^1(U(n, \xi), \Theta) = 3g - 3 ,$$

*except when* $g = 2$, $n = 3$.

It is likely that b) is also true in the exceptional case mentioned in the theorem*. Our proof does not apply to this case, as well as to the cases $n = 2$, $g = 2, 3$. In the latter cases, however, it turns out that the theorem is still valid, as the explicit computations of $\chi(U(n, \xi), \Theta)$, via the Riemann-Roch-Hirzebruch theorem show ([12], [15]).

The essential computation involved in Theorem 1 is that of $H^0$ and $H^1$, since the vanishing of the higher cohomologies follows from the Akizuki-Nakano theorem and the ampleness of the anti-canonical class of $U(n, \xi)$ ([15]). The determination of $H^0$ and $H^1$ is a consequence of Theorem 2 stated below.

It is known (since $n$ is relatively prime to deg $\xi$) that there is a universal (Poincaré) bundle $W$ on $U(n, \xi) \times X$ such that for $t \in U(n, \xi)$ the bundle $W|_{t \times X}$ is in the isomorphism class $t$. Let ad $W$ denote the bundle of endomorphisms of $W$ of *trace zero*. Let $x \in X$ and denote by $\text{ad}_x W$ the bundle on $U(n, \xi)$ obtained by restricting ad $W$ to $U(n, \xi) \times x$.

* The exceptions in Theorems 1 and 3 have since been removed (see [21]).

THEOREM 2.

a) *The infinitesimal deformation map* $T_x \to H^1(U(n, \xi), \mathrm{ad}_x W)$ *of the bundle* $W$, *considered as a family of bundles on* $U(n, \xi)$ *parametrised by* $X$, *is injective.*

b) *For any* $x \in X$, $H^0(U(n, \xi), \mathrm{ad}_x W) = 0$. *Moreover,*

$$\dim H^1(U(n, \xi), \mathrm{ad}_x W) = 1, \text{ except (possibly) when } g = 2, n = 2$$
$$H^2(U(n, \xi), \mathrm{ad}_x W) = 0, \text{ except (possibly) when } g = 2, n = 2, 3$$
$$g = 3, n = 2 .$$

Theorem 2 implies that the family $(W_x)_{x \in X}$ of vector bundles on $U(n, \xi)$ is complete and infinitesimally injective. In fact, it can be easily deduced from the results of this paper that this is also injectively parametrised. Thus the curve $X$ can be recovered as the deformation space of a vector bundle on $U(n, \xi)$.

Another interesting application of Theorem 2 is

THEOREM 3. *The third integral cohomology group of* $U(n, \xi)$ *is free, of rank* $2g$ *(except when* $g=2$, $n=3$) *and the canonically polarised intermediary Jacobian of* $U(n, \xi)$ *corresponding to the third cohomology group is naturally isomorphic to the canonically polarised Jacobian of* $X$.

For $n = 2$ this result is proved in [6]; see also [19] in this connection.

The proof of Theorem 2 yields in fact more general results, e.g., the vanishing of some of the higher cohomologies of $\mathrm{ad}_x(W)$ and results similar to those of Theorems 1 and 2 when $n$ and $\deg \xi$ are not necessarily coprime.

To deduce Theorem 1 from Theorem 2 we first observe the following: if we consider $W$ as a family of vector bundles on $U(n, \xi)$ (resp. $X$) parametrised by $X$ (resp. $U(n, \xi)$), it follows from Theorem 2 (resp. [15, Lemma 2.6]) that the infinitesimal deformation map is an isomorphism of the tangent sheaf of $X$ (resp. tangent sheaf of $U(n, \xi)$) onto the first direct image of the sheaf ad $W$ on $X$ (resp. on $U(n \ \xi)$). Then the Leray spectral sequence together with the vanishing given in Theorem 2 enables one to compare $H^i(X, \Theta)$ and $H^i(U(n, \xi), \Theta)$ to prove Theorem 1.

We shall now give an outline of the proof of Theorem 2. Let $x \in X$. For any family $V$ of vector bundles on $X$ parametrised by a space $T$, let $\pi_V$: $Q(V) \to T$ be the projective bundle on $T$ associated to the vector bundle $V^* \mid T \times (x)$, where $V^*$ denotes the dual bundle of $V$. One can construct a family $K(V)$ of vector bundles on $X$ parametrised by $Q(V)$, by interpreting a point $q$ of $Q(V)$ as a homomorphism of $\underline{V}_t = \underline{V} \mid \pi_V(q) \times X$ into the structure sheaf $\mathcal{O}/\mathfrak{m}_x$ of the point $x$ and associating to $q$ the dual of the kernel of this homomorphism. We iterate this procedure and obtain $\pi_{KV}$: $Q(KV) \to Q(V)$ and a vector bundle $K^2V$ on $Q(KV) \times X$. We apply this construction when $T = U(n, \xi)$ and $V$ is a Poincaré bundle $W$ on $U(n, \xi) \times X$. Let $\Omega$ be the open set of points $\omega \in Q(KW)$ such that $K^2W|_{\omega \times X}$ is a stable bundle on $X$. Then $\det K^2W|_{\omega \times X} \approx \xi$ and we have a classifying morphism $\varphi: \Omega \to U(n, \xi)$. We show that $\varphi$ can be lifted into an involution $\iota: \Omega \to \Omega$ (i.e., $\pi_W \circ \pi_{KW} \circ \iota = \varphi$ and $\iota^2 = \mathrm{Id}_\Omega$). The proof of Theorem 2 is then done in two steps. The first crucial step is to show that $\iota^*\Theta_{\pi_{KW}}$ is *dual* to the bundle $\pi_{KW}^*\Theta_{\pi_W}$, where $\Theta_{\pi_W}$ (resp. $\Theta_{\pi_{KW}}$) denotes the tangent bundle along the fibres of $\pi_W$ (resp. $\pi_{KW}$). This result together with the well-known theorem (of Hartogs-type) concerning extendability of cohomology classes, shows that, for $i+2 \leqq \operatorname{codim}(Q(KW) - \Omega)$, $H^i(U(n, \xi), \mathrm{ad}_x W) \approx H^{i-1}(U(n, \xi), \mathcal{O})$ (see § 7.1) while

$$\dim H^{i-1}(U(n, \xi), \mathcal{O}) = \begin{cases} 0 & \text{if } i \neq 1 \\ 1 & \text{if } i = 1 \end{cases}$$

since $U(n, \xi)$ is unirational. The second step consists in computing the codimension of $Q(KW) - \Omega$ in $Q(KW)$ and the exceptions are necessitated by this.

The correspondence variety $Q(W)$ and the family $K(W)$ which defines a correspondence between the spaces of vector bundles of degree $d$ and $(d-1)$ were introduced in [8] in the case of vector bundles of rank 2, suggested by the geometry of a quadratic complex of lines in $\mathbf{P}^3$. It turned out, somewhat surprisingly, that this correspondence can be looked upon as the geometric analogue of the correspondence by which the Hecke operator at a place in a function field in one variable (over a finite field) is defined (see [20, § 22]). We became aware of this in conversations with A. Weil and G. Harder. For further remarks on this, one may refer to Section 8. We are grateful to the referee for a careful reading of the manuscript and for pointing out a technical gap in the proof of Proposition 5.10.

## 2. Preliminaries

If $f: Y_1 \to Y_2$ is a morphism, $f^*$ and $f_*$ will denote respectively the inverse and direct images of $\mathcal{O}$-modules; also $R^q f_*$ will denote the $q^{\text{th}}$ direct image of sheaves. If $\mathcal{F}$ is a sheaf of $\mathcal{O}$-modules on $Y_2$, then the induced mapping $H^0(Y_2, \mathcal{F}) \to H^0(Y_1, f^*\mathcal{F})$ will also be denoted by $f^*$.

Let $E$ be a vector bundle on $Y$. Then there exist a projective bundle $p: \mathbf{P}(E) \to Y$, a line bundle $\tau_E$ (called the tautological hyperplane bundle) and a tautological exact sequence of vector bundles on $\mathbf{P}(E)$:

$$\text{Taut } E: \qquad 0 \longrightarrow \Theta_p^* \otimes \tau_E \longrightarrow p^*E^* \xrightarrow{\alpha_E} \tau_E \longrightarrow 0\,,$$

where $\Theta_p$ denotes the tangent bundle along the fibres of $p$. This has the following characteristic property, viz. if $f: Y' \to Y$ is a morphism, and $L$ a line subbundle of $f^*E$, then there exists a unique morphism $\tilde{f}: Y' \to \mathbf{P}(E)$ so that (i) $p \circ \tilde{f} = f$ and (ii) the canonical isomorphism of $\tilde{f}^* p^* E$ with $f^*E$ induces an isomorphism of $f^*$ Taut $E$ with the exact sequence

$$0 \longrightarrow (f^*(E)/L)^* \longrightarrow f^*E^* \longrightarrow L^* \longrightarrow 0 .$$

In particular we have

LEMMA 2.1. *If $L$ is a line subbundle of $E$, it defines a section $\sigma$ of $\mathbf{P}(E)$ and*

$$\sigma^*(\Theta_p^*) \approx L \otimes (E/L)^* .$$

LEMMA 2.2. *Let $0 \to E' \to E \to E'' \to 0$ be an exact sequence of vector bundles on $Y$. Identifying $\mathbf{P}(E')$ with a subvariety of $\mathbf{P}(E)$, we have $\alpha_E \mid \mathbf{P}(E') = \alpha_{E'} \circ p^*(j)$ where $j$ is the transpose of the inclusion $E' \to E$.*

*Proof.* Clear.

LEMMA 2.3.

(i) *$R^i(p)_*(\tau_E) = 0$ for $i \geqq 1$ and $p_*(\tau_E)$ is canonically isomorphic to $E^*$;*

(ii) *$R^i(p)_*(\Theta_p) = 0$ for $i \geqq 1$ and $p_*(\Theta_p) \approx$* ad *$E$;*

(iii) *$R^i(p)_*(\Theta_p^*) = 0$ for $i \neq 1$ and $R^1(p)_*(\Theta_p^*)$ is a trivial line bundle if* rk$E \geqq 2$.

*Proof.* These assertions are well-known and are simple consequences of the computations of cohomologies on projective spaces.

By a family $V = (V_t)_{t \in T}$ of vector bundles on a variety $Y$, parametrised by a space $T$, we mean a vector bundle $V$ on $T \times Y$, with $i_t^* V \approx V_t$, where $i_t: Y \to T \times Y$ is defined by $i_t(y) = (t, y)$. We do not require $T$ to be irreducible, but will assume that $V_t$ has the same rank for all $t \in T$. If $f: T' \to T$ is a morphism, then the pull-back of $V$ by the map $(f \times \mathrm{Id}_Y)$ defines a family of bundles on $Y$ parametrised by $T'$. This will be denoted $f^* V$.

Let us now specialize $Y$ to be the curve $X$. If, for all $t \in T$, $V_t$ is stable, rank $V_t = n$ and det $V_t \approx \xi$, then we have a classifying *morphism* $\theta_V: T \to U(n, \xi)$ which assigns to $t \in T$ the isomorphism class of $V_t$([17]). We will assume that $(n, \deg \xi) = 1$, so that there exists a Poincaré bundle $W$ on $U(n, \xi) \times X$ (i.e., such that $\theta_W = \mathrm{Id}_{U(n,\xi)}$) ([18]). Moreover $\theta_V^* W \approx V \otimes p_T^* L$, for some line bundle $L$ on $T$ ([15, Lemma 2.5]).

Let $M(n, d)$ denote the non-singular, quasi-projective variety of isomorphism classes of stable bundles of rank $n$ and degree $d$. This variety has dimension $n^2(g-1)+1$. We will use the notation above, also in this case. While there is no Poincaré family when $n$ and $d$ are not coprime, we nevertheless have

PROPOSITION 2.4. *Let $n$ and $d$ be integers, $n > 1$. Then there exist a non-singular variety $M'$ (with finitely many irreducible components and of dimension $n^2(g-1)+1$) and a family $V = (V_m)_{m \in M'}$ of stable vector bundles of rank $n$ and degree $d$, such that the classifying map $\theta_V: M' \to M(n, d)$ is étale and surjective.*

For the proof we require the following

LEMMA 2.5. *There exist a principal fibration $\pi: B \to M(n, d)$ by the projective linear group and a family $V' = \{V'_b\}_{b \in B}$ of stable bundles on $X$ parametrised by $B$ such that $\theta_{V'} = \pi$.*

*Proof.* This is more or less implicit in the construction of the variety $M(n, d)$. See [17]. In fact, after tensorization by a suitable line bundle, there exist a trivial vector bundle $E$ and a polynomial $P$ such that all stable bundles of given rank and degree occur in the universal quotient family $F = (F_q)_{q \in Q}$ on $X$ parametrised by $Q = \text{Quot}(E/P)$. Moreover, we may assume that two bundles $F_{q_1}$, $F_{q_2}$ are isomorphic if and only if $q_1$ and $q_2$ are in the same orbit under the action of Aut $E$ on $Q$. If we denote by $B$ the open set of points $q \in Q$ such that $F_q$ is stable, then clearly the classifying morphism $\theta_F: B \to M(n, d)$ is a PGL-fibration.

*Proof of Proposition* 2.4. Consider $\pi: B \to M(n, d)$ and the family $V'$ as in Lemma 2.5. Since $\pi$ is locally isotrivial, there exist a variety $M'$ and a morphism $f: M' \to B$ such that $\pi \circ f$ is étale and surjective. Now $V = f^* V'$ satisfies the condition of Proposition 2.4, since $\theta_V = \theta_{f^*V'} = \theta_{V'} \circ f = \pi \circ f$.

The following proposition states essentially that any vector bundle on $X$ can be 'approximated' by stable bundles.

PROPOSITION 2.6. *Let $V = \{V_t\}_{t \in T}$ be a family of vector bundles of rank $n$ and degree $d$, parametrised by $T$. Then there exists an irreducible, non-singular variety $T'$ parametrising a family $V' = \{V'_t\}_{t \in T'}$ of bundles on $X$, containing all bundles $V_t$, $t \in T$, and all stable bundles of rank $n$ and degree $d$. The variety $T'$ can be chosen so that $T'_\xi = \{t \in T \mid \det V_t \approx \xi\}$ is irreducible and non-singular for all $\xi \in J^d$. In particular, the open set $\{t' \in T' \mid V_{t'}$ is stable$\}$ (resp. the open set $\{t' \in T'_\xi \mid V_{t'}$ is stable$\}$) is non-empty and dense in $T'$ (resp. $T'_\xi$).*

*Proof.* On tensoring by a fixed line bundle, we may assume that bundles of the family $V = \{V_t\}_{t \in T}$ and all the stable bundles of the required type contain a trivial bundle of rank $(n-1)$. Let $P_d$ be the Poincaré family of line bundles of degree $d$ parametrised by the Jacobian of $X$. Then the family of all extensions of line bundles of degree $d$ by the trivial bundle of rank

$(n-1)$ (constructed using $P_d$) includes the bundles $V_t$ and all stable bundles of degree $d$. In this case the parameter variety is a vector bundle over the Jacobian, and has all the required properties. See [10, § 3] and [7, Proposition 3.1].

### 3. Some remarks on the infinitesimal deformation map

Let $N$, $Y$ be two irreducible non-singular varieties, and $V$ a vector bundle on $N \times Y$. Then we have an exact sequence on $N \times Y$,

$$0 \longrightarrow \text{End } V \longrightarrow \mathcal{G} \longrightarrow p_N^* \Theta_N \longrightarrow 0 \tag{3.1}$$

which we may look upon as the Atiyah sequence along the fibres of $p_Y$. We recall that this gives rise to a map

$$\rho\colon \Theta_N \longrightarrow R^1(p_N)_*(\text{End } V)$$

called the infinitesimal deformation map ([5, p. 370]). The induced map of the tangent space $T_0 = T_{n_0}$ at $n_0$ into $H^1(Y, \text{End } V|_{n_0 \times Y})$ is referred to as the infinitesimal deformation map at $n_0$.

If $V$ is of the form $p_Y^* E$, then it is clear that the sequence (3.1) splits canonically. Now let $V|_{n_0 \times Y} = E$ and suppose given an open covering $(U_i)_{i \in I}$ and isomorphisms $f_i\colon p_Y^* E \to V$ over $N \times U_i$ coinciding with identity at $n_0$. Then via $f_i$, we get splittings of the sequence (3.1) over $N \times U_i$. Using these splittings, one can compute the infinitesimal deformation map. Thus we obtain

LEMMA 3.2. *Let $(U_i)_{i \in I}$ be an open covering of $Y$ and assume that there exist isomorphisms $f_i\colon p_Y^* E \to V$ over $N \times U_i$, coinciding with identity on $n_0 \times Y$. Then the section $m_{ij} = f_i^{-1} f_j$ of $p_Y^*$ (Aut $E$) over $N \times U_i \cap U_j$ may be considered as a map $N \to \text{Aut}(E \mid U_i \cap U_j)$. Then $c_{ij} = (dm_{ij})_0$ is a map $T_0 \to H^0(U_i \cap U_j, \text{End } E)$. This forms a cocycle and the induced map $T_0 \to H^1(Y, \text{End } E)$ gives the infinitesimal deformation map at $n_0$.*

Now we have the following

PROPOSITION 3.3. *Let $N$ and $Y$ be irreducible, non-singular varieties, with $Y$ complete. Let $V$, $V''$ be vector bundles on $Y$ and suppose given an exact sequence*

$$0 \longrightarrow V' \longrightarrow p_Y^* V \xrightarrow{\ \pi\ } p_Y^* V'' \longrightarrow 0$$

*on $N \times Y$. For $n \in N$, let $T_n$ denote the tangent space at $n$. Considering $\pi$ as a function $\pi\colon N \to H^0(Y, \text{Hom}(V, V''))$, we have the differential map*

$$(d\pi)_n\colon T_n \longrightarrow H^0(Y, \text{Hom}(V, V''))$$

*and hence by restriction a map*

$$\zeta\colon T_n \longrightarrow H^0(Y, \operatorname{Hom}(V'|_{n\times Y}, V'')) .$$

*Then the following diagram is commutative:*

$$\begin{array}{ccc} T_n & \xrightarrow{\zeta} & H^0(Y, \operatorname{Hom}(V'|_{n\times Y}, V'')) \\ & {\scriptstyle -\rho_n}\searrow \quad \swarrow{\scriptstyle \delta} & \\ & H^1(Y, \operatorname{End} V'|_{n\times Y}) & \end{array}$$

*where $\rho_n$ is the infinitesimal deformation map at $n$ of the family $V'$ and $\delta$ is the connecting homomorphism obtained from the exact sequence*

$$0 \longrightarrow \operatorname{End} V'|_{n\times Y} \longrightarrow \operatorname{Hom}(V'|_{n\times Y}, V) \longrightarrow \operatorname{Hom}(V'|_{n\times Y}, V'') \longrightarrow 0 .$$

*Proof**. It is enough to prove the proposition in the case when $N$ is a suitable open subset of $H^0(X, \operatorname{Hom}(V, V''))$, consisting of surjections and $V'$ the family of kernels. We shall compute the infinitesimal deformation map at $n_0 \in N$. Let $(U_i)_{i\in I}$ be a finite open covering of $Y$ and $\sigma_i$ splittings over $U_i$ of the sequence

$$0 \longrightarrow V_0' \xrightarrow{\varphi} V \xrightarrow{n_0} V'' \longrightarrow 0 .$$

We may replace $N$ by a neighborhood of $n_0$ and assume that $n\circ\sigma_i$ is an automorphism of $V''$ over $U_i$ for all $n \in N$ and all $i$.

Now let $g_i$ be the endomorphism $\operatorname{Id} - \sigma_i\circ(n\circ\sigma_i)^{-1}\circ(n - n_0)$ of $V$ over $U_i$. We may assume that $g_i$ is invertible for all $n \in N$. Clearly $f_i = g_i\circ\varphi$ is an isomorphism of $V_0'$ with $V_n'$. Since $m_{ij} = f_i^{-1}\circ f_j = g_i^{-1}\circ f_j$, we have

$$(dm_{ij})_0 = df_j - dg_i\circ\varphi = (dg_j - dg_i)\circ\varphi .$$

But $(dg_i)_0 = -\sigma_i\circ dn$. This shows that $(dm_{ij})_0$ is the image under the connecting homomorphism of the cocycle $-dn\circ\varphi$, proving the proposition.

## 4. Constructions $Q(V)$ and $K(V)$

If $E$ is a vector bundle on $X$, then any non-zero element of $E_x^*$ (the fibre of $E^*$ at $x \in X$) can be interpreted as a surjective homomorphism of the sheaf $\underline{E}$ into $\mathcal{O}_x$, the structure sheaf of $x$. The kernel depends only on the one dimensional subspace of $E_x^*$ generated by the given element and one can construct in this way a family of vector bundles on $X$ parametrised by $\mathbf{P}(E_x^*)$. We will now do this a little more formally, also allowing $E$ to vary through any given family.

Let $x \in X$. To every family $V = \{V_t\}_{t\in T}$ of vector bundles of rank $n$ ($\geq 2$) on $X$ parametrised by $T$ we associate the following:

a) a projective bundle $\pi_V\colon Q(V) \to T$ associated to a vector bundle $E_V$ on $T$;

* Our thanks are due to R. R. Simha for this proof.

b) a family $H(V)$ of vector bundles on $X$ parametrised by $Q(V)$, i.e., a vector bundle $H(V)$ on $Q(V) \times X$;

c) an exact sequence of sheaves on $Q(V) \times X$:

$$A(V)\colon 0 \longrightarrow \underline{H}(V) \longrightarrow \pi_V^\# \underline{V} \xrightarrow{\beta_V} p_X^*(\mathcal{O}_x) \otimes p_{Q(V)}^*\tau_V \longrightarrow 0 ,$$

where $p_X$ and $p_{Q(V)}$ are the projections from $Q(V) \times X$ onto $X$ and $Q(V)$ respectively and $\tau_V$ is the tautological hyperplane bundle $\tau_{E_V}$ on $Q(V)$.

The construction goes as follows. Let $i_x^T$ (or simply $i_x$) be the map $T \to T \times X$ defined by $t \mapsto (t, x)$. Define $E_V$ to be the vector bundle $i_x^*(V^*)$ and $\pi_V\colon Q(V) \to T$ the projective bundle associated to $E_V$. To define $H(V)$ we now construct a surjective homomorphism $\beta_V\colon \pi_V^\# \underline{V} \to p_X^*(\mathcal{O}_x) \otimes p_{Q(V)}^*\tau_V$ or, what is the same, an element of

$$H^0(Q(V) \times X, \operatorname{Hom}(\pi_V^\# \underline{V}, p_X^*\mathcal{O}_x \otimes p_{Q(V)}^*\tau_V)) .$$

This space is mapped isomorphically by $i_x^*$ onto

$$\begin{aligned} &H^0(Q(V), i_x^*(\operatorname{Hom}(\pi_V^\# \underline{V}, p_X^*\mathcal{O}_x \otimes p_{Q(V)}^*\tau_V))) \\ &\approx H^0(Q(V), \operatorname{Hom}(\pi_V^* i_x^* V, \tau_V)) \\ &\approx H^0(Q(V), \operatorname{Hom}(\pi_V^* E_V^*, \tau_V)) . \end{aligned}$$

But then there is a canonical homomorphism $\alpha_{E_V}\colon \pi_V^* E_V^* \to \tau_V$ (see § 2). We define $\beta_V$ by setting $i_x^*\beta_V = \alpha_{E_V}$. It is easy to see that the homomorphism $\beta_V\colon \pi_V^\# \underline{V} \to p_X^*(\mathcal{O}_x) \otimes p_{Q(V)}^*\tau_V$ thus defined is surjective. Since $p_X^*\mathcal{O}_x \otimes p_{Q(V)}^*\tau_V$ has a locally free resolution of length one, we see that the kernel $\underline{H}(V)$ of this homomorphism is locally free. This gives the sequence $A(V)$. Compare [8, § 7].

We denote the dual of $H(V)$ by $K(V)$. From the functorial nature of the construction we have the following

LEMMA 4.1. *Let $f\colon T_1 \to T_2$ be a morphism and $V$ a vector bundle on $T_2 \times X$. We then have a Cartesian diagram*

$$\begin{array}{ccc} Q(f^\# V) & \xrightarrow{Q(f)} & Q(V) \\ \downarrow & & \downarrow \\ T_1 & \xrightarrow{\quad f \quad} & T_2 \end{array}$$

*and an isomorphism $Q(f)^\# A(V) \approx A(f^\# V)$. In particular $K(f^\# V) \approx Q(f)^\#(KV)$.*

We denote by $K^2 V$ the bundle $K(K(V))$ obtained from $K(V)$ by iterating the above procedure.

LEMMA 4.2. *We have*

a) $$\det K(V) \approx \det(\pi_V^\# V^*) \otimes p_X^* L_x$$

*and*

b) $$\det(K^2 V) \approx (\pi_V \circ \pi_{KV})^\# \det V\,,$$

*where $L_x$ is the line bundle on $X$ associated to the divisor $x$.*

*Proof.* From the sequence $A(V)$ we see that

$$\det K(V)^* = \det H(V) \approx \det(\pi_V^\# \underline{V}) \otimes \det(p_X^* \mathcal{O}_x \otimes p_{Q(V)}^* \tau_V)^{-1}\,.$$

Using the exact sequence

$$0 \longrightarrow \underline{L}_x^{-1} \longrightarrow \mathcal{O}_X \longrightarrow \mathcal{O}_x \longrightarrow 0$$

on $X$, we see that

$$\det(p_X^* \mathcal{O}_x \otimes p_{Q(V)}^* \tau_V) \approx p_X^*(\underline{L}_x)\,.$$

This proves a). Assertion b) is an immediate consequence of a).

LEMMA 4.3. *We have a natural isomorphism*

$$(\pi_V \times 1_X)_*(K(V))^* \approx p_X^* L_x^{-1} \otimes V\,.$$

*Proof.* Consider the direct image of the exact sequence $A(V)$ by the map $(\pi_V \times 1_X)$. Notice that the direct image of $\pi_V^\# \underline{V}$ is $\underline{V}$ and that of $p_X^* \mathcal{O}_x \otimes p_{Q(V)}^* \tau_V$ is naturally isomorphic to $p_X^* \mathcal{O}_x \otimes \underline{V}$. Moreover $\beta_V$ induces the natural restriction $\underline{V} \rightarrow p_X^* \mathcal{O}_x \otimes \underline{V}$ and hence the kernel is isomorphic to $p_X^* L_x^{-1} \otimes V$. Now the lemma follows from the direct image exact sequence.

PROPOSITION 4.4. *If $K(V)$ is considered as a family of vector bundles on $Q(V)$ parametrised by $X$, then the infinitesimal deformation map of this family at $x \in X$ is injective.*

*Proof.* Since the infinitesimal deformation map commutes with restrictions, it is enough to consider the case when $T$ is a point, i.e., when $V$ is a vector bundle over $X$. Let $N$ be a neighborhood of $x$ in $X$ on which $V$ is trivial. We identify $V \mid N$ with the trivial bundle with fibre $V_x = E_V^*$. The homomorphism

$$p_{Q(V)}^* \alpha_{E_V} \colon p_{Q(V)}^* \pi_V^* E_V^* = \pi_V^\# V \longrightarrow p_{Q(V)}^* \tau$$

maps the subsheaf $\underline{K(V)}^*$ of $\pi_V^\# \underline{V}$ into $p_{Q(V)}^* \tau \otimes p_X^* L_x^{-1}$ since $i_x^* p_{Q(V)}^* \alpha_{E_V} = \alpha_{E_V} = i_x^* \beta_V$. Now we have a commutative diagram on $Q(V) \times N$:

$$
\begin{array}{ccccccccc}
 & & & & 0 & & 0 & & \\
 & & & & \downarrow & & \downarrow & & \\
 & & & & p^*_{Q(V)}\tau \otimes p^*_X L_x^{-1} & = & p^*_{Q(V)}\tau \otimes p^*_X L_x^{-1} & & \\
 & & & & \downarrow & & \downarrow & & \\
0 & \longrightarrow & K(V)^* & \longrightarrow & p^*_{Q(V)}\tau \otimes p^*_X L_x^{-1} \oplus \pi^\#_V V & \xrightarrow{\eta} & p^*_{Q(V)}\tau & \longrightarrow & 0 \\
 & & \| & & \downarrow & & \downarrow & & \\
0 & \longrightarrow & K(V)^* & \longrightarrow & \pi^\#_V V & \longrightarrow & p^*_X \mathcal{O}_x \otimes p^*_{Q(V)}\tau & \longrightarrow & 0 \\
 & & & & \downarrow & & \downarrow & & \\
 & & & & 0 & & 0 & &
\end{array}
\tag{4.5}
$$

We will now compute the infinitesimal deformation map of $K(V)^*$. With the notation of Proposition 3.3, we have only to compute the element $\zeta$.

LEMMA 4.6. *The restriction of $p^*_{Q(V)}\alpha_{E_V}$ to $K(V)^*$ gives rise to a homomorphism $\underline{K(V)}^* \to p^*_{Q(V)}\tau \otimes p^*_X \underline{L}_x^{-1}$ which in turn yields a homomorphism*

$$\zeta' : T_x \longrightarrow H^0(Q(V), \operatorname{Hom}(K(V)^* \mid_{Q(V)\times x}, \tau))$$

*since $L_x^{-1}|_x \approx T_x^*$. We then have $\zeta = -\zeta'$.*

*Proof.* We have to interpret the map $\eta$ in diagram (4.5) as a function on $N$ with values in

$$H^0(Q(V), \operatorname{Hom}(\tau \oplus V, \tau)) = H^0(Q(V), \operatorname{End}\tau) \oplus H^0(Q(V), \operatorname{Hom}(V, \tau)) .$$

With our identifications, it is clear that the second component of $\eta$ is constant, while the first component is a local parameter at $x$. Hence its differential at $x$ is $(f, 0)$ where $f$ is the homomorphism $t \mapsto t\sigma_0$, $t \in T_x$ and $\sigma_0$ is the identity endomorphism of $\tau$. This proves the lemma.

To complete the proof of Proposition 4.4, we restrict the diagram (4.5) to $Q(V) \times x$ and note that the map $\tau \otimes L_x^{-1}|_x \to \tau$ is the zero map. Thus we obtain

a) $K(V)^* |_{Q(V)\times x} \approx \tau \otimes L_x^{-1}|_x \oplus \Omega^1_{Q(V)} \otimes \tau$,
b) the composite

$$T_x \xrightarrow{\zeta} H^0(Q(V), \operatorname{Hom}(K(V)^*_x, \tau)) \longrightarrow T_x \otimes H^0(Q(V), \operatorname{End}\tau)$$

is the homomorphism $t \mapsto t \otimes \sigma_0$, where $\sigma_0$ is the identity endomorphism of $\tau$.

Now consider the commutative diagram

$$
\begin{array}{ccc}
H^0(Q(V), \operatorname{Hom}(K(V)^*_x, \tau)) & \longrightarrow & H^1(Q(V), \operatorname{End} K(V)^*_x) \\
\downarrow & & \downarrow \\
T_x \otimes H^0(Q(V), \operatorname{Hom}(\tau, \tau)) & \longrightarrow & T_x \otimes H^1(Q(V), \operatorname{Hom}(\tau, K(V)^*_x))
\end{array}
$$

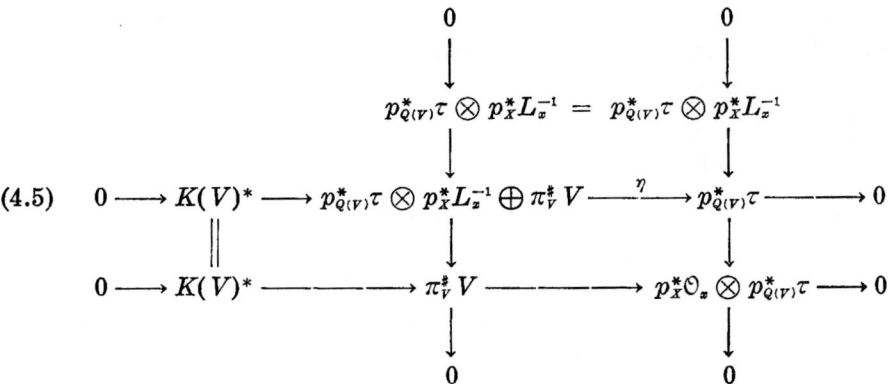

where $K(V)_x = K(V)|_{Q(V)\times x}$. Notice that the lower horizontal arrow is an isomorphism. Thus Proposition 4.4 follows from assertion b) above.

*Remark* 4.7. If $V$ is a vector bundle on $X$, we have also shown above that $K(V)|_{Q(V)\times x} \approx \tau^{-1} \oplus \tau^{-1} \otimes \Theta$. On the other hand, from the sequence $A(V)$, it is clear that $K(V)|_{Q(V)\times y}$ is trivial for $y \neq x$.

PROPOSITION 4.8. *There exists an exact sequence of vector bundles*

$$\Sigma(V): \quad 0 \longrightarrow T_x^* \otimes \tau_V \xrightarrow{\gamma_V} E_{KV} \longrightarrow \pi_V^*(E_V^*) \longrightarrow \tau_V \longrightarrow 0$$

*on $Q(V)$. Here $\pi_V^*(E_V^*) \to \tau_V$ is the tautological map (see § 2) and $T_x^*$ denotes the trivial line bundle with fibre $T_x^*$, the cotangent space at $x$ to $X$.*

*Proof.* Take the inverse image of the exact sequence $A(V)$ by the morphism $i_x: Q(V) \to Q(V) \times X$. Recalling that $K(V) = H(V)^*$, $E_{K(V)} = i_x^* H(V)$ and $E_V = i_x^* V^*$, we obtain an exact sequence

$$\Sigma(V): 0 \longrightarrow \mathfrak{L} \longrightarrow \underline{E}_{KV} \longrightarrow \pi_V^* \underline{E}_V^* \longrightarrow \tau_V \longrightarrow 0$$

on $Q(V)$, where

$$\mathfrak{L} = \mathcal{T}or_1^{\mathcal{O}_{Q(V)\times X}}\left(p_{Q(V)}^*(\tau_V) \otimes_{\mathcal{O}_{Q(V)\times X}} \mathcal{O}_{Q(V)}, \mathcal{O}_{Q(V)}\right),$$

$\mathcal{O}_{Q(V)}$ being considered as a $\mathcal{O}_{Q(V)\times X}$-module by means of $i_x$. Using the resolution

$$0 \longrightarrow \underline{L}_x^{-1} \longrightarrow \mathcal{O}_X \longrightarrow \mathcal{O}_x \longrightarrow 0,$$

we obtain the exact sequence

$$\begin{aligned} 0 \longrightarrow i_x^*(\mathfrak{L}) \longrightarrow p_{Q(V)}^*\tau \otimes p_X^*(\underline{L}_x^{-1} \otimes \mathcal{O}_x) &\longrightarrow p_{Q(V)}^*(\tau) \otimes \mathcal{O}_{Q(V)\times X} \\ &\longrightarrow p_{Q(V)}^*\tau \otimes \mathcal{O}_{Q(V)\times x} \longrightarrow 0 . \end{aligned}$$

This shows that $\mathfrak{L} \approx \tau \otimes T_x^*$, as $\underline{L}_x^{-1} \otimes \mathcal{O}_x \approx T_x^*$. From the definition of the sequence $A(V)$, it is clear that the map $\pi_V^* E_V^* \to \tau_V$ in $\Sigma(V)$ is the tautological map. This completes the proof of Proposition 4.8.

*Remark* 4.9. Dualizing the exact sequence $A(V)$ i.e., applying Hom $(\cdot, \mathcal{O})$, we obtain the exact sequence

$$A^*(V): \quad 0 \longrightarrow \pi_V^* \underline{V}^* \longrightarrow \underline{K}(V) \xrightarrow{\delta} \mathcal{E}xt^1(p_X^*\mathcal{O}_x \otimes p_{Q(V)}^*\tau, \mathcal{O}_{Q(V)}) \longrightarrow 0 .$$

Taking the inverse image of $A^*(V)$ by $i_x: Q(V) \to Q(V) \times X$, we note that the homomorphism $\pi_V^* E_V \to E_{KV}^*$ so obtained is the transpose of the map $E_{KV} \to \pi_V^* E_V^*$ in the sequence $\Sigma(V)$. Hence the inverse image of $A^*(V)$ by $i_x$ yields the exact sequence

$$0 \longrightarrow \tau_V^{-1} \longrightarrow \pi_V^* E_V \longrightarrow E_{KV}^* \longrightarrow T_x \otimes \tau_V^{-1} \longrightarrow 0$$

which is the dual of $\Sigma(V)$. With this identification, the map $i_x^*\delta: E_{KV}^* \to T_x \otimes$

$\tau^{-1}$ coincides, with $\gamma_V^*\colon E_{KV}^* \to T_x \otimes \tau^{-1}$, where $\gamma_V^*$ is the transpose of the map $\gamma_V$ in $\Sigma(V)$.

Now the line subbundle $T_x^* \otimes \tau_V$ of $E_{KV}$, given by $\Sigma(V)$ in Proposition 4.8, defines a section $\sigma_V$ of the projective bundle $Q(KV)$ over $Q(V)$. (See 2.1.)

LEMMA 4.10. *We have (with the notation of* 4.9)

$$\sigma_V^\sharp(A(KV)) \approx A^*(V)\ .$$

*Proof.* Recall that $A(KV)$ is the exact sequence

$$0 \longrightarrow \underline{(K^2V)^*} \longrightarrow \pi_{KV}^\sharp \underline{KV} \xrightarrow{\beta_{KV}} p_X^* \mathcal{O}_x \otimes p_{Q(KV)}^* \tau_{KV} \longrightarrow 0$$

on $Q(KV) \times X$. We clearly have $\sigma_V^\sharp \pi_V^\sharp KV \approx KV$. We shall now compute $(\sigma_V \times \mathrm{Id}_X)^* \beta_{KV}$. But this is determined by

$$i_x^*(\sigma_V \times \mathrm{Id}_X)^* \beta_{KV} = ((\sigma_V \times \mathrm{Id}_X) \circ i_x)^* \beta_{KV} = (i_x \circ \sigma_V)^* \beta_{KV} = \sigma_V^* i_x^* \beta_{KV}\ .$$

Now by the definition of the sequence $A(KV)$, the map $i_x^* \beta_{KV}$ is the tautological surjection $\alpha_{E_{KV}}\colon E_{KV}^* \to \tau_{KV}$. Hence we have

$$\sigma_V^* i_x^* \beta_{KV} = \sigma_V^* \alpha_{E_{KV}} = \gamma_V^* = i_x^* \delta$$

by Lemma 2.2 and Remark 4.9. (For the definition of $\alpha$ see the sequence 'Taut' in § 2.) Hence it follows that $(\sigma_V \times \mathrm{Id}_X)^* \beta_{KV} = \delta$. In particular $\sigma_V^\sharp \ker \beta_{KV} \approx \ker \delta$. This proves Lemma 4.10.

PROPOSITION 4.11. *We have*

a) $\sigma_V^\sharp K^2 V \approx \pi_V^\sharp V$, *the isomorphism being canonical.*

b) *The inverse image by $\sigma_V$ of the exact sequence*

$$\Sigma(KV)\colon\ 0 \longrightarrow T_x^* \otimes \tau_{KV} \longrightarrow E_{K^2V} \longrightarrow \pi_{KV}^* E_{KV} \longrightarrow \tau_{KV} \longrightarrow 0$$

*can be canonically identified with the sequence*

$$\Sigma^*(V)\colon\ 0 \longrightarrow \tau_V^{-1} \longrightarrow \pi_V^* E_V \longrightarrow E_{KV}^* \longrightarrow T_x \otimes \tau_V^{-1} \longrightarrow 0$$

*which is the dual of the sequence* $\Sigma(V)$.

*Proof.* As a part of the assertion of Lemma 4.10 we have $\sigma_V^\sharp (K^2V)^* \approx \pi_V^\sharp V^*$, which proves a). The assertion b) follows again from Lemma 4.10 by taking inverse images by $i_x$ of the sequences $\sigma_V^\sharp A(KV)$ and $A^*(V)$ and remarking that $i_x^* A^*(V)$ gives rise to the sequence $\Sigma^*(V)$.

PROPOSITION 4.12. *Let $\sigma_V\colon Q(V) \to Q(KV)$ be the section defined above. Then*

$$\Theta_{\pi_V} \approx \sigma_V^*(\Theta_{\pi_{KV}}^*) \otimes T_x\ ,$$

*where $\Theta_\pi$ denotes the tangent bundle along the fibres of $\pi$ and $T_x$ the trivial line bundle on $Q(V)$ with fibre $T_x$, the tangent space to $X$ at $x$.*

*Proof.* In view of the sequence $\Sigma(V)$ (see Proposition 4.8) we have

$$E_{KV}/\gamma(T_x^* \otimes \tau_V) \approx \ker(\pi_V^* E_V^* \longrightarrow \tau_V) \approx \tau_V \otimes \Theta^* \pi_V ,$$

by Lemma 2.1. Again by Lemma 2.1 it follows that

$$\begin{aligned} \sigma_V^*(\Theta_{\pi_{KV}}^*) &\approx (\tau_V^* \otimes \Theta_{\pi_V}) \otimes (T_x^* \otimes \tau_V) \\ &\approx \Theta_{\pi_V} \otimes T_x^* , \end{aligned}$$

which proves the proposition.

## 5. The involution $\iota$

In this article we apply the results of the last section to a universal (Poincaré) bundle $W$ on $U(n, \xi) \times X$ with $(n, \deg \xi) = 1$. Let $\Omega$ be the open set of points $\omega$ in $Q(KW)$ such that $K^2 W|_{\omega \times X}$ is stable. Since $\sigma_W^\#(K^2 W) \approx \pi_W^\# W$ by Proposition 4.11, we see that $\sigma_W(Q(W)) \subset \Omega$. Also, $\det K^2 W|_{\omega\times X} \approx \xi$, by Lemma 4.2. Hence there exists a morphism $\varphi\colon \Omega \longrightarrow U(n, \xi)$, which associates to $\omega \in \Omega$ the isomorphism class of $K^2 W|_{\omega \times X}$. We have then $\varphi^\#(W) \approx K^2 W \otimes p_\Omega^*(L)$ for some line bundle $L$ on $\Omega$, where $p_\Omega$ is the projection $\Omega \times X$ to $\Omega$ ([15, Lemma 2.5]).

Now we assert that $\varphi\colon \Omega \longrightarrow U(n, \xi)$ has a natural lift $\psi\colon \Omega \longrightarrow Q(W)$. We shall define $\psi$ by giving a section of the pull-back by $\varphi$ of the fibration $\pi_W\colon Q(W) \longrightarrow U(n, \xi)$. Since $Q(W)$ is the projective bundle associated to the vector bundle $E_W$ and $\varphi^*(E_W) \approx E_{K^2W} \otimes L^*$, we see that the pull-back of the fibration $\pi_W$ by $\varphi$ is isomorphic to the fibration $\pi_{K^2W}$ over $\Omega$; in other words, there exists a morphism $\tilde{\varphi}\colon \pi_{K^2W}^{-1}(\Omega) \longrightarrow Q(W)$ such that the diagram

(5.1)
$$\begin{array}{ccc} \pi_{K^2W}^{-1}(\Omega) & \xrightarrow{\tilde{\varphi}} & Q(W) \\ \downarrow{\scriptstyle \pi_{K^2W}} & & \downarrow{\scriptstyle \pi_W} \\ \Omega & \xrightarrow{\varphi} & U(n, \xi) \end{array}$$

is commutative. But there exists a canonical section $\sigma_{KW} \circ j\colon \Omega \longrightarrow \pi_{K^2W}^{-1}(\Omega)$, where $j\colon \Omega \longrightarrow Q(KW)$ is the inclusion map. We define $\psi$ by setting

$$\psi = \tilde{\varphi} \circ \sigma_{KW} \circ j .$$

LEMMA 5.2. *We have*
(i) $\psi^\# A(W) \approx j^\# A^*(KW) \otimes p_\Omega^*(L)$.
(ii) $\psi^\# A^*(W) \approx j^\# A(KW) \otimes p_\Omega^*(L^*)$.

*Proof.*

$$\begin{aligned} \psi^\# A(W) &= (\tilde{\varphi} \circ \sigma_{KW} \circ j)^\# A(W) \\ &\approx j^\# \circ \sigma_{KW}^\# A(\varphi^\# W) \qquad \text{(by functoriality; Lemma 4.1)} \\ &\approx j^\# \circ \sigma_{KW}^\# A(K^2 W \otimes p_\Omega^* L) \\ &\approx j^\# A^*(KW) \otimes p_\Omega^* L . \end{aligned}$$

(In the last step we have applied Lemma 4.10 to the section $\sigma_{KW}$.) This proves the lemma.

PROPOSITION 5.3. *With the notation of Proposition* 4.12, *the restriction of* $\Theta_{\pi_{KW}}$ *to* $\Omega$ *is isomorphic to* $\psi^*(\Theta^*_{\pi_W}) \otimes T_x$.

*Proof.* By Proposition 4.12, we have

$$\Theta_{\pi_{KW}} \approx \sigma^*_{KW}(\Theta^*_{\pi_{K^2W}}) \otimes T_x .$$

If $j: \Omega \to Q(KW)$ denotes the inclusion map, we have

$$\begin{aligned} j^*(\Theta_{\pi_{KW}}) &\approx (\sigma_{KW} \circ j)^*(\Theta^*_{\pi_{K^2W}}) \otimes T_x \\ &\approx (\sigma_{KW} \circ j)^*(\tilde{\varphi}^*\Theta^*_{\pi_W}) \otimes T_x \\ &\approx \psi^*(\Theta^*_{\pi_W}) \otimes T_x . \end{aligned}$$

This proves the proposition.

We now proceed to define a map $\iota: \Omega \to Q(KW)$ which lifts $\psi: \Omega \to Q(W)$. We start from the Cartesian diagrams

$$\begin{array}{ccc} \psi^*(Q(KW)) & \xrightarrow{\tilde{\psi}} & Q(KW) \\ {\scriptstyle \psi^*\pi_{KW}}\downarrow & & \downarrow{\scriptstyle \pi_{KW}} \\ \Omega & \xrightarrow{\psi} & Q(W) \end{array}$$

and

$$\begin{array}{ccc} \pi^*_{KW}Q(KW) & \xrightarrow{\tilde{\pi}_{KW}} & Q(KW) \\ {\scriptstyle \pi^*_{KW}\pi_{KW}}\downarrow & & \downarrow{\scriptstyle \pi_{KW}} \\ Q(KW) & \xrightarrow{\pi_{KW}} & Q(W) . \end{array} \tag{5.4}$$

Let $Q(\lambda): j^*\pi^*_{KW}Q(KW) \longrightarrow \psi^*Q(KW)$ be the isomorphism induced by the isomorphism $\lambda: \psi^*KW \longrightarrow \pi^*_{KW}(KW) \otimes p^*_\Omega(L^*)$ (over $\Omega \times X$) given by Lemma 5.2. We then have a commutative diagram

$$\begin{array}{ccc} j^*\pi^*_{KW}Q(KW) & \xrightarrow{Q(\lambda)} & \psi^*Q(KW) \\ & {\scriptstyle \pi^*_{KW}\pi_{KW}}\searrow \quad \swarrow{\scriptstyle \psi^*\pi_{KW}} & \\ & \Omega . & \end{array}$$

Now, by diagram (5.4), the section $\sigma_W: Q(W) \to Q(KW)$ defines a section $\pi^*_{KW}\sigma_W: Q(KW) \to \pi^*_{KW}Q(KW)$ such that $\tilde{\pi}_{KW} \circ \pi^*_{KW}\sigma_W = \sigma_W \circ \pi_{KW}$. We define $\iota: \Omega \to Q(KW)$ by

$$\iota = \tilde{\psi} \circ Q(\lambda) \circ \pi^*_{KW}\sigma_W \circ j .$$

We have clearly $\pi_{KW} \circ \iota = \psi$, so that $\iota$ lifts $\psi$.

LEMMA 5.5. $Q(\lambda)^{\#}\tilde{\psi}^{\#}A(KW) \approx A(j^{\#}\pi^{\#}_{KW}(KW)) \otimes p^{*}_{\Omega}(L)^{*}$.

*Proof.* Note that $\tilde{\psi}^{\#}A(KW) \approx A(\psi^{\#}KW)$ by Lemma 4.1 and $\lambda$ is the isomorphism $\psi^{\#}KW \approx j^{\#}\pi^{\#}_{KW}(KW) \otimes p^{*}_{\Omega}(L)^{*}$ given by Lemma 5.2.

PROPOSITION 5.6. *We have*
(i) $\iota^{\#}A(KW) \approx j^{\#}\pi^{\#}_{KW}(A^{*}(W)) \otimes p^{*}_{\Omega}(L)^{*}$,
(ii) $\iota^{\#}A^{*}(KW) \approx j^{\#}\pi^{\#}_{KW}(A(W)) \otimes p^{*}_{\Omega}(L)$.

*Proof.* By the functorial nature of $A$ (Lemma 4.1) we have $\tilde{\psi}^{\#}A(KW) = A(\psi^{\#}KW)$. By Lemma 5.5, it follows that

$$Q(\lambda)^{\#}\tilde{\psi}^{\#}A(KW) \approx A(j^{\#}\pi^{\#}_{KW}K(W)) \otimes p^{*}_{\Omega}(L)^{*}$$
$$\approx (\tilde{\pi}_{KW} \circ j)^{\#}A(KW) \otimes p^{*}_{\Omega}(L)^{*} .$$

Hence

$$p^{*}_{\Omega}(L) \otimes \iota^{\#}A(KW) \approx j^{\#}\widetilde{(\pi^{*}_{KW}\sigma_{W})}^{\#}(\tilde{\pi}_{KW} \circ j)^{\#}A(KW)$$
$$\approx (\pi_{KW} \circ j \circ \pi^{*}_{KW}\sigma_{W} \circ j)^{\#}A(KW)$$
$$\approx (\sigma_{W} \circ \pi_{KW} \circ j)^{\#}A(KW)$$
$$\approx j^{\#}\pi^{\#}_{KW}(A^{*}(W)) \text{ , by Lemma 4.10.}$$

This proves the first part of the proposition. The second part follows from (i).

COROLLARY 5.7. $\iota^{\#}K^{2}(W) \approx (\pi_{W} \circ \pi_{KW} \circ j)^{\#}(W) \otimes p^{*}_{\Omega}(L)$.

COROLLARY 5.8. *$\iota$ maps $\Omega$ into $\Omega$.*

*Proof.* Clearly the inverse image of $\Omega$ by $\iota\colon \Omega \to Q(KW)$ is the set of points corresponding to stable bundles in the family $\iota^{\#}K^{2}(W)$. Since $W$ consists entirely of stable bundles, Corollary 5.8 follows from Corollary 5.7.

We now proceed to prove that $\iota\colon \Omega \to \Omega$ is actually an involution.

PROPOSITION 5.9. *We have*
(i) $\varphi \circ \iota = \pi_{W} \circ \pi_{KW} \circ j$,
(ii) $\iota^{*}L \approx L^{*}$.

*Proof.* For $\omega \in \Omega$, $(\varphi \circ \iota)\omega$ is the isomorphism class of the bundle $\iota^{\#}K^{2}(W)|_{\omega \times X}$. But by Corollary 5.7, $\iota^{\#}K^{2}W|_{\omega \times X}$ is isomorphic to the bundle $(\pi_{W} \circ \pi_{KW})^{\#}W|_{\omega \times W}$, whose isomorphism class is clearly $\pi_{W} \circ \pi_{KW}(\omega)$. This proves (i).

We have

$$(\varphi \circ \iota)^{\#} \det W \approx \det(\iota^{\#}\varphi^{\#}W) \approx \det \iota^{\#}(K^{2}W \otimes p^{*}_{\Omega}L)$$
$$\approx (\pi_{W} \circ \pi_{KW} \circ j)^{\#} \det W \otimes p^{*}_{\Omega}(\iota^{*} L \otimes L)^{n}$$

by Corollary 5.7. On the other hand by (i), this is isomorphic to $(\pi_{W} \circ \pi_{KW} \circ j)^{\#} \det W$. Hence $(\iota^{*}L \otimes L)^{n}$ is trivial. Now the complement of $\Omega$

in $Q(KW)$ is of codim$\geq 2$, as is proved in Proposition 6.8. Hence $\operatorname{Pic} Q(KW) \to \operatorname{Pic} \Omega$ is an isomorphism, and $\operatorname{Pic} Q(KW)$ is torsion-free since $\operatorname{Pic} U(n, \xi)$ is torsion-free. (See for instance [15, Proposition 3.4].) It follows that $\iota^* L \otimes L$ is trivial.

PROPOSITION 5.10.

$$\iota^2 = \textit{Identity map of } \Omega .$$

*Proof.* We first show that $\psi \circ \iota = \pi_{KW} \circ j$. Since the composites of these two maps with $\pi_W$ are the same by Proposition 5.9 (i), we could use the criterion of Section 2 to prove this. Firstly there are isomorphisms

$$\begin{aligned} (\psi \circ \iota)^\# A(W) &\approx \iota^\# \psi^\# A(W). \\ &\approx \iota^\# A^*(KW) \otimes p_\Omega^* \iota^*(L) \text{ , by Lemma 5.2, (i).} \\ &\approx j^\# \pi_{KW}^\# A(W) \otimes p_\Omega^*(L \otimes \iota^* L) \text{ by Proposition 5.6, (ii).} \\ &\approx j^\# \pi_{KW}^\# A(W) \text{ by Proposition 5.10, (ii).} \end{aligned}$$

Clearly this induces also an isomorphism

$$(\psi \circ \iota)^* \operatorname{Taut} E_W \approx (\pi_{KW} \circ j)^\# \operatorname{Taut} E_W .$$

It remains to check that this isomorphism can be modified so as to induce the canonical isomorphism

$$\varphi^* E_W \approx (\pi_W \circ \pi_{KW} \circ j)^* E_W .$$

Since the canonical isomorphism as well as the isomorphism defined above are restrictions of isomorphisms

$$\varphi^\# W^* \approx (\pi_W \circ \pi_{KW} \circ j)^\# W^* ,$$

our assertion follows from the fact that the endomorphisms of $\varphi^\# W \approx K^2 W \otimes L$, are simply scalar functions on $\Omega$. The latter is a consequence of the isomorphisms

$$H^0(\Omega \times X, \operatorname{End} K^2 W) \approx H^0(\Omega, (p_\Omega)_* \operatorname{End} K^2 W) = H^0(\Omega, \mathcal{O}_\Omega)$$

since $K^2 W|_{\omega \times X}$ is simple for each $\omega \in \Omega$.

To prove $\iota^2 = \mathrm{Id}$, we again use the criterion of Section 2. We see that

$$\begin{aligned} (\iota^2)^\# A(KW) &\approx \iota^\# \pi_{KW}^\# A^*(W) \otimes p_\Omega^* \iota^* L && \text{by Proposition 5.6, (i)} \\ &\approx (\pi_{KW} \circ \iota)^\# A^*(W) \otimes p_\Omega^* \iota^* L \\ &\approx \psi^\# A^*(W) \otimes p_\Omega^* \iota^* L \\ &\approx j^\# A(KW) && \text{by Lemma 5.2, (ii) and Proposition 5.10.} \end{aligned}$$

The proof is completed as above, and we note that the set of stable bundles of the family $\psi^\# KW$ is non-empty, as will be shown in Corollary 6.9.

PROPOSITION 5.11. *We have*

$$\iota^*(\Theta_{\pi_{KW}}) \approx j^* \pi_{KW}^*(\Theta_{\pi_W}^*) \otimes T_x ,$$

*where $\Theta_\pi$ denotes the tangent bundle along the fibres of $\pi$.*

*Proof.* In fact,

$$\begin{aligned} j^*\pi^*_{KW}(\Theta^*_{\pi_W}) \otimes T_x &\approx (\pi_{KW} \circ j)^*(\Theta^*_{\pi_W}) \otimes T_x \\ &\approx (\psi \circ \iota)^*(\Theta^*_{\pi_W}) \otimes T_x \text{ , by Proposition 5.10,} \\ &\approx \iota^*\psi^*(\Theta^*_{\pi_W}) \otimes T_x \\ &\approx \iota^*(\Theta_{\pi_{KW}}) \text{ , by Proposition 5.3.} \end{aligned}$$

## 6. Codimension of $Q(KW) - \Omega$

We now wish to compute the codimension of $Q(KW) - \Omega$ in $Q(KW)$. (This computation is independent of Section 5.) This essentially means that we have to investigate the following question. Let $E_1$ be a stable vector bundle of rank $n$ and degree $d$, and let $N$ be obtained by the exact sequence

$$0 \longrightarrow \underline{N}^* \longrightarrow \underline{E}_1 \longrightarrow \mathcal{O}_x \longrightarrow 0 .$$

If $E_2$ is another vector bundle defined by the exact sequence

$$0 \longrightarrow \underline{E}_2^* \longrightarrow \underline{N} \longrightarrow \mathcal{O}_x \longrightarrow 0$$

then, under what condition is $E_2$ non-stable? Assume that $E_2$ is not stable and $E$ is a proper subbundle of $E_2^*$ with $\mu(E) \geqq \mu(E_2^*) = -\mu(E_1)$, where $\mu(E) = \deg E/\operatorname{rk} E$. Let $E'$ be the subbundle of $N$ generated by $E$ and $F'$ be the subbundle of $E_1$ generated by the subbundle $F = (E')^\perp$ of $N^*$ orthogonal to $E'$. Then we have

$$\begin{aligned} \mu(E_1) \operatorname{rk} F' > \deg F' \geqq \deg F &= -(\deg N - \deg E') \\ &= d - 1 + \deg E' \geqq (d-1) + \deg E \\ &\geqq d - 1 - d/n \operatorname{rk} E = -1 + d/n \operatorname{rk} F \\ &= -1 + \mu(E_1) \operatorname{rk} F' . \end{aligned}$$

Hence it follows that $\deg F = \deg F'$ and $\deg E = \deg E'$. In other words $F = F'$ and $E = E'$. Thus we have proved the second part of

LEMMA 6.1. *Let $N$, $E_1$, $E_2$ be vector bundles on $X$ and*

$$\begin{aligned} 0 \longrightarrow \underline{N}^* \longrightarrow \underline{E}_1 \longrightarrow \mathcal{O}_x \longrightarrow 0 , \\ 0 \longrightarrow \underline{E}_2^* \longrightarrow \underline{N} \longrightarrow \mathcal{O}_x \longrightarrow 0 \end{aligned}$$

*be two exact sequences of sheaves on $X$. If there exists a proper subbundle $F$ of $E_1$ with*

(i) $\mu(F) + 1/\operatorname{rk} F \geqq \mu(E_1)$;

(ii) *the restriction to $F$ of the map $\underline{E}_1 \longrightarrow \mathcal{O}_x$ is zero, so that $F$ is a subbundle of $N^*$;*

(iii) *the restriction to $F^\perp \subset N$ of the map $\underline{N} \longrightarrow \mathcal{O}_x$ is also the zero map; then $E_2$ is not stable.*

*Conversely, if $E_1$ is stable, but $E_2$ is not, then there exists a proper subbundle $F$ of $E_1$ satisfying* (i), (ii), (iii).

*Proof.* For the first part, note that by (iii), $F^\perp$ may be viewed as a subbundle of $E_2^*$ and we have

$$\deg(F^\perp) = -(\deg N^* - \deg F) = \deg F + \deg E_2^* + 1 \geqq \operatorname{rk} F\, d/n - d\,.$$

Hence $\mu(F^\perp) \geqq \mu(E_2^*)$, so that $E_2^*$ (and hence $E_2$) is not stable.

Lemma 6.1 shows that if $q \in Q(KW) - \Omega$, then $\pi_W \pi_{KW}(q)$ represents a stable bundle which admits a subbundle of a certain type. Hence we shall investigate the constructions $K(V)$, $K^2(V)$ in the following set-up. Let

$$(\mathfrak{S}) \qquad 0 \longrightarrow V' \longrightarrow V \xrightarrow{\;k\;} V'' \longrightarrow 0$$

be an exact sequence of vector bundles on $T \times X$. Since $i_x^*(V'')^* \subset i_x^*(V)^*$, we have an inclusion $Q(V'') \subset Q(V)$.

LEMMA 6.2. *In the above situation, we have a commutative diagram of sheaves on $Q(V'') \times X$ with exact rows and columns:*

$$\begin{array}{llcccccccc}
 & & & 0 & & 0 & & & & \\
 & & & \downarrow & & \downarrow & & & & \\
 & & & \pi_V^\# \underline{V}' & = & \pi_V^\# \underline{V}' & & & & \\
 & & & \downarrow & & \downarrow & & & & \\
A(V)|_{Q(V'')\times X}: & 0 & \longrightarrow & \underline{(KV)^*} & \longrightarrow & \pi_V^\# \underline{V} & \xrightarrow{\beta_V} & p_X^* \mathcal{O}_x \otimes p_{Q(V)}^* \tau_V & \longrightarrow & 0 \\
 & & & \downarrow & & \downarrow & & \| & & \\
A(V''): & 0 & \longrightarrow & \underline{(KV'')^*} & \longrightarrow & \pi_V^\# \underline{V}'' & \xrightarrow{\beta_{V''}} & p_X^* \mathcal{O}_x \otimes p_{Q(V'')}^* \tau_{V''} & \longrightarrow & 0 \\
 & & & \downarrow & & \downarrow & & & & \\
 & & & 0 & & 0 & & & &
\end{array}$$

*Proof.* We have only to show that $\beta_V|_{Q(V'')\times X} = \beta_{V''} \circ \pi_V^\#(k)$. But this follows from the definition of $\beta_V$ and Lemma 2.2.

LEMMA 6.3. (i) *With the notation of Lemma* 6.2, *the following sequence on $Q(V'') \times X$ is exact:*

$$0 \longrightarrow KV'' \longrightarrow KV \longrightarrow \pi_V^\#(V')^* \longrightarrow 0\,.$$

(ii) *Identifying $Q(\pi_V^\# V'^*)$ with a closed subvariety of $Q(KV)$, we have the exact sequence on $Q(\pi_V^\# V'^*) \times X$:*

$$0 \longrightarrow \pi_V^\# K((V')^*) \longrightarrow K^2 V \longrightarrow \pi_{KV}^\#(KV'')^* \longrightarrow 0\,.$$

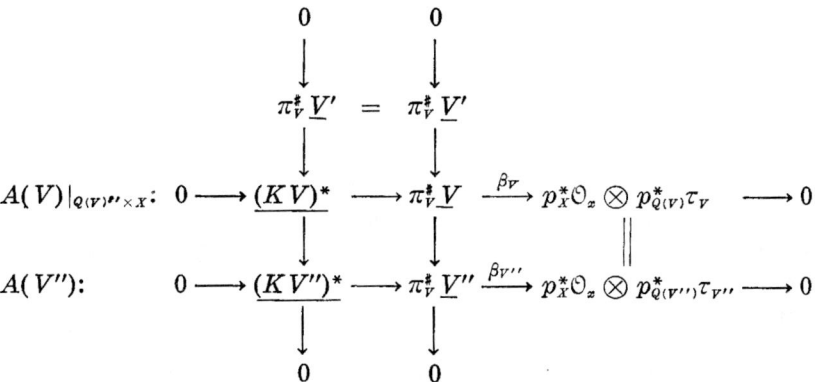

*Proof.* In fact, the sequence in (i) is the dual of the first vertical line in the diagram 6.2. The sequence in (ii) is obtained on applying (i) again to the sequence

$$0 \longrightarrow KV'' \longrightarrow KV \longrightarrow \pi_V^{\#}(V')^* \longrightarrow 0 .$$

*Remark* 6.4. We will denote the closed subvariety $Q(\pi_V^{\#} V'^*) \subset Q(KV)$ by $Z_{\mathfrak{S}}$. We have clearly

$$\begin{aligned} \dim Z_{\mathfrak{S}} &= (\operatorname{rk} V' - 1) + \dim Q(V'') = \operatorname{rk} V' - 1 + \operatorname{rk} V'' - 1 + \dim T \\ &= \dim T + \operatorname{rk} V - 2 . \end{aligned}$$

LEMMA 6.5. *If in the family $V'$, all bundles satisfy* $(\deg V'_t + 1)/\operatorname{rk} V' \geqq (\deg V_t)/\operatorname{rk} V$, *then all bundles of $K^2 V$ corresponding to points of $Z_{\mathfrak{S}}$ are non-stable.*

*Proof.* This follows from the first part of Lemma 6.1. For if $z \in Z_{\mathfrak{S}}$ and $t = \pi_V \pi_{KV} z$, then the subbundle $V'_t$ of $V_t$ satisfies conditions (ii) and (iii) of that lemma, while condition (i) is verified by hypothesis.

We have now the following converse to Lemma 6.5.

LEMMA 6.6. *Let $V$ be a stable vector bundle on $X$ of rank $n$ and degree $d$. Considering this as a trivial family (denoted $V$), let $q$ be a point of $Q(KV)$ with $K^2 V|_{q \times X}$ not stable. Then there exists an exact sequence*

$$(\mathfrak{S}) \qquad 0 \longrightarrow V' \longrightarrow V \longrightarrow V'' \longrightarrow 0$$

*with* (i) $\big(\deg(V') + 1\big)/\operatorname{rk} V' \geqq \mu(V)$, *and*

(ii) $q \in Z_{\mathfrak{S}}$.

*Proof.* Since $K^2 V|_{q \times X}$ is not stable, there exists a subbundle $V'$ satisfying (i), (ii), and (iii) of Lemma 6.1, giving rise to the sequence $\mathfrak{S}$. We have only to show that $q \in Z_{\mathfrak{S}}$. First we note that $\pi_{KV} q \in Q(V'')$, in view of condition (ii) in Lemma 6.1. Now $\pi_{KV} q$ gives rise to an exact sequence

$$0 \longrightarrow \underline{N}^* \longrightarrow \underline{V} \longrightarrow \mathcal{O}_x \longrightarrow 0$$

which can be imbedded in the diagram

$$\begin{array}{ccccccccc} & & 0 & & 0 & & & & \\ & & \downarrow & & \downarrow & & & & \\ & & \underline{V}' & = & \underline{V}' & & & & \\ & & \downarrow & & \downarrow & & & & \\ 0 & \longrightarrow & \underline{N}^* & \longrightarrow & \underline{V} & \longrightarrow & \mathcal{O}_x & \longrightarrow & 0 \\ & & \downarrow & & \downarrow & & & & \\ 0 & \longrightarrow & \underline{N}^*/\underline{V}' & \longrightarrow & \underline{V}'' & \longrightarrow & \mathcal{O}_x & \longrightarrow & 0 \\ & & \downarrow & & \downarrow & & & & \\ & & 0 & & 0 & & & & \end{array}$$

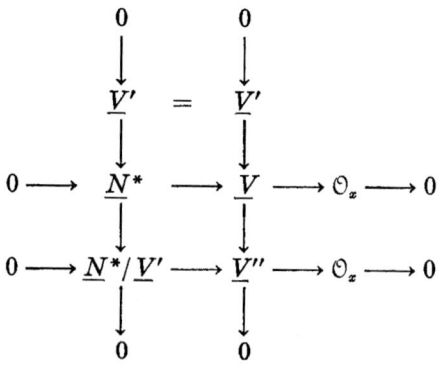

This is simply the diagram of Lemma 6.2 restricted to $\pi_{KV}q \times X$. Thus $KV'' \mid \pi_{KV}q \times X$ can be identified with the subbundle of $KV \mid \pi_{KV}q \times X$ orthogonal to $V'$. Now condition (iii) means that $q \in Z_{\mathcal{S}}$. Q.E.D.

Let $(\delta, \nu)$ be a pair of integers with $0 < \nu < n$, and $(\delta + 1)/\nu \geqq d/n > \delta/\nu$. For each such pair let $E = (E_r)_{r \in R}$ (resp. $F = (F_s)_{s \in S}$) be a family of vector bundles of rank $\nu$ (resp. $(n - \nu)$) and degree $\delta$ (resp. $(d - \delta)$). Let $G \subset R \times S$ be the open subset consisting of points $(r, s) \in R \times S$ with $H^0(X, \operatorname{Hom}(F_s, E_r)) = 0$. Then $\dim H^1(X, \operatorname{Hom}(F_s, E_r))$ is independent of $(r, s) \in G$ and hence the first direct image on $G$ of the sheaf $\operatorname{Hom}(F, E)$ is locally free. Let $\pi: P \to G$ be the associated projective bundle. By [15, Lemma 2.4] there exists an exact sequence of vector bundles on $P \times X$:

$$0 \longrightarrow V' \longrightarrow V \longrightarrow V'' \longrightarrow 0$$

where $V' = \pi^{\#} p_R^{\#} E \otimes p_P^{\#} \tau$ and $V'' = \pi^{\#} p_S^{\#} F$. Let $P' \subset P$ be the closed subvariety of $P$ consisting of points $p \in P$ such that $\det V|_{p \times X} \approx \xi$. Also let $P''$ be the open subvariety of $P'$ consisting of points $p \in P'$ with $V|_{p \times X}$ stable. The restriction of the above sequence to $P''$ will be denoted $\mathcal{S}(E, F)$. The classifying map $\theta_V: P'' \to U(n, \xi)$ lifts into a map $\Phi_{E,F}: Q(KV) \to Q(KW)$. By Lemma 4.1, $\Phi_{E,F}^{\#} K^2 W \approx K^2 V$ up to tensorisation with a line bundle on $Q(KV)$ and hence by Lemma 6.5, we have $\Phi_{E,F}(Z_{\mathcal{S}(E,F)}) \subset Q(KW) - \Omega$, in view of the choice of $\delta$ and $\nu$.

LEMMA 6.7. *For each pair of integers $(\delta, \nu)$ satisfying $0 < \nu < n$, $(\delta + 1)/\nu \geqq d/n > \delta/\nu$, let $E_{\delta,\nu}$ (resp. $F_{\delta,\nu}$) be fixed families (see Proposition 2.4) containing all stable bundles of rank $\nu$ (resp. $(n - \nu)$) and degree $\delta$ (resp. $(d - \delta)$) with parameter spaces $R$ (resp. $S$) of dimension $\nu^2(g - 1) + 1$ (resp. $(n - \nu)^2(g - 1) + 1$). Set $Z_{\delta,\nu} = Z_{\mathcal{S}(E_{\delta,\nu}, F_{\delta,\nu})}$, $\Phi_{\delta,\nu} = \Phi_{E_{\delta,\nu}, F_{\delta,\nu}}$. Then $\bigcup \Phi_{\delta,\nu}(Z_{\delta,\nu})$ is dense in $Q(KW) - \Omega$, where $(\delta, \nu)$ are allowed to run through pairs of integers as above.*

*Proof.* Let $q \in Q(KW) - \Omega$, and let $V$ belong to the isomorphism class $\pi_W \pi_{KW} q$. By restricting ourselves to the fibre of $\pi_W \pi_{KW}$ over $V$, (which is simply $Q(KV)$) and applying Lemma 6.6, we see that there exists an exact sequence

$$(\Sigma): \quad 0 \longrightarrow V' \longrightarrow V \longrightarrow V'' \longrightarrow 0$$

with $q \in Z_{\Sigma}$, and $(\delta + 1)/\nu \geqq d/n > \delta/\nu$, where $\delta = \deg V'$, $\nu = \operatorname{rank} V'$. Let $E'$, $F'$ (parametrised by varieties $R'$, $S'$ respectively) be families containing $V'$, $V''$ respectively and satisfying the conditions of Proposition 2.6. Let $R''$, $S''$ be the dense set of stable points of $R'$, $S'$ respectively. By [7, Lemma 2.1] we have $R'' \times S'' \subset G$ with the above notation. It is easy to see that

$\pi^{-1}(R'' \times S'') \cap P'$ is dense in $P'$. It follows that $\pi^{-1}(R'' \times S'') \cap P''$ is also dense in $P''$, since $P''$ is itself an open set in $P'$. Therefore $Z = Z_{\mathfrak{S}(E',F')} \cap \pi^{-1}(R'' \times S'')$ is non-empty and dense in $Z_{\mathfrak{S}(E',F')}$. It follows that $\Phi_{E',F'}(Z)$ contains $q$ in its closure. On the other hand, it is clear that $\Phi_{E',F'}(Z) \subset \Phi_{\delta,\nu}(Z_{\delta,\nu})$. Thus Lemma 6.7 is proved.

PROPOSITION 6.8. *The codimension of* $Q(KW) - \Omega$ *in* $Q(KW)$ *is greater than or equal to* $\inf \{(g-1)\nu(n-\nu) + n + (n\delta - \nu d)\}$ *where the infimum is taken over pairs of integers* $(\nu, \delta)$ *such that* $0 < \nu < n$ *and* $0 > n\delta - \nu d \geqq -n$. *In particular,*

$$\operatorname{Codim}(Q(KW) - \Omega) \begin{cases} \geqq n \geqq 2 \\ \geqq 3 \text{ except when } g = 2,\ n = 2 \\ \geqq 4 \text{ except when } g = 2,\ n = 2, 3 \text{ or } g = 3,\ n = 2\,. \end{cases}$$

*Proof.* We note that $\dim Q(KW) = (n^2 - 1)(g - 1) + 2(n - 1)$ and we will now compute, for each pair $(\delta, \nu)$ as above, $\dim Z_{\delta,\nu}$ with the notation of Lemma 6.7. By Remark 6.4, we have $\dim Z_{\delta,\nu} = (n - 2) + \dim P'$, while

$$\begin{aligned}\dim P' = \dim P - g = \dim R &+ \dim S \\ &+ (\dim H^1(X, \operatorname{Hom}(F_s, E_r)) - 1) - g\,.\end{aligned}$$

On the other hand, by the Riemann-Roch theorem,

$$\dim H^1(X, \operatorname{Hom}(F_s, E_r)) = -(n-\nu)\delta + \nu(d - \delta) - \nu(n - \nu)(1 - g)$$

and $\dim R$, $\dim S$ are respectively $\nu^2(g - 1) + 1$ and $((n - \nu)^2(g - 1) + 1)$. Finally, since $\dim \overline{\Phi_{\delta,\nu}(Z_{\delta,\nu})} \leqq \dim Z_{\delta,\nu}$ and $\dim Q(KW) = (n^2 - 1)(g - 1) + 2(n - 1)$, we get, using Lemma 6.7, the estimate on the codimension, as stated in the proposition.

To prove the second part we note first that since $(n, d) = 1$, the condition $n\delta - \nu d \geqq -n$ actually implies that $n\delta - \nu d \geqq -n + 1$. Hence in any case the required codimension $\geqq (g - 1)\nu(n - \nu) + 1 \geqq n$ always and $\geqq 3$ except when $g = 2$, $n = 2$. Also this codim $\geqq 4$ except when $g = 2$, $n = 2, 3$ or $g = 3$, $n = 2$.

COROLLARY 6.9. *The set of stable bundles in the family* $KW$ *is non-empty.*

*Proof.* In fact, in view of Proposition 6.8, it is enough to show that if $KW$ consisted entirely of non-stable bundles, then every fibre of $\pi_{KW}$ would intersect $Q(KW) - \Omega$. Let $E$ be a non-stable bundle of the family $KW$, and $F$ a subbundle of $E$ with $\mu(F) \geqq \mu(E)$. Then any homomorphism $E \to \mathcal{O}_x$ vanishing on $F_x$ has non-stable kernel. Hence $\pi_{KW}^{-1}(E) \not\subset \Omega$.

## 7. Proofs of the main theorems

7.1. *Proof of Theorem* 2.

a) With the notation of Section 5, we have the commutative diagram

$$\begin{array}{ccccc} & & H^1(Q(KW), \operatorname{End}(K^2W)_x) & & \\ & \nearrow & & \searrow & \\ T_x & & & & H^1(\Omega, \operatorname{End}(K^2W)_x) \\ & \searrow & & \nearrow & \\ & & H^1(U(n, \xi), \operatorname{ad}_x W) & & \end{array}$$

where the map $H^1(U(n, \xi), \operatorname{ad}_x W) \to H^1(\Omega, \operatorname{End}(K^2W)_x)$ is induced by $\varphi$ and the map $T_x \to H^1(Q(KW), \operatorname{End}(K^2W)_x)$ is the infinitesimal deformation map of $K^2W$ (considered as a family of bundles on $Q(KW)$). The latter is injective by Proposition 4.4. The map

$$H^1(Q(KW), \operatorname{End}(K^2W)_x) \longrightarrow H^1(\Omega, \operatorname{End}(K^2W)_x)$$

is injective by [3, Expose III; 16, § 5], in view of Proposition 6.8. Hence it follows that the map $T_x \to H^1(U(n, \xi), \operatorname{ad}_x W)$ is injective.

b) We have, by Lemma 2.3 and the Leray spectral sequence,

$$H^i(U(n, \xi), \operatorname{ad}_x W) \approx H^i(Q(W), \Theta_{\pi_W}) .$$

On the other hand ([4, Theorem 23.2.3]),

$$H^i(Q(W), \Theta_{\pi_W}) \approx H^i(Q(KW), \pi_{KW}^* \Theta_{\pi_W}) .$$

Suppose that $i + 2 \leqq \operatorname{codim}(Q(KW) - \Omega)$. Then by [3, Exp. III; 16, § 5] we see that

$$H^i(Q(KW), \pi_{KW}^* \Theta_{\pi_W}) \approx H^i(\Omega, j^* \pi_{KW}^* \Theta_{\pi_W})$$

where $j \colon \Omega \to Q(KW)$ is the inclusion map. Since $\iota$ is an involution and $\iota^* j^* \pi_{KW}^* (\Theta_{\pi_W}) \approx \Theta_{\pi_{KW}}^*$ by Proposition 5.10, we have

$$H^i(\Omega, j^* \pi_{KW}^* \Theta_{\pi_W}) \approx H^i(\Omega, \Theta_{\pi_{KW}}^*) .$$

By applying again the theorem on the extendability of cohomology classes ([3], [16]), we have

$$H^i(\Omega, \Theta_{\pi_{KW}}^*) \approx H^i(Q(KW), \Theta_{\pi_{KW}}^*) .$$

Since $R^i(\pi_{KW})_* \Theta_{\pi_{KW}}^* = 0$ for $i \neq 1$ and $R^1(\pi_{KW})_* \Theta_{\pi_{KW}}^* \approx \mathcal{O}_{Q(W)}$, by Lemma 2.3, we have

$$\begin{aligned} H^i(Q(KW), \Theta_{\pi_{KW}}^*) &\approx H^{i-1}(Q(W), \mathcal{O}) \\ &\approx H^{i-1}(U(n, \xi), \mathcal{O}) . \end{aligned}$$

Since $U(n, \xi)$ is unirational ([8]), we have

$$\begin{aligned} H^{i-1}(U(n, \xi), \mathcal{O}) &= 0 \quad \text{for } i \neq 1 \\ &\approx \mathbf{C} \quad \text{for } i = 1 . \end{aligned}$$

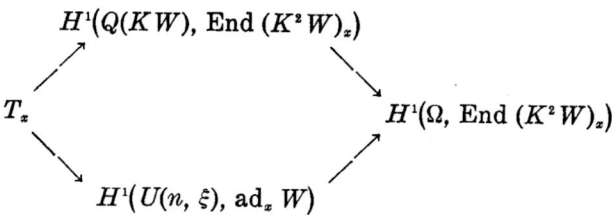

Now b) of Theorem 2 follows from Proposition 6.8.

*Remark* 7.2. It seems likely that all the cohomology groups $H^i(U(n, \xi), \mathrm{ad}_x W) = 0$ for $i \geqq 2$. We prove this in Proposition 7.7, when $n = 2$ and $g \geqq 3$.

**7.3.** *Proof of Theorem* 1. Since the canonical class of $U(n, \xi)$ is negative ([15]), it follows from the theorem of Akizuki-Nakano ([1]) that $H^i(U(n, \xi), \Theta) = 0$ for $i \geqq 2$. If $W$ is the universal bundle on $U(n, \xi) \times X$ we have two spectral sequences

$$H^p(X, R^q(p_X)_*(\mathrm{ad}\, W)) \Longrightarrow H^*(U(n, \xi) \times X, \mathrm{ad}\, W)\,,$$
$$H^p(U(n, \xi), R^q(p_U)_*(\mathrm{ad}\, W)) \Longrightarrow H^*(U(n, \xi) \times X, \mathrm{ad}\, W)\,.$$

But $R^q(p_U)_*(\mathrm{ad}\, W) = 0$ for $i \neq 1$ and $R^1(p_U)_*\, \mathrm{ad}\, W$ is locally free and is isomorphic to the tangent sheaf of $U(n, \xi)$. On the other hand, it follows from Theorem 2 that $(p_X)_*\, \mathrm{ad}\, W = 0$ and that $R^1(p_X)_*\, \mathrm{ad}\, W \approx \Theta_X$, except when $g = 2$, $n = 2$. Hence $H^0(U(n, \xi), \Theta) \approx H^0(X, \Theta_X) = 0$. (This is also known to be true for $g = 2$, $n = 2$ ([12]).) Now, again by Theorem 2, $R^2(p_X)_*(\mathrm{ad}\, W) = 0$, except when $g = 2$, $n = 2, 3$ and $g = 3$, $n = 2$, so that in all other cases we have the isomorphism

$$H^1(U(n, \xi), \Theta) \approx H^1(X, \Theta)\,,$$

and $\dim H^1(X, \Theta) = 3g - 3$. For $g = 2$, $n = 2$ this result is seen to be still valid by applying the Riemann-Roch-Hirzebruch theorem (see [11], [12]). Also for $g = 3$, $n = 2$, it has been shown ([15]) that $\chi(U(n, \xi), \Theta) = -6$, while we have just shown that $H^i(U(n, \xi), \Theta) = 0$ for $i \neq 1$. (There is a misprint in Theorem 4 of [15], where $\chi(U(n, \xi), \Theta)$ is stated to be $-3$.)

Since the dual of the canonical line bundle is ample ([15]), the group of automorphisms of $U(n, \xi)$ is an algebraic group. Since $H^0(U(n, \xi), \Theta) = 0$ it follows that this group is finite. This completes the proof of Theorem 1.

For the proof of Theorem 3, we need the following

PROPOSITION 7.4. *Let $\omega$ be a holomorphic differential on $X$ with simple zeros $x_1, \cdots, x_{2g-2}$. We then have an exact sequence of vector bundles*

$$0 \longrightarrow \Omega^1 \longrightarrow \bigoplus_{i=1}^{2g-2} (\mathrm{ad})_{x_i}(W) \longrightarrow \Theta \longrightarrow 0$$

*on $U(n, \xi)$.*

*Proof.* The differential $\omega$ gives rise to a positive divisor $D$ on $X$ and the following exact sequence of vector bundles on $U(n, \xi) \times X$:

$$0 \longrightarrow p_X^* K^{-1} \longrightarrow \mathcal{O} \longrightarrow \mathcal{O} \mid p_X^{-1} D \longrightarrow 0\,.$$

Tensoring this with $p_X^* K \otimes \mathrm{ad}\, W$, we get the exact sequence

$$0 \longrightarrow \operatorname{ad} W \longrightarrow p_X^* K \otimes \operatorname{ad} W \longrightarrow p_X^* K \otimes \operatorname{ad} W \otimes \mathcal{O}_{p_X^{-1}D} \longrightarrow 0 .$$

The $0^{\text{th}}$ (resp. first) direct image of ad $W$ by $p_{U(n,\xi)}$ is canonically dual to the first (resp. $0^{\text{th}}$) direct image of $p_X^* K \otimes \operatorname{ad} W$. Hence we have the exact sequence on $U(n, \xi)$,

$$0 \longrightarrow \Omega^1 \longrightarrow (p_{U(n,\xi)})_* (p_X^* K \otimes \operatorname{ad} W \otimes \mathcal{O}_{p_X^{-1}D}) \longrightarrow \Theta \longrightarrow 0$$

and it is clear that the middle term here is isomorphic to $\bigoplus_{i=1}^{2g-2} \operatorname{ad}_{x_i}(W)$.

*Remark* 7.5. To be canonical, the middle term of this sequence is actually $\bigoplus_{k=1}^{2g-2} K_{x_k} \otimes \operatorname{ad}_{x_k} W$, where $K_{x_i}$ is the trivial line bundle on $U(n, \xi)$ with fibre $K_{x_i}$. If $D$ has multiplicities, say $D = \sum_{k=1}^{r} n_k x_k$, we have an exact sequence

$$0 \longrightarrow \Omega^1 \longrightarrow \bigoplus_{k=1}^{r} V_{x_k} \longrightarrow \Theta \longrightarrow 0$$

where each $V_{x_k}$ is a successive extension of $\operatorname{ad}_{x_k} W$, $n_k$ times.

**7.6.** *Proof of Theorem* 3. It is enough to show that $\dim H^2(U(n, \xi), \Omega^1) = g$. For, $U(n, \xi)$ being rational ([14]), it would follow a) that the third Betti number is $2g$ and b) that $H^3(U(n, \xi), \mathbf{Z})$ is torsion-free ([2, Proposition 1]). Thus $H^3(U(n, \xi), \mathbf{Z})$ would be a free abelian group of rank $2g$ and the second part of the assertion of Theorem 3 can then be proved as in [15, Section 4], using [15, Lemma 2.1].

For $n = 2$, the Betti numbers of $U(n, \xi)$ have been calculated by P. E. Newstead ([11]) and in particular, the third Betti number is $2g$. Thus we may assume that $n \geqq 3$. If $\omega$ is a non-zero differential with simple zeros (which clearly exists) we may apply the exact sequence of Proposition 7.4. We then have the cohomology exact sequence (since $H^0(U(n, \xi), \Theta) = 0$)

$$\begin{aligned} 0 \longrightarrow H^1(U, \Omega^1) &\longrightarrow \bigoplus_{i=1}^{2g-2} H^1(U, \operatorname{ad}_{x_i} W) \longrightarrow H^1(U, \Theta) \\ \longrightarrow H^2(U, \Omega^1) &\longrightarrow \bigoplus_{i=1}^{2g-2} H^2(U, \operatorname{ad}_{x_i} W) \longrightarrow \cdots \end{aligned}$$

where $U = U(n, \xi)$. Now (except in the case $g = 2$, $n = 3$), we have by Theorems 1 and 2, $\dim H^1(U, \operatorname{ad}_{x_i} W) = 1$, $H^2(U, \operatorname{ad}_{x_i} W) = 0$, $\dim H^1(U, \Theta) = 3g - 3$. On the other hand, we have $\dim H^1(U, \Omega^1) = 1$ (see [15, Prop. 3.4]). Hence it follows that $\dim H^2(U, \Omega^1) = g$.

PROPOSITION 7.7. *Let* $U = U(2, \xi)$, *with* $\deg \xi = 1$. *For any* $x \in X$, *we have*

a) $H^i(U, \operatorname{ad}_x W) = 0$ *for* $i \geqq 3$.

b) $H^2(U, \operatorname{ad}_x W) = 0$ *except possibly when* $g = 2$.

*Proof.* a) Since $g \geqq 2$, there exists a non-zero differential $\omega$ vanishing at $x$. From [13], it follows that $H^i(U, \Omega^1) = 0$ for $i \geqq 3$. On the other hand, by Theorem 1, we have $H^i(U, \Theta) = 0$ for $i \geqq 3$. Hence from the exact sequence in Remark 7.5 corresponding to $\omega$, we obtain $H^i(U, V_x) = 0$ for

$i \geqq 3$. If $x$ were a simple zero of $\omega$, this would prove a). In the general case, we have by Remark 7.5 an exact sequence

$$0 \longrightarrow V' \longrightarrow V_x \longrightarrow \mathrm{ad}_x W \longrightarrow 0$$

where $V'$ is obtained as extensions of $\mathrm{ad}_x W$. Let now

$$i_0 = \sup \{i \mid H^i(U, \mathrm{ad}_x W) \neq 0\} .$$

If $i_0$ were $\geqq 3$, we would have $H^{i_0+1}(U, V') = 0$ and $H^{i_0}(U, V_x) = 0$ as has been just proved. It would follow that $H^{i_0}(U, \mathrm{ad}_x W) = 0$ contradicting the definition of $i_0$. Thus $i_0 < 3$, proving a).

b) Let $g \geqq 3$. If $\omega$ is a non-zero holomorphic differential, we get from 7.5 the exact sequence

$$\begin{aligned} 0 \longrightarrow H^1(U, \Omega^1) &\longrightarrow \bigoplus H^1(U, V_{x_k}) \longrightarrow H^1(U, \Theta) \\ \longrightarrow H^2(U, \Omega^1) &\longrightarrow \bigoplus H^2(U, V_{x_k}) \longrightarrow 0 \end{aligned}$$

since $H^0(U, \Theta) = 0 = H^2(U, \Theta)$. By Theorem 1, we have $\dim H^1(U, \Theta) = 3g-3$. Moreover, $\dim H^1(U, \Omega^1) = 1$ by [15] and $\dim H^2(U, \Omega^1) = g$. Now if all the $x_k$ occur with multiplicity 1, then $V_{x_k} \approx \mathrm{ad}_{x_k} W$ and hence, by Theorem 2, $\dim \bigoplus H^1(U, V_{x_k}) = 2g - 2$. We claim that this is true even if the zeros of $\omega$ are not simple. By the semicontinuity theorem, we see that $\dim \bigoplus H^1(U, V_{x_k}) \geqq 2g - 2$. On the other hand, since $V_{x_k}$ are obtained as successive extensions of $\mathrm{ad}_{x_k} W$, it follows that $\dim H^1(U, V_{x_k}) \leqq n_k$. Since $\sum n_k = 2g - 2$, we have $\dim \bigoplus H^1(U, V_{x_k}) = 2g - 2$ for all $\omega$. From the exact sequence above, we get $H^2(U, V_{x_k}) = 0$ for all $k$. Now the argument in a) gives $H^2(U, \mathrm{ad}_{x_k} W) = 0$. This proves b).

*Remark* 7.8. The assertion b) of 7.7 is obtained in Theorem 2 for $g \geqq 4$, and the proof given above is meant to cover the case $g = 3$.

## 8. Remarks on Hecke cycles

If $E$ is a vector bundle ($\neq 0$) on $X$, and $k \in \mathbf{Z}$, let $\mu_k(E)$ denote the rational number $(\deg E + k)/\mathrm{rk}\, E$.

DEFINITION 8.1. *A vector bundle $E$ on $X$ is said to be $(k, l)$-stable if for every proper subbundle $F$ of $E$, we have*

$$\mu_k(F) < \mu_{-l}(E/F) .$$

Let $M'$ be the locally closed subset of $U(n, 1 - d)$ consisting of $(0, 1)$-stable bundles whose determinants are of the form $\xi^{-1} \otimes L_x$ for some $x \in X$. If $(n, d) = 1$, $M'$ can be shown to be non-empty. If $V \in M'$, then it is easily seen that $K(V)$ consists entirely of stable bundles of rank $n$ and degree $d$. Hence we get a classifying morphism $\theta_{K(V)}: Q(V) \to U(n, \xi)$. Moreover, this is an imbedding of $Q(V)$ and the image cycle is denoted $Z_V$. Thus the (non-singular) variety $M'$ parametrises a family of cycles (projective spaces) in $U(n, \xi)$. It can be proved that

(i) this family is injectively parametrised,

(ii) if $\nu_V$ is the normal bundle of $Z_V$ in $U(n, \xi)$, then $H^i(Z_V, \nu_V) = 0$ for $i \geqq 1$, and

(iii) the infinitesimal deformation map $T_V \rightarrow H^0(Z_V, \nu_V)$ is an isomorphism (where $T_V$ is the tangent space of $M'$ at $V$).

In particular, $M'$ can be identified with an open subset of a suitable Chow component.

Moreover, if $E \in U(n, \xi)$ is a (1, 1)-stable vector bundle then the cycles $Z_V$ passing through $E$ can be canonically identified with the projective bundle $P(E)^*$. In particular, the curve $X$ and the projective Poincaré bundle can be recovered from this family of cycles. It should be remarked that (1, 1)-stable bundles form an open subset of $U(n, \xi)$ which is non-empty except when $g = 2$ and $d \equiv \pm 1 \pmod n$. For a systematic discussion of these and related questions and also proofs of the above assertions, see [9].

TATA INSTITUTE OF FUNDAMENTAL RESEARCH, BOMBAY

REFERENCES

[1] Y. AKIZUKI and S. NAKANO, Note on Kodaira-Spencer's proof of Lefschetz theorems, Proc. Jap. Acad. **30** (1954), 266-272.

[2] M. ARTIN and D. MUMFORD, Some elementary examples of unirational varieties which are not rational, Proc. Lond. Math. Soc. III Series, **25** (1972), 75-95.

[3] A. GROTHENDIECK, *Cohomologie locale des faisceaux coherents et Théorèmes de Lefschetz locaux et globaux* (SGA2), North-Holland, 1968.

[4] F. HIRZEBRUCH, *Topological Methods in Algebraic Geometry*, Springer-Verlag, 1966.

[5] K. KODAIRA and D. C. SPENCER, On deformations of complex analytic structures I-II, Ann. of Math. **67** (1968), 328-466.

[6] D. MUMFORD and P. E. NEWSTEAD, Periods of a moduli space of bundles on curves, Amer. J. Math. **90** (1968), 1201-1208.

[7] M. S. NARASIMHAN and S. RAMANAN, Moduli of vector bundles on a compact Riemann surface, Ann. of Math. **89** (1969), 19-51.

[8] ———, Vector bundles on curves, *Proceedings of the Bombay Colloquium on Algebraic Geometry*, 335-346, Oxford University Press, 1969.

[9] ———, Geometry of Hecke cycles, to appear.

[10] M. S. NARASIMHAN and C. S. SESHADRI, Stable and unitary vector bundles on a compact Riemann surface, Ann. of Math. **82** (1965), 540-567.

[11] P. E. NEWSTEAD, Topological properties of some spaces of stable bundles, Topology **6** (1967), 241-262.

[12] ———, Stable bundles of rank 2 and odd degree over a curve of genus 2, Topology **7** (1968), 205-215.

[13] ———, Characteristic classes of stable bundles of rank 2 over an algebraic curve, Trans. Amer. Math. Soc. **169** (1972), 337-345.

[14] ———, Rationality of moduli spaces of stable bundles, to appear.

[15] S. RAMANAN, The moduli spaces of vector bundles over an algebraic curve, Math. An-

nalen **200** (1973), 69–84.

[16] G. Scheja, Riemannsche Hebbarkeitssatze für Cohomologieklassen, Math. Ann. **144** (1961), 345–360.

[17] C. S. Seshadri, Space of unitary vector bundles on a compact Riemann surface, Ann. of Math. **85** (1967), 303–336.

[18] ———, Moduli of $\pi$-vector bundles over an algebraic curve, *Questions on Algebraic Varieties*, Roma, 1970.

[19] A. Tjurin, Analogs of Torelli's theorem for multi-dimensional vector bundles over an algebraic curve. Izv. Akad. Nauk. SSSR, Ser. Mat. Tom. **34** (2) (1970), 338–365, Math. USSR Izvestija Vol. **4** (2), 343–370.

[20] A. Weil, *Dirichlet Series and Automorphic forms*, Lecture notes in Mathematics **189**, Springer-Verlag, 1971.

[21] M. S. Narasimhan and S. Ramanan, Generalized Prym varieties as fixed points, Jour. Indian Math. Soc. (to appear).

(Received July 24, 1973)
(Revised November 12, 1974)

*Journal of the Indian Math. Soc.*
39 (1975) 1–19

# GENERALISED PRYM VARIETIES AS FIXED POINTS

*By* M. S. NARASIMHAN AND S. RAMANAN

*To Professor Laurent Schwartz on his sixtieth birthday.*

[Received September 17, 1973; revised October 31, 1974]

§ 1. **Introduction.** Let $X$ be a nonsingular, projective curve of genus $g \geqslant 2$. Then the elements of order $r$ of the Jacobian of $X$ act in a natural way on the moduli space $M(r, \xi)$ of stable vector bundles on $X$ of rank $r (\geqslant 2)$ whose determinants are isomorphic to a given line bundle $\xi$ of degree $d$. We shall assume that $r$ and $d$ are coprime and show that the fixed point variety corresponding to any element $\mu$ (strictly) of order $r$ is an abelian variety. In fact, this fixed point variety is isomorphic to the generalised Prym variety (See Remark 3.7) associated to $\mu$. We then apply the Atiyah-Singer fixed point theorem to compute the alternating sum

$$\sum (-1)^i \, Tr \, \mu \mid H^i (M(r, \xi), \Omega^p).$$

The computation of the invariants involved in the formula is fairly long and takes up the whole of § 4.

It is known that the action of $\mu$ on the complex cohomology groups of $M(r, \xi)$ is trivial. See [2, Theorem 1]. Thus we obtain a simple formula for the $\chi_y$-genus of $M(r, \xi)$. (See Theorem 1 in § 4). In particular, we have $\chi_y = (1-y)^{g-1}(1+y)^{2g-2}$ for $r = 2$.

§ 2. **Galois coverings and direct images.**

LEMMA 2.1. *Let $\pi: Y \to X$ be a finite, unramified Galois covering with Galois group $G$, over a complete variety $X$. Let $F$ be a vector bundle on $Y$ and $E$ its direct image on $X$. Then*
*(i) for any character $\chi$ on $G$, we have a canonical isomorphism $L_\chi \otimes E \to E$, where $L_\chi$ is the line bundle associated to $\chi$,*

(*ii*) $\pi^*E \approx \bigoplus_{s \in G} s^*F$; *moreover the isomorphism can be chosen to be compatible with the canonical G-action on* $\pi^*E$ *and the natural G-action (by shifting summands on* $\bigoplus_{s \in G} s^*F$*), and*

(*iii*) *If F and F′ are indecomposable vector bundles on Y with* $\pi_* F \approx \pi_* F'$, *then* $F \approx s^*F'$ *for some* $s \in G$.

PROOF. (i) Let $t_\chi : \mathbf{C} \to \pi^*(L_\chi)$ be the canonical trivialisation. Then we have canonical isomorphisms

$$E = \pi_*(F) \xrightarrow{\pi_*(t_\chi \otimes Id)} \pi_*(\pi^* L_\chi \otimes F) \to L_\chi \otimes \pi_*(F) = L_\chi \otimes E.$$

(ii) Notice first that the direct image of $s^*F$ is canonically isomorphic to $E$. We then have a natural evaluation map $\pi^*E \to \bigoplus_{s \in G} s^*F$. It is easy to check that this is a $G$-isomorphism.

(iii) This follows from the theorem of Krull-Remak-Schmidt and (ii).

Let $\pi: Y \to X$ be an unramified Galois covering with Galois group $G$, and $E$ a vector bundle on $X$ such that $L_\chi \otimes E \approx E$ for all characters $\chi$ of $G$, where $L_\chi$ is the line bundle on $X$ associated to $\chi$. We now define a group $H$ (which depends on $E$) as follows. An element of $H$ consists of a pair $(\chi, \phi)$ where $\chi \in \hat{G}$ and $\phi: L_\chi \otimes E \to E$ is an isomorphism. Define

$$(\chi_1, \phi_1) \cdot (\chi_2, \phi_2) = (\chi_1 \chi_2, \phi_1 (1 \otimes \phi_2)).$$

We then have an extension

$$(e) \to \mathrm{Aut}\ E \to H \xrightarrow{\eta} \hat{G} \to (e) \tag{2.2}$$

where $\eta$ is defined by $\eta(\chi, \phi) = \chi$.

Now the group $H$ operates on $\pi^*E$ as follows: Notice first that there is a canonical trivialisation $t_\chi : \mathbf{C} \to \pi^* L_\chi$. Moreover these trivialisations are compatible with tensor products of $L_\chi$ and multiplication of characters. If $(\chi, \phi) \in H$ and $\pi^*\phi : \pi^* L_\chi \otimes \pi^*E \to \pi^*E$ is the induced map, then $\pi^*\phi \circ (t_\chi \otimes 1) : \pi^*E \to \pi^*E$ is an isomorphism. It is easy to check that this gives a group action of $H$ on $\pi^*E$, the action being identity on the base and Aut $E$ acting naturally on $\pi^*E$.

LEMMA 2.3. *If F is a vector bundle on Y and* $E = \pi_*(F)$, *then*

(*i*) *the extension* (2.2) *splits canonically, and*

(*ii*) *writing* $\pi^*E = \bigoplus_{s \in G} s^*F$, *as in Lemma* 2.1, (*ii*), *the action of* $\hat{G}$ *on* $\pi^*E$ *through the splitting can be described as follows:*
$\hat{G}$ *leaves each* $s^*F$ *invariant and operates on it by the character* $\bar{s}$ *on* $\hat{G}$ *given by* $s$.

PROOF. (i) In fact, the isomorphism $\phi : L_\chi \otimes E \to E$ given by Lemma 2.1, (i) assigns to each $\chi \in \hat{G}$, an element $(\chi, \phi) \in H$. It is easy to check that this gives a splitting of (2.2).
(ii) The isomorphism $\phi: L_\chi \otimes E \to E$ gives rise to the isomorphism $\sum_{s \in G} s^*(\pi^* L_\chi \otimes F) \to \sum_{s \in G} s^*F$, where the induced map in each component is $s^*((t_\chi \otimes 1)^{-1})$. In other words, the map from $\pi^*L_\chi \otimes \bigoplus_{s \in G} s^*F \to \bigoplus_{s \in G} s^*F$ takes the component $\pi^*L_\chi \otimes s^*F$ to $s^*F$ and induces the map $s^*(t_\chi)^{-1}$. But $s^*(t_\chi)^{-1} = \chi(s) . (t_\chi)^{-1}$. This proves the assertion.

We seek to prove a converse to Lemma 2.3. We shall assume as before that $E$ is a vector bundle on $X$ such that $L_\chi \otimes E \approx E$ for all $\chi \in \hat{G}$. Now the group $G$ acts on $\pi^*E$. But this does not commute with the action of $H$ on $\pi^*E$. In fact we have, for $h \in H$ and $s \in G$,

$$s\, h\, s^{-1} = < s, \eta(h) > h \tag{2.4}$$

where $<,>$ denotes the pairing between $G$ and $\hat{G}$. For, let $h = (\chi, \phi)$. Then if $(y, v) \in \pi^*E$, we have

$$\begin{aligned} hs^{-1}(y, v) &= \pi^*\phi(t_\chi \otimes 1)\, s^{-1}(y, v) \\ &= \pi^*\phi((t_\chi)\, s^{-1}\, y\, (1) \otimes (s^{-1}y, v)) \\ &= \pi^*\phi(\chi(s)\, s^{-1}\, (t_\chi)_y\, (1) \otimes (y, v)) \\ &= \chi(s)\, s^{-1}\, \pi^*\phi(t_\chi \otimes 1)\, (y, v) \\ &= < s, \eta h > s^{-1}\, h(y, v). \end{aligned}$$

We now have

LEMMA 2.5. *Let* $Y \to X$ *be an unramified abelian covering over a complete variety, with Galois group* $G$. *Let* $E$ *be an indecomposable vector bundle on* $X$ *such that* $L_\chi \otimes E \approx E$ *for all* $\chi \in \hat{G}$. *Assume that the corresponding extension* (2.2) *splits. Then* $E$ *is isomorphic to the direct image of a vector bundle* $F$ *on* $Y$.

PROOF. Under the assumption, $\hat{G}$ acts on $\pi^*E$, identically on the base. It is easy to see that the isotypical components of the $\hat{G}$-action on the fibres of any given type $\rho$ build a subbundle $F_\rho$ of $\pi^*E$. Now by the commutation relation (2.4), we see that $sF_\rho = F_{\rho \otimes \tilde{s}^{-1}}$, where $\tilde{s}$ is the character on $\hat{G}$ induced by $s \in G$. Since $\hat{G}$ is abelian, each $\rho$ is one-dimensional and as $G$ is abelian, $\rho \otimes \tilde{s}^{-1}$ runs through all the characters of $G$. In other words, $G$ acts simply transitively on the components $F_\rho$. Hence all the $F_\rho$ have isomorphic direct images on $X$, say $\pi_*(F_\rho) \approx E'$. Then $\pi_*(\pi^*E) \approx E \otimes \pi_*(\mathbf{C})$ on the one hand and $\pi_*(\pi^*E) = \oplus\, \pi_*(F_\rho)$ is isomorphic to direct sum of some copies of $E'$ on the other. Hence $E$ is isomorphic to $E'$, by the Krull-Remak-Schmidt theorem.

In particular, we have

PROPOSITION 2.6. *Let $Y \to X$ be an unramified cyclic covering over a complete variety $X$ with Galois group $G$. Then any simple vector bundle $E$ on $X$ (i.e. a vector bundle such that the only endomorphisms are scalars) with $L_\chi \otimes E \approx E$ for all $\chi \in \hat{G}$, is isomorphic to the direct image of a vector bundle $F$ on $Y$.*

PROOF. Since a simple vector bundle is indecomposable, we have only to show (in view of Lemma 2.5) that the extension (2.2) splits in this case. In this case, (2.2) takes the form of a central extension

$$(e) \to \mathbf{C}^* \to H \to \hat{G} \to (e).$$

If $h \in H$ such that $\eta(h)$ generates $\hat{G}$, then $h^r \in \mathbf{C}^*$ and if $\lambda \in \mathbf{C}^*$ with $\lambda^r = h^r$, then $\lambda^{-1}h$ generates a subgroup of $H$ mapped isomorphically by $\eta$ onto $\hat{G}$.

## § 3. Determination of the fixed point set.

PROPOSITION 3.1. *Let $\pi: Y \to X$ be a finite, unramified, Galois covering over a complete, nonsingular, algebraic curve, and let $r$ be the order of the Galois group $G$. If $F$ is a vector bundle over $Y$ and $\pi_*F = E$, then we have*

(i) *$\deg E = \deg F$ and rank $E = r$.rank $F$,*

(ii) *$E$ is semistable if and only if $F$ is semistable,*

(iii) *$E$ is stable if and only if $F$ is stable and no two of the bundles $s^*F$, $s \in G$, are isomorphic, and*

*Here $\alpha$ is the albanese or the norm map composed with tensorisation by $L_\beta$, where $\beta$ is the determinant of the regular representation of $G$. Moreover, if $\xi \in J_d^X$, then $U \cap \theta^{-1}(M^X(mr, \xi)) = U_\xi$ has $r$ connected components.*

PROOF. The albanese map can be described as follows. If $L$ is a line bundle on $Y$, consider $\bigotimes_{s \in G} s^*L$ and the natural $G$-action on it. This defines by descent a line bundle alb$(L)$ on $X$. If $F$ is a vector bundle on $Y$, by Lemma 2.1, (ii), $\pi_*F$ is obtained by 'descending' the bundle $\bigoplus_{s \in G} s^*F$ by the natural $G$-action. Hence det $\pi_*F$ is obtained by 'descending' the line bundle $\det(\bigoplus s^*F) = \bigotimes s^* \det F$, where now the $G$-action on $\bigotimes_{s \in G} s^* \det F$ is the natural one, tensored with the determinant of the regular representation. This proves the first assertion. Since alb$^{-1}(\xi)$ has $r$ components, it would follow that $\theta^{-1}(M^X(mr, \xi))$ also has $r$ components, if we could show that det $U$ intersects every component of $\alpha^{-1}(\xi)$. Actually, the map det: $U \to J_d^Y$ is even surjective if $m \geqslant 2$. If $m = 1$, $U$ is the complement in $J_d^Y$ of $\bigcup \pi_i^* J_{d/r_i}^{X_i}$, where $r_i$ is a divisor of $d$ with $1 < r_i \leqslant r$ and $\pi_i$: $Y \to X_i$ is the quotient of $Y$ by the subgroup of $H$ of order $r_i$. Now the fact that the connected component of 1 in the kernel of alb: $J^Y \to J^X$ is not contained in $\pi_i^* J^{X_i}$ implies that $U$ intersects every component of $\alpha^{-1}(\xi)$.

PROPOSITION 3.5. *With the notation of Proposition* 3.3 *and Lemma* 3.4, *the quotient variety $U_\xi/G$ has $l$ components, where $l$ is the greatest common divisor of $d$ and $r$, $0 < l \leqslant r$.*

PROOF. In view of Lemma 3.4 and its proof, we have only to prove that $\alpha^{-1}(\xi)/G$ has $l$ components. Let $H$ be the unique subgroup of order $l$ of the cyclic group $G$. Denote by $\pi_1$, $\pi_2$ the coverings $Y \to Y/H$ and $Y/H \to X$. If $L$ is a line bundle of degree $d/l$ on $Y/H$, then $\pi_1^* L$ has degree $d$ and is fixed by $H$. It is clear that by a suitable choice of $L$, we may assume that $\pi_1^* L \in \alpha^{-1}(\xi)$. Now the component of $\alpha^{-1}(\xi)$ containing $\pi_1^* L$ is clearly left invariant by $H$. If $H' \supset H$ leaves this component invariant then it is easy to see that $G$ leaves invariant a component $C$ in $J_{d.[G:H']}$ of the fibre of the map Alb; for instance the component containing $\bigotimes_{g \in G/H'} g(\pi_1^* L)$. Now we claim that $G$ fixes an element of $C$. In fact let $\xi$ be any point of $C$ and $g$ be a generator of $G$. Then $g\xi \otimes \xi^{-1}$ belongs to the connected component $P$ of 1 in the kernel of Alb: $J_Y \to J_X$ and hence can be written (easily verified) as $gj \otimes j^{-1}$ for $j \in P$. Now clearly $\xi \otimes j^{-1}$ is a fixed point for $G$ and hence descends to a line bundle on $X$. This implies that $r$ divides $d[G:H']$ and hence that ord $(H')$ divides $l$. Hence $H = H'$. Thus there is one component of $\alpha^{-1}(\xi)$ left invariant precisely by $H$. Now since the kernel of the albanese map $J(Y) \to J(X)$ acts transitively on $\alpha^{-1}(\xi)$ and $G$ leaves all components of this kernel invariant, it follows that all components of $\alpha^{-1}(\xi)$ are left invariant by $H$. This proves the assertion.

COROLLARY 3.6. *If $d = \deg \xi$ and $r$ are coprime, then the fixed point variety under the $\hat{G}$-action on $M(nr, \xi)$ is isomorphic to the connected, non-singular variety $M^Y(r, P) \subset M^Y(r, d)$ consisting of bundles whose determinants are in a fixed component $P$ of $\alpha^{-1}(\xi)$.*

REMARK 3.7. When $d$ and $r$ are coprime, the fixed point variety for the $\hat{G}$-action on $M(\mathrm{r}, \xi)$ is $P$ itself. The abelian variety $P$ may be called the *generalised Prym variety* corresponding to the covering $Y \to X$. When the covering is 2-sheeted, this is the classical Prym variety. If $\mu$ is an element of order $r$ in the Jacobian of $X$ there exists a unique $r$-sheeted cyclic covering $\pi: Y \to X$, such that $\pi^*\mu$ is trivial. The connected component of 1 in the kernel of the albanese map $JY \to JX$ may be defined as the *generalised Prym variety associated to* $\mu$. Its dimension is $(r-1)(g-1)$.

## § 4. Application of the fixed point theorem.

Let $\pi: Y \to X$ be a Galois covering with cyclic Galois group $G$ of order $r$. Let $J = J^d(Y)$, $M = M^X(r, d)$ and $M_\xi = M^X(r, \xi)$. We assume that $(r, d) = 1$. Let $D$ be a Poincaré bundle on $J \times Y$, i.e., a line bundle whose restriction to $\alpha \times Y$, $\alpha \in J$, is in the class $\alpha$ and is trivial on $J \times Y$ for some $y \in Y$. Let $E$ be the direct image of $D$ on $J \times X$ by $Id \times \pi$. Let $P$ be a connected component of the set $\{j \in J: \det E|_{j \times X} \approx \xi\}$. Then by Corollary 3.6, the classifying map $\theta_E$ gives an isomorphism of $P$ with the fixed point variety of the action of $\hat{G}$ on $M_\xi$. We are interested in applying the Atiyah-Singer fixed point theorem [1, Formula 4.4] to this situation. Since $P$ is an abelian variety, its Todd class is 1 and hence the fixed point theorem takes the following form. Let $F$ be a vector bundle on $M_\xi$ to which the action of $\hat{G}$ lifts. We then have (if $\mu$ is a generator of $\hat{G}$),

$$\sum (-1)^i \, Tr(\mu \mid H^i(M_\xi, F)) = \widetilde{ch}\, F(\mu).(\prod_{x \neq 1} ch_{-<\mu^{-1}, x>} T_x^* )^{-1}[P]. \qquad (4.1)$$

In this formula, $\widetilde{ch}\, E$ is interpreted as an element of $H^*(P) \otimes R(\hat{G})$ where $R(\hat{G})$ is the representation ring of $\hat{G}$, and evaluation on $\mu$ gives rise to an element of $H^*(P)$. For $x \in G$, $T_x$ denotes the isotypical subbundle of the tangent bundle $T$ of $M_\xi$ restricted to $P$, corresponding to the character $\tilde{x}$ on $\hat{G}$ determined by $x \in G$. Also, if $x_i$ are the virtual Chern classes of $T_x$, $ch_t\, T_x^*$ denotes the cohomology class $\Pi(1 + te^{-x_i})$.

PROPOSITION 4.2. *There exists an element* $\Theta \in H^2(P, \mathbf{C})$ *such that*

$$\prod_{x \neq 1} ch_{-<\mu, x>} (T_x^* ) = r^{r(g-1)} . e^{-r\Theta} .$$

The proof of Proposition 4.2 will be completed in § 4.15. The element $\Theta$ in this proposition is in fact the restriction to $P$ of the principal polarisation on $J = J^d(Y)$. But we shall not need this fact. Before we proceed to the proof, we need some preliminaries. We first observe that there is a canonical isomorphism $a \longmapsto \tilde{a}$ of $H^1(Y, \mathbf{C})$ with $H^1(J, \mathbf{C})$. Let $(a_j)$ be a base of $H^1(Y, \mathbf{C})$ and $(b_i)$ the dual base with respect to the Poincaré form, i.e. we have $(a_i b_j)[Y] = \delta_{ij}$, or, what is the same, $a_i b_j = \delta_{ij} \gamma_Y$, where $\gamma_Y$ is the canonical generator of $H^2(Y, \mathbf{C})$. Then $\frac{1}{2} \sum \tilde{a}_i \tilde{b}_i \in H^2(J, \mathbf{C})$ is independent of the base $(a_i)$ and is defined to be the class $\Theta$. Its restriction to $P$ will also be denoted $\Theta$.

We will now choose a base $(a_i)$ of $H^1(Y, \mathbf{C})$ adapted to the action of $G$ on it as follows. Let $H^1(Y, \mathbf{C}) \approx \oplus H_\chi$, where $H_\chi$ is the isotypical component of type $\chi$. Since the Poincaré pairing $H^1(Y, \mathbf{C}) \times H^1(Y, \mathbf{C}) \to H^2(Y, \mathbf{C})$ is invariant under $G$ and the $G$-action on $H^2(Y, \mathbf{C})$ is trivial, we have: $H_{\chi_1}$ and $H_{\chi_2}$ are orthogonal if and only if $\chi_1 \chi_2 \neq 1$. We will choose $(a_i)$ by choosing a base $(a_i^\chi)$ for each $H_\chi$. Then in the dual base $(b_i)$, we have $b_i^\chi \in H_{\chi^{-1}}$.

LEMMA 4.3. *If*

$$c_\chi = \sum_i \tilde{a}_i^\chi \otimes b_i^\chi \in H^2(J \times Y) \text{ and } \zeta_\chi = \sum_i \tilde{a}_i^\chi . \tilde{b}_i^\chi \in H^2(J),$$

*then we have (i)* $c_{\chi_1} c_{\chi_2} = 0$ *for* $\chi_1 \chi_2 \neq 1$, *and* (ii) $c_\chi \, c_{\chi^{-1}} = -\zeta_\chi \otimes \gamma_Y$

PROOF. (i) follows from remarks above. As for (ii), we notice that if $(b_i)$ is the dual basis for $(a_i)$, then $(-a_i)$ is the dual basis for $(b_i)$. Hence

$$\begin{aligned} c_\chi \, c_{\chi^{-1}} &= \Big(\sum_i \tilde{a}_i^\chi \otimes b_i^\chi\Big)\Big(\sum_i \tilde{b}_i^\chi \otimes (-a_i^\chi)\Big) \\ &= \sum_{i,j} \tilde{a}_i^\chi \, \tilde{b}_j^\chi \otimes b_i^\chi \, a_j^\chi \\ &= -\zeta_\chi \otimes \gamma_Y. \end{aligned}$$

LEMMA 4.4. *If* $\chi \neq 1$, *dim* $H_\chi = 2(g-1)$ *and dim* $H_\chi = 2g$ *for* $\chi = 1$.

PROOF. Since $G$ acts freely on $Y$, we have, for all $s \neq 1 \in G$, using Lefschetz fixed point formula,

$$Tr\, s \mid H^\circ(Y, \mathbf{C}) - Tr\, s \mid H^1(Y, \mathbf{C}) + Tr. s \mid H^2(Y, \mathbf{C}) = 0.$$

But $G$ acts trivially on $H^\circ(Y, \mathbf{C})$ and $H^2(Y, \mathbf{C})$. Hence $Tr\, s \mid H^1(Y, \mathbf{C}) = 2$ for all $s \neq 1$, while on the other hand $Tr\, s \mid H^1(Y, \mathbf{C}) = 2(r-1)(g-1) + + 2g$ for $s = 1$, by the Riemann-Hurwitz theorem. This proves the lemma.

LEMMA 4.5. *If we denote by* $\zeta_\chi$ *also the restriction of* $\zeta_\chi$ *to* $H^2(P, \mathbf{C})$, *then for any* $a_\chi \in \mathbf{C}$, *we have*

$$\Big(\sum_{\chi \neq 1} a_\chi \zeta_\chi\Big)^{(r-1)(g-1)} [P] = C\big(\prod_{\chi \neq 1} a_\chi\big)^{g-1}$$

*for a constant C.*

PROOF. Note that the cohomology algebra of $P$ is canonically isomorphic to the exterior algebra of $\bigoplus_{\chi \neq 1} H_\chi$. But $\Lambda(\bigoplus_{\chi \neq 1} H_\chi) = \bigotimes_{\chi \neq 1} \Lambda(H_\chi)$. Since $\zeta_\chi \in \Lambda^2 H_\chi$, we have $(\zeta_\chi)^i = 0$ if $i > g-1$ by Lemma 4.4. This shows that any monomial $\Pi \zeta_\chi^{n_\chi}$ of degree $(r-1)(g-1)$ is zero unless $n_\chi = (g-1)$ for all $\chi$. Since in the L. H. S. of the lemma, the only nonvanishing monomial, viz. $\Pi\zeta_\chi^{(g-1)}$, has coefficient $C'(\Pi a_\chi)^{g-1}$, this proves Lemma 4.5.

REMARK 4.6. It is not difficult to compute directly the constant $C$ in Lemma 4.5. In fact we will need this later. However, since this can be deduced (See Corollary 4.16) from Proposition 4.2, we do not compute it here.

Let $W$ be a Poincaré bundle on $M_\xi \times X$. (For definition, see [5, §2]). The group $\hat{G}$ operates on $M_\xi$ by tensorisation and hence also on $M_\xi \times X$. This action lifts canonically to the bundle ad $W$ as follows.

LEMMA 4.7. *For each* $\alpha \in \hat{G}$, *we have* $\alpha^* W \simeq W \otimes p_X^* L_\alpha$.

PROOF. For every $\omega \in M_\xi$, $\alpha^* W|_{\omega\times X}$ and $W|_{\omega\times X} \otimes L_\alpha$ are clearly isomorphic. Hence by [6, Lemma 2.5], there exists a line bundle $L'$ on $M_\xi$ such that $\alpha^* W \approx W \otimes p_X^* L_\alpha \otimes p_{M_\xi}^* L'$. This implies in particular that $\alpha^*$ det $W \approx$ det $W \otimes (p_{M_\xi}^* L')^r$. But if det $W \approx p_{M_\xi}^* L'' \otimes p_X^* \xi$, then $\alpha^*$ det $W \approx p_{M_\xi}^* \alpha^* L'' \otimes p_X^* \xi$, so that we may conclude $\alpha^* L'' \approx L'' \otimes (L')^r$. However, since Pic $M_\xi \simeq \mathbf{Z}$ by [6, Proposition 3.4], it follows that $\alpha^* L'' \approx L''$ and hence that $L'$ is trivial. This proves the lemma.

Now any isomorphism $\alpha^* W \to W \otimes p_X^* L_\alpha$ gives rise to an isomorphism $\alpha^*$ ad $W \to$ ad $W$. Moreover since $W|_{\omega\times X}$ is simple for all $\omega \in M_\xi$, the bundle $W$ is also simple and hence this isomorphism is canonical. It is easy to check that this gives an action of $\hat{G}$ on ad $W$ lifting its action on $M_\xi$.

On the one hand, $\hat{G}$ operates on the tangent sheaf $T(M_\xi)$ of $M_\xi$ by differentiation, and on the other, the $\hat{G}$-action on ad $W$ induces an action on $R^1(p_M)_*(\text{ad } W)$. In view of its functorial nature, the infinitesimal deformation map, which gives an isomorphism between these two sheaves, is compatible with these $\hat{G}$-actions. In particular, by restriction to $P$, we get an isomorphism $T(M_\xi) \mid P \simeq R^1(p_p)_*(\text{ad } W \mid P \times X)$ compatible with $\hat{G}$-actions (now acting trivially on the base). Moreover it is clear that if

$V'_x \subset \text{ad } W \mid P$ is the isotypical subbundle corresponding to the character induced by $x$ on $\hat{G}$, then $R^1(p_P)_*(V'_x)$ is the isotypical subbundle of $R^1(p_P)_*(\text{ad } W)$ of type $x$ and thus we obtain,

LEMMA 4.8. *For any $x \in G$, let $T_x$ be the isotypical subbundle of type $x$ of $T(M_\xi) \mid P$. Then there is an isomorphism $T_x \approx R^1(p_P)_*(V'_x)$.*

By [6, Lemma 2.5], we have $W \mid P \times X \approx (E \mid P \times X) \otimes p_P^* L$ for some line bundle $L$ on $P$. This gives an isomorphism $\text{End } W \mid P \times X \approx \text{End } E \mid P \times X$. Moreover the $\hat{G}$-action on $\text{End } E \mid P$ under this isomorphism may be described as follows. For $\alpha \in \hat{G}$, the isomorphism $E \to E \otimes p_X^* L_\alpha$, given by Lemma 2.1, (i), induces an automorphism of End $E$. This is the action of $\alpha$ on End $E$. The action of $\hat{G}$ on $(1 \times \pi)^*$ End $E$ is even easier to determine. In fact, by Lemma 2.1, (ii),

$$(1 \times \pi)^* \text{ End } E \approx (\bigoplus_{s \in G} s^* D) \otimes (\bigoplus_{s \in G} s^* D^*)$$

$$\approx \bigoplus_{x \in G} (\bigoplus_{h \in G} (h^* x^* D \otimes h^* D^*)).$$

The canonical isomorphisms $E \to E \otimes p_X^* L_\alpha$ give rise to an action of $\hat{G}$ on $(1 \times \pi)^* E$. This action has $x^* D$ as isotypical component of type $x$ by Lemma 2.3, (ii). Thus we get

LEMMA 4.9. *If under the above isomorphism $\text{ad } W|_P \approx \text{ad } E|_P$, the isotypical component $V'_x$ of type $x$ is mapped onto $V_x$, then*

$$(1 \times \pi)^* V_x \approx \bigoplus_{h \in G} (h^* x^* D \otimes h^* D^*), \textit{ for all } x \neq 1 \in G.$$

Now Lemmas 4.8 and 4.9 give a means of computing $ch\ T_x$ which is the essential content of Proposition 4.2.

LEMMA 4.10. *If $(a_i)$ is a base of $H^1(Y, \mathbf{C})$ and $(b_i)$ the dual base with respect to the Poincaré form, then*

$$c_1(D) = d(1 \otimes \gamma_Y) - \Sigma\, \tilde{a}_i \otimes b_i.$$

*If $(a_i)$ is adapted to the G-action, then this may be written $c_1(D) = d(1 \times \gamma_Y) - \sum_{\chi \in G} c_\chi$, with the notation of 4.3.*

PROOF. This is well-known. (See [4, Formula 2.6]).

LEMMA 4.11. *For* $x \neq 1$, *ch* $V_x = r - \frac{1}{2} \sum_{\chi \neq 1} (2 - \chi(x) - \chi(x)^{-1}) \zeta_\chi \otimes \Upsilon_X$ *with the notation of* 4.3.

PROOF. By Lemma 4.9, we have for $x \neq 1$,

$$(1 \times \pi)^* \operatorname{ch} V_x = \sum_{h \in G} \operatorname{ch}(h^* x^* D) \operatorname{ch}(h^* D)^{-1}$$

$$= \sum_{h \in G} h^* \exp(x^* c_1(D) - c_1(D)).$$

But $x^* c_1(D) - c_1(D) = \sum_\chi (c_\chi - x^* c_\chi) = \sum_\chi (1 - \chi(x)^{-1}) c_\chi$. Hence

$$(1 \times \pi)^* \operatorname{ch} V_x = r + \sum_{h \in G} \sum_\chi (1 - \chi(x)^{-1}) h^* c_\chi$$

$$- \sum_\chi \frac{r}{2} (1 - \chi(x)) (1 - \chi(x^{-1})) \zeta_\chi \otimes \Upsilon_Y$$

using Lemma 4.3, (i) and (ii). Again it is clear that $h^* c_\chi = \chi(h)^{-1} c_\chi$ and hence

$$\sum_{h \in G} \sum_\chi (1 - \chi(x))^{-1} h^* c_\chi = \sum_\chi \sum_h (\chi(h)^{-1} - \chi(hx)^{-1}) c_\chi = 0,$$

by the orthogonality relations. Thus $(1 \times \pi)^*$ *ch* $V_x$ reduces to

$$r - r/2 \sum_\chi (1 - \chi(x)) (1 - \chi(x)^{-1}) \zeta_\chi \otimes \Upsilon_Y =$$

$$= (1 \times \pi)^* \left\{ r - \tfrac{1}{2} \sum_\chi (1 - \chi(x)) (1 - \chi(x)^{-1}) \zeta_\chi \otimes \Upsilon_X \right\}$$

using $\pi^*(1) = 1$ and $\pi^*(\Upsilon_X) = r \Upsilon_Y$. Since $(1 \times \pi)^*$ is injective, the lemma is proved.

LEMMA 4.12. *For* $x \neq 1$, *we have*

$$ch\ T_x = r(g-1) + \tfrac{1}{2} \sum_\chi (2 - \chi(x) - \chi(x)^{-1}) \zeta_\chi.$$

PROOF. By Lemma 4.8, the first direct image of $V_x$ is isomorphic to

$T_x$. Since $V_x \subset \operatorname{ad} E$, it is clear that the other direct images of $V_x$ are all zero. Hence

$$\operatorname{ch} T_x = -\operatorname{ch}(p_P)_!(V_x) = -(p_P)_* \operatorname{ch}(V_x).(1+(1-g)\gamma_X)$$

by the Grothendieck-Riemann-Roch formula. In view of 4.11, this proves the lemma.

A feature of the expression for the Chern character of $T_x$ is that the terms of order $> 2$ are all absent. In this situation we have the following

LEMMA 4.13. *Let $V$ be a vector bundle of rank $n$ with chern character $ch(V) = n + c_1(V)$. If $x_i$ are the virtual Chern classes of $V$, then we have*
*(i) $\Pi(1+tx_i) = e^{tc_1(V)}$, and*
*(ii) if $x_1^{r_1} \ldots x_n^{r_n}$ is any monomial with one of the $r_i's \geq 2$, then its symmetric function is zero.*

PROOF. (i) is a simple application of Newton's formulae, while (ii) can be easily proved by induction on the number of non-zero $r_i$'s occurring in the monomial.

LEMMA 4.14. *For $x \neq 1$, we have*

$$ch_t(T_x^*) = (1+t)^{r(g-1)} \exp\left\{\frac{-t}{2(1+t)} \sum_{\chi} (2 - \chi(x) - \chi(x)^{-1}) \zeta_\chi\right\}.$$

PROOF. If $x_i$ are the virtual Chern classes of $T_x$, then

$$\operatorname{ch}_t(T_x^*) = \Pi(1 + t \exp(-x_i)) = \Pi(1 + t - tx_i)$$

by Lemma 4.13, (ii). Hence

$$\operatorname{ch}_t(T_x^*) = (1+t)^{r(g-1)} \Pi\left(1 - \frac{t}{t+1} x_i\right)$$

$$= (1+t)^{r(g-1)} \exp\left(\frac{-t}{1+t} \cdot c_1(T_x)\right).$$

by Lemma 4.13, (i). Substituting from Lemma 4.12, we get the required expression.

4.15 PROOF OF PROPOSITION 4.2. If $\mu$ is a generator of $G$, then $\omega = \langle \mu, x \rangle$, $x \neq 1$ runs through the non-trivial $r^{th}$ roots of unity. We will denote by $\zeta_d$ the cohomology class $\zeta_{\mu^d}$. On substitution from 4.14, the L.H.S. of Proposition 4.2 becomes

$$\Pi\,(1-\omega)^{r(g-1)}\exp\left\{\tfrac{1}{2}\sum_{0<d<r}\left(\sum_{\omega}\frac{2\omega}{1-\omega}-\frac{\omega^{d+1}}{1-\omega}-\frac{\omega^{r-d+1}}{1-\omega}\right)\zeta_d\right\}.$$

Since $\omega$ runs through the non-trivial $r^{th}$ roots of unity, we obviously have $\prod_{\omega}(1-\omega)=r$. To compute the term in the exponent, we let

$a_d=\sum\omega^{d+1}/(1-\omega)$. Then we have $a_0-a_d=\sum_{k=1}^{d}(a_{k-1}-a_k)$

$=\sum_{k=1}^{d}\left(\sum_{\omega}\omega^k\right)=-d$ since $0<d<r$. Hence $2a_0-a_d-a_{r-d}=-r$.

Thus the L.H.S. is

$$r^{r(g-1)}\exp\left(\tfrac{1}{2}\sum_{d}(-r)\zeta_d\right)=r^{r(g-1)}\exp(-r\Theta)$$

as claimed.

COROLLARY 4.16. *(i)* $\Theta^{(r-1)(g-1)}[P]=\{(r-1)(g-1)\}!\,r^{g-1}$

*(ii) The constant C in Lemma* 4.5 *is* $2^{(r-1)(g-1)}.r^{g-1}.\{(r-1)(g-1)\}!$

PROOF. (i) Apply the Atiyah-Singer fixed point theorem in the form 4.1 to the trivial line bundle. Since $M_\xi$ is unirational (even rational), we have $H^i(M_\xi,\mathcal{O})=0$ for $i>0$. Hence the L.H.S. of 4.1 is 1. On the right side, we have

$$\frac{1}{r^{r(g-1)}}\,\frac{(r\Theta)^{(r-1)(g-1)}}{\{(r-1)(g-1)\}!}\,[P].$$

This proves (i).

(ii) Lemma 4.5, applied to $2\Theta=\sum_{\chi}\zeta_\chi$ yields

$$C=2^{(r-1)(g-1)}.r^{g-1}\{(r-1)(g-1)\}!.$$

REMARK 4.17. From 4.16, (i), we conclude that $\Theta$ is not a multiple of a principal polarisation of $P$ if $r>2$. For, if $\Theta=\lambda\Phi$, $\Phi$ principal, then we would have $r^{g-1}=\dfrac{\Theta^{(r-1)(g-1)}[P]}{\{(r-1)(g-1)\}!}=\lambda^{(r-1)(g-1)}$, or $r=\lambda^{r-1}$, which is impossible if $r>2$. If $r=2$, it is well-known (and, in fact, can be deduced from above) that $\Theta$ is twice the canonical polarisation on the Prym variety, which is principal.

We now wish to apply the fixed point theorem to the bundle $\lambda_y(T^*)=\Sigma\,y^p\Omega^p$, where $\Omega^p$ is the bundle of $p$-exterior forms. We have

$$\widetilde{\mathrm{ch}}\, \lambda_y(T^*) = \widetilde{\mathrm{ch}} \prod_{x \in G} \lambda_y(T^*_x) = \prod_{x \in G} \widetilde{\mathrm{ch}}\, \lambda_{x^{-1}y}(T^*_x).$$

If $x = 1$, then $\mathrm{ch}\, \lambda_{x^{-1}y}(T^*_x) = (1+y)^{rk\, T^*_x} = (1+y)^{(r-1)(g-1)}$. On the other hand, if $x \neq 1$, denoting the virtual Chern classes of $T_x$ by $x_i$, we get

$$\begin{aligned}\widetilde{\mathrm{ch}}\, \lambda_{x^{-1}y}(T^*_x) &= \prod_i (1 + x^{-1}ye^{-x_i})\\ &= \prod_i (1 + x^{-1}y - x^{-1}yx_i), \text{ (by 4.13, ii)}\\ &= (1+x^{-1}y)^{r(g-1)} \prod_i \left(1 - \frac{x^{-1}y}{1+x^{-1}y} x_i\right)\\ &= (1+x^{-1}y)^{r(g-1)} \exp\left(\frac{x^{-1}y}{1+x^{-1}y} \cdot \tfrac{1}{2} \sum_{\chi} (2 - \chi(x) - \chi(x^{-1})\zeta_\chi\right)\end{aligned}$$

by 4.12 and 4.13, (i).

Hence we have

$$\widetilde{\mathrm{ch}}\, \lambda_{-y}(T^*) = (1-y)^{(r-1)(g-1)} \cdot \prod_{x \neq 1} (1 - x^{-1}y)^{r(g-1)} \cdot \exp(\Sigma b_\chi \zeta_\chi)$$

where $$b_\chi = \tfrac{1}{2} \sum_{x \neq 1} (2 - \chi(x) - \chi(x)^{-1}) \frac{x^{-1}y}{1 - x^{-1}y}.$$

Let now $\mu$ be a generator of $\hat{G}$ and if $\chi = \mu^d$, we write $b_d$, $\zeta_d$ for $b_\chi(\mu)$, $\zeta_\chi$ respectively. In this notation we have

$$\widetilde{\mathrm{ch}}\, \lambda_{-y}(T^*)(\mu) = (1-y)^{(r-1)(g-1)} \cdot \prod (1 - \omega^{-1}y)^{r(g-1)} \cdot \exp\left(\sum_{0<d<r} b_d \zeta_d\right) \tag{4.18}$$

Here $\omega$ runs through the non-trivial $r^{th}$ root of unity.

Lemma 4.19. *(i)* $\prod(1 - \omega y) = (1 - y^r)/(1-y)$

*(ii)* $\sum_\omega \omega^{d+1}/(1-\omega y) = ry^{r-d-1}/(1-y^r) - 1/(1-y)$,

*for* $0 < d < r$. *In (i), (ii),* $\omega$ *runs through the non-trivial r-th roots of unity.*

Proof. (i) is obvious. To prove (ii), set $a_d = \sum_\omega \omega^{d+1}/(1-\omega y)$.

Then $a_{d+1} - ya_d = \sum \omega^d = -1$ so that

$$a_0 - y^d a_d = \sum_{k=1}^{d} y^{k-1}(a_{k-1} - ya_k) = -(1-y^d)/(1-y),$$

To compute $a_0$, differentiate logarithmically with respect to $y$ the equation (i). Then we obtain

$$-a_0 = \sum_{\omega} \frac{-\omega}{1-\omega y} = \frac{-ry^{r-1}}{1-y^r} + \frac{1}{1-y}$$

Hence $a_d = \{a_0 + (1-y^d)/(1-y)\}/y^d = ry^{r-d-1}/(1-y^r) - 1/(1-y)$, proving Lemma 4.19.

In view of Lemma 4.19, the constants $b_d$ in the equation 4.18 become

$$b_d = \tfrac{1}{2} \sum_{\omega \neq 1} \left( \frac{2\omega y}{1-\omega y} - \frac{\omega^{d+1} y}{1-\omega y} - \frac{\omega^{r-d+1}\, y}{1-\omega y} \right) = y/2\ (2a_0 - a_d - a_{r-d}) \quad (4.20)$$

with the notation of Lemma 4.19.

Now substituting 4.18 and Proposition 4.2 in the fixed point formula 4.1, and writing $\chi'_y = \Sigma\,(-1)^i\,(\Sigma\, y^p\, Tr\mu \mid H^i(M_\xi, \Omega^p))$, we obtain

$$\chi'_{-y} = \frac{1}{r^{r(g-1)}} \frac{(1-y^r)^{r(g-1)}}{(1-y)^{(g-1)}} \exp\left(\sum b_d\, \zeta_d + r\,\Theta\right)[P]$$

$$= \frac{1}{\{(r-1)(g-1)\}!} \frac{1}{r^{r(g-1)}} \frac{(1-y^r)^{r(g-1)}}{(1-y)^{(g-1)}} \Pi \left(b_d + \frac{r}{2}\right)^{g-1} . C$$

by Lemma 4.5. It remains to compute $\Pi \left(b_d + \frac{r}{2}\right)$.

Using 4.20 and 4.19, we get

$$b_d + \frac{r}{2} = \tfrac{1}{2}\,(2a_0 y - a_d y - a_{r-d}\, y + r)$$

$$= \tfrac{1}{2}\{2ry^r/(1-y^r) - ry^{r-d}/(1-y^r) - ry^d/(1-y^r) + r\}$$

$$= \frac{r}{2} \cdot \frac{(1-y^d)\,(1-y^{r-d})}{(1-y^r)}.$$

Now substituting for $C$ from Corollary 4.16, (ii), we get

$$\chi'_{-y} = \frac{2^{(r-1)(g-1)}}{r^{(r-1)(g-1)}} \cdot \frac{(1-y^r)^{r(g-1)}}{(1-y)^{(g-1)}} \times$$

$$\times \left\{ \frac{r^{(r-1)}}{2^{(r-1)}\,(1-y^r)^{(r-1)}} \prod_{0<d<r} (1-y^d)(1-y^{r-d}) \right\}^{(g-1)}$$

$$= \left\{ \frac{(1-y^r)}{(1-y)} \prod_{0<d<r} (1-y^d)^2 \right\}^{(g-1)} .$$

Since the action of $\mu$ on $H^*(M_\xi, C)$ is trivial [2, Theorem 1], so is its action on the spaces $H^q(M_\xi, \Omega^p)$. Hence we have

THEOREM 1. *Let $X$ be a nonsingular projective curve of genus $g \geqslant 2$ and $\xi$ a line bundle on $X$. Let $r \geqslant 2$ be an integer coprime to degree $\xi$ and*

*$M_\xi$ denote the nonsingular projective variety of isomorphism classes of stable vector bundles on X of rank r and determinant isomorphic to ξ. Then*

$$\chi_{-y}(M_\xi) = \sum (-1)^{p+q} \dim H^q(M_\xi, \Omega^p) y^p = \left\{ \frac{(1-y^r)}{(1-y)} \prod_{0<d<r} (1-y^d)^2 \right\}^{(g-1)}.$$

COROLLARY. *The signature and Euler characteristic of $M_\xi$ are zero.*

PROOF. In fact, these are respectively $\chi_1$ and $\chi_{-1}$.

We now deduce a somewhat amusing consequence of Theorem 1.

THEOREM 2. *If N is the dimension of $M_\xi$, then $c_{N-1}(M_\xi) = 0$ except when $g=2$ and $r=2$.*

PROOF. Under our hypothesis, namely $g \geqslant 3$ or $r \geqslant 3$, we see, by Theorem 1, that $(1+y)^3$ divides $\chi_y$. Hence the coefficient of $z^2$ in $\chi_{z-1}(M_\xi)$ is zero. On the other hand, by the Hirzebruch-Riemann-Roch theorem [3, § 21.3], we have denoting by $x_i$ the virtual Chern classes of $M_\xi$,

$$\begin{aligned}\chi_{z-1} &= \prod_{i=1}^{N} \frac{x_i}{1-e^{-x_i}} \prod_{i=1}^{N} (1+(z-1)e^{-x_i})\,[M_\xi] \\ &= \prod_{i=1}^{N} \left( x_i + \frac{x_i z}{e^{x_i}-1} \right) [M_\xi] \\ &= \prod_{i=1}^{N} \left( x_i + z\left(1 - \tfrac{1}{2}x_i + \frac{1}{12}x_i^2 \ldots\right)\right) [M_\xi],\end{aligned}$$

Equating coefficients of $z^2$ on both sides we get

$$\sum_{j,k} \left( \left(\prod_{i \neq j,k} x_i\right)\left(1 - \tfrac{1}{2}x_j + \frac{1}{12}x_j^2\right)\left(1 - \tfrac{1}{2}x_k + \frac{1}{12}x_k^2\right)\right) [M_\xi] = 0.$$

This implies that

$$\sum_{j,k} \left( \left(\prod_{i \neq j,k} x_i\right) \left( \tfrac{1}{4}x_j x_k + \frac{1}{12}x_j^2 + \frac{1}{12}x_k^2 \right)\right) [M_\xi] = 0,$$

or, what is the same

$$\tfrac{1}{4}\binom{N}{2} c_N + \frac{1}{12}(c_{N-1}c_1 - Nc_N) = 0.$$

Since the Euler characteristic of $M$ is zero, we have $c_N = 0$ which implies that $c_{N-1}\, c_1 = 0$. Now by [6, Theorem 1], we have $c_1 \neq 0$ and by [6, Proposition 3.4] the second Betti number of $M$ is 1. By duality, $c_{N-1}\, c_1 = 0$ implies $c_{N-1} = 0$.

REMARK 4.21. One may apply the fixed point theorem to the tangent sheaf $\Theta$ and get (since $H^i(M_\xi, \Theta) = 0$ for $i \geqslant 2$).

$$Tr\ \mu \mid H^0(M_\xi, \Theta) - Tr\ \mu \mid H^1(M_\xi, \Theta) = -(3g-3).$$

This agrees with the result of [5, Theorem 1]. It is somewhat easier to prove that $H^0(M_\xi, \Theta) = 0$, and thus one would also get an alternative proof of Theorem 1 of [5] if one could show directly that the action of $\mu$ on $H^1(M_\xi, \Theta)$ is trivial.

We take this opportunity to remove the exception in [5, Theorem 1]. We exclude the case $g = 2$, $n = 2$ where the theorem is known to be true,

Since $H^0(M_\xi, \Theta) = 0$, we get from [5, Proposition 7.4] the exact sequence

$$0 \to H^1(M_\xi, \Omega^1) \to \bigoplus_{i=1}^{2g-2} H^1(M_\xi, ad_{x_i} W) \to H^1(M_\xi, \Theta) \to H^2(M_\xi, \Omega^1).$$

Now by [5, Theorem 2b] dim $H^1(M_\xi, ad_{x_i} W) = 1$ and by [2, Theorem 3], dim $H^2(M_\xi, \Omega^1) = g$. Hence we have dim $H^1(M_\xi, \Theta) \leqslant 3g - 3$. But by the spectral sequence in [5, §7.3] we have dim $H^1(M_\xi, \Theta) \geqslant$ dim $H^1(X, \Theta) = 3g - 3$. Hence dim $H^1(M_\xi, \Theta) = 3g - 3$.

## REFERENCES


1. M.F. ATIYAH AND I.M. SINGER, The Index of Elliptic Operators III, *Ann. of Math.* 87 (1968), 546–604.
2. G. HARDER AND M.S. NARASIMHAN, On the cohomology groups of moduli spaces of vector bundles on curves, *Math. Ann.* 212 (1975) 215–248.
3. F. HIRZEBRUCH, *Topological methods in Algebraic geometry*, Springer, 1966.
4. K. KODAIRA, Characteristic linear systems of complete continuous systems, *Amer. J. of Math.*, 78 (1956), 716–744.
5. M.S. NARASIMHAN AND S. RAMANAN, Deformations of the moduli space of vector bundles over an algebraic curve. *Ann of Math.* 101 (1975) 391–417.
6. S. RAMANAN, The moduli spaces of vector bundles over an algebraic curve, *Math. Ann* 200, (1973), 69–84.


Tata Institute of Fundamental Research
Bombay 400 005

# GEOMETRY OF HECKE CYCLES – I

*By* M. S. NARASIMHAN and S. RAMANAN

**1. Introduction.** In the course of our study of vector bundles over a smooth projective curve we considered [5] a correspondence between the space vector bundles of rank $n$ and degree $d$ on the one hand and that of vector bundles of rank $n$ and degree $(1-d)$ on the other. This is the geometric analogue of the Hecke Correspondence which is defined in the case of a curve defined over a finite field. Let $\xi$ be a line bundle on the curve $X$ and $U(n,\xi)$ (resp. $U(n,\xi^{-1}X)$) be the moduli space of vector bundles on $X$ of rank $n$ with determinant isomorphic to $\xi$ (resp. isomorphic to $\xi^{-1}\otimes L_x, x\in X), n\geqslant 2$. For a general vector bundle $E\in U(n,\xi)$, the subscheme corresponding to $E$ in $U(n, \xi^{-1}X)$ under this correspondence is isomorphic to the projective bundle $P(E^*)$. We call these subschemes of $U(n, \xi^{-1}X)$ good Hecke cycles. This identifies a suitable open subset of $U(n,\xi)$ with an open subset of the Hilbert scheme of $U(n, \xi^{-1}X)$. Results of this type were announced in [5, § 8] and are proved here in § 5, which can be read independently of the rest of the paper.

In the case $n=2$, and $\xi$ is trivial we prove that the irreducible component (which we shall call for convenience the Hecke component) of the Hilbert scheme of $U(2, X)$ containing the good Hecke cycles is smooth and provides a non-singular model for $U(2,\xi)$. This is the main result of the paper (Theorem 8.14).

Now the Kummer variety associated to the Jacobian $J$ of $X$ is the set of non-stable (and even singular if the genus $g\geqslant 3$) points of $U(2,\xi)$. The possible elements of the Hecke component corresponding to points of the Kummer variety can be listed (§7) and they all turn out to be conic bundles over $X$. The fibre in the non-singular model over a non-nodal point $k$ of the Kummer variety is isomorphic to $PH^1(X,j^2)\times PH^1(X,j^{-2})$, where $j\in J$ lies above $k$, and the corresponding subschemes in $U(2,X)$ can be described as the union of two projective line bundles on $X$ corresponding to non-trivial extensions

$$0 \longrightarrow j^{-1} \longrightarrow E \longrightarrow j \longrightarrow 0$$

$$0 \longrightarrow j \longrightarrow E' \longrightarrow j^{-1} \longrightarrow 0$$

identified along the sections given by $j^{-1}$ and $j$ respectively. Over a node of the Kummer variety, the elements are trivial conic bundles contained in $X \times P(H^1(X, \mathcal{O}))_t$ (imbedded in $U(2,X)$), where $P(H^1 (X,\mathcal{O}))_t$ is the thickening of $PH^1 (X,\mathcal{O})$ corresponding to the universal quotient bundle $Q$ of $PH^1 (X, \mathcal{O})$. (See 4. 4 iii). Thus we get here not only conics contained in $PH^1 (X,\mathcal{O})$ but also lines in it which are thickened within this $Q^*$-thickening. (The latter will be referred to as 'outside thickenings'). It is proved that the Hilbert scheme is itself smooth at all these points except at 1) a pair of intersecting lines in $PH^1(X, \mathcal{O})$, 2) double lines contained schematically in $PH^1(X,\mathcal{O})$ (§8). At points of the above type another component of the Hilbert scheme hits the Hecke component. We show that there is a natural morphism of the union of these two components into the Jacobian of $X$, which is constant on the Hecke component. (This morphism should be thought of as playing the role of the Weil morphism into the intermediary Jacobian). By studying the differential of this morphism we show that the Hecke component is smooth also at these points. (See Lemmas 8.10, 8.11).

The conics contained schematically in $PH^1 (X,\mathcal{O})$ is a $\mathbf{P}^5$ bundle over the Grassmannian of planes in $PH^1 (X, \mathcal{O})$. It will be shown in a later paper that the non-singular model considered above can be blown down along these fibrations (one for each node) to another non-singular model.

It turns out that all the subschemes described above consist of bundles which are non-trivial extensions of line bundles of a particular kind. Such (triangular) bundles are parametrised by a projective bundle over $X \times J$. We have a morphism from this space into $U(2,\mathrm{X})$, which may also be looked upon as a family $\{P(D_j)\}_{j\in J}$ of smooth subvarieties of $U(2,X)$ parametrised by $J$, where each $P(D_j)$ is a projective bundle over $X$ of dimension $(g-1)$. One of the essential points in the proof is to study the first and second order differentials of this morphism and in particular to compute its

Hessian at the critical points (§6). The necessary preliminaries for this are discussed in §2. This study is necessary to prove the non-singularity of the Hilbert scheme at a point given by an 'outside thickening'. For this we need information about the conormal sheaf of the thickening $X \times PH^1(X,\mathcal{O})_t$ of $X \times PH^1(X\mathcal{O})$ in $U(2,X)$. Now this thickening is a special case of thickenings which arise in the study of a family of smooth subvarieties (in our case, the family $j \longmapsto P(D_j)$ mentioned above). In this situation the conormal sheaf of the thickened scheme can be described in terms of the Hessian (see Lemma 3.7 and Remark 3.9).

A different approach for the desingularisation of $U(2,\xi)$ has been found by C. S. Seshadri [10].

NOTATION. All schemes will be of finite type over an algebraically closed field $k$ of characteristic $\neq 2$. From § 5 on, $X$ will denote a smooth projective curve of genus $g \geqslant 2$. If $S$ is a subscheme of Pic $X$, $U(n, S)$ will denote the subscheme of the moduli scheme $U(n, d)$ of $S$-equivalence classes of semistable vector bundles of rank $n$ of degree $d$ obtained as the inverse image of $S$ by the morphism det: $U(n, d) \to$ Pic. The Jacobian of $X$ will be denoted $J$. If $E$ is a family of semistable vector bundles over $X$, then there is a canonical morphisms $\theta_E$ of the parameter space into the moduli space.

If $X$ is a subscheme of $Y$, we denote by $\check{N}_{X,Y}$ the conormal sheaf of $X$ in $Y$. In good cases, e.g. when $X$ is regularly imbedded in $Y$, the sheaf $\check{N}_{X,Y}$ is locally free and its dual is the normal bundle denoted $N_{X,Y}$ so that in that case, $\check{N}_{X,Y} = N^*_{X,Y}$. If $\pi: X \to Y$ is a smooth morphism, we denote by $T_\pi$. the tangent bundle along the fibres of $\pi$. Its dual is sometimes denoted by $\Omega^1_\pi$ as well. If $x$ is a (closed) point of $X$, the tangent space at $x$ is denoted $T_x$.

If $D$ is a Cartier divisor in a scheme $X$, then $L_D$ will denote the line bundle defined by it. If $L$ is a line bundle generically generated by sections, then the quotient sheaf of $H^0(X, L)^*_X$ by the subsheaf $L^*$ will be called the quotient sheaf of the linear system defined by $L$. If $\mathfrak{F}$ is a coherent sheaf on a closed subscheme $i: Y \to X$, the sheaf $i_*(F)$ will be denoted by $\tilde{\mathfrak{F}}$.

If $\pi: X \to Y$ is a projective morphism with a relatively ample sheaf then Hilb $(X, Y, P)$ will denote the relative Hilbert scheme over $Y$ with Hilbert polynomial $P$. If $Y = \operatorname{Spec} k$, we simply write Hilb $(X, P)$.

The projections from $X \times Y \times Z$ to the component schemes will be denoted $p_1, p_2, p_3, p_{12}, p_{23}, p_{1,3}$. When we are dealing with closed subschemes or closed imbeddings we usually omit the word "closed".

**2. Deformations of Principal bundles.** We wish to recall certain facts concerning deformations of locally trivial principal $G$-bundles, where $G$ is an algebraic group, and in particular study the second order infinitesimal deformations of such bundles.

DEFINITION 2.1. *Let $P$ be a principal $G$-bundle on $X$ and $(S, s_0)$ a pointed scheme. A deformation of $P$ parametrised by $S$ is a principal $G$-bundle $Q$ over $X \times S$, and an ismorphism $\Psi$ of $Q \mid X \times s_0$ with $P$. Two deformations $Q_1$, $Q_2$ are said to be equivalent if for every $s \in S$, there exists a neighbourhood $U$ and an isomorphism $f: Q_1 \mid X \times U \to Q_2 \mid X \times U$ such that the diagram*

$$\begin{array}{ccc} Q_1 & \xrightarrow{\Psi_1} & P \\ {\scriptstyle f \mid X \times s_0} \downarrow & \nearrow {\scriptstyle \Psi_2} & \\ Q_2 & & \end{array}$$

*commutes if $s_0 \in U$'.*

REMARK 2.2. (i) If the only automorphisms of $P$ are given by morphisms of $X$ into the centre of $G$, then it is clear that the equivalence class of a deformation $Q$ of $P$ is independent of $\Psi$.

(ii) If $P$ is a principal $G$-bundle on $X$, then we have obviously the deformation functor $\delta_P$ on the category of Artinian local algebras which associates to $A$, the equivalence class deformations of $P$ parametrised by $A$. This functor can be seen to satisfy the conditions $H_1$ and $H_2$ of Schlessinger [7, Theorem 2.11]. If $X$ is proper $k$, it satisfies $H_3$ as well. Moreover, if $P$ is such that $H^0(X, Z_X) \to H^0(X, \operatorname{Ad} P)$ is an isomorphism, where $Z$ denotes the centre of the Lie algebra of $G$, then $\delta_P$ also satisfies $H_4$ and hence is prorepresentable. We proceed to describe this functor a little more explicitly.

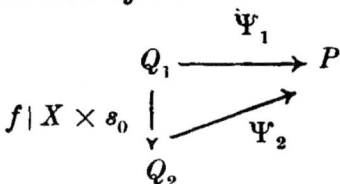

PROPOSITION 2.3. *Let $S$ be an Artinian local algebra, and $s_0$ its closed point. Then $\delta_P(S)$ is canonically bijective with the set $H^1(X,\underline{N})$ where $\underline{N}$ is the sheaf associated to the group scheme* $N = \ker p_* p^* G(\overline{P}) \to G(\overline{P})$, *where this map is given by restriction to $X \times s_0$, $G(P)$ is the group scheme* Aut $P$ *and $p: X \times S \to X$ is the projection.*

PROOF. The set of principal $G$-bundles on $X \times S$ are in (1,1) correspondence with the set $H^1(X \times S, p^* G(P)) \simeq H^1(X, p_* p^* G(P))$. Thus any deformation of $P$ gives rise to an element in $H^1(X, p_* p^* G(P)$ which is in the 'kernel' of $H^1(X, p_* p^* G(P)) \to H^1(X, G(P))$. But the prescription of $\Psi$ allows one to describe the deformation as a 1-cocycle for $p_* p^* G(P)$ and a 0-cochain for $G(P)$ of which its image in $G(P)$ is the coboundary. This proves the assertion.

PROPOSITION 2.4. *Let $(A, \mathbf{m})$ be an Artinian local algebra with $\mathbf{m}^2 = 0$ (resp. $\mathbf{m}^3 = 0$). Then the group scheme $N$ of 2.3 is isomorphic to the vector bundle* Ad $P \otimes \mathbf{m}$ *(resp. the group scheme* Ad $P \otimes \mathbf{m}$, *the group structure being defined by $(X, Y) \to X + Y + \frac{1}{2}[X, Y]$).*

PROOF. The exponential map yields a bijection of Ad $P \otimes \mathfrak{m}$ onto $N$, and the transfer of the group structure on $N$ to Ad $P \otimes \mathbf{m}$ is as given above, by virtue of the Campbell-Hausdorff formula [9, *SGA* 3, Expose VII, 3.1].

REMARK 2.5. If the field $k$ is of characteristic 0, we have an obvious generalisation of 2.4 to all Artinian local algebras using the Campbell-Hausdorff formula.

EXAMPLE 2.6. Let $A$ be the algebra $k[\epsilon_1, \epsilon_2]/(\epsilon_1^2, \epsilon_2^2)$. Then $H^1(X, \text{Ad } P \otimes \mathbf{m})$ can be described with respect to a suitable open covering $(U_i)$ of $X$ as follows. Any 1-cochain can be described by $f_{ij}\epsilon_1 + g_{ij}\epsilon_2 + h_{ij}\epsilon_1\epsilon_2$ where $f_{ij}, g_{ij}, h_{ij}$ belong to $H^0(U_i \cap U_j, \text{Ad } P)$. The cocycle condition with respect to the above group structure on Ad $P \otimes \mathfrak{m}$ is that, for every $i,j,k$, we have

$$f_{ik}\epsilon_1 + g_{ik}\epsilon_2 + h_{ik}\epsilon_1\epsilon_2 = (f_{ij}\epsilon_1 + g_{ij}\epsilon_2 + h_{ij}\epsilon_1\epsilon_2)(f_{jk}\epsilon_1 + g_{jk}\epsilon_2 + h_{jk}\epsilon_1\epsilon_2)$$
$$= (f_{ij} + f_{jk})\epsilon_1 + (g_{ij} + g_{jk})\epsilon_2 + \{\tfrac{1}{2}([f_{ij}, g_{jk}] + [g_{ij}, f_{jk}]) + h_{ij} + h_{jk}\}\epsilon_1\epsilon_2$$

or, what is the same, $f_{ij}, g_{ij}$ are 1-cocycles of Ad $P$, and $h_{ij}$ satisfies

$$h_{ij} + h_{jk} = h_{ik} - \tfrac{1}{2}([f_{ij}, g_{jk}] + [g_{ij}, f_{jk}]).$$

On the other hand, two cocycles $(f_{ij}, g_{ij}, h_{ij})$ and $(f'_{ij}, g'_{ij}, h'_{ij})$ are cohomologous if and only if there exist sections $(a_i, b_i, c_i)$ over $U_i$ of ad $P$ with

$$(f_{ji}\epsilon_1+g_{ij}\epsilon_2+h_{ij}\epsilon_1\epsilon_2)(a_j\epsilon_1+b_j\epsilon_2+c_j\epsilon_1\epsilon_2) = (a_i\epsilon_1+b_i\epsilon_2+c_i\epsilon_1\epsilon_2)\times$$
$$\times(f'_{ij}\epsilon_1+g'_{ij}\epsilon_2+h'_{ij}\epsilon_1\epsilon_2);$$

namely, $f'_{ij}-f_{ij}$ is the coboundary of $(a_i)$ in Ad $P$, $g'_{ij}-g_{ij}$ is the coboundary of $(b_i)$, and

$$h'_{ij}-h_{ij}=c_j-c_i+\tfrac{1}{2}\left([f_{ij}, b_j]+[g_{ij}, a_j]-[b_i, f'_{ij}]-[a_i, g'_{ij}]\right) \qquad (2.8)$$

2.9 HESSIAN. Let $f$ be a morphism of a smooth scheme $X$ into a scheme $Y$ and $df\colon T_x\to T_y, y=f(x)$ be its differential. Then the *Hessian* $h(f)$ of $f$ is defined to be a map ker $df\otimes T_x\to$ coker $df$. It is more convenient to define the dual map $\check{h}(f)\colon \ker(\check{df})\otimes T_x\to \operatorname{coker}\check{df}$. Let $(\mathcal{O}_x, \mathbf{m}_x)$, $(\mathcal{O}_y, \mathbf{m}_y)$ be the local rings at $x$, $y$. If $a\in\mathbf{m}_y$, with $a\circ f\in\mathbf{m}_x^2$ and $t\in T_x$, then we get an element of $T_y$, by contracting with $t$ the element in $S^2(\mathbf{m}_x/\mathbf{m}_x^2)=\mathbf{m}_x^2/\mathbf{m}_x^3$ given rise to by $a\circ f$. Its image in coker $df$ depends only on the class of $a$ modulo $\mathfrak{m}_y^2$ and is defined to be $\check{h}(f)(a, t)$.

2.10. FUNCTORIAL DESCRIPTION OF HESSIAN. In terms of $A$-valued points, the Hessian can be described as follows. Consider the ring $A=k[\epsilon_1, \epsilon_2]/(\epsilon_1^2, \epsilon_2^2)$ and the quotient rings $B_1$, $B_2$ and $C$ given by the ideals $(\epsilon_2)$, $(\epsilon_1)$ and $(\epsilon_1\epsilon_2)$. Giving vectors $a$, $t$ in ker $df$ and $T_x$ respectively is the same as giving a $C$-valued point $p$ of $X$ at $x$ such that the corresponding $B_1$-valued point is mapped by $f$ on the 0-vector at $y$. Let $q$ be an $A$-valued point extending $p$; there exists one such since $X$ is smooth. By assumption, the image $f(q)$ actually yields a $(k+k\epsilon_2+k\epsilon_1\epsilon_2)$-valued point at $y\in Y$. By restriction to Spec $(k+k\epsilon_1\epsilon_2)$, we get a vector at $y$. Its image modulo Image $(df)$ is independent of the extension $q$ of $p$ and gives $h(f)(a, t)$.

REMARK 2.11. The above definition of Hessian in terms of $A$-valued points enables one to define the Hessian of a morphism of a smooth, prorepresentable functor on the category of Artinian local algebras into another prorepresentable functor. In particular, if $P$ is a principal $G$-bundle with $H^2(X, \operatorname{Ad} P) = 0$, then it is easy to check that the functor $\delta_P$ defined in 2.2, (ii) is smooth. Hence if $P$ satisfies in addition the conditions of 2.2, (ii), then for any homomorphism of $G$ into another algebraic group $H$, the notion of Hessian of the morphism $\delta_P \to \delta_Q$ makes sense, where $Q$ is the principal $H$-bundle associated to $P$. We proceed to compute this in the case when $G \subset H$.

Proposition 2.12. (i) *Let $P$ be a principal $G$-bundle on a scheme $X$, proper over $k$. For any homomorphism $G \to H$, the induced morphism $e_{G,H}\colon \delta_P \to \delta_Q$ of the deformation functors has as differential, the natural map $H^1(X, \operatorname{Ad} P) \to H^1(X, \operatorname{Ad} Q)$, where $Q$ is the principal $H$-bundle associated to $P$.*

(ii) *If $H^2(X, \operatorname{Ad} P) = 0$ and $G \subset H$, then the cokernel of $de_{G,H}$ is canonically isomorphic to $H^1(X, E)$ where $E$ is the vector bundle associated to $P$ for the linear isotropy action of $G$ on the tangent space at $e$ of $H/G$.*

(iii) *Assume that $Q$ (and hence $P$) satisfies the condition in 2.2, (ii). Under the identification in (ii), the Hessian of $e_{G,H}$ can be described as follows. Let $t_1 \in \ker de_{G,H}$ and $t_2 \in H^1(X, \operatorname{Ad} P)$. Then $t_1$ is in the image under the boundary homomorphism of an element $s_1 \in H^0(X, E)$ associated to the exact sequence*

$$0 \to \operatorname{Ad} P \to \operatorname{Ad} Q \to E \to 0.$$

*Then $h(t_1, t_2) = [s_1, t_2]$, where the bracket is the cup product associated to the natural action of Ad $P$ on $E$.*

Proof. (i) is obvious from 2.4.

(ii) is a trivial consequence of (i) and the cohomology sequence of $0 \to \operatorname{Ad} P \to \operatorname{Ad} Q \to E \to 0$.

(iii) We will use the notation of 2.10. By the description of the Hessian given there and of the functors $\delta_P$, $\delta_Q$ given in 2.4, $h(t_1, t_2)$ is obtained as follows. The vectors $t_1$, $t_2$ give an element of $H^1(X, \operatorname{Ad} P \otimes \mathfrak{m}/(\epsilon_1, \epsilon_2))$ where $\mathfrak{m}$ is the maximal ideal of $A$. This is the image of an element in $H^1(X, \operatorname{Ad} P \otimes \mathfrak{m})$, the group structure in $\operatorname{Ad} P \otimes \mathfrak{m}$ being given in 2·4. Its image in $H^1(X, \operatorname{Ad} Q \otimes \mathfrak{m})$ goes to zero in $H^1(X, \mathrm{Ad}\, Q \otimes \mathfrak{m}/(\epsilon_2))$ and hence comes from an element in $H^1(X, \mathrm{Ad}\, Q \otimes (\epsilon_2))$. The image of this element in $H^1(X, E \otimes (\epsilon_2)/(k\epsilon_2))$ is $h(t_1, t_2)$. In terms of cocycles for a suitable covering $(U_i)$ of $X$ (as in 2.6) this means the following. Let $f_{ij}, g_{ij}$ be cocycles representing $t_1, t_2$, in $H^1(X, \mathrm{Ad}\, P)$. Then there exists $h_{ij} \in H^0(U_i \cap U_j, \mathrm{Ad}\, P)$ such that $(f_{ij}, g_{ij}, h_{ij})$ satisfy 2.7. Reading these as sections over $U_i \cap U_j$ of Ad $Q$, we get a corresponding element of $H^1(X, \mathrm{Ad}\, Q \otimes \mathfrak{m})$. We are given that $f_{ij}$ is a coboundary for Ad $Q$, namely, there exist $\lambda_i \in H^0(U_i, \mathrm{Ad}\, Q)$ with $f_{ij} = \lambda_j - \lambda_i$. Then the cocycle $(f_{ij}, g_{ij}, h_{ij})$ is cohomologous by 2.8 to $(0, g_{ij}, h_{ij} + \frac{1}{2}([\lambda_i, g_{ij}] - [g_{ij}, \lambda_j]))$, taking $(a_i, b_i, c_i) = (\lambda_i, 0, 0)$. Thus the element in $H^1(X, \mathrm{Ad}\, Q \otimes (k\epsilon_2 + k\epsilon_1\epsilon_2))$ is given by the cocycle $g_{ij}\epsilon_2 + h_{ji} + \frac{1}{2}([\lambda_1, g_{ij}] - [g_{ij}, \lambda_j])\epsilon_1\epsilon_2$. Finally, the Hessian $h(t_1, t_2)$ is given by the cocycle $\frac{1}{2}([\lambda_i, g_{ij}] - [g_{ij}, \lambda_j)])$ of $E$, since $h_{ij}$ is a section of Ad $P$. Notice that since $\lambda_i - \lambda_j$ is a section of Ad $P$, $\lambda_i = \lambda_j$ as sections of $E$ on $U_i \cap U_j$ and hence $(\lambda_i)$ determines a section $\lambda$ of $E$. Clearly $\lambda$ maps on $(f_{ij})$ under the boundary homomorphism and the cocycle $\frac{1}{2}([\lambda, g_{ij}] - [g_{ij}, \lambda]) = [\lambda, g_{ij}]$ represents the cup product of $\lambda$ and the class of $g_{ij}$ as was to be proved.

REMARKS 2.13. (i) We will apply Proposition 2.12 to the case when $G$ is $2 \times 2$ triangular (Borel) subgroup of $H = GL(2, k)$, and $P$ is a principal bundle over the curve $X$. Then $P$ is described by an exact sequence

$$0 \to L_1 \to W \to L_2 \to 0,$$

where $L_1$ and $L_2$ are line bundles. If $W$ is simple, then the conditions in Prop. 2.12 are satisfied. The bundle Ad $P$ is the bundle $\Delta(W)$ of endomorphisms of $W$ leaving $L_1$ invariant, Ad $Q$ = End $W$, and the vector bundle $E$ can be identified with Hom $(L_1, L_2)$. Moreover, the bundle Ad $P$ consisting of endomorphisms of trace 0 is isomorphic to Hom $(L_2, W)$. In particular, we have a map $\eta$ of Ad $P$ onto the sheaf $\mathcal{O}_X$, obtained by composing with the map $W \to L_2$. With these identifications, the action of Ad $P$ on $E$ is simply multiplication by its image in $\mathcal{O}$. Thus if $t_1 \in \ker H^1(X, \mathrm{Ad}\, P) \to H^1(X, \mathrm{Ad}\, Q)$, and $t_2 \in H^1(X, \mathrm{Ad}\, P)$, then $h(t_1, t_2) \in H^1(X, E)$ is the cup product of $\eta t_2 \in H^1(X, \mathcal{O})$ and the element $s_1 \in H^0(X, \mathrm{Hom}\,(L_1, L_2))$ of which $t_1$ is the image. In other words, $h(t_1, t_2)$ is simply the image of $\eta\, t_2 \in H^1(X, \mathcal{O})$ in $H^1(X, \text{Hom}\,(L_1, L_2))$ by the map $\mathcal{O} \to \text{Hom}\,(L_1, L_2)$ given by $t_1$.

(ii) If global fine moduli schemes for principal $G$ and $H$ bundles exist, then clearly Proposition 2.12 enables one to compute the Hessian of the morphism $e_{G,H}$ induced on the moduli schemes by extension of structure group.

**3. Thickenings.** Let $X$ be a scheme and $\mathcal{F}$ a coherent sheaf of $\mathcal{O}$-Modules on $X$. Let $\varphi : \mathcal{F} \to \Omega^1_X$ be a surjective $\mathcal{O}_X$-homomorphism of $\mathcal{O}_X$-Modules. Then one can define a ring structure on the subsheaf of $\mathcal{O} \oplus \mathcal{F}$ consisting of $\{(f, x) : df = \varphi x\}$. This is a subsheaf $\mathcal{O}_\varphi$ of rings if we consider $\mathcal{O} \oplus \mathcal{F}$ as an $\mathcal{O}$-Algebra with $\mathcal{F}^2 = 0$.

LEMMA 3.1. *$(X, \mathcal{O}_\varphi)$ is a scheme.*

PROOF. If $X = \text{Spec}\, A$ is affine, then $(X, \mathcal{O}_\varphi) \simeq \text{Spec}\,(M +_\varphi \Omega^1)$ where $\widetilde{M} = \mathcal{F}$. This shows in general that $(X, \mathcal{O}_\varphi)$ is a prescheme and since $(X, \mathcal{O}_\varphi)_{\text{red}} = (X, \mathcal{O})_{\text{red}}$ is a scheme, the lemma is proved.

If $I = \ker \varphi$, then we have an exact sequence of $\mathcal{O}_\varphi$-Modules ($I$ being considered as an $\mathcal{O}_\varphi$-Module)

$$0 \to I \to \mathcal{O}_\varphi \to \mathcal{O} \to 0.$$

LEMMA 3.2. *Let $X$ be a smooth irreducible scheme. The scheme $X_\varphi = (X, \mathcal{O}_\varphi)$ is Cohen-Macaulay if and only if $I$ is a locally free $\mathcal{O}_X$-Module. In this case, the dualising sheaf of $X_\varphi$ restricts to $X$ as $\underline{\text{Hom}}\,(I, \omega_X)$. Moreover, $X_\varphi$ is a local complete intersection if and only if $I$ is locally free of rank* 1 *or* 0.

PROOF. The problem being local, we may assume that $X = \text{Spec}\, A$ and $X_\varphi = \text{Spec}\, A_\varphi$, and since $\Omega^1_X$ is locally free, we may also assume that $A_\varphi = A \oplus I$, with $I^2 = 0$. In the local case, an $A$-sequence is an $A_\varphi$-sequence if and only if it is an $I$-sequence as well. Hence the first assertion. Now, if $I$ is free, then $A_\varphi = A \otimes (k \oplus V)$ with $V^2 = 0$, where $V$ is a finite dimensional $k$-vector space. Thus, our assertion on the dualising sheaf needs only to be proved when $X_\varphi = \text{Spec}\,(k \oplus V)$. In this case, the dualising sheaf is seen to be actually $k \oplus V^*$, in which $V$ operates trivially on $k$ and by duality on $V^*$. To prove the last assertion, we note that if $A_\varphi$ is a complete intersection, then so is $k \oplus V$, and hence $k \oplus V^*$ is free over $k \oplus V$. This clearly implies that dim $V \leqslant 1$, while, on the other hand, it is obvious that dim $V = 1$ implies $k \oplus V$ is a complete intersection.

**Remark 3.3.** Let $X$ be a smooth subscheme of a scheme $Y$. Then we have an exact sequence

$$0 \to \check{N}_{X,Y} \to \Omega^1_Y \mid_X \to \Omega^1_X \to 0.$$

If $\check{N}_{X,Y} \to I$ is an $\mathcal{O}_X$ - homomorphism into an $\mathcal{O}_X$ - Module $I$, then by the push-out construction, we obtain a sheaf $\mathcal{F}$ and a surjective homomorphism $\varphi : \mathcal{F} \to \Omega^1_X$. In particular, this gives rise to the scheme $X_\varphi$. Moreover, there is a natural morphism $X_\varphi \to Y$ making the diagram

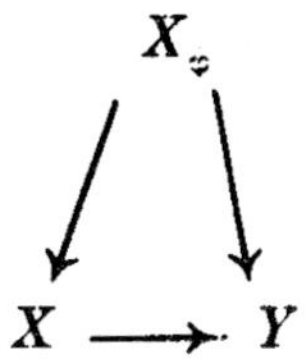

commutative. It is easy to see that this morphism $X_\varphi \to Y$ is a closed immersion if and only if $\check{N}_{X,Y} \to I$ is surjective (In this case, the thickened subscheme $X_\varphi$ of $Y$ will be denoted by $X_I$). In other words, all the thickenings of $X$ 'inside' $Y$ are given by Quot $\check{N}_{X,Y}$. In the case when $Y = Z_\psi$, with $Z$ smooth and $\psi$ a surjective homomorphism $\mathcal{F} \to \Omega^1_Z$, then $\check{N}_{X,Y}$ is the kernel of the map $\mathcal{F} \mid X \to \Omega^1_X$, obtained as the composite of $\psi \mid X$ and the map $\Omega^1_Z \mid X \to \Omega^1_X$.

**Lemma 3.4.** *Let $X$ be a smooth subscheme of a smooth scheme $Z$, and let $Q$ be a locally free quotient of $N^*_{X,Z}$. If $\eta : N^*_{X,Z} \to Q$ is the canonical map, then the restriction of $N_{X_Q,Z}$ to $X$ is locally free and fits in the exact sequence*

$$0 \to S^2(Q) \to N_{X_Q,Z} \mid X \to \ker \eta \to 0.$$

**Proof.** Let $I$ (resp. $J$) be the ideal sheaf of $X$ (resp. $X_Q$) in $Z$. We may choose local parameters $(x_1, \ldots\ldots x_l, y_1, \ldots\ldots, y_m, z_1, \ldots\ldots, z_n)$ at a point of $X$ so that $I = \{x_1, \ldots\ldots, x_l, y_1, \ldots\ldots y_m\}$ and $J = \{x_i\, x_j, y_1, \ldots\ldots, y_n\}$, $i, j, = 1, \ldots l$. From this it is clear that $I^2/IJ$

is locally free and the natural map $S^2(Q) = S^2(I/J) \to I^2/IJ$ is an isomorphism. On the other hand $N_{X_Q, Z}|X \simeq J|IJ$ fits in the exact sequence

$$0 \to I^2/IJ \to J|IJ \to J/I^2 \to 0.$$

Now $N^*_{X,Z} \simeq I/I^2$ and the map $\eta$ is the natural projection $I/I^2 \to I/J$ and hence $\ker \eta = J|I^2$, proving the lemma.

The following functorial remark on the extension 3.4 is an immediate consequence of the definition.

LEMMA 3.5. *If $X$ is a smooth subscheme of a smooth scheme $Z$ and $\eta_1 : N^*_{X,Z} \to Q_1$, $\zeta : Q_1 \to Q_2$ are surjections, then the extensions* 3.4 *corresponding to $X_{Q_1}$ and $X_{Q_2}$ fit in a commutative diagram*

$$\begin{array}{ccccccccc} 0 & \to & S^2(Q_1) & \to & N_{X_{Q_1}, Z}|X & \to & \ker \eta_1 & \to & 0 \\ & & \downarrow & & \downarrow & & \downarrow & & \\ 0 & \to & S^2(Q_2) & \to & \check{N}_{X_{Q_2}, Z}|X & \to & \ker \zeta \circ \eta_1 & \to & 0 \end{array}$$

REMARK 3.6. We will now determine the extension involved in 3.4 in the following situation. Let $\varphi : Y \to Z$ be a morphism of smooth schemes and $X$ a smooth subscheme of $Y$ on which $\varphi$ is an imbedding. Then the Hessian (see § 2.9) of the map $\varphi$ along $X$ goes down to a map $K \otimes N_{X,Y} \to \operatorname{Coker} d\varphi$ where $K = \ker d\varphi$. Now $d\varphi$ induces a map $N_{X,Y} \to N_{X,Z}$ and let $Q$ be the image of its transpose. We wish to consider the thickening of $X$ in $Z$ given by $\eta : N^*_{X,Z} \to Q$. It is easily seen that this thickening is in fact the schematic image (by $\varphi$) of the total normal thickening of $X$ in $Y$.

LEMMA 3.7. *In the situation of* 3.6, *we further assume that $K = \ker d\varphi$ has rank* 1 *at all points of $X$. Then the extension in Lemma* 3.4 *is the pull back by means of the transpose of the Hessian* : $\ker \eta \to K^* \otimes N^*_{X,Y}$ *of the exact sequence*

$$0 \to S^2(Q) \to S^2(N^*_{X,Y}) \to K^* \otimes N^*_{X,Y} \to 0$$

*obtained by symmetrising the exact sequence*

$$0 \to Q \to N^*_{X,Y} \to K^* \to 0.$$

PROOF. We use the notation of 3.4. Let $I'$ be the ideal of $X$ in $Y$. Then notice that $f \in I \Rightarrow f \circ \varphi \in I'$ and $f \in J \Rightarrow f \circ \varphi \in I'^2$. Hence we have a map $J/_{I,I} \to I'^2/I'^3 = S^2(N^*_{X,Y})$ induced by $\varphi$. The resulting map $J/I^2 = (N/Q)^* \to K^* \otimes N^*_{X,Y}$ is easily seen to be the transpose of the Hessian.

Putting together Lemmas 3.5 and 3.7, we obtain

PROPOSITION 3.8. *In the situation of 3.7, let $L$ be a locally free quotient of $Q$. Then the pullback of the extension 3.4 for $X_L$ by the inclusion* $\ker \eta \to \ker(N^*_{X,Z} \to L)$ *is isomorphic to the pullback of the sequence*

$$0 \to S^2(L) \to S^2(E) \to K^* \otimes E \to 0$$

*by the map $K^* \otimes N^*_{X,Y} \to K^* \otimes E$, where this latter extension is obtained by symmetrising the sequence*

$$0 \to L \to E \to K^* \to 0,$$

*$E$ being the push-out $N^*_{X,Y} \amalg_Q L$.*

PROOF. From Lemma 3.5, we obtain the commutative diagram

$$\begin{array}{ccccccccc} 0 & \to S^2(L) & \to & \check{N}_{X_L,Z} \mid X & \to & \ker(N^*_{X,Z} \to L) & \to 0 \\ & \uparrow & & \uparrow & & \uparrow & \\ 0 & \to S^2(Q) & \to & \check{N}_{X_Q,Z} \mid X & \to & \ker \eta & \longrightarrow 0 \end{array}$$

This shows that the required pullback is the push-out of the sequence

$$0 \to S^2(Q) \to \check{N}_{X_Q,Z} \mid X \to \ker \eta \to 0$$

by the map $S^2(Q) \to S^2(L)$. Now the proposition follows from the description of this sequence in Lemma 3.7 and the obvious commutative diagram

$$\begin{array}{ccccccc} 0 \to & S^2(L) & \to & S^2(E) & \to & K^* \otimes E & \to 0 \\ & \uparrow & & \uparrow & & \uparrow & \\ 0 \to & S^2(Q) & \to & S^2(N^*_{X,Y}) & \to & K^* \otimes N^*_{X,Y} & \to 0 \end{array}$$

REMARK 3.9. In our applications, we will be only considering the special case of (3.6), where we have $\psi : Y \to T$ is a proper smooth morphism and $\varphi$ realises the fibres $\{X\}$ of $\psi$ as a family of smooth subschemes of $Z$.

## 4. Conics.

DEFINITION 4.1. *A scheme $C$ together with a very ample line bundle $h$ is said to be a* conic *if its Hilbert polynomial is* $(2m + 1)$.

*A scheme $C$ over $T$ is said to be a* conic over $T$ *if with respect to a line bundle on $C$ its fibres are conics and the morphism $C \to T$ is faithfully flat.*

We will show in 4.2 and 4.3 that this coincides with the usual notion of a conic.

LEMMA 4.2 *If $C$ is a reduced scheme and $h$ a very ample line bundle on $C$ with Hilbert polynomial of the form $2m+\delta$, $\delta \leqslant 1$, then $\delta = 1$, $C$ is isomorphic to a (reduced) conic in $\mathbf{P}^2$, and $h$ is the restriction to $C$ of the hyperplane bundle on $\mathbf{P}^2$.*

PROOF. It is obvious that we may assume without loss of generality that $C$ is pure 1-dimensional. Clearly then, either $C$ is irreducible, or $C$ has two components $C_1$, $C_2$ with Hilbert polynomials $m+d_1$, $m+d_2$ with $d_1+d_2-l=\delta$, where $l$ is the length of the intersection $C_1 \cap C_2$. In the former case the normalisation of $C$ has to be rational; for, otherwise, a linear system of degree 2 is at most one dimensional and cannot imbed $C$. Moreover, if $C$ is not normal, the linear system of $h$ on $C$ will have dimension $\leqslant 1$, which cannot imbed $C$. Thus if $C$ is irreducible, it must be isomorphic to $\mathbf{P}^1$ and the bundle $h$ is the square of the hyperplane bundle on $\mathbf{P}^1$. In the case when $C$ has two components $C_1$ and $C_2$, we conclude as above that $C_1$, $C_2$ are both isomorphic to $\mathbf{P}^1$ and $h$ restricts to each of these as the hyperplane bundle. Thus we have $d_1 = d_2 = 1$. From the nature of the Hilbert polynomial, it is clear that $C_1$ and $C_2$ intersect. But dim $H^0(C, h) = 4 - 1 \geqslant 3$ since $h$ is very ample. Hence $l = 1$, *i.e.* $C$ is a pair of intersecting lines in $\mathbf{P}^2$.

LEMMA 4.3. *Let $C$ be a nonreduced scheme and $h$ an ample bundle on $C$ with Hilbert polynomial of the form $2m+\delta$, $\delta \leqslant 1$. If $h|C_{red}$ is very ample, then $C$ contains a unique subscheme $C'$ which is obtained by thickening $\mathbf{P}^1$ by a line bundle $L$ of degree $\leqslant -1$. If degree $L = -1$, then $L$ must be the dual of the hyperplane bundle and $C = C'$. In otherwords, $C$ is isomorphic to the total thickening of a line in $\mathbf{P}^2$. In particular, a nonreduced conic (in the sense of 4.1) is such a thickened line.*

PROOF. Clearly, we may assume $C$ has no zero-dimensional components. Consider the exact sequence

$$0 \to I \to \mathcal{O} \to \mathcal{O}_{red} \to 0$$

on $C$, where $I$ is the ideal of nilpotents in $\mathcal{O}$. If $I$ has finite support, the Hilbert polynomial of $\mathcal{O}_{red}$ is of the same form and hence by 4.2, must be $2m+1$. This shows that $\mathcal{O} = \mathcal{O}_{red}$, contrary to assumption. Thus the support of $I$ is 1-dimensional. It follows that the Hilbert polynomials of $I$ and $\mathcal{O}_{red}$ are of the form $m+d_1$, $m+d_2$. Since $h|C_{red}$ is very ample, we conclude as in 4.2 that $C_{red}$ is isomorphic to $\mathbf{P}^1$, and $h$ restricts to $\mathbf{P}^1$ as the hyperplane bundle. Now consider the filtration

$$\mathcal{O} \supset I \supset I^2 \ldots \supset I^n = (0).$$

Since the Hilbert polynomial $m+d_1$ of $I$ is the sum of the Hilbert polynomials of the sheaves $I^k/I^{k+1}$ on $\mathbf{P}^1$, it follows that $rk\, I/I^2 = 1$ and $I^k/I^{k+1}$ are all torsion sheaves for $k \geqslant 2$. If $L$ is the line bundle on $\mathbf{P}^1$ obtained as the free quotient of $I/I^2$, we get an $L$-thickening $C'$ of $\mathbf{P}^1$ as a subscheme of $C$. Since now the Hilbert polynomial of $\mathcal{O}_{red}$ is $m+1$, the Hilbert polynomial of $I$ is $m+\delta-1$ and hence that of $L$ is $m+d$, with $d \leqslant 0$. In particular, the degree of $L \geqslant -1$. If now deg $L = -1$, then $L$ is the dual of the hyperplane bundle on $\mathbf{P}^1$ whose Hilbert polynomial is $m$, and hence that of $C'$ is $2m+1$. This proves that $\delta = 1$ and $C = C'$. It is clear that the $L$-thickening of $\mathbf{P}^1$ is the normal thickening of a line in $\mathbf{P}^2$. Finally, if $h$ is already very ample on $C$, then from the exact sequence

$$0 \to L \to \mathcal{O}_{C'} \to \mathcal{O}_{red} \to 0$$

tensored with $h$, we obtain

$$0 \to H^0(\mathbf{P}^1, L \otimes h) \to H^0(C', h) \to H^0(\mathbf{P}^1, h).$$

Since $h$ is very ample on $C'$, we have dim $H^0(C', h) \geqslant 3$, and $H^0(\mathbf{P}^1, h)$ being 2-dimensional, it follows that $H^0(\mathbf{P}^1, L \otimes h) \neq 0$, *i.e.* deg $(L \otimes h) \geqslant 0$, proving that deg $L \geqslant -1$.

REMARKS 4.4 (i) From what we have seen above, it is clear that if $(C, h)$ is a conic as in 4.1, then the linear system of $h$ in $C$ is 2-dimensional and imbeds $C$ as a conic in $\mathbf{P}^2$. In particular, if $C$ is a subscheme of $\mathbf{P}^n$ and $C$ is a conic with respect to the restriction of the hyperplane bundle, then $C$ is contained in a unique plane in $\mathbf{P}^n$ as a conic.

(ii) Also, if $C$ is a subscheme of the one-point union of two projective spaces, and $C$ is a conic with respect to the restriction of the natural very ample (hyperplane) bundle on this union, then $C$ is either a conic in one of the projective spaces, or is a pair of lines, one in each of these spaces passing through their point of intersection.

(iii) Let $Y$ be the thickening of a projective space $Z$ by a surjection $\varphi : \mathcal{G} \to \Omega^1$, with ker $\varphi$ isomorphic to the dual of the universal quotient bundle $Q$ on $Z$. In this case, the hyperplane bundle on $Z$ extends uniquely to a line bundle $h$ on $Y$ which is very ample. If $C$ is a conic subscheme (with respect to $h$) of $Y$, then either (a) $C$ is a subscheme of $Z$ or (b) $C_{red}$ is a line $l$ in $Z$ and $C$ is obtained as a $\tau^{-1}$-thickening of this line in $Y$. By 3.3, the latter is obtained as a quotient of $N^*_{l,Z}$ which is isomorphic to $\tau^{-1}$. Consider the diagram on $l$:

$$
\begin{array}{ccccccccc}
 & & & & 0 & & 0 & & \\
 & & & & \downarrow & & \downarrow & & \\
0 & \to & Q & \to & \mathcal{G}' & \to & \tau^{-1} \otimes \text{ trivial} & \to & 0 \\
 & & \| & & \downarrow & & \downarrow & & \\
0 & \to & Q & \to & \mathcal{G} & \to & \Omega^1_Z & \to & 0 \\
 & & & & \downarrow & & \downarrow & & \\
 & & & & \Omega^1_l & = & \Omega^1_l & & \\
 & & & & \downarrow & & \downarrow & & \\
 & & & & 0 & & 0 & &
\end{array}
$$

Now $H^0(l, \text{Hom}\,(Q, \tau^{-1}))$ is 1-dimensional, since $Q|l \simeq \tau^{-1} +$ trivial. Hence any map $\mathcal{G}' \to \tau^{-1}$ when restricted to $Q$ must coincide (upto a scalar factor) with the natural map $Q|l \to \tau^{-1}$. If this map is zero, then the map $\mathcal{G}' \to \tau^{-1}$ factors through ($\tau^{-1} \otimes$ trivial) and in this case, $C$ is actually contained in $Z$ itself. All other $\tau^{-1}$ thickenings in $Y$ are parametrised by an affine space of dim $=$ codim of $l$ in $Z$.

(iv) If $C$ is a conic over $T$, by [9, SGA 6, VII], the morphism $\pi : C \to T$ is a complete intersection. Let $\omega$ be the relative dualising sheaf over $T$. Then clearly $C$ is imbedded as a $T$-scheme in the projective plane bundle $S$ over $T$ associated to $E = R^1 (\pi)_* \omega^2$. Moreover, if $C$ is imbedded as a $T$-scheme in a projective bundle $P$ over $T$ such that the fibres of $\pi$ are imbedded as conics, then the inclusion $C \hookrightarrow P$ factors through an imbedding $S \subset P$, linear on fibres. On $E$, there is on everywhere nontrivial quadratic form with values in a line bundle $L$ such that $C$ is the associated conic bundle over $T$.

**5. Good Hecke cycles.** If $E$ is a vector bundle ($\neq 0$) on $X$ and $k \in \mathbf{Z}$, we denote by $\mu_k(E)$ the rational number $(\deg E + k)/rk\, E$.

DEFINITION 5.1. *A vector bundle $E$ on $X$ is said to be $(k, l)$-stable (resp. $(k, l)$-semistable) if, for every proper subbundle $F$ of $E$, we have*

$$\mu_k (F) < \mu_{-l} (E/F) \text{ (resp. } \mu_k (F) \leqslant \mu_{-l} (E/F)).$$

REMARK 5.2. (i) The condition 5.1 is equivalent to $\mu_k(F) < \mu_{k-l}(E)$ or $\mu_{k-l}(E) < \mu_{-l}(E/F)$.

(ii) If $E$ is $(k, l)$-stable and $L$ any line bundle, then $E \otimes L$ is also $(k, l)$-stable.

(iii) If $E$ is $(k, l)$-stable, then $E^*$ is $(l, k)$-stable.

(iv) A vector bundle of degree 0 is stable if and only if it is $(0, 1)$-stable.

(v) A vector bundles of degree 1 is stable if and only if it is $(0, 1)$-semi-stable.

PROPOSITION 5.3. In any family $E$ of vector bundles on $X$, parametrised by $T$, the set of $(k, l)$-stable points is open in $T$.

PROOF. Consider the Quot-scheme of $E$ over $T$. The set of non-$(k, l)$-stable points is characterised as the union of the images of the components of the Quot scheme whose Hilbert polynomials satisfy an inequality which constrain them to vary over a finite set. Now our assertion follows from the properness of the Hilbert scheme over $T$, with a fixed Hilbert polynomial.

PROPOSITION 5.4. (i) *Except when $g=2$, $n=2$ and $d$ odd, there always exist* (0, 1)-*stable (and* (1, 0)-*stable) bundles of rank $n$ and degree $d$.*

(ii) *There exist* (1, 1)-*stable bundles of rank $n$ and degree $d$, except in the following cases.*

(a) $g=3$, *and $d$ both even*

(b) $g=2$, $d\equiv 0, \pm 1 \pmod{n}$

(c) $g=2$, $n=4$, $d\equiv 2 \pmod{4}$.

PROOF. It is enough to estimate the dimension of the subvariety of $U(n, d)$, consisting of non-(0,1)-stable (resp. non-(1,1)-stable) points and prove that it is a proper subvariety. Clearly any such bundle $E$ contains a subbundle $F$ satisfying the inequality

$$\frac{\deg F}{rk\ F} \geqslant \frac{\deg E - 1}{rk\ E}$$

(resp.) $$\frac{\deg F + 1}{rk\ F} \geqslant \frac{\deg E}{rk\ E}$$

By using [5, Proposition 2.6] as in [5, Lemma 6.7] we may as well assume that $F$ and $E/F$ are stable and compute the dimension of such bundles $E$. The dimension of a component corresponding to a fixed rank $r$ and degree $\delta$ of $F$, is majorised by $\dim U(r, \delta) + \dim U(n-r, d-\delta) + \dim H^1(X, \mathrm{Hom}(E/F, F)) - 1 = (g-1)(n^2 - nr + r^2) + 1 + (dr - \delta n)$. In order to show this is $< \dim U(n, d) = n^2(g-1) + 1$ we have only to verify that $(g-1)\, r(n-r) > dr - \delta n$. But this is a simple consequence of the inequality $dr - \delta n \leqslant r$ in the first case and $\leqslant n$ in the second case, taking into account the exceptions mentioned in the proposition.

LEMMA 5.5. *Let $x \in X$ and $0 \to \underline{E}' \to \underline{E} \to \mathcal{O}_X \to 0$ be an exact sequence of sheaves with $\underline{E}$, $\underline{E}'$ locally free. If $E$ is $(k, l)$-stable then $E'$ is $(k, l-1)$ stable. In particular, if $E$ is* (0, 1)-*stable then $E'$ is stable. Similar statements are valid when stable is replaced by semistable.*

PROOF. Let $F$ be a subbundle on $E'$ and $F'$ the subbundle of $E$ generated by the map $F \to E$. Then $F \to F'$ is of maximal rank and hence deg $F' \geqslant$ deg $F$. Now $\mu_k(F) \leqslant \mu_k(F') < \mu_{k-l}(E) = \mu_{k-l+1}(E')$ as deg $E$ = deg $E' + 1$. This proves that $E'$ is $(k, l-1)$ stable.

**Lemma** 5.6. (i) *Let $E$ be a (0, 1)-stable vector bundle of rank $n$ and $E'$ a stable vector bundle of rank $n$ and determinant isomorphic to* det $E \otimes L_x^{-1}$. *If $f: E' \to E$ is a non-zero homomorphism, we have an exact sequence*

$$0 \to \underline{E}' \to \underline{E} \to \mathcal{O}_x \to 0.$$

(ii) *Moreover* $\dim H^0(X, \mathrm{Hom}(E', E)) \leqslant 1$.

**Proof**. The map $f$ is of maximal rank. For, otherwise, let

$$\begin{array}{ccccc} E' & \to & G' & \to & 0 \\ & & \downarrow & & \\ E & \leftarrow & G & \leftarrow & 0 \end{array}$$

be the canonical factorisation of $f$, where $rk\, G' < n$ and $G' \to G$ is of maximal rank. We then have $\mu(G) \geqslant \mu(G') > \mu(E') = \mu_{-1}(E)$ which contradicts the (0, 1)-stability of $E$. Now the induced map $\overset{n}{\wedge} f : \overset{n}{\wedge} E' \to \overset{n}{\wedge} E$ is non-zero and can vanish only at $x$ (with multiplicity 1). Thus $f$ is of maximal rank at all points except $x$ and is of rank $(n-1)$ at $x$. This proves the first part.

If $f$ and $g$ are two linearly independent homomorphisms $E' \to E$, then for $y \neq x$, we can find a suitable (nontrivial) linear combination of $f$ and $g$ which is singular at $y$, see [4, Lemma 7.1]. This contradicts i) and completes the proof of Lemma 5.6.

Let now $W \to X \times S$ be a family of (0, 1)-stable vector bundles on $X$, with fixed determinant $\xi$, parametrised by $S$. Then we may construct a family $H(W)$ of vector bundles on $X$, parametrised by the associated projective bundle $\pi: P(W^*) \to X \times S$ as follows. It is easy to construct as in [5, § 4] a canonical surjection of the bundle $p_2^* \pi^* W$ on $X \times P(W^*)$ onto $p_2^* \tau \otimes \mathcal{O}_{P(W)^*}$, where $P(W)^*$ is considered a divisor in $X \times P(W^*)$, by the inclusion $(p_1 \circ \pi, Id)$. The kernel of this homomorphism is the vector bundle $H(W)$. Let $K(W)$ be its dual.

**Remark** 5.7. This construction is the same as that in [5], except that here we have allowed $x$ also to vary on the curve. From this point of view, the families constructed in [5] may properly be denoted $H_x(V)$, $K_x(V)$, etc.

By Lemma 5.5, $K(W)$ is a family of stable vector bundles on $X$ with determinant of the form $\xi^{-1} \otimes L_x$, $x \in X$. More precisely, we have a morphism $\theta_{K(W)}: P(W^*) \to U(n, \xi^{-1} X)$ with a commutative diagram

$$\begin{array}{ccc} P(W^*) & \xrightarrow{(\theta_{K(W)},\, p_2 \circ \pi)} & U(n, \xi^{-1} X) \times S \\ \downarrow & & \downarrow (\alpha, Id) \\ X \times S & \xrightarrow{Id} & X \times S \end{array} \tag{5.8}$$

where $\alpha$: $U(n, \xi^{-1} X) \to X$ is given by $L_{\alpha(E)} = (\det E) \otimes \xi^{-1}$ for $E \in U(n, \xi^{-1} X)$.

**Lemma 5.9.** *The morphism $(\theta_{K(W)}, p_2 \circ \pi)$ of $P(W)^*$ in 5.8 is a closed immersion.*

To prove this we will need the following

**Lemma 5.10.** *Let $X$ be a scheme proper/$k$ and $\mathscr{F}$ a coherent sheaf of $\mathcal{O}_X$-Modules. Let $S$ be a scheme and*

$$0 \to \mathscr{H} \to p_1^* \mathscr{F} \to \mathscr{G} \to 0$$

*be an exact sequence of coherent sheaves over $X \times S$ with $\mathscr{G}$ flat over $S$. Then $\mathscr{H}$ is a flat family of $\mathcal{O}_X$-Modules parametrised by $S$ and its infinitesimal deformation map at $s_0 \in S$, is the negative of the composite of the natural map $T_{s_0} \to H^0(X, \text{Hom}(\mathscr{H}_0, \mathscr{G}_0))$ (the differential of the morphism $S \to$Quot) and the boundary homomorphism $H^0(X, \text{Hom}(\mathscr{H}_0, \mathscr{G}_0)) \to \text{Ext}^1(X, \mathscr{H}_0, \mathscr{H}_0)$ associated to the sequence $0 \to \mathscr{H}_0 \to \mathscr{F} \to \mathscr{G}_0 \to 0$. Here $\mathscr{H}_0$, $\mathscr{G}_0$ denote the restrictions of $\mathscr{H}$, $\mathscr{G}$ to $X \times s_0$.*

**Proof.** It is clearly sufficient to consider the case $S = \text{Spec } k[\epsilon]$. To give a sheaf $\mathscr{L}$ on $X \times S$ flat over $S$ which extends a given sheaf $\mathscr{L}_0$ on $X \times s_0$, is the same as giving an extension

$$0 \to \epsilon \mathscr{L}_0 \to \mathscr{L} \to \mathscr{L}_0 \to 0$$

on $X$, where $\epsilon \mathscr{L}_0$ is another copy of $\mathscr{L}_0$. This identifies the space of infinitesimal deformations of $\mathscr{L}_0$ with $\text{Ext}^1(X, \mathscr{L}_0, \mathscr{L}_0)$. In particular,

$p_1^* \mathscr{F}$ is given by $\epsilon\, \mathscr{F} \oplus \mathscr{F}$. Now the map $p_1^*\mathscr{F} \to \mathscr{G}$ gives rise to a map $\mathscr{H}_0 \to \mathscr{F} \to \epsilon\, \mathscr{F} \oplus \mathscr{F} \to \mathscr{G}$. This map goes into $\epsilon\, \mathscr{G}_0$ and hence induces an element $t$ of $H^0(X, \mathrm{Hom}\,(\mathscr{H}_0, \mathscr{G}_0))$ and thus identifies this space with the fibre of $\mathrm{Quot}_{\mathscr{F}}(k[\epsilon]) \to \mathrm{Quot}_{\mathscr{F}}(k)$ over $\mathscr{G}_0$. We have a map $\mathscr{H} \to \epsilon\, \mathscr{F}$ obtained by composing $\mathscr{H} \to p_1^*\, \mathscr{F}$ with the projection $p_1^*\mathscr{F} \to \epsilon\, \mathscr{F}$. This fits in a commutative diagram

$$\begin{array}{ccccccccc} 0 & \to & \epsilon\, \mathscr{H}_0 & \to & \mathscr{H} & \to & \mathscr{H}_0 & \to & 0 \\ & & \downarrow & & \downarrow & & \downarrow & & \\ 0 & \to & \epsilon\, \mathscr{H}_0 & \to & \epsilon\, \mathscr{F} & \to & \epsilon\, \mathscr{G}_0 & \to & 0 \end{array}$$

The last vertical map in the diagram is easily checked to be $-t$. Hence the element of $\mathrm{Ext}^1\,(X, \mathscr{H}_0, \mathscr{H}_0)$ giving the extension $\mathscr{H}$ is the image of $-t$ under the connecting homomorphism $\mathrm{Hom}(X, \mathscr{H}_0, \mathscr{G}_0) \to \mathrm{Ext}^1\,(X, \mathscr{H}_0, \mathscr{H}_0)$.

REMARK 5.11. Proposition 3.3 of [5] is a particular case of 5.10, where $\mathscr{F}$ and $\mathscr{G}$ are locally free, in which case $\mathrm{Ext}^1\,(X, \mathscr{H}_0, \mathscr{H}_0)$ is the same as $H^1\,(X, \mathrm{End}\, \mathscr{H}_0)$.

PROOF OF LEMMA 5.9. In view of the diagram 5.8, we may assume that $T$ is a point. Thus we have a bundle $W$ on $X$ and the exact sequence

$$0 \to H\,(W) \to p_1^*\, W \to p_2^*\, \tau \otimes \mathcal{O}_{P(W^*)} \to 0$$

on $X \times P(W^*)$. Now it is easy to see that $P(W^*)$ is a component of Quot $E$ and hence its tangent bundle can be identified as in Lemma 5.10, with $(p_2)_*\,(\mathrm{Hom}\,(H(W), \tau \otimes \mathcal{O}_{P(W^*)}) = \mathrm{Hom}\,(H(W)|P(W^*), \tau)$. Moreover, by Lemma 5.10, the infinitesimal deformation map for the family $H(W)$ of bundles on $X$ is given, upto sign, by the connecting homomorphism $\mathrm{Hom}\,(H(W)|\, P(W^*), \tau) \to R^1(p_2)_*\, \mathrm{End}\, H(W)$. But this fits in the exact sequence

$$\begin{gathered} 0 \to (p_2)_*\,(\mathrm{End}\, H(W)) \to (p_2)_*\,(\mathrm{Hom}\,(H(W), p_1{}^*\, W)) \to \\ \mathrm{Hom}\,(H(W)\mid P(W^*), \tau) \to R^1(p_2)_*\, \mathrm{End}\, H(W). \end{gathered}$$

Now since $H(W)$ is a family of stable bundles,

$$(p_2)_*\,(\mathrm{End}\,(H(W)) \simeq \mathcal{O}_{P(W^*)}$$

and by Lemma 5.6, (ii), the same is true of

$$(p_2)_*\,(\mathrm{Hom}\,(H(W), p_2^*\, W)).$$

This proves that the differential of the map $\theta_{K(E)} : P(W^*) \to U(n, \xi^{-1} X)$ is injective. On the other hand, $\theta_{K(E)}$ is itself injective by Lemma 5.6, ii), thus proving Lemma 5.9.

Now, for each family of (0, 1)-stable bundles we have a morphism of the parameter space into $U(n, \xi^{-1} X)$, by Lemma 5.9. Since these morphisms are clearly functorial we get a morphism $\Phi$ on the open subscheme of $U(n, \xi)$ consisting of (0, 1)-stable points into the Hilbert scheme Hilb $(U(n, \xi^{-1} X))$.

Definition 5.12. *The cycles in* $U(n, \xi^{-1} X)$ *corresponding to* (0, 1)-*stable points of* $U(n, \xi)$ *will be called* good Hecke cycles. *Any subscheme in the same irreducible component of the Hilbert scheme will be called a* Hecke cycle.

Theorem 5.13. *Let $\xi$ be a line bundle on $X$. The open subscheme of $U(n, \xi)$ consisting of* (0, 1)-*stable points is isomorphic (by means of the map associating to a bundle $W$ the corresponding good Hecke cycle $\Phi(W)$) to an open subscheme of the Hilbert scheme of $U(n, \xi^{-1} X)$.*

Proof. To prove the injectivity, we will check the stronger assertion that for $W, W' \in U(n, \xi)$, both (0, 1)-stable, $\theta_{K_x(W)}(P(W_x^*)) = \theta_{K_x(W')}(P(W_x'^*))$ if and only if $W$ and $W'$ are isomorphic. Indeed, by [6, Lemma 2.5], such an assumption would yield an isomorphism $\varphi: P(W_x^*) \to P(W_x'^*)$ such that $\varphi^* K_x(W') \simeq K_x(W) \otimes p_2^* L$ on $X \times P(W_x^*)$, for some line bundle $L$ on $P(W_x^*)$. Now by [5, Remark 4.7], we see, on restricting this isomorphism to $P(W_x^*) \times (y), y \neq x$, that $L$ is trivial. But then, by [5, Lemma 4.3] we have $(p_1)_* \varphi^* K_x(W')^* \simeq (p_1)_* K_x(W')^* \simeq L_x^{-1} \otimes W'$ on the one hand, and $(p_1)_* K_x(W')^* \simeq L_x^{-1} \otimes W$ on the other. This proves that $W \simeq W'$.

Next we proceed to show that, if $W \in U(n, \xi)$ is (0, 1)-stable, the infinitesimal deformation map at $W$ of the family of good Hecke cycles from $T_W \simeq H^1(X, \text{Ad } W)$ to $H^0(P(W^*), N)$ is an isomorphism, where $N$ is the normal bundle of $P(W^*)$ in $U(n, \xi^{-1} X)$. This would prove that the Hilbert scheme is smooth at $\Phi(W)$ and that $\Phi$ is an isomorphism onto to an open subset, as claimed. To complete the proof of the theorem we need two lemmas.

**Lemma 5.14.** *Let $\pi_1 : P_1 \to U(n, \xi) \times X$ be the dual of the projective Poincaré bundle on $U(n, \xi) \times X$. Let $\pi_2 : P_2 \to U(n, \xi^{-1} X)$ be the restriction of the dual projective Poincaré bundle on $U(n, \xi^{-1} X) \times X$ to the divisor $U(n, \xi^{-1} X)$. Let $\Omega_1 \subset P_1$ and $\Omega_2 \subset P_2$ be the open subsets of points at which the families $K$ are stable. We then have an isomorphism $\widetilde{\theta} : \Omega_1 \to \Omega_2$ such that $\pi_2 \circ \widetilde{\theta} = \theta$.*

**Proof.** A point $p$ of $\Omega_1$ is described by a (0,1)-stable vector bundle in $U(n, \xi)$, a point $x \in X$ and an element of $P(E^*)_x$. Now this can be looked upon as a map $E \to \mathcal{O}_x$ with kernel $F$, where $F^*$ belongs to $U(n, \xi^{-1} X)$. By looking at the map $F_x \to E_x$ of the fibres at $x$, we can define a map $\Omega_1 \to P_2$ by associating the ker $F_x$ to $p$. It is easy to see that this lift of $\theta$ maps $\Omega_1$ isomorphically onto $\Omega_2$. In fact this map is induced by the section $\sigma$ [5, p. 402].

**Lemma 5.15.** *Let $W \in U(n, \xi)$ be (0,1)-stable. Then the normal bundle $N = N_{P(W^*), U(n, \xi^{-1} X)}$ of $P(W^*)$ in $U(n, \xi^{-1} X)$ fits into an exact sequence*

$$0 \to p_X^* T_X \otimes T_\pi^* \to P(W^*) \times T_W \to N \to 0.$$

**Proof.** Since $W$ is (0, 1)-stable, $P(W^*) = (p_2 \circ \pi_1)^{-1} W$ is contained in $\Omega_1$ and $N_{P(W^*), \Omega_1} \simeq P(W^*) \times T_W$. Moreover $N_{P(W^*), \Omega_1} \simeq N_{\widetilde{\theta}\, P(W^*), \Omega_2}$. Since $\pi_2 \circ \widetilde{\theta} = \theta$, and $\theta \mid P(W^*)$ is an imbedding (Lemma 5.9), we see that $\widetilde{\theta}(P(W^*))$ is transversal to the fibres of $\pi_2$. Hence we have an exact sequence (on $P(W^*)$)

$$0 \to \widetilde{\theta}^* T_{\pi_2} \to N_{\widetilde{\theta}(P(W^*)), \Omega_2} \to N_{(\theta(P(W^*)), U(n, \xi^{-1} X)} \to 0.$$

But by Proposition 4.12 in [5], we have $\widetilde{\theta}^* T_{\pi_2} \simeq (p_2 \circ \pi_1)^* T_X \otimes T_\pi^*$. This proves the Lemma.

**Completion of the proof of Theorem 5.13.** Taking the direct image of the exact sequence in Lemma 5.15, and noting that, $H^0(X, T_X) = 0$, we see that the natural map $T_W \to H^0(P(W^*), N)$ is an isomorphism. But this is clearly the differential of $\Phi$ at $W$.

Taking determinants in the exact sequence 5.15, we obtain

**Corollary 5.16.** *Let $P(W^*)$ be a good Hecke cycle in $U(n, \xi^{-1} X)$. Then the line bundle $K_{\det}$ on $U_X$ restricts to $P(W^*)$ as $\pi^* K_X^{-(n-1)} \otimes K_\pi^2$.*

REMARKS 5.17. (i) Although the Hilbert scheme in Theorem 5.11 is smooth, $H^1(P(W^*), N) \neq 0$ unlike the case discussed below.

(ii) Let us now fix $x \in X$. The open subset $\Omega_0$ of points in the projective dual Poincaré bundle over $U(n, \xi) \times (x)$, corresponding to the family $K_x$, is isomorphic to a similar open subset of the projective Poincaré bundle over $U(n, \eta) \times (x)$ where $\xi \otimes \eta = L_x$. Thus we have two maps

$$\begin{array}{ccc} & \Omega_0 & \\ {\scriptstyle \pi_1} \swarrow & & \searrow {\scriptstyle \pi_2} \\ U(n, \xi) & & U(n, \eta) \end{array}$$

which are projective fibrations over the open subsets of (0,1)-stable points of $U(n, \xi)$, $U(n, \eta)$ respectively. Now the content of [5, Proposition 4.12] is that $T^*_{\pi_1} \simeq T_{\pi_2}$. (See also [8a, Ch 5, §1]). If $W$ is a (0, 1)-stable point of $U(n, \xi)$, then $\pi_1^{-1}(W)$ is imbedded by $\pi_2$, thus giving a family of cycles in $U(n, \eta)$ parametrised by the (0, 1)-stable points of $U(n, \xi^{-1} X)$. For the normal bundle $N$ of the cycle $\Phi(W)$ corresponding to $W$ we have the exact sequence

$$0 \to \Omega^1_{P(W_x^*)} \to H^1(X \text{ Ad } W)_{P(W^*)} \to N \to 0.$$

This shows that $H^i(\Phi(W), N) = 0$ for $i \geqslant 1$ and $\dim H^0(\Phi(W), N) = \dim H^1(X, \text{Ad } W) + 1$ As in Theorem 5.13 we see that the open subset of (0, 1)- stable points of $U(n, \xi^{-1} X)$ can be identified with an open subset of a component of the Hilbert scheme of $U(n, \xi)$.

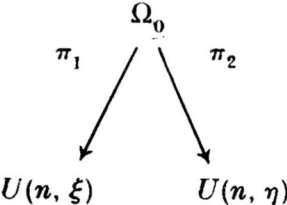

(iii) If $W$ is a (1, 1)-stable bundle in $U(n, \xi)$, the corresponding cycle $\Phi(W)$ in $U(n, \xi^{-1} X)$ which is isomorphic to $P(W^*)$, is contained entirely in the set of (0, 1)-stable points (Lemma 5.5 and Remark 5.2, iii)). By the reverse construction in ii), $P(W^*)$ is contained in the Hilbert scheme of $U(n, \xi)$ consisting of good Hecke cycles (this time, isomorphic to a projective space) passing through $W$. Thus $P(W^*)$ can be identified with this subscheme of the Hilbert scheme. From this one may conclude that any continuous group of automorphism of $U(n, \xi)$ lifts to the projective Poincaré bundle over some open subset of $U(n, \xi)$ consisting of (1,1) stable points. Thus, if (1,1)-stable points exist, then the automorphism group of $U(n, \xi)$ is finite. This was our original approach to the Theorems 1, 2 of [5].

**6. Triangular bundles.** We will deal with the case of rank 2 bundles in the rest of the paper. We will denote $U(2, \xi)$ (resp. $U(2, L_x)$) by $U_\xi$ (resp. $U_x$). When $\xi$ is trivial we write $U_\xi = U_0$.

We wish to study the limits of good Hecke cycles, i.e., points in the component of the Hilbert scheme, containing good Hecke cycles. It will turn out (§ 7) that these cycles consist of bundles which are extensions of line bundles of a particular kind. These bundles have been studied in [6] and [8] and the computations made in this article result from a closer analysis of the situation.

Let $R$ be a Poincaré bundle on $X \times J$, where $J$ is the Jacobian of $X$, Consider on $X \times X \times J$ the bundle $p_{13}^* R^2 \otimes p_{12}^* L_\Delta^{-1}$. where $\Delta$ is the diagonal in $X \times X$. Then the first direct image $R^1 (p_{23})_* (p_{13}^* R^2 \otimes p_{12}^* L_\Delta^{-1})$ on $X \times J$ is clearly a vector bundle $D$. For each $j \in J$, the restriction of $D$ to $X \times j$ gives rise to a bundle $D_j$ on $X$.

Lemma 6.1 *If $j^2 \neq 1$, then the bundle $D_j$ fits in an exact sequence*

$$0 \to j^2 \to D_j \to H^1(X, j^2)_X \to 0$$

*given by the identity element of $H^1(X\ \mathrm{Hom}\,(H^1(X, j^2)_X, j^2))$. If $j^2 = 1$, then $D_j$ is isomorphic, upto tensorisation by a line bundle, to $H^1(X, \mathcal{O})_X$. In particular, $D_j$ is a semistable vector bundle of degree 0 and any nonzero subbundle of $D_j$ of degree 0 is the inverse image of a trivial subbundle of $H^1(X, j^2)_X$.*

Proof. By the base change theorem, the bundle $D_j$ is isomorphic to $R^1(p_2)_* (p_1^* j^2 \otimes L_\Delta^{-1})$, where $p_1, p_2$ are the projections $X \times X \to X$. Hence it is dual to $(p_2)_* (p_1^* (K \otimes j^{-2}) \otimes L_\Delta)$. Consider on $X \times X$ the exact sequence

$$0 \to p_1^* (K \otimes j^{-2}) \to p_1^* (K \otimes j^{-2}) \otimes L_\Delta \to \widetilde{j}^{-2} \to 0, \quad (6.2)$$

where the sheaf $\widetilde{L}$ denotes the extension of a line bundle $L$ on $\Delta$ to $X \times X$ and we have used the fact that $L_\Delta|_\Delta \simeq K_X^{-1}$. We will show that by applying $(p_2)_*$ to this sequence and dualising, we get the canonical extension in 6.1. Now the pullback by $p_2^* j^{-2} \to \tilde{j}^{-2}$ of this sequence yields an extension of $p_2^* j^{-2}$ by $p_1^*(K \otimes j^{-2})$. The corresponding element in $H^1(X \times X, p_2^* j^2 \otimes p_1^*(K \otimes j^{-2})) = H^0(X, K \otimes j^{-2}) \otimes H^1(X, j^2)$ (since $j^2 \neq 1$) is the canonical element given by the duality theorem. Thus it is clear that this pulled-back extension is also the push-out by the evaluation map $H^0(K \otimes j^{-2})_{X \times X} \to p_1^*(K \otimes j^{-2})$ of the extension

$$0 \to H^0(X, K \otimes j^{-2})_{X \times X} \to \ldots \to p_2^* j^{-2} \to 0$$

given by the canonical element of $H^1(X \times X, p_2^* j^2 \otimes H^0(X, K \otimes j^{-2}))$. Hence the exact sequence obtained by taking direct image by $p_2$, namely,

$$0 \to H^0(X, K \otimes j^{-2}) \to \ldots \to j^{-2} \to 0$$

is again given by the canonical element. But this is also obtained by taking the direct image of the sequence (6.2), thus proving the first part of Lemma 6.1. If $j^2 = 1$, then $D_j = R^1(p_2)_*(L_\Delta^{-1}) \simeq H^1(X, \mathcal{O})_X$ as is seen from the exact sequence

$$0 \to L_\Delta^{-1} \to \mathcal{O} \to \tilde{\mathcal{O}} \to 0.$$

If $F$ is a subbundle of degree 0, then $F \cap j^2 = 0$ or $F \supset j^2$. In either case, the image of $F$ in $H^1(X, j^2)_X$ is a subbundle of degree 0 and hence trivial. If $F \cap j^2 = 0$, this would imply that the extension $0 \to j^2 \to D_j \to H^1(X, j^2)_X \to 0$ splits over a trivial subbundle of $H^1(X, j^2)_X$ which is seen to be not possible, in view of our description of this extension.

LEMMA 6.3. *Let* $C = \{(x, j) \in X \times J : j^{-2} \otimes L_x \in X\}$. *Then*

(i) *the bundle* $P(D)$ *is trivial on* $C$.

(ii) $X \times j \subset C$ *if and only if* $j$ *is an element of order* 2.

PROOF. (i) On $X \times C$, the bundle $(p_{13})^* R^2 \otimes (p_{12})^* L_\Delta^{-1}$ is isomorpic to $p_C^* \zeta \otimes (1 \times \alpha)^* L_\Delta^{-1}$ where $\zeta$ is some line bundle on $C$ and $\alpha : C \to X$ is the map defined by $L_{\alpha(x,j)} = j^{-2} \otimes L_x$. Hence $D|C$ is isomorphic (by base change theorem) to $\zeta \otimes \alpha^*(R^1(p_2)_* L_\Delta^{-1})$. But clearly $R^1(p_2)_* L_\Delta^{-1} \simeq H^1(X, \mathcal{O})_X$.

(ii) If $j$ is an element of order 2, clearly $X \times j \subset C$. On the other hand, if $X \times j \subset C$, then $D_j = D|_{X \times j} \simeq (\zeta \mid X \times j) \otimes$ trivial bundle, and hence by Lemma 6.1, this is possible only if $j$ is an element of order 2.

We now have a family $E$ of (triangular) vector bundles on $X \times P(D)$ given by [6, Lemma 2,4]:

$$0 \to (1\times\pi)^* p_{13}^* R \otimes p_2^* \tau \to E \to (1\times\pi)^* p_{13}^* R^{-1} \otimes (1\times\pi)^* p_{12}^* L_\Delta \to 0 \quad (6.4)$$

where $\pi$: $P(D) \to X \times J$ is the projective fibration associated to $D$.

The bundle $E$ on $X \times P(D)$ is a family of stable vector bundles [3, Lemma 10.1 (i)] on $X$ of rank 2 and determinants of the form $L_x$, $x \in X$. Thus there is an induced classifying morphism $\theta_E$: $P(D) \to U_X$ fitting in an obvious commutative diagram

$$\begin{array}{ccc} P(D) & \xrightarrow{\theta_E} & U_X \\ {\scriptstyle p_1\circ\pi}\downarrow & & \downarrow{\scriptstyle \det} \\ X & \xrightarrow{\mathrm{Id}} & X \end{array} \quad (6.5)$$

We now wish to study the morphism $\theta_E$ and, in particular, its differential. Clearly $E$ is a family of bundles with the triangular group as structure group. As such, there is an infinitesimal deformation map $T_{P(D)} \to R^1(p_2)_* (\Delta(E))$ where $\Delta(E)$ is the bundle of endomorphisms of $E$ preserving the exact sequence (6.4). This is because the associated bundle with the Lie algebra of the triangular group as fibre is $\Delta(E)$. (Remark 2,13, i).

Lemma 6.6. *The above infinitesimal deformation map induces an isomorphism of $T_{p_1\circ\pi}$ with the kernel of $R^1(p_2)_* (\Delta(E)) \to R^1(p_2)_*(\mathcal{O})$ given by the trace map $\Delta(E) \to \mathcal{O}$ and hence with $R^1(p_2)_* (S\Delta(E))$, where $S\Delta(E)$ is the subbundle of $\Delta(E)$ consisting of endomorphisms of trace* 0.

Proof. In view of the diagram 6.5 it is clear that $T_{p_1\circ\pi}$ is mapped into the kernel. From the diagram of exact rows (on $P(D)$),

$$\begin{array}{ccccccc} 0 \to & T_\pi & \longrightarrow & T_{p_1\circ\pi} & \longrightarrow & \pi^* p_2^* T_J & \to 0 \\ & \downarrow & & \downarrow & & \downarrow & \\ 0 \to & R^1(p_2)_* (E \otimes (1\times\pi)^* (p_{13}^* R \otimes p_{12}^* L_\Delta^{-1})) & \to & R^1(p_2)_*(\Delta(E)) & \to & H^1(X, \mathcal{O})_{P(D)} & \to 0 \end{array}$$

we see that it is enough to show that the infinitesimal deformation map maps $T_\pi$ isomorphically on $R^1(p_2)_*(E\otimes R\otimes L_\Delta^{-1})$, since it is obvious that the last vertical map is an isomorphism. Indeed, we have

LEMMA 6.7. *Let $V$, $W$ be two simple vector bundles on $X$. Let $P = PH^1(X, \text{Hom}(W, V))$, and*

$$0 \to p_1^* V \otimes p_2^* \tau \to E \to p_1^* W \to 0$$

*be the universal exact sequence on $X \times P$ (see [6, Lemma 2.3]). Then the infinitesimal deformation map of the family $E$ of bundles with the evident parabolic group as structure group, maps $T_P$ isomorphically onto the kernel of $R^1(p_2)_*(E\otimes p_1^* W^*) \to R^1(p_2)_*(End\ W)$.*

PROOF. From the definition of the universal extension, it follows that the map $\mathcal{O}_P = (p_2)_*(W\otimes W^*) \to R^1(p_2)_*(V\otimes\tau\otimes W^*) = H^1(X, V\otimes W^*)\otimes\tau$ is the natural inclusion. Hence its cokernel is isomorphic to the tangent bundle $T_P$ of $P$. From the cohomology exact sequence obtained from the given sequence tensored with $W^*$, we obtain an isomorphism of $T_P$ with the kernel of $R^1(p_2)_*(E\otimes p_1^* W^*) \to R^1(p_2)_*(\text{End } W)$. We have to check that this identification is the infinitesimal deformation map. This verification can be done by choosing splittings for the given sequence over $U_i\times P$ where $(U_i)$ is an open covering of $X$ and expressing the infinitesimal deformation map in terms of Čech cocycles with respect to this covering.

PROPOSITION 6.8. *Recall that $C = \{(x, j) \in X \times J : j^{-2}\otimes L_x \in X\}$. The differential $d\theta_E$ is injective outside $\pi^{-1}(C)$ and* ker *$d\theta_E$ is a line bundle on $\pi^{-1}(C)$.*

PROOF. Consider the exact sequence obtained from 6.4:

$$0 \to S\Delta(E) \to \text{Ad } E \to (1\times\pi)^*(p_{13}^* R^{-2}\otimes p_{12}^* L_\Delta)\otimes p_2^*\tau^{-1} \to 0. \quad (6.9)$$

Take the direct image on $P(D)$. The (zeroth) direct image of the bundle on the right is zero since it is clearly torsion free and zero outside $\pi^{-1}(C)$. Thus we get a short exact sequence (using Lemma 6.6)

$$0 \to T_{p_1\circ\pi} \to R^1(p_2)_*(\text{Ad } E) \to R^1(p_2)_*(R^{-2}\otimes L_\Delta)\otimes\tau^{-1} \to 0.$$

From [6, Lemma 2.6], we have $R^1(p_2)_*(\text{Ad } E) = \theta_E^* T_{\det}$ and it is easy to see that the first map is $d\theta_E$. Moreover, outside $\pi^{-1}(C)$, the last term in this sequence is clearly locally free of rank $(g-2)$ while its restriction to $\pi^{-1}(C)$ is also locally free of rank $(g-1)$. This proves the proposition.

Identifying $\pi^{-1}(C)$ with $C \times P$ where $P = PH^1(X, \mathcal{O})$, the restriction of $E$ to $X \times C \times P$ can be described as follows.

**Lemma** 6.10.(i) *There is a natural isomorphism* $H^1(X \times C \times P,$ $p_3^* \tau \otimes p_{12}^*(1 \times \alpha)^* L_\Delta^{-1})$ *with* $H^1(X, \mathcal{O}) \otimes H^1(X, \mathcal{O})^*$.

(ii) *The family* $E \otimes p_{13}^* R^{-1}$ *restricts to* $X \times C \times P$ *as an extension*

$$0 \to p_3^* \tau \to E \otimes p_{13}^* R^{-1} \to p_{12}^*(1 \times \alpha)^* L_\Delta \to 0$$

*given by the canonical element in* $H^1(X, \mathcal{O}) \otimes H^1(X, \mathcal{O})^*$, *using the isomorphism in* (i).

**Proof.** Note that $H^1(X \times C, (1 \times \alpha)^* L_\Delta^{-1}) \simeq H^0(C, R^1(p_2)_* (1 \times \alpha)^* L_\Delta^{-1}) \simeq H^0(C, \alpha^* R^1(p_2)_* L_\Delta^{-1}) \simeq H^0(C, \alpha^* R^1(p_2)_* \mathcal{O}_{X \times X}) \simeq H^1(X, \mathcal{O})_C$. This proves (i). That the canonical element gives the required extension follows from the definition of $E$ and the base change theorem.

**Proposition** 6.11. *Let* $0 \to p_2^* \tau \to F' \to \mathcal{O} \to 0$ *be the universal extension of* $\mathcal{O}$ *by* $\tau$ *on* $X \times PH^1(X, \mathcal{O})$. *Then the restriction of* $T_{p_1 \circ \pi}$ *to* $\pi^{-1}(C) = C \times P$ *is isomorphic to the quotient of* $F' \otimes H^1(X, \mathcal{O})$ *by the trivial subbundle contained in* $p_2^* \tau \otimes H^1(X, \mathcal{O})$.

**Proof.** We will apply Lemma 6.6, and use the description of $E \mid X \times C \times P$ given in Lemma 6.10. Thus we have the extension

$$0 \to p_3^* \tau \otimes p_{12}^*(1 \times \alpha)^* L_\Delta^{-1} \to S\Delta(E) \to \mathcal{O} \to 0.$$

This may be embedded in the commutative diagram

$$\begin{array}{ccccccc}
 & 0 & & 0 & & & \\
 & \downarrow & & \downarrow & & & \\
0 \to & p_3^* \tau \otimes (1 \times \alpha)^* L_\Delta^{-1} & \to & S\Delta(E) & \to & \mathcal{O} & \to 0 \\
 & \downarrow & & \downarrow & & \| & \\
0 \to & p_3^* \tau & \longrightarrow & F'' & \to & \mathcal{O} & \to 0 \\
 & \downarrow & & \downarrow & & & \\
 & p_3^* \tau \otimes \mathcal{O}_{\mathrm{Graph}\,\alpha} & = & p_3^* \tau \otimes \mathcal{O}_{\mathrm{Graph}\,\alpha} & & & \\
 & \downarrow & & \downarrow & & & \\
 & 0 & & 0 & & &
\end{array} \qquad (6.12)$$

where the middle line is obtained from the top line as the push-out by means of the map $p_3^* \tau \otimes (1 \times \alpha)^* L_\Delta^{-1} \to p_3^* \tau$. By Lemma 6.6,

$T_{p_1 \mathcal{O}_\pi} | C \times P$ is isomorphic to the first direct image on $C \times P$ of $S\Delta$ $(E)$. The middle vertical line of the above diagram yields an isomorphism of this with the first direct image of $F''$. For, $R^1(p_{23})_*(p_3^* \tau \otimes \mathcal{O}_{Graph\alpha})$ is clearly 0. Since $(p_{23})_*(p_3^* \tau) \to (p_{23})_*(\tau \otimes \mathcal{O}_{Graph\alpha})$ is clearly an isomorphism, it follows that $(p_{23})_* F'' \to (p_{23})_*(p_3^* \tau \otimes \mathcal{O}_{Graph\alpha})$ is surjective, proving our assertion above. It remains to compute $R^1(p_{23})_*(F'')$. From the description of the extension in Lemma 6.10 and hence that in the top horizontal line of the diagram 6.12, it can be checked that the element in $H^1(X \times C \times P, p_3^* \tau) = H^1(X, \mathcal{O}) \otimes H^1(X, \mathcal{O})^* \oplus H^1(C, \mathcal{O}) \otimes H^1(X, \mathcal{O})^*$ given by the extension $F''$ is the element $(Id, -\alpha^*(Id))$. Indeed, this follows on remarking that the map $H^1(X, \mathcal{O}) \simeq H^1(X \times X, L_\Delta^{-1} \Delta) \to H^1(X \times X, \mathcal{O}) \to H^1(X, \mathcal{O}) \oplus H^1(X, \mathcal{O})$ is given by $(Id, -Id)$. Now our assertion follows from

LEMMA 6.13. *Consider on $X \times X \times P$, the exact sequence*

$$0 \to p_3^* \tau \to F_0 \to \mathcal{O} \to 0$$

*given by the element $(Id, -Id)$ in $H^1(X \times X, \mathcal{O}) \otimes H^0(P, \tau) \simeq V \otimes V^* \oplus V \otimes V^*$, with $V = H^1(X, \mathcal{O})$. The direct image on $X \times P$ by $p_{23}$ of this sequence fits in a commutative diagram*

$$\begin{array}{ccccccc} 0 \to p_2^* T_P & \longrightarrow & R^1(p_{23})_* F_0 & \to & H^1(X, \mathcal{O})_{X \times P} & \to 0 \\ \uparrow & & \uparrow & & \uparrow & \\ 0 \to p_2^* \tau \otimes H^1(X, \mathcal{O}) & \to & F' \otimes H^1(X, \mathcal{O}) & \to & H^1(X, \mathcal{O})_{X \times P} & \to 0 \end{array}$$

*where $0 \to p_2^* \tau \to F' \to \mathcal{O} \to 0$ is the universal extension on $X \times P$ and the first vertical map is induced by the natural map $\tau \otimes H^1(X, \mathcal{O}) \to T_P$ on $P$.*

PROOF. In the following Lemma, take $L = K_X$, $L' = \mathcal{O}$, $T = X \times P$, and apply duality to prove the Lemma.

LEMMA 6.14. *Let*

$$0 \to p_1^* L \to M \to p_1^* L \otimes p_2^* L' \to 0$$

*be an exact sequence of vector bundles on the variety $X \times T$. Assume that* dim $H^0(X \times t, M)$ *is independent of $t$. Then the extension*

$$0 \to H^0(X, L)_T \to (p_T)_* M \to \ker \delta \to 0,$$

*where* $\delta$ *is the connecting homomorphism* $H^0(X, L) \otimes L' \to H^1(X,L)$, *is given by the image under the natural map* $H^1(T, L'^{-1}) \to H^1(T, (\ker\delta)^* \otimes H^0(X, L))$ *of the Künneth component in* $H^1(T, L'^{-1})$ *of the element of* $H^1(X \times T, \text{Hom}(L, L) \otimes L'^{-1})$ *determined by the extension* $M$.

**Proof.** Let $x_1, \ldots\ldots, x_N$ be points of $X$ such that the evaluation map $H^0(X, L) \to \underset{\oplus}{\Sigma} L_{x_i}$ is an isomorphism. This fits in a commutative diagram

$$\begin{array}{ccccccccc} 0 & \to & H^0(X, L)_T & \to & (p_T)_* M & \to & \ker\delta & \to & 0 \\ & & \downarrow & & \downarrow & & \downarrow & & \\ 0 & \to & \Sigma L_{x_i} & \to & \Sigma M_{x_i \times T} & \to & L' \otimes \Sigma L_{x_i} & \to & 0 \end{array}$$

The latter sequence is clearly defined by the required element in $H^1(T, L'^{-1})$.

**Proposition 6.15.** *The differential* $d\theta_E$ *on* $\pi^{-1} C = C \times P$ *fits in an exact sequence*

$$0 \to \tau^{-1} \to T_{p_1 \circ \pi} | \pi^{-1} C \xrightarrow{d\theta_E} \theta_E^* T_{\det} \to \tau^{-1} \otimes \pi^* \alpha^* F \to 0$$

*where* $F$ *is the vector bundle on* $X$ *defined by the exact sequence*

$$0 \to T_X \to H^1(X, \mathcal{O}) \to F \to 0$$

*corresponding to the canonical linear system and* $\alpha : C \to X$ *is the map defined by* $j^{-2} \otimes L_r = L_{\alpha(j,x)}$.

**Proof.** Note that if $\Delta$ denotes the diagonal divisor in $X \times X$, then $(p_2)_*(L_\Delta) \simeq \mathcal{O}_X$ and $R^1(p_2)_*(L_\Delta) \simeq F$. In fact, the first of these isomorphisms is obvious and the second follows from the exact sequence (on $X \times X$)

$$0 \to \mathcal{O}_{X \times X} \to L_\Delta \to \widetilde{T}_X \to 0$$

where $\widetilde{T}_X$ is the sheaf on $X \times X$ supported on $\Delta$ and inducing $T_X$ on it. Restricting 6.9 to $X \times C \times P$, we get the sequence

$$0 \to S\Delta(E) \to \text{Ad } E \to p_3^* \tau^{-1} \otimes p_{12}^* (1 \times \alpha)^* L_\Delta \to 0.$$

The direct image of this sequence, using base change for $\alpha$, yields the lemma, in view of Lemma 6.6.

We will now compute the Hessian of the map $\theta_E$ restricted to $\pi^{-1}C = C \times P$. By Proposition 6.8, $\pi^{-1}C$ is the critical set for $\theta_E$. Moreover, ker $d\theta_E$ on $C \times P$ is isomorphic to $\tau^{-1}$ by Proposition 6.15. The Hessian is thus a map $\tau^{-1} \otimes N_{C \times P; P(D)} \to \text{coker } d\theta_E | C \times P$. (See Remark 3.6). Now $N_{C \times P, P(D)} \simeq \pi^* N_{C, X \times J} \simeq \pi^* \alpha^* N_{X, J^1} \simeq \pi^* \alpha^* F$ where $F$ is the bundle defined in Proposition 6.15. On the other hand, by Proposition 6.15, coker $d\theta_E | C \times P$ is isomorphic to $\tau^{-1} \otimes \pi^* \alpha^* F$. With these identifications we have

**Proposition 6.16.** *The Hessian on $\pi^{-1} C = C \times P$ of the map $\theta_E$ considered as a map $\tau^{-1} \otimes \pi^* \alpha^* F \to \tau^{-1} \otimes \pi^* \alpha^* F$ is the identity map.*

**Proof.** By Lemma 6.6, the computation of the required Hessian may be made using Remark 2.13, i). From the description of the family $E$ over $X \times \pi^{-1} C$ given in Lemma 6.10, we see that the Hessian of $\theta_E$ is simply the map

$(p_{23})_* (\tau^{-1} \otimes \alpha^* L_\Delta) \otimes R^1 (p_{23})_* (\alpha^* L_\Delta^{-1} \otimes E \otimes R^{-1}) \to R^1(p_{23})_* (\tau^{-1} \otimes \alpha^* L_\Delta)$ given by the cup product for the natural map

$$(\tau^{-1} \otimes \alpha^* L_\Delta) \otimes (\alpha^* L_\Delta^{-1} \otimes E \otimes R^{-1}) \to \tau^{-1} \otimes E \otimes R^{-1} \to \tau^{-1} \otimes \alpha^* L_\Delta.$$

From this it is clear that this map is the identity on $\tau^{-1}$ tensored with the induced map $(p_{23})_* (\alpha^* L_\Delta) \otimes R^1 (p_{23})_* (\mathcal{O}) \to R^1 (p_{23})_* (\alpha^* L_\Delta)$. Now since $(p_{23})_* (\alpha^* L_\Delta)$ is canonically trivial. this is simply the inverse image by $\alpha \circ p_1$ of the map $H^1 (X, \mathcal{O})_X \to R^1 (p_2)_* (L_\Delta)$. This is clearly the natural map $H^1 (X, \mathcal{O})_X \to F$, proving our assertion.

**Remark 6.17.** From Proposition 6.16, we see that $\pi^{-1}C$ is a 'nondegenerate' critical manifold for $\theta_E$.

**Proposition 6.18.** *For every $j \in J$, the map $\theta_E : \pi^{-1} (X \times j) \to U_X$ is an imbedding.*

**Proof.** We first remark that in view of 6.5, it is enough to prove that $\theta_E : \pi^{-1}(x, j) \to U_X$ is an imbedding, for every $x \in X$ and $j \in J$. Secondly, from [3, Lemma 10.1], it follows that this map is injective. Moreover, if $(x, j) \in C$, then by Proposition 6.8, $\theta_E : \pi^{-1} (x, j) \to U_x$ is an imbedding. Finally, let $(x, j) \in C$. In order to show that $d\theta_E | T_\pi$ is injective, it is enough to prove that the composite of the inclusion $\tau^1 \to T_{p_1 \circ \pi}|_{C \times P}$ with the differential $d\pi : T_{p_1 \circ \pi} \to \pi^* T_{p_1} = H^1(X, \mathcal{O})_{C \times P}$ is injective. Consider the commutative diagram (on $\pi^{-1} C$)

$$\begin{array}{ccccccc} 0 \to & S\Delta(E) \to & \Delta(E) & \to & (1 \times \pi)^* (p_{13}^* R^{-2} \otimes p_{12}^* L_\Delta) \otimes p_2^* \tau^{-1} & \to 0 \\ & \downarrow & \downarrow & & \| & \\ 0 \to & \mathcal{O} & \to E \otimes (1 \times \pi)^* p_{13}^* R^{-1} \otimes p_2^* \tau^{-1} & \to & ,, & \to 0 \end{array}$$

From this we conclude that this composite is given by the boundary homomorphism of the lower sequence. Now our assertion follows from the definition of the extension $E$.

The nonsingular subvarieties $\theta_E(\pi^{-1}(X \times j)) \subset U_X$ will also be denoted $P(D_j)$.

**Proposition 6.19.** *If $j_1 \neq j_2$ or $j_2^{-1}$, then $P(D_{j_1})$ intersects $P(D_{j_2})$ in only finitely many points. If $j^2 \neq 1$, then $P(D_j)$ and $P(D_{j^{-1}})$ intersect transversally along the sections given by the exact sequence in Lemma* 6.1.

**Proof.** Let $j_1 \neq j_2, j_2^{-1}$ and $W \in P(D_{j_1}) \cap P(D_{j_2})$. Then $W$ is given by an extension

$$0 \to j_1 \to W \to j_1^{-1} \otimes L_x \to 0.$$

and $W$ contains $j_2$ also. The existence of a nonzero homomorphism $j_2 \to j_1^{-1} \otimes L_x$ implies that $j_1 \otimes j_2 = L_x \otimes L_y^{-1}$ for some $y \in X$. This proves that there are (if at all) only finitely many solutions for $x$ and $y$. If $(x, y)$ is one such, then $E$ has to be in the kernel of the map $H^1(j_1^2 \otimes L_x^{-1}) \to H^1(j_1^2 \otimes L_x^{-1} \otimes L_y)$ which is at most 1-dimensional. Thus $W$ is uniquely determined by such a choice of $x$ and $y$. This proves our first assertion.

Now let $W \in P(D_j) \cap P(D_{j^{-1}})$, $j^2 \neq 1$. As above, we see that $W$ is the unique bundle given by the kernel of $H^1(X, j^2 \otimes L_x^{-1}) \to H^1(X, j^2)$. Clearly, this is the kernel defined in Lemma 6.1. It remains to prove the transversality of the intersection of $P(D_j)$ and $P(D_{j^{-1}})$ Now $W$ is given simultaneously by two exact sequences

$$0 \to j \to W \to j^{-1} \otimes L_x \to 0$$

and

$$0 \to j^{-1} \to W \to j \otimes L_x \to 0.$$

Consider the map $j \oplus j^{-1} \to W$. Since $j \neq j^{-1}$, this map is a generic isomorphism and hence fits in an exact sequence

$$0 \to j \oplus j^{-1} \to W \to \mathcal{O}_x \to 0.$$

Let $M \subset \text{Ad } W$ be the bundle of endomorphisms (with trace 0) of $W$ taking $j$ into $j^{-1}$. Then we have the exact sequence

$$0 \to M \to \text{Ad } W \to L_x \to 0. \tag{6.20}$$

On the other hand, both Hom $(j^{-1} \otimes L_x, j) = j^2 \otimes L_x^{-1}$ and Hom $(j \otimes L_x, j^{-1}) = j^{-2} \otimes L_x^{-1}$ are clearly subbundles of $M$. At any $y \neq x$, we have $W_y = j_y \oplus j_y^{-1}$ and these two subspaces of $M_y$ may be represented in matrix form by $\begin{pmatrix} 0 & 0 \\ * & 0 \end{pmatrix}$ and $\begin{pmatrix} 0 & * \\ 0 & 0 \end{pmatrix}$ and hence are linearly independent. In other words, the map $j^2 \otimes L_x^{-1} \oplus j^{-2} \otimes L_x^{-1} \to M$ fails to be of maximal rank only at $x$. Since det $M = L_x^{-1}$, we have the exact sequence

$$0 \to j^2 \otimes L_x^{-1} \oplus j^{-2} \otimes L_x^{-1} \to M \to \mathcal{O}_x \to 0.$$

In particular, the natural map $H^1(j^2 \otimes L_x^{-1}) \oplus H^1(j^{-2} \otimes L_x^{-1}) \to H^1(M)$ is surjective. From (6.20), we conclude that the map $H^1(M) \to H^1(\text{Ad } W)$ has as cokernel $H^1(X, L_x)$ which is of dimension $(g-1)$. Thus the images of $H^1(j^2 \otimes L_x^{-1})$, $H^1(j^{-2} \otimes L_x^{-1})$ in $H^1(\text{Ad} W)$ by the natural inclusions have sum of dimension $(2g-2)$. But the image of $H^1(j^2 \otimes L_x^{-1})$ in $H^1(\text{Ad } W)$ is the tangent space at $W$ to the cycle $P(W_x^*)$ in $U_x$. This proves the transversality, since dim $U_x = 3g-3$.

Lemma 6.21. *Let $j \in J$ with $j^2 \neq 1$. Then identifying $X$ with the intersection of $P(D_j)$ and $P(D_{j^{-1}})$, we have the exact sequence*

$$0 \to T_{\pi_j} \oplus T_{\pi_{j^{-1}}} \to T_{\det} \mid X \to F \to 0$$

*where $T_{\pi_j}$ is the restriction to $X$ of the tangent bundle along the fibres of $P(D_j)$ and $F$ is the quotient bundle associated to the canonical linear system on $X$.*

Proof. The exact sequence 6.20 when $x$ is also varied, yields a sequence on $X \times X$

$$0 \to M \to \text{Ad } W \to L_\Delta \to 0.$$

Taking direct image on $X$, we get

$$R^1(p_1)_* M \to T_{\det} \mid X \to F \to 0.$$

On the other hand, the map $R^1(p_1)_*(j^2 \otimes L_\Delta^{-1}) \oplus R^1(p_1)_*(j^{-2} \otimes L_\Delta^{-1}) \to R^1(p_1)_* M$ has been proved to be surjective. As in Proposition 6.19, we see that the image of $R^1(p_1)_* (j^2 \otimes L_\Delta{}^{-1})$ is $T_{\pi_j}$, proving our assertion.

**Lemma** 6.22. (i) *If $j^2 \neq 1$, then we have the exact sequence on $P(D_j)$:*

$$0 \to H^1(X, \mathcal{O})_{P(D_j)} \to N_{P(D_j), U_X} \to \tau^{-1} \otimes \pi^* F(j) \to 0$$

*where $F(j)$ is the quotient sheaf defined by the linear system $K \otimes j^2$.*

(ii) *If $j^2 = 1$, then we have the exact sequence on $P(D_j) = X \times P$ with $P = PH^1(X, \mathcal{O})$*

$$0 \to \operatorname{Im} d\theta_E \mid P(D_j) \to T_{\det} \mid P(D_j) \to \tau^{-1} \otimes \pi^* F \to 0$$

*where $F$ is the quotient bundle on $X$ defined by the canonical linear system. Moreover, we have a commutative diagram on $P(D_j)$:*

$$\begin{array}{ccccccccc} 0 \to & \tau \otimes H^1(X, \mathcal{O}) & \to & F' \otimes H^1(X, \mathcal{O}) & \to & H^1(X, \mathcal{O}) & \to 0 \\ & \downarrow & & \downarrow & & \downarrow & \\ 0 \to & p_2^* T_P & \to & \operatorname{Im} d\theta_E & \to & \tau^{-1} \otimes p_2^* T_P & \to 0 \end{array}$$

*where the extreme vertical maps are the natural ones and the first extension is obtained by tensoring the universal extension on $X \times P$ with $H^1(X, \mathcal{O})$.*

**Proof.** (i) This follows from the sequence (6.9) and its direct image sequence noting that since $X \times j \not\subset C$ by Lemma 6.3. ii) and Proposition 6.8, the map $T_{p_1 \circ \pi} \to T_{U_X}$ restricted to $P(D_j)$ is still (sheaf theoretically) injective. The cokernel is clearly the tensor product of $\tau^{-1}$ and the sheaf $R^1(p_1)_*(p_2^* j^{-2} \otimes L_\Delta)$. From the exact sequence associated to

$$0 \to p_2^* j^{-2} \to p_2^* j^{-2} \otimes L_\Delta \to j^{-2} \widetilde{\otimes} K_X^* \to 0$$

we identify this sheaf with $F(j)$.

(ii) This is an immediate consequence of Propositions 6.15 and 6.11.

Corollary 6.23. For any $j \in J$, we have

$$K^*_{\det} \mid P(D_j) \simeq \tau^2 \otimes \pi^*(j^4 \otimes K).$$

Proof. From the exact sequence

$$0 \to K^* \otimes j^{-2} \to H^1(X, j^{-2})_X \to F(j) \to 0$$

we see that det $F(j) = K \otimes j^2$ and now from Lemma 6.22, i) we get det $T_{\det} \mid P(D_j) = \tau^{-(g-2)} \otimes \det F(j) \otimes \det T_\pi = \tau^{-(g-2)} \otimes K \otimes j^2 \otimes \tau^g \otimes j^2 = \tau^2 \otimes j^4 \otimes K$, as claimed. The case $j^2 = 1$ follows similarly from 6.22, ii).

**7. Limits of good Hecke cycles.** In this article, we study the limits of good Hecke cycles and their normal bundles in $U_X$. We first wish to fix an ample line bundle $\mathcal{O}(1)$ on $U_X$.

Lemma 7.1. *If $K_{\det}$ denotes the canonical line bundle along the fibres of the fibration* det: $U_X \to X$, *then the line bundle* $\mathcal{O}(1) = K^*_{\det} \otimes (\det)^* K_X$ *is an ample bundle on* $U_X$.

Proof. Fix $\alpha \in J^1$. Consider the map $f: J \to J^1$ given by $f(j) = j^2 \otimes \alpha$. Then we have a commutative diagram

$$\begin{array}{ccc} f^*U(2,1) & \rightarrow & U(2,1) \\ \downarrow & & \downarrow \\ J & \xrightarrow{f} & J^1 \end{array}$$

Now $f^*(U(2,1)) \to J$ is isomorphic to the product $U(2, \alpha) \times J \to J$. This gives in turn a commutative diagram

$$\begin{array}{ccc} U(2, \alpha) \times \widetilde{X} & \xrightarrow{\widetilde{f}} & U_X \\ \downarrow & & \downarrow \\ \widetilde{X} = f^{-1}(X) & \xrightarrow{f} & X \end{array}$$

Since $f$ is étale surjective, $X$ is nonsingular and $f^* K_X = K_X$. On the other hand, $f^* K_{\det} \simeq p_1^* K_{U_\alpha}$, so that our assertion follows from [2, Proposition 4.4] and the ampleness of $K^*$ on $U_\alpha$ (See [6, Theorem 1]) and of $K$ on $X$.

Lemma 7.2. *Let $Z$ be a good Hecke cycle. Then the polynomial $P(m, n) = \chi(Z, K_{\det}^{-m} \otimes dp^* K_X^n)$ is given by $(4m+1)(2m+2n-1)\times(g-1)$. In particular, the Hilbert polynomial of a good Hecke cycle (with respect to $\mathcal{O}(1)$) is $P(n) = (4n+1)(4n-1)(g-1)$.*

Proof. If $W$ is the vector bundle to which the Hecke cycle $Z$ corresponds, then by 5.16, we have $(K_{\det}^{-m} \otimes \det^* K_X^n)|_Z \simeq K_\pi^{-2m} \otimes \pi^* K_X^{m+n}$. Now

$$\pi_*(K_\pi^{-2m} \otimes \pi^* K_X^{m+n}) \simeq (\det W)^{2m} \otimes S^{4m}(W^*) \otimes K_X^{m+n}.$$

Since the higher direct images are zero for $m \geqslant 0$, we have

$$P(m, n) = \chi(X, (\det W)^{2m} \otimes S^{4m}(W^*) \otimes K_X^{m+n})$$

and the latter is computed to be as claimed, using the Riemann-Roch theorem.

Let $p : H \to U_X$ be the restriction of the dual Poincaré bundle on $X \times U_X$ to $U_X$ considered as a divisor by the map (det, Id). By 5.2, v) and Lemma 5.5, we obtain a map $h : H \to U_0$ which is a projective fibration over the set of stable points of $U_0$ by 5.2, iv) and Lemma 5.14. The morphism $\theta_E : P(D) \to U_X$ (in the notation of § 6) lifts to a map $\tilde{\theta}_E : P(D) \to H$ since $\theta_E^* H \simeq P(E)^*$ and $E^*$ has a natural line subbundle by construction. Let $\mathcal{K}$ be the variety of nonstable points of $U_0$, namely the Kummer variety $\mathcal{K} = J/i$, where $i$ is the involution $j \longmapsto j^{-1}$.

Lemma 7.3. *We have a commutative diagram*

$$\begin{array}{ccc} P(D) & \xrightarrow{\quad \theta_E \quad} & h^{-1}(\mathcal{K}) \subset H \\ \Big\downarrow {\scriptstyle p_2 \circ \pi} & & \Big\downarrow {\scriptstyle h} \\ J & \xrightarrow{\qquad\qquad} & \mathcal{K} \end{array}$$

*Moreover $\tilde{\theta}_E$ is onto $h^{-1}(\mathcal{K})$.*

Proof. A point of $P(D)$ is represented by an exact sequence

$$0 \to j \to E \to j^{-1} \otimes L_x \to 0$$

Its image in $H$ is given by the element $E \in U_X$ and the one-dimensional space of linear forms on $E_x$ vanishing on $j_x$. Now the construction $K$ on $H$ which defines the morphism $h$ associates to it the bundle $F$ obtained by

$$0 \to F \to E \to \mathcal{O}_x \to 0$$

given by the above linear form. It is obvious $j$ is a subbundle of $F$ and hence $F$ is $S$-equivalent to $j \oplus j^{-1}$ proving the commutativity of the diagram. Now a point of $h^{-1}(\mathscr{K})$ is given by a stable bundle $E \in U_x$, for some $x \in X$, and a linear form on $E_x$ such that the bundle $F$ defined by the sequence

$$0 \to F \to E \to \mathcal{O}_x \to 0$$

is non-stable. Let $j \subset F$. Clearly then $j$ is also contained in $E$, for otherwise some line bundle of the form $j \otimes L_D$, $D > 0$ will be contained in $E$ contradicting its stability. Thus $E$ can be written as an extension

$$0 \to j \to E \to j^{-1} \otimes L_x \to 0$$

proving the lemma.

LEMMA 7.4. (i) *If $k \in \mathscr{K}$ is not a node, then the reduced fibre of $h$ over $k$ is the push-out of $P(D_j)$ and $P(D_{j^{-1}})$ along the natural section $X$, where $j, j^{-1}$ are points in $J$ over $k$.*

(ii) *If $k \in \mathscr{K}$ is the image of an element $j \in J$ of order 2, then the reduced fibre of $h$ over $k$ is $X \times P$ where $P = PH^1(X, \mathcal{O})$. Moreover the schematic fibre of $h$ contains as a subscheme $X \times P_t$ where $P_t$ is the thickening of $P$ by $0 \to Q^* \to \mathscr{F} \to \Omega^1_P \to 0$, where $Q$ is the universal, quotient bundle of $P$.*

PROOF. (i) follows from diagram 7.3 and the fact that images by $\widetilde{\theta}_E$ of $(P(D_j))$ and $P(D_{j^{-1}})$ intersect transversally along $X$ since transversatility is true of $\theta_E$ (Lemma 6.19).

(ii) Clearly the total thickening $Z$ of $P(D_j)$ in $P(D)$ is in the fibre over $k$ of the morphism $h \circ \widetilde{\theta}_E$. Thus we have only to check that $\theta_E$ induces an epimorphism of $Z$ onto $X \times P_t$ and that the latter is a subscheme of $H$. But this follows from Remark 3.6.

**Remark** 7.5. The reduced scheme defined in 7.4, (i) will be denoted $F_k$, if $k$ is not the image of an element of order 2. If it is, then $F_k$ will denote the subscheme $X \times PH^1(X, \mathcal{O})_t$ of $U_X$. These are contained in the fibres of $H \to U_0$. Probably, these are precisely the schematic fibres, and some of the technical difficulties in this article can be obviated if one could prove this directly.

**Lemma** 7.6. *The natural morphism of the relative Hilbert scheme* Hilb $(H, U_0, P(n))$ *into* Hilb $(U_X, P(n))$ *is injective.*

**Proof.** Since $P(n)$ is of degree 2 in $n$, any element of Hilb $(U_X, P(n))$ represents a subscheme of $U_X$ of dimension 2. Hence our assertion follows from Proposition 6.19.

Let Hilb $(H, U_0, P(m, n))$ be the relative Hilbert scheme of the map $H \to U_0$ of cycles $Z$ satisfying $\chi(Z, p^*(K_{\det}^{-m} \otimes \det^* K_X^n) = P(m, n)$. If $H_2$ is the image of Hilb $(H, U_0, P(m, n))$ in Hilb $(U_X, P(n))$ then $H_2$ contains good Hecke cycles. Now if $Z \in H_2$ is not a good Hecke cycle, then $Z_{\mathrm{red}} \subset F_k$ for a unique $k \in \mathcal{K}$. For a fixed $k \in \mathcal{K}$ the set of elements of $H_2$ which are contained set theoretically in $F_k$ is denoted $H_{2,k}$.

7.7. **Definition of the varieties** $Q_j$, $Q_k$, $R_k$. Now we will define for $k$ not a node, three subvarieties $Q_j$, $Q_{j-1}$ and $R_k$ (with $j, j^{-1} \in J$ lying over $k$) of $H_{2,k}$. Consider the Grassmannian of lines in $PH^1(X, j^2)$; the symmetric product of the projective line bundle over it will be denoted $Q_j$. Also, we denote by $R_k$ the product of $PH^1(X, j^2)$ and $P(H^1(X, j^{-2}))$ for $j$ lying over $k \in \mathcal{K}$. Now points of $Q_j$, $Q_{j-1}$ and $R_k$ can be considered as subschemes of $F_k$. In fact, consider the exact sequences

$$0 \to j^2 \to D_j \to H^1(X, j^2)_X \to 0$$

$$0 \to j^{-2} \to D_{j-1} \to H^1(X, j^{-2})_X \to 0.$$

Any line in $PH^1(X, j^2)$ gives rise to a plane subbundle of $P(D_j)$ over $X$. A point of $Q_j$ gives rise a pair of projective line subbundles (possibly identical) of this plane bundle. Thus we obtain a subscheme of $F_k$. Similarly a point of $R_k$ gives rise to projective line subbundles of $P(D_j)$, $P(D_{j-1})$ containing the sections of $P(D_j)$, $P(D_{j-1})$ given by $j^2$ and $j^{-2}$ respectively. To compute the Hilbert polynomials of these schemes, we may assume without loss of generality that the scheme is obtained by gluing two projective line bundles $P(F_1)$, $P(F_2)$ on $X$ where $F_1$ and $F_2$ are of degree zero and contain either $j^2$ or $j^{-2}$ as line subbundles, along the sections given by these subbundles. Now, from Corollary 6.22, it follows that $\mathcal{O}(1)$ restricts to $P(D_j)$ as $K_X^2 \otimes j^4 \otimes \tau^2$. Its restriction to the section of $P(D_j)$ is therefore seen to be $K_X^2$. Hence the Hilbert polynomial of the subschemes corresponding to points of $Q_j$, $R_k$ is, on using the Mayer-Vietoris sequence,

$$2\chi(X, K_X^{m+n} \otimes S^{2m}(F^*)) - \chi(X, K_X^{m+n}) = (4m+1)(2m+2n-1)(g-1).$$

Thus, we have identified the sets $Q_j$, $Q_{j-1}$, $R_k$ with subsets of $H_{2,k}$.

If $k$ is a node, $Q_k$ is defined to be the variety of subschemes of the form $X \times C$, where $C$ is a conic in $P = PH^1(X, \mathcal{O})$, while $R_k$ denotes those of the form $X \times C$ where $C$ is a line thickened by $\tau^{-1}$ contained in $P_t$ but not in $P$. As above $Q_k$, $R_k$ are checked to be subsets of $H_{2,k}$.

PROPOSITION 7.8. *If $k$ is not a node of $\mathcal{X}$, and $j, j^{-1} \in J$ are points over $k$, then $H_{2,k} = Q_j \cup Q_{j-1} \cup R_k$. If $k$ is a node of $\mathcal{X}$, then $H_{2,k} = Q_k \cup R_k$.*

The rest of this section will be devoted to the proof of Proposition 7.8.

LEMMA 7.9. *If $Z \in H_2$, then the map $p: Z \to X$ is surjective and for generic $x \in X$, the fibre $Z_x$ has $(2m+1)$ as the Hilbert polynomial with respect to the ample generator $h$ of* Pic $U_x$.

PROOF. Since the polynomial $P(m, n)$ is not independent of $n$, it follows that the map $p: Z \to X$ is surjective. For large $m$, we have

$$P(m, n) = \chi(X, p_*(K_{\det}^{-m}) \otimes K_X^n).$$

In other words, $P(m, n)$ is the Hilbert polynomial of $p_*(K_{\det}^{-m})$ with respect to the ample line bundle $K_X$ on $X$. Hence the rank of the sheaf $p_*(K_{\det}^{-m})$ is $\frac{1}{\deg K}$ (the coefficient of $n$ in $P(m, n)$), namely $\frac{1}{2g-2}(8m+2)(g-1) = 4m+1$. But $K_{\det}^*$ restricts to $U_x$ as $K_{U_x}^*$, which is twice the ample generator of Pic $U_x$. Hence the lemma.

**Lemma 7.10.** *Let $S$ be a subscheme of $U_x$ having $2m+1$ as Hilbert polynomial. If $S_{red} \subset (F_k)_{red}$, for some $k \in \mathscr{K}$, then $S$ is a conic contained schematically in $F_k$.*

**Remark 7.11.** The proof of Lemma 7.10 given below can be simplified if one knew a priori that $h|S$ is very ample. This would follow for instance if we could show that $h$ itself is very ample which is very likely to be true and indeed so at least if $X$ is hyperelliptic [1, 5.10, II]. However, since the irreducible components of $(F_k)_{red} \cap U_x$ are projective spaces, the restriction of $h$ to these is very ample, as ample bundles on a projective space are very ample. From this it is easy to see that $h|(F_k)_{red} \cap U_x$ is very ample and hence $h|S_{red}$ is very ample. This fact will be used in the proof.

**Proof of Lemma 7.10.** Let $S$ be a subscheme of $U_x$ having $2m+1$ as Hilbert polynomial. If $S$ is reduced, our assertion follows from Lemma 4.2 and Remark 4.4. If $S$ is not reduced, we shall use Lemma 4.3. Let $S'$ be the thickening of $S_{red} = \mathbf{P}^1$ by the line bundle $L$ mentioned in that lemma. We first show that deg $L \geqslant -1$. Since $S'$ is a subscheme of $U_x$, the bundle $L$ is a quotient of the conormal bundle of $S_{red}$ in $U_x$. Now we have the exact sequences (Lemma 6.21)

$$0 \to \tau \otimes \text{ trivial} \to N^*_{P(D)(j,x), U_x} \to \text{trivial} \to 0$$

and

$$0 \to \tau \otimes \text{ trivial} \to N^*_{P(D)(j,x), U_x} \to \tau \otimes \Omega^1 \to 0$$

according as $j^2 \neq 1$ or $j^2 = 1$. On the other hand, we have the exact sequence

$$0 \to N^*_{P(D)_{(j,x)}, U_x} \to N^*_{\mathbf{P}^1, U_x} \to \tau^{-1} \otimes \text{ trivial} \to 0.$$

Since $\tau \otimes \Omega^1$ restricts to $\mathbf{P}^1$ as $\tau^{-1} \oplus$ trivial, it follows that any quotient line bundle of $N^*_{\mathbf{P}^1, U_x}$ is of deg $\geqslant -1$. Now if $j^2 \neq 1$, any quotient of $N^*_{\mathbf{P}^1, U_x}$ isomorphic to $\tau^{-1}$ is actually a quotient of $N^*_{\mathbf{P}^1, P(D)_{(j,x)}}$ so that $S \subset P(D)_{(j,x)}$. If $j^2 = 1$, we similarly conclude that $S \subset F_k = PH^1(X, \mathcal{O})_t$. This proves the lemma.

**Lemma 7.12.** *If $Z \in H_{2,k}$, then the morphism $p: Z \to X$ is flat.*

**Proof.** Using generic flatness, we see that on an open subset of $X$, $p$ itself defines a section of a suitable relative Hilbert scheme for the morphism $Z \to X$. This extends to a section over the whole of $X$ and thus gives a closed subscheme $Z'$ of $Z$ flat over $X$. Now from Lemma 7.9, it follows that *all* the fibres $Z'_x$ of $Z'$ have Hilbert polynomial $2m+1$ with respect to $h$. Now by Lemma 7.10, it follows that $Z'_x \subset F_k$ for each $x \in X$. Since the map $p: Z' \to X$ is flat, it follows that $Z' \subset F_k$. Now suppose we show that $K^m_{\det} \mid F_k$ has a direct image $V$ on $X$ of the form $K_X^{4m} \otimes$ (a semistable vector bundle of degree 0), at least for large $m$. Then we would have $p_*(K^m_{\det} \mid Z')$ is a quotient of $V$ for large $m$ and hence would be a vector bundle of rank $4m+1$ and degree $\geqslant 2m(g-1)(4m+1)$. Hence

$$\begin{aligned}\chi(Z', K^m_{\det} \otimes p^* K^n_X) &= \chi(X, p^*(K^m_{\det}) \otimes K^n_X) \\ &\geqslant (4m+1)(2m+2n-1)(g-1) = P(m, n)\end{aligned}$$

by Riemann-Roch theorem. On the other hand, if $I$ is the sheaf of ideals defining $Z'$ in $Z$, then from the exact sequence

$$0 \to I \to \mathcal{O}_Z \to \mathcal{O}_{Z'} \to 0,$$

we conclude that $\chi(Z, \mathcal{O}(n)) \geqslant \chi(Z', \mathcal{O}(n))$ for large $n$, and that equality holds only if $Z = Z'$. It remains to prove

**Lemma 7.13.** *The direct image of $K^m_{\det} \mid F_K$ on $X$ is of the form $K^m_X \otimes$ (a semistable vector bundle of degree 0).*

**Proof.** First assume that $k$ is not a node. Then $F_k = P(D)_j \cup P(D)_{j-1}$ with $P(D)_j \cap (PD)_{j-1} = X$. The restriction of $K_{\det}$ to $P(D)_j$ (resp. $X$) is isomorphic to $K_X \otimes j^4 \otimes \tau^2_{D_j}$ (resp. $K_X$) (Corollary 6.22). Now the direct image of $j^4 \otimes \tau^2_{D_j}$ is isomorphic to $j^4 \otimes S^2(D^*_j)$, and the restriction map to the direct image of $j^4 \otimes \tau^2_{D_j}$ on $X$ (namely, the trivial bundle) is induced by the natural map $S^2(D^*) \to j^{-4}$, given by the inclusion $j^2 \to D_j$. Hence, by the Mayer-Vietoris sequence, we see that $p^*(K^m_{\det} \mid F_k)$ is the tensor product of $K^m_X$ and the fibre product of $j^{4m} \otimes S^{2m}(D^*_j)$ and $j^{-4m} \otimes S^{2m}(D^*_j)$ over the trivial bundle. Since $S^{2m}(D^*_j)$ is obtainable (6.1) as successive extension of line bundles of degree 0, it is semistable of degree 0. Hence so is the fibre product mentioned above proving Lemma 7.13 in this case.

Now let $k$ be a node. Then we have Pic $(X \times PH^1(X, \mathcal{O})_t) \to$ Pic $(X \times PH^1(X, \mathcal{O}))$ is an isomorphism. In particular from 6.22, we see that the restriction of $K^m_{\det}$ to $X \times PH^1(X, \mathcal{O})_t$ is of the form $K^m_X \otimes$ a line bundle coming from $PH^1(X, \mathcal{O})_t$. Hence its direct image is of the form $K^m_X \otimes$ a trivial bundle, completing the proof of Lemma 7.13.

Proof of Proposition 7.9. Let $Z \in H_{2,k}$. By Lemma 7.12, the map $p: Z \to X$ is flat, and by Lemma 7.10, $Z$ is schematically contained in $F_k$ and all the fibres are conics. Let us first consider the case when $Z$ is a subscheme of $P(D)_j$ for some $j$ over $k$. By (4.4, iv) there exists a vector subbundle $E$ of $D_j$ of rank 3, such that $Z \subset P(E)$. Now $Z$ defines a divisor in $P(E)$ with $L_Z \simeq \tau^2_E \otimes L$ for some line bundle $L$ on $X$. We have the exact sequence

$$0 \to \tau_E^{-2} \otimes L^* \otimes K^{-m}_{\det}|_{P(E)} \otimes K^n_X \to K^{-m}_{\det}|_{P(E)} \otimes K^n_X \to K^{-m}_{\det}|_Z \otimes K^n_X \to 0.$$

Substituting $K^*_{\det}|_{P(E)} \simeq \tau^2_E \otimes K_X \otimes j^4$, and taking direct image on $X$, we obtain

$$\chi(Z, K^{-m}_{\det} \otimes K^n_X) = \chi(X, K^{m+n}_X \otimes S^{2m}(E^*)) - \chi(X, L^* \otimes K^{m+n}_X \otimes S^{2m-2}(E^*)).$$

The right side can be computed in terms of the degrees of $L$ and $E$ by the Riemann-Roch theorem. But the left side is given to be $P(m, n) = (4m+1)(2m+2n-1)(g-1)$. Equating coefficients of these two polynomials, we check that deg $L = 0$ and deg $E = 0$. In other words, when $Z \subset P(D)_j$, we have shown that $Z$ is a divisor in $P(E)$ given by a nonzero quadratic form $E \to E^* \otimes L$, where $L$ is of degree 0 and $E$ a subbundle of $D_j$, also of degree 0. Now, if $k$ is a node, clearly $E$ is also trivial since $D_j$ is trivial in this case. Moreover the quadratic form is a nonzero section of $S^2(D_j^*) \otimes L$, with deg $L = 0$, and hence it follows (a) that $L$ is trivial and (b) that the quadratic form is a constant. Thus if $Z \subset P(D)_j$ with $j^2 = 1$, then $Z \in R_k$. On the other hand, if $k$ is not a node, by §6.1, $E$ is the inverse image of a trivial vector subbundle of $H^1(X, j^2)_X$ of rank 2. Notice first that the map $E \to E^* \otimes L$ cannot be an isomorphism since $E^* \otimes L$ contains a direct sum of two line bundles of degree 0 while $E$ does not. Moreover the kernel of $E \to E^* \otimes L$ is a proper subbundle of degree 0 and, again by

§6.1, it must contain $j^2$. Thus the quadratic form on $E$ is induced by an $L$-valued quadratic form on the trivial subbundle $E/_{j^2} \subset H^1(X, j^2)_X$ viz. a nonzero section of $S^2((E/j^2)^* \otimes L$. As before, since $L$ is of degree 0, it follows (a) that $L$ is trivial, and (b) that the section of $S^2(E/j^2)^*$ is constant. This proves that if $Z$ is contained in $P(D)_j$ then $Z \in Q_j$.

It remains to consider the case when $Z \not\subset P(D)_j$ for any $j$ over $k$. Clearly if $k$ is not a node, $Z$ consists of two irreducible components $Z_1$ and $Z_2$ with $Z_1 \subset P(D)_j$, $Z_2 \subset P(D)_{j-1}$. Moreover, $Z_1$ (resp. $Z_2$) defines a subbundle $E_1$ (resp. $E_2$) of $D_j$ (resp. $D_{j-1}$) of rank 2, containing $j^2$ (resp. $j^{-2}$). Let their degrees be $d_1$, $d_2$. By the Mayer-Vietoris sequence, we see as before,

$$P(m, n) = \chi(Z, K_{\det}^{-m} \otimes K_X^n) = \chi(X, S^{2m}(E_1^*) \otimes K_X^{m+n} \otimes j^{4m})$$

$$+ \chi(X, S^{2m}(E_2^*) \otimes K_X^{m+n} \otimes j^{-4m})$$

$$- \chi(X, K_X^{m+n}).$$

Computing the right side by the Riemann-Roch theorem, and substituting for $P(m, n)$, we obtain $d_1 = d_2 = 0$. Thus $E_1$, $E_2$ are inverse images of line subbundle of the trivial bundles $H^1(X, j^2)_X$, $H^1(X, j^{-2})_X$ and since these subbundles are of degree 0, they must themselves be trivial. In other words, $Z \in R_k$.

Finally let $k$ be a node and $Z \in H_{2,k}$, $Z \not\subset P(D)_j = (F_k)_{\rm red}$. By 7.10, $Z \subset F_k$. In this case, $Z_{\rm red}$ is a projective bundle associated to a rank 2 vector subbundle $E$ of $D_j = H^1(X, \mathcal{O})_X$. Moreover $Z$ is given by thickening within $X \times F_k$ by a line bundle of the form $\tau_E^{-1} \otimes L$, where $L$ is a line bundle on $X$. From the exact sequence

$$0 \to \tau_E^{-1} \otimes L \to \mathcal{O}_Z \to \mathcal{O}_{Z_{\rm red}} \to 0$$

and the fact that $K_{\det} \mid X \times PH^1(X, \mathcal{O}) \simeq \tau^2 \otimes K_X$ we obtain

$$\chi(Z, K_{\det}^{-m} \otimes K_X^n) = \chi(X, S^{2m}(E^*) \otimes K_X^{m+n}) + \chi(X, S^{2m-1}(E^*) \otimes K_X^{m+n} \otimes L).$$

Equating this expression with $P(m, n)$, we obtain $\deg E = 0$, $\deg L = 0$. Since $E$ is contained in the trivial bundle $H^1(X, \mathcal{O})_X$, $E$

itself is trivial. Now we claim that if $L$ is nontrivial, Hom $(\tau_E \otimes L^{-1}, N_{X \times P(E), U_X}) = 0$. This follows from sequence $0 \to N_{X \times P(E), X \times P} \to N_{X \times P(E), UX} \to N_{X \times P, U_X}\big|_{X \times P(E)} \to 0$ and the sequences in 6.21, ii), since

(a) $H^0(\text{Hom}\,(\tau_E \otimes L^{-1}, N_{X \times P(E), X \times P})) = H^0(\text{Hom}(\tau_E \otimes L^{-1}, \tau \otimes \text{trivial})) = 0.$

(b) $H^0(\text{Hom}\,(\tau_E \otimes L^{-1}, \tau^{-1} \otimes F)) = 0.$

(c) $H^0(\text{Hom}\,(\tau_E \otimes L^{-1}, \tau^{-1} \otimes T_p)) = 0$, and

(d) $H^0(\text{Hom}\,(\tau_E \otimes L^{-1}, T_p)) = 0.$

It follows that $L$ is trivial, since $\tau^{-1} \otimes L \subset N_{X \times P(E), U_X}$. Thus the scheme $Z$ is of the form $X \times C$, where $C$ is a conic contained in $P_t$, and hence $Z \in R_k$. This completes the proof of Proposition 7.9.

**8. Nonsingularity of the Hecke component.** We will now prove that $H_0$, the irreducible component of Hilb$(U_X)$ containing Hecke cycles, is nonsingular. Since $H_0$ is a variety of dimension $3g-3$, it is enough to show that the Zariski tangent space at any point of $H_0$ is of dimension $3g-3$. This is done in Propositions 8.1, 8.5, 8.9 and 8.12. From the description in § 7 of the nature of the subschemes of $U_X$ occurring in $H_2$ (and hence $H_0$), it follows that all these schemes are local complete intersections and hence the Zariski tangent space at any $Z \in H_0$ to the Hilbert scheme itself is given by $H^0(Z, N_{Z, U_X})$ [9a]. Thus the nonsingularity of $H_0$ at some $Z \in H_0$ would follow if dim $H^0(Z, N_{Z, U_X}) \leqslant 3g-3$. However, it turns out that there are points in $H_0$ at which the Hilbert scheme itself is not smooth. We will first compute $H^0(Z, N_{Z, U_X})$ for $Z \in H_0$.

We have to discuss several cases.

PROPOSITION 8.1. *Let $Z \in R_k$, $k$ not a node and $Z \not\subset P(D)_j$, for any $j$ over $k$. Then the Hilbert scheme is smooth at $Z$.*

The proof is completed in 8.4.

By the definition of $R_k$, there exist points $p$, $p'$ in $PH^1(X, j^2)$, $P(H^1(X, j^{-2}))$ such that if $E, E'$ are the inverse images in $D_j, D_{j-1}$ of the trivial line bundles corresponding to these two points, then $Z = P(E) \cap P(E')$. Recall also that $P(E) \cap P(E') = X$ imbedded in $P(D_j)$, $P(D_{j-1})$ by means of the subbundles $j^2$, $j^{-2}$ of $D_j$, $D_{j-1}$ respectively.

**Lemma 8.2.** *Let $U$ be a nonsingular variety and $l_1$, $l_2$ be nonsingular subvarieties intersecting (schematically) in a nonsingular subvariety $X$ which is a divisor in both $l_1$ and $l_2$.*

(i) *Then the reduced scheme $l = l_1 \cup l_2$ is a local complete intersection.*

(ii) *Moreover, we have the exact sequence*

$$0 \to N^*_{l,U} \to \tilde{N}_1 \oplus \tilde{N}_2 \to N^*_{l,U}|X \to 0$$

*where $N_1$, $N_2$ are the restrictions of $N^*_{l,U}$ to $l_1$, $l_2$ respectively.*

(iii) *$N_1$ fits in an exact sequence*

$$0 \to N_1 \to N^*_{l_1,U} \to N^*_{X,l_2} \to 0, \text{ while}$$

(iv) *$N_{l,U}|X$ fits in the sequence*

$$0 \to N_{X,U}/_{N_{X,l_1} \oplus N_{X,l_2}} \to N_{l,U}|X \to N_{X,l_1} \otimes N_{X,l_2} \to 0.$$

**Proof.** At any point of $l$, we can choose a regular system of parameters $(x_1, \ldots x_r, y_1, y_2, z_1, \ldots, z_s)$ for $\mathcal{O}_U$ with $(x_1, \ldots, x_r, y_1) = I_1$ and $(x_1, \ldots, x_r, y_2) = I_2$ being the ideals defining $l_1$, $l_2$ respectively. Since $l$ is then defined by $I_1 \cap I_2 = (x_1, \ldots, x_r, y_1, y_2)$, it follows that $l$ is a complete intersection. The sequence (ii) is obtained by tensoring with $N^*_{l,U}$ the basic exact sequence

$$0 \to \mathcal{O}_l \to \tilde{\mathcal{O}}_{l_1} \oplus \tilde{\mathcal{O}}_{l_2} \to \tilde{\mathcal{O}}_X \to 0.$$

We have now the isomorphism $N_l = I_1 \cap I_2/_{(I_1 \cap I_2)^2} \simeq I_1 \cap I_2/_{I_1(I_1 \cap I_2)}$ and an exact sequence

$$0 \to I_1 \cap I_2/_{I_1(I_1 \cap I_2)} \to I_1/_{I_1^2} \to I_1/_{I_1^2 + I_1 \cap I_2} \to 0.$$

But $I_1/_{I_1^2+I_1 \cap I_2} \simeq \dfrac{I_1/_{I_1 \cap I_2}}{(I_1/I_1 \cap I_2)^2} \simeq \dfrac{I_1 + I_2/_{I_2}}{(I_1 + I_2/_{I_2})^2} \simeq N^*_{X,l_2}$ proving (iii).

Finally, restrict (iii) to $X$ to get the four term sequence

$$0 \to \underline{\mathrm{Tor}^1_{l_1}} (N^*_{X,l_2}, \mathcal{O}_X) \to N^*_{l,U}|X \to N^*_{l_1,U} | X \to N^*_{X,l_2} \to 0.$$

From the resolution

$$0 \to L_X^{-1} \to \mathcal{O}_{l_1} \to \mathcal{O}_X \to 0$$

and the isomorphism $L_X^{-1} | X \simeq N^*_{X, l_1}$, we evaluate the Tor-term to be $N^*_{X, l_2} \otimes N^*_{X, l_1}$. On the other hand, it is clear that $N^*_{l_1, U} | X \to N^*_{X, l_2}$ has kernel $\simeq (N_{X, U}/N_{X, l_1} \otimes N_{X, l_2})^*$. Dualising, we obtain (iv).

We will apply 8.2 to the case $U = U_X$, $l_1 = P(E)$, $l_2 = P(E')$ and $X = l_1 \cap l_2$. In this case we have $N_{X, l_2} \simeq \operatorname{Hom}(j^{-2}, \mathcal{O}) = j^2$ from the exact sequence

$$0 \to j^{-2} \to E' \to \mathcal{O} \to 0.$$

From the definition of $N_1$, we get $\det N_1 \simeq \det N^*_{l_1, U_X} \otimes (\det \widetilde{N}^*_{X, l_2})^{-1}$.

But $\det \widetilde{N}^*_{X, l_2} \simeq L_X$, where $L_X$ is the line bundle on $l_1$ associated to the divisor $X$, as is seen from the fact that the bundle $j^{-2}$ on $X$ extends to $l_1$ and from the resolution

$$0 \to L_X^{-1} \to \mathcal{O}_{l_1} \to \mathcal{O}_X \to 0$$

for the sheaf $\mathcal{O}_X$ on $l_1$. But $L_X$ is easily seen to be $\tau_E$. Thus if $\omega^l$ is the dualising sheaf on $l$, we get $\omega_l | l_1 \simeq \omega_{U_X} \otimes \det N_1^* \simeq \omega_{U_X} \otimes \det N_{l_1, U_X} \otimes (\det N^*_{X, l_2}) \simeq \omega_{l_1} \otimes \tau_E$. But $\omega_{l_1} \simeq K_X \otimes K_\pi$, where $K_\pi$ is the canonical bundle along the fibres of $P(E) \to X$ and hence $\simeq \tau_E^{-2} \otimes \det E^* \simeq \tau_E^{-2} \otimes j^{-2}$. This proves

**Lemma 8.3.** *The dualising sheaf $\omega_l$ of $l = l_1 \cup l_2$ restricts to $l_1$ as $\tau_E^{-1} \otimes j^2 \otimes K_X$. In particular, $\omega_l|_X \simeq K_X$.*

**Proof of Proposition 8.1.** We wish to compute $H^2(l, \omega_l \otimes N^*_{l, U_X})$ which is dual to the space $H^0(l, N_{l, U_X})$ required. To this end we first compute $H^2(l_1, \omega_l \otimes N_l)$. From the exact sequence $0 \to N_1 \to N^*_{l_1, U_X} \to \tilde{j}^{-2} \to 0$, we obtain $H^2(l_1\, \omega_l \otimes N_1) \simeq H^2(l_1, \omega_l \otimes N^*_{l_1, U_X})$ since $H^i(X, K_X \otimes j^{-2}) = 0$ for $i = 1, 2$. But $H^2(l_1, \omega_l \otimes N^*_{l_1, U_X}) = H^2(l_1, \omega_{l_1} \otimes \tau \otimes N^*_{l_1, U_X})$ by Lemma 8.3, which in turn is dual to $H^2(l_1, \tau^{-1} \otimes N_{l_1, U_X})$. To compute this, we use the exact sequence

$$0 \to N_{l_1, P(D_j)} \to N_{l_1, U_X} \to N_{P(D_j), U_X}|_{l_1} \to 0$$

and observe that $N_{l_1, P(D_j)} \simeq \tau \otimes V'$, where $V'$ is a vector space of rank $(g - 2)$. Hence $\dim H^0(l_1, \tau^{-1} \otimes N_{l_1, P(D_j)}) = g - 2$. To compute $H^0(l_1, \tau^{-1} \otimes N_{P(D_j), U_X})$, we appeal to Lemma 6.22, 1). Since $H^0(l_1, \tau^{-1} \otimes H^1(X, \mathcal{O})) = 0$ and $H^0(l_1, \tau^{-2} \otimes \pi^* F(j))$ is also seen to be zero, we get $H^0(l_1, \tau^{-1} \otimes N_{P(D_j), U_X}) = 0$ and hence dim $H^2(l_1, \omega_l \otimes N_1) = g - 2$. Now Proposition 8.1 will follow from the exact sequence in 8.2 tensored with $\omega_l$, if we prove

LEMMA 8.4. dim $H^1(X, N^*_{l,U_X} \otimes \omega_l) \leq g + 1$.

PROOF. Since $\omega_l|_X \simeq K_X$ by Lemma 8.3, we have only to show by duality, that dim $H^0(X, N_{l,U_X}) \leq g + 1$. For this, we will use Lemma 8.2, (iv). In fact, $N_{X,l_1} \simeq j^{-2}$ and $N_{X,l_2} \simeq j^2$ and hence $H^0(X, N_{X,l_2} \otimes N_{X,l_2})$ is of dimension 1. To compute $H^0(X, N_{X,U_X}/N_{X,l_1} \oplus N_{X,l_2})$, we use the exact sequence in 6.21, namely

$$0 \to N_{X,P(Dj)} \oplus N_{X,P(Dj-1)} \to N_{X,U_X} \to F \to 0.$$

Thus we get

$$0 \to (N_{l_1, P(Dj)} \oplus N_{l_2, P(Dj-1)})|X \to N_{X, U_X}/N_{X,l_1} \oplus N_{X,l_2} \to F \to 0.$$

But $N_{l_1, P(Dj)}|X \simeq \tau \otimes (D_j/l_1) \mid X \simeq j^{-2} \otimes$ trivial. Hence $H^0$ of the first term above is zero while $H^0(X, F)$ is clearly of dimension $g$. This proves Lemma 8.4, and hence completes the proof of Proposition 8.1.

PROPOSITION 8.5. *Let $Z \in Q_k$, $k$ a node. Assume that $Z = X \times C$, where $C$ is a conic in $P = PH^1(X, \mathcal{O})$. Then* dim $H^0(Z, N_{Z,U_X}) \leq$ $3g - 3$, $3g - 2$ *or* $3g - 1$ *according as $C$ is nondegenerate, is a pair of intersecting lines, or is a double line. In particular, if $C$ is nondegenerate, the Hilbert scheme is smooth at $Z$.*

PROOF. We first recall [9, SGA 6, VII, Proposition 1.7] that if $X_1 \subset X_2 \subset X_3$ with $X_2$, $X_3$ smooth and $X_1$ a local complete intersection, then by pushing out the restriction to $X_1$ of the exact sequence

$$0 \to T_{X_2} \to T_{X_3} \mid X_2 \to N_{X_2,X_3}, \to 0$$

by means of the map $T_{X_2} \mid X_1 \to N_{X_1,X_2}$, we obtain the exact sequence

$$0 \to N_{X_1,X_2} \to X_{X_1,X_3} \to N_{X_2,X_3} \mid X_1 \to 0.$$

We wish to apply this to the case $X_1 = Z$, $X_2 = X \times P$ and $X_3 = U_X$, and use the description of $T_{U_X}|X \times P$ given in 6.22,(ii). Then we see that $N_{Z,U_X}$ fits in exact sequence

$$0 \to N' \to N_{Z,U_X} \to \tau^{-1} \otimes F \mid Z \to 0 \qquad (8.6)$$

where $N'$ is obtained as the push-out of the sequence

$$0 \to T_P \mid Z \to \operatorname{Im} d\theta_E \mid Z \to \tau^{-1} \otimes T_P \mid Z \to 0$$

by means of the map $T_P \mid Z \to N_{Z,\, X\times P} = N_{C,\, P}$. Since the direct image of $\tau^{-1} \otimes F \mid Z$ on $X$ is zero, it follows that $H^0(Z, \tau^{-1} \otimes F) = 0$ and hence that $H^0(Z, N') \to H^0(Z, N_{Z,\, U_X})$ is an isomorphism. Again by 6.22, (ii), we have the commutative diagram (on $Z$)

$$\begin{array}{ccccccc} 0 \to & \tau \otimes H^1(X, \mathcal{O}) & \to & F' \otimes H^1(X, \mathcal{O}) & \to & H^1(X, \mathcal{O})_Z & \to 0 \\ & \downarrow & & \downarrow & & \downarrow & \\ 0 \longrightarrow & N_{C,\, P} & \longrightarrow & N' & \longrightarrow & \tau^{-1} \otimes T_P & \to 0 \end{array} \qquad (8.7)$$

where the top sequence is obtained by tensoring the universal extension $F'$ on $X \times P$ with $H^1(X, \mathcal{O})$. Now $H^0(N_{C,\, P})$ is easily computed to be $3g-4$, for instance by imbedding $C$ as a divisor in a plane $\varpi$ and proving dim $H^0(C, N_{C, \tilde{\omega}}) = 5$ and dim $H^0(C, N_{\tilde{\omega} P}) = 3(g-3)$. It remains therefore to compute the kernel of the boundary homomorphism $H^0(X \times C, \tau^{-1} \otimes T_P) \to H^1(X \times C, N_{C,\, P})$. But now it is easy to see that $H^0(P, \tau^{-1} \otimes T_P) \to H^0(C, \tau^{-1} \otimes T_P)$ is surjective by chekcing $H^0(P, \tau^{-1} \otimes T^P) \to H^0(\varpi, \tau^{-1} \otimes T_P)$ and $H^0(\varpi, \tau^{-1} \otimes T_P) \to H^0(C, \tau^{-1} \otimes T_P)$ are both surjective. Thus from (8.7), we see that the required boundary homomorphism is the composite of that of the top sequence $H^1(X, \mathcal{O}) \to H^1(X \times P,\ H^1(X, \mathcal{O}) \otimes \tau)$ and the natural map $H^1(X \times P,\ H^1(X, \mathcal{O}) \otimes \tau) \to H^1(C, N_{C,\, P})$. Again we have the diagram (on $Z$).

$$\begin{array}{cccccccc} 0 \to & \tau \otimes H^1(X, \mathcal{O}) & \to & F' \otimes H^1(X, \mathcal{O}) & \to & H^1(X, \mathcal{O})_Z & \to 0 & \\ & & & & & & \searrow & \\ & \Big\downarrow & & \Big\downarrow & & \Big\downarrow & & \tau^{-1} \otimes T_P \\ & & & & & & \swarrow & \\ 0 \to & N_{C,\, P} & \to & F' \otimes \tau^{-1} \otimes N_{C,\, P} & \to & \tau^{-1} \otimes N_{C,\, P} & \to 0 & \end{array} \qquad (8.8)$$

where the lower sequence is obtained as the tensor product on $X \times C$ of the universal extension $F'$ by $\tau^{-1} \otimes N_{C,\, P}$. From (8.8) we conclude that the required map is the composite of the natural map $H^0(C, \tau^{-1} \otimes T_P) \to H^0(C, \tau^{-1} \otimes N_{C,\, P})$ and the boundary homomorphism $H^0(C, \tau^{-1} \otimes N_{C,\, P}) \to H^1(X \times C, N_{C,\, P})$. From the definition of the universal extension, we conclude that the latter map is injective. Thus finally we have

$$\dim H^0(X\times C, N_{Z,\,U_X}) = \dim H^0(X\times C, N')$$
$$= 3g-4+\dim\ker H^0(C, \tau^{-1}\otimes T_P)\to H^0(C, \tau^{-1}\otimes N_{C,P})$$

so that Proposition 8.5 would follow from

LEMMA 8.9. *If $C$ is a conic in a projective space $P$, then* ker $H^0(C, \tau^{-1}\otimes T_P)\to H^0(C, \tau^{-1}\otimes N_{C,P})$ *is of dimension* 1, 2, *or* 3 *according as $C$ is nondegenerate, is a pair of lines, or is a double line.*

PROOF. By duality, and noting that $\tau^{-1}|C$ is the dualising sheaf, we have to compute the dimension of the cokernel of $H^1(C, N^*_{C,P}) \to H^1(C, \Omega^1_P)$. From the exact sequence

$$N^*_{C,P} \to \Omega^1_P|_C \to \Omega^1_C \to 0$$

we see that this cokernel is isomorphic to $H^1(C,\Omega^1_C)$ since $H^2(C,\mathscr{F})=0$ for any coherent sheaf $\mathscr{F}$ on $C$. For a nondegenerate conic, its dimension is clearly 1. On the other hand, if $C$ is a pair of lines $l_1$, $l_2$, we have a surjection $\Omega^1_C \to \widetilde{\Omega}^1_{l_1} \oplus \widetilde{\Omega}^1_{l_2}$ with the kernel sheaf supported at the point of interesection of $l_1$ and $l_2$. Hence $\dim H^1(\Omega^1_C)=2$. If $C$ is the $\tau^{-1}$-thickening of the projective line $l$, then we also have a projection $p\colon C \to l$. From this we get the exact sequence

$$0 \to p^*\,\Omega^1_l \to \Omega^1_C \to \Omega^1_p \to 0.$$

Now $\Omega^1_p$ is easily seen to be $i_*(\tau^{-1})$ where $i{:}l \to C$ is the inclusion, and hence $H^i(C, \Omega^1_p) = 0$ for all $i$. On the other hand, $H^1(C, p^*\,\Omega^1_l) \simeq H^1(l, \Omega^1_l) \oplus H^1(l, \tau^{-1}\otimes\Omega^1_l)$ is of dimension 3. This proves Lemma 8.9.

We will in fact see that if $Z\in H_{2,k}$ corresponds to a degenerate conic $C \subset P = PH^1(X, \mathcal{O})$, then Hilb $(P(n))$ is not smooth at $Z$.

To show that $H_0$ is nonsingular at such a $Z$, we proceed as follows. Let $\widetilde{Z} \subset H_2\times U_X$ be the universal subscheme representing $H_2$. Then the map $\widetilde{Z} \to H_2\times X$ given by $(p_{H_2}, \det)$ is flat. Since the fibres are conics, it follows that this is a morphism of complete intersection [9, SGA 6, VII]. Moreover, the direct image on $H_2\times X$ of the dual of the relative dualising sheaf is locally free of rank 3.

The determinant of this direct image is a line bundle and hence induces a map $\varphi: H_2 \to \text{Pic } X$. If $Z = P(E)$, is a good Hecke cycle, its image under $\varphi$ is given as follows: If $\pi: P(E) \to X$ is the projective fibration, $\varphi(Z) = \det \pi_*(T_\pi) \simeq \det \text{Ad} E$ is the trivial bundle. Hence $\varphi$ is a constant on $H_0$. On the other hand, if $Z \in Q_j$ is given by (rank 2) subbundles $F_1$, $F_2$ of $D_j$ containing $j^2$ with $F_1/_{j^2}$, $F_2/_{j^2}$ trivial, then $\omega_\pi^* \simeq \tau_E \otimes (F_1 \cap F_2)\cdot|_Z$ where $E = F_1 + F_2$. Now from the exact sequence (on $P(E)$)

$$0 \to \tau_E^{-1} \otimes F_1 \cap F_2 \to \tau_E \otimes F_1 \cap F_2 \to \omega_\pi^* \to 0$$

we get, on taking direct images on $X$,

$$\begin{aligned}\varphi(Z) &= \det \pi_* (\tau_E \otimes F_1 \cap F_2) = \det (E^* \otimes F_1 \cap F_2)\\ &= (F_1 \cap F_2)^3 \otimes \det E^* = j^4.\end{aligned}$$

since $F_1 \cap F_2 = j^2$ and $\det E = j^2$. Thus we have

Lemma 8.10. *If $Z \in H_0$, then $\varphi(Z)$ is trivial, while if $Z \in Q_j$ is given by two subbundles $F_1$, $F_2$ of $D_j$ containing $j^2$, then $\varphi(Z) = j^4$.*

Now in order to prove that $H_0$ is smooth at points $Z \in Q_k$, $k$ a node, we will show

Lemma 8.11. *$\text{Im}(d\varphi)_Z$ has dimension $\geqslant 1$ (resp. $\geqslant 2$) if $Z$ is represented by a pair of lines (resp. a double line).*

Assume the Lemma for the moment. Since $\varphi$ is a constant on $H_0$, it follows that $(d\varphi)_Z$ is zero on $T_Z(H_0)$. But by Proposition 8.5, $\dim T_Z(H_2) \leqslant 3g - 2$ (resp. $3g - 1$) in these cases. Now the map $(d\varphi)_Z : T_Z(H_2)/T_Z(H_0) \to T_{\varphi(Z)}(J)$ has image of dimension $\geqslant 1$ (resp. $\geqslant 2$). Hence $\dim T_Z(H_0) \leqslant (3g-3)$ in both these cases, proving $H_0$ is nonsingular at these points.

Proof of Lemma 8.11. Let $P$ be a Poincaré bundle on $J \times X$. The sheaf $\mathscr{F} = R^1(p_j)_*(P^2)$ on $J$ is locally free, outside elements of order 2, of rank $(g-1)$. It is easily seen that $\text{Grass}_{g-1}(\mathscr{F})$ is isomorphic to the blow up $\pi: \tilde{J} \to J$ at all elements of order 2. Hence on $\tilde{J}$ we have a surjection $\pi^*\mathscr{F} \to Q \to 0$ where $Q$ is the tautological quotient bundle or rank $(g-1)$. On the other hand, on $X \times J$ we have a surjection of the vector bundle $D$ onto $p_J^* \mathscr{F}$. This gives rise to a surjection $(1 \times \pi)^* D \to p^*_{\tilde{J}} Q$ (on $X \times \tilde{J}$) where now $Q$ is also locally free. Thus we get a family of subschemes of $P(D)$, parametrised by $P(Q)$. These subschemes are projective bundles associated to subbundles of $D_j$ of rank 2 containing the kernel of $D_j \to Q_j$ given by the surjection above. Thus $P(Q) \times_{\tilde{J}} P(Q)$ parametrises two families of projective line subbundles. It is easy to see that there is a flat family of schemes over $P(Q) \times_{\tilde{J}} P(Q)$ which is obtained as the union of these two schemes and that this family is a family of subschemes of $U_X$ with Hilbert polynomial $P(n)$. Thus we have a morphism $P(Q) \times_{\tilde{J}} P(Q) \to H_2$. By Lemma 8.10, we have the commutative diagram

$$\begin{array}{ccc} P(Q) \times_{\tilde{J}} P(Q) & \longrightarrow & H_2 \\ \downarrow & & \Big| \\ \tilde{J} & & \Big| \\ \downarrow & & \downarrow \\ J & \xrightarrow{\;\;4_J\;\;} & J \end{array}$$

where $4_J$ is the map $j \longmapsto j^4$ of $J$. If $Z \in H_2$ is represented by $X \times$ a pair of lines in $PH^1(X, \mathcal{O})$, then it is the image of some point in $P(Q) \times_{\tilde{J}} P(Q)$ over an element of order 2 of $J$. Since the map $\tilde{J} \to J$ has nonzero differential at any point, and since $P(Q) \times_{\tilde{J}} P(Q) \to \tilde{J}$ is a fibration, it follows that Im $(d\varphi)_Z$ has dimension $\geqslant 1$. If $Z$ is represented by $X \times$ a double line in $PH^1(X, \mathcal{O})$, then $Z$ can be obtained as the image of two points of $P(Q) \times_{\tilde{J}} P(Q)$ whose images $x$, $x'$ in $\tilde{J}$ are two different points over the given node. In fact, $x$ and $x'$ could be taken as any point on the line in $PH^1(X, \mathcal{O})$, to which $Z$ corresponds. Hence

$$\dim \operatorname{Im} (d\varphi)_Z \geqslant \dim (\operatorname{Im} (d\pi)_x + \operatorname{Im} (d\pi)_{x'}) \geqslant 2.$$

This completes the proof of 8.11 and hence the nonsingularity of $H_0$ at $Z \in Q_k$, $k$ a node.

It remains to consider the case $Z \in R_k$, $k$ a node.

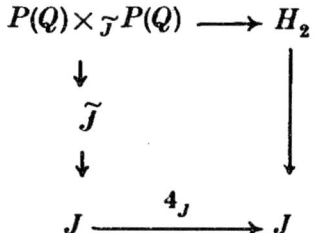

**Proposition** 8.12. *Let $Z \in R_k$, $k$ a node. Then* $\dim H^0(Z, N_{Z,U_X}) \leqslant 3g-3$.

**Proof.** Let $Z = X \times l_L$, where $l$ is a projective line in $PH^1(X, \mathcal{O})$ and $L$ is a subbundle of $N = N_{X \times l, U_X}$ isomorphic to the hyperplane bundle on $l$. Then we have the exact sequence

$$0 \to L^{-1} \otimes N_{Z,U_X}|X \times l \to N_{Z,U_X} \to N_{Z,U_X}|X \times l \to 0.$$

Now Proposition 8.12 follows from

**Lemma** 8.13. (i) $\dim H^0(X \times l, N_{Z,U_X}) \leqslant 2g-1$.

(ii) $\dim H^0(X \times l, L^{-1} \otimes N_{Z,U_X}) \leqslant g-2$.

**Proof of** (i). By § 3.4 we have the exact sequence

$$0 \to N/L \to N_{Z,U_X}|_{X \times l} \to L^2 \to 0. \tag{8.14}$$

Since $H^0(X \times l, L^2) \simeq H^0(l, \tau^2)$ is of dimension 3, (i) will be proved if we show that $\dim H^0(X \times l, N/L) \leqslant 2(g-2)$. Now for this computation, we need the exact sequences (Lemma 6.22. (ii))

$$0 \to \tau \otimes V/W \to N'/L \to V/W \to 0 \tag{8.15}$$

$$0 \to N'/L \to N/L \to \tau^{-1} \otimes F \to 0 \tag{8.16}$$

where $V = H^1(X, \mathcal{O})$, $W$ is the two dimensional subspace of $V$ corresponding to $l$, and $N'$ is the kernel of the surjection $N \to \tau^{-1} \otimes F$. From 8.15, we conclude that $H^0(X \times l, N'/L) \simeq H^0(X \times l, N/L)$, since $H^0(X \times l, \tau^{-1} \otimes F) = 0$. To compute $H^0(X \times l, N'/L)$, we note that $\dim H^0(X \times l, \tau \otimes V/W) = 2(g-2)$ and hence i), Lemma 8.13 will be proved if we can show that the boundary homomorphism $H^0(V/W) \to H^1(\tau \otimes V/W)$ is injective. From 6.22 (ii), we have the commutative diagram

$$\begin{array}{ccc} V/W \simeq H^0(V/W) & \to & H^1(\tau \otimes V/W) \simeq V \otimes V^* \otimes V/W \\ \uparrow & & \uparrow \\ V \simeq H^0(V) & \to & H^1(\tau \otimes V) \simeq V \otimes V^* \otimes V \end{array}$$

where the lower map is the boundary homomorphism given by the universal extension $F'$, and hence is the map $v \longmapsto \mathrm{Id}_V \otimes v$. From this it is easy to conclude that the top horizontal map is injective. This proves (i).

PROOF OF (ii). We proceed as in (i). By tensoring 8.14 with $L^{-1}$, we are reduced to computing a) $H^0(X \times l,\ L^{-1} \otimes N/L)$ and b) the boundary homomorphism $H^0(X \times l,\ L) \to H^1(X \times l,\ L^{-1} \otimes N/L)$. As for a), we have from 8.16, $H^0(L^{-1} \otimes N/L) \simeq H^0(L^{-1} \otimes N'/L)$ and from 8.15, $H^0(L^{-1} \otimes N'/L) \simeq H^0(V/W)$ which has dimension $g-2$. Now (ii) will be proved if we can show that the boundary homomorphism b) is injective. To prove this, we need a description of the extension 8.14 (which is the dual of the extension 3.4 for $l_L$). In other words, for $N_{X\times P, U_X}$ we have the exact sequence (6.22. (ii))

$$0 \to \tau^{-1} \otimes T_P \to N_{X\times P, U_X} \to \tau^{-1} \otimes F \to 0.$$

Now $L = \tau^{-1} \otimes T_l \subset \tau^{-1} \otimes T_P \,|\, X \times l$. Any line subbundle of $N_{X\times l, U_X}$ mapping isomorphically on $L$ gives rise to a thickening $X \times l_L$. From Proposition 3.8, we get the following information about 8.14. The map $d\theta_E \,|\, X \times l$ induces a map $N_{X\times P, P(D)} = H^1(X, \mathcal{O})_{X\times P} \to \tau^{-1} \otimes T_P$. Let $W$ be the inverse image of $L = \tau^{-1} \otimes T_l$ so that we have the exact sequence (on $X \times l$)

$$0 \to \tau^{-1} \to W \to \tau^{-1} \otimes T_l = L \to 0.$$

Of course, $W$ is actually the trivial bundle with fibre = the subspace of $P$ corresponding to $l$. Symmetrising this, we get the sequence

$$0 \to \tau^{-1} \otimes W \to S^2(W) \to L^2 \to 0. \tag{8.17}$$

Taking into account the description of the Hessian of $\theta_E|X \times P$, we see that the push-out of (8.14) by the map $N/L \to N_{X\times P, U_X}/\tau^{-1} \otimes T_P = \tau^{-1} \otimes F$ is isomorphic to the push-out of (8.17) by the natural map $\tau^{-1} \otimes W \to \tau^{-1} \otimes H^1(X, \mathcal{O}) \to \tau^{-1} \otimes F$. In view of this, we have a commutative diagram

$$\begin{array}{ccccc}
H^0(X\times l, L) & \to H^1(X\times l, L^{-1}\otimes\tau^{-1}\otimes W) \to & H^1(X\times l, L^{-1}\otimes\tau^{-1}\otimes H^1(X,\mathcal{O})) \\
\downarrow & & \downarrow \\
H^1(X\times l, L^{-1}\otimes N/L) & \longrightarrow & H^1(X\times l, L^{-1}\otimes \tau^{-1}\otimes F)
\end{array}$$

We have to show that the left vertical map is injective. The second top horizontal map is clearly injective, while the right vertical map is even an isomorphism. Thus our assertion will be proved if the map

$$H^0(l, L) \to H^1(l, L^{-1} \otimes \tau^{-1} \otimes W),$$

associated to (8.17) tensored with $L^{-1}$, is injective. But this is clear since $H^0(l, L^{-1} \otimes S^2(W)) = 0$, thus proving Lemma 8.13 and hence Proposition 8.12.

We now have

THEOREM 8.14. *There is a natural morphism of the Hecke component $H_0$ into the moduli space $U_0$ giving a non-singular model of $U_0$, which is an isomorphism over the set of stable points. The reduced fibre over a non-nodal point of the Kummer variety is isomorphic to $\mathbf{P}^{g-2} \times \mathbf{P}^{g-2}$ while that over the node is isomorphic to the union of a $\mathbf{P}^5$ bundle over the Grassmannian of planes in $\mathbf{P}^{g-1}$ and a $\mathbf{P}^{g-2}$ bundle over the Grassmannian of lines in $\mathbf{P}^{g-1}$.*

PROOF. The morphism $\pi$: Hilb $(H, U_0, P(m, n)) \to U_0$ is an isomorphism over the open set $U$ of stable points. Moreover the restriction of the morphism Hilb $(H, U_0, P(m, n)) \to$ Hilb $(U_X, P(n))$ to the (schematic) closure of $p^{-1}(U)$ is an injective (Lemma 7.6) and birational (Theorem 5.13) morphism onto $H_0$. Since $H_0$ is smooth, this morphism is an isomorphism onto $H_0$. This proves the first part of the theorem.

The fibre over a non-nodal point of $\mathscr{K}$ corresponding to $j \in J, j^2 \neq 1$, is isomorphic to $PH^1(X, j^2) \times PH^1(X, j^2)$ (Proposition 7.8). The fibre over a node is isomorphic to the union of the space of conics which are schematically contained in $PH^1(X, \mathcal{O})$ and the space of $\tau^{-1}$-thickenings of lines in $PH^1(X, \mathcal{O})$ which are contained in the thickening $PH^1(X, \mathcal{O})_t$ (Proposition 7.8). The first variety is a $\mathbf{P}^5$-bundle over the Grassmannian of planes in $PH^1(X, \mathcal{O})$ while the second is a $\mathbf{P}^{g-2}$ bundle over the Grassmannian of lines in $PH^1(X, \mathcal{O})$ (See Remark 4.4. iii).

## REFERENCES


1. U. V. Desale and S. Ramanan: Classification of vector bundles of rank 2 on hyperelliptic curves, *Inventiones Math.*, 38 (1976) 161-185.

2. R. Hartshorne: Ample subvarieties of algebraic varieties, *Springer Lecture Notes in Mathematics*, No. 156.

3. M. S. Narasimhan and S. Ramanan: Moduli of vector bundles on a compact Riemann surface, *Ann. of Math.*, (89) (1969) 19-51.

4. M. S. Narasimhan and S. Ramanan: Vector bundles on curves, *Proceedings of the Bombay Colloquium on Algebraic Geometry*, 335-346, Oxford University Press, 1969.

5. M. S. Narasimhan and S. Ramanan: Deformations of the moduli space of vector bundles over an algebraic curve, *Ann of Math.*, 101 (1975), 391-417.

6. S. Ramanan: The moduli spaces of vector bundles over an algebraic curve, *Math. Annalen*, 200 (1973), 69-84.

7. M. Schelessinger: Functors on Artin rings, *Trans. Amer. Math. Soc.*, 130 (1968), 208-222.

8. A. Tjurin: Analog of Torelli's theorem for two dimensional bundles over algebraic curves of arbitrary genus, Iz. Akad. Nauk, SSSR, Ser Mat Tom 33, (1869) 1149-1170, *Math. U.S.S.R. Izvestija*, Vol. 3 (1969) 1081-1101.

8.(a) A. N. Tjurin: The geometry of moduli of vector bundles, *Uspekhi Mat. Nauk*, 29 : 6 (1974), 59-88, *Russian Math. Surveys* 29 : 6 (1974), 57-88.

9. Séminaire de Géométrie Algébrique (SGA).

9.(a) Fondements de la Géométrie Algébrique (FGA).

10. C. S. Seshadri: Desingularisation of moduli varieties of vector bundles on curves, to appear in the *Proceedings of the Kyoto Conference on Algebraic Geometry*, 1977.

Commun. Math. Phys. 67, 121–136 (1979)

Communications in
Mathematical
Physics


# Geometry of $SU(2)$ Gauge Fields

M. S. Narasimhan and T. R. Ramadas

Tata Institute of Fundamental Research, Bombay 400005, India

**Abstract.** We study $SU(2)$ Yang-Mills theory on $S^3 \times \mathbb{R}$ from the canonical view-point. We use topological and differential geometric techniques, identifying the "true" configuration space as the base-space of a principal bundle with the gauge-group as structure group.

## 1. Introduction

We study in this paper the space of connections on the trivial $SU(2)$ bundle on $S^3$ and the action of the gauge-group on this space. Let $\mathscr{C} = \mathscr{C}^k$ denote the space of connections belonging to Sobolev class $(k)$, $k \geqq 3$. We introduce the groups Aut, $\mathrm{Aut}^o$ (see Sect. 2) of gauge transformations belonging to the Sobolev class $(k+1)$. We then define the space $\mathscr{C}_o$ of generic connections, which are the connections whose holonomy coincides with the whole group $SU(2)$, and prove that the above groups act properly on $\mathscr{C}$ (Proposition 2.4) and that $\mathscr{C}_o$ in a principal Aut (or $\mathrm{Aut}^o$) bundle (Propisition 4.3). The proof involves deriving estimates for certain elliptic operators whose coefficients belong to Sobolev spaces and are not necessarily $C^\infty$. We define the groups $\mathrm{Aut}_e$, $\mathrm{Aut}_e^o$ (Sect. 4b)) and show that the Aut (resp. $\mathrm{Aut}^o$) bundle cannot be reduced to the subgroup $\mathrm{Aut}_e$ [resp. $\mathrm{Aut}_e^o$ (Theorem 5.1)]. In particular gauge-fixing is not possible. This result is proved by looking at left-invariant differential forms on $S^3 = SU(2)$ with values in the Lie algebra of $SU(2)$ and by showing essentially that the principal $SO(3)$ bundle obtained by the action on $3 \times 3$ real matrices of rank $\geqq 2$, by multiplication on the left, is nontrivial (Theorem 6.2).

In Sect. 7 we introduce the Coulomb connection. We show (Theorem 7.5) that, in case we use the biinvariant metric on $S^3 = SU(2)$, the values of the curvature form of this connection at the point $\omega/2 \in \mathscr{C}_o$, where $\omega$ is the Maurer-Cartan form, span a dense subspace in the gauge algebra.

The study was motivated by the following physical considerations, taking Dirac's theory [1] of singular Lagrangians as starting point. We may recall that the Faddeev-Popov procedure was derived [2] by an extension of Dirac's

0010-3616/79/0067/0121/$03.20

constraint analysis programme. With the realisation due to Gribov [3] that the Coulomb gauge has ambiguities in the case of non-abelian theories, it has become necessary to examine anew the quantisation of such theories.

The $SU(2)$ Yang-Mills theory without matter-fields is described by the action

$$-\tfrac{1}{4}\int (F_{\mu\nu}F^{\mu\nu})d^4x$$

where $F_{\mu\nu}=\partial_\mu A_\nu-\partial_\nu A_\mu+[A_\mu, A_\nu]$. We assume that the fields $A_\mu(x)$ fall off fast enough at space-like infinity, that they can be mapped into fields on $S^3\times\mathbb{R}$, $\mathbb{R}$ representing the time-co-ordinate. Because of gauge-invariance, the Lagrangian is singular and the problem as is well-known, reduces to the following.

Consider the phase-space $\{A_i, \pi_i\}$ $(i, 1, 2, 3)$ of the space-components. There is a constraint on this space, usually expressed as $\partial_i\pi_i+[A_i, \pi_i]=0$. On this constrained space a "Hamiltonian" is defined.

$$\int (\pi_i\pi_i+1/2F_{ij}F_{ij})d^3x\,.$$

The constrained space, however is not a symplectic manifold. The "true" configuration space, and its phase space are obtained by factoring out by the "time-independent" gauge transformations. More precisely, time-independent gauge-transformations act on the space of fields $\mathscr{C}=\{A_i(x)\}$. The gauge-invariant configuration space $\boldsymbol{\mathscr{C}}$ is the quotient by this action, and the gauge-invariant phase-space, the corresponding phase-space. In terms of diagrams:

$$\begin{array}{ccccc}
T^*(\mathscr{C}) & \hookleftarrow & \mathscr{I} & \longrightarrow & \mathscr{C} \\
 & & \downarrow & & \downarrow \text{ gauge-group} \\
 & & T^*(\boldsymbol{\mathscr{C}}) & \xrightarrow[p]{} & \boldsymbol{\mathscr{C}}
\end{array}$$

Here $p$ is the projection from $T^*(\boldsymbol{\mathscr{C}})$, and $\mathscr{I}$, the fibre product over $\boldsymbol{\mathscr{C}}$ of $\mathscr{C}$ and $T^*(\boldsymbol{\mathscr{C}})$, is precisely the *constrained phase-space.*

The "Hamiltonian" given above goes down to $T^*(\boldsymbol{\mathscr{C}})$ and becomes a true Hamiltonian there. Correspondingly there is a well-defined, non-singular Lagrangian on $\boldsymbol{\mathscr{C}}$.

Faddeev [2] quantises by identifying $T^*(\boldsymbol{\mathscr{C}})$ with a section of the bundle $\mathscr{I}\to T^*(\boldsymbol{\mathscr{C}})$, this section representing the subsidiary constraint, which together with the first, forms a second-class system. Since $T^*(\boldsymbol{\mathscr{C}})\to\boldsymbol{\mathscr{C}}$ admits the zero section, it is clear that the existence of such a section is equivalent to the existence of a section for $\mathscr{C}\to\boldsymbol{\mathscr{C}}$.

The Lagrangian on $\boldsymbol{\mathscr{C}}$ can be obtained directly by the simple procedure of letting $A^o=0$ in the original Lagrangian, thus getting

$$\int (\dot{A}_i\dot{A}_i-\tfrac{1}{2}F_{ij}F_{ij})d^3x\,.$$

This Lagrangian has "time-independent gauge-transformations" as a symmetry, and gives rise to a Lagrangian on $\boldsymbol{\mathscr{C}}$ in a natural way. This involves defining a "horizontal space" at each point $A_i$ of $\mathscr{C}$: this is the space of tangent vectors $\dot{A}_i$ that satisfy

$$\partial_i\dot{A}_i+[A_i, \dot{A}_i]=0\,.$$

Note that in the abelian case the horizontal spaces form an integrable distribution, and the Coulomb gauge corresponds to taking a maximal integral manifold as the section $\boldsymbol{\mathscr{C}}\to\mathscr{C}$. Note also that in general this definition of horizontal spaces gives a connection in the bundle $\mathscr{C}\to\boldsymbol{\mathscr{C}}$. We call this the Coulomb connection.

In the absence of a section, a conceptually simple, although in practice difficult, path-integral procedure suggests itself. Suppose we consider transition amplitude between points **A**, **B**, in $\boldsymbol{\mathscr{C}}$. This involves integrating over all paths from **A** to **B**, using the Lagrangian in $\boldsymbol{\mathscr{C}}$. But for a given smooth path, the action is the same as the one given by lifting the path to a horizontal one in $\mathscr{C}$ (to a path satisfying $\partial_i \dot{A}_i + [A_i, \dot{A}_i] = 0$), between points A and B above. Thus the holonomy of the connection on $\mathscr{C}$ is clearly relevant. We calculate the holonomy for a special choice of a metric in $S^3$ and find that it is dense in the gauge-group. In other words, if we fix A on the fibre above **A**, a dense set of points above **B** can be joined to A by horizontal paths. (Thus the ambiguity is in some sense maximal.) Schematically:

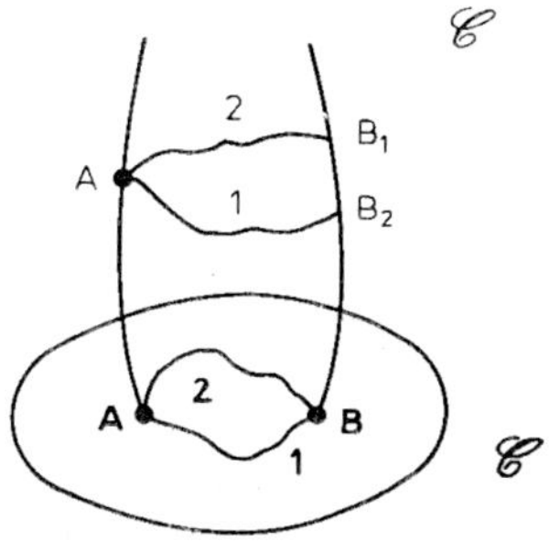


Note that in the abelian case the holonomy is trivial, and a horizontal path in $\mathscr{C}$ starting from a point in the Coulomb gauge always stays within the Coulomb gauge. In particular all paths from **A** to **B** below, when lifted through A, end in the same point B above **B**.

Results on gauge-fixing, applicable when the base-space is $S^3$ or $S^4$, and the structure-group is a general compact semi-simple Lie group [in particular $SU(N)$], have been announced by Singer [4]. The present work was done independently and our approach is different. In the particular case that we consider, our first main result (Theorem 5.1) is stronger than the nonexistence of a section for the action of the group of gauge-transformations. The second main result (Theorem 7.5) of this paper, on the holonomy of the "Coulomb connection", is new.

## 2. The Space of Connections and the Action of the Gauge Group

We shall consider connections on the trivial $SU(2)$ bundle over $S^3$. We identify the set of connections with the set of 1-forms with coefficients in the Lie Algebra, $\mathfrak{SU}(2)$, of $SU(2)$ by means of the map $\alpha \mapsto \sigma^*(\alpha)$ where $\sigma$ is the canonical section of the trivial $SU(2)$ bundle. We shall use connections which belong to the Sobolev class $(k)$ with $k \geqq 3$. We denote the space of such connections by $\mathscr{C}^k$ or simply $\mathscr{C}$ when once we have fixed $k \geqq 3$. Let *Aut denote the gauge group consisting of maps from $S^3$ to $SU(2)$ which belong to the Sobolev class $(k+1)$. *Aut$^o$ will refer to the subgroup of *Aut consisting of maps which are homotopic to the constant map $S^3 \to$ Identity.

In the rest of the paper, we will only occasionally need to distinguish between the groups *Aut and *Aut$^o$. We will let *G denote either one of them.

We will need

**Lemma 2.1.** *For* $i \geqq 2$,

i) *The Sobolev space* $H^i$ *of functions from* $S^3$ *to* $\mathbb{C}$ *of class* $(i)$ *forms a Banach algebra under pointwise multiplication.*

ii) *The multiplication* $H^i \times H^i \to H^i$ *is smooth.*

iii) *If we denote by* $\mathcal{M}$ *mappings of* $S^3$ *into* $M(2, \mathbb{C})$ *(complex* $2 \times 2$ *matrices) which are of Sobolov class* $(i)$ *then the group* *G *is a closed* $C^\infty$ *submanifold of* $\mathcal{M}$.

*Proof.* For a proof of (i) see [5], Theorem (5.23). Bilinearity and (i) imply (ii) and (iii) follows from [6, p. 78].

We have an action of *G on $\mathscr{C}$ given by

$$(\alpha, \varphi) \mapsto \varphi^{-1} \alpha \varphi + \varphi^{-1} d\varphi \equiv \alpha \circ \varphi \quad \text{for} \quad \alpha \in \mathscr{C}, \varphi \in {}^*\mathrm{G}.$$

We see from Lemma 1 that *G is a Lie group and that the above action is smooth.

The Lie algebra $\mathscr{G}$ of *G is identified with the Lie algebra of maps from $S^3$ to $\mathfrak{SU}(2)$ which are of Sobolev class $(k+1)$.

**Lemma 2.2.** *The isotropy of* *G *at any point of* $\mathscr{C}$ *is compact. In fact the isotropy group is isomorphic to the centraliser of the holonomy group in* $SU(2)$.

*Proof.* If $\varphi$ belongs to the isotropy group at $\alpha \in \mathscr{C}$ then $\varphi^{-1}\alpha\varphi + \varphi^{-1} d\varphi = \alpha$ or $d\varphi + [\alpha, \varphi] = 0$. Thus $\varphi$ is invariant under parallel translation, considered as a section of the bundle with $M(2, \mathbb{C})$ as fibre. Thus $\varphi$ is determined by $\varphi(e)$ and $\varphi(e)$ commutes with the elements of the holonomy group.

*Remark 2.3.* The group of constant functions with values in the centre of $SU(2)$ acts trivially on $\mathscr{C}$. The isotropy group of *G at $\alpha \in \mathscr{C}$ coincides with this subgroup if and only if the holonomy group is $SU(2)$; this condition in turn is easily seen (e.g. by Schur's lemma) to be equivalent to the condition: if $\beta$ is a 1-form with values in $\mathfrak{SU}(2)$, and $d\beta + [\alpha, \beta] = 0$ then $\beta = 0$. We call such connections, whose holonomy is the whole group $SU(2)$, *generic* and denote the set of generic connections by $\mathscr{C}_o$. Note that the gauge group *G acts on $\mathscr{C}_o$ and *G/($\mathbb{Z}/(2)$) acts freely. We will denote *G/($\mathbb{Z}/(2)$) by G.

**Proposition 2.4.** *The action of* *G *on* $\mathscr{C}$ *is proper.*

*Proof.* It is enough [7] to show that the map $\mu: \mathscr{C} \times {}^*\mathrm{G} \to \mathscr{C} \times \mathscr{C}$, $(\alpha, \varphi) \mapsto (\alpha \circ \varphi, \alpha)$ is closed and that the inverse image of each point by $\mu$ is compact. Lemma 2.2 shows that the inverse image of any point is compact. That $\mu$ is closed follows from

**Lemma 2.5.** *Let* $(\alpha_n, \varphi_n) \in \mathscr{C} \times {}^*\mathrm{G}$ *be a sequence such that* $\alpha_n \to \alpha$ *and* $\alpha_n \circ \varphi_n \equiv \beta_n \to \beta$ *in* $\mathscr{C}$. *Then there exists a subsequence* $\{\varphi_l\}$ *of* $\{\varphi_n\}$ *which tends to a limit* $\varphi$ *(so that* $\alpha \circ \varphi = \beta$*).*

Lemma 2.5 will follow from Lemmas 2.6–2.8. In these lemmas we use the notation of Lemma 2.5.

**Lemma 2.6.** *Let $U$ be an open co-ordinate cell in $S^3$ and $p$ a point of $U$. If there exists a subsequence $\{\varphi_l\}$ of $\{\varphi_n\}$ so that $\varphi_l(p)$ tends to a limit $g$ in $SU(2)$, then $\varphi_l$ tends uniformly on compact sets to a limit $\varphi: U \to M(2, \mathbb{C})$.*

*Proof.* We have $d\varphi_l = \varphi_l \beta_l - \alpha_l \varphi_l$, $\varphi_l(p) \to g$. *Define* $\hat{\varphi}_l = \varphi_l \varphi_l^{-1}(p)$, $\hat{\beta}_l = \varphi_l(p) \beta_l \varphi_l^{-1}(p)$. Then $d\hat{\varphi}_l = \hat{\varphi}_l \hat{\beta}_l - \alpha_l \hat{\varphi}_l$ and $\hat{\varphi}_l(p) = \text{Identity}$.

Introduce co-ordinates $(x_i)$ on $U$ with $x_i(p) = 0$, $U$ being mapped onto $\mathbb{R}^3$ by $(x_i)$. Denote by $\alpha_i(x)$ the components of a connection $\alpha \in \mathscr{C}$ in this co-ordinate system. When $x \in U$, $\alpha$, $\beta \in \mathscr{C}$ consider the system [with $t \in \mathbb{R}$, $y \in M(2, \mathbb{C})$]

$$\frac{dy_{(\alpha,\beta,x)}(t)}{dt} = \Big(\sum_i x_i \beta_i(xt)\Big) y(t) - y(t) \Big(\sum_i x_i \alpha_i(xt)\Big)$$

$$y_{(\alpha,\beta,x)}(0) = \text{Identity}.$$

This is an ordinary linear differential equation in $y(t)$ with $\alpha$, $\beta \in \mathscr{C}$ and $x \in U$ as parameters and a fixed initial condition. Then by (10.7.2.) of [8] the system has a unique solution $y_{(\alpha,\beta,x)}(t)$, defined for all $t$, $x$, $\alpha$, $\beta$, which is continuous in all four variables and differentiable in the first.

Now, note that by uniqueness $y_{(\alpha_l,\hat{\beta}_l,x)}(t) = \hat{\varphi}_l(x,t)$ and $\hat{\varphi}_l(x) = y_{(\alpha_l,\hat{\beta}_l,x)}(1)$. The lemma follows easily by continuity in $(\alpha, \beta)$.

**Lemma 2.7.** *There exists a subsequence $\{\varphi_l\}$ of $\{\varphi_n\}$ which tends uniformly on $S^3$ to a limit $\varphi$ which is a continuous map $\varphi: S^3 \to SU(2)$.*

*Proof.* Cover $S^3$ by two open cells $U$ and $U'$, choose a point $p$ in their intersection. Since $SU(2)$ is compact there exists a subsequence $\{\varphi_l\}$ such that $\varphi_l(p)$ tends to a limit. Let $V$, $V'$ be compact sets in $U$ and $U'$ respectively which also cover $S^3$. Then by Lemma 2.6, $\varphi_l$ converges uniformly on both $V$ and $V'$ and hence on $S^3$ to a continuous function $\varphi: S^3 \to M(2, \mathbb{C})$. Since $SU(2)$ is closed in $M(2, \mathbb{C})$, $\varphi$ has values in $SU(2)$.

**Lemma 2.8.** *If a subsequence $\{\varphi_l\}$ of $\{\varphi_n\}$ tends uniformly to a continuous function $\varphi: S^3 \to SU(2)$, then $\varphi$ is of class $(k+1)$ and $\varphi_l \to \varphi$ in* *G.

*Proof.* We have $\alpha_l \to \alpha$ and $\beta_l \to \beta$ in $H^k$ and $\varphi_l \to \varphi$ in $C^o$. But $d\varphi_l = \varphi_l \beta_l - \alpha_l \varphi_l \to \varphi\beta - \alpha\varphi$ in $C^o$ by Sobolev lemma; this implies that $\varphi$ is in $C^1$ and $\varphi_l \to \varphi$ in $C^1$. Similarly $\varphi_l \to \varphi$ in $C^2$ topology as $\alpha_l$, $\beta_l \in C^1$ by Sobolev. In particular $\varphi_l \to \varphi$ in the Sobolev space $H^2$. Now, since $d$ is an elliptic operator with injective symbol (on 0-forms) we see that $\varphi_l \to \varphi$ in $H^3$. We conclude by induction that $\varphi_l \to \varphi$ in $H^{k+1}$.

## 3. Some Estimates

In this section we shall derive some estimates connected with elliptic operators (whose coefficients are not necessarily $C^\infty$) arising from connections belonging to Sobolev class $(k)$. These will be needed in the rest of the paper and in particular to prove that the set $\mathscr{C}_o$ of connections whose holonomy is the whole group is a principal G-bundle.

Consider the tangent space to $\mathscr{C}$ at any point $\alpha$. This can be identified with $\mathscr{C}$ itself. Given a metric on $S^3$, we can define an inner product on $\mathscr{C}$ by

$$(\gamma, \beta) = -\int \mathrm{Tr}(\gamma \wedge {}^*\beta).$$

This gives rise to a (weak) Riemannian metric on $\mathscr{C}$.

Let $\mathscr{G}^i$ denote sections of Sobolev class $(i)$ of the adjoint bundle. For any $\alpha \in \mathscr{C}$ define $\partial_\alpha : \mathscr{C}^k \to \mathscr{G}^{k-1}$ by $(\beta, d_\alpha \Gamma) = (\partial_\alpha \beta, \Gamma)$ for $\beta \in \mathscr{C}^k$, $\Gamma \in \mathscr{G}^{k+1}$. Note that if $e(\alpha)$ denotes exterior multiplication by $\alpha$ with respect to the Lie algebra multiplication in $\mathfrak{SU}(2)$ and $i(\alpha)$ is the adjoint of $e(\alpha)$, then $d_\alpha = d + e(\alpha)$ and $\partial_\alpha = \partial + i(\alpha)$. [Note that $i(\alpha)$ is the interior multiplication by $\alpha$ defined using the metric and the Lie algebra multiplication in $\mathfrak{SU}(2)$]. Then $\Delta_\alpha = \partial_\alpha d_\alpha$ takes $\mathscr{G}^{k+1}$ to $\mathscr{G}^{k-1}$.

Note that for $\Gamma \in \mathscr{G}^{k+1}$, $\partial_\alpha d_\alpha \Gamma = 0$ if and only if $d_\alpha \Gamma = 0$ so that if $\alpha \in \mathscr{C}_o$, $\Delta_\alpha$ is injective, by Remark 2.3.

We now prove two lemmas which we will need in the proof of the next proposition.

**Lemma 3.1.** *If $v \in H^k$, $k \geqq 3$, then $v$ is a multiplier in $H^m$ for $-k \leqq m \leqq k$.*

*Proof.* It is enough to show that for $u \in H^m$, $0 \leqq m \leqq k$, $vu \in H^m$ and $u \to vu$ is continuous, for then we can define $vT$ for $T \in H^{-m}$ by duality: $\langle vT, u\rangle = \langle T, vu\rangle$. For $m \geqq 2$ this follows from the fact that $H^m$, $m \geqq 2$ forms a Banach algebra. Since $k \geqq 3$, $\varphi$ is $C^1$ (by Sobolev) and it is easy to show that $\varphi$ is a multiplier in $H^o$ and $H^1$ also.

**Lemma 3.2.** *Let $\alpha \in \mathscr{C}$, If $\Delta_\alpha u = 0$ and $u \in \mathscr{G}^{-(k-1)}$, then $u \in \mathscr{G}^{k+1}$.*

*Proof.* Write $\Delta_\alpha = \partial_\alpha d_\alpha = \Delta + \mathrm{B}$ where $\Delta = \partial d$ and $B = \partial e(\alpha) + i(\alpha) d + i(\alpha) e(\alpha)$. Then if $\Delta_\alpha u = 0$, $\Delta u = -Bu$. Since $u \in \mathscr{G}^{-(k-1)}$ we have by Lemma 3.1, $Bu \in \mathscr{G}^{-k}$. Since $\Delta$ has $C^\infty$ coefficients, we have $u \in \mathscr{G}^{-k+2}$. We see by induction that $u \in \mathscr{G}^{k+1}$.

**Proposition 3.3.** i) *Let $\alpha \in \mathscr{C}$. Then $\Delta_\alpha : \mathscr{G}^{k+1} \to \mathscr{G}^{k-1}$ is a quasi-monomorphism, i.e., its kernel is finite-dimensional and $\Delta_\alpha(\mathscr{G}^{k+1})$ is closed in $\mathscr{G}^{k-1}$. For $u \in \mathscr{G}^{k+1}$, we have*

$$|u|_{k+1} \leqq C\{|\Delta_\alpha u|_{k-1} + |u|_o\}$$

*for some constant $C$. Here $|u|_i$ denotes sum of the $L^2$-norms of partial derivatives of order $i$.*

ii) *Let $\alpha \in \mathscr{C}_o$, so that $\Delta_\alpha$ is injective. Then $\Delta_\alpha$ is actually an isomorphism.*

*Proof.* Write, as in the proof of Lemma 3.2, $\Delta_\alpha = \Delta + B$. Since $\Delta$ has smooth coefficients, we have as is well-known,

$$\begin{aligned} |u|_{k+1} &\leqq C\{|-Bu + \Delta_\alpha u|_{k-1} + |u|_o\} \\ &\leqq C\{|Bu|_{k-1} + |\Delta_\alpha u|_{k-1} + |u|_o\} \end{aligned}$$

for some constant $C$. Also

$$\begin{aligned} |Bu|_{k-1} &\leqq |(i)\alpha) d + \partial e(\alpha))u|_{k-1} + |i(\alpha) e(\alpha) u|_{k-1} \\ &\leqq C'\{|u|_k + |u|_{k-1}\} \end{aligned}$$

for some constant $C'$. On the other hand

$$|u|_l \leqq \varepsilon |u|_{k+1} + C(\varepsilon)|u|_o \quad \text{for} \quad 0 < l < k+1 \quad \text{for } \varepsilon > 0 \text{ and some function } C(\varepsilon).$$

Thus we see that, with a suitable constant $C$, we have

$$|u|_{k+1} \leqq C\{|\Delta_\alpha u|_{k-1} + |u|_o\}. \tag{1}$$

By Rellich lemma it follows that the kernel of $\Delta_\alpha$ is locally compact in the $L^2$-norm and hence is finite dimensional.

To see that $\Delta_\alpha$ has closed image, let $\mathscr{H}$ be the kernel and $W$ a topological supplement of $\mathscr{H}$. We see using the above estimate and Rellich lemma that there exists a constant $C''$ such that

$$|u|_o \leqq C''|\Delta_\alpha u|_{k-1} \quad \text{for} \quad u \in W \tag{2}$$

(see, for example [9, p. 456]). From (1) and (2) it is clear that $\operatorname{Im}\Delta_\alpha$ is closed.

ii) By i) it suffices to show that $\Delta_\alpha(\mathscr{G}^{k+1})$ is dense in $\mathscr{G}^{k-1}$, if $\alpha \in \mathscr{C}_o$. Let therefore $T \in \mathscr{G}^{-(k-1)}$ such that $T$ is zero on $\Delta_\alpha(\mathscr{G}^{k+1})$. Then we have $\Delta_\alpha * T = 0$, which implies by Lemma 3.2 that $*T \in \mathscr{G}^{k+1}$ so that since $\alpha \in \mathscr{C}_o$, $T = 0$.

We can now prove

**Proposition 3.4.** *For any $\alpha \in \mathscr{C}_o$ we have*

$$\mathscr{C} = d_\alpha(\mathscr{G}^{k+1}) \oplus (\ker \partial_\alpha)$$

*where $d_\alpha(\mathscr{G}^{k+1})$ and* $\ker \partial_\alpha$ *are closed subspaces.*

*Proof.* Let $G_\alpha = (\Delta_\alpha)^{-1} : \mathscr{G}^{k-1} \to \mathscr{G}^{k+1}$. $G_\alpha$ is continuous by Proposition 3.3, ii). Then we have $d_\alpha(\mathscr{G}^{k+1}) = \ker(1 - d_\alpha G_\alpha \partial_\alpha)$ and both spaces, being kernels of continuous operators, are closed. Since $\partial_\alpha d_\alpha \Gamma = 0$ if and only if $d_\alpha \Gamma = 0$, the sum is direct. Finally, if $\beta \in \mathscr{C}$, we have

$$\beta = d_\alpha G_\alpha \partial_\alpha \beta + (\beta - d_\alpha G_\alpha \partial_\alpha \beta)$$

with $\partial_\alpha(\beta - d_\alpha G_\alpha \beta) = 0$.

*Remark 3.5.* The above direct sum decomposition of $\mathscr{C}$ holds even if $\alpha \notin \mathscr{C}_o$, as can be seen by suitably defining $G_\alpha$.

## 4. The Space of Connections as a Principal Bundle

### *a) The Generic Connections*

**Lemma 4.1.** *The space $\mathscr{C}_o$ of generic connections is open in $\mathscr{C}$.*

*Proof.* By Remark 2.3 an element $\alpha_o$ of $\mathscr{C}$ belongs to $\mathscr{C}_o$ if and only if $d_{\alpha_o}: \mathscr{G}^{k+1} \to \mathscr{C}$ is injective. By Proposition 3.3 the image of $d_{\alpha_o}$ is also closed. Moreover, for $\alpha \in \mathscr{C}$, $\alpha \mapsto d_\alpha$ is a continuous map when we put the strong topology on the space of continuous linear maps from $\mathscr{G}^{k+1}$ to $\mathscr{C}$. From this it follows that $d_\alpha$ is injective in a neighbourhood of $\alpha_o$.

**Lemma 4.2.** *For every $\alpha \in \mathscr{C}_o$ the map* $\mathrm{G} \to \mathscr{C}$ *given by $\varphi \mapsto \alpha \circ \varphi$ is an injective immersion.*

*Proof.* The differential of the map at any point $\beta$ in the orbit is $d_\beta : \mathcal{G}^{k+1} \to \mathcal{C}$. By Proposition 3.4 the image is closed and admits a topological supplement, so that the lemma follows.

**Proposition 4.3.** *The action of* G *makes* $\mathcal{C}_0$ *a principal* G*-bundle.*

*Proof.* This proposition follows from Proposition 2.4 and Lemmas 4.12 and 4.2, using (6.2.3) of [10].

*b) The Groups* $\mathrm{Aut}_e$, $\mathrm{Aut}_e^o$

We now define the groups $\mathrm{Aut}_e$, $\mathrm{Aut}_e^o$. $\mathrm{Aut}_e$ is the (normal) subgroup of *Aut consisting of those elements $\varphi \in$ *Aut which take the value identity at a fixed point $e$ on $S^3$. (Note that as $k \geqq 3$, by Sobolev lemma $\varphi$ is of class $C^1$). Let $\mathrm{Aut}_e^o = \mathrm{Aut}_e \cap {}^*\mathrm{Aut}^o$. We will let $\mathrm{G}_e$ denote either $\mathrm{Aut}_e$ or $\mathrm{Aut}_e^o$.

Note that the groups $\mathrm{Aut}_e$, $\mathrm{Aut}_e^o$ act freely on $\mathcal{C}$. The Lie algebra $\mathcal{G}_e$ consists of elements of $\mathcal{G}$ which vanish at $e$. As in the case of G, $\mathrm{G}_e$ operates properly on $\mathcal{C}$. Also

**Lemma 4.4.** *For* $\alpha \in \mathcal{C}$ *the map* $\mathrm{G}_e \to \mathcal{C}$ *given by* $\alpha \mapsto \alpha \circ \varphi$ *is an injective immersion.*

*Proof.* Note that $\mathcal{G}_e$ is of finite co-dimension in $\mathcal{G}$ and we can write $\mathcal{G} = \mathcal{G}_e \oplus F$, where $F$ is a finite-dimensional space. The differential of the map $\mathrm{G}_e \to \mathcal{C}$ at any point $\beta$ in the orbit is $d_\beta : \mathcal{G}_e \to \mathcal{C}$. This is easily seen to be injective. By Remark (3.5), $\mathcal{C} = d_\beta(\mathcal{G}^{k+1}) \oplus \ker \partial_\beta = d_\beta(\mathcal{G}_e^{k+1}) \oplus d_\beta(F) \oplus \ker \partial_\beta$.

Finally, we have

**Proposition 4.5.** *The action of* $\mathrm{G}_e$ *makes* $\mathcal{C}$ *a principal* $\mathrm{G}_e$ *bundle.*

*Proof.* Same as Proposition 4.3.

## 5. Nonexistence of a Continuous Gauge

**Theorem 5.1.** *The* $\mathrm{Aut}^o$ *(resp.* Aut*) bundle* $\mathcal{C}_o$ *cannot be reduced to* $\mathrm{Aut}_e^o$ *(resp.* $\mathrm{Aut}_e$*). In particular these bundles do not admit sections.*

The rest of this section, and the next will be devoted to a proof of this theorem. But first we make the following

*Remark.* The $\mathrm{Aut}_e^o$ bundle is not trivial. In fact $\mathcal{C}$ is contractible while $\pi_i$ (third loop space of $SU(2)) = \pi_{i+3}(SU(2))$. But $\pi_4(S^3) = \mathbb{Z}_2$.

If the bundle were trivial, $\mathcal{C}$ would be homeomorphic to the product of $\mathrm{Aut}_e^o$ and some other topological space and $\pi_1$ ($\mathcal{C}$) would be different from zero. Nor is the $\mathrm{Aut}_e$ bundle trivial, for $\mathcal{C}$ is connected and $\mathrm{Aut}_e/\mathrm{Aut}_e^o$ discrete.

We identify $S^3$ with $SU(2)$, and the point $e$ on $S^3$ (used in the definition of $\mathrm{Aut}_e$, $\mathrm{Aut}_e^o$) with the identity in $SU(2)$.

The argument uses in a critical way, the space of left invariant forms on $SU(2)$ with values in $\mathfrak{SU}(2)$, the Lie algebra. Fix a basis of left-invariant vector fields $X_a$ such that $[X_a, X_b] = \varepsilon_{abc} X_c$ where $\varepsilon_{abc}$ is defined by $\varepsilon_{123} = +1$ and complete antisymmetry in the indices, and the corresponding dual basis of one-forms given by

$$\omega^a(X_b) = \delta_{ab}.$$

We also define a metric on $\mathfrak{SU}(2)$ by

$$(X_a, X_b) = \delta_{ab}.$$

A left-invariant Lie-algebra valued from $\varrho$ can be written as

$$\varrho = \sum_a L_\alpha \omega^a$$

where $L_a$ are elements of the Lie-algebra of $SU(2)$ (linear combinations of $X_a$). Of particular interest is the Maurer-Cartan form

$$\omega = \sum_a X_a \omega^a$$

which satisfies

$$d\omega + \tfrac{1}{2}[\omega, \omega] = 0.$$

A left-invariant form $\varrho = \sum_a L_a \omega^a$ gives a mapping $X_a \mapsto L_a$ and there is a one-one correspondence with $3 \times 3$ matrices $M_\varrho$ [which represent vector space endomorphism of the Lie algebra $\mathfrak{SU}(2)$]:

$$\varrho \leftrightarrow M_\varrho \quad \text{by} \quad L_a = M_\varrho X_a.$$

The Maurer-Cartan form corresponds to the identity homomorphism: $M_\omega = \text{Identity}$.

The curvature is a Lie-algebra valued two form:

$$F = d\varrho + \tfrac{1}{2}[\varrho, \varrho].$$

On left invariant vector fields $X$, $Y$, we have

$$F(X, Y) = [M_\varrho(X), M_\varrho(Y)] - M_\varrho([X, Y])$$

so that $F = 0$ if and only if $M_\varrho$ represents a Lie algebra homomorphism. With respect to our earlier choice of basis of left invariant vector fields, this means that $F = 0$ if and only if either $M_\varrho = 0$ or $M_\varrho \in SO(3)$ (with respect to the Lie algebra metric given earlier).

We will need the following lemmas

**Lemmas 5.2.** *Let $N$ denote the space of left-invariant forms $\varrho$ such that rank of $M_\varrho$ is $\geqq 2$, and $M_\varrho \notin SO(3)$. Then* G *acts freely at any point in $N$. There are no equivalences in $N$ under* $G_e$.

*Proof.* Let $\varrho = \sum_a L_a \omega^a$. If a gauge-transformation $g$ fixes $\varrho$

$$g^{-1} F g = F.$$

By hypothesis $F \neq 0$. Consider the image of $F$ at any point $x \in S^3$. If $\operatorname{Im} F$ is of dimension $\geqq 2$, $g(x) = \text{Identity}$; and if $\operatorname{Im} F$ is in the one-dimensional subspace $\mathfrak{h}$ of $\mathfrak{SU}(2)$, $g(x)$ is in the corresponding one-parameter subgroup $H$. By left invariance of $F$ we have thus two possibilities. Either $g(x) = \pm \text{Identity}\ \forall x \in S^3$ or, $\operatorname{Im} F \subset \mathfrak{h}$ and $g(x) \in H\ \forall x \in S^3$. In the second case.

$$g^{-1} L_a g + (g^{-1} dg)_a = L_a$$

which implies $g^{-1}L_a g - L_a \in \mathfrak{h}$. But $g^{-1}L_a g - L_a$ is orthogonal to $\mathfrak{h}$, and hence zero. Thus $L_a \in \mathfrak{h}$ for each $a$, and $\varrho \notin N$.

Now suppose that $g \in G_e$ takes $\varrho = \sum_a L_a \omega^a$ to $\varrho' = \sum L'_a \omega^a$, $\varrho' \neq \varrho$. Then since $g(e) = \text{Identity}$ we have $F = F'$ and again $g$ takes values in a one-parameter subgroup $H$ of $SU(2)$. We also have

$$g^{-1}L_a g + (g^{-1}dg)_a = L_a$$

which implies $L_a - L_a = (g^{-1}dg)_a(e) \in \mathfrak{h}$

$$g(x)L'_a g^{-1}(x) - L'_a \in \mathfrak{h}$$

so that again $\varrho, \varrho' \notin N$.

*Proof of Theorem 5.1.* As in Lemma 5.2, let $N$ denote the space of left invariant forms $\varrho$ such that rank $M_\varrho \geqq 2$ and $\varrho$ is not in the adjoint orbit of the Maurer-Cartan form. Consider the map $\eta : N \to \mathscr{C}_o/G_e$ induced by the canonical map $\mathscr{C}_o \to \mathscr{C}_o/G_e$. By Lemma 5.2 this map is injective. Let $N' = \eta(N)$

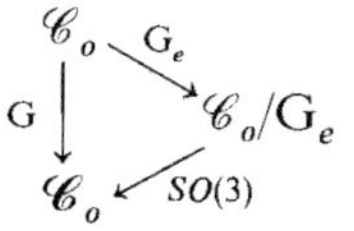


If the G bundle $\mathscr{C}_o$ could be reduced to the normal subgroup $G_e$, the $G/G_e = SO(3)$ fibration $\mathscr{C}_o/G_e \to \mathscr{C}_o/G$ would admit a (continuous) section. Note that the action of $SO(3)$ on $w'$ and the action of $SU(2)/\mathbb{Z}_2 \approx SO(3)$ on $N$ by $M_\varrho \to g M_\varrho g^{-1}$ commute. Hence the $SO(3)$ bundle $N$ would be trivial. But this is not the case as will be proved in the next section (Theorem 6.2).

## 6. Nontriviality of the "Three-Body" Bundle

Let $M(3)$ denote the vector space of $3 \times 3$ real matrices. Consider the right action of $SO(3)$ on $M(3)$ by $(B, g) \to g^{-1}B$, $g \in SO(3)$, $B \in M(3)$.

*Remark.* If we identify $M(3)$ with $(\mathbb{R}^3)^3$ by means of the map $B \mapsto (Be_1, Be_2, Be_3)$ where $\{e_1, e_2, e_3\}$ is the canonical basis in $\mathbb{R}^3$, the above action goes over to the diagonal action $((f_1, f_2, f_3), g) = (g^{-1}f_1, g^{-1}f_2, g^{-1}f_3)$, $f_i \in \mathbb{R}^3$. Hence the term "Three-Body Bundle".

**Lemma 6.1.** *The action of* $SO(3)$ *on* $M(3)$ *is free exactly at the set of matrices of rank* $\geqq 2$. *The isotropy group of a matrix of rank* 1 *is isomorphic to* $SO(2)$.

*Proof.* If $g^{-1}B = B$, every point of the image of $B$, considered as a linear map, is fixed by $g^{-1}$. Therefore if rank $B \geqq 2$, $g = \text{Identity}$ and if rank $B = 1$, the isotropy group is isomorphic to the special orthogonal group of the orthogonal complement of the image of $B$.

**Theorem 6.2.** *Let* $M_o$ *denote the manifold of* $3 \times 3$ *real matrices of rank* $k \geqq 2$. *The principal* $SO(3)$ *bundle* $M_o$ *(with the action* $(B, g) \to g^{-1}B$*) is not trivial on the complement of any point in* $M_o/SO(3) = \mathbf{M}_o$.

We first prove

**Lemma 6.3.** *The orbit space $M_0/SO(3)$ is homeomorphic to $\mathbb{R}\times(S^5-P)$ where $P$ is a submanifold of $S^5$ homeomorphic to the projective plane $\mathbb{P}^2(\mathbb{R})$.*

*Proof.* Consider first the action of $O(3)$ on $M(3)$ by multiplication on the left by $g^{-1}$, $g\in O(3)$. The quotient space is homeomorphic to the space of positive semidefinite matrices. This follows from remarking that if $B\in M(3)$ the non-negative square root $\sqrt{B'B}$ of $B'B$ is equivalent to $B$ under this action. This fact is well-known for nonsingular $B$; if $B$ is singular, let $B_n\to B$ with $B_n$ nonsingular, so that $B_n=g_n\sqrt{B_n'B_n}$ with $g_n\in O(3)$. Choosing a subsequence of $g_n$ tending to $g\in O(3)$ we see that $B=g\sqrt{B'B}$.

Now consider the diagram

$$\begin{array}{ccc} M(3) & \xrightarrow{SO(3)} & \mathscr{P}' \\ {\scriptstyle O(3)}\downarrow & \swarrow{\scriptstyle \mathbb{Z}_2} & \\ \mathscr{P} & & \end{array}$$

where $\mathscr{P}'$ is a (ramified) two sheeted covering of $\mathscr{P}$. We claim that there is ramification precisely over the set of positive semidefinite matrices, which are not definite. This follows from the following fact: If $B$ is a singular matrix there exists $g\in O(3)$ with $\det g=-1$ and $g^{-1}B=B$. (To see this it is sufficient to consider the case $B=P$ is singular and positive semidefinite. If $V\neq 0$ is the null space of $P$ and $h^{-1}P=P$ $(h^{-1}\in O(3))$, $h^{-1}$ leaves $V$ and $V_\perp=\operatorname{Im}P$ invariant. We can multiply $h^{-1}|_V$ by a suitable constant, without changing $h^{-1}|_{V_\perp}$, to get $g\in O(3)$ with $\det g=-1$ and $g^{-1}P=P$).

Now let $R$ denote the image of $\mathscr{P}-0$ (0 denoting the zero matrix) in the five-dimensional projective space associated with the vector space of $3\times 3$ symmetric matrics, $R'$ will denote the subset of $R$ corresponding to non-positive-definite matrices. The pair $(R,R')$ is homeomorphic to $(D^5,S^4)$ where $D^5$ denotes the closed 5-dimensional disc[1]. In fact, consider, in the space of symmetric matrices, the hyperplane $S$, consisting of elements $B$ such that $\operatorname{Tr}B=1$. Then $S\cap\mathscr{P}$ is mapped homeomorphically into $R$, mapping positive semidefinite matrices onto $R'$. It is clear that $S\cap\mathscr{P}$ is convex and compact and its interior is $S\cap\mathscr{P}_+$ where $\mathscr{P}_+$ denotes the set of positive definite matrices (Compactness is immediate since any element of $S\cap\mathscr{P}$ can be transformed by inner conjugation by $O(3)$ into a diagonal matrix $[\lambda_1,\lambda_2,\lambda_3]$ with $\lambda_i\geq 0,\sum_i\lambda_i=1$). It follows then, as is well known, that $(S\cap\mathscr{P},\operatorname{bd}S\cap\mathscr{P})$ is homeomorphic to $(D^5,S^4)$, (See, for example [11, p. 51]).

Now $(M(3)-0)/SO(3)$ is homeomorphic to the product of $\mathbb{R}$ and the space obtained by doubling $R$ along $R'$. This follows from the nature of the ramification locus of the map $\mathscr{P}'\to\mathscr{P}$. Since $(R,R')$ is homeomorphic to $(D^5,S^4)$ the corresponding double is homeomorphic to $S^5$. Hence $(M(3)-0)/SO(3)$ is homeomorphic to $\mathbb{R}\times S^5$ (and $M(3)/SO(3)$ to $\mathbb{R}^6$).

The subspace of $R'$ corresponding to quadratic forms of rank 1 is homeomorphic to $\mathbb{P}^2(\mathbb{R})$. In fact if $Q$ is a (positive semidefinite) quadratic form of rank 1, $Q$ defines a positive-definite quadratic form on the 1-dimensional space $Q$/(Nullity of

[1] This fact pointed out to us by R. R. Simha, who also supplied the proof

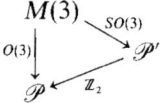

$Q$) and in a one-dimensional space there is, upto a scalar multiple, a unique positive definite quadratic form Thus the above space is homeomorphic to the projective space of 1-dimensional quotient subspaces of $\mathbb{R}^3$. (A similar interpretation of quadratic forms of rank 2 gives a decomposition of $S^4 \simeq R'$ into $\mathbb{P}^2$ and a disc bundle over $\mathbb{P}^2$).

Thus $M_o = M_o/SO(3)$ is homeomorphic to $\mathbb{R} \times (S^5 - \mathbb{P}^2)$.

*Proof of Theorem 6.2.* We first compute some homology groups of $S^5 - \mathbb{P}^2$. We have the exact sequence

$$H^{i-1}(S^5) \to H^{i-1}(\mathbb{P}^2) \to H^i(S^5, \mathbb{P}^2) \to H^i(S^5)$$

and isomorphisms

$$H^i(S^5, \mathbb{P}^2) \simeq H^i_c(S^5 - \mathbb{P}^2) = H_{5-i}(S^5 - \mathbb{P}^2)$$

where $H^i_c$ denotes cohomology with compact supports, the second isomorphism is given by Poincaré duality and the coefficient group is $\mathbb{Z}$. This gives, in particular.

$$H_2(S^5 - \mathbb{P}^2, \mathbb{Z}) \simeq H^2(\mathbb{P}^2, \mathbb{Z}) \simeq \mathbb{Z}/(2) \quad \text{(Alexander duality).}$$

Now $M(3) - 0$ is homeomorphic to $\mathbb{R} \times S^8$, and the space of rank 1 matrices is homeomorphic to $\mathbb{R} \times E$ where $E$ is a 4-dimensional subspace of $S^8$, so that $M_o$ is homeomorphic to $\mathbb{R} \times (S^8 - E)$.

From the exact sequence

$$\begin{array}{c} H^{i-1}(S^8) \to H^{i-1}(E) \to H^i(S^8, E) \to H^i(S^8) \\ \wr\wr \\ H_{8-i}(S^8 - E) \end{array}$$

We see that $H_2(S^8 - E) \approx H^5(E) = 0$. Thus we have $H_2(M_o) = \mathbb{Z}/(2)$ and $H_2(M_o) = 0$. It now follows that the bundle $M_o$ does not admit a section, for, if $\tilde{\sigma}$ were a section, the composite map $\pi_* \circ \tilde{\sigma}_*$ below would be the identity:

$$H_2(M_o) \xrightarrow[\tilde{\sigma}^*]{} H_2(M_o) \xrightarrow[\pi^*]{} H_2(M_o),$$

while $H_2(M_o) \neq 0$ and $H_2(M_o) = 0$.

Now if $p \in M_o/SO(3)$, $\pi^{-1}(p)$ is a submanifold of $M_o$ of co-dimension 6 and $p$ is a point in a 6-dimensional manifold. It follows, as the co-dimension is $\geqq 4$, that $H_2(M_o - p) \simeq H_2(M_o) = \mathbb{Z}_2$ and $H_2(M_o - \pi^{-1}(p)) \simeq H_2(M_o) = 0$. (See [12, p. 41]). The theorem then follows, as above.

*Remarks.* 1) The $SO(3)$ bundle $M_o$ cannot be reduced to any Lie subgroup of $SO(3)$. Any (connected) Lie subgroup of $SO(3) \neq \{e\}$, is isomorphic to $SO(2)$ and if there were a reduction, the corresponding complex line bundle would have Chern class zero as $H^2(M_o) = 0$. Hence the line-bundle would be trivial – but this would imply that $M_o$ itself is trivial.

2) A similar proof shows the following: The $SO(n)$ bundle of $n \times n$ matrices $(n \geqq 2)$ $B$ with rank $B \geqq n-1$ is nontrivial.

3) A simpler proof of Theorem 6.2, which however does not give information about the structure of $M_o$, can be given as follows. For $p \in M_o$, the codimension of $M(3) - M_o - \pi^{-1}(p)$ in $M(3)$ is greater than or equal to 3. Hence

$\pi_1(M_o-\pi^{-1}(p))=\pi_1(M(3))=0$. If the bundles were trivial, $M_o-\pi^{-1}(p)$ would be isomorphic to $SO(3)\times(M_o-p)$ and since $\pi_1(SO(3)\approx\mathbb{Z}/2$, $M_o-\pi^{-1}(p)$ could not be simply connected. The proof works for all $n$.

## 7. The Coulomb Connection, Its Curvature, and Holonomy

We now define a connection on the bundle $\mathscr{C}_o$: we take the horizontal space at $\alpha\in\mathscr{C}_o$ to be the space $H_\alpha=\{\beta\in\mathscr{C}\,|\,\partial_\alpha\beta=0\}$. The horizontal space is easily seen to be invariant under the group action, as the metric is invariant.

**Lemma 7.1.** *The above definition of horizontal space gives a connection on $\mathscr{C}_o$. The connection form at $\alpha\in\mathscr{C}_o$ is given by $G_\alpha\partial_\alpha$ where $G_\alpha$ is the inverse of $\Delta_\alpha=\partial_\alpha d_\alpha:\mathscr{G}^{k+1}\to\mathscr{G}^{k-1}$.*

*Proof.* By Proposition 3.3 (ii), $G_\alpha$ is well-defined. On a vertical vector $\beta=d_\alpha\Gamma$, $\Gamma\in\mathscr{G}^{k+1}$, we have $G_\alpha\partial_\alpha d_\alpha\Gamma=\Gamma$. On horizontal vectors $G_\alpha\partial_\alpha$ is zero by definition. Since $\Delta_\alpha:\mathscr{G}^{k+1}\to\mathscr{G}^{k-1}$ is a family of isomorphisms depending smoothly on $\alpha$ it follows that the inverse $G_\alpha$ also depends smoothly on $\alpha$.

We will denote by $\hat{\omega}$ the above connection form, and call this the *Coulomb connection* on $\mathscr{C}_o$.

**Lemma 7.2.** *Let $\beta_1,\beta_2$ be horizontal vectors at $\alpha\in\mathscr{C}_o$. If $\Omega$ is the curvature form corresponding to $\hat{\omega}$, we have*

$$\Omega(\beta_1,\beta_2)=G_\alpha(i(\beta_1)\beta_2-i(\beta_2)\beta_1)$$

*where $i(\beta)$ denotes interior product with respect to $\beta$.*

*Proof.* Consider $\beta_i(i=1,2)$ to be the infinitesimal generator of the one-parameter group of transformations $\alpha\mapsto t_i\beta_i+\alpha$. Then the vector fields $\beta_i$ satisfy $[\beta_1,\beta_2]=0$. Then

$$\begin{aligned}\Omega(\beta_1,\beta_2)&=d\hat{\omega}(\beta_1,\beta_2)\\&=\frac{\partial}{\partial t_1}\hat{\omega}_{(t_1,0)}(\beta_2)\Big|_{t_1=0}-\frac{\partial}{\partial t_2}\hat{\omega}_{(0,t_2)}(\beta_1)\Big|_{t_2=0}\\&=\frac{\partial}{\partial t_1}(G_{\alpha+t_1\beta_1}\partial_{\alpha+t_1\beta_1}(\beta_2))\Big|_{t_1=0}-(t_1\leftrightarrow t_2)\\&=G_\alpha(i(\beta_1)\beta_2-i(\beta_2)\beta_1)\quad\text{since}\quad\frac{\partial}{\partial t_1}\partial_{\alpha+t_1\beta_1}=i(\beta_1).\end{aligned}$$

We now calculate the 'holonomy group' of $\hat{\omega}$. *From now on we use as the metric on $S^3=SU(2)$. a biinvariant metric on $SU(2)$.*

Let $\omega$ be the Maurer-Cartan form and $\omega'=\omega/2$. Note that for left-invariant vector fields $X$ and $Y$, $F_{\omega'}(X,Y)=-\frac{1}{2}[X,Y]$ so that $\omega'\in\mathscr{C}_o$. Then we have

**Proposition 7.3.** *The linear subspace generated by elements of the form $i(\beta_1)\beta_2-i(\beta_2)\beta_1$ where $\beta_1,\beta_2$ are smooth horizontal vectors at $\omega'$ coincides with the space of smooth $\mathfrak{SU}(2)$-valued functions (which we will denote by $\mathscr{G}^\infty$).*

*Proof.* Note that $\mathscr{G}^\infty$ can be identified with the space of smooth 1-forms by $X_a \leftrightarrow \omega^a$. We shall prove that under the above identification, we can cover all smooth 1-forms.

We first construct 'enough' horizontal vectors. Note that with respect to a biinvariant metric any left-invariant form $\gamma$ satisfies $\partial\gamma = 0$ [By left-invariance of metric, $\gamma$ is a constant function, so that we have a linear map from the space of left-invariant forms into $\mathbb{R}$. Also, by right-invariance of metric $\partial R_g^* \gamma = R_g^* \partial\gamma = \partial\gamma$, so that this is a homomorphism of the adjoint representation of $SU(2)$ into the trivial representation. By Schur's. Lemma, $\partial\gamma = 0$]. Therefore, since $i(\omega)\omega = 0$,

$$\partial_{\omega'}\omega = 0.$$

Also, if $\zeta$ is a closed one-form (with values in $\mathbb{R}$) then $d_{\omega'}(\zeta \wedge \omega) = d\zeta \wedge \omega + \zeta \wedge d_{\omega'}\omega = 0$ since $d_{\omega'}\omega = d\omega + \frac{1}{2}[\omega, \omega] = 0$. Therefore

$$\partial_{\omega'} * (\zeta \wedge \omega) = 0.$$

If $\beta_i = \sum_a \beta_{ia}\omega^a (i = 1, 2)$ we have $i(\beta_1)\beta_2 - i(\beta_2)\beta_1 = \sum_a [\beta_{1a}, \beta_{2a}]$. Now

i) Let $\zeta$ be a closed 1-form. Take $\beta_1 = *(\zeta \wedge \omega)$, $\beta_2 = \omega'$. Then

$$\beta_1 = \sum_{a<b} [\zeta_a X_b - \zeta_b X_a] * (\omega^a \wedge \omega^b)$$

$$= \sum_{a<b} [\zeta_a X_b - \zeta_b X_a]\, \varepsilon_{abc}\omega^c = \sum \varepsilon_{abc}\zeta_a X_b \omega^c$$

$$i(\beta_1)\beta_2 - i(\beta_2)\beta_1 = \tfrac{1}{2}\sum \varepsilon_{abc}\zeta_a [X_b, X_c]$$

$$= \tfrac{1}{2}\sum_a \varepsilon_{abc}\varepsilon_{dbc}\zeta_a X_d$$

$$= \sum \zeta_a X_a \leftrightarrow \sum_a \zeta_a \omega^a = \zeta.$$

where $\leftrightarrow$ denotes the above-mentioned identification.

ii) Take $\beta_1 = *(\zeta_1 \wedge \omega)$, $\beta_2 = *(\zeta_2 \wedge \omega)$ with $\zeta_1$, $\zeta_2$ closed 1-forms. Then

$$i(\beta_1)\beta_2 - i(\beta_2)\beta_1 = \sum [\varepsilon_{abc}\zeta_{1b}X_c, \varepsilon_{ade}\zeta_{2d}X_e]$$

$$= \sum (\zeta_{1b}\zeta_{2b}[X_c, X_c] - \zeta_{1b}\zeta_{2c}[X_c, X_b]$$

$$= \sum \varepsilon_{abc}\zeta_{1b}\zeta_{2c}X_a \leftrightarrow *(\zeta_1 \wedge \zeta_2).$$

Thus closed 1-forms are clearly covered, and also co-closed 1-forms of the type $*(\beta_1 \wedge \beta_2)$ where $\beta_1$ and $\beta_2$ are closed. The next lemma completes the proof of the proposition.

**Lemma 7.4.** *Any smooth co-closed 1-form $\beta$ can be written as a finite sum $*\sum_p \beta_{1p} \wedge \beta_{2p}$ with $\beta_{1p}, \beta_{2p}$ closed, smooth 1-forms.*

*Proof.* It is enough to show that any smooth closed 2-form $\eta$ can be written as $\sum_p \beta_{1p} \wedge \beta_{2p}$ with $\beta_{1p}, \beta_{2p}$ smooth and closed. To see this, write $\eta = d\psi$ where $\psi$ is some smooth 1-form. By embedding $S^3$ in $\mathbb{R}^4$ and using, for instance, the retraction $\mathbb{R}^3-0\rightarrow S^3$ it is clear that $\psi$ can be written as

$$\psi=\sum_{p=1}^{4}\varphi_p dx_p$$

where $\varphi_p$ are smooth functions on $S^3$. Then $\eta=\sum_{p=1}^{4}d\varphi_p\wedge dx_p$ and the lemma is proved.

Now we can prove

**Theorem 7.5.** *Let $\omega'=\frac{\omega}{2}$ where $\omega$ is the Maurer-Cartan form. Then $\omega'\in\mathscr{C}_o$ and the set of values of the curvature form $\Omega$ of the Coulomb connection (defined using a biinvariant metric on $SU(2)$) at $\omega'$ is dense in the gauge algebra.*

*Proof.* Note that $\mathscr{G}^{\infty}$ is dense in $\mathscr{G}^{k-1}$. Then the theorem follows from Proposition 7.3 and the fact that $G_{\omega'}$ is an isomorphism.

*Note.* For the purposes of the present paper, the restricted holonomy group at a point of $\mathscr{C}_o$ is defined as in the finite-dimensional case. It is a differentiably arcwise connected subgroup of G.

**Lemma 7.6.** *Let $\beta_1,\beta_2\in H_\alpha$, $\alpha\in\mathscr{C}_o$. Then $\Omega(\beta_1,\beta_2)$ is the tangent vector to a curve in the restricted holonomy group at $\alpha$.*

*Proof.* This follows from the well-known geometric interpretation of curvature (see, e.g. [13, p. 75]).

**Proposition 7.7.** *Let* L *be a connected Banach Lie group,* $\mathrm{L}_o$ *a differentiably arcwise connected subgroup of* L. *Let $\mathscr{L}_o$ denote the subset of $\mathscr{L}$, the Lie algebra of* L, *consisting of tangent vectors to (piecewise smooth) curves in* $\mathrm{L}_o$ *through the Identity. If $\mathscr{L}_o$ is dense in $\mathscr{L}$, then* $\mathrm{L}_o$ *is dense in* L.

*Proof.* It is easily checked that $\mathscr{L}_o$ is a subalgebra of $\mathscr{L}$. Let then $X\in\mathscr{L}_o$. We shall show that $\exp X\in\bar{\mathrm{L}}_o$. Let $\gamma(t)$ be a curve in $\mathrm{L}_o$ with $\gamma(0)=e$ and $\dot\gamma(0)=X$. For small $t$, $\gamma(t)=\exp Z(t)$, $Z(t)\in\mathscr{L}$. Then $\frac{Z(t)}{t}\rightarrow\dot Z(0)=X$. Thus

$$\begin{aligned}\exp X&=\lim_{n\to\infty}\exp\left(nZ\left(\frac{1}{n}\right)\right)\\&=\lim_{n\to\infty}\left[\exp\left(Z\left(\frac{1}{n}\right)\right)\right]^n\\&=\lim_{n\to\infty}\left[\gamma\left(\frac{1}{n}\right)\right]^n.\end{aligned}$$

As $\left[\gamma\left(\frac{1}{n}\right)\right]^n\in\mathrm{L}_o$, it follows that $\exp X\in\bar{\mathrm{L}}_o$. Let $\mathscr{U}$(resp. $U$) be a neighbourhood of 0 (resp. $e$) in $\mathscr{L}$ (resp. $L$) such that $\exp:\mathscr{U}\rightarrow\mathrm{U}$ is a diffeomorphism. Now $\mathscr{L}_o\cap\mathscr{U}$ is dense in $\mathscr{L}\cap\mathscr{U}$ and $\exp(\mathscr{L}_o\cap\mathscr{U})=\bar{\mathrm{L}}_o\cap L$; hence $\bar{\mathrm{L}}_o\cap U=U$ so that $\mathrm{L}_o\cap U$ is dense in $U$. Since $U$ generates L, the lemma follows.

**Lemma 7.8.** *Let P be a principal bundle with structure group* L, *with both* $P$, L *connected. Let there be a connection on P with holonomy group* $\mathrm{L}_o$, *such that* $\bar{\mathrm{L}}_o = \mathrm{L}$. *Then if* $x \in P$, *the set of points in P which can be joined to x by horizontal paths is dense in P.*

*Proof.* Note that $P/\mathrm{L}$ is connected. Then the lemma follows from the fact that a dense set of points in the fibre through $x$ can be joined to $x$ by horizontal paths.

If we let $^*\mathscr{C}_o$ denote the connected component of $\mathscr{C}_o$ containing $\omega'$, it is clear that $^*\mathscr{C}_o$ is a principal $\mathrm{Aut}^o$ bundle. From Theorem 7.5, Lemma 7.6 and Proposition 7.7 it follows that a dense set of points in $^*\mathscr{C}_o$ can be connected to $\omega'$ by horizontal paths.

*Acknowledgement.* Our warmest thanks are due to P. P. Divakaran for encouragement and several illuminating conversations. We are also grateful to M. K. V. Murthy, M. V. Nori, M. S. Raghunathan, R. R. Simha and G. A. Swarup for many fruitful discussions.

## References

1. Dirac, P.A.M.: Lectures on quantum mechanics. New York: Belfer Graduate School Science, Yeshiva University 1964
2. Faddeev, L.D.: The Feynman integral for singular Lagrangians. Theor. Math. Phys. **1**, 3–18 (1963)
3. Gribov, V.N.: Instability of non-abelian gauge theories and impossibility of choice of Coulomb gauge. SLAC Translation **176**, (1977)
4. Singer, I.M.: Some remarks on the Gribov ambiguity. Commun. Math. Phys. **60**, 7–12 (1978)
5. Adams, R.A.: Sobolev spaces. New York, San Francisco, London: Academic Press 1975
6. Eels,Jr., J.: A setting for global analysis. Bull. Am. Math. Soc. **72**, 751–807 (1966)
7. Bourbaki, N.: Topologie générale, Chapt. 3–4. Paris: Hermann 1960
8. Dieudonné, J.: Foundations of modern analysis, Vol. 1. New York, London: Academic Press 1969
9. Kodaira, K., Nirenberg, L., Spencer, D.C.: On the existence of deformations of complex analytic structures. Ann. Math. **68**, 450–459 (1958)
10. Bourbaki, N.: Variétés différentielles et analytiques (Fascicule de resultats), Paragraphes 1 à 7. Paris: Hermann 1967
11. Seifert, H., Threlfall, W.: Lehrbuch der Topologie. New York: Chelsea 1947
12. Milnor, J.: Lectures on the *h*-cobordism theorem. Princeton: Princeton University Press 1965
13. Koszul, J.L.: Lectures on fibre bundles and differential geometry. Bombay: Tata Institute of Fundamental Research 1960

Communicated by J. Glimm

Received November 8, 1978

Proc. Indian Acad. Sci. (Math. Sci.), Vol. 90, Number 2, April 1981, pp. 125–128. 125


# Polarisations on an abelian variety

*By*

M S NARASIMHAN and M V NORI

## 1. Statement of the theorem and preliminaries

We shall prove the following.

*Theorem* 1.1. Let $X$ be an abelian variety over an algebraically closed field $k$ and let Aut $(X)$ denote the automorphism group of $X$. If, for any integer $d \neq 0$, $P_d$ denotes the subset of the Neron-Severi group of $X$ consisting of elements of degree $d$, then there are only finitely many orbits under the natural action of Aut $(X)$ on $P_d$.

For the definition of degree, see § 1.6.

In particular, if $P$ denotes the set of principal polarisations on $X$, there are only finitely many orbits under the natural action of Aut $(X)$ on $P$. This result was conjectured in the case $k = \mathbf{C}$, by Martens ([3], p. 47).

On applying Torelli's theorem we have

*Corollary* 1.2. There are, upto isomorphism, only finitely many smooth irreducible curves over $k$, having a given abelian variety as the Jacobian.

In the proof of theorem 1.1, we will use

*Theorem* 1.3. (Borel and Harish-Chandra) Let $G$ be a reductive algebraic group defined over $\mathbf{Q}$ and let $\Gamma$ be an arithmetic subgroup of $G$ [2]. Let $\rho : G \to GL(V)$ be a rational representation defined over $\mathbf{Q}$ and $L$ a lattice of $V$ invariant under $\Gamma$. If $O$ is a closed orbit of $G$, then $O \cap V$ consists of a finite number of orbits of $\Gamma$.

For a proof see ([1], theorem in § 9.11) or ([2], theorem 6.9).

1.4. *Aut* $(X)$ *as an arithmetic group*. Let $B$ denote the ring of endomorphisms of $X$ and let $B_K = B \otimes K$ for any field $K$ containing $\mathbf{Q}$. Then $B_{\mathbf{Q}}$ is a finite dimensional semi-simple algebra which contains $B$ as a lattice ([4], theorem 3 on p. 176; also p. 178). Moreover there is a homogeneous polynomial function, with rational coefficients, $\phi$ defined over $B$ such that $\phi(ab) = \phi(a)\phi(b)$ for $a, b \in B$ and $\phi(a) = \deg a$, for $a \in B$, where deg $a$ denotes the degree of the endomorphism $a$ ([4], theorem 2, p. 174; also p. 63).

*Lemma* 1.5. Let $\alpha \in B_K$, with $\phi(\alpha) \neq 0$. Then $\alpha$ is a unit of $B_K$.

*Proof.* For $\alpha \in B$, $\phi(\alpha) = \deg \alpha = n \neq 0$, there is an element $\beta$ of $B$ such that $\alpha\beta = \beta\alpha = n_X$ ([4], p. 169) proving that $\alpha$ is a unit of $B_{\mathbf{Q}}$.

Because $\phi$ is homogeneous, this proves the lemma for $K = Q$. The general case follows from this. In fact, let $B = \Pi S_i$ where $S_i$ are simple algebras over

**Q**. It then follows that $\phi = \Pi N_i^{k_i}$ with $k_i > 0$ where $N_i$ is the reduced norm in $S_i$. Since the determinant of the regular representation of $S_i \otimes_{\mathbf{Q}} K$ is a positive power of $N_{i,K}$ (the extension of $N_i$ to $S_i \otimes_{\mathbf{Q}} K$), we see that an element $a$ of $S_i \otimes K$ is a unit if and only if $N_{i,k}^{k_i}(a) \neq 0$. Thus $a$ is a unit if and only if $\phi(a) \neq 0$.

Putting

$$G_K = \{a \in B_K \mid \phi(a) = 1\}$$

for all fields $K$ containing **Q**, one gets a reductive algebraic group $G$ defined over **Q**, as $B_{\mathbf{Q}}$ is a semi-simple algebra. Moreover $G$ contains

$$\text{Aut}\ (X) = \{a \in B \mid a \text{ is unit}\} = \{a \in B \mid \phi(a) = \deg a = 1\}$$

as an arithmetic group.

1.6. *Rosati involution and the action of Aut (X) on the Neron-Severi group*

In this section, we construct a representation of $G$ defined over **Q**, which when restricted to the arithmetic subgroup Aut $(X)$ becomes the natural representation of Aut $(X)$ on the Neron-Severi group.

Let $\quad NS(X) = \text{Pic}\ X/\text{Pic}^\circ X$

denote the Neron-Severi group of $X$. Let $n \in NS(X)$ and $L$ a line bundle representing $n$. The homomorphism

$$\phi_L : X \to \text{Pic}^\circ X = \hat{X} \text{ (the dual abelian variety)}$$

defined by $\phi_L(x) = T_x^*(L) \otimes L^{-1}$ depends only on $n$ and will be denoted by $\phi_n$. We define

$$\deg n = \begin{cases} \deg \phi_n & \text{if } \phi_n \text{ is an isogeny} \\ 0 & \text{otherwise.} \end{cases}$$

Fix, once for all, an element $L_0$ in $NS(X)$ corresponding to an ample line bundle. Let $d_0 = \deg L_0$. For any endomorphism $f$ of $X$, put

$$f' = \phi_{L_0}^{-1} \circ \hat{f} \circ \phi_{L_0} \in B_{\mathbf{Q}},$$

where $\hat{f} : \hat{X} \to \hat{X}$ in the transpose of $f$. The map $f \mapsto f'$ extends to an involution $\theta_{\mathbf{Q}}$ of $B_{\mathbf{Q}}$, i.e., $\theta_{\mathbf{Q}}$ is a linear isomorphism of $B_{\mathbf{Q}}, \theta_{\mathbf{Q}}^2 = \text{identity}$, $\theta_{\mathbf{Q}}(ab) = \theta_{\mathbf{Q}}(b)\,\theta_{\mathbf{Q}}(a)$, and $\theta_{\mathbf{Q}}(a) = a'$ for $a \in B$ ($\theta_{\mathbf{Q}}$ is called the Rosati involution associated to $L_0$). Let $\theta_K = \theta_{\mathbf{Q}} \otimes Id_K$; then $\theta_K$ is an involution of $B_K$ and the $\theta_K : G_K \to G_K$, $\mathbf{Q} \subset K$, define a morphism $\theta : G \to G$.

If $\quad n \in NS(X)$, put $\rho(n) = \phi_{L_0}^{-1} \circ \phi_n$.

Then $\rho$ gives an injection of $NS(X)$ into $B$ and $\rho(NS(X)) = F$ is a lattice in

$$\mathbf{Q} \cdot F = S_{\mathbf{Q}} = \{a \in B_{\mathbf{Q}} \mid \theta_{\mathbf{Q}}(a) = a\} \text{ ([4], § 20, p. 190).}$$

Let $\quad S_K = \{b \in B_K \mid \theta_K(b) = b\}$.

Consider the following representation $\pi$ of $G$ on $S$ defined over $\mathbf{Q}$ : for $g \in G_K$ and $s \in S_K$ put $\pi(g^{-1}) s = \theta(g) sg$. For $n \in NS(X)$ and $g \in \mathrm{Aut}(X)$, let $g^*(n)$ denote the pull-back of $n$ by $g$. We then have

$$\theta(g) \rho(n) g = \phi_{L_0}^{-1} \hat{g} \phi_{L_0} \circ \phi_{L_0}^{-1} \phi_n \circ g = \phi_{L_0}^{-1} \hat{g} \phi_n g = \phi_{L_0}^{-1} \phi_{g^* n} = \rho(g^* n).$$

Thus $\pi(\mathrm{Aut}\, X)$ leaves the lattice $\rho(NS(X))$ in $B_{\mathbf{Q}}$ invariant and the action $\pi$ of $\mathrm{Aut}(X)$ on $\rho(NS(X))$ coincides with the natural action of $\mathrm{Aut}\, X$ on $NS(X)$.

## 2. Algebras with involution

In this section, we study algebras with involution over an algebraically closed field $K$ of characteristic zero. All the involutions encountered will be denoted by $\theta$.

Let $V$, $W$ be vector spaces of dimension $n$ over $K$ and $(,) : V \times W \to k$ be a non-degenerate bilinear pairing. If $A : V \to V$ is an endomorphism, denote by $A'$ the endomorphism of $W$ defined by $(AX, Y) = (X, A'Y)$ for $X \in V$, $Y \in W$. Then $A \mapsto A'$ gives an isomorphism of End $V$ onto End $W$. Put $C_n = \mathrm{End}\, V \times \mathrm{End}\, W$ and

$$\theta(A, B) = (B', A').$$

Now if $V = W$ and $A'' = A$ for all $A$ in End $V$, then the bilinear form $(,)$ is either symmetric or skew-symmetric. The algebra End $V$ with the involution $\theta(A) = A'$ will be denoted by $A_n$ in the first case and by $B_n$ in the second case.

*Lemma* 2.1. Let $a \in A_n$, $B_n$ or $C_n$ be a unit such that $\theta(a) = a$; then we may write $a = \beta\theta(j) j$ where $\beta$ is in the centre and $j$ satisfies:

(i) if $a \in A_n$ or $B_n$, $\det(j) = 1$,

(ii) if $a \in C_n$, $j = (j_1, j_2)$, then $\det j_1 = \det j_2 = 1$.

*Proof* : If $a \in C_n$, $a$ is of the form $(A, A')$; put $A = \beta B$ with $\beta \in K^*$, $\det B = 1$, and let $j = (B, 1)$. Then $a = \beta^2 \theta(j) j$.

If $a \in A_n$ (resp. $B_n$), the form $(aX, Y)$ is a non-degenerate bilinear symmetric (resp. antisymmetric) form on $V$, and so is $(X, Y)$.

Hence there exists $g \in \mathrm{Aut}\, V$ such that $(aX, Y) = (gX, gY)$ for all $X, Y \in V$. Again, putting $g = \beta j$ with $\beta \in K^*$ and $\det j = 1$, we have $a = \beta^2 \theta(j) j$.

*Lemma* 2.2. Let $B$ be a semi-simple algebra with involution $\theta$ over $K$. Let $a$ be a unit of $B$ with $\theta(a) = a$. Then we may write $a = \beta\theta(j) j$, where $\beta$ is in the centre and $N(j) = 1$ for any homogeneous polynomial function $N$ on $B$ satisfying $N(ab) = N(a) N(b)$ for all $a, b \in B$ and $N(1) = 1$.

*Proof* : It is well known that any such algebra with involution is a product of algebras with involution of the type $A_n$, $B_n$ or $C_n$ ([5], p. 595]. Write $B$ as such a product. Then $N$ is a product of powers of the determinant functions in the various factors. The lemma now follows from lemma 2.1.

## 3. Proof of theorem 1.1

In view of theorem 1.3 and the observations in § 1.6, the theorem will be a consequence of

*Lemma* 3.1. For any algebraically closed field $K$ of characteristic zero, there are only finitely many $G_K$-orbits in the closed set

$$S_d = \{a \in S_K \mid \phi(a) = d/d_0\},$$

where $\quad S_K = S_Q \otimes_Q K$

and $\quad S_Q = (a \in B_Q / \theta_Q(a) = a).$

Moreover, each of these orbits is closed.

*Proof*: Let $Z'$ be the centre of $B_K$ and $Z = Z' \cap S_K$, $R = G_K \cap Z$. If $a \in S_K$ with $\phi(a) = d/d_0$, $(d \neq 0)$, $a$ is a unit by lemma 1.5. Using lemma 2.2, we may write $a = \theta(j)\beta j$ with $\beta \in Z$ and $\phi(j) = 1$.

So, in order to show that there are finitely many $G_K$-orbits in $S_d$, it suffices to show that there exist only finitely many $R$-orbits in $S_d \cap Z$. If $a \in S_d \cap Z$, $a$ is a unit by lemma 1.5 and so $S_d \cap Z = aR$, from which it follows that the $R$-orbits in $S_d \cap Z$ can be put in bijective correspondence with $R/R^2$. Now $R$ is the kernel of $\phi \mid Z^*$ so that $R$ is the product of a torus and a finite abelian group, proving the first part of the lemma.

To prove that the orbits are closed it is enough to show that they are equidimensional or, what is the same, that the isotropy groups in question have the same dimension. For this, it is enough, by the preceding considerations, to look at the isotropy groups of points in $S_d \cap Z$. But here any isotropy group clearly consists of elements $g \in G_K$ such that $\theta(g)g =$ identity.

## References

[1] Borel A 1969 *Introduction aux groupes arithmetiques* (Paris : Hermann)

[2] Borel A and Harish-Chandra 1962 Arithmetic subgroups of algebraic groups; *Ann. Math.* **75** 485–535

[3] Martens H Riemann matrices with many polarisations, *Complex analysis and its applications.* Vol. III (ICTP Vienna: Summer School, Int. Atomic Energy Agency), pp. 35–48

[4] Mumford D 1969 *Abelian varieties, T.I.F.R. studies in mathematics* (Oxford University Press)

[5] Weil A 1960 Algebras with involution and classical groups; *J. Indian Math. Soc.* **24** 591

Tata Institute of Fundamental Research,
Bombay 400 005

# FIBRES DE 't HOOFT SPECIAUX ET APPLICATIONS

A. HIRSCHOWITZ et M.S. NARASIMHAN

On s'intéresse aux fibrés vectoriels algébriques de rang deux sur un espace projectif de dimension deux ou trois sur un corps de base algébriquement clos.

Les fibrés de 't Hooft ( [ 6 ] § 3.1.1) sont les fibrés de rang deux sur $\mathbb{P}^3$ associés aux réunions disjointes de droites. Nous introduisons les fibrés de 't Hooft spéciaux ('tHS) : ce sont les fibrés associés aux réunions de droites situées sur une même quadrique non-singulière. L'étude de ces fibrés nous conduit aux résultats suivants concernant l'espace des modules $M(c_1,c_2)$ des fibrés stables (de rang deux) de classes de Chern $c_1$ et $c_2$ :

THÉORÈME 1. *L'espace $M(0,2)$ des modules de fibrés instantons sur $\mathbb{P}^3$ avec $c_1 = 0$, $c_2 = 2$ est rationnel.*

THÉORÈME 2. *Si k est pair il n'existe pas de fibré universel algébrique (fibré de Poincaré) sur $\mathbb{P}^3 \times M_{\mathbb{P}^3}(0,k)$.*

Pour simplifier l'écriture, nous normalisons nos fibrés par $c_1=2$. Rappelons que $M(0,k)$ est isomorphe à $M(2,k+1)$. On sait ( [ 6 ] ) que tout fibré E de $M(2,3)$ s'insère dans une suite exacte $S(E)$ :

$$0 \longrightarrow \mathcal{O}\oplus\mathcal{O} \longrightarrow E \longrightarrow \tilde{\mathcal{F}}_E \longrightarrow 0$$

où $\tilde{\mathcal{F}}_E$ est le prolongement par zéro d'un fibré $\mathcal{F}_E$ de rang 1 et bidegré $\{2,-1\}$ sur une quadrique non singulière $Q_E$. Le faisceau $\tilde{\mathcal{G}}_E := \mathcal{E}xt^1(\tilde{\mathcal{F}}_E,\mathcal{O})$ est le prolongement par zéro d'un fibré $\mathcal{G}_E$ de rang 1 et bidegré $\{0,3\}$ sur $Q_E$ et le foncteur $\mathcal{H}om(\cdot,\mathcal{O})$ appliqué à $S(E)$ fournit deux sections qui engendrent $\mathcal{G}_E$ et donc un pinceau sans point de base de diviseurs de bidegré $\{0,3\}$ sur $Q_E$. Cette construction permet d'identifier M comme variété des pinceaux sans point de base de bidegré $\{0,3\}$ sur une quadrique non singulière. Nous donnons au § 4 une démonstration détaillée de cette identification qui fait de M(2,3) un fibré sur la variété $\tilde{\mathcal{K}}$ des quadriques non singulières de $\mathbb{P}^3$ munies d'un système de génératrices. Revêtement à deux feuillets de la variété $\mathcal{K}$ des quadriques non singulières, $\tilde{\mathcal{K}}$ est une variété rationnelle (cf § 3), et si C est le fibré en courbes rationnelles tautologique sur $\tilde{\mathcal{K}}$, alors M(2,3) est le fibré sur $\tilde{\mathcal{K}}$ des pinceaux sans point de base de degré 3 sur les fibres de C. C'est donc un ouvert d'un fibré en variétés grassmanniennes. Ce fibré en grassmanniennes n'est pas localement trivial dans la topologie de Zariski, mais il admet un sous-fibré ouvert localement trivial dans cette topologie (cf §2). Par suite M(2,3) est rationnel et p: M(2,3) $\longrightarrow \tilde{\mathcal{K}}$ admet des sections rationnelles. La suite exacte duale de S(E) :

$$0 \longrightarrow E^{V} \longrightarrow \mathcal{O} \oplus \mathcal{O} \longrightarrow \tilde{\mathcal{G}}_{E} \longrightarrow 0$$

permet de voir que s'il existe un fibré universel, alors $p^{*}C \longrightarrow M(2,3)$ provient d'un fibré vectoriel (Prop. 4.8). Ce n'est pas le cas parce que $C \longrightarrow \tilde{\mathcal{K}}$ ne provient pas d'un fibré vectoriel (§3) et que p admet une section rationnelle.

Les fibrés 'tHS sont les fibrés de M(2,2k+1) auxquels les raisonnements précédents s'étendent sans changement. Ils permettent de prouver le théorème 2.

La rationalité des modules de fibrés stables est connue dans certains cas sur les courbes ( [11] ) et sur $\mathbb{P}^2$ ( [1] [4] [7] ). Quant à la question du fibré universel, elle a d'abord été étudiée sur les courbes ( [10] [15] ). Maruyama [9] a ensuite donné en général une condition suffisante d'existence ; dans les cas qui nous préoccupent ici, cette condition est vérifiée pour $c_1=0$ et $c_2$ impair. Enfin Le Potier [8] a prouvé que pour les fibrés de rang deux sur $\mathbb{P}^2$, cette condition est aussi nécessaire. Nous expliquons rapidement au § 5 comment notre méthode permet de retrouver ce résultat de Le Potier dans le cas algébrique. Signalons aussi que Newstead [13] a étudié plus spécialement $M_{\mathbb{P}^3}(0,2)$ : la lecture de [13] est d'ailleurs à l'origine du présent travail.

Le Potier a démontré en même temps que nous le théorème 1 et le théorème 2 dans le cas particulier où k=2. Inspiré par les idées de Le Potier, Newstead démontre dans [14] les résultats de notre § 2.

## 1. FIBRÉS PROJECTIFS ET FIBRÉS EN CONIQUES.

Dans ce qui suit, un fibré projectif est un morphisme $\pi : C \longrightarrow B$ propre et plat de variétés algébriques (irréductibles) lisses dont les fibres sont isomorphe à l'espace projectif $\mathbb{P}^m$ de dimension m. Quand m=1, les fibres sont des courbes rationnelles lisses et on dit que $\pi$ est un fibré en coniques.
Le résultat suivant est bien connu [17, Prop.18]

PROPOSITION 1.1. *Soit $\pi: C \longrightarrow B$ un fibré projectif. Les conditions suivantes sont équivalentes :*

*1) Le fibré projectif est isomorphe au fibré projectif associé à un fibré vectoriel sur B.*

*2) $\pi : C \longrightarrow B$ admet une section rationnelle.*

*3) Il existe un fibré en droites sur C qui vaut $\mathcal{O}(1)$ sur chaque fibre.*

On dira qu'un fibré vérifiant ces conditions est banal.

REMARQUE 1.2. Supposons que S est une sous-variété lisse de B tel que $\pi^{-1}(S) \longrightarrow S$ ne soit pas banal. Alors si U est un ouvert non-vide de B, $\pi^{-1}(U) \longrightarrow U$ n'est pas banal. S'il l'était, $\pi : C \longrightarrow B$ serait banal par la proposition 1.1. et donc $\pi^{-1}(S) \longrightarrow S$ aussi.

REMARQUE 1.3. Soit $\pi : C \longrightarrow B$ un fibré en coniques. Le fibré tangent relatif $T_\pi$ est un fibré en droites qui est de degré 2 sur chaque fibre. Il s'ensuit que le fibré $\pi$ est banal s'il existe un fibré en droites sur C qui est de degré k sur chaque fibre, avec k impair.

Nous avons le résultat suivant assurant qu'un fibré en coniques n'est pas banal. Ce résultat remonte à Morin [16] pp 47-48. Voir aussi [[12] , Th. 2] .

PROPOSITION 1.4. *Soit $F : C' \longrightarrow B'$ un morphisme propre et plat de variétés lisses. On suppose qu'en dehors d'une sous variété lisse D de codimension 1 dans B', F est un fibré en coniques et que sur D les fibres sont des coniques dégénérées réduites (i.e., réunion de deux courbes rationnelles lisses se coupant en un point). Soit $B = B' - D$, $C = F^{-1}(B)$ et $\pi = F|C$. Si $F^{-1}(D)$ est irréductible alors le fibré en coniques $F : C \longrightarrow B$ n'est pas banal.*

DEMONSTRATION. Remarquons d'abord que F définit un revêtement $\tilde{D}$ à deux feuillets de D (on peut définir $\tilde{D}$ comme le schéma de Hilbert relatif des droites contenues dans les fibres sur D) ; de plus il existe un morphisme naturel $\eta : F^{-1}(D) - \mathcal{S} \longrightarrow \tilde{D}$, ou $\mathcal{S}$ est l'ensemble des points singuliers des fibres.

Supposons que $C \longrightarrow B$ est banal. Soit $\tau : U' \longrightarrow C$ une section sur un ouvert non-vide U de B. Comme F est propre, $\tau$ s'étend comme une section $\sigma : U \longrightarrow C'$ où U est un ouvert de B' avec codim $(B'-U) \geq 2$. Si on pose $W = U \cap D$, alors W est un ouvert non-vide de D. Si $w \in W$, $\sigma(w)$ ne peut pasêtre le point singulier de $F^{-1}(w)$ puisque F est lisse en $\sigma(w)$. En composant $\sigma|W$ avec $\eta$ , on aura une section de $\tilde{D} \longrightarrow D$ sur W. Puisque D est lisse, cette section s'étend en une section de $\tilde{D} \longrightarrow D$ sur D. Ainsi le revêtement $\tilde{D} \longrightarrow D$ se scinde, ce qui démontre que $F^{-1}(D)$ est réductible.

EXEMPLE 1.5. Soit B le schéma de Hilbert des coniques lisses dans $\mathbb{P}^2$ et $C \longrightarrow B$ le fibré en coniques tautologique. Alors $C \longrightarrow B$ n'est pas banal. En effet, soit $F : C' \longrightarrow B'$ la famille tautologique des coniques réduites dans $\mathbb{P}^2$ et $D \subset B'$ le diviseur paramétrant les coniques singulières. Alors $F^{-1}(D)$ est irréductible ; par exemple, parce que PGL(3) opère transitivement sur le revêtement $\tilde{D}$.

## 2. FIBRÉ EN GRASSMANNIENNES ASSOCIÉ A UN FIBRÉ EN CONIQUES.

Soit $\pi : C \longrightarrow B$ un fibré en coniques. Soit k un entier $\geq 1$. On désigne par $C^{(k)}$ le schéma de Hilbert $\mathrm{Hilb}^k\, C/B$ des diviseurs positifs de degré k sur les fibres de $C \longrightarrow B$. Alors $C^{(k)} \longrightarrow B$ est un fibré projectif (de fibre $\mathbb{P}^k$).

REMARQUE 2.1. Si k est pair, le fibré projectif $C^{(k)} \longrightarrow B$ est banal. Il est associé à l'image directe du faisceau $\otimes^{k/2} T_\pi$, où $T_\pi$ est le fibré tangent relatif.

On désigne par GC(d,k) la grassmannienne relative des sous-espaces projectifs de dimension d des fibres de $C^{(k)}$ $(0 \leq d \leq k)$. On note M(d,k) l'ouvert de GC(d,k) formé des systèmes linéaires de dimension d sans point de base.

PROPOSITION 2.2. *Supposons que $k$ est pair ou que $k$ et $d$ sont tous les deux impairs. Soit $G(d,k)$ la grassmannienne des sous espaces de dimension $d$ de l'espace projectif $\mathbb{P}^k$. Alors il existe un ouvert non-vide $U$ de $B$, un ouvert affine non-vide $A$ de $G(d,k)$ et un plongement ouvert de $U \times A$ dans $GC(d,k)$ au dessus de $B$.*

En particulier si B est rationnelle, GC(1,k) et M(1,k) sont rationnelles.

On va démontrer d'abord des lemmes.

LEMME 2.3. *Tout fibré de fibre l'espace affine $A^n$ de groupe structural le groupe affine est localement trivial dans la topologie de Zariski.*

DÉMONSTRATION. On va démontrer que chaque fibré principal de groupe structural le groupe affine est localement trivial dans la topologie de Zariski. D'après [17, Théorème 2, § 4.3] , il suffit de démontrer que le groupe affine est spécial i.e., qu'il existe un plongement du groupe affine H dans GL(N) comme sous groupe fermé tel que $GL(N) \longrightarrow GL(N)/H$ admette une section rationnelle. Mais le groupe affine est le sous groupe de GL(n+1) formé par les matrices de la forme

$$\begin{pmatrix} T & * \\ 0 & 1 \end{pmatrix} \quad \text{où } T \in GL(n)$$

Le sous-espace de GL(n+1) des matrices de la forme

$$\begin{pmatrix} I & 0 \\ * & \mu \end{pmatrix}$$

où I est la n x n matrice identique et $\mu \neq 0$, donne une section rationnelle.

LEMME 2.4. *Soit $V$ un sous-espace projectif de dimension $(k-d-1)$ dans $\mathbb{P}^k$, et $G^V(d,k)$ l'ouvert de la grassmannienne $G(d,k)$ des sous espaces projectifs de dimension $d$ disjoints de $V$. Alors $G^V(d,k)$ admet une structure d'espace affine pour laquelle l'action sur $G^V(d,k)$ du sous groupe* *de PGL(k+1) qui stabilise V est affine.*

DÉMONSTRATION. Soit $\mathbb{P}^k = \mathbb{P}(F)$ où F est un espace vectoriel de dimension k+1 et $\underline{V}$ le sous espace vectoriel de F correspondant à V. Alors $G^V(d,k)$ s'identifie à l'espace des scindages de la suite exacte d'espaces vectoriels

$$0 \longrightarrow \underline{V} \longrightarrow F \longrightarrow F/\underline{V} \longrightarrow 0,$$

qui est un espace affine (sous l'action de $\mathrm{Hom}(F/\underline{V},\underline{V})$. Soit $\underline{W}$ un supplémentaire de $\underline{V}$. Alors $G^V(d,k)$ s'identifie à $\mathrm{Hom}(\underline{W},\underline{V})$. Les éléments de GL(k+1) qui stabilisent V s'écrivent

$$\begin{pmatrix} f & g \\ o & h \end{pmatrix}$$

On vérifie que pour x dans $\mathrm{Hom}(\underline{W},\underline{V})$ $\begin{pmatrix} f & g \\ o & h \end{pmatrix}(x)$ est égal à $fxh^{-1} + gh^{-1}$, ce qui est bien affine en x.

LEMME 2.5. *Supposons que $\ell$ et k ont même parité. Alors $GC(\ell,k)$ admet une section rationnelle.*

DÉMONSTRATION. Comme $(k-\ell)$ est pair, $C^{(k-\ell)} \longrightarrow B$ admet une section rationnelle $\sigma$ (Remarque 2.1). On a un morphisme naturel $\varphi : C^{(k-\ell)} \times_B C^{(\ell)} \longrightarrow C^{(k)}$ ("addition des diviseurs"). Soit $(pr_1,\varphi) : C^{(k-\ell)} \times_B C^{\ell} \longrightarrow C^{(k-\ell)} \times C^{(k)}$ le morphisme associé. C'est un plongement dont l'image est propre et plate sur $C^{(k-\ell)}$. Il définit un B-morphisme $\psi : C^{(k-\ell)} \longrightarrow GC(\ell,k)$. Alors $\tau = \psi o \sigma$ est la section voulue.

DÉMONSTRATION DE LA PROPOSITION 2.2.

Si k est pair le résultat est évident, car GC(d,k) est localement trivial (dans la topologie de Zariski), $C^{(k)} \longrightarrow B$ étant localement trivial (Remarque 2.1). On va supposer k et d impairs. D'après le lemme 2.5, GC(k-d-1,k) admet une section $\sigma$ sur un ouvert non-vide U' de B. Soit M le sous-groupe de PGL(k+1) qui stabilise un (k-d-1) plan V dans $\mathbb{P}^k$. Si P est le fibré principal de groupe PGL(k+1) associé au fibré projectif $C^{(k)} \longrightarrow B$, la section $\sigma$ permet de définir une réduction du groupe structural de $P|U'$ à $M$. Soit $R$ le fibré principal sur $U'$ avec groupe $M$ ainsi obtenu. Avec la notation du Lemme 2.4, soit $E$ le fibré associé à $R$ avec fibre $G^V(d,k)$, à partir de l'action de $M$ sur $G^V(d,k)$. Alors $E$ est un fibré affine (Lemme 2.4) et il y a un plongement ouvert de $E$ dans $GC(d,k)$ au dessus de $U'$. D'après le Lemme 2.3, le fibré affine $E$ est localement trivial sur $U'$ (dans la topologie de Zariski), ce qui démontre la proposition.

PROPOSITION 2.6. *Soit $p : M(1,k) \longrightarrow B$ la projection canonique. Si le fibré projectif $C \longrightarrow B$ n'est pas banal, alors le fibré projectif $p^*(C) \longrightarrow M(1,k)$, image réciproque de $C \longrightarrow B$ par $p$, n'est pas banal non plus.*

DÉMONSTRATION. Avec la notation de la proposition 2.2, soit $\Omega \subset GC(1,k)$ l'image de $U \times A$. Comme $GC(1,k)$ est irréductible il existe $\xi \in \Omega \cap M(1,k)$. Soit $\sigma$ une section rationnelle de $\Omega \longrightarrow B$ passant par $\xi$ (il en existe puisque $\Omega \approx U \times A$). Alors pour un voisinage $W$ de $p(\xi)$ dans $B$, on aura $\sigma(W) \in M(1,k)$. Si $p^*(C) \longrightarrow M(1,k)$ était banal, $C$ serait banal sur $W$ et donc sur $B$ (Proposition 1.1).

## 3. LE REVÊTEMENT A DEUX FEUILLES DE L'ESPACE DES QUADRIQUES LISSES DANS $\mathbb{P}^3$ ET FIBRÉS ASSOCIÉS.

Soit $\mathcal{H}$ le schéma de Hilbert des quadriques lisses dans $\mathbb{P}^3$. Soient $\tilde{\mathcal{H}}$ le revêtement à deux feuilles de $\mathcal{H}$ correspondant au choix d'un système de génératrices sur les quadriques et $C \longrightarrow \tilde{\mathcal{H}}$ le fibré projectif dont la fibre en $\xi \in \tilde{\mathcal{H}}$ est la droite projective des génératrices correspondant à $\xi$ [3, § 2.7] .

On a le diagramme

$$\begin{array}{ccccc} C & \xleftarrow{\;g\;} & \tilde{Q} & \hookrightarrow & \tilde{\mathcal{H}} \times \mathbb{P}^3 \\ & {\scriptstyle p}\searrow & & \swarrow & \\ & & \tilde{\mathcal{H}} & & \end{array}$$

où $\tilde{Q}$ est l'image réciproque de la quadrique universelle $Q \longrightarrow \mathcal{H}$ .

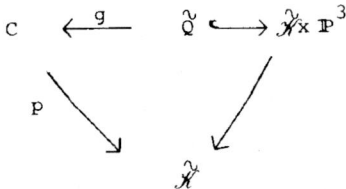

DEFINITION 3.1. Soit $C \longrightarrow \tilde{\mathcal{X}}$ le fibré projectif défini ci-dessus. On désigne par $M_k$ l'ouvert de $GC(1,k)$ formé par les systèmes linéaires de dimension 1 sans points de base, où $GC(1,k)$ est la grassmannienne relative des sous-espaces projectifs de dimension 1 des fibres de $C^{(k)}$. [$C^{(k)}$ est le fibré projectif $\mathrm{Hilb}^k(C/\tilde{\mathcal{X}})$, voir § 2.] . On va noter p la projection canonique $M_k \longrightarrow \tilde{\mathcal{X}}$.

REMARQUE 3.2. Soit $\pi : Q_S \longrightarrow S$, $Q_S \subset S \times \mathbb{P}^3$, une famille de quadriques lisses sur une variété S. Alors

1) - Si F est un fibré de rang un et de bidegré $\{0,k\}$, $k \geq 1$, sur $Q_S$, $(Q_S,F)$ définit un morphisme de S dans $\tilde{\mathcal{X}}$.

2) - Si D est un fibré de rang 2 sur S et $\delta : \pi^*(D) \longrightarrow F$ est un morphisme surjectif, $\delta$ définit un morphisme, noté $[\delta]$ , de S dans $M_k$.

3) - Soient F', D' et $\delta' : \pi^*(D') \longrightarrow F'$ vérifiant les mêmes hypothèses avec F et F' isomorphes. Si $[\delta] = [\delta']$ , alors $\delta$ et $\delta'$ sont isomorphes.

LEMME 3.3. *Soit S une variété et $\pi : Q_S \longrightarrow S$, $Q_S \subset S \times \mathbb{P}^3$, une famille lisse de quadriques sur S. Soit F un fibré en droites de bidegré $\{0,k\}$ , $k \geq 1$, sur $Q_S$ et $\varphi : S \longrightarrow M_k$ un morphisme tel que $p \circ \varphi : S \longrightarrow \tilde{\mathcal{X}}$ soit le morphisme associé a $(Q_S, F)$. Alors il existe un sous fibré (unique) D de rank 2 de $\pi_* F$ et un morphisme $\delta : \pi^*(D) \longrightarrow F$ tel que le morphisme $[\delta]$ associé vérifie $[\delta] = \varphi$ .*

DEMONSTRATION. On considère le diagramme commutatif

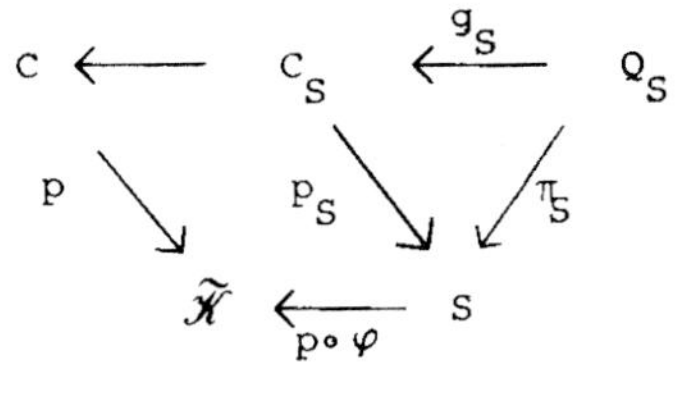


Alors $(g_S)_* F$ est un fibré en droites de degré relatif k sur $C_S$.
Or $(p_S)_*((g_S)_* F)$ est un fibré vectoriel E de rang k+1 sur S et le fibré projectif $\mathbb{P}(E)$ est canoniquement isomorphe au fibré projectif $C_S^{(k)}$. Cet isomorphisme induit un isomorphisme $\tau$ de $GC_S(1,k)$ (formé à partir de $C_S$) sur la variété grassmannienne relative GE(2) des sous-espaces vectoriels de dimension 2 des fibres de E. Soit W le (sous) fibré tau-

tologique de rang 2 sur GE(2) et $\tilde{\varphi}$ la section de $GC_S(1,k)$ induite par $\varphi$. Alors $D=(\tau \circ \tilde{\varphi})^*W$ est le fibré cherché. ($\pi^*D \longrightarrow L$ est surjectif car pour $s \in S$, $\varphi(s)$ est un système linéaire sans point de base).

THEOREME 3.4.

I) *Le fibré projectif* $C \longrightarrow \tilde{\mathcal{X}}$ *n'est pas banal.*

II) $\tilde{\mathcal{X}}$ *est une variété rationnelle.*

III) $M_k$ *(définition 3.1) est une variété rationnelle.*

IV) *Le fibré projectif* $p^*C \longrightarrow M_k$ *n'est pas banal (ici,* $p : M_k \longrightarrow \tilde{\mathcal{X}}$ *est la projection naturelle).*

DEMONSTRATION DE 3.4. I). Le résultat découle du lemme suivant en utilisant le fait qu'il existe des fibrés en coniques qui ne sont banals sur aucun ouvert non-vide de la base (Exemple 1.5 et Proposition 1.1).

LEMME 3.5. *Soit* $P \longrightarrow S$ *un fibré en coniques. Pour chaque* $s_0 \in S$ *il existe un voisinage* $U$ *de* $s_0$ *et un morphisme* $\varphi : U \longrightarrow \tilde{\mathcal{X}}$ *tel que* $P|U$ *soit isomorphe à l'image réciproque de* $C \longrightarrow \tilde{\mathcal{X}}$ *par* $\varphi$.

DEMONSTRATION. Il suffit de démontrer qu'on peut choisir un fibré vectoriel W de rang 4 sur S et une famille de quadriques lisses $Q \longrightarrow S$, $Q \subset \mathbb{P}(W)$, munie d'un système de génératrices, tel que le fibré P soit isomorphe au fibré en coniques défini par les génératrices choisies. Car, en trivialisant W au voisinage de $s_0$, on aura l'application $\varphi$ cherchée, vu la propriété universelle de $\tilde{\mathcal{X}}$.

Soit T le fibré principal avec groupe structural PGL(2) associé à P. On considère l'action diagonale de PGL(2) sur la quadrique $\mathbb{P}^1 \times \mathbb{P}^{1\vee} \subset \mathbb{P}(V \otimes V^\vee)$ où $\mathbb{P}^1 = \mathbb{P}(V)$, V étant un espace vectoriel de dimension 2. L'action de PGL(2) sur l'espace des génératrices paramétré par $\mathbb{P}^1$ correspond à l'action naturelle de $\mathbb{P}$GL(2) sur $\mathbb{P}^1$. Soit W le fibré vectoriel associé à T par l'action linéaire de PGL(2) sur $V \otimes V^\vee$. Si $Q \longrightarrow S$, est le fibré en quadriques associé à T par l'action de PGL(2) sur $\mathbb{P}^1 \times \mathbb{P}^{1\vee}$, on a $Q \subset \mathbb{P}(W)$ et Q vient avec le choix d'un système de génératrices sur chaque fibre. On vérifie que Q a la propriété voulue.

REMARQUE. W est en fait l'algèbre d'Azumaya associée au fibré projectif P.

Pour la clarté de l'exposé on va démontrer d'abord un résultat analogue à 3.4.II, pour le revêtement à deux feuilles de l'espace des formes quadratiques non-dégénérées, et puis on va indiquer les modifications nécessaires dans le cas de l'espace des quadriques non-singulières.

PROPOSITION 3.6. *Soit $\mathcal{B}$ l'espace des formes quadratiques non-dégénérées sur $V = k^4$ (k un corps algébriquement clos). Soit $\tilde{\mathcal{B}}$ le revêtement à deux feuilles associé au choix d'un système de sous espaces totalement singuliers maximaux. Alors $\tilde{\mathcal{B}}$ est une variété rationnelle.*

Avant de démontrer cette proposition nous rappelons ci-dessous la description explicite de la $k(\mathcal{B})$-algèbre $k(\tilde{\mathcal{B}})$ puis nous donnons une condition suffisante pour qu'une extension d'ordre deux d'une extension transcendante pure de k soit une extension transcendante pure de k. La démonstration de 3.6 consistera à vérifier cette condition, ce qui est immédiat si car $k \neq 2$.

RAPPEL 3.7.

1°) - (cf [3] Prop.2.8). Soit Q une forme quadratique non dégénérée sur $K^4$, C(Q) son algèbre de Clifford. Le centre Z de la partie paire de C(Q) s'identifie à l'anneau des fonctions rationnelles du K-schéma des systèmes de sous-espaces totalement singuliers maximaux de Q.

2°) - (cf [2] § 9 n° 4 Remarque 2 et Exercice 9) : Si K est un corps, Q une forme quadratique non dégénérée sur $K^4$, C(Q) son algèbre de Clifford et Z le centre de la partie paire de C(Q), alors:

A) Si car $K \neq 2$, il existe y dans Z tel que $\{1,y\}$ engendre Z sur K et que $y^2$ soit le discriminant de Q.

B) Si car K = 2, soit $(e_1,e_2,e_3,e_4)$ une base symplectique pour la forme bilinéaire (alternée!) associée à Q ; alors l'élément $y=e_1e_2+e_3e_4$ de C(Q) est dans Z, vérifie $y^2+y=Q(e_1)Q(e_2)+Q(e_3)Q(e_4)$, et est tel que $\{1,y\}$ engendre Z sur K.

LEMME 3.8. *Soit $k$ un corps et $K = k(x_1,\ldots,x_N)$ une extension transcendante pure de $k$ avec $\{x_1,\ldots,x_N\}$ algébriquement libre sur $k$. Soit $P$ un polynôme de $x_1,\ldots,x_N$ de la forme $P = x_1 T+R$ où $T$ et $R$ sont des polynômes de $x_2,\ldots,x_N$ et $1\neq 0$. Soit $Z$ une algèbre commutative sur $K$ admettant une base $\{1,y\}$ où $y$ vérifie l'équation $y^2+\lambda y = P$, $\lambda \in k$. Alors $Z$ est un corps et c'est une extension transcendante pure de $k$.*

DÉMONSTRATION. On vérifie que $P$ n'est pas de la forme $\xi^2+\lambda\xi = P$ avec $\xi \in K$, d'où $Z$ est un corps. Comme $x_1 = \frac{P-R}{T}$ on a $K = k(P,x_2,\ldots,x_N)$ et $\{P,x_2,\ldots,x_N\}$ est algébriquement libre sur $k$. Alors $Z = k(y, x_2,\ldots,x_N)$ avec $\{y,x_2,\ldots,x_N\}$ algébriquement libre sur $k$.

DEMONSTRATION DE LA PROPOSITION 3.6.*

Le cas où car $k \neq 2$ résulte immédiatement de 3.7 et 3.8. On suppose dans la suite que car $k = 2$.

Soit $(v_1,\ldots,v_4)$ la base canonique de $V = k^4$. Si $L$ est un surcorps de $k$ on va noter aussi $(v_1,\ldots,v_4)$ la base $(v_1\otimes 1_L,\ldots,v_4\otimes 1_L)$ de $V \otimes_k L$.

Soit $K$ le corps des fonctions rationnelles sur $\mathscr{B}$.
Si $\{x_{ij}\}_{i \leq j}$, $1 \leq i \leq 4$, $1 \leq j \leq 4$ sont des indéterminées sur $k$ on a $K = k(x_{ij})$ et la forme quadratique universelle $Q$ sur $K$ est donnée par :

$$Q\left(\sum_{i=1}^{4} \xi_i v_i\right) = \sum_{i=1}^{4} x_{ii}\,\xi_i^2 + \sum_{i<j} x_{ij}\,\xi_i\,\xi_j,\quad \xi_i \in K.$$

Considérons le sous-corps $K'$ de $K$ engendré par les $x_{ij}$ avec $i < j$ [ $K'$ est le corps de fonctions de l'espace des formes binéaires alternées sur $k^4$ ]. La forme alternée universelle $A$ (définie sur $K'$) est donnée par la matrice $(a_{ij})$, avec $a_{ij} = x_{ij}$ pour $i < j$, $a_{ji} = x_{ij}$ pour $j > i$ et $a_{ii} = 0$. De plus $A$ est la forme bilinéaire alternée associée à $Q$. Choisissons une base symplectique $(e_1,\ldots,e_4)$ pour $A$ dans $V \otimes_k K'$.

* Nous remercions M.V. Nori et A. Ramanathan de l'aide qu'ils nous ont apportée pour cette démonstration.

En utilisant 3.7.1, 3.7.2 B et 3.8, on peut terminer la démonstration de la proposition si on montre que $\{Q(e_1), Q(e_2), Q(e_3), Q(e_4)\}$ forme une base transcendante pure de K sur K' (car on aura alors $K = k(x_{ij}, Q(e_k))$, $i < j$, $1 \leq k \leq 4$). Pour le montrer il suffit de prouver que le sous-corps de K engendré sur K' par les $Q(e_i)$ est identique à K, puisque $K = K'(x_{11},\ldots,x_{44})$. Ecrivons

$$v_i = \sum_j \xi_{ij}\, e_j,\quad \xi_{ij} \in K'.$$

On a alors

$$x_{ii} = Q(v_i) = \sum_{j<k} \xi_{ij}\,\xi_{ik}\, A(e_j,e_k) + \sum_j \xi_{ij}^2\, Q(e_j).$$

Comme $\xi_{ij} \in K'$ et $A(e_j,e_k) \in k$, on voit que $K'(Q(e_i)) = K'(x_{ii})$.
Ceci complète la demonstration de 3.6.

DÉMONSTRATION DE 3.4.II.

La méthode de la démonstration de 3.4.II est analogue à celle de 3.8. Soit K le corps des fonctions rationnelles sur $\mathscr{K}$. L'idée est de choisir une section rationnelle de $\mathscr{B} \longrightarrow \mathscr{K}$ et de travailler avec la forme quadratique induite de Q sur K par cette section. On prendra la section donnée par les formes quadratiques q sur $k^4$ avec $q(v_1)=1$ i.e. définie par $x_{11} = 1$.

Avec la notation de la démonstration 3.6, le corps K des fonctions rationnelles sur $\mathscr{K}$ est isomorphe à $k(x_{22},x_{33},x_{44},x_{ij})$, $i < j$.
Considérons la forme quadratique Q sur K :

$$Q\Big(\sum_{i=1}^{4} \xi_i\, v_i\Big) = \xi_1^2 + \sum_{j=2}^{4} x_{jj}\,\xi_j^2 + \sum_{i<j} x_{ij}\,\xi_i\,\xi_j .$$

Quand car $k \neq 2$, en utilisant 3.7.1, 3.7.2.A et 3.8, on voit, comme dans la démonstration de 3.6, que $\mathscr{K}$ est rationnelle. Supposons donc car $k = 2$. Dans ce cas, K contient K' et la forme bilinéaire alternée associée à Q est A. Choisissons une base symplectique $(e_1,\ldots,e_4)$ dans $V \otimes_k K'$ avec $e_1 = v_1$. On voit comme dans la démonstration de 3.6 que $\{Q(e_2), Q(e_3), Q(e_4)\}$ est une base pure de K sur K'.
On termine en utilisant 3.8.

REMARQUE 3.9. On peut démontrer 3.4. II, en montrant, au moins en caractéristique nulle, que $\tilde{\mathcal{H}}$ est isomorphe à l'ouvert de la grassmannienne des plans dans $\mathbb{P}^5$, constitué par les plans qui coupent la quadrique de Klein suivant une conique lisse. Mais la démonstration donnée plus haut s'étend facilement au cas des quadriques de $\mathbb{P}^{2m+1}$, $m \geq 1$.

DÉMONSTRATION DE 3.4., III. C'est une conséquence immédiate de 3.4.II et de la proposition 2.2.

DÉMONSTRATION DE 3.4. IV. 3.4, IV résulte de 3.4.I et de la proposition 2.6.

## 4. FIBRÉS DE 't HOOFT SPÉCIAUX ('tHS).

DEFINITION 4.1. Un fibré vectoriel E sur $\mathbb{P}^3$ est un fibré 'tHS (fibré de 't Hooft spécial) s'il existe une quadrique non-singulière $Q_E$ et un fibré $\mathcal{F}_E$ de rang un sur $Q_E$ tel que E soit extension du prolongement $\tilde{\mathcal{F}}_E$ de $\mathcal{F}_E$ à $\mathbb{P}^3$ par un fibré trivial de rang 2.

REMARQUE 4.2. Les fibrés de 't Hooft sont les fibrés admettant une section dont le schéma des zéros est une réunion disjointes de droites (réduites). On peut montrer en caractéristique nulle que les fibrés 'tHS sont les fibrés de 't Hooft correspondant aux réunions disjointes de droites contenues dans une même quadrique non-singulière.

PROPOSITION 4.3. *Soit E un fibré 'tHS avec $c_2 \geq 3$. Alors $Q_E$ et $\mathcal{F}_E$ sont uniquement déterminés (à isomorphisme près pour $\mathcal{F}_E$). De plus, $H^0(\mathbb{P}^3, E)$ est de dimension deux et $\tilde{\mathcal{F}}_E$ est le conoyau de l'injection naturelle*

$$j_E = H^0(\mathbb{P}^3, E) \otimes \mathcal{O} \longrightarrow E.$$

*Le bidegré de $\mathcal{F}_E$ est $\{2, 2-c_2\}$.*

DÉMONSTRATION. Soit $\{a,b\}$ le bidegré de $\mathcal{F}_E$. Le polynôme de Hilbert de $\tilde{\mathcal{F}}_E$ est donc

$$\chi(t) = (t + a + 1)(t + b + 1)$$

et la formule de Riemann-Roch donne

$c_1(\tilde{\mathscr{F}}_E)=2$, $c_2(\tilde{\mathscr{F}}_E) = 4 - a - b$, $c_3(\tilde{\mathscr{F}}_E) = 2(a-2)(b-2)$.

Comme ce sont aussi les classes de Chern de E on doit avoir $c_3(\tilde{\mathscr{F}}_E)=0$ donc par exemple a=2 et la condition $c_2 \geq 3$ devient $b < 0$. On voit alors que $\tilde{\mathscr{F}}_E$ n'a pas de section globale, ce qui prouve que le morphisme donné $\mathcal{O}\oplus\mathcal{O} \longrightarrow E$ s'identifie au morphisme $j_E$ de l'énoncé : $\tilde{\mathscr{F}}_E$ s'identifie donc au conoyau de j et $Q_E$ est le support de ce conoyau.

PROPOSITION 4.4. *Soit S une variété, $\pi$ la projection $S \times \mathbb{P}^3 \longrightarrow S$ et $\mathscr{E}$ un fibré sur $S\times\mathbb{P}^3$ tel que pour tout s dans S, $\mathscr{E}(s)$ soit 'tHS avec $c_2 \geq 3$. On note $j_{\mathscr{E}}$ le morphisme naturel de $\pi^* \pi_* \mathscr{E}$ dans $\mathscr{E}$. Alors les quadriques $Q_{\mathscr{E}(s)}$ s'organisent en une famille plate notée $Q_{\mathscr{E}}$ et le conoyau de $j_{\mathscr{E}}$ est le prolongement par zéro d'un fibré de bidegré $\{2, 2-c_2\}$ sur $Q_{\mathscr{E}}$.*

DEMONSTRATION. Nous allons faire apparaitre $Q_{\mathscr{E}}$ comme support du conoyau de det $j_{\mathscr{E}}$.

D'après le théorème de changement de base, $\pi_*(\mathscr{E})$ est un fibré de rang deux ainsi que $\pi^* \pi_* \mathscr{E}$ et $\pi^* \pi_* \mathscr{E}(s)$ s'identifie à $H^0(\mathbb{P}^3, \mathscr{E}(s)) \otimes \mathcal{O}$, cependant que $j_{\mathscr{E}}(s)$ s'identifie au morphisme d'évaluation :

$$j_{\mathscr{E}}(s) = j_{\mathscr{E}(s)} : H^0(\mathbb{P}^3, \mathscr{E}(s)) \otimes \mathcal{O} \longrightarrow \mathscr{E}(s).$$

Comme, pour les fibrés, l'élévation à une puissance extérieure et la restriction commutent, ces identifications s'étendent au morphisme

$$\det j_{\mathscr{E}} : \Lambda^2 \pi^* \pi_* \mathscr{E} \longrightarrow \Lambda^2 \mathscr{E}$$

de sorte que det $j_{\mathscr{E}}(s)$ s'identifie à

$$\det j_{\mathscr{E}(s)} : \Lambda^2 H^0(\mathbb{P}^3, \mathscr{E}(s)) \otimes \mathcal{O} \longrightarrow \Lambda^2 \mathscr{E}(s).$$

Or ce dernier morphisme est injectif pour tout s. Par suite, si H désigne le conoyau de det $j_{\mathscr{E}}$, on voit que le polynôme de Hilbert de H(s) est indépendant de s. De ce fait H est S-plat [5 , Th III, 9.9 et sa preuve] , et $H \otimes \Lambda^2 \mathscr{E}$ est le faisceau structural de la famille $Q_{\mathscr{E}}$ cherchée. Nous allons maintenant voir que le conoyau coker $j_{\mathscr{E}}$ est le prolongement par zéro de sa restriction à $Q_{\mathscr{E}}$. Cela résulte du

LEMME 4.5. *Soit A et B deux fibrés de même rang sur une variété X et $j : A \longrightarrow B$ un morphisme injectif de faisceaux. Alors l'idéal défini par det j annule coker j, autrement dit : si Q est le sous-schéma défini par cet idéal, coker j s'identifie au prolongement par zero de sa restriction à Q.*

DÉMONSTRATION. L'énoncé étant local, on peut supposer que A et B sont triviaux. Les formules de Cramer donnent alors un morphisme $Cof_j : B \longrightarrow A$ vérifiant jo $Cof_j$ = det j. $Id_B$ ce qui prouve que $det_j(B)$ est dans l'image de A.

Pour terminer la démonstration de la Proposition 4.4., il nous suffit maintenant de prouver que la restriction à $Q_{\mathcal{E}}$ de $j_{\mathcal{E}}$ a pour conoyau un fibré de rang un (il aura le bon bidegré d'après la Proposition 4.3).

Comme $Q_{\mathcal{E}}$ est lisse, det $j_{\mathcal{E}}$ s'annule à l'ordre un sur $Q_{\mathcal{E}}$. Par suite $j_{\mathcal{E}}(x)$ est de rang un en tout point x de $Q_{\mathcal{E}}$. Pour conclure, il nous suffit d'appliquer le résultat général suivant : soient A et B deux fibres de rangs $r_A$ et $r_B$ sur la variété X et $j : A \longrightarrow B$ un morphisme de fibrés de rang constant r. ALors ker j, Im j, coker j sont des fibrés de rangs $r_A - r$ et $r_B - r$, dont la construction commute avec les changements de base.

Dans la suite de ce paragraphe nous fixons $c_2 \geq 3$ et nous étudions le foncteur 'tHS$(c_2)$, ou plus simplement 'tHS, sur la catégorie des variétés, qui à S associe l'ensemble des classes d'isomorphisme de fibrés sur $S \times \mathbb{P}^3$ verifiant l'hypothèse de la Proposition 4.4. Notons que cette proposition nous permettrait d'étendre le foncteur 'tHS à la catégorie des schémas, mais cela n'est pas utile ici.

Nous voulons montrer que la variété $M = M_{c_2}$ considérée en 3.1 est un module grossier pour 'tHS. Pour cela il nous faut le

LEMME 4.6. *Si F est le prolongement par zéro d'un fibré de rang un et bidegré {a, b} sur une famille plate de quadriques non-singulières Q sur S alors $\mathcal{E}xt^1(F, \mathcal{O})$ est le prolongement par zero d'un fibré de rang un sur Q de bidegré {2-a, 2-b}.*

DÉMONSTRATION. L'affirmation étant locale dans S on peut supposer que Q admet une équation

$$0 \longrightarrow \mathcal{O}(-2) \xrightarrow{q} \mathcal{O} \longrightarrow \mathcal{O}_Q \longrightarrow 0.$$

D'où

$$0 \longrightarrow \mathcal{O}(a-2) \xrightarrow{q} \mathcal{O}(a) \longrightarrow \mathcal{O}_Q(a) \longrightarrow 0$$

et en dualisant

$$\mathcal{O}(-a) \xrightarrow{q} \mathcal{O}(2-a) \longrightarrow \mathcal{E}xt^1(\mathcal{O}_Q(a),\mathcal{O}) \longrightarrow 0$$

Ceci prouve que $\mathcal{E}xt^1(\mathcal{O}_Q(a),\mathcal{O})$ est le prolongement par zéro d'un fibré de rang un sur Q de bidegré $\{2-a,2-a\}$. Il en résulte que $\mathcal{E}xt^1(F,\mathcal{O})$ est le prolongement par zéro d'un fibré de rang un sur Q. Le bidegré $\{2-a, 2-b\}$ de ce fibré se calcule en considérant des ouverts (non-affines) de $S \times \mathbb{P}^3$ où F est isomorphe à $\mathcal{O}_Q(a)$ où à $\mathcal{O}_Q(b)$.

DÉFINITION 4.7. *Soit $\mathcal{E}$ comme en 4.4 et $\delta : (\pi^* \pi_* \mathcal{E})^\vee \longrightarrow \mathcal{E}xt^1(\tilde{\mathcal{F}}_{\mathcal{E}},\mathcal{O})$ le conoyau de $j^\vee$. D'après le lemme 4.6 et la remarque 3.2, $[\delta]$ définit un morphisme de foncteurs, noté mod, de 'tHS dans le foncteur $\hat{M}$ représenté par M, où $M = M_{c_2}$ (Définition 3.1).*

L'image du morphisme mod est caractérisé par l'énoncé suivant.

PROPOSITION 4.8. Pour qu'un morphisme $\varphi : S \longrightarrow M$ soit dans l'image de mod (i.e. provienne d'une famille de fibrés 'tHS) il faut et il suffit qu'il existe sur le fibré $\varphi^* p^* \tilde{C}$ un fibré de rang un et de degré relatif $c_2$, où $p : M \longrightarrow \mathcal{K}$ est la projection canonique.

DEMONSTRATION. La condition est nécessaire ; en effet si $\mathcal{E}$ est un fibré 'tHS sur $S \times \mathbb{P}^3$ tel que mod $\mathcal{E} = \varphi$, alors d'après le lemme 4.6 $\mathcal{E}xt^1(\tilde{\mathcal{F}}_{\mathcal{E}},\mathcal{O})$ provient d'un fibré de rang un et de degré $c_2$ sur $\varphi^* p^* C$.

La condition est suffisante : Soit L un fibré de rang un et degré relatif $c_2$ sur $C_S = \varphi^* p^* C$. On considère le diagramme déduit du diagramme universel (§ 3),

$$\begin{array}{ccccc} C_S & \xleftarrow{g_S} & Q_S & \xrightarrow{i_S} & S \times \mathbb{P}^3 \\ & {\scriptstyle p_S}\searrow & \downarrow{\scriptstyle q_S} & \swarrow{\scriptstyle \pi_S} & \\ & & S & & \end{array}$$

D'après le lemme 3.3 (et sa preuve) $\varphi$ définit un sous fibré de rang 2, noté D, de $p_{S_*} L$ auquel correspond un morphisme

$$\delta' : p_S^* D \longrightarrow L$$

vérifiant $[\delta'] = \varphi$ .

Ce morphisme induit un morphisme surjectif

$$\delta : \pi_S^* D \longrightarrow i_{S_*} g_S^* L.$$

On pose $\mathcal{E} = (\ker \delta)^{\vee}$ et on montre que $\mathcal{E}$ est un fibré 'tHS vérifiant mod $\mathcal{E} = \varphi$ . D'abord Ker $\delta$ est un fibré parce que $i_{S_*} g_S^* L$ est de dimension homologique un. En dualisant la suite

$$0 \longrightarrow \mathcal{E}^{\vee} \xrightarrow{\alpha} \pi_S^* D \longrightarrow i_{S_*} g_S^* L \longrightarrow 0$$

on obtient

$$0 \longrightarrow \pi_S^* D^{\vee} \xrightarrow{\alpha^{\vee}} \mathcal{E} \longrightarrow \mathcal{E}xt^1(i_{S_*} g_S^* L, \mathcal{O}) \longrightarrow 0$$

ce qui prouve que $\mathcal{E}$ est un fibré 'tHS. On a alors par définition mod $\mathcal{E} = [\text{coker } \alpha^{\vee\vee}]$ et comme $\alpha^{\vee\vee} = \alpha$ , mod $\mathcal{E} = [\delta]$. Or $[\delta] = [\delta']$.

Pour montrer que M est un module grossier nous utiliserons le

LEMME 4.9. Soit F un foncteur sur la catégorie des variétés et mod : F $\longrightarrow \hat{M}$ un morphisme de F vers le foncteur représenté par M. On suppose

1) - pour toute variété S, si $\alpha$ et $\beta$ sont deux éléments de F(S) vérifiant mod $\alpha$ = mod $\beta$ , alors $\alpha$ et $\beta$ sont localement égaux dans S.

2) - il existe T et u dans F(T) tel que mod u soit propre et lisse.

Alors mod fait de M un module grossier.

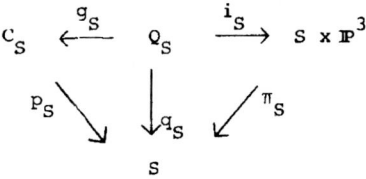

DEMONSTRATION. Observons que $T \times_M T$ est une variété et que les deux éléments $u_1$ et $u_2$ de $F(T\times_M T)$ induits par u sont localement égaux, d'après la première hypothèse. Par suite, si $n: F \to \hat{N}$ est un autre morphisme de foncteurs, alors $n(u_1)$ et $n(u_2)$ sont égaux. Comme $T \to M$ est un morphisme de descente, cela prouve que $n(u) : T \to N$ se factorise à travers M. Soit $f : M \to N$ le morphisme correspondant. Soit maintenant $v \in F(S)$ et montrons $n(v) = f \circ \mathrm{mod}\ v$. On observe que $S \times_M T$ est une variété et que v et u induisent des éléments v' et u' de $F(S\times_M T)$ localement égaux d'après 1). On a donc $n(v') = f \circ \mathrm{mod}\ v'$. Et le fait que $S\times_M T \to S$ est un épimorphisme permet de conclure.

Rappelons que si M est un module grossier, un élément de F(M) est un objet de Poincaré, si le morphisme associé est l'identité de M.

THEOREME 4.10. *Le morphisme de foncteurs mod (Définition 4.7) fait de $M_{c_2}$ (Définition 3.1) un module grossier pour 'tHS, qui admet un objet de Poincaré si et seulement si $c_2$ est pair. De plus M est rationnel.*

DEMONSTRATION. Pour appliquer le lemme 4.9 vérifions en les hypothèses. 1) Si $\mathscr{E}'$ et $\mathscr{E}''$ sont deux fibrés 'tHS sur $S \times \mathbb{P}^3$ ayant même morphisme modulaire, on a deux suites exactes

$$0 \longrightarrow \mathscr{E}'^{\vee} \longrightarrow (\pi^*\pi_*\mathscr{E}')^{\vee} \xrightarrow{\alpha'} \mathscr{E}xt^1(\tilde{\mathscr{F}}_{\mathscr{E}'},\mathcal{O}) \longrightarrow 0$$

$$0 \longrightarrow \mathscr{E}''^{\vee} \longrightarrow (\pi^*\pi_*\mathscr{E}'')^{\vee} \xrightarrow{\alpha''} \mathscr{E}xt^1(\tilde{\mathscr{F}}_{\mathscr{E}''},\mathcal{O}) \longrightarrow 0.$$

Le faisceau $\pi_*\mathscr{H}om\,(\mathscr{E}xt^1(\tilde{\mathscr{F}}_{\mathscr{E}},\mathcal{O}),\ \mathscr{E}xt^1(\tilde{\mathscr{F}}_{\mathscr{E}''},\mathcal{O}))$ est localement libre de rang un d'après le théorème de changement de base. On peut donc, quitte à restreindre S, supposer que $\mathscr{E}xt^1(\tilde{\mathscr{F}}_{\mathscr{E}},\mathcal{O})$ et $\mathscr{E}xt^1(\tilde{\mathscr{F}}_{\mathscr{E}''},\mathcal{O})$ sont isomorphes. On a $[\alpha'] = [\alpha'']$ et la remarque 3.2.3 permet de conclure que $\mathscr{E}'^{\vee}$ et $\mathscr{E}''^{\vee}$ (et par suite $\mathscr{E}'$ et $\mathscr{E}''$) sont isomorphes.

2) Il suffit d'observer que si $C \longrightarrow B$ est un fibré en coniques, alors $C\times_B C \to C$ est un fibré en coniques admettant des fibrés de rang un en tout degré relatif.

Ainsi M est un module grossier. L'affirmation concernant l'objet de Poincaré découle du fait que l'identité de M vérifie la condition de la proposition 4.8 si et seulement si $c_2$ est pair (On utilise la Proposition 1.1 (3), la Remarque 1.2 et le Théorème 3,4, IV).

La rationalité de M est démontrée dans le Théorème 3.4.,III.

COROLLAIRE 4.11. Le module grossier M(2,3) est rationnel.

DÉMONSTRATION. D'après [6] (Th. 9.7 et sa preuve) tout fibré stable avec $c_1 = 2$, $c_2 = 3$ est un fibré 'tHS.

COROLLAIRE 4.12. Il existe un objet de Poincaré pour $M(2,c_2)$ si et seulement si $c_2$ est pair.

DÉMONSTRATION. Si $c_2$ est pair il existe un objet de Poincaré d'après un résultat général de Maruyama [9] (Théorème 6.11). Inversement s'il existe un objet de Poincaré pour $M(2,c_2)$ alors il en existe un pour le module 'tHS correspondant et $c_2$ est pair d'après le théorème 4.10.

REMARQUE 4.13. En utilisant la Remarque 1.2 et le fait qu'un fibré de 't Hooft correspond à un point lisse dans $M(2,c_2)$ (cf. [6] ), on démontre qu'il n'y a pas de fibré de Poincaré sur $U \times \mathbb{P}^3$ où U est un ouvert non-vide de $M(2,c_2)$ dans la composante irréductible contenant les fibrés 'tHS, si $c_2$ est impair.

## 5. LE CAS DE $\mathbb{P}^2$.

Dans ce paragraphe nous indiquons brièvement comment notre méthode permet de traiter le cas de $\mathbb{P}^2$.

THÉORÈME 5.1. (Maruyama-Le Potier). *Pour qu'il existe un fibré de Poincaré pour $M(2,c_2)$, il faut et il suffit que $c_2$ soit pair.*

DEMONSTRATION. La condition suffisante est un cas particulier du résultat général de Maruyama utilisé plus haut.

Pour la condition nécessaire on suppose d'abord que $c_2 \geq 5$. On dit qu'un fibré E (de rang 2) sur $\mathbb{P}^2$ est un fibré de Hulsbergen spécial (HS) s'il existe une conique non-singulière $Q_E$ et un fibré $\mathcal{F}_E$ de rang un sur $Q_E$ tel que E soit l'extension du prolongement $\tilde{\mathcal{F}}_E$ de $\mathcal{F}_E$ à $\mathbb{P}^2$ par un fibré trivial de rang 2. On montre comme dans la proposition 4.3 que $Q_E$ et $\tilde{\mathcal{F}}_E$ sont uniquement déterminés, que $H^0(\mathbb{P}^2,E)$ est de dimension 2, que $\tilde{\mathcal{F}}_E$ est le conoyau de $H^0(\mathbb{P}^2,E) \otimes \mathcal{O} \longrightarrow E$ et que le degré de $\mathcal{F}_E$ est $4-c_2$. On montre ensuite comme dans la proposition 4.4 que cette construction passe bien aux familles de fibrés HS. On observe aussi que $\mathcal{E}xt^1(\tilde{\mathcal{F}}_E,\mathcal{O})$ est le prolongement par zéro d'un fibré de rang un et degré $c_2$ sur la conique. On définit alors, comme en 4.7, le morphisme mod du foncteur des fibrés HS dans le foncteur représenté par M', variété des $g^1_{c_2}$ sans point de base sur la conique universelle de $\mathbb{P}^2$. On montre comme en 4.8 qu'un morphisme $\varphi : S \longrightarrow M'$ est dans l'image de mod si et seulement s'il existe sur la conique universelle remontée à S un fibré de rang un et degré relatif $c_2$. On en déduit (comme dans le théorème 4.10) que M' est un module grossier qui admet un fibré de Poincaré si et seulement si $c_2$ est pair, en utilisant l'exemple 1.5 et la Proposition 2.6.

Comme dans la démonstration de Le Potier, le cas $c_2 = 3$ réclame un traitement spécial. Dans ce cas on sait que le diviseur des droites de saut est une conique de $\mathbb{P}^{2*}$ et que, p et q désignant les projections du diagramme standard

$$\begin{array}{ccc} \mathbb{P}^{2*} & \xleftarrow{\;q\;} & F \\ & & \big\downarrow p \\ & & \mathbb{P}^2 \end{array}$$

$R^1q_* \, p^* E(-1)$ est le prolongement par zéro d'un fibré de rang un et degré un sur cette conique (cf [1], § 6.2). Enfin le module grossier est la variété des coniques non-dégénérées de $\mathbb{P}^{2*}$. On conclut qu'il n'y a pas de fibré de Poincaré parce qu'il n'y a pas de fibré de rang un et degré relatif impair sur la conique universelle.

Remerciements : Ce travail a été effectué à l'Université de Nice. Le second auteur veut remercier l'Université de Nice pour un séjour agréable et d'excellentes conditions de travail.

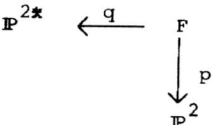

BIBLIOGRAPHIE


[1] W. BARTH — Moduli of vector bundles on the projective plane, Invent. Math. 42 (1977), 63-91.

[2] N. BOURBAKI — Eléments de Mathématiques, Formes sesquilinéaires et formes quadratiques, Hermann, Paris, 1959.

[3] P. DELIGNE — Quadriques, Exposé XII, Groupe de monodromie en géométrie algébrique, S.G.A. VII, Springer Lecture Notes 340 (1977).

[4] G. ELLINGSRUD and S.A. STROMME — On the moduli space of vector bundles of rank two on $\mathbb{P}_3$ with $c_1$ odd. Preprint Oslo 1979.

[5] R. HARTSHORNE — Algebraic Geometry, Graduate texts in mathematics 52, Springer 1977.

[6] R. HARTSHORNE — Stable vector bundles of rank 2 on $\mathbb{P}^3$, Math. Ann. 238 (1978), 229-280.

[7] K. HULEK — Stable Rank-2 Vector Bundles on $\mathbb{P}_2$ with $c_1$ Odd, Math. Ann. 242, 241-266 (1979).

[8] J. LE POTIER — Fibrés stables de rang 2 sur $\mathbb{P}_2(\mathbb{C})$, Math. Ann. 241 (1979), 217-256.

[9] M. MARUYAMA — Moduli of stable sheaves II, J. Math. Kyoto Univ. 18 (1978), 557-614.

[10] M.S. NARASIMHAN and S. RAMANAN — Vector bundles on curves, In algebraic Geometry (papers presented in the Bombay Colloquim 1968) (O.U.P. India 1969), 335-346.

[11] P.E. NEWSTEAD Rationality of moduli spaces of stable bundles, Math. Ann. 215 (1975), 251-268. Correction : Math. Ann. 249 (1980), 281-282.

[12] P.E. NEWSTEAD Comparison theorem for conic bundles, Math. Proc. Camb. Phil. Soc. 90 (1981), 21-31.

[13] P.E. NEWSTEAD On the cohomology and the Picard group of a moduli space of bundles on $\mathbb{P}^3$, soumis au Quarterly J. Math.

[14] P.E. NEWSTEAD Pencils on conic bundles. Preprint Liverpool 1982.

[15] S. RAMANAN The moduli spaces of vector bundles over an algebraic curve, Math. Ann. 200 (1973), 69-84.

[16] L. ROTH Algebraic threefolds, Springer, 1955.

[17] J.P. SERRE Espaces fibrés algébriques, Exposé 1, Séminaire Chevalley, 1958.

A. HIRSCHOWITZ
Département de Mathématiques
I.M.S.P.
Parc Valrose - 06034 NICE CEDEX

M.S. NARASIMHAN
Tata Institute of Fundamental Research,
BOMBAY 400 005 INDE.

# Projective bundles on a complex torus

By *G. Elencwajg* at Nice and *M. S. Narasimhan* at Bombay

---

Let $X$ be a complex manifold. If $P$ is a holomorphic projective bundle on $X$, we have an obstruction, which is a torsion element in $H^2(X, \mathcal{O}^*)$, for $P$ to be of the form $\mathbb{P}(E)$, where $E$ is a holomorphic vector bundle on $X$. This obstruction class is obtained by means of the connecting homomorphism

$$H^1(X, PGL(r, \mathcal{O})) \to H^2(X, \mathcal{O}^*)$$

associated with the exact sequence of sheaves of groups on $X$:

$$1 \to \mathcal{O}^* \to GL(r, \mathcal{O}) \to PGL(r, \mathcal{O}) \to 1.$$

A. Grothendieck has asked, in the algebraic case, whether every torsion element in $H^2_{et}(X, \mathcal{O}^*)$ is the obstruction class of a projective bundle ([4], p. 76). We deal with the analogue of this question in the case of a complex torus.

**Theorem 1.** *Let $X$ be a complex torus. Then every torsion element in $H^2(X, \mathcal{O}^*)$ is the obstruction class of a holomorphic projective bundle on $X$. In fact, the projective bundle may be chosen to be a flat bundle.*

For results of this type on abelian varieties we refer to [1] and [5].

It follows from the theorem that there exists, on any complex torus of complex dimension $\geqq 2$, a holomorphic projective bundle which is not associated to any holomorphic vector bundle (see Proposition 4. 1).

The proof of the theorem is purely algebraic. Let $\Gamma$ be the fundamental group of $X$. Using the fact that $X$ is a $K(\Gamma, 1)$ space we reduce the problem to one of showing that every element of $H^2(\Gamma, \mu_r)$ (where $\mu_r$ denotes the group of $r^{\text{th}}$ roots of unity) is the obstruction for lifting a projective representation of $\Gamma$ into a representation of $\Gamma$ into the special linear group. The latter problem is solved by explicit construction.

## 2. Projective representations of $\Gamma$

We shall denote by $SL(r)$ (resp. $PGL(r)$) the group $SL(r, \mathbb{C})$ (resp. $PGL(r, \mathbb{C})$) and by $g$ the complex dimension of $X$. Consider the exact sequence

$$1 \to \mu_r \to SL(r) \to PGL(r) \to 1.$$

0075-4102/83/0340-0001$02.00

Let $\rho: \Gamma \to PGL(r)$ be a homomorphism. The obstruction for lifting $\rho$ into a homomorphism of $\Gamma$ into $SL(r)$ is an element $\delta(\rho)$ in $H^2(\Gamma, \mu_r)$.

Note that there is a natural homomorphism $PGL(r) \times PGL(r') \to PGL(rr')$ induced by the homomorphism $SL(r) \times SL(r') \to SL(rr')$ given by

$$(A, B) \to A \otimes B, \quad A \in SL(r), \quad B \in SL(r').$$

If

$$\rho: \Gamma \to PGL(r) \quad \text{and} \quad \rho': \Gamma \to PGL(r')$$

are two homomorphisms, we thus have a natural homomorphism denoted by $\rho \otimes \rho'$, of $\Gamma$ into $PGL(rr')$. We have

$$\delta(\rho \otimes \rho') = \Phi_1(\delta(\rho)) \cdot \Phi_2(\delta(\rho'))$$

where $\Phi_1 : H^2(\Gamma, \mu_r) \to H^2(\Gamma, \mu_{rr'})$ and $\Phi_2 : H^2(\Gamma, \mu_{r'}) \to H^2(\Gamma, \mu_{rr'})$ are induced respectively by the inclusions of $\mu_r$ and $\mu_{r'}$ in $\mu_{rr'}$.

**Remark.** For several reasons it is better to think of the obstruction as an element of $H^2(\Gamma, \mu_\infty)$ where $\mu_\infty$ denotes the group of all roots of unity. But we shall not do so here.

**Proposition 2. 1.** *Every element $\xi$ in $H^2(\Gamma, \mu_r)$ occurs as the obstruction class of a projective representation $\rho: \Gamma \to PGL(r^g)$. (More precisely the image of $\xi$ in $H^2(\Gamma, \mu_{r^g})$ is the obstruction class of $\rho$.)*

**Remark.** We do not claim that $\xi$ is the obstruction class of a homomorphism of $\Gamma$ into $PGL(r)$.

For the proof of the proposition we need some lemmas.

**Lemma 2. 2.** *Let $\xi \in H^2(\Gamma, \mu_r)$. Then we can choose a basis $(e_1, \ldots, e_{2g})$ of $\Gamma$ with the following property: If $\Gamma_i$ denotes the subgroup $\mathbb{Z}e_{2i-1} + \mathbb{Z}e_{2i}$, there exists an element $\xi_i \in H^2(\Gamma_i, \mu_r)$, $1 \leqq i \leqq g$, such that*

$$\xi = \sum_{i=1}^{g} \tilde{\xi}_i$$

*where $\tilde{\xi}_i \in H^2(\Gamma, \mu_r)$ is the image of $\xi_i$ under the homomorphism $H^2(\Gamma_i, \mu_r) \to H^2(\Gamma, \mu_r)$ induced by the projection $\Gamma \to \Gamma_i$ (with respect to the direct sum decomposition $\Gamma = \bigoplus_i \Gamma_i$).*

*Proof.* Note first that, since $X$ is a $K(\Gamma, 1)$ space, we have $H^i(\Gamma, \mathbb{Z}) \xrightarrow{\sim} H^i(X, \mathbb{Z})$, $i \geqq 0$. [3], Ch. XVI, § 9. In particular we see that $H^3(\Gamma, \mathbb{Z})$ is torsion free. Consider the exact sequence

$$0 \longrightarrow \mathbb{Z} \longrightarrow \mathbb{Z} \xrightarrow{e^{\frac{2\pi i}{(r)}(.)}} \mu_r \longrightarrow 1$$

and the corresponding exact cohomology sequence

$$H^2(\Gamma, \mathbb{Z}) \to H^2(\Gamma, \mu_r) \to H^3(\Gamma, \mathbb{Z}) \to H^3(\Gamma, \mathbb{Z}).$$

Since $H^3(\Gamma, \mathbb{Z})$ is torsion-free, the map $H^2(\Gamma, \mathbb{Z}) \to H^2(\Gamma, \mu_r)$ is surjective. Let $\eta$ be an element of $H^2(\Gamma, \mathbb{Z})$ mapping into $\xi$.

Let $Z^2(\Gamma, \mathbb{Z})$ be the space of 2-cocycles of $\Gamma$ with values in $\mathbb{Z}$ and $\mathrm{Alt}^2(\Gamma, \mathbb{Z})$ the space of alternating 2-forms on $\Gamma$ with values in $\mathbb{Z}$. One knows that the mapping $Z^2(\Gamma, \mathbb{Z}) \to \mathrm{Alt}^2(\Gamma, \mathbb{Z})$ mapping a cocycle $f$ into the alternating 2-form

$$\varphi(x, y) = f(x, y) - f(y, x)$$

induces an isomorphism of $H^2(\Gamma, \mathbb{Z})$ onto $\mathrm{Alt}^2(\Gamma, \mathbb{Z})$ ([6], Ch. I, § 2, Lemma on page 16). Let $\Phi$ be the alternating 2-form corresponding to $\eta$. By a theorem of Frobenius ([2], § 5, Th. 1) we can choose a basis $(e_1, \ldots, e_{2g})$ of $\Gamma$ and integers $d_1, \ldots, d_g \geqq 0$ such that

1) $\Phi(e_{2i-1}, e_{2i}) = d_i \quad (i = 1, \ldots, g)$,

2) $\Phi(e_i, e_j) = 0$ for all other $(i, j)$ with $i \leqq j$.

Let $\eta_i \in H^2(\Gamma_i, \mathbb{Z})$ be the element which maps into the alternating form $\Phi_i$ on $\Gamma_i$ given by $\Phi_i(e_{2i-1}, e_{2i}) = d_i$. If $\xi_i \in H^2(\Gamma_i, \mu_r)$ is the image of $\eta_i$ under the canonical map

$$H^2(\Gamma_i, \mathbb{Z}) \to H^2(\Gamma_i, \mu_r),$$

then $\{\xi_i\}$ satisfy the required conditions.

**Lemma 2. 3.** *Let $\Gamma_0$ be a free abelian group on two generators and $\psi$ be an element of $H^2(\Gamma_0, \mu_r)$. Then there exists a projective representation $\rho_0\colon \Gamma_0 \to PGL(r)$ whose obstruction class is $\psi$.*

*Proof.* Let $(e_1, e_2)$ be a basis of $\Gamma_0$ and let $\psi$ correspond to the (multiplicative) alternating form $\varphi(e_1, e_2) = \lambda \in \mu_r$. Let $A$ and $B$ be two elements of $SL(r)$ such that $ABA^{-1}B^{-1} = \lambda I$.

[For example, if $(f_1, \ldots, f_r)$ is the canonical basis in $\mathbb{C}^r$ let $\tilde{A}$ be the linear transformation in $\mathbb{C}^r$ defined by $\tilde{A}(f_i) = \lambda^{i-1} f_i$ $(i = 1, \ldots, r)$ and let $\tilde{B}$ be defined by $\tilde{B} f_i = f_{i+1}$ for $i < r$ and $\tilde{B} f_r = f_1$. Take

$$A = \frac{\tilde{A}}{\sqrt[r]{\det \tilde{A}}} \quad \text{and} \quad B = \frac{\tilde{B}}{\sqrt[r]{\det \tilde{B}}}\Bigg].$$

If $\underline{A}$ and $\underline{B}$ be the images of $A$ and $B$ in $PGL(r)$, then $\underline{A}$ and $\underline{B}$ commute in $PGL(r)$ and let $\rho_0\colon \Gamma \to PGL(r)$ be the homomorphism such that $\rho_0(e_1) = \underline{A}$ and $\rho_0(e_2) = \underline{B}$. We verify easily that the obstruction class of $\rho_0$ is $\psi$.

*Proof of Proposition* 2. 1. Let $\xi \in H^2(\Gamma, \mu_r)$. Choose $\xi_i \in H^2(\Gamma_i, \mu_r)$ as in Lemma 2. 2. By Lemma 2. 3, we can find a homomorphism $\rho_i\colon \Gamma_i \to PGL(r)$ with $\delta(\rho_i) = \xi_i$. Let $\tilde{\rho}_i = \rho_i \circ p_i$ where $p_i\colon \Gamma \to \Gamma_i$ is the projection with respect to the decomposition $\Gamma = \oplus \Gamma_i$. Define $\rho\colon \Gamma \to PGL(r^g)$ to be $\tilde{\rho}_1 \otimes \cdots \otimes \tilde{\rho}_g$. Then $\rho$ has the required property.

## 3. Proof of Theorem 1

Consider the Kummer exact sequence

$$1 \longrightarrow \mu_r \longrightarrow \mathcal{O}^* \xrightarrow{(\cdot)^r} \mathcal{O}^* \longrightarrow 1$$

and the corresponding exact cohomology sequence

$$H^2(X, \mu_r) \longrightarrow H^2(X, \mathcal{O}^*) \xrightarrow{(\cdot)^r} H^2(X, \mathcal{O}^*).$$

From this we see that every $r$-torsion element of $H^2(X, \mathcal{O}^*)$ is the image of an element in $H^2(X, \mu_r)$. Now $X$ being a $K(\Gamma, 1)$ space, the natural homomorphism

$$H^2(\Gamma, \mu_r) \longrightarrow H^2(X, \mu_r)$$

is an isomorphism.

If $\chi$ is an $r$-torsion element in $H^2(X, \mathcal{O}^*)$, choose $\xi \in H^2(\Gamma, \mu_r)$ mapping into $\chi$. Let $\xi'$ be the image of $\xi$ in $H^2(\Gamma, \mu_{r^g})$ and $\rho: \Gamma \to PGL(r^g)$ a homomorphism whose obstruction class is $\xi'$ (such a $\rho$ exists by Proposition 2. 2). Let $P_\rho$ be the flat holomorphic projective bundle associated to $\rho$. Using the commutativity of the diagram

$$\begin{array}{ccc} H^2(\Gamma, \mu_r) & \longrightarrow & H^2(X, \mathcal{O}^*) \\ & \searrow \qquad \nearrow & \\ & H^2(\Gamma, \mu_{r^g}) & \end{array}$$

we see that the obstruction class of $P_\rho$ is $\chi$.

## 4. An application

**Proposition 4. 1.** *On any complex torus of complex dimension* $\geqq 2$, *there exists a holomorphic projective bundle which does not arise from a holomorphic vector bundle.*

*Proof.* Applying Theorem 1, it suffices to show that $H^2(X, \mathcal{O}^*)_{tors} \neq 1$. We see, using the Kummer sequence, that we have an exact sequence

$$1 \to \frac{NS(X)}{r \cdot NS(X)} \to H^2(X, \mu_r) \to H^2(X, \mathcal{O}^*)_{r\text{-}tors} \to 1,$$

where $NS(X)$ is the Neron-Severi group of $X$. On the other hand we have

1) $$\# \frac{NS(X)}{r \cdot NS(X)} = r^s,$$

where $s$ is the rank of the Neron-Severi group and

2) $$\# H^2(X, \mu_r) = r^{\binom{2g}{2}}.$$

Since $s \leqq h^{1,1}(X) = g^2$ we have $\binom{2g}{2} > g^2 \geqq s$ für $g \geqq 2$.

Hence $H^2(X, \mathcal{O}^*)_{r\text{-}tors} \neq 1$ for $r > 1$, $g \geqq 2$.

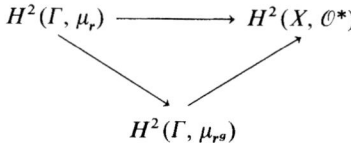

## References

[1] *V. G. Berkovich*, The Brauer group of abelian varieties, Funct. Anal. appl. **6** (1972), 180—184, translated from Funktional 'nyi Analiz i Ego Prilozheniya **6** (1972).
[2] *N. Bourbaki*, Elements de mathématique, Formes Sesquilinéaires et formes quadratiques, Paris 1959.
[3] *H. Cartan* and *S. Eilenberg*, Homological Algebra, Princeton 1956.
[4] *A. Grothendieck*, Le groupe de Brauer, Dix exposés sur la cohomologie des schémas, Amsterdam 1968.
[5] *R. Hoobler*, Brauer groups of abelian schemes, Ann. Sc. de l'E.N.S. (4) **5** (1972), 45—70.
[6] *D. Mumford*, Abelian varieties, T.I.F.R, Studies in Mathematics, London 1970.

Institut de Mathématiques, Université de Nice, Parc Valrose, F-06034 Nice

Tata Institute of Fundamental Research, Bombay 400 005, India.

Eingegangen 31. März 1982

Math. Ann. 266, 55–72 (1983)

Mathematische Annalen


# Maximal Subbundles of Rank Two Vector Bundles on Curves

H. Lange[1] and M. S. Narasimhan[2]

1 Mathematisches Institut der Universität, Bismarckstrasse $1^1/_2$, D-8520 Erlangen, Federal Republic of Germany

2 Tata Institute of Fundamental Research, Homi Bhabha Road, Bombay 400 005, India

Let $X$ be a complete non-singular irreducible curve over an algebraically closed field of characteristic zero. For a vector bundle $E$ of rank two over $X$ define

$$s(E) = \deg(E) - 2\max\deg(L),$$

where the maximum is taken over all line subbundles $L$ of $E$. Note that $s(E) \equiv \deg(E) \bmod 2$ and that $E$ is stable (resp. semi-stable) if and only if $s(E) \geqq 1$ (resp. $s(E) \geqq 0$). Nagata proved in [8] that $s(E) \leqq g$, where $g$ is the genus of $X$.

A line subbundle of $L$ of $E$ of maximal degree will be called a maximal subbundle. Let $M(E)$ be the subscheme of $\mathrm{Pic}(X)$ formed by the maximal line subbundles of $E$. Maruyama [6] proved the following results:

1) If $s(E) = g$, then $\dim M(E) = 1$.

2) For a general bundle $E$ with $s(E) \leqq (g-1)$, there are only finitely many maximal line subbundles.

In Sects. 3 and 4 we rederive the results of Nagata and Maruyama, adding certain complements.

Suggested by 2), Maruyama conjectured that, if $E$ is a bundle which is not of the form $L \oplus L$ and $s(E) \leqq (g-1)$, the number of maximal line subbundles of $E$ is finite [6, p. 58]. This problem is also raised in Nagata [9]. The main contribution of the present paper is a study of this question. It turns out that the conjecture is not in general valid.

When $s=2$ (and $g \geqq 3$) we determine completely the curves and bundles $E$ for which $M(E)$ is not finite (Theorem 5.1). In particular such bundles exist on any curve of genus 3; for such a bundle to exist on a curve of genus $\geqq 4$, the curve has to be a two sheeted covering of an elliptic curve and for each line bundle of degree $g$ on the elliptic curve we can construct such a bundle. The method of dealing with this question is to translate the problem into one of projective geometry; namely, to determine curves of genus $g$ and degree $2g$ in the projective space $\mathbb{P}^g$ and points in $\mathbb{P}^g$, not on the curve, through which infinitely many secant lines of the curve pass. The results regarding the secant spaces, which are used throughout the paper, are developed in Sects. 1 and 2. (The ideas of Sect. 1 go back to Atiyah (cp. [12]).) In particular we obtain, in passing, a criterion for an extension of line bundles to be stable [and more generally to have a given value of $s(E)$] in terms of the secant spaces.

In Sect. 6, we show that for any $g \geqq 4$ there is a curve $X$ of genus $g$ and a vector bundle $E$ on $X$ with $s(E)=3$ and $\dim M(E)=1$. By projection one is reduced to finding a curve of genus $g$ in $\mathbb{P}^g$ of degree $(2g+1)$ having infinitely many trisecant lines. Such curves are constructed by studying linear systems of curves on certain elliptic scrolls.

In the last section we construct examples with $\dim M(E)=1$ for arbitrary $s$ for $g$ sufficiently large compared with $s$. However this method does not yield as complete results as those in Sect. 5.

The authors would like to thank the University of Nice where this work was begun. The first named author would like to thank the Tata Institute of Fundamental Research for hospitality and excellent working conditions.

## 1. The Invariant $s$ and Sectant Varieties

Let $X$ denote a complete nonsingular curve of genus $g$ over an algebraically closed field $k$ of characteristic 0. For a vector bundle $E$ of rank 2 on $X$ we define the invariant $s(E)$ by

$$s(E)=c_1(E)-2\max c_1(L)$$

the maximum being taken over all line subbundles $L$ of $E$. It is easy to see that if $P(E)$ denotes the associated ruled suface, then $s(E)$ is the minimum of the self-intersection numbers of all sections of $P(E)$ over $X$ [5]. By definition, $E$ is stable (resp. semi-stable) if and only if $s(E)>0$ (resp. $\geqq 0$). If $T$ is any algebraic variety over $k$ and $\mathscr{E}$ a vector bundle of rank $Z$ on $X\times T$ then the function $s: T\to\mathbb{Z}$ defined by $s(t)=s(\mathscr{E}|(X\times t))$ is lower semicontinuous. The proof is the same as the usual proof for the openness of stability condition.

Now suppose we are given a fixed line sub-bundle of $E$ which is not a direct summand. Since we are interested in $s(E)$ and since $s(E\otimes L)=s(E)$ for every line bundle $L$ on $X$ we may assume the given line sub-bundle to be the trivial one, i.e. suppose we are given a non-split exact sequence

$$0\to\mathcal{O}_X\to E\to L\to 0. \tag{e}$$

The points of $\mathbb{P}=P(H^0(K_X\otimes L))$ are the codimension 1 vector subspaces of $H^0(K_X\otimes L)$ or equivalently the classes of nonzero linear forms on $H^0(K_X\otimes L)$ modulo the canonical operation of $k^*$. Hence using Serre duality the class of (e) may be considered as an element of $\mathbb{P}$. As such it will be denoted also by e. Consider the rational map $\varphi=\varphi_{K_X\otimes L}: X\to\mathbb{P}$, $x\mapsto\{\sigma\in H^0(K_X\otimes L)|\sigma(x)=0\}$ corresponding to the complete linear system $|K_X\otimes L|$.

We want to analyse the geometric meaning of $s(E)$ in relation to the map $\varphi\cdot X\to\mathbb{P}$ and the point $e\in\mathbb{P}$. In order to avoid trivial cases, in which the statements have to be slightly modified, we assume in the rest of the section that $d=c_1(L)\geqq 2$ (resp. 3 or 4) if $g\geqq 2$ (resp. 1 or 0). In particular $\dim\mathbb{P}\geqq 2$ and $\varphi$ is a morphism. Moreover $2-d\leqq s(E)\leqq d$ since (e) is not split.

If $\varphi$ is an embedding and $D=\sum_{i=1}^{t} n_i x_i$ an effective divisor of $X$ we define *span*$(D)$ to be the projective linear subspace of $\mathbb{P}$ spaned by the $n_i^{\text{th}}$ osculating spaces of $\varphi(x_i)$ in $\mathbb{P}$ with respect to $\varphi(X)$ for $i=1,\dots,t$. If $\varphi$ is not an embedding, then necessarily $c_1(L)=2$ and $g \geqq 2$ and we need the notation span$(D)$ only for divisors of degree 1, in which case it has an obvious meaning. For a fixed $j$ with $1 \leqq j \leqq d-1$ the union of all linear spaces span $D$ of $\mathbb{P}$ for all effective divisors $D$ of $X$ of degree $j$ is denoted by $\mathrm{Sec}_j(X)$. It is called the $j^{\text{th}}$ *secant variety of $X$ with respect to* $\varphi$. A consequence of the proposition below and the semicontinuity of $s$ is that $\mathrm{Sec}_j(X)$ is a closed subvariety of $\mathbb{P}$.

**Proposition 1.1.** *Let the notation be as above. Then for any integer* $s \equiv d \bmod 2$ *with* $s \leqq d$ *and* $s \geqq 4-d$ *(resp.* $6-d$, $8-d$ *if* $g=1,0$*) the following conditions are equivalent*

(i) $s(E) \geqq s$,

(ii) $e \notin \mathrm{Sec}_{\frac{1}{2}(d+s-2)}(X)$.

*Proof.* (i) $\Rightarrow$ (ii). Suppose $e \in \mathrm{span}(D)$ with an effective divisor $D$ of degree $\frac{1}{2}(d+s-2)$. According to the definitions this means that the exact sequence (e) is contained in the kernel of the canonical map

$$H^0(K_X \otimes L)^* \to H^0(K_X \otimes L(-D))^* .$$

If $\psi: L(-D) \to L$ denotes the canonical inclusion this means that the upper exact sequence of the pullback diagram splits:

$$\begin{array}{ccccccccc} 0 \to & \mathcal{O}_X & \to & E \times_L L(-D) & \to & L(-D) & \to 0 \\ & \downarrow & & \downarrow & \swarrow & \downarrow \psi & \\ 0 \to & \mathcal{O}_X & \longrightarrow & E & \longrightarrow & L & \longrightarrow 0 \end{array}$$

In particular $L(-D)$ is a subsheaf of $E$. It follows that $s(E) \leqq c_1(E) - 2c_1(L(-D)) = d - 2d + 2\deg D = s-2$ contradicting (i).

(ii) $\Rightarrow$ (i). Suppose $s(E) \leqq s-2$. Then there is an invertible subsheaf $M$ of $E$ of degree $\frac{1}{2}(d-s+2)$. Since $\deg M > 0$ the composed homomorphism $\psi: M \to L$ is injective which implies that $M = L(-D)$ with an effective divisor $D$ of degree $\frac{1}{2}(d+s-2)$. This means that $e \in \mathrm{span}(D)$ contradicting (ii). This completes the proof of Proposition 1.1.

Consider the "flag" of secant varieties

$$X = \mathrm{Sec}_1(X) \subsetneqq \mathrm{Sec}_2(X) \subseteqq \dots \subseteqq \mathrm{Sec}_{d-2}(X) \subseteqq \mathrm{Sec}_{d-1}(X) \subseteqq \mathbb{P} .$$

Proposition 1.1 means that if $e$ is a point in $\mathbb{P}$ and $E$ the corresponding vector bundle of rank 2, then $s(E)$ is determined by the number $j$ such that $e$ is contained in $\mathrm{Sec}_j(X)$ but not in $\mathrm{Sec}_{j-1}(X)$. Also we get as a special case: $E$ is stable if and only if $e \notin \mathrm{Sec}_{d/2}(X)$.

*Remark.* If $X$ is the projective line, then $\varphi(X)$ is the rational normal curve in $\mathbb{P}$. In this case Proposition 1.1 has been proved by Mülich in [7] using explicit transition matrices for $E$ and by Barth (unpublished) using cohomological methods.

In Sect. 3 we need

**Corollary 1.2.** *Let $g \geqq 2$ and $L_1$ and $L_2$ be line bundles on $X$ with $0 < c_1(L_2) - c_1(L_1) = s \leqq g$. The there is an extension*

$$0 \to L_1 \to E \to L_2 \to 0$$

*with $s(E) = s$.*

*Proof.* We may assume $s \geqq 2$ the case $s = 1$ being obvious. For $L = L_2 \otimes L_1^*$ let $\varphi = \varphi_{K_X \otimes L} : X \to \mathbb{P} = P(H^0(K_X \otimes L))$ be as above. Since $\dim \mathrm{Sec}_{s-2}(X) \leqq 2s - 3 < 2s - 2 \leqq g + s - 2 = \dim \mathbb{P}$, there is an $e \in \mathbb{P} - \mathrm{Sec}_{s-2}(X)$. Let (e) $: 0 \to \mathcal{O}_X \to E \to L \to 0$ denote a corresponding extension. Proposition 1.1 implies that $s(E) = s$. Now the extension (e)$\otimes L_1$ has the required property.

## 2. Maximal Subbundles

Given a vector bundle $E$ of rank 2 on $X$ we wish to determine the dimension of the space of line sub-bundles of maximal degree of $E$, called *maximal subbundles* in the sequel. In this section the problem will be translated into a geometric one along the lines of Sect. 1. We need the following lemma.

**Lemma 2.1.** *If $E$ is indecomposable, the natural map from the set of maximal line bundles of $E$ into* Pic$(X)$ *is injective.*

*Proof.* We have to show that if $i_1 : L_1 \to E$ and $i_2 : L_2 \to E$ are two maximal subbundles of $E$ with $L_1 \simeq L_2$, then $i_1 = \lambda i_2$ for a nonzero constant $\lambda$. Suppose $i_1 \neq \lambda i_2$ for all $\lambda$, then the map $i_1 + j i_2 : L_1 \oplus L_1 \to E$, where $j : L_1 \to L_2$ is an isomorphism, is of rank 2 and hence injective. Since it is not an isomorphism, there is a point $x \in X$ with $(i_1 + j i_2)(x) = 0$ which implies that $L_1 \otimes \mathcal{O}_X(x)$ is subsheaf of $E$ contradicting the maximality of $L_1$.

**Lemma 2.2** *Let $E$ denote an indecomposable vector bundle of* rank 2 *on $X$. The set $M(E)$ of maximal subbundles of $E$ considered as a subset of* Pic$(X)$ *is closed in* Pic$(X)$.

*Proof.* By definition $M(E)$ is contained in $P = \mathrm{Pic}^{\frac{1}{2}(c_1(E) - s(E))}(X)$. Let $\mathcal{P}$ denote a Poincaré bundle on $X \times P$ and consider $\mathcal{P}^* \otimes p_1^* E$ where $p_1 : X \times P \to X$ denotes the projection. The semi-continuity theorem implies that the set

$$\{p \in P \mid h^0(X \times \{p\}, \mathcal{P}^* \otimes p_1^* E \mid X \times \{p\}) \geqq 1\}$$

is closed in $P$. But this is just $M(E)$.

We may assume $s(E) \geqq 1$ and $g \geqq 2$, the problem being obvious for $s(E) \leqq 0$ and wellknown for $g \leqq 1$. Suppose we are given a maximal subbundle of $E$. Since the situation does not change under twistings with line bundles, we may assume moreover the given line subbundle to be $\mathcal{O}_X$. Then we have

**Lemma 2.3.** *Given an exact sequence*

$$0 \to \mathcal{O}_X \to E \to L \to 0 \qquad \text{(e)}$$

*with a line bundle $L$ of degree $d=s(E)\geqq 1$ on $X$ with $g\geqq 2$. For a line bundle $M$ of degree 0 on $X$ the following conditions are equivalent*

(i) *$M$ is a subbundle of $E$ (necessarily or maximal degree) and $M\neq \mathcal{O}_X$.*

(ii) *$L\otimes M^{-1}\simeq \mathcal{O}_X(D)$ with an effective divisor $D$ of degree $d$ and* (e) *is in the kernel of the canonical map $H^0(K_X\otimes L)^*\to H^0(K_X\otimes L(-D))^*$.*

*Proof.* (i) $\Rightarrow$ (ii). Let $M\to E$ denote the canonical injection. Then the composition $\psi: M\to E\to L$ is nonzero. Hence $L\otimes M^{-1}\simeq \mathcal{O}_X(D)$ with $D$ effective of degree $d$ and the upper line of the pullback diagram splits

$$\begin{array}{ccccccccc} 0 & \to & \mathcal{O}_X & \to & E\times_L M & \to & M & \to & 0 \\ & & \downarrow & & \downarrow & \swarrow & \downarrow{\scriptstyle \psi} & & \\ 0 & \to & \mathcal{O}_X & \to & E & \to & L & \to & 0 \end{array}$$

which means (e)$\in \operatorname{Ker}(H^1(L^{-1})\to H^1(M^{-1}))$. Now Serre duality implies (ii).

(ii) $\Rightarrow$ (i). Since $D$ is effective there is an injection $\psi: M\to L$ given by $D$ and (ii) means that the upper line of the above diagram splits, that is, that $M$ is a subbundle of $E$. Lemma 2.1 implies $M\neq \mathcal{O}_X$.

*Remark.* In other terms the proof is given by noting that the diagram

$$\begin{array}{ccccc} \operatorname{Hom}(M,E) & \to & \operatorname{Hom}(M,L) & \to & H^1(M^{-1}) \\ & & \uparrow & & \uparrow \\ & & \operatorname{Hom}(L,L) & \to & H^1(L^{-1}) \end{array}$$

is commutative and the upper sequence is exact.

As in Sect. 1 consider the projective space $\mathbb{P}=P(H^0(K_X\otimes L))$, the map $\varphi=\varphi_{K_X\otimes L}: X\to \mathbb{P}$, and the sequence (e) as a point $e$ of $\mathbb{P}$. If $D$ is any effective divisor of degree $d$ of $X$, condition (ii) of Lemma 2.3 means that the vector subspace $H^0(K_X\otimes L(-D))$ of $H^0(K_X\otimes L)$ is contained in the vector subspace corresponding to the point $e$. To express this geometrically it is convenient to introduce the following definition. A line bundle $N$ of degree $d\geqq 1$ on $X$ is called a *$d$-secant line bundle with respect to $\varphi=\varphi_{K_X\otimes L}: X\to \mathbb{P}$*, if $N=\mathcal{O}_X(D)$ with an effective divisor $D$ of degree $d$ on $X$ such that $\dim \operatorname{span}(\varphi(D))=d-1$. Note that $\deg L=d$. Hence $\operatorname{span}(\varphi(D))$ has been defined in Sect. 1 if $d\geqq 3$, $\varphi$ being a embedding. If $d=2$, the definition is the same except when $D=x_1+x_2$ with $x_1\neq x_2$ but $\varphi(x_1)=\varphi(x_2)$ or $D=2x$ with $\varphi(x)$ being cusp of $\varphi(X)$ in which case $\operatorname{span}(\varphi(D))$ has to be considered as zero-dimensional (see Lemma 5.2 below). If $d=1$ and $D=x$, then $\operatorname{span}\varphi(D)$ has a meaning and $\dim \operatorname{span}(\varphi(D))=0$ if and only if $\varphi$ is defined at $x$.

According to Proposition 1.1 the condition $s(E)=d$ implies for a divisor $D$ satisfying condition (ii) of Lemma 2.3 that $\dim \operatorname{span}(D)=d-1$ that is that $\mathcal{O}_X(D)$ is a $d$-secant line bundle, with respect to $\varphi$, for otherwise $e$ would be contained in the linear span of a divisor of degree $d-1$. Thus Lemma 2.3 can be expressed as

**Proposition 2.4.** *Let $g\geqq 2$. Given an exact sequence* (e) *with $s(E)=c_1(L)=d\geqq 1$, let $\varphi=\varphi_{K_X\otimes L}: X\to \mathbb{P}=PH^0(K_X\otimes L)$ denote the map associated to $K_X\otimes L$ and consider*

*the extension* (e) *as a point e of* $\mathbb{P}$. *Then there is a canonical bijection between the following sets*

(i) *maximal subbundles of E different from* $\mathcal{O}_X$,

(ii) *d secant line bundles N of X with respect to* $\varphi$ *such that e is contained in a corresponding d-secant of* $\varphi(X)$.

## 3. Stratification of the Moduli Space According to the Invariant *s*

Let $M(d)$ denote the moduli space of stable vector bundles of rank 2 and degree $d$ on a curve $X$ of genus $g \geqq 2$. The function $s: M(d) \to \mathbb{Z}$ defined by $E \mapsto s(E)$ is lower semicontinuous and this gives a stratification of $M(d)$ into locally closed subsets $M(d, s)$ according to the value of $s$. The following proposition gives us the dimension of $M(d, s)$.

**Proposition 3.1.** *Let* $1 \leqq s \leqq g$ *and* $s \equiv d \bmod 2$. *Then* $M(d, s)$ *is an irreducible algebraic variety of dimension* $3g + s - 2$ *(resp.* $4g - 3$*) if* $s \leqq g - 2$ *(resp.* $s = g - 1$ *or* $g$*).*

*Remark.* Because of the semicontinuity of $s$ an immediate consequence of Proposition 3.1 is Nagata's theorem that $s(E) \leqq g$ for every vector bundle $E$ of rank 2 on $X$. Proofs of this result were given by Nagata [8], Gunning [1], Stuhler [11], and probably first by Segre [10].

*Proof.* Let $d_1$ and $d_2$ denote the integers such that $d_1 + d_2 = d$ and $d_2 - d_1 = s$. Let $J_i$ denote the Jacobian variety of degree $d_i$ and $\mathcal{P}_i$ a Poincaré bundle on $X \times J_i$ for $i = 1, 2$. If $p_{ij}$ denote the projections of $X \times J_1 \times J_2$ then $V = R^1_{p_{12*}}(p_{02}^* \mathcal{P}_2^* \otimes p_{01}^* \mathcal{P}_1)$ is locally free on $J_1 \times J_2$ of rank $g + s - 1$ by Riemann-Roch. Let $P = P(V^*)$ and $\pi: X \times P \to X \times J_1 \times J_2$ denote the canonical projection. According to [4, Corollary 4.5] there is an extension on $X \times P$

$$0 \to \pi^* p_{01}^* \mathcal{P}_1 \otimes \mathcal{O}_P(1) \to \mathcal{E} \to \pi^* p_{02}^* \mathcal{P}_2 \to 0 \tag{1}$$

which is universal for the classes of nonsplit extensions

$$0 \to L_1 \to E \to L_2 \to 0 \tag{2}$$

with $L_i \in J_i$, modulo the canonical operation of $k^*$. In particular $(1)|X \times \{p\}$ is an extension representing $p$ for every $p \in P$. Corollary 1.2 and semicontinuity imply that there is an open dense subset $U_s \subset P$ such that $s(\mathcal{E}(X \times p)) = s$ if and only if $p \in U_s$. The universal property of $M(d)$ yields a morphism

$$f_s: U_s \to M(d)$$

such that if (2) corresponds to $p \in U_s$ then $f_s(p)$ is just the point of $M(d)$ representing $E$. This means that $f_s(U_s) = M(d, s)$ and hence $M(d, s)$ is irreducible being the image of an irreducible variety under a morphism.

In [5, Lemma 4.2] it is proved in a more general setting that for a general point $p \in U_s$ we have, if (2) is the corresponding extension:

$$\dim f_s^{-1}(E) \leqq h^0(L_1^* \otimes L_2) = \begin{cases} 0 & \text{if } s \leqq g - 1 \\ 1 & \text{if } s = g \end{cases} \tag{3}$$

the last equality being true since $L_1$, $L_2$ are general. It follows, for $s \leqq g-1$,

$$\dim M(d,s) = \dim f_s(U_s) = \dim U_s = 3g+s-2$$

and for $s=g$

$$4g-3 = \dim M(d,g) \geqq \dim \operatorname{Im} f_s(U_s) \geqq \dim P - \dim f_s^{-1}(E) = 4g-2-1 = 4g-3$$

which completes the proof of the proposition.

**Corollary 3.2** (Maruyama [6]). (i) *If* $d \equiv g-1 \bmod 2$ *there is an open dense set* $V_{g-1} \subset M(d)$ *such that every* $E \in V_{g-1}$ *admits only a finite number of maximal subbundles.*

(ii) *If* $d \equiv g \bmod 2$ *we have* $\dim M(E) \geqq 1$ *for every* $E \in M(d,g)$ *and there is an open dense set* $V_g \subset M(d,g)$ *such that for every* $E \in V_g$ *we have* $\dim M(E) = 1$.

*Proof.* Since obviously $\dim M(E) = \dim f_s^{-1}(E)$ the corollary is a consequence of (3) and (4).

Corollary 3.2 (ii) will be improved in Sect. 4 (Corollary 4.7) in the form that for every $E \in M(d,g)$ we have $\dim M(E) = 1$. An analogous statement for the case of Corollary 3.2 (i) is false as we shall see in the subsequent sections. For $1 \leqq s \leqq g-2$ there is a result more precise in another direction, namely we have

**Proposition 3.3.** *Suppose* $1 \leqq s \leqq g-2$ *and* $d \equiv s \bmod 2$. *There is an open dense set* $V_s \subset M(d,s)$ *such that every vector bundle E of* $V_s$ *has exactly one maximal subbundle.*

*Proof.* Let the notation be as in the proof of Proposition 3.1. Applying (3) we get: there is an open dense set $V' \subseteq M(d,s)$ such that for every $E \in V'$ the number $\# f_s^{-1}(E)$ is finite. Using [EGA IV, 15.5.1] there is an open dense set $V'' \subset V'$ such that the function $V'' \to \mathbb{Z}$, $E \mapsto \# f_s^{-1}(E)$ is lower semicontinuous. Hence the set $\{E \in V'' \mid \# f_s^{-1}(E) = 1\}$ is constructibel in $M(d,s)$. Since the number of maximal subbundles does not change under twisting by line bundles it suffices to show that for every $L \in \operatorname{Pic}^s(X)$ there is an open dense set $V_L \subset \mathbb{P} = P(H^0(K_X \otimes L))$ such that for every $e \in V_L$ we have $\# f_s^{-1}(f_s(e)) = 1$.

Since $s \leqq g-2$ we have

$$\dim \operatorname{Sec}_s(X) \leqq 2s-1 \leqq g-3+s < g-2+s = \dim \mathbb{P}.$$

Hence $V_L = \mathbb{P} - \operatorname{Sec}_s(X)$ is open and dense in $\mathbb{P}$ and for every $e \in V_L$ there is no $s$-secant of $X$ in $\mathbb{P}$ passing through $e$. By Proposition 2.4 this means that the only maximal subbundle of $E = f_s(e)$ is the given subbundle $\mathcal{O}_X$ which is clearly maximal according to Proposition 1.1. This completes the proof of Proposition 3.3.

## 4. The Cases $s=1$ and $s=g$

In this section we want to analyse the dimension of the space of maximal subbundles of a vector bundle $E$ with $s=1$ and $s=g$ on a curve $X$ of genus $g \geqq 2$. For the case $s=1$ we need a lemma.

**Lemma 4.1.** *Let $L$ denote a line bundle of degree 1 on $X$ and $\varphi$ the rational map corresponding to the complete linear system $|K_X \otimes L|$. Then we have:*

$$\deg\varphi \leqq \begin{cases} 3 & \text{if} \quad g=2 \\ 2 & \text{if} \quad g \geqq 3. \end{cases}$$

*Proof.* $2g-1=\deg(K_X\otimes L)=\deg\varphi\cdot\deg\varphi(X)\geqq\deg\varphi\cdot(g-1)$ since $\varphi(X)$ is a nondegenerate curve in $\mathbb{P}^{g-1}$. This implies the assertion.

Let

$$0\to\mathcal{O}_X\to E\to L\to 0 \tag{e}$$

denote a nonsplit exact sequence, $L$ being a line bundle of degree 1 on $X$. Let $\varphi=\varphi_{K_X\otimes L}: X\to\mathbb{P}=P(H^0(K_X\otimes L))$ be the associated rational map. Since for every nonsplit exact sequence (e) we have $s(E)=1$ and since a 1-secant line bundle of $X$ with respect to $\varphi$ is just a point bundle $\mathcal{O}_X(x)$ such that $\varphi$ is defined at the point $x\in X$, we get the result immediately using Proposition 2.4 and Lemma 4.1.

**Proposition 4.2.** *Let $E$ denote a vector bundle of* rank 2 *with $s(E)=1$ on $X$. Then the number of maximal sub bundles of $E$ is $\leqq 4$ if $g=2$ and $\leqq 3$ if $g\geqq 3$.*

*Remark.* (a) Using a more refined version of Lemma 4.1 (analogous to Lemma 5.2) it is easy to say precisely when this number is 1, 2, 3, or 4. It is in general 4 if $g=2$ and 1 in general if $g\geqq 3$ (which we know already from Proposition 3.3).

(b) In [6, Theorem 3.17 (iii)] it was proved that if $s(E)=1$, the number of maximal sub line bundles is $\leqq 2g$.

The rest of this section is devoted to the computation of $\dim M(E)$ if $s(E)=g$, $g\geqq 2$. For this it is convenient to use the corresponding projective bundle $P(E)$ and the notion of an elementary transformation. If $p$ is a point of $P(E)$, the proper transform of the fibre over $X$ passing through $p$ of the blow up of $p$ is contractible. Its contraction is again a projective bundle denoted by $\mathrm{elm}_p(P(E))$ and called the *elementary transformation of $P(E)$ in $p$*. The line sub-bundles of $E$ are in canonical one to one correspondence with the sections of $P(E)$ over $X$ and it is easy to see (see [5]) that $s(E)=\min\sigma^2$, the minimum being taken over all sections $\sigma$ of $P(E)$ over $X$. By a *minimal section* of $P(E)$ we understand a section with minimal self intersection number. Thus $M(E)$ can be identified canonically with the space of all minimal sections of $P(E)$.

**Lemma 4.3.** *Let $P(E')=\mathrm{elm}_p(P(E))$.*

(i) *If no minimal section of $P(E)$ passes through $p$, then*

$$s(E')=s(E)+1.$$

(ii) *If a minimal section of $P(E)$ passes through $p$, then*

$$s(E')=s(E)-1$$

*and the minimal sections of $P(E')$ are exactly the proper transforms of the minimal sections of $P(E)$ passing through $p$.*

The proof is easy and will be omitted.

**Lemma 4.4.** *If* $\dim M(E) \geqq 1$, *then for every point* $p \in P(E)$ *there is a minimal section* $\sigma$ *of* $P(E)$ *with* $p \in \operatorname{supp} \sigma$.

*Proof.* Let $C \subset M(E)$ be a curve and consider $C$ as a one dimensional family of minimal sections of $P(E)$. According to Lemma 2.2 we may suppose $C$ to be proper. Consider the morphism

$$\varphi : X \times C \to P(E)$$
$$(x, \sigma) \mapsto \sigma(x).$$

Certainly $\varphi$ is of rank 2. Since $\varphi$ is proper it is surjective which was to be shown.

**Proposition 4.5.** *Let E denote a vector bundle of* rank 2 *on X. Then there is no point p of* $P(E)$ *which admits a one dimensional family of minimal sections* $\sigma$ *of* $P(E)$ *with* $p \in \operatorname{supp} \sigma$.

*Proof.* Suppose $C$ is a curve of minimal sections of $P(E)$ passing through $p \in P(E)$. If $P(E') = \operatorname{elm}_p(P(E))$, Lemma 4.3 (ii) says that $s(E') = s(E) - 1$ and $\dim M(E') \geqq 1$. Combining Lemma 4.4 and Lemma 4.3 (ii) this implies: for every elementary transformation $P(E'') = \operatorname{elm}_q(P(E'))$ we have $s(E'') = s(E) - 2$. But this cannot be true, since for $P(E') = \operatorname{elm}_p(P(E))$ there is an inverse elementary transformation, say $P(E) = \operatorname{elm}_{p'}(P(E'))$ for which we would have $s(E) = s(E') - 1 = s(E) - 2$.

**Corollary 4.6.** *For any vector bundle E on X we have* $\dim M(E) \leqq 1$.

*Proof.* Otherwise there would be a point $p \in P(E)$ with $p \in \operatorname{supp} \sigma$ for infinitely many minimal sections $\sigma$ of $P(E)$.

Since $\dim M(E) \geqq 1$ according to Corollary 3.2 (ii), we obtain

**Corollary 4.7** (Maruyama [6, Proposition 3.14]). *If E is a vector bundle of* rank 2 *with* $s(E) = g$ *on X, then* $\dim M(E) = 1$.

## 5. The Case $s = 2$, $g \geqq 3$

In the following theorem, which is the main result of this section, a *double elliptic curve* means a curve which admits a morphism of degree 2 onto an elliptic curve. A *trigonality* is a morphism of degree 3 onto a curve of genus 0.

**Theorem 5.1.** *Let* $d \equiv 0 \bmod 2$ *and* $g(X) \geqq 3$. *Then we have*

(i) $\dim M(E) = 0$ *for every* $E \in M(d, 2)$ *if and only if* $g(X) \geqq 4$ *and X is not double elliptic.*

(ii) *If X is double elliptic, every double cover of an elliptic curve yields a g dimensional subvariety of* $M(d, 2)$ *all of whose E have* $\dim M(E) = 1$. *If X is of genus 3 (not necessarily double elliptic), every trigonality of X yields a 3-dimensional subvariety of* $M(d, 2)$ *all of whose E have* $\dim M(E) = 1$. *For every E of* $M(d, 2)$ *outside these subvarieties we have* $\dim M(E) = 0$.

The two steps of the proof are first to construct the "exceptional" sets of $M(d, 2)$ with $\dim M(E) = 1$ and then to show that these are the only ones with this property.

Since the dimension of the space of maximal subbundles does not change under twisting with line bundles it is enough to consider bundles $E$ which admit an exact sequence

$$0 \to \mathcal{O}_X \to E \to L \to 0 \tag{e}$$

with a line bundle $L$ of degree 2 on $X$. We need the following auxiliary result for the morphism $\varphi = \varphi_{K_X \otimes L} : X \to \mathbb{P}^g = P(H^0(K_X \otimes L))$ associated to $K_X \otimes L$.

**Lemma 5.2.** *Let $X$ be a smooth projective curve of genus $g \geqq 2$ over $k$ and $L$ a line bundle of degree 2 on $X$. If $\varphi : X \to \mathbb{P}^g$ denotes the morphism given by the complete linear system $|K_X \otimes L|$, then we have*

(i) *If $h^0(L) = 0$, then $\varphi$ is an embedding.*

(ii) *If $h^0(L) = 1$, then $\varphi$ is birational and* $\operatorname{Im} \varphi$ *has exactly one double point.*

(iii) *If $h^0(L) \geqq 2$, then $\varphi$ is hyperelliptic and $\varphi$ is a double cover of the rational normal curve in $\mathbb{P}^g$.*

The proof is an easy exercise using the criterion that a line bundle $M$ on $X$ is very ample if and only if for every two points $x_1, x_2 \in X$ we have $h^0(M(-x_1 - x_2)) = h^0(M) - 2$.

**Proposition 5.3.** *Let $\pi : X \to Z$ be a double cover of an elliptic curve of genus $g(X) \geqq 3$. To every $M \in \operatorname{Pic}^g(Z)$ one can associate in a canonical way a vector bundle $E$ on $X \in M(2, 2)$ with* $\dim M(E) = 1$.

*Remarks.* (a) $E$ turns out to be of determinant $\pi^* M \otimes K_X^{-1}$. Hence twisting by a line bundle of degree $\frac{d-2}{2}$ on $X$ we get an element $\tilde{E} \in M(d, 2)$ with $\dim M(\tilde{E}) = 1$. Varying $M$ in $\operatorname{Pic}^g(Z)$ and $N$ in $\operatorname{Pic}^{\frac{d-2}{2}}(X)$ and "factoring" by the one dimensional families of maximal subbundles we get an algebraic family of such bundles in $M(d, 2)$ of dimension $g$ as stated in Theorem 5.1.

(b) If one is only interested in finding examples $E \in M(d, 2)$ with $\dim M(E) = 1$ and not in classifying them, (that is in proving Theorem 5.1) the translation into geometry given below is not necessary. In fact this is easily done as follows: Start with a bundle $F$ on the elliptic curve $Z$ with $s(F) = 1$. Then $s(\pi^* F) = 2$ and the one dimensional family of maximal subbundles of $F$ pulls back under $\pi : X \to Z$ to give a one dimensional family of maximal subbundles of $E = \pi^* F$. This point of view is exploited in Sect. 7 to construct vector bundles $E$ on some curves $X$ with $\dim M(E) = 1$ and higher invariant $s < g - 1$.

*Proof. of Proposition 5.3.* Let $i : Z \to \mathbb{P}^{g-1} = P(H^0(Z, M))$ denote the embedding associated to $M$. If $\pi^* M = K_X \otimes L$, then $L$ is of degree 2 on $X$. Denote by $\varphi : X \to \mathbb{P}^g = P(H^0(K_X \otimes L))$ the morphism associated to $K_X \otimes L$. The map $\pi : X \to Z$ yields an embedding

$$H^0(Z, M) \to H^0(X, K_X \otimes L).$$

Since $H^0(Z, M)$ is of codimension 1 in $H^0(X, K_X \otimes L)$, it defines a point $e \in \mathbb{P}^g = P(H^0(K_X \otimes L))$. Let $p$ denote the linear projection of $\mathbb{P}^g$ with centre $e$. By definition the following diagram is commutative

$$\begin{array}{ccc} \mathbb{P}^g - \{e\} & \xrightarrow{p} & \mathbb{P}^{g-1} \\ \uparrow & & \uparrow i \\ X - \{\varphi^{-1}(e)\} & \xrightarrow{\pi} & Z \end{array} \qquad (*)$$

Using Lemma 5.2 we see that $\varphi$ is of degree 1, since otherwise restricting $p$ to the image $Y$ of $X$ under $\varphi$ we would get a rational map of a rational curve onto an elliptic curve. It follows $\deg(p|Y)=2$. On the other hand $Y=\varphi(X)$ is of degree $2g$ and $i(Z)$ is of degree $g$, which imply that $e$ is not on $Y$.

Now $e$ may be interpreted, upto a nonzero constant as a nonsplit exact sequence $0\to\mathcal{O}_X\to E\to L\to 0$. According to Proposition 1.1, $e\notin Y$ means $s(E)=2$. The geometric meaning of $p|Y$ is that the points of $i(Z)-p(y_0)$, $y_0$ denoting the possible singular point of $Y$, parametrize the 2-secants of $X$ with respect to $\varphi$ passing through $e$. This means, according to Proposition 2.4, that $i(Z)-p(y_0)$ parametrizes the maximal subbundles of $E$ different from $\mathcal{O}_X$, since obviously the corresponding line bundles are non isomorphic. This complete the proof of Proposition 5.3.

**Proposition 5.4.** *To every threefold covering $\pi:X\to\mathbb{P}^1$ of a curve $X$ of genus* 3 *we can associate in a canonical way a vector bundle $E\in M(2,2)$ with* $\dim M(E)=1$.

*Remark.* $E$ turns out to be of determinant $\pi^*(\mathcal{O}_{\mathbb{P}_1}(2))\otimes K_X^{-1}$. Hence twisting by a line bundle $N$ of degree $\frac{d-2}{2}$ on $X$ we get an element $\tilde{E}\in M(d,2)$ with $\dim M(\tilde{E})=1$. Varying $\pi$ (there is a one dimensional family of trigonalities on every curve of genus 3) and $N$ in $\mathrm{Pic}^{\frac{d-2}{2}}(X)$ and factoring by the one-dimensional family of maximal subbundles we get a 3-dimensional algebraic family of such bundles as stated in Theorem 5.1.

*Proof.* Let $i:\mathbb{P}^1\to\mathbb{P}^2$ denote the embedding given by the complete linear system $|\mathcal{O}_{\mathbb{P}^1}(2)|$. If we write $\pi^*\mathcal{O}_{\mathbb{P}^1}(2)=K_X\otimes L$, then $L$ is of degree 2 on $X$. Denote by $\varphi:X\to\mathbb{P}^3=P(H^0(K_X\otimes L))$ the morphism associated to $K_X\otimes L$. As in the proof of Proposition 5.3, $H^0(\mathbb{P}^1,\mathcal{O}_{\mathbb{P}^1}(2))$ is of codimension 1 in $H^0(X,K_X\otimes L)$ and therefore defines a point $e$ of $\mathbb{P}^3=P(H^0(K_X\otimes L))$. Let $p$ denote the linear projection of $\mathbb{P}^3$ with centre $e$. Then again we have the commutative diagram (*) as in the proof of Proposition 5.3, now with $g=3$ and $\mathbb{P}^1$ instead of $Z$.

Since $\deg(p\circ\varphi)=\deg\pi=3$, $\varphi$ cannot be of degree 2 which according to Lemma 5.2 means $\deg\varphi=1$. It follows that the degree of $p$ restricted to $Y=\mathrm{Im}\,\varphi$ is 3. On the other hand $Y$ is of degree 6 in $\mathbb{P}^3$ and $i(\mathbb{P}^1)$ of degree 2 in $\mathbb{P}^2$. This implies that $e$ is not on $Y$.

Now $e$ may be interpreted upto a nonzero constant as a nonsplit exact sequence $0\to\mathcal{O}_X\to E\to L\to 0$. According to Proposition 1.1 $e\notin Y$ means $s(E)=2$.

The geometric meaning of $p|Y$ is that the points of $i(\mathbb{P}_1)$ parametrize the trisecant lines of $Y$ in $\mathbb{P}^3$ passing through $e$. Every trisecant line $\overline{y_1y_2y_3}$ of $Y$ in $\mathbb{P}^3$ with $y_i$ different from each other and from the possible singular point of $Y$ yields three 2-secant line bundles with respect to $\varphi$ (using the terminology of Sect. 2) namely $\mathcal{O}_X(x_1+x_2)$, $\mathcal{O}_X(x_1+x_3)$, and $\mathcal{O}_X(x_2+x_3)$ if $y_i=\varphi(x_i)$. Moreover if $\overline{y_1y_2y_3}$ passes through $e$, so do $\overline{y_1y_2}$, $\overline{y_1y_3}$, and $\overline{y_2y_3}$.

According to Proposition 2.4 every such 2-secant line bundle yields a maximal subbundle of $E$. We claim that no two of them define isomorphic line subbundles. Suppose $\mathcal{O}_X(x_1+x_2)\simeq\mathcal{O}_X(x_1'+x_2')$ and let $\mathcal{O}_X(x_1+x_2+x_3)$ and $\mathcal{O}_X(x_1'+x_2'+x_3')$ denote the corresponding trisecant line bundles of $X$ with respect to $\varphi$. Since $i(\mathbb{P}^1)$

is of genus 0 we have $x_1+x_2+x_3$ linearly equivalent to $x_1'+x_2'+x_3'$ and hence $x_3$ linearly equivalent to $x_3'$ which means $x_3=x_3'$. But this implies $x_1+x_2=x_1'+x_2'$ as divisors. It follows that $\dim M(E)=1$.

**Proposition 5.5.** *Let $X$ be of genus $g\geqq 3$ and suppose given an exact sequence* (e) *as above with* $s(E)=2$ *and* $\dim M(E)=1$. *Then the pair* $(X,E)$ *is of the type of Propositions 5.3 or 5.4.*

*Proof.* Consider again the morphism $\varphi=\varphi_{K_X\otimes L}:X\to\mathbb{P}^g=P(H^0(K_X\otimes L))$ associated to $K_X\otimes L$ and (e) as a point $e$ of $\mathbb{P}^g$. According to Proposition 1.1 $e$ is not on the curve $X'=\varphi(X)$. Applying Proposition 2.4 we get a one dimensional family of secant lines of $X'$ passing through $e$. Consider the linear projection $p:\mathbb{P}^g-e\to\mathbb{P}^{g-1}$ with centre $e$. Let $Z'=\mathrm{Im}(p(X'))$ and $i:Z\to Z'$ its normalization. Then we have the commutative diagram

$$\begin{array}{ccc} \mathbb{P}^g-e & \xrightarrow{p} & \mathbb{P}^{g-1} \\ \cup\uparrow & & \cup\uparrow \\ X' & \xrightarrow{q} & Z' \\ \uparrow & & \uparrow i \\ X & \xrightarrow{\pi} & Z \end{array},$$

where $q=p|X'$ and $\pi$ denotes the normalization of $q$. $q$ is of degree $\geqq 2$, since there are infinitely many secant lines of $X'$ passing through $e$ and $Z'$ cannot have infinitely many singularities. Suppose $i$ is given by the linear system $\ell\subset|M|$ with a line bundle $M$ on $Z$. The commutativity of the diagram implies that $\pi^*M\simeq K_X\otimes L$. Suppose $\ell\subset|M|$ is not a complete linear system. Since $\dim\ell=g-1$ we would have $h^0(M)\geqq g+1$. On the other hand: for any line bundle $N\in\mathrm{Pic}(Z)$ of positive degree we have $h^0(N)\leqq\deg N$, which means for $M$: $h^0(M)\leqq\frac{2g}{\deg q}\leqq g$, a contradiction. Hence $\ell$ is the complete linear system $|M|$ itself. According to Lemma 5.2 we have

$$\deg\varphi=\begin{cases}1 & \text{if} \quad h^0(L)\leqq 1\\ 2 & \text{if} \quad h^0(L)=2\end{cases}$$

which implies that

$$\deg Z'\cdot\deg q=\deg X'=\begin{cases}2g & \text{if} \quad h^0(L)\leqq 1\\ g & \text{if} \quad h^0(L)=2.\end{cases} \tag{1}$$

Since $Z'$ is a nondegenerate curve in $\mathbb{P}^{g-1}$ we have

$$\deg Z'\geqq g-1. \tag{2}$$

If $h^0(L)=2$ we get from (1) and (2), $(g-1)\deg q\leqq g$ which means $2\leqq\deg q\leqq\frac{g}{g-1}<2$, a contradiction. If follows that $h^0(L)\leqq 1$ and (1) and (2) yield:

$$\deg q\leqq\frac{2g}{g-1}\quad\text{which means}\quad \deg q\leqq\begin{cases}3 & \text{if} \quad g=3\\ 2 & \text{if} \quad g\geqq 4.\end{cases}$$

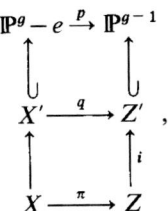

Thus we are led to consider the following two cases:

*Case (a).* $\deg Z' = g$, $\deg q = 2$, $g \geqq 3$.

Denote by $g'$ the genus of $Z$. If $M$ is a special line bundle, we have $g = \deg M \leqq 2g' - 2$. But the Riemann-Hurwitz formula implies $g \geqq 2g' - 1$, a contradiction. Hence $M$ is non-special and therefore

$$g = h^0(Z, M) = \deg M - g' + 1 = g - g' + 1$$

which implies $g' = 1$. It follows that $X$ is double elliptic and that we are in the situation of Proposition 5.3.

*Case (b).* $\deg Z' = 2$, $\deg q = 3$, $g = 3$.

Again denote by $g'$ the genus of $Z$. If $M$ is special, $\deg M = 2$ implies $g' = 2$ contradicting the Riemann-Hurwitz formula. Hence $M$ is non-special and therefore

$$3 = h^0(Z, M) = 2 - g' + 1$$

whence $g' = 0$ and we are in the situation of Proposition 5.4. This completes the proof of Proposition 5.5.

*Proof of Theorem 5.1.* Propositions 5.3–5.5 together with the remark following Propositions 5.3 and 5.4 complete the proof of Theorem 5.1.

Consider the special case $g = 3$ and let $\pi: X \to Z$ denote the double elliptic curve of Proposition 5.3 (resp. $\pi: X \to \mathbb{P}^1$ denote the trigonal cover of Proposition 5.4). If $M \in \operatorname{Pic}^3(Z)$ (resp. $M = \mathcal{O}_{\mathbb{P}^1}(2)$) and $\pi^* M = K_X \otimes L$ we have the following commutative diagram of the proof of Proposition 5.3 (resp. Proposition 5.4 with $\mathbb{P}^1$ instead of $Z$ and $i(\mathbb{P}^1)$ instead of $Z'$):

$$\begin{array}{ccc}
\mathbb{P}^3 & \xrightarrow{-e} & \mathbb{P}^2 \\
\uparrow & & \uparrow \\
Y & \longrightarrow & Z' \\
\uparrow & & \uparrow{\scriptstyle i} \\
X & \xrightarrow{\pi} & Z
\end{array}$$

The question is: What is the curve $Y$ explicitly and can one "see" the infinitely many secants of $Y$ passing through $e$?

**Proposition 5.6.** (i) *In the case of double elliptic curve $Y$ is a complete intersection with one ordinary double point (i.e. $p_a(Y) = 4$) of a cubic cone with centre $e$ and a quadric not passing through $e$.*

(ii) *In the case of trigonal covering $Y$ is a complete intersection with one ordinary double point (i.e. $p_a(Y) = 4$) of a quadric cone with centre $e$ and a cubic surface not passing through $e$.*

*Proof.* (i) Considering $\mathbb{P}^2$ as a plane in $\mathbb{P}^3$ not containing $e$, it is obvious from the above diagram that $Y$ is contained in a cubic cone $K_1$ passing through $e$. Consider the exact sequence

$$0 \to H^0(I_Y(3)) \to H^0(\mathcal{O}_{\mathbb{P}^3}(3)) \to H^0(\mathcal{O}_Y(3)).$$

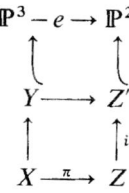

Since $h^0(\mathcal{O}_{\mathbb{P}^3}(3))=20$ and $h^0(\mathcal{O}_Y(3))=16$, we see that there is a smooth cubic surface $K_2 \neq K_1$ containing $Y$. Let $C=K_1 \cdot K_2$ denote the complete intersection of $K_1$ and $K_2$. Then $\deg C=9$ and $p_a(C)=10$. It follows that $C=Y \cup Y'$ with a (not necessarily irreducible) curve $Y'$ in $\mathbb{P}^3$ of degree 3 and hence $p_a(Y') \leqq 1$. From [2, V, Exercise 1.3] we get

$$10=p_a(C)=p_a(Y)+p_a(Y')+6-1 \leqq p_a(Y)+6$$

which means $p_a(Y) \geqq 4$. On the other hand Castelnuovo's theorem [2, IV, Theorem 6.4] gives $p_a(Y) \leqq 4$, so that in fact $p_a(Y)=4$. Moreover Castelnuovo's theorem says that if the Castelnuovo's inequality is actually an equality, than $Y$ lies on quadric surface $Q$. Since $6=\deg Y \leqq \deg(K_1 \cdot Q)=6$, $Y$ is the complete intersection of $K_1$ and $Q$.

(ii) Here it is obvious that $Y$ is contained in a quadric cone $Q$ passing through $e$. As above we see that $Y$ is contained in a cubic surface $K$. Comparison of the degrees of $Y$ and the complete intersection of $Q$ and $K$ completes the proof.

*Remark.* Conversely it is clear that every complete intersection as in Proposition 5.6 is of the type of Propositions 5.3 or 5.4.

Suppose that $Y$ is in particular a complete intersection of a quadric cone with centre $e$ and a cubic cone with centre $e'$ with $e \neq e'$. Then the generators of the quadric cone (resp. cubic cone) give a 1-parameter family of trisecants (resp. bisecants) of $Y$ passing through $e$ (resp. $e'$). Thus $e$ and $e'$ give bundles with $s=2$ and infinitely many maximal bundles on the normalization of $Y$.

## 6. The Case $s=3$, $g \geqq 4$

The aim of this section is to prove

**Theorem 6.1.** *For every $g \geqq 4$ there is a curve $X$ of genus $g$ over $k$ which admits a vector bundle $E$ of* rank 2 *with $s(E)=3$ and* $\dim M(E)=1$.

According to Proposition 2.4 we have to construct a curve $X$ of genus $g$ and degree $2g+1$ in $\mathbb{P}^{g+1}$ such that there is a one dimensional family of (non-equivalent) 3-secant planes of $X$ passing through a point $e$ not in $\mathrm{Sec}_2(X)$. Projecting from $e$ this would give a curve $Y \subset \mathbb{P}^g$ isomorphic to $X$ with infinitely many trisecant lines. We first construct $Y$. For

$$n=\left[\frac{g+1}{2}\right] \quad \text{and} \quad e=2n-g-2=\begin{cases}-1 & \text{if } g \text{ even} \\ 0 & \text{if } g \text{ odd}\end{cases},$$

let $F_{e,n}$ denote an elliptic ruled surface with minimal section $C_0$ with $C_0^2=-e$ and fibre $f$, embedded by means of $C_0+nf$ as a scroll of degree $g+1$ in $\mathbb{P}^g$. First we need a technical lemma

**Lemma 6.2.** *For every divisor $D$ of $F_{e,n}$ numerically equivalent to $(3n-g-2)f$ we have for $i \geqq 1$*

$$h^i(F_{e,n}, \mathcal{O}_{F_{e,n}}(aC_0+D-kf))=0 \quad \textit{for} \quad (a,k)=(3,0),(2,0), \textit{ and } (3,1).$$

*Proof.* It is enough to show that $(a+2)C_0+(5n-2g-(k+3)f)$ is ample, since then Kodaira's vanishing theorem can be applied and yields just what we want. To prove ampleness, the criteria [2, V, Propositions 2.20 and 2.21] apply.

**Lemma 6.3.** *$F_{e,n}$ admits a smooth irreducible curve $Y$ numerically equivalent to $3C_0+(3n-g-2)f$. $Y$ is of genus and degree $2g+1$.*

*Proof.* Let $D$ be a divisor numerically equivalent to $(3n-g-2)f$. We want to show that the linear system $|3C_0+D|$ contains a smooth irreducible curve. From Riemann-Roth theorem and Lemma 6.2 we see that $h^0(F_{e,n}, \mathcal{O}(3C_0+D)) > \max(h^0(F_{e,n}, \mathcal{O}(2C_0+D)), h^0(F_{e,n}, \mathcal{O}(3C_0+D-f))$. This implies that $|3C_0+D|$ contains no fixed component. Suppose that $x_0$ is a base point of $|3C_0+D|$. Then either $x_0 \in \operatorname{Supp}(C_0)$ or $x_0$ is contained in a fibre, contained as a component in $|3C_0+D|$. Suppose $x_0 \in \operatorname{Supp}(C_0)$ (the argument in the other case is analogous). Then the restriction of $|3C_0+D|$ to $C_0$ has $x_0$ as a base point. But according to Lemma 6.2 the restriction map $H^0(F_{e,n}, \mathcal{O}_{F_{e,n}}(3C_0+D)) \to H^0(C_0, \mathcal{O}_{C_0}(3C_0+D))$ is surjective and a linear system of degree $\geqq 2$ on an elliptic curve is base point free, a contradiction.

It follows that the map $\varphi: F_{e,n} \to \mathbb{P}^N$ corresponding to the complete linear system $|3C_0+D|$ is a morphism. We claim that $\dim \operatorname{Im}\varphi = 2$. If $g \geqq 7$, then $D$ varies already with fixed $3C_0$ in $|3C_0+D|$, so this is clear, $C_0$ not being a fixed component. Suppose $g=4$, 5 or 6 and $\dim \operatorname{Im}\varphi = 1$. Then $\varphi: F_{e,n} \to \operatorname{Im}\varphi$ is a fibration, which implies $(3C_0+D)^2=0$. But $(3C_0+D)^2 = 9$, 12, or 18 if $g=4, 5$, or 6, respectively, a contradiction.

Hence the theorem of Bertini can be applied and a general element $Y$ of $|3C_0+D|$ is smooth and irreducible. The computations of the degree and genus of $Y$ are obvious.

*Proof of Theorem 6.1.* Let $Y \subset \mathbb{P}^g$ be the curve of Lemma 6.3. $Y$ is of genus $g$ and degree $2g+1$ in $\mathbb{P}^g$, hence $h^0(Y, \mathcal{O}_Y(1)) = g+2$. Let $\varphi: Y \hookrightarrow X \subset \mathbb{P}^{g+1}$ denote the corresponding embedding. The linear system defining $Y$ in $\mathbb{P}^g$ is of codimension 1 in $|\mathcal{O}_Y(1)|$. This means that there is a point $e \in \mathbb{P}^{g+1}$ such that following diagram is commutative:

$$\begin{array}{ccc} X & \hookrightarrow & \mathbb{P}^{g+1} - \{e\} \\ {\scriptstyle \pi = \varphi^{-1}} \downarrow & & \downarrow {\scriptstyle p_e} \\ Y & \hookrightarrow & \mathbb{P}^g \end{array}$$

The point $e$ is not on the secant variety $\operatorname{Sec}_2(X)$, since $\pi$ is an isomorphism. Let $Z$ denote the elliptic curve corresponding to the ruling of $F_{e,n}$ and $(H_z)_{z \in Z}$ the system of planes in $\mathbb{P}^{g+1}$ passing through $e$, which are given by the ruling of the scroll $F_{e,n}$.

Suppose that $\dim \overline{H_z \cap X} = 1$ for infinitely many $z \in Z$. Then this is true for all $z \in Z$, since the condition of three points being linearly dependent is an algebraic condition.

Let $\tilde{F}$ denote the surface spanned by the lines $\overline{H_z \cap X}$ in $\mathbb{P}^{g+1}$. Then it is easy to see that the restriction of the linear projection $p_e$ to $\tilde{F}$ is an isomorphism of $\tilde{F}$ onto $F_{e,n}$. But this contradicts the fact that $F_{e,n}$ is embedded in $\mathbb{P}^g$ by a complete linear system whereas $\tilde{F}$ is not contained in a hyperplane of $\mathbb{P}^{g+1}$, since $X$ is nondegenerate. It follows that $\dim \overline{H_z \cap X} = 2$ for almost all $z \in Z$. This means that for all but a finite number of $z \in Z$, $H_z$ is a proper 3-secant plane of $X$ passing through $e$.

By construction $X$ is a smooth irreducible curve of genus $g \geqq 4$ embedded by a complete linear system say $|K_X \otimes L|$ with $\deg L = 3$ in $\mathbb{P}^{g+1}$ such that there is a point $e \in \mathbb{P}^{g+1}$ not on $\mathrm{Sec}_2(X)$ such that there is a one dimensional family of 3-secant planes of $X$ passing through $e$. Again $e$ can be considered as a class of extensions $0 \to \mathcal{O}_X \to E \to L \to 0$ modulo the canonical operation of $k^*$. By Proposition 1.1, $s(E) = 3$ and by Proposition 2.4 $\dim M(E) = 1$. This completes the proof of Theorem 6.1.

## 7. Examples for Arbitrary $s$ and $g \gg 0$

It seems difficult to construct vector bundles on curves with high $s < g$ and infinitely many maximal subbundles using the method of Proposition 2.4, that is, roughly by constructing curves in high dimensional projective spaces with degenerate $s$-secant behaviour. However there is a method which serves to construct such bundles directly which we shall outline in this section. The idea is to start with a vector bundle on a curve $Y$ with infinitely many maximal subbundles and to study its pullback under some covering $X$ of $Y$. Notice that now Proposition 2.4 can be applied to give curves with more $s$-secants than generally expected.

Let $Y$ denote a smooth projective curve of genus $g' \geqq 1$ which admits a vector bundle $F$ with $1 \leqq s(F) = s' \leqq g'$ and infinitely many maximal subbundles one of which is $\mathcal{O}_Y$. Then there is an exact sequence

$$0 \to \mathcal{O}_Y \to F \to M \to 0 \tag{1}$$

with $c_1(M) = s'$. For $s' = g'$ every vector bundle on $Y$ (suitably twisted by a line bundle) has these properties according to Corollary 4.7.

Let $n \geqq 2$ and $ns' = s \leqq g - 1$.

**Lemma 7.1.** *There is an n-fold (ramified) covering $\pi: X \to Y$ of genus g which does not admit a factorization via a covering of Y of degree $d (2 < d < n)$ whenever $g > ng' - n + 1$. This condition holds if $s' = g'$.*

*Proof.* The first assertion is a consequence of [3, Satz I], using an additional argument if $g' = 1$. The second assertion follows since $g \geqq ng' + 1 \geqq ng' - n + 1$.

Now define $E = \pi^* F$. Pulling back (1) we get

$$0 \to \mathcal{O}_X \to E \to \pi^* M \to 0. \tag{2}$$

**Lemma 7.2.** *If $g > (g' - 2)n + 2 + \frac{s}{2}(n-1)$, then $s(E) = s$.*

*Proof.* Obviously $s(E) \leqq c_1(E) = s$. Suppose $s(E) < s$. Then there is an exact sequence $0 \to L_1 \to E \to L_2 \to 0$ with $c_1(L_2) < s$. By the universal property of the projective bundle $p: P(F) \to Y$ there is a morphism $g: X \to P(F)$ over $Y$ such that $g^* \mathcal{O}_{P(F)}(1) \simeq L_2$.

Let $\tilde{X}=\operatorname{Im} g$. Since $\pi$ does not admit a non trivial factorization we have $\deg(g:X\to\tilde{X})=1$ or $=n$. Suppose $\deg g=n$, then $\tilde{X}$ is a section of $p$, $\sigma: Y\to\tilde{X}\subseteqq P(E)$ and hence $g=\sigma\circ\pi$. It follows that $L_2=\pi^*(\sigma^*\mathcal{O}_{P(F)}(1))$ and $\sigma^*(\mathcal{O}_{P(F)}(1))$ is a quotient of $F$ of degree $<s'$ contradicting $s(F)=1$. Thus we have $\deg g=1$.

Let $Y_0$ resp. $f$ denote a divisor in the class of $\mathcal{O}_{P(F)}(1)$ resp. $\mathcal{O}_{P(F)}(\text{fibre})$ in $\operatorname{Num} P(F)$. Since $Y_0$ and $f$ form a basis of $\operatorname{Num} P(F)$ over $\mathbb{Z}$ we have $X\equiv\alpha Y_0+\beta f$ with $\alpha,\beta\in\mathbb{Z}$. Now on the one hand

$$\begin{aligned}\alpha&=(\alpha Y_0+\beta f)\cdot f=\tilde{X}\cdot f=\mathcal{O}_{P(F)}(\tilde{X})\cdot p^*\mathcal{O}_Y(\text{point})\\&=c_1(g^*p^*\mathcal{O}_Y(\text{point}))=c_1(\pi^*\mathcal{O}_Y(\text{point}))=n\end{aligned}$$

and on the other hand

$$\begin{aligned}c_1(E)+\beta&=ns'+\beta=nY_0^2+\beta \quad (\text{since } Y_0^2=s')\\&=(\alpha Y_0+\beta f)Y_0=\tilde{X}\cdot Y_0\\&=\mathcal{O}_{P(F)}(\tilde{X})\cdot\mathcal{O}_{P(F)}(1)=c_1(g^*\mathcal{O}_{P(F)}(1))\\&=c_1(L_2)=c_1(E)-c_1(L_1).\end{aligned}$$

It follows that $X\equiv nY_0-c_1(L)f$ in $\operatorname{Num} P(F)$. By the adjunction formula

$$\begin{aligned}p_a(\tilde{X})&=\tfrac{1}{2}(nY_0-c_1(L_1)f)((n-2)Y_0+(2g'-2+s'-c_1(L_1))f)+1\\&=(1-n)(c_1(L_1)+1)+\frac{n^2s'}{2}-\frac{ns'}{2}+ng'\\&\leqq 2-2n+\frac{ns}{2}-\frac{s}{2}+ng' \quad \text{since} \quad c_1(L_1)>1\,.\end{aligned}$$

It follows that

$$g\leqq p_a(\tilde{X})\leqq(g'-2)n+2+\frac{ns}{2}-\frac{s}{2},$$

contradicting the hypothesis.

**Proposition 7.3.** *Suppose* $n\geqq 2$, $g'\geqq 1$, $s=ng'$, *and* $g>\operatorname{Max}\left(s,\frac{s}{2}(n+1)-2n+2\right)$. *Then for every n-fold covering* $\pi: X\to Y$ *with smooth projective curves X of genus g and Y of genus g′ as in Lemma 7.1 there is a vector bundle E of* rank 2 *on X with* $s(E)=s(<g)$ *which admits infinitely many maximal subbundles.*

*Proof.* According to Corollary 4.7 every vector bundle $F$ of rank 2 on $Y$ with $s(F)=s'=g'$ admits infinitely many maximal subbundles. According to Lemma 7.2 we have for $E=\pi^*F$, $s(E)=s$, since $g>\frac{s}{2}(n+1)-2n+2=(g'-2)n+2+\frac{ns}{2}-\frac{s}{2}$. Hence pulling back a maximal subbundle of $F$ we get a maximal subbundle of $E$. Since the canonical map $\operatorname{Pic}(Y)\xrightarrow{\pi^*}\operatorname{Pic}(X)$ has finite kernel, this implies the proposition.

*Remark.* (a) Starting with $g'=1$ we thus get examples of curves of genus $g$ which admit a vector bundle $E$ of rank 2 with $s(E)=s<g$ with infinitely many maximal

subbundles for $g \geqq \max(s+1, \frac{1}{2}(s^2-3s+6))$ i.e. for

| | | | | | | | | | |
|---|---|---|---|---|---|---|---|---|---|
| and | $s=2$ | 3 | 4 | 5 | 6 | 7 | 8 | 9 | ... |
| | $g \geqq 3$ | 4 | 5 | 8 | 12 | 17 | 23 | 30 | .... |

Starting with $g'=2$ we get examples for $g \geqq n^2-n+3$ i.e. for

| | | | | |
|---|---|---|---|---|
| and | $s=4$ | 6 | 8 | ... |
| | $g \geqq 5$ | 9 | 15 | .... |

Especially for $s \geqq 6$ we get examples in a range which is not covered by starting with $g'=1$. Similarly starting with $g'=3$ we get for $s \geqq 6$ examples in a range which is not covered by starting with $g'=1$ and 2.

*Remark.* (b) One may also start with a pair $(Y, F)$ with $g(Y)=g' \geqq 2$ and $s(F)=s'<g'$ such that $\dim M(F)=1$. By the same argument as in Proposition 7.3 this gives bundles $E$ on $X$ with $s(E)<g(X)$ and infinitely many maximal subbundles. They are certainly different from the examples above. But there seem to be no examples for pairs $(g, s)$ for which Proposition 7.3 does not apply. For example we do not know an example for $g=6$, $s=5$.

*Remark.* (c) Using Proposition 2.4 we get examples of curves $X$ of genus $g$ in $\mathbb{P}^{g+s-2}$ of degree $2g+s-2$ with exceptional $s$-secants passing through a point $e \notin \mathrm{Sec}^{s-1}(X)$ for the pairs $(g, s)$ as above.

## References


1. Gunning, R.C.: On the divisor order of vector bundles of rank 2 on a Riemann surface. Bull. Acad. Sinica **6**, 295–302 (1978)
2. Hartshorne, R.: Algebraic geometry. Graduate Texts in Mathematics, Vol. 52. Berlin, Heidelberg, New York: Springer 1977
3. Lange, H.: Kurven mit rationaler Abbildung. Crelles J. **295**, 80–115 (1977)
4. Lange, H.: Universal families of extensions. J. Algebra **83**, 101–112 (1983)
5. Lange. H.: Zur Klassifikation von Regelmannigfaltigkeiten. Math. Ann. **262**, 447–459 (1983)
6. Maruyama, M.: On classification of ruled surfaces. Lectures in Mathematics. Kyoto Univ. No. 3, Tokyo (1970)
7. Mülich, G.: Familien holomorpher Vektorraumbündel über der projektiven Geraden und unzerlegbare holomorphe 2-Bündel über der projektiven Ebene. Dissertation, Göttingen (1974)
8. Nagata, M.: On self intersection number of vector bundles of rank 2 on a Riemann surface. Nagoya Math. J. **37**, 191–196 (1970)
9. Nagata, M.: On ruled surfaces. Publ. Sém. Math. Sup. **46**, 27–48 (1971)
10. Segre, C.: Recherches générales sur les courbes et les surfaces réglés algébriques. II. Math. Ann. **34**, 1–25 (1889)
11. Stuhler, U.: Unterbündel maximalen Grades von Vektorbündeln auf algebraischen Kurven. Manuscripta Math. **27**, 313–321 (1979)
12. Atiyah, M.F.: Complex fibre bundles and ruled surfaces. Proc. London Math. Soc. **5**, 407–434 (1955)


Received May 3, 1983

Abdus Salam International Centre for Theoretical Physics, Trieste, 2005

THE COLLECTED PAPERS OF

# M.S. Narasimhan

# THE COLLECTED PAPERS OF
# M.S. Narasimhan

Volume II

1985 - 2001

**Nitin Nitsure**
Editor

Published for the
**Tata Institute of Fundamental Research**
by

HINDUSTAN
BOOK AGENCY

International Distribution by
**American Mathematical Society, USA**

HINDUSTAN BOOK AGENCY (INDIA)
P 19 Green Park Extension, New Delhi 110 016

email: hba@vsnl.com
www.hindbook.com



ISBN - 13: 978-81-85931-77-7 (2 volume set)
ISBN - 10: 81-85931-77-1 (2 volume set)

# Contents

## Volume I


Editorial Preface xiii

Curriculum Vitae xv

*A Brief Summary of the Work of M.S. Narasimhan* xvii

*M.S. Narasimhan and his Work* by C.S. Seshadri xx

Acknowledgements xxvii

1. *The problem of limits on a Riemannian manifold*, J. Indian Math. Soc. (N.S.) **20** (1956), 291–297. 1
2. *The identity of the weak and strong extensions of a linear elliptic differential operator*, Proc. Nat. Acad. Sci. U.S.A. **43** (1957), 513–514. 8
3. *The identity of the weak and strong extensions of a linear elliptic differential operator. II*, Proc. Nat. Acad. Sci. U.S.A. **43** (1957), 620. 10
4. *The type and the Green's kernel of an open Riemann surface*, Ann. Inst. Fourier. Grenoble **10** (1960), 285–296. 11
5. *Variations of complex structures on an open Riemann surface*, Ann. Inst. Fourier Grenoble **11** (1961), 493–514, XVI–XVII. 23
6. *Existence of universal connections* (with S. Ramanan), Amer. J. Math. **83** (1961), 563–572. 45
7. *Existence of universal connections, II* (with S. Ramanan), Amer. J. Math. **85** (1963), 223–231. 55
8. *Regularity theorems for fractional powers of a linear elliptic operator* (with Takeshi Kotake), Bull. Soc. Math. France **90** (1962), 449–471. 64
9. *Stable bundles and unitary bundles on a compact Riemann surface* (with C.S. Seshadri), Proc. Nat. Acad. Sci. U.S.A. **52** (1964), 207–211. 87
10. *Holomorphic vector bundles on a compact Riemann surface* (with C.S. Seshadri), Math. Ann. **155** (1964), 69–80. 92
11. *Stable and unitary vector bundles on a compact Riemann surface* (with C.S. Seshadri), Ann. of Math. (2) **82** (1965), 540–567. 104
12. *Manifolds with ample canonical class* (with R.R. Simha), Invent. Math. **5** (1968), 120–128. 132

13. *Vector bundles on curves* (with S. Ramanan), Algebraic Geometry (Internat. Colloq., Tata Inst. Fund. Res., Bombay, 1968) pp. 335–346 Oxford Univ. Press, London, 1968. 141

14. *Moduli of vector bundles on a compact Riemann surface* (with S. Ramanan), Ann. of Math. (2) **89** (1969), 14–51. 153

15. *An analogue of the Borel-Weil-Bott theorem for hermitian symmetric pairs of non-compact type* (with K. Okamoto), Ann. of Math. (2) **91** (1970), 486–511. 191

16. *Geometry of moduli spaces of vector bundles*, Actes du Congres International des Mathematiciens (Nice, 1970), Tome 2, pp. 199–201. Gauthier-Villars, Paris, 1971. 217

17. *On the cohomology groups of moduli spaces of vector bundles on curves* (with G. Harder), Math. Ann. **212** (1975), 215–248. 220

18. *Deformations of the moduli space of vector bundles over an algebraic curve* (with S. Ramanan), Ann. of Math. (2) **101** (1975), 391–417. 254

19. *Generalised Prym varieties as fixed points* (with S. Ramanan), J. Indian Math. Soc. (N.S.) **39** (1975), 1–19. 281

20. *Geometry of Hecke cycles, I* (with S. Ramanan), C.P. Ramanujam—a tribute, pp. 291–345, Tata Inst. Fund. Res. Studies in Math., **8** Springer, Berlin-New York, 1978. 300

21. *Geometry of* SU(2) *gauge fields* (with T.R. Ramadas), Comm. Math. Phys. **67** (1979), no. 2, 121–136. 355

22. *Polarisations on an abelian variety* (with M.V. Nori), Proc. Indian Acad. Sci. Math. Sci. **90** (1981), no. 2, 125–128. 371

23. *Fibres de 't Hooft speciaux et applications* (with A. Hirschowitz), (French) [Special 't Hooft bundles and applications] Enumerative geometry and classical algebraic geometry (Nice, 1981), pp. 143–164, Progr. Math., **24** Birkhauser, Boston, Mass., 1982. 375

24. *Projective bundles on a complex torus* (with G. Elencwajg), J. Reine Angew. Math. **340** (1983), 1–5. 397

25. *Maximal subbundles of rank two vector bundles on curves* (with H. Lange), Math. Ann. **266** (1983), no. 1, 55–72. 402

# Volume II


26. *Survey of vector bundles on curves*, Singularities, representation of algebras, and vector bundles (Lambrecht, 1985), 1–8, Lecture Notes in Math., **1273** Springer, Berlin, 1987. 420

27. *2θ-linear systems on abelian varieties* (with S. Ramanan), Vector bundles on algebraic varieties (Bombay, 1984), 415–427, Tata Inst. Fund. Res. Stud. Math., **11** Tata Inst. Fund. Res., Bombay, 1987. 428

28. *Squares of ample line bundles on abelian varieties* (with H. Lange), Exposition. Math. **7** (1989), no. 3, 275–287. 441

29. *Spectral curves and the generalised theta divisor* (with Arnaud Beauville and S. Ramanan), J. Reine Angew. Math. **398** (1989), 169–179. 454

30. *Groupe de Picard des varieties de modules de fibres semi-stables sur les courbes algebriques* (with J.-M. Drezet), (French) [The Picard group of moduli varieties of semistable bundles over algebraic curves] Invent. Math. **97** (1989), no. 1, 53–94. 465

31. *Compactification of $M_{P_3}(0,2)$ and Poncelet pairs of conics* (with G. Trautmann), Pacific J. Math. **145** (1990), no. 2, 255–365. 507

32. *Rank 2 vector bundles on $P_4$ with $c_1$ odd and contact curves* (with W. Decker and F.-O. Schreyer), Math. Z. **205** (1990), no. 1, 123–136. 618

33. *The Picard group of the compactification of $M_{P_3}(0,2)$* (with G. Trautmann), J. Reine Angew. Math. **422** (1991), 21–44. 632

34. *Factorisation of generalised theta functions, I* (with T.R. Ramadas), Invent. Math. **114** (1993), no. 3, 565–623. 656

35. *Vector bundles as direct images of line bundles* (with A. Hirschowitz), K. G. Ramanathan memorial issue. Proc. Indian Acad. Sci. Math. Sci. **104** (1994), no. 1, 191–200. 715

36. *Infinite Grassmannians and moduli spaces of G-bundles* (with Shrawan Kumar and A. Ramanathan), Math. Ann. **300** (1994), no. 1, 41–75. 725

37. *Picard group of the moduli spaces of G-bundles* (with Shrawan Kumar), Math. Ann. **308** (1997), no. 1, 155–173. 760

38. *Hodge classes of moduli spaces of parabolic bundles over the general curve* (with Indranil Biswas), J. Algebraic Geom. **6** (1997), no. 4, 697–715. 779

39. *Hermitian-Einstein metrics on parabolic stable bundles* (with Jia Yu Li), Acta Math. Sin. (Engl. Ser.) **15** (1999), no. 1, 93–114. 798

40. *A note on Hermitian-Einstein metrics on parabolic stable bundles* (with Jia Yu Li), Acta Math. Sin. (Engl. Ser.) **17** (2001), no. 1, 77–80. 820

41. *A generalisation of Nagata's theorem on ruled surfaces* (with Yogish I. Holla), Compositio Math. **127** (2001), no. 3, 321–332. 824

Acknowledgements 837

Volume II

1985 - 2001

# SURVEY OF VECTOR BUNDLES ON CURVES

M.S. Narasimhan
School of Mathematics
Tata Institute of Fundamental Research
Homi Bhabha Road, Bombay 400 005
INDIA

## 1. Moduli problem for vector bundles on curves.

Let $X$ be a compact Riemann surface, or what is the same, a smooth projective irreducible algebraic curve over $\mathbb{C}$. It is well known that the set of isomorphism classes of holomorphic (or algebraic) line bundles of degree $d$ has a natural structure of a smooth projective variety; if $d = 0$ we obtain an abelian variety, the Jacobian of $X$. Moreover any line bundle of degree $0$ on $X$ is associated to a (unitary) character of the fundamental group of $X$ [9].

The corresponding 'moduli problem' for (algebraic) vector bundles of higher rank on $X$ was first envisaged by A. Weil in 1938 in a famous paper [30]. Naively formulated, the question is whether there is a natural structure of a variety on the set of isomorphism classes of vector bundles of a given rank and degree on $X$. However it is easy to see that even 'locally' around certain bundles of rank $\geq 2$ one can not have a structure of a variety, for example due to the jump-phenomenon [see 17, p. 126]. This suggests that one can expect a structure of variety only on a suitable subset of the isomorphism classes of vector bundles.

In what follows we shall assume that the genus $g$ of $X$ is $\geq 2$. It is well known that any vector bundle on a curve of genus $0$ is a direct sum of line bundles. Vector bundles on curves of genus $1$ were investigated in detail by M.F. Atiyah [1].

## 2. Flat bundles and a theorem of A. Weil.

We shall consider in this section a class of vector bundles on $X$, namely flat bundles, which play an important role in the moduli problem for vector bundles on curves.

Let $\tilde{X}$ be a universal covering of $X$ and let the fundamental group $\pi = \pi_1(X)$ act on $\tilde{X}$ on the right. If $\rho$ is a homomorphism of $\pi$ into the full linear group $GL(r,\mathbb{C})$, we can construct a holomorphic vector bundle $W_\rho = \tilde{X} \times_\pi \mathbb{C}^r$ on $X$; the bundle $W_\rho$ is the quotient of $\tilde{X} \times \mathbb{C}^r$ under the action of $\pi$ given by :

$$(\tilde{x},v)\gamma = (\tilde{x}\gamma,\ \rho(\gamma)^{-1}v),\ \tilde{x} \in \tilde{X},\ v \in \mathbb{C}^r,\ \gamma \in \pi\ .$$

We say that a holomorphic vector bundle $V$ on $X$ arises from a representation (resp. unitary representation) of $\pi$, if $V$ is isomorphic to $W_\rho$ where $\rho$ is a homomorphism of $\pi$ into $GL(r,\mathbb{C})$ (resp. into to the unitary group $U(r)$).

Among other results A. Weil proved in [30]

Theorem 2.1. A vector bundle on $X$ arises from a representation of the fundamental group of $X$ if and only if each of its indecomposable components is of degree zero.

A. Weil also expected that bundles which arise from unitary representations would play an important role.

3. Stable and semi-stable bundles.

The crucial step in the progress of moduli problem for vector bundles on curves was the introduction by David Mumford of the all important notion of stable vector bundles. This concept was motivated by the geometric invariant theory [6].

Definition 3.1. A vector bundle $V$ on $X$ is said to be stable (resp. semi-stable) if for every proper subbundle $W$ of $V$ we have

$$\frac{\deg W}{\text{rank } W} < \frac{\deg V}{\text{rank } V} \text{ (resp. } \frac{\deg W}{\text{rank } W} \leq \frac{\deg V}{\text{rank } V}) ,$$

where $\deg (V) = C_1(V)[X]$, $C_1(V)$ denoting the first chern class of $V$.

Observe that a semi-stable bundle is automatically stable, if its rank and degree are coprime.

D. Mumford proved [7]

Theorem 3.1. The set of isomorphism classes of stable vector bundles on $X$ of rank $r$ and degree $d$ has a natural structure of a smooth quasi-projective variety of dimension $r^2(g-1)+1$.

4. Stable bundles and unitary bundles.

The following basic theorem was proved by M.S. Narasimhan and C.S. Seshadri [15,16].

Theorem 4.1. A vector bundle on $X$ of degree $0$ is stable if and only if it arises from an irreducible unitary representation of the fundamental group of $X$.

As a consequence one sees that a vector bundle arises from a unitary representation of $\pi_1(X)$ if and only if each of its indecomposable components is of degree $0$ and stable. Moreover it is easy to show that two vector bundles arising from unitary representations are isomorphic if and only if the representations are equivalent.

In the same paper a characterisation similar to Theorem 4.1 was also given for stable bundles of arbitrary degree in terms of irreducible unitary representations of a certain Fuchsian group. This result implies that, if $(r,d) = 1$, the space of isomorphism classes of stable bundles of rank $r$ and degree $d$ is compact and is hence a smooth projective variety.

5. The moduli space of semi-stable bundles.

The results stated in § 4 suggest a natural compactification of the space of stable bundles of degree 0, namely the space of equivalence classes of unitary representations (not necessarily irreducible) of a given rank. C.S. Seshadri proved that this compactification is a projective variety [23]. Before stating his result precisely we will introduce an equivalence relation among semi-stable bundles.

Let $V$ be a semi-stable vector bundle on $X$. Then $V$ has a strictly decreasing filtration by subbundles

$$V = V_0 \supset V_1 \supset \cdots \supset V_k = (0)$$

such that for $1 \leq i \leq k$ the bundle $W_i = V_i/V_{i-1}$ is stable and satisfies $\frac{\deg W_i}{\operatorname{rank} W_i} = \frac{\deg V}{\operatorname{rank} V}$. Moreover the bundle $Gr(V) = \bigoplus_{i=1}^{k} (V_i/V_{i-1})$ is uniquely determined by $V$ upto isomorphism (Jordan-Hölder theorem). We say that two semi-stable bundles $V_1$ and $V_2$ are S-equivalent if $Gr(V_1)$ is isomorphic to $Gr(V_2)$. Observe that two stable bundles are S-equivalent if and only if they are isomorphic. It is clear, using Theorem 4.1, that the set of equivalence classes of unitary representations is in canonical bijective correspondence with the set of S-equivalence classes of semi-stable vector bundles of degree 0.

C.S. Seshadri, using geometric invariant theory, proved [23]

Theorem 5.1. There is a unique structure of a normal projective (irreducible) variety $U(r,d)$ on the set of S-equivalence classes of semi-stable vector bundles on $X$ of rank $r$ and degree $d$ such that the following property holds : if $\{V_t\}_{t \in T}$ is an algebraic family of semi-stable bundles on $X$ of rank $r$ and degree $d$ parametrised by an algebraic variety $T$, then the map $T \longrightarrow U(r,d)$, sending $t \in T$ to the S-equivalence class of $V_t$, is a morphism.

We shall call the variety given by Theorem 5.1 the moduli space of (semi-stable) vector bundles of rank $r$ and degree and denote it by $U(r,d)$.

Theorem 5.1 is also valid for a curve $X$ over an algebraically closed field of arbitrary characteristic [25].

6. Singularities of $U(r,d)$.

The set of singular points of $U(r,d)$ has been determined by M.S. Narasimhan and S. Ramanan [11].

Theorem 6.1. The set of non-singular points of $U(r,d)$ is precisely the set of stable points in $U(r,d)$ except when $g = 2$, $r = 2$ and $d$ even. In this exceptional case $U(r,d)$ is smooth.

Explicit desingularisations of $U(2,0)$ have been given by M.S.Narasimhan-S. Ramanan [14] and by C.S. Seshadri [24].

## 7. Poincaré bundles.

Let $U_S(r,d)$ denote the open set of stable points in $U(r,d)$.

Definition 7.1. Let $\Omega$ be a non-empty open subset of $U_S(r,d)$. A Poincaré family of vector bundles parametrised by $\Omega$ is an algebraic vector bundle P on $\Omega \times X$ such that for any $\omega \in \Omega$ the bundle on $X$ obtained by restricting $P$ to $\omega \times X$ is in the isomorphism class $\omega$.

It is not hard to see that when $(r,d) = 1$ there is a Poincaré bundle on $U(r,d) \times X$ [8]. S. Ramanan proved [22]

Theorem 7.1. If $r$ and $d$ are not coprime there is no Poincaré family on $X$ parametrised by any non-empty open subset of $U_S(r,d)$.

The special case of this theorem where $g = 2$, $r = 2$, $d$ even, was proved earlier in [10].

## 8. The variety $S(r,d)$.

In order to study the varieties $U(r,d)$ let us fix a line bundle $L$ of degree $d$ and consider the subvariety of $U(r,d)$ corresponding to semi-stable bundles $V$ with $\Lambda^r V \simeq L$. We will denote this variety by $S_X^L(r,d)$ or simply by $S(r,d)$. The varieties $S(r,d)$ have been studied intensively by G. Harder, M.S. Narasimhan, P.E. Newstead, S. Ramanan and A. Tjurin, especially in the case $(r,d) = 1$. The results obtained pertain to the computation of numerical invariants like the Betti numbers, questions concerning the rationality of these varieties, the relation between the moduli of curves and the moduli of the varieties $S(r,d)$ and the explicit determination of these varieties in low genus or rank.

## 9. Betti numbers of $S(r,d)$.

The Betti numbers of $S(r,d)$ were first calculated by P.E. Newstead in the case $r = 2$, $d = 1$, by topological methods using the results of § 4 [18]. Based on these results G. Harder verified the Weil conjecture for $S(2,1)$ in the case of a curve defined over a finite field [4], at a time when the Weil conjecture was not proved in general. Harder observed in turn that the Betti numbers of $S(2,1)$ can be computed by arithmetical methods on the basis of the Weil conjecture.

Harder's method was generalised by him and M.S. Narasimhan to bundles of arbitrary rank [5]. It was shown, in the case $(r,d) = 1$, that the $\zeta$-function of $S(r,d)$ can be calculated from the $\zeta$-function of $X$. This result, together with Weil conjecture proved by P. Deligne, gives a method for computing the Betti numbers of $S(r,d)$ when $(r,d) = 1$.

As special cases of results proved in [5] we have

Theorem 9.1. 1) The second Betti number of $S(r,d)$ is 1 and the third Betti number is $2g$, when $(r,d) = 1$.

2) Let $d$ and $d'$ be such that $(r,d) = (r,d') = 1$, and $d \not\equiv d' (\mathrm{mod}\ r)$. Then $S(r,d)$ and $S(r,d')$ are not homeomorphic.

Another result proved in [5] is the following :

Theorem 9.2. Let $(r,d) = 1$ and $J_r$ denote the subgroup of Pic (X) consisting of line bundles of order r. Consider the action of $J_r$ on $S(r,d)$ given by

$$(\xi, V) \mapsto \xi \times V,\ \xi \in J_r,\ V \in S(r,d).$$

Then the induced action of $J_r$ on the real cohomology groups of $S(r,d)$ is trivial.

Another method to compute the Betti numbers of $S(r,d)$ was given recently by M.F. Atiyah and R. Bott [2,29]. They prove also that the integral cohomology is torsion-free, if $(r,d) = 1$.

Using Theorem 8.2 and the Atiyah-Singer fixed point theorem for the action of $J_r$ on $S(r,d)$, M.S. Narasimhan and S. Ramanan computed explicitly the Euler characteristic $\chi(S(r,d),\Omega^p)$ of the sheaf $\Omega^p$ [13]. In particular the Euler - characteristic and the index of $S(r,d)$ are zero $[(r,d) = 1]$.

10. Rationality of $S(r,d)$.

It is easy to see that the varieties $S(r,d)$ are unirational [10, p.337]. It is also known that $S(r,d)$ is rational if $d \equiv \pm 1 \ (\mathrm{mod}\ r)$ [21]; in particular $S(2,1)$ is rational. However it is not known whether $S(r,d)$ is rational in general. It is not even known whether $S(2,0)$ is rational when $g \geq 3$.

11. Moduli of curves and moduli of $S(r,d)$.

In this section we shall assume $(r,d) = 1$.

Theorem 11.1. The canonically polarised intermediate jacobian of $S(r,d)$ corresponding to the third cohomology group is naturally isomorphic to the canonically polarised jacobian of X.

This theorem was proved in the case $r = 2$ by D. Mumford and P.E. Newstead [8] and in general by M.S. Narasimhan and S. Ramanan [12]. As a consequence one sees that if $X_1$ and $X_2$ are two curves such that the corresponding moduli spaces $S_{X_1}(r,d)$ and $S_{X_2}(r,d)$ are isomorphic, then $X_1$ and $X_2$ are isomorphic. This result was also proved by A. Tjurin [26,27].

The following result proved by M.S. Narasimhan and S. Ramanan [12] implies that any small deformation of $S_X(r,d)$ is of the form $S_{X_t}(r,d)$ for a deformation

$X_t$ of X.

Theorem 11.2. 1) The group of automorphisms of S(r,d) is finite and

$$H^i(S(r,d),\Theta) = 0 \quad \text{for} \quad i \neq 1 ,$$

where $\Theta$ is the tangent sheaf.

2) $\dim H^1(S(r,d),\Theta) = 3g-3$ .

12. Explicit determination of S(r,d) in special cases.

Theorem 12.1. Let $g = 2$. Then

1) S(2,0) isomorphic to $\mathbb{P}^3$.

2) S(2,1) is isomorphic to a smooth intersection of two (smooth) quadrics in $\mathbb{P}^5$.

The proof of Theorem 12.1 given in [11] exploits the surprising connection between the moduli spaces of bundles of rank 2 on a curve of genus 2 and the classical theory of Kummer surfaces and quadratic complexes. The second part of the theorem was also proved by P.E. Newstead [19].

The second part of the theorem was generalised by U.V. Desale and S. Ramanan [3] as follows :

Theorem 12.2. Let X be an hyperelliptic curve of genus $g \geq 2$ and let $\lambda_0,\ldots,\lambda_{2g+1}$ be the branch points of X in $\mathbb{P}^1$. Then S(2,1) is isomorphic to the space of all (g-2) dimensional linear subspaces of $\mathbb{P}^{2g+1}$ which are contained in the intersection of the two quadrics

$$\sum_{i=0}^{2g+1} X_i^2 = \sum_{i=0}^{2g+1} \lambda_i X_i^2 = 0 .$$

## BIBLIOGRAPHY

1. M.F. Atiyah : Vector bundles over an elliptic curve. Proc. London Math. Soc. 7 (1957), 414-452.

2. M.F. Atiyah and R. Bott : The Yang-Mills equations on a Riemann surface. Phil. Trans. Royal Soc. Lond. A 308 (1982), 523-621.

3. U.V. Desale and S. Ramanan : Classification of vector bundles of rank 2 on hyper-elliptic curves, Inventiones Math., 38 (1976), 161-185.

4. G. Harder : Eine Bemurkung zu einer Arbeit von P.E. Newstead. Jour. für. Math. 242 (1970), 16-25.

5. G. Harder and M.S. Narasimhan : On the cohomology groups of moduli spaces of vector bundles on curves, Math. Annalen 212 (1975), 215-248.

6. D. Mumford : Geometric Invariant Theory, Springer Verlag, 1965.

7. D. Mumford : Projective invariants of projective structures and applications, Proc. International Congress of Math., (Stockholm 1962), 526-530.

8. D. Mumford and P.E. Newstead : Periods of a moduli space of bundles on curves. Amer. J. Math. 90 (1968), 1200-1208.

9. M.S. Narasimhan : Vector bundles on compact Riemann surfaces, in Complex Analysis and its Applications, Vol. III, International Atomic Energy Agency, Vienna, 1976.

10. M.S. Narasimhan and S. Ramanan : Vector bundles on curves, in Algebraic Geometry, Bombay Colloquium 1968, (O.U.P. 1969), 335-346.

11. M.S. Narasimhan and S. Ramanan : Moduli of vector bundles on a compact Riemann surface, Ann. of Math. 89 (1969), 14-51.

12. M.S. Narasimhan and S. Ramanan : Deformations of the moduli space of vector bundles on an algebraic curve, Ann. of Math. 101 (1975), 391-417.

13. M.S. Narasimhan and S. Ramanan : Generalized prym varieties as fixed points, J. Indian Math. Soc. 39 (1975), 1-19.

14. M.S. Narasimhan and S. Ramanan : Geometry of Hecke Cycles I, In C.P. Ramanujan - A Tribute (T.I.F.R. Studies in Math.) Springer-Verlag, 1978, 291-345.

15. M.S. Narasimhan and C.S. Seshadri : Holomorphic vector bundles on a compact Riemann surface, Math. Annalen 155 (1964), 69-80.

16. M.S. Narasimhan and C.S. Seshadri : Stable and unitary vector bundles on a compact Riemann surface, Ann. of Math. 82 (1965), 540-567.

17. P.E. Newstead : Introduction to moduli problems and orbit spaces, T.I.F.R. Lecture Notes, Springer-Verlag, 1978.

18. P.E. Newstead : Topological properties of some spaces of stable bundles,Topology 6 (1967), 241-262.

19. P.E. Newstead : Stable bundles of rank 2 and odd degree over a curve of genus 2, Topology 7 (1968), 205-215.

20. P.E. Newstead : Characteristic classes of stable bundles of rank 2 over an algebraic curve, Trans. Amer. Math. Soc. 169 (1972), 337-345.

21. P.E. Newstead : Rationality of moduli spaces of stable bundles, Math. Annalen 215 (1975), 251-268, Correction in 249 (1980) 281.

22. S. Ramanan : The moduli spaces of vector bundles over an algebraic curve, Math. Annalen 200 (1973), 69-84.

23. C.S. Seshadri : Space of unitary vector bundles on a compact Riemann surface, Ann. of Math. 85 (1967), 303-336.

24. C.S. Seshadri : Desingularisation of the moduli variety of vector bundles on curves, Proc. Int. Symp. Alg. Geom. Kyoto 1977, Kinokuniya, Tokyo, 1978.

25. C.S. Seshadri : Fibrés vectoriels sur les courbes algébriques (rédigé par J.M. Drezet), Astérisque (96), 1982.

26. A. Tjurin : Analogue of Torelli's theorem for two dimensional bundles over algebraic curves of arbitrary genus Izv. Akad. Nauk. SSSR. Ser. Mat. 30 (1966), 1353-1356, English trans., Math. U.S.S.R. Izv 3 (1969), 1081-1101.

27. A. Tjurin : Analogues of Torelli's theorem for multidimensional vector bundles over an arbitrary algebraic curve, Izv. Akad. Nauk,SSSR Ser Mat. 34 (1970), 338-365, English trans., Math. U.S.S.R. - Izv 4 (1970), 343-370.

28. A. Tjurin : The geometry of moduli of vector bundles, Uspehi Mat. Nauk 29 : 6 (1974) 59-88, English trans., Russian Math. Surveys 29 : 6 (1974), 57-88.

29. J.-L. Verdier and J. Le Potier (Editors) : Module des fibrés stables sur les courbes algébriques, Progress in Maths. vol. 54, Birkhäuser, 1985.

30. A. Weil : Généralisation des functions abéliennes, J. Math. Pures Appl. 17 (1938), 47-87.

# $2\theta$-LINEAR SYSTEMS ON ABELIAN VARIETIES

*By* M. S. NARASIMHAN and S. RAMANAN

## §1. Introduction and Statement of Main Theorem

We would like to consider the $2\theta$-linear system on an abelian variety with a principal polarisation $\theta$. In the case when the abelian variety is the Jacobian of a projective non-singular curve $X$, there is a close relationship between semistable vector bundles of rank 2 and trivial determinant on $X$ and the $2\theta$-linear system. On the one hand, one can describe the moduli $SU_X(2)$ of such bundles on $X$ in terms of the $2\theta$-linear system. On the other, classical questions regarding Kummer varieties or the Schottky relation may be better understood in terms of this 'new' variety $SU_X(2)$.

We first considered this relationship some fifteen years ago in the case of genus 2 and proved

THEOREM 1 [5]. *The variety $SU_X$ (2) is canonically isomorphic to the projective space of divisors on the Jacobian, linearly equivalent to $2\theta$.*

This result was somewhat of a surprise for the following reason. For every line bundle $j$ on any curve $X$, consider the semistable bundle $j \oplus j^{-1}$. This imbeds in $SU_X(2)$, the Kummer variety $\mathcal{K} = J/i$, where $J = \mathrm{Pic}^0(X)$ is the Jacobian of $X$ and $i$ is the involution $x \to x^{-1}$ of $J$. It is easy to see that $SU_X(2)-\mathscr{K}$ is smooth. That $SU_X(2)$ is itself smooth in the case of genus 2 is surprising in view of the following

THEOREM 2 [5]. *The Kummer variety $\mathscr{K}$ is precisely the singular locus of $SU_X(2)$, if $g \geq 3$.*

Some of the ideas relating to Theorem 1 have been generalised [2] to hyperelliptic curves of arbitrary genus $g \geq 2$. In particular, we have

THEOREM 3 [2]. *If $X$ is hyperelliptic of genus 3 and $i$ the involution of $SU_X(2)$ induced by the hyperelliptic involution on $X$, then $SU_X(2)/i$ is a quadric in $\mathbb{P}^7$.*

The aim of this paper is the following generalisation of Theorem 1.

MAIN THEOREM. *If $X$ is non-hyperelliptic of genus 3, then $SU_X(2)$ is isomorphic to a quartic hypersurface in $\mathbb{P}^7$*

In particular, of course, the Kummer variety is imbedded in $\mathbb{P}^7$ and is the singular locus of a quartic hypersurface. Suppose $f$ is the quartic polynomial in the homogenous coordinates $(Z_1, \ldots, Z_8)$ defining $SU_X(2)$. Then Theorem 2 implies that $\mathscr{K}$ is defined by $\frac{\partial f}{\partial z_i} = 0$, $i = 1, \ldots 8$. Thus we have, as an application, the

COROLLARY. *$\mathscr{K}$ can be defined by cubic polynomials.*

Wirtinger [7] had shown that $\mathscr{K}$ can be defined by quartics and it was an open problem if cubics would suffice. (See Coble [1, p. 106]).

## §2. The relationship between vector bundles and 2θ-linear systems

Let us now make explicit the map of $SU_X(2)$ into the projective linear system $P$ of $2\theta$, which exists for any $g$. If we denote by $J^d$ the space of line bundles on $X$ of degree $d$, notice that the natural divisor $\theta$ lives only in $J^{g-1}$. Hence the linear system of $2\theta$ is a system of divisors in $J^{g-1}$. The map that we have in mind associates to a vector bundle $E$ of rank 2 with trivial determinant, the subset $D_E = \{\xi \in J^{g-1} : \Gamma(\xi \otimes E) \neq 0\}$. If $E$ is not semistable, then it has a line subbundle $L$ of positive degree. For every $\xi \in J^{g-1}$, we have $\Gamma(\xi \otimes L) \neq 0$ so that $D_E = J^{g-1}$ in this case. On the other hand, Raynaud [6] showed that if $E$ is semistable, the $D_E$ is a proper subset of $J^{g-1}$. Then it is easy to see that $D_E$ is the support of a divisor linearly equivalent to $2\theta$.

Actually, one can associate a divisor $D_E$ to $E$ (and not merely its support) as follows. Take $x_1, \ldots x_N \in X$, with $N$ sufficiently large. If $Z$ is the divisor $x_1 + \ldots + x_N$, tensorise the exact sequence

$$0 \to \mathcal{O}(-Z) \to \mathcal{O}_X \to \mathcal{O}_Z \to 0$$

with $\xi \otimes E$. If $N$ is large enough so that $\Gamma(\xi \otimes E \otimes \mathcal{O}(-Z)) = 0$, then clearly $\Gamma(\xi \otimes E)$ is the kernel of the connecting homomorphism

$$\sum_{i=1}^{N} (\xi \otimes E)_{x_i} \to H^1 (\xi \otimes E \otimes \mathcal{O}(-Z))$$

We will now allow $\xi$ to vary over $J^{g-1}$. In others words, taking $\xi$ to be the universal line bundle of degree $g$-i parametrised by $J^{g-1}$ we get two vector bundles on $J^{g-1}$, both of rank $2N$ and a homomorphism. This defines a section of a line bundle, namely the difference between their determinants. It is easy to compute this line bundle and show that it is isomorphic to $2\theta$.

## § 3. Generalities on polarised abelian varieties

Let $A$ be an abelian variety and $\tau$ a principal polarisation on 4. Consider $B$ = Pic $^{\tau}$ $A$, namely the space of isomorphism classes of line bundles on $A$ with Chern class $\tau$. Then $A$ acts on $B$ by $a.\ \xi$ = translation by $a$ of $\xi$. All line bundles $\xi \epsilon B$ are ample and hence $H^i(\xi) = 0$ for $i \geq 1$ [3, p. 150]. On the other hand, $\Gamma(\xi)$ is 1-dimensional, since the polarisation is principal. Hence the line bundles $\xi \epsilon B$ may be identified with their divisors in $A$. Let $\theta \subset B$ be defined by

$$\theta = \{ \xi \epsilon B : \text{divisor of } \xi \text{ passes through } 0 \}$$

This belongs to the principal polarisation $\tau$ in the sense that, for any choice of a point in $B$, the natural identification $A \to B$ leads to a divisor in $A$ whose class is $\tau$. Moreover, $A \amalg B$ can be made into a group, using the involution $i : B \to B$ given by $\xi \to i^* \xi$ where $i$ is the morphism $x \to -x$ of $A$. We have then a natural exact sequence

$$0 \to A \to A \amalg B \to \mathbb{Z}/2 \to 0.$$

Of course, this sequence splits, but the point is that there is no canonical splitting. Elements of $A_2 = \{a \in A : 2a = 0\}$ are called *period characteristics* and elements of $B_2 = \{b \in B : 2b = 0\}$ are called *theta characteristics*. As a group $A_2$ is isomorphic to $\mathbb{F}_2^{2g}$ and the principal polarisation on $A$ gives a nondegenerate alternating form $\tau$ on $A_2$. Moreover, $B_2$ can be identified with quadratic forms on $A_2$ whose associated alternating form is $\tau$. The Arf invariant then distinguishes between odd and even theta characteristics. We have the well-known [4]

THEOREM. *The parity of the multiplicity of the divisor $\theta$ at $b \in B_2$ determines the parity of the theta characteristic.*

On $A \times B$, we may consider the divisor obtained by pulling back $\theta$ by the natural group action $A \times B \to B$. This bundle serves as a Poincaré bundle for line bundles on $A$ with Chern class $\tau$, and also as one for line bundles on $B$ with Chern class $\tau$. Indeed, the space $B$ and the divisor $\theta$ are characterized by this property. It then follows that if $A$ is the Jacobian of a curve $X$ of genus $g$, then $B$ may be identified with $J^{g-1}$

Although there is no canonical $\theta$-divisor on $A$, there is a divisor class $\Phi$ with Chern class $2\tau$. For every $b \in B$, $T_b^* \theta$, is the divisor in $A$, corresponding to $b$. The linear equivalence class of $T_b^* \theta + T_{-b}^* \theta$ is independent of $b \in B$ and this defines $\Phi$. More elegantly, this may be constructed as follows. Consider the map $\mu : A \times B \to B \times B$ given by $(a, b) \mapsto (a+b, a-b)$. The pull back of tensor product of $p_1^*\mathcal{O}(\theta) \otimes p_2^*\mathcal{O}(\theta)$ on $B \times B$ to $A \times B$ may be looked upon as a family of line bundles on $B$ parametrised by $A$. But then for any a, this gives the isomorphism class of $\mathcal{O}(2\theta)$. Hence there is a line bundle $\Phi$ on $A$ such that

$$p_1^* \phi \otimes p_2^* \mathcal{O}(2\theta) \approx \mu^*(p_1^*\mathcal{O}(\theta) \otimes p_2^* \mathcal{O}(\theta)).$$

Moreover, the situation comes with a section of the line $p_1^*\Phi \otimes p_2^* \mathcal{O}(2\theta)$, so that we have a canonical element of $\Gamma(A, \Phi) \otimes \Gamma(B, \mathcal{O}(2\theta))$. This can be proved to give a perfect pairing so that we have

3.1 THEOREM. *There is a canonical duality between* $\Gamma(A,\Phi)$ *and* $\Gamma(B,\mathcal{O}(2\theta))$.

We have a natural morphism $A \to P\Gamma(A,\Phi)^*$ given by linear system of $\Phi$. On the other hand, for $a \in A$, we have the divisor $T_a\theta + T_{-a}\theta$ linearly equivalent to $2\theta$. In other words, we have a morphism $A \to P\Gamma(B, \mathcal{O}(2\theta))$. Under the duality of Theorem 3.1, these two morphism are compatible.

When $A$ is the Jacobian and $B = J^{g-1}$ of the curve $X$, we may interpret the morphism $A \to P\Gamma(B, \mathcal{O}(2\theta))$ defined above as follows. For any $a \in A$, consider the semistable vector bundle $E = a \oplus (a)^*$. Then $E$ is of rank 2 and trivial determinant so that there is a divisor $D_E$ linearly equivalent to $2\theta$ defined in §2. It is clear that this is the same as the map given above. Thus one may say that the morphism $SU_X(2) \to P\Gamma(J^{g-1}, \mathcal{O}(2\theta))$ given by $E \mapsto D_E$ restricts to the Kummer variety $\mathcal{K}$ as the morphism given by the linear system of $\Phi$. It has been shown by Andreotti and Meyer over $\mathbb{C}$ that for $\theta$ irreducible, $\Phi$ imbeds the Kummer variety. This does not seem to have been proved in the literature for characteristic $p > 0$. It is easy enough to show that if $\theta$ is irreducible, then the Kummer variety is mapped injectively. We can prove (following ideas of Wirtinger [7]) over any field of characteristic $\neq 2$, the following

THEOREM. *Let $A$ be any abelian variety, $\tau$ a principal polarisation on it and $B = \mathrm{Pic}^{\tau} A$. If no even theta characteristic lies on the canonical theta divisor in $B$, then $\mathcal{K}$ is imbedded by the canonical divisor class $\Phi$ in $A$ with Chern class $2\tau$.*

## §4. Proof of the Main Theorem

We will sketch here the alinea of proof of our main theorem. The detailed proofs will appear elsewhere.

4.1 LEMMA. *Let $\xi$ be a line bundle of degree 1 on $X$. Then $X$ may be imbedded in $J$ by $x \to \xi^{-1} \otimes \mathcal{O}(x)$. The induced map $\Gamma(J, \Phi) \to \Gamma(X, K \otimes \xi^2)$ is onto.*

PROOF. Any divisor linearly equivalent to $K \otimes \xi^2$ may be written as a sum of two divisors of degree $g$ each. Let these belong to classes $K \otimes \xi \otimes \alpha^{-1}$ and $\xi \otimes \alpha$ for $\alpha \in J^{g-1}$. It is easy to see that 'most' sections in $\Gamma(K \otimes \xi^2)$ can be thus split up in which $\Gamma(K \otimes \xi \otimes \alpha^{-1})$ and $\Gamma(\xi \otimes \alpha)$ are of dimension one each. It is clear then that these divisors are images of elements in $\Gamma(\Phi)$ given by $\alpha\theta + \alpha^{-1}\theta \sim \Phi$.

Consider the injective map

$$H^1(X, \xi^{-2}) \xrightarrow{\sim} \Gamma(X, K \otimes \xi^2)^* \to \Gamma(J, \Phi)^* \xrightarrow{\sim} \Gamma(J^{g-1}, \mathcal{O}(2\theta)).$$

We wish now to interpret the map $PH^1(X, \xi^{-2}) \to P\Gamma(J^{g-1}, \mathcal{O}(2\theta))$.

4.2. LEMMA. *Let $\xi$ be a line bundle on X, of degree 1. Any point of $PH^1(X, \xi^{-2})$ gives a nontrivial extension $0 \to \xi^{-1} \to E \to \xi \to 0$, where E is a semi-stable vector bundle of rank 2 and trivial determinant.*

PROOF. See [5, Lemma 5.1].

By allowing $\xi$ to vary over $J^1$, one can thus construct a bundle $V$ on $J^1$ whose fibre over $\xi$ is identifiable with $H^1(X, \xi^{-2})$, and an exact sequence

$$0 \to V \to \Gamma(J^{g-1}, \mathcal{O}(2\theta))_{J^1} \to W \to 0$$

on $J^1$ It can be shown that the fibre $V_\xi$ is the subspace of $\Gamma(J^{g-1}, \mathcal{O}(2\theta))$ consisting of sections which vanish on $\xi X$.

4.3 LEMMA. *The map $PH^1(X, \xi^{-2}) \to P\Gamma(J^{g-1}, (2\theta))$ can be interpreted as follows. If E is the vector bundle associated to $v \in PH^1(X, \xi^{-2})$, then the image of v is the divisor $D_E$.*

4.4 LEMMA. *Let X be of genus 3. If $E \in SU_X(2)$ is stable and $\xi X \subset D_E$ for some $\xi \in J^1$, then $\Gamma(\xi \otimes E) \neq 0$.*

These two lemmas ensure that the morphism $E \mapsto D_E$ is injective. Firstly, it is easy to show (see 5.4 below for a slightly stronger statement) that there exists $\xi \in J^1$ with $\Gamma(\xi \otimes E) \neq 0$. This implies of course that $\xi X \subset D_E$. If then $D_E = D_{E'}$ it follows from 4.4 that $\Gamma(\xi \otimes E') \neq 0$ as well, which implies that both $E$ and $E'$ occur as extensions of $\xi$ by $\xi^{-1}$. Now 4.3 implies that $E \mapsto D_E$ is injective.

We have also to show that the morphism $E \to D_E$ is an imbedding. From the injectivity, we see that the image is a hypersurface in $\mathbf{P}^7 = \mathbf{P}\,\Gamma(J^{g-1}\ \mathcal{O}(2\theta))$. We will compute the differential of the morphism $P(V) \to \mathbf{P}^7$. Let $E \in SU_X(2)$ be a stable bundle. Then we will show that in the diagram

$$\begin{array}{ccc} P(V) & \longrightarrow & \mathbf{P}^7 \\ & \searrow \quad \nearrow & \\ & SU_X(2) & \end{array}$$

there exists a point of $P(V)$ lying over $E$ at which the differential of the map $P(V) \to \mathbf{P}^7$ is injective. It would follow that the hypersurface in question is normal, since the set of nonstable points is of codim 3 in $SU_X(2)$. Hence by Zariski's Main Theorem, we are through.

## §5. Computation of differential

But the computation of the differential of the morphism $P(V) \to SU_X(2)$ turns out to be somewhat hard. The result is

5.1 LEMMA. *The differential is injective at every point of* $PH^1(\xi^{-2})$ *corresponding to stable* $E$, *if* $\xi^2 \notin \theta$.

Since the bundle $V$ is induced by a map $\varphi$ of the Jacobian $J^1$ into the Grassmannian of 4 dimensional subspaces of

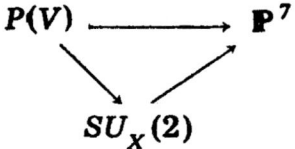

$\Gamma(J^2, \mathcal{O}(2\theta))$, Lemma 5.1 can be proved by computing the differential of this map and a simple computation of the differential of the natural map of the projective bundle associated to the universal subbundle on the Grassmannian, into the projective space. Thus it is easy to see that Lemma 5.1 would follow from the following statement regarding the differential of $\varphi$ which is a map $H^1(X, \mathcal{O}) \to \mathrm{Hom}(V_\xi, \Gamma(K^2 \otimes \xi^{-2}))$.

5.2 LEMMA. *Let $v \in T_\xi(J^1)$, $\xi^2 \notin \theta$. a) if $v$ does not belong to any 1-dimensional subspace of $H^1(X, \mathcal{O})$ corresponding to a point of $X$ in its canonical imbedding, then the image of $v$ under the differential of $J^1 \to \mathrm{Gr}_4(\Gamma(J^2, \mathcal{O}(2\theta))$ is an injective map $V_\xi \to \Gamma(K^2 \otimes \xi^{-2})$. b) If $v$ belongs to such an 1-dimensional subspace, then the image has 1-dimensional kernel and this, as a point of $P(V_\xi)$, corresponds to a nonstable extension.*

The space $V_\xi$ can also be identified with the kernel of $\Gamma(J^2, \mathcal{O}(2\theta)) \to \Gamma(X, K^2 \otimes \xi^{-2})$ obtained by imbedding $X$ as $\xi X \subset J^2$. In other words, $V_\xi = \Gamma(J^2, I_X \otimes \mathcal{O}(2\theta))$, where $I_X$ is the sheaf of ideals of $\xi X$ in $J^2$. From the exact sequence

$$0 \to I_X^2 \to I_X \to N^* \to 0$$

we obtain a natural map

$V_\xi \to \Gamma(N^* \otimes K^2 \otimes \xi^{-2})$. It can be shown that the required differential factors as follows.

$$\begin{array}{ccc} H^1(X, \mathcal{O}) & \to & \mathrm{Hom}(V_\xi, \Gamma(K^2 \otimes \xi^{-2})) \\ & \searrow & \downarrow \\ & & \mathrm{Hom}(\Gamma(N^* \otimes K^2 \otimes \xi^{-2}), \Gamma(K^2 \otimes \xi^{-2})) \end{array}$$

The vertical map is induced by the map $V_{\xi} \rightarrow \Gamma(N^* \otimes K^2 \otimes \xi^{-2})$ mentioned above. The map $H^1(X, \mathcal{O}) \rightarrow \text{Hom}(\Gamma(N^* \otimes K^2 \xi^{-2}), \Gamma(K^2 \otimes \xi^{-2}))$ can be determined explicitly by using the natural inclusion $N^* \rightarrow \Gamma(X, K)_X$. In fact, it is the composite of the natural map

$$H^1(X, \mathcal{O}) \rightarrow \text{Hom}(\Gamma(X, K) \otimes \Gamma(K^2 \otimes \xi^{-2}), \Gamma(K^2 \otimes \xi^{-2}))$$

and the map $\text{Hom}(\Gamma(X, K) \otimes \Gamma(K^2 \otimes \xi^{-2}), \Gamma(K^2 \otimes \xi^{-2})) \rightarrow \text{Hom}(\Gamma(N^* \otimes K^2 \otimes \xi^{-2}), \Gamma(K^2 \otimes \xi^{-2}))$ induced by $N^* \rightarrow \Gamma(X,K)_X$. In view of this, Lemma 5.2 will follow from

5.3 LEMMA. *The natural map* $V_\xi \rightarrow \Gamma(N^* \otimes K^2 \otimes \xi^{-2})$ *is an isomorphism if* $\xi^2 \notin$ supp $\theta$. *Geometrically speaking, there is no divisor linearly equivalent to* $2\theta$ *which vanishes on* $\xi X$ *with multiplicity* 2, *if* $\xi^2 \notin$ Supp $\theta$.

The proof that $E \longmapsto D_E$ is an imbedding is completed by proving

5.4 LEMMA. *Every stable bundle* $E$ *can be obtained as an extension* $0 \rightarrow \xi^{-1} \rightarrow E \rightarrow \xi \rightarrow 0$, $\xi \in J^1$ *and* $\xi^2 \notin$ Supp $\theta$.

To prove that the image is a quartic surface, we first observe that the map $\mathbf{P}(V) \rightarrow \mathbf{P}^7$ is generically finite and is of degree 8. If $\tau_V$ is the tautological hyperplane bundle along the fibres it suffices to show that $[c_1(\tau_V)]^6\ [P(V)] = 32$. This follows from

5.5 LEMMA. *Let* $c_i$ $(i = 1, 2, 3)$ *denote the Chern classes of the bundle* $V$ *and let* $c(\theta) \in H^2(J^1, \mathbb{Z})$ *be the cohomology class defined by a* $\theta$*-divisor in* $J^1$. Then we have

1) $c_1(V) = -4c(\theta)$, $c_2(V) = 8\,[c(\theta)]^2$ *and*

$$c_3(V) = -\frac{16}{3}[c(\theta)]^3$$

2) $[c_1(\tau_V)]^6\,[P(V)] = (-c_1^3 + 2c_1\,c_2 - c_3)\,[J^1]$.

## REFERENCES


1. A. B. Coble. Algebraic geometry and theta functions, *A.M.S. Colloquium Publications* X, 1929.

2. U. V. Desale and S. Ramanan . Classification of vector bundles of rank 2 on hyperelliptic curves, *Inventiones Math.* 38 (1976) 161-185.

3. D. Mumford . Abelian varieties, *T.I.F.R. Studies in Maths.* O.U.P , 1974.

4. D. Mumford . On the equations defining abelian varieties, *Inventiones Math.* 1 (1966) 287-354.

5. M. S. Narasimhan and S. Ramanan Moduli of vector bundles on a compact Riemann surface. Ann. of Math. 89 (1969) 19-51.

6. M. Raynaud . Sections des fibrés vectoriels sur une courbe, *Bull. Soc. Math. France* 110 (1982) 103-125.

7. W. Wirtinger Untersuchungen über Thetafunktionen, *Teubner*, 1895.

School of Mathematics,
Tata Institute of Fundamental Research,
Homi Bhabha Road, Colaba,
Bombay - 400 005.
INDIA.

EXPOSITIONES
MATHEMATICAE

Expo. Math. 7 (1989), 275–287


Mathematical Notes

# Squares of ample line bundles on abelian varieties

H. Lange and M.S. Narasimhan[1]

Let $X$ denote an abelian variety of dimension $g$ over the field of complex numbers and $L$ an ample line bundle. According to a classical theorem of Lefschetz the line bundle $L^n$ is very ample for $n \geq 3$, that is the map $\varphi_{L^n} : X \to P(H^0(L^n))$ associated to the complete linear system $|L^n|$ is an embedding. It is the aim of this note to give an account of the corresponding property for the line bundle $L^2$, that is of the behaviour of the map

$$\varphi = \varphi_{L^2} : X \to \mathbf{P}_N = P(H^0(L^2))$$

associated to $|L^2|$. Most of the results are known.

To state the main theorem let

$$(X, L) = (X_1, L_1) \times \cdots \times (X_s, L_s)$$

denote the decomposition of the polarized abelian variety $(X, L)$ into a product of irreducible polarized abelian varieties. Suppose that $(X_v, L_v)$ for $v = 1, \ldots, r$ are principally and for $v = r+1, \ldots, s$ are not principally polarized varieties. For $v = 1, \ldots, r$ let $K_v = X_v/\pm\,\mathrm{id}$ denote the Kummer-variety of $X_v$ and $p_v : X_v \to K_v$ the canonical projection.

Define $K = K_1 \times \cdots \times K_r \times X_{r+1} \times \cdots \times X_s$ and $p = p_1 \times \cdots \times p_r \times id_{X_{r+1}} \times \cdots id_{X_s}$ : $X \to K$. With these notations $\varphi = \varphi_{L^2}$ factors as

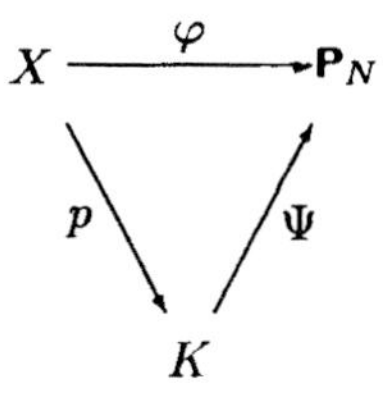


with a holomorphic mapping $\Psi$. The main result is

**Theorem:** $\Psi$ is an embedding.

[1]Supported by DFG

0723-0869/89/030275-13$4.10

Thus $\varphi$ is of degree $2^r$ onto its image. In particular if none of the components $(X_i, L_i)$ is principally polarized, $\varphi$ is an embedding. On the other hand in the case of an irreducible principally polarized abelian variety the result says that $\varphi$ embeds the Kummer variety. The main contributors to this topic are (to the best of our knowledge) Wirtinger, Andreotti-Mayer, Sasaki, Ramanan, and Ohbuchi.

# 1 Notations and Preliminaries

Let $\pi : V \to X$ denote the universal covering of $X$. $V \simeq C^g$ may be considered as the tangent space $\mathcal{T}_{X,0}$ of $X$ at 0 and $\pi$ as the exponential map. $\Lambda = \pi^{-1}(0)$ is a lattice of maximal rank in $V$ such that $X \simeq V/\Lambda$. Since $\pi^*L$ is trivial on $V$, the global sections of $L$ and $L^2$ can and will be considered as theta functions on $V$. If $\mathbf{P}_N$ denotes the projective space of hyperplane sections $\varphi = \varphi_{L^2}$ is induced by the map

$$\begin{array}{ccc} V & \longrightarrow & \mathbf{P}_N = P(H^0(L^2)) \\ v & \mapsto & \{\theta \in H^0(L^2), \theta(v) = 0\} \end{array}$$

also denoted by $\varphi$. It is well-known and easy to see that $L^2$ is generated by its global sections, that is $\varphi$ is everywhere defined (see [M, p.60]).

We recall some basic facts about invertible sheaves and their sections on abelian varieties.

## 1.1 First Chern class

The first Chern class map is the connecting homomorphism $c_1 : H^1(X, \mathcal{O}_X^*) \to H^2(X, \mathbf{Z})$ associated to the exponential sequence $0 \to \mathbf{Z} \to \mathcal{O}_X \to \mathcal{O}_X^* \to 0$. Since $H^1(X, \mathcal{O}_X^*)$ is the group of line bundles $\mathrm{Pic}(X)$ on $X$ and since there is a canonical identification of $H^2(X, \mathbf{Z})$ with the group of alternating 2-forms $\mathrm{Alt}^2(\Lambda, \mathbf{Z})$ with values in $\mathbf{Z}$ on the lattice $\Lambda$ (see [M, p.3]), we may consider the first Chern class as a map

$$c_1 : \mathrm{Pic}(X) \to \mathrm{Alt}^2(\Lambda, \mathbf{Z}).$$

According to a theorem of Frobenius for any line bundle $M$ on $X$ there is a basis of $\Lambda$ such that with respect to this basis the 2-form $c_1(M)$ is given by a matrix

$$c_1(M) = \begin{pmatrix} 0 & D \\ -D & 0 \end{pmatrix}$$

where $D = \mathrm{diag}(d_1, \ldots, d_g)$ with nonnegative integers $d_i$ such that $d_i$ divides $d_{i+1}$ for $i = 1, \ldots, g-1$. We call the $g$-tupel $(d_0, \ldots, d_g)$ the *type of the line bundle* $M$.

## 1.2 Dual abelian variety

The subgroup $\mathrm{Pic}^0(X)$ of $\mathrm{Pic}(X)$ of line bundles of type $(0,\ldots,0)$ admits itself the structure of an abelian variety called the *dual abelian variety* and denoted by $X^*$. Given a line bundle $M$ on $X$, the map

$$\Phi_M : \begin{cases} X & \longrightarrow & X^* \\ x & \longmapsto & T_x^* M \otimes M^{-1} \end{cases}$$

is a homomorphism of abelian varieties. Here $T_x$ denotes the translation map

$$T_x : X \longrightarrow X, \quad y \mapsto y + x.$$

$\Phi_M$ is an isogeny if and only if $c_1(M)$ is a nondegenerate alternating form. The connected component $K_0(M)$ of its kernel is an abelian subvariety and in fact the maximal abelian subvariety of $X$ such that there is a line bundle $N$ on $X/K_0(M)$ with $M \simeq p^*N$, where $p : X \longrightarrow X/K_0(M)$ denotes the natural projection map (see [M, p.231]). Moreover $c_1(N)$ is a nondegenerate alternating form and $h^0(M) = h^0(N)$.

## 1.3 Riemann–Roch

Let $M$ denote a line bundle of type $(d_1,\ldots,d_g)$ on $X$. Riemann–Roch's theorem for the Euler–Poincaré characteristic $\chi(M)$ in terms of the selfintersection number $(M^g)$ of $M$ says

$$\chi(M) = \frac{(M^g)}{g!}$$

Another form of this theorem is

$$\chi(M)^2 = \deg\Phi_M = (d_1\ldots d_g)^2.$$

## 1.4 Polarization

A *polarization* on $X$ is by definition the first Chern class $c_1(M)$ of an ample line bundle $M$ on $X$. By abuse of notation we consider the line bundle $M$ itself as a polarization. The type of $M$ is called the *type of the polarization.*

It is easy to check whether a given line bundle is ample: $M \in \mathrm{Pic}(X)$ is ample if and only if $h^0(M) \neq 0$ and $c_1(M)$ is a nondegenerate form. Since for an ample line bundle $M$ on $X$ all higher cohonology groups vanish, Riemann–Roch says in this case

$$h^0(M) = \frac{(M^g)}{g!} = \sqrt{\deg\Phi_M} = d_1\ldots d_g.$$

## 1.5 Gauss–map

Let $D$ denote an irreducible reduced divisor on $X$, $L = \mathcal{O}_X(D)$ the corresponding line bundle, and $\theta$ a corresponding theta function under a suitable isomorphism of $\pi^*L$ with the trivial linebundle on $V$. Let $D_s$ denote the smooth part of $D$. For every $x \in D_s$ the tangent space $\mathcal{T}_{D,x}$ is a $(g-1)$-dimensional subspace and its translation to zero is a well defined $(g-1)$-dimensional vector subspace of $\mathcal{T}_{X,0} = V$. The onedimensional subspace of the dual vector space $V^*$ determined by $\mathcal{T}_{D,x}$ is generated by $(d\theta)_x$ and the *Gauss-map* $G$ of $D$ is defined as

$$G : \begin{cases} D_s & \longrightarrow \quad \mathbf{P}_{g-1} = P(V) \\ x & \longmapsto \quad [(d\theta)_x] \end{cases}$$

We will use the following property of $G$, for the proof of which we refer to [Ra, Cor.II, 11, p.466]:
*If $D$ is an irreducible (reduced) ample divisor, then $G$ is dominant.*

## 1.6 Sums of translations of divisors

The following statement is well known and in fact the main ingredient of theorem of Lefschetz mentioned above (see [M, p.163]):
*Let $x_1, \ldots, x_m \in X$ with $\sum_{i=1}^m x_i = 0$ and $D$ a divisor on $X$. Then*

$$\sum_{i=1}^{m} T_{x_i}^* D \quad \sim mD$$

An immediate consequence is the fact that *if $D$ is effective, a general element of the linear system $|D|$ has no multiple components.*

The following lemma is also well known.

**Lemma 1.7** *Let $L \in Pic(X)$ be ample. Then for an open dense set of divisors $D$ in $|L|$ $T_x^*D = D$ holds only for $x = 0$.*

Proof: Suppose $0 \neq x \in X$ and $T_x^*D = D$ for all $D \in |L|$. In particular $T_x^*L = L$ and $x$ is contained in the finite gruop $\mathrm{Ker}\Phi_L$. $x$ generates a finite subgroup $G$, say of order $n$, of $X$. Let $Y = X/G$ and $p : X \to Y$ the canonical projection map. The assumption implies that for any $D \in |L|$ there is a divisor $E$ on $Y$ such that $D = \pi^*E$. Clearly all such divisors $E$ are linearly equivalent and belong to a line bundle $M$ on $Y$ with $\pi^*M = L$. Since $M$ is ample if and only if $L$ is ample, Riemann–Roch gives the contradiction

$$h^0(M) = h^0(L) = \deg p\, h^0(M) > h^0(M).$$

Hence for a fixed $x \neq 0$ the set of $D \in |L|$ with $T_x^* D \neq D$ is open and dense in $|L|$. This completes the proof of Lemma 1.6, since there are only finitely many possibilities for $x$. ■

# 2 Decomposition of the map $\varphi$

Our aim in this section is to show that it is sufficient to prove the theorem in the case of a polarization without fixed component and in the case of an irreducible principal polarization. For $i = 1,2$ let $M_i$ be a line bundle on the abelian variety $X$ with $h^0(M_i) \geq 1$. According to 1.2 there is a unique maximal abelian subvariety $K_0(M_i)$ such that $M_i = p_i^* N_i$ for some line bundle $N_i$ on $X_{M_i} = X/K_0(M_i)$ where $p_i : X \to X_{M_i}$ denotes the natural projection.

**Lemma 2.1** *Suppose $M_1 \otimes M_2$ is ample and the homomorphism*

$$(p,q) : X \to X_{M_1} \times X_{M_2}$$

*is not surjective and has finite kernel. Then*

$$h^0(M_1 \otimes M_2) \geq h^0(M_1) + h^0(M_2).$$

Proof: Write $g_i = \dim X_{M_i}$ for $i = 1,2$. Since $(p_1,p_2)$ has finite kernel, $g \leq g_1 + g_2$ and we have

$$\frac{1}{g!}(M_1 \otimes M_2)^g = \frac{1}{g!}(p_1^* N_1 \otimes p_2^* N_2)^g = \sum_{v=g-g_2}^{g_1} \frac{(p_1^* N_1)^v}{v!} \cdot \frac{(p_2^* N_2)^{g-v}}{(g-v)!} \tag{1}$$

since $(p_1^* N_1)^v = 0$ for $v > g_1$ and $(p_1^* N_2)^{g-v} = 0$ for $v > g_2$. For the summand with index $v = g_1$ we have

$$\begin{aligned}
\frac{(p_1^* N_1)^{g_1}}{g_1!} \cdot \frac{(p_2^* N_2)^{g-g_1}}{(g-g_1)!} &= \frac{(p_1^* N_1)^{g_1}}{g_1!} \frac{1}{(g-g_1)!} \left\{ (p_2^* N_2)^{g-g_1} + \textstyle\sum_{\mu=1}^{g-g_1} \binom{g-g_1}{\mu} (p_1^* N_1)^{\mu} (p_2^* N_2)^{g-g_1-\mu} \right\} \\
&= \frac{(p_1^* N_1)^{g_1}}{g_1!} \frac{(L)^{g-g_1}}{(g-g_1)!} \\
&= h^0(N_1) \frac{p_1^*(\text{point}) \cdot (L)^{g-g_1}}{(g-g_1)!} \quad \text{(since } N_1 \text{ is ample, see 1.2 and 1.3)} \\
&= h^0(N_1) \frac{(L|_{K_0(M_1)})^{g-g_1}}{(g-g_1)!} \\
&= h^0(N_1) \cdot h^0(L|K_0(M_1)) \quad \text{(since } L \text{ is ample)} \\
&\geq h^0(N_1) \\
&= h^0(M_1) \qquad \text{(by 1.2)}
\end{aligned}$$

Similarly we have for the summand with $v = g - g_2$

$$\frac{(p_1^*N_1)^{g-g_2}}{(g-g_2)!} \cdot \frac{(p_2^*N_2)^{g_2}}{g_2!} \geq h^0(M_2)$$

Now by our assumption $g_1 \neq g - g_2$. Since all summands on the right hand side of (1) are nonnegative this implies the assertion. ∎

Now let $L$ denote an ample line bundle on $X$ and

$$L = M \otimes F$$

a decomposition of $L$ such that $h^0(L) = h^0(M)$ and $h^0(F) = 1$. Let $p : X \to X_M = X/K_0(M)$ and $q : X \to X_F = X/K_0(F)$ be defined as above and $M = p^*N$, $F = q^*G$. Under these hypothesis we have

**Proposition 2.2** *If $K_0(M) \cap K_0(F)$ is finite, then $(p,q) : X \to X_M \times X_F$ is an isomorphism.*

Proof: $(p,q)$ is an isogeny, since otherwise according to Lemma 2.1

$$h^0(L) \geq h^0(M) + h^0(F) = h^0(L) + 1$$

a contradiction.
Let now $p_1$ and $p_2$ denote the projections of $X_M \times X_F$ onto its factors. We have $L = (p,q)^*(p_1^*N \otimes p_2^*G)$ and hence applying Riemann-Roch and the Künneth-formula

$$\begin{aligned}(L^g) &= \deg(p,q) \cdot (p_1^*N \otimes p_2^*G)^g \\ &= \deg(p,q) \cdot g!h^0(p_1^*N \otimes p_2^*G) \\ &= \deg(p,q) \cdot g!h^0(N) \cdot h^0(G) \\ &= \deg(p,q)(L)^g \text{ which implies } \deg(p,q) = 1.\end{aligned}$$

∎

Proposition 2.2 will be applied twice.

**Corollary 2.3** *Let $L \in Pic(X)$ be ample and $L = M \otimes F$ the unique decomposition such that $|M|$ is base component free and $|F|$ the fixed part of the linear system $|L|$. Then the canonical map*

$$(p,q) : X \to X_M \times X_F$$

*is an isomorphism of polarized abelian varieties $(X, L)$ and $(X_F, q_1^*N \otimes q_2^*G)$.*

Here by abuse of notation we denote by $q_1$ and $q_2$ also the projections of $X_M \times X_F$ onto its factors. Note that according to 1.2 and 1.4 $N$ is a polarization on $X_M$ and $G$ a principal polarization on $X_F$. As $K_0(M) \cap K_0(F) \subset \mathrm{Ker}\Phi_L$, which is finite, since $L$ is ample, Corollary 2.3 is a direct consequence of Proposition 2.2.

**2.4** Secondly let $G \in \mathrm{Pic}(X)$ denote a principal polarization. Let

$$G = G_1 \otimes \cdots \otimes G_r$$

be the unique decomposition of $G$ such that $|G| = |G_1| + \cdots + |G_r|$ is the decomposition of the divisor $|G|$ into its irreducible components $|G_i|$. For every $i = 1, \ldots, r$ we have $G_i = q_i^* H_i$ with an irreducible principal polarization $H_i$ on $X_{G_i} = X/K_0(G_i)$ and $q_i : X \to X_{G_i}$. By abuse of notation denote with $q_i$ also the $i$-th projection of $X_{G_1} \times \cdots \times X_{G_r}$. By 1.6 the divisors $|G_i|$ are pairwise different from each other. This implies that Proposition 2.2 can succesively applied $r - 2$ time to give

**Corollary 2.5** *With the notations as in* 2.4

$$(q_1, \ldots, q_r) : X \to X_{G_1} \times \cdots \times X_{G_r}$$

*is an isomorphism of polarized abelian varieties* $(X, G)$ *and* $(X_{G_1} \times \cdots \times X_{G_r}, q_1^* H_1 \otimes \cdots \otimes q_r^* H_r)$.

**2.6 Conclusion** Combining Corollaries 2.3 and 2.5 we obtain the following result:

*Let* $L \in \mathrm{Pic}(X)$ *be a polarization on the abelian variety* $X$. *There is a unique decomposition*

$$L = M \otimes G_1 \otimes \cdots \otimes G_r$$

*such that* $|M|$ *is free of fixed components,* $|G_1 \otimes \cdots \otimes G_r|$ *is the fixed part of the linear system* $|L|$, *and* $|G_1 \otimes \cdots \otimes G_r| = |G_1| + \cdots + |G_r|$ *its decomposition into irreducible components.*

*As above we have* $M = p^* N$ *and* $G_i = q_i^* H_i$ *for* $i = 1, \ldots, r$ *with a polarization* $N$ *on* $X_M = X/K_0(M)$ *and irreducible principal polarizations* $H_i$ *on* $X_{G_i} = X/K_0(G_i)$, *where* $p : X \to X_M$ *and* $q : X \to X_{G_i}$ *denote the natural projections. We will also denote by* $p$ *and* $q_i$ *the projections of* $X_M \times X_{G_1} \times \cdots \times X_{G_r}$ *onto its factors.*

From Corollaries 2.3 and 2.5 we conclude:

*The map*

$$(p, q_1, \ldots, q_r) : X \to X_M \times X_{G_1} \times \cdots \times X_{G_r}$$

*is an isomorphism of polarized abelian varieties* $(X, L)$ *and* $(X_M \times X_{G_1} \times \cdots \times X_{G_r}, p^* N \otimes q_1^* H_1 \otimes \cdots \otimes q_r^* H_r)$.

Thus we can decompose the map $\varphi = \varphi_{L^2} : X \to \mathbf{P}_N$ as follows: Let $\varphi_0 = \varphi_{M^2} : X_M \to \mathbf{P}_{N_0}$ and for $i = 1, \ldots, r$ $\varphi_i = \varphi_{G_i^2} : X_{G_i} \hookrightarrow \mathbf{P}_{N_i}$ denote the maps associated to $M^2$ and $G_i^2$ and $\Psi : \mathbf{P}_{N_0} \times \cdots \times \mathbf{P}_{N_r} \to \mathbf{P}_N$ the Segre embedding. Then

$$\varphi = \Psi \circ (\varphi_0 \times \cdots \times \varphi_r)$$

Thus for the investigation of the map $\varphi_{L^2}$ we are reduced to the following to cases

I. $L \in \mathrm{Pic}(X)$ a polarization without fixed components

II. $L \in \mathrm{Pic}(X)$ an irreducible principal polarization

and they will be studied seperately in the next two sections.

# 3 $|L|$ without fixed components

For the whole section let $L$ denote an ample line bundle without fixed components on an abelian variety $X$ of dimension $g$. From Bertini's theorem we would get that a general member of the linear system $|L|$ is irreducible if only the dimension of the image of the map $\varphi_L$ were greater than 1. The following lemma says that this is true even if this dimension is one.

**Lemma 3.1** *A general divisor of the linear system $|L|$ is irreducible.*

Proof: Suppose this is not the case that is

$$L = M \otimes N$$

with $h^0(M) \geq 2$, $h^0(N) \geq 2$ and the map $|M| \times |N| \to |L|$, $(D_1, D_2) \to D_1 + D_2$ is surjective.

As in section 2 there are ample line bundles $M'$ on $X_M = X/K_0(M)$ and $N'$ on $X_N = X/K_0(N)$ such that $M = p^*M'$ and $N = q^*N'$ where $p : X \to X_M$ and $q : X \to X_N$ denote the natural projections. Consider the map

$$(p, q) : X \to X_M \times X_N$$

$(p, q)$ has finite kernel, since $K_0(M) \cap K_0(N) \subseteq \ker\Phi_L$ and $L$ is ample.

Suppose first that $(p, q)$ is not surjective. Applying Lemma 2.1 and the fact that $\dim|M| + \dim|N| \geq \dim|L|$, we get

$$h^0(L) \geq h^0(M) + h^0(N) \geq h^0(L) + 1,$$

a contradiction.

Hence $(p, q)$ is a isogeny. Denoting by $p_1$ and $p_2$ the projections of $X_M \times X_N$ and applying Künneth's formula we get

$$\begin{aligned} h^0(M) + h^0(N) - 1 &\geq h^0(L) \\ &= \frac{1}{g!}(L^g) \\ &= \frac{1}{g!}\deg(p,q)\cdot(p_1^*M' \otimes p_2^*N')^g \\ &= \deg(p,q)\cdot h^0(p_1^*M' + p_2^*N') \\ &= \deg(p,q)\cdot h^0(M)\cdot h^0(N) \\ &\geq h^0(M)\cdot h^0(N) \end{aligned}$$

and hence

$$1 \leq (h^0(M)-1)(h^0(N)-1) = h^0(M)\cdot h^0(N) - h^0(M) - h^0(N) + 1 \leq 0,$$

a contradiction. ∎

Now we are in a position to prove the main result of this section. It was proven by Ramanan (see [R]) for a generic abelian variety and by Ohbuchi (see[O]) in general.

**Theorem:** *If $L \in Pic(X)$ is ample without fixed components, then $L^2$ is very ample.*

Proof: Step I: *Injectivity of $\varphi = \varphi_{L^2} : X \to \mathbf{P}_N$:*

Suppose this is not the case. Then there are points $p \neq q$ in $X$ such that for all divisors $D$ in $|L^2|$ we have $p \in D$ if and only if $q \in D$.

According to Lemma 1.6 and Lemma 3.1 there is an irreducible reduced divisor $D_1 \in |L|$ such that $T_x^* D_1 = D_1$ only if $x = 0$.

Fixing $D_1$ there is an irreducible reduced divisor $D_2 \in |L|$ with $D_2 \neq (-1)^* T_{p+q}^* D_1$.

For $x \in T_p^* D_1$ we have $p \in T_x^* D_1 \subset T_x^* D_1 + T_{-x}^* D_2 \in |L^2|$. By assumption this implies $q \in T_x^* D_1 + T_{-x}^* D_2$ which in turn is equivalent to $x \in T_q^* D_1 + (-1)^* T_q^* D_2$. Since this holds for every $x \in T_p^* D_1$, we get

$$T_p^* D_1 \subset T_q^* D_1 + (-1)^* T_q^* D_2$$

and thus

$$D_1 = T_{-p}^* T_p^* D_1 \subset T_{q-p}^* D_1 + (-1)^* T_{p+q}^* D_2$$

Since $D_1$ and $D_2$ are irreducible and $D_1 \neq (-1)^* T_{p+q}^* D_2$, we finally get $D_1 = T_{q-p}^* D_1$ implying $q - p = 0$, a contradiction.

Step II: *Injectivity by the differential $d\varphi_p$ for every $p \in X$:*

Without loss of generality we may assume $p = 0$, since otherwise consider $T_p^* L$ instead of $L$.

Suppose $0 \neq \xi \in T_0X = V$. We have to find a divisor $D \in |L^2|$ with $0 \in D$ and $\xi$ not tangent to $D$.

For a general divisor $D_1 \in |L|$ we have $D_1 \neq (-1)^*D_1$. Otherwise the divisors $T_x^*D_1 + T_{-x}^*((-1)^*D_1)$ span the linear system $|L^2|$ which would contradict the injectivity of $\varphi$.

According to Lemma 3.1 a general divisor $D_1 \in |L|$ is irreducible and reduced and by assumption ample. Choose a divisor $D_1 \in |L|$ with these properties.

Since by 1.5 the Gauss map $G : D_{1,s} \longrightarrow P(V)$ is dominant, there is an element $y \in D_{1,s} - (-1)^*D_1$ such that $\xi$ is not tangent to $D_1$ at $y$.

Then

$$D = T_y^*D_1 + T_{-y}^*D_1 \in |L^2|$$

and we have

$$0 = y - y \in T_y^*D_1 \subset D$$

and

$$0 \notin T_{-y}^*D_1 \text{ since otherwise } y \in (-1)^*D_1.$$

Since $y \in D_{1,s}$, $0 = y - y$ is contained in the smooth part of $T_y^*D_1$ and hence $D$ is smooth at 0 and $\xi$ is not tangential to $D$ at 0. ∎

# 4 Kummer–Varieties

In this section let $L$ denote an ample line bundle defining an irreducible principal polarization. Let $K_X = X/(\pm 1)$ denote the *Kummer variety* associated to $X$ that is the quotient of $X$ by the inversion $(-1)$ of $X$. $K_X$ is a variety of dimension $g$ over $\mathbf{C}$ smooth apart from the $2^{2g}$ singular points, the images of the 2–division points of $X$ under the natural map $p : X \longrightarrow K_X$. Again we consider the global sections of $L$ and $L^2$ as theta function on the universal cover $V$ of $X$. Since the map $\varphi = \varphi_{L^2}$ does not depend on the point 0 of the abelian variety $X$, we may choose 0 in such a way that $L$ is symmetric that is $(-1)^*L \simeq L$. It is easy to see that then every element of $H^0(L^2)$ is an even theta function on $V$. Hence there is a map $\Psi = \Psi_{L^2} : K_X \longrightarrow \mathbf{P}^{2^g-1}$ such that the following diagram commutes

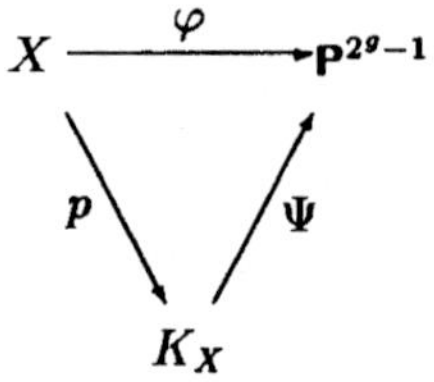

**Theorem 4.1** *If $L \in Pic(X)$ is symmetric and defines an irreducible principal polarization on $X$, $\Psi : K_X \longrightarrow \mathbf{P}^{2^g-1}$ is an embedding.*

Wirtinger [W] proved the injectivity of $\Psi$, Andreotti and Mayer [AM] showed that $\Psi$ is an embedding outside the singular points. Sasaki [S] finally showed that $\Psi$ is an embedding also at the singular points. Our proof uses the Gauss map $G$ of 1.5 .

Proof: For abbreviation we write $\Theta = |L|$ and $\bar{a} = p(a)$ for every $a \in X$.

Step I: *Injectivity of $\Psi$:*

Suppose $a, b \in X$ with $a \neq \pm b$. We have to show that there is a divisor $D \in |L^2|$ with $a \in D$ and $b \notin D$.

By Riemann–Roch the homomorphism $\Phi_L : X \to X^*$ is an isomorphism. This implies that $\Theta \neq T^*_{b-a}\Theta$ and $\Theta \neq T^*_{-(a+b)}\Theta$. Hence there is an element $x \in \Theta$ with $x \notin T^*_{b-a}\Theta \cup T^*_{-(a+b)}\Theta$. For

$$D := T^*_{x-a}\Theta + T^*_{a-x}\Theta \in |L^2|$$

we have

$$a = x + (a - x) \in T^*_{x-a}\Theta \subset D,$$
$$b \notin T^*_{x-a}\Theta, \text{ since otherwise } x \in T^*_{b-a}\Theta, \text{ and}$$
$$b \notin T^*_{a-x}\Theta, \text{ since otherwise } x \in \left(T^*_{-(a+b)}(-1)^*\Theta\right) = T^*_{-(a+b)}\Theta.$$

and hence $b \notin D$.

Step II: *Injectivity of the differential $d\Psi_{\bar{a}}$ for $a \in X$ with $2a \neq 0$:*

Suppose $0 \neq \xi \in \mathcal{T}_a X$. We have to find a divisor $D \in |L^2|$ with $a \in D$ and $\xi$ not tangential to $D$ at $a$.

Since the Gauss map $G$ is dominant by 1.5, there is an element $x \in \Theta_s - T^*_{-2a}\Theta$ such that $\xi$ is not tangent to $\Theta$ at $x$. For

$$D := T^*_{x-a}\Theta + T^*_{a-x}\Theta \in |L^2|$$

we have

$$a = x + (a - x) \in T^*_{x-a}\Theta \subset D$$

and

$$a \notin T^*_{a-x}\Theta, \text{ since otherwise } x \in T^*_{-2a}(-1)^*\Theta = T^*_{-2a}\Theta.$$

Since $x \in \Theta_s$ also $a = x + (a - x) \in \left(T^*_{-(a-x)}\Theta\right)_s$ and $D$ is smooth at $a$. Thus $\xi$ is not tangential to $D$ at $a$.

Step III: *Injectivity of the differential $d\Psi_{\bar{a}}$ for $a \in X$ with $2a = 0$:*

Without loss of generality we may assume $a = 0$. Since we identified the tangent space $\mathcal{T}_{X,0}$ with $V$, the tangent space of $K_X$ at $\bar{0}$ can be identified with the symmetric product $S^2V$: $\mathcal{T}_{K_X,\bar{0}} = S^2V$. The tangent space of $\mathbf{P}^{2^g-1} = P(H^0(L^2))$ at a point $y$ (considered as a hyperplane in $H^0(L^2)$) is $\mathcal{T}_{P^{2^g-1},y} = \text{Hom}(y, H^0(L^2)/y)$ and we have to show that the map

$$d\Psi_{\bar{0}} : S^2(V) \longrightarrow \text{Hom}(\varphi(0), H^0(L^2)/\varphi(0))$$

is injective. It is defined as follows: Choose an isomorphism $H^0(L^2)/\varphi(0) \simeq \mathbf{C}$ such that $\text{Hom}(\varphi(0), H^0(L^2)/\varphi(0)) = \varphi(0)^*$.

If $z_1, \ldots, z_g$ are a basis of $V$, the partial derivatives $\frac{\partial}{\partial z_1}, \ldots, \frac{\partial}{\partial z_g}$ are a basis of $\mathcal{T}_{X,0}$ identified with $V$. Hence any element of $S^2V$ is of the form $\sum_{i,j} \alpha_{ij} \frac{\partial}{\partial z_i}\frac{\partial}{\partial z_j}$ with constants $\alpha_{ij} \in \mathbf{C}$. Recall that by definition $\varphi(0) = \{s \in H^0(L^2) | s(0) = 0\}$. Then we have

$$d\Psi_{\bar{0}}\left(\sum_{i,j} \alpha_{ij}\frac{\partial}{\partial z_i}\frac{\partial}{\partial z_j}\right) = \begin{cases} \varphi(0) \to \mathbf{C} \\ s \mapsto \sum_{i,j} \alpha_{ij} \left(\frac{\partial^2 s}{\partial z_i \partial z_i}\right)_0 \end{cases}$$

and we have to show: Given $0 \neq \sum \alpha_{ij} \frac{\partial}{\partial z_i}\frac{\partial}{\partial z_j} \in S^2V$, there is a section $s \in H^0(V)$ with $s(0) = 0$ and $\sum_{i,j} \alpha_{ij} \left(\frac{\partial^2 s}{\partial z_i \partial z_j}\right)_0 \neq 0$.

Let $Q$ denote the quadric defined by $\sum_{i,j} \alpha_{ij} \frac{\partial}{\partial z_i}\frac{\partial}{\partial z_j} = 0$ in $P(V)$. Since $G$ is dominant there is an element $a \in \Theta_s$ such that $G(a) \notin Q$. In other words

$$\sum_{i,j} \alpha_{ij} \left(\frac{\partial \theta}{\partial z_i}\right)_a \left(\frac{\partial \theta}{\partial z_j}\right)_a \neq 0$$

where $\theta$ denotes a theta function associated to $\Theta$.

Now consider

$$s_a := \theta(z + a) \cdot \theta(z - a) \in H^0(L^2)$$

we have

$$s_a(0) = \theta(a) \cdot \theta(-a) = 0$$

since $a \in \Theta$ and

$$\begin{aligned}
\left(\frac{\partial^2 s_a}{\partial z_i \partial z_j}\right)_0 &= \left(\frac{\partial\theta(z+a)}{\partial z_i} \cdot \frac{\partial\theta(z-a)}{\partial z_j}\right)_0 + \left(\frac{\partial\theta(z-a)}{\partial z_i} \cdot \frac{\partial\theta(z+a)}{\partial z_j}\right)_0 \\
&= \left(\frac{\partial\theta}{\partial z_i}\right)_a \cdot \left(\frac{\partial\theta}{\partial z_j}\right)_{-a} + \left(\frac{\partial\theta}{\partial z_i}\right)_{-a} \cdot \left(\frac{\partial\theta}{\partial z_j}\right)_a \\
&= -2\left(\frac{\partial\theta}{\partial z_i}\right)_a \cdot \left(\frac{\partial\theta}{\partial z_j}\right)_a,
\end{aligned}$$

since $\frac{\partial\theta}{\partial z_i}$ and $\frac{\partial\theta}{\partial z_j}$ are odd functions. Hence

$$\sum_{i,j} \alpha_{ij} \left(\frac{\partial^2 s_a}{\partial z_i \partial z_j}\right)_0 = -2\sum_{i,j} \alpha_{ij} \left(\frac{\partial\theta}{\partial z_i}\right)_a \cdot \left(\frac{\partial\theta}{\partial z_j}\right)_a \neq 0$$

which completes the proof of the theorem. ■

## References

[AM] **A. Andreotti, A.L. Mayer:** On period relations for abelian integrals on algebraic curves, Ann. Sc. Norm. Sup. Pisa 21 (1967), 189–238

[M] **D. Mumford:** Abelian varieties, Oxford Univ. Press (1970)

[O] **A. Ohbuchi:** Some remarks on ample line bundles on abelian varieties, Manuscr. Math. 57 (1987), 225–238

[R] **S. Ramanan:** Ample divisors on abelian surfaces, Proc. London. Math. Soc. 57, (1985), 231–245

[Ra] **Z. Ran:** On subvarieties of abelian varieties, Inv. Math. 62 (1981), 459–479

[S] **R. Sasaki:** Modular forms vanishing at the reducible points of the Siegel upper-half space, Journ. reine angew. Math. 345 (1983), 111–121

[W] **W. Wirtinger:** Untersuchungen über Thetafunktionen, Teubner, Leipzig (1895)

Received: 27.07.88

H. Lange
Mathem. Institut der Universität
Bismarkstr. 1½
D-8520 Erlangen

M.S. Narasimhan
Tata Institute of Fundamental Research
Homi Bhabha Road
Bombay 400 005

J. reine angew. Math. **398** (1989), 169—179

Journal für die reine und angewandte Mathematik


# Spectral curves and the generalised theta divisor

By *Arnaud Beauville* at Paris, *M. S. Narasimhan*[1]) and *S. Ramanan* at Bombay

## § 1. Statement of results

Let $X$ be a smooth, irreducible, projective curve over an algebraically closed field of characteristic 0 and let $g = g_X \geqq 2$ be its genus. We show in this paper that a generic vector bundle on $X$ of rank $n$ and any fixed degree can be obtained as the direct image of a line bundle on an $n$-sheeted (ramified) covering of $X$. From this we deduce results concerning linear systems on the moduli space of vector bundles of rank $n$ and degree 0.

Let $\mathscr{U}_X(n, d)$, or simply $\mathscr{U}(n, d)$ (resp. $\mathscr{S}\mathscr{U}(n, \zeta)$) denote the moduli space of semistable vector bundles of rank $n$ and degree $d$ (resp. with determinant isomorphic to a fixed line bundle $\zeta$) on $X$. When $\zeta$ is the trivial line bundle, we simply write $\mathscr{S}\mathscr{U}(n)$ for $\mathscr{S}\mathscr{U}(n, \zeta)$. When we deal with $\mathscr{S}\mathscr{U}(n, \zeta)$, but the particular $\zeta$ that we fix is not of importance, we denote it by $\mathscr{S}\mathscr{U}(n, d)$. Let $Y$ be any smooth curve and $\pi\colon Y \to X$ a finite morphism. We can then associate to any line bundle $\xi$ on $Y$ a vector bundle, namely the direct image $\pi_*(\xi)$, on $X$. This yields a morphism of the open set $\mathscr{T}_Y^\delta$ of $J_Y^\delta$ into $\mathscr{U}(n, d)$, where a) $J_Y^\delta$ is the space of line bundles of degree $\delta$ on $Y$, b) $\mathscr{T}_Y^\delta$ is the set of those $\xi$ for which $\pi_*(\xi)$ is semistable, and c) $d$ and $\delta$ are connected by the relation $d = \delta + \deg(\pi_*(\mathcal{O}_Y))$.

**Theorem 1.** *For any pair of integers n and d, there exists an n-sheeted covering* $\pi\colon Y \to X$ *with Y smooth and irreducible, such that the morphism* $\pi_*\colon \mathscr{T}_Y^\delta \to \mathscr{U}_X(n, d)$ *is dominant with* $\delta = d - \deg \pi_*(\mathcal{O}_Y)$.

Before stating the next result, we need to introduce the notion of the *generalised theta divisor.* It is easy to define a divisor $\Theta$ in $\mathscr{U}(n, d)$ with $d = n(g-1)$ supported on the set of bundles $E$ with $\Gamma(E) \neq 0$. This defines an invertible sheaf [2] on $\mathscr{U}(n, n(g-1))$, which we may denote by $\mathcal{O}(\Theta)$. In the case $n = 1$, this divisor is the wellknown Riemann theta divisor on the variety $J^{g-1}$.

**Theorem 2.** *We have* $\dim \Gamma(\mathscr{U}(n, n(g-1)), \mathcal{O}(\Theta)) = 1$.

This theorem is proved by considering the dominant morphism $\pi_*\colon \mathscr{T}_Y^\delta \to \mathscr{U}(n, d)$ whose existence is asserted in Theorem 1, with $d = n(g-1)$. One can determine the pullback of $\Theta$ by the morphism $\pi_*$. In fact this turns out to be the Riemann theta

[1]) Supported by DFG.

divisor. Now one shows that the complement of $\mathcal{T}_Y^\delta$ in $J_Y^\delta$ is of codimension at least 2 so that we get an injection of $\Gamma(\mathcal{U}(n, d), \Theta)$ into $\Gamma(\mathcal{T}_Y^\delta, \Theta) = \Gamma(J_Y, \Theta)$, proving our assertion.

There is a close relationship [9] between the moduli space $\mathcal{S}\mathcal{U}(n)$ and the linear system of the invertible sheaf $\mathcal{O}(n\Theta)$ on $J^{g-1}$. Indeed, associate to a general vector bundle $E$, the divisor $D_E$ given by $\{\xi \in J^{g-1}: \Gamma(E \otimes \xi) \neq 0\}$. This gives a rational map of $\mathcal{S}\mathcal{U}(n)$ into the linear system of $n\Theta$ on $J^{g-1}$. On the other hand, let $\xi$ be an element of $J^{g-1}$, and let $\Delta_\xi = \{E \in \mathcal{S}\mathcal{U}(n): \Gamma(E \otimes \xi) \neq 0\}$. One can show [2] that $\Delta_\xi$ defines a Cartier divisor and the corresponding line bundle $h$ generates $\mathrm{Pic}(\mathcal{S}\mathcal{U}(n))$.

**Theorem 3.** *The (rational) map, which associates to a general point $e$ of $\mathcal{S}\mathcal{U}(n)$ the divisor $D_E$ in $J^{g-1}$, where $E$ is a semistable bundle in the class of $e$, induces an isomorphism of $\Gamma(J^{g-1}, \mathcal{O}(n\Theta))$ with $\Gamma(\mathcal{S}\mathcal{U}(n), h)^*$, and the rational map is defined by this linear system. In particular, $\Gamma(\mathcal{S}\mathcal{U}(n), h)$ has dimension $n^g$.*

The proof of this theorem is similar to that of Theorem 2, except that the Jacobian of $Y$ is in this case replaced by the Prym variety associated to the covering.

In the case of vector bundles of rank 2, these results have been proved by A. Beauville [1]. In the rank $n$ case, vector bundles arising from line bundles on an étale covering have been studied by Narasimhan and Ramanan [8]. Ideas of Hitchin [4], and the existence of 'very stable' vector bundles proved by Drinfeld and Laumon [5] have been useful to us.

## § 2. Preliminaries

**2. 1. Twisted endomorphisms.** Let $E$ be a vector bundle of rank $n$, $L$ a line bundle on $X$ and $\varphi: E \to L \otimes E$ a homomorphism. Then one can define its trace as an element of $\Gamma(L)$ by interpreting it to be a map $E \otimes E^* \to L$ and taking the image of the identity section of $E \otimes E^*$. More generally, its characteristic coefficients $a_i \in \Gamma(X, L^i)$, for $0 \leq i \leq n$, may be defined by setting $a_i = (-1)^i \mathrm{Tr}\, \Lambda^i \varphi$. The Cayley-Hamilton theorem then asserts that $\varphi$ satisfies its characteristic equation. This means that $\sum a_i \varphi^{n-i}$, interpreted as a homomorphism $E \to L^n \otimes E$, is zero. In particular, if all $a_i$ are zero for $1 \leq i \leq n$, then $\varphi^n = 0$, and we may say that $\varphi$ is *nilpotent*.

**2. 2. Generalised theta divisor.** Let $\mathcal{U}(n, n(g-1))$ be the moduli space of vector bundles of rank $n$ and degree $n(g-1)$. Then by the Riemann-Roch theorem, we see that $\chi(E) = 0$ for all $E$ in $\mathcal{U}(n, n(g-1))$. Let $\{E_t\}_{t \in T}$ be a family of vector bundles, given by a vector bundle $E$ on $T \times X$, with $E_t = E|_{t \times X}$ semistable for all $t \in T$. Let $p: T \times X \to T$ denote the first projection. The first direct image $R^1 p_* E$ is wellbehaved under base change and is supported on the set $\{t \in T: \Gamma(E_t) \neq 0\}$. Then $(\det R^1 p_* E) \otimes (\det p_* E)^{-1}$ is defined to be the invertible sheaf $\mathcal{O}(\Theta_T)$. It can be shown [2] that there exists an invertible sheaf $\mathcal{O}(\Theta)$ on $\mathcal{U}(n, d)$ such that $\mathcal{O}(\Theta_T)$ is the pullback of $\mathcal{O}(\Theta)$ by the canonical classifying morphism $T \to \mathcal{U}(n, d)$.

**2. 3. Remarks on principally polarised Abelian varieties.** Suppose $(A, \Theta)$ is a principally polarised Abelian variety and $N$ an Abelian subvariety of $A$. Consider the subgroup $P$ of $A$ consisting of all $a \in A$ such that $T_a^* \mathcal{O}(\Theta)$ and $\mathcal{O}(\Theta)$ restrict to isomorphic line bundles on $N$. We call $P$ the *orthogonal* of $N$. It is an Abelian subvariety of $A$ such that the maps $P \to A/N$ and $N \to A/P$ are isogenies. Let $\mathcal{O}(\Theta_N)$

and $\mathcal{O}(\Theta_P)$ be the restrictions of $\mathcal{O}(\Theta)$ to $N$ and $P$ respectively. One can then define a rational map $\psi$ of $P$ into $|\Theta_N|$ by setting $\psi(p)=T_p^*(\Theta)|N$. On the other hand, we also have a rational map $\varphi: P \longrightarrow |\Theta_P|^*$, defined by the linear system $|\Theta_P|$.

**2. 4. Proposition.** *There is a canonical isomorphism* $\iota: |\Theta_P|^* \longrightarrow |\Theta_N|$ *such that* $\iota \circ \varphi = \psi$.

*Proof.* Consider the isogeny $\pi: N \times P \longrightarrow A$ sending $(n, p)$ to $n+p$. The pullback $\pi^* \mathcal{O}(\Theta)$ is isomorphic to $p_N^* \mathcal{O}(\Theta_N) \otimes p_P^* \mathcal{O}(\Theta_P)$, by the theorem of the square. The canonical section of $\mathcal{O}(\Theta)$ pulls back to an element of $\Gamma(N, \mathcal{O}(\Theta_N)) \otimes \Gamma(P, \mathcal{O}(\Theta_P))$. One can now check that the induced map $\iota: \Gamma(P, \mathcal{O}(\Theta_P))^* \longrightarrow \Gamma(N, \mathcal{O}(\Theta_N))$ is an isomorphism. This follows from the fact that the theta group of $\pi^* \mathcal{O}(\Theta)$ is the natural pushout of the direct product of the theta groups of $\mathcal{O}(\Theta_N)$ and $\mathcal{O}(\Theta_P)$ with respect to the imbedding of the multiplicative group $\mathbf{G}_m$ by $\lambda \mapsto (\lambda, \lambda^{-1})$ and that the irreducible theta action is given by that on $\Gamma(N, \mathcal{O}(\Theta_N)) \otimes \Gamma(P, \mathcal{O}(\Theta_P))$. It is easy to identify the unique 1-dimensional subspace fixed by the maximal level subgroup $\mathrm{Ker}(\pi)$, and see that it gives a nondegenerate element as claimed. The rational map $\varphi$ of the variety $P$ into $|\Theta_P|^*$ is characterised by its equivariance under the action of the theta group. Now one can check that $\iota^{-1} \circ \psi$ does have this property, thus proving the equality $\iota \circ \varphi = \psi$.

**2. 5. Polarisation on the Prym variety of a covering.** Let $\pi: Y \longrightarrow X$ be a finite morphism of smooth, irreducible curves. Let us denote by $\mathfrak{d}$ the line bundle $\det(\pi_*(\mathcal{O}_Y))^{-1}$. The relative duality theorem gives an isomorphism of

$$\pi_*(\mathcal{O}_Y)^* \text{ with } \pi_*(K_Y \otimes \pi^* K_X^{-1}).$$

The ramification divisor $\Delta$ is given by a section of $\pi^* K_X^{-1} \otimes K_Y$, so that we have $\mathfrak{d} = \det \pi_*(\mathcal{O}(\Delta))$. For any line bundle $L$ on $Y$, we may associate the sheaf $\pi_*(L)$ on $X$. It is actually a vector bundle of rank $n$ if $\pi$ is $n$-sheeted. Then one can show easily that $\det \pi_*(L) = \mathrm{Nm}(L) \otimes \mathfrak{d}^{-1}$, where Nm is the norm map associated to the covering $\pi$. In particular, $\det \pi_*(\mathcal{O}(\Delta)) = \mathrm{Nm}(\mathcal{O}(\Delta)) \otimes \mathfrak{d}^{-1}$ and hence $\mathrm{Nm}\, \mathcal{O}(\Delta) = \mathfrak{d}^2$.

Suppose that the induced homomorphism $\pi^*$ of $J_X$ into $J_Y$ is *injective*. Then we may set $N = \pi^* J_X$ and $A = J_Y$ and apply the considerations of 2. 3 and 2. 4. The norm homomorphism $\mathrm{Nm}: J_Y \longrightarrow J_X$ can be identified with the transpose of $\pi^*$ and consequently has connected kernel [7]. This is the Prym variety of the map $\pi$ and we would like to take for $P$ this kernel. However, in order to avoid making too many choices, we will consider the varieties $N' = \pi^* J_X^{g-1}$, $A' = J_Y^{h-1}$ and $P' = \mathrm{Nm}^{-1}(\mathfrak{d})$, where $g = g_X$, $h = g_Y$ and Nm is the norm map $J_Y^{\deg(\mathfrak{d})} \longrightarrow J_X^{\deg(\mathfrak{d})}$. Then we have a natural addition map $N' \times P' \longrightarrow A'$ (namely tensor product). Up to translation, this is the same as the map defined in 2. 3. Hence the pullback of the line bundle $\mathcal{O}(\Theta_Y)$ by this map is the tensor product of a line bundle on $N'$ and one on $P'$. In fact, under our assumption, $N'$ is isomorphic to $J_X^{g-1}$ on which is defined Riemann's theta divisor $\Theta_X$. Now it is easy to deduce that the restriction of $\mathcal{O}(\Theta_Y)$ to $N' \times \{\alpha\}$ is isomorphic to $\mathcal{O}(n\Theta)$ for any $\alpha \in P'$. Hence there is a natural line bundle $\tau$ on $P'$ such that the pullback mentioned above is $p_1^* \mathcal{O}(n\Theta_X) \otimes p_2^* \tau$. Now the pullback of the natural section of $\mathcal{O}(\Theta_Y)$ gives rise, in view of our remarks above, to a nondegenerate element of $\Gamma(P', \tau) \otimes \Gamma(J_X^{g-1}, \mathcal{O}(n\Theta))$, that is to say there is a natural isomorphism of $\Gamma(P', \tau)^*$ with $\Gamma(J_X^{g-1}, \mathcal{O}(n\Theta))$. Thus we have

**2. 6. Proposition.** *Let* $\pi: Y \longrightarrow X$ *be a finite morphism of projective nonsingular curves such that the induced morphism* $\pi^*: J_X \longrightarrow J_Y$ *is injective. Then with the notation*

*above, there is a natural isomorphism of* $\Gamma(J_X^{g-1}, \mathcal{O}(n\Theta))$ *with* $\Gamma(P', \tau)^*$ *such that the following diagram is commutative.*

$$\begin{array}{ccc} P' & \longrightarrow & P\Gamma(P',\tau)^* \\ & \searrow & \downarrow \approx \\ & & P\Gamma(J_X^{g-1}, \mathcal{O}(n\Theta)) \end{array}$$

Here the rational map $P' \longrightarrow P\Gamma(P', \tau)^*$ is given by the linear system of $\tau$, while the rational map $P' \longrightarrow P\Gamma(J_X^{g-1}, \mathcal{O}(n\Theta))$ is given by $p \mapsto T_p^* \Theta_Y \cap \pi^* J_X^{g-1}$.

**2. 7. Remark.** One can show in fact that under the assumption that $\pi^*: J_X \longrightarrow J_Y$ is injective, the polarisation on $P$ (and on $P'$ as well) is of type $(1, 1, \ldots, n, n, \ldots)$ where the $n$'s are repeated $g$ times. In particular its Pfaffian is $n^g$ which is of course the dimension of $\Gamma(J^{g-1}, \mathcal{O}(n\Theta))$. It is easy to show that $\tau$ is a primitive element in the group of algebraic equivalence classes.

## § 3. Spectral curves

Spectral curves have been treated in recent years by several authors, in connection with some special kinds of nonlinear differential equations. In our context it was used by Hitchin [4]. Let $L$ be a line bundle on $X$ and $s = (s_k)$ be sections of $L^k$ for $k = 1, 2, \ldots, n$. Then we will construct a scheme $X_s$ and a finite morphism $\pi: X_s \longrightarrow X$. Let $p: P = \mathbf{P}(\mathcal{O} \oplus L^*) \longrightarrow X$ be the natural projection and $\mathcal{O}(1)$ the relatively ample bundle, or what may be called "the hyperplane bundle along the fibres". Then $p_*(\mathcal{O}(1))$ is naturally isomorphic to $\mathcal{O} \oplus L^*$ which has a canonical section, namely, the constant section 1 of $\mathcal{O}$. This gives a section of $\mathcal{O}(1)$ over $P$ which we denote by $y$. On the other hand, $p_*(p^* L \otimes \mathcal{O}(1))$ is isomorphic, by the projection formula, to $L \otimes (\mathcal{O} \oplus L^*) = L \oplus \mathcal{O}$. Hence it also has a canonical section and we denote the corresponding section of $p^* L \otimes \mathcal{O}(1)$ by $x$. Note that the two sections $x$ and $y$ have as their zero schemes the natural subvarieties $\mathbf{P}(\mathcal{O})$ and $\mathbf{P}(L^*)$ of $P$. Now consider the section

$$x^n + (p^* s_1) \cdot y \cdot x^{n-1} + \cdots + (p^* s_n) \cdot y^n$$

of $p^* L^n \otimes \mathcal{O}(n)$. Its zero scheme is the scheme $X_s$ we have in mind. It is clear that the restriction $\pi$ of $p$ to $X_s$ is finite and that at any point $v$ of $X$ the fibre of $\pi$ is the subscheme of $\mathbf{P}^1$ given by

$$x^n + a_1 y x^{n-1} + \cdots + a_n y^n = 0,$$

where $(x, y)$ is a homogeneous coordinate system, and $a_i$ is the value of $s_i$ at $v$ on identifying the fibre of $L$ at $P$ with the residue field at $v$.

**3. 1. Remark.** It is easily seen that the set of all $(s) \in \Gamma(L) \oplus \Gamma(L^2) \oplus \cdots \oplus \Gamma(L^n)$ for which the scheme $X_s$ is integral (i.e., reduced and irreducible), is open. Also, it is easy to see that it is nonempty, whenever $L^n$ admits a section whose divisor is not of the form $mD$ for some integer $m$ dividing $n$. For, in that case we can take all $s_i$ for $i \leqq n-1$ to be 0 and $s_n$ to be a section as postulated.

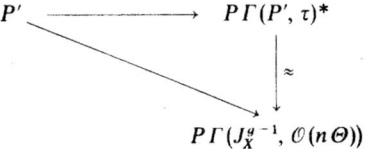

Hereafter we will assume that $(s)$ has been so chosen that $X_s$ is integral. Since the zeros of $x$ and $y$ are disjoint in $P$, it follows that the restriction of $y$ to $X_s$ is everywhere nonzero and hence the restriction of $\mathcal{O}(1)$ to $X_s$ is trivial. The restriction of $x$ to $X_s$ can therefore be considered as a section of $\pi^*(L)$. An alternative description of $X_s$ is given as follows. The direct image $\pi_*(\mathcal{O})$ is naturally isomorphic to $\mathcal{O} \oplus L^{-1} \oplus \cdots \oplus L^{-(n-1)}$. Its algebra structure is better understood by interpreting this to be $\mathrm{Sym}(L^{-1})/\mathcal{I}$ where $\mathcal{I}$ is the ideal sheaf generated by the image of the homomorphism $u\colon L^{-n} \to \mathrm{Sym}(L^{-1})$ given as the sum of the imbeddings $L^{-n} \to L^{-(n-i)}$ defined by $s_i$. In terms of this sheaf of algebras, $X_s$ is simply $\mathrm{Spec}(\mathrm{Sym}(L^{-1})/\mathcal{I})$.

**3. 2. Remark.** The genus $g'$ of $X_s$ may be computed from the above interpretation. In fact, $1-g'$ is given by

$$\begin{aligned}\chi(X_s, \mathcal{O}) = \chi(X, \pi_*(\mathcal{O})) = \sum \chi(X, L^i)&\\ &= (\deg L)\cdot(0-1-\cdots-(n-1)) + n(1-g)\\ &= -(\deg L)\cdot n(n-1)/2 + n(1-g)\end{aligned}$$

and hence $g' = (\deg L)\cdot n(n-1)/2 + n(g-1) + 1$.

**3. 3. Remark.** Assume that $X_s$ is integral. Then consider the resultant of the polynomial

$$x^n + s_1 y x^{n-1} + s_2 y^2 x^{n-2} + \cdots + s_n y^n,$$

and its derivative

$$n x^{n-1} + (n-1) s_1 y x^{n-2} + \cdots + s_{n-1} y^{n-1}.$$

This can be interpreted to be a section of $(p^* L \otimes \mathcal{O}(1))^{n(n-1)/2}$. One easily verifies that the map $\pi\colon X_s \to X$ is étale at a point if and only if this resultant does not vanish at that point. In other words, the resultant gives the ramification divisor of the map $\pi$.

**3. 4. Example.** Let $L$ be a line bundle with $L^n$ isomorphic to $\mathcal{O}$. We could then take $s_i$ to be 0 for $i \leqq n-1$, and $s_n$ to be any nonzero section of $L^n$. In this case one can conclude from 3. 3 that the map $\pi$ is actually étale. However, the scheme $X_s$ is connected (and therefore irreducible) if and only if the section $s_n$ is not a power of a section of $L^r$ for some $r$ which divides $n$. This would happen if and only if $L^r$ is trivial. In other words, $X_s$ is irreducible if and only if $L$ is of order $n$ in the Jacobian of $X$.

**3. 5 Remark.** Let us assume that $X_s$ is integral. Then using the Jacobian criterion it is not difficult to prove that $X_s$ is smooth if and only if at every point which is a multiple zero of the section $s_n$, the section $s_{n-1}$ is nonzero. In particular, the set of sections $(s)$ such that the scheme $X_s$ is smooth, is open on the one hand, and is nonempty if $L^n$ admits a section without multiple zeros, on the other.

**3. 6. Proposition.** *Let $X$ be a curve and $L$ any line bundle on it. Let $s = (s_i)$ be a set of sections of $L^i$ for $1 \leqq i \leqq n$. Assume that the corresponding scheme $X_s$ is integral. Then there is a bijective correspondence between isomorphism classes of torsion free* *sheaves of rank 1 on $X_s$ and isomorphism classes of pairs $(E, \varphi)$ where $E$ is a vector bundle of rank $n$ and $\varphi: E \to L \otimes E$ a homomorphism with characteristic coefficients $s_i$. The correspondence is given by associating to any line bundle $M$ on $X_s$ the sheaf $\pi_*(M)$ on $X$ and the natural isomorphism*

$$\pi_*(M) \to L \otimes \pi_*(M) \approx \pi_*(\pi^* L \otimes M)$$

*given by the section $x$ of $\pi^*(L)$.*

*Proof.* If $M$ is a torsion free sheaf of rank 1 on $X_s$, then the sheaf $\pi_* M$ is a vector bundle of rank $n$ on $X$, endowed with a $\pi_*(\mathcal{O}_{X_s})$-module structure. Conversely, since $\pi$ is affine, a vector bundle $E$ of rank $n$ with a $\pi_*\mathcal{O}$-module structure defines a sheaf $M$ with $\pi_*(M)$ isomorphic to $E$. Now $M$ is torsion free of rank 1 because $X_s$ is integral.

The $\mathcal{O}_X$-Algebra $\pi_*\mathcal{O}_{X_s}$ is isomorphic to $\mathrm{Sym}(L^{-1})/\mathcal{I}$, so that the data of a $\pi_*(\mathcal{O}_{X_s})$-module structure on $E$ is equivalent to an algebra homomorphism

$$\mathrm{Sym}(L^{-1})/\mathcal{I} \to \mathcal{E}nd(E),$$

that is to say an $\mathcal{O}_X$-homomorphism $\varphi: E \to L \otimes E$ such that $P_s(\varphi) = 0$, where $P_s$ is the polynomial determined by $(s)$. Since $X_s$ is integral, $P_s$ is irreducible over the function field of $X$, and is hence the characteristic polynomial of $\varphi$. Conversely, if $E$ is a vector bundle of rank $n$ on $X$, and $\varphi: E \to L \otimes E$ a linear homomorphism with characteristic coefficients $s_i$, then $P_s(\varphi) = 0$ by the Cayley-Hamilton theorem. This proves the proposition.

**3. 7. Remark.** The sequence

$$0 \longrightarrow M(-\Delta) \longrightarrow \pi^* E \xrightarrow{\pi^*\varphi - x} \pi^*(L \otimes E) \longrightarrow \pi^* L \otimes M \longrightarrow 0$$

is exact. In fact one could actually define the line bundle $M$ corresponding to $(E, \varphi)$ by this exact sequence. This is essentially the point of view in Hitchin [4].

**3. 8. Remark.** When $X_s$ is nonsingular, then 'torsion free sheaves of rank 1' may be replaced by 'line bundles' in Proposition 3. 6.

**3. 9. Example.** Let us get back to Example 3. 5. If $L$ is a line bundle of order $n$ in the Jacobian of $X$, then according to Proposition 3. 6, there is a natural bijection between line bundles on $X_s$ and pairs $(E, \varphi)$. Now there exists a map $\varphi: E \to L \otimes E$ with the given characteristic polynomial if and only if $L \otimes E$ is isomorphic to $E$. In this particular case this map was studied and indeed many of the computations in this paper were carried out in [8].

**3. 10. Remark.** If $M$ is a nontrivial line bundle on $X$, and $\pi^* M$ is trivial on $X_s$, then it follows that $\pi_*(\pi^* M) \cong M \otimes \pi_*(\mathcal{O}) \cong \sum M \otimes L^{-i}$ admits a nontrivial section. In particular this implies that the degree of $L$ is 0. But since $s_i \in \Gamma(L^i)$, we must have for each $i$, either $s_i = 0$, or $L^i$ is trivial. Thus we see that except in this case (which includes Example 3. 9), the induced map $\pi^*: J_X \to J_{X_s}$ is injective.

## § 4. Applications to the moduli space of vector bundles

In what follows, we will take for the line bundle $L$ the canonical line bundle $K$ of a smooth, projective curve $X$ of genus $g \geqq 2$. Denoting by $W$ the direct sum $\bigoplus_{k=1}^{k=n} \Gamma(X, K^k)$, we see immediately from the Riemann-Roch theorem that

$$\dim W = n^2(g-1)+1,$$

which is also the dimension of $\mathscr{S}\mathscr{U}(n, d)$. *Let* $s=(s_i)$ be an element of $W$ such that the corresponding curve $X_s$ is smooth and irreducible. By Remark 3. 2, the genus of $X_s$ is $n^2(g-1)+1$. Moreover, for any line bundle $\xi$ on $X_s$ we have

$$\det \pi_*(\xi) = \mathrm{Nm}(\xi) \otimes \det \pi_*(\mathcal{O}) = \mathrm{Nm}(\xi) \otimes K^{-n(n-1)/2}$$

where $\mathrm{Nm}\colon \mathrm{Pic}(X_s) \to \mathrm{Pic}(X)$ is the norm map associated to $\pi$. In particular, the degree of $\pi_*(\xi)$ is $\deg(\xi) - n(n-1)(g-1)$.

Let $\mathscr{U}$ be the moduli space of stable vector bundles of rank $n$ and degree $d = \deg(\xi) - n(n-1)(g-1)$. Let $\Omega = \Omega^1$ denote its cotangent bundle. If $E$ is a stable bundle of rank $n$ and degree $d$, then the tangent space to $\mathscr{U}$ at the corresponding point can be canonically identified with $H^1(X, \mathscr{E}nd\, E)$. By duality, the cotangent space may be identified with $\Gamma(X, K \otimes \mathscr{E}nd\, E)$. Therefore a point of $\Omega$ can be seen as a pair $(E, \varphi)$, where $E$ is a vector bundle in $\mathscr{U}$ and $\varphi\colon E \to K \otimes E$ is a homomorphism. By associating to $(E, \varphi)$ the characteristic polynomial of $\varphi$, we define a morphism $H\colon \Omega \to W$. We will call this map the Hitchin map [4].

**4. 1.** *Proof of the Theorem* 1. Consider the map $\lambda\colon \Omega \to \mathscr{U} \times W$ where the first factor of $\lambda$ is the projection onto $\mathscr{U}$ and the second factor is the Hitchin map $H$. By a theorem of Drinfeld and Laumon [5], there exists a stable bundle $E_0$ of rank $n$ and degree $d$ such that every nilpotent homomorphism $E_0 \to K \otimes E_0$ is zero. This implies that the inverse image by $\lambda$ of $(E_0, 0)$ in $\mathscr{U} \times W$ consists of a single point, namely the zero cotangent vector at $E_0 \in \mathscr{U}$. Now $\dim(\Omega) = \dim(\mathscr{U} \times W) = 2(n^2(g-1)+1)$. By the dimension theorem, the morphism $\lambda$ is dominant and moreover its differential is an isomorphism at the generic point of $\Omega$. It follows that the projection $H^{-1}(s) \to \mathscr{U}$ is dominant for all point $s$ in an open set in $W$. Moreover we may assume that for $s$ in this open set the spectral curve $X_s$ is smooth and irreducible in view of Remark 3. 5. Now by Proposition 3. 6, $H^{-1}(s)$ consists of pairs $(E, \varphi)$ where $E$ is a stable bundle obtained as the direct image of a line bundle $\xi$ on $X_s$ and $\varphi\colon E \to K \otimes E$ is the homomorphism induced by the canonical section of $\pi^*(K)$ on $X_s$. This proves Theorem 1.

## § 5. The basic linear system on $\mathscr{S}\mathscr{U}(n)$

We apply the considerations of the previous section to the case when $d = n(g-1)$. Let us choose an $n$-sheeted covering $Y$ of $X$ such that most vector bundles of rank $n$ and degree $d$ are direct images of line bundles of degree $\delta$ with $\delta = g_Y - 1 = n^2(g-1)$. We will denote the Riemann theta divisor in $J_X^{g-1}$ (whose support consists of effective divisor classes of degree $(g-1)$) by $\Theta_X$. Let $\mathcal{T}_Y^\delta$ denote the open subset of the component of the Jacobian consisting of $\xi$ such that $\pi_*(\xi)$ is semistable. Thus there is a natural morphism $\pi_* : \mathcal{T}_Y^\delta \to \mathcal{U}(n, n(g-1))$.

**5. 1. Proposition.** a) *For every point $\xi$ belonging to $J_Y^\delta - \Theta_Y$, the bundle $\pi_*(\xi)$ is semi-stable.*

(b) *There is a point $\xi$ on the theta divisor $\Theta_Y$ such that $\pi_*(\xi)$ is semistable.*

c) *The complement of $\mathcal{T}_Y^\delta$ in $J_Y^\delta$ is of codimension greater than* 1.

d) *There is a point $\xi$ in $\Theta_Y$ such that $\pi_*(\xi)$ is stable and the map $\pi_*$ has maximal rank at $\xi$.*

*Proof.* a) By our assumption, $\Gamma(Y, \xi) = 0$. Hence $\Gamma(X, \pi_* \xi) = 0$. If $\pi_*(\xi)$ were not semistable, it would admit a subbundle $F$ of degree greater than $\mathrm{rk}(F) \cdot (g-1)$. Now the Riemann-Roch theorem implies that $\Gamma(X, F) \neq 0$, which is a contradiction.

b) If $\alpha \in J_X$, then the projection formula implies that $\pi^*(\alpha) \otimes \xi$ has semistable direct image for every $\xi$ as in a). Since $\Theta$ is ample and $\pi^*: J_X \to J_Y$ is a finite map, it follows that it intersects the positive dimensional variety $\pi^*(J_X)\xi$.

c) Since the complement of $\mathcal{T}_Y^\delta$ in $J_Y^\delta$ is contained properly in the irreducible divisor $\Theta$, it follows that it is of codimension at least 2.

d) We have assumed that the map $\pi_*$ is dominant so that there exists some $\eta \in J_Y^\delta$ such that $\pi_*(\eta)$ is stable and $\pi_*$ is of maximal rank at $\eta$. Now as in b) above, we can translate $\eta$ by $\pi^*(\alpha)$ for an $\alpha \in J_X$ so that $\xi = \eta \otimes \pi^* \alpha$ belongs to $\Theta_Y$. It is clear from the commutativity of the diagram

$$\begin{array}{ccc} \mathcal{T}_Y^\delta & \longrightarrow & \mathcal{U}(n, d) \\ \downarrow {\scriptstyle \otimes \pi^* \alpha} & & \downarrow {\scriptstyle \otimes \alpha} \\ \mathcal{T}_Y^\delta & \longrightarrow & \mathcal{U}(n, d) \end{array}$$

that $\pi_*(\xi)$ is stable and that the map $\pi_*$ is of maximal rank at $\xi$ also.

**5. 2. Remark.** The complement of the set $S$ of points $\xi$ of $J_Y^\delta$ such that $\pi_* \xi$ is stable can be shown to have codimension at least $2g-2$, if $n \geqq 3$. In fact suppose that $Z$ is the set of points $\xi$ in $J_Y^\delta$ such that $\pi_* \xi$ has a quotient line bundle of degree $\leqq d/n$. We will estimate the codimension of $Z$ in $J_Y^\delta$. By the duality theorem applied to the map $\pi$, the existence of a nonzero homomorphism of $\pi_* \xi$ into $M$, where $M$ is a line bundle on $X$, is equivalent to the existence of a nonzero homomorphism of $\xi$ into $\pi^* M \otimes \omega_\pi$ where $\omega_\pi$ is the relative dualising sheaf, namely $K_Y \otimes \pi^* K_X^{-1}$. Thus every element of $Z$ is of the form $\pi^* M \otimes \omega_\pi(-D)$ where $M$ is a line bundle on $X$ with $\deg(M) \leqq d/n$, and $D$ is an effective divisor on $Y$. We have

$$\deg(D) = n \cdot \deg M + \deg \omega_\pi - \delta \leqq d - \delta + 2\{g_Y - 1 - n(g_X - 1)\}.$$

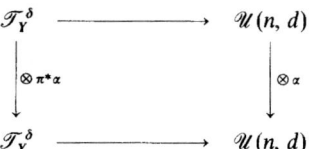

It follows that $\operatorname{codim}(Z) \geqq (n-1)(g-1) \geqq 2(g-1)$ if $n \geqq 3$. If $E$ is a general stable bundle on $X$, then by [10], Lemma 2. 1, for every subbundle $F$ of $E$ we have

$$\deg(F)/\operatorname{rk}(F) \leqq \{\deg(E)-(\operatorname{rk}(E)-\operatorname{rk}(F))(g-1)\}/\operatorname{rk}(E).$$

Let now $L$ be a general element of $J_Y^{\delta+(2g-3)}$ such that $\pi_* L = E$ satisfies the condition given above. If $D$ is an effective divisor of degree $2g-3$ on $Y$, the bundle $\pi_*(L(-D))$ is of rank $n$ and degree $d$ and is a subsheaf of $\pi_* L$. If $F$ is a subbundle of $\pi_*(L(-D))$ with $\operatorname{rk}(F) \leqq n-2$, it follows that $\deg(F)/\operatorname{rk}(F) < d/n$.

On the other hand, as $D$ varies, the line bundles $L(-D)$ form a subvariety $W_L$ of $J_Y^\delta$ of dimension $(2g-3)$. Changing $L$ amounts to applying a translation on $W_L$. Since the variety $Z$ defined earlier is of codimension at least $2g-2$, we can choose $L$ such that $W_L$ does not intersect $Z$. For such a choice of $L$, $W_L$ is contained in $S$. Now $W_L$ is cohomologous to $\Theta^k/k!$ with $k = \operatorname{codim}(W_L)$ in $J_Y^\delta$. Hence $W_L$ intersects every subvariety of $J_Y^\delta$ of codimension $2g-3$. This proves our assertion.

**5. 3. Remark.** For $n=2$, the codimension of $(J_Y^\delta - S)$ is at least $g-1$. Coupled with Remark 5. 2, this implies that with the exception of the case $g=r=2$, this codimension is at least 2. Consider the Hitchin map $H: \Omega \to W$. The restriction of the natural symplectic form $\omega$ to the fibre $H^{-1}(a)$ over a general point $a \in W$ is the boundary of a holomorphic 1-form. On the other hand, $H^{-1}(a)$ can be identified with the complement in $J_Y^\delta$ of a subvariety of codimension at least 2 and hence this 1-form can be extended to the whole of $J_Y^\delta$. It is therefore closed and so $\omega$ restricts to 0 on $H^{-1}(a)$. In other words, the map $H$ defines a completely integrable Hamiltonian system. Moreover the Hamiltonian vector fields given by components (with respect to a basis of $W$) of $H$ also extend to the Jacobian and thus the system is linearised on the variety $J_Y^\delta$. This gives another proof of the results of Hitchin [4].

**5. 4. Remark.** *The map* $\pi_*: \mathcal{T}_Y^\delta \to \mathcal{U}(n, d)$ *is of degree* $2^{3g-3} \cdot 3^{5g-5} \ldots n^{(2n-1)(g-1)}$.

*Proof.* Indeed, the required degree is equal to the degree of the map $H_E: \Omega_E^1 \to W$, namely the restriction of $H$ to the cotangent space to the moduli space at a sufficiently general point $E$. Since this is clearly defined by $3g-3$ polynomials of degree two, $5g-5$ polynomials of degree three, ... the assertion follows from Bezout's theorem.

**5. 5.** *Proof of Theorem* 2. Let $d = n(g-1)$ and $\delta = n^2(g-1)$. We will compute on $\mathcal{T}_Y^\delta$ the pullback of the theta divisor on $\mathcal{U}(n, d)$. Indeed, the vector bundle on $\mathcal{T}_Y^\delta \times X$ which induces the map $\pi_*: \mathcal{T}_Y^\delta \to \mathcal{U}(n, d)$ is easy to describe. Consider a Poincaré bundle $P$ on $\mathcal{T}_Y^\delta \times Y$. Then the required bundle is $U = (1 \times \pi)_*(P)$. In order to compute the pullback of $\mathcal{O}(\Theta)$, we need to compute only $(p_1)_*(U)$ and $R^1(p_1)_*(U)$. Since $(1 \times \pi)$ is an affine morphism these are simply the sheaves $(p_1)_*(P)$ and $R^1(p_1)_*(P)$. But the sheaf $(p_1)_*(P)$ is 0 and $\det(R^1(p_1)_*(P))$ is the Riemann theta divisor on $J_Y^\delta$. Hence the pullback of $\mathcal{O}(\Theta)$ by $\pi_*$ is isomorphic to the restriction of $\mathcal{O}(\Theta)$ to $\mathcal{T}_Y^\delta$. Since the complement of $\mathcal{T}_Y^\delta$ in $J_Y^\delta$ is of codimension $\geqq 2$ it follows that $\pi$ induces an injective map $\Gamma(\mathcal{U}(n, d), \mathcal{O}(\Theta)) \to \Gamma(J_Y^\delta, \mathcal{O}(\Theta))$. This proves that the dimension of $\Gamma(\mathcal{U}(n, d), \mathcal{O}(\Theta))$ is at most 1. But it is clearly given by an effective divisor so that Theorem 2 is proved.

**5. 6.** *Proof of Theorem* 3. Take $d=0$ and $\delta = (n^2-n)(g-1)$ and choose a covering $\pi: Y \to X$ as in Theorem 1. We will use here the notation of 2. 3—2. 6. Firstly the rational map $J_Y^\delta \to \mathcal{U}(n, 0)$ is dominant.

**5. 7. Proposition.** *The direct image map induces a dominant rational map $P' \to \mathcal{SU}(n)$. Moreover, the complement of the open set $\mathcal{T}_Y^{\delta} \cap P'$ is of codimension at least 2.*

*Proof.* If $\pi_*(\alpha)$ is semistable, so is $\pi_*(\pi^* L \otimes \alpha)$ for any line bundle $L$ on $X$ and we can therefore conclude that there is an induced dominant rational map $P' \to \mathcal{SU}(n)$. The rest of the assertions is also proved in the same way, using Proposition 5. 1.

Now, from the commutative diagram

$$\begin{array}{ccc} P' \times J_X^{g_X-1} & \longrightarrow & J_Y^{g_Y-1} \\ \big\downarrow & & \big\downarrow \\ \mathcal{SU}(n) \times J_X^{g_X-1} & \longrightarrow & \mathcal{U}(n, n(g-1)) \end{array}$$

we conclude that the pullback of the line bundle $\mathcal{O}(\Theta)$ to $\mathcal{SU}(n) \times J_X^{g-1}$ is of the form $p_1^* M \otimes p_2^* \mathcal{O}(n\Theta)$ for some line bundle $M$. This implies that the pullback of $M$ by the map $P' \to \mathcal{SU}(n)$ is $\tau$. But $\tau$ is a primitive element in the group of algebraic equivalence classes of divisors in $P'$. Hence the same is true of $M$. In other words, the line bundle is the same as the ample generator $h$ of the Picard group, which is isomorphic to **Z** [2]. From Remark 4. 9, Proposition 2. 6 and the codimension computation in Proposition 5. 1, we deduce, as in the proof of Theorem 2, that the map $\Gamma(\mathcal{SU}(n), h) \to \Gamma(P', \tau)$ is injective. Now the group of $n$ division points of the Jacobian of $X$ acts on $\mathcal{SU}(n)$ as well as on $P(\Gamma(P', \tau))$ and it is easy to see that the above map is compatible with these actions. Now it is easily seen that $\Gamma(\mathcal{SU}(n), h) \neq 0$ so that we conclude that this map is bijective and in particular that $\dim \Gamma(\mathcal{SU}(n), h) = n^g$, since the action of the theta group on $\Gamma(P', \tau)$ is irreducible [6]. Finally, it remains to use the isomorphism of $\Gamma(P', \tau)^*$ with $\Gamma(J_X^{g-1}, \mathcal{O}(n\Theta))$, and identify the induced rational map with $E \mapsto D_E$. In fact, this only amounts to computing the rational map of $P'$ into $P\Gamma(J_X^{g-1}, \mathcal{O}(n\Theta))$. But, in view of Proposition 2. 6, this means that we have to verify that for all $p$ belonging to an open set in $P'$, the divisor $T_p^* \Theta \cap J^{g-1}$ is the same as $D_{\pi_* p}$. The latter is given by $\{\xi \in J^{g-1} : \Gamma(\xi \otimes \pi_* p) \neq 0\}$. The projection formula implies that this consists of those $\xi$ with $\Gamma(\pi^* \xi \otimes p) \neq 0$. This is then the divisor $\{\xi \in J^{g-1} : T_p \pi^* \xi \in \Theta_Y\} = T_p^* \Theta \cap J^{g-1}$, as was to be proved.

**5. 8. Remark.** On the one hand, we have a natural rational map $\Delta: J^{g-1} \to P\Gamma(\mathcal{SU}(n), h)$ given by mapping any $\xi \in J^{g-1}$ to the pullback of the theta divisor on $\mathcal{U}(n, n(g-1))$ by the morphism $E \mapsto E \otimes \xi$ of $\mathcal{SU}(n)$ into $\mathcal{U}(n, n(g-1))$. On the other, the linear system of $n\Theta$ gives a rational map of $J^{g-1}$ into $P\Gamma(J^{g-1}, \mathcal{O}(n\Theta))^*$. These maps fit into a commutative diagram

$$\begin{array}{ccc} J^{g-1} & \xrightarrow{\ \Delta\ } & P\Gamma(\mathcal{SU}(n), h) \\ & \searrow & \big\downarrow \\ & & P\Gamma(J^{g-1}, \mathcal{O}(n\Theta))^* \end{array}$$

in which the vertical map is given by the duality in Theorem 3.

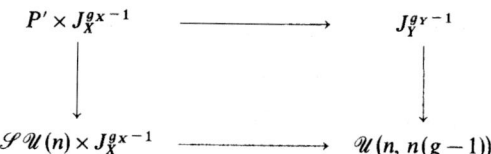

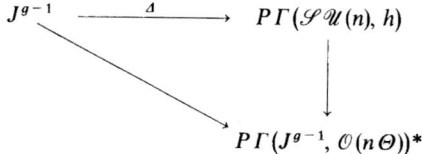

**5. 9. Remark.** The linear system of $h$ in $\mathcal{SU}(n)$ is base point free for $n=2$. On the other hand, there are examples when it does have base points. Indeed, by 5. 8., it is spanned by the image of $J^{g-1}$ under $\Delta$. Therefore a point $E \in \mathcal{SU}(n)$ is a base point if and only if all the divisors in the image of $\Delta$ pass through $E$. This condition means that for all $\xi$ in $J^{g-1}$, one has $\Gamma(X, E \otimes \xi) \neq 0$. There are examples of bundles with this property ([11], § 3).

## References


[1] *A. Beauville*, Fibrés de rang 2 sur une courbe, fibré déterminant et fonctions thêta, Bull. Soc. Math. France **116** (1988).

[2] *J. M. Drezet* and *M. S. Narasimhan*, Groupes de Picard des variétés de modules de fibrés semistables sur les courbes algébriques, to appear.

[3] *R. Hartshorne*, Algebraic Geometry, Berlin-Heidelberg-New York 1977.

[4] *N. Hitchin*, Stable bundles and integrable systems, Duke Math. J. **54** (1987), 91—114.

[5] *G. Laumon*, Un analogue global du cône nilpotent, Duke Math. J. **57** (1988), 667—671.

[6] *D. Mumford*, Equations defining abelian varieties, Invent. Math. **1** (1966), 287—354.

[7] *D. Mumford*, Prym Varieties I, In Contributions to Analysis, London-New York-San Francisco 1974, 325—350.

[8] *M. S. Narasimhan* and *S. Ramanan*, Generalised Prym Varieties as fixed points, J. Ind. Math. Soc. **39** (1975), 1—19.

[9] *M. S. Narasimhan* and *S. Ramanan*, 2$\Theta$-linear systems, Proc. of the Int. Colloq. on Vector Bundles, Bombay 1984, Oxford (1988).

[10] *H. Lange*, Zur Klassifikation von Regelmannigfaltigkeiten, Math. Ann. **262** (1983), 447—459.

[11] *M. Raynaud*, Sections des fibrés vectoriels sur une courbe, Bull. Soc. Math. France **110** (1982), 103—125.



Mathématiques — Bât. 425, Université Paris-Sud, 91405 Orsay Cedex, France

Tata Institute of Fundamental Research, Homi Bhabha Road, Bombay 400005, India


Eingegangen 20. September 1988

Invent. math. 97, 53–94 (1989)

*Inventiones mathematicae*



# Groupe de Picard des variétés de modules de fibrés semi-stables sur les courbes algébriques

J.-M. Drezet [1] et M.S. Narasimhan [2]

[1] Université Paris VII, UER de Mathématiques, Aile 45-55, 5ème étage, 2, place Jussieu; F-75251 Paris Cedex 05, France

[2] Tata Institute of Fundamental Research, Homi Bhabha Road, 400005 Bombay, India

## 0. Introduction

Soient $X$ une courbe algébrique projective lisse de genre $g \geqq 2$ sur $\mathbb{C}$, $r$, $d$ des entiers, avec $r \geqq 2$. On note $U(r, d)$ la variété de modules des fibrés algébriques semi-stables sur $X$ de rang $r$ et de degré $d$, $U_s(r, d)$ l'ouvert de $U(r, d)$ correspondant aux fibrés stables. On sait que $U(r, d)$ est une variété algébrique projective irréductible et normale. Si $r$ et $d$ ne sont pas premiers entre eux, $U(r, d)$ n'est pas lisse, sauf dans une exception: le cas où $g=2$, $r=2$, et $d$ est pair (cf. [17]). On supposera par la suite qu'on n'est pas dans ce cas. On a alors $\operatorname{codim}_{U(r,d)}(U(r, d) \setminus U_s(r, d)) \geqq 2$, et $U(r, d) \setminus U_s(r, d)$ est le lieu des points singuliers de $U(r, d)$. Si $L$ est un fibré en droites de degré $d$ sur $X$, on note $U(r, L)$ (resp. $U_s(r, L)$) la sous-variété fermée de $U(r, d)$ (resp. $U_s(r, d)$) correspondant aux fibrés vectoriels de déterminant isomorphe à $L$. Le but de ce travail est l'étude de $\operatorname{Pic}(U(r, d))$ et $\operatorname{Pic}(U(r, L))$.

### 0.1. *Factorialité de $U(r, d)$ et $U(r, L)$*

Le premier résultat est le

**Théorème A.** *Les variétés $U(r, d)$ et $U(r, L)$ sont localement factorielles.*

Donnons une idée de la démonstration de ce théorème. Pour toute variété algébrique $Y$ on note $CL(Y)$ le groupe des classes d'équivalence linéaire de diviseurs de Weil de $Y$. On a un diagramme commutatif:

$$\begin{array}{ccc} \operatorname{Pic}(U(r, d)) & \xrightarrow{\varphi} & \operatorname{Cl}(U(r, d)) \\ \downarrow{\scriptstyle r_1} & & \downarrow{\scriptstyle r_2} \\ \operatorname{Pic}(U_s(r, d)) & \xrightarrow{\varphi_s} & \operatorname{Cl}(U_s(r, d)) \end{array}$$

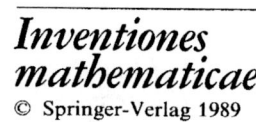

les flèches horizontales étant les morphismes canoniques, les verticales étant les restrictions. Le morphisme $\varphi_s$ est bijectif car $U_s(r, d)$ est lisse, et il en est de même de $r_2$ car $\operatorname{codim}_{U(r,d)}(U(r, d)\setminus U_s(r, d)) \geqq 2$. Nous verrons que $U(r, d)$ est localement factorielle si et seulement si $\varphi$ est un isomorphisme, et ceci se produit si et seulement si $r_1$ en est un. L'injectivité de $r_1$ découle de la normalité de $U(r, d)$. Nous devrons donc essentiellement prouver que $r_1$ est surjectif.

Pour cela, on considère la construction habituelle de $U(r, d)$ comme bon quotient d'un ouvert lisse $R^{ss}$ d'un schéma de Grothendieck par un groupe du type $PGL(q)$ (cf. 1.1.4). Soient

$$\pi: R^{ss} \to U(r, d)$$

le morphisme quotient, $R^s = \pi^{-1}(U_s(r, d))$, et $\pi_s: R^s \to U_s(r, d)$ la restriction de $\pi$. On note $\operatorname{Pic}^G(R^{ss})$ le groupe des classes d'isomorphisme de $PGL(q)$-fibrés en droites sur $R^{ss}$ (un $PGL(q)$-fibré en droites sur $R^{ss}$ étant un fibré en droites algébrique sur $R^{ss}$ muni d'une action linéaire algébrique de $PGL(q)$ au dessus de l'action de ce groupe sur $R^{ss}$). On définit de même $\operatorname{Pic}^G(R^s)$.

Du fait notamment que $PGL(q)$ agit librement sur $R^s$, nous verrons qu'on a un isomorphisme naturel

$$\operatorname{Pic}(U_s(r, d)) \simeq \operatorname{Pic}^G(R^s)$$

(si $L$ est un fibré en droites sur $U_s(r, d)$, on lui associe le $PGL(q)$-fibré en droites $\pi^* L$ sur $R^s$, et réciproquement, si $L'$ est un $PGL(q)$-fibré en droites sur $R^s$, il existe un bon quotient $L'/PGL(q)$, qui a une structure naturelle de fibré en droites algébrique sur $U_s(r, d)$, et c'est ce fibré qui est associé à $L'$).

Ensuite, on montre que le morphisme de restriction

$$\operatorname{Pic}^G(R^{ss}) \to \operatorname{Pic}^G(R^s)$$

est surjectif. Partant d'un fibré en droites $L$ sur $U_s(r, d)$ nous pouvons donc définir un $PGL(q)$-fibré en droites $L'$ sur $R^{ss}$ tel que $\pi_s^* L \simeq L'|_{R^s}$.

Pour continuer nous utiliserons un «lemme de descente» donnant des conditions suffisantes pour qu'il existe un fibré en droites $\bar{L}$ sur $U(r, d)$ tel que $\pi^* \bar{L} \simeq L'$. Ce lemme de descente, dans sa version finale, est dû à Kempf. Nous avions précédemment une version plus compliquée (qui nécessitait plus de vérifications sur $L'$), inspirée de ([1], Prop. 2.3). La condition à vérifier sur $L'$ est la suivante: pour tout point fermé y de $R^{ss}$ tel que l'orbite de y dans $R^{ss}$ soit fermée, le stabilisateur de y dans $PGL(q)$ agit trivialement sur la fibre $L'_y$. Nous verrons que ceci est vrai, d'où l'existence de $\bar{L}$.

Ce fibré en droites $\bar{L}$ obtenu sur $U(r, d)$ est le prolongement voulu de $L$. On obtient aussi en cours de démonstration un ismorphisme

$$\operatorname{Pic}(U(r, d)) \simeq \operatorname{Pic}^G(R^{ss}).$$

En ce qui concerne $U(r, L)$, il suffit de faire la même démonstration, avec $\pi^{-1}(U(r, L))$ à la place de $R^{ss}$.

*Remarque.* On voit facilement que les complétés des anneaux locaux des points singuliers de $U(r, d)$ ne sont pas nécessairement factoriels (cf. [3]).

*0.2. Description de* Pic$(U(r, d))$ *et* Pic$(U(r, L))$

Nous avons deux descriptions de Pic$(U(r, d))$. La première fait intervenir une généralisation aux variétés de modules de fibrés semi-stables de la notion de «diviseur théta» des jacobiens. La seconde utilise un sous-groupe du groupe de Grothendieck $K(X)$ de $X$.

*0.2.1. Diviseurs théta généralisés.* Posons $n = PGCD(r, d)$. Soit $F$ un fibré vectoriel sur $X$ tel que

$$\deg(F) = \frac{-d + r(g-1)}{n}, \qquad rg(F) = \frac{r}{n}.$$

Ces conditions entrainent que $\chi(E \otimes F) = 0$ pour tout fibré vectoriel $E$ sur $X$ de rang $r$ et de degré $d$. D'après [8], on peut trouver un tel $F$ tel qu'il existe un fibré stable $E$ sur $X$, de rang $r$ et de degré $d$, tel que

$$H^0(X, E \otimes F) = H^1(X, E \otimes F) = 0.$$

Dans ces conditions soit $\Theta_F^s$ (resp. $\Theta_{F,L}^s$) l'ensemble des points de $U_s(r, d)$ (resp. $U_s(r, L)$) correspondant aux fibrés stables $E$ tels que

$$H^0(X, E \otimes F) \neq 0.$$

Nous verrons que c'est une hypersurface de $U_s(r, d)$ (resp. $U_s(r, L)$). Soit $\Theta_F$ (resp. $\Theta_{F,L}$) l'adhérence de $\Theta_F^s$ (resp. $\Theta_{F,L}^s$) dans $U(r, d)$ (resp. $U(r, L)$). Les hypersurfaces $\Theta_F$ et $\Theta_{F,L}$ sont appelées *diviseurs théta.*

D'après le Théorème A, le faisceau d'idéaux de $\Theta_F$ (resp. $\Theta_{F,L}$) est un fibré en droites $\mathcal{O}(\Theta_F)$ (resp. $\mathcal{O}(\Theta_{F,L})$). On peut aussi définir ces fibrés par une propriété universelle: par exemple $\mathcal{O}(\Theta_F)$ est entièrement déterminé par la propriété suivante: pour toute variété algébrique $S$ et toute famille $E$ de fibrés vectoriels semi-stables sur $X$, de rang $r$ et de degré $d$ paramétrée par $S$, on peut définir un «sous-schéma de saut» $Z$ de $S$, dont les points fermés sont les points $s$ tels que $H^0(E_s \otimes F) \neq 0$, et dont le faisceau d'idéaux $\mathcal{O}(Z)$ est localement libre. Alors, si $f_E: S \to U(r, d)$ est le morphisme canonique déduit de $E$, on a un isomorphisme

$$f_E^* \mathcal{O}(\Theta_F) \simeq \mathcal{O}(Z).$$

On a une propriété universelle analogue pour définir $\mathcal{O}(\Theta_{F,L})$ (ces faits ne seront pas utilisés dans la suite). On démontrera le

**Théorème B.** (a) *Le fibré en droites* $\mathcal{O}(\Theta_{F,L})$ *est indépendant du choix de* $F$.

(b) *Le groupe* Pic$(U(r, L))$ *est isomorphe à* $\mathbb{Z}$, *et est engendré par* $\mathcal{O}(\Theta_{F,L})$.

L'isomorphisme Pic$(U(r, L)) \simeq \mathbb{Z}$ dans le cas $n = 1$ a été prouvé d'une autre façon par Seshadri (cf. [20]).

Examinons maintenant $\mathrm{Pic}(U(r, d))$. Soit $J^{(d)}$ le jacobien des fibrés en droites de degré $d$ sur $X$. On a un morphisme canonique

$$\det: U(r, d) \to J^{(d)}$$

qui permet de considérer $\mathrm{Pic}(J^{(d)})$ comme un sous-groupe de $\mathrm{Pic}(U(r, d))$. On a alors le

**Théorème C.** *Les inclusions* $\mathrm{Pic}(J^{(d)}) \subset \mathrm{Pic}(U(r, d))$ *et* $\mathbb{Z}.\,\mathcal{O}(\Theta_F) \subset \mathrm{Pic}(U(r, d))$ *induisent un isomorphisme*

$$\mathrm{Pic}(U(r, d)) \simeq \mathrm{Pic}(J^{(d)}) \oplus \mathbb{Z}.$$

Nous verrons dans l'autre description que nous donnons de $\mathrm{Pic}(U(r, d))$ quelle est la dépendance de $\mathcal{O}(\Theta_F)$ en $F$.

*0.2.2. Groupe de Grothendieck de X et* $\mathrm{Pic}(U(r, d))$. Soit $S$ une variété algébrique. Une *famille de fibrés de* $U(r, d)$ *paramétrée par* $S$ est un fibré vectoriel $E$ sur $S \times X$, plat sur $S$, tel que pour tout point fermé $s$ de $S$, $E_s = E|_{\{s\} \times X}$ soit semi-stable de rang $r$ et de degré $d$. On note $p_S$, $p_X$ les projections $S \times X \to S$ et $S \times X \to X$ respectivement. Deux telles familles $E$, $E'$ paramétrées par $S$ sont dites *equivalentes* s'il existe un fibré en droites $L$ sur $S$ et un isomorphisme $E' \simeq E \otimes p_S^* L$.

On note $F(r, d)$ le foncteur

$$\text{Variétés algébriques} \to \text{Ensembles}$$

associant à $S$ l'ensemble des classes d'équivalence de familles de fibrés de $U(r, d)$ paramétrées par $S$. Si $f: S' \to S$ est un morphisme de variétés algébriques, et si $E$ est une famille de fibrés de $F(r, d)$ paramétrée par $S$, l'image par $F(r, d)(f)$ de la classe de $E$ est celle de $f^* E$. On sait qu'on a un morphisme canonique de foncteurs

$$F(r, d) \to \mathrm{Hom}(-, U(r, d)).$$

Autrement dit, à toute famille $E$ de fibrés de $U(r, d)$ pramétrée par $S$ on associe un morphisme

$$f_E: S \to U(r, d)$$

ne dépendant que de la classe d'équivalence de $E$.

Si $Y$ est une variété algébrique, on note $K(Y)$ le groupe de Grothendieck de $Y$, et si $E$ est un faisceau cohérent sur $Y$, $[E]$ désigne la classe de $E$ dans $K(Y)$. Soient $\eta$ la classe dans $K(X)$ d'un fibré vectoriel de rang $r$ et de degré $d$, $H$ le noyau du morphisme

$$K(X) \to \mathbb{Z}$$
$$\alpha \mapsto \chi(\alpha \otimes \eta).$$

En fait, $H$ ne dépend que de $r$ et $d$, c'est pourquoi on notera $H = H(r, d)$. La classe du fibré $F$ considéré dans 0.2.1 est par exemple un élément de $H(r, d)$. On peut voir $\mathrm{Pic}^0(X)$ comme un sous-groupe de $H(r, d)$ (cf. 3.3), et on a

$$H(r, d) \simeq \mathrm{Pic}^0(X) \oplus \mathbb{Z} \cdot [F].$$

Soit $E$ une famille de fibrés de $U(r, d)$ paramétrée par une variété lisse $S$, et $\alpha$ un élément de $H(r, d)$. On définit successivement les éléments

$$\begin{array}{rl} E \otimes p_X^* \alpha & \text{de } K(S \times X), \\ p_{S!}(E \otimes p_X^* \alpha) & \text{de } K(S), \\ \gamma_E(\alpha) = \det(p_{S!}(E \otimes p_X^* \alpha)) & \text{de } \operatorname{Pic}(S) \end{array}$$

(en fait ceci est possible même si $S$ est seulement intègre). Il est aisé de voir que puisque $\alpha$ est dans $H(r, d)$, $\gamma_E(\alpha)$ ne dépend que de la classe d'équivalence de $E$. Les $\gamma_E(\alpha)$ dépendent fonctoriellement de $E$ et nous verrons qu'on définit ainsi un élément de $\operatorname{Pic}(F(r, d))$, le *groupe de Picard du foncteur* $F(r, d)$ (cf. § 3).

Rappelons que $\operatorname{Pic}^0(X)$ peut être vu naturellement comme un sous-groupe de $\operatorname{Pic}(J^{(d)})$ ou $H(r, d)$. On prouvera le

**Théorème D.** (a) *Soit $\alpha$ un élément de $H(r, d)$. Il existe un unique $\mathbb{L}(\alpha)$ dans $\operatorname{Pic}(U(r, d))$ tel que pour toute variété lisse $S$ et toute famille $E$ de fibrés de $U(r, d)$ paramétrée par $S$, on ait*

$$\gamma_E(\alpha) \simeq f_E^* \mathbb{L}(\alpha).$$

(b) *Le morphisme de groupes $\mathbb{L}: H(r, d) \to \operatorname{Pic}(U(r, d))$ est injectif.*

(c) *On a $\operatorname{Pic}(U(r, d)) = \operatorname{Im}(\mathbb{L}) + \operatorname{Pic}(J^{(d)})$, le diagramme suivant est commutatif*

$$\begin{array}{ccc} \operatorname{Pic}^0(X) \simeq Z & \longrightarrow & H(r, d) \\ \downarrow & & \swarrow \\ \operatorname{Pic}(J^{(d)}) & & \\ \downarrow \det^* & \swarrow \mathbb{L} & \\ \operatorname{Pic}(U(r, d)) & & \end{array}$$

*et on a*

$$\operatorname{Im}(\mathbb{L}) \cap \operatorname{Pic}(J^{(d)}) = \mathbb{L}(Z) = \det^*(\operatorname{Pic}^0(X)).$$

On a une inclusion naturelle $i: \operatorname{Pic}(U(r, d)) \to \operatorname{Pic}(F(r, d))$, et on a défini un morphisme de groupes $\gamma: H(r, d) \to \operatorname{Pic}(F(r, d))$. La partie (a) du théorème D dit simplement que l'image de $\gamma$ est contenue dans celle de $i$. Nous montrerons qu'on a en fait $\operatorname{Pic}(U(r, d)) = \operatorname{Pic}(F(r, d))$.

Pour faire le lien entre les deux descriptions, notons qu'on a

$$\mathcal{O}(\Theta_F) = \mathbb{L}(-[F]).$$

On déduit de (c) que si $F'$ est un autre fibré sur $X$ ayant les mêmes propriétés que $F$, on a

$$\mathcal{O}(\Theta_{F'}) = \mathcal{O}(\Theta_F) \otimes \det^*(\det(F') \otimes \det(F)^{-1})$$

($\det(F') \otimes \det(F)^{-1}$ étant vu comme un élément de $\operatorname{Pic}^0(X) \subset \operatorname{Pic}(J^{(d)})$).

Donnons maintenant une idée des démonstrations des théorèmes B, C, D. On utilise une construction due à Seshadri. On peut toujours supposer $d$ aussi grand qu'on le veut, et dans ce cas tout fibré semi-stable $E$ de rang $r$ et de degré $d$ peut s'écrire comme extension

$$0 \to \mathcal{O}_X \otimes \mathbb{C}^{r-1} \to E \to L_0 \to 0,$$

$L_0$ étant un fibré en droites isomorphe à $\det(E)$. La famille de toutes ces extensions est un fibré en espaces projectifs $\mathbb{P}$ sur $J^{(d)}$ (cf. §7). Soit $\mathbb{P}^s$ l'ouvert des extensions dont le terme du milieu est stable. On peut déduire $\mathrm{Pic}(U_s(r, d))$ de $\mathrm{Pic}(\mathbb{P}^s)$, qui on le verra est isomorphe à $\mathrm{Pic}(\mathbb{P})$, par une étude précise du morphisme canonique $\mathbb{P}^s \to U_s(r, d)$.

### *0.3. Faisceau dualisant et faisceau canonique*

On peut calculer le faisceau dualisant de $U(r, d)$ ou $U(r, L)$. Soit $T_U$ le faisceau tangent de $U(r, d)$. Puisqu'il est localement libre en dehors d'un fermé de codimension au moins 2, on peut définir son déterminant, qui est un fibré en droites $\Delta$ sur $U(r, d)$. On note $\omega$ le dual de $\Delta$. On définit de même le fibré en droites $\omega_L$ sur $U(r, L)$.

**Théorème E.** (a) *Soit $F_0$ un fibré vectoriel sur $X$ de rang $2r$ et de degré $2(-d+r(g-1))$. Alors on a*

$$\omega \simeq \mathbb{L}([F_0]) \otimes \det^* \Lambda,$$

*où $\Lambda$ est le fibré en droites sur $J^{(d)}$*

$$\Lambda = [\det(p_{J!}[L]) \otimes \det(p_{J!}[L^*])]^{r-1} \otimes \det(p_{J!}([L \otimes p_X^* F_0]))^{-1},$$

*$L$ désignant un fibré de Poincaré sur $J^{(d)} \times X$, et $p_J$, $p_X$ les projections $J^{(d)} \times X \to J^{(d)}$ et $J^{(d)} \times X \to X$.*

(b) *Le faisceau dualisant de $U(r, d)$ est isomorphe à $\omega$.*

Rappelons que $n = PGCD(r, d)$. On a alors le

**Théorème F.** (a) *On a $\omega_L \simeq \mathcal{O}(-2n\,\Theta_{F,L})$.*

(b) *Le faisceau dualisant de $U(r, L)$ est isomorphe à $\omega_L$.*

### *0.4. Résultats annexes*

Soit $U$ un ouvert non vide de $U_s(r, d)$. On appelle *fibré de Poincaré sur $U$* un fibré vectoriel $E$ sur $U \times X$ tel que pour tout point fermé $u$ de $U$, $E_u$ soit stable de rang $r$ et de degré $d$, et que $u$ soit le point de $U_s(r, d)$ correspondant à $E_u$. Cela revient à dire que le morphisme canonique $f_E$: $U \to U(r, d)$ est l'inclusion. Nous donnons une autre démonstration d'un résultat de Ramanan [20]: si $n > 1$, il n'existe pas de fibré de Poincaré sur $U$ (théorème 5.5).

Ce résultat découle de l'étude des $GL(q)$-fibrés en droites sur $R^{ss}$. Si $L$ en est un, il existe un entier $k$ tel qu'en tout point $y$ de $R^{ss}$. l'action d'un élément $t$ de $\mathbb{C}^* \subset GL(q)$ soit la multiplication par $t^k$. Nous verrons que $k$ est toujours multiple de $n$ (proposition 5.1), et que l'existence d'un fibré de Poincaré sur $U$ entrainerait celle d'un $L$ pour lequel on aurait $k=1$.

Notre méthode s'applique aussi aux variétés de modules de faisceaux semi-stables sur $\mathbb{P}_2(\mathbb{C})$ (et sans doute à d'autres cas aussi)! Plus précisément, soient $r \geqq 2$, $c_1, c_2$ des entiers, $M(r, c_1, c_2)$ la variété de modules des faisceaux semi-stables sur $\mathbb{P}_2(\mathbb{C})$ de rang $r$ et de classes de Chern $c_1, c_2$, $M_s(r, c_1, c_2)$ l'ouvert de $M(r, c_1, c_2)$ correspondant aux faisceaux stables. On suppose que $\dim(M(r, c_1, c_2)) > 0$. Posons

$$\chi = r - c_2 + \frac{c_1(c_1+3)}{2}$$

(c'est la caractéristique d'Euler-Poincaré des faisceaux de rang $r$ et de classes de Chern $c_1, c_2$ sur $\mathbb{P}_2(\mathbb{C})$). Si $r, c_1$ et $\chi$ sont premiers entre eux, il existe un faisceau de Poincaré sur $M_s(r, c_1, c_2)$ (Maruyama [13], theorem 6.11). Dans les autres cas nous obtenons

**Théorème G.** *Si $r, c_1$ et $\chi$ ne sont pas premiers entre eux, et si $U$ est un ouvert non vide de $M_s(r, c_1, c_2)$, il n'existe pas de faisceau de Poincaré sur $U$.*

*0.5. Le cas $g=2$, $r=2$ et $d$ pair*

Dans ce cas il existe une description explicite de $U(r, d)$ et $U(r, L)$ permettant de prouver les théorèmes A à F (cf. [18]).

Le premier auteur remercie le National Board for Higher Mathematics pour l'avoir invité à passer quelques semaines en Inde, durant lesquelles ce travail a été commencé. Le second auteur remercie le *DFG* et l'université de Kaiserslautern pour leur hospitalité pendant la réalisation de cet article.

*Définitions et notations*

Si $X_1, \ldots, X_p$ sont des ensembles, on notera $p_{X_i}$ la projection

$$X_1 \times \ldots \times X_p \rightarrow X_i.$$

Si $f: S' \rightarrow S$ est un morphisme de variétés algébriques, et $F$ un faisceau cohérent sur $S \times X$, on posera

$$f^{\#} F = (f \times I_X)^* F.$$

Soit $G$ un groupe algébrique et $Y$ une variété algébrique sur laquelle $G$ opère algébriquement. Un $G$-fibré sur $Y$ est un fibré vectoriel algébrique sur $Y$ sur lequel $G$ opère linéairement et algébriquement, au dessus de l'action de $G$ sur $Y$.

Si $E=Y\times\mathbb{C}^r$ est un fibré trivial, l'action de $G$ sur $E$ triviale sur les fibres est par définition l'action produit sur $Y\times\mathbb{C}^r$, $\mathbb{C}^r$ étant muni de l'action triviale de $G$.

Soit $E$ un faisceau cohérent sur une variété algébrique projective. S'il n'y a pas d'ambigüité, on posera

$$H^i(E)=H^i(Y,E),$$

et

$$h^i(E)=\dim_{\mathbb{C}}(H^i(Y,E)) \quad \text{pour tout entier } i.$$

Soit $m$ un entier tel que $m\geqq 1$. On notera $m\cdot E$ le faisceau $E\oplus\ldots\oplus E$, somme directe de $m$ copies de $E$.

Soit $Y$ une variété algébrique et $H$ un sous-groupe de $\operatorname{Pic}(Y)$. Soit $L$ un fibré en droites sur $Y$. Nous dirons pour simplifier que $\operatorname{Pic}(Y)$ est isomorphe à $H\oplus\mathbb{Z}\cdot L$ si le morphisme

$$i\colon \mathbb{Z}\to\operatorname{Pic}(Y)$$
$$m\mapsto L^m$$

est injectif, et si on a $\operatorname{Pic}(Y)=H\oplus\operatorname{Im}(i)$.

## 1. Préliminaires

### *1.1. Variétés de modules de fibrés semi-stables*

*1.1.1. Fibrés stables et fibrés semi-stables.* Soit $E$ un fibré vectoriel non nul sur $X$. On pose

$$\mu(E)=\frac{\deg(E)}{rg(E)},$$

qu'on appelle la *pente* de $E$. On dit que $E$ est *semi-stable* (resp. *stable*) si pour tout sous-fibré propre $F$ de $E$ on a

$$\mu(F)\leqq\mu(E) \quad (\text{resp.} <).$$

*1.1.2. Variétés de modules.* Soient $r, d$ des entiers, avec $r\geqq 1$. On a défini dans l'Introduction les familles de fibrés semi-stables de rang $r$ et de degré $d$ paramétrées par une variété algébrique $S$, ainsi que la notion d'équivalence de telles familles, et le foncteur $F(r,d)$. La variété de modules $U(r,d)$ est définie (à isomorphisme près) par les deux propriétés suivantes:

(i) Il existe un morphisme de foncteurs:

$$\Psi\colon F(r,d)\to\operatorname{Hom}(-,U(r,d)),$$

donc à toute famille $E$ de fibrés semi-stables sur $X$ de rang $r$ et de degré $d$ paramétrée par $S$ on associe un morphisme

$$f_E\colon S\to U(r,d),$$

et $f_E$ ne dépend que de la classe d'équivalence de $E$.

(ii) Si $M$ est une variété algébrique, et

$$\Psi' \colon F(r, d) \to \mathrm{Hom}(-, M)$$

un morphisme de foncteurs, il existe un unique morphisme

$$f \colon U(r, d) \to M$$

tel que le diagramme suivant soit commutatif:

$$\begin{array}{ccc} F(r,d) & \xrightarrow{\Psi} & \mathrm{Hom}(-, U(r,d)) \\ & \searrow^{\Psi'} & \downarrow{\scriptstyle \mathrm{Hom}(-,f)} \\ & & \mathrm{Hom}(-, M). \end{array}$$

La variété $U(r, d)$ est projective, normale et irréductible.

*1.1.3. Filtration de Jordan-Hölder.* Soit $E$ un fibré semi-stable sur $X$. Il existe une filtration de $E$ par des sous-fibrés vectoriels

$$0 = E_0 \subset E_1 \subset \ldots \subset E_p = E$$

telle que pour $1 \leqq i \leqq p$, $E_i/E_{i-1}$ soit stable et de même pente que $E$. Une telle filtration n'est pas en général unique, mais la classe d'isomorphisme du gradué $\bigoplus_{1 \leqq i \leqq p} E_i/E_{i-1}$ l'est. La filtration précédente est une *filtration de Jordan-Hölder* de $E$, et on note

$$Gr(E) = \bigoplus_{1 \leqq i \leqq p} E_i/E_{i-1}.$$

Deux fibrés semi-stables $E$, $E'$ sont dits *équivalents* si $Gr(E) \simeq Gr(E')$. Les points fermés de $U(r, d)$ s'identifient naturellement aux classe d'équivalence de fibrés semi-stables sur $X$ de rang $r$ et de degré $d$. En particulier, les classes d'isomorphisme de fibrés stables de rang $r$ et de degré $d$ sur $X$ constituent un ouvert $U_s(r, d)$ de $U(r, d)$. Cet ouvert est non vide et lisse. Si $r$ et $d$ sont premiers entre eux, les notions de stabilité et de semi-stabilité sont équivalentes, donc $U(r, d) = U_s(r, d)$ est une variété lisse. Si $r$ et $d$ ne sont pas premiers entre eux, l'ouvert des points lisses de $U(r, d)$ est réduit à $U_s(r, d)$ sauf dans l'exception mentionnée dans l'Introduction.

*1.1.4. Construction de $U(r, d)$.* Soit $\mathcal{O}_X(1)$ un fibré en droites très ample sur $X$. Il existe un entier $m_0$ tel que pour tout entier $m \geqq m_0$ et tout fibré semi-stable $E$ de rang $r$ et de degré $d$ sur $X$, $E(m) = E \otimes \mathcal{O}_X(m)$ soit engendré par ses sections, et $h^1(E(m)) = 0$. On pose, si $m \geqq m_0$,

$$q = h^0(E(m)) = d + r \cdot \deg(\mathcal{O}_X(1)) \cdot m + r \cdot (1 - g),$$

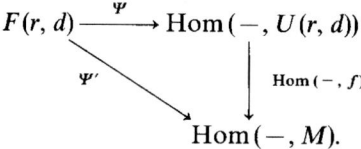

et soit $P$ le polynôme

$$d+r(1-g)+r\cdot\deg(\mathcal{O}_X(1))\cdot T.$$

Soit

$$R=\mathrm{Quot}_P(\mathcal{O}_X(-m)\otimes\mathbb{C}^q).$$

Rappelons que c'est la variété projective représentant le foncteur

$$\Psi_0: \text{Variétés algébriques} \to \text{Ensembles}$$

associant à $S$ l'ensembles des classes d'isomorphisme de morphismes surjectifs de faisceaux sur $S\times X$

$$p_X^*(\mathcal{O}_X(-m)\otimes\mathbb{C}^q)\to E,$$

où $E$ est plat sur $S$, et où pour tout point fermé $s$ de $S$, $E_s$ a pour polynôme de Hilbert $P$ relativement à $\mathcal{O}_X(1)$ (deux tels morphismes $f$: $p_X^*(\mathcal{O}_X(-m)\otimes\mathbb{C}^q)\to E$ et $f'$: $p_X^*(\mathcal{O}_X(-m)\otimes\mathbb{C}^q)\to E'$ sont isomorphes s'il existe un isomorphisme $\varphi$: $F\to F'$ tel qu'on ait $f'=\varphi\circ f$). Il existe un morphisme surjectif «universel»

$$\theta: p_X^*(\mathcal{O}_X(-m)\otimes\mathbb{C}^q)\to\mathbb{F}_o$$

sur $R\times X$ (cf. [6]). On note $R^{ss}$ (resp. $R^s$) l'ouvert des points $y$ de $R$ tels que

$$\theta_y: \mathcal{O}_X(-m)\otimes\mathbb{C}^q\to\mathbb{F}_{0y}$$

induise un isomorphisme

$$\mathbb{C}^q\simeq H^0(\mathbb{F}_{0y}(m)),$$

et que $\mathbb{F}_{0y}$ soit semi-stable (resp. stable). C'est un ouvert lisse de $R$. De la restriction de $\mathbb{F}_0$ à $R^{ss}\times X$ on déduit un morphisme

$$\pi=f_{\mathbb{F}_0}: R^{ss}\to U(r,d).$$

Le groupe $PGL(q)$ agit de façon naturelle sur $R^{ss}$, et on peut montrer que si $m$ est assez grand, $\pi$ est un bon quotient de $R^{ss}$ par $PGL(q)$, tandis que la restriction de $\pi$: $R^s\to U_s(r,d)$ est un quotient géométrique (cf. par exemple [23]). On supposera toujours dans la suite que $m$ est assez grand pour que les propriétés précédentes soient vérifiées.

De l'action de $PGL(q)$ sur $R^{ss}$ on déduit une action de $GL(q)$. Si $y$ est un point fermé de $R^{ss}$, le stabilisateur de $y$ dans $GL(q)$ s'identifie naturellement au groupe des automorphismes de $\mathbb{F}_{oy}$.

## 2. Descente de fibrés vectoriels

Soit $Y$ une variété algébrique intègre sur laquelle opère algébriquement un groupe algébrique réductif $G$. On suppose qu'il existe un bon quotient

$$\pi: Y\to M$$

(cf. [15, 19]).

**Lemme 2.1.** *Soit y un point fermé de Y. Il existe une unique orbite fermée de Y contenue dans* $\overline{Gy}$. *Cette orbite fermée est aussi la seule qui soit contenue dans* $\pi^{-1}(\pi(y))$.

*Existence.* Il suffit de prendre une orbite de dimension minimale dans $\overline{Gy}$.
Unicité: Soient $\Gamma_1$, $\Gamma_2$ deux orbites fermées de $Y$ contenues dans $\pi^{-1}(\pi(y))$. Alors on a $\pi(\Gamma_1)=\pi(\Gamma_2)=\{\pi(y)\}$, donc puisque $\pi$ est un bon quotient, et $\Gamma_1$, $\Gamma_2$ fermées, $\Gamma_1 \cap \Gamma_2$ est non vide. Donc $\Gamma_1=\Gamma_2$. Ceci démontre le lemme 2.1.

Pour tout fibré vectoriel algébrique $E$ sur $M$, le fibré $\pi^* E$ est muni d'une structure naturelle de $G$-fibré vectoriel sur $Y$. Si $F$ est un $G$-fibré vectoriel sur $Y$, on dit que $F$ *descend à* $M$ s'il existe un fibré vectoriel $E$ sur $M$ tel que les $G$-fibrés $F$ et $\pi^* E$ soient isomorphes.

**Lemme 2.2.** *Soit F un G-fibré vectoriel de rang r sur Y. Alors F descend à M si et seulement si pour tout point fermé m de M il existe un voisinage U de m et un G-isomorphisme*

$$F|_{\pi^{-1}(U)} \simeq r \cdot \mathcal{O}_{\pi^{-1}(U)},$$

*l'action de G sur* $r \cdot \mathcal{O}_{\pi^{-1}(U)}$ *étant triviale sur les fibres.*

Nécessité: On prend pour $U$ un voisinage de $m$ tel que $E|_U$ soit trivial.
Suffisance: Supposons la condition du lemme réalisée. Il existe alors un recouvrement ouvert $(U_i)_{i\in I}$ de $M$ tel que pour tout $i$ dans $I$ il existe un $G$-isomorphisme

$$f_i: F|_{\pi^{-1}(U_i)} \xrightarrow{\simeq} r \cdot \mathcal{O}_{\pi^{-1}(U_i)}.$$

Posons

$$g_{ij}=f_j \circ f_i^{-1}: r \cdot \mathcal{O}_{\pi^{-1}(U_i \cap U_j)} \xrightarrow{\simeq} r \cdot \mathcal{O}_{\pi^{-1}(U_i \cap U_j)}.$$

Les $g_{ij}$ sont des isomorphismes $G$-invariants et définissent donc des isomorphismes

$$\bar{g}_{ij}: r \cdot \mathcal{O}_{U_i \cap U_j} \to r \cdot \mathcal{O}_{U_i \cap U_j}.$$

Les $\bar{g}_{ij}$ constituent une famille de cocycles définissant un fibré vectoriel $E$ sur $M$ tel que $\pi^* E$ soit isomorphe à $F$. Ceci démontre le lemme 2.2.

**Théorème 2.3** («Lemme de descente»). *Soit F un G-fibré vectoriel sur Y. Alors E descend à M si et seulement si pour tout point fermé y de Y tel que Gy soit fermée, le stabilisateur de y dans G agit trivialement sur* $F_y$.

Ce résultat est dû à Kempf. Nous en avions précédemment une version plus compliquée, limitée au cas où $F$ est de rang 1, avec plus de conditions à vérifier sur le fibré $F$. Cette version s'inspirait de [1], prop. (2.3).

Si $F$ descend à $M$, il est immédiat que le stabilisateur de tout point de $Y$ agit trivialement sur la fibre de $F$ en ce point.

Réciproquement, supposons que $F$ possède la propriété du théorème 2.3. Soit $y$ un point fermé de $Y$. D'après le lemme 2.2, il faut montrer qu'il existe un ouvert $U$ de $M$ contenant $\pi(y)$ et un isomorphisme

$$s: r\cdot\mathcal{O}_{\pi^{-1}(U)} \xrightarrow{\simeq} F|_{\pi^{-1}(U)}$$

$G$-invariant, c'est à dire tel que pour tout point $y'$ de $\pi^{-1}(U)$ et tous $v$ dans $r\cdot\mathcal{O}_{\pi^{-1}(U),y'}$, $g$ dans $G$, on ait $s(gv)=gs(v)$. Il revient au même de trouver $r$ sections $G$-invariantes

$$s_i: \mathcal{O}_{\pi^{-1}(U)} \to F|_{\pi^{-1}(U)}$$

qui engendrent $F|_{\pi^{-1}(U)}$. Soit $C=Gz$ l'unique orbite fermée de $Y$ contenue dans $\pi^{-1}(\pi(y))$. Montrons qu'il existe $r$ sections $G$-invariantes

$$\sigma_i: \mathcal{O}_C \to F|_C$$

engendrant $F|_C$: soient $u_1, \ldots, u_r$ une base de $F_z$, et $G_z$ le stabilisateur de $z$ dans $G$. On a un diagramme commutatif

$$\begin{array}{cccc}
g & G & \xrightarrow{g\mapsto gu_i} & F|_C \\
\Big\downarrow & \Big\downarrow & & \Big\downarrow \text{projection} \\
gz & C & = & C
\end{array}$$

Puisque $G_z$ agit trivialement sur $F_z$, on en déduit un diagramme commutatif

$$\begin{array}{ccc}
G/G_z & \xrightarrow{\psi} & F|_C \\
\varphi\Big\downarrow & & \Big\downarrow \\
C & = & C
\end{array}$$

Le morphisme $\varphi$ est un isomorphisme. On en déduit

$$\psi\circ\varphi^{-1}: C \to F|_C$$

qui définit la section $G$-invariante $\sigma_i$ de $F|_C$. Il est immédiat que ces sections engendrent $F|_C$.

Soit $V$ un ouvert affine de $M$ contenant $\pi(y)$. Puisque $\pi$ est un bon quotient, $\pi^{-1}(V)$ est un ouvert affine de $Y$, qui contient $C$. On considère l'action de $G$ sur $H^0(\pi^{-1}(V), F)$:

$$(g, s)\mapsto gs,$$

où $gs(y')=gs(g^{-1}y')$, pour tout $y'$ dans $\pi^{-1}(V)$. Les éléments $G$-invariants de $H^0(\pi^{-1}(V), F)$ sont précisément les sections $G$-invariantes de $F|_{\pi^{-1}(V)}$.

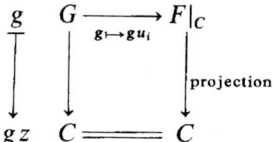

Montrons qu'il existe un opérateur de Reynolds

$$R\colon H^0(\pi^{-1}(V), F) \to H^0(\pi^{-1}(V), F)^G$$

bien que $H^0(\pi^{-1}(V), F)$ ne soit pas de dimension finie. Pour cela il suffit de vérifier que dans $H^0(\pi^{-1}(V), F)$, chaque $G$-orbite est contenue dans un sous-espace vectoriel de dimension finie. Pour cela, on considère un ouvert dense affine $V_0$ de $\pi^{-1}(V)$ sur lequel $F$ est trivial. Soit $s$ un élément de $H^0(\pi^{-1}(V), F)$. Le morphisme

$$G \times V_0 \to F|_{V_0} \simeq r \cdot \mathcal{O}_{V_0}$$
$$(g, v_0) \to g s(g^{-1} v_0)$$

peut être vu comme une fonction régulière

$$f\colon G \times V_0 \to \mathbb{C}^r.$$

Puisque $\mathbb{C}[G \times V_0] \simeq \mathbb{C}[G] \otimes \mathbb{C}[V_0]$, on peut mettre $f$ sous la forme

$$f = \sum_{1 \leqq i \leqq p} \varphi_i \otimes \psi_i,$$

où pour $1 \leqq i \leqq p$, $\varphi_i\colon V \to \mathbb{C}^r$, $\psi_i\colon G \to \mathbb{C}$ sont des morphismes. Alors $Gs|_{V_0}$ est contenu dans le sous-espace vectoriel de $H^0(V_0, F)$ engendré par $\psi_1, \ldots, \psi_p$, et $Gs$ est contenu dans l'intersection de ce sous-espace avec $H^0(\pi^{-1}(V), F)$. On a donc prouvé l'existence de $R$.

On a de même un opérateur de Reynolds

$$R'\colon H^0(C, F) \to H^0(C, F)^G.$$

Soit $\alpha\colon r \cdot \mathcal{O}_C \to F|_C$ un isomorphisme $G$-invariant, dont l'existence a été prouvée précédemment. Soient

$$\alpha_i\colon \mathcal{O}_C \to F|_C, \qquad 1 \leqq i \leqq r,$$

les restrictions de $\alpha$ aux facteurs directs de $r \cdot \mathcal{O}_C$. Puisque $\pi^{-1}(C)$ est une sous-variété fermée de la variété affine $\pi^{-1}(V)$, $\alpha_i$ admet un prolongement

$$\bar{\alpha}_i\colon \mathcal{O}_{\pi^{-1}(V)} \to F|_{\pi^{-1}(V)},$$

qui est une section non nécessairement $G$-invariante. Mais $R(\bar{\alpha}_i)$ l'est, et

$$R(\bar{\alpha}_i)|_C = \alpha_i$$

par fonctorialité de l'opérateur de Reynolds. On pose

$$\bar{\alpha} = (R(\bar{\alpha}_1), \ldots, R(\bar{\alpha}_p))\colon r \cdot \mathcal{O}_{\pi^{-1}(V)} \to F|_{\pi^{-1}(V)},$$

qui est un morphisme $G$-invariant qui prolonge $\alpha$.

Il reste à prouver que l'ouvert $W$ de $\pi^{-1}(V)$ des points au dessus desquels $\bar{\alpha}$ est un isomorphisme contient un ouvert de la forme $\pi^{-1}(U)$, $U$ étant un ouvert de $M$ contenant $\pi(y)$. Les fermés $\pi^{-1}(V) \setminus W$ et $C$ de $\pi^{-1}(V)$ sont $G$-

invariants et disjoints, donc puisque la restriction de $\pi: \pi^{-1}(V) \to V$ est un bon quotient, $\pi(\pi^{-1}(V) \setminus W)$ et $\pi(C) = \{\pi(y)\}$ sont des fermés disjoints de $V$. Il suffit de prendre

$$U = V \setminus \pi(\pi^{-1}(V) \setminus W).$$

Ceci achève la démonstration du théorème 2.3.

*Remarque.* Le fibré vectoriel $E$ sur $M$ tel que les $G$-fibrés vectoriels $\pi^* E$ et $F$ soient isomorphes est unique à isomorphisme près, car il découle aisément du lemme 2.2 que la projection

$$\pi^* E \to E$$

est un bon quotient par $PGL(q)$. On a donc $E = F/PGL(q)$.

## 3. Groupe de Picard de $F(r, d)$ et $PGL(q)$-fibrés en droites sur $R^{ss}$

### 3.1. *Définitions*

**Définition 1.** Un *fibré en droites* $L$ sur $F(r, d)$ est défini par la donnée de

(i) Pour toute famille $F$ de fibrés de $U(r, d)$ paramétrée par une variété lisse $S$, d'un fibré en droites $L_F$ sur $S$, ne dépendant que de la classe d'équivalence de $F$.

(ii) Pour tout morphisme $f: S' \to S$ de variétés lisses, d'un isomorphisme

$$\alpha_F^L(f): L_{f^*F} \xrightarrow{\simeq} f^* L_F$$

ne dépendant que de la classe d'équivalence de $F$, tel que si $g: S'' \to S'$ est un autre morphisme de variétés lisses, on ait

$$\alpha_F^L(f \circ g) = \alpha_{f^*F}^L(g) \circ g^*(\alpha_F^L(f)),$$

(en particulier $\alpha_F^L(I_S) = I_{L_F}$).

**Définition 2.** Soient $L, L'$ des fibrés en droites sur $F(r, d)$. Un *isomorphisme* $L \simeq L'$ est la donnée, pour toute famille $F$ de fibrés de $U(r, d)$ paramétrée par une variété lisse $S$, d'un isomorphisme

$$\sigma_F: L_F \xrightarrow{\simeq} L'_F$$

ne dépendant que de la classe d'équivalence de $F$, tel que si $f: S' \to S$ est un morphisme de variétés lisses, on ait

$$\sigma_{f^*F} = \alpha_F^{L'}(f) \circ f^* \sigma_F \circ \alpha_F^L(f)^{-1}.$$

**Définition 3.** Les classes d'isomorphisme de fibrés en droites sur $F(r, d)$ constituent de façon évidente un groupe commutatif, appelé *groupe de Picard* de $F(r, d)$, et noté $\mathrm{Pic}(F(r, d))$.

Soit $L_0$ un fibré en droites sur $U(r, d)$. On en déduit un fibré en droites $L$ sur $F(r, d)$ défini par

$$L_F = f_F^* L_0,$$

pour toute famille $F$ de fibrés de $U(r, d)$ paramétrée par une variété lisse, les $\alpha_F^L(f)$ étant définis de manière évidente. On obtient ainsi un morphisme de groupes

$$i\colon \operatorname{Pic}(U(r, d)) \to \operatorname{Pic}(F(r, d)).$$

**Definition 4.** On dit qu'un élément de $\operatorname{Pic}(F(r, d))$ *provient* de $\operatorname{Pic}(U(r, d))$ s'il est dans l'image de $i$.

*3.2. Groupe de Picard de $F(r, d)$ et $PGL(q)$-fibrés en droites sur $R^{ss}$ et $R^s$*

Soit $\eta\colon R^{ss} \times PGL(q) \to R^{ss} \times PGL(q)$

$$(y, g) \mapsto (g y, g).$$

**Lemme 3.1.** *Les familles $\eta^\# p_{R^{ss}}^\# \mathbb{F}_0$ et $p_{R^{ss}}^\# \mathbb{F}_0$ de fibrés de $U(r, d)$ paramétrées par $R^{ss} \times PGL(q)$ sont équivalentes.*

Il est clair que pour tout point $y$ de $R^{ss}$ et tout $g$ dans $PGL(q)$, on a un isomorphisme

$$\eta^\# p_{R^{ss}}^\# \mathbb{F}_0|_{\{(y,g)\}\times X} \simeq p_{R^{ss}}^\# \mathbb{F}_0|_{\{(y,g)\}\times X}.$$

Il en découle, par simplicité des fibrés stables, que

$$p_{(R^{ss}\times PGL(q))*}(\underline{\mathrm{Hom}}(\eta^\# p_{R^{ss}}^\# \mathbb{F}_0, p_{R^{ss}}^\# \mathbb{F}_0))|_{R^s \times PGL(q)}$$

est un fibré en droites $L$ sur $R^s \times PGL(q)$. On a donc un isomorphisme

$$p_{R^{ss}}^\# \mathbb{F}_0|_{R^s\times PGL(q)\times X} \simeq p_{R^s}^* L \otimes \eta^\# p_{R^{ss}}^\# \mathbb{F}_0|_{R^s\times PGL(q)\times X}.$$

Puisque $R^{ss}$ est lisse, $L$ se prolonge en un fibré en droites sur $R^{ss} \times PGL(q)$, aussi noté $L$. Puisque $\operatorname{codim}_{R^{ss}}(R^{ss} \setminus R^s) \geqq 2$, l'isomorphisme précédent se prolonge à $R^{ss} \times PGL(q) \times X$. Ceci démontre le lemme 3.1.

Soit $L$ un fibré en droites sur $F(r, d)$. On déduit du lemme 3.1 un isomorphisme

$$\eta^* p_{R^{ss}}^* L_{\mathbb{F}_0} \simeq p_{R^{ss}}^* L_{\mathbb{F}_0}.$$

Cet isomorphisme définit une structure de $PGL(q)$-fibré en droites sur $L_{\mathbb{F}_0}$ (cela découle de la définition 1(ii)).

On note $\operatorname{Pic}^G(R^{ss})$ le groupe des classes d'isomorphisme de $PGL(q)$-fibrés en droites sur $R^{ss}$. On a donc obtenu un morphisme de groupes

$$\operatorname{Pic}(F(r, d)) \to \operatorname{Pic}^G(R^{ss}).$$

**Lemme 3.2.** *Le morphisme d'oubli* $\operatorname{Pic}^G(R^{ss}) \to \operatorname{Pic}(R^{ss})$ *est injectif.*

Soit $L$ un $PGL(q)$-fibré en droites, trivial comme fibré en droites. Il faut montrer qu'il est aussi trivial comme $PGL(q)$-fibré. Fixons un isomorphisme $L \simeq \mathcal{O}_{R^{ss}}$. Montrons que l'action de $PGL(q)$ sur $L$ est triviale. Une action de $PGL(q)$ sur $R^{ss} \times \mathbb{C}$ provient d'un «morphisme croisé», c'est à dire d'un morphisme

$$\chi: PGL(q) \times R^{ss} \to \mathbb{C}^*$$

tel que

$$(*) \qquad \chi(g g', y) = \chi(g, g' y) \cdot \chi(g', y)$$

pour tous $g, g'$ dans $G$ et $y$ dans $R^{ss}$ (on a alors $g \cdot (y, t) = (g y, \chi(g, y) t)$). Les seules fonctions régulières inversibles sur $PGL(q)$ sont les constantes, donc $\chi(g, y)$ est une fonction de $y$ seulement:

$$\chi(g, y) = \chi_0(y).$$

La relation $(*)$ s'écrit alors

$$\chi_0(y) = \chi_0(g' y) \cdot \chi_0(y),$$

d'où $\chi_0 = 1$, et $\chi = 1$. L'action de $PGL(q)$ sur $L$ est donc bien triviale. Ceci démontre le lemme 3.2.

**Proposition 3.3.** *Le morphisme de groupes*

$$\mathrm{Pic}(F(r, d)) \to \mathrm{Pic}(R^{ss})$$
$$L \mapsto L_{\mathbb{F}_0}$$

*est injectif.*

Soit $L$ un fibré en droites sur $F(r, d)$ tel que $L_{\mathbb{F}_0} \simeq \mathcal{O}_{R^{ss}}$. Soit $S$ une variété lisse, et $F$ une famille de fibrés de $U(r, d)$ paramétrée par $S$. Montrons qu'on a un isomorphisme canonique

$$u_F: L_F \xrightarrow{\simeq} \mathcal{O}_S.$$

Le faisceau cohérent $W = p_{S*}(F \otimes p_X^* \mathcal{O}_X(m))$ est localement libre de rang $q$. Soit $z$ un point de $S$ et $U_z$ un voisinage de $z$ tel qu'on ait une trivialisation

$$\beta_z: W|_{U_z} \xrightarrow{\simeq} \mathcal{O}_{U_z} \otimes \mathbb{C}^q.$$

On a un morphisme canonique surjectif

$$p_S^* W \otimes p_X^* \mathcal{O}_X(-m) \to F,$$

et en utilisant la trivialisation $\beta_z$, on obtient un morphisme surjectif

$$p_X^*(\mathcal{O}_X(-m) \otimes \mathbb{C}^q) \to F|_{U_z}.$$

On en déduit un morphisme

$$f_z: U_z \to R^{ss},$$

et un isomorphisme $f_z^* \mathbb{F}_0 \simeq F|_{U_z}$. On obtient alors le diagramme commutatif

$$\begin{array}{ccc}
L_F|_{U_z} \simeq L_{(F|_{U_z \times X})} & \xrightarrow{\alpha_{F|U_z \times X}} & f_z^* L_{\mathbb{F}_0} \\
& \searrow^{\lambda_z} & \downarrow \simeq \\
& & f_z^* \mathcal{O}_{R^{ss}} \\
& & \| \\
& & \mathcal{O}_{U_z}
\end{array}$$

On va montrer que $\lambda_z$ est indépendant de la trivialisation $\beta_z$. Supposons qu'on ait une autre trivialisation

$$\beta_z' \colon W|_{U_z} \to \mathcal{O}_{U_z} \otimes \mathbb{C}^q.$$

On en déduit

$$f_z' \colon U_z \to R^{ss} \quad \text{et} \quad \lambda_z' \colon L_F|_{U_z} \to \mathcal{O}_{U_z}.$$

La trivialisation $\beta_z'$ est la multiplication de $\beta_z$ et d'un morphisme $\varepsilon \colon U_z \to PGL(q)$: pour tout $y'$ dans $U_z$ on a

$$\beta'_{z,y'} = \varepsilon(y') \circ \beta_{z,y'}.$$

Il en découle que

$$f_z'(y') = \varepsilon(y') \cdot f_z(y').$$

On a, pour tout $t$ dans $L_{F,y'}$,

$$\lambda_z'(t) = \varepsilon(y) \cdot \lambda_z(t),$$

(où $\lambda_z(t)$, $\lambda_z'(t)$ sont vus respectivement comme éléments de $L_{\mathbb{F}_0, f_z(y')}$, $L_{\mathbb{F}_0, f_z'(y')}$). Comme l'action de $PGL(q)$ sur $L_{\mathbb{F}_0}$ est triviale d'après le lemme 3.2, on a $\lambda_z'(t) = \lambda_z(t)$ (comme nombres complexes). Il en découle que les isomorphismes $\lambda_z$ se recollent et en définissent un

$$u_F \colon L_F \to \mathcal{O}_S.$$

Il reste à voir que ces isomorphismes sont compatibles avec les $\alpha_F^L(f)$, c'est à dire que si $f \colon S' \to S$ est un morphisme de variétés lisses, on a un diagramme commutatif

$$\begin{array}{ccc}
L_{f^*F} & \xrightarrow{u_{f^*F}} & \mathcal{O}_{S'} \\
{\scriptstyle \alpha_F^L(f)} \downarrow & & \downarrow \simeq \\
f^* L_F & \xrightarrow{f^* u_F} & f^* \mathcal{O}_S.
\end{array}$$

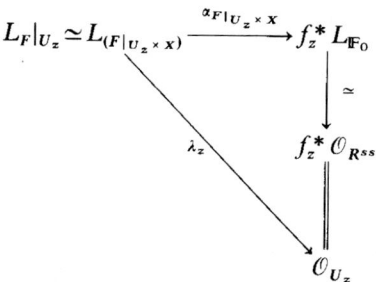

Pour cela, il suffit de montrer que tout point de $S$ possède un voisinage $U$ tel que ce qui précède soit vrai si on remplace $S$, $S'$ par $U$, $f^{-1}(U)$ respectivement, et $F$ par sa restriction à $U$. On choisit $U = U_z$, avec les notations précédentes. Si

$$W' = p_{S'*}(f^* F \otimes p_X^* \mathcal{O}_X(m)),$$

on déduit de $\beta_z$ une trivialisation de $W'|_{f^{-1}(U)}$ et un morphisme

$$\lambda' : f^{-1}(U) \to R^{ss}$$

tel qu'on ait un diagramme commutatif

$$\begin{array}{ccc} f^{-1}(U) & \xrightarrow{\lambda'} & R^{ss} \\ {\scriptstyle f}\downarrow & & \| \\ U & \xrightarrow{\lambda} & R^{ss}, \end{array}$$

d'où, d'après la définition de $u_F$, $u_{f^* F}$, un diagramme commutatif

$$\begin{array}{ccc} L_{f^\# F|_{f^{-1}(U)}} & \xrightarrow{u_{f^\# F|_{f^{-1}(U)}}} & \mathcal{O}_{f^{-1}(U)} \\ \downarrow {\scriptstyle \alpha_F^L(f)|_{f^{-1}(U)}} & & \| \\ f^* L_{F|_{f^{-1}(U)}} & \xrightarrow[f^* u_F|_{f^{-1}(U)}]{} & \mathcal{O}_{f^{-1}(U)}. \end{array}$$

Ceci démontre la proposition 3.3.

**Corollaire 3.4.** *Le morphisme de groupes*

$$\begin{array}{c} \mathrm{Pic}(F(r, d)) \to \mathrm{Pic}(R^s) \\ L \mapsto L_{\mathbb{F}_0|R^s} \end{array}$$

*est injectif.*

Cela découle immédiatement de la proposition 3.3 et du fait que $\mathrm{codim}_{R^{ss}}(R^{ss} \setminus R^s) \geqq 2$.

Rappelons (cf. §2) que si $L$ est un $PGL(q)$-fibré en droites sur $R^{ss}$, on dit que $L$ descend à $U(r, d)$ s'il existe un fibré en droites $L_0$ sur $U(r, d)$ tel que les $PGL(q)$-fibrés $L$ et $\pi^* L_0$ soient isomorphes. Voici une autre conséquence immédiate de la proposition 3.3:

**Corollaire 3.5.** *Soit $L$ un élément de* Pic$(F(r, d))$. *Alors $L$ provient de* Pic$(U(r, d))$ *si et seulement si $L_{\mathbb{F}_0}$ descend à $U(r, d)$. Si $L_0$ est l'unique élément de* Pic$(U(r, d))$ *tel que $L_{\mathbb{F}_0} \simeq \pi^* L_0$, on a $L = i(L_0)$.*

*3.3. Exemples fondamentaux d'éléments de* Pic$(F(r, d))$

Soit $K(X)$ le groupe de Grothendieck de $X$. On sait que le morphisme de groupes

$$K(X) \to \mathrm{Pic}(X) \oplus \mathbb{Z}$$
$$\alpha \to (\det(\alpha), rg(\alpha))$$

est un isomorphisme. On note comme dans l'Introduction $H(r, d)$ le sous-groupe de $K(X)$ constitué des $\alpha$ tels que $\chi([\mathrm{E}] \otimes \alpha) = 0$, pour un fibré $E$ de rang $r$ et de degré $d$ sur $X$. Cela signifie que

$$d \cdot rg(\alpha) + r \cdot \deg(\alpha) + r \cdot rg(\alpha) \cdot (1 - g) = 0.$$

En particulier, $H(r, d)$ contient tous les $\alpha$ de rang et degré nuls. Ceux-ci constituent un sous-groupe $Z$ de $K(X)$, isomorphe à $\mathrm{Pic}^{(0)}(X)$, l'isomorphisme étant

$$\mathrm{Pic}^{(0)}(X) \to Z$$
$$L \mapsto [L_0 \otimes L] - [L_0],$$

($L_0$ étant un fibré en droites sur $X$ fixé, l'isomorphisme ci-dessus est indépendant du choix de $L_0$).

Soit $\alpha$ un élément de $H(r, d)$. On va en déduire un élément $\gamma(\alpha)$ de Pic$(F(r, d))$. Si $S$ est une variété algébrique lisse, et $F$ une famille de fibrés de $U(r, d)$ paramétrée par $S$, on aura

$$\gamma(\alpha)_F = \det(p_{S!}(F \otimes p_X^* \alpha)).$$

Vérifions tout de suite que $\gamma(\alpha)_F$ ne dépend que de la classe d'équivalence de $F$. Soit $L$ un fibré en droites sur $S$. Alors on a

$$\gamma(\alpha)\, F \otimes p_S^* L = \gamma(\alpha)_F \otimes L^{\chi(F_s \otimes \alpha)},$$

($s$ étant un point fermé quelconque de $S$). Puisque $\alpha$ est dans $H(r, d)$, on a $\chi(F_s \otimes \alpha) = 0$, donc $\gamma(\alpha)\, F \otimes p_S^* L = \gamma(\alpha)_F$.

On ne peut pas déduire de $\alpha$ un fibré en droites sur $F(r, d)$, mais un élément de Pic$(F(r, d))$. Pour obtenir un fibré en droites $L$ sur $F(r, d)$, on considère une représentation de $\alpha$ sous la forme

$$\alpha = [E_1] - [E_2],$$

$E_1$, $E_2$ étant des fibrés vectoriels sur $X$. On pose

$$L_F = \det(p_{S!}(F \otimes p_X^* E_1)) \otimes \det(p_{S!}(F \otimes p_X^* E_2))^{-1}.$$

Les $\alpha_F^L(f)$ sont définis de manière évidente, et il est aisé de voir que pour des choix différents de $E_1$, $E_2$ (donnant le même $\alpha$), on obtient un fibré en droites sur $F(r, d)$ isomorphe à $L$. Par définition $\gamma(\alpha)$ est la classe d'isomorphisme de $L$.

*Remarque.* Avec les notations précédentes, supposons $S$ seulement intègre. Alors il est facile de voir que $p_{S!}(F \otimes p_X^* \alpha)$ est contenu dans le sous-groupe de $K(S)$

engendré par les classes des faisceaux localement libres. En utilisant ce fait, on peut donner une définition de $\gamma(\alpha)_F$ dans ce cas aussi. Nous nous sommes limités au cas des variétés lisses pour alléger l'exposé.

*Exemple.* Le cas $r=1$.

Les définitions précédentes s'appliquent aussi dans le cas où $r=1$. Pour tout $\alpha$ dans $Z$, $\gamma(\alpha)$ provient de $\mathrm{Pic}(U(1,d))=\mathrm{Pic}(J^{(d)})$, et on peut ainsi considérer $Z=\mathrm{Pic}^{(0)}(X)$ comme un sous-groupe de $\mathrm{Pic}(J^{(d)})$ (cf. [12]).

### 3.4. *Effets de la torsion par un fibré en droites*

Soit $k$ un entier et $L_0$ un fibré en droites sur $X$ de degré $k$. On a un isomorphisme

$$U(r,d)\xrightarrow{\varphi} U(r,d+kr)$$

associant à la classe d'équivalence de $[E]$ celle de $[E\otimes L_0]$. Pour tout fibré en droites $L$ de degré $d$ sur $X$, cet isomorphisme en induit un

$$U(r,L)\rightarrow U(r,L\otimes L_0^r).$$

L'isomorphisme $\varphi$ est compatible avec un isomorphisme évident de foncteurs

$$F(r,d)\rightarrow F(r,d+kr).$$

On a un automorphisme de groupes

$$\begin{aligned} K(X)&\rightarrow K(X)\\ \alpha&\rightarrow \alpha\otimes[L_0^{-1}], \end{aligned}$$

induisant un isomorphisme $H(r,d)\rightarrow H(r,d+kr)$. Il est immédiat qu'on a un diagramme commutatif

$$\begin{array}{ccc} H(r,d) & \xrightarrow{\gamma} & \mathrm{Pic}(F(r,d)) \\ \big\downarrow & & \big\downarrow \\ H(r,d+kr) & \xrightarrow{\gamma} & \mathrm{Pic}(F(r,d+kr)). \end{array}$$

Il en découle que pour prouver tous les résultats de l'Introduction, on peut supposer $d$ arbitrairement grand.

## 4. Étude des $PGL(q)$-fibrés en droites sur $R^{ss}$

Le résultat suivant permet, à l'aide du lemme de descente (§ 2), de prouver que tout $PGL(q)$-fibré en droites sur $R^{ss}$ descend à $U(r,d)$:

**Proposition 4.1.** *Soit $L$ un $PGL(q)$-fibré en droites sur $R^{ss}$, et $y$ un point fermé de $R^{ss}$ tel que l'orbite $Gy$ soit fermée. Alors le stabilisateur de $y$ dans $PGL(q)$ agit trivialement sur $L_y$.*

**Lemme 4.2.** *Soit $y$ un point fermé de $R^{ss}$. Alors l'orbite $Gy$ est fermée si et seulement si le fibré $\mathbb{F}_{0y}$ est isomorphe à une somme directe de fibrés stables.*

Soit $y$ un point de $R^{ss}$ tel qu'on ait une suite exacte

$$0 \to E_1 \to \mathbb{F}_{0y} \to E_2 \to 0,$$

$E_1$, $E_2$ étant des fibrés vectoriels de pente $d/r$. En utilisant des extensions de $E_2$ par $E_1$ on construit une famille $E$ de fibrés de $U(r, d)$ paramétrée par $\mathbb{C}$ telle que $E_t \simeq \mathbb{F}_{0y}$ si $t \neq 0$, et $E_0 \simeq E_1 \oplus E_2$. On en déduit que dans $\overline{Gy}$ il existe un point $z$ tel que $\mathbb{F}_{0y} \simeq E_1 \oplus E_2$.

En raisonnant par récurrence on montre que dans l'adhérence de toute orbite il existe un point $z$ tel que $\mathbb{F}_{0z}$ soit isomorphe à une somme directe de fibrés stables. Le lemme 4.2 en découle aisément.

Démontrons maintenant la proposition 4.1. Soit $y$ un point de $R^{ss}$ tel que $Gy$ soit fermée. D'après le lemme 4.1, $\mathbb{F}_{0y}$ est isomorphe à une somme directe de fibrés stables:

$$\mathbb{F}_{0y} \simeq m_1 E_1 \oplus \ldots \oplus m_p E_p,$$

où pour $1 \leqq i \leqq p$, $m_i$ est un entier, $m_i \geqq 1$, $E_i$ un fibré stable, et $E_i$ n'est pas isomorphe à $E_j$ si $i \neq j$.

Pour tout point $y'$ de $R^{ss}$, on note $G_{y'}$ le stabilisateur de $y'$ dans $GL(q)$. Alors $G_y$ est isomorphe à $GL(m_1) \times \ldots \times GL(m_p)$. L'action de $G_y$ sur $L_y$ est de la forme

$$G_y \times L_y \to L_y$$
$$(g, u) \to \lambda(g)\, u,$$

$\lambda$ étant un caractère de $G$, défini par des entiers $n_1, \ldots, n_p$ tels que

$$(*) \qquad m_1 n_1 + \ldots + m_p n_p = 0;$$

si $g = (g_1, \ldots, g_p)$ est un élément de $G_y = GL(m_1) \times \ldots \times GL(m_p)$, on a

$$\lambda(g) = \prod_{1 \leqq i \leqq p} \det(g_i)\, n_i,$$

(l'équation $(*)$ découle du fait que le sous-groupe des homothéties de $GL(q)$ agit trivialement sur $L$). Si $p = 1$, c'est à dire si $\mathbb{F}_{0y} \simeq m_1 E_1$, alors la proposition 4.1 est vraie, car le caractère $\lambda$ est trivial.

Il suffit de prouver l'assertion suivante: il existe une variété algébrique intègre $R_0$ et un morphisme $\varphi\colon R_0 \to R^{ss}$ tels que:

(i) L'image de $\varphi$ contient $y$.

(ii) L'image de $\varphi$ contient un point $y_0$ tel qu'on ait un isomorphisme

$$\mathbb{F}_{0 y_0} \simeq nE,$$

$E$ étant un fibré stable.

(iii) Pour tout point $y'$ de l'image de $\varphi$, on a $G_y \subset G_{y'}$.

En effet, supposons cela vérifié. Alors les propriétés (i), (ii), (iii) sont encore vraies si on remplace l'image de $\varphi$ par son adhérence $S$ qui est une sous-variété irréductible de $R^{ss}$, sur laquelle $G_y$ agit trivialement. A priori, l'action de $G_y$ sur $L|_S$ s'écrit

$$\begin{aligned} G_y \times L|_S &\to L|_S \\ (g, u) &\to \lambda_{\alpha(u)}(g)\, u, \end{aligned}$$

où $\alpha: L|_S \to S$ est la projection, et pour tout point $s$ de $S$, $\lambda_s$ est un caractère de $G_y$.

Montrons que pour tout point $s$ de $S$, on a $\lambda_s = \lambda$. Soit $g$ un élément de $G_y$. En prenant des trivialisations locales de $L|_S$, on voit que

$$\begin{aligned} \Psi: G_y \times L|_S &\to \mathbb{C}^* \\ (g, u) &\to \lambda_{\alpha(u)}(g) \end{aligned}$$

est un morphisme. Puisque le groupe des caractères de $G_y$ est dénombrable, la restriction de $\Psi$

$$\begin{aligned} L|_S &\to \mathbb{C}^* \\ u &\to \lambda_{\alpha(u)}(g) \end{aligned}$$

prend une quantité dénombrable de valeurs, donc est constante. Le caractère $\lambda_s$ ne dépend donc pas du point $s$ de $S$, donc $\lambda_s = \lambda$.

On a vu que $G_{y_0}$ agit trivialement sur $L_{y_0}$, donc il en est de même de $G_y$ d'après (iii). Donc $\lambda$ est trivial, et $G_y$ agit trivialement sur $L_y$, ce qui démontre la proposition 4.1.

Il reste à trouver le morphisme $\varphi: R_0 \to R^{ss}$. Un tel morphisme équivaut à la donnée d'un morphisme surjectif de faisceaux sur $R_0 \times X$:

$$p_X^* \mathcal{O}_X(-m) \otimes \mathbb{C}^q \xrightarrow{\theta'} \mathbb{E},$$

où $\mathbb{E}$ est une famille de fibrés de $U(r, d)$ paramétrée par $R_0$, tel qu'en tout point fermé $z$ de $R_0$, $\theta'_z$ induise un isomorphisme $\mathbb{C}^q \simeq H^0(X, \mathbb{E}_z(m))$. La condition (iii) équivaut à la suivante: pour tout $g$ dans $G_y$ on a un diagramme commutatif

$$\begin{array}{ccc} p_X^* \mathcal{O}_X(-m) \otimes \mathbb{C}^q & \xrightarrow{\theta'} & \mathbb{E} \\ \downarrow g & & \downarrow \simeq \\ p_X^* \mathcal{O}_X(-m) \otimes \mathbb{C}^q & \xrightarrow{\theta'} & \mathbb{E} \end{array}$$

($g$ étant vu comme élément de $GL(q)$).

On considère, pour $1 \leqq i \leqq p$, l'ouvert $R_i^{ss}$ du schéma de Grothendieck correspondant à $E_i$,

$$\theta_i \colon p_X^* \mathcal{O}_X(-m) \otimes \mathbb{C}^{q_i} \to \mathbb{F}_{0i}$$

le morphisme surjectif universel sur $R_i^{ss} \times Y$. Fixons des isomorphismes

$$\mathbb{C}^{q_i} \simeq H^0(E_i(m)) \qquad \text{pour } 1 \leqq i \leqq p.$$

Alors du point $y$ on déduit un isomorphisme

$$\mathbb{C}^q \simeq \bigoplus_{1 \leqq i \leqq p} (\mathbb{C}^{q_i})^{m_i} \simeq \bigoplus_{1 \leqq i \leqq p} \mathbb{C}^{m_i} \otimes \mathbb{C}^{q_i}.$$

On prend maintenant $R_0 = R_1^{ss} \times \ldots \times R_p^{ss}$, et $\theta'$ est la composée

$$p_X^* \mathcal{O}_X(-m) \otimes \mathbb{C}^q \simeq \bigoplus_{1 \leqq i \leqq p} p_X^* \mathcal{O}_X(-m) \otimes (\mathbb{C}^{q_i})^{m_i} \to \bigoplus_{1 \leqq i \leqq p} (p_{R_i^{ss}}^{\#} \mathbb{F}_{0i})^{m_i},$$

le second morphisme étant un produit des $\theta_i$. Il est immédiat que $\theta'$ vérifie bien les conditions requises.

La proposition 4.1 est donc démontrée.

**Corollaire 4.3.** *Le morphisme canonique*

$$i \colon \mathrm{Pic}(U(r, d)) \to \mathrm{Pic}(F(r, d))$$

*est un isomorphisme.*

Cela découle immédiatement du corollaire 3.5, de la proposition 4.1 et du lemme de descente (théorème 2.3).

## 5. Étude des $GL(q)$-fibrés en droites sur $R^s$

Rappelons que $n = PGCD(r, d)$, et qu'un $GL(q)$-fibré en droites sur $R^s$ est un fibré en droites algébrique sur $R^s$ muni d'une action algébrique linéaire de $GL(q)$ au dessus de l'action de $PGL(q)$ sur $R^s$.

Soit $L$ un $GL(q)$-fibré en droites sur $R^s$. Alors il existe un entier $p$ tel que pour tout point $y$ de $R^s$ et tout $t$ dans $\mathbb{C}^* \subset GL(q)$, l'action de $t$ sur $L_y$ soit la multiplication par $t^p$. On posera

$$e(L) = p.$$

Si $e(L) = 0$, $L$ est en fait un $PGL(q)$-fibré en droites sur $R^s$.

**Proposition 5.1.** *Soit $p$ un entier. Alors il existe un $GL(q)$-fibré en droites $L$ sur $R^s$ tel que $e(L) = p$ si et seulement si $p$ est multiple de $n$.*

Soit $p$ un multiple de $b$: $p = kn$. Montrons l'existence d'un $GL(q)$-fibré en droites $L$ sur $R^s$ tel que $e(L) = p$. Il suffit de traiter le cas où $k = 1$, c'est à dire $p = n$: car si $L_0$ est un $GL(q)$-fibré en droites sur $R^s$ tel que $e(L_0) = n$, il suffit de prendre $L = L_0^k$.

Considérons l'action de $GL(q)$ sur $\mathbb{F}_0$. L'action d'un scalaire $t$ sur une fibre de $\mathbb{F}_0$ est la multiplication par $t$. Il en est donc de même en ce qui concerne les $GL(q)$-fibrés sur $R^s$

$$E_1 = p_{S*}(\mathbb{F}_0 \otimes p_X^* \mathcal{O}_X(m))$$
$$E_2 = p_{S*}(\mathbb{F}_0 \otimes p_X^* (\mathcal{O}_X(m) \otimes \Lambda)),$$

($\Lambda$ étant un fibré en droites de degré 1 sur $X$). On a

$$rg(E_1) = d + r(1 - g + m \cdot \deg(\mathcal{O}_X(1))),$$
$$rg(E_2) = rg(E_1) + r,$$

donc $PGCD(rg(E_1), rg(E_2)) = n$. Il existe donc des entiers $a$, $b$ tels que

$$a \cdot rg(E_1) + b \cdot rg(E_2) = n,$$

et il suffit de prendre

$$L_0 = \det(E_1)^a \otimes \det(E_2)^b.$$

Réciproquement, soit $L$ un $GL(q)$-fibré en droites sur $R^s$. Il faut prouver que $e(L)$ est multiple de $n$.

**Lemme 5.2.** *Tout $GL(q)$-fibré en droites sur $R^s$ peut être prolongé en un $GL(q)$-fibré en droites sur $R^{ss}$.*

Il en découle que tout $PGL(q)$-fibré en droites sur $R^s$ peut aussi être prolongé en un $PGL(q)$-fibré en droites sur $R^{ss}$ (nous utiliserons ce résultat pour démontrer le théorème A dans le § 6).

Soit $L$ un $GL(q)$-fibré en droites sur $R^s$. Puisque $R^{ss}$ est lisse, le fibré en droites $L$ se prolonge en un fibré en droites sur $R^{ss}$, aussi noté $L$. Il faut montrer que la structure de $GL(q)$-fibré sur $L$ peut aussi être prolongée. Il suffit de montrer que le morphisme

$$\varphi: GL(q) \times L|_{R^s} \to L|_{R^s}$$

(définissant l'action de $GL(q)$) peut être prolongé en un morphisme

$$GL(q) \times L \to L,$$

car les propriétés que doit vérifier ce morphisme pour qu'il définisse une structure de $GL(q)$-fibré sur $L$ découleront de celles de $\varphi$, par continuité. Soit $(U_i)_{i \in I}$ un recouvrement ouvert affine de $U(r, d)$. Alors $(\pi^{-1}(U_i))_{i \in I}$ est un recouvrement ouvert affine $PGL(q)$-invariant de $R^{ss}$. Il suffit de prouver que le morphisme

$$\varphi_i: GL(q) \times L|_{R^s \cap \pi^{-1}(U_i)} \to L|_{R^s \cap \pi^{-1}(U_i)}$$

restriction de $\varphi$ peut être prolongé en un morphisme

$$GL(q) \times L|_{\pi^{-1}(U_i)} \to L|_{\pi^{-1}(U_i)},$$

car puisque $\operatorname{codim}_{R^{ss}}(R^{ss}\backslash R^s)\geqq 2$, tous ces morphismes se recolleront. Du morphisme $\varphi_i$ on déduit le morphisme d'anneaux

$$\mathbb{C}[L|_{R^s\cap\pi^{-1}(U_i)}]\to\mathbb{C}[GL(q)\times L|_{R^s\cap\pi^{-1}(U_i)}].$$

Puisque $\operatorname{codim}_{R^{ss}}(R^{ss}\backslash R^s)\geqq 2$, on a

$$\mathbb{C}[L|_{R^s\cap\pi^{-1}(U_i)}]=\mathbb{C}[L|_{\pi^{-1}(U_i)}]$$

et

$$\mathbb{C}[GL(q)\times L|_{R^s\cap\pi^{-1}(U_i)}]=\mathbb{C}[GL(q)\times L|_{\pi^{-1}(U_i)}].$$

On a donc un morphisme

$$\mathbb{C}[L|_{\pi^{-1}(U_i)}]\to\mathbb{C}[GL(q)\times L|_{\pi^{-1}(U_i)}].$$

Puisque les variétés $L|_{\pi^{-1}(U_i)}$ et $GL(q)\times L|_{\pi^{-1}(U_i)}$ sont affines, de ce morphisme d'anneaux on déduit un morphisme

$$GL(q)\times L|_{\pi^{-1}(U_i)}\to L|_{\pi^{-1}(U_i)},$$

qui est le prolongement recherché de $\varphi_i$. Ceci achève la démonstration du lemme 5.2.

Reprenons maintenant la démonstration de la proposition 5.1. Soit $y$ un point de $R^{ss}$ tel que $\mathbb{F}_{0y}$ soit isomorphe à la somme directe de $n$ copies d'un fibré stable:

$$\mathbb{F}_{0y}\simeq n\cdot E.$$

On note $L$ l'extension à $R^{ss}$ du $GL(q)$-fibré en droites $L$ sur $R^s$. Alors le stabilisateur $G_y$ de $y$ dans $GL(q)$ s'identifie à $GL(n)$. Par conséquent, de l'action de $GL(q)$ sur $L$ on déduit une action de $GL(n)$ sur $L_y$, qui est de la forme

$$\begin{aligned} GL(n)\times L_y &\to L_y\\ (\varphi,u)&\to \det(\varphi)^a\cdot u, \end{aligned}$$

avec a entier. L'action du sous-groupe des homothéties est donc

$$\begin{aligned} \mathbb{C}^*\times L_y &\to L_y\\ (t,u)&\to t^{an}\cdot u. \end{aligned}$$

On a donc

$$e(L)=an,$$

ce qui démontre la proposition 5.1.

La proposition 5.1 sera utilisée dans le §7, mais on en donne ici d'autres conséquences. Les résultats qui suivent dans ce chapitre sont inutiles à la démonstration des théorèmes A à F.

*5.1. Fibrés universels*

Soit $U$ un ouvert non vide de $U_s(r, d)$. Rappelons qu'un fibré de Poincaré sur $U$ est une famille $E$ de fibrés de $U(r, d)$ paramétrée par $U$ telle que le morphisme canonique $f_E\colon U \to U(r, d)$ soit l'inclusion $U \subset U(r, d)$.

**Lemme 5.3.** *Il existe un fibré de Poincaré sur U si et seulement s'il en existe un sur $U_s(r, d)$.*

Il faut montrer que s'il existe un fibré de Poincaré sur $U$ il en existe aussi un sur $U_s(r, d)$. On définit de faҫn évidente la notion de «fibré projectif universel» sur $U$: c'est un morphisme lisse

$$\varphi\colon W \to U \times X$$

tel que pour tout point $u$ de $U$, $\varphi^{-1}(u)$ soit un fibré en espaces projectifs sur $X$:

$$\varphi^{-1}(u) \simeq \mathbb{P}(E), \qquad \text{(droites de } E\text{)},$$

$E$ étant un fibré stable de rang $r$ et de degré $d$ sur $X$ dont le point associé de $U(r, d)$ soit $u$.

Il existe un fibré projectif universel sur $U_s(r, d)$: c'est $\mathbb{P}(\mathbb{F}_0)/PGL(q) = \mathbb{P}_0$.

Soit $V$ un fibré de Poincaré sur $U$. Alors, puisque tout fibré stable sur $X$ est simple,

$$\Lambda = p_{\pi^{-1}(U)*}(\underline{\mathrm{Hom}}(\mathbb{F}_0, \pi^{\#} V))$$

est un fibré en droites sur $\pi^{-1}(U)$. Il en découle un isomorphisme

$$\Lambda \otimes \mathbb{F}_0|_{\pi^{-1}(U) \times X} \simeq \pi^{\#} V,$$

d'où un isomorphisme

$$\mathbb{P}(V) \simeq \mathbb{P}_0|_U,$$

c'est à dire que $\mathbb{P}_0|_U$ est banal. Mais la banalité d'un fibré projectif est un problème birationnel (cf. [9]). Donc si $\mathbb{P}_0|_U$ est banal, il en est de même de $\mathbb{P}_0$, c'est à dire qu'il existe un fibré de Poincaré sur $U_s(r, d)$. Ceci démontre le lemme 5.3.

**Lemme 5.4.** *Les assertions suivantes sont équivalentes:*

(i) *Il existe un $GL(q)$-fibré en droites $L$ sur $R^s$ tel que $e(L) = 1$.*
(ii) *Il existe un fibré de Poincaré sur $U_s(r, d)$.*

Supposons que $L$ existe. Il suffit de prendre pour fibré de Poincaré le quotient

$$\mathbb{F}_0 \otimes p_X^* L^{-1}/PGL(q).$$

Réciproquement, si $V$ est un fibré de Poincaré sur $U_s(r, d)$,

$$L = p_{R^s_*}(\underline{\mathrm{Hom}}(\pi^{\#} V, \mathbb{F}_0))$$

est un $GL(q)$-fibré en droites et $e(L) = 1$.

De la proposition 5.1 et des deux lemmes précédents on déduit le

**Théorème 5.5.** *Si $r$ et $d$ ne sont pas premiers entre eux, et si $U$ est un ouvert non vide de $U_s(r, d)$, il n'existe pas de fibré de Poincaré sur $U$.*

C'est le résultat de Ramanan [20].

*Remarque.* Cette démonstration ne marche pas si $g=2$, $r=2$ et $d$ est pair, car alors $\operatorname{codim}_{R^{ss}}(R^{ss}\setminus R^s) = 1$. Mais dans ce cas on dispose d'une description très précise de $U(r, d)$ permettant de faire une démonstration directe du théorème 5.5 (cf. [18]).

*5.2. Faisceaux de Poincaré sur les variétés de modules de faisceaux semi-stables sur $\mathbb{P}_2(\mathbb{C})$*

La démonstration du théorème G suit pas à pas celle du théorème 5.5, compte tenu des deux faits suivants:

– $M(r, c_1, c_2)$ s'obtient aussi comme quotient d'un ouvert lisse irréductible d'un schéma de Grothendieck par un groupe du type $PGL(q)$ (cf. [13]).

– Le nombre maximal de termes d'une somme directe de faisceaux cohérents sur $\mathbb{P}_2(\mathbb{C})$ qui est un faisceau semi-stable de rang $r$ et de classes de Chern $c_1, c_2$ est exactement $PGCD(r, c_1, \chi)$.

Evidemment, ce qui précède n'est valable que si

$$\operatorname{codim}_{M(r, c_1, c_2)}(M(r, c_1, c_2)\setminus M_s(r, c_1, c_2)) \geqq 2.$$

Dans le cas constraire, on peut montrer que $M(r, c_1, c_2)$ s'identifie à $\mathbb{P}_5$ (espace des coniques de $\mathbb{P}_2(\mathbb{C})$), $M_s(r, c_1, c_2)$ étant l'ouvert des coniques non dégénérées (cf. [2]). Sans entrer dans les détails, disons que pour montrer qu'il n'existe pas de fibré de Poincaré sur $M_s(r, c_1, c_2)$, on utilise le fait qu'il n'existe pas de section de la projection

$$\mathbb{P}_5 \times \mathbb{P}_2 \supset Q \to \mathbb{P}_5,$$

$Q$ étant la conique universelle.

## 6. Démonstration du théorème A

On veut prouver que $U(r, d)$ et $U(r, L)$ sont localement factorielles. On ne traitera que le cas de $U(r, d)$, celui de $U(r, L)$ étant analogue.

D'après [7], proposition 6.2 et [14], p. 141, $U(r, d)$ est localement factorielle si et seulement si tout diviseur de $U(r, d)$ est localement principal, c'est à dire si et seulement si l'idéal de toute hypersurface de $U(r, d)$ est localement libre.

Cette condition équivaut à la suivante, $U(r, d)$ étant normale: le morphisme canonique

$$\mathrm{Pic}(U(r, d)) \to Cl(U(r, d))$$

est un isomorphisme.

Nous avons vu dans l'Introduction qu'il suffisait de montrer que le morphisme de restriction

$$\mathrm{Pic}(U(r, d)) \to \mathrm{Pic}(U_s(r, d))$$

est surjectif. Soit $L$ un fibré en droites sur $U_s(r, d)$. Alors $\pi_s^* L$ est un $PGL(q)$-fibré en droites sur $R^s$. D'après le lemme 5.2, ce $PGL(q)$-fibré en droites s'étend en un $PGL(q)$-fibré en droites $\bar{L}'$ sur $R^{ss}$. D'après la proposition 4.1, $\bar{L}'$ vérifie les hypothèses du lemme de descente (théorème 2.3). Il existe donc un fibré en droites $\bar{L}$ sur $U(r, d)$ tel que les $PGL(q)$-fibrés en droites $\bar{L}'$ et $\pi^* \bar{L}$ soient isomorphes. On a alors

$$\bar{L}|_{R^s} = L'/PGL(q) = L,$$

donc $\bar{L}$ est l'extension voulue de $L$.

Le théorème A est donc démontré.

## 7. Description de Pic ($U(r, d)$) et Pic ($U(r, L)$)

### *7.1. Fibrés engendrés par leurs sections*

**Lemme 7.1.** *Soit $E$ un fibré vectoriel de rang $r$ sur $X$, engendré par ses sections globales. Alors il existe un sous-espace vectoriel $H$ de $H^0(E)$ de dimension $r-1$, tel que la restriction du morphisme canonique*

$$H \otimes \mathcal{O}_X \to E$$

*soit un morphisme injectif de fibrés vectoriels.*

Soit $x$ un point de $X$. Alors le morphisme canonique $H^0(E) \to E_x$ est surjectif. Il en découle que le sous-espace vectoriel de $H^0(E)$ des sections qui s'annulent en $x$ est de codimension $r$. Par conséquent la sous-variété homogène de $H^0(E)$ constitutée des sections s'annulant en au moins un point de $X$ est de codimension $r-1$. Le lemme 7.1 en découle immédiatement.

Il existe un entier $k_0$ tel que pour tout entier $k \geqq k_0$, tout fibré semi-stable de rang $r$ et de degré $d+kr$ soit engendré par ses sections (cf. [23]). D'après 3.4, on peut donc supposer que tout fibré semi-stable de rang $r$ et de degré $d$ est engendré par ses sections. Soit $E$ un tel fibré. On a donc d'après le lemme 7.1 une suite exacte

$$0 \to \mathcal{O}_X \otimes \mathbb{C}^{r-1} \to E \to \det(E) \to 0.$$

*7.2. Construction d'une grande famille de fibrés stables*

Soient $J^{(d)}$ la jacobienne des fibrés en droites de degré $d$ sur $X$, et $D$ un fibré de Poincaré sur $J^{(d)} \times X$. Posons

$$V = R^1 p_{J*}(D^* \otimes \mathbb{C}^{r-1}).$$

C'est un fibré vectoriel sur $J^{(d)}$ de rang $(r-1)(g-1+d)$. Soit $\mathbb{P}$ le fibré en espaces projectifs associé à $V$. D'après [23], App. II proposition 2, il existe un fibré vectoriel $\mathbb{E}$ sur $\mathbb{P} \times X$ et une suite exacte

$$(*) \qquad 0 \to \mathcal{O} \otimes \mathbb{C}^{r-1} \to \mathbb{E} \to \pi_0^\# D \otimes p_{\mathbb{P}*} \mathcal{O}_{\mathbb{P}}(-1) \to 0$$

($\pi_0$ désignant la projection $\mathbb{P} \to J^{(d)}$) telle que pour tout point $y$ de $\mathbb{P}$, si $\alpha = \pi_0(y)$, la restriction de $(*)$ à $\{y\} \times X$:

$$0 \to \mathcal{O}_X \otimes \mathbb{C}^{r-1} \to \mathbb{E}_y \to D_\alpha \otimes y \to 0$$

soit associée à l'inclusion

$$y \to \operatorname{Ext}^1(D_\alpha, \mathcal{O}_X \otimes \mathbb{C}^{r-1}) = V_\alpha,$$

(vue comme élément de $\operatorname{Ext}^1(D_\alpha \otimes y, \mathcal{O}_X \otimes \mathbb{C}^{r-1})$).

On notera $\mathbb{P}^s$ l'ouvert de $\mathbb{P}$ constitué des points $y$ tels que $\mathbb{E}_y$ soit stable. La restriction de $\mathbb{E}$ à $\mathbb{P}^s \times X$, aussi notée $\mathbb{E}$, est une famille de fibrés stables de rang $r$ et de degré $d$ sur $X$. D'après le lemme 7.1, le morphisme canonique

$$f_{\mathbb{E}}: \mathbb{P}^s \to U_s(r, d)$$

est surjectif.

*7.3. Autre construction de la même famille*

*7.3.1.* Soit $\mathbb{F}_0$ le fibré canonique sur $R^s \times X$ (cf. 1.1). Alors le faisceau $F_0 = p_{R^s_*} \mathbb{F}_0$ est localement libre de rang $k = d + r(1-g)$. Soit $Gr_0$ le fibré en grassmanniennes des sous-espaces de dimension $r-1$ de $F_0$.

Il existe un bon quotient $Gr_0/PGL(q)$: on considère pour cela l'action de $GL(q)$ sur $R^s$. Le groupe $GL(q)$ agit canoniquement sur $\mathbb{F}_0$, donc sur $F_0$ et $\Lambda^{r-1} F_0$. La projection $\Lambda^{r-1} F_0 \to R^s$ est un $GL(q)$-morphisme affine. Il en découle d'après Ramanathan ([21], lemma 4.1) qu'il existe un bon quotient $\Lambda^{r-1} F_0/GL(q)$. Par conséquent il existe un bon quotient $Gr_0/PGL(q)$. C'est un quotient géométrique. On notera $\Gamma_0 = Gr_0/PGL(q)$.

Soit $p_0: \Gamma_0 \to U_s(r, d)$ le morphisme déduit de la projection $Gr_0 \to R^s$. Soit $y$ un point de $R^s$. Alors il existe un isomorphisme canonique

$$p_0^{-1}(\pi(y)) \simeq Gr^{r-1}(H^0(\mathbb{F}_{0y})).$$

On note $\pi_{Gr}: Gr_0 \to \Gamma_0$ le morphisme quotient.

*7.3.2. Groupe de Picard de $\Gamma_0$.* Soient $Q_{Gr}$ le fibré quotient canonique relatif sur $Gr_0$ (si $y$ est un point de $R^s$, et $H \subset H^0(\mathbb{F}_{0y})$ est de dimension $r-1$, on a $Q_{Gr,H} = H^0(\mathbb{F}_{0y})/H$), et

$$\mathcal{O}_{Gr}(1) = \Lambda^{k-r+1} Q_{Gr}.$$

On sait que

$$\operatorname{Pic}(Gr_0) \simeq \operatorname{Pic}(R^s) \otimes \mathbb{Z} \cdot \mathcal{O}_{Gr}(1).$$

Le fibré $\mathcal{O}_{Gr}(1)$ est muni de l'action canonique de $GL(q)$, et on a

$$e(\mathcal{O}_{Gr}(1)) = k - r + 1$$

(notation analogue à celle du § 5).

**Proposition 7.2.** (a) *Il existe un fibré en droites sur $\Gamma_0$, noté $\mathcal{O}_{\Gamma_0}(n)$, tel qu'on ait, pour tout point y de $R^s$*

$$\mathcal{O}_{\Gamma_0}(n)|_{p_0^{-1}(\pi(y))} \simeq \mathcal{O}_{Gr^{r-1}(H^0(\mathbb{F}_{0y}))}(n).$$

(b) *On a un isomorphisme canonique*

$$\operatorname{Pic}(\Gamma_0) \simeq \operatorname{Pic}(U_s(r,d)) \oplus \mathbb{Z} \cdot \mathcal{O}_{\Gamma_0}(n).$$

Construisons d'abord $\mathcal{O}_{\Gamma_0}(n)$. Le fibré $F_0$ est de rang $k = d + r(1-g)$, et si $x$ est un point de $X$, $\mathbb{F}_{0x} = \mathbb{F}_0|_{R^s \times x}$ est de rang $r$. On a $PGCD(r,k) = n$, donc il existe des entiers $a$, $b$ tels que

$$ak + br = -n(k-1+r).$$

Donc, si $L = \det(F_0)^a \otimes \det(\mathbb{F}_{0x})^b$, on a $e(L) = -n(k-1+r)$. Par conséquent on a

$$e(L \otimes \mathcal{O}_{Gr_0}(n)) = 0,$$

et $L \otimes \mathcal{O}_{Gr_0}(n)$ est en fait muni d'une action de $PGL(q)$. Puisque $\pi_{Gr}$ est un quotient géométrique, il existe un fibré en droites $\mathcal{O}_{\Gamma_0}(n)$ sur $\Gamma_0$ tel que

$$\pi_{Gr}^*(\mathcal{O}_{\Gamma_0}(n)) \simeq \mathcal{O}_{Gr_0}(n) \otimes L.$$

Il est immédiat que $\mathcal{O}_{\Gamma_0}(n)$ possède bien la propriété requise. Ceci démontre (a).

Prouvons (b). Si $L$ est un fibré en droites sur $U_s(r,d)$, on a $p_{0*}(p_0^* L) \simeq L$, donc $p_0^*$: $\operatorname{Pic}(U_s(r,d)) \to \operatorname{Pic}(\Gamma_0)$ est injectif. On en déduit avec (a) un morphisme injectif

$$\operatorname{Pic}(U_s(r,d)) \oplus \mathbb{Z} \cdot \mathcal{O}_{\Gamma_0}(n) \to \operatorname{Pic}(\Gamma_0).$$

Il reste à montrer qu'il est surjectif. Soit $\Delta$ un fibré en droites sur $\Gamma_0$. Alors, puisque $\operatorname{Pic}(Gr_0) \simeq \operatorname{Pic}(R^s) \oplus \mathbb{Z} \cdot \mathcal{O}_{Gr_0}(1)$, il existe un entier $m$ et un fibré en droites $\Delta'$ sur $R^s$ tels que

$$\pi_{Gr}^* \Delta \simeq p_1^* \Delta' \otimes \mathcal{O}_{Gr_0}(-m),$$

$p_1$ désignant la projection $Gr_0 \to R^s$. Le fibré $\Lambda'$ est muni d'une action de $GL(q)$, et on a

$$e(\Lambda') = m(k-r+1).$$

D'après la proposition 5.1, $e(\Lambda')$ est un multiple de $n$, et comme $n$ et $k-r+1$ sont premiers entre eux, $n$ divise $m$: posons $m = -an$. Alors $p_0^*(\Delta \otimes \mathcal{O}_{\Gamma_0}(n)^a) = p_1^* \Lambda''$, $\Lambda''$ étant un fibré en droites sur $R^s$ avec $e(\Lambda'') = 0$. Il en découle que $\Lambda''$ provient de $R^s$. Donc $\Delta \otimes \mathcal{O}_{\Gamma_0}(n)^a$ aussi. Ceci prouve (b) et achève la démonstration de la proposition 7.2.

*7.3.3.* Soit $Gr'_0$ l'ouvert de $Gr_0$ constitué des points $H$ tels que le morphisme canonique de fibrés vectoriels

$$\mathcal{O}_X \otimes H \to \mathbb{F}_{0y}$$

soit injectif ($y$ désignant l'image de $H$ dans $R^{ss}$). C'est un ouvert $PGL(q)$-invariant, et $\Gamma'_0 = \pi_{Gr}(Gr'_0)$ est un ouvert de $\Gamma_0$.

**Lemma 7.3.** *Pour tout point $y$ de $R^s$, $p_0^{-1}(\pi(y)) \cap (\Gamma_0 \setminus \Gamma'_0)$ est une hypersurface irréductible de $p_0^{-1}(\pi(y))$.*

Cela signifie la chose suivante: si $E$ est un fibré stable de rang $r$ et de degré $d$ sur $X$, la sous-variété fermée $Y$ de $Gr^{r-1}(H^0(e))$ constituée des sous-espaces $H$ tels que le morphisme de fibrés vectoriels

$$\mathcal{O}_X \otimes H \to E$$

soit non injectif est une hypersurface irréductible.

Soit $W$ la sous-variété fermée de $X \times Gr^{r-1}(H^0(E))$ constituée des couples $(x, H)$ tels que l'évaluation en $x: H \to E_x$ ne soit pas injective. Si $X \times Gr^{r-1}(H^0(E)) \to Gr^{r-1}(H^0(E))$ est la projection, on a $p_G(W) = Y$. Il suffit donc de prouver que $W$ est irréductible de codimension 2, et que les fibres de $p_G$: $W \to Y$ au dessus d'un point général de $Y$ sont finies. Pour démontrer la première assertion, on remarque que pour tout point $x$ de $X$, $p_X^{-1}(x)$ est une sous-variété fermée de $Gr^{r-1}(H^0(E))$ qui est de codimension 2: elle est constituée des sous-espaces $H$ tels que

$$H \cap \mathrm{Ker}(H^0(E) \to E_x) \neq \{0\}.$$

Il reste à prouver qu'il existe un point $H$ de $Y$ tel que $\mathcal{O}_x \otimes H \to E$ soit non injectif en seulement un nombre fini de points de $X$. Pour cela, on choisit un point $x$ de $X$ et un sous-espace vectoriel $H$ de dimension $r-1$ de $H^0(E)$ tel que la restriction de l'évaluation $H \to E_x$ soit injective (cela est possible car $H^0(E) \to E_x$ est surjective). Ceci démontre le lemme 7.3.

**Corollaire 7.4.** *La sous-variété fermée $\Gamma_0 \setminus \Gamma'_0$ de $\Gamma_0$ est une hypersurface irréductible.*

La restriction de $p_0$: $\Gamma_0 \setminus \Gamma'_0 \to U_s(r, d)$ est surjective, et ses fibres sont d'après le lemme 7.3 irréductibles et de même dimension. Il en découle d'après [24] (théorème 8, p. 61) que $\Gamma_0 \setminus \Gamma'_0$ est irréductible. Il est immédiat que c'est une hypersurface.

**Lemme 7.5.** *Le faisceau d'idéaux de $\Gamma_0 \setminus \Gamma'_0$ est de la forme $p_0^* \Lambda \otimes \mathcal{O}_{\Gamma_0}(n)^{-d/n}$, avec $\Lambda$ dans* $\mathrm{Pic}(U_s(r, d))$.

Il suffit de montrer que pour tout fibré stable $E$ de rang $r$ et de degré $d$ sur $X$, l'hypersurface $Y$ de $Gr^{r-1}(H^0(E))$ constituée des sous-espaces $H$ tels que le morphisme de fibrés $\mathcal{O}_X \otimes H \to E$ soit non injectif est de degré $d$. Pour cela, considérons le morphisme canonique de fibrés vectoriels sur $X \times Gr^{r-1}(H^0(E))$

$$p_G^* U \xrightarrow{\Phi} p_X^* E,$$

$U$ désignant le sous-fibré universel de $\mathcal{O}_{Gr^{r-1}(H^0(E))} \otimes H^0(E)$, $p_G$ la projection $X \times Gr^{r-1}(H^0(E)) \to Gr^{r-1}(H^0(E))$. Soit $W$ la sous-variété de $X \times Gr^{r-1}(H^0(E))$ constituée des couples $(x, H)$ tels que l'évaluation $H \to E_x$ ne soit pas injective. Alors $W$ est le lieu des points où $\Phi$ n'est pas injectif. D'après la formule de Porteous (cf. [4]), $c_2(p_X^* E - p_G^* U)$ est un multiple entier de $[W]$ dans $A^2(X \times Gr^{r-1}(H^0(E)))$. On a

$$c_2(p_X^* E - p_G^* U) = \alpha_1^2 - \alpha_2 - \alpha_1 \cdot \det(E),$$

$\alpha_1$, $\alpha_2$ étant respectivement la première et seconde classe de Chern de $U$. Puisque $c_2(p_X^* E - p_G^* U)$ n'est pas divisible dans $A^2(X \times Gr^{r-1}(H^0(E)))$, on a

$$[W] = \alpha_1^2 - \alpha_2 - \alpha_1 \cdot \det(E).$$

On a alors dans $A^1(Gr^{r-1}(H^0(E)))$

$$[Y] = p_G^*[W] = -\deg(E)\,\alpha_1,$$

donc $Y$ est de degré $d$. Ceci démontre le lemme 7.5.

**Proposition 7.6.** *On a une suite exacte*

$$0 \to \operatorname{Pic}(U_s(r, d)) \to \operatorname{Pic}(\Gamma_0') \to \mathbb{Z}/(d/n)\mathbb{Z} \to 0.$$

Montrons d'abord que $\Phi = p_0^*$: $\operatorname{Pic}(U_s(r, d)) \to \operatorname{Pic}(\Gamma_0')$ est injectif. Soit $L_0$ un fibré en droites sur $U_s(r, d)$ tel que $p_0^* L_0$ soit trivial sur $\Gamma_0'$. On a une suite exacte de groupes

$$\mathbb{Z} \xrightarrow{i} \operatorname{Pic}(\Gamma_0) \xrightarrow{\text{restriction}} \operatorname{Pic}(\Gamma_0') \to 0$$

([7], proposition II.6.5), le morphisme $i$ associant à $m$ la puissance mième du faisceau d'idéaux de $\Gamma_0 \backslash \Gamma_0'$. Il en découle qu'on a, dans $\operatorname{Pic}(\Gamma_0)$,

$$p_0^* L_0 \simeq p_0^* \Lambda^m \otimes \mathcal{O}_{\Gamma_0}(n)^{-dm/n},$$

avec $m$ entier, $p_0^* \Lambda \otimes \mathcal{O}_{\Gamma_0}(n)^{-d/n}$ étant le faisceau d'idéaux de $\Gamma_0 \backslash \Gamma_0'$, d'après le lemme 7.5. D'après la proposition 7.2(b), on a $m=0$, d'où $p_0^* L_0 \simeq \mathcal{O}_{\Gamma_0}$, et on en déduit, toujours à l'aide de la même proposition, que $L_0$ est trivial. Donc $\Phi$ est injectif.

On a un diagramme commutatif de groupes, avec lignes et colonnes exactes

$$\begin{array}{ccccccccc}
 & & & & 0 & & 0 & & \\
 & & & & \downarrow & & \downarrow & & \\
 & & & & \mathbb{Z} & = & \mathbb{Z} & & \\
 & & & & \downarrow{\scriptstyle i} & & \downarrow{\scriptstyle j} & & \\
0 & \longrightarrow & \mathrm{Pic}(U_s(r,d)) & \xrightarrow{p_0^*} & \mathrm{Pic}(\Gamma_0) & \longrightarrow & \mathbb{Z} & \longrightarrow & 0 \\
 & & \| & & \downarrow & & & & \\
0 & \longrightarrow & \mathrm{Pic}(U_s(r,d)) & \xrightarrow{\Phi} & \mathrm{Pic}(\Gamma_0') & & & & \\
 & & & & \downarrow & & & & \\
 & & & & 0 & & & &
\end{array}$$

La première ligne exacte découle de la proposition 7.2(b) et $j$ est la multiplication par $d/n$. On a

$$\begin{aligned}
\mathrm{Pic}(\Gamma_0')/\mathrm{Pic}(U_s(r,d)) &= (\mathrm{Pic}(\Gamma_0)/i(\mathbb{Z}))/\mathrm{Pic}(U_s(r,d)) \\
&= \mathrm{Pic}(\Gamma_0)/(i(\mathbb{Z}) + \mathrm{Pic}(U_s(r,d))) \\
&= (\mathrm{Pic}(\Gamma_0)/\mathrm{Pic}(U_s(r,d)))/j(\mathbb{Z}) \\
&= \mathbb{Z}/j(\mathbb{Z}) \simeq \mathbb{Z}/(d/n)\mathbb{Z}.
\end{aligned}$$

Ceci achève la démonstration de la proposition 7.6.

*7.3.4.* Soit $U_{Gr}$ le sous-fibré universel relatif de $\mathcal{O}_{Gr_0} \otimes F_0$. Pour tout point $y$ de $R^{ss}$ et tout $H$ dans $Gr^{r-1}(H^0(\mathbb{F}_{0y}))$, on a $U_{Gr,H} = H$. Soit

$$T = \underline{\mathrm{Hom}}(\mathcal{O}_{Gr_0} \otimes \mathbb{C}^{r-1}, U_{Gr}),$$

fibré vectoriel sur $Gr_0$ muni d'une action évidente de $GL(q)$. On pose

$$\mathbb{T} = T/GL(q) = \mathbb{P}(T)/PGL(q), \qquad T' = \underline{\mathrm{Isom}}(\mathcal{O}_{Gr_0} \otimes \mathbb{C}^{r-1}, U_{Gr}).$$

**Lemme 7.7.** *Il existe sur* $\mathbb{T}$ *une structure de fibré en espaces projectifs localement trivial sur* $\Gamma_0$.

Si $\mathcal{O}_{Gr_0}(1) = \Lambda^{r-1} U_{Gr}$, muni de l'action de $GL(q)$ déduite de celle de $GL(q)$ su $U_{Gr}$, on a

$$e(\mathcal{O}_{Gr_0}(1)) = r - 1.$$

D'autre part, il existe d'après la proposition 5.1 un fibre en droites $L$ sur $Gr_0$, provenant de $R^{ss}$, tel que $e(L) = n$. Puisque $r-1$ et $n$ sont premiers entre eux, il existe des entiers $a, b$ tels que $an + b(r-1) = 1$. Alors, si $L_0 = L^a \otimes \mathcal{O}_{Gr_0}(b)$, on a $e(L_0) = 1$. Il en découle que $T \otimes L_0^{-1}$ est en fait muni d'une action de $PGL(q)$.

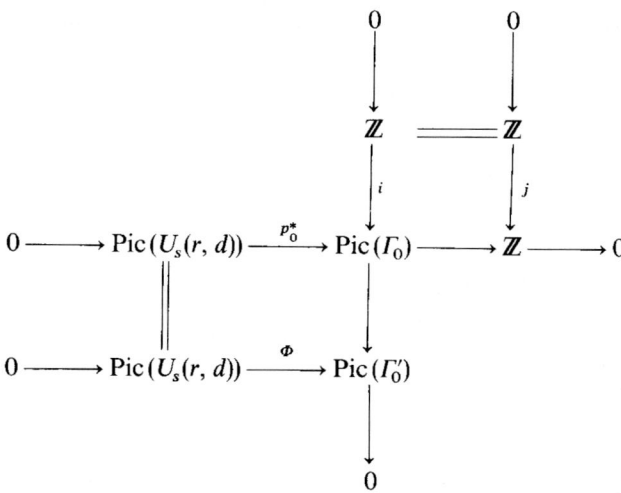

Le quotient $(T \otimes L_0^{-1})/PGL(q)$ est un fibré vectoriel sur $\Gamma_0$ dont le fibré en espaces projectifs associé est $\mathbb{T}$. Ceci démontre le lemme 7.7.

Soit $p'_0: \mathbb{T} \to \Gamma_0$ le morphisme canonique. Pour tout point $y$ de $R^{ss}$ et tout $H$ dans $p_0^{-1}(\pi(y)) = Gr^{r-1}(H^0(\mathbb{F}_{0y}))$, on a $p_0'^{-1}(H) = \mathbb{P}(\mathrm{Hom}(\mathbb{C}^{r-1}, H))$. Soit $\mathbb{T}'$ l'ouvert de $\mathbb{T}$ des points au dessus de $\Gamma'_0$ correspondant aux isomorphismes. On a

$$p_0'^{-1}(H) \cap \mathbb{T}' = \mathbb{P}(\mathrm{Isom}(\mathbb{C}^{r-1}, H)).$$

**Proposition 7.8.** *On a une suite exacte*

$$0 \to \mathrm{Pic}(\Gamma'_0) \to \mathrm{Pic}(\mathbb{T}') \to \mathbb{Z}/(r-1)\mathbb{Z} \to 0.$$

La démonstration est analogue à celle de la proposition 7.6, compte tenu du fait que le faisceau d'idéaux de $\mathbb{T} \backslash \mathbb{T}'$ est de la forme

$$\mathcal{O}_{\mathbb{T}}(r-1) \otimes \varphi^* L,$$

$L$ étant un fibré en droites sur $\Gamma'_0$ et $\varphi: \mathbb{T}' \to \Gamma'_0$ la projection.

*7.3.5. Conclusion*

**Proposition 7.9** (Seshadri). *On a un isomorphisme* $\mathbb{P}^s \simeq \mathbb{T}'$.

Soit $y$ un point de $\mathbb{P}^s$. On a donc une suite exacte

$$0 \to \mathcal{O}_X \otimes \mathbb{C}^{r-1} \xrightarrow{i} \mathbb{E}_y \to D_\alpha \otimes y \to 0,$$

avec $\alpha = \pi_0(y)$ (cf. 7.2). De $\mathbb{E}_y$ on déduit un point $z$ de $U_s(r, d)$, de $\mathrm{Im}(H^0(i)) \subset H^0(\mathbb{E}_y)$ un point $H$ de $\Gamma'_0$ au dessus de $z$, puis de l'isomorphisme $\mathbb{C}^{r-1} \simeq \mathrm{Im}(H^0(i))$ déduit de $i$ un point de $\mathbb{T}'$ au dessus de $H$. Pour montrer qu'on définit ainsi un morphisme $\Phi: \mathbb{P}^s \to \mathbb{T}'$, on considère pour tout $y$ de $\mathbb{P}^s$ un isomorphisme sur un voisinage $U$ de $y$:

$$\mathcal{O}_U \otimes \mathbb{C}^q \simeq p_{\mathbb{P}^s *}(\mathbb{E} \otimes p_X^* \mathcal{O}_X(m))$$

permettant de factoriser $\Phi|_U$ à travers $T$.

L'isomorphisme réciproque est défini de la façon suivante: soient $y$ un point de $R^{ss}$, $\gamma$ un point de $T$ au dessus de $y$. On en déduit une suite exacte

$$0 \to \mathcal{O}_X \otimes \mathbb{C}^{r-1} \to \mathbb{F}_{0y} \to \det(\mathbb{F}_{0y}) \to 0.$$

Si $\alpha$ est le point de $J^{(d)}$ cprrespondant à $\det(\mathbb{F}_{0y})$, la suite exacte précédente définit un point de $\mathbb{P}^s$ au dessus de $\alpha$. On obtient ainsi un morphisme $GL(q)$-invariant $T' \to \mathbb{P}^s$ qui donne par passage au quotient un morphisme $\mathbb{T}' \to \mathbb{P}^s$ inverse de $\Phi$. Ceci démontre la proposition 7.9.

*Remarques.* 1. Soient $L$ un fibré en droites de degré $d$ sur $X$,

$$\mathbb{P}_L = \pi_0^{-1}(L), \quad \mathbb{P}_L^s = \mathbb{P}_L \cap \mathbb{P}^s,$$

$\Gamma'_{0L}$, $\mathbb{T}'_L$ les sous-variétés fermées de $\Gamma'_0$, $\mathbb{T}'$ respectivement, images réciproques de $U_s(r, L)$. L'isomorphisme $\Phi$ en induit un $\mathbb{P}^s_L \simeq \mathbb{T}'_L$, et on a des suites exactes

$$0 \to \operatorname{Pic}(U_s(r, L)) \to \operatorname{Pic}(\Gamma'_{0L}) \to \mathbb{Z}/(d/n)\mathbb{Z} \to 0,$$

$$0 \to \operatorname{Pic}(\Gamma'_{0L}) \to \operatorname{Pic}(\mathbb{P}^s_L) \to \mathbb{Z}/(r-1)\mathbb{Z} \to 0.$$

Remarquons que $\mathbb{P}_L$ est un espace projectif.

2. Soit $p_1: Gr'_0 \to R^s$ la projection. On a vu dans le lemme 7.7 qu'il existe un $GL(q)$-fibré en droites $L$ sur $Gr'_0$ tel que $e(L)=1$. Alors $V = p_1^{\#} \mathbb{F}_0|_{R^s \times X} \otimes L^{-1}$ est en fait un $PGL(q)$-fibré vectoriel sur $Gr'_0 \times X$. Le quotient

$$\mathbb{E}' = V/PGL(q)$$

est une famille de fibrés de $U(r, d)$ paramétrée par $\Gamma'_0$. Le morphisme canonique

$$f_{\mathbb{E}'}: \Gamma'_0 \to U(r, d)$$

n'est autre que la restriction de $p_0$, défini en 7.3.1. Si

$$p': \mathbb{P}^s = \mathbb{T}' \to \Gamma'_0$$

est la projection, il existe un fibré en droites $L_0$ sur $\mathbb{P}^s$ et un isomorphisme

$$p'^* \mathbb{E}' \otimes L_0 \simeq \mathbb{E}$$

(cela découle aisément de la simplicité des fibrés stables).

### 7.4. *Applications à l'étude de* Pic$(U(r, d))$ *et* Pic$(U(r, L))$

*7.4.1. Etude de l'image de* $\operatorname{Pic}(J^{(d)})$ *dans* $\operatorname{Pic}(U_s(r, d))$. Soit $L_0$ un élément de $\operatorname{Pic}^{(0)}(X)$, qu'on peut voir comme un élément de $\operatorname{Pic}^{(0)}(J^{(d)})$, ou de $Z \subset H(r, d)$ (cf. § 3). Soit

$$\det: U(r, d) \to J^{(d)}$$

le morphisme canonique. On a défini dans le § 3 un morphisme de groupes

$$\gamma: H(r, d) \to \operatorname{Pic}(F(r, d)) = \operatorname{Pic}(U(r, d)).$$

**Proposition 7.10.** *On a* $\gamma(L_0) = \det^* L_0$.

**Lemme 7.11.** *Le morphisme*

$$\begin{aligned} \operatorname{Pic}(F(r, d)) &\to \operatorname{Pic}(\mathbb{P}^s) \\ L &\to L_{\mathbb{E}} \end{aligned}$$

*est injectif.*

On a un diagramme commutatif

$$\begin{array}{ccc} \mathbb{T}'=\mathbb{P}^s & \xrightarrow{f_{\mathbb{E}}} & U_s(r,d) \\ \uparrow{\scriptstyle \pi'} & & \uparrow{\scriptstyle \pi} \\ T' & \xrightarrow{p} & R^{ss}, \end{array}$$

$\pi'$ étant le morphisme quotient, et $p$ la projection canonique. Si $L$ est élément de $\mathrm{Pic}(F(r,d))$ tel que $L_{\mathbb{E}}$ soit trivial, $\pi'^* L_{\mathbb{E}} = L_{\pi'^*\mathbb{E}}$ l'est aussi. Mais d'après la remarque 2 de 7.3.4, la famille $\pi'^*\mathbb{E}$ est équivalente à $p^*\mathbb{F}_0|_{R^s\times X}$. Donc

$$L_{p^*\mathbb{F}_0|_{R^s\times X}} \simeq p^* L_{\mathbb{F}_0|_{R^s\times X}}$$

est aussi trivial. Le morphisme

$$p^*: \mathrm{Pic}(R^{ss}) \to \mathrm{Pic}(T')$$

est injectif: la démonstration est analogue à celle de l'injectivité de $\mathrm{Pic}(U_s(r,d)) \to \mathrm{Pic}(\Gamma_0')$ dans la proposition 7.6. Il en découle que $L_{\mathbb{F}_0|_{R^s\times X}}$ est trivial. D'après le corollaire 3.4, $L$ est aussi trivial. Ceci démontre le lemme 7.11.

Pour démontrer la proposition 7.10 il suffit donc de prouver que

$$\gamma(L_0)_{\mathbb{E}} \simeq f_{\mathbb{E}}^* \det{}^* L_0.$$

Soit $L_1$ un fibré en droites de degré 0 sur $X$. L'élément de $Z$ correspondant à $L_0$ est $[L_0\otimes L_1]-[L_1]$, et l'élément de $\mathrm{Pic}^{(0)}(J^{(d)})$ correspondant à $L_0$ est aussi celui provenant de l'élément précédent de $Z$ (cf. 3.3). D'autre part, le morphisme composé

$$\mathbb{P}^s \to U_s(r,d) \xrightarrow{\det} J^{(d)}$$

est celui qui est défini par la famille $\det(\mathbb{E})$ de fibrés en droites de degré $d$ sur $X$. Il suffit donc de prouver le résultat suivant: pour tout fibré en droites $L_1$ de degré 0 sur $X$, on a un isomorphisme

$$(*) \qquad \det(p_{\mathbf{P}^s*}(\mathbb{E}\otimes p_X^* L_1)) \simeq \det(p_{\mathbf{P}^s*}(\det(\mathbb{E})\otimes p_X^* L_1)),$$

et

$$R^1 p_{\mathbf{P}^s*}(\mathbb{E}\otimes p_X^* L_1)) = R^1 p_{\mathbf{P}^s*}(\det(\mathbb{E})\otimes p_X^* L_1) = 0.$$

Les égalités précédentes sont vérifiées car $d$ est positif. Il reste donc à prouver l'isomorphisme $(*)$. D'après la suite exacte

$$0 \to \mathcal{O}_{\mathbf{P}^s\times X}\otimes \mathbb{C}^{r-1} \to \mathbb{E} \to \pi_0^{\#} D \otimes p_{\mathbf{P}^s*}\mathcal{O}_{\mathbf{P}}(-1) \to 0$$

on a

$$\det(\mathbb{E}) \simeq \pi_0^{\#} D \otimes p_{\mathbf{P}^s*}\mathcal{O}_{\mathbf{P}}(-1).$$

On a donc une suite exacte

$$0 \to \mathcal{O}_{\mathbb{P}^s \times X} \otimes \mathbb{C}^{r-1} \to \mathbb{E} \to \det(\mathbb{E}) \to 0,$$

et l'isomorphisme (∗) en découle immédiatement. Ceci démontre la proposition 7.10.

*7.4.2. Démonstration du théorème B.* Soient

$$r' = \frac{r}{n}, \qquad d' = \frac{-d + r(g-1)}{n}.$$

Il existe d'après [8], § 4.6, un fibré vectoriel $F$ sur $X$ de rang $r'$ et de degré $d'$ tel qu'il existe un fibré stable de rang $r$ et de degré $d$ tel que

$$h^0(E \otimes F) = 0.$$

On note comme dans l'introduction $\Theta^s_{F,L}$ le sous-ensemble de $U_s(r, L)$ constitué des points correspondant aux fibrés stables $E$ tels que $h^0(E \otimes F) \neq 0$.

**Lemme 7.12.** *Le groupe* $\mathrm{Pic}(U(r, L))$ *est isomorphe à* $\mathbb{Z}$.

Puisque $\mathrm{codim}_{U(r,L)}(U(r, L) \setminus U_s(r, L)) \geqq 2$, et que $U(r, L)$ est localement factorielle, il suffit de montrer que $\mathrm{Pic}(U_s(r, L)) \simeq \mathbb{Z}$. Le groupe de Picard de $\mathbb{P}^s_L$ est un quotient de $\mathbb{Z}$, car $\mathbb{P}^s_L$ est un ouvert d'un espace projectif. Puisque $\mathrm{codim}_{U(r,L)}(U(r, L) \setminus U_s(r, L)) \geqq 2$, $\mathrm{Pic}(U_s(r, L))$ possède au moins un élément qui n'est pas d'ordre fini, et comme $\mathrm{Pic}(U_s(r, L)) \subset \mathrm{Pic}(\mathbb{P}^s_L)$ d'après la remarque 1 de 7.3.5, on a $\mathrm{Pic}(U_s(r, L)) \simeq \mathbb{Z} \simeq \mathrm{Pic}(\mathbb{P}^s_L)$, ce qui démontre le lemme 7.12.

**Proposition 7.13.** *L'image de* $\mathrm{Pic}(U_s(r, L))$ *dans* $\mathrm{Pic}(\mathbb{P}^s_L) \simeq \mathbb{Z}$ *est le sous-groupe* $\mathbb{Z}/(r-1)\dfrac{d}{n} \cdot \mathbb{Z}$.

On a d'après 7.3.5 des suites exactes

$$0 \to \mathrm{Pic}(U_s(r, L)) \to \mathrm{Pic}(\Gamma'_{0L}) \to \mathbb{Z}/(d/n)\mathbb{Z} \to 0,$$
$$0 \to \mathrm{Pic}(\Gamma'_{0L}) \to \mathrm{Pic}(\mathbb{P}^s_L) \to \mathbb{Z}/(r-1)\mathbb{Z} \to 0,$$

d'où une suite exacte

$$0 \to \mathbb{Z}/(r-1)\mathbb{Z} \to \mathrm{Pic}(\mathbb{P}^s_L)/\mathrm{Pic}(U_s(r, L)) \to \mathbb{Z}/(d/n)\mathbb{Z} \to 0.$$

On en déduit que $\mathrm{Pic}(\mathbb{P}^s_L)/\mathrm{Pic}(U_s(r, L))$ possède $(r-1)\dfrac{d}{n}$ éléments, ce qui démontre la proposition 7.13.

D'après le corollaire 4.3, le morphisme canonique

$$i: \mathrm{Pic}(U(r, d)) \to \mathrm{Pic}(F(r, d))$$

est un isomorphisme. Posons

$$\mathbb{L} = i^{-1} \circ \gamma : H(r, d) \to \mathrm{Pic}(U(r, d))$$

(cf. § 3). On note $\mathbb{E}_L$ la restriction de $\mathbb{E}$ à $\mathbb{P}^s_L \times X$, et pour tout entier $m$ on pose

$$\mathcal{O}_{\mathbb{P}^s_L}(m) = \mathcal{O}_{\mathbb{P}_L}(m)|_{\mathbb{P}^s_L}.$$

**Lemma 7.14.** *On a* $\gamma([F])_{\mathbb{E}_L} = \mathcal{O}_{\mathbb{P}^s_L}\left(-(r-1)\,\dfrac{d}{n}\right)$.

Considérons la suite exacte

$$0 \to \mathcal{O}_{\mathbb{P}^s_L \times X} \otimes \mathbb{C}^{r-1} \to \mathbb{E}_L \to p^*_{\mathbb{P}^s_L}\,\mathcal{O}_{\mathbb{P}^s_L}(-1) \otimes L \to 0.$$

On en déduit la suite exacte

$$0 \to \mathcal{O}_{\mathbb{P}^s L} \otimes H^0(\mathbb{C}^{r-1} \otimes F) \to p_{\mathbb{P}^s_L *}(\mathbb{E}_L \otimes F) \to \mathcal{O}_{\mathbb{P}^s_L}(-1) \otimes H^0(F \otimes L) \to \dots$$
$$\to \mathcal{O}_{\mathbb{P}^s_L} \otimes H^1(\mathbb{C}^{r-1} \otimes F) \to R^1 p_{\mathbb{P}^s_L *}(\mathbb{E}_L \otimes F) \to \mathcal{O}_{\mathbb{P}^s_L}(-1) \otimes H^1(F \otimes L) \to 0,$$

et l'isomorphisme

$$\det(p_{\mathbb{P}^s_L !}(\mathbb{E}_L \otimes F)) \simeq \mathcal{O}_{\mathbb{P}^s_L}(-\chi(F \otimes L)).$$

Le lemme 7.14 découle du fait que $\chi(F \otimes L) = (r-1)\,\dfrac{d}{n}$.

Pour démontrer le théorème B il reste à prouver que $\Theta^s_{F,L}$ est une hypersurface de $U_s(r, L)$ et que

$$\mathcal{O}(-\theta^s_{F,L}) \simeq \mathbb{L}([F])|_{U_s(r,L)}.$$

Considérons le morphisme

$$\alpha\colon \mathcal{O}_{\mathbb{P}^s_L}(-1) \otimes H^0(F \otimes L) \to \mathcal{O}_{\mathbb{P}^s_L} \otimes H^1(\mathbb{C}^{r-1} \otimes F)$$

du lemme 7.14. Puisque $p_{\mathbb{P}^s_L *}(\mathbb{E}_L \otimes F)$ est de torsion, $\alpha$ est un morphisme génériquement bijectif, dont le schéma des zéros est exactement $f^{-1}_{\mathbb{E}_L}(\Theta^s_{F,L})$. Il en découle que $f^{-1}_{\mathbb{E}_L}(\Theta^s_{F,L})$ est une hypersurface de $\mathbb{P}^s_L$. Les fibres de $f_{\mathbb{E}_L}$ étant de dimension constante, $\Theta^s_{F,L}$ est bien une hypersurface de $U_s(r, L)$.

Le fibré $\gamma(-[F])_{\mathbb{E}_L}$ est isomorphe au faisceau d'idéaux du schéma des zéros de $\alpha$, et engendre l'image de $\operatorname{Pic}(U_s(r,d))$ dans $\operatorname{Pic}(\mathbb{P}^s_L)$ d'après la proposition 7.13. Le fibré $\mathcal{O}(f^{-1}_{\mathbb{E}_L}(\Theta^s_{F,L}))$ est une puissance de $\gamma([F])_{\mathbb{E}_L}$. On doit donc avoir en fait

$$\mathcal{O}(f^{-1}_{\mathbb{E}_L}(\Theta^s_{F,L})) = \gamma(-[F])_{\mathbb{E}_L},$$

d'où

$$\mathcal{O}(-\Theta^s_{F,L}) = \mathbb{L}([F])|_{U_s(r,L)}.$$

Ceci achève la démonstration du théorème B.

*7.4.3. Démonstration du théorème C.* On démontre d'abord de la même façon que pour $U(r, L)$ que l'image de $\operatorname{Pic}(U_s(r,d))$ dans $\operatorname{Pic}(\mathbb{P}^s)$ est le sous-groupe engendré par $\mathrm{Pic}(J^{(d)})$ et $\mathcal{O}_{\mathbb{P}^s}\left((r-1)\dfrac{d}{n}\right)$. Ensuite, comme dans les démonstrations précédentes, on montre que

$$f_{\mathbb{E}}^* \mathcal{O}(-\Theta_F^s) = \gamma([F])_{\mathbb{E}},$$

et que ce fibré se met sous la forme $\mathcal{O}_{\mathbb{P}^s}\left((r-1)\dfrac{d}{n}\right) \otimes L$, le fibré en droites $L$ provenant de $\mathrm{Pic}(J^{(d)})$. Compte tenu du théorème A, le théorème C en découle immédiatement. On obtient aussi que

$$\mathbb{L}([F]) = \mathcal{O}(-\Theta_F),$$

ce qui montre comme indiqué dans l'introduction que si $F'$ est un fibré vectoriel sur $X$ analogue à $F$, on a

$$\mathcal{O}(\Theta_{F'}) = \mathcal{O}(\Theta_F) \otimes \det^*(\det(F') \otimes \det(F)^{-1}),$$

$(\det(F') \otimes \det(F)^{-1}$ étant vu comme un élément de $\mathrm{Pic}(J^{(d)}))$.

*7.4.4. Démonstration du théorème D.* Remarquons que

$$H(r,d) = Z + \mathbb{Z} \cdot [F],$$

$Z$ désignant le sous-groupe de $H(r,d)$ isomorphe à $\mathrm{Pic}^{(0)}(X)$ (cf. § 3). La partie (a) du théorème D découle immédiatement du corollaire 4.3.

Prouvons (b). Il suffit de montrer que la restriction de

$$\mathbb{L}: H(r,d) \to \mathrm{Pic}(U(r,d))$$

à $Z$ est injective. En effet, la composée

$$H(r,d) \xrightarrow{\mathbb{L}} \mathrm{Pic}(U(r,d)) \to \mathrm{Pic}(U(r,L))$$

est triviale sur $Z$ et injective sur $\mathbb{Z} \cdot [F]$ d'après 7.4.2. Donc si la restriction de $\mathbb{L}$ à $Z$ est injective, il en est de même de $\mathbb{L}$. D'après la proposition 7.10, il suffit, pour prouver l'injectivité de $\mathbb{L}|_Z$, de prouver celle de

$$\det^*: \mathrm{Pic}^{(0)}(X) \subset \mathrm{Pic}(J^{(d)}) \to \mathrm{Pic}(U(r,d)).$$

Si $L_0$ est un fibré en droites sur $J^{(d)}$, on a, les fibres de $\det: U(r,d) \to J^{(d)}$ étant projectives et irréductibles

$$\det_*(\det^* L_0) \simeq L_0,$$

donc $\det^*$ est injective. Ceci prouve (b).

Du théorème C et de l'égalité $\mathbb{L}([F]) = \mathcal{O}(-\Theta_F)$, on déduit

$$\mathrm{Pic}(U(r,d)) = \mathrm{Im}(\mathbb{L}) + \mathrm{Pic}(J^{(d)}).$$

L'égalité

$$\mathrm{Im}(\mathbb{L}) \cap \mathrm{Pic}(J^{(d)}) = \mathrm{Pic}^{(0)}(X)$$

est encore une conséquence du théorème C et de la proposition 7.10.

Le théorème D est donc démontré.

### 7.5. *Faisceau canonique et faisceau dualisant*

On démontre ici les théorèmes *E* et *F*. On ne démontrera que le théorème E, l'autre étant analogue. Prouvons d'abord (a). On note $T_U$ le fibré tangent de $U_s(r, d)$. On a un isomorphisme

$$f_{\mathbb{E}}^* T_U \simeq R^1 p_{\mathbb{P}^s *}(\mathbb{E}^* \otimes \mathbb{E}),$$

et il suffit de calculer le déterminant de ce dernier faisceau dans $\mathrm{Pic}(\mathbb{P}^s)$. On a une suite exacte

$$(**) \qquad 0 \to \mathcal{O}_{\mathbb{P}^s \times X} \otimes \mathbb{C}^{r-1} \to \mathbb{E} \to \pi_0^{\#} D \otimes p_{\mathbb{P}^s}^* \mathcal{O}_{\mathbb{P}^s}(-1) \to 0,$$

d'où une suite exacte

$$\begin{aligned} 0 \to \mathcal{O}_{\mathbb{P}^s} \to \mathcal{O}_{\mathbb{P}^s}(-1) \otimes p_{\mathbb{P}^s *}(\mathbb{E}^* \otimes \pi_0^{\#} D) \to R^1 p_{\mathbb{P}^s *}(\mathbb{E}^* \otimes \mathbb{C}^{r-1}) \to \dots \\ \to R^1 p_{\mathbb{P}^s *}(\mathbb{E}^* \otimes \mathbb{E}) \to 0 \end{aligned}$$

(car $d \gg 0$). Il en découle que

$$\det(R^1 p_{\mathbb{P}^s *}(\mathbb{E}^* \otimes \mathbb{E})) \simeq \det(R^1 p_{\mathbb{P}^s *} \mathbb{E}^*)^{r-1} \otimes \det(\mathcal{O}_{\mathbb{P}^s}(-1) \otimes p_{\mathbb{P}^s *}(\mathbb{E}^* \otimes \pi_0^{\#} D))^{-1}.$$

En utilisant la suite exacte duale de $(**)$ on trouve la suite exacte

$$\begin{aligned} 0 \to \mathcal{O}_{\mathbb{P}^s} \otimes \mathbb{C}^{r-1} \to \mathcal{O}_{\mathbb{P}^s}(1) \otimes R^1 p_{\mathbb{P}^s *}(\pi_0^{\#} D^*) \to \dots \\ \to R^1 p_{\mathbb{P}^s *} \mathbb{E}^* \to \mathcal{O}_{\mathbb{P}^s} \otimes \mathbb{C}^{(r-1)g} \to 0, \end{aligned}$$

d'où un isomorphisme

$$\det(R^1 p_{\mathbb{P}^s *} \mathbb{E}^*) \simeq \det(\mathcal{O}_{\mathbb{P}^s}(1) \otimes R^1 p_{\mathbb{P}^s *}(\pi_0^{\#} D^*)).$$

De même, en utilisant la suite exacte duale de $(**)$ on trouve l'isomorphisme

$$\det(\mathcal{O}_{\mathbb{P}^s}(-1) \otimes p_{\mathbb{P}^s *}(\mathbb{E}^* \otimes \pi_0^{\#} D)) \simeq \det(p_{\mathbb{P}^s *}(\pi_0^{\#} D) \otimes \mathbb{C}^{r-1} \otimes \mathcal{O}_{\mathbb{P}^s}(-1)).$$

Si $L$ est un fibré en droites de degré $d$ sur $X$, on a

$$h^0(L) = \chi(L) = d + 1 - g, \qquad h^1(L^*) = \chi(L^*) = d + g - 1,$$

donc

$$\det(R^1 p_{\mathbb{P}^s *} \mathbb{E}^*) \simeq \mathcal{O}_{\mathbb{P}^s}(d + g - 1) \otimes \det(R^1 p_{\mathbb{P}^s *}(\pi_0^{\#} D)),$$

et

$$\det(\mathcal{O}_{\mathbb{P}^s}(-1) \otimes p_{\mathbb{P}^s *}(\mathbb{E}^* \otimes \pi_0^{\#} D)) \simeq \mathcal{O}_{\mathbb{P}^s}((r-1)(g-d-1)) \otimes \det(p_{\mathbb{P}^s *}(\pi_0^{\#} D))^{r-1}.$$

Finalement, on a

$$\det(R^1 p_{\mathbb{P}^s*}(\mathbb{E}^*\otimes\mathbb{E}))=\mathcal{O}_{\mathbb{P}^s}(2(r-1)d)\otimes\Delta^{r-1},$$

avec

$$\Delta=\det(R^1 p_{\mathbb{P}^s*}(\pi_0^\# D^*))\otimes\det(p_{\mathbb{P}^s*}(\pi_0^\# D))^{-1}.$$

Soit $F_0$ un fibré vectoriel de rang $2r$ et de degré $2(-d+r(g-1))$. En utilisant la suite exacte (**) on voit que

$$\gamma([F_0])_{\mathbb{E}}\simeq\mathcal{O}_{\mathbb{P}^s}(-2d(r-1))\otimes\det^*(p_{J(d)*}(D\otimes p_X^* F_0)),$$

d'où

$$\det(R^1 p_{\mathbb{P}^s*}(\mathbb{E}^*\otimes\mathbb{E}))\simeq\gamma([F_0])_{\mathbb{E}}^{-1}\otimes\det^*(p_{J(d)*}(D\otimes p_X^* F_0))\otimes\Delta^{r-1},$$

ce qui démontre (a) (la formulation du théorème E est légèrement différente pour être valable même si on n'a pas $d\gg 0$).

Prouvons (b). D'après [22] le faisceau dualisant d'une variété normale est divisoriel, c'est à dire qu'il provient d'un diviseur de Weil de cette variété. Le faisceau dualisant de $U(r, d)$ est donc divisoriel. Mais cette variété est localement factorielle, donc tout diviseur de Weil de $U(r, d)$ est de Cartier, et par conséquent le faisceau dualisant de $U(r, d)$ est localement libre. Il est isomorphe à $\omega$, puisqu' il coïncide avec $\omega$ sur $U_s(r, d)$ et que $\operatorname{codim}_{U(r,d)}(U(r, d)\setminus U_s(r, d))\geqq 2$. Ceci achève la démonstration du théorème E.

*Remarque.* Voici une autre démonstration du fait que le faisceau dualisant de $U(r, d)$ est localement libre: du fait que $R^{ss}$ est lisse, $U(r, d)$ est une variété de Cohen-Macaulay (cela découle du théorème de Hochster-Roberts, [15], p. 153, [10, 11]). Et une variété de Cohen-Macaulay localement factorielle est de Gorenstein, c'est à dire que son faisceau dualisant est inversible (cf. [16]).

## Bibliographie


1. Drezet, J.—M.: Groupe de Picard des variétés de modules de faisceaux semi-stables sur $\mathbb{P}_2(\mathbb{C})$. A paraitre aux Annales de l'institut Fourier
2. Drezet, J.—M.: Fibrés exceptionnels et variétés de modules de faisceaux semi-stables sur $\mathbb{P}_2(\mathbb{C})$. Z. Angew. Math. Mech. **380**, 14–58 (1987)
3. Drezet, J.-M.: Groupe de Picard des variétés de modules de faisceaux semi-stables sur $\mathbb{P}_2$. Singularities, representation of algebras, and vector bundles. Proc. Lambrecht 1985. (Lect. Notes Math., Vol. 1273). Berlin-Heidelberg-New York: Springer 1987
4. Fulton, W.: Intersection theory. (Ergebnisse der Mathematik und ihre Grenzgebiete). Berlin-Heidelberg-New York: Springer 1984
5. Gieseker, D.: On the moduli of vector bundles on an algebraic surface. Ann. Math. **106**, 45–60 (1977)
6. Grothendieck, A.: Technique de descente et théorèmes d'existence en géométrie algébrique. IV Les schémas de Hilbert. Séminaire Bourbaki **221**, (1960/61)
7. Hartshorne, R.: Algebraic geometry. (Grad. Texts in Math., Vol. 52). Berlin-Heidelberg-New York: Springer 1977
8. Hirschowitz, A.: Problèmes de Brill-Noether en rang supérieur. Preprint
9. Hirschowitz, A., Narasimhan, M.S.: Fibrés de t'Hooft spéciaux et applications. Enumerative geometry and Classical algebraic geometry. Progr. Math., Boston **24**, (1982)

10. Hochster, M., Roberts, J.: Rings of invariants of reductive groups acting on regular rings are Cohen-Macaulay. Adv. Math. **13**, 115 (1974)
11. Kempf, G.: Hochster-Roberts theorem in invariant theory. Mich. Math. J. **26**, 19 (1979)
12. Lang, S.: Abelian varieties. New York: Interscience Publ. 1959
13. Maruyama, M.: Moduli of stable sheaves II. J. Math. Kyoto Univ. **18**, 557–614 (1978)
14. Matsumura, H.: Commutative algebra. New York: W.A. Benjamin Co. 1970
15. Mumford, D., Fogarty, J.: Geometric invariant theory. (Ergebnisse der Mathematikund ihre Grenzgebiete). Berlin-Heidelberg-New York: Springer 1984
16. Murthy, M.P.: A note on factorial rings. Arch. Math. **15**, 418–420 (1964)
17. Narasimhan, M.S., Ramanan, S.: Moduli of vector bundles on a compact Riemann surface. Ann. Math. **89**, 14–51 (1969)
18. Narasimhan, M.S., Ramanan, S.: Vector bundles on curves. (Proc. of Bombay Coll. of Algebraic Geometry, pp. 335–346.) Oxford Univ. Press 1969
19. Newstead, P.E.: Introduction to moduli problems and orbit spaces. TIFR Lect. Notes **51**, (1978)
20. Ramanan, S.: The moduli spaces of vector bundles on an algebraic curve. Math. Ann. **200**, 69–84 (1973)
21. Ramanathan, A.: Stable principal bundles on a compact Riemann surface, construction of moduli space. Ph.D. Thesis. Univ. of Bombay (1976)
22. Reid, M.: Canonical 3-folds. Journées de Géométrie Algébrique d'Angers, edité par A. Beauville (1979) 273–310
23. Seshadri, C.S.: Fibrés vectoriels sur les courbes algébriques. Astérisque **96**, (1982)
24. Shafarevitch, I.R.: Basic algebraic Geometry. (Grundlehren, Bd. 213). Berlin-Heidelberg-New York: Springer 1974

Oblatum 18-VI-1988

PACIFIC JOURNAL OF MATHEMATICS
Vol. 145, No. 2, 1990

# COMPACTIFICATION OF $M_{\mathbb{P}_3}(0, 2)$ AND PONCELET PAIRS OF CONICS

M. S. NARASIMHAN AND G. TRAUTMANN

**Let $M(0, 2)$ denote the quasi-projective variety of isomorphism classes of stable rank 2 vector bundles on $\mathbb{P}_3(C)$ with $c_1 = 0$ and $c_2 = 2$. In this paper we study a natural (irreducible) compactification of $M(0, 2)$ and describe explicitly the sheaves on $\mathbb{P}_3$ which occur in the closure of $M(0, 2)$ in the moduli space of semi-stable sheaves on $\mathbb{P}_3$ with $c_1 = 0$, $c_2 = 2$ and $c_3 = 0$.**

**Introduction.** The space $M(0, 2)$ of stable rank 2 vector bundles on $\mathbb{P}_3$ with $c_1 = 0$, $c_2 = 2$ was investigated in detail by Hartshorne [**Ha2**]. (See also [**Au-Dou**].) He proved that $M(0, 2)$ has the structure of a fibre space over the 9-dimensional variety $R$ of reguli, the fibre being an open subset of a smooth quadric in $\mathbb{P}_5$. (A regulus is a smooth quadric in $\mathbb{P}_3$ with a distinguished system of generating lines.) If $S$ is the smooth conic in the Grassmannian $\mathbb{G}$ of lines in $\mathbb{P}_3$ given by the generators of a regulus $\rho$, then the fibre over $\rho$ consists of smooth conics $C$ such that $S$ and $C$ are Poncelet related with $S$ as the inner conic, i.e. a triangle can be inscribed in $C$ which circumscribes $S$.

To obtain a natural compactification of $M(0, 2)$, we first compactify the fibres over $R$ by taking all conics $S$, smooth or not, which are Poncelet related to $S$; the fibre over $\rho = S$ is then a smooth quadric in $\mathbb{P}_5$. We then take as the compactification of the space $R$ of reguli the *Hilbert Scheme* $C(\mathbb{G})$ of all conics contained in the Grassmannian $\mathbb{G}$. The quadric bundle over $R$ extends to a bundle over $C(\mathbb{G})$, namely the Poncelet quadric bundle associated to the tautological conic bundle over $C(\mathbb{G})$; it is constructed by considering also the space of conics which are Poncelet related to singular conics, such that the fibre of this quadric bundle is a pair of hyperplanes in $\mathbb{P}_5$ in the case of a pair of lines and a double hyperplane in $\mathbb{P}_5$ in the case of a double line. This Poncelet quadric bundle $Q$, which is a normal projective variety, is the compactification of $M(0, 2)$ we study.

The space $Q$ essentially parametrises a family of semi-stable sheaves of rank 2 with $c_1 = c_3 = 0$, $c_2 = 2$. More precisely it is shown that $Q$ is a G.I.T. quotient of a space $X^{ss}$ by SL(2) and that

$X^{ss}$ parametrises a *flat* family of *semi-stable* sheaves with $c_1 = c_3 = 0$, $c_2 = 2$ invariant under the action of SL(2) (see 8.1, 8.2). The smooth points of $Q$ correspond exactly to stable sheaves. We describe in §§9, 10 explicitly the sheaves occurring in the family parametrised by $X^{ss}$.

Let $\overline{M(0,2)}$ by the (schematic) closure of $M(0,2)$ in the Maruyama scheme of semi-stable sheaves on $\mathbb{P}_3$ with $c_1 = 0$, $c_2 = 2$, $c_3 = 0$. We investigate the canonical morphism $Q \to \overline{M(0,2)}$ defined by the family parametrised by $X^{ss}$ and prove (Theorem 4.4) that the normalisation $\widetilde{M(0,2)}$ of $\overline{M(0,2)}$ is isomorphic to the variety obtained by blowing down $Q$ along the fibres of a $\mathbb{P}_1$-fibration (see 4.2) on a codimension 5 subvariety contained in the singular locus of $Q$. Moreover the canonical map $\widetilde{M(0,2)} \to \overline{M(0,2)}$ is bijective and the smooth points of $\widetilde{M(0,2)}$ are precisely the stable sheaves.

We now briefly describe the contents of the different sections of the paper.

In §2 we mainly review the theory of $M(0,2)$ from the point of view of monads, jumping lines and Poncelet conics. It is in particular shown that the set of second order jumping lines of a bundle $\mathcal{E} \in M(0,2)$ is the conic $S^0 \subset \mathbb{G}$ "conjugate" to the conic $S$ defined by the regulus associated with $\mathcal{E}$. This result will be generalized in 7.6 to the case of sheaves which are limits of elements in $M(0,2)$.

We deal with the Hilbert scheme $C(\mathbb{G})$ of conics in $\mathbb{G}$ and the associated Poncelet quadric bundle $Q \to C(\mathbb{G})$ in §3. It is shown that $C(\mathbb{G})$ is smooth (3.8) and that $Q$ is a normal variety (3.13). We determine the singularities of $Q$ in terms of Poncelet pairs $(S, C^\vee)$ (3.12).

In §4 we define 4 irreducible Weil divisors $Q_0$, $Q_\alpha$, $Q_\beta$, $Q_e$ on $Q$ and the complement $M$ of the union of these divisors consists of Poncelet related pairs $(S, C^\vee)$ where $S$ is a regular cut of $\mathbb{G}$ by a plane in $\mathbb{P}_5$ and $C^\vee$ is smooth (i.e. corresponds to $M(0,2)$). Let $\mathrm{Sing}(Q)$ be the singular set of $Q$ and let $Q_{\mathrm{exc}}$ be the elements of $\mathrm{Sing}(Q)$ lying over the space of double lines in $C(\mathbb{G})$. It is shown in 4.2 that $Q_{\mathrm{exc}}$ is fibred naturally into a $\mathbb{P}_1$-bundle, the fibres $\mathbb{P}_1$ being the spaces of double structures on a line contained in $\mathbb{G}$.

The main theorem comparing $Q$ and $\overline{M(0,2)}$ is stated in 4.4. Assuming certain results that are proved in the later sections, it is proved in 4.5 that $Q$ can be blown down to a (normal) variety along the $\mathbb{P}_1$-fibration of $Q_{\mathrm{exc}}$ and that the canonical map $Q \to \overline{M(0,2)}$ induces an isomorphism of this blown down variety onto the normalisation of $\overline{M(0,2)}$.

A geometric invariant theoretic (G.I.T.) description of $C(\mathbb{G})$ is given in §5: $C(\mathbb{G}) = Y^{ss}//\mathrm{SL}(2)$ where $Y^{ss}$ is the space of semistable points for a linearised action of SL(2) on a space $Y$. In this section we also give a criterion for a point of $\mathrm{Gr}(U \otimes W)$ to be stable (resp. semi-stable) for the action of $\mathrm{SL}(U)$, where $U$ and $W$ are finite dimensional spaces and $\mathrm{Gr}_q$ denotes the Grassmannian of $q$-dimensional subspaces, Prop. 5.1.1.

In §6 a similar G.I.T. parametrisation of $Q = X^{ss}//\mathrm{SL}(2)$ is given for the Poncelet bundle $Q$.

We construct in §7 a *flat* family $\{\mathcal{N}_y\}$ of sheaves (of rank 4 on $\mathbb{P}_3$) parametrised by $y \in Y^{ss}$. These will correspond to kernel sheaves in the monad description of sheaves which are limits of elements in $M(0, 2)$. The proof of the flatness of the family, which involves, among other things, the use of the Eagon-Northcott complex, is given in Proposition 7.1. If $y \in Y$ and $S$ is the corresponding conic in $C(\mathbb{G})$, it is shown in 7.6 that the space of "second-order" jumping lines of $\mathcal{N}_y$ (defined as the support of the sheaf $R^1\mathcal{N}_y$ on $\mathbb{G}$) is the "conjugate" conic $S^0$. This result is of importance in the investigation of the map $Q \to \overline{M(0, 2)}$.

In §8 we construct a flat family $\{\mathcal{F}_x\}$, $x \in X^{ss}$, of rank 2 sheaves on $\mathbb{P}_3$ with $c_1 = c_3 = 0$, $c_2 = 2$ parametrised by $X^{ss}$. In fact a family of monads parametrised by $X^{ss}$ is constructed; these monads are not necessarily self-dual as the sheaves are not self-dual. We calculate some cohomology groups of $\mathcal{F}_x$.

In the last two sections we give explicit descriptions of the sheaves $\mathcal{F}_x$, essentially in terms of the configuration in $\mathbb{P}_3$ defined by a Poncelet pair $(S, C^\vee)$. For instance sheaves in $Q_e \setminus Q_\alpha \cup Q_\beta \cup Q_0$ are given by suitable elementary transformations of a null-correlation bundle or of the trivial bundle of rank 2 (9.1). A detailed study of all these sheaves is carried out to prove their stability (resp. semi-stability).

**0. Notation and conventions.** All vector spaces and varieties will be over a fixed algebraically closed field $k$ of characteristic $0$.

$G_m V$ denotes the Grassmannian of $m$-dimensional subspaces of the vector space $V$, $\mathbb{P}_n = \mathbb{P}V = G_1 V$ the projective space, $\dim V = n + 1$.

The invertible sheaf of degree $d$ on $\mathbb{P}V$ is $\mathcal{O}(d)$, s.t. $V^\vee = \Gamma(\mathbb{P}V, \mathcal{O}(1))$. For an $\mathcal{O}_{\mathbb{P}V}$-module $\mathcal{F}$ we use the abbreviations $\mathcal{F}(d) = \mathcal{F} \otimes \mathcal{O}(d)$ and $h^i\mathcal{F}(d)$ for the dimension of $H^i\mathcal{F}(d) = H^i(\mathbb{P}V, \mathcal{F}(d))$. The sheaf of the trivial vector bundle with fibre $F$ is denoted by $F \otimes \mathcal{O}$.

0.1. The evaluation map $V^\vee \otimes \mathcal{O} \to \mathcal{O}(1)$ gives rise to the Koszul complex homomorphisms $\bigwedge^{p+1} V^\vee \otimes \mathcal{O}(-1) \to \bigwedge^p V^\vee \otimes \mathcal{O}$ defined as the composition $\bigwedge^{p+1} V^\vee \otimes \mathcal{O}(-1) \to \bigwedge^p V^\vee \otimes V^\vee \otimes \mathcal{O}(-1) \to \bigwedge^p V^\vee \otimes \mathcal{O}$. The image is identified with $\Omega^p(p)$, the sheaf of $p$-differentials in twist $p$. In particular $\Omega^n(n) = \bigwedge^{n+1} \otimes \mathcal{O}(-1)$ and $\Gamma\Omega^p(p+1) = \bigwedge^{p+1} V^\vee$. The Koszul homomorphism with respect to the fibres over $\langle x\rangle \in \mathbb{P}V$ is contraction with $x$, $\bigwedge^{p+1} V^\vee \otimes \langle x\rangle \to \bigwedge^p V^\vee$.

We frequently use isomorphisms $\bigwedge^{n-p} V \simeq \bigwedge^{p+1} V^\vee$ based on a fixed isomorphism $\bigwedge^{n+1} V \simeq \mathbb{k}$. Then the Koszul homomorphism for the fibres is $\wedge x$ (up to sign) and we have the commutative diagram

$$\begin{array}{ccccc} \bigwedge^{p+1} \otimes \langle x\rangle & \longrightarrow & \Omega^p(p)(\langle x\rangle) & \subset & \bigwedge^p V^\vee \\ \wr\| & & \wr\| & & \wr\| \\ \bigwedge^{n-p} V \otimes \langle x\rangle & \longrightarrow & \bigwedge^{n-p} V \wedge x & \subset & \bigwedge^{n-p+1} V \end{array}$$

Here $\mathcal{F}(\text{pt})$ denotes the fibre $\mathcal{F}_{\text{pt}}/m_{\text{pt}}\mathcal{F}_{\text{pt}}$.

Using the Koszul complex it is standard to verify that there are natural isomorphisms

$$\bigwedge^k V \underset{\approx}{\longrightarrow} \operatorname{Hom}(\mathbb{P}V, \Omega^{p+k}(p+k), \Omega^p(p))$$

for any $k, p \geq 0$. The homomorphism corresponding to $a \in \bigwedge^k V$ is contraction on the fibres or wedging:

$$\bigwedge^{n-p-k} V \wedge x \underset{a\wedge}{\longrightarrow} \bigwedge^{n-p} V \wedge x$$

and it extends to the Koszul complex. Under these isomorphisms composition of homomorphisms corresponds to the wedge product up to signs. More generally, if $E$ and $F$ are vector spaces, we have canonical isomorphisms

$$\operatorname{Hom}\left(E, F \otimes \bigwedge^k V\right) \simeq \operatorname{Hom}(E \otimes \Omega^{p+k}(p+k), F \otimes \Omega^p(p))$$

for any $p, k \geq 0$. Given an operator of the left side the homomorphism of the sheaves is uniquely induced by the diagram

$$\begin{array}{ccc} E \otimes \bigwedge^{n-p-k} V \otimes \mathcal{O}(-1) & \longrightarrow & F \otimes \bigwedge^{n-p} V \otimes \mathcal{O}(-1) \\ \downarrow & & \downarrow \\ E \otimes \Omega^{p+k}(p+k) & \longrightarrow & F \otimes \Omega^p(p) \end{array}$$

where we use $\bigwedge^{n-p} V \simeq \bigwedge^{q+1} V^\vee$. Finally, if we choose bases of the vector spaces, a homomorphism

$$\mathscr{k}^m \otimes \Omega^{p+k}(p+k) \to \mathscr{k}^n \otimes \Omega^p(p)$$

is considered as an $m \times n$-matrix $(a_{ij})$ of elements $a_{ij} \in \bigwedge^k V$ in such a way that $\mathscr{k}^m \to \mathscr{k}^m \otimes \bigwedge^k V$ is described by $(c_1, \dots, c_m) \to (c_1, \dots, c_m) \cdot (a_{ij})$. It is sometimes convenient, to consider $\mathscr{k}^m \otimes \bigwedge^k V^\vee \to \mathscr{k}^n$ instead.

As a special case we mention:

0.2. LEMMA. *Let $B \subset \mathscr{k}^m \otimes V$ and let $\mathscr{k}^m \otimes \Omega^1(1) \xrightarrow{b} B^\vee \otimes \mathscr{O}$ be the homomorphism induced by $\mathscr{k}^m \otimes V^\vee \to B^\vee$. Then $b$ is an epimorphism iff $(\mathscr{k}^m \otimes v) \cap B = 0$ for any $v \in V$.*

*Proof*. Consider $B$ as a matrix $\mathscr{k}^p \to \mathscr{k}^m \otimes V$. Then $b$ is an epimorphism iff $b^\vee$ is a subbundle, i.e. $\mathscr{k}^p \otimes x \xrightarrow[B]{} \mathscr{k}^m \otimes V \wedge x$ is injective for any $x \in V$. Since $\lambda \circ B \wedge x = 0$ is equivalent to $\lambda \circ B = c \otimes x$ for some $c \in \mathscr{k}^m$, the lemma follows.

**0.3. Incidence transformation.** From now on $\dim V = 4$, $\mathbb{P}_3 = \mathbb{P}V$, and $\mathbb{G} = G_2 V \subset \mathbb{P}\bigwedge^2 V$. We consider the flag manifold $\mathbb{F} \subset \mathbb{P}_3 \times \mathbb{G}$ of pairs $(x, 1)$ with $x \in 1$ and let $\mathbb{P}_3 \xleftarrow[p]{} \mathbb{F} \xrightarrow[q]{} \mathbb{G}$ denote the projections, which is a $\mathbb{P}_2$ (resp. $\mathbb{P}_1$) bundle. Since $p^*$ is exact, the functor $R^\cdot = R^\cdot q_* p_*$ is a cohomology functor. Some of the standard direct images are:

$$R^0\mathscr{O}_{\mathbb{P}_3} = \mathscr{O}_{\mathbb{G}}, \quad R^1\Omega^1 = \mathscr{O}_{\mathbb{G}}, \quad R^0\Omega^1(1) = Q^\vee,$$

$$R^1\mathscr{O}_{\mathbb{P}_3}(-m-2) = S^m S \otimes \bigwedge^2 S = S^m S \otimes \mathscr{O}_{\mathbb{G}}(-1),$$

where $S$, $Q$ denote the universal sub-, quotient bundles on $\mathbb{G}$, and $S^m$ denotes the symmetric power, $m \geq 0$.

**0.4. Conics in $\mathbb{G}$ and reguli.** We denote by $C(\mathbb{G})$ the Hilbert scheme of conics in $\mathbb{G}$. This is a smooth variety of dimension 9, see 3.8. Each conic $S \subset \mathbb{G}$ defines a plane $P \subset \mathbb{P}\bigwedge^2 V$, such that $S \subset \mathbb{G} \cap P$. If $P$ is not contained in $\mathbb{G}$ (as an $\alpha$-plane, i.e. a plane consisting of all lines through a point in $\mathbb{P}_3$, or as a $\beta$-plane, i.e. a plane consisting of all lines in a plane in $\mathbb{P}_3$) then $S = \mathbb{G} \cap P$. The system of lines in $\mathbb{P}_3$ parametrised by a given conic $S \subset \mathbb{G}$ can be visualised as a "complete" regulus. This is a quadric $Q \subset \mathbb{P}_3$ with $pq^{-1}(S)$ as its underlying set together with the system of lines on it given by $S$. We give below a list of all types of complete reguli in $\mathbb{P}_3$, which arise in this way from conics in $\mathbb{G}$. The complete reguli obtained by the configuration of the dual lines in $\mathbb{P}_3^\vee$ are also given and denoted by $Q^\vee$.

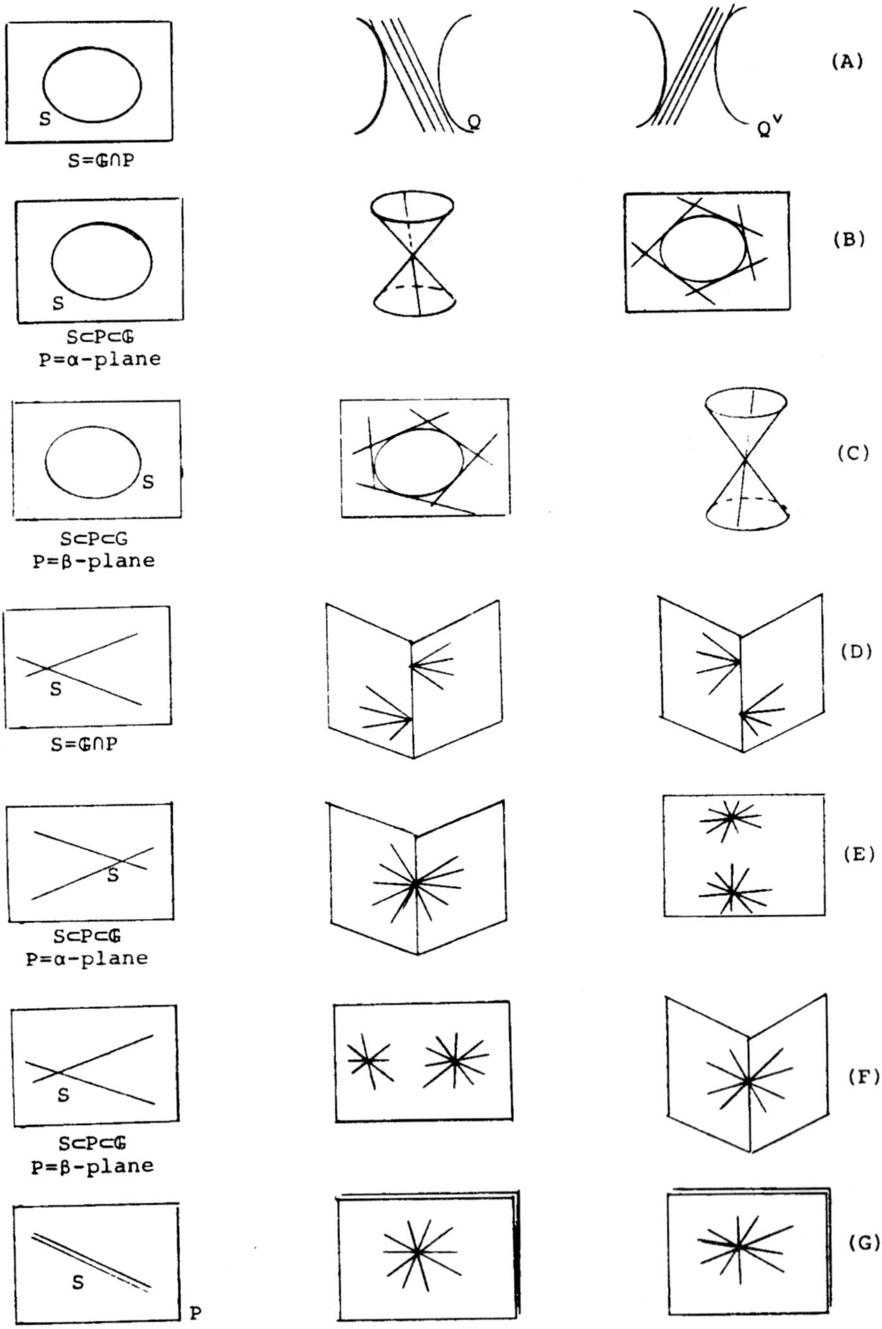

Note that in the last case (G) the double regulus cannot remember the plane $P$ of the conic $S$. In other words, while the reduced line $S$ can be recovered from the configuration of this regulus in $\mathbb{P}_3$, the double structure on the line is not determined by it, see 3.9. If $S = G \cap P$ is a regular conic section the quadric $Q$ spanned by the lines is regular and has two systems of lines. The conic $S$ is isomorphic to any line of the second system.

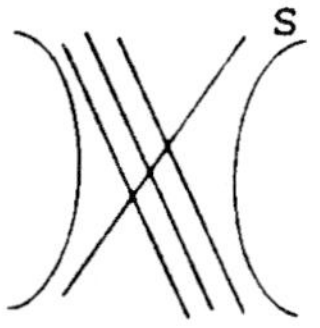


Since $Q \simeq \mathbb{P}_1 \times \mathbb{P}_1$ we can identify $S$ with the first factor, s.t.

$$\Gamma\mathcal{O}_S(3) = \Gamma\mathcal{O}_Q(3, 0).$$

The second factor parametrises a second (the conjugate) conic $S^0 \subset \mathbb{G}$ which is the intersection $S^0 = \mathbb{G} \cap P^0$, where $P^0$ is the plane orthogonal to $P$ with respect to the quadratic form of $\mathbb{G}$.

We denote by $C^0(\mathbb{G})$ the open part of the Hilbert scheme of regular plane sections.

**1. Conics and kernel bundles.**

1.1. *Standard resolution of* $\mathcal{O}_Q(3, 0)$. Let $\mathcal{O}_Q(3, 0)$ be defined by the conic $S \in C^0(\mathbb{G})$. We can choose a basis $e_0, \ldots, e_3 \in V$ with dual basis $z_0, \ldots, z_3 \in V^\vee$, s.t. $Q$ has the equation $z_0z_3 - z_1z_2 = 0$ and is the image of the standard Segre imbedding $z_0 = s_0t_0$, $z_1 = s_0t_1$, $z_2 = s_1t_0$, $z_3 = s_1t_1$. Then $\mathcal{O}_Q(3, 0)$ is generated by the liftings of $s_0^3$, $s_0^2s_1$, $s_0s_1^2$, $s_1^3 \in \Gamma\mathcal{O}_{\mathbb{P}_1}(3)$. It is then straightforward to verify that the sequence

$$0 \to k^2 \otimes \mathcal{O}(-2) \xrightarrow[B]{} k^6 \otimes \mathcal{O}(-1) \xrightarrow[A]{} k^4 \otimes \mathcal{O} \to \mathcal{O}_Q(3, 0) \to 0$$

with

$$B = \left(\begin{array}{ccc|ccc} -z_3 & z_1 & & z_2 & -z_0 & \\ & -z_3 & z_1 & & z_2 & -z_0 \end{array}\right), \quad A = \left(\begin{array}{cccc} -z_2 & z_0 & & \\ & -z_2 & z_0 & \\ & & -z_2 & z_0 \\ \hline -z_3 & z_1 & & \\ & -z_3 & z_1 & \\ & & -z_3 & z_1 \end{array}\right),$$

is a resolution in $\mathbb{P}_3$.

We also have the exact sequence

$$0 \to k^2 \xrightarrow[N^*]{} k^2 \otimes V \xrightarrow[B]{} k^6 \to 0$$

where $N^*$ is the matrix

$$N^* = \boxed{\begin{matrix} e_0 & e_2 \\ e_1 & e_3 \end{matrix}}\,.$$

REMARK. $\det N^* = e_0e_3 - e_1e_2$ is the equation of the dual quadric $Q^\vee \subset \mathbb{P}V^\vee$ as can be easily verified.

1.2. If $\mathscr{K}$ denotes the kernel of $B^\vee(-1)$ we obtain the exact diagram

$$\begin{array}{ccccccccc}
 & & 0 & & 0 & & & & \\
 & & \downarrow & & \downarrow & & & & \\
0 & \longrightarrow & \mathscr{K} & \longrightarrow & k^6 \otimes \mathscr{O} & \xrightarrow{B^\vee} & k^2 \otimes \mathscr{O}(1) & \longrightarrow & 0 \\
 & & \downarrow & & \downarrow{\scriptstyle B^\vee} & & \| & & \\
0 & \longrightarrow & k^2 \otimes \Omega^1(1) & \longrightarrow & k^2 \otimes V^\vee \otimes \mathscr{O} & \longrightarrow & k^2 \otimes \mathscr{O}(1) & \longrightarrow & 0 \\
 & & \downarrow{\scriptstyle N^\vee} & & \downarrow{\scriptstyle N^\vee} & & & & \\
 & & k^2 \otimes \mathscr{O} & = & k^2 \otimes \mathscr{O} & & & & \\
 & & \downarrow & & \downarrow & & & & \\
 & & 0 & & 0 & & & &
\end{array} \tag{1}$$

By 0.2 $N^\vee$ is an epimorphism and thus $\mathscr{K}$ is locally free. Of course by the resolution above we also have the exact sequence

$$0 \to \mathscr{K}^\vee(-1) \to \Gamma^\vee \otimes \mathscr{O} \to \mathscr{O}_Q(3\,,\,0) \to 0\,, \tag{2}$$

where $\Gamma^\vee = \Gamma\mathscr{O}_Q(3\,,\,0)$, and we obtain dually

$$0 \to \Gamma \otimes \Omega^3(4) \to \mathscr{K}(1) \to \mathscr{O}_Q(-1\,,\,2) \to 0\,, \tag{$2^\vee$}$$

since $\mathrm{Ext}^1_{\mathscr{O}}(\mathscr{O}_Q(a\,,\,b)\,,\,\mathscr{O}) = \mathscr{O}_Q(2-a\,,\,2-b)$, which follows since the dualizing sheaf $\omega_Q = \mathscr{O}_Q(-2\,,\,-2)$. We are going to investigate the sections of $\mathscr{K}(1)$. By the first column of (1) we are given the diagram

$$\begin{array}{ccccccccc}
0 & \to & \Gamma\mathscr{K}(1) & \to & k^2 \otimes \Gamma\Omega^1(2) & \to & k^2 \otimes \Gamma\mathscr{O}(1) & \to & 0 \\
 & & \| & & \| & & \| & & \\
 & & \Gamma \otimes \Lambda^4 V^\vee & & k^2 \otimes \Lambda^2 V^\vee & & k^2 \otimes V^\vee & & \\
 & & \wr\| & & \wr\| & & \wr\| & & \\
0 & \to & \Gamma & \to & k^2 \otimes \Lambda^2 V & \xrightarrow[\wedge N^\vee]{} & k^2 \otimes \Lambda^3 V & \to & 0
\end{array}$$

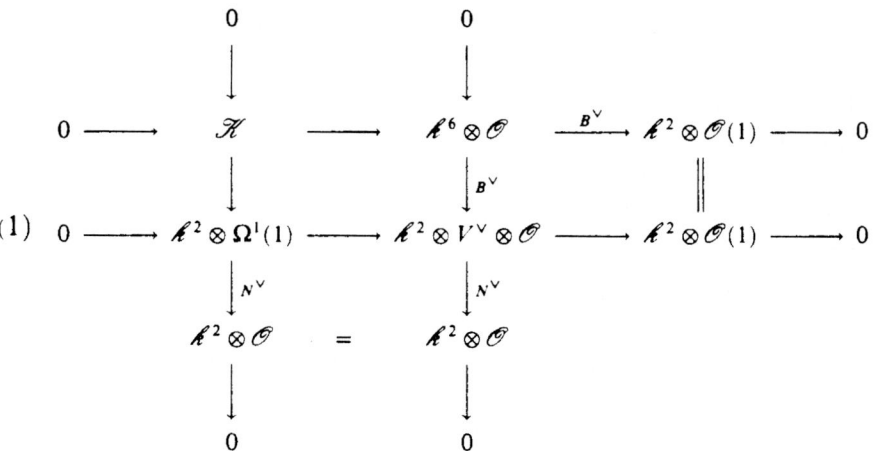

A direct calculation shows that $\dim \Gamma = 4$ and that $\Gamma$ is presented by the matrix $\Gamma^*: \mathcal{O}^4 \to \mathcal{O}^2 \otimes \bigwedge^2 V$

$$\Gamma^* = \begin{bmatrix} \xi & 0 \\ \omega & \xi \\ \eta & \omega \\ 0 & \eta \end{bmatrix}$$

with $\xi = e_0 \wedge e_1$, $\omega = e_0 \wedge e_3 - e_1 \wedge e_2$, $\eta = e_2 \wedge e_3$. In particular $H^1\mathscr{K}(1) = 0$.

1.3. LEMMA. (1) *The conic* $S$ *is parametrised by* $s^2\xi + st\omega + t^2\eta$.

(2) *If the zero scheme* $Z(\gamma)$ *of a section* $\gamma \in \Gamma \simeq \Gamma\mathscr{K}(1)$ *is not empty, it is a line* $l \in S$ *and*

(3) $\qquad \gamma = (s, t) \otimes (s^2\xi + st\omega + t^2\eta) = (s^3, s^2t, st^2, t^3) \circ \Gamma^*.$

*Proof*. (1) is immediate from 1.1 by looking at the embedding of the first factor of $\mathbb{P}_1 \times \mathbb{P}_1$.

(2) If $\gamma \in \Gamma$ then $\gamma$ vanishes in $\langle x \rangle$ iff $\gamma \wedge x = 0$, see 0.1. If $\gamma = (\alpha_0, \dots, \alpha_3) \circ \Gamma^*$ this means that

$$\alpha_0\xi \wedge x + \alpha_1\omega \wedge x + \alpha_2\eta \wedge x = 0,$$
$$\alpha_1\xi \wedge x + \alpha_2\omega \wedge x + \alpha_3\eta \wedge x = 0.$$

However by the definition of $\xi$, $\omega$, $\eta$ the vectors $\xi \wedge x$, $\eta \wedge x$ are linearly independent, and there is at most one relation of the vectors, i.e.

$$\operatorname{rank} \begin{bmatrix} \alpha_0 & \alpha_1 & \alpha_2 \\ \alpha_1 & \alpha_2 & \alpha_3 \end{bmatrix} = 1.$$

But it is well known that then

$$(\alpha_0, \dots, \alpha_3) = (s^3, s^2t, st^2, t^3)$$

which proves (3), and thus $Z(\gamma) = s^2\xi + st\omega' + t^2\eta \in S$.

1.3.1. COROLLARY. *The correspondence* $\mathscr{K} \leftrightarrow S$ *is* 1:1 *between the kernels* $\mathscr{K}$ *of regular* $N$*'s (with four independent entries) and the regular conics* $S \in C^0(\mathbb{G})$.

1.3.2. COROLLARY. *If* $\mathscr{K}$ *is defined by a regular* $N$, *we have the exact sequence*

$$0 \to \Gamma(\mathscr{K}(1)) \otimes \mathscr{O} \to \mathscr{K}(1) \to \mathscr{O}_Q(-1, 2) \to 0.$$

*Proof*. Clearly $\mathscr{K}$ is defined by $S \in C^0(\mathbb{G})$. By the lemma conversely $S$ is defined by $\mathscr{K}$. Given an arbitrary regular $N$ and $\mathscr{K}$, a conic $S$ is defined by $s^2\xi + st\omega + t^2\eta$ which gives $\mathscr{K}$. Then the corollary follows from $(2^\vee)$ or the lemma.

1.4. Remark. Let $G_2^0(k^2 \otimes V)$ denote the open set of the Grassmannian of all 2-dimensional subspaces $N \subset k^2 \otimes V$ which are presented by matrices with 4 independent vectors. The map $N \mapsto S$ is a morphism

$$G_2^0(k^2 \otimes V) \to C^0(\mathbb{G})$$

onto $C^0(\mathbb{G})$. It is invariant under the action of SL(2) given by $(g, N) \to (f \otimes \mathrm{id})(N)$, and thus factorizes into an isomorphism

$$G_2^0(k^2 \otimes V)/\,\mathrm{SL}(2) \simeq C^0(\mathbb{G}).$$

The transposition map $\begin{pmatrix} x & x' \\ y & y' \end{pmatrix} \to \begin{pmatrix} x' & y \\ x' & y' \end{pmatrix}$ induces the involution $S \to S^0$.

We finally state two further beautiful geometric properties of a bundle $\mathscr{K}$.

1.5. Proposition. *Let $\mathscr{K}$ be defined as above and let $S$ resp. $Q$ be the associated conic resp. quadric. Then*

(i) *the dual quadric $Q^\vee \subset \mathbb{P}_3^\vee$ is the set of jumping planes of $\mathscr{K}$, i.e. of all planes $P \subset \mathbb{P}_3$ with $h^0(\mathscr{K}|P) \neq 0$.*

(ii) $R^1\mathscr{K} = \mathscr{O}_{S^0}(1)$, *where $R^1$ is the first incidence transform,* 0.3.

*Proof*. (i) Let $H = \{f = 0\}$ with $f \in V^\vee$. There is a splitting of

$$0 \to f \otimes \mathscr{O}_H \to \Omega^1(1)|H \to \Omega_H^1(1) \to 0$$

which is induced from the Koszul-complex. Therefore we obtain the exact sequence

$$\begin{array}{ccccc} 0 \to \Gamma(\mathscr{K}|H) & \to & k^2 \otimes \Gamma(\Omega^1(1)|H) & \to & k^2 \otimes \Gamma\mathscr{O}_H \\ & & \wr\| & & \wr\| \\ & & k^2 \otimes f & \xrightarrow[N^{\bullet\vee}]{} & k^2 \end{array}$$

Hence $h^0(\mathscr{K}|H) \neq 0$ iff $\det = f(e_0)f(e_3) - f(e_1)f(e_2) = 0$, i.e. iff $f \in Q^\vee$.

(ii) If we apply $R^1$ to $(2^\vee)$ we obtain $R^1\mathscr{K} = R^1\mathscr{O}_Q(-2, 1)$. Since $S$ is the first factor, we find $H^1\mathscr{O}_Q(-2, 1)\otimes\mathscr{O}_l = 0$ except for $l \in S^0$. Now $R^1\mathscr{O}_Q(-2, 1)$ being supported on $S^0$, we can obtain it as the simple direct image under $Q \to S^0$, which is the second projection. Therefore $R^1\mathscr{O}_Q(-2, 1) = \mathscr{O}_{S^0}(1)$.

**2. Review of** $M(0, 2)$. The bundles $\mathscr{E} \in M(0, 2)$ can be constructed in two different ways: from a linear system on a conic $S \subset \mathbb{G}$ as mentioned in the introduction and from monads, see **[Ha2]**. We summarize both in the following

2.1. THEOREM. *A rank-2 bundle* $\mathscr{E}$ *on* $\mathbb{P}_3$ *belongs to* $M(0, 2)$ *if and only if it is a member of one of the following exact diagrams (displays). These can be derived from each other.*

$$(\mathrm{D}_1)\qquad \begin{array}{ccccccccc}
 & & 0 & & 0 & & & & \\
 & & \downarrow & & \downarrow & & & & \\
0 & \longrightarrow & \mathscr{E}^\vee(-1) & \longrightarrow & L\otimes\mathscr{O} & \longrightarrow & \mathscr{O}_Q(3, 0) & \longrightarrow & 0 \\
 & & \downarrow & & \downarrow & & \| & & \\
0 & \longrightarrow & \mathscr{K}^\vee(-1) & \longrightarrow & \Gamma^\vee\otimes\mathscr{O} & \longrightarrow & \mathscr{O}_Q(3, 0) & \longrightarrow & 0 \\
 & & \downarrow & & \downarrow & & & & \\
 & & M^\vee\otimes\mathscr{O} & = & M^\vee\otimes\mathscr{O} & & & & \\
 & & \downarrow & & \downarrow & & & & \\
 & & 0 & & 0 & & & &
\end{array}$$

$$(\mathrm{D}_1^\vee)\qquad \begin{array}{ccccccccc}
 & & 0 & & 0 & & & & \\
 & & \uparrow & & \uparrow & & & & \\
0 & \longrightarrow & L^\vee\otimes\Omega^3(4) & \longrightarrow & \mathscr{E}(1) & \longrightarrow & \mathscr{O}_Q(-1, 2) & \longrightarrow & 0 \\
 & & \uparrow & & \uparrow & & \| & & \\
0 & \longrightarrow & \Gamma\otimes\Omega^3(4) & \longrightarrow & \mathscr{K}(1) & \longrightarrow & \mathscr{O}_Q(-1, 2) & \longrightarrow & 0 \\
 & & \uparrow & & \uparrow & & & & \\
 & & M\otimes\Omega^3(4) & = & M\otimes\Omega^3(4) & & & & \\
 & & \uparrow & & \uparrow & & & & \\
 & & 0 & & 0 & & & &
\end{array}$$

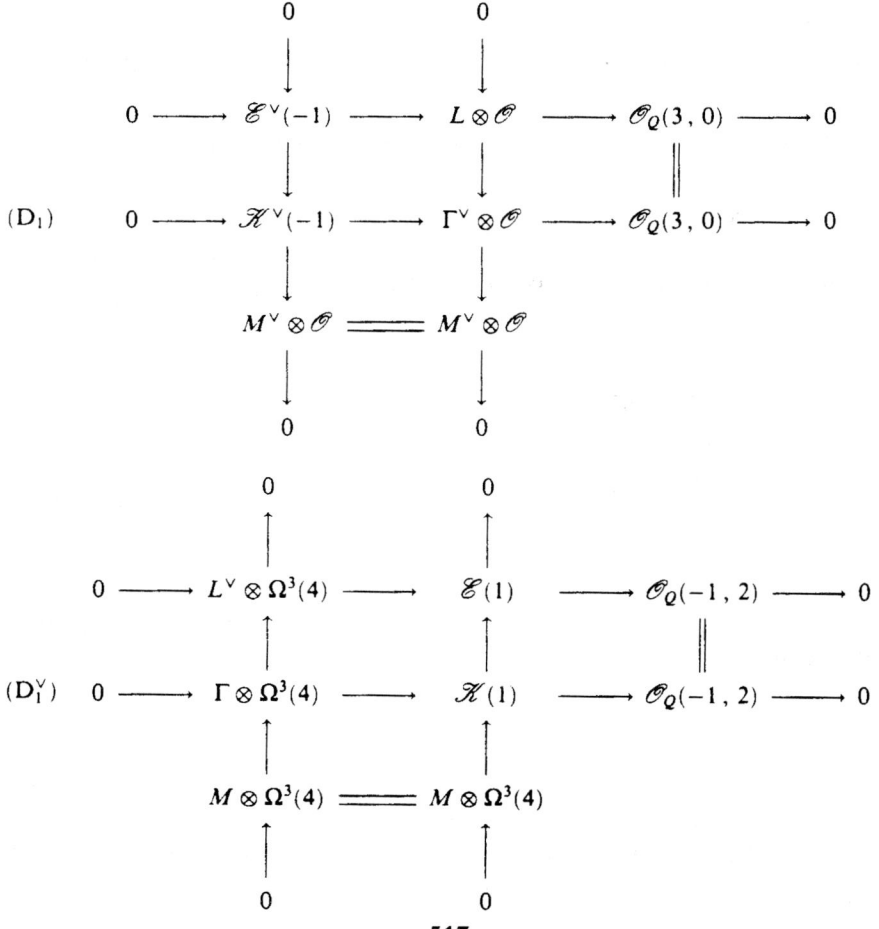

$$
(\mathrm{D}_2)\qquad
\begin{array}{ccccccccc}
 & & & & 0 & & 0 & & \\
 & & & & \downarrow & & \downarrow & & \\
0 & \longrightarrow & M\otimes\Omega^3(3) & \longrightarrow & \mathscr{K} & \longrightarrow & \mathscr{E} & \longrightarrow & 0 \\
 & & \| & & \downarrow & & \downarrow & & \\
0 & \longrightarrow & M\otimes\Omega^3(3) & \longrightarrow & \mathscr{k}^2\otimes\Omega^1(1) & \longrightarrow & \mathscr{M} & \longrightarrow & 0 \\
 & & & & \downarrow & & \downarrow & & \\
 & & & & N^\vee\otimes\mathscr{O} & = & N^\vee\otimes\mathscr{O} & & \\
 & & & & \downarrow & & \downarrow & & \\
 & & & & 0 & & 0 & &
\end{array}
$$

*Explanation of* $(\mathrm{D}_1)$. $Q$ is the regulus of a regular conic section $S\subset\mathbb{G}$, s.t. $\Gamma\mathscr{O}_S(3)=\Gamma\mathscr{O}_Q(3,0)=\Gamma^\vee$ and $L\subset\Gamma^\vee$ is a 2-dimensional subspace without base points. $L\otimes\mathscr{O}\to\mathscr{O}_Q(3,0)$ and $\Gamma^\vee\otimes\mathscr{O}\to\mathscr{O}_Q(3,0)$ are the induced epimorphisms, see 1.2.

*Explanation of* $(\mathrm{D}_1^\vee)$. This is obtained by applying $\mathscr{H}om(\cdot,\mathscr{O})$ to $(\mathrm{D}_1)$, where we use $\mathscr{O}^\vee=\Omega^3(4)$ by formal reasons and

$$\mathscr{E}xt^1_{\mathscr{O}}(\mathscr{O}_Q(a,b),\mathscr{O})\simeq\mathscr{O}_Q(2-a,2-b).$$

The latter follows by using the dualizing sheaf $\omega_Q=\mathscr{O}_Q(-2,-2)$.

*Explanation of* $(\mathrm{D}_2)$. $M\subset\mathscr{k}^2\otimes\bigwedge^2V$ and $N\subset\mathscr{k}^2\otimes V$ are 2-dimensional subspaces such that $M$ is contained in the kernel of the composed operator $\mathscr{k}^2\otimes\bigwedge^2V\to N^\vee\otimes V\otimes\bigwedge^2V\to N^\vee\otimes\bigwedge^3V$. By 0.1 we obtain a complex

$$M\otimes\Omega^3(3)\xrightarrow[\mu]{}\mathscr{k}^2\otimes\Omega^1(1)\xrightarrow[\nu]{}N^\vee\otimes\mathscr{O},$$

and we suppose that $\mu$ is a subbundle and $\nu$ an epimorphism. Such a complex is called *monad* and the display $(\mathrm{D}_2)$ is called the display of the monad.

*Proof*. (1) If $\mathscr{E}$ is defined by $(\mathrm{D}_1)$, it must be a rank-2 bundle since $\mathscr{K}^\vee(-1)$ is locally free by 1.2 and $\mathscr{E}=\mathscr{E}^\vee(-1)^\vee(-1)$. Furthermore its Chern classes must be $c_1=0$, $c_2=2$, and $h^0\mathscr{E}=0$ since $h^0\mathscr{K}=0$. Hence it is stable and a member of $M(0,2)$. The same can be proved if $\mathscr{E}$ is defined by $(\mathrm{D}_2)$.

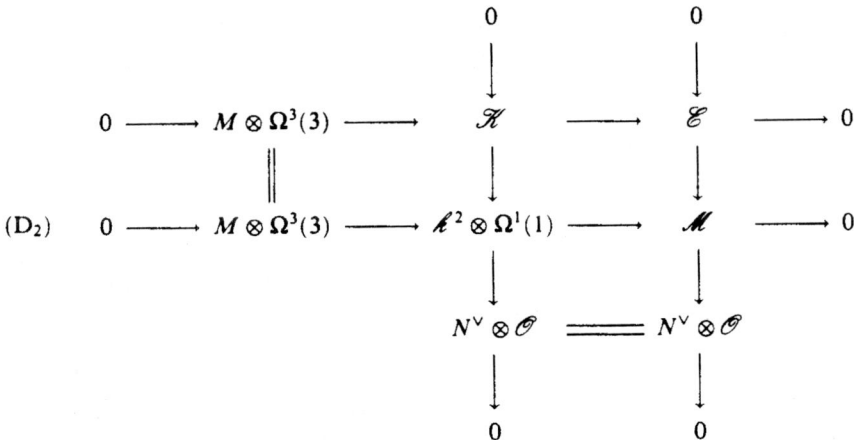

(2) It was shown in **[Ha2]** that $h^1\mathscr{E}(-2) = 0$ for any $\mathscr{E} \in M(0, 2)$. Then the Beilinson spectral sequence, see **[OSS]**, of $\mathscr{E}$ degenerates, and its $E_2$ level yields the monad in $(\mathrm{D}_2)$:

$$H^2(\mathscr{E}(-3)) \otimes \Omega^3(3) \to H^1(\mathscr{E}(-1)) \otimes \Omega^1(1) \to H^1(\mathscr{E}) \otimes \mathscr{O}$$

so that $M = H^2\mathscr{E}(-3)$, $k^2 = H^1\mathscr{E}(-1)$, $H^1\mathscr{E} = N^\vee$. Then automatically $\mu$ and $\nu$ are sub- resp. quotient bundles.

(3) Clearly the displays $(\mathrm{D}_1)$ and $(\mathrm{D}_1^\vee)$ are dual to each other. If $(\mathrm{D}_1)$ is given we get $(\mathrm{D}_2)$ from the results of 1.2, for there it was shown that $\mathscr{K}$ is a kernel as in the column in the middle of $(\mathrm{D}_2)$. We also can derive $(\mathrm{D}_1^\vee)$ from $(\mathrm{D}_2)$ as follows. By 0.2, $\nu$ is an epimorphism iff $(k^2 \otimes v) \cap N = 0$ for any $v \in V$. Let now $k^2 \otimes V^\vee \to N^\vee$ be given by the matrix $N^*: k^2 \to k^2 \otimes V$. It is elementary to derive that the condition for $N$ is satisfied iff $N^*$ is one of the matrices

$$\begin{bmatrix} e_0 & e_1 \\ e_2 & e_3 \end{bmatrix} \quad \text{or} \quad \begin{bmatrix} e_0 & e_1 \\ e_2 & e_0 \end{bmatrix}$$

where $e_0, \dots, e_3 \in V$ is a basis. In the first case $\mathscr{K}$ is an extension as in $(\mathrm{D}_1^\vee)$ by 1.2, and hence $(\mathrm{D}_1^\vee)$ follows from $(\mathrm{D}_2^\vee)$. We show now that the second case cannot occur: As in 1.2 we find that the kernel $\Gamma$ of $k^2 \otimes \bigwedge^2 V \xrightarrow[\wedge N^*]{} k^2 \otimes \bigwedge^3 V$ is generated by the matrix $(e_{ij} = e_i \wedge e_j)$

$$\Gamma^* = \begin{bmatrix} e_{01} & 0 \\ e_{21} & e_{01} \\ e_{20} & e_{21} \\ 0 & e_{20} \end{bmatrix}.$$

Since $M \subset \Gamma$, the matrix $M^*: k^2 \to k^2 \otimes \bigwedge^2 V$ representing $M$ must be a product $M^* = A \circ \Gamma^*$ with a usual $2 \times 4$ matrix $A$. It follows that the entries $a_{ij}$ of $M^*$ are contained in the span of $e_0 \wedge e_1$, $e_0 \wedge e_2$, $e_1 \wedge e_2$. Therefore for any $z \in V$

$$a_{ij} \wedge z = \alpha_{ij}(z) e_0 \wedge e_1 \wedge e_2 + z_3 \tilde{a}_{ij}$$

where $\alpha_{ij}$ are linear in the coordinates $z_0$, $z_1$, $z_2$ only and $\tilde{a}_{ij} \in \bigwedge^3 V$. Hence, if $z_3 = 0$,

$$M^* \wedge z = (a_{ij} \wedge z) = (\alpha_{ij}(z) e_0 \wedge e_1 \wedge e_2)$$

and we see that this matrix is degenerate on the conic $z_0 = 0$, $\det(\alpha_{ij}(z)) = 0$. This shows that $k^2 \otimes \Omega^3(3) \xrightarrow[M^*]{} k^2 \otimes \Omega^1(1)$ is not a subbundle along this conic.

2.1.1. Remark. Assume that $\nu$ is regular in $(\mathrm{D}_2)$ and $\mu$ injective but not necessarily a subbundle. Then $L = (\Gamma/M)^\vee \subset \Gamma^\vee$ is base point free iff $\mu$ is a subbundle.

*Proof*. We still obtain diagram $(\mathrm{D}_1^\vee)$, but in $(\mathrm{D}_1)$ there might occur the cokernel $\mathrm{Ext}^1(\mathscr{E}(1), \mathscr{O})$ in the top row and left column. Now both conditions are satisfied iff $\mathrm{Ext}^1(\mathscr{E}(1), \mathscr{O}) = 0$.

2.1.2. Remark. If $\nu$ is regular in $\mathrm{D}_2$, i.e. coming from a conic $S \in C^0(G)$, and $\mu$ injective, the sheaf $\mathscr{E}$ is still stable, of rank 2, $h^0\mathscr{E} = 0$, $c_1 = 0$, $c_2 = 2$, $c_3 = 0$. These sheaves occur as kernels in sequences

$$0 \to \mathscr{E} \to \mathscr{E}' \to \mathscr{O}_l(1) \to 0$$

where $\mathscr{E}' \in \overline{M(0, 1)}$ and $l \subset \mathbb{P}_3$ is a line, see 9.1.

2.1.3. Remark. The monad in $(\mathrm{D}_2)$ is determined by $\mathscr{E}$ up to equivalence. This means that if $\mathscr{E}$ and $\mathscr{E}'$ are given by $(M, N)$ and $(M', N')$ then $\mathscr{E} \simeq \mathscr{E}'$ iff there exists $g \in \mathrm{GL}(2, k)$ s.t.

$$\begin{array}{ccc}
M \subset k^2 \otimes \Lambda^2 V & & k^2 \otimes V \supset N \\
\downarrow \quad \big\downarrow g\otimes \mathrm{id} & \text{and} & \big\uparrow g^\vee \otimes \mathrm{id} \quad \uparrow \\
M' \subset k^2 \otimes \Lambda^2 V & & k^2 \otimes V \supset N'
\end{array}$$

The proof follows easily from the identifications of $M$, $k^2$, $N^\vee$ with $H^2\mathscr{E}(-3)$, $H^1\mathscr{E}(-1)$, $H^1\mathscr{E}$ respectively.

2.2. *Sections of* $\mathscr{E}(1)$. If $\lambda \in L^\vee \simeq \Gamma\mathscr{E}(1)$ is a section of $\mathscr{E}(1)$ we obtain the exact diagram

$$\begin{array}{ccccccc}
 & & 0 & & 0 & & \\
 & & \uparrow & & \uparrow & & \\
\mathscr{E}^\vee(-1) & \longrightarrow & \langle\lambda\rangle^\vee \otimes \mathscr{O} & \longrightarrow & \mathscr{O}_Z & \longrightarrow & 0 \\
\| & & \uparrow & & \uparrow & & \\
\mathscr{E}^\vee(-1) & \longrightarrow & L \otimes \mathscr{O} & \longrightarrow & \mathscr{O}_Q(3, 0) & \longrightarrow & 0 \\
 & & \uparrow & & \uparrow & & \\
 & & \langle f \rangle \otimes \mathscr{O} & \simeq & \mathscr{O} & & \\
 & & \uparrow & & & & \\
 & & 0 & & & &
\end{array}$$

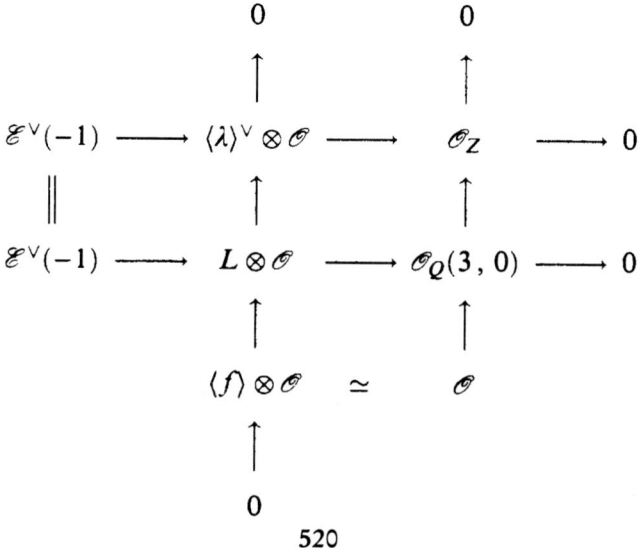

where $Z$ denotes the zero scheme of the section, and $f$ spans the kernel. Then $f$ as an element of $\Gamma^\vee = \Gamma\mathcal{O}_S(3)$ has three zeros on $S$ with the sequence

$$0 \to \mathcal{O}_S \to \mathcal{O}_S(3) \to \mathcal{O}_y \to 0.$$

If $Q \xrightarrow[\pi]{} S$ is the projection to the first factor, $\pi^*$ yields the right-hand column and thus $Z = \pi^{-1}(y)$ consists of three lines of the system $S$.

Note that we have an isomorphism $\mathbb{P}L^\vee \simeq \mathbb{P}L$ since $L$ is 2-dimensional.

2.2.1. Lemma. *Let $\gamma \in \Gamma \simeq \Gamma\mathcal{K}(1)$ and $\lambda \in L^\vee \simeq \Gamma\mathcal{E}(1)$ be two sections with $\varnothing \neq Z(\gamma) \subset Z(\lambda)$. Then $\gamma$ maps into $\langle\lambda\rangle \subset L^\vee$ under $\Gamma \to L^\vee$.*

*Proof*. Let $\lambda'$ be the image of $\gamma$. Obviously also $Z(\gamma) \subset Z(\lambda')$. If $f'$, $f$ are the polynomials in $L \subset \Gamma^\vee = \Gamma\mathcal{O}_S(3)$ corresponding to $\lambda'$, $\lambda$ they must have a common zero. If $\lambda'$, $\lambda$ were independent, also $f'$, $f$ would be independent, contradicting the assumption that $L^\vee$ is base point free.

2.3. *Jumping lines of $\mathcal{E}$*. Let $(M, N)$ be a monad defining a bundle $\mathcal{E} \in M(0, 2)$. Using the isomorphism $\bigwedge^2 V \simeq \bigwedge^2 V^\vee$ we can consider a representing matrix $M^*$ of $M$ as a matrix of linear forms on $\bigwedge^2 V$. If $\xi \in \bigwedge^2 V$ we write

$$M^*(\xi): k^2 \xrightarrow[*]{} k^2 \otimes \bigwedge^2 V \xrightarrow[\wedge\xi]{} k^2 \otimes \bigwedge^4 V \simeq k^2$$

or $M^*(\xi) = M^* \wedge \xi$. The equation $\det M^*(\xi) = 0$ is then uniquely determined by $\mathcal{E}$ up to a scalar. If we apply the incidence transformation $R^1$ to the monad ($\mathrm{D}_2$) we obtain $R^1\mathcal{E}(-1) = R^1\mathcal{M}(-1)$ and

$$0 \to k^2 \otimes \mathcal{O}_G(-1) \xrightarrow[R^1\mu]{} k^2 \otimes \mathcal{O}_G \to R^1\mathcal{E}(-1) \to 0$$

such that $R^1\mu$ is induced by the matrix $M^*(\xi)$. Similarly if we apply $\otimes\mathcal{O}_l(-1)$ to the monad we get

$$\begin{array}{ccccccc} k^2 \otimes H^1\Omega^3(2) \otimes \mathcal{O}_l & \longrightarrow & k^2 \otimes H^1\Omega^1 \otimes \mathcal{O}_l & \longrightarrow & H^1\mathcal{M}|_l(-1) & \longrightarrow & 0 \\ \| & & \| & & & & \\ k^2 & \xrightarrow[M^*(l)]{} & k^2 & & & & \end{array}$$

and we obtain, that for a line $l$

$$(4) \qquad \mathcal{E}|_l \simeq \mathcal{O}_l(-i) \otimes \mathcal{O}_l(i) \quad \text{iff} \quad \operatorname{rk} M^*(l) = 2 - i.$$

As a consequence, if $\tilde{J} = \{\det M^* = 0\}$ is the hypersurface in $\mathbb{P}\bigwedge^2 V$, the hypersurface

$$J = \tilde{J} \cap \mathbb{G} = \operatorname{Supp} R^1\mathcal{E}(-1)$$

is the set of jumping lines of $\mathcal{E}$. If $W^\perp \subset \bigwedge^2 V$ is the orthogonal of $W = \langle \xi, \omega, \eta \rangle$, we have $\xi|W^\perp = \omega|W^\perp = \eta|W^\perp = 0$ and thus $M^* = 0$ on $\mathbb{P}W^\perp$. Since $S^0 = \mathbb{G} \cap PW^\perp$ we find that $J$ is singular along $S^0$. We even have

$$S^0 = \operatorname{Sing} J,$$

since $M^* \neq 0$ away from $W^\perp$. By (4) this is the set of jumping lines of order 2.

Let finally $C = \tilde{J} \cap \mathbb{P}W$ be the conic cut out by $\mathbb{P}W$. Since $\mathbb{P}W^\perp \cap \mathbb{P}W \neq \emptyset$ and $\mathbb{P}W^\perp$ is exactly the singular locus of $\det M^*$, $\tilde{J}$ is the cone over $C$ with vertex $\mathbb{P}W^\perp$. Note that $C$ must be smooth, since otherwise $M^*$ could be given the form $\left(\begin{smallmatrix}\alpha & 0\\ \beta & \gamma\end{smallmatrix}\right)$ with $\alpha \in \mathbb{G}$, and then $\mu$ would be degenerate in $\alpha$.

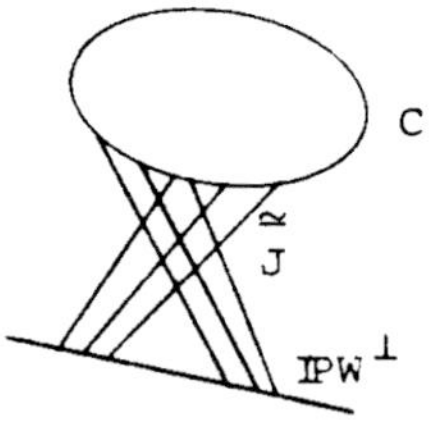


2.4. *The associated Poncelet pair.* Let $\mathcal{E} \in M(0, 2)$ and let $S$, $C$ be the conics associated with $\mathcal{E}$, see 2.2, 2.3. These are conics in the same plane $P \subset \mathbb{P}\bigwedge^2 V$ with $S = P \cap \mathbb{G}$, $C = P \cap \tilde{J}$.

2.4.1. PROPOSITION (*Hartshorne* **[Ha2]**). *The conic $C$ is Poncelet related to $S$ with respect to the pencil $\mathbb{P}L \subset |\mathcal{O}_S(3)|$, i.e. the tangents to $S$ in the points of any divisor of the pencil meet on $C$.*

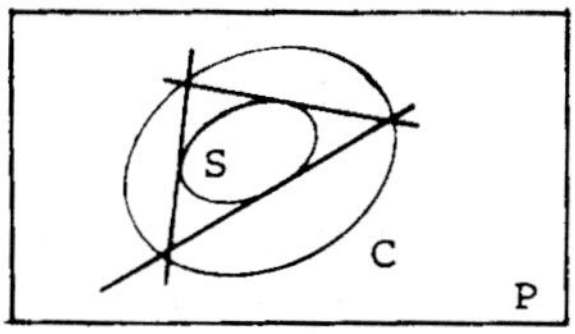


*Proof.* $L \subset \Gamma^\vee = \Gamma\mathscr{O}_S(3)$ and let $\Gamma \subset \mathscr{k}^2 \otimes \bigwedge^2 V$ be presented by a matrix $\Gamma^*: \mathscr{k}^4 \to \mathscr{k}^2 \otimes \bigwedge^2 V$ as in 1.2, see also display $(D_1)$. Then $S$ is the conic parametrised by $s^2\xi + st\omega + t^2\eta$, and $P = \mathbb{P}W$, $W = \langle \xi, \omega, \eta \rangle$. By Lemma 1.3 a section $\gamma \in \Gamma \simeq \Gamma\mathscr{H}(1)$ with the zero line $l = s^2\xi + st\omega + t^2\eta$ is given by

$$\gamma = (s^3, s^2t, st^2, t^3) \circ \Gamma^* = (s, t) \otimes l.$$

Let now $\lambda \in L^\vee$ and $f \in L \subset \Gamma^\vee$ the corresponding polynomial having $Z(\lambda)$ as its zeros, 2.2. If

$$l_i = s_i^2\xi + s_it_i\omega + t_i^2\eta$$

are two zeros of $f$, $i = 0, 1$, let $\gamma_i = (s_i, t_i) \otimes l_i$ be the corresponding sections of $\mathscr{H}(1)$. By Lemma 2.2.1 both $\gamma_0$ and $\gamma_1$ map into $\langle\lambda\rangle \subset L^\vee$ under $\Gamma \simeq \Gamma\mathscr{H}(1) \to \Gamma\mathscr{E}(1) \simeq L^\vee$. Therefore $M$, $\gamma_1$, $\gamma_2$ are in a 3-dimensional subspace of $\Gamma$. If $M$ is given by $M^* = A \circ \Gamma^*: \mathscr{k}^2 \xrightarrow{A} \mathscr{k}^4 \xrightarrow{\Gamma^*} \mathscr{k}^2 \otimes \bigwedge^2 V$ there must be a nontrivial relation

$$bA + a_0(s_0^3, s_0^2t_0, s_0t_0^2, t_0^3) + a_1(s_1^3, \dots, t_1^3) = 0$$

(with $b \neq 0$ since the two rows are independent).

Now it is elementary to verify that the equations of the tangents to $S$ in $l_i$ in the plane $\mathbb{P}W$ are

$$s_i^2\xi(p) + s_it_i\omega(p) + t_i^2\eta(p) = 0.$$

If $\langle p \rangle \in \mathbb{P}W$ is the intersection point, we therefore find

$$(s_i^3, s_i^2t_i, s_it_i^2, t_i^3) \circ \Gamma^*(p) = 0$$

by the shape of $\Gamma^*$. But then the above relation implies $b \circ M^*(p) = b \circ A \circ \Gamma^*(p) = 0$, i.e. $\det M^*(p) = 0$.

2.4.2. COROLLARY. *There is a bijection* $[\mathscr{E}] \leftrightarrow (S, C)$ *between* $M(0, 2)$ *and the set of Poncelet pairs with* $S \in C^0(\mathbb{G})$ *and* $C$ *smooth.*

*Proof.* If $(S, C)$ is given, the conic $C$ determines a pencil on $S$, since by Poncelet's closure theorem of **[Gr]**, **[Gr-Ha]**, through any point of $C$ there is a triangle tangent to $S$. By $(D_1)$, $S$ and the pencil determine a bundle in $M(0, 2)$.

2.5. REMARKS. (1) This corollary has been generalized to arbitrary instanton bundles with $h^0\mathscr{E}(1) = 2$ in **[Bö-Tr]**, where $C$ becomes a curve of $\deg = c_2$.

(2) The pair $(C, S)$ of conics reflects the two components of the monad $(M, N)$. The Poncelet relation between $S$, $C$ is the geometric expression for the monad to form a complex.

2.6. Summarizing the results of 1.5, 2.2, 2.3, 2.4, we have: If $\mathscr{E} \in M(0, 2)$ there is a conic $S \in C^0(\mathbb{G})$ with quadric $Q$ and conjugate conic $S^0$, and a smooth Poncelet conic $C$ in the plane $P$ of $S$, s.t.

(1) The pencil describing the Poncelet relation is the pencil of zero lines of sections of $\mathscr{E}(1)$.

(2) The conic $S^0$ is the set of jumping lines of $\mathscr{E}$ of order 2 and $S^0 = \operatorname{Supp} R^1\mathscr{E}$ $(= \operatorname{Supp} R^1\mathscr{K})$.

(3) If $\tilde{J}$ is the cone over $C$ with vertex $P^\perp$ then $J = \tilde{J} \cap \mathbb{G}$ is the hypersurface of all jumping lines of $\mathscr{E}$ and $J = \operatorname{Supp} R^1\mathscr{E}(-1)$ $(= \operatorname{Supp} R^1\mathscr{M}(-1))$.

(4) $S^0 = \operatorname{Sing} J$.

(5) The dual quadric $Q^\vee$ is the set of all jumping planes of $\mathscr{E}$, i.e. of all planes $H$ with $H^0(\mathscr{E}|H) \neq 0$.

## 3. Quadric bundles of Poncelet conics.

3.1. Given two conics $S$ and $C$ in the projective plane $\mathbb{P}_2$ one can try to inscribe a triangle in $C$ which is circumscribed about $S$. Poncelet's theorem states that, if there is one such triangle, one can start from any point on $C$ to construct such a triangle, see also **[Gr-Ha]**, **[Gr]**. If $C^\vee$ is the polar dual of $C$ in the dual plane with respect to $S$, the Poncelet condition simply says that there are three points on $S$, such that the dual triangle in $\mathbb{P}_2^\vee$ has its vertices on $C^\vee$. This condition is now symmetric in $S$, $C^\vee$. If it is satisfied we also call $S$, $C^\vee$ a Poncelet pair of conics and $C^\vee$ a Poncelet conic with respect to $S$. For a given regular $S$ the set of all regular Poncelet conics $C^\vee$

with respect to $S$ is open in a quadric in the $\mathbb{P}_5$ of all conics with $\mathbb{P}_2^\vee$. We need an explicit description of this quadric, which doesn't seem to exist in the literature. There is a formula for $S$, $C$ to be a Poncelet pair, first derived by Cayley, see **[Sa]**, p. 342, and **[Gr-Ha]**. This however is not symmetric in $S$, $C$ and one has to transform it for a pair $S$, $C^\vee$. In 3.2 we give both a functorial and an explicit description of it.

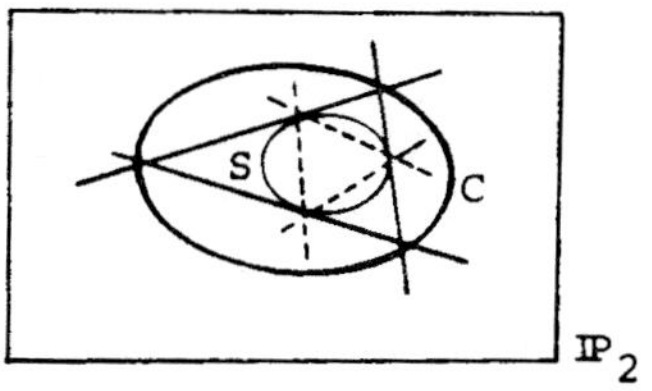


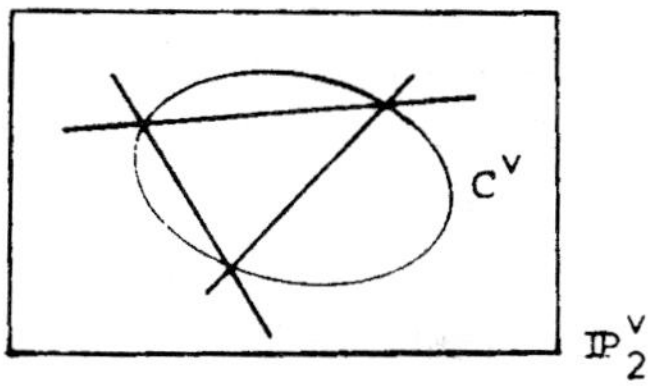


3.2. Let $W$ be a 3-dimensional vectorspace and the conic $S \subset \mathbb{P}W$ be given by $\sigma \in (S^2 W)^\vee$, i.e. by a symmetric bilinear form $W \times W \underset{\sigma}{\longrightarrow} k$ or $W \underset{\sigma}{\longrightarrow} W^\vee$. To $\sigma$ we associate the two canonical forms:

$$s^2\sigma : s^2 W \to S^2 W^\vee \underset{\approx}{\to} (S^2 W)^\vee$$
$$\sigma \cdot \sigma : S^2 W \to (S^2 W)^\vee .$$

The first is the functorial map $S^2\sigma$ followed by the canonical isomorphism, the second is defined by

$$x \cdot y \mapsto \sigma(x, y)\sigma.$$

Now we define

$$Q\sigma = S^2\sigma - \tfrac{1}{2}\sigma \cdot \sigma.$$

Thus $Q\sigma : S^2 W \to (S^2 W)^\vee$ is a symmetric bilinear form on $S^2 W$ and defines a quadric in $\mathbb{P}S^2 W$. We can define $Q\sigma$ as well by

$$Q\sigma(xy, x'y') = \tfrac{1}{2}[\sigma(x, x')\sigma(y, y') + \sigma(x, y')\sigma(y, x') - \sigma(x, y)\sigma(x', y')].$$

3.3. *The matrix representation of* $Q\sigma$. Let $e_0$, $e_1$, $e_2$ be a basis of $W$ and $z_0$, $z_1$, $z_2$ the dual basis of $W^\vee$. The given form $\sigma$ will be expressed by

$$(\sigma(e_i, e_j)) = \begin{pmatrix} 2s_{00} & 2_{01} & s_{02} \\ s_{01} & 2s_{11} & s_{12} \\ s_{02} & s_{12} & 2s_{22} \end{pmatrix}$$

such that the conic $S$ is given by the equation $\sum_{i\leq j} s_{ij}z_iz_j = 0$. Then the symmetric matrix $Q(\sigma) = (Q\sigma(e_ie_j, e_ke_l))$ is

$Q(\sigma) =$

| | 00 | 01 | 02 | 11 | 12 | 22 |
|---|---|---|---|---|---|---|
| 00 | $2s_{00}^2$ | $s_{00}s_{01}$ | $s_{00}s_{02}$ | $s_{01}^2 - 2s_{00}s_{11}$ | $s_{01}x_{02} - s_{00}s_{12}$ | $s_{01}s_{02} - s_{00}s_{22}$ |
| 01 | | $2s_{00}a_{11}$ | $s_{00}s_{12}$ | $s_{01}s_{11}$ | $s_{02}s_{11}$ | $s_{02}s_{12} - s_{01}s_{22}$ |
| 02 | | | $2s_{00}s_{22}$ | $s_{01}s_{12} - s_{02}s_{11}$ | $s_{01}s_{22}$ | $s_{02}s_{22}$ |
| 11 | | | | $2s_{11}^2$ | $s_{11}s_{12}$ | $s_{12}^2 - 2s_{11}s_{22}$ |
| 12 | | | | | $2s_{11}s_{22}$ | $s_{12}s_{22}$ |
| 22 | | | | | | $2s_{22}^2$ |

When the conic $C^\vee$ is given by the equation $\sum_{i\leq j} c_{ij}e_ie_j = 0$ in $\mathbb{P}_2^\vee$ and $c$ denotes the column of the coefficients ordered as above, we get

$$Q\sigma(c, c) = c^t \circ Q(s) \circ c =: 2Q(s, c).$$

Explicitly we have for $Q(s, c)$ the expression

$$\begin{aligned} s_{00}^2c_{00}^2 &+ s_{00}s_{01}c_{00}c_{01} + s_{00}s_{02}c_{00}c_{02} + (s_{01}^2 - 2s_{00}s_{11})s_{00}s_{11} \\ &+ (s_{01}s_{02} - s_{00}s_{12})c_{00}c_{12} + (s_{02}^2 - 2s_{00}s_{22})c_{00}c_{22} \\ &+ s_{00}s_{11}c_{01}^2 + \cdots . \end{aligned}$$

REMARK. When ordered by the products $s_{ij}s_{kl}$ the coefficients as functions in the $c_{ij}$ are the same as in the $s_{ij}$. This proves that $Q(s, c) = Q(c, s)$ and that the condition $Q(s, c) = 0$ is symmetric in $s$ and $c$.

3.4. PROPOSITION. *Let $S \subset \mathbb{P}W$ and $C^\vee \subset \mathbb{P}W^\vee$ be the regular conics with the equations*

$$\sum_{i\leq j} s_{ij}z_iz_j = 0 \quad \textit{resp.} \sum_{i\leq j} c_{ij}e_ie_j = 0.$$

*Then $(S, C^\vee)$ is a Poncelet pair if and only if $Q(s, c) = 0$.*

*Proof*. Let $A$, $B$ be two $3 \times 3$ matrices, not necessarily symmetric and define $\theta(A, B)$, $\theta'(A, B)$ by

$$\det(\lambda A + B) = \lambda^3 \det A + \lambda^2\theta(A, B) + \lambda\theta'(A, B) + \det B.$$

Then for any matrix $M$ we have $\theta(M\circ A, M\circ B) = \det(M)\cdot\theta(A, B)$.

The formula of *Cayley*, [**Sa**], says that if $S$, $T$ are symmetric matrices representing two conics $S, T \subset \mathbb{P}U$ then $S$, $T$ is a Poncelet pair (with $S$ as the "inner" conic) if and only if

$$\theta(S, T)^2 = 4\det(S)\theta(T, S).$$

Now let

$$S = \begin{pmatrix} 2s_{00} & s_{01} & s_{02} \\ s_{01} & 2s_{11} & s_{12} \\ s_{02} & s_{12}2 & s_{22} \end{pmatrix}, \quad \text{resp. } C = \begin{pmatrix} 2c_{00} & c_{01} & c_{02} \\ c_{01} & 2c_{11} & c_{12} \\ c_{02} & c_{12} & 2c_{21} \end{pmatrix},$$

be the matrices of the given conics. The polar dual $C$ of $C^\vee$ with respect to $S$ then has the matrix $T = S \circ C \circ S$. Applying the formula of Cayley we get

$$\theta(S, S \circ C \circ S)^2 = 4\det(S)\theta(S \circ C \circ S, S),$$

which is equivalent to

$$\theta(I, C \circ S)^2 = 4\theta(C \circ S, I).$$

Now by a rather lengthy calculation this condition is equivalent to $Q(s, c) = 0$.

3.5. The quadratic form $Q(s, c)$ determines a *quadric bundle* $Q \subset \mathbb{P}S^2W^\vee \times \mathbb{P}S^2W$ over $\mathbb{P}S^2W^\vee$ with fibres

$$Q_s = \{\langle c\rangle \in \mathbb{P}S^2W | Q(s, c) = 0\}.$$

By using the homogeneous coordinates $s_{00}, \dots, s_{22}$ and $c_{00}, \dots, c_{22}$ we can easily determine the singular locus of $Q$. If the conic $S \subset \mathbb{P}W$ given by $\langle s\rangle \in \mathbb{P}S^2W^\vee$ is non-degenerate, then obviously $Q_s$ is smooth, and therefore $Q$ is smooth over the open set of regular conics in $\mathbb{P}S^2W^\vee$.

3.6. *Singularities of $Q$ over $\mathbb{P}S^2W^\vee$*. Since $\mathrm{Sing}(Q)$ is contained in the inverse image of the discriminant locus of $\mathbb{P}S^2W^\vee$ of degenerate conics, it is enough to describe $Q_s \cap \mathrm{Sing}(Q)$ for $s$ degenerate. We consider the two cases, where $S$ is a pair of distinct lines or a double line.

*Case* 1. If the conic $S$ given by $s$ consists of a *pair* of *distinct lines*, we can choose a basis of $W$ in such a way that $S$ is given by $z_1z_2 = 0$, i.e. $s_{ij} = 0$ for $(i, j) \neq (1, 2)$. Then $Q(s, c) = c_{11}c_{22}$ and $Q_s$ is a pair of distinct 4-planes in $\mathbb{P}S^2W$.

Calculating in addition all the partial derivatives

$$\frac{\partial Q}{\partial s_{ij}}, \quad \frac{\partial Q}{\partial c_{ij}}$$

in $(s, c)$, we find that

$$Q_s \cap \operatorname{Sing}(Q) = \{\langle c\rangle \in \mathbb{P}S^2W \mid c_{11} = c_{22} = 0,\ c_{01}c_{02} - c_{00}c_{12} = 0\}.$$

This is a regular quadric in the 3-dimensional intersection $c_{11} = c_{22} = 0$ of the two components of $Q_s$. In order to illustrate the points of $Q_s \cap \operatorname{Sing}(Q)$ as conics in $\mathbb{P}W^\vee$ let as before $C^\vee$ denote the conic $\sum_{i\le j} c_{ij}e_ie_j = 0$. The condition $c_{11} = c_{22} = 0$ means that $C^\vee$ is a conic through both $z_1, z_2 \in \mathbb{P}W^\vee$. The possible cases of $C^\vee$ are:

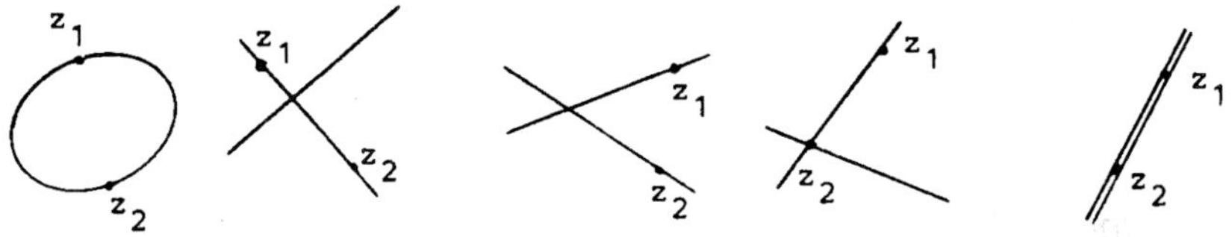

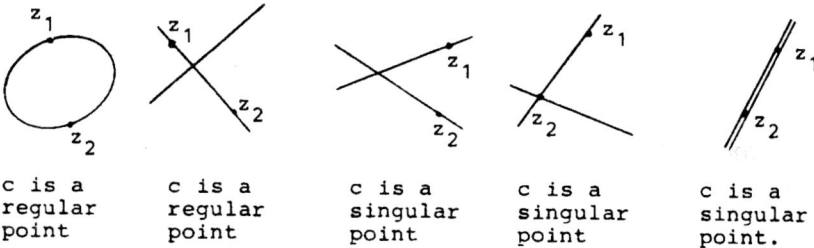

c is a regular point | c is a regular point | c is a singular point | c is a singular point | c is a singular point.

These cases can be checked easily by the above description of $Q_s \cap \operatorname{Sing}(Q)$. Note that in the second case $C^\vee$ is singular whereas $(s, c)$ is a regular point of $Q$.

If $c_{11} = 0$ but $c_{22} \neq 0$ the pair $(s, c)$ is in one component of the fibre $Q_s$ which consists of all conics $C^\vee$ passing through $z_1$ but not through $z_2$:

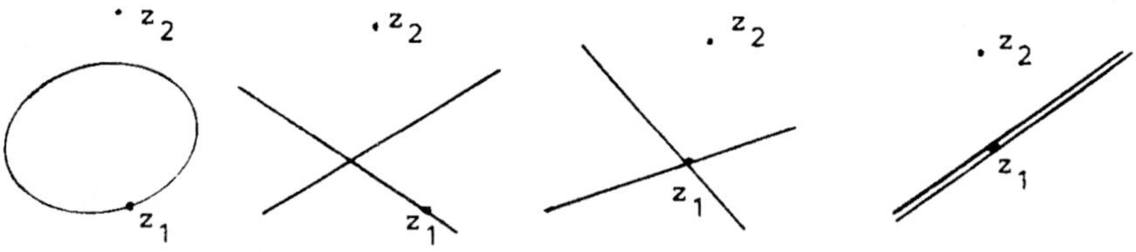


In each of these cases $(s, c)$ is a regular point.

*Case* 2. If the conic $S$ given by $s$ consists of a *double line*, we can choose a basis of $W$ in such a way that $S$ is given by $z_2^2 = 0$, i.e. $s_{ij} = 0$ for $(i, j) \neq (2, 2)$. In this case $Q(s, c) = c_{22}^2$ and $Q_s$ is a double 4-plane. Again by looking at the partial derivatives we find that

$$Q_s \cap \operatorname{Sing}(Q) = \{\langle c\rangle \in \mathbb{P}S^2W \mid c_{22} = c_{02} = c_{12} = 0\},$$

which is a 2-plane in the 4-plane $Q_s$.

The condition $c_{22} = 0$ means that any $C^\vee \in Q_s$ passes through $z_2$. The additional conditions $c_{02} = c_{12} = 0$ say that $C^\vee$ is a conic

$$c_{00}e_0^2 + c_{01}e_0e_1 + c_{11}e_1^2 = 0.$$

Any such is degenerate with vertex $z_2$ and conversely. Thus the list of possible conics $C^\vee$ in the fibre $Q_s$ is:

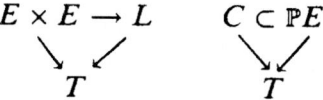

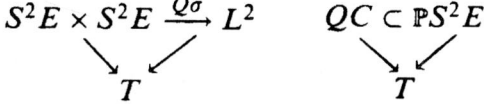

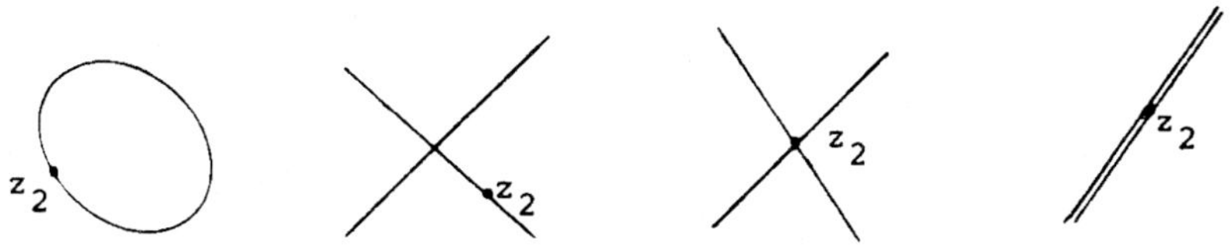


c is a regular point | c is a regular point | c is a singular point | c is a singular point.

3.7. *Associated quadric bundles.* The functor $Q$ of 3.2 can be applied to any scheme of quadrics. In our case we need it only for schemes of conics. Let $E \to T$ be a rank 3 vectorbundle over a scheme together with a quadratic form $\sigma$ as a morphism $E \times E \xrightarrow[\sigma]{} L$ to a line bundle over $T$. Then $\sigma$ defines the scheme of zeros $C \subset \mathbb{P}E$ in the projectified bundle, which we call a scheme of conics:

$$\begin{array}{ccc} E \times E \to L & \qquad & C \subset \mathbb{P}E \\ \searrow \ \swarrow & & \searrow \ \swarrow \\ T & & T \end{array}$$

It is constructed in such a way that for geometric points $t \in T$ we have a conic $C_t \subset \mathbb{P}E_t$. We need this only in the case where $T$ is a reduced variety.

REMARK. In **[Na-Ra]** Narasimhan-Ramanan give a more abstract definition and prove that any conic bundle is of the above form.

Now we can apply the functor $Q$ to obtain the quadric bundle $QC$ as the zero scheme of $Q\sigma$:

$$\begin{array}{ccc} S^2E \times S^2E \xrightarrow{Q\sigma} L^2 & \qquad & QC \subset \mathbb{P}S^2E \\ \searrow \ \swarrow & & \searrow \ \swarrow \\ T & & T \end{array}$$

We call $QC$ the *Poncelet quadric bundle* associated to $C$. We shall need this construction only in the case of the universal conic over the Hilbert scheme of conics contained in the Grassmannian $\mathbb{G} \subset \mathbb{P}\bigwedge^2 V$.

3.8. *The Hilbert scheme of conics in* $\mathbb{G} \subset \mathbb{P}\bigwedge^2 V$. Let $C(\mathbb{G})$, resp. $C(\mathbb{P}\bigwedge^2 V)$, denote the Hilbert schemes of conics in $\mathbb{G}$, resp. $\mathbb{P}\bigwedge^2 V$. There is a natural embedding $C(\mathbb{G}) \subset C(\mathbb{P}\bigwedge^2 V)$ as the subscheme of conics $S \subset \mathbb{P}\bigwedge^2 V$ contained in $\mathbb{G}$ schematically. Any conic $S \in C(\mathbb{G})$ defines a plane $P \subset \mathbb{P}\bigwedge^2 V$ such that $S \subset \mathbb{G} \cap P$ as a subscheme. If $P \not\subset \mathbb{G}$ then $S = \mathbb{G} \cap P$ is a plane section. We distinguish the following exceptional sets of $C(\mathbb{G})$:

$$\Sigma_0 = \text{set of singular conics in } C(\mathbb{G}),$$
$$\Sigma_\alpha = \text{set of conics contained in an } \alpha\text{-plane } P \subset \mathbb{G},$$
$$\Sigma_\beta = \text{set of conics contained in a } \beta\text{-plane } P \subset \mathbb{G}.$$

Then $C^0(\mathbb{G}) = C(\mathbb{G}) \setminus \Sigma_0 \cup \Sigma_\alpha \cup \Sigma_\beta$ is the open part of regular plane conic sections.

3.8.1. PROPOSITION. *$C(\mathbb{G})$ is a smooth, irreducible variety of dimension 9 and $\Sigma_0$, $\Sigma_\alpha$, $\Sigma_\beta$ are irreducible divisors in this manifold.*

For the proof we will make use of the following lemma, see SGA, VII, Prop. 1.7.

LEMMA. *Let $X_1 \subset X_2 \subset X_3$ be schemes with $X_2$, $X_3$ smooth and $X_1$ a locally complete intersection. Then there is an exact sequence (on $X_1$)*

$$0 \to N_{X_1, X_2} \to N_{X_1, X_3} \to N_{X_2, X_3}|X_1 \to 0$$

*where $N_{X_i, X_j}$ denotes the normal bundle of $X_i$ in $X_j$.*

*Proof*. (1) Using the differential criterion for the smoothness for Hilbert schemes, the smoothness of $C(\mathbb{G})$ and its dimension will follow if we have proved $h^1(S, N_{S,\mathbb{G}}) = 0$, $h^0(S, N_{S,\mathbb{G}}) = 9$ for any conic $S \subset \mathbb{G}$.

(2) If $S \subset P \subset \mathbb{P}_5 = \mathbb{P}\bigwedge^2 V$ is a conic in a plane $P$ in $\mathbb{P}_5$ it is immediate to see that $h^1(S, N_{S,P}) = 0$, $h^0(S, N_{S,P}) = 5$, and using the lemma, that also $h^1(S, N_{S,\mathbb{P}_5}) = 0$, $h^0(S, N_{S,\mathbb{P}_5}) = 14$, where one can use that $N_{P,\mathbb{P}_5} = 3\mathcal{O}_P(1)$.

(3) Let $P \subset \mathbb{G}$. Then $h^1(P, N_{P,\mathbb{G}}) = 0$, $h^1(P, N_{P,\mathbb{G}}(-2)) = 1$ and $h^0(P, N_{P,\mathbb{G}}(-2)) = h^2(P, N_{P,\mathbb{G}}(-2)) = 0$, whereas $h^0(P, N_{P,\mathbb{G}}) = 3$. This can be proved by applying the lemma to $P \subset \mathbb{G} \subset \mathbb{P}_5$, so that

we have the exact sequence

$$0 \to N_{P,\mathbb{G}} \to 3\mathcal{O}_P(1) \to \mathcal{O}_P(2) \to 0,$$

since $N_{\mathbb{G},\mathbf{P}_5} = \mathcal{O}_{\mathbb{G}}(2)$. Dualizing this and then tensoring by $\mathcal{O}_P(1)$ we obtain

$$0 \to \mathcal{O}_P(-1) \to 3\mathcal{O}_P \to N^{\vee}_{P,\mathbb{G}} \otimes \mathcal{O}_P(1) \to 0.$$

But $N_{P,\mathbb{G}}$ is of rank 2 and its determinant bundle is $\mathcal{O}_P(1)$, so that $N^{\vee}_{P,\mathbb{G}} \otimes \mathcal{O}_P(1) \cong N_{P,\mathbb{G}}$. Thus we have the exact sequence

$$0 \to \mathcal{O}_P(1) \to 3\mathcal{O}_P \to N_{P,\mathbb{G}} \to 0,$$

and the statements in (3) follow immediately.

(4) To prove (1) we distinguish *case* 1: $P \subset \mathbb{G}$ and *case* 2: $P \not\subset \mathbb{G}$. In *case* 1 we have an exact sequence for $S \subset P$

$$0 \to N_{S,P} \to N_{S,\mathbb{G}} \to N_{P,\mathbb{G}}|S \to 0.$$

From the exact sequence

$$0 \to N_{P,\mathbb{G}}(-2) \to N_{P,\mathbb{G}} \to N_{P,\mathbb{G}}|S \to 0$$

we see, using (3), that

$$h^1(S, N_{P,\mathbb{G}}|S) = 0 \quad \text{and} \quad h^0(S, N_{P,\mathbb{G}}|S) = 4.$$

Using (2) we now conclude that indeed $h^1(S, N_{S,\mathbb{G}}) = 0$ and $h^0(S, N_{S,\mathbb{G}}) = 9$. In *case* 2 the conic $S$ is the (schematic) intersection $S = \mathbb{G} \cap P$. Let us consider the diagram

$$\begin{array}{ccccccccc}
 & & & & 0 & & & & \\
 & & & & \downarrow & & & & \\
 & & & & N_{S,P} & \searrow & & & \\
 & & & & \downarrow & & & & \\
0 & \longrightarrow & N_{S,\mathbb{G}} & \longrightarrow & N_{S,\mathbf{P}_5} & \longrightarrow & N_{\mathbb{G},\mathbf{P}_5}|S & \longrightarrow & 0 \\
 & & & & \downarrow & & & & \\
 & & & & N_{P,\mathbf{P}_5}|S & & & & \\
 & & & & \downarrow & & & & \\
 & & & & 0 & & & &
\end{array}$$

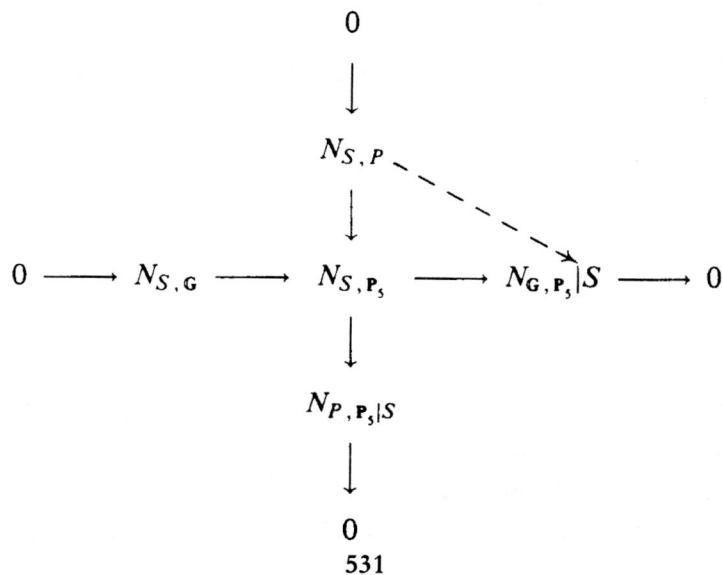

We claim that the vertical sequence is split; in fact the natural map $N_{S,P} \to N_{\mathbb{G},\mathbb{P}_5}|S$ is an isomorphism. This follows from the fact that the line bundle on $P$ determined by the ideal sheaf of $S$ in $P$ is the pull back of the line bundle determined by the ideal sheaf of $\mathbb{G}$ in $\mathbb{P}_5$, by the inclusion $P \to \mathbb{P}_5$. Since the vertical sequence is split, the map

$$H^1(S, N_{S,\mathbb{G}}) \to H^1(S, N_{S,\mathbb{P}_5})$$

is an injection while $H^1(S, N_{S,\mathbb{P}_5}) = 0$ by (2). Thus $H^1(S, N_{S,\mathbb{G}}) = 0$. We also have

$$h^0(S, N_{S,\mathbb{G}}) = h^0(S, N_{S,\mathbb{P}_5}) - h^-(S, N_{\mathbb{G},\mathbb{P}_5}|S) = 14 - 5 = 9.$$

This completes the proof of the smoothness.

(5) By their definition $\Sigma_0$, $\Sigma_\alpha$, $\Sigma_\beta$ are subvarieties of $C(\mathbb{G})$. Clearly $\Sigma_0$ is 1-codimensional and $\Sigma_\alpha$, $\Sigma_\beta$ are $\mathbb{P}_5$-bundles over a $\mathbb{P}_3$ (as space of all $\alpha$- resp. $\beta$-planes) and hence also 1-codimensional. Being bundles $\Sigma_\alpha$, $\Sigma_\beta$ are already irreducible. So we are left to show that also $\Sigma_0$ is irreducible. To do this let $\Omega \subset \Sigma_0$ be the open set of conics consisting of two different lines and which are not in $\Sigma_\alpha \cup \Sigma_\beta$. It suffices to show that $\Omega$ is irreducible (in fact $\dim(\Sigma_0 \setminus \Omega) < 8$ since $\dim \Sigma_0 \cap \Sigma_\alpha$, $\dim \Sigma_0 \cap \Sigma_\beta = 7$ and the subvariety $\Sigma_0' \subset \Sigma_0$ of double lines is of dimension 6, see 3.9.1). Now PGL($V$) acts transitively on $\Omega$ as can be seen from the configuration (D) in 0.4: A conic $S \in \Omega$ corresponds to a pair of planes and two points in their intersection in $\mathbb{P}V$ and thus is determined by a pair $(p, p_1)$, $(q, q_1)$ of pairs of points in $\mathbb{P}_3$ such that the four points are independent. It is now immediate that, given two such configurations in $\mathbb{P}_3$, one can be taken into another by a linear transformation in PGL($V$).

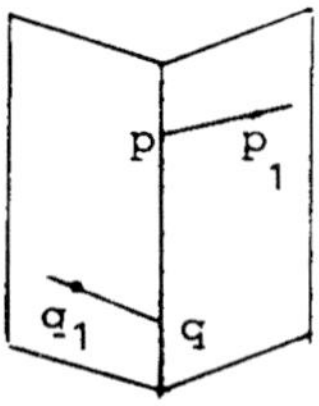


(6) Finally to prove the irreducibility of $C(\mathbb{G})$, it is sufficient to prove that the open dense subset $C^0(\mathbb{G}) = C(\mathbb{G}) \setminus \Sigma_0 \cup \Sigma_\alpha \cup \Sigma_\beta$ is irreducible. But this is isomorphic to the open set $G_3^0 \Lambda^2 V$ in the Grassmannian of 2-planes in $\mathbb{P}\Lambda^2 V$ cutting $\mathbb{G}$ in a regular conic,

see the morphism $C(\mathbb{G}) \to G_3 \bigwedge^2 V$ in 3.10 below. But the latter is irreducible.

3.9. *The double lines in* $C(\mathbb{G})$. If $l \subset \mathbb{G}$ is a line in $\mathbb{P}\bigwedge^2 V$ contained in $\mathbb{G}$, there is a unique $\alpha$-plane $P_\alpha$ and a unique $\beta$-plane $P_\beta$ with $l \subset P_\alpha, P_\beta$. The pencil $\mathrm{Pen}(l)$ of planes spanned by these is the unique pencil such that $l = \mathbb{G} \cap P$ for $P \in \mathrm{Pen}(l)$, $P \neq P_\alpha, P_\beta$. This can be proved easily by choosing a basis of $\bigwedge^2 V$ containing a basis of $l$.

If now $S \in C(\mathbb{G})$ is a double line, the plane $P$ with $S \subset \mathbb{G} \cap P$, which is determined by $S$, must be a member of $\mathrm{Pen}(S_{\mathrm{red}})$, and conversely any $P \in \mathrm{Pen}(S_{\mathrm{red}})$ determines a conic structure $S$ on $S_{\mathrm{red}}$ with $S \subset \mathbb{G} \cap P$ schematically. Therefore the conics $S \in C(\mathbb{G})$ supported by a line $l \subset \mathbb{G}$ are in 1:1 correspondence with the planes $P \in \mathrm{Pen}(l)$. We even have

3.9.1. LEMMA. *Let* $\Sigma_0' \subset \Sigma_0 \subset C(\mathbb{G})$ *be the subvariety of double lines in* $C(G)$. *Then* $\Sigma_0'$ *is a* $\mathbb{P}_1$*-bundle over the Hilbert scheme* $\mathscr{L}(\mathbb{G})$ *of lines in* $\mathbb{G}$, *and* $\dim \Sigma_0' = 6$.

*Proof.* $S \to S_{\mathrm{red}}$ is a morphism $\Sigma_0' \to \mathscr{L}(\mathbb{G})$ whose fibres are the pencils $\mathrm{Pen}(l)$. It is left to the reader to verify that this is a $\mathbb{P}_1$-bundle. It follows that $\dim \Sigma_0' = 6$.

3.10. *The modification* $C(\mathbb{G}) \to G_3 \bigwedge^2 V$. Let $Z = G_3 \bigwedge^2 V$ be the Grassmannian of 2-planes in $\mathbb{P}\bigwedge^2 V$. We denote by $\mathbf{W} \to Z$ the tautological 3-bundle and by $W_z$ its fibre over $z$. The projective bundle $\mathbb{P}S^2\mathbf{W}^\vee$ of quadratic forms in the fibres of $\mathbf{W}$ can be considered as the Hilbert scheme of conics in $\mathbb{P}\bigwedge^2 V$. Therefore we have an embedding

$$\begin{array}{ccc} C(\mathbb{G}) & \hookrightarrow & \mathbb{P}S^2\mathbf{W}^\vee \\ & {\scriptstyle\mu}\searrow \quad \swarrow{\scriptstyle\pi} & \\ & Z & \end{array}$$

and the composed morphism is a modification of the Grassmannian $G_3 \bigwedge^2 V$ with exceptional divisors $\Sigma_\alpha$, $\Sigma_\beta$. Thus the modification consists in putting in all conics in $\alpha$- or $\beta$-planes.

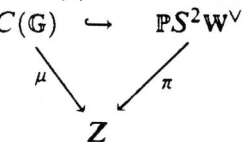

3.11. *The quadric bundle* $Q \to C(\mathbb{G})$. The universal conic over $\mathbb{P}S^2\mathbf{W}^\vee$ can be constructed as follows. Let $\pi^*\mathbf{W}$ be the pull back of the tautological bundle to $\mathbb{P}S^2\mathbf{W}^\vee$. There is a universal quadratic form $\sigma$ on $\pi^*\mathbf{W}$ which has values in the relative hyperplane bundle $\mathscr{L} = \mathscr{O}_{\mathbb{P}S^2\mathbf{W}^\vee}(1)$ such that we have

$$(5)\qquad \begin{array}{ccc} \pi^*\mathbf{W}\otimes\pi^*\mathbf{W} \xrightarrow{\sigma} \mathscr{L} & & C \to \mathbb{P}\pi^*\mathbf{W} \\ \searrow \quad \swarrow & & \searrow \quad \swarrow \\ \mathbb{P}S^2\mathbf{W}^\vee & & \mathbb{P}S^2\mathbf{W}^\vee \end{array}$$

The universal conic $C$ over $\mathbb{P}S^2\mathbf{W}^\vee$ is the zero locus of the form $\sigma$. Clearly the restriction of the conic bundle $C$ to $C(\mathbb{G}) \subset \mathbb{P}S^2\mathbf{W}^\vee$ is the universal conic bundle over the Hilbert scheme $C(\mathbb{G})$.

Now we apply the Poncelet functor $Q$ to the universal conic bundle over $C(\mathbb{G}) \subset \mathbb{P}S^2\mathbf{W}^\vee$. Thus from (5) we obtain the quadric bundle $QC$ and its restriction $Q$ to $C(\mathbb{G})$:

$$(6)\qquad \begin{array}{ccccc} Q & \hookrightarrow & QC & \subset & \pi^*\mathbb{P}S^2\mathbf{W} \\ \downarrow & & \downarrow & & \\ C(\mathbb{G}) & \hookrightarrow & \mathbb{P}S^2\mathbf{W}^\vee . & & \end{array}$$

If we consider $\pi^*\mathbb{P}S^2\mathbf{W}$ as a fibre product we have

$$(7)\qquad \begin{array}{ccccc} Q & \hookrightarrow & QC & \hookrightarrow & \mathbb{P}S^2\mathbf{W}^\vee \times_Z \mathbb{P}S^2\mathbf{W} \\ & \searrow & \downarrow & \swarrow & \\ & & Z & & \end{array}$$

Then for any $z \in Z$ the fibre

$$(QC)_z \subset \mathbb{P}S^2W_z^\vee \times \mathbb{P}S^2W_z$$

is the Poncelet hypersurface of bidegree 2 considered in 3.5. It is at the same time the quadric bundle over $\mathbb{P}S^2W_z^\vee$, or $QC|\mathbb{P}S^2W_z^\vee$.

3.12. *Singularities of* $Q$. If the conic $S \in C(\mathbb{G})$ is regular then the fibre $Q_S$ of $Q$ over $S$ is non-degenerate and therefore $Q$ is smooth over the open part of regular conics. If $S$ is singular, then $Q_S \cap \operatorname{Sing} Q$ has the same description as in 3.6. The proof consists in using local coordinates (derived from $C(\mathbb{G}) \subset \mathbb{P}S^2\mathbf{W}^\vee$ for example) and then in calculating partial derivative as in 3.6. Thus

(i) If $S \in C(\mathbb{G})$ consists of two different lines $e, f$, the pair $(S, C^\vee)$ is a singular point iff $C^\vee$ is singular and passes through both $e$ and $f$.

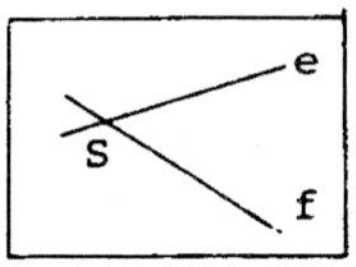

P

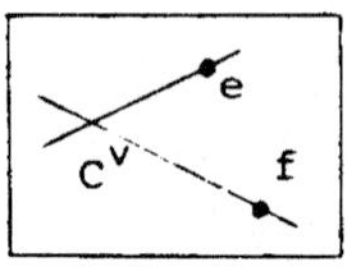

P∨

(ii) If $S \in C()$ is a double line with $S_{\text{red}} = e$, then $(S, C^\vee)$ is a singular point iff $C^\vee$ is singular and $e \in \operatorname{Sing} C^\vee$.

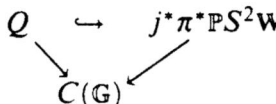

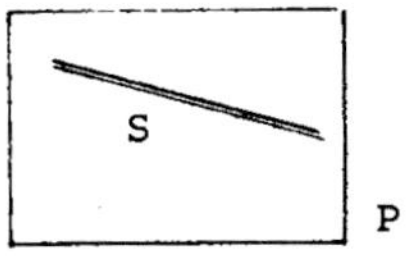


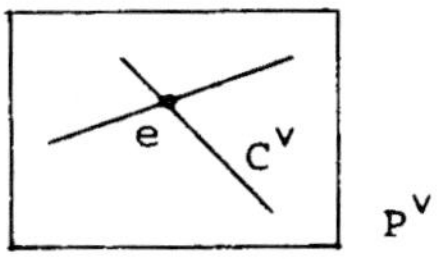


3.12.1. Corollary. *The codimension of* $\operatorname{Sing} Q$ *is* 3.

3.13. Proposition. *The quadric bundle $Q$ is a normal irreducible variety.*

*Proof.* Let $C(\mathbb{G}) \underset{j}{\hookrightarrow} \mathbb{P}S\mathrm{W}^\vee$ be the embedding of 3.9. From diagram (6) we obtain the diagram

$$\begin{array}{ccc} Q & \hookrightarrow & j^*\pi^*\mathbb{P}S^2\mathrm{W} \\ & \searrow \quad \swarrow & \\ & C(\mathbb{G}) & \end{array}$$

where $j^*\pi^*\mathbb{S}P^2\mathrm{W}$ is the bundle of all conics in the dual plane defined by the universal conic over $C(\mathbb{G})$. The embedding of $Q$ into this bundle is regular, since any fibre of $Q$ consists of the Poncelet quadric in the corresponding fibre $\mathbb{P}S^2W$ of $j^*\pi^*\mathbb{P}S^2\mathrm{W}$. Since $C(\mathbb{G})$ is smooth by 3.8.1, it follows that $Q$ is a local complete intersection, see SGA 6, VIII, Prop. 1.5. On the other hand, the codimension of the singular set of $Q$ is $\geq 2$ by 3.12.1. It follows from [**Ha1**, Prop. 8.23] that $Q$ is normal. Since $Q$ is equidimensional of dimension 13, $Q|C^0(\mathbb{G})$ is irreducible and $\dim(Q \setminus Q|C^0(\mathbb{G})) < 13$, $Q$ must also be irreducible.

**4. Boundary components of $Q$ and Main Theorem.** We define 4 (positive) divisors on $Q$ as follows. Let $Q_0$, $Q_\alpha$, $Q_\beta$ be respectively the inverse images of $\Sigma_0$, $\Sigma_\alpha$, $\Sigma_\beta$ (3.8) by the canonical projection $Q \to C(\mathbb{G})$, and let $Q_e$ be the subvariety of pairs $(S, C^\vee)$ with $C^\vee$ singular.

4.1. Proposition. *The divisors $Q_0$, $Q_\alpha$, $Q_\beta$ and $Q_e$ are irreducible.*

*Proof.* Since the Poncelet bundle associated to the space of conics in $\mathbb{P}_2$ is irreducible, we see that $Q_\alpha$ and $Q_\beta$ are irreducible. To show that $Q_e$ is irreducible, let $\Omega'$ be the open subset of $Q_e$ consisting of $(S, C^\vee)$ where $S$ is a regular cut of $\mathbb{G}$ by a plane in $\mathbb{P}_5$. Then $\Omega'$ is irreducible and $\Omega'$ is dense in $Q_e$ and $Q_e$ is of pure dimension 12 and $\dim(Q_3 - \Omega') \leq 11$.

To prove that $Q_0$ is irreducible consider the diagram, see (7) in 3.11,

$$\begin{array}{ccccccc} Q_0 & \hookrightarrow & Q & \hookrightarrow & PS^2\mathbf{W}^\vee \times_Z PS^2\mathbf{W} & \rightarrow & PS^2\mathbf{W} \\ \downarrow & & \downarrow & & \downarrow & \swarrow & \\ \Sigma_0 & \overset{j}{\hookrightarrow} & C(\mathbb{G}) & \overset{\mu}{\longrightarrow} & Z & & \end{array}$$

and the induced map $\varphi$

$$\begin{array}{ccc} Q_0 & \overset{\varphi}{\longrightarrow} & j^*\mu^*\mathbb{P}S^2(\mathbf{W}) \\ & \searrow \quad \swarrow & \\ & \Sigma_0 & \end{array}$$

Let $R \subset j^*\mu^*\mathbb{P}S^2(\mathbf{W})$ be the space of smooth conics. Since $\Sigma_0$ is irreducible we see that $R$ is irreducible. For $C \in R$, $\varphi^{-1}(C)$ consists of pairs $(S, C)$ where $S$ is a singular conic in the dual plane one of whose components touches the dual conic $C^\vee$.

Thus $\varphi^{-1}(C)$ is irreducible of dimension 3, being the image of $\mathbb{P}_1 \times \mathbb{P}_2$ by a finite map. Hence $\varphi^{-1}(R)$ is irreducible. Now $\varphi^{-1}(R)$ is open and dense in $Q_0$, since, in the space of conics in $\mathbb{P}_2$ through a point in $\mathbb{P}_2$, the subspace consisting of smooth conics is dense.

4.2. *$\mathbb{P}_1$-fibration on the exceptional set $Q_{\text{exc}}$*. Recall, 3.12, that the singular set, $\operatorname{Sing} Q$, of $Q$ is contained in $Q_0 \cap Q_0$ and consists of pairs $(S, C^\vee)$ where $S$ and $C^\vee$ are both singular having a position as in (i), (ii), 3.12. Let $\Sigma_0' \subset \Sigma_0$ be the space of double lines and let

$$Q_{\text{exc}} = \operatorname{Sing} Q \cap \pi^{-1}(\Sigma_0'), \quad \pi : Q \rightarrow C(\mathbb{G}).$$

We claim that $Q_{\text{exc}}$ has a natural structure of a $\mathbb{P}_1$-bundle, a fibre being the pencil $\mathbb{P}_1$ of double structures on a line contained in $\mathbb{G}$, see 3.9.1. Let $L(\mathbb{G})$ be the Hilbert scheme of lines contained in $\mathbb{G}$ and $D \rightarrow L(\mathbb{G})$ the tautological $\mathbb{P}_1$-bundle. Let $S^2(D) \rightarrow L(\mathbb{G})$ be the $\mathbb{P}_2$-bundle which is the relative Hilbert scheme of pairs of points on the fibres of $D \rightarrow L(\mathbb{G})$.

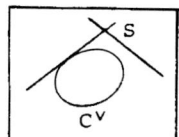

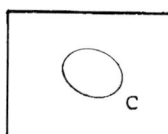

A point of $Q_{\text{exc}}$ consists of a pair $(S, C^\vee)$ where $S$ is a line with a double structure in $\mathbb{G}$ and $C^\vee$ is a singular conic in the dual plane having a singularity at the point in the dual plane determined by $S$. Let $l$ be the reduced line associated to $S$. We may view $C^\vee$ as a pair of points on $l$. Thus we obtain a map $Q_{\text{exc}} \to S^2(D)$, whose fibre at $(l, p, q) \in S^2(D)$, where $l$ is a line in $\mathbb{G}$ and $p, q \in l$, is the pencil $\text{Pen}(l)$ of double structures on $l$ contained in $\mathbb{G}$.

Note that we have the diagram of $\mathbb{P}_1$-fibrations

$$\begin{array}{ccc} Q_{\text{exc}} & \longrightarrow & S^2(D) \\ \downarrow & & \downarrow \\ \Sigma_0' & \longrightarrow & L(\mathbb{G}) \end{array}.$$

4.3. *The component* $\overline{M(0,2)}$ *of the Maruyama scheme containing* $M(0,2)$. Let $\overline{M}(2;0,2,0)$ be the Maruyama scheme of all semistable coherent rank 2 sheaves on $\mathbb{P}_3$ with Chern classes $c_1 = 0$, $c_2 = 2$, $c_3 = 0$. The moduli space of vector bundles $M(0,2)$ is a smooth connected open subset of $\overline{M}(2;0,2,0)$; we denote by $\overline{M(0,2)}$ the (reduced) schematic closure of $M(0,2)$ in $\overline{M}(0;0,2,0)$, and by $\widetilde{M(0,2)} \xrightarrow[\nu]{} \overline{M(0,2)}$ its normalisation.

4.4. THEOREM. (1) *The variety* $Q$ (*see* 3.11) *can be blown down along the* $\mathbb{P}_1$*-fibration* $Q_{\text{exc}} \to S^2(D)$ (*defined in* 4.2) *to a normal variety* $\underset{\sim}{Q}$, *i.e. the push-out* $\underset{\sim}{Q}$ *of the diagram*

$$\begin{array}{ccc} Q_{\text{exc}} & \longrightarrow & Q \\ \downarrow & & \\ S^2(D) & & \end{array}$$

*exists in the category of varieties over* $k$.

(2) *There exists a canonical morphism*

$$Q \xrightarrow[\varphi]{} \overline{M(0,2)}$$

*which induces an isomorphism of the blown down variety* $\underset{\sim}{Q}$ *onto the normalisation* $\widetilde{M(0,2)}$ *of* $\overline{M(0,2)}$.

(3) *Let $Q^0$ denote $Q \setminus Q_e \cup Q_\alpha \cup Q_\beta \cup Q_0$ (see 4.1). The restriction of $\varphi$ to $Q^0$ maps $Q^0$ isomorphically onto $M(0, 2)$. The inverse of this isomorphism is the map of Corollary 2.4.2 which associates to a bundle the corresponding Poncelet pair $(S, C)$ of smooth conics. Moreover the "boundary" $\overline{M(0, 2)} \setminus M(0, 2)$ is the union of the four Weil divisors which are the images by $\varphi$ of $Q_e$, $Q_\alpha$, $Q_\beta$ and $Q_0$.*

(4) *The normalisation map $\widetilde{M(0, 2)} \xrightarrow{\nu} \overline{M(0, 2)}$ is bijective and the smooth points of $M(0, 2)$ correspond precisely to the stable sheaves in $\overline{M(0, 2)}$.*

REMARK 1. In the formulation of the theorem in [**Na-Tr**] the blow down of the $\mathbb{P}_1$-fibration of $Q_{exc}$ had been overlooked.

REMARK 2. Under $X^{ss}//SL(2) \cong Q$ the stable points of $X^{ss}$ under the SL(2)-action correspond precisely to the smooth points of $Q$, see 6.7.1.

REMARK 3. The sheaves or their equivalence classes in the 4 boundary components can be characterised geometrically by the Poncelet pairs $(S, C^\vee)$ by which they are defined. This can be found in §§9, 10.

In particular the generic points of the divisor $Q_e$ are the sheaves $\mathcal{F}$, which are obtained by the elementary transformations

$$0 \to \mathcal{F} \to \mathcal{E}' \to \mathcal{O}_L(1) \to 0,$$

where $\mathcal{E}'$ is a bundle in $M(0, 1)$ and $L$ is a line in $\mathbb{P}_3$, see 9.1.

REMARK 4. The semi-stable but non-stable sheaves in the boundary are characterised as extensions

$$0 \to \mathcal{J}_{L \cup q} \to \mathcal{F} \to \mathcal{J}_{K \cup p} \to 0,$$

where $L$, $K$ are lines and $p$, $q$ points in $\mathbb{P}_3$, $\mathcal{J}_{L\cup q}$ is the ideal sheaf of the union $L \cup \{q\}$ (with a simple multiple structure in $q$ if $q \in L$), see Theorem 10.5. The blow down of the $\mathbb{P}_1$-fibration of $Q_{exc}$ is explained in terms of the sheaves in 10.6.

REMARK 5. For any $[\mathcal{F}] \in \overline{M(0, 2)}$ let $S^0 = \operatorname{Supp} R^1\mathcal{F}$, $J = \operatorname{Supp} R^1\mathcal{F}(-1)$, where $R^1$ is the incidence transformation. These sets are the generalised sets of jumping lines of order $\geqq 2$ resp. of all jumping lines of $\mathcal{F}$. In the proof of 8.3 it is shown that $S^0$ and $J$ only depend on the equivalence class $[\mathcal{F}]$. However, which is more important, we have the

COROLLARY. *Any* $[\mathcal{F}] \in \overline{M(0,2)}$ *is already determined by the pair* $(S^0, J)$.

*Proof*. By 8.3 (d) the pair $(S^0, J)$ determines a quadric hypersurface $\tilde{J} \subset \mathbb{P}\bigwedge^2 V$ such that $J = \mathbb{G} \cap \tilde{J}$ and $\tilde{J}$ is singular along $S^0 \subset \operatorname{Sing} J$. If $[\mathcal{F}] \notin \varphi(Q_{\mathrm{exc}})$ then $(S^0, \tilde{J})$ determines the plane $\mathbb{P}W$ or $\mathbb{P}W^\perp$ by 8.3 (e), (f). If $\mathbb{P}W^\perp \subset \mathbb{P}\bigwedge^2 V$ is any splitting of $\mathbb{P}\bigwedge^2 V \cong \mathbb{P}\bigwedge^2 V^\vee \to \mathbb{P}W^\vee$, see 8.3 (b), then $C^\vee = \mathbb{P}W^\vee \cap \tilde{J}$, 8.3 (d), and hence $\mathcal{F}$ is determined by $(S^0, J)$ through $(S, C^\vee)$. If however $[\mathcal{F}] \in \varphi(Q_{\mathrm{exc}})$, then we can choose any plane $\mathbb{P}W$ in the pencil $\operatorname{Pen}(S_{\mathrm{red}})$ and define $C^\vee = \mathbb{P}W^\vee \cap \tilde{J}$. Then $(S, C^\vee)$ determines the class $[\mathcal{F}]$ independently of the choice of $\mathbb{P}W$ by (1) of the theorem, i.e. $(S^0, J)$ determines $[\mathcal{F}]$ in this case, too.

4.5. *Proof of Theorem* 4.4. (a) We first prove the existence of the canonical map $\varphi$ and part (3) of the theorem. In §6 we will construct a projective variety $X$ with an SL(2)-action such that, if $X^{ss}$ is the open subset of semi-stable points for this action, then the good quotient $X^{ss}//\mathrm{SL}(2)$ is isomorphic to $Q$. Moreover we construct in §8 a flat family $\{\mathcal{F}_x\}$, $x \in X^{ss}$, of rank 2 coherent sheaves on $\mathbb{P}_3$ with $c_1 = 0$, $c_2 = 2$, $c_3 = 0$. It is proved in §§9, 10 that the sheaves $\mathcal{F}_x$ are semi-stable. Moreover if $x$ and $x'$ are on the same SL(2)-orbit then $\mathcal{F}_x \simeq \mathcal{F}_{x'}$. Hence there is a canonical morphism from $X^{ss}$ to the Maruyama scheme $\overline{M}(2; 0, 2, 0)$, and this induces a morphism $Q \xrightarrow[\Phi]{} \overline{M}(2; 0, 2, 0)$.

(b) We will now prove that $\Phi$ maps $Q$ onto $\overline{M(0,2)}$, so that we will obtain a morphism $Q \xrightarrow[\varphi]{} \overline{M(0,2)}$. We first show that $\Phi$ maps $Q^0$ isomorphically onto $M(0,2) \subset \overline{M}(2; 0, 2, 0)$. Let $X^0$ be the inverse image of $Q^0$ in $X^{ss}$. Now each sheaf $\mathcal{F}_x$, $x \in X^{ss}$, comes with a monad display (22) in §8. If $x \in X^0$ this is a monad of a bundle by 6.8 with $\mathcal{N}_y = \mathcal{K}_N$ (see 7.1.2) and $\mathcal{A}_y = N^\vee \otimes \mathcal{O}$. So, $X^0$ and hence its quotient $Q^0$ are mapped into $M(0,2)$ under $\Phi$. The Poncelet pair $(S, C^\vee)$ of smooth conics associated to $x$ is the Poncelet pair associated to the bundle in 2.4.2 with $C^\vee$ the polar dual of $C$, see 6.8. Now by Corollary 2.4.2 $\Phi|Q^0$ is a bijection from $Q^0$ onto $M(0,2)$. Since both varieties are smooth $\Phi|Q^0$ is an isomorphism. Since $Q$ is irreducible by Proposition 3.13 and $Q^0$ is dense in $Q$, we now see that $\Phi$ maps $Q$ onto $\overline{M(0,2)}$. We thus have the canonical morphism $Q \xrightarrow[\varphi]{} \overline{M(0,2)}$. Observe that we have proved already the first parts of (2) and (3) of the theorem.

(c) We now proceed to prove (1) and (2). It is proved in 8.4 with preparations in §7 that $Q \xrightarrow[\varphi]{} \overline{M(0,2)}$ is injective on $Q \setminus Q_{\mathrm{exc}}$, constant on the fibres of the $\mathbb{P}_1$-fibration on $Q_{\mathrm{exc}}$ and induces an injective map $S^2(D) \to \overline{M(0,2)}$. Since $Q$ is normal (Proposition 3.13) and $\varphi$ is onto, $\varphi$ lifts to a surjective morphism $Q \xrightarrow{\tilde{\varphi}} \widetilde{M(0,2)}$. Since the fibres of $\varphi$ are connected, the normalization map $\widetilde{M(0,2)} \xrightarrow[\nu]{} \overline{M(0,2)}$ is bijective. Using the Stein factorization of $\tilde{\varphi}$ and Zariski's main theorem, we see that

$$\tilde{\varphi}_* \mathscr{O}_Q = \mathscr{O}_{\widetilde{M(0,2)}}.$$

If $Q_{\mathrm{exc}} \xrightarrow[\eta]{} S^2(D)$ is the $\mathbb{P}_1$-fibration we also have $\eta_* \mathscr{O}_{Q_{\mathrm{exc}}} = \mathscr{O}_{S^2(D)}$, so that the map $S^2(D) \xrightarrow[\psi]{} \widetilde{M(0,2)}$ induced by $S^2(D) \to \overline{M(0,2)}$ is a morphism. We thus have a commutative diagram of morphisms

$$\begin{array}{ccc} Q_{\mathrm{exc}} & \longrightarrow & Q \\ \downarrow{\scriptstyle \eta} & & \downarrow{\scriptstyle \tilde{\varphi}} \\ S^2(D) & \longrightarrow & \widetilde{M(0,2)} \end{array}$$

with $\tilde{\varphi}_* \mathscr{O}_Q = \mathscr{O}_{\widetilde{M(0,2)}}$. It is easy to verify from this that $\widetilde{M(0,2)}$ is the required push out. Thus we have proved (1) and (2) of the theorem, and also that the normalization map $\nu$ is bijective.

(d) From the above we know that

$$Q \setminus Q_{\mathrm{exc}} \to \widetilde{M(0,2)} \setminus \tilde{\varphi}(Q_{\mathrm{exc}})$$

is an isomorphism. Under this the divisors $Q_e$, $Q_\alpha$, $Q_\beta$, $Q_0$ are mapped to divisors, since $Q_{\mathrm{exc}} \subset Q_0$ and $\dim Q_{\mathrm{exc}} = 8$ (as a $\mathbb{P}_2$-bundle over $\Sigma_0' \subset C(\mathbb{C})$). Hence they are also mapped to divisors in $\widetilde{M(0,2)}$ or $\overline{M(0,2)}$. This proves the second part of (3).

(e) To complete the proof of (4), observe that $Q_{\mathrm{exc}}$ is contained in the singular locus $Q_{\mathrm{Sing}} \subset Q_0$ of $Q$. Now $Q_{\mathrm{Sing}} \setminus Q_{\mathrm{exc}}$ is dense in $Q_{\mathrm{Sing}}$. In fact by 3.12, a Poncelet pair $(S, C^\vee)$ in $Q_{\mathrm{Sing}}$ corresponds to a singular conic $S$ with a pair of points one on each component. We can approximate a double line with a pair of (eventually coincident) points on it by a singular conic consisting of distinct lines along with a point on each component. Since $\tilde{\varphi}$ is an isomorphism on $Q \setminus Q_{\mathrm{exc}}$, we see that $\tilde{\varphi}(Q_{\mathrm{Sing}}) = \widetilde{M(0,2)}_{\mathrm{Sing}}$. On the other hand by Theorem

10.5, a point in $Q$ corresponds to a stable sheaf precisely when it is a non-singular point of $Q$.

**5. Geometric invariant parametrisation of** $C(\mathbb{G})$. In order to define a universal family of sheaves we have to construct a suitable parametrisation of the quadric bundle $Q \to C(\mathbb{G})$, since a universal family does not even exist over $M(0, 2)$ **[Hi-Na]**. This will be done in such a way that we construct a morphism $X \to Y$ of projective varieties, acted on by $\mathrm{SL}(2, \mathscr{k})$, which is equivariant, and such that the induced morphism $X^{ss}//\mathrm{SL}(2) \to Y^{ss}//\mathrm{SL}(2)$ on the good quotients is the quadric bundle.

As a first step we construct $Y$ in this section. Recall that $C^0(\mathbb{G}) = G_2^0(\mathscr{k}^2 \otimes V)/\mathrm{SL}(2)$, where $G_2^0(\mathscr{k}^2 \otimes V)$ is the open part of the Grassmannian consisting of regular subspaces $N \subset \mathscr{k}^2 \otimes V$ defining the right parts $\mathscr{k}^2 \otimes \Omega^1(1) \xrightarrow[\nu]{} N^\vee \otimes \mathscr{O}$ of bundle monads, 1.4. The space $G_2^{ss}(\mathscr{k}^2 \otimes V)$ of semi-stable points of the Grassmannian however does not parametrise the complete Hilbert-scheme, see Remark 5.9. The parameter space $Y$ is constructed in such a way that its semi-stable points form a modification of $G_2^{ss}(\mathscr{k}^2 \otimes V)$. It is essentially the set of $\Gamma \in G_4(\mathscr{k}^2 \otimes \bigwedge^2 V)$ such that

$$\Gamma \subset \mathrm{Ker}\left(\mathscr{k}^2 \otimes \bigwedge^2 V \to N^\vee \otimes \bigwedge^3 V\right)$$

for some $N \in G_2(\mathscr{k}^2 \otimes V)$, and only if $N$ is regular we have an exact sequence

$$0 \to \Gamma \to \mathscr{k}^2 \otimes \bigwedge^2 V \to N^\vee \otimes \bigwedge^3 V \to 0. \tag{8}$$

It turned out that in the degenerate cases of $N$ the subspaces $\Gamma$ provide us with the necessary information in the limit cases: They determine the degenerate conics in $C(\mathbb{G})$ and we have $\Gamma \simeq \Gamma\mathscr{N}(1)$, where $\mathscr{N}$ is the corresponding kernel sheaf, thereby generalizing the results of §1, whereas the degenerate spaces $N$ do not determine them. The result is the space $Y \subset G_4(\mathscr{k}^2 \otimes W)$ constructed in 5.7. In 5.8 we show that $N$ is determined by $\Gamma$ for semi-stable $\Gamma$.

As a preparation we prove a stability criterion for points in Grassmannians, of the above type, which is essential for our constructions.

5.1. Let $U$, $W$ be finite dimensional vector spaces and let $\mathrm{SL}(U)$ act on $G_q(U \otimes W)$ by $L \mapsto (g \otimes \mathrm{id})(L)$. This action is induced from the linear action of SL($U$) on $\bigwedge^q(U\otimes W)$ via the Plücker embedding $G_q(U\otimes W)\subset \mathbb{P}\bigwedge^q(U\otimes W)$. Therefore it makes sense to characterize the stable and semi-stable points of this action in the sense of Mumford, **[Mu-Fo]**, see also **[Ne1]**. The result is

5.1.1. PROPOSITION (Stability criterion). *Let* SL($U$) *act on the Grassmannian* $G_q(U\otimes W)$ *as above. Then a point* $L\in G_q(U\otimes W)$ *is stable* (*semi-stable*) *if and only if*

$$\dim L\cap (U'\otimes W) \underset{(\leq)}{<} \frac{\dim U'}{\dim U}\dim L$$

*for any proper subspace* $0\neq U'\subsetneq U$.

*Proof*. Let $\lambda$ be a (non-trivial) 1-parameter subgroup of SL($U$). In a basis $(e_0, \dots, e_l)$ of $U$, $\lambda$ is given by the diagonal matrix $\lambda(t)=[t^{\alpha_0}, \dots, t^{\alpha_l}]$, $\alpha_0\geq\alpha_1\geq\cdots\geq\alpha_l$, $\sum_k\alpha_k=0$.

Let $v_0, \dots, v_n$ be any basis of $W$. Consider the basis $\{e_i\otimes v_j\}$ of $U\otimes W$. Since the action of SL($U$) on $U\otimes W$ is given by $g(u\otimes w)=gu\otimes w$ for $u\in U$, $w\in W$, the action of $\lambda$ on $U\otimes W$ is given by the diagonal matrix

$$[t^{\alpha_0}, \dots, t^{\alpha_0}, t^{\alpha_1}, \dots, t^{\alpha_1}, \dots, t^{\alpha_l}, \dots, t^{\alpha_l}]$$

(each $t^{\alpha_j}$ occurring $(n+1)$ times), with respect to the basis $f_0=e_0\otimes v_0, \dots, f_n=e_0\otimes v_n$; $f_{n+1}=e_1\otimes v_0, \dots, f_{2n+1}=e_1\otimes v_n$; $\dots$; $f_{ln+1}=e_l\otimes v_0, \dots, f_{(l+1)n+l}=e_l\otimes v_n$.

We now use **[Mu-Fo**, p. 88, $(***)_N$ or **Ne1**, p. 121**]** with $N=1$ to get

$$\mu(L, \lambda)=-q\cdot r_{(l+1)n+l}+\sum_{i=0}^{(l+1)n+1}\dim(L\cap L_i)(r_{i+1}-r_i)$$

where $L_i$ is the subspace spanned by $(f_0, \dots, f_i)$. In our case

$$r_0=\cdots=r_n=\alpha_0;\quad r_{n+1}=\cdots=r_{2n+1}=\alpha_1;$$
$$r_{ln+1}=\cdots=r_{(l+1)n+l}=\alpha_l.$$

Hence

$$(*)\qquad \mu(L, \lambda)=-q\alpha_l+\sum_{k=0}^{l-1}\dim(L\cap L_{(k+1)n_k})(\alpha_{k+1}-\alpha_k).$$

As in [**Ne1**, p. 121] we consider the cases

$$\alpha_0 = \cdots = \alpha_p = l-p; \quad \alpha_{p+1} = \cdots = \alpha_l = -(p+1) \quad \text{for } 0 \le p < l-1.$$

Here $\mu(L, \lambda) = q(p+1) - \dim(L \cap L_{(p+1)n+p})(l+1)$. But $L_{(p+1)n+p} = U'_p \otimes W$, where $U'_p$ is the span (in $U$) of $e_0, \dots, e_p$.

Thus $\mu(L, \lambda) = q(p+1) - \dim(L \cap (U'_p \otimes W))(l+1)$.

Thus $\mu(L, \lambda) > 0$ (resp. $\ge 0$) if and only if

$$\dim(L \cap (U'_p \otimes W)) < \frac{l+1}{p+1} \cdot q \qquad (\text{resp. } \le)$$

for $0 \le p < l$. Since every $(p+1)$-dimensional subspace of $U$ is conjugate under $\mathrm{SL}(U)$ to $U'_p$ the result follows. In 6.3 we need

5.1.2. **Proposition.** *Let $U = k^2$, $\dim W = 3$ and let $M \in G_2(k^2 \otimes W)$ and $\Gamma \in G_4(k^2 \otimes W)$ with $M \subset \Gamma$. Then*

(1) *The pair $(M, \Gamma) \in G_2 \times G_4$ is semi-stable if and only if $M$ and $\Gamma$ are semi-stable in $G_2$ and $G_4$ respectively.*

(2) *If $(M, \Gamma)$ is semi-stable and one of the components is stable, then $(M, \Gamma)$ is stable.*

*Proof*. (1) Let $\lambda(t) = \begin{pmatrix} t^{\alpha_0} & \\ & t^{\alpha_1} \end{pmatrix}$ be a 1-parameter subgroup with $\alpha_0 + \alpha_1 = 0$, $\alpha_1 < 0$. As in the proof of 5.1.1 we have

$$\mu(M, \lambda) = \alpha_1(-\dim M + 2\dim M \cap (e_0 \otimes W)),$$
$$\mu(\Gamma, \lambda) = \alpha_1(-\dim \Gamma + 2\dim \Gamma \cap (e_0 \otimes W)).$$

Since the action on the product is linearised via the tensor-product $\bigwedge^2(k^2 \otimes W) \otimes \bigwedge^4(k^2 \otimes W)$ we must have

$$(*) \qquad \mu(M, \Gamma, \lambda) = \mu(M, \lambda) + \mu(\Gamma, \lambda).$$

From this it is clear that $(M, \Gamma)$ is semi-stable if both components are semi-stable. To prove the converse, let $(M, \Gamma)$ be semi-stable. Then we have

$$\dim M \cap (e_0 \otimes W) \dim \Gamma \cap (e_0 \otimes W) \le \tfrac{1}{2}(\dim M + \dim \Gamma) \le 3.$$

If $M$ were not semi-stable, $\dim M \cap e_0 \otimes W = 2$ or $M \subset e_0 \otimes W$. By the inclusion $M \subset \Gamma$ also $\dim \Gamma \cap (e-0 \otimes W) \ge 2$, but by the previous inequality this should be $\le 1$. Similarly if $\Gamma$ were not semi-stable, $\dim \Gamma \cap (e_0 \otimes W) \ge 3$ and then $M \cap (e_0 \otimes W) = 0$. However this is not possible since $\dim M = 2$.

(2) Follows directly from the formula $(*)$.

From now on $W$ will be 3-dimensional and $\mathbb{P}_2 = \mathbb{P}W$.

5.2. *The conic* $S(\Gamma)$. Let $\Gamma \subset k^2 \otimes W$ be a 4-dimensional subspace, $\dim W = 3$. The description of the conic $S(\Gamma)$ in 1.3 motivates the following definition

$$S(\Gamma) := \{\langle w\rangle \in \mathbb{P}W \mid u \otimes w \in \Gamma \text{ for some } u \neq 0 \text{ in } k^2\}.$$

For arbitrary $\Gamma$ this set could be the whole plane, for example if $u \otimes W \subset \Gamma$ for some $u$.

5.2.1. Proposition. (1) *For* $\Gamma \in G_4(k^2 \otimes W)$ *the following conditions are equivalent*:

(a) $\Gamma$ *is semi-stable.*
(b) $S(\Gamma)$ *is a conic.*
(c) $S(\Gamma) \neq \mathbb{P}W$.

(2) $\Gamma$ *is stable iff the conic* $S(\Gamma)$ *is regular.*

*Proof.* (1) If $\Gamma$ is not semi-stable, by 5.1.1 there is some $0 \neq u \in k^2$ with $u \otimes W \subset \Gamma$ $(\dim(u \otimes W) \cap \Gamma \geq 3)$ and hence $S(\Gamma) = \mathbb{P}W$. This proves (c) $\Rightarrow$ (a). It remains to prove (a) $\Rightarrow$ (b). We consider the 2-dimensional kernel $\Sigma$ in

$$0 \to \Sigma \to k^2 \otimes W^\vee \to \Gamma^\vee \to 0.$$

Since $G_4(k^2 \otimes W) \to G_2(k^2 \otimes W^\vee)$ is an SL(2)-equivariant isomorphism $\Sigma$ is stable (semi-stable) iff $\Gamma$ is stable (semi-stable). By the criterion 5.1.1 this means $\dim \Sigma \cap (u \otimes W^\vee) < 1$ $(\leq 1)$ for any $u \neq 0$. If $\Sigma$ is the image of the matrix $\Sigma^* = \begin{pmatrix} z & w \\ z' & w' \end{pmatrix} : k^2 \to k^2 \otimes W^\vee$, we find that $\Sigma$ is semi-stable iff the determinant $zw' - z'w \neq 0$ in $S^2 W^\vee$, and this then defines a conic. To see that this is $S(\Gamma)$, we consider the diagram on $\mathbb{P}W$

$$
(9)\qquad
\begin{array}{ccccccccc}
 & & & & 0 & & & & \\
 & & & & \downarrow & & & & \\
 & & 0 & \longrightarrow & k^2 \otimes \Omega^1(1) & \longrightarrow & \Gamma^\vee \otimes \mathcal{O} & \longrightarrow & \mathcal{C} \longrightarrow \mathcal{O} \\
 & & \downarrow & & \downarrow & & \| & & \\
0 & \longrightarrow & \Sigma \otimes \mathcal{O} & \longrightarrow & k^2 \otimes W^\vee \otimes \mathcal{O} & \longrightarrow & \Gamma^\vee \otimes \mathcal{O} & \longrightarrow & 0 \\
 & & \downarrow & & \downarrow & & & & \\
 & & k^2 \otimes \mathcal{O}(1) & = & k^2 \otimes \mathcal{O}(1) & & & & \\
 & & \downarrow & & \downarrow & & & & \\
 & & \mathcal{C} & & 0 & & & & \\
 & & \downarrow & & & & & & \\
 & & 0 & & & & & &
\end{array}
$$

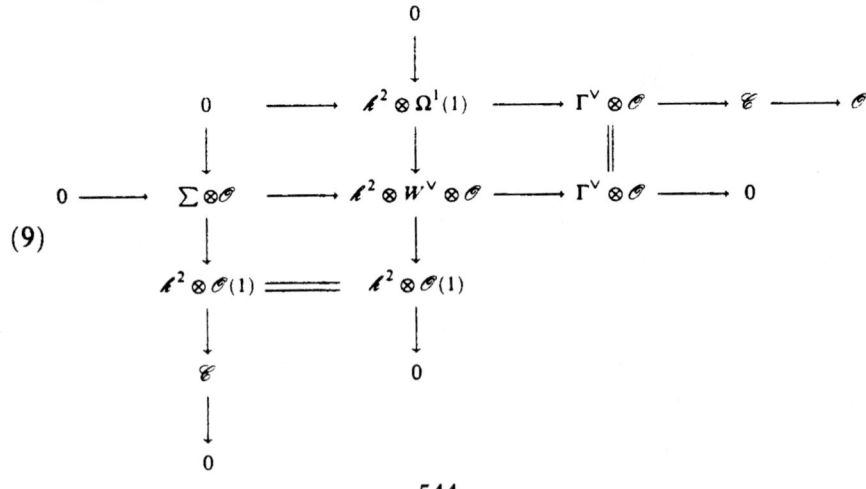

in which the two cokernels identify. The homomorphism $\Sigma \otimes \mathscr{O} \to \mathscr{k}^2 \otimes \mathscr{O}(1)$ is injective in the semi-stable case. By the fibre description of $\mathscr{k}^2 \otimes \Omega^1(1) \to \Gamma^\vee \otimes \mathscr{O}$, see 0.1, we find that $\operatorname{supp} \mathscr{C} = S(\Gamma)$, and the left column shows that this is a conic given by $\det \Sigma^* = 0$.

(2) Now it is easy to see that $S(\Gamma)$ is regular iff $\Sigma^*$ has no zero as any entry in any equivalent representation. This is equivalent to $\Sigma \cap (u \otimes W^\vee) = 0$ for any $u \neq 0$, i.e. $\Sigma$ stable.

5.2.2. **Remark.** (1) If $S(\Gamma)$ is regular then $\mathscr{C} \simeq \mathscr{O}_S(3)$ of degree 3 on $S$. In the following $S(\Gamma)$ shall always be given the structure of the equation $\det \Sigma^* = 0$.

(2) $S(\Gamma)$ is nothing but the determinantal variety of the homomorphism of the top row.

5.3. *Normal forms.* In the following tableau the matrices are $\Sigma^*$, $\Gamma^*$ defining $\Sigma$, $\Gamma$ as images of $\mathscr{k}^2 \to \mathscr{k}^2 \otimes W^\vee$, $\mathscr{k}^4 \to \mathscr{k}^2 \otimes W$ respectively, and $e_0, e_1, e_2 \in W$, $z_0, z_1, z_2 \in W^\vee$ are dual bases.

S(Γ)

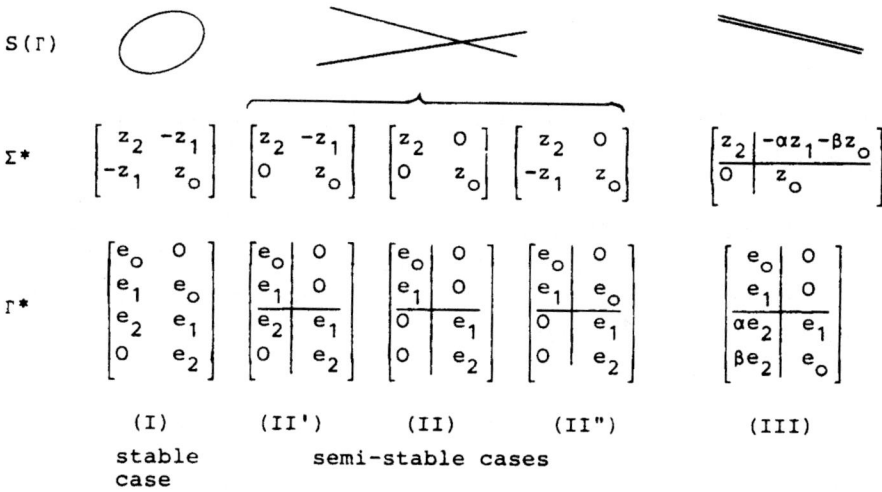

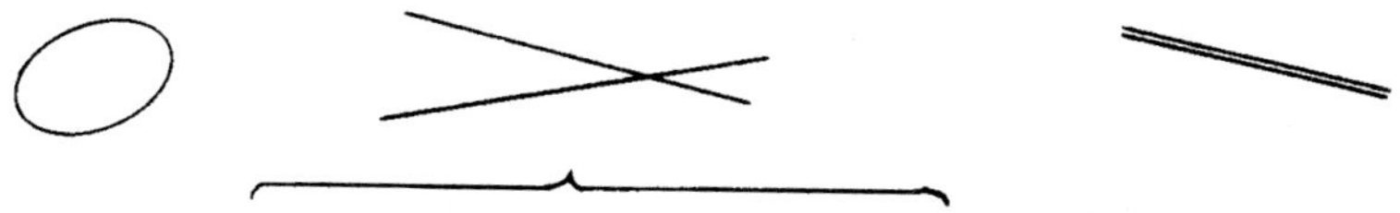

| | (I) | (II') | (II) | (II") | (III) |
|---|---|---|---|---|---|
| Σ* | $\begin{bmatrix} z_2 & -z_1 \\ -z_1 & z_0 \end{bmatrix}$ | $\begin{bmatrix} z_2 & -z_1 \\ 0 & z_0 \end{bmatrix}$ | $\begin{bmatrix} z_2 & 0 \\ 0 & z_0 \end{bmatrix}$ | $\begin{bmatrix} z_2 & 0 \\ -z_1 & z_0 \end{bmatrix}$ | $\begin{bmatrix} z_2 & -\alpha z_1 - \beta z_0 \\ 0 & z_0 \end{bmatrix}$ |
| Γ* | $\begin{bmatrix} e_0 & 0 \\ e_1 & e_0 \\ e_2 & e_1 \\ 0 & e_2 \end{bmatrix}$ | $\begin{bmatrix} e_0 & 0 \\ e_1 & 0 \\ e_2 & e_1 \\ 0 & e_2 \end{bmatrix}$ | $\begin{bmatrix} e_0 & 0 \\ e_1 & 0 \\ 0 & e_1 \\ 0 & e_2 \end{bmatrix}$ | $\begin{bmatrix} e_0 & 0 \\ e_1 & e_0 \\ 0 & e_1 \\ 0 & e_2 \end{bmatrix}$ | $\begin{bmatrix} e_0 & 0 \\ e_1 & 0 \\ \alpha e_2 & e_1 \\ \beta e_2 & e_0 \end{bmatrix}$ |
| | stable case | semi-stable cases | | | |

5.3.1. **Lemma.** *If $\Gamma \in G^4(\mathscr{k}^2 \otimes W)$ is semi-stable it can be presented by one of the normal forms in the above tableau.*

The proof follows immediately if we choose the bases of $W^\vee$ so that $S(\Gamma)$ is defined by the matrices $\Sigma^*$ of the form given.

5.4. **Remark.** Note that the normal forms of type II', II, II" give the same conic although they are not on the same orbits under SL(2).

However the orbit of $\Gamma$ of type II is in the closure of that of type II′ or II″ in the tableau. For if we consider the 1-parameter subgroup $\begin{pmatrix} \alpha & \\ & \alpha^{-1} \end{pmatrix}$, its action on the $\Gamma$ of type II′ is given by

$$\operatorname{Im}\begin{bmatrix} e_0 & \\ e_1 & \\ e_2 & e_1 \\ & e_2 \end{bmatrix}\begin{bmatrix} \alpha & \\ & \alpha^{-1} \end{bmatrix} = \operatorname{Im}\begin{bmatrix} \alpha e_0 & \\ \alpha e_1 & \\ \alpha e_2 & \alpha^{-1} e_1 \\ & \alpha^{-1} e_2 \end{bmatrix} = \operatorname{Im}\begin{bmatrix} e_0 & \\ e_1 & \\ \alpha^2 e_2 & e_1 \\ & e_2 \end{bmatrix}$$

and the latter tends to the direct sum as $\alpha \to 0$.

5.5. *An unusual parametrisation of* $\mathbb{P}_5$. By the previous results we have a morphism $G_4(\mathscr{k}^2 \otimes W)^{ss} \to \mathbb{P}S^2W^\vee$ given by $\Gamma \to \langle \det \Sigma^* \rangle$, where $\Sigma^*$ is any matrix defining $\Sigma$. This morphism factors through the good quotient, **[Mu-Fo]**, **[Ne1]**, $G_4(\mathscr{k}^2 \otimes W)^{ss}//\mathrm{SL}(2) \to \mathbb{P}S^2W^\vee$ and we have the

5.5.1. Proposition. *$G_4(\mathscr{k}^2 \otimes W)^{ss}//\mathrm{SL}(2) \simeq \mathbb{P}S^2W^\vee$ is an isomorphism.*

*Proof.* Indeed this is a bijective morphism, which follows from the listing of the normal forms above. Since $\mathbb{P}DS^2W^\vee$ is smooth and the quotient is irreducible and reduced, it must be an isomorphism by Zariski's main theorem.

5.6. In order to obtain a similar *parametrisation of the Hilbert scheme* of all conics in $\mathbb{P}_5 = \mathbb{P}\bigwedge^2 V$ and later in $\mathbb{G} = G_2V \subset \mathbb{P}\bigwedge^2 V$ we use the above parametrisation for each plane $P \subset \mathbb{P}_5$ and let the planes vary in the Grassmannian $G_3 \bigwedge^2 V$. So we consider the tautological bundle

$$\mathbf{W} \to G_3 \bigwedge^2 V = Z.$$

We denote by $W = W_z \subset \bigwedge^2 V$ the 3-dimensional subspace given by $z \in Z$, see 3.10. As in the absolute case there is the induced action of $\mathrm{SL}(2)$ on the Grassmann bundle

$$G_4(\mathscr{k}^2 \otimes \mathbf{W}) \to Z.$$

This can be linearised as follows. Since $\mathbf{W} \subset Z \times \bigwedge^2 V$ we get the embedding

$$G_4(\mathscr{k}^2 \otimes \mathbf{W}) \subset Z \times G_4\left(\mathscr{k}^2 \otimes \bigwedge^2 V\right) \subset Z \times \mathbb{P}\bigwedge^4\left(\mathscr{k}^2 \otimes \bigwedge^2 V\right)$$

as the subvariety of pairs $(z, \Gamma)$ satisfying $\Gamma \subset k^2 \otimes \bigwedge^2 V$. Since $Z$ is not affected by the action, it is enough to consider the linear action on $\bigwedge^4(k^2 \otimes \bigwedge^2 V)$. Therefore the relative statement on stability reads the same way as that in 5.1.1, 5.1.2.

5.6.1. **Proposition.** *For* $(z, \Gamma) \in G_4(k^2 \otimes \mathrm{W})$ *the following conditions are equivalent:*

(i) $(z, \Gamma)$ *is stable (semi-stable).*
(ii) $\dim \Gamma \cap (u \otimes W_z) \leq 1$ $(\leq 2)$ *for any* $0 \neq u \in k^2$.
(iii) $S(\Gamma) \subset \mathbb{P} W_z$ *is a regular conic (conic).*

Also we obtain analogously

5.6.2. **Proposition.** *The map* $(z, \Gamma) \to S(\Gamma)$ *induces an isomorphism* $G_4(k^2 \otimes \mathrm{W})^{ss}//\mathrm{SL}(2) \simeq \mathbb{P} S^2 \mathrm{W}^\vee$ *of the good quotient with the Hilbert scheme* $C(\mathbb{P} \bigwedge^2 V) = \mathbb{P} S^2 \mathrm{W}^\vee$ *of conics in* $\mathbb{P} \bigwedge^2 V$.

5.7. *Parametrisation of the Hilbert scheme* $C(\mathbb{G})$. Since we are only interested in conics contained in the Plücker quadric $\mathbb{G} = G_2 V \subset \mathbb{P} \bigwedge^2 V$, we have to characterise those $(z, \Gamma)$ for which the conic $S(\Gamma) \subset \mathbb{G} \cap \mathbb{P} W_z$ scheme-theoretically. Note that by this, the case where $\mathbb{G} \cap \mathbb{P} W_z$ is a pair of lines and $S(\Gamma)$ is a double line with $S(\Gamma)_{\mathrm{red}}$ being one of the lines, is excluded. To do this we consider the quadratic forms:

$$\sigma(\Gamma) = \det \Sigma^* \in S^2 W_z \text{ determined up to a scalar},$$
$$\rho(z)k = \text{quadratic form } \in S^2 W_z \text{ of } \mathbb{G} \cap \mathbb{P} W_z.$$

Note that $\sigma(\Gamma) = 0$ if $\Gamma$ is not semi-stable, and $\rho(z) = 0$ if $\mathbb{P} W_z \subset \mathbb{G}$. The condition $S(\Gamma) \subset \mathbb{G} \cap \mathbb{P} W_z$ is now expressed by $\rho(z) \wedge \sigma(\Gamma) = 0$, which is well defined. Next we consider the open subset

$$G_2^0(k^2 \otimes V)$$

of the Grassmannian of right monads $N$ as in §1, s.t. the morphism $k^2 \otimes \Omega^1(1) \to N^\vee \otimes \mathcal{O}$ is surjective, see 1.2. Recall that then $N$ defines a 4-dimensional kernel $\Gamma$ by (8) and a regular conic section $S(\Gamma) = \mathbb{G} \cap \mathbb{P} W_z$ by 1.3. Therefore we have an equivariant morphism

$$G_2^0(k^2 \otimes V) \xrightarrow{\varepsilon} G_4(k^2 \otimes \mathrm{W})^s.$$

Now we define

$$Y \subset G_4(\mathcal{A}^2 \otimes \mathbf{W})$$

as the closure of the image of $\varepsilon$, which is an imbedding. Then $Y$ is 12-dimensional and irreducible. Also $Y$ is invariant under SL(2) since the image of $\varepsilon$ is. Therefore we have the induced action and linearisation and

$$Y^{ss} = Y \cap G_4(\mathcal{A}^2 \otimes \mathbf{W})^{ss}.$$

Since $Y$ is defined as the closure, the condition $\sigma(\Gamma) \wedge \rho(z) = 0$ is satisfied for any $(z, \Gamma) \in Y^{ss}$ and therefore the conic $S(\Gamma)$ is contained in $\mathbb{G} \cap \mathbb{P}W_z$ as a subscheme. Therefore there is the morphism

$$Y^{ss} \to C(\mathbb{G}) \subset \mathbb{P}S^2\mathbf{W}^\vee$$

given by $\sigma(\Gamma)$. This factorizes through the good quotient $Y^{ss}//\mathrm{SL}(2)$.

5.7.1. Proposition. *$Y^{ss}//\mathrm{SL}(2) \to C(\mathbb{G})$ is an isomorphism.*

*Proof*. Clearly the morphism is surjective, since its image contains that of $G_2^0(\mathcal{A}^2 \otimes V)$, which is dense, and since the quotient is projective. It is also injective: $\mathbb{P}W_z$ is determined by $S(\Gamma)$ and also the equivalence class of $\Gamma \subset \mathcal{A}^2 \otimes W_z$ by the normal forms, see 5.4. Since the quotient is also integral, **[Mu-Fo]**, and $C(\mathbb{G})$ is smooth, it must be an isomorphism by Zariski's main theorem.

5.7.2. Remark. One can even show that $Y^{ss}$ is smooth, whereas $Y$ is singular. To do this, consider the subvariety $Y' \subset G_4(\mathcal{A}^2 \otimes \mathbf{W})$ defined by $\sigma(\Gamma) \wedge \rho(z) = 0$. We have $Y \subset Y'$ and $Y^{ss} \subset Y'^{ss}$, $Y'$ being also invariant. Now one can show that $\dim T_p Y'^{ss} = 12$ for any $p \in Y'^{ss}$. One has to consider the different types of points, choose appropriate bases of $\Gamma$ and the bundle $W$, and to use local coordinates of the Grassmannian $G_3 \bigwedge^2 V$. It turns out that the Jacobian matrix of the equations of $Y'$ always has rank 5 in the 17-dimensional manifold $G_4(\mathcal{A}^2 \otimes \mathbf{W})^{ss}$. Therefore for any $p \in Y^{ss}$

$$12 = \dim_p Y^{ss} \leq \dim_p Y'^{ss} \leq \dim T_p Y'^{ss} = 12.$$

Hence $Y^{ss}$ is smooth and defines a component of $Y'$. However we don't need this result.

5.8. *The morphism* $Y^{ss} \to G_2^{ss}(k^2 \otimes V)$. If $\Gamma \in G_4(k^2 \otimes \bigwedge^2 V)$ we can consider the induced homomorphism

$$(10)\qquad \begin{array}{ccccc} k^2 \otimes \bigwedge^2 V^\vee & \to k^2 \otimes \bigwedge^2 V^\vee \otimes V^\vee \to & & \Gamma^\vee \otimes V^\vee \\ \wr\| & & & \wr\| \\ k^2 \otimes V & \xrightarrow[h(\Gamma)]{\qquad\qquad} & & \Gamma^\vee \otimes \bigwedge^3 V \end{array}$$

5.8.1. LEMMA. *If* $(z, \Gamma) \in Y^{ss}$ *then* $N = \operatorname{Ker} h(\Gamma)$ *is 2-dimensional and semi-stable.*

By this we obtain a morphism $Y^{ss} \xrightarrow{\nu} G_2^{ss}(k^2 \otimes V)$. It is now easy to see that the morphism $\varepsilon$ of 5.7 is a *section* of $\nu$ over $G_2^0(k^2 \otimes V)$, since for regular $N$ the space $\Gamma$ defined through (8) is stable with $N \subset \operatorname{Ker} h(\Gamma)$ and thus $N = \operatorname{Ker} h(\Gamma)$.

*Proof of the Lemma.* (a) If the conic $S(\Gamma)$ is regular, then $\Gamma$ is presented by a matrix

$$k^4 \xrightarrow[\begin{bmatrix} \xi & 0 \\ \omega & \xi \\ \eta & \omega \\ 0 & \eta \end{bmatrix}]{\qquad\qquad} k^2 \otimes W_z, \qquad W_z \subset \bigwedge^2 V,$$

such that each vector $s^2\xi + st\omega + t^2\eta \in \bigwedge^2 V$ is decomposable, since $S(\Gamma) \subset \mathbb{G}$. This is equivalent to $\xi \wedge \xi = 0$, $\eta \wedge \eta = 0$, $\xi \wedge \omega = 0$, $\eta \wedge \omega = 0$ and $\omega \wedge \omega + 2\xi \wedge \eta = 0$.

If $S(\Gamma) = \mathbb{G} \cap \mathbb{P}W_z$ is a regular conic section, there is nothing to prove, because then $N$ must be given as in 1.2. If $\mathbb{P}W_z \subset \mathbb{G}$ is an $\alpha$-plane, then there is a vector $x \in V$ with $\xi = x \wedge x'$, $\eta = x \wedge y$, $\omega = x \wedge y'$ for some $x', y, y' \in V$. These vectors must form a basis, since $\xi$, $\omega$, $\eta$ are independent. Now we see that $N$ must be presented by

$$k^2 \xrightarrow[\begin{bmatrix} x & 0 \\ 0 & x \end{bmatrix}]{\qquad\qquad} k^2 \otimes V,$$

which is semi-stable.

(b) If $\mathbb{P}W_z \subset \mathbb{G}$ is a $\beta$-plane, we must have $\xi, \omega, \eta \in \bigwedge^2 H$, $H \subset V$, s.t. also $\omega \wedge \omega = 0$ and hence all products are zero. Since $\xi$, $\omega$, $\eta$ are independent, we find a basis $x$, $x'$, $y$ of $H$ s.t. $\xi = x \wedge x'$, $\omega = -x' \wedge y$, $\eta = y \wedge x$. In this case the kernel is represented by

$$k^2 \xrightarrow[\begin{bmatrix} x & y \\ x' & x \end{bmatrix}]{} k^2 \otimes V$$

which is stable in this case.

(c) If $S(\Gamma) = \mathbb{G} \cap \mathbb{P}W_z$ is a pair of lines, $\Gamma$ must be presented in normal form, see 5.3,

$$\begin{bmatrix} \xi & 0 \\ \omega & 0 \\ \eta & \omega \\ 0 & \eta \end{bmatrix},$$

where the lines in $\mathbb{G}$ are parametrised by $\langle s\xi + t\omega \rangle$ and $\langle s\omega + t\eta \rangle$. Then $\xi = x \wedge x'$, $\omega = x \wedge y'$, $\eta = y \wedge y'$ for some vectors $x$, $y$, $y'$, which are independent. Then the kernel is represented by

$$k^2 \xrightarrow[\begin{bmatrix} x & 0 \\ y & y' \end{bmatrix}]{} k^2 \otimes V$$

which again is semi-stable.

(d) All other cases are treated analogously.

5.9. Remark. One can consider $Y^{ss} \to G_2^{ss}(k^2 \otimes V)$ as a Schubert-type blow up by filling in the 4-dimensional subspaces

$$\Gamma \subset \operatorname{Ker}\left(k^2 \otimes \bigwedge^2 V \to N^\vee \otimes \bigwedge^3 V\right).$$

If we go to the quotients we get a modification

$$\begin{array}{ccc} Y^{ss}//\mathrm{SL}(2) & \longrightarrow & G_2^{ss}(k^2 \otimes V)//\mathrm{SL}(2) \\ \| & & \| \\ C(\mathbb{G}) & \longrightarrow & R \end{array}$$

It is not difficult to see that $R$ is a ramified cover of the space $\mathbb{P}S^2V$ of all quadrics in $\mathbb{P}V^\vee$ by looking at $[N] \to \langle \det N^* \rangle$. Away from the singular quadrics this is the 2-sheeted cover of regular reguli by distinguishing a system of lines in a quadric, see Remark 1.4. The transposition

$$\begin{bmatrix} x & x' \\ y & y' \end{bmatrix} \to \begin{bmatrix} x & y \\ x' & y' \end{bmatrix}$$

of the parametrising matrices defines the decktransformation. The space $R$ is the minimal completion of this covering. By the above modification $C(\mathbb{G}) \to R$ it is possible to extend the quadric bundle of Poncelet conics.

**6. Geometric invariant parametrisation of the quadric bundle $Q \to C(\mathbb{G})$.**

6.1. As in 5.6 the Grassmann bundle $G_2(\mathcal{k}^2 \otimes \mathrm{W}) \to Z = G_3 \bigwedge^2 V$ can be described as the flag variety of pairs $(z, M) \in Z \times G_2(\mathcal{k}^2 \otimes \bigwedge^2 V)$ with $M \subset \mathcal{k}^2 \otimes W_z$. Analogously the induced group action of SL(2) can be linearised through the Plücker embedding by the action of $\bigwedge^2(\mathcal{k}^2 \otimes \bigwedge^2 V)$. Since $Z$ is not affected, $(z, M)$ is (semi-)stable iff $M \in G_2(\mathcal{k}^2 \otimes \bigwedge^2 V)$ is (semi-)stable. To each $M$ we can also associate a quadratic form $\det M^* \in S^2 W_z$ where $M^* : \mathcal{k}^2 \to \mathcal{k} \otimes W_z$ represents $M$. As in 4.6 we obtain the

6.1.1. PROPOSITION. (1) *For each* $(z, M) \in G_2(\mathcal{k}^2 \otimes \mathrm{W})$ *the following are equivalent*:

(i) $(z, M)$ *is semi-stable* (*stable*).

(ii) $\dim M \cap (u \otimes W_z) \leq 1$ $(= 0)$ *for any* $u \neq 0$.

(iii) $\det M^* \neq 0$ *in* $S^2 W_z$ ($\det M^* = 0$ *is the equation of a regular conic*).

6.2. By this result we obtain a morphism $(z, M) \to \langle \det M^* \rangle$

$$G_2^{ss}(\mathcal{k}^2 \otimes \mathrm{W}) // \mathrm{SL}(2) \to \mathbb{P} S^2 \mathrm{W}.$$

This morphism is bijective and hence an isomorphism.

6.3. Let now $X \subset G_2(\mathcal{k}^2 \otimes \mathrm{W}) \times_Z G_4(\mathcal{k}^2 \otimes \mathrm{W})$ be the flag subvariety of the product bundle, defined as the set of all $(z, M, \Gamma)$ with $(z, \Gamma) \in Y$ and $M \subset \Gamma$. If $Y \xrightarrow[\gamma]{} G_4(\mathcal{k}^2 \otimes \bigwedge^2 V)$ is the canonical composition

$$Y \hookrightarrow G_4(\mathcal{k}^2 \otimes \mathrm{W}) \hookrightarrow Z \times G_4\left(\mathcal{k}^2 \otimes \bigwedge^2 V\right) \to G_4\left(\mathcal{k}^2 \otimes \bigwedge^2 V\right),$$

and if $\mathbb{T}$ is the tautological subbundle on the Grassmannian $G_4$, we obviously have $X = G_2\gamma^*\mathbb{T}$. Thus $X \to Y$ is a Grassmann bundle, $\dim X = 16$. The induced SL(2)-action on the product bundle leaves $X$ invariant, and the action is again linearised via the embeddings

$$\begin{aligned} X &\subset G_2(\mathscr{k}^2 \otimes \mathbf{W}) \times_Z G_4(\mathscr{k}^2 \otimes \mathbf{W}) \\ &\subset Z \times G_2\left(\mathscr{k}^2 \otimes \bigwedge^2 V\right) \times G_4\left(\mathscr{k}^2 \otimes \bigwedge^2 V\right) \\ &\subset Z \times \mathbb{P}\bigwedge^2\left(\mathscr{k}^2 \otimes \bigwedge^2 V\right) \times \mathbb{P}\bigwedge^4\left(\mathscr{k}^2 \otimes \bigwedge^2 V\right) \\ &\subset Z \times \mathbb{P}\left(\bigwedge^2\left(\mathscr{k}^2 \otimes \bigwedge^2 V\right) \otimes \bigwedge^4\left(\mathscr{k}^2 \otimes \bigwedge^2 V\right)\right). \end{aligned}$$

Thus $(z, M, \Gamma) \in X$ is (semi-)stable iff $(M, \Gamma) \in G_2(\mathscr{k}^2 \otimes W_z) \times G_4(\mathscr{k}^2 \otimes W_z)$ is (semi-)stable. Fortunately we have the

6.3.1. Lemma. (1) *$(z, M, \Gamma) \in X$ is semi-stable iff each component is semi-stable in $G_2(\mathscr{k}^2 \otimes W_z)$, $G_4(\mathscr{k}^2 \otimes W_z)$ respectively.*

(2) *$(z, M, \Gamma) \in X^{ss}$ is stable if one of the components is stable.*

(3) *There are stable pairs $(z, M, \Gamma) \in X^s$ without $M$ and $\Gamma$ being stable, see Remark* 6.7.2.

*Proof*. (1) and (2) had been proved in 5.1.2. In this proof it is possible that $\mu(M, \lambda)$ or $\mu(\Gamma, \lambda')$ are zero for different $\lambda$'s but never simultaneously. In such a case $(M, \Gamma)$ is stable but not $M$ and $\mathbb{T}$. An example is provided by the pair $M \subset \Gamma$, defined as the images of the matrices

$$\begin{bmatrix} e_0 + e_2 & e_1 \\ 0 & e_2 \end{bmatrix} : \mathscr{k}^2 \to \mathscr{k}^2 \otimes W, \qquad \begin{bmatrix} e_0 & 0 \\ e_1 & 0 \\ e_2 & e_1 \\ 0 & e_2 \end{bmatrix} : \mathscr{k}^4 \to \mathscr{k}^2 \otimes W$$

respectively. Then $(M, \Gamma)$ is stable but neither $M$ nor $\Gamma$.

6.4. By the above lemma the projection $X \xrightarrow[\pi]{} Y$ maps $X^{ss} \to Y^{ss}$ and $X^{ss} \subset \pi^{-1}(Y^{ss})$. Since $X^{ss}$ is open and $\pi^{-1}(Y^{ss})$ is irreducible as a bundle over $Y^{ss}$, also $X^{ss}$ is irreducible and smooth.

6.5. **Proposition.** *If* $(z, M, \Gamma) \in X$ *is semi-stable, the pair* $(\langle\sigma(\Gamma)\rangle, \langle\det M^*\rangle) \in \mathbb{P}S^2W_z^\vee \times \mathbb{P}S^2W_z$ *is a Poncelet pair and determines a pair of Poncelet conics* $S(\Gamma) \subset \mathbb{P}W_z$, $C^\vee(M) \subset \mathbb{P}W_z^\vee$.

*Proof*. For the proof we use the normal forms of the spaces $\Gamma \subset k^2 \otimes W_z$ given in 5.3. Let $e_0, e_1, e_2 \in W_2$ and $z_0, z_1, z_2 \in W_z^\vee$ be dual bases, as in 5.3.

*Case* 1. $S(\Gamma)$ is regular and its form is $\sigma(\Gamma) = z_0z_2 - z_1^2$. Then with the notations of 3.3 the Poncelet quadric

$$Q_{\sigma(\Gamma)} = \{c_{00}c_{22} - c_{01}c_{12} + c_{02}c_{11} + c_{11}^2 = 0\} \subset \mathbb{P}S^2W_z.$$

We have to verify that $\det M^* = \sum_{i\leq j} c_{ij}e_ie_j$ satisfies this condition. As $M \subset \Gamma$ is a 2-dim. subspace, we have

$$M^* = \begin{bmatrix} \alpha_1 & \alpha_2 & \alpha_3 & \alpha_4 \\ \beta_1 & \beta_2 & \beta_3 & \beta_4 \end{bmatrix} \begin{bmatrix} e_0 & 0 \\ e_1 & e_0 \\ e_2 & e_1 \\ 0 & e_2 \end{bmatrix},$$

where we use the normal form of $\Gamma^*$. When $a_{ij} = \alpha_i\beta_j - \alpha_j\beta_i$ we get $c_{00} = a_{12}$, $c_{01} = a_{13}$, $c_{02} = a_{14} - a_{23}$, $c_{11} = a_{23}$, $c_{12} = a_{24}$, $c_{22} = a_{34}$. Inserted into the formula:

$$c_{00}c_{22} - c_{01}c_{12} + c_{02}c_{11} + c_{11}^2 = a_{12}a_{34} - a_{13}a_{24} + (a_{14} - a_{23})a_{23} + a_{23}^2 = 0$$

this expression vanishes because of the Plücker relation of the $a_{ij}$.

*Case* 2. $S(\Gamma)$ is a pair of lines, $z_0z_2 = 0$, and the matrix $\Gamma^*$ has the form (II$'$) say. Now by 3.3

$$Q_{\sigma(\Gamma)} = \{c_{00}c_{22} = 0\} \subset \mathbb{P}S^2W_z$$

and from

$$M^* = \begin{bmatrix} \alpha_1 & \alpha_2 & \alpha_3 & \alpha_4 \\ \beta_1 & \beta_2 & \beta_3 & \beta_4 \end{bmatrix} \begin{bmatrix} e_0 & \\ e_1 & \\ e_2 & e_1 \\ & e_2 \end{bmatrix}$$

we obtain $c_{00} = 0$. This shows that again $\langle\det M^*\rangle \in Q_{\sigma(\Gamma)}$. If $\Gamma^*$ has the form (II$''$) we would get $c_{22} = 0$.

*Case* 3. $S(\Gamma)$ is a double line; this can be treated as case 2.

6.6. By the last proposition the morphism

$$X^{ss} \to \mathbb{P}S^2\mathbf{W}^\vee \times_Z \mathbb{P}S^2\mathbf{W},$$
$$(z, M, \Gamma) \to (\langle\sigma(\Gamma)\rangle, \langle\det M^*\rangle)$$

has its image contained in the Poncelet quadric bundle

$$\begin{array}{ccc} Q & \subset \mathbb{P}S^2\mathbf{W}^\vee \times_Z \mathbb{P}S^2\mathbf{W} \\ \downarrow & \\ C(\mathbb{G}) & \end{array}$$

such that for the fibres over a point $z$ we have the Poncelet bundle

$$Q_z \subset \mathbb{P}S^2W_z^\vee \times \mathbb{P}S^2W_z$$

of the plane $\mathbb{P}W_z$, see 3.3. Again we have

6.6.1. PROPOSITION. *The induced morphism* $\varphi$

$$\begin{array}{ccc} X^{ss}//\mathrm{SL}(2) & \xrightarrow[\approx]{\varphi} & Q \\ \downarrow & & \downarrow \\ Y^{ss}//\mathrm{SL}(2) & \xrightarrow[\approx]{} & C(\mathbb{G}) \end{array}$$

*is an isomorphism.*

*Proof*. Clearly the morphism factorizes through the good quotient. The surjectivity can be shown directly by constructing $M \subset \Gamma$ for a pair of conics using the normal forms, or simply by remarking that it must have a dense image and that $X^{ss}//\mathrm{SL}(2)$ is projective, whereas $Q$ is irreducible.

The morphism is also *injective*. Since we know this already for $Y^{ss}//\mathrm{SL}(2) \to C(\mathbb{G})$, we have to show injectivity in the fibres for fixed $\sigma(\Gamma)$. If $M_1, M_2 \subset \Gamma$ and $\langle\det M_1^*\rangle = \langle\det M_2^*\rangle$, we can assume

$$M_\nu^* = A_\nu \circ \Gamma^*$$

with the same $\Gamma^*$. When $a_{ij}^\nu$ are the Plücker coordinates of $A_\nu$ and $c_{ij}^\nu$ the coefficients of $\det M_\nu^*$, the formulas of the proof of 6.5 show that $a_{ij}^2 = \lambda a_{ij}^1$ in the different cases of $S(\Gamma)$. But this proves that the matrices $A_1, A_2$ span the same subspaces, i.e. $M_1 = M_2$.

Finally as in the previous cases $X^{ss}//\mathrm{SL}(2) \to Q$ must be an isomorphism, since the quotient is integral and $Q$ is normal.

6.7. By the general theory of good quotients there are open sets $Q^s \subset Q$ and $C(\mathbb{G})^s \subset C(\mathbb{G})$ such that their inverse images in the parameter spaces are $X^s$ and $Y^s$ respectively. (However $X^s$ is not mapped necessarily into $Y^s$, nor $Q^s$ into $C(\mathbb{G})^s$.) But we know from Lemma 6.3.1 that $Q|C(\mathbb{G})^s \subset Q^s$, since the stable conics are the regular ones in $C(\mathbb{G})$. Therefore the semi-stable but non-stable points can only lie over $C(\mathbb{G}) \setminus C(\mathbb{G})^s$, i.e. in the fibres over the degenerate conics in $G$. It turns out that we even have:

6.7.1. **Theorem.** *The non-stable points of $Q$ are exactly the singular points of $Q$.*

The proof follows easily from the description of the singular points $(S, C^\vee) \in Q$ in 3.12 on the one hand, and from the characterisation of the semi-stable points $(z, M, \Gamma) \in X$ in 6.3.1 on the other hand. However one has to take special care of those points $(z, M, \Gamma)$ which are stable without $M, \Gamma$ being stable.

6.7.2. **Remark.** The stable points $(z, M, \Gamma)$ for which neither $M$ nor $\Gamma$ are stable correspond exactly to those pairs $(S, C^\vee)$ which are smooth points of $Q$ but with both conics $S$ and $C^\vee$ singular. The corresponding sheaves in $\overline{M(0, 2)}$ are stable and will be treated in 10.4, 10.7, see also 10.5.

6.8. *Points in $X^{ss}$ parametrising bundle monads.* The projective variety $X$ has been constructed in such a way that it completes the space of monads $(D_2)$ of bundles in 2 and simultaneously serves as a parameter space of $Q$. We are going to identify the part of $X^{ss}$ which consists of monads for bundles. Let $Q^0$ be the complement of the 4 Weil divisors $Q_e$, $Q_\alpha$, $Q_\beta$, $Q_0$ in $Q$ as defined in §4. Then $Q^0$ consists entirely of pairs of regular conics and maps onto $C^0(\mathbb{G})$. Let $X^0 \subset X^s$ resp. $Y^0 \subset Y^s$ be the inverse images of $Q^0$ resp. $C^0(\mathbb{G})$ in $X$ resp. $Y$. We then have the diagram of SL(2)-quotients

$$\begin{array}{ccc} X^0 & \longrightarrow & Q^0 \\ \downarrow & & \downarrow \\ Y^0 & \longrightarrow & C^0(\mathbb{G}) \end{array}$$

6.8.1. **Lemma.** (i) *The open and dense set* $X^0 \subset X^{ss}$ *is the set of monads of bundles in* $M(0, 2)$.

(ii) *The open and dense set* $Y^0 \subset Y^{ss}$ *is the set of right arrows of such monads, and is isomorphic to* $G_2^0(k^2 \otimes V)$, *see Remark* 1.4.

*Proof.* If $(M, N)$ denotes a monad $(D_2)$ for a bundle, we have $N \in G_2^0(k^2 \otimes V)$ and $M \subset \Gamma = \mathrm{Ker}(k^2 \otimes \bigwedge^2 V \to N^\vee \otimes \bigwedge^2 V)$. The pair of conics $(S, C)$ associated to the bundle and thus to $(M, N)$ had been described in 2.4, see also 1.3, as $S = S(\Gamma) = \mathbb{G} \cap \mathbb{P}W$ and $C = \{\det M^* = 0\} \cap \mathbb{P}W$, where $\mathbb{P}W = \mathbb{P}W_z$ is the plane of $S$. Then $x = (z, M, \Gamma)$ is a point of $X$ belonging to $X^0$, since $S$ and the polar dual $C^\vee$ of $C$ form by this description exactly the pair $(S, C^\vee)$ associated to $x$. If on the other hand $x = (z, M, \Gamma) \in X^0$ is given, then $y = (z, \Gamma) \in Y^0$ defines a regular plane conic section $S(\Gamma) = \mathbb{G} \cap \mathbb{P}W_z$. By the proof of 5.8.1 we find that there is an $N \in G_2^0(k^0 \otimes V)$ defining $\Gamma$ as its kernel. (In fact $Y^0$ is the image of the imbedding $\varepsilon$ of 5.7.) Now the pair $(M, N)$ is a bundle monad $(D_2)$ by 2.4.2, since both conics $S(\Gamma)$, $C^\vee(M)$ are regular.

**7. The universal kernel sheaf over** $Y^{ss}$. By the construction of $Y$ we have got the following diagram of morphisms

$$(11)\qquad \begin{array}{ccccc} & & Y^{ss} \hookrightarrow G_4^{ss}(k^2 \otimes W) \hookrightarrow Z \times G_4^{ss}(k^2 \otimes \bigwedge^2 V) & & \\ & \overset{\nu}{\swarrow} & \overset{\gamma}{\searrow} \qquad\qquad \overset{\mathrm{pr}}{\longleftarrow} & & \\ G_2^{ss}(k^2 \otimes V) & & G_4^{ss}(k^2 \otimes \bigwedge^2 V) & & \end{array}$$

Let $\mathbb{N}$ resp. $\mathbb{T}$ denote the tautological subbundles on the Grassmannians respectively. Let furthermore $p$ and $q$ denote the first and second projection of $\mathbb{P}_3 \times T$ for any second space $T$. We define the sheaves $\mathcal{N}$, $\mathcal{A}$ and $\mathcal{C}$ as kernel, image and cokernel of the composed homomorphism

$$k^2 \otimes p^*\Omega^1(1) \hookrightarrow k^2 \otimes V^\vee \otimes \mathcal{O}_{P_3 \times Y^{ss}} \to q^*\nu^*\mathbb{N}^\vee,$$

which is derived from the imbedding $\Omega^1(1) \subset V^\vee \otimes \mathcal{O}_{\mathbb{P}_3}$ and the canonical epimorphism $k^2 \otimes V^\vee \otimes \mathcal{O}_{G_2} \to \mathbb{N}^\vee$. Therefore we have the exact sequence

$$(12)\qquad 0 \to \mathcal{N} \to k^2 \otimes p^*\Omega^1(1) \xrightarrow{\mathcal{A}} q^*\nu^*\mathbb{N}^\vee \to \mathcal{C} \to 0.$$

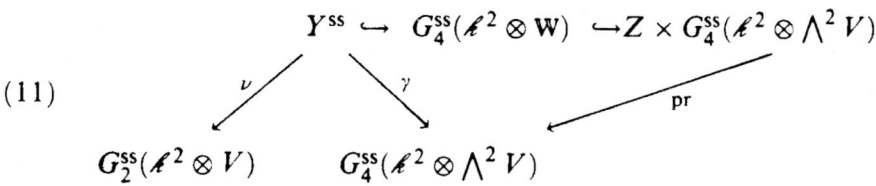

We can transform this sequence into the equivalent sequence (12′) by the same construction as in 5.2 (9), where $\mathbf{A}$ is the universal quotient on $G_2(\mathscr{k}^2 \otimes V)$,

(12′) $$0 \to \mathscr{N} \to q^*\nu^*\mathbf{A}^\vee \to \mathscr{k}^2 \otimes p^*\mathscr{O}_{\mathbf{P}_3}(1) \to \mathscr{C} \to 0.$$

If $y \in Y^{ss}$, we denote by

$$\mathscr{N}_y, \quad \mathscr{A}_y, \quad \mathscr{C}_y$$

the sheaves induced on the fibre $\mathbf{P}_3 \times \{y\} \simeq \mathbf{P}_3$ and call them the fibres of the sheaves respectively.

7.1. PROPOSITION. (i) *The sheaves* $\mathscr{A}$ *and* $\mathscr{N}$ *are flat over* $Y^{ss}$.
(ii) $q_*\mathscr{N}(1) = \gamma^*\mathbf{T} \otimes \bigwedge^4 V^\vee$ *is locally free, and for a point* $y = (z, \Gamma) \in Y^{ss}$ *we have* $\Gamma \otimes \bigwedge^4 V^\vee = \Gamma\mathscr{N}_y(1)$.
(iii) $\operatorname{Supp}\mathscr{C}$ *is finite over the exceptional set* $E \subset Y^s$, *where* $E = \nu^{-1}(E_0)$, $E_0 = G_2^{ss}(\mathscr{k}^2 \otimes V) \setminus G_2^s(\mathscr{k}^2 \otimes V)$.

7.1.1. REMARK. One could do the same construction of $\mathscr{N}$ and $\mathscr{A}$ over $G_2^{ss}(\mathscr{k}^2 \otimes V)$. However in this case $\mathscr{N}$ would not be flat. The modification $Y^{ss} \xrightarrow[\nu]{} G_2^{ss}$ is necessary to obtain a flat sheaf, and indeed our $\mathscr{N}$ can be considered the flattening of the corresponding sheaf over $G_2^{ss}$. On the other hand the modification $\nu$ was necessary in order to extend the quadric bundle of Poncelet conics across the boundary of the completion $R$, see 5.9.

*Proof.* (a) First we are going to show (iii). If $N = \nu(y)$ we have an exact sequence

$$0 \to \mathscr{K}_N \to \mathscr{k}^2 \otimes \Omega^1(1) \to N^\vee \otimes \mathscr{O}_{\mathbf{P}_3} \to \mathscr{C}_N \to 0$$

with $\mathscr{C}_N = \mathscr{C}_y$ and the

7.1.2. LEMMA. ($\alpha$) $N$ *is stable iff* $\mathscr{C}_N = 0$.
($\beta$) *If* $N \in E_0$ *then* $\mathscr{C}_N$ *is a skyscraper sheaf* $\mathscr{k}_x$ *or* $\mathscr{k}_x \oplus \mathscr{k}_y$ *or an extension* $0 \to \mathscr{k}_x \to \mathscr{C}_N \to \mathscr{k}_x \to 0$ *where* $\mathscr{k}_x = \mathscr{O}/m(x)$ *on* $\mathbf{P}_3$.

*Proof.* ($\alpha$) It follows from the stability criterion 5.1.1 that $N$ is stable iff $N \cap (\xi \otimes V) = 0$ for any $\xi \in \mathscr{k}^2$. On the other hand $\mathscr{C}_N = 0$ iff $N \cap (\mathscr{k}^2 \otimes z) = 0$ for any $z \in V$, see proof of Theorem 2.1 and 0.2. It is immediate to see that these two conditions are equivalent.

($\beta$) If $N$ is only semi-stable it must be represented by a matrix $N^*: \mathscr{k}^2 \to \mathscr{k}^2 \otimes V$ of the form

$$\begin{bmatrix} x & 0 \\ y & y' \end{bmatrix} \quad \text{or} \quad \begin{bmatrix} x & 0 \\ 0 & y' \end{bmatrix}$$

up to equivalence, with $y \notin \operatorname{Span}(x, y')$. Now it is easy to verify that the cokernel of

$$(13) \qquad k^2 \otimes \Omega^1(1) \xrightarrow[N^{*\vee}]{} k^2 \otimes \mathcal{O} \to \mathcal{C}_N \to 0$$

is $k_x$ or $k_x \oplus k_y$, or an extension of $k_x$ by itself, by using the equivalent presentation

$$(13') \qquad k^6 \otimes \mathcal{O} \xrightarrow[A^{*\vee}]{} k^2 \otimes \mathcal{O}(1) \to \mathcal{C}_N \to 0,$$

where $A^{*\vee}$ can be determined as the kernel of $N^{*\vee}: k^2 \otimes V^\vee \to k^2$.

Now (iii) follows directly from this lemma.

(b) To prove (ii) we first remark that $q_* p^* \Omega^1(2) = \bigwedge^2 V^\vee \otimes \mathcal{O}_{Y^{ss}}$ and

$$q_*(p^*\mathcal{O}_{\mathbb{P}_3}(1) \otimes q^*\nu^*\mathrm{N}^\vee) = q_*p^*\mathcal{O}_{\mathbb{P}_3}(1) \otimes \nu^*\mathrm{N}^\vee = V^\vee \otimes \nu^*\mathrm{N}^\vee$$

by the projection formula. Therefore we obtain the exact sequence

$$0 \to q_*\mathcal{N}(1) \to k^2 \otimes \bigwedge^2 V^\vee \otimes \mathcal{O}_{Y^{ss}} \to V^\vee \otimes \nu^*\mathrm{N}^\vee.$$

On the other hand by the definition of $\nu$ we have

$$\Gamma \in \operatorname{Ker}\left(k^2 \otimes \bigwedge^2 V \to N^\vee \otimes \bigwedge^3 V\right) \quad \text{if } N = \nu(z, \Gamma),$$

and therefore, using $\bigwedge^i V^\vee \simeq \bigwedge^{4-i} V$,

$$\gamma^*\mathrm{T} \otimes \bigwedge^4 V^\vee \subset q_*\mathcal{N}(1).$$

The quotient sheaf $\mathcal{R} = q_*\mathcal{N}(1)/\gamma^*\mathrm{T} \otimes \bigwedge^4 V^\vee$ is supported on $E$, since for $N \notin E_0$ the space $\Gamma$ is the kernel. But $\mathcal{R}$ is also a subsheaf of the quotient bundle $k^2 \otimes \bigwedge^2 V^\vee \otimes \mathcal{O}_{Y^{ss}}/\gamma^*\mathrm{T} \otimes \bigwedge^4 V^\vee$. Since $Y^{ss}$ is irreducible, $\mathcal{R} = 0$.

(Remark. If we knew flatness already then the base change homomorphism $\Gamma \otimes \bigwedge^4 V^\vee \to \Gamma\mathcal{N}_y(1)$ would be an isomorphism already. We are going to prove this directly, which then implies flatness.)

(c) Lemma. *For any point* $y \in Y^{ss}$, $H^1(\mathbb{P}_3 \times U, m(y)\mathcal{N}(1)) = 0$ *for sufficiently small neighborhoods* $U(y) \subset Y^{ss}$ (*for the proof see* (e)).

Using this lemma we find that $\Gamma\otimes\bigwedge^4 V^\vee \to \Gamma\mathcal{N}_y(1)$ is onto by the diagram

$$\begin{array}{ccccc}
\Gamma(U\,,\, q_*\mathcal{N}(1)) & \longrightarrow & \Gamma\mathcal{N}_y(1) & \longrightarrow & 0 \\
\| & & \uparrow & & \\
\Gamma(U\,,\, y^*\Gamma\otimes\bigwedge^4 V^\vee) & \longrightarrow & \Gamma\otimes\bigwedge^4 V^\vee & \longrightarrow & 0
\end{array} .$$

On the other hand from

$$q^*y^*\Gamma\otimes\bigwedge^4 V^\vee \underset{\approx}{\longrightarrow} q^*q_*\mathcal{N}(1)\to\mathcal{N}(1)$$

we get the diagram

$$\begin{array}{ccc}
(q^*y^*\Gamma\otimes\bigwedge^4 V^\vee)_y & \longrightarrow & \mathcal{N}_y(1) \\
\| & & \uparrow \\
\Gamma\otimes\bigwedge^4 V^\vee\otimes\mathcal{O}_{\mathbb{P}_3} & \longrightarrow & \ell^2\otimes\Omega^1(2)
\end{array} ,$$

which induces the diagram on sections

$$\begin{array}{ccc}
 & \nearrow & \Gamma\mathcal{N}_y(1) \\
\Gamma\otimes\bigwedge^4 V^\vee & & \downarrow \\
\| & \searrow & \ell^2\otimes\Gamma\Omega^1(2) \\
\Gamma & & \| \\
 & \hookrightarrow & \ell^2\otimes\bigwedge^2 V.
\end{array}$$

This proves that $\Gamma\otimes\bigwedge^4 V^\vee\to\Gamma\mathcal{N}_y(1)$ is an isomorphism.

(d) *Proof of the flatness.* We put $\mathcal{T}or_i(\mathcal{F}\,,\,y) = \mathcal{T}or_i(\mathcal{F}\,,\,\mathcal{O}_y/m(y))$ for any $\mathcal{O}_{\mathbb{P}_3\times Y}$-module. From (12) we get the exact sequences, $N = \nu(y)$, on $\mathbb{P}_3$

$$0\longrightarrow\mathcal{T}or_1(\mathcal{E}\,,\,y)\longrightarrow\mathcal{A}_y(1)\longrightarrow N^\vee\otimes\mathcal{O}(1)\longrightarrow\mathcal{E}_N\longrightarrow 0,$$
$$0\longrightarrow\mathcal{T}or_1(\mathcal{A}(1)\,,\,y)\longrightarrow\mathcal{N}_y(1)\longrightarrow\ell^2\otimes\Omega^1(2)\longrightarrow\mathcal{A}_y(1)\longrightarrow 0.$$

By the previous diagram $\Gamma\mathcal{N}_y(1)\to\ell^2\otimes\Gamma\Omega^1(2)$ is *injective*, and hence $\mathcal{T}or_1(\mathcal{A}(1)\,,\,y)$ has no sections. But since $\mathcal{E}_N$ is a sky-scraper sheaf, also $\mathcal{T}or_1(\mathcal{A}(1)\,,\,y)$ is a sky-scraper and hence must be zero. This proves that $\mathcal{A}$ is flat over $Y^{ss}$. Now also $\mathcal{N}$ must be flat over $Y^{ss}$ since $p^*\Omega^1(1)$ is a bundle.

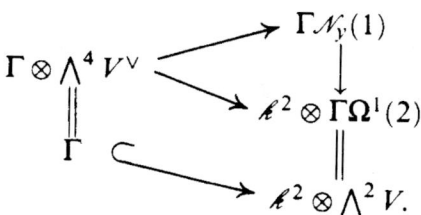

(e) *Proof of Lemma* (c). Since $\operatorname{Supp}\mathscr{C} \to Y^{ss}$ is finite we have $H^i(\mathbb{P}_3 \times U, \mathscr{C}) = 0$ for $i > 0$, and the same is true for any coherent sheaf $\mathscr{F}$ with $\operatorname{Supp}\mathscr{F} \subset \operatorname{Supp}\mathscr{C}$. If $E \to F$ is a homomorphism of vector bundles there is the Eagon-Northcott complex

$$\cdots \to \bigwedge^{f+3} E \otimes S^2 F^\vee \to \bigwedge^{f+2} E \otimes F^\vee \to \bigwedge^{f+1} E \to E \otimes \bigwedge^{f} F \to F \otimes \bigwedge^{f} F \to 0$$

which is exact wherever $E \to F$ is onto. We consider this complex in the case $q^*\nu^*\mathbb{A}^\vee(-1) \to \mathscr{k}^2 \otimes \mathscr{O}_{\mathbb{P}_3 \times Y^{ss}}$, see (12′), which has $\mathscr{N}(-1)$ as kernel.

Putting $q\mathscr{O}(d) = \mathscr{k}^q \otimes p^*\mathscr{O}_{\mathbb{P}_3}(d)$ for the moment, the Eagon-Northcott complex is locally over $Y$ of the form

$$\xrightarrow{\alpha_4} q_3\mathscr{O}(-4) \xrightarrow{\alpha_3} q_2\mathscr{O}(-3) \xrightarrow{\alpha_2} 6\mathscr{O}(-1) \xrightarrow{\alpha_1} 2\mathscr{O} \longrightarrow \mathscr{C} \longrightarrow 0.$$

Let $\mathscr{Z}_i = \operatorname{Ker}\alpha_i$, $\mathscr{B}_i = \operatorname{Im}\alpha_{i+1}$, $\mathscr{C}_i = \mathscr{Z}_i/\mathscr{B}_i$, and let us write $H^j\mathscr{F}$ for $H^j(\mathbb{P}_3 \times U, \mathscr{F})$. Since the complex is exact away from $\operatorname{Supp}\mathscr{C}$, we have $\operatorname{Supp}\mathscr{C}_i \subset \operatorname{Supp}\mathscr{C}$ and $H^j\mathscr{C}_i = 0$ for $j > 0$.

Since $H^4\mathscr{F} = 0$ for any $\mathscr{F}$ and $H^3\mathscr{O}(-3) = 0$, we get the following chain of vanishings.

$$H^3\mathscr{Z}_3(2) = H^3\mathscr{B}_3(2) = 0, \quad H^2\mathscr{Z}_2(2) = H^2\mathscr{B}_2(2) = 0,$$
$$H^1\mathscr{B}_1(2) = 0, \quad H^1\mathscr{N}(1) = H^1\mathscr{Z}_1(2) = 0.$$

By the same method we even get

$$H^i(\mathbb{P}_3 \times U, \mathscr{N}(d)) = 0 \quad \text{for } d \geq 2 - i, \; i > 0.$$

To get the vanishing of the lemma we consider a local resolution

$$m_2\mathscr{O}_y \xrightarrow{\beta_2} m_1\mathscr{O}_y \xrightarrow{\beta_1} m_0\mathscr{O}_y \xrightarrow{\beta_0} \mathfrak{m}(y) \longrightarrow 0$$

of the maximal ideal on $U \subset Y^{ss}$ and put $\mathscr{K}_i = \operatorname{Ker}\beta_i$. We obtain the exact sequence

$$0 \longrightarrow \mathscr{N}_0 \longrightarrow \mathfrak{m}(y) \otimes \mathscr{N}(1) \longrightarrow \mathfrak{m}(y)\mathscr{N}(1) \longrightarrow 0$$
$$0 \longrightarrow \mathscr{N}_1 \longrightarrow \mathscr{K}_0 \otimes \mathscr{N}(1) \longrightarrow m_0\mathscr{N}(1) \longrightarrow \mathfrak{m}(y) \otimes \mathscr{N}(1) \longrightarrow 0$$
$$0 \longrightarrow \mathscr{N}_2 \longrightarrow \mathscr{K}_1 \otimes \mathscr{N}(1) \longrightarrow m_1\mathscr{N}(1) \longrightarrow \mathscr{K}_0 \otimes \mathscr{N}(1) \longrightarrow 0,$$

where the $\mathscr{N}_i$ are $\mathscr{T}\!or_1(\mathscr{N}(1), \mathscr{K}_{i-1})$. Since $\mathscr{N}$ is locally free outside $\operatorname{Supp}\mathscr{C}$, $\operatorname{Supp}\mathscr{N}_i \subset \operatorname{Supp}\mathscr{C}$ and again $H^j\mathscr{N}_i = 0$. As in the previous part we get from $H^i\mathcal{N}(1) = 0$ the desired vanishing

$$H^1(\mathbb{P}_3 \times U, m(y)\mathcal{N}(1)) = 0.$$

This *completes the proof* of the proposition. By the same calculation we even get

$$H^i(\mathbb{P}_3 \times U, m(y)\mathcal{N}(d)) = 0 \quad \text{for } i > 0, \ e \geq 3 - i.$$

Hence we have the

7.2. COROLLARY. *For any* $y \in Y^{ss}$, $H^i\mathcal{N}_y(d) = 0$ *for* $i > 0$, $d \geq 2 - i$.

*Proof*. We have

$$H^i(\mathbb{P}_3 \times U, \mathcal{N}(d)) \to H^i\mathcal{N}_y(d) \to H^{i+1}(\mathbb{P}_3 \times U, m(y)\mathcal{N}(d)).$$

7.3. *Evaluation map and the sheaf* $\mathcal{G}$. The isomorphism $\gamma^*\Gamma \otimes \bigwedge^4 V^\vee = q_*\mathcal{N}(1)$ induces the homomorphism (called evaluation map)

$$0 \to q^*\gamma^*\mathbb{T} \otimes \bigwedge^4 V^\vee \to \mathcal{N}(1) \to \mathcal{G}(1) \to 0$$

with cokernel $\mathcal{G}(1)$.

7.3.1. LEMMA. *The evaluation map is injective and* $\mathcal{G}$ *is flat over* $Y^{ss}$.

*Proof*. If $y \in Y^0 \simeq G_2^0(\ell^2 \otimes V) =$ set of regular bundle epimorphism, then $\mathcal{N}_y$ is a bundle and $\mathcal{G}_y$ is a line bundle on a quadric in $\mathbb{P}_3$. This proves that $\operatorname{rank}\mathcal{G}_y = 0$. Since $\operatorname{rank}\Gamma = 4 = \operatorname{rank}\mathcal{N}$, the kernel must have rank $= 0$ and thus is $0$. To show that $\mathcal{G}$ is flat we consider the sequence

$$\begin{array}{c} 0 \to \mathcal{T}or_1(\mathcal{G}, y) \to \Gamma \otimes \bigwedge^4 V^\vee \otimes \mathcal{O} \to \mathcal{N}_y(1) \to \mathcal{G}_y(1) \to 0 \\ \qquad\qquad\qquad\qquad\qquad\qquad \searrow^{\varphi} \\ \qquad\qquad\qquad\qquad\qquad\qquad\qquad \ell^2 \otimes \Omega^1(2), \end{array}$$

where we have put $\mathcal{T}or_1(\mathcal{G}, y) = \mathcal{T}or_1(\mathcal{G}, \mathcal{O}_y/m(y))$, as in 7.1.2, (d). Since $\varphi$ is injective, the sheaf $\mathcal{T}or_1 = 0$. This proves flatness.

7.4. *The display of* $\mathcal{N}_y$. If $y \in Y^{ss}$ we write $\mathcal{T}_y = \mathcal{T}or_1(\mathcal{C}, y)$. Clearly $\operatorname{Supp}\mathcal{T}_y \subset \operatorname{Supp}\mathcal{C}_y$. From the defining sequence (12) we obtain the exact diagram for the fibre sheaves on $\mathbb{P}_3$:

$$
\begin{array}{ccccccccccc}
 & & & & & & 0 & & & & \\
 & & & & & & \downarrow & & & & \\
 & & 0 & & & & \mathcal{T}_y & & & & \\
 & & \downarrow & & & & \downarrow & & & & \\
0 & \longrightarrow & \mathcal{N}_y & \longrightarrow & k^2 \otimes \Omega^1(1) & \longrightarrow & \mathcal{A}_y & \longrightarrow & 0 & & \\
 & & \downarrow & & \| & & \downarrow & & & & \\
0 & \longrightarrow & \mathcal{K}_n & \longrightarrow & k^2 \otimes \Omega^1(1) & \longrightarrow & N^\vee \otimes \mathcal{O} & \longrightarrow & \mathcal{C}_n & \longrightarrow & 0 \\
 & & \downarrow & & & & \downarrow & \diagup\!\!\diagup & & & \\
 & & \mathcal{T}_y & & & & \mathcal{C}_y & & & & \\
 & & \downarrow & & & & \downarrow & & & & \\
 & & 0 & & & & 0 & & & &
\end{array} \tag{14}
$$

By the flatness of $\mathcal{A}$, $\chi\mathcal{A}_y(m) = 2\chi\mathcal{O}(m)$ is constant and we obtain $h^0\mathcal{T}_y = h^0\mathcal{C}_y$. In particular the skyscraper $\mathcal{T}_y \neq 0$ iff $\mathcal{C}_y \neq 0$.

7.5. *Zero sets of section of* $\mathcal{N}_y(1)$. We are now able to generalise the results for kernels of bundle monads in Lemma 1.3 to any of the sheaves $\mathcal{N}_y$.

7.5.1. PROPOSITION. (1) *Let* $y \in Y^{ss}$ *and* $s \in \Gamma\mathcal{N}_y(1)$. *If the zero scheme* $Z(s)$ *is neither empty nor a point, it is a line belonging to the conic* $S(\Gamma)$.

(2) *The conic* $S(\Gamma)$ *is exactly the set of all zero lines of sections of* $\mathcal{N}_y(1)$.

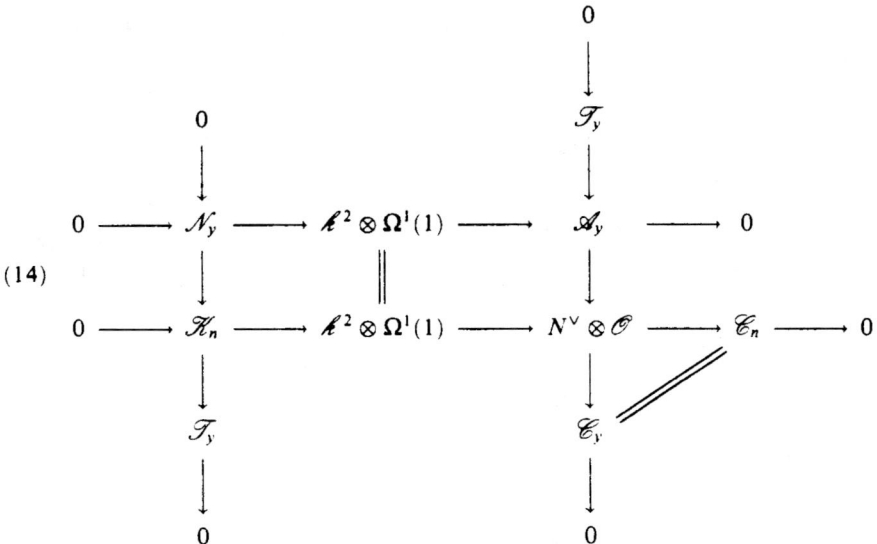

*Proof*. Since supports of $\mathcal{C}_y$, $\mathcal{T}_y$ are 0-dimensional we find $\mathcal{E}xt^1(\mathcal{A}_y, \mathcal{O}) = 0$. It follows that $s \in \Gamma\mathcal{N}_y(1)$ has the same zero scheme as a section of $\mathcal{N}_y(1)$ and as a section of $k^2 \otimes \Omega^1(2)$, since $k^2 \otimes \Omega^1(1)^\vee \to \mathcal{N}_y(1)^\vee$ is onto. Let now $y \in \Gamma \subset k^2 \otimes \bigwedge^2 V$, $y = (\xi, \eta)$, correspond to $s$. A point $\langle z \rangle \in \mathbb{P}V$ is a zero of $s$ iff $(\xi \wedge z, \eta \wedge z) = 0$. If now $Z(s)$ is not 0-dimensional, then $\xi$ and $\eta$ must be linearly dependent and by the definition of the conic $S(\Gamma)$ in 4.2 define a point of $S(\Gamma)$. Conversely any point of $S(\Gamma)$ comes from some $\gamma \in \Gamma$ with a line as zero locus.

7.6. *Conjugate conic* $S^0$ *and* $R^1\mathcal{N}$. If $S \in C(\mathbb{G})$ is a conic we define the "conjugate" conic $S^0 \in C(\mathbb{G})$ as follows. If $S = \mathbb{G} \cap \mathbb{P}W$ we let $S^0 = \mathbb{G} \cap \mathbb{P}W^\perp$ where $W^\perp$ is the orthogonal of $W$ with respect to the quadratic form of $\mathbb{G}$. Then $S$ is regular iff $S^0$ is regular. If $S$ is a double line then also $S^0$ is a double line with the same reduced line but with a different plane. If however $S \subset \mathbb{P}W \subset \mathbb{G}$ we define $S^0 = S$. It can be shown that this map $S \mapsto S^0$ is an involutive morphism of $C(\mathbb{G})$.

REMARK. $\Gamma \to \Gamma^0$ can be defined by continuity.

We are now going to generalise Proposition 1.5, (ii) to any kernel sheaf of our construction:

7.6.1. PROPOSITION. *Let* $y \in Y^{ss}$ *and* $S = S(\Gamma)$. *Then* $S^0 = \operatorname{Supp} R^1\mathcal{N}_y = \operatorname{Supp} R^1\mathcal{G}_y$ *as reduced schemes.*

*Proof*. Since it seems complicated to show that the family $R_1\mathcal{N}_y$ and their supports form a flat family, we proceed to calculate $R^1\mathcal{G}_y$ in the different cases of $S(\Gamma)$, which also gives a beautiful insight into the structure of those sheaves.

*Case* 1. $y \in Y^0$ *and defines a regular conic* $S(\Gamma) \in C^0(\mathbb{G})$ was treated in 1.5.

*Case* 2. $y$ *defines a regular conic* $S(\Gamma) \subset \mathbb{P}W_z\mathbb{G}$ *in a* $\beta$*-plane.* Since the entries of $\Gamma$ are decomposed we can choose a basis $e_0, e_1, e_2 \in W_z$ s.t. $\Gamma$ is represented by the matrix

$$\Gamma^* = \begin{bmatrix} e_{01} & \\ e_{12} & e_{01} \\ e_{02} & e_{12} \\ & e_{02} \end{bmatrix}, \quad N^{\vee *} = \begin{bmatrix} e_0 & e_1 \\ -e_2 & e_0 \end{bmatrix},$$

where $e_{ij} = e_i \wedge e_j$, and then $N^\vee$ is represented by the matrix $N^{\vee *}$ in the above form, see (10) in 5.8. By 0.3 the homomorphism induced is surjective, i.e. $\mathcal{E}_y = 0$ and we have the exact sequence

$$0 \to \mathcal{N}_y \to \mathcal{k}^2 \otimes \Omega^1(1) \xrightarrow[N^{\vee *}]{} \mathcal{k}^2 \otimes \mathcal{O} \to 0.$$

If we apply $R^{\cdot}$ we get, see 0.4,

$$k^2 \otimes Q^\vee \xrightarrow[N^{\vee *}]{} k^2 \otimes \mathcal{O}_{\mathbb{G}} \to R^1\mathcal{N}_y \to 0.$$

We can now calculate this homomorphism as follows. Let $p_{ij}$ be dual to $e_{ij}$, i.e. the homogeneous Plücker-coordinates of $\mathbb{P}\bigwedge^2 V$, and let $\mathbb{G}_{ij} = \{p_{ij} \neq 0\}$. If $x = \sum x_i e_i$ defines the homomorphism $Q^\vee \xrightarrow[x]{} \mathcal{O}_{\mathbb{G}}$ and if $Q^\vee | G_{ij} \simeq k^2 \otimes \mathcal{O}_{\mathbb{G}}$ is trivialised, this homomorphism can be expressed by the matrix

$$\begin{bmatrix} x_k - \dfrac{P_{ik}}{P_{ij}} x_j + \dfrac{P_{jk}}{P_{ij}} x_i \\ x_l - \dfrac{P_{il}}{P_{ij}} l_j + \dfrac{P_{jl}}{P_{ij}} x_i \end{bmatrix} : 2\mathcal{O}_{\mathbb{G}} \to \mathcal{O}_{\mathbb{G}}.$$

This follows immediately if we choose a basis of $(V/U)^\vee$, the fibre of $Q^\vee$ at $U \in \mathbb{G}_{ij}$, where $p_{\mu\nu}$ are the Plücker coordinates of $U$, and $k$, $l$ are complementary to $i$, $j$. If we choose for example $\mathbb{G}_{01}$, we get the homomorphism

$$k^2 \otimes k^2 \otimes \mathcal{O}_{\mathbb{G}} \to k^2 \otimes \mathcal{O}_{\mathbb{G}} \to R^1\mathcal{N}_y \to 0,$$

$$\left[\begin{array}{c|c} p_{12} & -p_{02} \\ p_{13} & -p_{03} \\ \hline -1 & p_{12} \\ 0 & p_{13} \end{array}\right] \cdot p_{01}^{-1}.$$

The Fitting ideal of $R^1\mathcal{N}_y | \mathbb{G}_{01}$ therefore is generated by $p_{13}$, $p_{03}$, $p_{12}^2 - p_{02}$ which is the ideal of the conic $S(\Gamma)$ parametrised by $s^2 e_{01} + st e_{12} + t^2 e_{02}$. Since $S = S^0$ here, this settles Case 2.

*Case* 3. $y$ defines a regular conic $S(\Gamma) \subset \mathbb{P}W_z \subset \mathbb{G}$ in an $\alpha$-plane. Again upon choosing a basis of $V$ we can assume that $\Gamma$, $N^\vee$ are represented by the matrices

$$\Gamma^* = \begin{bmatrix} e_{01} & \\ e_{03} & e_{01} \\ e_{02} & e_{02} \\ & e_{03} \end{bmatrix}, \quad N^{\vee *} = \begin{bmatrix} e_0 & \\ & e_0 \end{bmatrix}.$$

In this case $\mathcal{K}_N = k^2 \otimes \mathcal{Z}$ where $\mathcal{Z}$ is the kernel in

(15) $$0 \to \mathcal{Z} \to \Omega^1(1) \xrightarrow[e_0]{} \mathcal{O} \to k_{e_0} \to 0$$

and $\mathcal{N}_y$ is the kernel in

$$0 \to \mathcal{N}_y \to k^2 \otimes \mathcal{Z} \to k^2 \otimes k_{e_0} \to 0 \tag{16}$$

where we use that here $\mathcal{T}_{e_0} = k^2 \otimes k_{e_0}$.

To proceed further we have to digress into computations for the sheaf $\mathcal{Z}$. If we write $x$ for $e_0$ and apply the same transformation as in 5.2 (9), we find that $\mathcal{Z}$ is the first syzygy of $m(x)(1)$, i.e. we have exact sequences

$$\begin{gathered} 0 \to \mathcal{Z} \to (V/x)^\vee \otimes \mathcal{O} \to \mathcal{M}(1) \to 0, \\ 0 \to \bigwedge^3 (V/x)^\vee \otimes \mathcal{O}(-2) \to \bigwedge^2 (V/x)^\vee \otimes \mathcal{O}(-1) \to \mathcal{Z} \to 0. \end{gathered} \tag{17}$$

From (17) we get $\Gamma\mathcal{Z}(1) = \bigwedge^2(V/x)^\vee$ and that

$$\mathrm{Hom}(\mathcal{Z}, k_x) = \bigwedge^2 (V/x).$$

Hence if $\xi \in \bigwedge^2(V/x)$ induces $\mathcal{Z} \xrightarrow[\xi]{} k_x$, the same element gives the induced homomorphism $\Gamma\mathcal{Z}(1) \xrightarrow[\xi]{} k$. Moreover, if we apply $R^0$ we get

$$\begin{array}{ccc} R^0\mathcal{Z} & \xrightarrow{R^0\xi} & R^0 k_x \\ \| & & \| \\ \bigwedge^3(V/x)^\vee \otimes \mathcal{O}_{\mathrm{G}}(-1) & \longrightarrow & \mathcal{O}_{P_x} \\ \downarrow & \swarrow \xi & \\ \bigwedge^3(V/x)^\vee \otimes \mathcal{O}_{P_x} & & \end{array},$$

where $P_x = \mathbb{P}(V/x)$ is the $\alpha$-plane of $x$ and $\xi$ is identified with an element of $(V/x)^\vee = \bigwedge^2(V/x)$.

Now we are able to calculate $R^1\mathcal{N}_y$ in Case 3. First we determine the homomorphism in (16) by the induced sequence

$$\begin{array}{ccccccccc} 0 & \longrightarrow & \Gamma\mathcal{N}_y(1) & \longrightarrow & k^2 \otimes \Gamma\mathcal{Z}(1) & \longrightarrow & k^2 & \longrightarrow & 0 \\ & & \| & & \| & & \| & & \\ 0 & \longrightarrow & k^4 \otimes \bigwedge^4 V^\vee & \xrightarrow[\Gamma^\cdot]{} & k^2 \otimes \bigwedge^2(V/x)^\vee & \xrightarrow[A]{} & k^2 & \longrightarrow & 0 \end{array}.$$

It follows that $A$ is the matrix

$$A = \begin{bmatrix} e_{12} & e_{13} \\ -e_{23} & e_{12} \end{bmatrix}.$$

Passing now to $R^0\mathscr{Z}$ we get the diagram

$$\begin{array}{ccccccc}
k^2 \otimes R^0\mathscr{Z} & \longrightarrow & k^2 \otimes R^0 k_{e_0} & \longrightarrow & R^1\mathscr{N}_y & \longrightarrow & 0 \\
\downarrow & & \| & & \| & & \\
k^2 \otimes \mathcal{O}_{P_{e_0}}(-1) & \xrightarrow[A]{} & k^2 \otimes \mathcal{O}_{P_{e_0}} & \longrightarrow & R^1\mathscr{N}_y & \longrightarrow & 0
\end{array}$$

Now under $\bigwedge^2(V/e_0) \simeq (V/e_0)^\vee$ we have $e_{12} \leftrightarrow e_3^\vee$, $e_{13} \leftrightarrow -e_2^\vee$, $e_{23} \leftrightarrow e_1^\vee$ and hence

$$\det A = e_3^{\vee 2} - e_1^\vee e_2^\vee.$$

But this is exactly the equation of the conic $S(\Gamma)$ in the $\alpha$-plane $P_{e_0} = \mathbb{P}(V/e_0) = \mathbb{P}\langle e_{01}, e_{03}, e_{02}\rangle$ which is given by $s^2e_{01} + ste_{03} + t^2e_{02}$. Since here also $S = S^0$, this proves Case 3.

*Case* 4. $y$ defines a pair of lines $S(\Gamma)$. We assume that $S(\Gamma)$ is a plane section as in (D), 0.4, since the other situations are only special cases of this. If $S$ is the union of the two lines $e$, $f$ which define the pencil of lines in $E$ through $p$, $F$ through $q$ respectively, then $S^0$ consists of the lines $e_0$, $f_0$ which describe the pencil of lines in $E$ through $q$, $F$ through $p$, respectively.

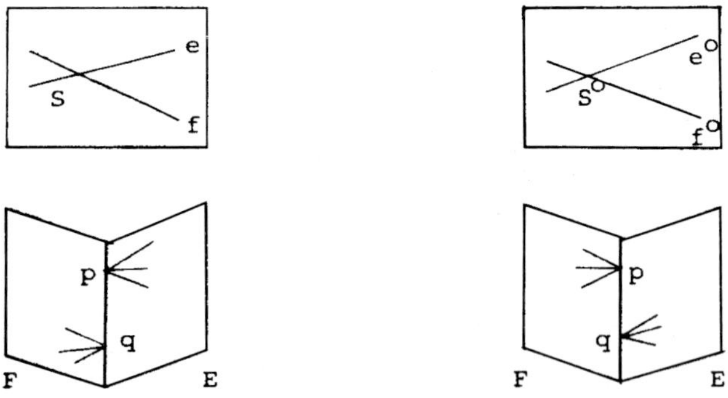


Let $\mathscr{G}_y$ be the quotient of $\mathscr{N}_y$ s.t. $\operatorname{Supp}\mathscr{G}_y = E \cup F$ and $R^1\mathscr{N}_y = R^1\mathscr{G}_y$. It will be shown in 10.2 that $\mathscr{G}_y$ is an extension

$$0 \to m_E(q)(-1) \to \mathscr{G}_y \to m_F(p)(-1) \to 0$$

where $m_F(p) \subset \mathscr{O}_F$ is the ideal sheaf of $p$ in the plane $F$. Choosing two generating sections of $m_F(p)$ we obtain the resolution

$$(18) \qquad 0 \to \mathscr{O}_F(-3) \xrightarrow[(a,b)]{} k^2 \otimes \mathscr{O}_F(-2) \to m_F(p)(-1) \to 0.$$

Since $R^1\mathscr{O}(-m-2) = S^2S \otimes \bigwedge^2 S$, where $S$ is the tautological subbundle on $\mathbb{G}$, it is easy to derive that $R^1\mathscr{O}_F(-m-2)$ is the restriction of $R^1\mathscr{O}(-m-2)$ to the $\beta$-plane $P_F$ of all lines in $F$, and that the homomorphism $\mathscr{O}_F(-3) \to \mathscr{O}_F(-2)$ becomes contraction with $a\colon S \otimes \bigwedge^2 S|P_F \to \bigwedge^2 S|P_F$. Hence from (18) we get the sequence

$$(19) \quad 0 \to S \otimes \mathscr{O}_{P_F}(-1) \xrightarrow[(a,b)]{} k^2 \otimes \mathscr{O}_{P_F}(-1) \to R^1(m_F(x)(-1)) \to 0.$$

This shows that the support of $R^1 m_F(p)(-1)$ is the line $f^0$ and that the homomorphism $(a, b)$ must be injective. Indeed $R^1 m_F(p)(-1) = \mathscr{O}_{f^0}$, since $S|P_F \simeq \Omega^1_{P_F}(1)$ and (19) can be transformed into

$$0 \to \mathscr{O}_{P_F}(-1) \xrightarrow[\alpha]{} \mathscr{O}_{P_F} \to R^1 m_F(p)(-1) \to 0$$

as in 5.2 (9), where $\alpha$ is the equation of $f^0 \subset P_F$.

Similarly we obtain $R^1 m_E(q)(-1) = \mathscr{O}_{e^0}$.

Now we consider the resolution of $\mathscr{G}$ which can be constructed from the resolution of the ends.

$$\begin{array}{ccccccccc}
 & & 0 & & 0 & & 0 & & \\
 & & \downarrow & & \downarrow & & \downarrow & & \\
0 & \to & \mathscr{O}_E(-3) & \to & \mathscr{O}_E(-3) \oplus \mathscr{O}_F(-3) & \to & \mathscr{O}_F(-3) & \to & 0 \\
 & & \downarrow & & \downarrow & & \downarrow & & \\
0 & \to & k^2 \otimes \mathscr{O}_E(-2) & \to & k^2 \otimes \mathscr{O}_E(-2) \oplus k^2 \otimes \mathscr{O}_F(-2) & \to & k^2 \otimes \mathscr{O}_F(-2) & \to & 0 \\
 & & \downarrow & & \downarrow & & \downarrow & & \\
0 & \to & m_E(q)(-1) & \to & \mathscr{G}_y & \to & m_F(p)(-1) & \to & 0 \\
 & & \downarrow & & \downarrow & & \downarrow & & \\
 & & 0 & & 0 & & 0 & &
\end{array}$$

If we apply $R^1$ we find that the sequence

$$0 \to R^1 m_E(q)(-1) \to R^1\mathscr{G}_y \to R^1 m_F(p)(-1) \to 0$$

is exact because the left-hand side of (19) is injective. This proves that $R^1\mathscr{G}_y$ is an extension of $\mathscr{O}_{f^0}$ by $\mathscr{O}_{e^0}$, and that $\operatorname{Supp} R^1\mathscr{N}_y = S^0$.

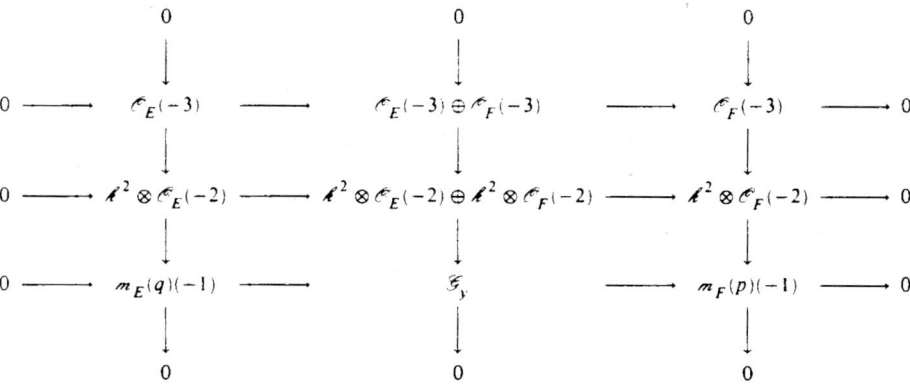

**8. The universal sheaf $\mathscr{F}$ over $X^{ss}$.** Recall that in 6.3 we have defined $X$ to be the flag subvariety

$$X \subset Z \times G_2\left(k^2 \otimes \bigwedge^2 V\right) \times G_4\left(k^2 \otimes \bigwedge^2 V\right)$$

of all triples $(z, M, \Gamma)$ satisfying $(z, \Gamma) \in Y$ and $M \subset \Gamma \subset k^2 \otimes W_z$. Let

$$\begin{array}{ccccc} & X & \xrightarrow{\pi} & & Y \\ \mu \swarrow & & \searrow \gamma & & \downarrow \gamma_Y \\ G_2(k^2 \otimes \bigwedge^2 V) & & & & G_4(k^2 \otimes \bigwedge^2 V) \end{array}$$

be the projections in $\mathbb{M}$, $\mathbb{T}$ the tautological subbundles respectively. By the definition of $X$ we get the exact sequence

$$0 \to \mu^*\mathbb{M} \to \gamma^*\mathbb{T} \to \mathbb{B} \to 0$$

on $X$ where $\mathbb{B}$ is defined as the quotient bundle. As before we denote by $p$, resp. $q$, the first and second projection of $\mathbb{P}_3 \times X^{ss}$, and we denote $(\mathrm{id} \times \pi)^*\mathcal{N}$ again by $\mathcal{N}$, so that we have $\mathcal{N}_x = \mathcal{N}_{\pi(x)}$ by abuse of notation and similarly for $\mathscr{G}$. Since we had the inclusion $q^*\gamma_y^*\mathbb{T} \hookrightarrow \mathcal{N}(1)$ we obtain the exact diagram (up to the factor $\otimes \bigwedge^4 V^\vee$ in the top row)

$$\begin{array}{ccccccccc} & & & & 0 & & 0 & & \\ & & & & \downarrow & & \downarrow & & \\ 0 & \longrightarrow & q^*\mu^*\mathbb{M} & \longrightarrow & q^*\gamma^*\mathbb{T} & \longrightarrow & q^*\mathbb{B} & \longrightarrow & 0 \\ & & \| & & \downarrow & & \downarrow & & \\ 0 & \longrightarrow & q^*\mu^*\mathbb{M} & \longrightarrow & \mathcal{N}(1) & \longrightarrow & \mathscr{F}(1) & \longrightarrow & 0, \\ & & & & \downarrow & & \downarrow & & \\ & & & & \mathscr{G}(1) & = & \mathscr{G}(1) & & \\ & & & & \downarrow & & \downarrow & & \\ & & & & 0 & & 0 & & \end{array}$$

in which $\mathscr{F}$ is defined as cokernel. Since $\mathbb{B}$ is a bundle and $\mathscr{G}$ is flat we conclude that also $\mathscr{F}$ is flat over $X^{ss}$. Thus for any point

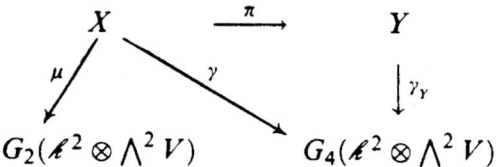

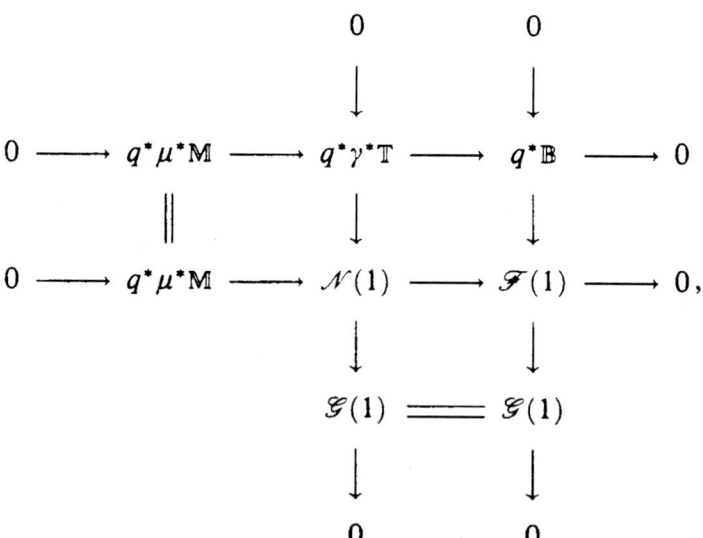

$x = (z, M, \Gamma)$ we have the diagram (up to $\otimes \bigwedge^4 V^\vee$)

$$
\begin{array}{ccccccccc}
 & & & & 0 & & 0 & & \\
 & & & & \downarrow & & \downarrow & & \\
0 & \longrightarrow & M \otimes \mathcal{O} & \longrightarrow & \Gamma \otimes \mathcal{O} & \longrightarrow & (\Gamma/M) \otimes \mathcal{O} & \longrightarrow & 0 \\
 & & \| & & \downarrow & & \downarrow & & \\
0 & \longrightarrow & M \otimes \mathcal{O} & \longrightarrow & \mathcal{N}_y(1) & \longrightarrow & \mathcal{F}_x(1) & \longrightarrow & 0 \\
 & & & & \downarrow & & \downarrow & & \\
 & & & & \mathcal{G}_y(1) & = & \mathcal{G}_y(1) & & \\
 & & & & \downarrow & & \downarrow & & \\
 & & & & 0 & & 0 & &
\end{array} \tag{21}
$$

on $\mathbb{P}_3$. We also recall that we have a monad display generalising $(D_2)$ of §2:

$$
\begin{array}{ccccccccc}
 & & & & 0 & & 0 & & \\
 & & & & \downarrow & & \downarrow & & \\
0 & \longrightarrow & M \otimes \Omega^3(3) & \longrightarrow & \mathcal{N}_x & \longrightarrow & \mathcal{F}_x & \longrightarrow & 0 \\
 & & \| & & \downarrow & & \downarrow & & \\
0 & \xrightarrow[A]{} & M \otimes \Omega^3(3) & \longrightarrow & \mathscr{K}^2 \otimes \Omega^1(1) & \longrightarrow & \mathcal{M}_x & \longrightarrow & 0 \\
 & & & & \downarrow & & \downarrow & & \\
 & & & & \mathcal{A}_y & = & \mathcal{A}_y & & \\
 & & & & \downarrow & & \downarrow & & \\
 & & & & 0 & & 0 & &
\end{array} \tag{22}
$$

where $\mathcal{M}_x$ is defined as the cokernel and $\mathcal{A}_y$ comes with the definition of $\mathcal{N}_y$, see 7.4.

8.1. **Proposition.** *For any* $x \in X^{ss}$ *the sheaf* $\mathcal{F}_x$ *is semi-stable of rank* 2 *with Chern classes* $c_1 = 0$, $c_2 = 2$, $c_3 = 0$ *on* $\mathbb{P}_3$.

Therefore the family $\mathscr{F}$ defines a morphism of $X^{ss}$ into the Maruyama scheme which will be discussed in 8.3.

*Proof*. We only prove here that $\mathscr{F}_x$ is torsionfree with the Chern classes indicated. Semi-stability will follow from the geometric description of the sheaves in §§9 and 10.

It is enough to prove that $\mathscr{M}_{\langle x\rangle}$ is torsionfree by diagram (22). Since depth $\mathscr{M}_{\langle x\rangle} \geq 2$ everywhere, it is enough to show that $\mathscr{M}$ is locally free outside a curve, see f.e. **[Si-Tr]**. If $M^*$ is a matrix representing $M$, and the homomorphism of the fibre over $\langle x\rangle \in \mathbb{P}V$,

$$\mathscr{k}^2 \otimes \langle x\rangle \xrightarrow{M\dot{\wedge} x} \mathscr{k}^2 \otimes \bigwedge^2 V \wedge x$$

is degenerate, i.e. has rank $< 2$, then for any $y \in V$ also the matrix $M^* \wedge x \wedge y$ has determinant zero and therefore vanishes on the $\alpha$-plane $P_{\langle x\rangle}$. If now $M^*$ is degenerate on a surface, its $\det M^*$ would be identically zero on the Grassmannian, which contradicts semi-stability of $M$, see Proposition 6.1.1. This proves that $\mathscr{F}_x$ is torsionfree. The calculation of rank and Chern classes follows immediately from the diagrams.

8.2. *Cohomology of* $\mathscr{F}_x$. The cohomology dimensions $h^i\mathscr{N}_y(d)$, $h^i\mathscr{F}_x(d)$ can be summarised in the following tables.

| $d$ | $h^0$ | $h^1$ | $h^2$ | $h^3$ |
|---|---|---|---|---|
| $\geq 2$ | • | | | |
| 1 | 4 | | | |
| 0 | | 2 | | |
| $-1$ | | 2 | | |
| $-2$ | | $t$ | $t$ | |
| $-3$ | | $t$ | $t$ | |
| $\leq -4$ | | $t$ | $t$ | • |

for $\mathscr{N}_y$

| $d$ | $h^0$ | $h^1$ | $h^2$ | $h^3$ |
|---|---|---|---|---|
| $\geq 2$ | • | | | |
| 1 | 2 | | | |
| 0 | | 2 | | |
| $-1$ | | 2 | | |
| $-2$ | | $t$ | $t$ | |
| $-3$ | | $t$ | $t+2$ | |
| $\leq -4$ | | $t$ | • | • |

for $\mathscr{F}_x$

Here $t = h^0\mathscr{T}_y = h^0\mathscr{C}_y \leq 2$, see 7.1.2.

*Proof*. We fix $x$ and $y$ and omit the index. It was shown in Corollary 7.2 that $h^i\mathscr{N}(d) = 0$ for $i > 0$ and $d \geq 2 - i$.

Since $H^i\mathscr{N}(d) = H^i\mathscr{F}(d)$ for $i > 0$ and $d \geq -2$, the same is true for $\mathscr{F}$. Next we show that also $0 = h^3\mathscr{N}(d) = h^3\mathscr{F}(d)$ for $d = -2, -3$, which settles the case $h^3$: From the display (14) of $\mathscr{N}_y = \mathscr{N}$ in 7.4 we obtain easily $0 = H^2\mathscr{A}(d) = H^3\mathscr{N}(d)$ for these $d$.

Next we state that $h^2\mathscr{N}(d) = t$ for $d \leq -2$, which also follows from the same display by $h^0\mathscr{C} = h^1\mathscr{A}(d) = h^2\mathscr{N}(d)$.

The case $h^2\mathscr{N}(-1) = 0$ is more subtle. To obtain this we note that $h^2\mathscr{N}(-1) = h^2\mathscr{K}(-1)$ and that for $\mathscr{K}$ we can replace the row of $\mathscr{K}$ in (14), 7.4 by the row

$$0 \to \mathscr{K} \to k^6 \otimes \mathscr{O} \xrightarrow[R]{} k^2 \otimes \mathscr{O}(1) \to \mathscr{C} \to 0,$$

where the matrix $R$ is the kernel of

$$0 \to k^6 \to k^2 \otimes V^\vee \xrightarrow[N^{*\vee}]{} k^2 \to 0,$$

see (9) in the case of a plane. Now it is easy to see that in the few cases of degenerate $N^{*\vee}$ the induced homomorphism $k^2 \otimes \Gamma\mathscr{O} \to \Gamma\mathscr{C}(-1)$ is onto which implies $H^2\mathscr{K}(-1) = H^1 \operatorname{Im}(R)(-1) = 0$. By this the case $h^2\mathscr{N}(d)$ is settled for all $d$.

For $d \geq -2$ we also have $H^2\mathscr{N}(d) = H^2\mathscr{F}(d)$, and for $d = -3$ the exact sequence

$$0 \to H^2\mathscr{N}(-3) \to H^2\mathscr{F}(-3) \to M \otimes H^3\mathscr{O}(-3) \to 0,$$

and hence $h^2\mathscr{F}(-3) = 2 + t$.

Finally $h^1\mathscr{N}(-1) = h^1\mathscr{F}(-1) = 2$ and $h^1\mathscr{N} = h^1\mathscr{F} = 2$ follows from Riemann-Roch and $h^0\mathscr{N} = h^0\mathscr{F} = 0$. Of course $h^0\mathscr{N}(1) = 4$ and $h^0\mathscr{F}(1) = 2$ by our earlier result.

8.3. *Morphism* $Q \xrightarrow[\varphi]{} \overline{M(0,2)}$. Let $\overline{M}(2; 0, 2, 0)$ be the Maruyama scheme of all semi-stable coherent rank 2 sheaves on $\mathbb{P}_3$ with Chern classes $c_1 = 0$, $c_2 = 2$, $c_3 = 0$ which contains $M(0, 2)$ as an open part. Let $\overline{M(0,2)}$ be its closure. The family $\mathscr{F}_x$, $x \in X^{ss}$, provides us with a morphism $X^{ss} \to \overline{M}(2; 0, 2, 0)$. Since by our construction $\mathscr{F}_x \cong \mathscr{F}_{x}$, if $O(x) = O(x')$, this morphism is SL(2)-equivariant and factors through the good quotient $Q = X^{ss}/\!/\,\mathrm{SL}(2)$. By the description of $M(0, 2)$ in 2.4.2 the open set $Q \setminus Q_0 \cup Q_\alpha \cup Q_\beta \cup Q_e$, see 4, maps isomorphically onto $M(0, 2)$. Therefore we have a surjective birational morphism $Q \xrightarrow[\varphi]{} \overline{M(0,2)}$. We are going to investigate how far it is from being an isomorphism.

Let $Q \xrightarrow{\pi} C(\mathbb{G})$ be the projection of the quadric bundle and let $\Sigma_0' \subset \Sigma_0 \subset C(\mathbb{G})$ be the subvarieties of all double lines, resp. of all singular conics. We write

$$Q_{\mathrm{exc}} = \pi^{-1}(\Sigma_0') \cap \operatorname{Sing} Q.$$

By 4.2 this is 2-dimensional over $\Sigma_0'$ and indeed a $\mathbb{P}_2$-bundle.

8.4. **Proposition.** (1) *$Q \setminus Q_{\mathrm{exc}} \underset{\varphi}{\longrightarrow} \overline{M(0,2)}$ is injective.*

(2) *The fibres of $Q_{\mathrm{exc}} \underset{\varphi}{\longrightarrow} \overline{M(0,2)}$ are in the $\mathbb{P}_1$'s of double structures of the conics in $\Sigma_0'$.*

*Proof.* (a) The injectivity on $Q \setminus Q_{\mathrm{exc}}$ will follow when we prove that the pair $(S, C^\vee)$ of conics given by $x \in X^{ss}$ is determined by the class $[\mathcal{F}_x]$ through $\operatorname{Supp} R^1\mathcal{F}_x$ and $\operatorname{Supp} R^1\mathcal{F}_x(-1)$. Since $R^1\mathcal{N}_y = R^1\mathcal{F}_x$ the reduced conic $S$ is already determined by $S^0 = \operatorname{Supp} R^1\mathcal{F}_x$, see 7.6.

(b) If $x = (z, M, \Gamma) \in X^{ss}$ denote $W = W_z$, $\mathcal{F} = \mathcal{F}_x$, $S = S(\Gamma)$, and $C^\vee = C^\vee(M)$. We consider the diagram

$$\begin{array}{ccccccccc} 0 & \longrightarrow & W^\perp & \longrightarrow & \bigwedge^2 V & \longrightarrow & W' & \longrightarrow & 0 \\ & & \downarrow \wr & & \| & & \|\wr & & \\ 0 & \longrightarrow & \left(\bigwedge^2 V/W\right)^\vee & \longrightarrow & \bigwedge^2 V^\vee & \longrightarrow & W^\vee & \longrightarrow & 0 \end{array}$$

where the vertical arrow in the middle is the quadratic form of the Grassmannian, which identifies the orthogonal $W^\perp$ with $(\bigwedge^2 V/W)^\vee$, and provides an isomorphism of the cokernel $W'$ with $W^\vee$. If we take any splitting of the first sequence we get a projection

$$\mathbb{P}\bigwedge^2 V \setminus \mathbb{P}W^\perp \to \mathbb{P}W^\perp \subset \mathbb{P}\bigwedge^2 V.$$

(c) From diagram (22) we get the exact sequence

$$0 \to R^0\mathcal{M}(-1) \to \mathcal{k}^2 \otimes \mathcal{O}_{\mathbb{G}}(-1) \xrightarrow{M^*} \mathcal{k}^2 \otimes \mathcal{O}_{\mathbb{G}} \to R^1\mathcal{M}(-1) \to 0,$$

where $M^*$ is a matrix representing $M$, which also is a homomorphism on the Grassmannian by $\mathcal{k}^2 \otimes \bigwedge^4 V^\vee \otimes \bigwedge^2 U \to \mathcal{k}^2$ for $U \in G_2 V$, and which we also denote by $M^*$. It follows that $R^0\mathcal{M}(-1) = 0$ and

$$\operatorname{Supp} R^1\mathcal{M}(-1) = \{\det M^* = 0\}.$$

Moreover from the display (14) we obtain that $R^0\mathscr{A}(-1) = R^0\mathscr{T}$, $R^1\mathscr{A}(-1) = R^0\mathscr{C}$, and from (22) the exact sequence

$$0 \to R^0\mathscr{T} \to R^1\mathscr{F}(-1) \to R^1\mathscr{M}(-1) \to R^0\mathscr{C} \to 0.$$

Since $\operatorname{Supp}\mathscr{T} = \operatorname{Supp}\mathscr{C}$, also $\operatorname{Supp} R^0\mathscr{T} = \operatorname{Supp} R^0\mathscr{C}$, and the sequences show that

$$J = \operatorname{Supp} R^1\mathscr{F}(-1) = \operatorname{Supp} R^1\mathscr{M}(-1) = \{\det M^* = 0\}.$$

REMARK. $M^*$ is determined by $R^1\mathscr{M}(-1)$ through $\mathscr{k}^2 \to \mathscr{k}^2 \otimes \Gamma\mathscr{O}_{\mathbb{G}}(1)$ up to equivalence.

(d) LEMMA. *There is a unique quadric hypersurface* $\tilde{J} \subset \mathbb{P}\bigwedge^2 V$ *such that* $J = \mathbb{G} \cap \tilde{J}$ *and* $\tilde{J}$ *is singular along* $S^0 \subset J$.

*Proof*. Let $f$ be the equation of any quadric hypersurface $\tilde{J}$ with $J = \mathbb{G} \cap \tilde{J}$ and let $q$ be the equation of $\mathbb{G}$. Since $S^0$ is contained in the singular locus $\operatorname{Sing}(J)$ (because $S^0 \subset \mathbb{P}W^\perp \cap \mathbb{G} \subset J$ and $M^*$ vanishes on $\mathbb{P}W^\perp$), for any $p \in S^0$ there is a unique $\lambda(p) \in \mathscr{k}$ such that

$$\frac{\partial f}{\partial p_{ij}}(p) = \lambda(p)\frac{\partial q}{\partial p_{ij}}(p)$$

for all derivatives with respect to the Plücker coordinates of $\mathbb{P}\bigwedge^2 V$. Because $\mathbb{G}$ is regular, $\lambda$ is a regular function on $S^0$ and hence constant. Then $\tilde{J} = \{f - \lambda q = 0\}$ is the unique hypersurface of the lemma.

Since however $\{\det M^* = 0\}$ has the properties of $\tilde{J}$ in the lemma, $\tilde{J} = \{\det M^* = 0\}$. On the other hand the conic $C^\vee(M)$ in $\mathbb{P}W^\vee$ has exactly the same equation, see 6.5. If $\mathbb{P}W^\vee \subset \mathbb{P}\bigwedge^2 V$ by some splitting in the diagram of (b), we obtain

$$C^\vee = \mathbb{P}W^\vee \cap \tilde{J}.$$

Therefore the injectivity of $\varphi$ on $Q \setminus Q_{\text{exc}}$ will be proved if the plane $\mathbb{P}W$ or $\mathbb{P}W^\perp$ can be determined by $\mathscr{F}$.

(e) If we consider $Q \setminus \pi^{-1}(\Sigma_0')$ clearly $\mathbb{P}W$ is determined by $S^0 = \operatorname{Supp} R^1\mathscr{F}$, since each $S$ and $S^0$ is a pair of distinct lines. In this case the injectivity follows if we show that $\operatorname{Supp} R^1\mathscr{F}$ and $\operatorname{Supp} R^1\mathscr{F}(-1)$ are invariants of the class $[\mathscr{F}] \in \overline{M(0,2)}$. If $\mathscr{F}$ is stable, there is nothing to prove. If $\mathscr{F}$ is semi-stable and non-stable then the pair

$(S, C^\vee)$ is semi-stable, see 3.12, 10.5 with $S$ and $C^\vee$ both degenerate. It is shown in 10.5 that in this case $\mathcal{F}$ is an extension of the type

$$0 \to \mathcal{I}_{L\cup q} \to \mathcal{F} \to \mathcal{I}_{K\cup p} \to 0,$$

where $\mathcal{I}_{L\cup q}$, $\mathcal{I}_{K\cup p}$ are ideal sheaves of a line and a point as indicated in the figure, which is determined by $(S, C^\vee)$.

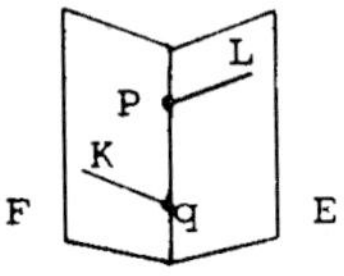


(If $q \in L$ then $\mathcal{I}_{L\cup q}$ is the ideal sheaf of the line $L$ with a multiple structure in $q$ with tangent vector in the plane $E$, similarly for $K\cup p$.) It follows first from 7.6.1, Case 4, that, if we consider the sheaf $\mathcal{N}$, $\operatorname{Supp} R^1\mathcal{N}$ is independent of the extension class; hence the same is true for $R^1\mathcal{F} = R^1\mathcal{N}$. Second we have $\mathcal{I}_{L\cup q}(-1) \subset \mathcal{I}_L(-1) \subset \mathcal{O}(-1)$ and hence $R^0\mathcal{I}_{L\cup q}(-1) = 0$. Therefore from the extension sequence of $\mathcal{F}$ we also obtain the short exact sequence

$$0 \to R^1\mathcal{I}_{L\cup q}(-1) \to R^1\mathcal{F}(-1) \to R^1\mathcal{I}_{K\cup p}(-1) \to 0,$$

which shows that the support of $R^1\mathcal{F}(-1)$ is independent of the extension. This proves injectivity of $\varphi|Q \setminus \pi^{-1}(\Sigma_0')$.

(f) Let us now consider the regular points over $\Sigma_0'$, i.e. $\pi^{-1}(\Sigma_0') \setminus Q_{\text{exc}}$. Because these correspond to stable points $x \in X^{ss}$ with $\mathcal{F}_x$ stable, see 1, the supports of $R^1\mathcal{F}_x$, $R^1\mathcal{F}_x(-1)$ are determined by $[\mathcal{F}_x]$. But here we have to show that $\mathcal{F}_x$ determines the plane $\mathbb{P}W$ or $\mathbb{P}W^\perp$.

Now in the case of stable pairs $(S, C^\vee)$ with $S$ a double line we only have two cases of $C^\vee$ as indicated in the picture, see 3.12.

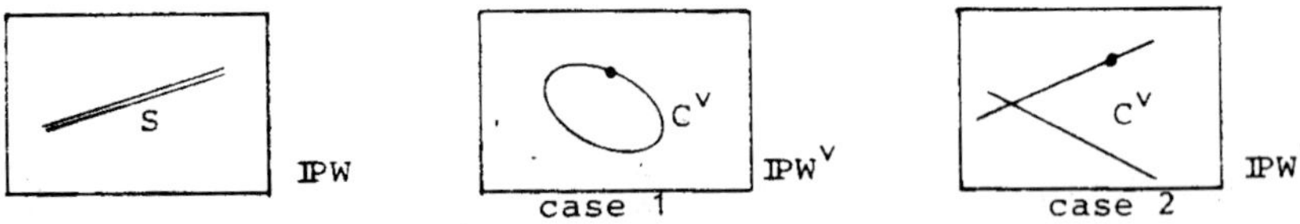


In Case 1 $\mathbb{P}W^\perp = \operatorname{Sing}\tilde{J}$ since $\mathbb{P}W^\perp \subset \operatorname{Sing}\tilde{J}$ and the latter is 2-dimensional. Therefore the plane is determined by $\mathcal{F}$ in this case.

In Case 2, $\tilde{J}$ has an equation $a\cdot b = 0$ such that one factor, $a$ say, is not in $S^0 = S \subset W \subset \bigwedge^2 V$ as a 2-space. This implies

$$S^\perp \not\subset a^\perp \cap b^\perp = \operatorname{Sing}\tilde{J}.$$

Because here $S^{\perp}$ and $\operatorname{Sing}\tilde{J}$ both are 3-dimensional and contain $\mathbb{P}W^{\perp}$

$$\mathbb{P}W^{\perp} = S^{\perp} \cap \operatorname{Sing}\tilde{J}.$$

This again proves that $\mathbb{P}W^{\perp}$ is determined by $\mathcal{F}$, since $S = S^0$ and $\tilde{J}$ are determined by $\mathcal{F}$.

(g) Finally we consider the restriction $\varphi | Q_{\mathrm{exc}}$. If $(S, C^{\vee}) \in Q_{\mathrm{exc}}$ the conics are of the type

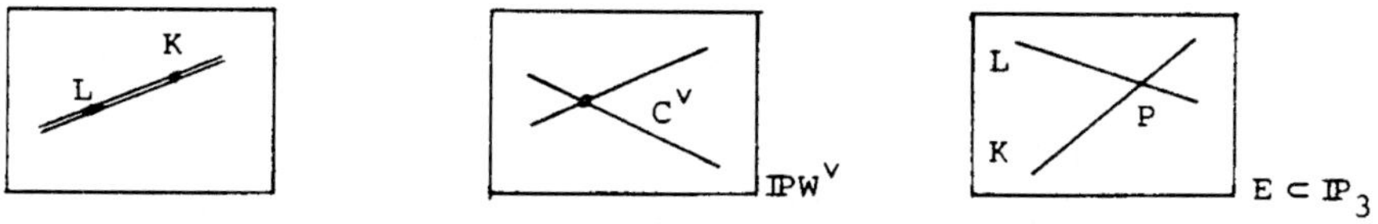


with the singular point of $C^{\vee}$ being the point $S \in \mathbb{P}W^{\vee}$ and $C^{\vee}$ determining two points $L, K \in S$. The triple $(S, L, K)$ determines a pair of lines in $\mathbb{P}_3$ with a plane $E$ and a point $p \in E$. The sheaf $\mathcal{F} = \mathcal{F}_x$ coming from a point $x$ defining $(S, C^{\vee})$ again is an extension

$$0 \to \mathcal{I}_{L \cup p} \to \mathcal{F} \to \mathcal{I}_{K \cup p} \to 0,$$

see (e), and $[\mathcal{F}] = [\mathcal{I}_{L\cup p} \oplus \mathcal{I}_{K \cup p}]$ in $\overline{M(0, 2)}$, where the double structure of $p$ in one of the lines shows in the direction of the plane $E$. But now the class $[\mathcal{F}]$ cannot remember the plane (whereas the extension class of $\mathcal{F}$ can, as can be shown easily). Thus $(S, C^{\vee}) \xrightarrow{\varphi} [\mathcal{F}]$ forgets the plane $\mathbb{P}W$, but $[\mathcal{F}]$ determines the triple $(S, L, K)$. This shows that $\varphi | Q_{\mathrm{exc}}$ blows down the $\mathbb{P}_1$ 's of double structures of the conics $S \in \Sigma_0'$, see 4.2, 3.9.

8.5. **Remark.** An "Orbit-Lemma" is true for $Q \setminus Q_{\mathrm{exc}}$: Let $x, x' \in X^{ss}$ with $q(x), q(x') \notin Q_{\mathrm{exc}}$. Then $\mathcal{F}_x \cong \mathcal{F}_{x'}$ iff $O(x) = O(x')$.

**9. Sheaves in the boundary with regular conic $S$.** In this section we start with the detailed geometric description of the sheaves representing boundary points of the moduli space. Since we fix a semi-stable parameter point $x = (z, M, \Gamma)$ in each case with $y = (z, \Gamma)$ and associated space $N$, we drop the indices and write

$$\mathcal{F} = \mathcal{F}_x, \quad \mathcal{N} = \mathcal{N}_y, \quad \mathcal{G} = \mathcal{G}_y, \quad \mathcal{K} = \mathcal{K}_N, \quad \mathcal{C} = \mathcal{C}_N, \quad \mathcal{T} = \mathcal{T}_y$$

and also

$$S = S(\Gamma), \quad C^{\vee} = C^{\vee}(M), \quad W = W_z.$$

In this section we assume that $S$ is a regular conic, which could be a regular plane section, or contained in a $\beta$- or $\alpha$-plane, see 9.1, 9.2,

9.3 respectively. It is convenient in this case to consider the Poncelet conic $C$ in the same plane $\mathbb{P}W$ of $S$, which is the polar dual of $C^\vee$ w.r.t. $S$. Since the Poncelet condition for degenerate $C$ just means that one of the lines of $C$ must be tangent to $S$, we have to consider the following cases:

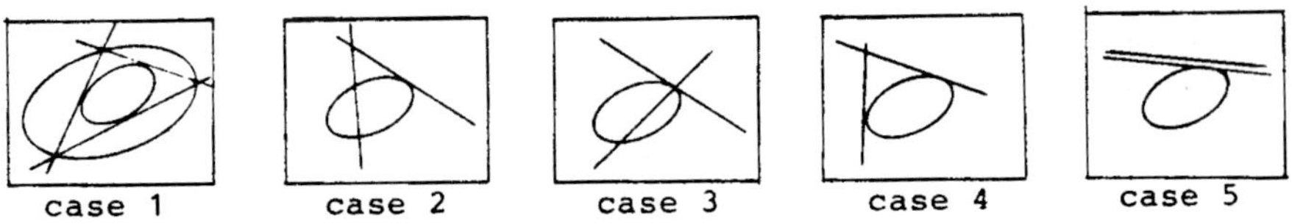


9.1. *The sheaves in* $Q_e \setminus Q_\alpha \cup Q_\beta \cup Q_0$. The pairs $(S, C)$ in this set are characterised by $S$ to be a regular plane section and $C$ to be singular. In such a case the homomorphism defined by $N$ is a regular epimorphism as in 1.1, 1.2, and we have $\mathscr{N} = \mathscr{K}$, $\mathscr{E} = \mathscr{T} = 0$, and $\mathscr{G} = \mathscr{O}_Q(-2, 1)$, where $Q$ here denotes the quadric defined by $S$.

9.1.1. Proposition. *The sheaves* $\mathscr{F}$ *in* Case 2/3 *are exactly those which can be obtained by an "elementary transformation"*

$$0 \to \mathscr{F} \to \mathscr{E}' \xrightarrow[\pi]{} \mathscr{O}_l(1) \to 0,$$

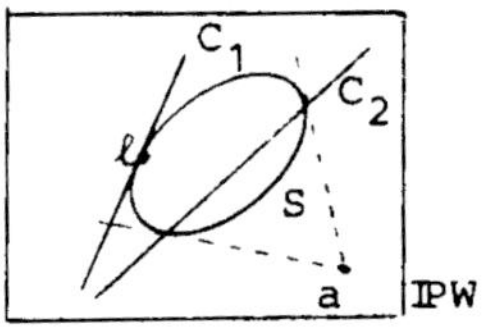


*where* $\mathscr{E}' \in M(0, 1)$ *is an instanton bundle with* $c_2\mathscr{E}' = 1$ *(i.e. a null-correlation bundle),* $l$ *is a line in* $\mathbb{P}_3$ *and* $\pi$ *is an epimorphism. The data* $(\mathscr{E}', l, \pi)$ *are in* 1:1 *correspondence with the pairs* $(S, C) \in Q_e \setminus Q_\alpha \cup Q_\beta \cup Q_e$ *as follows*

(i) $\mathscr{E}'$ *is the bundle in* $M(0, 1) = \mathbb{P}\Lambda^2 V \setminus G_2 V$ *determined by the pole* $a$ *of the component* $C_2$ *of* $C$ *in the plane* $\mathbb{P}W$,

(ii) $l$ *is the tangent point of the component* $C_1$,

(iii) *the epimorphism* $\pi$ *corresponds (in a way described in the proof) to the plane* $\mathbb{P}W$ *through* $a$, $l$ *and intersecting* $\mathbb{G}$ *regularly in* $S$.

*Cases* 2 *and* 3 *can be distinguished by* $\mathscr{E}'|l \simeq 2\mathscr{O}_l$ *and* $\mathscr{E}'|l \simeq \mathscr{O}_l(-1) \oplus \mathscr{O}_l(1)$.

Corollary 1. *Each such* $\mathscr{F}$ *is* $\mu$*-stable, since* $\mathscr{E}'$ *is* $\mu$*-stable.*

COROLLARY 2. *Let $B \to \mathbb{G} \times M(0, 1)$ be the projective bundle of homomorphisms $\mathcal{E}' \to \mathcal{O}_l(1)$ mod scalars for $(l, \mathcal{E}') \in \mathbb{G} \times M(0, 1)$ and let $B^0$ be the open part of epimorphisms. Then the elementary transformation gives us an isomorphism*

$$B^0 \hookrightarrow Q_e | C^0(\mathbb{G})$$

*onto the open part of the boundary component $Q_e$ defined by* 9.1, *Cases* 2 *and* 3.

*Proof*. (a) Let $\mathcal{F}$ be given. Since $S$ is regular we can choose a basis $e_i$ of $V$ such that the space $N$ can be presented by the matrix

$$N^* = \begin{bmatrix} e_0 & e_2 \\ e_1 & e_3 \end{bmatrix},$$

see 1.1, 1.2, and such that $l = e_{01} = e_0 \wedge e_1$. Moreover since $C^\vee(M)$ is a pair of lines, the matrix representing $M$ must have the shape

$$M^* = \begin{bmatrix} l & 0 \\ b & a \end{bmatrix},$$

such that $M^* \wedge N^{*t} = 0$. By our convention $\bigwedge^2 V \simeq \bigwedge^2 V^\vee$, the conic $C$ has the equation $l \circ a = 0$ in $\mathbb{P}W \simeq \mathbb{P}W^\vee$ (duality given by $\mathbb{G}$ or $S$). By this form of $M^*$ we obtain the exact diagram

$$\begin{array}{ccccccccc}
 & & 0 & & 0 & & 0 & & \\
 & & \downarrow & & \downarrow & & \downarrow & & \\
0 & \longrightarrow & \Omega^3(3) & \xrightarrow{\ l\ } & \Omega^1(1) & \longrightarrow & \mathcal{F}' & \longrightarrow & 0 \\
 & & \downarrow & & \downarrow & & \downarrow & & \\
0 & \longrightarrow & k^2 \otimes \Omega^3(3) & \xrightarrow[\binom{l}{b\ a}]{} & k^2 \otimes \Omega^1(1) & \longrightarrow & \mathcal{M} & \longrightarrow & 0 \\
 & & \downarrow & & \downarrow & & \downarrow & & \\
0 & \longrightarrow & \Omega^3(3) & \xrightarrow[a]{} & \Omega^1(1) & \longrightarrow & \mathcal{E}' & \longrightarrow & 0 \\
 & & \downarrow & & \downarrow & & \downarrow & & \\
 & & 0 & & 0 & & 0 & &
\end{array} \tag{23}$$

with cokernels $\mathcal{F}'$, $\mathcal{M}$, $\mathcal{E}'$ respectively. On the other hand the monad

(22) of $\mathcal{F}$ in 8. gives the mid row of the exact diagram

$$
\begin{array}{ccccccccc}
 & & & & 0 & & 0 & & \\
 & & & & \downarrow & & \downarrow & & \\
 & & & & \mathcal{F}' & = & \mathcal{F}' & & \\
 & & & & \downarrow & & \downarrow & & \\
0 & \longrightarrow & \mathcal{F} & \longrightarrow & \mathcal{M} & \longrightarrow & k^2 \otimes \mathcal{O} & \longrightarrow & 0. \\
 & & \| & & \downarrow & & \downarrow & & \\
0 & \longrightarrow & \mathcal{F} & \longrightarrow & \mathcal{E}' & \longrightarrow & \mathcal{C}oker & \longrightarrow & 0 \\
 & & & & \downarrow & & \downarrow & & \\
 & & & & 0 & & 0 & &
\end{array}
\tag{24}
$$

In this the composite $\mathcal{F}' \to k^2 \otimes \mathcal{O}$ is still injective, which follows from the upper right-hand square of (23) and from

$$
\begin{array}{ccc}
k^2 \otimes \Omega^1(1) & \longrightarrow & \mathcal{M} \\
 & \searrow^{\left[\begin{smallmatrix} e_0 & e_1 \\ e_2 & e_3 \end{smallmatrix}\right]} & \downarrow \\
 & & k^2 \otimes \mathcal{O}
\end{array}
,
$$

since $l = e_{01}$. This shows that (24) is exact. Since $a$ is indecomposable, $\mathcal{E}'$ is a typical bundle of $M(0, 1) = \mathbb{P}\bigwedge^2 V \setminus \mathbb{G}$, see **[Ha2]**. It remains to identify the cokernel. By the definition of $\mathcal{F}' \to k^2 \otimes \mathcal{O}$ we get the diagram

$$
\begin{array}{ccccccccc}
 & & & & & & 0 & & \\
 & & & & & & \downarrow & & \\
0 & \longrightarrow & \Omega^3(3) & \xrightarrow{\ l\ } & \Omega^1(1) & \longrightarrow & \mathcal{F}' & \longrightarrow & 0 \\
 & & & & \| & & \downarrow & & \\
 & & & & \Omega^1(1) & \xrightarrow[(e_0, e_1)]{} & k^2 \otimes \mathcal{O} & \longrightarrow & \mathcal{O}_l(1) & \longrightarrow & 0, \\
 & & & & & & \downarrow & & \\
 & & & & & & \mathcal{C}oker & & \\
 & & & & & & \downarrow & & \\
 & & & & & & 0 & &
\end{array}
$$

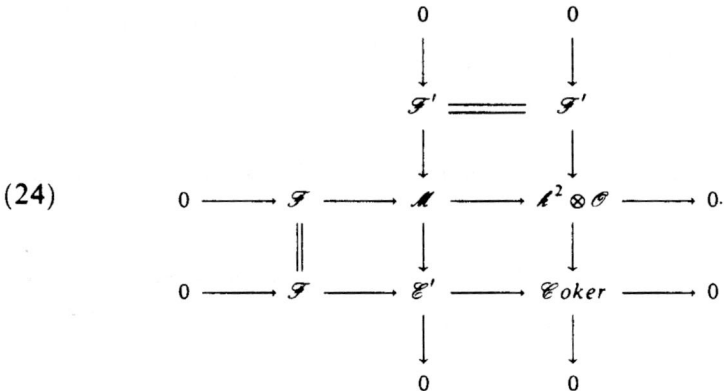

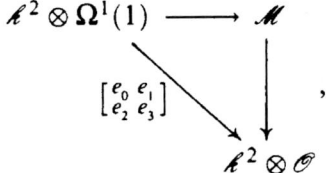

which shows that $\mathscr{C}oker = \mathscr{O}_l(1)$. (In particular we have obtained the two equivalent descriptions of $\mathscr{F}'$ which is a sheaf of the boundary of $M(0, 1)$.)

(b) It is also easy to verify that the subspace $W \subset \bigwedge^2 V$ is isomorphic under $\bigwedge^2 V \simeq \bigwedge^2 V^\vee$ to the kernel of the composed map

$$\bigwedge^2 V^\vee = \Gamma\Omega^1(2) \to \Gamma\mathscr{E}'(1) \to \Gamma\mathscr{O}_l(2),$$

using the above matrices. This shows that the plane $\mathbb{P}W$ is determined by $\pi$. Conversely we had just constructed $\pi$ from $l$, $a$ and the plane. Thus we have established (i), (ii), (iii) if $\mathscr{F}$ is given.

(c) Let now an elementary transformation be given. We can find a monad for $\mathscr{F}$ by going backwards in the diagrams (23) and (24). First we can determine $l$, $a$ and the plane $\mathbb{P}W$ by $\pi$ as in (b). Let $\mathscr{F}'$ be defined as the kernel of $k^2 \otimes \mathscr{O} \to \mathscr{O}_l(1)$, and define $\mathscr{M}$ as the pullback in diagram (24). Since $\mathscr{F}'$ and $\mathscr{E}'$ have the resolutions as in (23), we get the resolution of $\mathscr{M}$ by adding up. Then $0 \to \mathscr{F} \to \mathscr{M} \to k^2 \otimes \mathscr{O}$ and the resolution of $\mathscr{M}$ give us a monad. To see that this defines a pair $(S, C)$ of the above type, we consider the composed homomorphism

$$k^2 \otimes \Omega^1(1) \xrightarrow[N^{*\vee} = \begin{pmatrix} e_0 & e_1 \\ v & w \end{pmatrix}]{} k^2 \otimes \mathscr{O} \longrightarrow 0,$$

which must have the entries $e_0$, $e_1$ in its first row because of $l = e_{01}$ and $M^* \wedge N^{*\vee} = 0$, and which must be an epimorphism. By 0.2 we must have $\dim\langle e_0, e_1, v, w\rangle = 4$ or $= 3$ in a special configuration. If $\dim = 3$ it would follow that the entries $l$, $a$, $b$ of $M^*$ are contained in a $\beta$-plane, see 7.6, Case 2, and $a$ would be decomposable. Therefore $N^*$ defines a regular conic $S$ and $M^*$, by its shape, a degenerate conic $C$ as in Cases 2 or 3.

*Case* 4. If the conic $C$ consists of two tangents we get a degenerate case of the elementary transformation by replacing $\mathscr{E}'$ by a sheaf of the type $\mathscr{F}'$ considered above. Thus a sheaf $\mathscr{F}$ in 9.1, Case 4, is given as the kernel of an epimorphism $\pi$

$$0 \to \mathscr{F} \to 2\mathscr{O} \xrightarrow{\pi} \mathscr{O}_{l_1}(1) \oplus \mathscr{O}_{l_2}(1) \to 0,$$

where $\pi$ corresponds to the plane $\mathbb{P}W$ through the axis $\overline{l_1, l_2}$. The proof is a special case of the one just made. Note that by this we extend the morphism of elementary transformation to $\overline{M(0, 1)}$. If we compactify this along the direction of the epimorphisms we would leave the set of regular conics $S$.

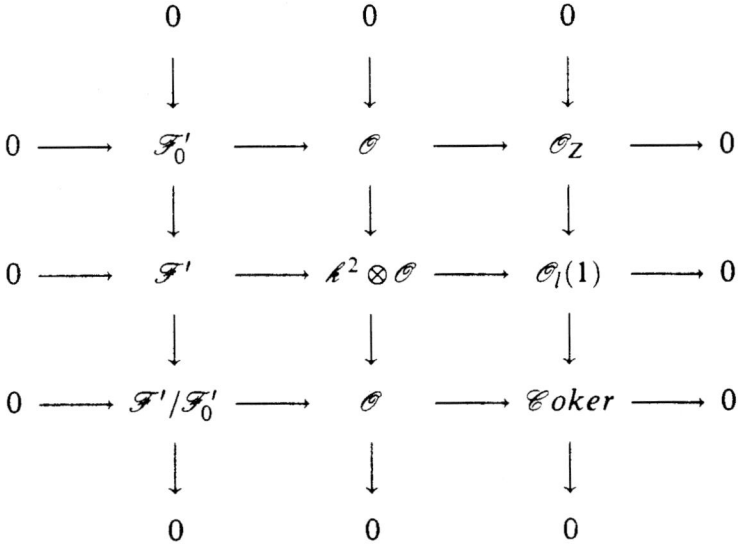

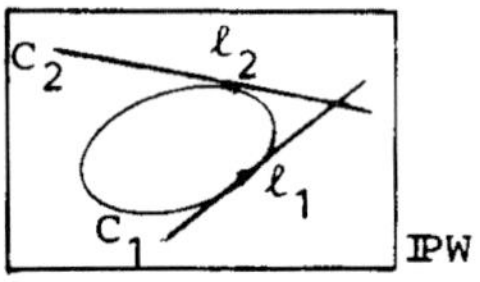


Note that also in this case $\mathcal{F}$ is stable, since it is easy to show that any sheaf $\mathcal{F}'$ as above is stable (but not $\mu$-stable any more): If $\mathcal{F}_0' \subset \mathcal{F}'$ is a sub-sheaf of rank 1 with $\mathcal{F}'/\mathcal{F}_0'$ torsionfree, we can assume $c_1\mathcal{F}_0' = 0$ and hence $\mathcal{F}_0' \subset \mathcal{O}$ an ideal sheaf. The diagram

$$\begin{array}{ccccccccc}
 & & 0 & & 0 & & 0 & & \\
 & & \downarrow & & \downarrow & & \downarrow & & \\
0 & \longrightarrow & \mathcal{F}_0' & \longrightarrow & \mathcal{O} & \longrightarrow & \mathcal{O}_Z & \longrightarrow & 0 \\
 & & \downarrow & & \downarrow & & \downarrow & & \\
0 & \longrightarrow & \mathcal{F}' & \longrightarrow & k^2 \otimes \mathcal{O} & \longrightarrow & \mathcal{O}_l(1) & \longrightarrow & 0 \\
 & & \downarrow & & \downarrow & & \downarrow & & \\
0 & \longrightarrow & \mathcal{F}'/\mathcal{F}_0' & \longrightarrow & \mathcal{O} & \longrightarrow & \mathcal{C}oker & \longrightarrow & 0 \\
 & & \downarrow & & \downarrow & & \downarrow & & \\
 & & 0 & & 0 & & 0 & &
\end{array}$$

shows that $\operatorname{Supp}(\mathcal{C}oker)$ is at most 0-dimensional if $Z \neq \varnothing$, and that then $\chi\mathcal{F}_0'(m) < \frac{1}{2}\chi\mathcal{F}'(m)$ for large $m$.

*Case* 5. Again this is a degeneration of Case 3 or Case 4. We obtain here an exact sequence

$$0 \to \mathcal{F} \to 2\mathcal{O} \xrightarrow{\pi} \mathcal{R} \to 0,$$

where $\mathcal{F}^{\vee\vee} = 2\mathcal{O}$ as in the previous case and where $\mathcal{R}$ is supported on $L$ as an $\mathcal{O}$-module extension

$$0 \to \mathcal{O}_l(1) \to \mathcal{R} \to \mathcal{O}_l(1) \to 0.$$

Here both the extension and the epimorphism depend on the plane $\mathbb{P}W$, but we omit further details. Again $\mathcal{F}$ is stable.

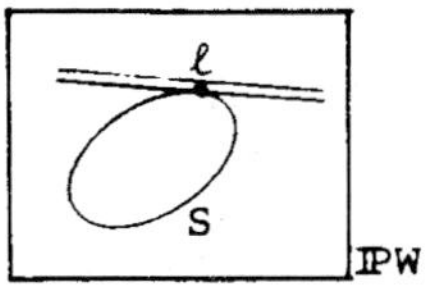


9.2. *The sheaves in* $Q_\beta \setminus Q_0$. These are the sheaves corresponding to a pair $(S, C^\vee)$ where $S$ is a regular conic in a $\beta$-plane. Since $\mathbb{P}W$ is a $\beta$-plane, we have $W = \bigwedge^2 U$ where $\mathbb{P}U \subset \mathbb{P}V$ is a plane, which can be considered now the dual of $\mathbb{P}W$ by $\bigwedge^2 U^\vee \simeq U$. The dual conic $S^\vee \subset \mathbb{P}U$ can now be considered as the base conic for the Poncelet property and we can also consider $C^\vee(M)$ as a conic in $\mathbb{P}U$ given by the equation

$$\det M^* = 0 \quad \text{in } \mathbb{P}U,$$

where the entries of $M^*$ are elements of $W = \bigwedge^2 U \simeq U^\vee$. Moreover we choose a basis $e_i$ of $V$ such that

$$U = \langle e_0, e_1, e_2 \rangle, \qquad U \oplus \langle e_3 \rangle = V,$$

and such that the matrices $\Gamma^*$, $N^*$ representing the spaces $\Gamma$, $N$ are given by

$$\Gamma^* = \begin{bmatrix} e_{01} & 0 \\ e_{12} & e_{01} \\ e_{02} & e_{12} \\ 0 & e_{02} \end{bmatrix}, \qquad N^{*\vee} = \begin{bmatrix} e_0 & e_1 \\ -e_2 & e_0 \end{bmatrix}$$

as in Case 2 of 7.6.

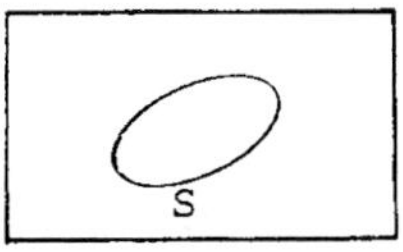


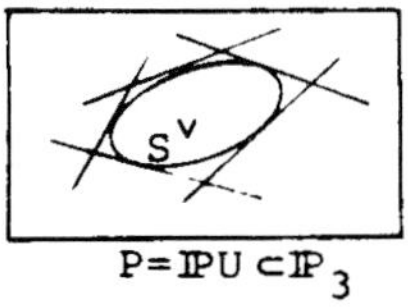

9.2.1. **Proposition.** (1) *The sheaves* $\mathcal{F}$ *in* $Q_\beta \setminus Q_0$ *are elementary transformations of the type*

$$0 \to \mathcal{F} \to k^2 \otimes \mathcal{O} \to \mathcal{R}(1) \to 0$$

*where* $\mathcal{R}$ *is supported by the conic* $C^\vee \subset P \subset \mathbb{P}_3$ *and is the cokernel of* $M^*$ :

$$0 \to k^2 \otimes \Omega_P^2(2) \xrightarrow[M^*]{} k^2 \otimes \mathcal{O}_P \to \mathcal{R} \to 0,$$

*where* $M^*$ *is the matrix of* $M$ *with entries in* $W = \bigwedge^2 U$ ($\mathcal{R}$ *is a Cohen-Macaulay module on* $C^\vee$).

(2) *The sheaf* $\mathcal{G}$ *is a stable rank-2 bundle on* $P$ *with* $c_1\mathcal{G}(1) = 0$, $c_2\mathcal{G}(1) = 2$ *and is the kernel of* $N^{*\vee}$,

$$0 \to \mathcal{G} \to k^2 \otimes \Omega_P^1(1) \xrightarrow{N^{*\vee}} k^2 \otimes \mathcal{O}_P \to 0$$

(*with entries of* $N^*$ *in* $U$). *Its jumping lines are the points of* $S$, *which at the same time are the zero loci of sections of* $\mathcal{N}(1)$.

**Remark.** These sheaves are well understood, see **[Ba]**.

(3) The Poncelet relation of $S^\vee$ with $C^\vee$ in $P$ has its expression in the exact sequence

$$0 \to \mathcal{G}(-1) \to (\Gamma/M) \otimes \Omega^2 P(2) \to \mathcal{R} \to 0$$

obtained in (25) of the proof.

(4) The restriction of $\mathcal{F}$ to $P$ splits into

$$\mathcal{F}|P = \mathcal{G} \oplus \mathcal{R}.$$

(5) Each such $\mathcal{F}$ is stable.

*Proof.* (a) We first remark that the homomorphism $\Omega^3(3) \xrightarrow[a]{} \Omega^1(1)$ defined by $a \in \bigwedge^2 U \subset \bigwedge^2 V$ splits into

$$\Omega_P^2(2) \xrightarrow[(a,0)]{} \mathcal{O}_P \oplus \Omega_P^1(1)$$

when restricted to the plane $P = \mathbb{P}U$, see 0.1.

(b) Let us consider now the diagram

$$\begin{array}{ccccccccc}
 & & & & 0 & & 0 & & \\
 & & & & \downarrow & & \downarrow & & \\
0 & \longrightarrow & k^4\otimes\Omega^3(3) & \longrightarrow & \mathcal{N} & \longrightarrow & \mathcal{G} & \longrightarrow & 0\\
 & & \| & & \downarrow & & \downarrow & & \\
0 & \longrightarrow & k^4\otimes\Omega^3(3) & \xrightarrow[\Gamma^*]{} & k^2\otimes\Omega^1(1) & \longrightarrow & \tilde{\mathcal{G}} & \longrightarrow & 0,\\
 & & & & \downarrow{\scriptstyle N^{*\vee}} & & \downarrow & & \\
 & & & & k^2\otimes\mathcal{O} & = & k^2\otimes\mathcal{O} & & \\
 & & & & \downarrow & & \downarrow & & \\
 & & & & 0 & & 0 & &
\end{array}$$

which is defined by $\Gamma^*$. From this it can be proved first that $z_3\mathcal{G} = 0$, i.e. $\mathcal{G}$ is an $\mathcal{O}_P$-module ($z_0, \dots, z_3 \in V^\vee$ are the dual coordinates). If we restrict this diagram to $P$ we obtain the splitting as indicated in the diagram, where we identify $\mathcal{T}or_1(\mathcal{G}, \mathcal{O}_P) = \mathcal{G}(-1) = \mathcal{T}or_1(\tilde{\mathcal{G}}, \mathcal{O}_P)$,

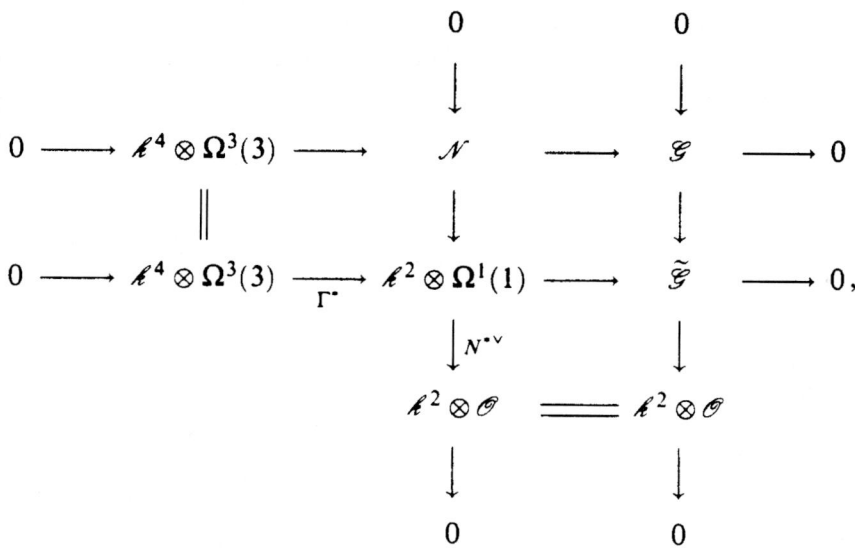

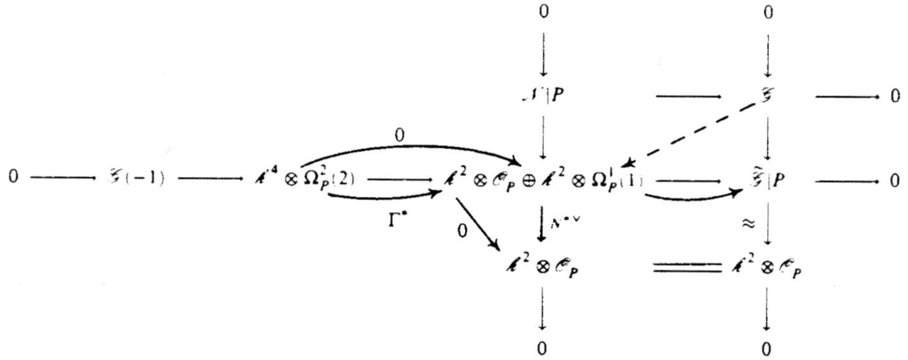


It follows that

$$\mathcal{N}|_P \simeq k^2\otimes\mathcal{O}_P\oplus\mathcal{G},$$

and that we obtain two different presentations of the sheaf $\mathscr{G}$ which are equivalent by using a transformation based on $\Omega^1_P(1) \subset U^\vee \otimes \mathscr{O}_P$ as in (9) of 5.2.

(c) Let now the sheaf $\mathscr{R}$ be defined as the cokernel of $M^*$ as homomorphism on $P$ and as in the proposition. Then we obtain the exact diagram

(25)

$$
\begin{array}{ccccccccc}
 & & & & p & & & & \\
 & & & & \downarrow & & & & \\
 & & 0 & \longrightarrow & M \otimes \Omega^2_P(2) & \longrightarrow & k^2 \otimes \mathscr{O}_P & \longrightarrow \mathscr{R} & \longrightarrow 0 \\
 & & \downarrow & & \downarrow & & \| & & \\
\mathscr{O} & \longrightarrow & \mathscr{G}(-1) & \longrightarrow & \Gamma \otimes \Omega^2_P(2) & \longrightarrow & k^2 \otimes \mathscr{O}_P & \longrightarrow 0 & \\
 & & \downarrow & & \downarrow & & & & \\
 & & (\Gamma/M) \otimes \Omega^2_P(2) & = & (\Gamma/M) \otimes \Omega^2_P(2) & & & & \\
 & & \downarrow & & \downarrow & & & & \\
 & & \mathscr{R} & & 0 & & & & \\
 & & \downarrow & & & & & & \\
 & & 0 & & & & & &
\end{array}
$$

and in particular the sequence (3) of the proposition.

(d) If we restrict display (22) of $\mathscr{F}$ to $P$ we obtain

$$
\begin{array}{ccccccccc}
 & & & & 0 & & 0 & & \\
 & & & & \downarrow & & \downarrow & & \\
0 & \longrightarrow & M \otimes \Omega^2_P(2) & \longrightarrow & k^2 \otimes \mathscr{O}_P \oplus \mathscr{G} & \longrightarrow & \mathscr{F}|_P & \longrightarrow & 0 \\
 & & \| & & \| \qquad \downarrow & & \downarrow & & \\
0 & \longrightarrow & M \otimes \Omega^2_P(2) & \longrightarrow & k^2 \otimes \mathscr{O}_P \oplus k^2 \otimes \Omega^1_P(1) & \longrightarrow & \mathscr{M}|_P & \longrightarrow & 0 \\
 & & & & {\scriptstyle 0}\searrow \quad \swarrow{\scriptstyle N^{*\vee}} & & \downarrow & & \\
 & & & & k^2 \otimes \mathscr{O}_P & = & k^2 \otimes \mathscr{O}_P & & \\
 & & & & \downarrow & & \downarrow & & \\
 & & & & 0 & & 0 & &
\end{array}
$$

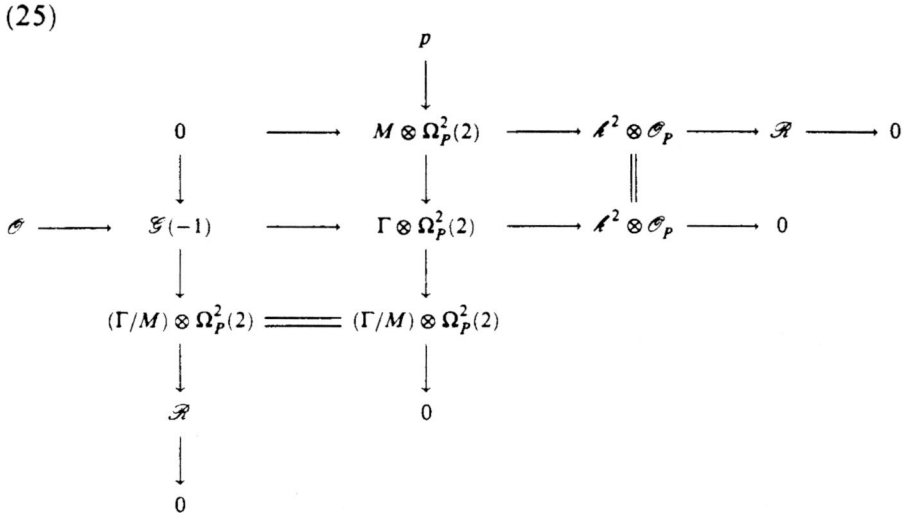

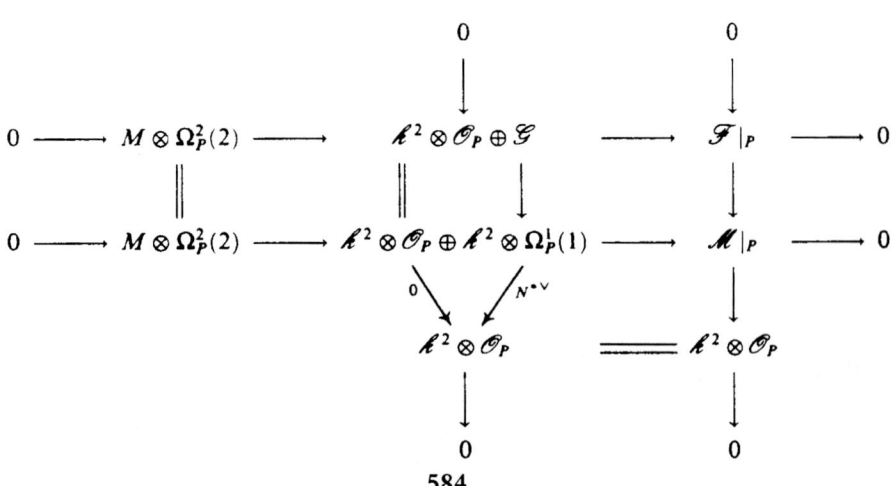

thereby obtaining the splitting

$$\mathscr{F}|_P = \mathscr{R} \oplus \mathscr{G}.$$

(e) Finally we consider the diagram

$$\begin{array}{ccccccccccc}
 & & & & 0 & & & & 0 & & & & \\
 & & & & \downarrow & & & & \downarrow & & & & \\
 & & 0 & \longrightarrow & \Gamma/M \otimes \Omega^3(2) & & \longrightarrow & & \mathscr{F}(-1) & \longrightarrow & \mathscr{G}(-1) & \longrightarrow & 0 \\
 & & & & \downarrow{\scriptstyle z_3} & & {\scriptstyle \alpha}\swarrow & & \downarrow{\scriptstyle z_3} & & \vdots\downarrow{\scriptstyle 0} & & \\
 & & 0 & \longrightarrow & \Gamma/M \otimes \Omega^3(3) & & \longrightarrow & & \mathscr{F} & \longrightarrow & \mathscr{G} & \longrightarrow & 0 \\
 & & & & \downarrow & & \searrow{\scriptstyle \beta} & & \downarrow & & \| & & \\
0 & \longrightarrow & \mathscr{G}(-1) & \longrightarrow & \Gamma/M \otimes \Omega^2_P(2) & & \longrightarrow & & \mathscr{R} \oplus \mathscr{G} & \longrightarrow & \mathscr{G} & \longrightarrow & 0 \\
 & & & & \downarrow & & & & \downarrow & & & & \\
 & & & & 0 & & & & 0 & & & &
\end{array}$$

Since $z_3\mathscr{G}(-1) = 0$ the multiplication by $z_3$ lifts to $\alpha$, and by the splitting of the bottom row we obtain the exact diagram

$$\begin{array}{ccccccccc}
 & & & & 0 & & 0 & & \\
 & & & & \downarrow & & \downarrow & & \\
0 & \longrightarrow & \Gamma/M \otimes \Omega^3(2) & \longrightarrow & \mathscr{F}(-1) & \longrightarrow & \mathscr{G}(-1) & \longrightarrow & 0 \\
 & & \| & & \downarrow{\scriptstyle \alpha} & & \downarrow & & \\
0 & \longrightarrow & \Gamma/M \otimes \Omega^3(2) & \longrightarrow & \Gamma/M \otimes \Omega^3(3) & \longrightarrow & \Gamma/M \otimes \Omega^2_P(2) & \longrightarrow & 0 \\
 & & & & \downarrow{\scriptstyle \beta} & & \downarrow & & \\
 & & & & \mathscr{R} & = & \mathscr{R} & & \\
 & & & & \downarrow & & \downarrow & & \\
 & & & & 0 & & 0 & &
\end{array}$$

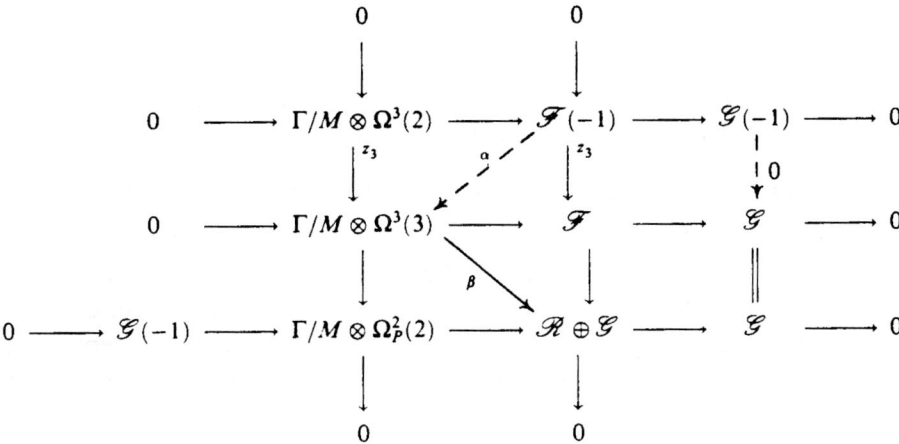

Altogether this proves (1), ..., (4) of the proposition. For the proof of stability we can take a subsheaf $\mathscr{F}' \subset \mathscr{F}$ with $\mathscr{F}/\mathscr{F}'$ torsionfree, with $c_1\mathscr{F}' = 0$, rank$\mathscr{F}' = 1$, since $2\mathscr{O}$ is $\mu$-semi-stable. Now a diagram analogous to the one of 9.1, Case 4, shows that the $\mathcal{C}oker$ is 0-dimensional and thus $\chi\mathcal{F}'(m) < \frac{1}{2}\chi\mathcal{F}(m)$ for large $m$.

REMARK 1. The sequence (3) of the proposition describes the Poncelet situation in terms of bundles in the plane $P$, see **[Ba]** and also **[Tr2]**. Since $\mathcal{E}xt^1(\mathcal{G}, \mathcal{O}) = \mathcal{G}^\vee(1)$, $\mathcal{G}^\vee = \mathcal{G}(2)$, and $\mathcal{E}xt^2_{\mathcal{O}}(\mathcal{R}, \mathcal{O}) = \mathcal{R}^\vee(2)$ (dual of $\mathcal{R}$ on its support), we also obtain the exact sequence

$$0 \to (\Gamma/M)^\vee \otimes \mathcal{O}_P \to \mathcal{G}(2) \to \mathcal{R}^\vee(1) \to 0,$$

where $\Gamma^\vee = \Gamma\mathcal{G}(2)$. This shows that we get all Poncelet curves if we vary the 2-dimensional subspaces of $\Gamma\mathcal{G}(2)$.

REMARK 2. We can also investigate the epimorphisms $k^2 \otimes \mathcal{O} \xrightarrow{\pi} \mathcal{R}(1) \to 0$ for a given conic $C^\vee$ in $P$ as in 9.1. The result is that the pencils $\subset \mathbb{P}\Gamma\mathcal{R}(1) = \mathbb{P}_3$ describe the 4-dimensional family of regular conics $S^\vee \subset P$ to which $C^\vee$ is Poncelet related. Thus also in Case 9.2 the epimorphisms $\pi$ correspond to the regular conics. Moreover the elementary transformations investigated here also extend the ones of 9.1 to the case of $\beta$-planes.

Now we can describe the different situations of the conic $C^\vee$.

*Case* 1: in which $C^\vee$ is regular. Then $\mathcal{R} = \mathcal{O}_{C^\vee}(1)$ is the line bundle of degree 1 on $C^\vee$.

*Cases* 2, 3: in which $C^\vee$ is a pair of lines. In this case the matrix $M^*$ cannot be split and defines $\mathcal{R}$ as a nontrivial extension

$$0 \to \mathcal{O}_L \to \mathcal{R} \to O_K \to 0,$$

where $L$, $K$ are the two lines of $C^\vee$ in the plane $P$.

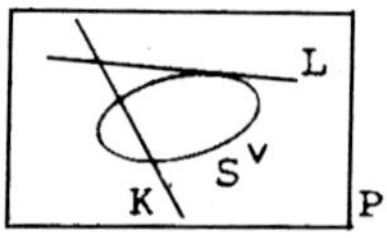


*Case* 4: in which $C^\vee$ consists of a pair of tangents. Here $\mathcal{R}$ is the direct sum $\mathcal{O}_L \oplus \mathcal{O}_K$.

*Case* 5: in which $C^\vee$ is a double tangent. Now $\mathcal{R}$ can be a nontrivial extension again depending on $M^*$, $0 \to \mathcal{O}_L \to \mathcal{R} \to \mathcal{O}_L \to 0$.

9.3. *The sheaves in* $Q_\alpha \backslash Q_0$. These are the sheaves corresponding to a pair $(S, C^\vee)$ where $S$ is regular in an $\alpha$-plane and thus determines a cone $Q \subset \mathbb{P}_3$. Any plane $P = \mathbb{P}U$ in $\mathbb{P}_3$ not passing through the vertex $e_0$ serves as a base of the cone which is isomorphic to the $\alpha$-plane $\mathbb{P}W$, and we assume

$$W = e_0 \wedge U.$$

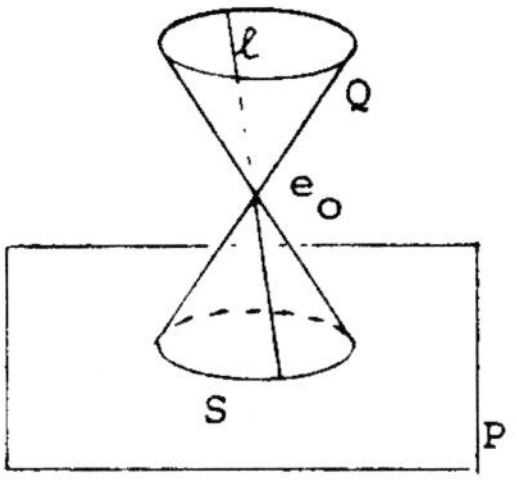


The conic $Q \cap P$ can be identified with the given conic $S$. Now we can choose a basis $e_0, \ldots, e_3$ such that

$$U = \langle e_1, e_2, e_3 \rangle \quad \text{and} \quad \Gamma^* = \begin{bmatrix} e_{01} & 0 \\ e_{03} & e_{01} \\ e_{02} & e_{03} \\ 0 & e_{02} \end{bmatrix},$$

see 7.6, Case 3. Then the equation of $S$ in $P$ is $z_2^2 - z_1 z_3 = 0$, where the $z_j$ denote the dual coordinates, and the matrix $N^*$ is necessarily a direct product

$$N^* = \begin{bmatrix} e_0 & \\ & e_0 \end{bmatrix}.$$

As shown in 7.6, Case 3, we have in this case $\mathscr{E} = \mathscr{T} = k^2 \otimes k_{e_0} =$ 2 times the skyscraper sheaf $k_{e_0} = \mathscr{O}/m(e_0)$, and

$$\mathscr{K} = k^2 \otimes \mathscr{Z},$$

where $\mathscr{Z}$ is the first syzygy of the ideal sheaf $m(e_0)(1)$, or equivalently

$$0 \to \mathscr{Z} \to \Omega^1(1) \xrightarrow[e_0]{} \mathscr{O} \to k_{e_0} \to 0.$$

We first investigate the sheaf $\mathcal{G}$, which of course by §7 is determined by the cone $Q$ alone. Let $\tilde{\mathcal{G}}$ be the cokernel in:

$$
\begin{array}{ccccccccc}
 & & & & 0 & & 0 & & \\
 & & & & \downarrow & & \downarrow & & \\
0 & \longrightarrow & \Gamma\otimes\Omega^3(3) & \longrightarrow & \mathcal{N} & \longrightarrow & \mathcal{G} & \longrightarrow & 0 \\
 & & \| & & \downarrow & & \downarrow & & \\
0 & \longrightarrow & \Gamma\otimes\Omega^3(3) & \longrightarrow & \mathcal{K} & \longrightarrow & \tilde{\mathcal{G}} & \longrightarrow & 0. \\
 & & & & \downarrow & & \downarrow & & \\
 & & & & \mathcal{T} & = & \mathcal{T} & & \\
 & & & & \downarrow & & \downarrow & & \\
 & & & & 0 & & 0 & &
\end{array}
$$

9.3.0. **Proposition.** *$\tilde{\mathcal{G}}$ is the ideal sheaf $\mathcal{J}_{l,Q} \subset \mathcal{O}_Q$ of any line of the cone and $\mathcal{G} = m(e_0)\mathcal{J}_{l,Q}$ ($\tilde{\mathcal{G}}$ is a reflexive Cohen-Macaulay module of the singularity $e_0$).*

Note that in this case, we have only $\mathcal{G}(2) \subset \tilde{\mathcal{G}}(2) \simeq \mathcal{E}xt^1_{\mathcal{O}}(\mathcal{G}, \mathcal{O})$.

*Proof.* We have $\Gamma\mathcal{K}(1) = k^2 \otimes W \subset k^2 \otimes \bigwedge^2 V$ and thus the diagram

$$
\begin{array}{ccccccccc}
 & & & & 0 & & 0 & & \\
 & & & & \downarrow & & \downarrow & & \\
 & & & & \Gamma\otimes\mathcal{O} & = & \Gamma\otimes\mathcal{O} & & \\
 & & & & \downarrow & & \downarrow & & \\
0 & \longrightarrow & k^2\otimes\mathcal{O}(-1) & \longrightarrow & k^2\otimes W\otimes\mathcal{O} & \longrightarrow & \mathcal{K}(1) & \longrightarrow & 0, \\
 & & \| & & \downarrow & & \downarrow & & \\
0 & \longrightarrow & k^2\otimes\mathcal{O}(-1) & \xrightarrow[\varphi]{} & k^2\otimes\mathcal{O} & \longrightarrow & \tilde{\mathcal{G}}(1) & \longrightarrow & 0 \\
 & & & & \downarrow & & \downarrow & & \\
 & & & & 0 & & 0 & &
\end{array}
$$

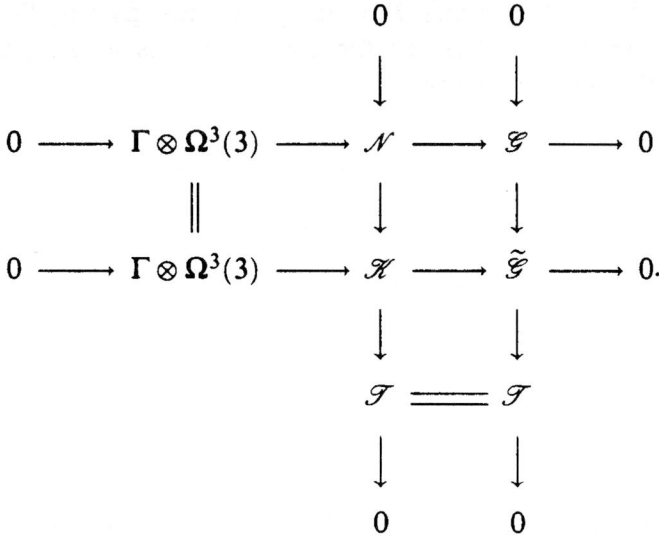

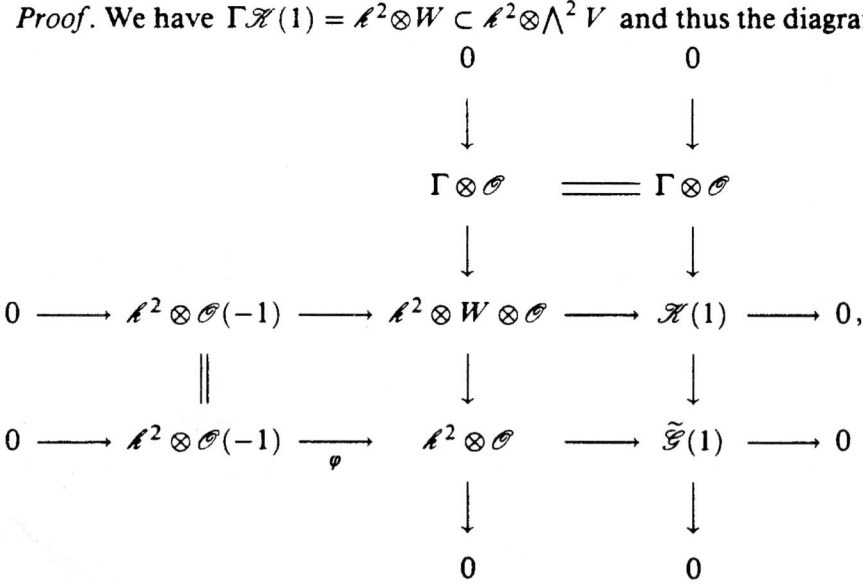

in which $\varphi$ becomes the matrix $\left(\begin{smallmatrix} z_3 & -z_2 \\ -z_1 & z_3 \end{smallmatrix}\right)$ by the entries of $\Gamma^*$ and the canonical resolution of $\mathscr{K}$ as $\ell^2 \otimes Z$. But such a $\varphi$ is exactly the resolution of an ideal $\mathscr{J}_{l,W}$ with $l = \{z_2 = z_3 = 0\}$, say. It follows that $\tilde{\mathscr{G}}/m(e_0)\tilde{\mathscr{G}}$ has dimension 2 and hence is isomorphic to $\mathscr{T}$. Therefore $\mathscr{G} \simeq m(e_0)\tilde{\mathscr{G}}$.

The different situations of $C$ or $C^\vee$ (we also identify $C$ with a conic in the plane $P \simeq \mathbb{P}W$) can now be interpreted by the structure of the bidual sheaf $\mathscr{F}^{\vee\vee}$. The situation is similar to that in 9.1 except that all sheaves are singular in the vertex and for elementary transformations we have to consider lines on the cone.

9.3.1. PROPOSITION. *Let $\mathscr{F}$ correspond to a Poncelet pair $(S, C^\vee)$ with $S$ a regular $\alpha$-conic. Then for $0 \to \mathscr{F} \to \mathscr{F}^{\vee\vee} \to \mathscr{R} \to 0$, we have,*

(i) *$\mathscr{F}^{\vee\vee}$ is locally free outside the vertex $e_0$ and $c_1\mathscr{F}^{\vee\vee} = 0$.*

(ii) *$\mathscr{F}^{\vee\vee} = i_*\pi^*(\mathscr{F}^{\vee\vee}|P)$, where $\mathbb{P}_3 \xleftarrow{i} \mathbb{P}_3 \setminus \{e_0\} \xrightarrow{\pi} P$, and thus $\mathscr{F}^{\vee\vee}$ is determined by the 2-bundle $\mathscr{F}^{\vee\vee}|P$ with Chern classes $c_1 = 0$, $0 \le c_2 \le 2$.*

(iii)

$$c_2\mathscr{F}^{\vee\vee} = \begin{cases} 2 & \text{if } C \text{ is regular}, \\ 1 & \text{if } C \text{ is a tangent and a secant}, \\ 0 & \text{if } C \text{ is a pair of tangents.} \end{cases}$$

(iv) *$\mathscr{F}$ is stable in each of the different cases of $C$, which will be described below.*

*Proof.* (a) The sheaf $\mathscr{Z}$ is also the kernel of $(V/e_0)^\vee \otimes \mathscr{O} \to m(e_0)(1)$ and $\mathscr{Z}|P = \Omega^1_P(1)$. It follows immediately that

$$i_*\pi^*(\mathscr{Z}|P) = \mathscr{Z}.$$

Moreover, the homomorphism

$$\mathscr{Z}^\vee \to \mathscr{O}(1) = i_*\pi^*(\Omega^1_P(2) \xrightarrow{a} \mathscr{O}_P(1))$$

can be described by

$$\begin{array}{ccc} (V/e_0) \otimes \mathscr{O} & \xrightarrow{e_0 \wedge a} e_0 \wedge \bigwedge^2 V \otimes \mathscr{O} \subset & \bigwedge^3 V \otimes \mathscr{O} \\ \downarrow & & \downarrow \\ \mathscr{Z} & \xrightarrow{\qquad\qquad} & \mathscr{O}(1) \end{array},$$

and the dual of any homomorphism $\Omega^3(3) \xrightarrow[e_0\wedge a]{} \mathscr{Z} \subset \Omega^1(1)$ is of this form.

(b) Let now $\widetilde{\mathscr{F}}$ denote the cokernel of

$$0 \to k^2 \otimes \Omega^3(3) \xrightarrow[M^*]{} k^2 \otimes \mathscr{Z} \to \widetilde{\mathscr{F}} \to 0$$
$$\cap$$
$$k^2 \otimes \Omega^1(1),$$

where as before $M^*$ represents $M$. Then from $0 \to \mathscr{F} \to \widetilde{\mathscr{F}} \to \mathscr{T} \to 0$ we get $\widetilde{\widetilde{\mathscr{F}}}^\vee = \mathscr{F}^\vee$. In our case moreover $M^* = e_0 \wedge A$ for some $k^2 \xrightarrow[A]{} k^2 \otimes U$ and thus we have the diagram

$$\begin{array}{ccccccc} & & & k^2 \otimes (V/e_0) \otimes \mathscr{O} & \xrightarrow{e_0\wedge A} & k^2 \otimes \bigwedge^3 V \otimes \mathscr{O} \\ (26) & & & \downarrow & & \downarrow & , \\ 0 \longrightarrow & \mathscr{F}^\vee & \longrightarrow & k^2 \otimes \mathscr{Z}^\vee & \longrightarrow & k^2 \otimes \mathscr{O}(1) \end{array}$$

which, after restriction to $P$, gives the exact sequence

$$(26_P) \qquad 0 \to \mathscr{F}^\vee|P \to k^2 \otimes \Omega^1_P(2) \xrightarrow{A'} k^2 \otimes \mathscr{O}_P(1) \to \Phi \to 0.$$

(c) This proves the proposition: Since $\mathscr{F}^\vee|P$ is reflexive on $P$, it is locally free. We must have

$$i_*\pi^*(\mathscr{F}^\vee|P) = \mathscr{F}^\vee$$

by (a) and diagram (26). Hence $\mathscr{F}^\vee$ is locally free on $\mathbb{P}_3\setminus\{e_0\}$. Taking the dual of this identity yields (i), (ii). From $(26_P)$ we get $c_1(\mathscr{F}^\vee|P) = 0$, $c_2(\mathscr{F}^\vee|P) = 2 - h^0\Phi$, where we note that $\Phi$ must have 0-dimensional support. This proves (iii), since it is shown below that $h^0\Phi = 0, 1, 2$ in the different cases of $C$.

We are going now to describe $\mathscr{F}^{\vee\vee} \to \mathscr{R}$ in the different cases of $C$. Note first that the conic $C^\vee \subset \mathbb{P}U^\vee \simeq PW^\vee$ has the equation $\det A = 0$, where as above $M^* = e_0 \wedge A$ with entries in $U$. This follows from our convention $\bigwedge^2 V \simeq \bigwedge^2 V^\vee$ and $\bigwedge^2 W \simeq W^\vee$, $W = e_0 \wedge U$.

*Case* 1: in which $C^\vee$ is regular. In this case the entries of $M^*$ or $A$ span the space $W$ or $U$ and the sheaf $\mathscr{M}$ in display (22) must be locally free except at $e_0$. Since $\operatorname{Supp}\mathscr{E} = \{e_0\}$, it follows from the same display that also $\mathscr{F}$ is locally free on $\mathbb{P}_3 \backslash \{e_0\}$. Moreover, if we consider ($26_P$) in this case, we see by the form of the matrix $A$ that $\Phi = 0$ and hence $\mathscr{F}^\vee | P$ and $\mathscr{F} | P$ are bundles with Chern-classes $c_1 = 0$, $c_2 = 2$. Its jumping lines are exactly the points of $C^\vee$ as can be calculated from its representing matrix $A$. Since $C^\vee$ is the polar dual of $C$ w.r.t. $S$, the jumping lines are the polars of points of $C$ w.r.t. $S$.

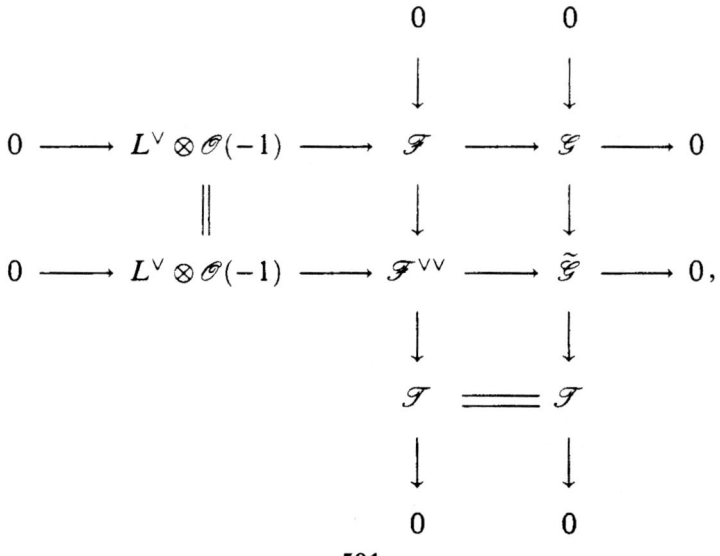

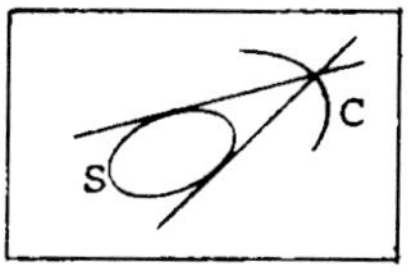


There is a unique subspace $L^\vee \subset \Gamma(\mathscr{F}|P)(1)$ s.t.

$$0 \to L^\vee \otimes \mathscr{O}_P(1) \to \mathscr{F}|P \to \mathscr{O}_S(-1) \to 0, \tag{27}$$

the cokernel of the evaluating homomorphism is $\mathscr{O}_S(-1)$, and this sequence is nothing but the restriction of the sequence

$$0 \to (\Gamma/M) \otimes \mathscr{O}(-1) \to \mathscr{F} \to \mathscr{G} \to 0.$$

If we start with (27) and apply $i_* \pi^*$ we obtain the diagram

$$\begin{array}{ccccccccc}
 & & & & 0 & & 0 & & \\
 & & & & \downarrow & & \downarrow & & \\
0 & \longrightarrow & L^\vee \otimes \mathscr{O}(-1) & \longrightarrow & \mathscr{F} & \longrightarrow & \mathscr{G} & \longrightarrow & 0 \\
 & & \| & & \downarrow & & \downarrow & & \\
0 & \longrightarrow & L^\vee \otimes \mathscr{O}(-1) & \longrightarrow & \mathscr{F}^{\vee\vee} & \longrightarrow & \tilde{\mathscr{G}} & \longrightarrow & 0, \\
 & & & & \downarrow & & \downarrow & & \\
 & & & & \mathscr{T} & = & \mathscr{T} & & \\
 & & & & \downarrow & & \downarrow & & \\
 & & & & 0 & & 0 & &
\end{array}$$

in which $\mathscr{F}$ is the pullback of $\mathscr{G} \subset \tilde{\mathscr{G}}$. Therefore $\mathscr{R} = \mathscr{T}$ in this case.

REMARK. If we pursue the point of view of elementary transformations $\mathscr{F}^{\vee\vee} \xrightarrow[h]{} \mathscr{T}$, the sheaves $\mathscr{F}^{\vee\vee}$ and $\mathscr{T}$ can be defined by $C \subset P$ and $e_0$, and then the epimorphism $h$ corresponds to a conic $S$ which is regular and in Poncelet relation with $C$ or $C^\vee$.

*Stability.* In Case 1 the sheaf $\mathscr{F}^{\vee\vee}$ has no non-zero section and thus $\mu$-stable, hence also $\mathscr{F}$ is $\mu$-stable.

*Cases* 2/3: in which $C$ is a pair of lines, a tangent and a secant of $S$. Let $L$ be the line joining the tangent point with the vertex and $K$ be the line joining the pole of the secant with the vertex. In this case the sheaf $\mathscr{F}$ can be described as follows:

(a) $\mathscr{R}$ is the structure sheaf

$$\mathscr{R} = \mathscr{O}_Q/\mathscr{G} = \mathscr{O}_Q/m(e_0)\mathscr{I}_{L,Q}$$

of the line $L$ with a multiple point in $e_0$ and we have the exact sequence

$$0 \to \mathscr{T} \to \mathscr{R} \to \mathscr{O}_L \to 0.$$

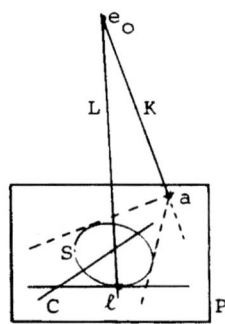

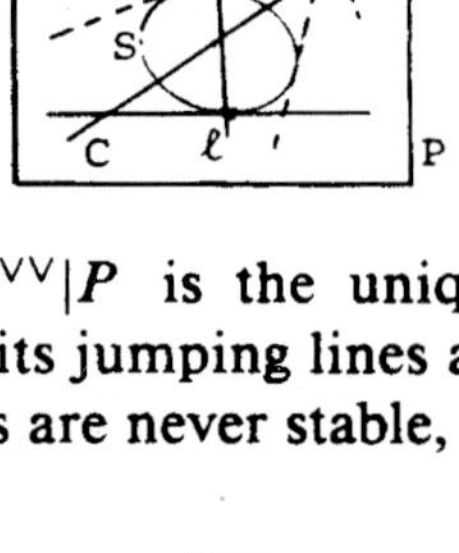


(b) The restriction $\mathscr{F}^{\vee\vee}|P$ is the unique 2-bundle on $P$ with $c_2(\mathscr{F}^{\vee\vee}|P) = 1$ such that its jumping lines are the lines in $P$ through the pole $a$. (Such bundles are never stable, since $h^0(\mathscr{F}^{\vee\vee}|P) = 1$, see **[Ba]**.)

(c) There is a unique subspace $L^\vee \subset \Gamma(\mathscr{F}^{\vee\vee}(1)|P)$ s.t. the evaluation map yields the sequence

$$0 \to L^\vee \otimes \mathscr{O}_P(-1) \to \mathscr{F}^{\vee\vee}|P \to \mathscr{O}_S \to 0.$$

Pulling this up via $i_* \pi^*$ we get the pullback diagram

$$\begin{array}{ccccccccc}
 & & & & 0 & & 0 & & \\
 & & & & \downarrow & & \downarrow & & \\
0 & \longrightarrow & L^\vee \otimes \mathscr{O}(-1) & \longrightarrow & \mathscr{F} & \longrightarrow & \mathscr{G} & \longrightarrow & \mathscr{O} \\
 & & \| & & \downarrow & & \downarrow & & \\
0 & \longrightarrow & L^\vee \otimes \mathscr{O}(-1) & \longrightarrow & \mathscr{F}^{\vee\vee} & \longrightarrow & \mathscr{O}_Q & \longrightarrow & 0. \\
 & & & & \downarrow & & \downarrow & & \\
 & & & & \mathscr{R} & = & \mathscr{R} & & \\
 & & & & \downarrow & & \downarrow & & \\
 & & & & 0 & & 0 & &
\end{array}$$

(d) $\mathscr{F}$ is *stable* (although $\mathscr{F}^{\vee\vee}$ is not semi-stable).

*Proof*. We first investigate the sheaf $\widetilde{\mathscr{F}}$ which was introduced as the cokernel of $M \otimes \Omega^3(3) \to \mathscr{k}^2 \otimes \mathscr{Z}$. Since now $M^* = e_0 \wedge A$ with (up to equivalence)

$$A = \begin{bmatrix} l & 0 \\ b & a \end{bmatrix},$$

and since we get the exact sequences

$$0 \to \Omega^3(3) \xrightarrow[e_0 \wedge a]{} \mathscr{Z} \to \mathscr{I}_K \to 0,$$
$$0 \to \Omega^3(3) \xrightarrow[e_0 \wedge l]{} \mathscr{Z} \to \mathscr{I}_L \to 0,$$

the sheaf $\widetilde{\mathscr{F}}$ must be an extension of the kind

$$0 \to \mathscr{I}_L \to \widetilde{\mathscr{F}} \to \mathscr{I}_K \to 0.$$

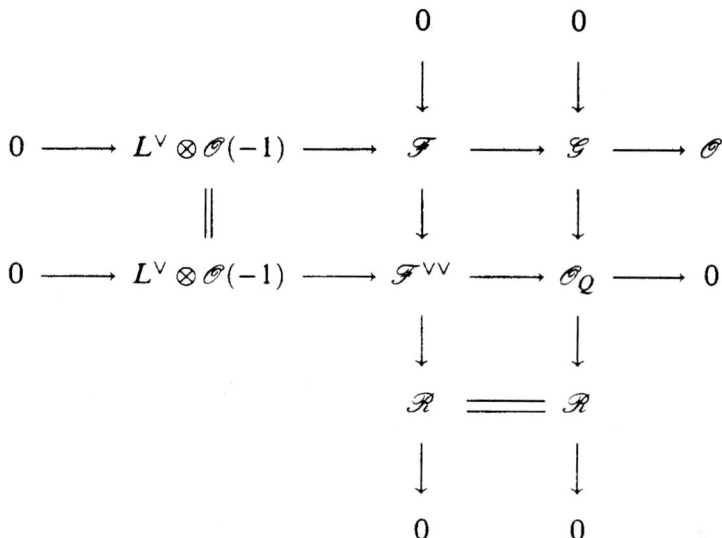

Taking the bidual gives us a diagram

$$
\begin{array}{ccccccccc}
 & & 0 & & 0 & & & & \\
 & & \downarrow & & \downarrow & & & & \\
0 & \longrightarrow & I_L & \longrightarrow & \widetilde{\mathscr{F}} & \longrightarrow & \mathscr{J}_K & \longrightarrow & 0 \\
 & & \downarrow & & \downarrow & & \| & & \\
0 & \longrightarrow & \mathscr{O} & \longrightarrow & \mathscr{F}^{\vee\vee} & \longrightarrow & \mathscr{J}_K & \longrightarrow & 0, \\
 & & \downarrow & & \downarrow & & & & \\
 & & \mathscr{O}_L & = & \mathscr{O}_L & & & & \\
 & & \downarrow & & \downarrow & & & & \\
 & & 0 & & 0 & & & &
\end{array}
\tag{28}
$$

as can be easily checked. Restricting the evaluating sequence $(\Gamma/M)\otimes\mathscr{O}(1)\to\mathscr{F}\to\mathscr{G}$ to $P$ we obtain the diagram

$$
\begin{array}{ccccccccc}
 & & & & 0 & & 0 & & \\
 & & & & \downarrow & & \downarrow & & \\
0 & \longrightarrow & (\Gamma/M)\otimes\mathscr{O}_P(-1) & \longrightarrow & \mathscr{F}|P & \longrightarrow & \mathscr{G}|P & \longrightarrow & 0 \\
 & & \| & & \downarrow & & \downarrow & & \\
0 & \longrightarrow & (\Gamma/M)\otimes\mathscr{O}_P(-1) & \longrightarrow & \mathscr{F}^{\vee\vee}|P & \longrightarrow & \mathscr{L} & \longrightarrow & 0, \\
 & & & & \downarrow & & \downarrow & & \\
 & & & & \mathscr{O}_L\otimes\mathscr{O}_P & = & \mathscr{O}_L\otimes\mathscr{O}_P & & \\
 & & & & \downarrow & & \downarrow & & \\
 & & & & 0 & & 0 & &
\end{array}
\tag{29}
$$

where $\mathscr{L}$ is defined as cokernel. Now $\mathscr{L}$ is supported on $S = Q\cap P$ and an $\mathscr{O}_S$-module, since $\mathscr{G}|P = \mathscr{O}_S(-1)$ and $\mathscr{O}_L\otimes\mathscr{O}_P = k_l$. Since by the middle row its depth is $=1$, it is a line bundle on $S$. But $h^0\mathscr{L} = 1$, indeed $h^0(\mathscr{F}^{\vee\vee}|P) = 1$. Therefore $\mathscr{L} = \mathscr{O}_S$. Now the proposition can be derived:

It is clear that $c_1(\mathscr{F}^{\vee\vee}|P) = 0$, $c_2(\mathscr{F}^{\vee\vee}|P) = 1$ by (29) and that $h^0(\mathscr{F}^{\vee\vee}|P) = 0$. The jumping lines are exactly those through $a$, which follows from $0 \to \mathcal{O} \to \mathcal{F}^{\vee\vee} \to \mathcal{I}_K \to 0$ by restricting to $P$ and by investigating the result of $\otimes\mathcal{O}_{L'}$ for a line $L' \subset P$. Finally (c) follows by pulling back the middle row of (29), which also gives the definition of $\mathcal{R}$. Since $\mathcal{O}_L = \mathcal{O}_Q/\widetilde{\mathcal{G}}$ we get (a).

To prove the stability of $\mathcal{F}$ we remark that for any nonzero section of $\mathcal{F}^{\vee\vee}$ the composed homomorphism must be onto $\mathcal{R}$:

$$\begin{array}{ccc}
\mathcal{O} & \longrightarrow & \mathcal{O}_Q \\
\downarrow & \searrow & \downarrow \\
\mathcal{F}^{\vee\vee} & \longrightarrow & \mathcal{R}
\end{array}.$$

If now $\mathcal{F}' \subset \mathcal{F}$ is a rank-1 subsheaf with $\mathcal{F}/\mathcal{F}' = \mathcal{F}''$ torsionfree, we can assume that $c_1\mathcal{F}' = 0$. Then we get a diagram

$$\begin{array}{ccccccccc}
 & & 0 & & 0 & & 0 & & \\
 & & \downarrow & & \downarrow & & \downarrow & & \\
0 & \longrightarrow & \mathcal{F}' & \longrightarrow & \mathcal{O} & \longrightarrow & \mathcal{O}_{Z'} & \longrightarrow & 0 \\
 & & \downarrow & & \downarrow & \overset{\psi}{\searrow} & \downarrow & & \\
0 & \longrightarrow & \mathcal{F} & \longrightarrow & \mathcal{F}^{\vee\vee} & \longrightarrow & \mathcal{R} & \longrightarrow & 0, \\
 & & \downarrow & & \downarrow & & \downarrow & & \\
0 & \longrightarrow & \mathcal{F}'' & \underset{\varphi''}{\longrightarrow} & \mathcal{I}_K & \longrightarrow & \mathcal{C}'' & \longrightarrow & 0 \\
 & & \downarrow & & \downarrow & & \downarrow & & \\
 & & 0 & & 0 & & 0 & &
\end{array}$$

which is exact, because $\varphi''$ is nonzero and injective, and because $\mathcal{F}''$ is torsionfree. Since $\psi$ must be surjective, we conclude that $\mathcal{C}'' = 0$ and $\mathcal{F}'' = \mathcal{I}_K$.

Now

$$\chi\mathcal{F}'(m) = \chi\mathcal{O}(m) - \chi\mathcal{R}(m) = \chi\mathcal{O}(m) - \chi\mathcal{O}_L(m) - 2,$$
$$\chi\mathcal{F}''(m) = \chi\mathcal{O}(m) - \chi\mathcal{O}_K(m).$$

This shows that $\chi\mathcal{F}'(m) < \frac{1}{2}\chi\mathcal{F}(m)$.

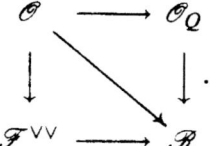

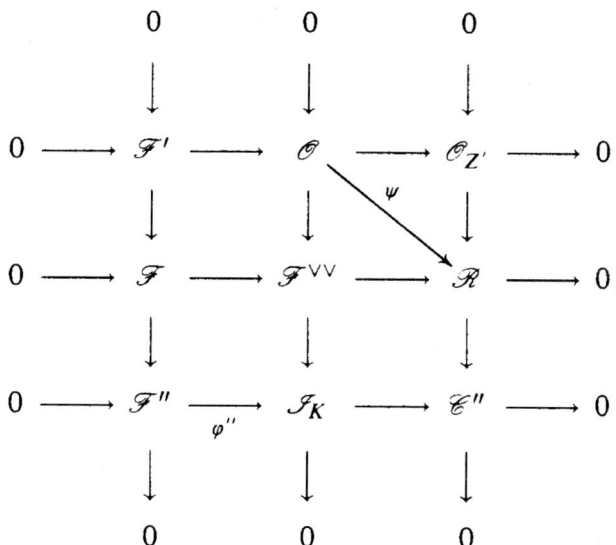

*Case* 4: in which $C$ is a pair of tangents. In this case the bidual $\mathscr{F}^{\vee\vee} = 2\mathscr{O}$, too, and we have the diagram

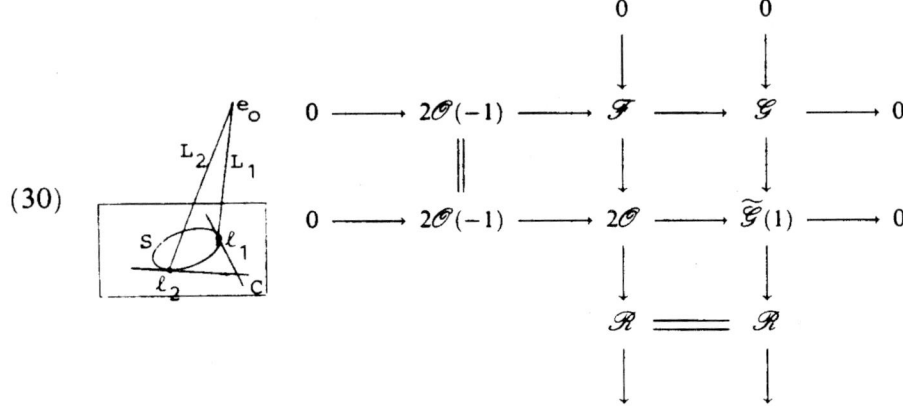

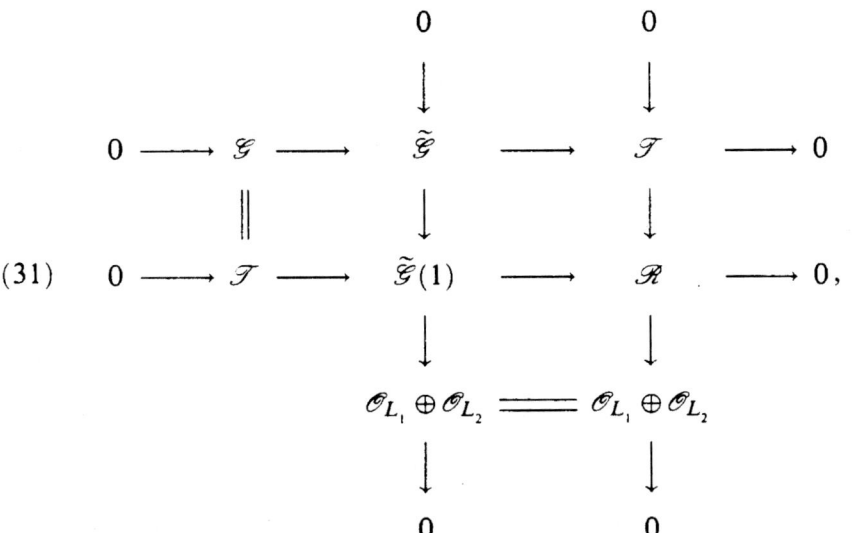

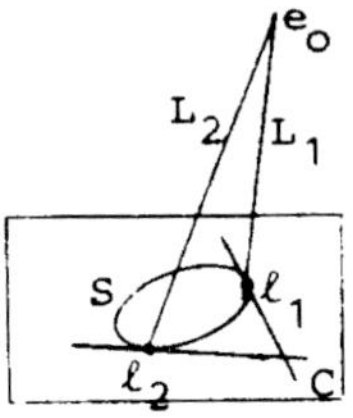


$$
(30)\qquad
\begin{array}{ccccccccc}
 & & & & 0 & & 0 & & \\
 & & & & \downarrow & & \downarrow & & \\
0 & \longrightarrow & 2\mathscr{O}(-1) & \longrightarrow & \mathscr{F} & \longrightarrow & \mathscr{G} & \longrightarrow & 0 \\
 & & \| & & \downarrow & & \downarrow & & \\
0 & \longrightarrow & 2\mathscr{O}(-1) & \longrightarrow & 2\mathscr{O} & \longrightarrow & \widetilde{\mathscr{G}}(1) & \longrightarrow & 0 \\
 & & & & \downarrow & & \downarrow & & \\
 & & & & \mathscr{R} & = & \mathscr{R} & & \\
 & & & & \downarrow & & \downarrow & &
\end{array}
$$

in which $\mathscr{R}$ is defined by the right-hand column, and thus determined by the cone. It fits into the diagram

$$
(31)\qquad
\begin{array}{ccccccccc}
 & & & & 0 & & 0 & & \\
 & & & & \downarrow & & \downarrow & & \\
0 & \longrightarrow & \mathscr{G} & \longrightarrow & \widetilde{\mathscr{G}} & \longrightarrow & \mathscr{T} & \longrightarrow & 0 \\
 & & \| & & \downarrow & & \downarrow & & \\
0 & \longrightarrow & \mathscr{T} & \longrightarrow & \widetilde{\mathscr{G}}(1) & \longrightarrow & \mathscr{R} & \longrightarrow & 0, \\
 & & & & \downarrow & & \downarrow & & \\
 & & & & \mathscr{O}_{L_1} \oplus \mathscr{O}_{L_2} & = & \mathscr{O}_{L_1} \oplus \mathscr{O}_{L_2} & & \\
 & & & & \downarrow & & \downarrow & & \\
 & & & & 0 & & 0 & &
\end{array}
$$

which is derived in the proof. Also in this case $\mathscr{F}$ is *stable.*

*Proof*. Because $C$ consists of two tangents we can choose the basis of $U$ so that $l_1 = e_1$, $l_2 = e_2$. By the shape of the matrix $\Gamma^*$ above we find that the only possibility of $M^* = e_0 \wedge A$ is the direct sum

$$A = \begin{bmatrix} e_1 & 0 \\ 0 & e_2 \end{bmatrix}.$$

It follows from the previous proof that $\widetilde{\mathscr{F}} = \mathscr{I}_{L_1} \oplus \mathscr{I}_{L_2}$ and therefore $\mathscr{F}^{\vee\vee} = \widetilde{\mathscr{F}}^{\vee\vee} = 2\mathscr{O}$. Furthermore the diagram (29) now becomes

$$
\begin{array}{ccccccccc}
 & & & & 0 & & 0 & & \\
 & & & & \downarrow & & \downarrow & & \\
0 & \longrightarrow & 2\mathscr{O}_P(-1) & \longrightarrow & \mathscr{F}|P & \longrightarrow & \mathscr{G}|P & \longrightarrow & 0 \\
 & & \| & & \downarrow & & \downarrow & & \\
0 & \longrightarrow & 2\mathscr{O}_P(-1) & \longrightarrow & 2\mathscr{O}_P & \longrightarrow & \mathscr{L} & \longrightarrow & 0, \\
 & & & & \downarrow & & \downarrow & & \\
 & & & & k_{l_1} \oplus k_{l_2} & = & k_{l_1} \oplus k_{l_2} & & \\
 & & & & \downarrow & & \downarrow & & \\
 & & & & 0 & & 0 & &
\end{array} \tag{29'}
$$

and we conclude that $\mathscr{L} = \mathscr{O}_S(1)$ as in Cases 2/3. Since $\mathscr{G}|P = \mathscr{O}_S(-1)$, the right-hand column becomes the top row of the next diagram

$$
\begin{array}{ccccccccc}
 & & 0 & & 0 & & 0 & & \\
 & & \uparrow & & \uparrow & & \uparrow & & \\
0 & \longrightarrow & \mathscr{O}_S(-1) & \xrightarrow[z_3|S]{} & \mathscr{O}_S(1) & \longrightarrow & k_{e_1} \oplus k_{e_2} & \longrightarrow & 0 \\
 & & \uparrow & & \uparrow & & \uparrow & & \\
0 & \longrightarrow & 2\mathscr{O}_P(-1) & \xrightarrow[z_3]{} & 2\mathscr{O}_P & \longrightarrow & 2\mathscr{O}_{\wp} & \longrightarrow & 0, \\
 & & \uparrow{\scriptstyle \varphi} & & \uparrow{\scriptstyle \varphi} & & \uparrow{\scriptstyle \varphi|\wp} & & \\
0 & \longrightarrow & 2\mathscr{O}_P(-2) & \xrightarrow[z_3]{} & 2\mathscr{O}_P(-1) & \longrightarrow & 2\mathscr{O}_{\wp}(-1) & \longrightarrow & 0 \\
 & & \uparrow & & \uparrow & & \uparrow & & \\
 & & 0 & & 0 & & 0 & &
\end{array} \tag{32}
$$

where $z_3$ is the equation of the line $g \subset P$ through $e_1$, $e_2$. The homomorphism $\varphi$ can be given (up to equivalence) by

$$\varphi = \begin{bmatrix} z_3 & -z_2 \\ -z_1 & z_3 \end{bmatrix},$$

s.t. $\det \varphi = z_3^2 - z_1 z_2$ is the equation of the conic $S$ (see definition of $\tilde{\mathscr{G}}$). From this we see that

$$\varphi|g = \begin{bmatrix} 0 & -z_2 \\ -z_1 & 0 \end{bmatrix}.$$

If we apply $i_* \pi^*$ to the last diagram (32) we obtain

$$\begin{array}{ccccccccc}
 & & 0 & & 0 & & 0 & & \\
 & & \uparrow & & \uparrow & & \uparrow & & \\
0 & \longrightarrow & \tilde{\mathscr{G}} & \longrightarrow & \tilde{\mathscr{G}}(1) & \longrightarrow & \mathscr{O}_{L_1} \oplus \mathscr{O}_{L_2} & \longrightarrow & 0 \\
 & & \uparrow & & \uparrow & & \uparrow & & \\
0 & \longrightarrow & 2\mathscr{O}(-1) & \xrightarrow[z_3]{} & 2\mathscr{O} & \longrightarrow & 2\mathscr{O}_E & \longrightarrow & 0, \\
 & & \uparrow \varphi & & \uparrow \varphi & & \uparrow \varphi|_E & & \\
0 & \longrightarrow & 2\mathscr{O}(-2) & \xrightarrow[z_3]{} & 2\mathscr{O}(-1) & \longrightarrow & 2\mathscr{O}_E(-1) & \longrightarrow & 0 \\
 & & \uparrow & & \uparrow & & \uparrow & & \\
 & & 0 & & 0 & & 0 & &
\end{array} \tag{33}$$

where now $E$ is the plane $z_3 = 0$, spanned by the two lines $L_1 \cup L_2 = Q \cap E$. The top row of (33) gives us the diagram (31) with the definition of $\mathscr{R}$. Diagram (30) follows from diagram (29′) by pulling back via $i_* \pi^*$ again, which first gives the corresponding diagram with $\widetilde{\mathscr{F}} = i_* \pi^*(\mathscr{F}|P)$ and $\tilde{\mathscr{G}}$, and then imbedding the sequence $0 \to 2\mathscr{O}(-1) \to \mathscr{F} \to \mathscr{G} \to 0$ into its first row.

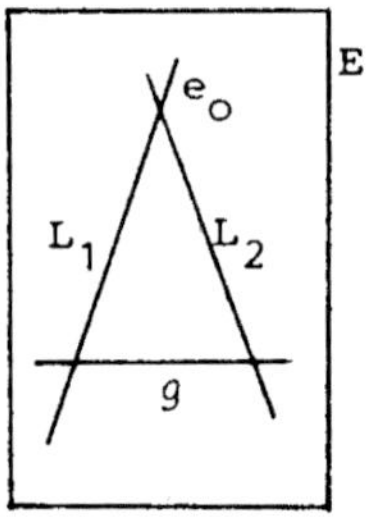

The proof of stability of $\mathscr{F}$ is reduced to that of Cases 2 and 3 as follows: We have the two diagrams

$$\begin{array}{ccccccccc}
 & & & & 0 & & 0 & & \\
 & & & & \downarrow & & \downarrow & & \\
0 & \longrightarrow & \tilde{\mathscr{G}} & \longrightarrow & \mathscr{O}_Q & \longrightarrow & \mathscr{O}_{L_1} & \longrightarrow & 0 \\
 & & \| & & \downarrow & & \downarrow & & \\
0 & \longrightarrow & \tilde{\mathscr{G}} & \longrightarrow & \tilde{\mathscr{G}}(1) & \longrightarrow & \mathscr{O}_{L_1} \oplus \mathscr{O}_{L_2} & \longrightarrow & 0 \\
 & & & & \downarrow & & \downarrow & & \\
 & & & & \mathscr{O}_{L_2} & = & \mathscr{O}_{L_2} & & \\
 & & & & \downarrow & & \downarrow & & \\
 & & & & 0 & & 0 & &
\end{array}$$

$$\begin{array}{ccccccccc}
 & & & & 0 & & 0 & & \\
 & & & & \downarrow & & \downarrow & & \\
0 & \longrightarrow & \mathscr{G} & \longrightarrow & \mathscr{O}_Q & \longrightarrow & \mathscr{R}_1 & \longrightarrow & 0 \\
 & & \| & & \downarrow & & \downarrow & & \\
0 & \longrightarrow & \mathscr{G} & \longrightarrow & \tilde{\mathscr{G}}(1) & \longrightarrow & \mathscr{R} & \longrightarrow & 0 \\
 & & & & \downarrow & & \downarrow & & \\
 & & & & \mathscr{O}_{L_2} & = & \mathscr{O}_{L_2} & & \\
 & & & & \downarrow & & \downarrow & & \\
 & & & & 0 & & 0 & &
\end{array}$$

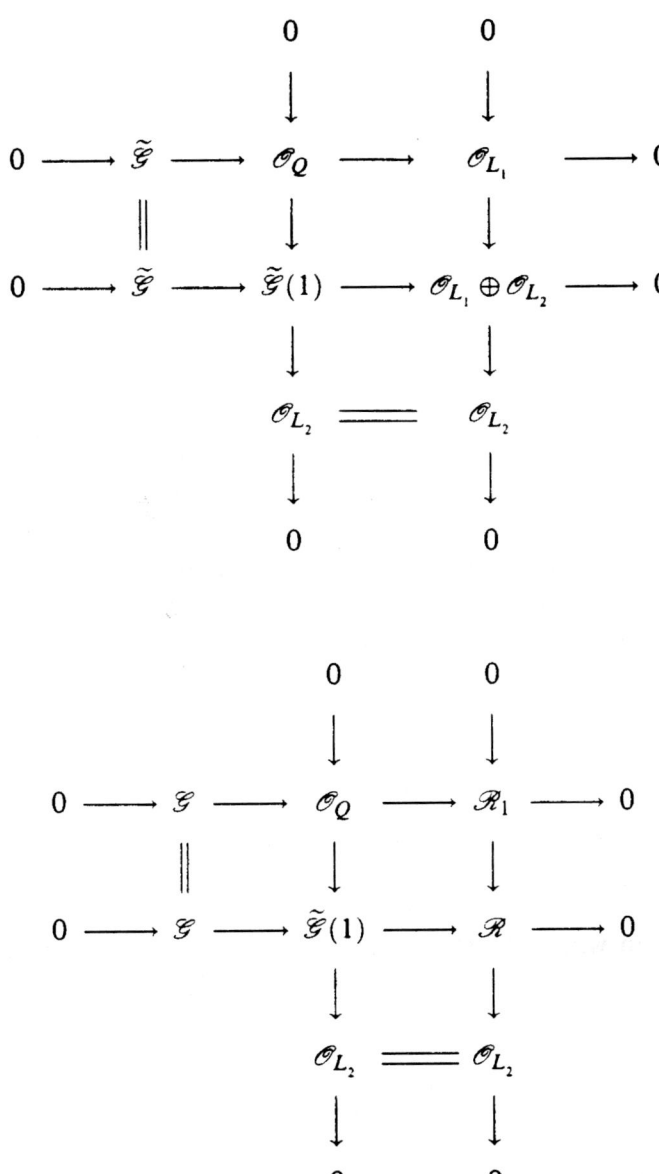

where we use in the first one, that $\mathscr{G} = \mathscr{I}_{L_1, Q}$, and where $\mathscr{R}_1$ in the second is the sheaf $\mathscr{R}$ of the Case 2/3 with $L = L_1$. From the right-hand column of the second we obtain the diagram

$$\begin{array}{ccccccccc}
 & & & & 0 & & 0 & & \\
 & & & & \downarrow & & \downarrow & & \\
0 & \longrightarrow & \mathscr{F} & \longrightarrow & \mathscr{F}_1 & \longrightarrow & \mathscr{R}_1 & \longrightarrow & 0 \\
 & & \| & & \downarrow & & \downarrow & & \\
0 & \longrightarrow & \mathscr{F} & \longrightarrow & 2\mathscr{O} & \longrightarrow & \mathscr{R} & \longrightarrow & 0 \\
 & & & & \downarrow & & \downarrow & & \\
 & & & & \mathscr{O}_{L_2} & = & \mathscr{O}_{L_2} & & \\
 & & & & \downarrow & & \downarrow & & \\
 & & & & 0 & & 0 & &
\end{array}$$

in which $\mathscr{F}_1$ is the pullback and must be isomorphic to $\mathscr{O} \oplus \mathscr{I}_{L_2}$. Now we can use the top row to proceed as in Cases 2/3 to prove stability of $\mathscr{F}$, because any non-zero section of $\mathscr{F}_1$ factorises through $\mathscr{O}_Q$:

$$\begin{array}{ccc}
\mathscr{O} & \longrightarrow & \mathscr{O}_Q \\
\downarrow & \searrow & \downarrow \\
\mathscr{F}_1 & \longrightarrow & \mathscr{R}_1
\end{array} .$$

*Case* 5: in which $C$ is a double tangent. This is a special case of

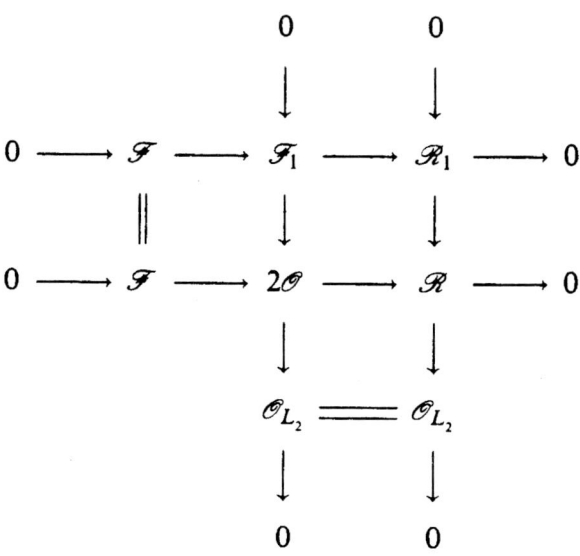

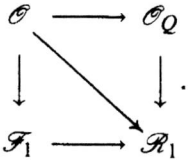

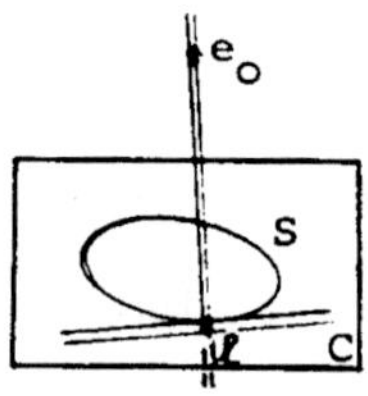


Case 4 and we obtain by the same method that $\mathscr{F}^{\vee\vee} = 2\mathscr{O}$ and the

diagrams

$$\begin{array}{ccccccccc}
 & & & & 0 & & 0 & & \\
 & & & & \downarrow & & \downarrow & & \\
0 & \longrightarrow & \mathcal{G} & \longrightarrow & \tilde{\mathcal{G}} & \longrightarrow & \mathcal{T} & \longrightarrow & 0 \\
 & & \| & & \downarrow & & \downarrow & & \\
0 & \longrightarrow & \mathcal{G} & \longrightarrow & \tilde{\mathcal{G}}(1) & \longrightarrow & \mathcal{R} & \longrightarrow & 0 \\
 & & & & \downarrow & & \downarrow & & \\
 & & & & \mathcal{O}_{\tilde{\mathcal{L}}} & = & \mathcal{O}_{\tilde{\mathcal{L}}} & & \\
 & & & & \downarrow & & \downarrow & & \\
 & & & & 0 & & 0 & &
\end{array}$$

$$\begin{array}{ccccccccc}
 & & & & 0 & & 0 & & \\
 & & & & \downarrow & & \downarrow & & \\
0 & \longrightarrow & 2\mathcal{O}(-1) & \longrightarrow & \mathcal{F} & \longrightarrow & \mathcal{G} & \longrightarrow & 0 \\
 & & \| & & \downarrow & & \downarrow & & \\
0 & \longrightarrow & 2\mathcal{O}(-1) & \longrightarrow & 2\mathcal{O} & \longrightarrow & \tilde{\mathcal{G}}(1) & \longrightarrow & 0, \\
 & & & & \downarrow & & \downarrow & & \\
 & & & & \mathcal{R} & = & \mathcal{R} & & \\
 & & & & \downarrow & & \downarrow & & \\
 & & & & 0 & & 0 & &
\end{array}$$

where now $\mathcal{O}_{\tilde{L}}$ denotes the double structure of $L$ in $Q$. Again by the same method the stability of $\mathcal{F}$ can be proved.

**10. Sheaves in the boundary with singular $S$.** If the conic $S$ is degenerate the sheaves $\mathcal{N}$ and $\mathcal{F}$ are of completely different nature. The sheaf $\mathcal{N}$ is always semi-stable and $S$ only determines the stable gradation of $\mathcal{N}$. We are going to describe this gradation first. As in

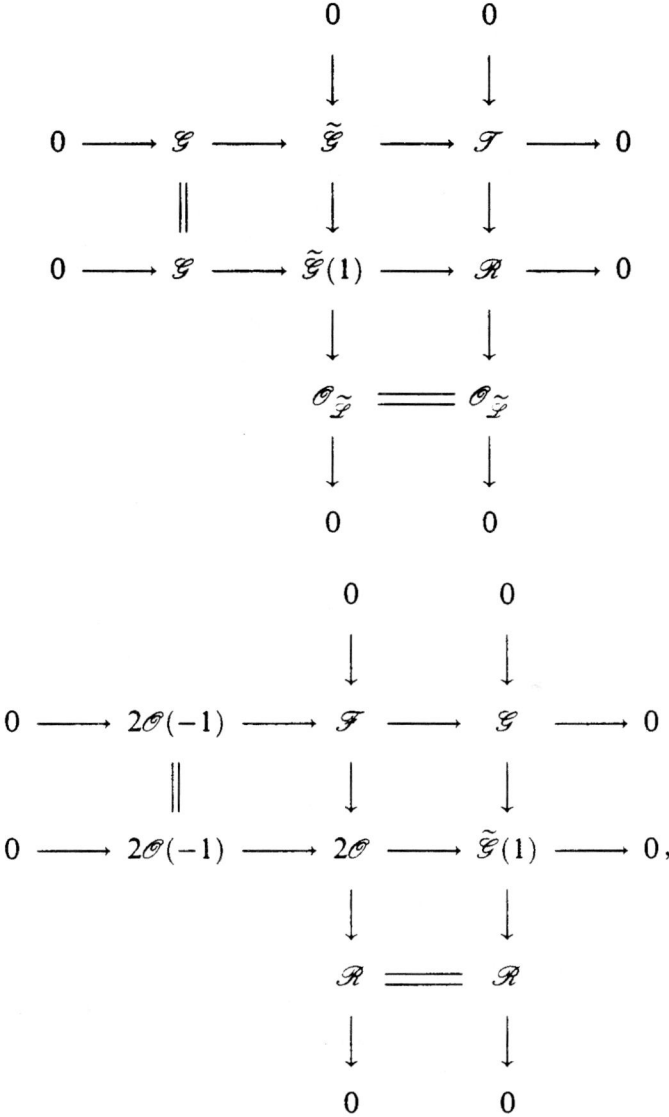

§9 we drop the indices of the sheaves and conics for a given point $x = (z, M, \Gamma) \in X^{ss}$, $y = (z, \Gamma)$.

If $\mathcal{F}$ is any coherent sheaf on $\mathbb{P}V$ we write as usual $\mathcal{P}(\mathcal{F})(d) = \chi\mathcal{F}(d)/\operatorname{rk}\mathcal{F}$. If $\mathcal{F}$ is semi-stable there is a filtration $0 = \mathcal{F}_0 \subset \mathcal{F}_1 \subset \cdots \subset \mathcal{F}_n = \mathcal{F}$ by coherent subsheaves, such that $\mathcal{F}_i/\mathcal{F}_{i-1}$ is stable for $1 \leq i \leq n$ and $\mathcal{P}(\mathcal{F}_i) = \mathcal{P}(\mathcal{F})$, **[Ma2]**. The direct sum $\operatorname{Gr}(\mathcal{F}) = \bigoplus \mathcal{F}_i/\mathcal{F}_{i-1}$ is unique up to isomorphisms and called the stable gradation. In order to describe the stable gradation of the $\mathcal{N}$'s for singular conic $S$, we consider the following rank-2 sheaves associated to planes in $\mathbb{P}_3$ together with an ordered pair of points in the plane.

10.1. Let $E \subset \mathbb{P}_3$ be a plane and $p, q \in E$. The sheaf $\mathcal{M}(p, E, q) = \mathcal{M}$ is defined by the exact diagram as follows:

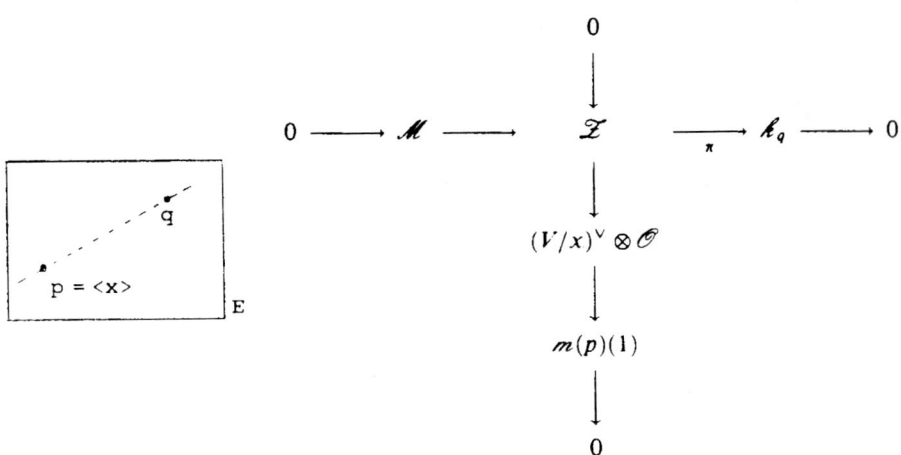

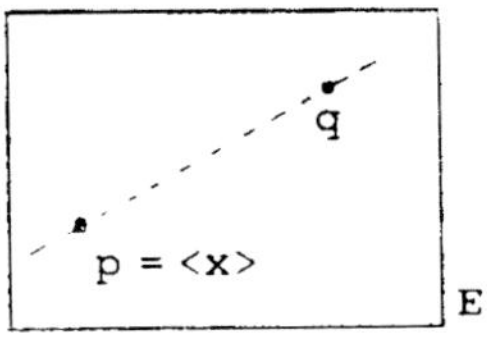


$$\begin{array}{ccccccccc}
 & & & & 0 & & & & \\
 & & & & \downarrow & & & & \\
0 & \longrightarrow & \mathcal{M} & \longrightarrow & \mathcal{Z} & \xrightarrow{\ \pi\ } & k_q & \longrightarrow & 0 \\
 & & & & \downarrow & & & & \\
 & & & & (V/x)^\vee \otimes \mathcal{O} & & & & \\
 & & & & \downarrow & & & & \\
 & & & & \mathfrak{m}(p)(1) & & & & \\
 & & & & \downarrow & & & & \\
 & & & & 0 & & & &
\end{array}$$

In this diagram $\mathfrak{m}(p)(1)$ is the ideal sheaf of $p$ in twist $1$, $\mathcal{Z}$ its first syzygy and $\pi$ the epimorphism defined by the plane $E$ by the

10.1.1. Lemma. *There is a* $1:1$ *correspondence between* $\mathbb{P}\operatorname{Hom}(\mathcal{Z}, k_q)$ *and the set of planes* $E$ *through* $p$, $q$ *(if* $p = q$ *the line* $\overline{p, q}$ *is replaced by a tangent direction in* $p$*).*

*Proof*. The Koszul resolution of $\mathcal{Z}$ and an epimorphism give rise to a diagram

$$\begin{array}{ccccccccc}
0 & \longrightarrow & \mathcal{O}(-2) & \xrightarrow{\ \alpha\ } & 3\mathcal{O}(-1) & \longrightarrow & \mathcal{Z} & \longrightarrow & 0 \\
 & & \downarrow & \searrow^{e} & \downarrow & & \downarrow & & \\
0 & \longrightarrow & \mathfrak{m}(q)(-1) & \longrightarrow & \mathcal{O}(-1) & \longrightarrow & k_q & \longrightarrow & 0
\end{array}$$

with $\alpha = (z_1, z_2, z_3)$ consisting of a basis of $(V/x)^\vee$, and with the linear form $e$ vanishing in $p = \langle x \rangle$ and $q$. Conversely any such $e$ factorises through $\alpha$ and $\mathcal{m}(q)(-1)$, thereby defining a non-zero homomorphism $\mathcal{Z} \to \mathcal{k}_q$.

10.1.2. **Proposition.** *The sheaf* $\mathcal{M}(p, E, q)$ *has the following properties*:

(i) $\operatorname{rk}\mathcal{M} = 2$ *and* $\mathcal{P}\mathcal{M}(d) = \frac{1}{12}(d+2)(d+3)(2d-1)$.

(ii) $\mathcal{M}$ *has Chern polynomial* $1 - h + h^2 - h^3$.

(iii) $\mathcal{M}$ *is* $\mu$*-stable.*

(iv) *The sections of* $\mathcal{M}(1)$ *are in one to one correspondence with the lines in* $E$ *through* $p$, *so that a line is the zero locus of the section.*

(v) $h^0\mathcal{M}(1) = 2$ *and* $\mathcal{M}$ *has the evaluation sequence*

$$0 \to 2\mathcal{O} \to \mathcal{M}(1) \to \mathcal{m}_E(q)(1) \to 0,$$

*where* $\mathcal{m}_E(q) \subset \mathcal{O}_E$ *denotes the ideal sheaf of* $q$ *in* $E$.

(vi) $h^1\mathcal{M}(d) = 0$ *for* $d \geq 1$.

*Proof*. (i) and (ii) follow directly from the defining diagram. Since $\mathcal{Z}$ is reflexive with $c_1\mathcal{Z} = -1$ and $h^0\mathcal{Z} = 0$, this sheaf is $\mu$-stable and then also $\mathcal{M}$. That $h^0\mathcal{M}(1) = 2$ and $h^1\mathcal{M}(d) = 0$ for $d \geq 1$ also follow easily from the definition and properties of $\mathcal{Z}$. To prove (v) we consider the diagram

$$\begin{array}{ccccccccc}
 & & & & 0 & & 0 & & \\
 & & & & \downarrow & & \downarrow & & \\
 & & 0 & \longrightarrow & 2\mathcal{O}(-1) & \longrightarrow & \mathcal{M} & \longrightarrow & \mathcal{C} \longrightarrow 0 \\
 & & \downarrow & & \downarrow & & \downarrow & & \\
0 & \longrightarrow & \mathcal{O}(-2) & \longrightarrow & 3\mathcal{O}(-1) & \longrightarrow & \mathcal{Z} & \longrightarrow & 0 \\
 & & {\scriptstyle e'}\downarrow & \overset{e}{\searrow} & \downarrow & & \downarrow & & \\
0 & \longrightarrow & \mathcal{m}(q)(-1) & \longrightarrow & \mathcal{O}(-1) & \longrightarrow & \mathcal{k}_q & \longrightarrow & 0 \\
 & & \downarrow & & \downarrow & & \downarrow & & \\
 & & \mathcal{m}_E(q)(-1) & & 0 & & 0 & & \\
 & & \downarrow & & & & & & \\
 & & 0 & & & & & &
\end{array}$$

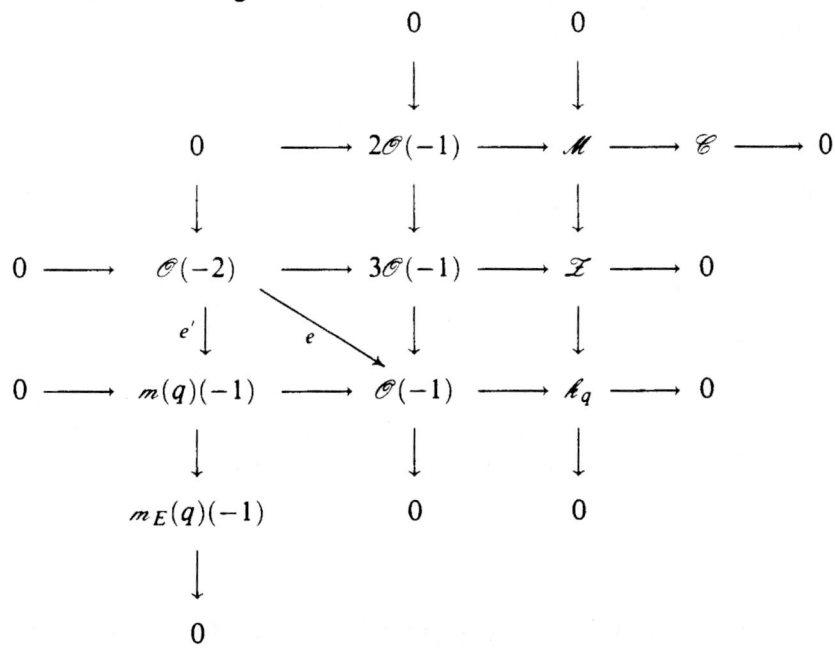

used already in the proof of the lemma. The cokernel of $e'$ is $m_E(q)(-1)$ and this is isomorphic to $\mathscr{C}$. From this we also see that any section of $\mathscr{M}(1)$ must have its zero locus in $E$. Finally to prove (iv) we note that any section of $\mathscr{Z}(1)$ vanishes exactly on a line $L \ni p$ and gives rise to a diagram

$$
\begin{array}{ccccccccc}
 & & & & 0 & & 0 & & \\
 & & & & \downarrow & & \downarrow & & \\
 & & & & \mathscr{O}(-1) & = & \mathscr{O}(-1) & & \\
 & & & & \downarrow & & \downarrow & & \\
0 & \longrightarrow & \mathscr{O}(-2) & \xrightarrow[\alpha]{} & 3\mathscr{O}(-1) & \longrightarrow & \mathscr{Z} & \longrightarrow & 0, \\
 & & \| & & \downarrow & & \downarrow & & \\
0 & \longrightarrow & \mathscr{O}(-2) & \xrightarrow[\beta]{} & 2\mathscr{O}(-1) & \longrightarrow & \mathscr{J}_L & \longrightarrow & 0 \\
 & & & & \downarrow & & \downarrow & & \\
 & & & & 0 & & 0 & &
\end{array}
$$

where $\beta$ consists of two independent linear forms with cokernel the ideal of the line they define. Therefore a section of $\mathscr{M}(1)$ must vanish on a line $L$ in $E$ through $p$, and gives rise to the diagram

$$
\begin{array}{ccccccccc}
 & & 0 & & 0 & & & & \\
 & & \downarrow & & \downarrow & & & & \\
 & & \mathscr{O}(-1) & = & \mathscr{O}(-1) & & & & \\
 & & \downarrow & & \downarrow & & & & \\
0 & \longrightarrow & \mathscr{M} & \longrightarrow & \mathscr{Z} & \longrightarrow & k_q & \longrightarrow & 0, \\
 & & \downarrow & & \downarrow & & \| & & \\
0 & \longrightarrow & \mathscr{J}_{L\cup q} & \longrightarrow & \mathscr{J}_L & \longrightarrow & k_q & \longrightarrow & 0 \\
 & & \downarrow & & \downarrow & & & & \\
 & & 0 & & 0 & & & &
\end{array}
$$

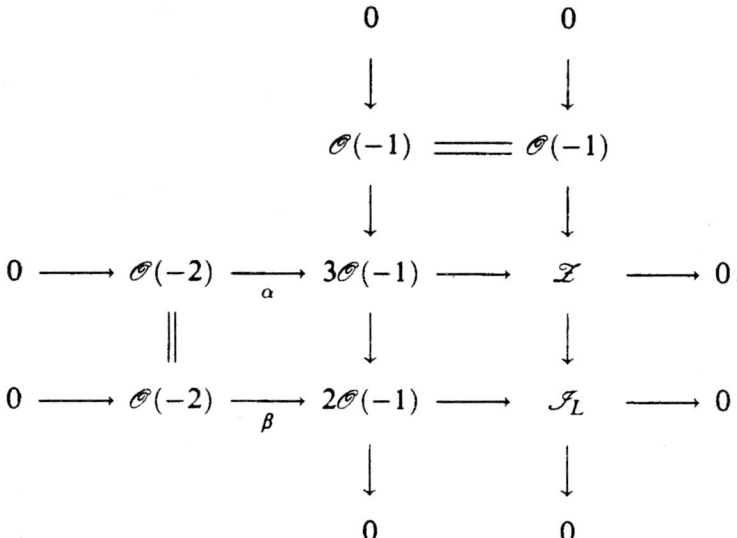

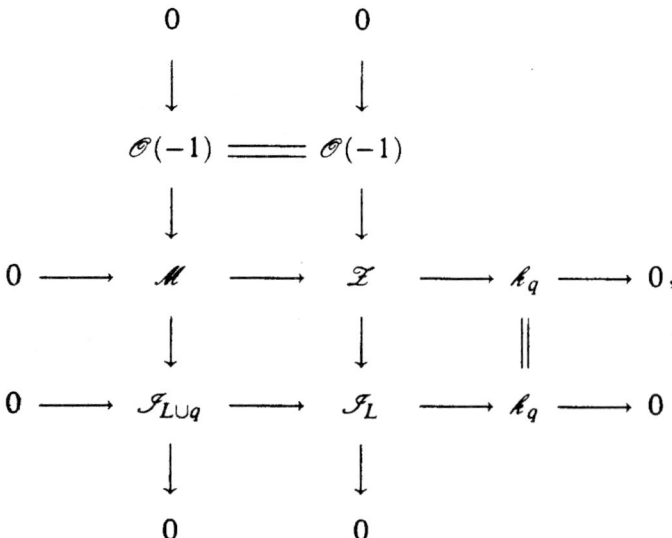

in which either $q \notin L$ or $\mathscr{J}_{L\cup q}$ is the ideal sheaf of $L$ with a multiple structure in $q$. Conversely given any such line, we can define the section by the last two diagrams.

10.1.3. **Corollary.** *For any line $L$ with $p \in L \subset E$ there is a diagram*

$$\begin{array}{ccccccccc}
 & & 0 & & 0 & & & & \\
 & & \downarrow & & \downarrow & & & & \\
 & 0 \longrightarrow & \mathcal{O}(-1) & \longrightarrow & 2\mathcal{O}(-1) & \longrightarrow & \mathcal{O}(-1) & \longrightarrow & 0 \\
 & & \downarrow & & \downarrow & & & & \\
 & & \mathcal{M} & = & \mathcal{M} & & & & \\
 & & \downarrow & & \downarrow & & & & \\
0 \longrightarrow \mathcal{O}(-1) & \longrightarrow & \mathcal{I}_{L\cup q} & \longrightarrow & m_E(q)(-1) & \longrightarrow & 0 & & \\
 & & \downarrow & & \downarrow & & & & \\
 & & 0 & & 0 & & & &
\end{array}$$

For later use we also need the

10.1.4. **Lemma.** *Any non-trivial extension of $m_E(q)(-1)$ by $\mathcal{O}(-1)$ is of the above form $\mathcal{I}_{L\cup q}$ for some line $L$ as above.*

Its proof can be derived from the equalities

$$\mathrm{Ext}^1_{\mathcal{O}}(m_E(q), \mathcal{O}) = \mathrm{Ext}^1_{\mathcal{O}}(\mathcal{O}_E, \mathcal{O}) = \Gamma\mathcal{O}_E(1)$$

and will be left to the reader.

10.2. *The sheaf $\mathcal{N}$ for singular conic $S$.* We consider first the generic case in which $S$ is the intersection of $G$ with a plane and consists of two different lines $e$, $f$. Then $S$ defines a regulus in $\mathbb{P}_3$ supported by two planes $E \cup F$.

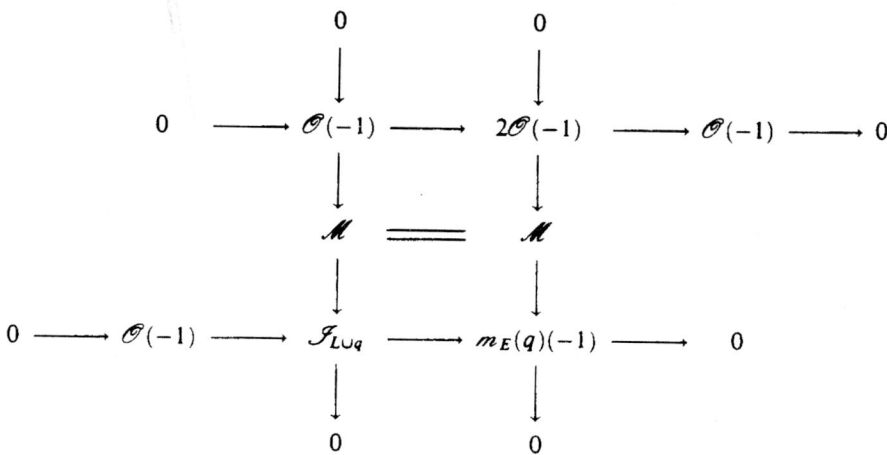

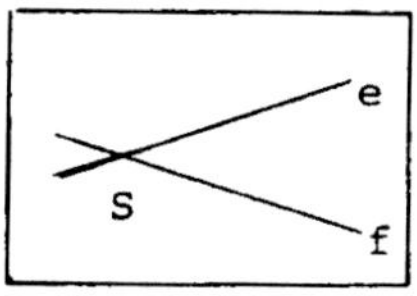


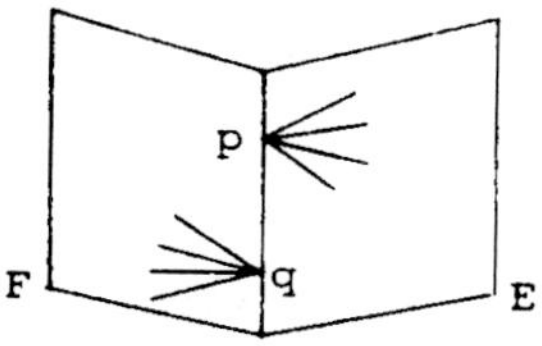

By the construction in 10.1 this configuration defines the sheaves

$$\mathcal{N}^e = \mathcal{M}(p, E, q) \quad \text{and} \quad \mathcal{N}^f = \mathcal{M}(q, F, p).$$

10.2.1. PROPOSITION. *Let* $\mathcal{N} = \mathcal{N}_y$ *be defined by* $y = (z, \Gamma)$ *with* $S = S(\Gamma)$ *as above. Then*

(i) $\mathcal{N}^e \oplus \mathcal{N}^f$ *is the stable gradation of* $\mathcal{N}$.

(ii) *Let* $\Gamma$ *be presented by one of the normal forms*

$$\text{(a)} \begin{pmatrix} \xi & 0 \\ \omega & 0 \\ \eta & \omega \\ 0 & \eta \end{pmatrix} \quad \text{(b)} \begin{pmatrix} \xi & \\ \omega & \\ & \omega \\ & \eta \end{pmatrix} \quad \text{(c)} \begin{pmatrix} \xi & \\ \omega & \xi \\ & \omega \\ & \eta \end{pmatrix} \quad \textit{see 5.3.}$$

*Then in these different cases* $\mathcal{N}$ *is an extension:*

(a) $0 \to \mathcal{N}^e \to \mathcal{N} \to \mathcal{N}^f \to 0$ (*non-trivial*),

(b) $\mathcal{N} = \mathcal{N}^e \oplus \mathcal{N}_f$,

(c) $0 \to \mathcal{N}^f \to \mathcal{N} \to \mathcal{N}^e \to 0$ (*non-trivial*).

(iii) *The different cases of* $\mathcal{N}$ *are distinguished by the singularities on* $\mathcal{N}$:

(a) $\operatorname{Sing}\mathcal{N} = \{p\}$, (b) $\operatorname{Sing}\mathcal{N} = \{p, q\}$, (c) $\operatorname{Sing}\mathcal{N} = \{q\}$.

10.2.2. REMARK. At the first glimpse it is a surprise that each of the sheaves $\mathcal{N}^e$, $\mathcal{N}^f$ has two singular points whereas $\mathcal{N}$ has only one in cases (a) and (c), but this is in accordance with the depths of the sheaves in these points.

*Proof*. By 5.3, $\Gamma$ has only three normal forms in (ii). Let $N$ be the space associated to $\Gamma$ by 5.8. Then $N^\vee$ is presented by a matrix of the type

$$\begin{pmatrix} x & 0 \\ y & y' \end{pmatrix}, \quad \begin{pmatrix} x & 0 \\ 0 & y' \end{pmatrix}, \quad \begin{pmatrix} x & x' \\ 0 & y' \end{pmatrix}$$

in the three different cases respectively, where $p = \langle x \rangle$, $q = \langle y' \rangle$. In the direct sum case (b), we then have

$$\mathcal{C} = k_p \oplus k_q = \mathcal{T},$$

and the display diagram (14) gives us $\mathcal{K} = \mathcal{Z}^p \oplus \mathcal{Z}^q$, where $\mathcal{Z}^p$

denotes the syzygy of $\mathscr{m}(p)(1)$. In the situation (a) we get the diagram

$$\begin{array}{ccccccccc}
 & & 0 & & 0 & & & & \\
 & & \downarrow & & \downarrow & & & & \\
0 & \longrightarrow & \mathcal{N}^e & \longrightarrow & \mathcal{N} & \longrightarrow & \mathcal{Z}^q & \longrightarrow & \mathscr{k}_p & \longrightarrow & 0 \\
 & & \downarrow & & \downarrow & & \| & & \| & & \\
0 & \longrightarrow & \mathcal{Z}^p & \longrightarrow & \mathcal{K} & \longrightarrow & \mathcal{Z}^q & \longrightarrow & k\mathscr{k}_p & \longrightarrow & 0 \\
 & & \downarrow & & \downarrow & & & & \\
 & & \mathscr{k}_q & = & \mathcal{T} & & & & \\
 & & \downarrow & & \downarrow & & & & \\
 & & 0 & & 0 & & & &
\end{array}$$

in which the middle row follows from the shape of $\left(\begin{smallmatrix} x & 0 \\ x & y' \end{smallmatrix}\right)$ and a corresponding extension diagram. Here $\mathcal{C} = \mathscr{k}_q$, too. In case (c) we get the analogous diagram with $p$, $q$ interchanged, and in the direct sum case (b) the diagram specialises to

$$\begin{array}{ccccccccc}
 & & 0 & & 0 & & 0 & & \\
 & & \downarrow & & \downarrow & & \downarrow & & \\
0 & \longrightarrow & \mathcal{N}^e & \longrightarrow & \mathcal{N} & \longrightarrow & \mathcal{N}^f & \longrightarrow & 0 \\
 & & \downarrow & & \downarrow & & \downarrow & & \\
0 & \longrightarrow & \mathcal{Z}^p & \longrightarrow & \mathcal{K} & \longrightarrow & \mathcal{Z}^q & \longrightarrow & 0. \\
 & & \downarrow & & \downarrow & & \downarrow & & \\
0 & \longrightarrow & \mathscr{k}_q & \longrightarrow & \mathcal{T} & \longrightarrow & \mathscr{k}_p & \longrightarrow & 0 \\
 & & \downarrow & & \downarrow & & \downarrow & & \\
 & & 0 & & 0 & & 0 & &
\end{array}$$

From these diagrams we easily derive (i), (ii), and also (iii) by looking at the local cohomology groups $H_{\{p\}}\mathcal{N}$, $H_{\{q\}}\mathcal{N}$ etc.

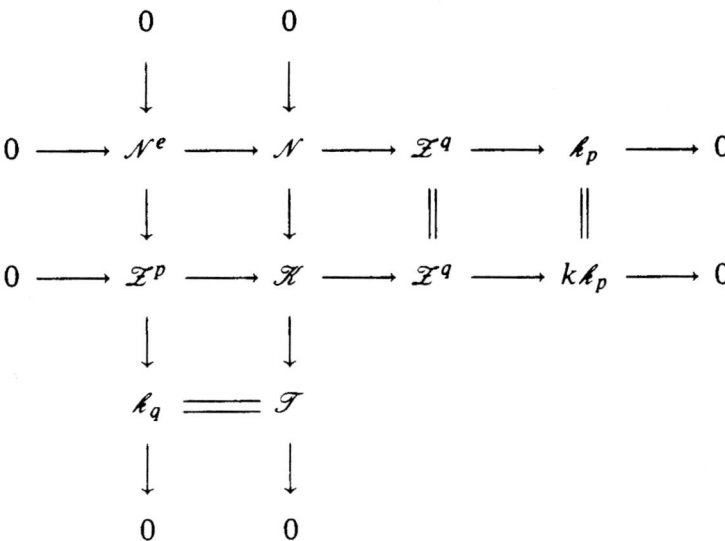

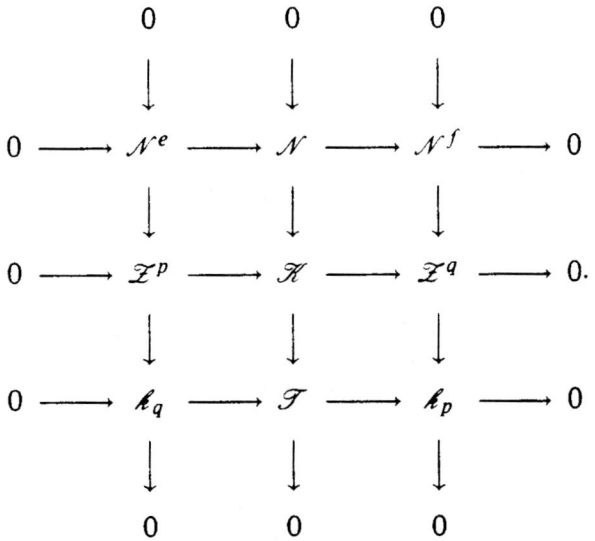

From the extensions in (ii) we can determine also the sheaf $\mathcal{G}$ by the diagram (in case (a) for example)

$$\begin{array}{ccccccccc}
 & & 0 & & 0 & & 0 & & \\
 & & \downarrow & & \downarrow & & \downarrow & & \\
0 & \longrightarrow & \Gamma' \oplus \Omega^3(3) & \longrightarrow & \Gamma \oplus \Omega^3(3) & \longrightarrow & \Gamma'' \oplus \Omega^3(3) & \longrightarrow & 0 \\
 & & \downarrow & & \downarrow & & \downarrow & & \\
0 & \longrightarrow & \mathcal{N}^e & \longrightarrow & \mathcal{N} & \longrightarrow & \mathcal{N}^f & \longrightarrow & 0, \\
 & & \downarrow & & \downarrow & & \downarrow & & \\
0 & \longrightarrow & m_E(q)(-1) & \longrightarrow & \mathcal{G} & \longrightarrow & m_F(p)(-1) & \longrightarrow & 0 \\
 & & \downarrow & & \downarrow & & \downarrow & & \\
 & & 0 & & 0 & & 0 & &
\end{array}$$

where $\Gamma'$ and $\Gamma''$ are defined by the shape of $\Gamma$, by which we have an exact sequence

$$\begin{array}{ccccccccc}
0 & \to & \Gamma' & \to & \Gamma & \longrightarrow & \Gamma'' & \to & 0. \\
 & & \cap & & \cap & & \cap & & \\
0 & \to & W_z & \to & k^2 \oplus W_z & \to & W_z & \to & 0.
\end{array}$$

10.3. LEMMA. *Let* $x = (z, M, \Gamma) \in X^{ss}$ *be a semi-stable point with* $\Gamma$ *as in* (a) *or* (b). *With the previous notation the following are equivalent*:

(i) $x$ *is stable,*

(ii) $M \cap \Gamma' = 0$,

(iii) *the pair* $(S, C^\vee(M))$ *is not singular, i.e.* $C^\vee(M)$ *is not a pair of lines passing through the two points* $e, f \in \mathbb{P}W^\vee$ *of* $S$, *see* 3.12, (i).

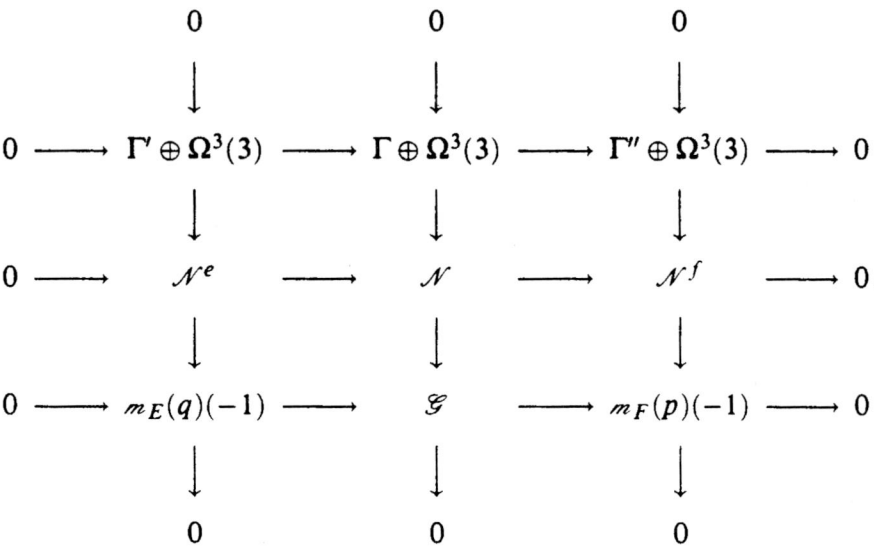

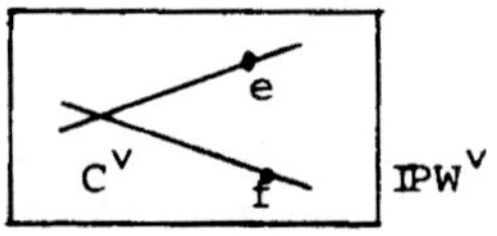

*Proof*. For example in case (a) we have: $M \cap \Gamma' \neq 0$ if and only if $M$ is presented by a matrix of the form

$$\begin{pmatrix} a\xi + b\omega & 0 \\ * & c\omega + d\eta \end{pmatrix},$$

and such $M$ are exactly those which define singular conics $C^\vee$ passing through both points $e$, $f$. The cases (b), (c) are proved similarly.

10.4. PROPOSITION. *Let* $x = (z, M, \Gamma)$ *be a* stable *point with* $\Gamma$ *of type* (a) *or* (b). *Then the sheaf* $\mathcal{F} = \mathcal{F}_x$ *is* stable *and fits into a diagram*

$$\begin{array}{ccccccccc}
 & & & & 0 & & 0 & & \\
 & & & & \downarrow & & \downarrow & & \\
 & & & & M \otimes \Omega^3(3) & \cong & 2\Omega^3(3) & & \\
 & & & & \downarrow & & \downarrow & & \\
0 & \longrightarrow & \mathcal{N}^e & \longrightarrow & \mathcal{N} & \longrightarrow & \mathcal{N}^f & \longrightarrow & 0 \\
 & & \| & & \downarrow & & \downarrow & & \\
0 & \longrightarrow & \mathcal{N}^e & \longrightarrow & \mathcal{F} & \longrightarrow & \mathfrak{m}_F(p)(-1) & \longrightarrow & 0 \\
 & & & & \downarrow & & \downarrow & & \\
 & & & & 0 & & 0 & &
\end{array}$$

*A similar statement holds in case* (c) *with* $e$, $f$ *interchanged.*

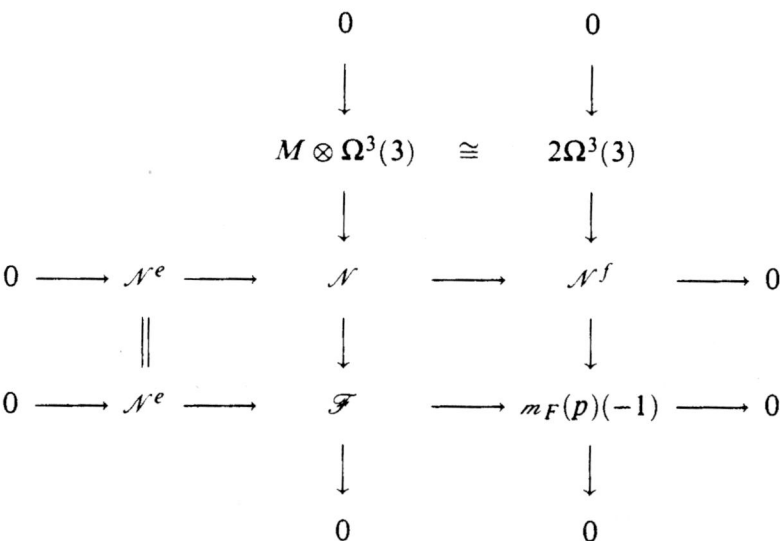

*Proof*. The diagram follows immediately from 10.3 because $M \cap \Gamma\mathcal{N}^e(1) = 0$, and from (v) of the proposition in 10.1. To prove stability, let $\mathcal{F}' \subset \mathcal{F}$ be any rank 1 subsheaf with torsionfree quotient $\mathcal{F}'' = \mathcal{F}/\mathcal{F}'$. We can assume $c_1\mathcal{F}' = 0$, otherwise $P(\mathcal{F}')(d) < \frac{1}{2}P(\mathcal{F})(d)$ would be trivially satisfied. If $\mathcal{N}' = \mathcal{N}^e \cap \mathcal{F}'$, we obtain an exact diagram

$$\begin{array}{ccccccccc}
 & & 0 & & 0 & & 0 & & \\
 & & \downarrow & & \downarrow & & \downarrow & & \\
0 & \longrightarrow & \mathcal{N}' & \longrightarrow & \mathcal{F}' & \longrightarrow & \mathcal{C}' & \longrightarrow & 0 \\
 & & \downarrow & & \downarrow & & \downarrow & & \\
0 & \longrightarrow & \mathcal{N}^e & \longrightarrow & \mathcal{F} & \longrightarrow & m_F(p)(-1) & \longrightarrow & 0. \\
 & & \downarrow & & \downarrow & & \downarrow & & \\
0 & \longrightarrow & \mathcal{N}'' & \longrightarrow & \mathcal{F}'' & \longrightarrow & \mathcal{C}'' & \longrightarrow & 0 \\
 & & \downarrow & & \downarrow & & \downarrow & & \\
 & & 0 & & 0 & & 0 & &
\end{array}$$

By definition we have $\chi m_F(p)(d-1) = \binom{d+1}{2} - 1$ and $\chi\mathcal{C}'(d) = \binom{d+1}{2} - 1 - l(d)$ with $l(d) = \chi\mathcal{C}''(d)$. We have to consider

$$\Delta(d) = \frac{1}{2}\chi\mathcal{F}(d) - \chi\mathcal{F}'(d) = \binom{d+3}{3} - (d+2) - \chi\mathcal{N}'(d) - \chi\mathcal{C}'(d)$$
$$= \binom{d+2}{3} + l(d) - \chi\mathcal{N}'(d).$$

Since $\operatorname{rk}\mathcal{N}' = 1$ we have $\mathcal{N}'^{\vee\vee} = \mathcal{O}(c)$ and by the definition of $\mathcal{N}^e$ there is a diagram

$$\begin{array}{ccccccccc}
0 & \longrightarrow & \mathcal{N}' & \longrightarrow & \mathcal{O}(c) & \longrightarrow & \mathcal{L}_q & \longrightarrow & 0 \\
 & & \cap & & \cap & & \cap & & , \\
0 & \longrightarrow & \mathcal{N}^e & \longrightarrow & \mathcal{I} & \longrightarrow & k_q & \longrightarrow & 0
\end{array}$$

s.t. $\chi\mathcal{N}'(d) = \chi\mathcal{O}(d+c) - \varepsilon$ with $\varepsilon = 0, 1$.

Now $\mathcal{N}^e$ is $\mu$-stable with $c_1 = -1$ and therefore $c = c_1\mathcal{N}' \leq -1$. If $c \leq -2$, we get immediately

$$\Delta(d) = \binom{d+2}{3} - \binom{d+c+3}{3} + \varepsilon + l(d) > l(d) \geq 0 \quad \text{for } d \gg 0.$$

If however $c = -1$, we only have

$$\Delta(d) = \varepsilon + l(d) \geq 0 \quad \text{for } d \gg 0.$$

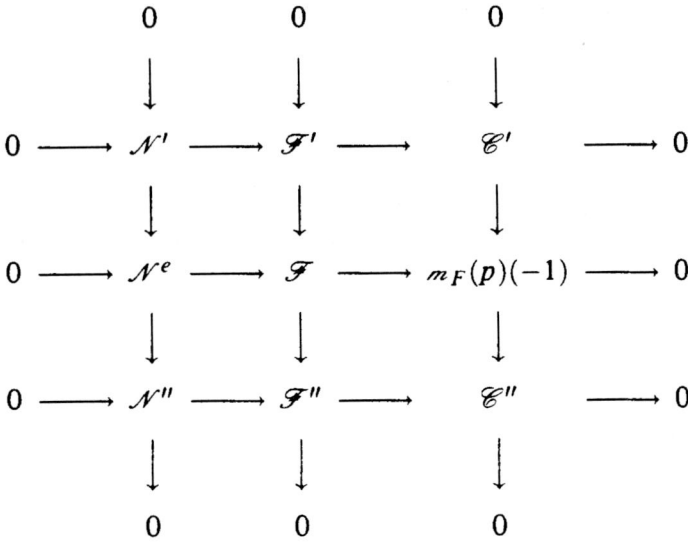

This proves that $\mathscr{F}$ is semi-stable, and that it is even stable if the case $\varepsilon = 0$, $\mathscr{C}'' = 0$ does not occur. But in this case $\mathscr{N}' = \mathscr{O}(-1)$, and by Corollary 10.1.3

$$\mathscr{F}'' = \mathscr{N}'' = \mathscr{I}_{L \cup q}$$

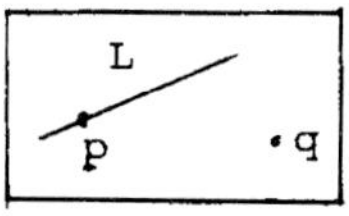


E

for some line $L$ in $E$ through $p$ and finally

$$\mathscr{F}' = \mathscr{I}_{K \cup p}$$

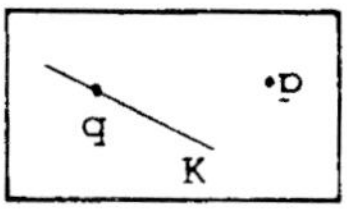


F

by Lemma 10.1.4 for some line $K$ in $F$ through $q$. Therefore we have proved that if $\mathscr{F}$ is not stable under the assumptions of the proposition, it must be an extension of the form

$$0 \to \mathscr{I}_{K \cup p} \to \mathscr{F} \to \mathscr{I}_{L \cup q} \to 0. \tag{34}$$

Now the proof will follow from Lemma 10.3 and from

10.4.1. LEMMA. *If $\mathscr{F}$ is an extension as in* (34) *then $C^\vee \subset \mathbb{P}W^\vee$ is a singular conic through $e$ and $f$.*

*Proof.* We use the incidence transform to show that $\operatorname{Supp} R^1\mathscr{F}(-1) = \mathbb{G} \cap \tilde{J}$ is given by a union $\tilde{J} = \tilde{H}_k \cup \tilde{H}_L$ of two hyperplanes in $\mathbb{P}\bigwedge^2 V$ and s.t. $\tilde{J} \cap \mathbb{P}W' = C^\vee$ passes through $e$ and $f$ (for notation see 8.4, (b)). This contradicts the assumption of the proposition by 10.3.

First we note that $R^0\mathscr{I}_{L\cup q}(-1) = 0$ since $R^0\mathscr{O}(-1) = 0$, and therefore we obtain the exact sequence

$$0 \to R^1\mathscr{I}_{K\cup p}(-1) \to R^1\mathscr{F}(-1) \to R^1\mathscr{I}_{L\cup q}(-1) \to 0.$$

Therefore it is enough to determine the supports of the ends of this sequence. Since we have the sequence

$$0 \to \mathscr{I}_{L\cup q} \to \mathscr{I}_L \to k_q \to 0,$$

we obtain the exact sequence

$$\begin{array}{ccccccccc} 0 & \longrightarrow & R^0 k_q & \longrightarrow & R^1\mathscr{I}_{L\cup q}(-1) & \longrightarrow & R^1\mathscr{I}_L(-1) & \longrightarrow & 0 \\ & & \| & & & & \| & & \\ & & \mathscr{O}_{P_q} & & & & R^0\mathscr{O}_L(-1) & & \end{array},$$

where $P_q$ is the $\alpha$-plane of $q$ in $\mathbb{G}$. Further from $0 \to \mathcal{O}_L(-1) \to \mathcal{O}_L \to k_a \to 0$ with some $a \in L$ we get

$$\begin{array}{ccccccccc} 0 & \longrightarrow & R^0\mathcal{O}_L(-1) & \longrightarrow & R^0\mathcal{O}_L & \longrightarrow & R^0 k_a & \longrightarrow & 0 \\ & & \| & & \| & & \| & & , \\ 0 & \longrightarrow & \mathcal{J}_{P_a, H_L} & \longrightarrow & \mathcal{O}_{H_L} & \longrightarrow & \mathcal{O}_{P_a} & \longrightarrow & 0 \end{array}$$

where $H_L \subset \mathbb{G}$ is the cone of all lines in $\mathbb{P}_3$ meeting $L$. It is a hyperplane section

$$H_L = \mathbb{G} \cap \widetilde{H}_L.$$

Therefore $\operatorname{Supp} R^1\mathcal{J}_L(-1) = H_L$ and hence

$$\operatorname{Supp} R^1\mathcal{J}_{L\cup q}(-1) = H_L \cup P_q.$$

Similarly

$$\operatorname{Supp} R^1\mathcal{J}_{K\cup p} = H_K \cup P_p.$$

But since $p \in L$ and $q \in K$ we have $P_q \subset H_K$ and $P_p \subset H_L$ and thus

$$\operatorname{Supp} R^1\mathcal{F}(-1) = H_L \cup H_K.$$

Finally it is easy to show that the unique hyperplanes $\widetilde{H}_L$, $\widetilde{H}_K$ intersecting $\mathbb{G}$ in $H_L$, $H_K$ pass through the points $e$ resp. $f$ if we choose an embedding of $\mathbb{P}W^\vee$ as in 8.4 (b). This proves the lemma.

Now we can prove the

10.5. THEOREM. *Let $x = (z, M, \Gamma) \in X^{ss}$ with $S = S(\Gamma)$ singular, and let $\mathcal{F} = \mathcal{F}_x$ be the corresponding sheaf. Then $\mathcal{F}$ is always semi-stable and the following conditions are equivalent:*

(i) *$(S, C^\vee)$ is a singular point of $Q$.*
(ii) *$x$ is not stable.*
(iii) *$\mathcal{F}$ is not stable.*
(iv) *$\mathcal{F}$ is an extension of the type*

$$0 \to \mathcal{J}_{K\cup p} \to \mathcal{F} \to \mathcal{J}_{L\cup q} \to 0,$$

*which also defines the stable filtration of $\mathcal{F}$.*

*Proof*. We restrict ourselves to the generic situation of a singular $S$, the proof in the other cases is the same. By the previous proof we obtain (iii) $\Rightarrow$ (ii). Since (iv) $\Rightarrow$ (iii) is obvious and (i) $\Leftrightarrow$ (ii) by 10.3, we only have to show that (ii) $\Rightarrow$ (iv). It is sufficient to consider only the case (a). By 10.3 we have $M \cap \Gamma\mathcal{N}^e(1) \neq 0$ if $x$ is not stable, and the dimension of this intersection cannot be 2. Therefore we obtain a diagram

$$\begin{array}{ccccccccc}
 & & 0 & & 0 & & 0 & & \\
 & & \downarrow & & \downarrow & & \downarrow & & \\
0 & \longrightarrow & M' \otimes \Omega^3(3) & \longrightarrow & M \otimes \Omega^3(3) & \longrightarrow & M'' \otimes \Omega^3(3) & \longrightarrow & 0 \\
 & & \downarrow & & \downarrow & & \downarrow & & \\
0 & \longrightarrow & \mathcal{N}^e & \longrightarrow & \mathcal{N} & \longrightarrow & \mathcal{N}^f & \longrightarrow & 0 \\
 & & \downarrow & & \downarrow & & \downarrow & & \\
0 & \longrightarrow & \mathcal{J}_{L\cup q} & \longrightarrow & \mathcal{F} & \longrightarrow & \mathcal{J}_{K\cup p} & \longrightarrow & 0 \\
 & & \downarrow & & \downarrow & & \downarrow & & \\
 & & 0 & & 0 & & 0 & &
\end{array}$$

where $M' = M \cap \Gamma\mathcal{N}^e(1)$, $M''$ is the image of $M$ under $\Gamma\mathcal{N}(1) \to \Gamma\mathcal{N}^f(1)$, and where the cokernels must be of the form $\mathcal{J}_{L\cup q}$ by 10.1.

The other cases of $(S, C^\vee)$ with singular $S$ are treated similarly; one only has to interpret $L\cup q$ in case $q \in L$ as a line $L$ with a double structure in the point $q$, the tangent plane of which determines the plane $E$, in which $L$ is contained.

10.6. Remark. If $S$ is a double line the non-stable pairs $(S, C^\vee)$ are given by two points $L, K \in S$ or two lines in a plane in $\mathbb{P}_3$.

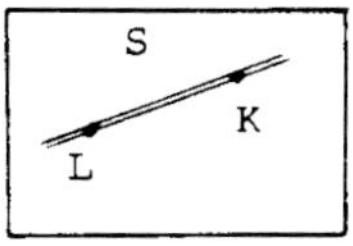


IPW

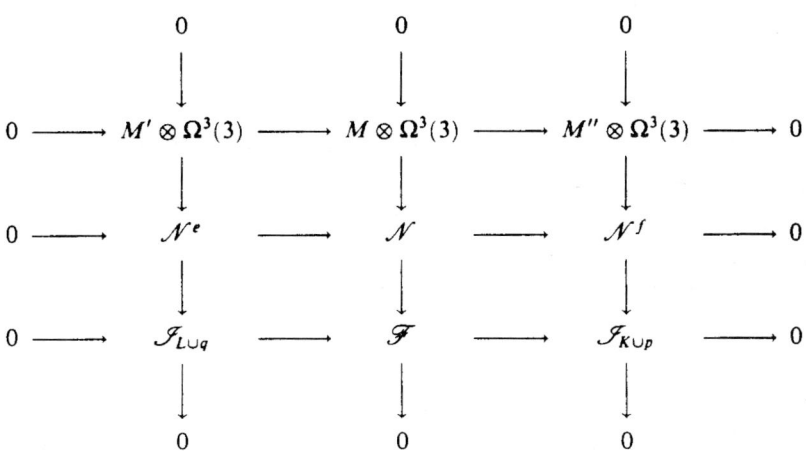

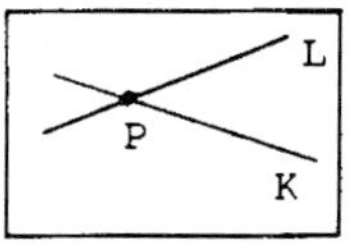


E = F

The corresponding sheaf is now an extension

$$0 \to \mathcal{J}_{L\cup p} \to \mathcal{F} \to \mathcal{J}_{K\cup p} \to 0.$$

The class $[\mathcal{F}] = [\mathcal{J}_{L\cup p} \oplus \mathcal{J}_{K\cup p}]$ cannot remember the plane $\mathbb{P}W$ belonging to the $\mathbb{P}_1$-fibration of $Q_{\text{exc}}$, which is blown down. If $L = K$ we obtain the most degenerate element $[\mathcal{J}_{L\cup p} \oplus \mathcal{J}_{L\cup p}]$ in $\overline{M(0, 2)}$.

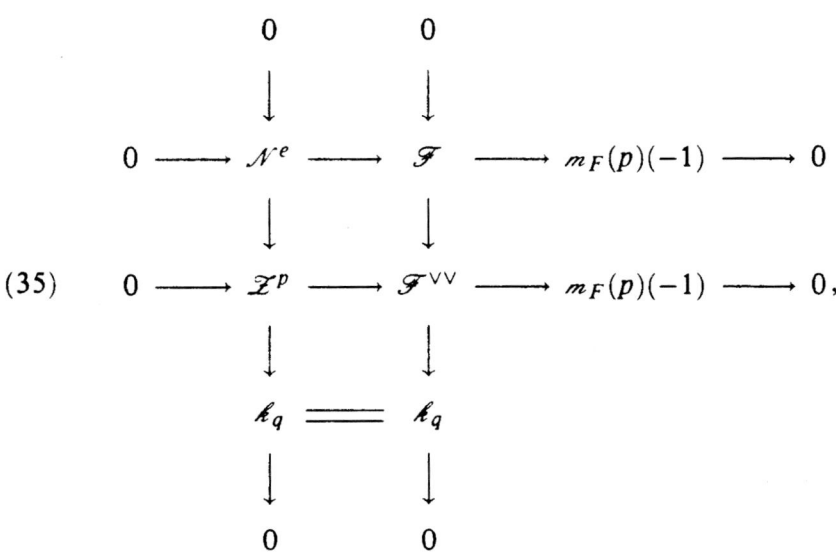

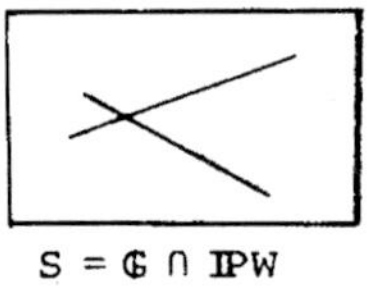


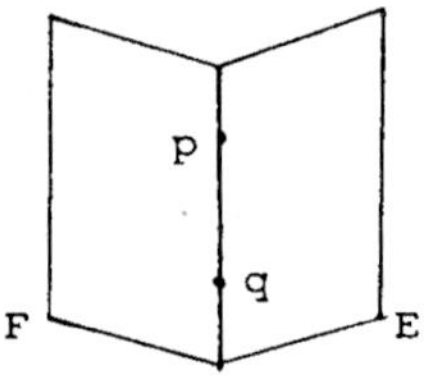


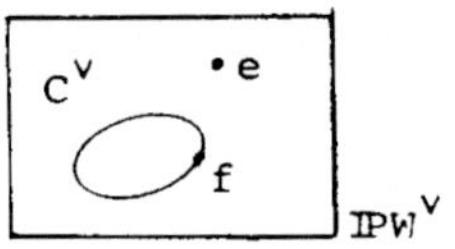


10.7. We close with some remarks on the embedding $\mathcal{F} \to \mathcal{F}^{\vee\vee}$ for a generic stable pair $(S, C^\vee) \in Q_0$.

We have shown in 10.4 that in such a case the sheaf $\mathcal{F}$ is stable and an extension of $m_F(p)(-1)$ by $\mathcal{N}^e$. Now we can show that for smooth $C^\vee$ through one of the points, say $f$, there is a diagram

$$\begin{array}{ccccccccc} & & 0 & & 0 & & & & \\ & & \downarrow & & \downarrow & & & & \\ 0 & \longrightarrow & \mathcal{N}^e & \longrightarrow & \mathcal{F} & \longrightarrow & m_F(p)(-1) & \longrightarrow & 0 \\ & & \downarrow & & \downarrow & & & & \\ 0 & \longrightarrow & \mathcal{Z}^p & \longrightarrow & \mathcal{F}^{\vee\vee} & \longrightarrow & m_F(p)(-1) & \longrightarrow & 0, \\ & & \downarrow & & \downarrow & & & & \\ & & k_q & = & k_q & & & & \\ & & \downarrow & & \downarrow & & & & \\ & & 0 & & 0 & & & & \end{array} \tag{35}$$

and that moreover $\mathcal{F}$ is locally free outside $q$. This implies that $c_3\mathcal{F}^{\vee\vee} = 2$, whereas $c_3\mathcal{F}^{\vee\vee} = 4$ in 9.2 for sheaves in $Q_\alpha$.

*Proof*. The first column is the definition of $\mathcal{N}^e$. Next we prove that $\mathcal{F}$ is locally free outside of $q$ in this case and that the cokernel of $\mathcal{F} \subset \mathcal{F}^{\vee\vee}$ is $k_q$. For this we consider the display diagram (22). One can show that $\mathcal{M}$ is locally free outside $q$ if $C^\vee$ is regular as in the picture above, and that $\mathcal{M}$ is reflexive. Moreover $\mathcal{C} = \mathcal{T} = k_q$ in that case, so that also $\mathcal{A}$ is locally free outside $q$ and therefore $\mathcal{F}$, too. Finally we have the diagram

$$
\begin{array}{ccccccccc}
 & & & & & & 0 & & \\
 & & & & & & \downarrow & & \\
 & & 0 & & & & \mathscr{k}_q & & \\
 & & \downarrow & & & & \downarrow & & \\
0 & \longrightarrow & \mathscr{F} & \longrightarrow & \mathscr{M} & \longrightarrow & \mathscr{A} & \longrightarrow & 0 \\
 & & \downarrow & & \| & & \downarrow & & \\
0 & \longrightarrow & \mathscr{F}^{\vee\vee} & \longrightarrow & \mathscr{M} & \longrightarrow & 2\mathscr{O} & & \\
 & & \downarrow & & & & & & \\
 & & \mathscr{R} & & & & & & \\
 & & \downarrow & & & & & & \\
 & & 0 & & & & & &
\end{array} \tag{36}
$$

from which it follows that $\mathscr{R} = \mathscr{k}_q$. Going back to (35) the induced homomorphism $\mathscr{k}_q \to \mathscr{k}_q$ must be nonzero and hence an isomorphism, for otherwise $\mathscr{k}_q$ would inject into $\mathscr{m}_F(p)(-1)$, which is not possible. This proves the diagram (35).

If in the previous example however $C^\vee$ is a smooth conic through both of the points $e$ and $f$, we see by the analogue of diagram (36) that now the kernel and cokernel of $\mathscr{A} \to 2\mathscr{O}$ is $k_p \oplus k_q$ and equals $\mathscr{R}$. Thus in this case $\mathscr{M}$ and $\mathscr{F}^{\vee\vee}$ are singular at $p$ and $q$ and $c_3\mathscr{F}^{\vee\vee} = 4$.

## References

[Au-Dou] H. Aupetit and A. Douady, *Fibrés stables de rang 2 sur $\mathbb{P}_3$ avec $c_1 = 0$, $c_2 = 2$*, Sem. E.N.S. 1977–78, Exp. X, Asterisque, (1980), 71–72.

[Ba] W. Barth, *Moduli of vector bundles on the projective plane*, Invent. Math., **42** (1977), 63–91.

[Ba-Hu] W. Barth and K. Hulek, *Monads and moduli of vector bundles*, Manuscripta Math., **25** (1978), 323–347.

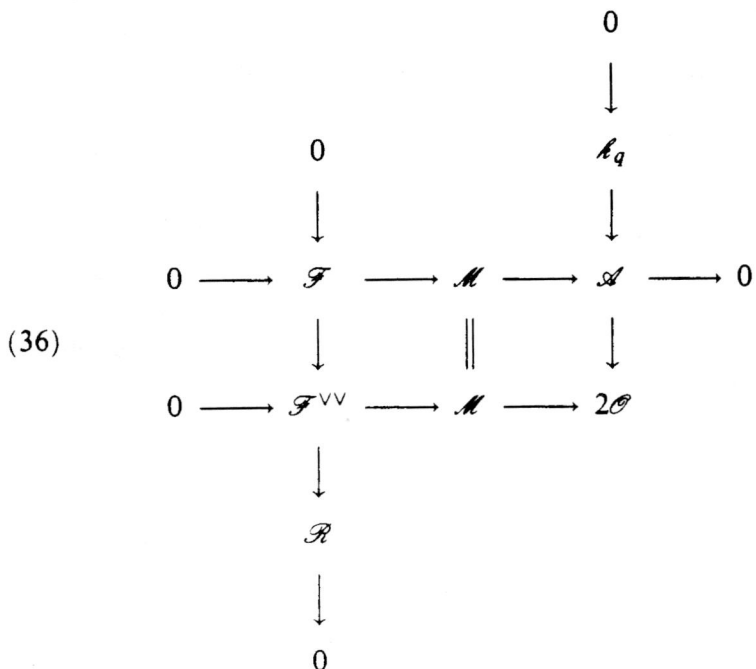

[Bö-Tr] W. Böhmer and G. Trautmann, *Special Instanton Bundles and Poncelet curves, in Singularities, Representation of Algebras and Vector Bundles*, Proc. Lambrecht 1985, 325–336, Springer L.N. 1273, 1987.

[De-Pr] C. Deconcini and C. Procesi, *Complete Symmetric Varieties*, in *Proceedings Montecatini* 1982, Invariant Theory, Springer Lecture Notes in Mathematics 996 (1983).

[Gr] Ph. Griffiths, *Complex analysis and algebraic geometry*, Bull. Amer. Math. Soc., **1** (1979), 595–626.

[Gr-Ha] Ph. Griffiths and J. Harris, *On Cayley's explicit solution to Poncelet's porism*, L'enseignement Mathématique, **24** (1978), 31–40.

[Gru-La-Pe] L. Gruson, R. Lazarsfeld and Ch. Peskine, *On a theorem of Castelnuovo, and the equations defining space curves*, Invent. Math., **72** (1983), 492–506.

[Ha 1] R. Hartshorne, *Algebraic Geometry*, Springer 1977.

[Ha 2] ——, *Stable vector bundles of rank 2 on $\mathbb{P}_3$*, Math. Ann., **238** (1978), 229–280.

[Hi] H. Hironaka, *Flattening theorem in complex analytic geometry*, Amer. J. Math., **97** (1975), 503–547.

[Hi-Na] A. Hirschowitz and M. S. Narasimhan, *Fibrés de 't Hooft speciaux et applications*, Proc. Nice Conf. 1981, Birkhäuser 1982.

[Ma 1] M. Maruyama, *Moduli of stable sheaves* I, J. Math. Kyoto Univ., **17** (1977), 91–126.

[Ma 2] ——, *Moduli of stable sheaves* II, J. Math. Kyoto Univ., **18** (1978), 557–614.

[Mu-Fo] D. Mumford and J. Fogarty, *Geometric Invariant Theory*, 2nd ed., Springer 1982.

[Na-Ra] M. S. Narasimhan and S. Ramanan, *Geometry of Hecke cycles* I, in C. P. Ramanujan, A Tribute, Tata Institute, Springer 1978.

[Na-Tr] M. S. Narasimhan and G. Trautmann, *Compactification of $M(0, 2)$*, in *Vector Bundles on Algebraic Varieties*, Bombay Colloquium 1984, 429–443, Oxford Univ. Press 1987.

[Ne 1] P. E. Newstead, *Introduction to moduli problems and orbit spaces*, Tata Institute Lectures 51, Springer 1978.

[Ne 2] ——, *On the cohomology and the Picard group of a moduli space of bundles on $\mathbb{P}_3$*, Quarterly J. Math., **33** (1983), 349–353.

[Ne 3] ——, *The fundamental group of a moduli space of bundles on $\mathbb{P}_3$*, Topology, **19** (1980), 419–426.

[OSS] Chr. Okonek, M. Schneider and H. Spindler, *Vector Bundles on Complex Projective Spaces*, Birkhäuser 1980.

[Sa] G. Salmon, *A treatise on conic sections*, 6th ed. Chelsea.

[Si-Tr] Y. T. Siu and G. Trautmann, *Gap Sheaves and Extension of Coherent Analytic Subsheaves*, Springer Lecture Notes in Mathematics 172 (1971).

[Tr 1] G. Trautmann, *Moduli for vectorbundles on $\mathbb{P}_n(\mathbb{G})$*, Math. Ann., **237** (1978), 167–186.

[Tr 2] ——, *Poncelet curves and associated theta characteristics*, Expositiones Math., **6** (1988), 29–64.

[Va] I. Vainsencher, *Schubert calculus for complete quadrics*, Proc. Nice Conf. 1981, Birkhäuser 1982.

[Ve] J. L. Verdier, *Instantons*, Sem. E.N.S. 1977–78, exp. VII, Asterisque, (1980), 71–72.

Received November 2, 1988 and in revised form May 18, 1989. Both authors were supported by the Deutsche Forschungsgemeinschaft (DFG) during preparation of the paper.

TATA INSTITUTE OF FUNDAMENTAL RESEARCH
HOMI BHABHA ROAD
BOMBAY, INDIA

AND

FB MATHEMATIK DER UNIVERSITÄT
6750 KAISERSLAUTERN
F. R. GERMANY

Math. Z. 205, 123–136 (1990)

**Mathematische Zeitschrift**



# Rank 2 vector bundles on $P_4$ with $c_1$ odd and contact curves

**W. Decker**[1], **M.S. Narasimhan**[2], **and F.-O. Schreyer**[3]

[1] Universität Kaiserslautern, Fachbereich Mathematik, Postfach 3049, D-6750 Kaiserslautern, Federal Republic of Germany
[2] Tata Institute of Fundamental Research, Homi Bhabha Road, 400005 Bombay, India
[3] Mathematisches Institut der Universität, Postfach 101251, D-8580 Bayreuth, Federal Republic of Germany

Received March 20, 1989; in final form June 2, 1989

## 0. Introduction

There have been great efforts to construct indecomposable rank 2 vector bundles on $\mathbf{P}_4 = \mathbf{P}_4(\mathbf{C})$. But so far only the Horrocks-Mumford-bundle $\mathscr{F}_{HM}$ [H–M] and its satellites are known. Barth [Ba2] suggested a construction principle for further 2-bundles on $\mathbf{P}_4$ with $c_1 = -1$:

**Theorem** (Barth). *Let $R \subset \mathbf{P}_3$ be a nonsingular curve satisfying*

(B1) *Set-theoretically $R$ is the complete intersection of two surfaces $S_1, S_2 \subset \mathbf{P}_3$, both of degree $n$.*

(B2) *$R$ is the curve of contact of $S_1$ and $S_2$, i.e. no point of $R$ is singular for both $S_1$ and $S_2$, and the intersection multiplicity $i_R(S_1, S_2) = 2$.*

(B3) *The sequence*

$$0 \to \mathscr{S} \to \mathscr{N}_{R/\mathbf{P}_3} \to \mathscr{Q} \to 0$$

*splits, where $\mathscr{S} = \mathscr{N}_{R/S_1} = \mathscr{N}_{R/S_2} \subset \mathscr{N}_{R/\mathbf{P}_3}$ is the common normal bundle with quotient $\mathscr{Q}$.*

(B4) $\mathscr{Q} \cong \mathscr{S}(-1)$

(B5) *$R$ is linearly normal.*

*Fix a point $p \in \mathbf{P}_4$. Then there exists a stable 2-bundle $\mathscr{F}$ on $\mathbf{P}_4$ with $c_1 = -1$, $c_2 = n$ such that:*

(i) *$\mathscr{F}(n)$ has a section vanishing with multiplicity 2 along $X_R = \sigma(\pi^{-1}(R))$, the cone over $R$ obtained by blowing up $\mathbf{P}_4$ in $p$:*

$$\begin{array}{ccc} \tilde{\mathbf{P}}_4 & \xrightarrow{\sigma} & \mathbf{P}_4 \\ {\scriptstyle\pi}\downarrow & & \\ \mathbf{P}_3 & & \end{array}$$

(ii) *$R$ is the variety of jumping lines of $\mathscr{F}$ through $p$.*

It is the aim of this paper to give some more details of Barth's result. We first show:

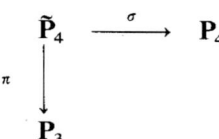

**Theorem.** (A) *Let* $\mathscr{F}$ *be a stable 2-bundle on* $\mathbf{P}_4$ *with* $c_1=-1$, $c_2=n$ *and* $p\in\mathbf{P}_4$ *a point.*

*Suppose that the variety* $R\subset\mathbf{P}_3$ *of jumping lines of* $\mathscr{F}$ *through* $p$ *is a curve. Then:*

(i) *$R$ is locally Cohen-Macaulay,* $\deg R=n^2/2$ *and* $p_a(R)=n^2(4n-13)/12+1$. *In particular $R$ is nondegenerate. Its Hartshorne-Rao-module is*

$$\bigoplus_{m\in\mathbf{Z}} H^1(\mathbf{P}_3,\mathscr{I}_R(m))\cong\bigoplus_{m\in\mathbf{Z}} H^2(\mathbf{P}_4,\mathscr{F}(m-n)),$$

*considered as* $\bigoplus_{m\in\mathbf{Z}}\Gamma(\mathbf{P}_3,\mathscr{O}(m))$*-modules. In particular $R$ is linearly normal iff* $H^2(\mathbf{P}_4,\mathscr{F}(n-5))=0$.

(ii) *Suppose moreover that there are at most finitely many jumping lines of* $\mathscr{F}$ *through $p$ of order* $\geqq 2$. *Then $R$ is directly self-linked by a complete intersection $C$ of two surfaces of degree $n$.*

(iii) *$R$ is a locally complete intersection iff there are no jumping lines of* $\mathscr{F}$ *through $p$ of order* $\geqq 2$. *In this case* $\mathscr{I}_R/\mathscr{I}_C$ *is an invertible* $\mathscr{O}_R$*-module,*

$$\mathscr{I}_R/\mathscr{I}_C\cong\omega_R(-2n+4).$$

*Moreover every hyperplane* $H\subset\mathbf{P}_4$, $H\not\ni p$ *induces a splitting*

$$\mathscr{I}_R/\mathscr{I}_R^2\cong\omega_R(-2n+4)\oplus\omega_R(-2n+5).$$

(B) *Let conversely* $R\subset\mathbf{P}_3$ *be a linearly normal locally complete intersection curve with splitting conormal bundle*

$$\mathscr{I}_R/\mathscr{I}_R^2\cong\omega_R(-2n+4)\oplus\omega_R(-2n+5).$$

*Then the double structure $C$ on $R$ induced by* $\omega_R(-2n+4)$ *is a complete intersection of two surfaces of degree $n$ iff*

$$h^0(\mathbf{P}_3,\mathscr{I}_C(n-1))=0.$$

*In this case $R$ is directly self-linked by $C$ and we obtain a bundle* $\mathscr{F}$ *via Barth's construction.*

*Remark:* (i) In (B) we drop the assumption that $R$ is smooth.

(ii) Our arguments work more generally over $\mathbf{P}_r$, $r\geqq 4$. But if $r\geqq 5$ then for every $p\in\mathbf{P}_r$ there are jumping lines of $\mathscr{F}$ through $p$ of order $\geqq 2$. □

To prove the theorem we proceed as follows. Let $\mathscr{F}$ be a stable 2-bundle on $\mathbf{P}_r$, $r\geqq 2$ with $c_1=-1$. Fix a point $p\in\mathbf{P}_r$ and blow up $\mathbf{P}_r$ in $p$:

$$\begin{array}{ccc}\tilde{\mathbf{P}}_r & \xrightarrow{\ \sigma\ } & \mathbf{P}_r\\ {\scriptstyle\pi}\big\downarrow & & \\ \mathbf{P}_{r-1} & & \end{array}$$

Then $\operatorname{supp} R^1\pi_*\sigma^*\mathscr{F}$ is the set of jumping lines of $\mathscr{F}$ through $p$. In Sect. 1 we recall from [G-P2] the definition and some properties of the correct scheme structure. The proof of the theorem is given in Sect. 2. The crucial point is that $\pi^{-1}(R)$ comes with a double structure $Y$ inducing double structures

$$C = Y \cap E \quad \text{and} \quad C_H = Y \cap H$$

on $R$. Here we identify the exceptional divisor $E \subset \tilde{\mathbf{P}}_r$ and a hyperplane $H \subset \mathbf{P}_r$, $H \not\ni p$ with $\mathbf{P}_{r-1}$ via $\pi$.

In Sect. 3 we consider the case of the Horrocks-Mumford-bundle. For the general point $p \in \mathbf{P}_4$ the variety $R = R_p(\mathscr{F}_{\mathrm{HM}})$ has the expected codimension [B-H-M1, Prop. 17]. We explain how to determine the equations of $R$ and end up with two explicit examples.

## 1. The variety of jumping lines through a point

Let $\mathscr{F}$ be a stable 2-bundle on $\mathbf{P}_r$, $r \geqq 2$, with Chern-classes $c_1 = -1$, $c_2 = n$.

For each line $L \subset \mathbf{P}_r$ there exists $i \geqq 0$ such that

$$\mathscr{F} | L \cong \mathcal{O}_L(i) \oplus \mathcal{O}_L(-i-1)$$

by Grothendieck's splitting theorem (cf. [O-S-S, I, 2.1]). For the generic line $i=0$ by Grauert-Mülich [Ba1]. $L$ is called an *$i^{\text{th}}$-order jumping line of $\mathscr{F}$* if $i > 0$. Let $R_p^i = R_p^i(\mathscr{F})$ denote the *variety of jumping lines of $\mathscr{F}$ of order $\geqq i$ through a fixed point* $p \in \mathbf{P}_r$. We recall its definition and some properties from [G-P2, Appendice B]. Blow up $\mathbf{P}_r$ in $p$:

$$\begin{array}{ccc} \tilde{\mathbf{P}}_r & \xrightarrow{\sigma} & \mathbf{P}_r \\ {\scriptstyle \pi}\downarrow & & \\ \mathbf{P}_{r-1} & & \end{array}$$

$R_p^i$ is defined by the $(i-1)^{\text{st}}$ Fitting ideal of $R^1 \pi_* \sigma^* \mathscr{F}$,

$$\mathscr{I}_{R_p^i} = F_{i-1}(R^1 \pi_* \sigma^* \mathscr{F}).$$

Alternatively

$$\mathscr{I}_{R_p^i} = F_0(R^1 \pi_* \sigma^* (\mathscr{F}(i-1))$$

since

$$F_j(R^1 \pi_* \sigma^* (\mathscr{F}(m))) = F_{j-1}(R^1 \pi_* \sigma^* (\mathscr{F}(m+1)))$$

[loc. cit., Prop. B2].

An explicit determinantal description can be obtained from a presentation

$$0 \to \mathscr{K}' \to q\mathcal{O}(-m) \to \mathscr{F} \to 0$$

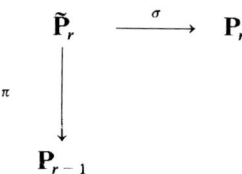

for $m \gg 0$ via $\pi_* \sigma^*$: In the induced sequence

$$0 \to \pi_* \sigma^*(\mathscr{F}(i-1)) \to \mathscr{K} \xrightarrow{\Phi} \mathscr{L} \to R^1 \pi_* \sigma^*(\mathscr{F}(i-1)) \to 0 \qquad (*)$$

$\mathscr{K}$ and $\mathscr{L}$ are locally free with ranks

$$l = \mathrm{rk}\, \mathscr{L} = \mathrm{rk}\, \mathscr{K} - 2i + 1.$$

$R_p^i$ is then defined by the $l \times l$-minors of $\Phi$, hence

$$\operatorname{codim} R_p^i \leqq 2i.$$

Moreover, if equality holds, then $R_p^i$ is locally Cohen-Macaulay, it has degree

$$\deg R_p^i = c_{2i}(\pi_* \sigma^*(\mathscr{F}(i-1)) - R^1 \pi_* \sigma^*(\mathscr{F}(i-1))$$

and it is a locally complete intersection iff there are no jumping lines of order $\geqq i+1$ through $p$. We abbreviate

$$R = R_p = R_p(\mathscr{F}) = R_p^1(\mathscr{F})$$

for the variety of (all) jumping lines of $\mathscr{F}$ through $p$.

**Proposition 1.1.** *Suppose that $R_p$ has expected codimension 2. Then:*

(i) $\pi_* \sigma^* \mathscr{F} \cong \mathscr{O}(-n)$
(ii) $\deg R = n^2/2$
(iii) $R^1 \pi_* \sigma^* \mathscr{F} \cong \omega_R(-n+r)$
(iv) *For $r \geqq 4$ and a general $\mathbf{P}_3 \subset \mathbf{P}_{r-1}$*

$$p_a(\mathbf{P}_3 \cap R) = n^2(4n-13)/12 + 1.$$

(v) *For $r \geqq 5$*

$$\deg R_p^2 = n^2(n+2)^2/24$$

*if $R_p^2$ has expected codimension 4.*

*Proof.* $\pi_* \sigma^* \mathscr{F}$ has rank 1 by Grauert-Mülich and it is reflexive, so it is a line bundle. Grothendieck-Riemann-Roch applied to $\pi: \tilde{\mathbf{P}}_r \to \mathbf{P}_{r-1}$ gives

$$\mathrm{ch}\,(\pi_* \sigma^* \mathscr{F} - R^1 \pi_* \sigma^* \mathscr{F}) = 1 - n H_{r-1} + n^2/12\, H_{r-1}^3 \\ + \text{higher terms}$$

($H_{r-1}$ is the hyperplane class on $\mathbf{P}_{r-1}$). This proves (i) and (ii) since $\operatorname{codim} R = 2$ implies

$$c_1(R^1 \pi_* \sigma^* \mathscr{F}) = 0.$$

We may identify the resolution

$$0 \longrightarrow \mathscr{O}(-n) \xrightarrow{\Psi} \mathscr{K} \xrightarrow{\Phi} \mathscr{L} \longrightarrow R^1 \pi_* \sigma^* \mathscr{F} \longrightarrow 0 \qquad (*)$$

with the Eagon-Northcott-complex associated to $\Phi$ [loc. cit.]:

$$0 \to \Lambda^{l+1}\mathcal{K} \otimes \Lambda^{l}\mathcal{L}^* \xrightarrow{\Lambda^l \Phi^t} \mathcal{K} \xrightarrow{\Phi} \mathcal{L} \to R^1\pi_* \sigma^* \mathcal{F} \to 0,$$

exactness follows from codim $R=2$. So $\Psi$ is given by the $l \times l$-minors of $\Phi$ and we obtain a resolution of $\mathcal{O}_R$ by dualizing:

$$0 \to \mathcal{L}^*(-n) \to \mathcal{K}^*(-n) \to \mathcal{O} \to \mathcal{O}_R \to 0,$$

exactness follows again from codim $R=2$. In particular

$$\omega_R \cong \mathcal{E}xt^2(\mathcal{O}_R, \mathcal{O}_{\mathbf{P}_{r-1}}(-r)) \cong (R^1\pi_* \sigma^* \mathcal{F})(n-r),$$

i.e. (iii).

To prove (iv) we may restrict ourselves to the case $r=4$ by base change. Hirzebruch-Riemann-Roch on $\mathbf{P}_3$ gives

$$\chi(R^1\pi_* \sigma^* \mathcal{F}) = n^2(-2n+11)/12.$$

Hence (iv) because

$$\chi(R^1\pi_* \sigma^* \mathcal{F}) = \chi(\omega_R(-n+4)) = -\chi(\mathcal{O}_R(n-4)) = n^2(n-4)/2 + 1 - p_a.$$

(v) follows again from Grothendieck-Riemann-Roch. □

**Proposition 1.2.** *Suppose that $R_p$ has codimension 2. Then for $i=2, \ldots, r-2$*

$$\bigoplus_{m \in \mathbf{Z}} H^i(\mathbf{P}_r, \mathcal{F}(m)) \cong \bigoplus_{m \in \mathbf{Z}} H^{r-i-1}(\mathbf{P}_{r-1}, \mathcal{J}_R(n-r-m))^*$$

*as $\bigoplus_{m \in \mathbf{Z}} \Gamma(\mathbf{P}_{r-1}, \mathcal{O}(m))$-modules.*

*Proof.* The ideal sheaf of the exceptional divisor $E \subset \tilde{\mathbf{P}}_r$ is

$$\mathcal{O}_{\tilde{P}_r}(-E) \cong \pi^* \mathcal{O}(1) \otimes \sigma^* \mathcal{O}(-1).$$

Consequently

$$H^i(\mathbf{P}_r, \mathcal{F}(m)) \cong H^i(\tilde{\mathbf{P}}_r, \sigma^* \mathcal{F}(m)) \cong H^i(\tilde{\mathbf{P}}_r, \pi^* \mathcal{O}(m) \otimes \sigma^* F)$$

for $i=2, \ldots, r-2$ and $m \in \mathbf{Z}$ since

$$H^i(\sigma^* \mathcal{F} \otimes \mathcal{O}_E(l)) = 0, \quad i=1, \ldots, r-2, \quad l \in \mathbf{Z}.$$

Moreover the $\bigoplus_{m \in \mathbf{Z}} \Gamma(\mathbf{P}_{r-1}, \mathcal{O}(m))$-module structure is just induced by the inclusion

$$\bigoplus_{m \in \mathbf{Z}} \Gamma(\mathbf{P}_{r-1}, \mathcal{O}(m)) \subset \bigoplus_{m \in \mathbf{Z}} \Gamma(\mathbf{P}_r, \mathcal{O}(m))$$

which corresponds to the projection $\mathbf{P}_r \setminus \{p\} \to \mathbf{P}_{r-1}$.

We compute the right hand side with Leray's spectral sequence

$$H^i(\mathbf{P}_{r-1}, (R^j\pi_* \sigma^* \mathcal{F})(m)) \Rightarrow H^{i+j}(\tilde{\mathbf{P}}_r, \pi^* \mathcal{O}(m) \otimes \sigma^* \mathcal{F}).$$

Identifying

$$R^0\pi_*\sigma^*\mathcal{F}\cong\mathcal{O}(-n) \quad \text{and} \quad R^1\pi_*\sigma^*\mathcal{F}\cong\omega_R(r-n)$$

we find

$$\begin{aligned}H^i(\mathbf{P}_r,\mathcal{F}(m))&\cong H^{i-1}(\mathbf{P}_{r-1},\omega_R(r-n+m))\\&\cong H^{r-i-2}(R,\mathcal{O}_R(n-r-m))^*\\&\cong H^{r-i-1}(\mathbf{P}_{r-1},\mathcal{J}_R(n-r-m))^*\end{aligned}$$

for $i=2,\ldots,r-3$ and

$$\begin{aligned}H^{r-2}(\mathbf{P}_r,\mathcal{F}(m))&\cong\ker(H^{r-3}(\mathbf{P}_{r-1},\omega_R(r-n+m))\to H^r(\mathbf{P}_{r-1}\mathcal{O}(m-n)))\\&\cong\operatorname{coker}(H^0(\mathbf{P}_{r-1},\mathcal{O}(n-r-m))\to H^0(R,\mathcal{O}_R(n-r-m))^*\\&\cong H^1(\mathbf{P}_{r-1},\mathcal{J}_R(n-m-r))^*.\ \square\end{aligned}$$

## 2. The double structure on $\pi^{-1}(R)$

**Lemma 2.1.** *Let* $\mathcal{E}$ *on* $\mathbf{P}_1\times\mathbf{C}^2\xrightarrow{q}\mathbf{C}^2$ *be the versal deformation of* $\mathcal{E}_0=\mathcal{O}(1)\oplus\mathcal{O}(-2)$ *on* $\mathbf{P}_2$. *Then the evaluation map*

$$\mathcal{O}\cong q^*q_*\mathcal{E}\to\mathcal{E}$$

*defines a double structure on* $\mathbf{P}_1\times\{0\}\subset\mathbf{P}_1\times\mathbf{C}^2$.

*Proof.* Let $s$, $t$ be homogenous coordinates on $\mathbf{P}_1$ and $a$, $b$ coordinates on

$$\mathbf{C}^2\cong\operatorname{Ext}^1(\mathcal{E}_0,\mathcal{E}_0).$$

Consider the covering $\{s\neq 0\}$, $\{t\neq 0\}$ of $\mathbf{P}_1\times\mathbf{C}^2$. The transition matrix

$$\begin{bmatrix}z^2 & a+bz\\ 0 & z^{-1}\end{bmatrix},\qquad z=s/t$$

gives the versal deformation $\mathcal{E}$ of $\mathcal{E}_0$. $q_*\mathcal{E}$ is reflexive of rank 1, so $q_*\mathcal{E}\cong\mathcal{O}_{\mathbf{C}^2}$. Indeed the section

$$\begin{bmatrix}a^2\\ az^{-1}-b\end{bmatrix}=\begin{bmatrix}z^2 & a+bz\\ 0 & z^{-1}\end{bmatrix}\begin{bmatrix}b^2\\ a-bz\end{bmatrix}$$

generates $H^0(\mathbf{P}_r\times\mathbf{C}^2,\mathcal{E})=H^0(\mathbf{C}^2,q_*\mathcal{E})$ as a $\mathbf{C}[a,b]$-module. The zeroes of this section give the double structure

$$(as-bt,a^2,ab,b^2)$$

on the line $\{a=b=0\}$. $\square$

We apply Lemma 2.1 in the situation of Sect. 1. We denote by $E\subset\tilde{\mathbf{P}}_r$ the exceptional divisor and identify $E$ and a given hyperplane $H\subset\mathbf{P}_r$, $H\not\ni p$ with $\mathbf{P}_{r-1}$ via $\pi$.

**Proposition 2.2.** *Suppose that* $R_p$ *has codimension* 2. *Then:*

(i) *If* codim $R_p^2\geq 3$ *then the zero locus* $Y$ *of the evaluation map*

$$\pi^*\pi_*\sigma^*\mathcal{F}\to\sigma^*\mathcal{F}$$

*is a double structure on* $\pi^{-1}(R)$. *The induced double structures on* $R$

$$C = Y \cap E \quad \text{and} \quad C_H = Y \cap H$$

*belong via Serre-correspondence to*

$$\sigma^* \mathscr{F} \otimes \mathcal{O}_E(n) = 2\mathcal{O}_E(n) \quad \text{and} \quad \mathscr{F}_H(n) = \sigma^* \mathscr{F} \otimes \mathcal{O}_H(n)$$

*resp. In particular* $C$ *is the complete intersection of two hypersurfaces of degree* $n$ *and* $R$ *is directly self-linked by* $C$.

(ii) *If* $R_p^2 = \emptyset$ *then* $\mathscr{J}_{\pi^{-1}(R)}/\mathscr{J}_Y$ *is an invertible* $\mathcal{O}_{\pi^{-1}(R)}$*-module and the projection*

$$\mathscr{J}_{\pi^{-1}(R)}/\mathscr{J}^2_{\pi^{-1}(R)} \to \mathscr{J}_{\pi^{-1}(R)}/\mathscr{J}_Y$$

*induces an isomorphism*

$$\mathscr{J}_R/\mathscr{J}_R^2 \cong \mathscr{J}_R/\mathscr{J}_C \oplus \mathscr{J}_R/\mathscr{J}_{C_H} \cong \omega_R(-2n+r) \oplus \omega_R(-2n+r+1).$$

*In particular*

$$\omega_R^{\otimes 3} \cong \mathcal{O}_R(4n-13).$$

*Proof*. (i) Clearly supp $Y = \operatorname{supp} \pi^{-1}(R)$. Let $z \in R$ be a point such that

$$\mathscr{F} | L \cong \mathcal{O}_L(1) \oplus \mathcal{O}_L(-2), \; L = \pi^{-1}(z).$$

From the versality of $\mathscr{E}$ we obtain a commutative diagramm

$$\begin{array}{ccc} \mathbf{P}_1 \times U & \xrightarrow{\Phi = id \times \varphi} & \mathbf{P}_1 \times \mathbf{C}^2 \\ \Big\downarrow{\scriptstyle \pi} & & \Big\downarrow{\scriptstyle q} \\ U & \xrightarrow[\varphi]{} & \mathbf{C}^2 \end{array} \quad , \sigma^* \mathscr{F} | \mathbf{P}_1 \times U \cong \Phi^* \mathscr{E}$$

over a convenient neighbourhood $U$ of $z$. The canonical morphism

$$\varphi^* q_* \mathscr{E} \to \pi_* \Phi^* \mathscr{E}$$

is an isomorphism over $U \setminus R$ by base change, hence it is an isomorphism since codim $R_p = 2$. So, over $\mathbf{P}_1 \times U$, the evaluation map

$$\pi^* \pi_* \sigma^* \mathscr{F} \to \sigma^* \mathscr{F}$$

is the pullback

$$\Phi^*(q^* q_* \mathscr{E}) \to \Phi^*(\mathscr{E}).$$

By Lemma 1 we have near $z: \pi^{-1}(R) \subset Y$ and $\mathscr{J}_{\pi^{-1}(R)}/\mathscr{J}_Y$ is a locally free $\mathcal{O}_{\pi^{-1}(R)}$-module of rank 1. Since $\pi^{-1}(R)$ and $Y$ are locally Cohen-Macaulay and codim $R_p^2 \geqq 3$ the inclusion $\pi^{-1}(R) \subset Y$ holds everywhere. Moreover

$$\mathscr{A}nn_{\mathcal{O}_{\mathbf{P}_r}}(\mathscr{J}_{\pi^{-1}(R)}/\mathscr{J}_Y) = \mathscr{J}_{\pi^{-1}(R)}.$$

In other words $Y$ is a double structure on $\pi^{-1}(R)$.

Serre-correspondence yields the exact sequence

$$0 \to \mathcal{O}_{\tilde{\mathbf{P}}_r}(-n, 0) \to \sigma^* \mathcal{F} \to \mathcal{O}_{\tilde{\mathbf{P}}_r}(n, -1) \to \mathcal{O}_Y(n, -1) \to 0 \tag{1}$$

$(\mathcal{O}_{\tilde{\mathbf{P}}_r}(a, b) = \pi^* \mathcal{O}_{\mathbf{P}_{r-1}}(a) \otimes \sigma^* \mathcal{O}_{\mathbf{P}_r}(b))$. Restricting to $E$ and $H$ resp. gives (i) since

$$\mathcal{A}nn_{\mathcal{O}_{\mathbf{P}_{r-1}}}(\mathcal{J}_R/\mathcal{J}_C) = \mathcal{A}nn_{\mathcal{O}_{\tilde{\mathbf{P}}_r}}(\mathcal{J}_{\pi^{-1}(R)}/\mathcal{J}_Y) \otimes \mathcal{O}_E = \mathcal{J}_R.$$

(ii) If furthermore $R_p^2 = \emptyset$ then $\mathcal{Q} = \mathcal{J}_{\pi^{-1}(R)}/\mathcal{J}_Y$ is invertible everywhere. The restrictions of the projection

$$\mathcal{J}_{\pi^{-1}(R)}/\mathcal{J}^2_{\pi^{-1}(R)} \to \mathcal{Q}$$

to $E$ and $H$ resp. add up to a map

$$\mathcal{J}_R/\mathcal{J}_R^2 \to \mathcal{Q}_E \oplus \mathcal{Q}_H.$$

Locally this is just the pullback of the splitting of the conormal bundle of $0 \in \mathbf{C}^2$ obtained from the points $0 = (1:0)$, $\infty = (0:1)$:

$$(a, b)/(a^2, ab, b^2) \to [a] \oplus [-b].$$

Since $E \cap H = \emptyset$ and $R_p^2 = \emptyset$ this glues to a global splitting. It remains to identify $\mathcal{Q}_E$ and $\mathcal{Q}_H$. Consider the commutative diagram

$$\begin{array}{ccccccccc}
 & & & & 0 & & 0 & & \\
 & & & & \downarrow & & \downarrow & & \\
 & & 0 & \longrightarrow & \mathcal{J}_Y & \longrightarrow & \mathcal{J}_{\pi^{-1}(R)} & \longrightarrow & \mathcal{Q} \longrightarrow 0 \\
 & & & & \downarrow & & \downarrow & & \\
 & & & & \mathcal{O}_{\tilde{\mathbf{P}}_r} & = & \mathcal{O}_{\tilde{\mathbf{P}}_r} & & \\
 & & & & \downarrow & & \downarrow & & \\
0 & \longrightarrow & \mathcal{J}_{\pi^{-1}(R)/Y} & \longrightarrow & \mathcal{O}_Y & \longrightarrow & \mathcal{O}_{\pi^{-1}(R)} & \longrightarrow & 0 \\
 & & & & \downarrow & & \downarrow & & \\
 & & & & 0 & & 0 & &
\end{array} \tag{2}$$

with exact rows and columns. By the snake lemma $\mathcal{Q} \cong \mathcal{J}_{\pi^{-1}(R)/Y}$. We apply $\mathcal{E}xt^2(-, \omega_{\tilde{\mathbf{P}}_r})$ and obtain the exact sequence

$$0 \to \omega_{\pi^{-1}(R)} \to \omega_Y \to \mathcal{Q}^* \otimes \omega_{\pi^{-1}(R)} \to 0 \tag{3}$$

from the bottom row of (2). Similarly from (1)

$$\begin{aligned}\omega_Y &\cong \mathcal{E}xt^2(\mathcal{O}_Y, \omega_{\tilde{\mathbf{P}}_r}) \cong \mathcal{E}xt^2(\mathcal{O}_Y, \mathcal{O}_{\tilde{\mathbf{P}}_r}(-r+1, -2)) \\ &\cong \mathcal{O}_Y(2n - r + 1, -3).\end{aligned}$$

Comparing (3) and the bottom row of (2) gives

$$\mathcal{Q} \cong \omega_{\pi^{-1}(R)}(-2n + r - 1, 3).$$

Our result follows since

$$\omega_R \cong \omega_{\pi^{-1}(R)}(E) \otimes \mathcal{O}_E \cong \omega_{\pi^{-1}(R)}(H) \otimes \mathcal{O}_H. \quad \square$$

## 3. Example: the Horrocks-Mumford-bundle

Recall:

**Lemma 3.1.** *Let $V$ be a $r+1$-dimensional $\mathbf{C}$-vector space and*

$$\begin{array}{ccc} \tilde{\mathbf{P}}_r & \xrightarrow{\sigma} & \mathbf{P}_r = \mathbf{P}(V) \\ \downarrow{\scriptstyle \pi} & & \\ \mathbf{P}_{r-1} = \mathbf{P}(V/\langle a \rangle) & & \end{array}$$

*the blow up of $\mathbf{P}(V)$ in a point $p = \langle a \rangle \in \mathbf{P}(V)$. Then*

$$\pi_* \sigma^* (\Omega^i_{\mathbf{P}_r}(i)) \cong \Omega^i_{\mathbf{P}_{r-1}}(i)$$

*and*

$$\begin{array}{ccc} \operatorname{Hom}(\Omega^i_{\mathbf{P}_r}(i), \Omega^j_{\mathbf{P}_r}(j)) & \xrightarrow{\pi_*\sigma^*} & \operatorname{Hom}(\Omega^i_{\mathbf{P}_{r-1}}(i), \Omega^j_{\mathbf{P}_{r-1}}(j)) \\ \| \wr & & \| \wr \\ \Lambda^{i-j} V & \xrightarrow[\text{projection}]{} & \Lambda^{i-j}(V/\langle a \rangle) \end{array}$$

*commutes.*

We consider the case $r=4$ and recall the definition of the Horrocks-Mumford-bundle $\mathscr{F}_{\mathrm{HM}}$ ([H-M] or [Ba2]): Fix a basis $e_0, \ldots, e_4$ of $V$ and define matrices $A = (a_{ij})$, ${}^*A = \begin{bmatrix} 0 & 1 \\ -1 & 0 \end{bmatrix} \cdot A^t$ by

$$\left. \begin{aligned} a^{i0} &= e_{i+2} \wedge e_{i+3} \\ a^{i1} &= e_{i+1} \wedge e_{i+4} \end{aligned} \right\} i \bmod 5.$$

Then

$$(M) \quad 5\mathcal{O}(-1) \xrightarrow{{}^*A} 2\Omega^2(2) \xrightarrow{A} 5\mathcal{O}$$

is a monad, its cohomology is the Horrocks-Mumford-bundle. $\mathscr{F}_{\mathrm{HM}}$ is a stable 2-bundle with Chern-classes $c_1 = -1$, $c_2 = 4$.

**Proposition 3.2.** *For every point $p \in \mathbf{P}_4$ the variety $R_p(\mathscr{F}_{\mathrm{HM}})$ is either the whole $\mathbf{P}_3$ or a nondegenerate, linearly normal curve of degree 8 and arithmetic genus 5. It is a smooth curve for the general point.*

*Proof.* Suppose that $R = R_p(\mathscr{F}_{\mathrm{HM}})$ has a 2-dimensional component. The same arguments as in the proof of Proposition 1.1 show that

$$\pi_* \sigma^* \mathscr{F}_{\mathrm{HM}} \cong \mathcal{O}(-4 + \mathrm{e}),$$

where $e$ denotes the degree of the 2-dimensional part of $R$. Evaluation gives a section in

$$H^0(\tilde{\mathbf{P}}_4, (\sigma^* \mathscr{F}_{\mathrm{HM}})(m, 0)) \to H^0(\tilde{\mathbf{P}}_4, (\sigma^* \mathscr{F}_{\mathrm{HM}})(0, m)) \cong H^0(\mathbf{P}_4, \mathscr{F}_{\mathrm{HM}}(m))$$

with $m = 4 - e$. The zero locus of this section is supported on a cone with vertex in $p$. But $H^0(\mathbf{P}_4, \mathscr{F}_{\mathrm{HM}}(m)) = 0$ for $m \leqq 2$ [H-M] and no section of $\mathscr{F}_{\mathrm{HM}}(3)$ has a zero locus supported on a cone [B-H-M2, Theorem 0.1]. So $R$ cannot have a 2-dimensional component.

If $R$ is a curve then it is linearly normal by our theorem since $H^2(\mathbf{P}_4, \mathscr{F}_{\mathrm{HM}}(-1)) = 0$ [H-M]. That it is a smooth curve for the general point has been proved in [B-H-M1, Proposition 17]. This also follows from our explicit example below. □

Let $p = \langle a \rangle \in \mathbf{P}_4 = \mathbf{P}(V)$ be a point such that $R = R_p(\mathscr{F}_{\mathrm{HM}})$ is a curve. We derive an explicit description of $R$ from the monad. Apply $\pi_* \sigma^*$ to the display

$$\begin{array}{ccccc}
5\mathcal{O}(-1) & \longrightarrow & \mathscr{K} & \longrightarrow & \mathscr{F}_{\mathrm{HM}} \\
\| & & \downarrow & & \downarrow \\
5\mathcal{O}(-1) & \xrightarrow{\;{}^*A\;} & 2\Omega^2(2) & \longrightarrow & \mathscr{Q} \\
 & & \downarrow{\scriptstyle A} & & \downarrow \\
 & & 5\mathcal{O} & = & 5\mathcal{O}
\end{array}$$

of $(M)$. Since

$$\pi_* \sigma^* \mathscr{K} \cong \pi_* \sigma^* \mathscr{F}_{\mathrm{HM}} \cong \mathcal{O}_{\mathbf{P}_3}(-4),$$
$$R^1 \pi_* \sigma^* \mathscr{K} \cong R^1 \pi_* \sigma^* \mathscr{F}_{\mathrm{HM}}$$

and

$$R^1 \pi_* \sigma^* (\Omega^2(2)) = 0$$

we obtain the exact sequence

$$0 \longrightarrow \mathcal{O}_{\mathbf{P}_3}(-4) \longrightarrow 2\Omega^2_{\mathbf{P}_3}(2) \xrightarrow{\;A'\;} 5\mathcal{O}_{\mathbf{P}_3} \longrightarrow R^1 \pi_* \sigma^* \mathscr{F}_{\mathrm{HM}} \longrightarrow 0. \qquad (*)$$

Here $A'$ is the image of $A$ under $\Lambda^2 V \to \Lambda^2(V/\langle a \rangle)$. $R$ is the determinantal variety defined by the $2 \times 2$-minors of $A'$. Dualizing $(*)$ gives the resolution

$$0 \longrightarrow 5\mathcal{O}_{\mathbf{P}_3}(-4) \xrightarrow{\;{}^tA'\;} 2\Omega^1_{\mathbf{P}_3}(-2) \longrightarrow \mathcal{O}_{\mathbf{P}_3} \longrightarrow \mathcal{O}_R \longrightarrow 0.$$

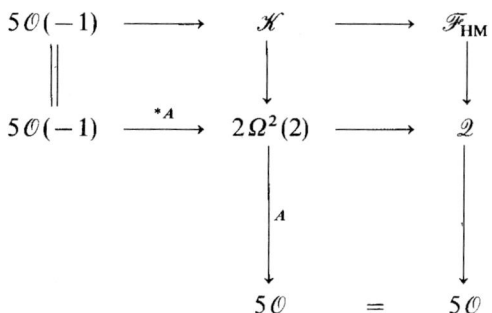

Hence the Koszul-resolution of $\Omega^1_{\mathbf{P}_3}(-2)$ induces the minimal free resolution of $\mathcal{O}_R$:

$$
\begin{array}{ccccccccccc}
 & & & & 0 & & 0 & & & & \\
 & & & & \downarrow & & \downarrow & & & & \\
 & & & & 2\mathcal{O}_{\mathbf{P}_3}(-6) & = & 2\mathcal{O}_{\mathbf{P}_3}(-6) & & & & \\
 & & & & \downarrow & & \downarrow{\scriptstyle \alpha_2} & & & & \\
 & & & & 8\mathcal{O}_{\mathbf{P}_3}(-5) & = & 8\mathcal{O}_{\mathbf{P}_3}(-5) & & & & \\
 & & & & \downarrow & & \downarrow{\scriptstyle \alpha_1} & & & & \\
0 & \longrightarrow & 5\mathcal{O}_{\mathbf{P}_3}(-4) & \xrightarrow{{}^tA'} & 12\mathcal{O}_{\mathbf{P}_3}(-4) & \longrightarrow & 7\mathcal{O}_{\mathbf{P}_3}(-4) & \longrightarrow & 0 & & \qquad (**) \\
 & & \| & & \downarrow & & \downarrow{\scriptstyle \alpha_0} & & & & \\
0 & \longrightarrow & 5\mathcal{O}_{\mathbf{P}_3}(-4) & \xrightarrow{{}^tA'} & 2\Omega^1_{\mathbf{P}_3}(-2) & \longrightarrow & \mathcal{O}_{\mathbf{P}_3} & \longrightarrow & \mathcal{O}_R & \longrightarrow & 0 \\
 & & & & \downarrow & & \downarrow & & & & \\
 & & & & 0 & & \mathcal{O}_R & & & & \\
 & & & & & & \downarrow & & & & \\
 & & & & & & 0 & & & &
\end{array}
$$

For a given point the computation of ${}^tA'$ and $\alpha_1$ is straightforward. The seven quartics defining $R$, i.e. the entries of $\alpha_0$ or ${}^t\alpha_0$ are the syzgies of ${}^t\alpha_1$.

Let us determine the two quartics in the ideal of $R$ which define the complete intersection $C$. Identifying

$$\sigma^*(\Omega^i(i)) \otimes \mathcal{O}_E \cong \Lambda^i(V/\langle a \rangle)^* \otimes_{\mathbf{C}} \mathcal{O}_E \cong \binom{4}{i} \mathcal{O}_{\mathbf{P}_3},$$

we may consider the transposed sequence (**) as part of the display

$$
\begin{array}{ccccc}
5\mathcal{O}_{\mathbf{P}_3} & \longrightarrow & 7\mathcal{O}_{\mathbf{P}_3} & \xrightarrow{\varphi} & 2\mathcal{O}_{\mathbf{P}_3} \\
\| & & \downarrow & & \downarrow \\
5\mathcal{O}_{\mathbf{P}_3} & \xrightarrow{{}^*A'} & 12\mathcal{O}_{\mathbf{P}_3} & \longrightarrow & 7\mathcal{O}_{\mathbf{P}_3} \\
 & & \downarrow{\scriptstyle A'} & & \downarrow \\
 & & 5\mathcal{O}_{\mathbf{P}_3} & = & 5\mathcal{O}_{\mathbf{P}_3}
\end{array}
$$

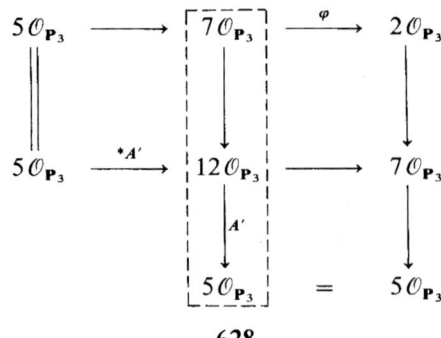

of $\sigma^*(M)\otimes\mathcal{O}_E$. Hence the composition

$$\mathcal{O}_{\mathbf{P}_3}(-4) \xrightarrow{{}^t\alpha_0} 7\mathcal{O}_{\mathbf{P}_3} \longrightarrow 2\mathcal{O}_{\mathbf{P}_3}$$

is the evaluation map

$$\pi^*\mathcal{O}_{\mathbf{P}_3}(-4) \longrightarrow \sigma^*\mathcal{F}_{\mathrm{HM}}$$

restricted to $E\cong\mathbf{P}_3$.

We give two examples, computed with the help of the computer program Macaulay [B-S]. Choose homogenous coordinates $(\tilde{v}:\tilde{w}:\tilde{x}:\tilde{y}:\tilde{z})$ on $\mathbf{P}_4$ and $(w:x:y:z)$ on $\mathbf{P}_3$.

*Example 3.3.* For the point $P=(2:1:1:1:1)$ we take

$$w=\tilde{w}-2\tilde{v},\ \ldots,\ z=\tilde{z}-2\tilde{v}.$$

Then $R_p(\mathcal{F}_{\mathrm{HM}})$ is the smooth curve defined by the following seven quartics:

$$w^2x^2+4wx^3+4w^3y-10w^2xy+4wx^2y+w^2y^2-10wx^2z+4w^2yz+20wxyz$$
$$-10wy^2z+4xy^2z+4y^3z+4wxz^2+x^2z^2-10xyz^2+y^2z^2+4xz^3$$

$$2w^4-2w^3x+3w^2x^2-wx^3-2w^3y+2wx^2y+5w^2y^2-5wxy^2+4x^2y^2$$
$$-4wy^3+4xy^3-2x^3z-2w^2xz-3wx^2z+x^3z+2w^2yz-10x^2yz-wy^2z$$
$$+xy^2z+10w^2z^2+4wxz^2+4x^2z^2-8wyz^2-2wz^3$$

$$-4w^4+4w^3x-6w^2x^2+2wx^3+5w^2xy-6wx^2y+2x^3y-11w^2y^2+12wxy^2$$
$$-10x^2y^2+2xy^3-2y^4+4w^3z+3w^2xz+2wx^2z-2x^3z+w^2yz-4wxyz$$
$$+18x^2yz+2wy^2z-4xy^2z+4y^3z-20w^2z^2+2wxz^2-8x^2z^2-2wyz^2$$
$$+3xyz^2-7y^2z^2-xz^3+9yz^3-4z^4$$

$$w^3x+4w^2x\,2-w^3y-3w^3xy+4wx^2y-2x^3y+3w^2y^2-2wxy^2+10x^2y^2$$
$$-2wy^3-2xy^3+2y^4-10w^2xz+2w^2yz-8x^2yz+2xy^2z-2y^3z$$
$$+4w^2z^2+wxz^2-5wyz^2-xyz^2+5y^2z^2+4wz^3-4yz^3$$

$$wx^2y+4x^3y+4w^2y^2-10wxy^2+4x^2y^2+wy^3-2w^3z-8w^2xz-wx^2z$$
$$-4x^3z+4w^2yz-5x^2yz-3wy^2z+2xy^2z-y^3z+10w^2z^2+2wxz^2$$
$$+5x^2z^2-2wyz^2+3y^2z^2-2wz^3-2xz^3-2yz^3+2z^4$$

$$-2w^4-11w^2x^2-4x^4+4w^3y+2w^2xy-11w^2y^2+8wxy^2-24x^2y^2$$
$$+8wy^3-4y^4+4w^3z+20w^2xz+6wx^2z+8x^3z-8w^2yz-4wxyz$$
$$+8x^2yz+6wy^2z-20w^2z^2-8wxz^2$$
$$-11x^2z^2+20wyz^2+2xyz^2-11y^2z^2+4wz^3+4xz^3-2z^4$$

$$4w^4-8w^3y+11w^2x^2-4wx^3+2x^4-6w^2xy+8wx^2y-4x^3y+11w^2y^2$$

$$-20wxy^2+20x^2y^2-4xy^3+2y^4-8w^2xz-2wx^2z+4wxyz-20x^2yz$$
$$-2wy^2z+8xy^2z-4y^3z+24w^2z^2+11x^2z^2-8wyz^2-6xyz^2+11y^2z^2$$
$$-8yz^3+4z^4$$

The last two equations give the Kummer surfaces [Ba2] defining the complete intersection $C$.

*Example 3.4.* If $p=(1:0:0:0:0)$, then $R=R_p(\mathscr{F}_{\mathrm{HM}})$ is not everywhere a locally complete intersection, it is nonreduced and reducible. Nevertheless the equations of $R$ are much simpler. In coordinates

$$w=\tilde{w}, \ldots, z=\tilde{z}$$

we obtain

$$I_R=(wxyz, wxy^2, w^2xz, wyz^2, x^2yz, w^2z^2, x^2y^2)$$

with primary decomposition

$$I_R=(w, x^2)\cap(x, z^2)\cap(z, y^2)\cap(y, w^2).$$

So $R$ consists of four double lines:

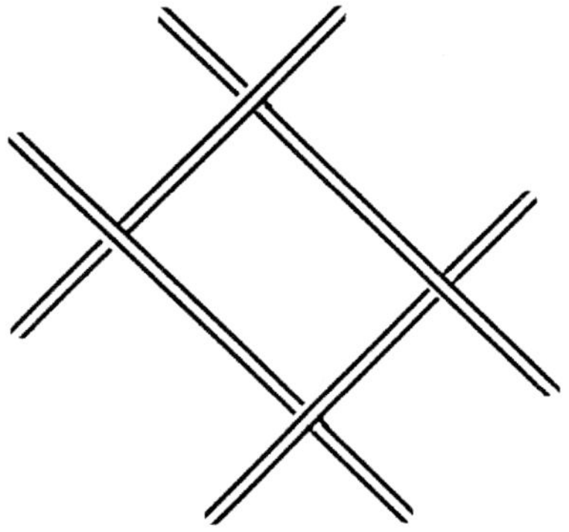

**Fig. 1**

The complete intersection $C$ is given by

$$I_C=(w^2y^2, x^2z^2).$$

Even the second double structure is not too complicated. For $H=\{\tilde{v}=0\}$ it has the primary decomposition

$$\begin{aligned}I_{C_H}&=(wxyz, wxy^4+x^4yz-w^3z^3, x^3y^3-w^4xz-wyz^4,\\&\qquad x^5yz-w^3xz^3, w^2yz^4, w^4xz^2, x^4y^2z)\\&=(w^2, wx, x^4, wz^4-x^3y^2)\\&\cap(x^2, xz, z^4, xy^4-z^3w^2)\\&\cap(z^2, yz, y^4, zw^4-y^3x^2)\\&\cap(y^2, yw, w^4, yx^4-w^3z^2).\quad\square\end{aligned}$$

## References


[Ba1] Barth, W.: Some properties of stable rank-2 vector bundles on $\mathbf{P}_n$. Math. Ann. **226**, 125–150 (1977)

[Ba2] Barth, W.: Kummer surfaces associated with the Horrocks-Mumford bundle. In: Journées de géométrie algébrique d'Angers 1979. Alphen aan den Rijn: Sijthoff and Noordhoff 1980

[B-H-M1] Barth, W., Hulek, K., Moore, R.: Shioda's modular surface $S(5)$ and the Horrocks-Mumford bundle. Proceedings of the Tata conference on algebraic vector bundles over algebraic varieties, Bombay 1984. Oxford: Oxford University Press 1987

[B-H-M2] Barth, W., Hulek, K., Moore, R.: Degenerations of Horrocks-Mumford surfaces. Math. Ann. **277**, 735–755 (1987)

[B-S] Bayer, D., Stillman, M.: Macaulay, a computer algebra system for algebraic geometry

[G-P1] Gruson, L., Peskine, C.: Genre des courbes de l'espace projectif. In: Algebraic geometry, proceedings, Tromsø 1977. Berlin Heidelberg New York: Springer 1978

[G-P2] Gruson, L., Peskine, C.: Courbes de l'espace projectif: Variétés de secantes. In: Enumerative geometry and classical algebraic geometry, Nice 1981. Boston Basel Stuttgart: Birkhäuser 1982

[H-M] Horrocks, G., Mumford, D.: A rank 2 vector bundle on $\mathbf{P}^4$ with 15000 symmetries. Topology **12**, 63–81 (1973)

[O-S-S] Okonek, C., Schneider, M., Spindler, H.: Vector bundles on complex projective spaces. Basel: Birkhäuser 1980

J. reine angew. Math. **422** (1991), 21—44

Journal für die reine und angewandte Mathematik


# The Picard group of the compactification of $M_{\mathbb{P}_3}(0,2)$

By *M.S. Narasimhan* at Bombay and *G. Trautmann* at Kaiserslautern

---

### Introduction and statement of the main theorem

Let $M(0,2)$ denote the moduli space of rank 2 vector bundles on $\mathbb{P}_3$, with $c_1 = 0$, $c_2 = 2$. We denote by $\overline{M(0,2)}$ the schematic closure of $M(0,2)$ in the Maruyama scheme of rank 2 semi-stable sheaves on $\mathbb{P}_3$ with $c_1 = 0$, $c_2 = 2$, $c_3 = 0$ and by $\tilde{M}(0,2)$ the normalisation of $\overline{M(0,2)}$. In [NT] we had constructed a normal projective variety $Q$ which is a quadric bundle ("Poncelet bundle") over the Hilbert scheme $C(G)$ of conics in the Grassmannian of projective lines in $\mathbb{P}_3$ and a birational surjective morphism $Q \to \tilde{M}(0,2)$ such that the exceptional set is of codimension $\geqq 2$ (in fact of codimension 5). From this it follows that $\mathrm{Cl}(Q) = \mathrm{Cl}\,\tilde{M}(0,2)$ (where 'Cl' denotes the set of classes of Weil divisors on a normal variety) and that $\tilde{M}(0,2)$ is not locally factorial.

We prove

**Theorem.** 1) *The variety $Q$ is locally factorial.*

2) *The varieties $Q$ and $\tilde{M}(0,2)$ have only rational singularities.*

3) $\mathrm{Pic}(Q) \approx \mathbb{Z}^4$ *and* $\mathrm{Pic}(\tilde{M}(0,2)) \approx \mathbb{Z}^3$, *where* Pic *denotes the Picard group.*

The local factoriality of $Q$ is proved in 3.6. The second part of the theorem is proved in 5.2 and the third part in 2.4 and 5.4. We also give explicit generators of the Picard groups.

To calculate $\mathrm{Pic}(\tilde{M}(0,2))$ we need to show that a line bundle on $Q$ which is trivial on the fibres of the morphism $\varphi: Q \to \tilde{M}(0,2)$ descends to a line bundle on $\tilde{M}(0,2)$; to prove part 2 of the theorem we have to show that $R^i\varphi_*(\mathcal{O}_Q) = 0$ for $i > 0$. These results are proved in § 4 and § 5, the essential point being the proof of the vanishing of the first cohomology of the sheaves $\mathcal{J}/\mathcal{J}^{m+1}$, $m > 0$, where $\mathcal{J}$ is the ideal sheaf in $Q$ of a fibre of $\varphi$.

In the last section we study an interesting Weil divisor on $\tilde{M}(0,2)$ which is not locally principal.

Since we know the components of the divisor $Q \setminus M(0,2)$, the explicit computation of $\mathrm{Pic}(Q)$ allows us to rederive the result of Newstead that $\mathrm{Pic}(M(0,2)) \approx \mathbb{Z}/6\mathbb{Z}$, see 2.6.

**Acknowledgement.** During the preparation of this work both authors were supported by Deutsche Forschungsgemeinschaft, DFG.

**Notation.** $k$ denotes a fixed algebraically closed field of characteristics 0. All varieties and vector-spaces will be over $k$.

If $V$ is a finite dimensional vector space, $PV$ denotes the projective space of 1-dimensional subspaces, $G_m V \subset P\Lambda^m V$ the Grassmannian of $m$-dimensional linear subspaces with its Plücker embedding, $S^m V$, $\Lambda^m V$ the symmetric resp. exterior $m$ th power.

As usual, if $D$ is a Cartier divisor on a variety $X$, $\mathcal{O}_X(D)$ is the associated line bundle. $D_1 \sim D_2$ means linear equivalence $\mathcal{O}_X(D_1) \simeq \mathcal{O}_X(D_2)$. We write $\mathcal{O}_{PV}(d)$ for $\mathcal{O}_{PV}(dH)$ where $H \subset PV$ is a hyperplane, $d \in \mathbb{Z}$.

A normal variety is locally factorial if any Weil-divisor is a Cartier-divisor, see [AK], i.e. if it is locally principal.

## 1. Description of the quadric bundle $Q$

**1.1. Poncelet quadrics.** Let $U$ be a 3-dimensional vector space and $\sigma \in (S^2 U)^\vee$, which we also consider as a symmetric bilinear form $U \times U \to k$ or $U \to U^\vee$. To $\sigma$ we associate the new symmetric bilinear form on $S^2 U$

$$Q\sigma := S^2\sigma - \frac{1}{2}\sigma\cdot\sigma\colon S^2 U \to (S^2 U)^\vee$$

where $S^2\sigma$ is the composition $S^2 U \to S^2 U^\vee \simeq (S^2 U)^\vee$ and $\sigma\cdot\sigma$ is defined by $x\cdot y \to \sigma(x,y)\sigma$. If $c \in S^2 U$ satisfies $Q\sigma(c,c) = 0$, then the conic $C^\vee = \{c = 0\} \subset PU^\vee$ is called Poncelet related to the conic $S = \{\sigma = 0\} \subset PU$. (It had been shown in [NT], § 3, that this is the correct definition of Poncelet related conics, which applies also to the degenerate cases, and generalizes the classical description for smooth conics in one plane $PU$, see also [T] for higher degree.)

The quadric $Q(\sigma) = \{c \mid Q\sigma(c,c) = 0\}$ in $PS^2 U$ is called the Poncelet quadric of $\sigma$. It is smooth if $S$ is smooth, consists of two distinct 4-planes if $S$ consists of two different lines (as the set of conics in $PU^\vee$ passing through one of the points given by $S$), and it is a double 4-plane if $S$ is a double line, see [NT], 3.6. We shall also consider bundles of Poncelet quadrics over families of conics.

**1.2.** Let from now on $V$ be a *4-dimensional vector space*, and let $W \to Z$ be the tautological subbundle on the Grassmannian $Z := G_3\Lambda^2 V$ of 3-dimensional subspaces of $\Lambda^2 V$. We write $W_z \subset \Lambda^2 V$ and $P_z = PW_z \subset P\Lambda^2 V$ for the linear and the associated projective subspace given by $z \in Z$. The projective bundle $PS^2 W^\vee$ can be considered as the

Hilbert scheme of conics in $P\Lambda^2 V$. Under this identification the projection $PS^2 W^\vee \xrightarrow[\varrho]{} Z$ associates to each conic in $P\Lambda^2 V$ the unique plane in which it is contained.

If $G = G^2 V \subset P\Lambda^2 V$ denotes the Grassmannian of lines in $PV$, we let $C(G) \subset PS^2 W^\vee$ be the subvariety of conics contained in $G$. This is the Hilbert scheme of curves in $G$ with Hilbertpolynomial $2m+1$. We are given the diagram

$$\begin{array}{ccc} C(G) & \overset{j}{\hookrightarrow} & PS^2 W^\vee \\ & {\scriptstyle\mu}\searrow \quad \swarrow{\scriptstyle\varrho} & \\ & Z & \end{array} .$$

It had been verified in [NT], 3.8, that $C(G)$ is smooth, connected and of dimension 9. Let $Z_\alpha, Z_\beta \subset Z$ be the subvarieties of $\alpha$- and $\beta$-planes, i.e. of planes in $P\Lambda^2 V$ which are contained in $G$. Then $Z_\alpha$ and $Z_\beta$ are both isomorphic to $\mathbb{P}_3$. Let $\mu = \varrho \circ j$ and $\Sigma_\alpha$, $\Sigma_\beta$ be the inverse images of $Z_\alpha, Z_\beta$. Then $\Sigma_\alpha, \Sigma_\beta$ are $\mathbb{P}_5$-bundles respectively over $Z_\alpha, Z_\beta$, the fibre over a plane being the space of all conics in this plane. Moreover

$$C(G)\backslash \Sigma_\alpha \cup \Sigma_\beta \xrightarrow[\mu]{} Z \backslash Z_\alpha \cup Z_\beta$$

is an isomorphism, [NT], 3.10.

Next let $\tilde{Z}$ be the blow up of $Z$ along $Z_\alpha \cup Z_\beta$. Since $C(G)$ is smooth, $\Sigma_\alpha$ and $\Sigma_\beta$ are locally principal (smooth) divisors. Thus by the universal property of the blow up we have a morphism $C(G) \to \tilde{Z}$ which is easily seen to be an isomorphism by Zariski's main theorem. We let $Z_0 \subset Z$ be the subvariety of those planes in $P\Lambda^2 V$ on which the quadratic form of $G$ has rank $\leqq 1$. This is irreducible of codimension 1 and $Z_\alpha, Z_\beta \subset Z_0$ as the loci of planes where the rank $= 0$. We let $\Sigma_0$ be the proper transform of $Z_0$ and obtain the picture:

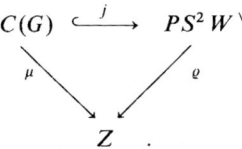

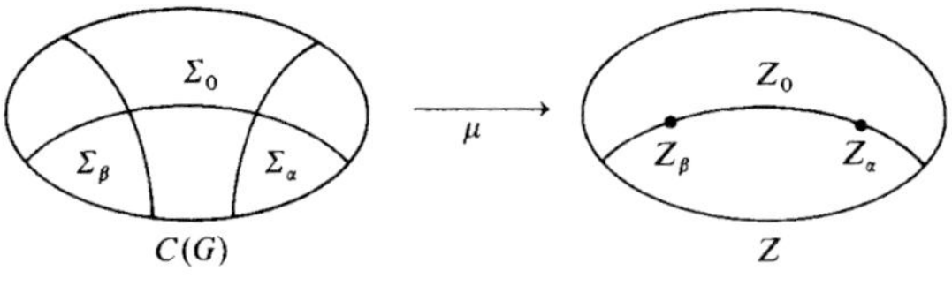


**1.3.** There is a canonical universal quadratic form $\varrho^* W \otimes \varrho^* W \xrightarrow[\sigma]{} \mathcal{O}_{PS^2W^\vee}(1)$ with values in the relative hyperplane bundle on $PS^2 W^\vee$. If we restrict this to $C(G)$ we obtain the associated Poncelet form

$$Q\sigma = S^2\sigma - \frac{1}{2}\sigma \cdot \sigma \colon \mu^* S^2 W \otimes \mu^* S^2 W \to j^* \mathcal{O}_{PS^2W^\vee}(2) .$$

Its scheme of zeros in $P\mu * S^2 W$ is a quadric bundle $Q$ over $C(G)$, such that each fibre $Q(S) \subset PS^2 W_z$ is the Poncelet quadric of $S$. We have the diagram

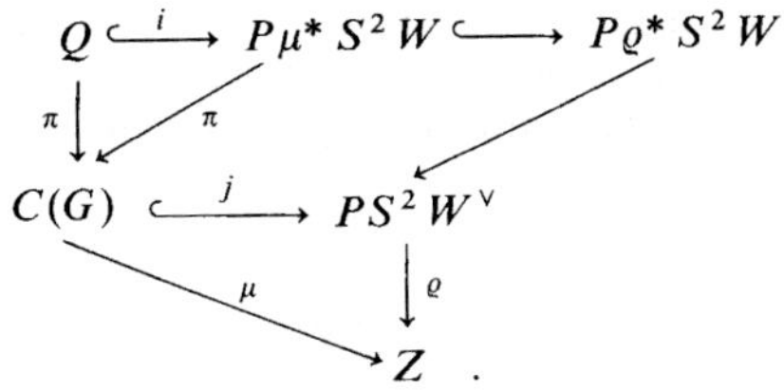


The variety $Q$ can be viewed as the set of pairs $(S, C^\vee)$ of conics, $S \subset G \cap P_z$, $z = \mu(S)$, $C^\vee \subset P_z^\vee$, which are Poncelet related. In $Q$ we are given the following subvarieties:

$$Q_0 = \pi^{-1}(\Sigma_0),\; Q_\alpha = \pi^{-1}(\Sigma_\alpha),\; Q_\beta = \pi^{-1}(\Sigma_\beta),$$

$$Q_e = \text{set of pairs } (S, C^\vee), \text{ where } C^\vee \text{ is singular.}$$

Clearly $Q_0$, $Q_\alpha$, $Q_\beta$ are locally principal divisors and irreducible. Also $Q_e$ is an irreducible subvariety of codimension 1, see [NT], prop. 4.1. We have $\operatorname{Sing} Q \subset Q_0$, since $Q \setminus Q_0 \to C(G) \setminus \Sigma_0$ is a smooth fibration.

By [NT], 3.12, the singular locus of $Q$ consists of the pairs $(S, C^\vee)$, for which both $S$ and $C^\vee$ are singular and such that $C^\vee$ contains both points corresponding to $S$ in the dual plane and such that each line of $C^\vee$ passes through one of these points. These are the cases

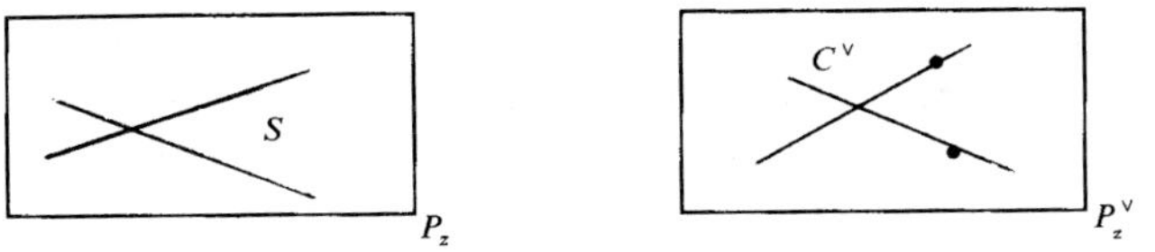


Figure 1

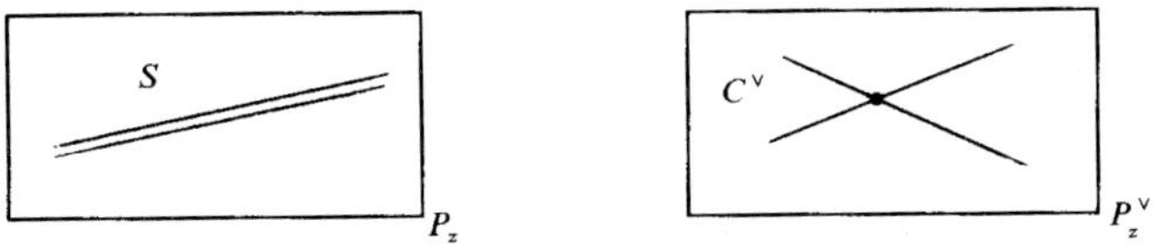


Figure 2

where $C^\vee$ might also be a double line. It follows that codim Sing $Q = 3$. By [NT], 3.13, $Q$ is normal and irreducible.

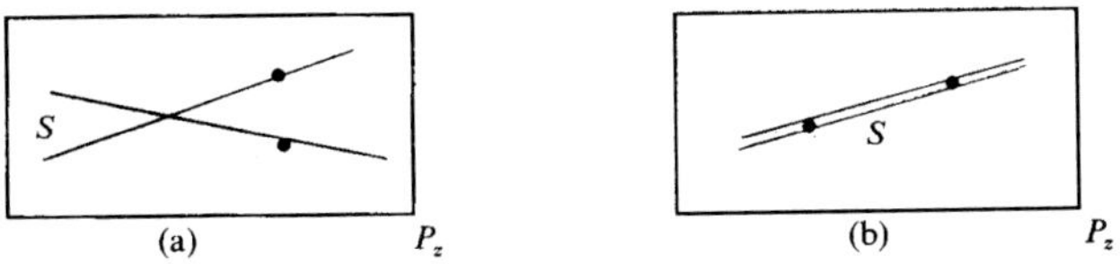


Figure 3

If we mark the two lines (or the double line) of $C^\vee$ as points (or a double point) on $S$, Sing $Q$ consists of the objects (a), (b), where it is allowed that the two points coincide.

**1.4. The $\mathbb{P}_1$-fibration.** We let $Q_{\text{exc}} \subset \operatorname{Sing} Q \subset Q_0$ be the set of objects as in figure 3, (b). If $\Sigma_0' \subset \Sigma_0$ is the subvariety of double lines in $C(G)$, we have $Q_{\text{exc}} = \pi^{-1}(\Sigma_0') \cap \operatorname{Sing} Q$. It follows from the next diagram that the subvariety $Q_{\text{exc}}$ has dimension 8. We have the diagram as follows:

$$\begin{array}{ccc} Q_{\text{exc}} & \xrightarrow{r} & \operatorname{Hilb}_2(D) \\ {\scriptstyle\pi}\downarrow & & \downarrow \\ \Sigma_0' & \xrightarrow{\bar{r}} & L(G) \end{array} .$$

Here $\pi$ is the restriction of the projection, which is a $\mathbb{P}_2$-bundle (consisting of pairs of points in each $S$). $L(G)$ denotes the Hilbert scheme of lines in $G$, $D$ its universal $\mathbb{P}_1$-bundle, and $\operatorname{Hilb}_2(D)$ the relative Hilbert scheme of two points in its fibres. The morphisms $r$ and $\bar{r}$ are defined by $S \mapsto S_{\text{red}}$. They are $\mathbb{P}_1$-bundles, since the double structure of each $S$ corresponds to the plane $P_z \supset S_{\text{red}}$, $S \subset G \cap P_z$, and these planes form a pencil for fixed $S_{\text{red}}$, which is spanned by the unique $\alpha$- and $\beta$-plane containing $S_{\text{red}}$, see [NT], 4.2, 3.9.1, 3.8.

**1.5.** The main result of [NT], §4, was that there exists a canonical morphism

$$Q \xrightarrow{\varphi} \tilde{M}(0,2)$$

which blows down $Q_{\text{exc}}$ to $\varphi(Q_{\text{exc}}) \simeq \operatorname{Hilb}_2(D)$ and such that $Q \setminus Q_{\text{exc}} \to \tilde{M}(0,2) \setminus \varphi(Q_{\text{exc}})$ is an isomorphism. Moreover $\varphi(Q_i)$, $i = 0, e, \alpha, \beta$, are irreducible of codimension 1 and form the boundary components of $\tilde{M}(0,2) \setminus M(0,2)$. Since $Q_{\text{exc}} \subset Q_0$ we have

$$Q \setminus Q_0 \cup Q_e \cup Q_\alpha \cup Q_\beta \simeq M(0,2) .$$

We are going to use the $\mathbb{P}_1$-fibration $Q_{\text{exc}}$ to calculate Pic $\tilde{M}(0,2)$ from Pic $Q$.

## 2. The Picard group of $Q$

**2.1.** Since $C(G) \xrightarrow[\mu]{} Z$ is a blow up with the disjoint exceptional divisors $\Sigma_\alpha$, $\Sigma_\beta$, we obtain the exact diagram, cf. [H1], p. 133, 145,

$$\begin{array}{ccccccc} 0 \longrightarrow & \mathbb{Z}^2 & \xrightarrow{\varepsilon} & \operatorname{Pic} C(G) & \longrightarrow & \operatorname{Pic}(C(G) \setminus \Sigma_\alpha \cup \Sigma_\beta) & \longrightarrow 0 \\ & & & \uparrow {\scriptstyle\mu^*} & & \uparrow \wr & \\ & & & \operatorname{Pic} Z & \xrightarrow[\approx]{} & \operatorname{Pic}(Z \setminus Z_\alpha \cup Z_\beta) . & \end{array}$$

$\varepsilon$ is given by $a\Sigma_\alpha + b\Sigma_\beta$ and injective and also $\mu^*$ is injective. It follows that

$$\operatorname{Pic} C(G) \simeq \mu^* \operatorname{Pic}(Z) \oplus \mathbb{Z}^2 \simeq \mathbb{Z}^3 .$$

To determine the third generator, we denote by $\sigma_{100}$ (in the notation of [GH] the standard Schubert cycle which is represented by the hyperplane section of $Z = G_3 \Lambda^2 V \subset P\Lambda^3(\Lambda^2 V)$ of 2-planes in $P\Lambda^2 V$ meeting a fixed 2-plane. It is the free generator of Pic $Z$. Let $\Sigma_1$ be the total transform of $\sigma_{100}$ in $C(G)$, i.e. $\mathcal{O}_{C(G)}(\Sigma_1) = \mu^* \mathcal{O}_Z(\sigma_{100})$. Then $\Sigma_1$, $\Sigma_\alpha$, $\Sigma_\beta$ are free generators of Pic $C(G)$.

**Lemma.** (i) $Z_0 \sim 2\sigma_{100}$.

(ii) $\Sigma_0 \sim 2\Sigma_1 - \Sigma_\alpha - \Sigma_\beta$.

*Proof.* (i) Since $Z_0$ is a divisor, $Z_0 \sim m\sigma_{100}$. Let $\sigma_{332}$ denote the complementary Schubert cycle of dimension 1. It is represented by any variety of planes $P \subset P\Lambda^2 V$, satisfying $P_1 \subset P \subset P_3$ for a pair $P_1 \subset P_3$ of a fixed line and a fixed 3-plane. If they are in general position $m = \#(Z_0 \cap \sigma_{332})$. Hence we can assume that $P_1 \cap G = \{p, q\}$ and that $P_3 \cap G$ is a smooth quadric. Then $P \in Z_0 \cap \sigma_{332}$ means that $P \cap G = P \cap (P_3 \cap G)$ consists of two lines each passing through one of the points, $p$, $q$. Since $P_3 \cap G$ is smooth, there are exactly two such planes $P$, i.e. $m = 2$.

(ii) Since $\mu^* \mathcal{O}(Z_0) = \mathcal{O}(\Sigma_0 + \Sigma_\alpha + \Sigma_\beta)$ as the total transform, we obtain $2\Sigma_1 \sim \Sigma_0 + \Sigma_\alpha + \Sigma_\beta$.

**2.2. Corollary.** (i) $\operatorname{Pic}(C(G) \setminus \Sigma_0 \cup \Sigma_\alpha \cup \Sigma_\beta) \simeq \mathbb{Z}/2$.

(ii) $\operatorname{Pic}(C(G) \setminus \Sigma_0) \simeq \mathbb{Z}^2$ *and is freely generated by* $\Sigma_\alpha \setminus \Sigma_0$, $\Sigma_1 \setminus \Sigma_0$.

*Proof.* (i) follows immediately from $C(G) \setminus \Sigma_0 \cup \Sigma_\alpha \cup \Sigma_\beta \simeq Z \setminus Z_0$ and the exact sequence

$$\begin{array}{ccccccccc} 0 & \longrightarrow & \mathbb{Z} & \longrightarrow & \operatorname{Pic} Z & \longrightarrow & \operatorname{Pic}(Z \setminus Z_0) & \longrightarrow & 0, \\ & & 1 & \longmapsto & Z_0 . & & & & \end{array}$$

(ii) We also have the exact diagram

$$\begin{array}{ccccccccc} 0 & \longrightarrow & \mathbb{Z} & \xrightarrow{\varepsilon} & \operatorname{Pic} C(G) & \xrightarrow{\varrho} & \operatorname{Pic}(C(G) \setminus \Sigma_0) & \longrightarrow & 0 \\ & & \| & & \| & & \| & & \\ 0 & \longrightarrow & \mathbb{Z} & \xrightarrow[(-1,-1,2)]{} & \mathbb{Z}^3 & \xrightarrow[\alpha]{} & \mathbb{Z}^2 & \longrightarrow & 0 \end{array}$$

in which $\varepsilon$ is given by $m \to m\Sigma_0$ and $\alpha$ by $(a, b, c) \to (a - b, c + 2b)$. Since $\Sigma_\alpha$, $\Sigma_1$, correspond to $(1, 0, 0)$, $(0, 0, 1)$ in $\mathbb{Z}^3$, we see that $\varrho(\Sigma_\alpha)$ and $\varrho(\Sigma_1)$ correspond to $(1, 0)$ and $(0, 1)$ in $\mathbb{Z}^2$.

**2.3.** In addition to the divisors $Q_e$, $Q_0$, $Q_1$, $Q_\alpha$, $Q_\beta$ which are (with the exception of $Q_e$) pullbacks of $\Sigma_0$, $\Sigma_1$, $\Sigma_\alpha$, $\Sigma_\beta$, we also consider the divisor $H$ defined by $\mathcal{O}_Q(H) = i^* \mathcal{O}_P(1)$, $P = P\mu^* S^2 W$, see diagram in 1.3.

It will be shown in § 3 that $Q$ is *locally factorial*. Assuming this here it is now standard to determine $\operatorname{Pic}(Q \setminus Q_0)$ and $\operatorname{Pic} Q$.

Firstly $Q \setminus Q_0 \xrightarrow{\pi} C(G) \setminus \Sigma_0$ is a smooth quadric bundle. If $F$ is one of its fibres, $F \subset PS^2 W_z$ is a smooth quadric. We consider the canonical sequence

$$\begin{array}{ccccccccc} 0 & \longrightarrow & \operatorname{Pic}(C(G)\setminus\Sigma_0) & \xrightarrow{\pi^*} & \operatorname{Pic}(Q\setminus Q_0) & \xrightarrow{i^*} & \operatorname{Pic}(F) & \longrightarrow & 0\,. \\ & & \wr\| & & & & \wr\| & & \\ & & \mathbb{Z}^2 & & & & \mathbb{Z} & & \end{array}$$

**2.3.1. Lemma.** *This sequence is exact,* $\operatorname{Pic}(Q \setminus Q_0) \simeq \mathbb{Z}^3$ *and is freely generated by the restrictions of* $H$, $Q_\alpha$, $Q_1$.

*Proof.* Clearly $i^* \circ \pi^* = 0$. The map $i^*$ is surjective, since $i^* \mathcal{O}_Q(H)$ is the restriction of the hyperplane bundle of $PS^2 W_z$ to the quadric $F$, and this generates $\operatorname{Pic}(F) \simeq \mathbb{Z}$. The map $\pi^*$ is injective: If $L$ is a line bundle on $C(G) \setminus \Sigma_0$, and $\pi^* L \simeq \mathcal{O}$ we obtain, since $\pi_* \mathcal{O}_Q = \mathcal{O}_{C(G)}$ that $L = \pi_* \pi^* L \simeq \pi_* \mathcal{O} \simeq \mathcal{O}$. Finally let $\mathcal{L}$ be any line bundle on $Q \setminus Q_0$ such that $\mathcal{L} | F$ is trivial for the one fibre. Then $\mathcal{L} | Q(S) \simeq \mathcal{O}_{Q(S)}$ for any fibre, since the first Chern class is constant. It follows that $L = \pi_* \mathcal{L}$ is locally free of rank 1 and that the canonical homomorphism

$$\pi^* L \simeq \pi^* \pi_* \mathcal{L} \to \mathcal{L}$$

must be an isomorphism. This proves exactness. The rest follows automatically since $\pi^* \mathcal{O}(\Sigma_i) = \mathcal{O}(Q_i)$ for $i = 1, \alpha$ and $i^* \mathcal{O}(H)$ is the generator of $\operatorname{Pic}(F)$.

Finally we consider the diagram

$$\begin{array}{ccccccccc} & & & & & & \mathbb{Z}^3 & & \\ & & & & & & \| & & \\ 0 & \longrightarrow & \mathbb{Z} & \xrightarrow{\hat{\varepsilon}} & \operatorname{Pic} Q & \xrightarrow{\hat{\varrho}} & \operatorname{Pic}(Q\setminus Q_0) & \longrightarrow & 0 \\ & & \| & & \uparrow{\scriptstyle \pi^*} & & \uparrow{\scriptstyle \pi^*} & & \\ 0 & \longrightarrow & \mathbb{Z} & \xrightarrow[\varepsilon]{} & \operatorname{Pic} C(G) & \xrightarrow[\sigma]{} & \operatorname{Pic}(C(G)\setminus\Sigma_0) & \longrightarrow & 0 \\ & & & & \mathbb{Z}^3 & \xrightarrow[\alpha]{\qquad} & \mathbb{Z}^2 & & \end{array}$$

where $\hat{\varepsilon}(1) = Q_0$, $\varepsilon(1) = \Sigma_0$. The bottom row is exact by the proof of Corollary 2.2, and the top row is exact, i.e. $\hat{\varrho}$ is surjective, since $Q$ is *locally factorial* and normal, [H 1], p. 133, 145.

The maps $\pi^*$ are injective by the previous proof and exactness. It follows from the diagram and Lemma 2.3.1 that $Q_0$, $Q_1$, $Q_\alpha$, $H$ generate $\operatorname{Pic} Q$ and are independent. Since

$$\Sigma_0 \sim 2\Sigma_1 - \Sigma_\alpha - \Sigma_\beta$$

we also obtain

$$Q_0 \sim 2Q_1 - Q_\alpha - Q_\beta,$$

and we can replace the generators by $Q_1$, $Q_\alpha$, $Q_\beta$, $H$. We have proved

**2.4. Proposition.** Pic $Q \simeq \mathbb{Z}^4$ *and is freely generated by* $Q_1$, $Q_\alpha$, $Q_\beta$, $H$.

In addition to the formula for $Q_0$ we are going to prove

**2.5. Lemma.** $Q_e \sim 3H - 2Q_1$.

*Proof.* The intersection of $Q_e$ with a fibre $Q(S)$ of $Q$ is the intersection of $Q(S)$ with the discriminant locus $\Delta_z \subset PS^2 W_z$, $z = \mu(S)$, of singular conics in $PW_z^\vee$. There is a universal discriminant locus $\Delta \subset PS^2 \mu^* W$, whose intersection with any fibre $(P\mu^* S^2 W)_S = PS^2 W_z$ is $\Delta_z$.

To see this, let $E \to X$ be any 3-bundle and consider $PS^2 E \xrightarrow[\pi]{} X$. As in 1.3 there is the universal bilinear form $\pi^* E^\vee \otimes \pi^* E^\vee \to \mathcal{O}_{PS^2E}(1)$. The determinant of $\pi^* E^\vee \to \pi^* E \otimes \mathcal{O}_{PS^2E}(1)$ gives us $\pi^* \Lambda^3 E^\vee \to \pi^* \Lambda^3 E \otimes \mathcal{O}_{PS^2E}(3)$ and therefore a section of $\pi^*(\Lambda^3 E)^{\otimes 2} \otimes \mathcal{O}_{PS^2E}(3)$. The zero locus $\Delta$ of this section is the discriminant, s.t. $\mathcal{O}_{PS^2E}(\Delta) \simeq \pi^*(\Lambda^3 E)^{\otimes 2} \otimes \mathcal{O}_{PS^2E}(3)$, and such that $\Delta \cap PS^2 E_x$ is the discriminant of $PS^2 E_x$. If we apply that to $\mu^* W \to C(G)$, we obtain the divisor $\Delta \subset P\mu^* S^2 W = P$, such that $Q_e = Q \cap \Delta$ and

$$\mathcal{O}_P(\Delta) = \pi^* \mu^* (\Lambda^3 W)^{\otimes 2} \otimes \mathcal{O}_P(3).$$

Since $i^* \mathcal{O}_P(1) = \mathcal{O}_Q(H)$ and $\mu^*(\Lambda^3 W) = \mu^* \mathcal{O}_Z(-\sigma_{100}) = \mathcal{O}_{C(G)}(-\Sigma_1)$, we obtain

$$\mathcal{O}_Q(Q_e) \simeq \mathcal{O}_Q(-2Q_1) \otimes \mathcal{O}_Q(3H),$$

which proves $Q_e \sim 3H - 2Q_1$.

**2.6. Corollary** (Newstead [N]). Pic $M(0,2) = \mathbb{Z}/6$.

*Proof.* We have $Q \backslash Q_0 \cup Q_\alpha \cup Q_\beta \cup Q_e \simeq M(0,2)$, and since $Q \backslash Q_0$ is smooth, the sequence

$$(*) \quad 0 \longrightarrow \mathbb{Z} \longrightarrow \operatorname{Pic}(Q \backslash Q_0 \cup Q_\alpha \cup Q_\beta) \longrightarrow \operatorname{Pic} M(0,2) \longrightarrow 0$$
$$1 \longrightarrow Q_e \cap$$

is exact. On the other hand, as in 2.3 the sequence

$$\begin{array}{ccccccccc} 0 & \longrightarrow & \operatorname{Pic}(C(G)\setminus\Sigma_0\cup\Sigma_\alpha\cap\Sigma_\beta) & \longrightarrow & \operatorname{Pic}(Q\setminus Q_0\cup Q_\alpha\cup Q_\beta) & \longrightarrow & \operatorname{Pic}(F) & \longrightarrow & 0 \\ & & \| & & & & \| & & \\ & & \mathbb{Z}/2 & & & & \mathbb{Z} & & \end{array}$$

is exact, such that the group in the middle is isomorphic to $\mathbb{Z}/2\oplus\mathbb{Z}$ and freely generated by

$$Q_1\cap \quad \text{and} \quad H\cap$$

with $2Q_1\cap\sim 0$. Since $Q_e\sim 3H-2Q_1$ the sequence (*) becomes

$$0\longrightarrow \mathbb{Z}\xrightarrow[(3,0)]{}\mathbb{Z}\oplus\mathbb{Z}/2\longrightarrow\mathbb{Z}/3\oplus\mathbb{Z}/2\longrightarrow 0,$$

and hence the result.

## 3. Local factoriality of $Q$

To prove that $Q$ is locally factorial we use the geometric invariant theoretical description $X^{ss}//\mathrm{SL}(2)=Q$ of [NT], § 6, and the method of [DN]. Before we recall briefly the construction of the parameter space $X$, we need the definition of the following quadratic forms.

**3.1.** Let $U$ be a 3-space and $\Gamma\subset k^2\otimes U$ a 4-dimensional subspace. If $k^2\otimes U\to\Sigma$ is the cokernel, we obtain a morphism $k^2\otimes\mathcal{O}_{PU}(-1)\to k^2\otimes U\otimes\mathcal{O}_{PU}\to\Sigma\otimes\mathcal{O}_{PU}$ and we let $\sigma(\Gamma)$ denote its determinant. It is a quadratic form which might be zero. If $\sigma(\Gamma)\neq 0$ we denote by $S(\Gamma)\subset PU$ its conic. Similarly if $M\subset k^2\otimes U$ is a 2-subspace, we obtain the morphism $M\otimes\mathcal{O}_{PU^\vee}\to k^2\otimes U^{\vee\vee}\otimes\mathcal{O}_{PU^\vee}\to k^2\otimes\mathcal{O}_{PU^\vee}(1)$. If the determinant is nonzero, we denote the conic defined by it by $C^\vee(M)\subset PU^\vee$.

**3.2.** Let now $X\subset Z\times G_2(k^2\otimes\Lambda^2V)\times G_4(k^2\otimes\Lambda^2V)$ be the subset of triples $(z,M,\Gamma)$ which satisfy

(i) $M\subset\Gamma\subset k^2\otimes W_z$,

(ii) the quadratic form $\sigma(\Gamma)$ on $PW_z$ is proportional to the quadratic form induced on $PW_z$ by the Plücker quadric $G\subset P\Lambda^2V$ (both may be zero).

Obviously $X$ is a closed subvariety (in the flag variety defined by (i)). There is an SL(2)-action on the above product induced by the natural action on $k^2$, and $X$ is invariant under this action. The action is linearised by Plücker and Segre embeddings. Therefore there is the notion of semi-stable (ss) and stable (s) points in the above SL(2)-varieties. By [NT], 6.3.1, $X^{ss}=X\cap(Z\times G_2(\ldots)^{ss}\times G_4(\ldots)^{ss})$, and by [NT], 5.7.2 and 6.4, $X^{ss}$ is smooth. Furthermore the map $(z,M,\Gamma)\to(S(\Gamma),C^\vee(M))$ (notations as in 3.1, if $M$ and $\Gamma$ are semi-stable then $C^\vee(M)$ and $S(\Gamma)$ are defined as conics, which are Poncelet related) is a morphism $X^{ss}\xrightarrow[q]{}Q$ which yields $X^{ss}//\mathrm{SL}(2)\simeq Q$ as a good quotient, [NT], 6.6.1. By loc. cit. 6.7.1 we have $X^{ss}\setminus X^s=q^{-1}(\mathrm{Sing}\,Q)$.

**3.3. Orbits.** For each $(z, M, \Gamma) \in X$ the spaces $M, \Gamma$ can be represented by matrices $k^2 \xrightarrow{*M} k^2 \otimes W_z$, $k^4 \xrightarrow{*\Gamma} k^2 \otimes W_z$ with entries in $W_z \subset \Lambda^2 V$. In particular if $q_0 \in Q$ is visualised by the figure A, see 1.3,

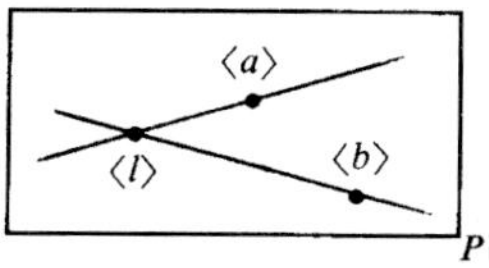


Figure A

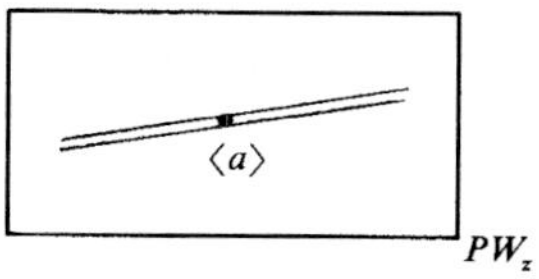


Figure B

where $W_z = \operatorname{span}(\ell, a, b)$, $x_0 = (z, M, \Gamma) \in q^{-1}(q_0)$ can be chosen so that

$$*M = \begin{pmatrix} a & 0 \\ 0 & b \end{pmatrix}, \qquad *\Gamma = \left(\begin{array}{c|c} a & 0 \\ \ell & 0 \\ \hline 0 & \ell \\ 0 & b \end{array}\right) \qquad \text{(type A)}\,.$$

Such an $x_0$ has minimal closed orbit in $q^{-1}(q_0)$ and its stabilizer subgroup $\mathrm{SL}(2)_{x_0}$ is the group $k^*$ of diagonal matrices $\operatorname{diag}(t, t^{-1})$. Moreover there are only two non-closed orbits in $q^{-1}(q_0)$, $o(z, M_1, \Gamma_1)$, $o(z, M_2, \Gamma_2)$, represented by

$$*M_1 = \begin{pmatrix} a & 0 \\ 0 & b \end{pmatrix}, \qquad *\Gamma_1 = \left(\begin{array}{c|c} a & 0 \\ \ell & 0 \\ \hline b & \ell \\ 0 & b \end{array}\right)$$

and

$$*M_2 = \begin{pmatrix} a & 0 \\ 0 & b \end{pmatrix}, \qquad *\Gamma_2 = \left(\begin{array}{c|c} a & 0 \\ \ell & a \\ \hline 0 & \ell \\ 0 & b \end{array}\right)$$

see [NT], 5.3.1. For such points the stabilizer is $\{\pm \mathrm{id}\}$. A point $q_0 \in Q$ with figure B can be presented by an $x_0 = (\operatorname{Span}(\ell, a, b), M, \Gamma)$ with

$$*M = \begin{pmatrix} a & 0 \\ 0 & a \end{pmatrix}, \qquad *\Gamma = \left(\begin{array}{c|c} a & 0 \\ \ell & 0 \\ \hline 0 & \ell \\ 0 & a \end{array}\right) \qquad \text{(type B)}$$

such that the orbit $O(x_0)$ is closed and minimal in $q^{-1}(q_0)$ with stabilizer the whole group $\mathrm{SL}(2)$.

**3.4. Lemma.** *Let $x_0 \in X^{ss} \setminus X^s$ be a point of type* A. *Then there is a rational irreducible curve $R \subset X^{ss} \setminus X^s$, contained in the union of closed orbits, such that*

(i) $x_0 \in R$,

(ii) *R contains a point $y_0$ of type* B,

(iii) $\ell^* \subset \mathrm{SL}(2)_y$ for any $y \in R$.

$$\begin{array}{l} \| \\ \mathrm{SL}(2)_{x_0} \end{array}$$

*Proof.* Since $\ell, a, b \in \Lambda^2 V$ are decomposable and $\ell \wedge a = 0, \ell \wedge b = 0$ ($\langle \ell \rangle, \langle a \rangle$ and $\langle \ell \rangle, \langle b \rangle$ belong to the same line in $G_2 V$) we can find vectors $x, y, x', y' \in V$ with $\ell = x \wedge y$, $a = x \wedge x'$, $b = y \wedge y'$. These vectors are linear independent. Let now

$$v(s,t) = sy + tx, \quad v'(s,t) = sy' + tx'$$

and $b(s,t) = v(s,t) \wedge v'(s,t)$. Then $b(1,0) = y \wedge y' = b$ and $b(0,1) = x \wedge x' = a$. The map $\mathbb{P}_1 \to Z$

$$\langle s, t \rangle \mapsto z(s,t) = \mathrm{span}\,(\ell, a, b(s,t))$$

is well-defined. The curve $\mathbb{P}_1 \to R \subset X^{ss} \setminus X^s$ is now defined by

$$\left(z(s,t)\,,\ \mathrm{span}\begin{pmatrix} a & 0 \\ 0 & b(s,t) \end{pmatrix},\ \mathrm{span}\left(\begin{array}{c|c} a & 0 \\ \ell & 0 \\ \hline 0 & \ell \\ 0 & b(s,t) \end{array}\right)\right).$$

By 3.3 any point $y \in R$ has a closed orbit, and for $\langle s, t \rangle \neq \langle 0, 1 \rangle$ the point is of type A, whereas $y_0 \in R$ for $\langle 1, 0 \rangle$ is of type B. This proves 3.4.

**3.5. Lemma.** $\operatorname{codim}(X^{ss} \setminus X^s) \geqq 2$.

*Proof.* Since $X^{ss} \setminus X^s = q^{-1}(\mathrm{Sing}\, Q)$ and $\operatorname{codim}(\mathrm{Sing}\, Q) = 3$, we have only to show that the general fibre $q^{-1}(q_0)$, $q_0 \in \mathrm{Sing}\, Q$, has dimension $\leqq 4$. But as pointed out in 3.3, $q^{-1}(q_0)$ is the union of three orbits, two of them non-closed of dimension 3 and one closed in the closure of the former. Hence $\dim q^{-1}(q_0) \leqq 3$ for general $q_0 \in \mathrm{Sing}\, Q$, whereas $\dim X^{ss} = 16$.

**3.6.** We can now prove that $Q$ is locally factorial as in [DN] using 3.4, 3.5. It is enough to show that any line bundle $L$ on $Q \setminus \mathrm{Sing}\, Q$ can be extended to a line bundle on $Q$, because then any irreducible codimension-1 subvariety is locally principal. Let $L$ be such a line bundle. The bundle $q^* L$ on $X^s$ can be extended to a *unique* line bundle $\mathbb{L}$ on $X^{ss}$ ($q^* L$ extends using Lemma 3.5 to a coherent algebraic sheaf $\mathcal{F}$ of rank 1 and $\mathcal{F}^{\vee\vee}$ is locally free being a reflexive rank 1 sheaf on the smooth variety $X^{ss}$). The bundle $q^* L$ comes with a natural structure of an SL(2)-bundle over $\mathrm{SL}(2) \times X^s \to X^s$. As in [DN], Lemma 5.2, one can prove using 3.5 and the smoothness of $X^{ss}$ that the action $\mathrm{SL}(2) \times q^* L \to q^* L$ over $\mathrm{SL}(2) \times X^s \to X^s$ extends to an action $\mathrm{SL}(2) \times \mathbb{L} \to \mathbb{L}$ over $\mathrm{SL}(2) \times X^{ss} \to X^{ss}$.

By the theorem of Kempf and Drezet-Narasimhan [DN], theorem 2.3, the bundle $\mathbb{L}$ descends to a bundle $L$ on $Q$ if the stabilizer $\mathrm{SL}(2)_x$ of the action on $X^{ss}$ acts trivially on $\mathbb{L}_x$ for any $x$ with a closed orbit $\mathrm{O}(x)$. This has only to be verified for points in $X^{ss}\setminus X^s$. If $x_0 \in X^{ss}\setminus X^s$ is not of type B (in that case $\mathrm{SL}(2)_x = \mathrm{SL}(2)$ and the representation of SL(2) on $\mathbb{L}_x$ is trivial) it is of type A and by 3.4 there is a rational curve $R \subset X^{ss}\setminus X^s$ meeting only closed orbits with (i), (ii), (iii). Since $k^* = \mathrm{SL}(2)_{x_0}$ acts on each $\mathbb{L}_y$ $y \in R$, by (iii) we get a family of 1-dimensional representations of $k^*$, i.e. a family $\chi_y\colon k^* \to k^*$ of characters. Since $\mathbb{L}|R$ is locally trivial, $y \to \chi_y$ must be continous. But the characters of $k^*$ form a discrete family, hence $\chi_y$ is constant. However $\chi_{y_0}$ is induced by the action of SL(2) on $\mathbb{L}_{y_0}$ and hence must be trivial. This proves that $\chi_{x_0} = 1$ and that means that $\mathrm{SL}(2)_{x_0}$ acts trivially on $\mathbb{L}_{x_0}$. By the theorem of descent mentioned above $\mathbb{L}$ descends to $Q$, $\bar{L} = \mathbb{L}//\mathrm{SL}(2)$ extends L.

## 4. Vanishing Lemma

The canonical morphism $Q \xrightarrow[\varphi]{} \tilde{M}(0,2)$ blows down the $\mathbb{P}_1$-fibration $Q_{\mathrm{exc}} \to \mathrm{Hilb}_2(D)$, 1.4, 1.5. We are going to show that $\mathrm{Pic}\,\tilde{M}(0,2) \subset \mathrm{Pic}\,Q$ is the subgroup of line bundles $\mathscr{L}$ on $Q$ which are trivial on the fibres $A$ of $\varphi$. To prove that for such bundles $\varphi_*\mathscr{L}$ is locally free, we need their triviality in infinitesimal neighborhoods $A_m$ of $A$ in $Q$, and this is proved by showing that $h^1(\mathcal{O}/\mathcal{J}_A^{m+1}) = 0$ for all $m$. For this we first give a precise description of such a fibre.

**4.1.** If $(g, p_1, p_2) \in \mathrm{Hilb}_2(D)$ is a line with two marked points, the fibre in $Q_{\mathrm{exc}}$ consists of pairs $(S_P, C^\vee)$, where $P$ varies in the pencil of planes through $g$, see 1.4, $S_P$ is the double structure on $g$ defined by $g \subset P$, and $C^\vee \subset P^\vee$ is the pair of lines given by the two points on $g$ (if $p_1 = p_2$ then $C^\vee$ is the double line in $P^\vee$ given by the pt.). To describe the fibre as a rational curve in $Q$ we choose two 2-dimensional subspaces

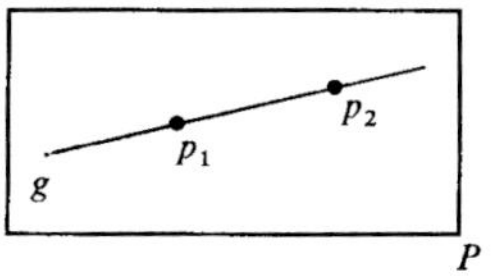


$$F, U \subset \Lambda^2 V \quad \text{and} \quad f_1, f_2 \in F$$

such that $F \cap U = 0$, $g = PF$, $p_i = \langle f_i \rangle$ and $P(F \oplus \langle u \rangle)$ are the planes of the pencil as $\langle u \rangle$ varies in $PU$. We write $S_{\langle u \rangle}$ for $S_{P(F\oplus\langle u \rangle)}$ and note that the equation of $C^\vee \subset P(F \oplus \langle u \rangle)^\vee$ is given by

$$f_1 f_2 \in S^2(F \oplus \langle u \rangle) = S^2 F \oplus (F \otimes \langle u \rangle) \oplus \langle u \rangle^2 .$$

Since $PS^2(F \oplus \langle u \rangle)$ is the fibre of the bundle $PS^2\mu^* W$ over the conic $S_{\langle u \rangle} \in C(G)$, see 1.3, we obtain the morphism

$$PU \xrightarrow{\ a\ } Q \subset PS^2\mu^* W$$

by $\langle u \rangle \to (S_{\langle u \rangle}, \langle f_1 f_2 \rangle_{\langle u \rangle})$. This is obviously an embedding and its image $A$ is the fibre $r^{-1}(g, p_1, p_2)$ in $Q_{\mathrm{exc}}$. We denote the images of $\pi \circ a$ and $\mu \circ \pi \circ a$ by $B$ and $C$ and thus get the diagram

$$\begin{array}{ccccccc}
 & & A & \stackrel{\alpha}{\hookrightarrow} & Q & \stackrel{i}{\hookrightarrow} & PS^2\mu^* W =: P \\
 & \nearrow^{\approx} & \downarrow\wr & & \downarrow \pi & & \\
\mathbb{P}U & \xrightarrow{\approx} & B & \stackrel{\beta}{\hookrightarrow} & C(G) & & \\
 & \searrow_{\approx} & \downarrow\wr & & \downarrow & & \\
 & & C & \stackrel{\gamma}{\hookrightarrow} & Z\,. & &
\end{array}$$

Thus $B$ is the image of the embedding $\langle u\rangle \to S_{\langle u\rangle}$ and $C$ is the image of the embedding $\langle u\rangle \to P(F \oplus \langle u\rangle)$.

**4.1.1. Remark.** a) Since $\mu$ is the blow up of $Z$ along $Z_\alpha$, $Z_\beta$, the curve $B$ can also be considered as the proper transform of $C$.

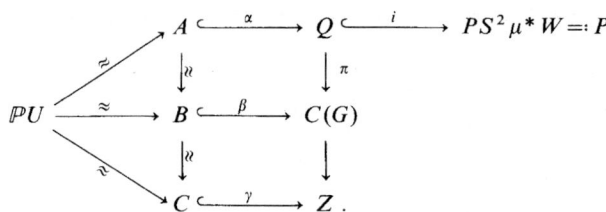

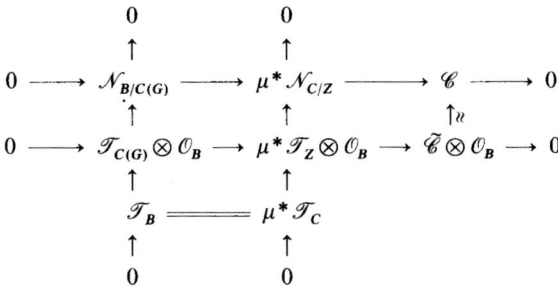

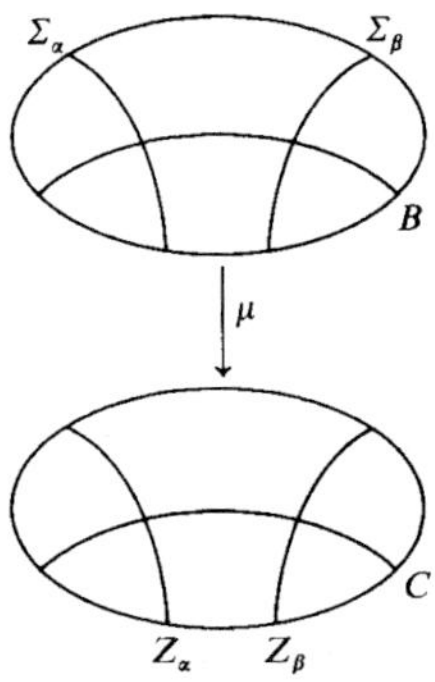


b) It can be easily verified that $C$ meets $Z_\alpha$ and $Z_\beta$ each in one point and that the intersection is transversal by choosing canonical coordinates of the Grassmannian in these points. Hence also $B$ intersects $\Sigma_\alpha$ and $\Sigma_\beta$ transversally each in a point.

c) We are given the canonical exact diagram of the normal and tangent sheaves over $B$.

$$\begin{array}{ccccccccc}
 & & 0 & & 0 & & & & \\
 & & \uparrow & & \uparrow & & & & \\
0 & \longrightarrow & \mathcal{N}_{B/C(G)} & \longrightarrow & \mu^*\mathcal{N}_{C/Z} & \longrightarrow & \mathscr{C} & \longrightarrow & 0 \\
 & & \uparrow & & \uparrow & & \uparrow\wr & & \\
0 & \longrightarrow & \mathcal{T}_{C(G)} \otimes \mathcal{O}_B & \longrightarrow & \mu^*\mathcal{T}_Z \otimes \mathcal{O}_B & \longrightarrow & \tilde{\mathscr{C}} \otimes \mathcal{O}_B & \longrightarrow & 0 \\
 & & \uparrow & & \uparrow & & & & \\
 & & \mathcal{T}_B & = & \mu^*\mathcal{T}_C & & & & \\
 & & \uparrow & & \uparrow & & & & \\
 & & 0 & & 0 & & & &
\end{array}$$

Here $\mathscr{C}$ is the canonical bundle of rank 5 on $\Sigma_\alpha \cup \Sigma_\beta$ given by the exact sequence

$$0 \to \mathcal{O}_{\Sigma_\alpha}(-1) \oplus \mathcal{O}_{\Sigma_\beta}(-1) \to \mu^*\mathcal{N}_{Z_\alpha/Z} \oplus \mu^*\mathcal{N}_{Z_\beta/Z} \to \tilde{\mathscr{C}} \to 0$$

arising from $\Sigma_\alpha = P(\mathcal{N}_{Z_\alpha/Z})$.

**4.2. Lemma.** (i) $\mathcal{O}_P(d)|A = \mathcal{O}_A$ *for any d, where* $P = PS^2\mu^* W$.

(ii) *The relative tangent bundle* $\mathcal{T}(P/C(G))$ *restricts to*

$$\mathcal{T}(P/C(G))|A \simeq 2\mathcal{O}_A \oplus 2\mathcal{O}_A(-1) \oplus \mathcal{O}_A(-2).$$

*Proof.* (i) Since $W$ is a subbundle of the trivial bundle $Z \times \Lambda^2 V$ the projective bundle $P = P\mu^* S^2 W$ is a subbundle of $C(G) \times PS^2\Lambda^2 V$. In particular $\mathcal{O}_P(1) = \tau^* \mathcal{O}_{PS^2\Lambda^2 V}(1)$ where $\tau$ is the composition

$$P \longrightarrow C(G) \times PS^2\Lambda^2 V \longrightarrow PS^2\Lambda^2 V.$$

By construction $A$ is contained in the fibre of $\tau$ over $\langle f_1, f_2\rangle$. Therefore $\mathcal{O}_P(1)$ is trivial on $A$.

(ii) On $P$ we have the relative tautological sequence

$$0 \longrightarrow \mathcal{O}_P(-1) \longrightarrow \pi^*\mu^* S^2 W \longrightarrow \mathcal{T}(P/C(G)) \otimes \mathcal{O}_P(-1) \longrightarrow 0.$$

If we restrict this to $A$ we obtain by (i)

$$0 \longrightarrow \mathcal{O}_A \xrightarrow{\ \varepsilon\ } \mathcal{E}_A \longrightarrow \mathcal{T}(P/C(G))|A \longrightarrow 0$$

where $\mathcal{E}_A = S^2 F \otimes \mathcal{O}_A \oplus F \otimes \mathcal{O}_A(-1) \oplus \mathcal{O}_A(-2)$ is the lift of $\mathcal{E}$ to $A$. In this sequence $\varepsilon$ is given by the constant vector $f_1 f_2 \in S^2 F$. If we cancel this, we obtain the isomorphism (ii).

**4.2.1. Corollary.** *If* $h^0\mathcal{N}_{B/C(G)}(-1) = 0$ *then* $h^0\mathcal{N}_{A/P}(-1) = 0$.

*Proof.* There is the canonical sequence of bundles

$$0 \longrightarrow \mathcal{T}(P/C(G))|A \longrightarrow \mathcal{N}_{A/P} \longrightarrow \mu_A^* \mathcal{N}_{B/C(G)} \longrightarrow 0.$$

Since $\pi|A$ is an isomorphism the result follows from (ii).

In order to prove $h^0\mathcal{N}_{B/C(G)}(-1) = 0$, we first show

**4.3. Lemma.** $\mathcal{N}_{C/Z} \simeq 4\mathcal{O}_C(1) \oplus 4\mathcal{O}_C$.

*Proof.* Let $U \subset E \subset \Lambda^2 V$ be maximal with $E \cap F = 0$. Then $E/\langle u\rangle \simeq \Lambda^2 V/F \oplus \langle u\rangle$. The tangent map at $\langle u\rangle$ of the linear embedding $\langle u\rangle \to F \oplus \langle u\rangle$, $PU \hookrightarrow G_3\Lambda^2 V = Z$ is then given by

$$\begin{array}{ccc} \mathrm{Hom}(\langle u\rangle, U/\langle u\rangle) & \longrightarrow & \mathrm{Hom}(F \oplus \langle u\rangle, \Lambda^2 V/F \oplus \langle u\rangle) \\ & & \| \\ & & \mathrm{Hom}(\langle u\rangle, E/\langle u\rangle) \oplus \mathrm{Hom}(F, E/\langle u\rangle) \end{array}$$

assigning to $h$ the pair $(\tilde{h}, 0)$, where $\tilde{h}\colon \langle u\rangle \to E/\langle u\rangle$ is the induced map. It follows that the cokernel of this map is the fibre $N_{\langle u\rangle}$ of the normal bundle $\mathcal{N}_{C/Z}$ at $\langle u\rangle$:

$$\operatorname{Hom}(\langle u\rangle, E/U) \oplus \operatorname{Hom}(F, E/\langle u\rangle) = N_{\langle u\rangle}.$$

To determine the bundle $\mathcal{N}_{C/Z}$ we note first that $\langle u\rangle$ resp. $U/\langle u\rangle$ are the fibres of $\mathcal{O}_C(-1)$ resp. $\mathcal{O}_C(1)$ at $\langle u\rangle$. If $\mathcal{A}$ is the bundle with fibres $E/\langle u\rangle$, the exact sequences $0 \to U/\langle u\rangle \to E/\langle u\rangle \to E/U \to 0$ give us an extension $0 \to \mathcal{O}_C(1) \to \mathcal{A} \to E/U \otimes \mathcal{O}_C \to 0$, which must be trivial. Hence $\mathcal{A} = \mathcal{O}_C(1) \oplus E/U \otimes \mathcal{O}_C$. Now it follows from the formula for $N_{\langle u\rangle}$ that

$$\mathcal{N}_{C/Z} = E/U \otimes \mathcal{O}_C(1) \oplus F^\vee \otimes \mathcal{O}_C(1) \oplus F^\vee \otimes E/U \otimes \mathcal{O}_C.$$

In our case $\dim E = 4$. This proves the Lemma.

**4.4. Lemma.** $h^0 \mathcal{N}_{B/C(G)}(-1) = 0$.

*Proof.* a) Since $\mu|B$ is an isomorphism onto $C$ we get $\mu^* \mathcal{N}_{C/Z} = 4\mathcal{O}_B(1) \oplus 4\mathcal{O}_B$. The canonical sequence in 4.1.1 c) gives us the exact sequence

$$0 \longrightarrow \mathcal{N}_{B/C(G)} \longrightarrow 4\mathcal{O}_B(1) \oplus 4\mathcal{O}_B \longrightarrow \mathscr{C} \longrightarrow 0$$

where $\mathscr{C}$ is supported on the two intersection points and $\mathscr{C}(p) = \mathscr{C}_p \otimes k(p) = k(p)^5$ for $p = x, y$.

b) Since $B$ is the projective line, the bundle $\mathcal{N}_{B/C(G)}$ splits into a direct sum of line bundles $\mathcal{O}_B(d)$, and by the injection all $d \leqq 1$. Thus $\mathcal{N}_{B/C(G)} = a\mathcal{O}_B(1) \oplus b\mathcal{O}_B \oplus \mathcal{L}$, where $\mathcal{L}$ has only negative factors.

c) Let $\Sigma_0' \subset C(G)$ be the subvariety of those conics which are double lines. By 1.4, $\Sigma_0'$ is a $\mathbb{P}_1$-bundle $\Sigma_0' \to L(G)$ and $B$ is one of its fibres. Therefore

$$\mathcal{N}_{B/\Sigma_0'} \simeq 5\mathcal{O}_B.$$

On the other hand there is the canonical sequence

$$0 \longrightarrow \mathcal{N}_{B/\Sigma_0'} \longrightarrow \mathcal{N}_{B/C(G)} \longrightarrow \mathcal{N}_{\Sigma_0'/C(G)}|B \longrightarrow 0,$$

which proves that $\mathcal{N}_{B/C(G)}$ must have $5\mathcal{O}_B$ as a subbundle. This must be contained in $a\mathcal{O}_B(1) \oplus b\mathcal{O}_B$. Hence $5 \leqq a + b$.

d) Using now the properties of $\mathscr{C}$ and the exact sequence

$$0 \longrightarrow a\mathcal{O}_B(1) \oplus b\mathcal{O}_B \oplus \mathcal{L} \longrightarrow 4\mathcal{O}_B(1) \oplus 4\mathcal{O}_B \longrightarrow \mathscr{C} \longrightarrow 0$$

we are going to exclude the case $a > 0$, which then proves the Lemma.

Since $a\mathcal{O}_B(1)$ is mapped injectively into $4\mathcal{O}_B(1)$, we must have $a \leqq 4$ and $a\mathcal{O}_B(1)$ can be cancelled from the sequence. Moreover if $m$ is the rank of $b\mathcal{O}_B \to 4\mathcal{O}_B$, then also $m\mathcal{O}_B$ can be cancelled. We are thus left with a resolution

$$0 \longrightarrow (b-m)\mathcal{O}_B \oplus \mathcal{L} \xrightarrow[\left(\begin{array}{c|c} L & O \\ \hline * & M \end{array}\right)]{} (4-a)\mathcal{O}_B(1) \oplus (4-m)\mathcal{O}_B \longrightarrow \mathcal{C} \longrightarrow 0$$

in which the block $L$ is linear. Since $\mathcal{C}(x) = k^5(x)$ we must have $(4-a)+(4-m) \geqq 5$ or

$$3 \geqq a+m.$$

Moreover $b-m \leqq 4-a$ since $L$ is injective. Hence also

$$5 \leqq a+b \leqq 4+m.$$

We are thus left with the cases

$$(a,b,m) = (2,4,1),\ (1,4,1),\ (1,4,2),\ (1,5,2).$$

In the first case $(2,4,1)$ we have an epimorphism $2\mathcal{O}_B(1) \oplus 3\mathcal{O}_B \to \mathcal{C} \to 0$ and it follows from $\mathcal{C}(p) = k(p)^5$ at $p = x, y$ that $L(x) = 0$, $L(y) = 0$, hence $L = 0$ because it is linear. This contradicts injectivity of $L$ or $b-m = 3$. By the same argument the cases $(1,4,2)$, $(1,5,2)$ can be excluded. In case $(1,4,1)$ we have the resolution

$$0 \longrightarrow 3\mathcal{O}_B \oplus \mathcal{L} \longrightarrow 3\mathcal{O}_B(1) \oplus 3\mathcal{O}_B \longrightarrow \mathcal{C} \longrightarrow 0.$$

and $L$ is a $3 \times 3$ square matrix. Again by $\mathcal{C}(\mathrm{p}) = k(p)^5$ for $p = x, y$ we obtain that

$$\operatorname{rk} L(p) \leqq 1 \quad \text{for} \quad p = x, y.$$

If $z$ is a local coordinate of $\mathbb{P}_1$ defined at $x, y$, we find that $\det L$ is divisible by $(z-x)^2$, $(z-y)^2$. But since $L$ is linear, $\det L$ is a form of degree 3, this is a contradiction. This finally proves the Lemma. By 4.2.1 we obtain

**4.4.1. Corollary.** $h^0\mathcal{N}_{A/P}(-1) = 0$.

**4.5. Lemma.** *Let* $\mathcal{I} = \mathcal{I}_A$ *be the ideal sheaf of* $A$ *in* $Q$. *Then*

$$h^1(\mathcal{I}^m/\mathcal{I}^{m+1}) = h^1(\mathcal{O}/\mathcal{I}^{m+1}) = h^1(\mathcal{I}/\mathcal{I}^{m+1}) = 0 \ \textit{for} \ m \geqq 0.$$

*Proof.* Let in addition $\mathcal{J} \subset \mathcal{O}_P$ be the ideal sheaf of $A$ in $P$. Then $\mathcal{J}^m/\mathcal{J}^{m+1} \simeq S^m\mathcal{N}_{A/P}^\vee$. Since $P$ is smooth $\mathcal{N}_{A/P}$ is a bundle and by 4.4.1 it is of the form $\bigoplus \mathcal{O}_A(d_i)$ with $d_i \leqq 0$. Hence $S^m\mathcal{N}_{A/P}^\vee$ contains only summands of degree $\geqq 0$ and therefore $h^1\mathcal{J}^m/\mathcal{J}^{m+1} = 0$ for all $m$.

We have now the exact diagram

$$\begin{array}{ccccccccc}
0 & \longrightarrow & \mathscr{J} & \longrightarrow & \mathcal{O}_P & \longrightarrow & \mathcal{O}_A & \longrightarrow & 0 \\
 & & \downarrow & & \downarrow & & \| & & \\
0 & \longrightarrow & \mathcal{J} & \longrightarrow & \mathcal{O}_Q & \longrightarrow & \mathcal{O}_A & \longrightarrow & 0 \\
 & & \downarrow & & \downarrow & & & & \\
 & & 0 & & 0 & & & &
\end{array}$$

and obtain from this the induced diagram of epimorphisms

$$\begin{array}{ccc}
\mathscr{J}^m & \twoheadrightarrow & \mathscr{J}^m/\mathscr{J}^{m+1} \\
\downarrow & & \downarrow \\
\mathcal{J}^m & \twoheadrightarrow & \mathcal{J}^m/\mathcal{J}^{m+1} .
\end{array}$$

Since $A$ is isomorphic to $\mathbb{P}_1$, $h^1\mathscr{J}^m/\mathscr{J}^{m+1} = 0$ implies $h^1\mathcal{J}^m/\mathcal{J}^{m+1} = 0$. If we look at the exact sequences

$$0 \longrightarrow \mathcal{J}^m/\mathcal{J}^{m+1} \longrightarrow \mathcal{J}/\mathcal{J}^{m+1} \longrightarrow \mathcal{J}/\mathcal{J}^m \longrightarrow 0$$

we find by induction that $h^1\mathcal{J}/\mathcal{J}^{m+1} = 0$ for $m \geqq 0$ and similarly from

$$0 \longrightarrow \mathcal{J}^m/\mathcal{J}^{m+1} \longrightarrow \mathcal{O}/\mathcal{J}^{m+1} \longrightarrow \mathcal{O}/\mathcal{J}^m \longrightarrow 0$$

that $h^1(\mathcal{O}/\mathcal{J}^{m+1}) = 0$ for $m \geqq 0$.

For later use we need

**4.6. Lemma.** $\mathcal{O}_Q(Q_i)|A \simeq \mathcal{O}_A(1)$ *for* $i = 1, \alpha, \beta$.

*Proof.* a) We still use the diagram of 4.1. Let $\mathcal{O}_Z(1)$ be given by the Plücker embedding. Then $\mathcal{O}_{C(G)}(\Sigma_1) = \mu^*\mathcal{O}_Z(1)$ and $\mathcal{O}_Q(Q_1) = \pi^*\mu^*\mathcal{O}_Z(1)$. If $A \underset{\varrho}{\longrightarrow} C$ denotes the isomorphism $A \underset{\approx}{\rightarrow} B \underset{\approx}{\rightarrow} C$, we have $\mathcal{O}_Q(Q_1)|A = \varrho^*(\mathcal{O}_Z(1)|C)$. But since $C \hookrightarrow Z \hookrightarrow P\Lambda^3\Lambda^2 V$ is linearly embedded $\mathcal{O}_Z(1)|C = \mathcal{O}_C(1)$. This gives the result for $Q_1$.

b) If $i = \alpha, \beta$ it is enough to show that $\mathcal{O}_{C(G)}(\Sigma_i)|B = \mathcal{O}_B(1)$. We had seen that $B \cap \Sigma_\alpha$ is one point and the intersection is transversal in the proof of 4.4 a). Hence the result for $Q_\alpha$, $Q_\beta$.

## 5. Descent and Pic $\tilde{M}(0,2)$

We consider now the morphism $Q \xrightarrow{\varphi} \tilde{M}(0,2)$ of the quadric bundle to the normalization, which is an isomorphism away from $Q_{\text{exc}}$ and which blows down the $\mathbb{P}_1$-fibration $Q_{\text{exc}} \underset{r}{\longrightarrow} \text{Hilb}_2(D)$. It follows that $\text{Pic}\,\tilde{M}(0,2) \xrightarrow{\varphi^*} \text{Pic}\,Q$ is injective. We can now prove

**5.1. Proposition.** *If* $\mathcal{L}$ *is a line bundle on* $Q$ *which is trivial on the fibres of* $\varphi$ *(i.e. trivial on the fibres of the* $\mathbb{P}_1$*-fibration of* $Q_{\text{exc}}$*) then* $\varphi_*\mathcal{L}$ *is locally free and* $\varphi^*\varphi_*\mathcal{L} = \mathcal{L}$.

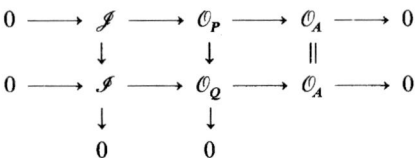

*Proof* (see also [ES]). Let $A_b$ be one of the fibres $\simeq \mathbb{P}_1$, $b \in \tilde{M} = \tilde{M}(0,2)$, and let $\mathscr{I}$ be its ideal sheaf in $\mathcal{O}_Q$. By a Theorem of Grothendieck, [H$_1$], III, 11.1,

$$(R^1 \varphi_* \mathscr{I}\mathscr{L})_b = \varprojlim_m H^1(A_b, \mathscr{I}\mathscr{L}/\mathscr{I}^{m+1}\mathscr{L}).$$

Since $\mathscr{L}|A$ is free, $\mathscr{I}/\mathscr{I}^{m+1} \otimes \mathscr{L} = \mathscr{I}\mathscr{L}/\mathscr{I}^{m+1}\mathscr{L}$ and it follows from Lemma 4.5 that

$$(R^1 \varphi_* \mathscr{I}\mathscr{L})_b = 0.$$

There is also an affine neighborhood $U(b) \subset \tilde{M}$ on which $R^1 \varphi_* \mathscr{I}\mathscr{L}$ is zero, because this sheaf is coherent. By Leray's spectral sequence

$$H^1(\tilde{U}, \mathscr{I}\mathscr{L}) = \Gamma(U, R^1 \varphi_* \mathscr{I}\mathscr{L}) = 0$$

where $\tilde{U} = \varphi^{-1}(U)$. Therefore the restriction map

$$\Gamma(\tilde{U}, \mathscr{L}) \longrightarrow \Gamma(A, \mathscr{L}|A)$$

is surjective. Let $s \in \Gamma(\tilde{U}, \mathscr{L})$ restrict to 1 in $\Gamma(A, \mathcal{O}_A)$. Then $s$ generates $\mathscr{L}$ in a neighborhood of $A$, i.e. after shrinking $U$ there is an epimorphism

$$\mathcal{O}|\tilde{U} \underset{\approx}{\longrightarrow} \mathscr{L}|\tilde{U}$$

which must be an isomorphism. On the other hand we know that $\varphi_* \mathcal{O}_Q = \mathcal{O}_{\tilde{M}}$ (the fibres of $\varphi$ are connected, see [NT], 4.5), so that $\varphi_* \mathscr{L}$ is locally free of rank 1. Then automatically $\varphi^* \varphi_* \mathscr{L} \approx \mathscr{L}$.

**5.2. Corollary.** *$Q$ and $\tilde{M}(0,2)$ only have rational singularities.*

*Proof.* Since $X^{ss}$ is smooth it follows from the result of Boutot, [B], that $Q$ has only rational singularities: If $\tilde{Q} \xrightarrow[p]{} Q$ is a desingularization then $p_* \mathcal{O}_{\tilde{Q}} = \mathcal{O}_Q$ and $R^i p_* \mathcal{O}_{\tilde{Q}} = 0$ for $i > 0$. Since $\varphi_* \mathcal{O}_Q = \mathcal{O}_{\tilde{M}}$ and $R^i \varphi_* \mathcal{O}_Q = 0$ (by the same proof as in 5.1) it follows that also $(\varphi \circ p)_* \mathcal{O}_{\tilde{Q}} = \mathcal{O}_M$ and $R^i (\varphi \circ p)_* \mathcal{O}_{\tilde{Q}} = 0$ for $i > 0$.

**5.3. Remark.** Whereas $Q$ is locally factorial, $\tilde{M}(0,2)$ is not. This follows f.e. from $\operatorname{Pic} \tilde{M}(0,2) \subsetneqq \operatorname{Pic} Q$, whereas $\operatorname{Cl} \tilde{M}(0,2) = \operatorname{Cl} Q$.

**5.4. Proposition.** $\operatorname{Pic} \tilde{M}(0,2) \simeq \mathbb{Z}^3$ *and is freely generated by*

$$\varphi_* \mathcal{O}_Q(H),\ \varphi_* \mathcal{O}_Q(Q_\alpha - Q_1),\ \varphi_* \mathcal{O}_Q(Q_\beta - Q_1).$$

*Proof.* By proposition 5.1, $\operatorname{Pic} \tilde{M}(0,2) \subset \operatorname{Pic} Q$ is the subgroup of those bundles $\mathscr{L}$ on $Q$ which are trivial on the fibres of $\varphi$. By proposition 2.4 $\operatorname{Pic} Q = \mathbb{Z}^4$ with free generators $H, Q_1, Q_\alpha, Q_\beta$. Now

$$\mathscr{L} = \mathcal{O}_Q(a_h H + a_1 Q_1 + a_\alpha Q_\alpha + a_\beta Q_\beta)$$

is trivial on such a fibre $A \simeq \mathbb{P}_1$ if and only if

$$a_1 + a_\alpha + a_\beta = 0$$

by Lemma 4.6. It follows that Pic $\tilde{M}(0,2)$ is isomorphic to $\mathbb{Z}^3$ as the kernel in the exact sequence

$$0 \longrightarrow \mathbb{Z}^3 \xrightarrow[\begin{pmatrix} 1 & 0 & 0 & 0 \\ 0 & -1 & -1 & 0 \\ 0 & 0 & -1 & 1 \end{pmatrix}]{} \mathbb{Z}^4 \xrightarrow[\begin{pmatrix} 0 \\ 1 \\ 1 \\ 1 \end{pmatrix}]{} \mathbb{Z} \longrightarrow 0 .$$

We see from these matrices that the bundles of the proposition form a basis of Pic $\tilde{M}(0,2)$.

## 6. The divisor $E$

There is an interesting divisor $E_M$ in $M(0,2)$, defined by the condition $\operatorname{Ext}^2(\mathscr{F}, \mathscr{F}(-2)) \neq 0$. It is the ramification divisor for the restriction map

$$\mathscr{F} \to \mathscr{F} \mid \text{Quadric}$$

for any smooth quadric in $\mathbb{P}_3$, see f.e. [P]. We let $E$ denote the closure of $E_M$ in $Q$, where we consider $M(0,2) \subset Q$ as an open subset by 1.5.

**6.1. Proposition.** *Let $\mathscr{F} \in M(0,2)$ correspond to the pair $(S, C^\vee)$ of (smooth) conics in $Q$, $S \subset PW_z$, $C^\vee \subset PW_z^\vee$, let $s \in s^2 W_z^\vee$, $c^\vee \in S^2 W_z$ be their equation and let $C \subset PW_z$ be the polar dual of $C^\vee$ w.r.t. $s$. Then the following are equivalent:*

(1) $\operatorname{Ext}^2(\mathscr{F}, \mathscr{F}(-2)) \neq 0$, (1') $\operatorname{Ext}^1(\mathscr{F}, \mathscr{F}(-2)) \neq 0$,

(2) $\ell(s, c^\vee) = 0$ *for the canonical pairing* $S^2 W^\vee \times S^2 W_z \xrightarrow{\ell} k$,

(3) $S \cap C = \{p_1, p_2, p_3, p_4\}$ *consists of four different points and*

$$\text{Crossratio}_S(p_{i_1}, p_{i_2}, p_{i_3}, p_{i_4}) \cdot \text{Crossratio}_C(p_{i_1} p_{i_2}, p_{i_3}, p_{i_4}) = 1$$

*for any permutation of the points (the points are then inparticular in "equianharmonic" position on each of the conics).*

We only sketch the proof:

a) The bundle $\mathscr{F}$ is the cohomology of the Beilinson monad

$$H^2(\mathscr{F}(-3)) \otimes \Omega^3(3) \xrightarrow{m} H^1(\mathscr{F}(-1)) \otimes \Omega^1(1) \xrightarrow{b} H^1(\mathscr{F}) \otimes \mathcal{O},$$

cf. [NT], 2.1. Writing down certain Ext sequences one arrives at the exact sequence

$$0 \to \operatorname{Ext}^1(\mathcal{F}, \mathcal{F}(-2)) \to H^2(\mathcal{F}(-3)) \otimes H^2(\mathcal{F}(-3))$$
$$\xrightarrow[\phi_m]{} H^1(\mathcal{F}(-1)) \otimes H^1(\mathcal{F}(-1)) \to \operatorname{Ext}^2(\mathcal{F}, \mathcal{F}(-2)) \to 0,$$

where $\phi_m$ depends only on $m$. Note that $m$ is in 1-1 correspondence with the "Kronecker"-module $H^2\mathcal{F}(-3) \xrightarrow{\tilde{m}} H^1(\mathcal{F}(-1)) \otimes \Lambda^2 V$, which is induced by the cup product

$$\Gamma\Omega^1(2) \otimes H^2\mathcal{F}(-3) \longrightarrow H^2(\Omega^1 \otimes \mathcal{F}(-1)) \simeq H^1\mathcal{F}(-1).$$

b) By [NT], 2.1, the pair $(S, C^\vee)$ can be represented by a point $x = (z, M, \Gamma)$, such that $M \subset \Gamma$ are represented by matrices

$$\underset{M}{\begin{pmatrix} a & b \\ b & c \end{pmatrix}} = \underset{A}{\begin{pmatrix} \alpha_0 & \alpha_1 & \alpha_2 & \alpha_3 \\ \alpha_1 & \alpha_2 & \alpha_3 & \alpha_4 \end{pmatrix}} \qquad \underset{\Gamma}{\begin{pmatrix} \xi & 0 \\ \omega & \xi \\ \eta & \omega \\ 0 & \eta \end{pmatrix}}$$

with the condition that

$$J = \det \begin{pmatrix} \alpha_0 & \alpha_1 & \alpha_2 \\ \alpha_1 & \alpha_2 & \alpha_3 \\ \alpha_2 & \alpha_3 & \alpha_4 \end{pmatrix} \neq 0,$$

where $z = \operatorname{span}(\xi, \omega, \eta) \subset \Lambda^2 V$, and such that $M: k^2 \to k^2 \otimes W_z \hookrightarrow k^2 \otimes \Lambda^2 V$ represents the Kronecker-module $\tilde{m}$. ($\tilde{m}$ has to be symmetric in the case of a bundle $\mathcal{F}$, and thus $A$ is persymmetric, and $J \neq 0$ is the condition for $C^\vee(M)$ to be smooth.) Now one verifies that the matrix

$$M \wedge M = \left(\begin{array}{c|c} a \wedge M & b \wedge M \\ \hline b \wedge M & c \wedge M \end{array}\right): k^4 \to k^4 \otimes \Lambda^4 V \simeq k^4$$

represents $\phi_m$. If we replace the 4-vectors by scalars via $\Lambda^4 V \simeq k$, one easily calculates

$$\det(M \wedge M) = 4IJ^2,$$

where $I = \alpha_0\alpha_4 - 4\alpha_1\alpha_3 + 3\alpha_2^2$ is the other invariant of the binary form $\Sigma \alpha_i s^{4-i} t^i$. Since $J \neq 0$, it follows that

$$\operatorname{Ext}^1(\mathcal{F}, \mathcal{F}(-2)) \neq 0 \quad \text{iff} \quad I = 0.$$

c) Let $x, w, y \in W_z^\vee$ be dual to the basis $\xi, \omega, \eta \in W_z$. The conic $S = S(\Gamma) \subset PW_z$ is parametrized by $\langle s, t\rangle \to \langle s^2\xi + st\omega + t^2\eta\rangle$ and has the equation $xy - w^2 = 0$, see [NT] and $C^\vee(M) \subset PW_z^\vee$ is given by the quadratic form $ac - b^2 = 0$. One can show now by elementary (but lengthy) calculations that (3) is equivalent to $I = 0$.

d) The pairing $S^2 W_z^\vee \times S^2 W_z \xrightarrow{\ell} k$ is defined by

$$\ell(z_1 z_2, \zeta_1 \zeta_2) = z_1(\zeta_1) z_2(\zeta_2) + z_2(\zeta_1) z_1(\zeta_2) .$$

Since $S$ has the equation $xy - w^2$ and $C^\vee$ the equation

$$ac - b^2 = (\alpha_0 \alpha_2 - \alpha_1^2)\xi^2 + \ldots + (\alpha_2 \alpha_4 - \alpha_3^2)\eta^2$$

we find

$$\ell(s, c^\vee) = \ell(xy - w^2, ac - b^2) = (\alpha_0 \alpha_4 - \alpha_1 \alpha_3) - 3(\alpha_1 \alpha_3 - \alpha_2^2) = I .$$

This proves (1) $\Leftrightarrow$ (2).

**6.2. Remark.** The set $\{[\mathscr{F}] \in \tilde{M}(0, 2) \text{ with } \operatorname{Ext}^2(\mathscr{F}, \mathscr{F}(-2)) \neq 0\}$ is bigger than the divisor $\varphi(E) = \bar{E}_M$.

**6.3.** We can interpret the condition (2) of 6.1 as the vanishing of a section of the bundle $\mathcal{O}_Q(H) \otimes \pi^* j^* \mathcal{O}_{PS^2W^\vee}(1)$, where as in 1.3 we consider the diagram:

$$\begin{array}{ccccc}
Q & \xhookrightarrow{i} & P\mu^* S^2 W & \hookrightarrow & P\varrho^* S^2 W \\
\downarrow{\scriptstyle \pi} & \swarrow{\scriptstyle \pi} & & \swarrow & \\
C(G) & \xhookrightarrow{j} & PS^2 W^\vee & & \\
& {\scriptstyle \mu}\searrow & \downarrow{\scriptstyle \varrho} & & \\
& & Z . & &
\end{array}$$

If $p \in P\varrho^* S^2 W$ is a point, $p = \langle c^\vee \rangle \in (P\varrho^* S^2 W)_{\langle \sigma_z \rangle} = PS^2 W_z$ where $\langle \sigma_z \rangle = \pi(p)$ and $z = \mu \langle \sigma_z \rangle$, $\sigma_z \in S^2 W_z^\vee$. The fibre of the bundle space of

$$\mathscr{L} = \mathcal{O}_{P\varrho^* S^2 W}(1) \otimes \pi^* \mathcal{O}_{PS^2 W^\vee}(1)$$

at $p$ is then the 1-dimensional space

$$\langle c^\vee \rangle^\vee \otimes \langle \sigma_z \rangle^\vee .$$

Now there is a canonical element $\lambda_p$ in that space, namely the homomorphism induced by

$$\langle \sigma_z \rangle \otimes \langle c^\vee \rangle \hookrightarrow S^2 W_z^\vee \otimes S^2 W_z \longrightarrow k .$$

We thus are given a canonical section $\lambda$ of the bundle $\mathscr{L}$, such that $\lambda(p) = 0$ is exactly the condition (2) in 6.1. We denote by the same letter the restriction of $\lambda$ to $Q$.

**6.3.1. Corollary.** *The subvariety $E$ is irreducible of codimension* 1 *and equal to the zero scheme* $(\lambda)$.

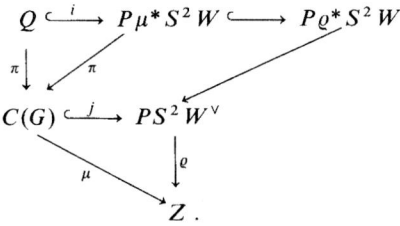

*Proof.* $(Q\setminus Q_0)\cap(\lambda)$ is irreducible of codimension 1, since $Q\setminus Q_0 \to C(G)\setminus\Sigma_0$ is a smooth fibration and $(\lambda)$ is the relative hyperplane in $P\mu^* S^2 W$. On the other hand one easily verifies that $(\lambda)\cap Q_0 \neq Q_0$ such that $(\lambda)$ has no other component in $Q_0$. Hence $(\lambda)$ is irreducible. Since $E_M \subset (\lambda)$ and $E_M = (\lambda)\cap M(0,2)$, it follows that $E = (\lambda)$ being the closure of $E_M$.

**6.4. Proposition.** $E \sim H - Q_\alpha - Q_\beta$.

**6.4.1. Remark.** $\tilde{E} = \varphi(E)$ is the closure of $E_M$ in $\tilde{M}(0,2)$ and this is an irreducible Weil-divisor, and $E = \varphi^{-1}(\tilde{E})$.

$\tilde{E}$ is *not locally principal*, for then

$$\varphi^* \mathcal{O}_{\tilde{M}}(\tilde{E}) = \mathcal{O}_Q(E) = \mathcal{O}_Q(H - Q_\alpha - Q_\beta)\,.$$

But $\mathcal{O}_Q(H - Q_\alpha - Q_\beta)|_A = \mathcal{O}_A(-1)$ by 4.6 for any fibre $A$ of the $\mathbb{P}_1$-fibration, contradicting $\varphi^*\mathcal{O}_{\tilde{M}}(\tilde{E})|_A = \mathcal{O}_A$.

*Proof of proposition* 6.4. a) By 6.3.1, $\mathcal{O}_Q(E) = \mathcal{O}_Q(H)\otimes \pi^* j^* \mathcal{O}_{PS^2W^\vee}(1)$ and it remains to prove that

$$j^*\mathcal{O}_{PS^2W^\vee}(1) = \mathcal{O}_{C(G)}(-\Sigma_\alpha - \Sigma_\beta)\,.$$

We shall use the abbreviations $\mathbb{P} = PS^2W^\vee$ and $\mathbb{P}_\alpha = \varrho^{-1}(Z_\alpha)$, $\mathbb{P}_\beta = \varrho^{-1}(Z_\beta)$, and we obtain the diagram

$$\begin{array}{ccc} C(G)\setminus\Sigma_\alpha\cup\Sigma_\beta & \overset{j}{\hookrightarrow} & \mathbb{P}\setminus\mathbb{P}_\alpha\cup\mathbb{P}_\beta \\ \uparrow\searrow & & \swarrow \\ & Z\setminus Z_\alpha\cup Z_\beta\,. & \end{array}$$

We are given the subbundle $\mathcal{O}_{\mathbb{P}}(-1)\subset \varrho^* S^2 W^\vee$ and the section $v$ gives us the subbundle

$$(*) \qquad (j\circ v)^*\mathcal{O}_{\mathbb{P}}(-1)\subset S^2 W^\vee\,.$$

If $\langle\sigma_z\rangle \in \mathbb{P}$ then the fibres are $\langle\sigma_z\rangle \subset S^2 W_z^\vee$, and since $j\circ v$ associates to $z$ the quadratic form $q\,|\,W_z$, where $\Lambda^2 V\otimes \Lambda^2 V \xrightarrow[q]{} \Lambda^4 V \simeq k$ is the form of $G$, the fibres of $(*)$ are

$$\langle q\,|\,W_z\rangle \subset S^2 W_z^\vee\,.$$

If we fix the isomorphism $\Lambda^4 V \simeq k$, we are thus given a section of $(j\circ v)^*\mathcal{O}_{\mathbb{P}}(1)$ without zeros, since $z\notin Z_\alpha\cup Z_\beta$. It follows that

$$(j\circ v)^*\mathcal{O}_{\mathbb{P}}(-1) = \mathcal{O}_{Z\setminus Z_\alpha\cup Z_\beta}$$

and hence (applying $\mu^*$)

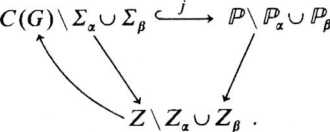

$$i^*\mathcal{O}_P(-1)_{|C(G)\setminus\Sigma_\alpha\cup\Sigma_\beta} = \mathcal{O}_{C(G)}.$$

b) Let now $j^*\mathcal{O}_P(1) = \mathcal{O}_{C(G)}(a_1\Sigma_1 + \alpha_\alpha\Sigma_\alpha + a_\beta\Sigma_\beta)$, 2.1. If we restrict this to $C(G)\setminus\Sigma_\alpha\cup\Sigma_\beta$ we obtain

$$\mathcal{O}_{C(G)}(-a_1\Sigma_1) = j^*\mathcal{O}_P(-1)_{|\ldots} = \mathcal{O}_{C(G)}.$$

But $\Sigma_1 \cap (C(G)\setminus\Sigma_\alpha\cup\Sigma_\beta)$ is a free generator, hence $a_1 = 0$.

c) To determine $a_\alpha$, $a_\beta$ we restrict $j^*\mathcal{O}_P(1)$ to $\Sigma_\alpha, \Sigma_\beta$. Now $\Sigma_\alpha \xrightarrow{\mu} Z_\alpha$ is a projective bundle with fibres $PS^2 W_z^\vee$ for $z \in Z_\alpha$. If we restrict $j^*\mathcal{O}_P(1)$ to such a fibre, we obtain by the definition of $\mathcal{O}_P(1)$

$$j^*\mathcal{O}_P(1)|PS^2 W_z^\vee = \mathcal{O}_{PS^2W_z^\vee}(1).$$

It follows that $j^*\mathcal{O}_P(1)|\Sigma_\alpha$ is the bundle $\mathcal{O}_{\Sigma_\alpha}(1)$ of the projective bundle $\Sigma_\alpha$.

On the other hand $\mathcal{O}_{C(G)}(a_\alpha\Sigma_\alpha + a_\beta\Sigma_\beta)|\Sigma_\alpha$ is the bundle

$$\mathcal{O}_{\Sigma_\alpha}(\Sigma_\alpha)^{\otimes a_\alpha}$$

i.e. the $a_\alpha$-th power of the normal bundle. But now we *claim* that

$$\mathcal{O}_{\Sigma_\alpha}(\Sigma_\alpha)|PS^2 W_z^\vee = \mathcal{O}_{PS^2W_z^\vee}(-1), \quad z \in Z_\alpha.$$

It follows that $-a_\alpha = 1$, and analogously $\alpha_\beta = -1$.

*Proof of the claim.* Let $X$ be a transversal slice to $Z_\alpha$ through $z$, $\dim X = 6$. Then $\mathcal{O}_{\Sigma_\alpha}(\Sigma_\alpha)|\mu^{-1}(X)$ is the normal bundle of the exceptional divisor $PS^2 W_z^\vee$ of the blow up of $X$ at $z$. But this normal bundle is $\mathcal{O}_{PS^2W_z^\vee}(-1)$. This completes the proof of proposition 6.4.

## References


[AK] *A Altman, St. Kleiman,* Introduction to Grothendieck duality theory, Lect. Notes Math. **146**, Berlin–Heidelberg–New York 1970.

[BT] *W. Böhmer, G. Trautmann,* Special instanton bundles and Poncelet curves, in Singularities, Representation of Algebras and Vector Bundles, Lect. Notes Math. **1273**, Berlin–Heidelberg–New York 1987.

[B] *J.F. Boutot,* Singularités rationelles et quotients par les groupes réductifs, Invent. math. **88** (1987), 65–68.

[DN] *J.M. Drezet, M.S. Narasimhan,* Groupe de Picard des varités de modules de fibrés semi-stables sur les courbes algébriques, Invent. math. **97** (1989), 53–94.

[ES] *G. Ellingsrud, S.A. Strømme,* Stable rank – 2 vector bundles on $\mathbb{P}^3$ with $c_1 = 0$, and $c_2 = 3$, Math. Ann. **255** (1981), 123–135.

[GH] *Ph. Griffiths, J. Harris,* Principles of Algebraic Geometry, New York 1978.

[H1] *R. Hartshorne,* Algebraic Geometry, Berlin–Heidelberg–New York 1977.

[H2] *R. Hartshorne,* Stable vector bundles of rank 2 on $\mathbb{P}_3$, Math. Ann. **238** (1978), 229–280.

[HN] *A Hirschowitz, M.S. Narasimhan,* Fibrés de 't Hooft speciaux et applications, Proc. Nice Conf. 1981, Basel–Stuttgart–Boston 1982.

[MF] *D. Mumford, J. Fogarty,* Geometric Invariant Theory, 2nd ed., Berlin–Heidelberg–New York 1982.

[NT] *M.S. Narasimhan, G. Trautmann,* Compactification of $M_{\mathbb{P}_3}(0, 2)$ and Poncelet pairs of conics, Pacific J. Math. **145** (1990), 255–365.
[N] *P.E. Newstead,* On the cohomology and the Picard group of a moduli space of bundles on $\mathbb{P}_3$, Quart. J. Math. **33** (1983), 349–353.
[P] *J. LePotier,* Sur l'espace de modules des fibrés de Yang et Mills, in Mathématique et Physique, Sém. Ecole Norm. Sup. 1979–1982, Basel–Stuttgart–Boston 1983.
[ST] *W. Singhof, G. Trautmann,* On the topology of the moduli space $M(0, 2)$ of stable bundles of rank 2 on $\mathbb{P}_3$, Quart. J. Math., to appear.
[T] *G. Trautmann,* Poncelet curves and associated theta characteristics, Exp. Math. **6** (1988), 29–64.

Tata Institute of Fundamental Research, Homi Bhabba Road, Bombay 400005, India
FB Mathematik der Universität, W-6750 Kaiserslautern, Germany

Eingegangen 11. Mai 1990

Invent. math. 114, 565–623 (1993)

*Inventiones mathematicae*


# Factorisation of generalised theta functions. I

**M.S. Narasimhan**[1] **and T.R. Ramadas**[2]

[1] International Centre for Theoretical Physics, P.O. Box 586, I-34100 Trieste, Italy, and School of Mathematics, TIFR, Homi Bhabha Road, Bombay 400005, India
[2] School of Mathematics, TIFR, Homi Bhabha Road, Bombay 400005, India

Oblatum 23-I-1992 & 21-VI-1993

**Abstract.** We prove a version of "factorisation", relating the space of sections of theta bundles on the moduli spaces of (parabolic, rank 2) vector bundles on curves of genus $g$ and $g-1$.

## 1. Introduction

*1a.* Let $X_1$ be a smooth projective irreducible curve over $\mathbf{C}$ of genus $g$. Let $\mathcal{U}_{X_1} = \mathcal{U}_{X_1}(d)$ be the moduli space of semistable vector bundles of rank 2 and degree $d$ on $X_1$. On $\mathcal{U}_{X_1}$ we have a natural (ample) line bundle, defined up to algebraic equivalence, which generalises the line bundle on the jacobian of $X_1$ defined by the Riemann theta divisor [D-N]. We call this the theta line bundle and denote it by $\theta_1$. A section of $\theta_1^k$ over $\mathcal{U}_{X_1}$ may be called a generalised theta function of order $k$.

We would like to study the space $H^0(\mathcal{U}_{X_1}, \theta_1^k)$ by relating it to the space of generalised theta functions associated with a smooth curve of genus $g-1$. Such a relationship has been suggested by conformal field theory under the name of "factorisation rule" or "glueing axiom".

From the point of view of algebraic geometry it is natural to study this relationship by degenerating $X_1$ into an irreducible curve $X = X_0$ which is smooth except for a single node, so that the normalisation $\tilde{X}$ of $X$ is a smooth curve of genus $g-1$. We can then consider the space of generalised theta functions on a suitable moduli space $\mathcal{U}_X$ associated to $X$ and then seek to relate this space with a space of generalised theta function associated with the normalisation $\tilde{X}$. The space $\mathcal{U}_X$ is the moduli space of semistable torsion-free sheaves of rank 2 and degree $d$ on $X$ and it carries a natural theta line bundle $\theta$. If moreover $H^1(\theta^k) = H^1(\theta_1^k) = 0$, one would have that $\dim H^0(\theta^k) = \dim H^0(\theta_1^k)$.

Let then $X$ be an irreducible curve over $\mathbf{C}$ of genus $g$, smooth except for one node $x_0$. We denote by $\tilde{X}$ the normalisation of $X$, $\tilde{g} = g-1$ the genus of $\tilde{X}$, and $\pi\colon \tilde{X} \to X$ the canonical map. Let $\{x_1, x_2\}$ be the inverse image of $x_0$ in $\tilde{X}$. The factorisation rule is:

$$H^0(\mathcal{U}_X, \theta^k) \sim \bigoplus_{\mu} H^0(\mathcal{U}_{\tilde{X}}^{\mu}, \theta_{\mu}),$$

where $\mu$ runs through a certain indexing set depending on $k$, $\mathscr{U}^{\mu}_{\tilde{X}}$ is the moduli space of parabolic vector bundles (of rank 2 and degree $d$) on $\tilde{X}$ with parabolic structures [M-S] at $x_1$ and $x_2$ (with weights depending on $\mu$) and $\theta_\mu$ is a line bundle on $\mathscr{U}^{\mu}_{\tilde{X}}$ (the generalised theta bundle).

It is clear that to carry through the induction on $g$ one has to start with moduli spaces of parabolic torsion-free sheaves of rank 2 on a nodal curve $X$ with parabolic structures at a finite number of smooth points and prove a factorisation rule for generalised theta functions on them, as well as a corresponding vanishing theorem for $H^1$. This is what is done in this paper.

### *1b. Statement of the main theorem*

First, some preliminaries:

(1) Let $X$ be an irreducible curve of genus $g$, smooth but for one node $x_0$. Choose a finite set $\{y_i\}_I$ of smooth points on $X$. Let $\tilde{X}$ be the normalisation of $X$, $\pi\colon \tilde{X} \to X$ the canonical map, and $\pi^{-1}(x_0) = \{x_1, x_2\}$.

(2) Fix integers $d, k > 0$, and also, for each $i \in I$ integers $0 \leqq \alpha_i < \beta_i < k$ satisfying the condition: $dk + \sum_i(\alpha_i + \beta_i)$ is even.

(3) Define "weights" $\{(a_i, b_i)\}_I$ by $a_i = \alpha_i/k$, $b_i = \beta_i/k$. We construct in the Appendix A the moduli space $\mathscr{U}_X = \mathscr{U}(X, d, \{(a_i, b_i)\}_I)$ of ($s$-equivalence classes of) parabolic torsion-free sheaves of rank 2 and degree $d$ on $X$, with parabolic structures at the $\{y_i\}_I$, semistable with respect to the weights $\{(a_i, b_i)\}_I$. The space $\mathscr{U}_{\tilde{X}} \equiv \mathscr{U}(\tilde{X}, d, \{(a_i, b_i)\}_I)$ is constructed similarly. The definitions can be extended to the case when $a_q = b_q$ for $q \in Q \subset I$ (§2c).

(4) For $\mu = (\alpha, \beta)$, $0 \leqq \alpha \leqq \beta < k$, let $\mathscr{U}^{\mu}_{\tilde{X}}$ be the moduli space of semistable parabolic bundles on $\tilde{X}$ with parabolic structures at the $\{y_i\}_I$ and weights $\{(a_i, b_i)\}_I$, and in addition, parabolic structures at $x_1$ and $x_2$, both of weight $(a, b) = (\alpha/k, \beta/k)$.

(5) We will define (§2), up to algebraic equivalence, a natural ample line bundle $\theta = \theta(d, k, \{(a_i, b_i)\}_I)$ on $\mathscr{U}_X$. Analogous bundles $\theta_\mu$ can be defined on the $\mathscr{U}^{\mu}_{\tilde{X}}$ (Definition 5.5).

We have then the

**Main theorem**

(A) *We have a (noncanonical) isomorphism*:

$$H^0(\mathscr{U}_X, \theta) \sim \bigoplus_{\mu} H^0(\mathscr{U}^{\mu}_{\tilde{X}}, \theta_\mu),$$

*where $\mu$ runs through the integers* $(\alpha, \beta)$, $0 \leqq \alpha \leqq \beta < k$.

(B) *Assume* $g \geqq 4$. $H^1(\mathscr{U}_X, \theta) = 0$.

The statement (A) is proved in §5b and (B) is a restatement of Theorem 7.

*1c.* We give in this sub-section a proof of factorisation in the case of rank 1 sheaves. There are few technical complications here, and the main ideas of the proof are best understood by studying this case.

If $X_t$ is a (flat) family of curves such that $X_0 = X$, and the $X_t$, for $t \neq 0$ are smooth, there exists, for every integer $d$, a corresponding family of jacobians $J^d_{X_t}$ (of degree $d$ line bundles) specialising to the *compactified jacobian* of $X$ (which we denote by $\bar{J}^d_X$). The latter parametrises rank 1 torsion-free sheaves on $X$, and is a compactification of $J^d_X$, the moduli space of line bundles of degree $d$ on $X$. In particular, consider $J^{g-1}_X$. This has a canonically defined ample line bundle on it – the theta bundle – which can be defined as Grothendieck's "determinant bundle of cohomology" [K-M] of *any* Poincaré bundle on $X_t \times J^{g-1}_{X_t}$. We shall from now on denote this bundle $\theta_t$, and set $\theta_0 = \theta$. Given a vanishing theorem for $H^1(\bar{J}^{g-1}_X, \theta^k)$, we can compute $\dim H^0(J^{g-1}_{X_t}, \theta^k_t)$ for generic $X_t$ by specialising to $t = 0$.

Giving a line bundle $N$ on $X$ is equivalent to giving one, say $L$, on $\tilde{X}$ together with an isomorphism between $L_{x_1}$ and $L_{x_2}$. To such an isomorphism we can associate its graph, a one-dimensional subspace $S$ of $L_{x_1} \oplus L_{x_2}$, and in turn, the quotient $Q$ by $S$, thought of as a point of the projective space of $L_{x_1} \oplus L_{x_2}$. This motivates the following well-known construction. Let $J^d_{\tilde{X}}$ denote the jacobian of degree $d$ line bundles on $\tilde{X}$. Given a Poincaré bundle $\mathscr{L}$ on $\tilde{X} \times J^{g-1}_{\tilde{X}}$, let $\mathbf{P}$ be the projective bundle on $J^{g-1}_{\tilde{X}}$ associated to the vector bundle (with an obvious notation) $\mathscr{L}_{x_1} \oplus \mathscr{L}_{x_2}$. We have on $\mathbf{P}$ the tautological exact sequence of bundles $0 \to \mathscr{S} \to \rho^*(\mathscr{L}_{x_1} \oplus \mathscr{L}_{x_2}) \to \mathscr{Q} \to 0$. Let $\pi_* \mathscr{L}$ denote the sheaf on $X \times \mathbf{P}$, got by taking the direct image of $\mathscr{L}$ by $\pi \times I_{J^{g-1}_{\tilde{X}}}$, and pulling back the resulting sheaf from $X \times J^{g-1}_{\tilde{X}}$. We can think of $\mathscr{Q}$ as a sheaf on $X \times \mathbf{P}$ supported on $\{x_0\} \times \mathbf{P}$. There is an obvious homomorphism $\pi_* \mathscr{L} \to \mathscr{Q}$ and we define $\tilde{\mathscr{N}}$ to be the kernel sheaf. Thus we have constructed a family of rank 1 torsion-free sheaves on $X$ parametrised by $\mathbf{P}$.

There is therefore a morphism $\phi \colon \mathbf{P} \to \bar{J}^{g-1}_X$ such that for any Pioncaré sheaf $\mathscr{N}$ on $X \times \bar{J}^{g-1}_X$ we have $(I_X \times \phi)^* \mathscr{N} = \tilde{\mathscr{N}}$ up to tensoring by a line bundle from $\mathbf{P}$:

$$\begin{array}{ccc} \mathbf{P} & \xrightarrow{\phi} & \bar{J}^{g-1}_X \\ {\scriptstyle \rho}\downarrow & & \\ J^{g-1}_{\tilde{X}} & & \end{array}$$

One can, by functoriality of the determinant bundle [L, VI, §1], compute the pull-back of $\theta$ to $\mathbf{P}$. Here it is important that we are working with line bundles of Euler characteristic 0:

$$\phi^* \theta = \rho^*(\det R\pi_{J^{g-1}_{\tilde{X}}} \mathscr{L}) \otimes \mathscr{Q}, \tag{1.1}$$

where we use the notation $\det R\pi_{Z_1} \mathscr{A}$ for the determinant bundle of cohomology of a family $\mathscr{A}$ of sheaves on $Z_1 \times Z_2$ parametrised by $Z_1$ (see 1f.(2)). One can check that this is independent of the choice of $\mathscr{L}$.

Let $\mathscr{D}_1, \mathscr{D}_2$ denote the two divisors in $\mathbf{P}$ given by $\mathscr{L}_{x_1}$ and $\mathscr{L}_{x_2}$, respectively. It is a fact that $\phi$ restricted to the complement of $\mathscr{D}_1 \cup \mathscr{D}_2$ is an isomorphism onto $J^{g-1}_X \subset \bar{J}^{g-1}_X$, and *each of the $\mathscr{D}_j$ maps isomorphically onto the singular locus $\mathscr{W}$ of* $\bar{J}^{g-1}_X$. Also, $\bar{J}^{g-1}_X$ is *seminormal* (see §4 below for the definition) and this allows us to write the exact sequence of $\mathscr{O}_{\bar{J}^{g-1}_X}$-modules:

$$0 \to \phi_* \mathscr{O}_{\mathbf{P}}(-\mathscr{D}_1 - \mathscr{D}_2) \to \mathscr{O}_{\bar{J}^{g-1}_X} \to \mathscr{O}_{\mathscr{W}} \to 0,$$

which yields

$$0 \to H^0(\phi^* \theta^k(-\mathscr{D}_1 - \mathscr{D}_2)) \to H^0(\theta^k) \to H^0(\theta^k|_{\mathscr{W}}). \tag{1.2}$$

We will argue below that the last map is a surjection. Note that $H^0(\theta^k|_{\mathscr{W}}) \sim H^0(\phi^*\theta^k|_{\mathscr{D}_1})$. Thus $H^0(\theta^k)$ is an extension:

$$0 \to H^0(\phi^*\theta^k(-\mathscr{D}_1 - \mathscr{D}_2)) \to H^0(\theta^k) \to H^0(\phi^*\theta^k|_{\mathscr{D}_1}) \to 0.$$

Now, each of the cohomology spaces on either side of the middle can be computed by taking direct images on $J^{g-1}_{\tilde{X}}$. Standard arguments, using the expression (1.1) and also $\mathcal{O}(\mathscr{D}_j) = \mathcal{O} \otimes \mathscr{L}^{-1}_{x_j}$, yield:

$$\begin{aligned} \rho_*(\phi^*\theta^k(-\mathscr{D}_1 - \mathscr{D}_2)) &= (\det R\pi_{J^{g-1}_{\tilde{X}}} \mathscr{L})^k \otimes \mathscr{L}_{x_1}\mathscr{L}_{x_2} \otimes \rho_* \mathscr{Q}^{(k-2)} \\ &= (\det R\pi_{J^{g-1}_{\tilde{X}}} \mathscr{L})^k \otimes \mathscr{L}_{x_1}\mathscr{L}_{x_2} \otimes S^{k-2}(\mathscr{L}_{x_1} \oplus \mathscr{L}_{x_2}) \\ &= (\det R\pi_{J^{g-1}_{\tilde{X}}} \mathscr{L})^k \otimes \mathscr{L}_{x_1}\mathscr{L}_{x_2} \otimes \left\{ \bigoplus_{l=0,\dots,k-2} \mathscr{L}^l_{x_1} \mathscr{L}^{k-2-l}_{x_2} \right\}, \end{aligned}$$

where $S^{k-2}$ denotes the $(k-2)$th symmetric product. Similarly,

$$(\rho|_{\mathscr{D}_1})_* \phi^*\theta^k = (\det R\pi_{J^{g-1}_{\tilde{X}}} \mathscr{L})^k \otimes \mathscr{L}^k_{x_2}.$$

We have thus found an expression $H^0(\bar{J}^{g-1}, \theta^k)$ in terms of line bundles on $J^{g-1}_{\tilde{X}}$.

We still need to show that the sequence (1.2) is exact on the right. For this it suffices to show that $H^1(\phi^*\theta^k(-\mathscr{D}_1 - \mathscr{D}_2)) = 0$. For this observe that $R^1\rho_*(\phi^*\theta^k(-\mathscr{D}_1 - \mathscr{D}_2)) = 0$ and $\rho_*\theta^k(-\mathscr{D}_1 - \mathscr{D}_2))$ is a direct sum of ample line bundles on $J^{g-1}_{\tilde{X}}$. A similar argument shows that $H^1(\bar{J}^{g-1}_{\tilde{X}}, \theta^k) = 0$.

As a simple exercise let us compute the dimension of $H^0(\bar{J}^{g-1}, \theta^k)$. Choose Poincaré bundle $\mathscr{L}$ which is trivial on (say $\{x_1\} \times J^{g-1}_{\tilde{X}}$. Then $\det R\pi_{J^{g-1}_{\tilde{X}}} \mathscr{L}$ is in the algebraic equivalence class of theta, and the $\mathscr{L}_{x_j}$ are algebraically equivalent to the trivial bundle. Thus

$$\dim H^0(J^{g-1}_X, \theta^k) = (k-1)k^{\tilde{g}} + k^{\tilde{g}} = k^g,$$

as expected.

*1d.* We describe briefly the main steps in the proof of the Main Theorem.

When comparing bundles on a singular curve $X$ and its normalisation $\tilde{X}$, we use a variant of a concept, due to Bhosle [B1], of a "generalised parabolic bundle on $\tilde{X}$ with a generalised parabolic structure over the divisor $\{x_1, x_2\}$". Such a bundle of rank 2 is given by a pair $(E, Q)$ where $E$ is a rank 2 vector bundle on $\tilde{X}$ and $Q$ a two-dimensional quotient of $E_{x_1} \oplus E_{x_2}$. Given a generalised parabolic bundle (GPB from now on) one obtains a torsion-free sheaf $F$ on $X$ which fits into the exact sequence: $0 \to F \to \pi_* E \to {}_{x_0}Q \to 0$, where ${}_{x_0}Q$ is the skyscraper sheaf on $X$ with support $x_0$ and fibre $Q$ (it is easy to show that degree $F$ = degree $E$). One can define the notion of a semistable GPB, and prove that $F$ is a semistable torsion-free sheaf iff $(E, Q)$ is a semistable GPB. All this goes through if there are additional parabolic structures at the $\{y_i\}_I$. There is therefore a morphism $\phi$: $\mathscr{P} \to \mathscr{U}_X$, where $\mathscr{P}$ denotes a suitable moduli space of generalised parabolic bundles on $\tilde{X}$. We will study this morphism in §4c, and see that it is in particular birational – in fact $\mathscr{P}$ is the normalisation of $\mathscr{U}_X$. (One has in fact to allow for torsion at the points $x_j$ so it is more appropriate to talk of generalised parabolic sheaves – this is done in the main body of the paper.)

We will consider a certain locally universal family (parametrised by a variety $\tilde{\mathscr{R}}_F$) of rank 2 vector bundles $E$ on $\tilde{X}$ with degree $E = d$, and parabolic structures at the $\{y_i\}_I$: $\mathscr{U}_{\tilde{X}}$ is a geometric invariant theory quotient of the semistable points of $\tilde{\mathscr{R}}_F$ with respect to the action of a suitable reductive group and a certain linearisation by a line bundle $\hat{\theta}$. *Let* $\rho$: $\tilde{\mathscr{R}}'_F \to \tilde{\mathscr{R}}_F$ denote the grassmannian bundle of two-dimensional quotients of $E_{x_1} \oplus E_{x_2}$ (the reason for the notation will become clear later). Using the results of §4 (namely, "seminormality" of $\mathscr{U}_X$,) we then (§5b) characterise the subspace

$$H^0(\mathscr{U}_X, \theta_{\mathscr{U}_X}) \subset H^0(\tilde{\mathscr{R}}'_F, \rho^* \hat{\theta} \otimes \mathbf{L})^{\text{inv}} = H^0(\tilde{\mathscr{R}}_F, \hat{\theta} \otimes \rho_* \mathbf{L})^{\text{inv}}, \tag{1.3}$$

where $\mathbf{L}$ is essentially the line bundle $O(k)$ along the fibres of the grassmannian bundle, and $\{.\}^{\text{inv}}$ denotes a space of invariants for the group action. The computation of $\rho_* \mathbf{L}$ amounts to the following problem when is easily solved. Let $Gr$ be the grassmannian of 2 dimensional subspaces of $\mathbf{C}^4$, $m$ a positive integer: decompose the representation of $GL(4, \mathbf{C})$ on $H^0(Gr, \mathcal{O}(m))$ into irreducible representations of $GL(2) \times GL(2) \subset GL(4)$. The decomposition (A) follows from this. (Note that the (1.3) refers to invariant sections on all of $\tilde{\mathscr{R}}_F$ and not just on the open subscheme of semistable points – this is because of Lemma 4.15 below.)

We turn next to the vanishing theorem for $H^1$. The map $\phi$: $\mathscr{P} \to \mathscr{U}_{\tilde{X}}$ is finite; and we will see that it suffices to prove the vanishing of $H^1$ for $\theta_{\mathscr{U}_X}$ pulled back to $\mathscr{P}$ and restricted to a "fixed determinant subvariety" $\mathscr{P}^L \subset \mathscr{P}$, $L \in J^d_{\tilde{X}}$. We will denote this pull-back bundle by $\theta_{\mathscr{P}}$. We consider a new set of data $(d, \bar{k}, \bar{\alpha}_i, \bar{\beta}_i)$ such that $\bar{k} = k + 4$, and $\bar{\beta} - \bar{\alpha} = \beta - \alpha + 2$. Let $\bar{\mathscr{P}}$ denote the corresponding moduli space of GPS's, we show that $H^1(\mathscr{P}, \theta_{\mathscr{P}}) = H^1(\bar{\mathscr{P}}, \theta_{\bar{\mathscr{P}}} \otimes \bar{\Omega})$ where $\theta_{\bar{\mathscr{P}}}$ is an ample line bundle on $\bar{\mathscr{P}}$ and $\bar{\Omega}$ is the dualising sheaf of $\bar{\mathscr{P}}$. (This would be the case, for example, if there is a common open set $\mathscr{P}_0$ in both $\mathscr{P}$ and $\bar{\mathscr{P}}$ such that the complement of $\mathscr{P}_0$ in each of them is of high codimension and such that $\theta_{\mathscr{P}}|_{\mathscr{P}_0} = \theta_{\bar{\mathscr{P}}} \otimes \bar{\Omega}|_{\mathscr{P}_0}$. Actually, we give a slightly different proof.) A Kodaira-type vanishing theorem for $\theta_{\bar{\mathscr{P}}} \otimes \bar{\Omega}$ now yields the desired vanishing theorem (§5b).

We introduce the moduli spaces of parabolic vector bundles and define the theta bundle in §2. In Appendices A and B we give a Geometric Invariant Theory construction of the moduli spaces of interest. The construction of moduli Simpson [Si]. The same method is used to construct the moduli space of generalised parabolic sheaves.

We prove in §3 that $\mathscr{U}_X$ is seminormal and in Appendix C that $\mathscr{P}$ is normal and has rational singularities. These properties are essentially used in the proof.

*1e.* In a subsequent work we will remove the restriction on genus in the statement of the Main Theorem (B). The results of this paper can then be used to give a proof of the "Verlinde Formula" for the dimension of generalised theta functions on the moduli space of (parabolic) bundles.

It should be mentioned that a factorisation rule for "conformal blocks", defined via representations of affine Lie algebras, has been proved in[T-U-Y].

### *1f. Notation*

(1) We will let $\det R\pi_{Z_1} \mathscr{A}$ denote the determinant bundle of a flat family $\mathscr{A}$ of sheaves parameterised by $Z_1$. A convenient reference for the determinant bundle of a family is [L] – *our definition of the determinant bundle is, however, the inverse of the*

*one used there.* For example, if $Z_2$ is a projective curve, $\mathscr{A}$ a coherent sheaf on $Z_1 \times Z_2$ flat over $Z_1$, and $x \in Z_1$, we have

$$\{\det R\pi_{Z_1}\mathscr{A}\}_x = \{\det H^0(Z_2, \mathscr{A}_x)\}^{-1} \otimes \{\det H^1(Z_2, \mathscr{A}_x)\}.$$

(2) Unless otherwise mentioned, $X$ will denote an irreducible curve of genus $g$, with one node $x_0$, $\{y_i\}_I$ a finite set of smooth points on $X$, and $y$ yet another smooth point. Let $\tilde{X}$ be the normalisation of $X, \pi\colon \tilde{X} \to X$ the canonical map, and $\pi^{-1}(x_0) = \{x_1, x_2\}$.

(3) We shall fix an integer $d$, the degree, another integer $k > 0$, and also, for each $i \in I$ integers $0 \leqq \alpha_i < \beta_i < k$. We define $n = d + 2(1-g)$ and let $l$ denote the number determined by

$$nk = 2k|I| + 2l - \sum_i (\alpha_i + \beta_i). \tag{1.4}$$

*We shall assume that the data are such that $l$ is an integer, i.e. that $dk + \sum_i (\alpha_i + \beta_i)$ is even.* Let $a_i = \alpha_i/k$, $b_i = \beta_i/k$, and set $\omega = \{(a_i, b_i)\}_I$. Finally, let $\tilde{n} = n + 2$, $\tilde{l} = l + k$.

(4) At a point $x \in X$ we let $\mathcal{O}_x$ denote the local ring and $\mathcal{M}_x$ the maximal ideal. Given a coherent sheaf $F$ on $X$, we mean by $F_x$ the vector space $F \otimes \mathcal{O}_x/\mathcal{M}_x$. The slight ambiguity of notation should not cause confusion. We let $Tor\, F$ denote the torsion subsheaf of $F$. By the degree of a torsion sheaf $\tau$ on $X$ we mean $\dim H^0(X, \tau)$. We let $h^r(F) \equiv \dim H^r(F)$.

(5) Given a vector space $\mathbf{W}$ we mean by ${}_x\mathbf{W}$ the "skyscraper sheaf" supported at the reduced point $x$, with fibre $\mathbf{W}$. Note $\mathbf{W} = H^0({}_x\mathbf{W})$. *We will often write simply* $\mathbf{W}$ *when we mean* ${}_x\mathbf{W}$.

(6) GIT is short for "geometric invariant theory". The GIT quotient of a $G$-variety $V$ is denoted by $V//G$. By a scheme we mean a (separated) scheme of finite type over $\mathbf{C}$. By a variety we mean a reduced scheme, which will be assumed irreducible unless otherwise mentioned.

## 2. The theta bundles

It will be clear that the results of this section continue to be valid if the number of nodes of $X$ is *any nonnegative integer* as long as $X$ is irreducible.

### *2a. Parabolic sheaves*

Let $F$ be a torsion-free sheaf of rank 2 and degree $d$ on $X$ – clearly such a sheaf is a vector bundle outside the node $x_0$.

**Definition 2.1a.** By a *quasi-parabolic structure* on $F$ at a smooth point $x \in X$ we mean a choice of a one-dimensional quotient $F_x \to Q \to 0$ of the fibre of $F$ at the point $x$. If in addition real numbers ("weights") $0 \leqq a < b < 1$ are given, this is a *parabolic structure.*

We shall consider sheaves with parabolic structures at the points $\{y_i\}_I$; the weights will be $\omega = \{(a_i, b_i)\}_I$ and shall denote by $Q_i$ the quotient at the point $y_i$. Such a sheaf will be called a "parabolic sheaf". The parabolic degree of a parabolic sheaf $F$ is by definition par degree $F = d + \sum_i (a_i + b_i)$; given a rank one subsheaf

$L \subset F$ such that $F/L$ is torsion-free, its parabolic degree is by definition par degree $L = \text{degree } L + \sum_{R^c} a_i + \sum_R b_i$ where $R \equiv R(L) \subset I$ is the subset where $L_{y_i} \subset \ker(F_{y_i} \to Q_i)$ and $R^c \equiv R^c(L)$ its complement. (We shall usually write simply $R$ when we mean $R(L)$ etc.)

Note that equation (1.4) can be rewritten:

$$\text{par degree } F = 2(|I| + l/k - 1 + g), \tag{2.1}$$

where the parabolic degree is defined with respect to the weights $\omega$.

**Definition 2.1b.** A parabolic sheaf $F$ is said to be *stable* (respectively, *semistable*) with respect to the weights $\{(a_i, b_i)\}_I$ if for every such subsheaf $L$ we have par degree $L <_{(\text{resp. } \leqq)} \frac{1}{2}$ (par degree $F$) – in other words, if

$$2 \text{ degree } L \underset{(\text{resp. } \leqq)}{<} d + \sum_{R^c} (b_i - a_i) - \sum_R (b_i - a_i). \tag{2.2}$$

By a family of rank 2 parabolic sheaves parametrised by a variety $T$ one means a sheaf $\mathscr{F}_T$ on $X \times T$, flat over $T$, and torsion-free (with rank 2 and degree $d$) on $X \times \{t\}$ for every point $t \in T$, together with, for each $y_i$, a quotient line bundle $\mathscr{Q}_{T,i}$ of $\mathscr{F}_T|_{\{y_i\} \times \mathscr{F}}$. The following theorem is proved in Appendix A.

**Theorem X1.** *There exists a (coarse) moduli space $\mathscr{U}^s(X, d, \omega)$ of stable parabolic sheaves $F$. We have an open immersion $\mathscr{U}^s(X, d, \omega) \subsetneq \mathscr{U}(X, d, \omega)$ where $\mathscr{U}(X, d, \omega)$ denotes the space of s-equivalence classes of semistable parabolic sheaves. The latter is a projective variety. If $X$ is smooth, then $\mathscr{U}$ is normal, with rational singularities.*

We will set $\mathscr{U}_X = \mathscr{U}(X, d, \omega)$ and $\mathscr{U}_X^s = \mathscr{U}^s(X, d, \omega)$.

*Remark 2.2.* If $M$ is a fixed line bundle on $X$, $F \mapsto F \otimes M$ takes (semi)stable sheaves to (semi)stable sheaves, and also preserves $s$-equivalence.

We begin by outlining the construction of the moduli space $\mathscr{U}(X, d, \omega)$ (see Appendix A for details). Take $d$ to be large; let $\mathbf{Q}$ denote the Quot scheme of coherent sheaves (of degree $d$ and rank 2) over $X$ which are quotients of $\mathcal{O}^n$, where $n = d + 2(1 - g)$. Thus there is on $X \times \mathbf{Q}$ a sheaf $\mathscr{F}_{\mathbf{Q}}$, flat over $\mathbf{Q}$, and an exact sequence $\mathcal{O}^n \xrightarrow{p} \mathscr{F}_{\mathbf{Q}} \to 0$. Let $\mathscr{F}_{y_i}$ be the sheaf on $\mathbf{Q}$ given by restricting $\mathscr{F}_{\mathbf{Q}}$ to $\{y_i\} \times \mathbf{Q}$, and let $\text{Flag}_{(1,2)}(\mathscr{F}_{y_i})$ be the relative flag scheme of locally-free quotients of $\mathscr{F}_{y_i}$ of rank $(1, 2)$ [EG A-I, 9.9.2]. Let $\mathscr{R}$ be the fibre product over $\mathbf{Q}$:

$$\mathscr{R} = \underset{i \in I}{\times_{\mathbf{Q}}} \text{Flag}_{(1,2)}(\mathscr{F}_{y_i})$$

Let $\mathscr{R}^s$ (respectively, $\mathscr{R}^{ss}$) denote the open subscheme of $\mathscr{R}$ corresponding to stable (respectively, semistable) parabolic sheaves such that $H^0(p)$ is an isomorphism. The variety $\mathscr{U}(X, d, \omega)$ is the "good quotient" [S1, Definitions 1.5, 1.6] of $\mathscr{R}^{ss}$ by the action of $SL(n)$ which, in fact, acts through $PSL(n)$. We will denote by $\psi$ the projection $\mathscr{R}^{ss} \to \mathscr{U}_X$.

Choose an ample line bundle of degree 1 on $X$, denoted by $\mathcal{O}(1)$ from now on. For large enough $m$ we have a $SL(n)$-equivariant embedding $\mathscr{R} \subsetneq \mathbf{G}$ where

$$\mathbf{G} \equiv \text{Grass}_{P(m)}(\mathbf{C}^n \otimes W) \times \underset{i}{\times} \{\text{Grass}_2(\mathbf{C}^n) \times \text{Grass}_1(\mathbf{C}^n)\},$$

$P(m) = n + 2m$, and $W \equiv H^0(X, \mathcal{O}(m))$. Each factor on the right has a canonical ample generator of the Picard group. We give **G** the polarisation (using the obvious notation):

$$\frac{l}{m} \times \times_i \{(k - \beta_i), (\beta_i - \alpha_i)\} \tag{2.3}$$

and take on $\mathcal{R}$ the induced polarisation. We show that the set of semistable points for the $SL(n)$ action on $\mathcal{R}$ is precisely $\mathcal{R}^{ss} \cdot \mathcal{R}^{ss}$ is reduced and irreducible and $\mathcal{U}_X$ is its GIT quotient. (The above polarisation is in general only rational since $l/m$ need not be an integer; we will see, however, that on $\mathcal{R}^{ss}$ it is indeed given by a line bundle.)

## 2b. The theta bundle

The following Theorem characterises the theta bundle.

**Theorem 1.** (A) *There is a unique line bundle $\theta_{\mathcal{U}_X} \equiv \theta(d, k, \alpha_i, \beta_i)$ on $\mathcal{U}_X$ such that given any family of semistable parabolic sheaves parametrised by a variety T, we have $\Phi_T^* \theta_{\mathcal{U}_X} = \theta_{\mathcal{F}_T}$ where*

$$\theta_{\mathcal{F}_T} \equiv (\det R\pi_T \mathcal{F}_T)^k \otimes \bigotimes_i \{(\mathcal{Q}_{T,i})^{\beta_i - \alpha_i} \otimes (\det(\mathcal{F}_T)_{y_i})^{k - \beta_i}\} \otimes (\det(\mathcal{F}_T)_y)^l \tag{2.4}$$

*and $\Phi_T$ is the induced map $T \to \mathcal{U}_X$.*

(B) *The bundle $\theta_{\mathcal{U}_X}$ is ample.*

*Proof of Theorem 1(A).* We claim that $\theta_{\mathcal{F}_{\mathcal{R}^{ss}}}$ descends to $\mathcal{U}_X$. To see this we use a result of Kempf [D-N] (Lemma 2.3 below).

The bundle $\theta_{\mathcal{F}_{\mathcal{R}^{ss}}}$ is a $PGL(n)$ bundle: given $\lambda \in \mathbf{C}^*$, its action on the fibre of $\theta_{\mathcal{F}_{\mathcal{R}^{ss}}}$ at $F$ is given by the character $\lambda \mapsto \lambda^{-kn + 2l + \sum_i(\beta_i - \alpha_i) + 2\sum_i(k - \beta_i)} = \lambda^0$ where we have used equation (1.4).

We apply Lemma 2.3 to our situation, taking $G = PGL(n)$. We first check the condition (*) of Lemma 2.3 for a stable point $F$. By an analogue of [N, Theorem 5.3(iv)] and [S2, Proposition 9(d)], the stabiliser of the $GL(n)$-action at such a point is just the centre $\mathbf{C}^* \subset GL(n)$, and the stabiliser of the $PGL(n)$ action therefore trivial.

We turn next to a semistable point $F$ such that the orbit through $F$ is closed. At such a point $F = L_1 \oplus L_2$ where the $L_i$ are rank one torsion-free sheaves, with

$$\text{par degree } L_i = \tfrac{1}{2}(\text{par degree } F) \tag{2.5}$$

Consider first the case when the (parabolic) line bundles $L_1$ and $L_2$ are not isomorphic (this is necessarily the case when $|I| > 0$). Up to $PGL(n)$ action we can write $\mathcal{O}^n = \mathcal{O}^{n_1} \oplus \mathcal{O}^{n_2}$ with $\mathcal{O}^{n_i} \sim H^0(L_i)$. The parabolic structure of $F$ at the $y_i$ is such that either

(1) $(L_1)_{y_i} \mapsto 0$ in $Q_i$, in which case the weights assigned to $(L_1)_{y_i}$ and $(L_2)_{y_i}$ are $b_i$ and $a_i$ respectively (we let $R_1 \subset I$ denote the set of such $i$), or

(2) $(L_2)_{y_i} \mapsto 0$ in $Q_i$, in which case the weights assigned to $(L_1)_{y_i}$ and $(L_1)_{y_i}$ are $a_i$ and $b_i$ respectively (we let $R_2 \subset I$ denote the set of such $i$).

(Note that $R_1 \cap R_2 = \emptyset$, $R_1 \cup R_2 = I$, par degree $L_1$ = degree $L_1 + \sum_{R_1} b_i + \sum_{R_2} a_i$ and par degree $L_2$ = degree $L_2 + \sum_{R_2} b_i + \sum_{R_1} a_i$.) Then by [S2, Proposition 25(ii)] the isotropy at $F$ of the $GL(n)$-action is $\mathbf{C}^* \times \mathbf{C}^* \subset GL(n_1) \times GL(n_2)$. Given $(\lambda_1, \lambda_2) \in \mathbf{C}^* \times \mathbf{C}^*$ its action on the fibre of $\theta_{\mathscr{F}_{\mathscr{R}^{ss}}}$ at $F$ is given by

$$\begin{aligned}&\lambda_1^{-kn_1 + l + k|I| - \sum_i \beta_i + \sum_{R_2}(\beta_i - \alpha_i)} \times \lambda_2^{-kn_2 + l + k|I| - \sum_i \beta_i + \sum_{R_1}(\beta_i - \alpha_i)}\\ &= \lambda_1^{-k(\text{par degree } L_1) - k(1-g) + l + k|I|} \times \lambda_2^{-k(\text{par degree } L_2) - k(1-g) + l + k|I|}\\ &= \lambda_1^0 \lambda_2^0,\end{aligned}$$

where we have used equations (2.1) and (2.5).

If $|I| = 0$ and the line bundles $L_j$ are isomorphic the isotropy subgroup for the $PGL(n)$-action is $PGL(2)$ which has no nontrivial characters so again we are done.

This finishes the proof of the claim.

Arguments similar to those in [D-N, §3] show that the line bundle $\theta_{\mathscr{U}_X}$, defined as the "descendant" of $\theta_{\mathscr{F}_{\mathscr{R}^{ss}}}$ to $\mathscr{U}_X$, has the universal properties asserted in Theorem 1(A). □

**Lemma 2.3.** (Theorem 2.3 of [D-N]). *Let $V$ be a variety with a $G$-action, where $G$ is a reductive algebraic group. Suppose a good quotient $\pi$: $V \to V//G$ exists. Let $E$ be a $G$-vector bundle on $V$. Then $E$ descends to $V//G$ iff the following condition holds:*

(∗) *For every point $y$ such that the oribit $Gy$ is closed, the stabiliser of $y$ acts trivially on $E_y$.*

*Remark 2.4.* If there exist semistable parabolic bundles which are not parabolic stable, and $|I| > 0$, then for $i \in I$

$$\tilde{\mathscr{L}}_i \equiv \mathscr{Q}^2_{\mathscr{R}^{ss},i} \otimes (\det(\mathscr{F}_{\mathscr{R}^{ss}})_{y_i})^{-1}$$

is a $PSL(n)$ line bundle which does not satisfy the condition (∗) of Lemma 2.3 at points with nontrivial isotropy. From this it follows that if $|I| > 0$, the genus $g$ is large enough, and there exist semistable parabolic bundles which are not parabolic stable, then the moduli space of semistable bundles is *not locally factorial*. To see this note that the restriction of $\tilde{\mathscr{L}}_i$ to $\mathscr{R}^s$, which we denote by $\tilde{\mathscr{L}}_i^s$, clearly descends to a line bundle $\mathscr{L}_i^s$ on $\mathscr{U}_X^s$; if $\mathscr{U}_X$ were locally factorial $\mathscr{L}_i^s$ would extend to $\mathscr{U}_X$ as a line bundle $\mathscr{L}_i$, and its pull-back to $\mathscr{R}^{ss}$, which we denote by $\tilde{\mathscr{L}}'_i$, would be an extension of $\tilde{\mathscr{L}}_i^s$ which does indeed satisfy (∗). For large enough $g$ codimensions are high and all the above extensions would be unique, so that $\tilde{\mathscr{L}}'_i = \tilde{\mathscr{L}}_i$ (as line bundles with $PSL(n)$-action). This yields a contradiction. (cf. [D-N, §7].)

*Remark 2.5.* (a) Note that if $\mathscr{F}'_T = \mathscr{F}_T \otimes \mathscr{N}$ and $\mathscr{Q}'_{T,i} = \mathscr{Q}_{T,i} \otimes \mathscr{N}$, with $\mathscr{N}$ a line bundle on $T$, we have, by Eq. (1.4) and elementary properties of the determinant bundle of family, a canonical isomorphism $\theta_{\mathscr{F}'} \sim \theta_{\mathscr{F}}$.

(b) When a Poincaré sheaf exists, formula (2.4) can be used to define $\theta_{\mathscr{U}_X}$.

(c) Different choices of $y$ give algebraically equivalent bundles. We sketch the proof: Let $X^{\text{reg}}$ denote the smooth points of $X$, and consider the quotient $\mathscr{R}^{ss} \times X^{\text{reg}} \to \mathscr{U} \times X^{\text{reg}}$. This is a good quotient by Lemma 2.6 below. Lemma 2.3, applied to a suitable line bundle on $\mathscr{R}^{ss} \times X^{reg}$, yields, as in the proof of Theorem 1(A), a line bundle on $\mathscr{U} \times X^{reg}$ that gives the desired algebraic equivalence.

(d) Similarly, given integers $v_i$ such that $0 \leqq \alpha_i + v_i < \beta_i + v_i < k$, $\mathscr{U}(X, d, \omega) = \mathscr{U}(X, d, a_i + v_i/k, b_i + v_i/k)$, and $\theta(d, k, \alpha_i, \beta_i)$ is algebraically equivalent to $\theta(d, k, \alpha_i + v_i, \beta_i + v_i)$.

(e) For $m \in \mathbf{Z}$, $F \mapsto F \otimes \mathscr{O}(\pm y)$ gives an isomorphism of $\mathscr{U}(X, d, \omega)$ and $\mathscr{U}(X, d \pm 2, \omega)$, such that $\theta(d \pm 2, k, \alpha_i, \beta_i)$ pulls back to $\theta(d, k, \alpha_i, \beta_i)$. Note that $l \mapsto l \pm k$.

(f) Suppose $|I| = 0$. Then Eq. (2.4) becomes: $\theta_{\mathscr{F}_T} \equiv (\det R\pi_T \mathscr{F}_T)^k \otimes \otimes (\det(\mathscr{F}_T)_y)^{\frac{1}{2}nk}$ where $n$ is the Euler characteristic of $\mathscr{F}_t$, for $t \in T$. Note that when $d$ is odd we have to take $k$ even. If $X$ is smooth the results of [D-N] show that the bundles $\theta(d, 1)$ (when $d$ is even) and $\theta(d, 2)$ (when $d$ is odd) are ample and in fact generate the Picard group of the moduli space of bundles with fixed determinant. (The first case is immediate; when $d$ is odd one has to deform the bundle $F$ of [D-N, p. 55] to the bundle $\mathscr{O} \oplus \mathscr{O}(-ny)$.)

**Lemma 2.6.** *Suppose $V \to V//G$ is a good quotient and $T$ is any variety with trivial G-action. Then $V \times T \to V//G \times T$ is a good quotient.*

*Proof.* By [N, Proposition 3.10(b)] we can assume $T$ and $V$ are affine. The result then follows from the fact ([M-F, Theorem 1.1]) that $V \to V//G$ is a universal categorical quotient (when the base field has characteristic zero.) □

*Proof of Theorem 1(B).* We will show that $\theta_{\mathscr{U}_X}$ is the descendant of the ample line bundle (A.4) on $\mathscr{R}$ used to linearise the action of $SL(n)$ (cf. [D, the proof of Proposition 5.4]) if the line bundle $\mathscr{O}(1)$ on $X$ is chosen to be $\mathscr{O}(y)$.

Note that the construction of Appendix A requires that for every semistable point the map $\mathbf{C}^n \to H^0(F)$ is an isomorphism. This implies that on $\mathscr{R}^{ss}$ we have (we will drop the suffix specifying the parameter space, which will be $\mathscr{R}^{ss}$ below)

$$\theta_{\mathscr{F}} = (\det \mathscr{O}^n)^{-k} \otimes \bigotimes_i \{\mathscr{Q}_i^{\beta_i - \alpha_i} \otimes (\det \mathscr{F}_{y_i})^{k - \beta_i}\} \otimes (\det \mathscr{F}_y)^l.$$

On the other hand one can compute the restriction of the polarisation (A.4) to $\mathscr{R}^{ss}$; this is

$$\theta_{\mathscr{F}} = (\det R\pi_{\mathscr{R}^{ss}} \mathscr{F}(my))^{l/m} \otimes \bigotimes_i \{\mathscr{Q}_i^{\beta_i - \alpha_i} \otimes (\det \mathscr{F}_{y_i})^{k - \beta_i}\}.$$

Using natural isomorphisms we see that this equals $\theta_{\mathscr{F}}$, upto tensoring by a power of the trivial line-bundle $\det \mathscr{O}^n$.

Now, some multiple of the polarisation (A.4) descends as an ample line bundle by general properties of GIT quotients. Thus some multiple of $\theta_{\mathscr{U}_X}$ is ample, and hence $\theta_{\mathscr{U}_X}$ itself. □

## 2c. *Parabolic weights*

We have required $0 \leqq \alpha_i < \beta_i < k$ so far, but the construction in Appendix A calls for $0 < \alpha_i < \beta_i \leqq k$. Also, in the statement of the decomposition theorems below we will need to consider the case $0 \leqq \alpha_i \leqq \beta_i < k$. We extend the range allowed in the Appendix ($0 < \alpha_i < \beta_i \leqq k$) to cover also $0 \leqq \alpha_i \leqq \beta_i < k$ as follows.

Suppose $\alpha_q = \beta_q$ for $q \in Q \subset I$. Denote by $\mathscr{U}^s(X, d, \omega)$ to moduli space of stable parabolic sheaves with parabolic structures at $\{y_i\}_{I \setminus Q}$, and parabolic weights $\{(a_i, b_i)\}_{I \setminus Q}$. A similar convention holds for $\mathscr{U}(X, d, \omega)$.

(2) Secondly if $\beta_i < k \forall i$ we define $\alpha_i^* = 1$, $\beta_i^* = \beta_i + 1$ whenever $\alpha_i = 0$. The corresponding change in weights does not alter the notion of (semi)stability, on the other hand it conforms to the convention used in the Appendix.

We need to be sure that the results above the theta bundle and its ampleness are unaffected by these redefinitions. This is true because of the following.

*Remark 2.7.* Suppose given smooth points $z_q$ indexed by $q \in Q$, and integers $l_q$, for $q \in Q$. Let $\theta(d, k, \alpha_i, \beta_i, z_q, l_q)$ be the line bundle given by the construction of Theorem 1(A), with $(\det(\mathscr{F}_T)_y)^l$ replaced by $\otimes_{q \in Q}(\det(\mathscr{F}_T)_{z_q})^{l_q} \otimes (\det(\mathscr{F}_T)_y)^{l + l_0}$ where $\sum_{q \in Q} l_q = -l_0$. (It is easy to check that the descent conditions are satisfied with this change.) It is clear (as in 2.5(c)) that these line bundles are all algebraically equivalent to $\theta(d, k, \alpha_i, \beta_i)$. Moreover, these line bundles are also ample, because they correspond to a different choice of the line bundle $\mathcal{O}(1)$ on the curve, the new choice being such that $\mathcal{O}(l) = \mathcal{O}(\sum_{q \in Q} l_q z_q + (l + l_0)_y)$.

## 3. Seminormality of $\mathscr{U}_X$

### *3a. Torsion-free sheaves on a nodal curve*

Note that a torsion-free sheaf $F$ on $X$ is actually free outside $x_0$, since $\dim X = 1$. Also, if rank $F = 2$ and if $F$ is not locally-free at $x_0$, we have [S2, p. 164], either $F \otimes \mathcal{O}_{x_0} \sim \mathcal{O}_{x_0} \oplus \mathscr{M}_{x_0}$ or $F \otimes \mathcal{O}_{x_0} \sim \mathscr{M}_{x_0} \oplus \mathscr{M}_{x_0}$. (We denote by $\mathscr{M}_x$ the maximal ideal at a point $x$.) This yields a decomposition of the space $\mathscr{R}^{ss}$: $\mathscr{R}^{ss} = \mathscr{R}_0 \cup \mathscr{R}_1 \cup \mathscr{R}_2$ where

**Notation 3.1.** $\mathscr{R}_a$ consists of semistable quotients $\mathcal{O}^n \to F \to 0$ satisfying

$$F \otimes \mathcal{O}_{x_0} = a\mathcal{O}_{x_0} \oplus (2 - a)\mathscr{M}_{x_0}. \tag{3.1}$$

By semicontinuity $\bigcup_{b \leq a} \mathscr{R}_b$ is closed in $\mathscr{R}^{ss}$. We will let $\hat{\mathscr{W}}$ denote the set $\bigcup_{b \leq 1} \mathscr{R}_b$, and $\hat{\mathscr{W}}'$ the set $\mathscr{R}_0$, each endowed with its reduced structure. The subschemes $\hat{\mathscr{W}}$ and $\hat{\mathscr{W}}'$ are $SL(n)$-invariant, and yield (by Lemma 4.14) closed reduced subschemes of $\mathscr{U}_X$, which we denote by $\mathscr{W}$ and $\mathscr{W}'$ respectively. Note that the $\mathscr{R}_a$ are *not* necessarily saturated sets for the quotient map, for the condition (3.1) need not be preserved by $s$-equivalence (see the 'Remarque' on p. 172 of [S2]).

We will prove that the spaces $\mathscr{U}_X$ and $\mathscr{W}$ are seminormal. This is a local property of a variety $V$, which implies in particular that any (algebraic) function on the normalisation $\hat{V}$ that is constant on the fibres descends to an algebraic function on $V$. The method of the proof is to show that the variety $\mathscr{R}^{ss}$, of which $\mathscr{U}_X$ is a GIT quoteint, is seminormal. A general property of GIT quotients then yields the desired result. The seminormality of $\mathscr{R}^{ss}$ in turn is proved using Seshadri's description of its local structure. A similar proof works for $\mathscr{W}$.

We summarise Seshadri's description in the following theorem. First we make a preliminary.

**Definition 3.2.** Given a scheme $Z$ and closed subschemes $Z_2 \subsetneq Z_1 \subsetneq Z$, we say that an analytical model at $p \in Z_2$ is given by schemes $Z'_2 \subsetneq Z'_1 \subsetneq Z'$ (with $(Z'_1$

and $Z'_2$ closed) and a point $q$ in $Z'_2$ if for some $r$ and some $s$, we have a diagram

$$\begin{array}{ccccc} (\hat{\mathcal{O}}_{Z_2})_p[[u_1,\dots,u_r]] & \longleftarrow & (\hat{\mathcal{O}}_{Z_1})_p[[u_1,\dots,u_r]] & \longleftarrow & (\hat{\mathcal{O}}_{Z})_p[[u_1,\dots,u_r]] \\ \sim\downarrow & & \sim\downarrow & & \sim\downarrow \\ (\hat{\mathcal{O}}_{Z'_2})_q[[v_1,\dots,v_s]] & \longleftarrow & (\hat{\mathcal{O}}_{Z'_1})_q[[v_1,\dots,v_s]] & \longleftarrow & (\hat{\mathcal{O}}_{Z'})_q[[v_1,\dots,v_s]] \end{array}$$

**Theorem 2.** (1) *$\mathcal{R}_2$ is a smooth variety.*

(2) *Let $p \in \mathcal{R}_1 \backslash \mathcal{R}_0$. The analytical local model for $\mathcal{R}_1 \subsetneq \mathcal{R}^{ss}$ at $p$ is* Spec $A/(u,v) \subsetneq$ Spec $A$ where $A = \mathbf{C}[u,v]/(uv)$.

(3) *Let $X = (X_{ij})$ and $Y = (Y_{lm})$ be $2\times 2$ matrices of indeterminates. Let* $A = \mathbf{C}[X,Y]/I$, $I = ((XY)_{ij}, (YX)_{lm})$. $J = (Y_{lm}, \det X) \cap (X_{ij}, \det Y)$. *Let $p \in \hat{\mathcal{W}}'$. An analytical local model for $\hat{\mathcal{W}}' \subsetneq \hat{\mathcal{W}} \subsetneq \mathcal{R}^{ss}$ at $p$ is* Spec $A/(X,Y) \subsetneq$ Spec $A/J \subsetneq$ Spec $A$.

*Proof.* This theorem follows from the results of [S2, Huitième Partie, III] and properties of smooth morphisms (see §4d). □

The following lemma is implicit in [B1].

**Lemma 3.3.** *Let $E'$ be a rank 2 (semi)stable parabolic bundle on $\tilde{X}$, of degree $d-2$. Then its direct image $F = \pi_* E'$ is a (semi)stable parabolic sheaf of degree $d$ on $X$, such that $F \otimes \mathcal{O}_{x_0} \sim \mathcal{M}_{x_0} \otimes \mathcal{M}_{x_0}$. We have $E' = \pi^* F/(T$ or $\pi^* F)$.*

*Proof.* That $E' \mapsto F \equiv \pi_* E'$, $F \mapsto E \equiv \pi^* F/(\mathrm{Tor}\, \pi^* F)$ gives a bijection between the set of isomorphism classes of rank 2 bundles $E'$ on $\hat{X}$ with degree $d-2$ and torsion-free sheaves $F$ on $X$ with degree $d$ and $F \otimes \mathcal{O}_{x_0} \sim \mathcal{M}_{x_0} \oplus \mathcal{M}_{x_0}$ is clear from [S2, Septième Partie, Proposition 10] (see also the proof of Lemma 4.6(4).)

We check that the (semi)stability of $E'$ implies that of $F$: Let $L$ be a torsion-free rank 1 quotient of $F$. One checks that $L \otimes \mathcal{O}_{x_0} \sim \mathcal{M}_{x_0}$. As in the last paragraph, we have $L = \pi_* L'$, with $L' = \pi^* L/(T$ or $\pi^* L)$ locally-free and degree $L' =$ degree $L - 1$. One checks that $L'$ is a quotient of $E'$ and this gives par degree $L' >_{(\text{resp. } \geq)}$, par degree $E'$ and rewriting we get par degree $L >_{(\text{resp. } \geq)}$, par degree $F$. The converse is similarly verified. □

## 3b. Seminormality

All rings considered in this section will be noetherian, with characteristic zero. The basic references are [T] and [Sw]. We recall from [Sw]:

**Definition 3.4.** An extension $A \subsetneq B$ of reduced rings is *subintegral* if

(1) $B$ is integral over $A$

(2) Spec $B \to$ Spec $A$ is a bijection

(3) $\forall \wp \in$ Spec $B$, $k_{A \cap \wp} \to k_\wp$ is an isomorphism, where $k_\wp = B_\wp/\wp B_\wp$

**Definition 3.5.** If $A \subsetneq B$, both rings reduced, we say $A$ is *seminormal in $B$* if there is no extension $A \subsetneq C \subsetneq B$ with $C \neq A$ and $A \subsetneq C$ subintegral. We say $A$ is seminormal if it is seminormal in its total ring of quotients.

We will use the following characterisation of seminormal rings ([Sw, Corollary 3.2]):

**Proposition 3.6.** *A reduced ring $A$ is seminormal if $\forall b, c \in A$ with $b^3 = c^2$ there is a unique $a \in A$ with $b = a^2$ and $c = a^3$.*

*Remark.* The uniqueness of $a$ depends only on the fact that $A$ is reduced, and can be seen as follows. Given $a_i$, $i \in \{1,2\}$ such that $b = a_i^2$ and $c = a_i^3$ we compute

$$\begin{aligned}(a_1 - a_2)^3 &= 3a_1a_2(a_1 - a_2)\\ &= 3/4\{(a_1 + a_2)^2 - (a_1 - a_2)^2\}(a_1 - a_2)\\ &= -3/4 \times (a_1 - a_2)^3,\end{aligned}$$

where we use $a_1^3 - a_2^3 = 0$, and $(a_1 + a_2)^2(a_1 - a_2) = (a_1 + a_2)(a_1^2 - a_2^2) = 0$. This shows, since $A$ is reduced, that $a_1 = a_2$.

Recall that given a variety $V$, with normalisation $\sigma: \tilde{V} \to V$, the *conductor* $\mathscr{C}$ is the $\mathcal{O}_V$-ideal defined as the annihilator of $\mathbf{D}$, with $\mathbf{D}$ being defined by the exact sequence of sheaves on $V$: $0 \to \mathcal{O}_V \to \sigma_*\mathcal{O}_{\tilde{V}} \to \mathbf{D} \to 0$. In fact $\mathscr{C}$ is a $\mathcal{O}_{\tilde{V}}$-ideal as well, and the biggest such. Also, the variety defined by $\mathscr{C}$ is the non-normal locus $W$ in $V$ [B, Chapter 5, §1.5, Corrollary 5]. Let $\hat{W}$ be the set-theoretic inverse image of $W$ in $\tilde{V}$.

We have then

**Lemma 3.7.** *If $V$ is seminormal, then $\mathscr{C}$ is the ideal of functions vanishing on $\tilde{W}$.*

*Proof.* Immediate from [T, Lemma 1.3]. □

Given a local ring $A$, let $\hat{A}$ denote its completion w.r.t. the maximal ideal.

**Lemma 3.8.** *Let $V$ be an variety. Assume that $\forall p \in V$, $\hat{\mathcal{O}}_p[[u_1, \ldots, u_n]]$ is seminormal for some $n$. Then $V$ is seminormal.*

*Proof.* It is enough [Sw, Theorem 1] to prove $A[u_1, \ldots, u_n]$ is seminormal (where $A$ denotes, as before, the ring of functions on $V$) and further, by [Sw, Proposition 4.7] that $A[u_1, \ldots, u_n]$, localised at any maximal ideal is of the form $\mathscr{I}_p + (u_1 - a_1, \ldots, u_n - a_n)$, where $\mathscr{I}_p$ is the ideal of functions vanishing at $p \in V$, and $a_i \in \mathbf{C}$. We can, without loss of generality, assume $a_i = 0$. The localisation of $A$ at such a maximal ideal is $(\mathcal{O}_p[u_1, \ldots, u_n])_{\mathscr{I}_p + (u_1, \ldots, u_n)}$ and its completion, by [A-M, Exercise 10.5], is $\hat{\mathcal{O}}_p[[u_1, \ldots, u_n]]$. The result now follows from the next lemma. □

**Lemma 3.9.** *LetA be a local domain, $\hat{A}$ its completion* w.r.t. *the maximal ideal. Then if $\hat{A}$ is seminormal so is $A$*

*Proof. Let $b, c \in A$* such that $b^3 = c^2$ (one can assume these are nonzero). Then $\exists \hat{a} \in \hat{A}$ such that $\hat{a}^3 = c$, $\hat{a}^2 = b$. Thus $\hat{a}b = c \in \hat{A}$, which implies, by faithful flatness, that $\exists\, a \in A$ such that $ab = c \in A$. One now computes: $b^2(a^2 - b) = c^2 - c^2 = 0$ which yields $b = a^2$, $c = a^3$. The uniqueness of $a$ is clear. □

**Lemma 3.10.** *Let $X = (X_{ij})$ and $Y = (Y_{lm})$* be $2 \times 2$ *matrices of indeterminates. Let $A = \mathbf{C}[X, Y]/I$, $I = ((XY)_{ij}, (YX)_{lm})$. Then $A$ is seminormal.*

*Proof.* We follow [S2, Theorem 30], where the proof, due to Cowsik, that $A$ is reduced is given. One finds $I = \wp_1 \cap \wp_2 \cap \wp_3$ where $\wp_1 = (X_{ij})$, $\wp_2 = (Y_{lm})$, and $\wp_3 = (I, \det X, \det Y)$, and one checks that these are prime ideals. We claim now that

(1) $\wp_1 \cap \wp_2$ is radical, and
(2) $\wp_1 \cap \wp_2 + \wp_3$ is radical

Granting this claim, Lemma 3.11 (below) finishes the proof..

We turn now to the claim. That $\wp_1 \cap \wp_2$ is radical is clear. On the other hand we now show $\wp_1 \cap \wp_2 + \wp_3 = J_1 \cap J_2$ where $J_1 = (X_{ij}, \det Y)$ and $J_2 = (Y_{lm}, \det X)$.

That $\wp_1 \cap \wp_2 + \wp_3 \subset J_1 \cap J_2$ is clear. Consider now an element in $J_1 \cap J_2$: $\alpha = \sum a_{ij} X_{ij} + b \det Y = \sum c_{ij} Y_{ij} + d \det X$. We write $\alpha = \{\sum a_{ij} X_{ij} - d \det X\} + \{d \det X + b \det Y\}$. The second term is in $\wp_3$, and the first term, which can also be written $\sum c_{ij} Y_{ij} - b \det Y$, is in $\wp_1 \cap \wp_2$. It remains to remark that $J_1$ and $J_2$ are prime – this is because $(\det X)$ is. $\square$

**Lemma 3.11.** *Let $I_1$ and $I_2$ be two radical ideals in a ring $A$ such that $I_1 + I_2$ is radical. Then if $A/I_i$ is seminormal for $i = 1, 2$ then so is $A/(I_1 \cap I_2)$.*

*Proof.* ([K-P, Lemma on p. 587]). Let $b, c \in A/(I_1 \cap I_2)$ such that $b^2 = c^3$. Then $\exists\, a_i \in A/I_i$, $i = 1, 2$ such that $b = a_i^3$, $c = a_i^2$ in $A/I_i$.

On the other hand, by the Remark following Proposition 3.6, we have $a_1 - a_2 = 0$ in $A/(I_1 + I_2)$ (since $A/(I_1 + I_2)$ is reduced). From the exact sequence of $A$-modules

$$0 \to A/I_1 \cap I_2 \to A/I_1 \oplus A/I_2 \to A/(I_1 + I_2) \to 0,$$

we see that in fact there exists an $a$ in $A/I_1 \cap I_2$ as required. $\square$

**Lemma 3.12.** *Let $A$ be as in the statement of Lemma 3.10, $\mathcal{M}$ the maximal ideal $(X_{ij}, Y_{lm})$, $\hat{A}$ the completion of $A_{\mathcal{M}}$ w.r.t. $\mathcal{M}A_{\mathcal{M}}$. Then $\hat{A}[[u_1, \ldots, u_n]]$ is seminormal for any $n$.*

*Proof.* The proof of Lemma 3.10 goes through almost word for word. The only-point to note is by [Z, Theorem 2] that the ideals $\wp_3$ and $(\det X)$ remain prime under completion, since each defines a normal variety. (That $\det X$ defines a normal variety is well-known; $\wp_3$, in Cowsik's description, defines the cone over $\mathbf{P}_1 \times \mathbf{P}_1 \times \mathbf{P}_1$, embedded in the complete linear system of $\mathcal{O}(1) \otimes \mathcal{O}(1) \otimes \mathcal{O}(1)$ where each $\mathcal{O}(1)$ comes from one of the factors. The projective normality of $\mathbf{P}_1 \times \mathbf{P}_1 \times \mathbf{P}_1$ is clear, yielding normality of the cone.) $\square$

By Theorem X1 of Appendix A, $\mathcal{U}_X$ is a variety. We can now prove (the notation of §4c is used below).

**Theorem 3.** *$\mathcal{U}_X$ is seminormal.*

*Proof.* By Lemma 3.13 below it suffices to show that $\mathcal{R}^{ss}$ is seminormal. We now use Theorem 2. The points in $\mathcal{R}_2$ are smooth and hence the local rings are seminormal. Using the Theorem 2, Lemma 3.8 and 3.12 we see that the local rings at points of $\mathcal{R}_1$ and $\mathcal{R}_0$ are seminormal as well. The Theorem now follows by [Sw, Proposition 3.7]. $\square$

**Lemma 3.13.** *A GIT quotient of a seminormal variety is seminormal*

*Proof.* The result to be proved is: Given a seminormal domain $A$ with a $G$-action, the ring of invariants (denoted $A^G$ below) is seminormal. One needs to show that if $a \in A$, with $a^2$ and $a^3$ in $A^G$, then $a \in A^G$. One can assume $a \neq 0$, for if $a = 0$ the result is trivially true. For any $g \in G$, $(a - a_g)(a + a_g) = a^2 - a_g^2 = a^2 - (a^2)_g = 0$, which yields $a = \pm a_g$. On the other hand $a^3 = a_g^3$ which rules out $a = -a_g$. □

By Proposition 3.15 below $\mathcal{W}$ is a variety. We have

**Proposition 3.14.** *The variety $\mathcal{W}$ is seminormal.*

*Proof.* The analysis proceeds as above. The local result to be proved is this: Let $X = (X_{ij})$ and $Y = (Y_{lm})$ be matrices of indeterminates. Let $A = \mathbf{C}[X, Y]/I$, $I = (Y_{lm}, \det X) \cap (X_{ij}, \det Y)$. Then $A$ is seminormal. But this is clear. □

**Proposition 3.15.** (1) *$\mathcal{W}$ is irreducible.*

(2) *$\mathcal{W}'$ is irreducible.*

(3) *$\mathcal{W}'$ is normal.*

(4) *$\mathcal{W}$ is the non-normal locus of $\mathcal{U}_X$.*

(5) *$\mathcal{W}'$ is the non-normal locus of $\mathcal{W}$.*

(6) *The map $E' \mapsto F = \pi_* E'$ gives a morphism $\mathcal{U}(\tilde{X}, d - 2, \omega) \to \mathcal{W}'$.*

*Proof.* (1–3) We will see below (Lemma 3.16) that the $\mathcal{R}_a$ $(a = 0, 1, 2)$ are irreducible. These statements are now easy consequences of Theorem 2, using general properties of GIT quotients.

(4 and 5) The proof will be given in §4, immediately following the proof of Proposition 4.11.

(6) By Lemma 3.3 there is a morphism $\mathcal{U}(\tilde{X}, d - 2, \omega) \to \mathcal{U}_X$, whose set-theoretic image is $\mathcal{W}'$. Since $\mathcal{U}(\tilde{X}, d, \omega)$ and $\mathcal{W}'$ are reduced this actually yields a morphism $\mathcal{U}(\tilde{X}, d - 2, \omega) \to \mathcal{W}'$. □

**Lemma 3.16.** *The $\mathcal{R}_a$ $(a = 0, 1, 2)$ are irreducible.*

*Proof.* In the course of the proof of Theorem X1 we show that $\mathcal{R}^{ss}$ is irreducible. Hence so is its open subset $\mathcal{R}_2$. The cases $a = 0, 1$ will be treated later, immediately following the proof of Proposition 4.11. □

## 4. Preliminaries

### *4a. Generalised parabolic sheaves*

**Definition 4.1a.** Let $E$ be a sheaf on $\tilde{X}$, torsion-free of rank 2 outside $\{x_1, x_2\}$. *A generalised parabolic structure on $E$ over the divisor $\{x_1, x_2\}$* is a two-dimensional quotient $Q$ of $E_{x_1} \oplus E_{x_2}$.

The pair $(E, Q)$ is said to be a "generalised parabolic sheaf" (GPS). We do not define a generalised quasiparabolic structure since a certain choice of "generalised weights" is assumed. We shall consider generalised parabolic sheaves $E$ with, in addition, parabolic structures at the $\{y_i\}_I$ (i.e. a one-dimensional quotient $E_{y_i} \to Q_i \to 0$ of the fibre of $E$ at each point $y_i$, and weights $0 \leq a_i < b_i < 1$ as before).

**Definition 4.1b.** A GPS $(E, Q)$ is said to be *stable* (respectively, *semistable*) with respect to the weights $\omega$ if for every nontrivial subsheaf $E'$ such that $E/E'$ is torsion-free outside the reduced points $\{x_1, x_2\}$, we have

$$\text{par degree}\, E' \underset{(\text{resp. } \leqq)}{<} \frac{\text{rank } E'}{2}(\text{par degree } E) - (\text{rank } E' - \dim Q^{E'}), \qquad (4.1)$$

where, for any subsheaf $E'$ we denote by $Q^{E'}$ the image of $E'_{x_1} \oplus E'_{x_2}$ in $Q$.

Note that in the above definition the parabolic degree of $E'$ needs to be defined. If $E'$ is torsion this is just its degree ( = length), otherwise $E'$ is actually a subbundle of $E$ outside $\{x_1, x_2\}$ and the earlier Definition (2.1a) extends in a clear way.

*Remark 4.2.* If $(E, Q)$ is a semistable GPS, $\operatorname{Tor} E$ is supported on the reduced subscheme $\{x_1, x_2\}$ and $(\operatorname{Tor} E)_{x_1} \oplus (\operatorname{Tor} E)_{x_2} \subsetneq Q$. This follows from (4.1).

**Theorem X2.** *There exists a (coarse) moduli space $\mathscr{P}^s(\tilde{X}, d, \omega)$ of stable GPS's on $\tilde{X}$. We have an open immersion $\mathscr{P}^s(\tilde{X}, d, \omega) \subsetneq \mathscr{P}(\tilde{X}, d, \omega)$ where $\mathscr{P}(\tilde{X}, d, \omega)$ denotes the space of s-equivalence classes of semistable GPSs. The former is a smooth variety; the latter a normal projective variety with rational singularities.*

This theorem is proved in Appendix B. The definition of $s$-equivalence is given there. We shall set $\mathscr{P}^s = \mathscr{P}^s(\tilde{X}, d, \omega)$ and $\mathscr{P} = \mathscr{P}(\tilde{X}, d, \omega)$.

We make explicit the notion of a family of GPSs parametrised by a variety $T$. This consists of

(1) a rank 2 sheaf $\mathscr{E}_T$ (on $\tilde{X} \times T$) flat over $T$ and locally free outside $\{x_1, x_2\} \times T$

(2) a locally-free rank 2 quotient $\mathscr{Q}_T$ (on $T$) of $(\mathscr{E}_T)_{x_1} \oplus (\mathscr{E}_T)_{x_2}$, and

(3) a locally-free rank 1 quotient $\mathscr{Q}_{T,i}$ (on $T$) of $(\mathscr{E}_T)_{y_i}$ for $i \in I$,

where we have set, for $x \in \tilde{X}$, $(\mathscr{E}_T)_x \equiv \mathscr{E}_T|_{\{x\} \times \mathscr{T}}$. (We will on occasion regard $\mathscr{Q}_T$ as a sheaf on $X \times T$ supported on $\{x_0\} \times T$.) Take now $T = \tilde{\mathscr{R}}'$, the parameter-space of the locally universal family of Appendix B:

$$\tilde{\mathscr{R}}' = \operatorname{Grass}_2(\mathscr{E}_{x_1} \oplus \mathscr{E}_{x_2}) \times_{\tilde{\mathrm{Q}}} \left\{ \underset{i \in I}{\times}{}_{\tilde{\mathrm{Q}}} \operatorname{Flag}_{(1,2)}(\mathscr{E}_{y_i}) \right\},$$

where $\tilde{\mathbf{Q}}$ is the Quot scheme of rank 2 degree $d$ quotients of $\mathscr{O}^{\tilde{n}}_{\tilde{X}}$. The degree $d$ is assumed large.(We have let $\mathscr{E} = \mathscr{E}_{\tilde{\mathscr{R}}'}$; we will similarly let $\mathscr{Q} \equiv \mathscr{Q}_{\tilde{\mathscr{R}}'}$.) The polarisation on $\tilde{\mathscr{R}}'$ is defined in Appendix B (equation B-2). The moduli space $\mathscr{P}$ is the GIT quotient of $\tilde{\mathscr{R}}'^{ss}$ by $SL(\tilde{n})$. (We have $SL(\tilde{n})$ rather than $SL(n)$ because we are considering bundles of degree $d$ on $\tilde{X}$ rather than on $X$.) We will denote by $\tilde{\psi}'$ the projection $\tilde{\mathscr{R}}'^{ss} \to \mathscr{P}$.

**Notation 4.3a.** Define $\mathscr{H}$ to be the set of (closed) points $(\mathscr{O}^{\tilde{n}} \to E \to 0, Q)$ in $\tilde{\mathscr{R}}'$, where $\mathbf{C}_{\tilde{n}} \to H^0(E)$ is an isomorphism, $H'(E(-x_1 - x_2 - x)) = 0$ for $x \in \tilde{X}$, and

(T) $\operatorname{Tor} E$ is supported on the reduced subscheme $\{x_1, x_2\}$ and $(\operatorname{Tor} E)_{x_1} \oplus (\operatorname{Tor} E)_{x_2} \subsetneq Q$.

Requiring that $H^1(E(-x_1 - x_2 - x)) = 0$ ensures that $H^1(E) = 0$, $E$ is generated by sections, $H^0(E) \to E_{x_1} \oplus E_{x_2}$ is onto, and $E(-x_1 - x_2)$ is generated by sections.

It will be clear from Appendices B and C that $\tilde{\mathscr{R}}'^{ss} \subsetneq_{\text{open}} \mathscr{H} \subsetneq_{\text{open}} \tilde{\mathscr{R}}'$.

**Notation 4.3b.** Define $\tilde{\mathbf{Q}}_F$ to be the open subscheme of $\tilde{\mathbf{Q}}$ consisting of locally-free quotients $\mathcal{O}\tilde{n} \to E \to 0$) such that
(1) $\mathbf{C}^{\tilde{n}} \to H^0(E)$ is an isomorphism, and
(2) $H^1(E(-x_1 - x_2 - x)) = 0$ for $x \in \tilde{X}$.

**Notation 4.3c.** Let $\tilde{\mathcal{R}}'_F$ be the inverse image of $\tilde{\mathbf{Q}}_F$ by the projection $\tilde{\mathcal{R}}' \to \tilde{\mathbf{Q}}$. This is a grassmannian bundle over $\tilde{\mathcal{R}}_F$, where

$$\tilde{\mathcal{R}}_F = \underset{i \in I}{\times_{\tilde{\mathbf{Q}}_F}} \mathrm{Flag}_{(1,2)}(\mathcal{E}_{y_i}).$$

We let $\rho$ denote the projection $\tilde{\mathcal{R}}'_F \to \tilde{\mathcal{R}}_F$. Note that $\tilde{\mathcal{R}}'_F \subset \mathcal{H}$. On $\tilde{\mathcal{R}}'_F$, consider the morphism of vector bundles $\mathcal{E}_{x_1} \to \mathcal{Q}$ given by the generalised parabolic structure. The zero scheme of this morphism is denoted by $\hat{\mathcal{V}}_{1,F}$($\mathcal{V}$ for "vertex"). The determinant of this map defines a subscheme which we denote $\hat{\mathcal{D}}_{1,F}$. The subschemes $\hat{\mathcal{V}}_{2,F}$ and $\hat{\mathcal{D}}_{2,F}$ are defined similarly. Clearly $\hat{\mathcal{V}}_{j,F} \subsetneq \hat{\mathcal{D}}_{j,F}, j = 1,2$. As a set, $\hat{\mathcal{D}}_{1,F}$ consists of pairs $(E, Q)$ such that the map $E_{x_1} \to Q$ is not of maximal rank and $\hat{\mathcal{V}}_{1,F}$ of pairs such that the map $E_{x_1} \to \mathcal{Q}$ is zero. Note that $\mathcal{O}(\hat{\mathcal{D}}_{j,F}) = (\det \mathcal{Q})(\det \mathcal{E}_{x_j})^{-1}$.

**Notation 4.3d.** The schematic closure of $\hat{\mathcal{D}}_{j,F}$ in $\mathcal{H}$ is denoted $\hat{\mathcal{D}}^f_j$. The $\hat{\mathcal{D}}_{j,F}$ are reduced and irreducible divisors and so the $\hat{\mathcal{D}}^f_j$ are also reduced prime divisors. The subscheme $\hat{\mathcal{V}}^f_j$ is defined as the schematic of $\hat{\mathcal{V}}_{j,F}$ in $\mathcal{H}$.

**Notation 4.3e.** We define $\hat{\mathcal{D}}^t_1$ to be component of $\mathcal{H} \backslash \mathcal{H}_F$ parametrising sheaves with *non-zero torsion at* $x_2$. We take $\hat{\mathcal{D}}^t_1$ to have its reduced structure. $\hat{\mathcal{D}}^t_2$ is defined similarly.

We quote from Appendix C the

**Proposition C.7.** (1) *The $\hat{\mathcal{D}}^f_j$ are reduced, irreducible, and normal.*
(2) *The $\hat{\mathcal{D}}^t_j$ are reduced, irreducible, and normal.*
(3) *The $\hat{\mathcal{V}}^f_j$ are smooth. We have $\hat{\mathcal{V}}^f_j \cap \{\hat{\mathcal{D}}^t_1 \cap \hat{\mathcal{D}}^t_2\} = \emptyset$.*
(4) *The closed orbits in $\hat{\mathcal{D}}^f_j$ and $\hat{\mathcal{D}}^t_j$ are contained in $\hat{\mathcal{D}}^f_j \cap \hat{\mathcal{D}}^t_j$.*

**Notation 4.3f.** The closed subschemes $\hat{\mathcal{D}}^f_j \cap \tilde{\mathcal{R}}'^{ss}$ and $\hat{\mathcal{V}}^f_j \cap \tilde{\mathcal{R}}'^{ss}$ are $SL(\tilde{n})$-invariant, and therefore yield (by Lemma 4.14 below) closed subschemes of $\mathcal{P}$ which we denote by $\mathcal{D}_j$ and $\mathcal{V}_j$ respectively.

Proposition C.7 has the following

**Corollary 4.4.** (1) *The $\mathcal{D}_j$ and the $\mathcal{V}_j$ are reduced, irreducible and normal.*
(2) $\mathcal{V}_j \cap \{\mathcal{D}_1 \cap \mathcal{D}_2\} = \emptyset$.
(3) *$\mathcal{D}_j$ is also the quotient of $(\hat{\mathcal{D}}^t_j)^{ss}$.*

### *4b. The map $\phi$*

Given a GPS on $\tilde{X}$ one obtains a sheaf $F$ on $X$ which fits into the exact sequence: $0 \to F \to \pi_* E \to {}_{x_0}Q \to 0$, where ${}_{x_0}Q$ is defined as in Notation 1f(5). (Note: $\pi_* E \otimes_{\mathcal{O}_x} k(x_0) = E_{x_1} \oplus E_{x_2}$). *We will often omit the subscript $x_0$ and simply write $Q$ when we mean ${}_{x_0}Q$.* The sheaf $F$ has, of course, a natural parabolic structure at the $\{y_i\}_I$.

*Remark 4.5.* Since $\pi$ is a finite morphism, $\chi(E) = \chi(\pi_*(E))$, and $\chi(F) = \chi(\pi_*(E)) - \chi({}_{x_0}Q) = \chi(E) - 2$, which, rewritten in terms of degrees, becomes degree $F + 2(1-g) =$ degree $E + 2(1-\tilde{g}) - 2$. Thus degree $F =$ degree $E$. Note that the computation also gives, for any coherent sheaf $E$ on $\tilde{X}$, degree $\pi_*E =$ degree $E$ + rank $E$).

**Lemma 4.6.** (1) *Let* $(E, Q)$ *be a GPS, and* $F$ *the associated sheaf on* $X$. $F$ *is torsion-free iff the condition* $(T)$ *of Notation* 4.3a *holds.*

(2) *If* $E$ *is a vector bundle and the maps* $E_{x_j} \to Q$ *isomorphisms, then the associated* $F$ *is a vector bundle. Otherwise* $F$ *is not locally free.*

(3) *If* $F$ *is a vector bundle on* $X$, *there is a unique GPS* $(E, Q)$ *which yields* $F$ *by the above construction. Infact* $E = \pi^*F$.

(4) *If* $F$ *is torsion-free but not locally free there is a GPS* $(E, Q)$ *that yields* $F$, *with* $E$ *a vector bundle and the map* $E_{x_2} \to Q$ *an isomorphism. The rank of the map* $E_{x_1} \to Q$ *is then*

(1) 1 *iff* $F \otimes \mathcal{O}_{x_0} \sim \mathcal{O}_{x_0} \oplus \mathcal{M}_{x_0}$, *and*

(2) 0 *iff* $F \otimes \mathcal{O}_{x_0} \sim \mathcal{M}_{x_0} \oplus \mathcal{M}_{x_0}$.

*The roles of* $x_1$ and $x_2$ can of course be reversed.

(5) *Every torsion-free rank* 2 *sheaf* $F$ *on* $X$ *comes from a pair* $(E, Q)$, *with* $E$ *a vector bundle.*

*Proof.* Many of these results are in [B1]. For completeness we sketch proofs. For any sheaf $A$ on $X$ define $Q_A$ by the exact sequence $A \xrightarrow{a} \pi_*\pi^*A \to {}_{x_0}Q_A \to 0$. (The map $a$ is generically an isomorphism and hence an injection when $A$ is torsion-free.)

(1) It is clear that the assumption (T) is equivalent to: $\mathrm{Tor}\,\pi_*E$ $(= \pi_*(\mathrm{Tor}\,E)) \subsetneq {}_{x_0}Q$.

(2) If the maps $E_{x_j} \to Q$ are isomorphisms, this gives an isomorphism between $E_{x_1}$ and $E_{x_2}$, which can be used to show that $F$ is locally free. That otherwise $F$ is not locally free follows from (3).

(3) We show next that if $F$ is a vector bundle the GPS$(E, Q)$ is uniquely determined. In fact $E$ is just $\pi^*F$ and $Q = Q_F$. To see this, consider

$$\begin{array}{ccccccccc} 0 & \longrightarrow & F & \xrightarrow{a} & \pi_*\pi^*F & \longrightarrow & {}_{x_0}(Q_F) & \longrightarrow & 0 \\ & & =\downarrow & & {}^{b}\downarrow & & {}^{c}\downarrow & & \\ 0 & \longrightarrow & F & \longrightarrow & \pi_*E & \longrightarrow & {}_{x_0}Q & \longrightarrow & 0 \end{array}$$

If $F$ is locally $\pi_*\pi^*F$ is torsion-free and the map $b$ is an injection. Thus $c$ is an injection and therefore an isomorphism because $\dim Q_F = 2 = \dim Q$. The Snake Lemma now yields the isomorphism $\pi_*\pi^*F = \pi_*E$ from which it easily follows that $E = \pi^*F$.

(4) Define the vector bundle $\tilde{E}$ by the exact sequence $0 \to \mathrm{Tor}\,\pi^*F \to \pi^*F \to \tilde{E} \to 0$. Consider the diagram

$$\begin{array}{ccccccccc} 0 & \longrightarrow & F & \longrightarrow & \pi_*\pi^*F & \longrightarrow & {}_{x_0}(Q_F) & \longrightarrow & 0 \\ & & =\downarrow & & \downarrow & & \downarrow & & \\ 0 & \longrightarrow & F & \xrightarrow{d} & \pi_*\tilde{E} & \longrightarrow & {}_{x_0}\tilde{Q} & \longrightarrow & 0 \end{array}$$

where $d$ is an injection (as in the above cases) because $F$ is torsion-free, and $\tilde{Q}$ is defined to make the second sequence exact. The vertical arrows are clearly surjections, so we see that ${}_{x_0}\tilde{Q} = {}_{x_0}(Q_F)/\{\pi_*(\mathrm{Tor}\,\pi^*F)\}$. Local computation show that in case (2) $\tilde{Q} = 0$, and increase (1) $\dim \tilde{Q} = 1$. In both cases it is easy to manufacture a GPS as required. We describe the case (2) which is less involved. In this case $F = \pi_* \tilde{E}$, with degree $E = d - 2$. Take $E = \tilde{E}(x_2)$, $Q = \tilde{E}_{x_2} \otimes (\Omega_{\tilde{X}})^{-1}_{x_2}$, and the maps $E_{x_j} \to Q$ as follows: the map is zero for $j = 1$ and the residue map for $j = 2$.

(5) This follows from (3) and (4) □

**Proposition 4.7.** (1) *If $F$ is semistable then $(E, Q)$ is semistable.*

(2) *If $F$ is a stable vector bundle the GPS $(E, Q)$ (which is unique by Lemma* 4.6(2) *is stable.*

(3) *If $(E, Q)$ is (semi)stable then $F$ is (semi)stable.*

*Proof* Given a subsheaf $E'$ of $E$ recall that we denote by $Q^{E'}$ the image of $E'_{x_1} \oplus E'_{x_2}$ in $Q$.

(1) Suppose $F$ is semistable. Given a sub-sheaf $E'$ of $E$ define the subsheaf $F'$ of $F$ via the commutative diagram

$$\begin{array}{ccccccccc} 0 & \longrightarrow & F & \longrightarrow & \pi^* F & \longrightarrow & {}_{x_0}Q & \longrightarrow & 0 \\ & & \uparrow & & \uparrow & & \uparrow & & \\ 0 & \longrightarrow & F' & \longrightarrow & \pi_* E' & \longrightarrow & {}_{x_0}(Q^{E'}) & \longrightarrow & 0 \end{array}$$

with the vertical arrows being inclusions. It is now easy to verify that the criterion (4.1) is satisfied and (1) is proved.

(2) It could happen in the above proof that $E'$ is a nontrivial subsheaf of $E$ but $F' = 0$ or $F' = F$. This is why stability of $F$ does not guarantee stability of $(E, Q)$, but only semistability. If $F$ were a vector bundle a nontrivial subsheaf $E'$ yields a nontrivial sub-sheaf $F'$, whence the claim in part (2) of the Proposition that $(E, Q)$ is stable if $F$ is a stable vector bundle.

(3) Suppose now that $(E, Q)$ is a (semi)stable GPS. Note that by Remark 4.2 $F$ is torsion-free. Let $L'$ be a rank 1 sub-sheaf of $F$ such that $F/L'$ is torsion-free. Define the sheaf $K_1$ to be the kernel of the composite map $\pi^* L'(\to \pi^* \pi_* E) \to E$; let $E'$ denote the image. Consider the commutative diagram of sheaves on $X$:

$$\begin{array}{ccccccccc} 0 & \longrightarrow & F & \xrightarrow{f} & \pi_* E & \longrightarrow & {}_{x_0}Q & \longrightarrow & 0 \\ & & \uparrow & & \uparrow & & \uparrow & & \\ 0 & \longrightarrow & L' & \longrightarrow & \pi_* \pi^* L' & \longrightarrow & {}_{x_0}(Q_{L'}) & \longrightarrow & 0 \end{array}$$

The second sequence is left exact since $L'$ is torsion-free. The first vertical arrow is an inclusion, and the quotient $F/L'$ is torsion-free. This yields, for the subsheaf $E'$ of $E$, the equality ${}_{x_0}(Q^{E'}) = {}_{x_0}(Q_{L'})/\{\pi_* K_1\}$. We have the following sequences of inequalities, each of which implies the next, and the first follows the semistability of $(E, Q)$:

$$\begin{aligned} 2(\text{par degree } E') &\leq \text{par degree } E - (2 - 2\dim Q^{E'}) \\ 2(\text{par degree } \pi^* L' - h^0(K_1)) &\leq \text{par degree } E - (2 - 2\dim Q^{E'}) \\ 2(\text{par degree } \pi_* \pi^* L' - 1 - \dim K_1) &\leq \text{par degree } E - (2 - 2\dim Q^{E'}) \\ 2(\text{par degree } L') &\leq \text{par degree } E + (2 - 2\dim Q^{E'} \\ &\quad + h^0(K_1) - \dim Q_{L'}) \\ &= \text{par degree } E \\ &= \text{par degree } F \end{aligned}$$

(In case $(E, Q)$ is stable all the inequalities are strict.) This proves (3). □

*Remark 4.8.* $\mathscr{P}^s$ is nonempty iff $\mathscr{U}_X^s$ is nonempty. (This follows from Proposition 4.7.) In this case $\dim \mathscr{P} = 4\tilde{g} + |I| + 1 = 4g + |I| - 3 = \dim \mathscr{U}_X$.

**Definition 4.9a.** We now define a morphism $\mathscr{P} \to \mathscr{U}_X$. For any family of GPSs as above we construct a family $\mathscr{F}_T$ of sheaves on $X$ parameterised by $T$: $\mathscr{F}_T$ is defined by the exact sequence

$$0 \to \mathscr{F}_T \to (\pi \times I_T)_* \mathscr{E}_T \to \mathscr{Q}_T \to 0. \tag{4.2}$$

where $\mathscr{Q}_T$ is regarded as a sheaf on $X \times T$ supported on $\{x_0\} \times T$. Now $(\pi \times T_T)_* \mathscr{E}_T$ is flat over $T$ since $\mathscr{E}_T$ is flat and $\pi$ is finite, $\mathscr{Q}_T$ is locally-free on $T$ and hence flat, and therefore so is $\mathscr{F}_T$. If, further, the family consists of semistable GPSs, by the above Lemma and the universal property of $\mathscr{U}_X$, we get a morphism $\phi_T\colon T \to \mathscr{U}_X$. This applies in particular to $T = \tilde{\mathscr{R}}'^{ss}$, and the resulting morphism clearly induces a morphism $\phi\colon \mathscr{P} \to \mathscr{U}_X$.

**Definition 4.9b.** Define on $\tilde{\mathscr{R}}'^{ss}$ a line bundle $\hat{\theta}'$ by

$$\hat{\theta}' \equiv (\det R\pi_{\tilde{\mathscr{R}}''} \mathscr{E})^k \otimes (\det \mathscr{Q})^k \otimes \bigotimes_i \{\mathscr{Q}_i^{\beta_i - \alpha_i} \otimes (\det \mathscr{E}_{y_i})^{k - \beta_i}\} \otimes (\det \mathscr{E}_y)^l.$$

As in §2 one can check that $\hat{\theta}'$ is the (restriction of) the ample bundle on $\tilde{\mathscr{R}}'$ used to linearise the action of $SL(\tilde{n})$, and that this descends to an (ample) line bundle $\theta_{\mathscr{P}}$ on $\mathscr{P}$.

**Definition 4.9c.** The variety $\tilde{\mathscr{R}}_F$ is a locally universal family of (ordinary) parabolic bundles on $\tilde{X}$. We let $\hat{\theta}$ be the line bundle on $\tilde{\mathscr{R}}_F$ defined by the data $(d, k, \alpha_i, \beta_i)$ as in §2b:

$$\hat{\theta} = (\det R\pi_{\tilde{\mathscr{R}}_F} \mathscr{E})^k \otimes \bigotimes_i \{(\mathscr{Q}_i)^{\beta_i - \alpha_i} \otimes (\det \mathscr{E}_{y_i})^{k - \beta_i}\} \otimes (\det \mathscr{E}_y)^{\tilde{l}},$$

where $\tilde{l} = l + k$.

Recall that $\tilde{\psi}'$ denotes the projection $\tilde{\mathscr{R}}'^{ss} \to \mathscr{P}$.

**Lemma 4.10.** (1) *Let* $\eta_x \equiv (\det \mathscr{Q})(\det \mathscr{E}_x)^{-1}$ *for a point* $x \in \tilde{X}$. *Then*

$$\hat{\theta}' = \rho^* \hat{\theta} \otimes \eta_y^k.$$

(2) $\theta_{\mathscr{P}} = \phi^* \theta_{\mathscr{U}_X}$.

*Proof.* The first claim is easily checked. From the exact sequence (4.2) we get

$$\begin{aligned}\det R\pi_T \mathscr{F}_T &= (\det R\pi_T(\pi_* \mathscr{E}_T)) \otimes (\det \mathscr{Q}_T)\\ &= (\det R\pi_T \mathscr{E}_T) \otimes (\det \mathscr{Q}_T).\end{aligned}$$

From this and (2.4) we see that $(\phi \circ \tilde{\psi}')^* \theta$ is equal to the restriction to $\tilde{\mathscr{R}}'^{ss}$ of $\hat{\theta}'$. This proves (2) □

Some of the notation of the next proposition is defined in §4a and §4b.

**Proposition 4.11.** (1) *The map* $\phi\colon \mathscr{P} \to \mathscr{U}_X$ *is finite and surjective.*

(2) *Each of the* $\mathscr{D}_j$ *maps onto* $\mathscr{W}$. *This is a finite map.*

(3) $\mathscr{P} \backslash (\mathscr{D}_1 \cup \mathscr{D}_2)$ *maps isomorphically to* $\mathscr{U}_X \backslash \mathscr{W}$.

(4) *Each of the* $\mathscr{V}_j$ *maps isomorphically onto* $\mathscr{W}'$.

(5) *$\mathscr{D}_1 \cap \mathscr{D}_2$ maps to $\mathscr{W}'$.*
(6) *Let $\mathscr{D}_j^0 = \mathscr{D}_j \backslash (\mathscr{V}_j \cup (\mathscr{D}_1 \cap \mathscr{D}_2))$. Then $\mathscr{D}_j^0$ maps isomorphically onto $\mathscr{W} \backslash \mathscr{W}'$.*
(7) *$\mathscr{W}$ is irreducible.*
(8) *$\mathscr{P}$ is the normalisation of $\mathscr{U}_X$.*
(9) *Each $\mathscr{D}_j$ is the normalisation of $\mathscr{W}$.*

*Proof.* (1) Finiteness follows from Lemma 4.10(2) and ampleness of $\theta_{\mathscr{U}_X}$ and $\theta_{\mathscr{P}}$. Surjectivity follows from Lemma 4.6(5) and Proposition 4.7(1).

(2) Consider the morphism $\phi_{\tilde{\mathscr{R}}'^{ss}}$. Using Lemma 4.6 and Proposition 4.7(3) we see that $\hat{\mathscr{D}}_{j,F} \cap \tilde{\mathscr{R}}'^{ss}$ maps onto $\mathscr{W}$ set-theoretically; hence so does $\hat{\mathscr{D}}_j^f \cap \tilde{\mathscr{R}}'^{ss}$. Thus $\mathscr{D}_j$ maps set-theoretically into $\mathscr{W}$. Since both schemes are reduced in fact this is a morphism. Finiteness now follows from (1).

(3) By Lemma 4.12(1) below and Corollary 4.4(3) $\phi(\mathscr{P} \backslash (\mathscr{D}_1 \cup \mathscr{D}_2)) = \mathscr{U}_X \backslash \mathscr{W}$. On the other hand $\phi|_{\mathscr{P} \backslash (\mathscr{D}_1 \cup \mathscr{D}_2)}$ has a section. To see this, note first that $\tilde{\psi}^{-1}(\mathscr{U}_X \backslash \mathscr{W}) \subset \mathscr{R}_2$. Now, given a vector bundle on $X$ the pull-back to $\tilde{X}$ has a canonical generalised parabolic structure which is semistable iff the bundle is semistable (Proposition 4.7(b)). This gives a map from $\tilde{\psi}^{-1}(\mathscr{U}_X \backslash \mathscr{W})$ to $\mathscr{P}$ which induces a section $(\mathscr{U}_X \backslash \mathscr{W}) \to \mathscr{P} \backslash (\mathscr{D}_1 \cup \mathscr{D}_2)$. Since $\mathscr{P}$ is irreducible, so is its open subset $\mathscr{P} \backslash (\mathscr{D}_1 \cap \mathscr{D}_2)$ and we conclude that $\phi|_{\mathscr{P} \backslash (\mathscr{D}_1 \cup \mathscr{D}_2)}$ is an isomorphism.

(4) One verifies as in part (2) that $\hat{\mathscr{V}}_j^f \cap \tilde{\mathscr{R}}'^{ss}$ maps onto $\mathscr{W}'$, inducing a morphism $\mathscr{V}_j \to \mathscr{W}'$. As in the proof of (3) we can see that this map has a section. (We use Lemma 4.13 below.)

(5) One checks as above that $(\hat{\mathscr{D}}_{1,F} \cap \hat{\mathscr{D}}_{2,F}) \cap \tilde{\mathscr{R}}'^{ss}$ maps to $\mathscr{W}'$. Now as in the proof of the irreducibility of $\mathscr{H}$ (Lemma C.2) it is possible to show that the $(\hat{\mathscr{D}}_{1,F} \cap \hat{\mathscr{D}}_{2,F})$ is dense in $(\hat{\mathscr{D}}_1 \cap \hat{\mathscr{D}}_2)$. This yields the result.

The proof of (6) is similar to that of statement (4), we use Lemma 4.12(2). The claim (7) follows from (2) and Proposition 4.4(1). The statements (8) and (9) are consequences of the normality of $\mathscr{P}$ and $\mathscr{D}_j$ and statements (1–3) and (6). □

*Proof of Lemma 3.16 (continued).* We have in the above proofs used the following facts:

(1) One can construct a family of torsion-free (but not locally free) semi-stable sheaves on $X$ parametrised by $\hat{\mathscr{D}}_{j,F} \cap \tilde{\mathscr{R}}'^{ss}$. This family contains every such sheaf.

(2) One can construct a family of torsion-free semi-stable sheaves $F$ on $X$ (with $F \otimes \mathscr{O}_{x_0} \sim \mathscr{M}_{x_0} \oplus \mathscr{M}_{x_0}$) parametrised by $\hat{\mathscr{V}}_{j,F} \cap \tilde{\mathscr{R}}'^{ss}$. This family contains every such sheaf.

The parameter spaces are in both cases reduced and irreducible. The irreducibility of $\mathscr{R}_a$ $(a = 0, 1)$ now follows by a standard argument. □

*Proof of Proposition 3.15 (4 and 5).* We prove (4) first. Consider the map $\phi$: $\mathscr{P} \to \mathscr{U}_X$. By Proposition 4.11(8) this is the normalisation map, and by Proposition 4.11(3) the non-normal locus of $\mathscr{U}_X$ is contained in $\mathscr{W}$. Since $\mathscr{W}$ is irreducible it suffices to show that the non-normal locus is nonempty (i.e. that the map $\phi$ is not an isomorphism) unless $\mathscr{W}$ is empty. Suppose then that $\mathscr{W}$ is nonempty. Then so too are the divisors $\mathscr{D}_j$ in $\mathscr{P}$ (by 4.11(2)). If $\mathscr{D}_1 \cap \mathscr{D}_2$ is nonempty, $\mathscr{W}'$ is nonempty and the proof of part (5) below shows that $\phi$ is not an isomorphism. If $\mathscr{D}_1 \cap \mathscr{D}_2 = \emptyset$ the inverse image of a point on $\mathscr{W}$ is not connected, and we are again through.

We turn to (5) next. Consider the map $\mathscr{D}_j \to \mathscr{W}$. This is the normalisation of $\mathscr{W}$ by Proposition 4.11(9), and an isomorphism outside $\mathscr{W}'$ by Proposition 4.11(6).

On the other hand by parts (4) and (5) of the same proposition, Corollary 4.4(2), and Zariski's Main Theorem it is clear from points on $\mathscr{W}'$ are *not* normal. □

**Lemma 4.12.** *Let $(E,Q)$ be a GPS, and $F$ the associated sheaf on $X$.*

(1) *If $F$ is s-equivalent to a non-locally free sheaf, then $(E,Q)$ is s-equivalent to a GPS $(E_1,Q_1)$ with $E_1$ not locally free.*

(2) *If $F$ is s-equivalent to a non-locally free sheaf $F_1$ with $F_1 \otimes \mathcal{O}_{x_0} \sim \mathcal{M}_{x_0} \oplus \mathcal{M}_{x_0}$, then $(E,Q)$ is s-equivalent to a GPS $(E_1,Q_1)$ with $E_1$ having a torsion subsheaf of degree 2.*

*Proof.* We consider (1) first. If $F$ is not locally free, either $E$ is not torsion-free and we are done, or $E$ is torsion-free and one of the maps $E_{x_j} \to Q$ is not in isomorphism. In the latter case we are again done by Proposition C.7(4). Suppose now that $F$ is locally free. Then we have the following situation:

**There is an exact sequence $0 \to L_1 \to F \to L_2 \to 0$, with $\mathbf{L}_q$ torsion-free, 2 par degree $L_q =$ par degree $F$ for $q = 1,2$, and neither $L_q$ locally free.

(One can check, by tensoring with $\mathcal{O}_{x_0}/\mathcal{M}_{x_0}$, that if one of the $L_q$ is not locally free then neither is.) It is clear that in case (2) also condition (**) holds so that we can now combine the two proofs.

Write $L_1 = \pi_* L'_1$ where $L'_1$ is a line bundle on $\tilde{X}$ with degree $L'_1 =$ degree $L_1 - 1$. There is a map of sheaves $L'_1 \to E$ on $\tilde{X}$ which is generically injective and hence everywhere injective since $L'_1$ is a line bundle. Let $L''_1$ be the quotient. Consider the commutative diagram:

$$
\begin{array}{ccccccccc}
 & & 0 & & 0 & & & & \\
 & & \uparrow & & \uparrow & & & & \\
0 & \longrightarrow & L_2 & \longrightarrow & \pi_* L''_1 & \longrightarrow & {}_{x_0}Q'' & \longrightarrow & 0 \\
 & & \uparrow & & \uparrow & & = \uparrow & & \\
0 & \longrightarrow & F & \longrightarrow & \pi_* E & \longrightarrow & {}_{x_0}Q & \longrightarrow & 0 \\
 & & \uparrow & & \uparrow & & & & \\
 & & L_1 & \xrightarrow{=} & \pi_* L'_1 & & & & \\
 & & \uparrow & & \uparrow & & & & \\
 & & 0 & & 0 & & & &
\end{array}
$$

It is easy to check (in the notation of Appendix Bb) that

$$\mu_G[(L'_1,0)] = \mu_G[(L''_1,Q'')] = \mu_G[(E,Q)].$$

Note that $L''_1$ is a rank one sheaf and $\dim Q = 2$. We leave it to the reader to check that such a semi-stable GPS must be $s$-equivalent to one with a torsion subsheaf of degree 2.

**Lemma 4.13.** *Let $T$ be a variety, $\mathscr{F}$ a sheaf on $X \times T$, flat over $T$, such that for $t \in T$ the sheaf $\mathscr{F}_t$ on $X$ is torsion-free of rank 2. Then $\mathscr{F}$ is torsion-free on $X \times T$. Suppose further that $\exists 0 \leq a \leq 2$ such that $\forall t \in T$ we have $\mathscr{F}_t \otimes \mathcal{O}_{x_0} \sim a\mathcal{O}_{x_0} \oplus (2-a)\mathcal{M}_{x_0}$. By "flat" we shall mean "flat over $T$". Then*

(1) *$(\pi \times I_T)^* \mathscr{F}$ is flat.*

(2) *If $a = 0$ there* exists a vector bundle *$\tilde{\mathscr{E}}$ on $\tilde{X} \times T$ such that $\mathscr{F} = (\pi \times I_{TR})_* \mathscr{E}$.*

(3) *If* $a = 1$ *there* exists a vector bundle $\tilde{\mathscr{E}}$ *on* $\tilde{X} \times T$ *and a line-bundle quotient* $\tilde{\mathscr{Q}}$ *of* $\tilde{\mathscr{E}}_{x_1} \oplus \tilde{\mathscr{E}}_{x_2}$ *such that the following sequence is exact:*

$$0 \to \mathscr{F} \to (\pi \times I_T)_* \tilde{\mathscr{E}} \to {}_{x0}\tilde{\mathscr{Q}} \to 0 .$$

*Proof.* It is possible to prove, as in [S2, Huitième Partie, pp. 180–182] that $\mathscr{F}$ is a subsheaf of a locally free sheaf. This implies it is torsion-free.

Consider now the sequence $\mathscr{F} \underset{i}{\to} (\pi \times I_T)^*(\pi \times I_T)^* \mathscr{F} \to \mathscr{Q}_1 \to 0$ which defines $\mathscr{Q}_1$. Since $i$ is generically an injection and $\mathscr{F}$ is torsion-free $i$ is an injection. Specialising, we see that $\dim h^0((\mathscr{Q}_1)_t) = 4 - a$ and hence constant. Since $T$ is reduced $\mathscr{Q}_1$ is flat. This show $(\pi \times I_T)^* \mathscr{F}$ is flat.

Next, consider the map $(\pi \times I_T)^* \mathscr{F} \to (\pi \times I_T)^* \mathscr{F} \otimes \mathscr{Q}_{\tilde{X}}(x_1 + x_2)$. By specialising as before one sees that the cokernel is flat, and hence also the image and kernel. Let $\tilde{\mathscr{E}}$ be the image. One can now show that $\tilde{\mathscr{E}}$ is a vector bundle, and we have an exact sequence of flat sheaves $0 \to T\,or\,(\pi \times I_T)^* \mathscr{F} \to (\pi \times I_T)^* \mathscr{F} \to \tilde{\mathscr{E}} \to 0$. We now repeat the construction of Lemma 4.6(4) "over" $T$ to prove (2) and (3). □

It is worth pointing out that the varieties $\mathscr{W}$ or $\mathscr{W}'$ could *à priori* be empty; also it could happen that $\mathscr{U}_X = \mathscr{W}$. In fact we always have $\emptyset \neq \mathscr{W} \neq \mathscr{U}_X$ (Remark 6.19).

### 4c. *Some general results*

We collect here some general statements needed elsewhere in the paper. The following fact about GIT quotients is standard.

**Lemma 4.14.** *Let* $V$ *be a projective scheme on which a reductive group* $G$ *acts,* $\tilde{\mathscr{L}}$ *an ample line bundle linearising the* $G$*-action, and* $V^{ss}$ *the open subscheme of semistable points. Let* $V'$ *be a* $G$*-invariant closed subscheme of* $V^{ss}$, $\bar{V}'$ *its schematic closure in* $V$. *Then*

(1) $\bar{V}'^{ss} = V'$, *and*

(2) $V'//G$ *is a closed subscheme of* $V^{ss}//G$.

*Proof.* (1) See the last paragraph of the proof of [M-F, Chapter 1, §5]. (2) Clearly we can take $V$ to be affine. Then this is a consequence of "algebraic fact number 3" on p. 29 of the same reference.

**Lemma 4.15.** *Suppose* $V$, $G$, *and* $V^{ss}$ *are as in the statement of the previous lemma. Let* $W$ *be an open* $G$*-invariant (irreducible) normal subscheme of* $V$ *containing* $V^{ss}$. *Then* $H^0(V^{ss}, \tilde{\mathscr{L}})^{inv} = H^0(W, \tilde{\mathscr{L}})^{inv}$ *where* $(\ )^{inv}$ *denotes the invariant subspace for an action of* $G$.

*Proof.* Assume first that $V$ is irreducible and normal. In this case we will show that any invariant section on $V^{ss}$ in fact extends to $V$ (cf. [S1, Theorem 4.1 (iii)). This is clear if $D = V \setminus V^{ss}$ has codimension $> 1$. Suppose otherwise and for simplicity assume there is only one irreducible component $D_1$. Consider an invariant section $s$ on $V^{ss}$, and assume it has a pole along $D_1$. By the definition of semistability there is an invariant section $s_1$ on $V$, vanishing on $D_1$. For some integers $l, m$ the section $s_1^l s^m$ will extend to $D_1$ and be nonvanishing there. This will contradict non-semistability of points on $D_1$. This shows that in fact $s$ extends to $V$; it is clearly $G$-invariant there. In case there are more than one component, we work by induction on the number of such components. Write $D = \bigcup D_q$. As above we can find an invariant section regular along $D_1$ and nonzero there. If this section is everywhere regular, we have the desired contradiction. If not the polar divisor of the new section has fewer components and induction is possible.

In general, we replace $V$ by the irreducible component $V_1$ containing $W$, and endow $V_1$ with its reduced structure. Using [M-F, Chapter 1, §5] (Theorem 1.19 and the remarks in the last paragraph) we see that $V_1^{ss} = V^{ss}$. The argument of the previous paragraph, applied to the normalisation of $V_1$ (again using the above results) finishes the proof. □

**Lemma 4.16.** *Let $V$ be a normal variety with a $G$-action, where $G$ is a reductive algebraic group. Suppose a good quotient $\pi$: $V \to U$ exists. Let $\tilde{\mathscr{L}}$ be a $G$-line bundle on $V$, and suppose it descends as a line bundle $\mathscr{L}$ on $U$. Let $V'' \subset V' \subset V$ be open $G$-invariant subvarieties of $V$, such that $V'$ maps onto $U$ and $V'' = \pi^{-1}(U'')$ for some nonempty open subset $U''$ of $U$. Then any invariant section of $\tilde{\mathscr{L}}$ on $V'$ extends to $V$.*

*Proof.* (cf. the proof of [Lu, Lumme 1.8].) Clearly we can assume $U$ and $V$ are affine, and $\mathscr{L}$ is trivial. A nowhere vanishing section of $\mathscr{L}$ pulls back to a $G$-invariant trivialisation of $\tilde{\mathscr{L}}$. Thus we can assume $\tilde{\mathscr{L}}$ is the trivial line bundle with the trivial action of $G$. Let $k[V]$ denote the ring of regular functions on a variety $V$. Suppose $f$ is an invariant regular function on $V'$ which does not extend to $V$. Then $f \in k[V'']^G = k[U'']$ ([N, Theorem 3.5(iii)]) and can therefore be written as $f = g/h$, with $g$, $h$ in $k[U] = k[V]^G$. Since $U$ is normal, there exists a codimension one subset $F \subset U$ such that $h|_F = 0$, and $g|_F \neq 0$. Let $y \in F$ such that $g(y) \neq 0$ and let $x \in V'$ such that $\pi(x) = y$. Then $0 \neq g(y) = g(x) = f(x)h(y) = 0$, which is a contradiction. □

The next result is from [Kn] – *we have retained the notation of that work, and there should be no confusion with notation used elsewhere in this paper.*

**Lemma 4.17.** *Let $X$ be a normal, Cohen–Macaulay variety on which a reductive group $G$ acts, such that a good quotient $\pi$: $X \to Y$ exists. Suppose that the action is generically free and that* $\dim G = \dim X - \dim Y$, *and further suppose that*

(1) *the subset where the action is not free has codimension* $\geqq 2$, *and*

(2) *for every prime divisor $D$ in $X$, $\pi(D)$ has codimension $\leqq 1$. Here $D$ need not be invariant.*

*Then $\omega_Y = (\pi_* \omega_X)^G$ where $\omega_X$, $\omega_Y$ are the respective dualising sheaves and the superscript $(\ )^G$ denotes the $G$-invariant direct image.*

*Proof.* This follows from Satz 5 of [Kn], noting (again in the notation of that paper) that condition (1) implies that $D_\mu = 0$, and condition (2) that $D_\pi = 0$. The result is stated in [Kn] for the case when $X$ is an affine variety, but this is not necessary, because under our hypothesis there is a canonical morphism $(\pi_* \omega_X)^G \to \omega_Y$. □

### *4d. Smooth morphisms*

We shall use the following device (cf. [S2, Huitième Partie]) to analyse singularities of a variety $V$. We shall find varieties $W$ and $V'$ and smooth morphisms $f$: $W \to V$ and $f'$: $W \to V'$, such that the singularities of $V'$ are easy to analyse. Recall that a *smooth morphism* of schemes $f$: $V \to W$ is one which is flat and has smooth scheme-theoretic fibres. Equivalently, for every $p \in \tilde{V}$, the completion of the local ring $\tilde{\mathcal{O}}_p$ is isomorphic, as a $\hat{\mathcal{O}}_{f(p)}$-algebra, to $\hat{\mathcal{O}}_{f(p)}[[x_1, \ldots, x_n]]$ for some $n$. There is a lifting property which characterises smooth morphisms; see [Mu, 2.1]. We have the following well-known result, see for example [Ma, Theorem 32.2 (i)]:

**Lemma 4.18.** *Let $f: W \to V$ be a smooth morphism. Then $W$ is reduced (respectively, normal, Cohen–Macaulay, Gorenstein) if and only if $V$ is.*

We will also need

**Proposition 4.19.** *Let $V_1$ and $V_2$ be varieties over $\mathbf{C}$ and, for $j = 1, 2$, let $v_j \in V_j$. Let $\mathcal{O}_j$ be the respective local rings. Suppose that the completions $\hat{\mathcal{O}}_j$ are isomorphic. Then if $V_1$ has rational singularities at $v_1$, then so does $V_2$ at $v_2$.*

*Proof.* Let $K_V$ denote the Grauert–Riemenschneider sheaf [G–R] on a variety $V$, obtained as the direct image of the canonical sheaf of a desingularisation of $V$ and let $\Omega_V$ denote the dualising sheaf of $V$. By [K] $V$ has rational singularities if and only if

(1) $V$ is Cohen–Macaulay, and
(2) the canonical map $i: K_V \to \Omega_V$ is an isomorphism.
Now, condition (2) is equivalent to:
(3) $i^{\text{an}}: K_V^{\text{an}} \to \Omega_V^{\text{an}}$ is an isomorphism,

where for a coherent $\mathcal{O}_V$ sheaf $F$, $F^{\text{an}}$ denotes the analytic sheaf obtained on the analytic space $V^{\text{an}}$ associated with $V$. Moreover for normal $V$, $K_V^{\text{an}}$ has an intrinsic characterisation in terms of $V^{\text{an}}$; in fact, it can be defined as the direct image of the presheaf of square-integrable holomorphic forms of top degree of the complement of the singular set [G-R, §2.2, p. 271].

Since $\hat{\mathcal{O}}_1 = \hat{\mathcal{O}}_2$ and $\mathcal{O}_1$ is Cohen–Macaulay and normal it follows that so is $\mathcal{O}_2$ [Z-S]. By [GAGA, §2, Proposition 3] $\hat{\mathcal{O}}_j = \widehat{\mathcal{O}_j^{\text{an}}}$. Since $\widehat{\mathcal{O}_1^{\text{an}}} = \widehat{\mathcal{O}_2^{\text{an}}}$ there are neighbourhoods of $v_1$ in $V_1$ and $v_2$ in $V_2$ which are analytically isomorphic [A, Corollary 1.6, p. 282]. Using the intrinsic characterisation of $K_V^{\text{an}}$ it follows that $i$ is an isomorphism. □

## 5. The decomposition theorem

We assume $k > 0$. Let $\mathcal{I}_Z(Z')$ denote the ideal sheaf on $Z$ of a subvariety $Z'$. (We omit the subscript $Z$ when it is superfluous.) When $Z'$ is of codimension one (not necessarily a Cartier divisor) we set $\mathcal{O}_Z(-Z') = \mathcal{I}_Z(Z')$.

### 5a. *A decomposition theorem on $\mathcal{P}$*

We prove first a decomposition theorem (Theorem 4) for $H^0(\mathcal{P}, \theta_{\mathcal{P}})$. This will be used in the proof of the vanishing theorem in §6; the results proved here will be of use in the next subsection as well.

For $j = 1, 2$ let $E_j$ be two-dimensional vector spaces. Let $Gr$ denote the grassmannian of two-dimensional quotients $E_1 \oplus E_2 \xrightarrow[P]{} Q$. We define two divisors $D_1$ and $D_2$ in $Gr$. Let $l_j$ denote the line bundle $(\det E_j)^{-1} \otimes \det Q$. This has a canonical section $\det P|_{E_j}$. Its zero-scheme is the divisor $D_j$; thus $l_j = \mathcal{O}(D_j)$. One checks easily that the divisors $D_j$ are reduced, irreducible and normal. As a set

$D_j = \{P \mid (\ker P) \cap E_j \neq \{0\}\}$. The action of $GL(E_1) \times GL(E_2)$ on $Gr$ lifts to the $l_j$'s, and (for $m \in \mathbf{Z}$) $H^0(l_j^m)$ and $H^0(l_1^m|_{D_1})$ are $GL(E_1) \times GL(E_2)$ modulus. We have then (with $\zeta$ denoting the one-dimensional representation $(\det E_1)^{-1} \otimes \det E_2$):

**Lemma 5.1.** *For $m \in \mathbf{Z}$ we have natural isomorphisms of $GL(E_1) \times GL(E_2)$ modulus:*

(1) $H^0(l_1^m|_{D_1 \cap D_2}) = S^m E_1^* \otimes S^m E_2$.

(2) $H^0(l_1^m|_{D_1}) = \bigoplus_{q=0,\dots,m} \zeta^{m-q} \otimes S^q E_1^* \otimes S^q E_2$.

(3) $H^0(l_1^m) = \bigoplus_{p=0,\dots,m} \left( \bigoplus_{q=0,\dots,p} \zeta^{p-q} \otimes S^q E_1^* \otimes S^q E_2 \right)$

(4) *All the corresponding first cohomology groups vanish for $m \geqq 0$.*

*Proof.* We use the notation $H^l(l_1^q|_{D_1}) \equiv \Delta_q^l$. We will use the following easy facts:

(a) The canonical bundle of $Gr$ is $l_1^{-4}\zeta^2$, $l_1$ is ample. Note that this gives

(b) $H^1(l_1^q) = \{H^3(l_1^{(-q-4)})\}^* = 0$ for $q > -4$.

(c) Also, $H^0(l_1) = \mathbf{C} \oplus \zeta \oplus E_1^* \otimes E_2$.

Consider the exact sequence:

$$0 \to l_1^{q-1} \to l_1^q \to l_1^q|_{D_1} \to 0 . \tag{5.1}$$

This, together with (b), shows:

(d) for $q > 0$ there is an exact sequence $0 \to H^0(l_1^{q-1}) \to H^0(l_1^q) \to H^0(l_1^q|_{D_1}) \to 0$.

Let $\Pi$ denote the product of two projective spaces corresponding to $E_1$ and $E_2$, and $\mathbf{q}_1, \mathbf{q}_2$ denote the respective tautological quotient bundles. Then $D_1 \cap D_2 \sim \Pi$ and $Q|_{D_1 \cap D_2} \sim \mathbf{q}_1 \oplus \mathbf{q}_2$. The assertion (1) of the Lemma follows.

Consider now, for any integer $q$, the exact sequence:

$$0 \to l_1^q|_{D_1}(-(D_1 \cap D_2)) \to l_1^q|_{D_1} \to l_1^q|_{D_1 \cap D_2} \to 0 .$$

We can rewrite this:

$$\text{(on } D_1\text{)} \qquad 0 \to \zeta \otimes l_1^{q-1} \to l_1^q \to (\det E_1)^{-q} \mathbf{q}_1^q \mathbf{q}_2^q \to 0 .$$

The long exact cohomology sequence now gives:

$$(n \geqq 0) \qquad 0 \to \zeta \otimes \Delta_{q-1}^0 \to \Delta_q^0 \xrightarrow{P} A^q E_1^* \otimes S^q E_2 \to \zeta \otimes \Delta_{q-1}^1 \Delta_q^1 \to 0 .$$

The map $P$ is trivially onto for $q \leqq 0$. From (c) and (d) it follows that $P$ is onto for $q = 1$, and therefore it is nonzero for all $q > 1$. Since $S^q E_1^* \otimes S^q E_2$ is an irreducible $GL(E_1) \times GL(E_2)$ module the map is onto (and in fact has a canonical splitting, because by induction $\Delta_{q-1}^0$ does not contain the representation $S^q E_1^* \otimes S^1 E_2$). This yields (2). We also see that for all $q$, we have $\Delta_{q-1}^1 \sim \Delta_q^1$, which yields $\Delta_m^1 = 0$. Together with (b) this proves (4).

Assertion (a), together with (1), now gives (3). □

Recall that $\rho$ denotes the projection $\tilde{\mathcal{R}}'_F \to \tilde{\mathcal{R}}_F$. The decomposition of $H^0(\mathcal{P}, \theta_{\mathcal{P}})$ is obtained by considering the projection $\rho$. We set, for $x \in \tilde{X}$, $\eta_x \equiv (\det \mathcal{Q})(\det \mathcal{E}_x)^{-1}$. Thus $\eta_{x_j} = (\hat{\mathcal{D}}_{j,F})$. We also set $(\det \mathcal{E}_{x_1})^{-1} \otimes (\det \mathcal{E}_{x_2}) \equiv \xi$ and $\xi_j = (\det \mathcal{E}_y)^{-1} \otimes (\det \mathcal{E}_{x_j})$.

**Lemma 5.2.** *Let $m$ be an integer. Then*

(1) *If $m \geqq 0$.*

$$\rho_*(\eta_{x_1}^m|_{D_{1,F}}) = \bigoplus_{q=0,\dots,m} \xi^{m-q} \otimes S^q \mathcal{E}_{x_1}^* \otimes S^q \mathcal{E}_{x_2} .$$

*Otherwise* $\rho_*(\eta^m_{x_1}|_{\hat{\mathscr{D}}_{1,F}}) = 0.$

(2) $R^1\rho_*(\eta^m_{x_1}|_{\hat{\mathscr{D}}_{1,F}}) = 0.$

(3) *If* $m \geqq 0.$

$$\rho_*\eta^m_{x_1} = \bigoplus_{p=0,\ldots,m}\left(\bigoplus_{q=0,\ldots,p} \xi^{p-q}\otimes S^q\mathscr{E}^*_{x_1}\otimes S^q\mathscr{E}_{x_2}\right).$$

*Otherwise* $\rho_*\eta^m_{x_1} = 0.$

(4) $R^1\rho_*\eta^m_{x_1} = 0.$

*Proof.* Immediate corollary of Lemma 5.1. □

**Lemma 5.3.** *The following maps are isomorphisms*:

(1) $H^0(\tilde{\mathscr{R}}'^{ss}, \hat{\theta}')^{inv} \to H^0(\tilde{\mathscr{R}}'^{ss}\cap\tilde{\mathscr{R}}'_F, \hat{\theta}')^{inv}$, *and*

(2) $H^0((\hat{\mathscr{D}}^f_1)^{ss}, \hat{\theta}')^{inv} \to H^0((\hat{\mathscr{D}}^f_1)^{ss}\cap\hat{\mathscr{D}}_{1,F}, \hat{\theta}')^{inv}.$

*Proof.* (1) We use Lemma 4.16 with the identification $V = \tilde{\mathscr{R}}'^{ss}$, $U = \mathscr{P}$, $\pi = \tilde{\psi}'$, $V' = \tilde{\mathscr{R}}'^{ss}\cap\tilde{\mathscr{R}}'_F$, and $U'' = \mathscr{P}\backslash(\mathscr{D}_1\cup\mathscr{D}_2)$. To show that $\mathscr{P}\backslash(\mathscr{D}_1\cup\mathscr{D}_2)$ is nonempty it suffices, by Proposition 4.11(3), to show that $\mathscr{U}_X\backslash\mathscr{W}$ is nonempty. This is true by Remark 6.19 below. To show that $\tilde{\mathscr{R}}'^{ss}\cap\tilde{\mathscr{R}}'_F$ maps onto $\mathscr{P}$ we use Corollary B.17.

(2) We use normality of $\hat{\mathscr{D}}_1$ (Proposition C.7(1)) and Remark C.5(e). □

**Proposition 5.4.** *There exists a canonical isomorphism*

$$H^0(\mathscr{P}, \theta_{\mathscr{P}}) \sim \bigoplus_{(p=0,\ldots,k)}\ \bigoplus_{(q=0,\ldots,p)} H^0(\tilde{\mathscr{R}}_F, \hat{\theta}\otimes\xi^k_1\otimes\xi^{p-q}\otimes S^q\mathscr{E}^*_{x_1}\otimes S^q\mathscr{E}_{x_2})^{inv}. \tag{5.2}$$

*Proof.* We have $H^0(\mathscr{P}, \theta_{\mathscr{P}}) = H^0(\tilde{\mathscr{R}}'^{ss}, \hat{\theta}')^{inv} = H^0(\tilde{\mathscr{R}}'^{ss}\cap\tilde{\mathscr{R}}'_F, \hat{\theta}')^{inv}$ where the second equality follows from Lemma 5.3(1). On the other hand, by Lemma 4.15 and C.3 $H^0(\tilde{\mathscr{R}}'^{ss}, \hat{\theta}')^{inv} = H^0(\mathscr{H}, \hat{\theta}')^{inv}$ so that we can write $H^0(\mathscr{P}, \theta_{\mathscr{P}}) = H^0(\tilde{\mathscr{R}}'_F, \hat{\theta}')^{inv}$.

Recall Eq. (4.4): on $\tilde{\mathscr{R}}'_F$ we have $\hat{\theta}' = \hat{\theta}\otimes\eta^k_y$. Taking direct images on $\tilde{\mathscr{R}}_F$ and using Lemma 5.2.(3) we get Eq. (5.2). □

**Definition 5.5.** For $\tilde{\mu} = (\alpha, \beta)$, $0 \leqq \alpha \leqq \beta \leqq k$, let $\mathscr{U}^{\tilde{\mu}}_{\tilde{X}}$ be the moduli space of semistable parabolic bundles on $\tilde{X}$ with parabolic structures at the $\{y_i\}_I$ and weights $\{(a_i, b_i)\}_I$, and in addition, parabolic structures at $x_1$ and $x_2$, both of weight $(a, b) = (\alpha/k, \beta/k)$. Let $\theta_{\tilde{\mu}} = \theta(d, k, \alpha_i, \beta_i, x_2, l_2)$ be the line bundle on $\mathscr{U}^{\tilde{\mu}}_{\tilde{X}}$ defined as in Remark 2.7, with $Q = \{2\}$ and $l_2 = -k + \beta + \alpha$.

**Theorem 4.** *We have a (canonical) isomorphism*:

$$H^0(\mathscr{P}, \theta_{\mathscr{P}}) \sim \bigoplus_{\tilde{\mu}} H^0(\mathscr{U}^{\tilde{\mu}}_{\tilde{X}}, \theta_{\tilde{\mu}}).$$

*where* $\tilde{\mu}$ *runs through the integers* $(\alpha, \beta)$, $0 \leqq \alpha \leqq \beta \leqq k$.

*Proof.* We first rewrite (5.2) as follows, substituting $p = \beta$, $q = \beta - \alpha$:

$$H^0(\mathscr{P}, \theta_{\mathscr{P}}) \sim \bigoplus_{(\beta=0,\ldots,k)}\ \bigoplus_{(\alpha=0,\ldots,\beta)} H^0(\tilde{\mathscr{R}}_F, \hat{\theta}\otimes\xi^k_1\otimes\xi^{\alpha}\otimes S^{\beta-\alpha}\mathscr{E}^*_{x_1}\otimes S^{\beta-\alpha}\mathscr{E}_{x_2})^{inv}. \tag{5.3}$$

Note now that the bundles $S^q\mathscr{E}_{x_j}$ are direct images of line bundles on the projective bundle $\mathbf{P}(\mathscr{E}_{x_j})$; thus the cohomology groups on the right hand side of (5.3) can be written as sections of suitable line bundles $\hat{\theta}_{\tilde{\mu}}$ on $\tilde{R}_F^+$ where

$$\tilde{\mathscr{R}}^+ \equiv \times_{\tilde{Q}}{}_{i\in I} \operatorname{Flag}_{(1,2)}(\mathscr{E}_{y_i}) \times \times_{\tilde{Q}}{}_{j=1,2} \operatorname{Flag}_{(1,2)}(\mathscr{E}_{x_j}) .$$

(Recall that for a 2-dimensional vector space $\operatorname{Flag}_{(1,2)}$ is just the projective space. Thus $\tilde{\mathscr{R}}_F^+$ is the fibre product of two $P^1$ bundles over $\tilde{\mathscr{R}}_F$.) In fact one checks easily that

$$\hat{\theta}_{\tilde{\mu}} = \hat{\theta} \otimes \mathcal{O}_{\mathbf{P}(\mathscr{E}_{x_1})}(\beta - \alpha) \otimes \mathcal{O}_{P(\mathscr{E}_{x2})}(\beta - \alpha)$$
$$\otimes (\det \mathscr{E}_{x_1})^{k-\beta} \otimes (\det \mathscr{E}_{x_2})^{+\alpha} \otimes (\det \mathscr{E}_y)^{-k} .$$

Each $\hat{\theta}_{\tilde{\mu}}$ is the restriction to $\tilde{\mathscr{R}}_F^+$ of a line bundle linearising the $SL(\tilde{n})$-action on the projective variety $\tilde{\mathscr{R}}^+$ where $\tilde{\mathscr{R}}^+$ is the analogue of $\mathscr{R}$, for parabolic bundles on $\tilde{X}$ with parabolic structures at $\{y_i\}_I \cup \{x_1, x_2\}$ and the moduli space $\mathscr{U}_{\tilde{X}}^{\tilde{\mu}}$ is the GIT quotient of the semistable points $(\tilde{\mathscr{R}}^+)^{ss} \subset \tilde{\mathscr{R}}_F^+$. There is a small point to be checked here, namely, that the integers $n, m$ involved in the GIT construction of $\mathscr{P}$ and $\mathscr{U}_X$ can be made to work for these additional moduli space as well. But this is clear since the index $\tilde{\mu}$ runs over a fixed finite set depending only on $k$.

The variety $\tilde{\mathscr{R}}_F^+$ is normal (in fact smooth) so that Lemma 4.15 applies, and we can conclude

$$HG^0(\mathscr{P}, \theta_{\mathscr{P}} \sim )\bigoplus_{\tilde{\mu}} H^0(\tilde{\mathscr{R}}_F^+, \hat{\theta}_{\tilde{\mu}})^{\text{inv}} \sim \bigoplus_{\tilde{\mu}} H^0(\mathscr{U}_{\tilde{X}}^{\tilde{\mu}}, \theta_{\tilde{\mu}}) . \tag{5.4}$$

This finishes the proof. □

We close this subsection with two results which will be used in the proof of Theorem 5.

**Proposition 5.6.** *Let $m \geqq 0$ be a integer. Consider the inclusions of sheaves:*

(1) *on* $\tilde{\mathscr{R}}'_F$, $\eta_{x_1}^m(-\hat{\mathscr{D}}_{1,F}) \to \eta_{x_1}^m$,

(2) *on* $\tilde{\mathscr{R}}'_F$, $\eta_{x_1}^m(-\hat{\mathscr{D}}_{1,F} - \hat{\mathscr{D}}_{2,F}) \to \eta_{x_1}^m$, *and*

(3) *on* $\hat{\mathscr{D}}_{1,F}$, $\eta_{x_1}^m \otimes \mathscr{I}_{\hat{\mathscr{D}}_{1,F}}(\hat{\mathscr{V}}_{1,E} \cup (\hat{\mathscr{D}}_{1,F} \cap \hat{\mathscr{D}}_{2,F})) \to \eta_{n_1}^m|_{\hat{\mathscr{D}}_{1,F}}$

*Each of the above sheaves has $R^1\rho_*(\cdot) = 0$. The induced inclusions of the zeroth direct images by $\rho$ (which are therefore* vector bundles) *have a $SL(\tilde{n})$-invariant splitting.*

*Proof.* The cases (1) and (2) are an immediate consequence of Lemma 5.2.

We turn next to (3). Considser

$$(\text{on } \hat{\mathscr{D}}_{1,F}) \qquad 0 \to \eta_{x_1}^m \otimes \mathscr{I}_{\hat{\mathscr{D}}_{1,F}}(\hat{\mathscr{V}}_{1,F}) \to \eta_{x_1}^m \to \eta_{x_1}^m|_{\hat{\mathscr{V}}_{1,F}} \to 0 .$$

We have, using the fact that on $\hat{\mathscr{V}}_{1,F}$, $\det \mathscr{Q} \sim \det \mathscr{E}_{x_2}$, the equality $\eta_{x_1}|_{\hat{\mathscr{V}}_{1,F}} = \xi|_{\hat{\mathscr{V}}_{1,F}}$. By Lemma 5.2 therefore

$$\rho_*(\eta_{x_1}^m \otimes \mathscr{I}_{\hat{\mathscr{D}}_{1,F}}(\hat{\mathscr{V}}_{1,F})) = \bigoplus_{n=1,\ldots,m} \xi^{m-n} \otimes S^n\mathscr{E}_{x_1}^* \otimes S^n\mathscr{E}_{x_2} ,$$

$$R^1\rho_*((\eta_{x_1}^m \otimes \mathscr{I}_{\hat{\mathscr{D}}_{1,F}}(\hat{\mathscr{V}}_{1,F}))) = 0 .$$

Consider next

(on $\hat{\mathscr{D}}_{1,F}$)

$$0 \to \eta_{x_1}^m \otimes \mathscr{I}_{\hat{\mathscr{D}}_{1,F}}(\hat{\mathscr{V}}_{1,F} \cup (\hat{\mathscr{D}}_{1,F} \cap \hat{D}_{2,F})) \to \eta_{x_1}^m \otimes \mathscr{I}_{\hat{\mathscr{D}}_{1,F}}(\hat{\mathscr{V}}_{1,F}) \to \eta_{x_1}^m|_{\hat{\mathscr{D}}_{1,F} \cap \hat{\mathscr{D}}_{2,F}} \to 0,$$

where we have used the fact that $(\hat{\mathscr{V}}_{1,F} \cap (\hat{\mathscr{D}}_{1,F} \cap \hat{\mathscr{D}}_{2,F})) = \emptyset$. As in the foregoing proofs we see that

$$\bigoplus_{n=1,\ldots,m-1} \xi^{m-n} \otimes S^n \mathscr{E}_{x_1}^* \otimes S^n \mathscr{E}_{x_2} = \rho_*(\eta_{x_1}^m \otimes \mathscr{I}_{\hat{\mathscr{D}}_{1,F}}(\hat{\mathscr{V}}_{1,F} \cup (\hat{\mathscr{D}}_{1,F} \cap \hat{\mathscr{D}}_{2,F})))$$

$$\to \rho_*(\eta_{x_1}^m = \bigoplus_{n=0,\ldots,m} \xi^{m-n} \otimes S^n \mathscr{E}_{x_1}^* \otimes S^n \mathscr{E}_{x_2},$$

splits, and $R^1\rho_*(\eta_{x_1}^m \otimes \mathscr{I}_{\hat{\mathscr{D}}^{1,F}}(\hat{\mathscr{V}}_{1,F} \cup (\hat{\mathscr{D}}_{1,F} \cap \hat{\mathscr{D}}_{2,F}))) = 0$. □

**Lemma 5.7.** *The following maps are surjections:*

(1) $H^0(\theta_{\mathscr{P}}) \to H^0(\theta_{\mathscr{P}}|_{\mathscr{D}_1})$,

(2) $H^0(\theta_{\mathscr{P}}) \to H^0(\theta_{\mathscr{P}}|_{\mathscr{D}_1 \cup \mathscr{D}_2})$,

(3) $H^0(\theta_{\mathscr{P}}|_{\mathscr{D}_1}) \to H^0(\theta_{\mathscr{P}}|_{\mathscr{V}_1 \cup \mathscr{D}_1 \cap \mathscr{D}_2})_2$.

*Proof.* Let us deal with (1) in detail. Consider the diagram:

$$\begin{array}{ccc} H^0(\tilde{\mathscr{R}}'^{ss}, \hat{\theta}')^{inv} & \xrightarrow{a} & H^0((\hat{\mathscr{D}}_1^f)^{ss}, \hat{\theta}')^{inv} \\ e\uparrow & & f\uparrow \\ H^0(\mathscr{H}, \hat{\theta}')^{inv} & \longrightarrow & H^0(\hat{\mathscr{D}}_1^f, \hat{\theta}')^{inv} \\ b\downarrow & & d\downarrow \\ H^0(\tilde{\mathscr{R}}'_F, \hat{\theta}')^{inv} & \xrightarrow{c} & H^0(\hat{\mathscr{D}}_{1,F}, \hat{\theta}')^{inv} \end{array}$$

We need to prove that a is surjective. The maps e, f are equalities because of Lemma 4.15. The map b is an isomorphism by Lemmas 5.3 and 4.15. The map d is similarly an isomorphism, so that the result follows by Proposition 5.6 which states that c is surjective.

The statements (2) and (3) are proved along similar lines. There is a complication in case (2) because $D_1 \cup D_2$ is not normal. In this case we have an analogous diagram, with $\hat{\mathscr{D}}_1^f$ replaced by $\hat{\mathscr{D}}_1^f \cup \hat{\mathscr{D}}_2^f$ etc. (We will continue to use the same letters to denote the maps.) We can no longer assert that f and d are equalities. But given a section $\sigma$ of $H^0((\hat{\mathscr{D}}_1^f)^{ss} \cup (\hat{\mathscr{D}}_2^f)^{ss}. \hat{\theta}')^{inv}$, it certainly extends to sections $\sigma_j$ on $\hat{\mathscr{D}}_j^f$ which are equal on $(\hat{\mathscr{D}}_1)^{ss} \cap (\hat{\mathscr{D}}_2)^{ss}$. By seminormality of $\hat{\mathscr{D}}_{1,F} \cup \hat{\mathscr{D}}_{2,F}$, a fact easily checked, this yields a section there. The rest of the proof goes through as before. □

### *5b. The decomposition theorem on $\mathscr{U}_X$*

We start with a general result relating sections of a line bundle on a semi-normal variety to those of the pull-back on the normalisation.

**Proposition 5.8.** *Suppose given a seminormal variety $V$, with normalisation $\sigma: \tilde{V} \to V$. Let the non-normal locus be $W$, endowed with its reduced structure. Let $\tilde{W}$ be the set-theoretic inverse image of $W$ in $\tilde{V}$, endowed with its reduced structure. Let $\mathscr{N}$ be a line bundle on $V$, and let $\tilde{\mathscr{N}}$ be its pull-back to $\tilde{V}$. Suppose $H^0(\tilde{V}, \tilde{\mathscr{N}}) \to H^0(\tilde{W}, \tilde{\mathscr{N}})$ is surjective. Then*

(1) *There is an exact sequence*

$$0 \to H^0(\tilde{V}, \tilde{\mathcal{N}} \otimes \mathcal{I}(\tilde{W})) \to H^0(V, \mathcal{N}) \to H^0(W, \mathcal{N}) \to 0 .$$

(2) *If* $H^1(W, \mathcal{N}) \to H^1(\tilde{W}, \tilde{\mathcal{N}})$ *is injective so is* $H^1(V, \mathcal{N}) \to H^1(\tilde{V}, \tilde{\mathcal{N}})$.

*Proof.* Consider the commutative diagram of sheaves on $V$:

$$\begin{array}{ccccccc} 0 \to & \mathcal{I}(W) & \to & \mathcal{O}_V & \to \mathcal{O}_W \to 0 \\ & = \downarrow & & \downarrow & \downarrow \\ 0 \to & \sigma_* \mathcal{I}(\tilde{W}) & \to & \sigma_* \mathcal{O}_{\tilde{V}} & \to \mathcal{O}_{\tilde{W}} \to 0 \end{array}$$

where the equality is a consequence of Lemma 3.7. Note that the vertical arrows are inclusions. Tensoring by $\mathcal{N}$ and using the projection formula we get

$$\begin{array}{ccccccc} 0 \to & \mathcal{N} \otimes \mathcal{I}(W) & \to & \mathcal{N} & \to \mathcal{N}|_W \to 0 \\ & = \downarrow & & \downarrow & \downarrow \\ 0 \to & \sigma_*(\tilde{\mathcal{N}} \otimes \mathcal{I}(\tilde{W}) & \to & \sigma_* \tilde{\mathcal{N}} & \to \tilde{\mathcal{N}}|_{\tilde{W}} \to 0 \end{array}$$

Taking cohomologies gives

$$\begin{array}{ccccccc} 0 \to & H^0(\mathcal{N} \otimes \mathcal{I}(W)) & \to & H^0(\mathcal{N}) & \to & H^0(\mathcal{N}|_W & \xrightarrow{a} H^1(\mathcal{N} \otimes \mathcal{I}(W)) \\ & = \downarrow & & \downarrow & & c\downarrow & = \downarrow \\ 0 \to & H^0(\tilde{\mathcal{N}} \otimes \mathcal{I}(\tilde{W}) & \to & H^0(\tilde{\mathcal{N}}) & \to & H^0(\tilde{\mathcal{N}}|_{\tilde{W}}) & \xrightarrow{b} H^1(\tilde{\mathcal{N}} \otimes \mathcal{I}(\tilde{W})) \end{array}$$

where we have used the fact that $\sigma$ is finite to identify $H^1(\sigma_*(\tilde{\mathcal{N}} \otimes \mathcal{I}(\tilde{W})))$ with $H^1(\tilde{\mathcal{N}} \otimes \mathcal{I}(\tilde{W}))$. By assumption b is zero. Since c is an injection we see that a is zero as well. This implies the first part of the Proposition.

Continuing with the two cohomology sequences and using the above results we also get

$$\begin{array}{ccccccc} 0 \to & H^1(\mathcal{N} \otimes \mathcal{I}(W)) & \to & H^1(\mathcal{N}) & \to & H^1(\mathcal{N}|_W) \to \\ & = \downarrow & & \downarrow & & \downarrow \\ 0 \to & H^1(\tilde{\mathcal{N}} \otimes \mathcal{I}(\tilde{W})) & \to & H^1(\tilde{\mathcal{N}}) & \to & H^1(\tilde{\mathcal{N}}|_{\tilde{W}}) \to \end{array}$$

This implies the second claim. □

The subvarieties $\mathcal{W}$ and $\mathcal{W}'$ of $\mathcal{U}_X$ are defined in §3a. The seminormality of $\mathcal{U}_X$ and $\mathcal{W}$ and in particular, its main consequence, as stated in Lemma 3.7, will be used repeatedly below. Recall also (Lemma 4.10(2)) that $\theta_{\mathcal{P}} = \phi^* \theta_{\mathcal{U}_X}$.

**Proposition 5.9.** *There exists a (noncanonical) isomorphism*:

$$H^0(\mathcal{U}_X, \theta_{\mathcal{U}_X}) \sim H^0(\theta_{\mathcal{P}}(-\mathcal{D}_1 - \mathcal{D}_2)) \oplus H^0(\mathcal{D}_1, \theta_{\mathcal{P}}(-\mathcal{D}_2)) .$$

*Proof.* We use Proposition 5.8(1). By seminormality of $\mathcal{U}_X$ and Proposition 3.15(4) we have an exact sequence

$$0 \to H^0(\theta_{\mathcal{P}}(-\mathcal{D}_1 - \mathcal{D}_2)) \to H^0(\theta_{\mathcal{U}_X}) \to H^0(\theta_{\mathcal{U}_X}|_{\mathcal{W}}) \to 0 . \tag{5.5a}$$

(Proposition 5.8(1) applies because of Lemma 5.7(2).) Again, by seminormality of $\mathcal{W}$ and Proposition 3.15(5) we get

$$0 \to H^0(\theta_{\mathcal{P}} \otimes \mathcal{I}_{\mathcal{D}_1}(\mathcal{V}_1 \cup (\mathcal{D}_1 \cap \mathcal{D}_2))) \to H^0(\mathcal{W}, \theta_{\mathcal{U}_X}) \to H^0(\mathcal{W}', \theta_{\mathcal{U}_X}) \to 0 . \tag{5.5b}$$

(Proposition 5.8(1) applies because of Lem,ma 5.7(3).

On the other hand, by Corollary 4.4(2) we have on $\mathscr{P}$ an exact sequence

$$0 \to H^0(\theta_{\mathscr{P}} \otimes \mathscr{I}_{\mathscr{D}_1}(\mathscr{V}_1 \cup (\mathscr{D}_1 \cap \mathscr{D}_2))) \to H^0(\mathscr{D}_1(-\mathscr{D}_2), \theta_{\mathscr{P}}) + H^0(\mathscr{V}, \theta_{\mathscr{P}}) \to 0\,, \tag{5.5c}$$

where the surjectivity on the right follows from Lemma 5.7(3), again using 4.4(2).

By Proposition 4.11(3) we have $H^0(\mathscr{V}, \theta_{\mathscr{P}}) = H^0(\mathscr{W}', \theta_{\mathscr{U}_X})$, and this, together with Eqs. (5.5) yields the desired result. □

We can now prove the

**Theorem 5.** *Let* $\eta_x \equiv (\det \mathscr{Q})(\det \mathscr{E}_x)^{-1}$ *for a point* $x \in X$, *and* $\xi_1 \equiv \eta_y \eta_{x_1}^{-1}$. *Then there exists a noncanonical isomorphism*

$$H^0(\mathscr{U}_X, \theta_{\mathscr{U}_X}) \sim \bigoplus_{(p=0,\ldots,k-1)} \bigoplus_{(q=0,\ldots,p)} H^0(\tilde{\mathscr{R}}_F, \hat{\theta} \otimes \xi_1^k \otimes \xi^{1+p-q} \otimes S^q \mathscr{E}_{x_1}^* \otimes S^q \mathscr{E}_{x_2})^{\mathrm{inv}}\,. \tag{5.6}$$

*Proof.* By Proposition 5.9 and Lemma 4.15

$$H^0(\mathscr{U}_X, \theta_{\mathscr{U}_X}) \sim H^0(\mathscr{H}, \hat{\theta}' \otimes \mathscr{O}(-\hat{\mathscr{D}}_1 - \hat{\mathscr{D}}_2))^{\mathrm{inv}} \oplus H^0(\hat{\mathscr{D}}_1^f, \hat{\theta}' \otimes \mathscr{O}(-(\hat{\mathscr{D}}_1^f \cap \hat{\mathscr{D}}_2^f)))^{\mathrm{inv}}\,.$$

We have applied Lemma 4.15 with the identification $W = \mathscr{H}$; note that the Lemma applies since, for example, sections of $\hat{\theta}' \otimes \mathscr{O}(-\hat{\mathscr{D}}_1 - \hat{\mathscr{D}}_2)$ are also sections of $\hat{\theta}'$. By Lemma 5.5 the sections on the right are determined by their restrictions to $\tilde{\mathscr{R}}'_F$ and $\hat{\mathscr{D}}_{1,F}$ respectively. Now use Lemma 5.2. □

*Proof of main Theorem (A).* This follows from Theorem 5 exactly as Theorem 4 follows from Proposition 5.4. □

*Remark 5.10.* For $j = 1, 2$, let $\mathscr{F}r_j$ denote the frame-bundle of $\mathscr{E}_{x_j}$, thought of as a principal $GL(2)$-bundle. The bundle $\tilde{\mathscr{R}}'_F \xrightarrow{\rho} \tilde{\mathscr{R}}_F$ can be regarded as associated to the principal $GL(2) \times GL(2)$ bundle $\mathscr{F}r_1 \times_{\tilde{\mathscr{R}}_F} \mathscr{F}r_2$. The various (zeroth-) direct image sheaves that we encounter can be thought of as vector bundles associated to representations of $GL(2) \times GL(2)$. In particular, equation (5.6) can be rewritten in terms of vector bundles associated to $\mathscr{F}r_1 \times_{\tilde{\mathscr{R}}_F} \mathscr{F}r_2$:

$$H^0(\mathscr{U}_X, \theta_{\mathscr{U}_X}) \sim \bigoplus_{\mu} H^0(\tilde{\mathscr{R}}_F, \hat{\theta} \otimes \xi_1^k \otimes \xi \otimes (\mathscr{E}_{x_1}^{\mu})^* \otimes \mathscr{E}_{x_2}^{\mu})^{\mathrm{inv}}\,,$$

where $\mu$ runs over (highest weights of) irreducible representations of $GL(2)$, $\mu = (\alpha, \beta)$, $0 \leqq \alpha \leqq \beta < k$, and $\mathscr{E}_{x_j}^{\mu}$ is the bundle associated to $\mathscr{F}r_j$ through the representation with highest weight $\mu$. (The representation corresponding to $(\alpha, \beta)$ is $(\det \varrho)^{\otimes \alpha}\,\mathrm{Symm}^{\beta-\alpha}(\varrho)$ where $\varrho$ is the defining representation of $GL(2)$.

## 6. The vanishing theorem

Consider now a family $X_t$ of smooth curves degenerating, as in the Introduction, to $X_0 = X$. Clearly, to be able to assert that $h^0(\mathscr{U}_{X_t}, \theta_{\mathscr{U}_{X_t}}) = h^0(\mathscr{U}_X, \theta_{\mathscr{U}_X})$ we need a vanishing theorem.

*6a. A vanishing theorem on* $\mathcal{U}_{\tilde{X}}$

We will first prove a vanishing Theorem for $\mathcal{U}_{\tilde{X}}$. This will (with the replacement $\tilde{X} \to X_t$) prove the constancy of $h^0(\mathcal{U}_{X_t}, \theta_{\mathcal{U}_X'})$ for $t \neq 0$. It will also be needed in the next subsection.

We begin by computing the dualising sheaf of $\mathcal{U}_{\tilde{X}}$ using Lemma 4.17. The space $\tilde{\mathcal{R}}_F$ is defined in Notation 4.3c; $\mathcal{U}_{\tilde{X}}$ is the good quotient of the open subset of semistable points $\tilde{\mathcal{R}}^{ss}$. We will denote by $\tilde{\psi}$ the projection $\tilde{\mathcal{R}}^{ss} \to \mathcal{U}_{\tilde{X}}$.

**Notation 6.1.** Let Det denote the morphism $\tilde{\mathcal{R}}_F \to J^d_{\tilde{X}}$ given by the determinant of the universal quotient bundle. This induces a morphism $\mathcal{U}_{\tilde{X}} \to J^d_{\tilde{X}}$, which will also be denoted det. Let $\mathcal{L}$ denote a Poincaré line bundle on $\tilde{X} \times J^d_{\tilde{X}}$ and let $\theta_y$ denote the line-bundle on $J^d_{\tilde{X}}$ defined by

$$\theta_y \equiv (\det R\pi_J \mathcal{L}) \otimes (\det \mathcal{L}_y)^{(d+1-\tilde{g})}. \tag{6.1}$$

**Proposition 6.2.** *Assume $\tilde{g} \geq 1$. Let $\Omega_{\tilde{X}}$ be the canonical bundle of $\tilde{X}$, and suppose $\Omega_{\tilde{X}} = \mathcal{O}(\sum_{q \in Q} z_q)$. Let $\Omega_{\tilde{\mathcal{R}}_F}$ denote the canonical bundle of $\tilde{\mathcal{R}}_F$. We have*

$$\Omega^{-1}_{\tilde{\mathcal{R}}_F} = (\det R\pi_{\tilde{\mathcal{R}}} \mathcal{E})^4 \otimes \bigotimes_i \{\mathcal{Q}_i^2 \otimes (\det \mathcal{E}_{y_i})^{-1}\}$$
$$\otimes \bigotimes_q \{(\det \mathcal{E}_{z_q})^{-1}\} \otimes (\det \mathcal{E}_y)^{2\tilde{n}+2\tilde{g}-2} \otimes (\mathrm{Det}^* \theta_y)^{-2}. \tag{6.2}$$

*Proof.* $\tilde{\mathcal{R}}_F$ is a fibre-product of $\mathbf{P}^1$-bundles over $\tilde{\mathbf{Q}}_F$, and we first need an expression for $\Omega_{\tilde{\mathbf{Q}}_F}$. (The spaces $\tilde{\mathbf{Q}}$ and $\tilde{\mathbf{Q}}_F$ are defined in §4.a.) On $\tilde{X} \times \tilde{\mathbf{Q}}_F$ we have an exact sequence of vector bundles $0 \to \mathcal{K} \to \mathcal{O}^{\tilde{n}} \to \mathcal{E} \to 0$, and the tangent space at a point $0 \to K \to \mathcal{O}^{\tilde{n}} \to E \to 0$ is $H^0(\tilde{X}, K^* \otimes E)$. From the properties of $\tilde{\mathbf{Q}}_F$ (the Notation 4.3b) it follows that

$$\Omega^{-1}_{\tilde{\mathcal{R}}_F} = \det R\pi_{\tilde{\mathcal{R}}_F}(\mathcal{E} \otimes \mathcal{E}^*) \otimes \bigotimes_i \{\mathcal{Q}_i^2 \otimes (\det \mathcal{E}_{y_i})^{-1}\}.$$

We now use a variant of the method of [D–N] to evaluate $\det R\pi_{\tilde{\mathcal{R}}_F}(\mathcal{E} \otimes \mathcal{E}^*)$. Consider on $\tilde{\mathcal{R}}_F$ the projective bundle $\mathbf{P}$ associated to the vector bundle $((\pi_{\tilde{\mathcal{R}}_F})_* \mathcal{E})^*$. We have on $\tilde{X} \times \mathbf{P}$ an injection of sheaves $0 \to \mathcal{O}_{\mathbf{P}}(-1)$. Let $D'$ denote the (reduced) subscheme where this section vanishes. Its projection to $\mathbf{P}$, which we will denote by $D$, is an irreducible divisor. (One sees this by intersecting with the fibre over a point of $\tilde{\mathcal{R}}_F$ – see [D-N, Lemma 7.3 and Corollaire 7.4]). Outside $D'$ we have an exact sequence of vector bundles $0 \to \mathcal{O}_{\mathbf{P}}(-1) \to \mathcal{E} \to \det \mathcal{E} \otimes \mathcal{O}_{\mathbf{P}}(+1) \to 0$, which yields, outside $D$,

(1) an isomorphism $\det R\pi_{\mathbf{P}} \mathcal{E} = \det R\pi_{\mathbf{P}} \mathcal{L} \otimes (\mathcal{L}^{-1} \det \mathcal{E})^{-(d+1-\tilde{g})} \otimes \mathcal{O}_{\mathbf{P}}(-d)$.

(2) an isomorphism $\det R\pi_{\mathbf{P}}(\mathcal{E} \otimes \mathcal{E}^*) = \det R\pi_{\mathbf{P}} \mathcal{E} \otimes \det R\pi_{\mathbf{P}} \mathcal{E}^* \otimes \mathcal{O}_{\mathbf{P}}(-2d)$.

(We have written $\mathrm{Det}^* \mathcal{L} = \mathcal{L}$).

By duality $\det R\pi_{\mathbf{P}} \mathcal{E}^* = \det R\pi_{\mathbf{P}}(\mathcal{E} \otimes \Omega_{\tilde{X}})$. From this and the exact sequence $0 \to \mathcal{E} \to \mathcal{E} \otimes \Omega_{\tilde{X}} \to \bigoplus_q \mathcal{E}_{z_q} \otimes (\Omega_{\tilde{X}})_{z_q} \to 0$ we get $\det R\pi_{\mathbf{P}} \mathcal{E}^* = \det R\pi_{\mathbf{P}} \mathcal{E} \otimes \bigotimes_q \{(\det \mathcal{E}_{z_q})^{-1}\}$.

Thus we have an isomorphism outside $D$:

$$\det R\pi_{\mathbf{P}}(\mathcal{E} \otimes \mathcal{E}^*) = (\det R\pi_{\mathbf{P}} \mathcal{E})^4 \otimes \bigotimes_q \{(\det \mathcal{E}_{z_q})^{-1}\}$$
$$\otimes (\det R\pi_{\mathbf{P}} \mathcal{L})^{-2} \otimes (\mathcal{L}^{-1} \otimes \det \mathcal{E})^{2(d+1-\tilde{g})}.$$

If we now use the fact that $\mathscr{L}^{-1} \otimes \det \mathscr{E}$ on $\tilde{X} \times \mathbf{P}$ is a line bundle pulled back from $\mathbf{P}$ we get (6.2) outside $D$. (That is, the line bundles on the two sides of (6.2), when pulled back to $\mathbf{P}$, are isomorphic outside $D$.)

We now claim that the map $\mathrm{Pic}(\tilde{\mathscr{R}}_F) \to \mathrm{Pic}(\mathbf{P} \backslash D)$ is injective; this will clearly finish the proof. To see the truth of the claim, one uses the fact (cf. [D-N, Lemma 7.3]) that each fibre of the morphism $\mathbf{P} \backslash D \to \tilde{\mathscr{R}}_F$ is the complement of an irreducible divisor on a projective space so that any nowhere-vanishing regular function on the fibre is a constant. (This shows that if the pull-back of a line bundle is trivial then the line bundle itself is trivial, for a nowhere-vanishing section of the pull-back descends to a nowhere-vanishing section of the original bundle.) □

**Lemma 6.3.** *Assume* $\tilde{g} \geqq 2$. *Then* $(\tilde{\psi}_* \Omega_{\tilde{\mathscr{R}}^{ss}})^{\mathrm{inv}} = \Omega_{\mathscr{U}_{\tilde{X}}}$ *where* $\Omega_{\mathscr{U}_{\tilde{X}}}$ *is the dualising sheaf of* $\mathscr{U}_{\tilde{X}}$.

*Proof.* Consider the action of $PSL(\tilde{n})$ on $\tilde{\mathscr{R}}^{ss}$. We will see (Lemma 6.14(1)) that if $\tilde{g} \geqq 2$ the complement of the set $\tilde{\mathscr{R}}^{s}$ of stable points has codimension $> 1$. Since $\tilde{\mathscr{R}}^{s} \to \mathscr{U}^{s}_{\tilde{X}}$ is an étale locally trivial $PSL(\tilde{n})$-bundle we see that the conditions of Lemma 4.17 are satisfied; and this implies the above result. □

We can now prove

**Theorem 6.** *Assume* $\tilde{g} \geq 3$. *Then* $H^1(\mathscr{U}_{\tilde{X}}, \theta_{\mathscr{U}_{\tilde{X}}}) = 0$.

*Proof.* We use the following device: we consider a new set of data $(d, \bar{k}, \bar{\alpha}_i, \bar{\beta}_i)$ such that $\bar{k} = k + 4$, and $\bar{\beta}_i - \bar{\alpha}_i = \beta_i - \alpha_i + 2$. Let $\bar{\omega}$ denote denote the new set of parabolic weights, $\hat{\theta}_{\bar{\omega}}$ the line bundle on $\tilde{\mathscr{R}}$ defined by the new data, $\mathscr{U}_{\tilde{X},\bar{\omega}}$ the corresponding moduli space, and $\theta_{\mathscr{U}_{\tilde{X},\bar{\omega}}}$ the "descendant" of $\hat{\theta}_{\bar{\omega}}$. Recall that the parabolic data do not quite suffice to define $\hat{\theta}_{\bar{\omega}}$, but a choice of degree 1 line bundle on $\tilde{X}$ is also needed (see Remark 2.7). We make this choice so that

$$\hat{\theta}_{\bar{\omega}} = (\det R\pi_{\tilde{\mathscr{R}}_F}\mathscr{E})^{k+4} \otimes \bigotimes_i \{(\mathscr{Q}_i)^{\bar{\beta}_i - \bar{\alpha}_i} \otimes (\det \mathscr{E}_{y_i})^{k-\beta_i}\}$$

$$\otimes \bigotimes_i \{(\det \mathscr{E}_{y_i})^{\bar{\beta}_i - \beta_i - 1}\} \otimes (\det \mathscr{E}_y)^{2\tilde{n} + 2\tilde{g} - 2 + \tilde{l}}.$$

We shall assume that the integers $n, m$ in the construction of $\mathscr{U}_{\tilde{X}}$ are chosen so that they work for $\mathscr{U}_{\tilde{X},\bar{\omega}}$ as well, so that $\mathscr{U}_{\tilde{X},\bar{\omega}}$ is the quotient of the semistable points $\tilde{\mathscr{R}}^{ss}_{\bar{\omega}}$ of $\tilde{\mathscr{R}}$ with respect to the new polarisation. Using (6.2) we see that on $\tilde{\mathscr{R}}_F$ we have

$$\hat{\theta} = \hat{\theta}_{\bar{\omega}} \otimes \Omega_{\tilde{\mathscr{R}}_F} \otimes (\mathrm{Det}^* \theta_y)^{-2}. \tag{6.3}$$

Since $\tilde{\mathscr{R}}^{ss}$ is a dense open subset of $\tilde{\mathscr{R}}_F$ this continues to hold in $\tilde{\mathscr{R}}^{ss}$.

We now write

$$\begin{aligned} H^1(\mathscr{U}_{\tilde{X}}, \theta_{\mathscr{U}_{\tilde{X}}})_1 &\underset{1}{=} H^1(\tilde{\mathscr{R}}^{ss}, \hat{\theta})^{\mathrm{inv}} \\ &\underset{2}{=} H^1(\tilde{\mathscr{R}}^{ss}_{\bar{\omega}}, \hat{\theta})^{\mathrm{inv}} \\ &\underset{3}{=} H^1(\tilde{\mathscr{R}}^{ss}_{\bar{\omega}}, \hat{\theta}_{\bar{\omega}} \otimes \Omega_{\tilde{\mathscr{R}}_F} \otimes (\mathrm{Det}^* \theta_y)^{-2})^{\mathrm{inv}} \\ &\underset{4}{=} H^1(\mathscr{U}_{\tilde{X},\bar{\omega}}, \theta_{\mathscr{U}_{\tilde{X},\bar{\omega}}} \otimes ((\tilde{\psi}_{\bar{\omega}})_* \Omega_{\tilde{\mathscr{R}}_F})^{\mathrm{inv}} \otimes (\mathrm{Det}^* \theta_y)^{-2}) \\ &\underset{5}{=} H^1(\mathscr{U}_{\tilde{X},\bar{\omega}}, \theta_{\mathscr{U}_{\tilde{X},\bar{\omega}}} \otimes \Omega_{\mathscr{U}_{\tilde{X},\bar{\omega}}} \otimes (\mathrm{Det}^* \theta_y)^{-2}), \end{aligned}$$

where $\Omega_{\mathscr{U}_{\tilde{X},\tilde{\omega}}}$ is the dualising sheaf of $\mathscr{U}_{\tilde{X},\tilde{\omega}}$ and $\tilde{\psi}_{\tilde{\omega}}$ the quotient map $\tilde{\mathscr{R}}^{ss}_{\tilde{\omega}} \to \mathscr{U}_{\tilde{X},\tilde{\omega}}$. The second equality holds because of the Lemma 6.14(2) below, using a Hartogs-type extension theorem for first cohomology. The third uses Equation 6.2, and the fourth Lemma 6.3. The first and fifth equalities follow from the fact that for good quotients the space of invariants of the cohomology of an invariant line bundle is the same as the cohomology of the invariant direct image. This fact is easily proved (as pointed out by J.M. Drezet) by taking an invariant affine cover and applying Reynold's operator to Cĕch cochains.

We will prove below (Lemma 6.4) that $\theta_{\mathscr{U}_{\tilde{X},\tilde{\omega}}} \otimes (\mathrm{Det}^* \theta_y)^{-2}$ is ample. Since $\mathscr{U}_{\tilde{X},\tilde{\omega}}$ has rational singularities a Kodaira-type vanishing theorem [S-S, Theorem 7.80(f)] now applies and we can conclude that $H^1(\mathscr{U}_{\tilde{X}}, \theta_{\mathscr{U}_{\tilde{X}}}) = 0$. □

**Lemma 6.4.** *$\theta_{\mathscr{U}_{\tilde{X}}} \otimes (\mathrm{Det}^* \theta_y)^{-2}$ is ample if $k > 4$.*

*Proof.* Consider the morphism Det: $\mathscr{U}_{\tilde{X}} \to J^d_{\tilde{X}}$, and let $\mathscr{U}^L_{\tilde{X}}$, and let $\mathscr{U}^L_{\tilde{X}}$ denote the fibre above $L$. One has a $2^{2\tilde{g}}$-fold covering $\mathscr{U}^L_{\tilde{X}} \times J^0_{\tilde{X}} \to \mathscr{U}_{\tilde{X}}$. We will show that $\theta_{\mathscr{U}_{\tilde{X}}} \otimes (\mathrm{Det}^* \theta_y)^{-2}$ is ample when pulled back to this finite cover.

One can show by a standard method (as for example, in [S2, p. 53]) that $\mathscr{U}^L_{\tilde{X}}$ is unirational. Hence its $\mathrm{Pic}_0$ is trivial, and the pull-back bundle is therefore a product of line bundles coming from the two factors. It suffices to check that the restriction to each factor is ample. The restriction to the first factor is $\theta$, and clearly ample.

Write the restriction to $J^0_{\tilde{X}}$ as $M_1 \otimes M_2$, where $M_1$ the pull-back of $\theta_{\mathscr{U}_{\tilde{X}}}$ and $M_2$ is the pull-back of $(\mathrm{Det}^* \theta_y)^{-2}$. Now $\theta_y$ is essentially the theta bundle on $J^d_{\tilde{X}}$, and ample. We will identify $J^d_{\tilde{X}}$ and $J^0_{\tilde{X}}$, and also work up to algebraic equivalence. One checks (using well-known properties of theta bundles on abelian varieties) that $M_2$ is algebraically equivalent to $\theta_y^{-8}$. Also, $M_2$ is algebraically equivalent to $\theta_y^{2k}$. (Consider a family $E \otimes \mathscr{L}$ of parabolic bundles, for $E$ a fixed parabolic bundle, and then deform $E$ to a bundle of the form $\mathscr{O}_X \oplus \mathscr{O}_X(\sum_{h=1,\ldots,d} x_h)$.) Clearly $M_1 \otimes M_2$ is ample if $k > 4$. □

*6b. Vanishing Theorem on $\mathscr{U}_X$*

We turn now to the vanishing theorem for $\mathscr{U}_X$.

**Theorem 7.** *Assume $g \geqq 4$. Then $H^1(\mathscr{U}_X, \theta) = 0$.*

*Proof.* This is a consequence of the next lemma and Theorem 8 below. □

**Lemma 6.5.** *$H^1(\mathscr{U}_X, \theta_{\mathscr{U}_X})$ injects into $H^1(\mathscr{P}, \theta_{\mathscr{P}})$.*

*Proof.* By Proposition 5.8(2) it suffices to prove that $H^1(\mathscr{W}, \theta_{\mathscr{U}_x})$ injects into $H^1(\mathscr{D}_1 \cup \mathscr{D}_2, \theta_{\mathscr{P}})$. For this it clearly suffices to show that $H^1(\mathscr{W}, \theta_{\mathscr{U}_x})$ injects into $H^1(\mathscr{D}_1, \theta_{\mathscr{P}})$. Again using the Proposition 5.8(2) we se that it is enough to show that $H^1(\mathscr{W}', \theta_{\mathscr{U}_x})$ injects into $H^1(\mathscr{V}_1 \cup (\mathscr{D}_1 \cap \mathscr{D}_2), \theta_{\mathscr{P}})$, and as above it is enough to show that $H^1(\mathscr{W}', \theta_{\mathscr{U}_x})$ injects into $H^1(\mathscr{V}_1, \theta_{\mathscr{P}})$. This is clear because the map $\phi$: $\mathscr{V}_1 \to \mathscr{W}'$ is an isomorphism. □

*6c. A vanishing theorem on $\mathscr{P}$*

We are left with the task of proving

**Theorem 8.** *Assume $\tilde{g} \geqq 3$. Then $H^1(\mathscr{P}, \theta_{\mathscr{P}}) = 0$.*

This in turn is proved along the lines of Theorem 6. There are complications, however. First, it takes more work to prove a formula for the dualising sheaf. Second, one cannot prove the analogue of Lemma 6.4.

**Proposition 6.6.** *Let $\Omega_{\tilde{X}}$ be the canonical bundle of $\tilde{X}$, and suppose $\Omega_{\tilde{X}} = \mathcal{O}(\sum_{q \in Q} z_q)$. Let $\Omega_{\tilde{\mathscr{R}}'_F}$ denote the canonical bundle of $\tilde{\mathscr{R}}'_F$. We have*

$$\Omega_{\tilde{\mathscr{R}}'_F}^{-1} = (\det R\pi_{\tilde{\mathscr{R}}'_F}\mathscr{E})^4 \otimes \bigotimes_i \{\mathscr{Q}_i^2 \otimes (\det \mathscr{E}_{yi})^{-1}\} \otimes (\det \mathscr{Q})^4 (\det \mathscr{E}_{x_1})^{-2} (\det \mathscr{E}_{x_2})^{-2}$$

$$\otimes \bigotimes_q \{(\det \mathscr{E}_{z_q})^{-1}\} \otimes (\det \mathscr{E}_y)^{2\tilde{n}+2\tilde{g}-2} \otimes (\mathrm{Det}^*\, \theta_y)^{-2}\,. \qquad (6.5)$$

*Proof.* $\tilde{\mathscr{R}}'_F$ is a grassmannian bundle over $\tilde{\mathscr{R}}_F$. Now use Proposition 6.2. □

We need an expression for the canonical bundle of $\mathscr{H}$. (By Proposition C.3 $\mathscr{H}$ is Gorenstein and has a canonical bundle). The idea is to find an extension of the right-hand side of (6.5) to $\mathscr{H}$ as a $PSL(\tilde{n})$ line-bundle, and then to argue that this gives the canonical bundle.

*Remark 6.7.* (a) We have, on $\tilde{X} \times \tilde{\mathscr{R}}'$ a surjection $\mathcal{O}^{\tilde{n}} \to \mathscr{E}_{\tilde{\mathscr{R}}'} \to 0$. The kernel $\mathscr{K}$ is flat over $\tilde{\mathscr{R}}'$, and since $\tilde{X}$ is smooth, it is locally free (this needs an argument using [N, Lemma 5.4]). On $\mathscr{H}$ we have the identity (for $x \in \tilde{X} \setminus \{x_1, x_2\}$):

$$\det \mathscr{K}_x \otimes \det \mathscr{E}_x = \det \mathcal{O}^{\tilde{n}} \sim \mathcal{O}\,.$$

(b) In the definition of $\theta'$ (4.9b) we can replace the term $(\det \mathscr{E}_y)^l$ by $\otimes_{q \in Q} (\det \mathscr{E}_{z_q})^{l_q} \otimes (\det \mathscr{E}_y)^{l+l_0}$ (cf., Remark 2.7) as long as for every $q \in Q$ we have $z_q \notin \{x_1, x_2\}$. Using (a), we can in fact replace any (or all) of the factors $(\det \mathscr{E}_{z_q})^{l_q}$ by $(\det \mathscr{K}_{z_q})^{-l_q}$, and, after this change, allow $z_q$ to be one of the points $\{x_1, x_2\}$. It is clear that all these choices give algebraically equivalent ample line bundles on $\mathscr{P}$.

**Proposition 6.8.** *Let $\Omega_{\mathscr{H}}$ denote the canonical bundle of $\mathscr{H}$. We have*

$$\Omega_{\mathscr{H}}^{-1} = (\det R\pi_{\mathscr{H}}\mathscr{E})^4 \otimes \bigotimes_i \{\mathscr{Q}_i^2 \otimes (\det \mathscr{E}_{y_i})^{-1}\} \otimes (\det \mathscr{Q})^4 (\det \mathscr{K}_{x_1})^2 (\det \mathscr{K}_{x_2})^2$$

$$\otimes \bigotimes_q \{(\det \mathscr{E}_{z_q})^{-1}\} \otimes (\det \mathscr{E}_y)^{2\tilde{n}+2\tilde{g}-2} \otimes (\mathrm{Det}^*\, \theta_y)^{-2}\,, \qquad (6.6)$$

*where the vector bundle $\mathscr{K}$ is defined in Remark 6.7(a) above.*

*Proof.* Let $\Omega^{-1}$ denote the RHS of (6.6). By Proposition 6.6 the isomorphism $\Omega = \Omega_{\mathscr{H}}$ holds outside the $\hat{\mathscr{D}}_j^t$. We will check that it extends to each $\hat{\mathscr{D}}_j^t$.

For definiteness take $j = 1$ and for simplicity of notation suppose there are no ordinary parabolic points. The proof will use the methods of Appendix C (to which we refer the reader for unexplained notation) to determine $\Omega_{\mathscr{H}}$ in a neighbourhood of a suitable point of $\hat{\mathscr{D}}_1^t$. Since $\hat{\mathscr{D}}_1^t$ is irreducible, it will be enough to show that the isomorphism (6.6) extends to one such neighbourhood.

Consider then a point $(\mathcal{O}^{\tilde{n}} \to E \to 0, Q)$ in $\mathscr{H}$ where

(1) $E$ has torsion at $x_2$ (i.e. the point lies on $\hat{\mathscr{D}}_1^t$),

(2) $E$ is locally free at $x_1$, and

(3) the maps $E_{x_j} \to Q$ are onto for both $j = 1, 2$.

Define the vector bundle $\tilde{E}$ to be the kernel of the map sequence $E \to {}_{x_1}Q \to 0$ ($\tilde{E}$ is a vector bundle because of condition (3) in the definition of $\mathscr{H}$). The conditions (2) and (3) will continue to hold in a neighbourhood $\mathbf{U}_1$. On $\tilde{X} \times \mathbf{U}_1$ one can define a locally free sheaf $\tilde{\mathscr{E}}$ by the exact sequence $0 \to \tilde{\mathscr{E}} \to \mathscr{E} \to {}_{x_1}\mathscr{Q} \to 0$ where (where ${}_{x_1}\mathscr{Q}$ is the sheaf on $\tilde{X} \times \tilde{\mathscr{R}}'$ got by pulling back $\mathscr{Q}$ from $\mathscr{R}'$ and then restricting to $\{x_1) \times \tilde{\mathscr{R}}'$). Suppose the vector bundle $\tilde{E}$ is stable (such points certainly exist). Then this will continue to hold in a open set $\mathbf{U}$, with $(E, Q) \in \mathbf{U} \subset \mathbf{U}_1 \subset \mathscr{H}$. Note that on $\mathbf{U}$ we have an isomorphism of vector bundles $\mathscr{E}_{x_1} \sim \mathscr{Q}$.

We construct another space $E$ as follows. For simplicity assume that the degree $d$ is odd so that a Poincaré bundle exists for stable bundles of degree $d-2$. (An argument with étale open sets is needed otherwise.) Denote this bundle by $\tilde{\mathscr{E}}'$: this is a vector bundle on $\tilde{X} \times \mathscr{U}_{\tilde{X}}(d-2)$. On $\tilde{X} \times \mathscr{U}(\tilde{X}, d-2)$ consider the bundle of extensions $\mathbf{E}$ whose fibre over $\tilde{E}'$ is $\mathrm{Ext}^1({}_{x_2}(\tilde{E}'_{x_1}), \tilde{E}')$. On $\tilde{X} \times \mathbf{E}$ there is an universal extension $0 \to \tilde{\mathscr{E}}' \to \mathscr{E}' \to {}_{x_2}(\tilde{\mathscr{E}}'_{x_1}) \to 0$.

There is a morphism $H\colon \mathbf{U} \to \mathbf{E}$ such that $H^*\tilde{\mathscr{E}}' = \tilde{\mathscr{E}} \otimes \mathscr{N}$, $H^*\mathscr{E}' = \mathscr{E} \otimes \mathscr{N}$. for some line bundle $\mathscr{N}$ on $\mathbf{U}$. One checks easily that

(1) $H$ is a submersion,

(2) the fibres of $H$ are $PSL(\tilde{n})$ orbits, and

(3) $PSL(\tilde{n})$ acts freely on $\mathbf{U}$.

From this it follows that $\Omega_{\mathbf{U}} = H^*\Omega_{\mathbf{E}}$.

We now proceed to check that $H^*\Omega_{\mathbf{E}} = \Omega$. One easily computes:

$$H^*\Omega_{\mathbf{E}}^{-1} \otimes \Omega = (\det \tilde{\mathscr{E}}_y)^{-4} \otimes (\det \mathscr{K}_{x_1})^2 \otimes (\det \mathscr{K}_{x_2})^2$$

$$= (\det \tilde{\mathscr{E}}_{x_2})^2 \otimes (\det \mathscr{K}_{x_2})^2.$$

We will now show that $\det \mathscr{K}_{x_2} = (\det \tilde{\mathscr{E}}_{x_2})^{-1}$. Consider the commutative diagram of sheaves on $\tilde{X} \times \mathbf{U}$:

$$\begin{array}{ccccccccc}
 & & & & 0 & & 0 & & \\
 & & & & \uparrow & & \uparrow & & \\
 & & & & {}_{x_2}\mathscr{Q} & = & {}_{x_2}\mathscr{Q} & & \\
 & & & & \uparrow & & \uparrow & & \\
0 & \to & \mathscr{K} & \to & \mathscr{O}^{\tilde{n}} & \to & \mathscr{E} & \to & 0 \quad \text{(a)} \\
 & & = \uparrow & & \uparrow & & \uparrow & & \\
0 & \to & \mathscr{K} & \to & \mathscr{K}' & \to & \tilde{\mathscr{E}} & \to & 0 \quad \text{(b)} \\
 & & & & \uparrow & & \uparrow & & \\
 & & & & 0 & & 0 & &
\end{array}$$

where the (b) is the pull-back of (a) by the inclusion $\tilde{\mathscr{E}} \to \mathscr{E}$ – this defines $\mathscr{K}'$. One sees easily that $\mathscr{K}'$ is a vector bundle. We have therefore the equality of line bundles on $\tilde{X} \times \mathbf{U}$: $\det \mathscr{K} \otimes \det \tilde{\mathscr{E}} = \det \mathscr{K}'$, which yields the equality of line bundles on $\mathbf{U}$: $(\det \mathscr{K})_{x_2} \otimes (\det \tilde{\mathscr{E}})_{x_2} = (\det \mathscr{K}')_{x_2}$. On the other hand we get from the exact sequence $0 \to \mathscr{K}' \to \mathscr{O}^{\tilde{n}} \to {}_{x_2}\mathscr{Q} \to 0$ the exact sequence of bundles on $\mathbf{U}$: $0 \to \mathscr{Q} \otimes (\Omega_{\tilde{X}})_{x_2} \to \mathscr{K}'_{x_2} \to \mathscr{O}^{\tilde{n}} \to \mathscr{Q} \to 0$. This shows that $(\det \mathscr{K}')_{x_2}$ is trivial. □

We next prove the analogue of Lemma 6.3:

**Lemma 6.9.** *Assume $\tilde{g} \geqq 2$. Then $(\tilde{\psi}'_*\Omega_{\mathscr{H}})^{\mathrm{inv}} = \Omega_{\mathscr{P}}$ where $\Omega_{\mathscr{P}}$ is the dualising sheaf of $\mathscr{P}$.*

*Proof.* We check that the conditions of Lemma 4.17 are satisfied. By Corollary 6.18 and Remark 6.19 there exists stable bundles on $X$. By Proposition 4.7(2), there exist stable generalised parabolic bundles on $\tilde{X}$. Thus there exist stable points in $\tilde{\mathscr{R}}'$, and the action of $PSL(\tilde{n})$ is therefore generically free. We now check conditions (1) and (2) of 4.17.

(1) By Lemma 6.15(1) below one sees that in $\tilde{\mathscr{R}}'^{ss} \backslash \hat{\mathscr{D}}_1 \cup \hat{\mathscr{D}}_2$ the nonstable locus has codimension $\geqq 2$. We next show that each of the $(\hat{\mathscr{D}}_j^f)^{ss}$ or $(\hat{\mathscr{D}}_j^t)^{ss}$ contains a GPS with no nontrivial automorphism. Take $j = 1$ for definiteness. Let $E'$ be a stable (parabolic) bundle on $\tilde{X}$ of degree $d - 2$, let $E = E' \otimes {}_{x_2}\mathbf{C}$ and define the GPS structure on $E$ as follows. We take $Q = \mathbf{C}^2$, the map $E_{x_2} \to Q$ to be the obvious projection, and the map $E_{x_1} \to Q$ any isomorphism. This yields, after an identification $H(E) \sim \mathbf{C}^{\tilde{n}}$, a point on $\hat{\mathscr{D}}_1^t$ as required. Next consider $E = E'(x_2)$, the GPS structure being given by taking $Q = E'_{x_2} \otimes (\Omega_{\tilde{X}})_{x_2}^{-1}$, the map $E_{x_1} \to Q$ being zero, and the map $E_{x_2} \to Q$ the residue. This yields a point on $\hat{\mathscr{D}}_1^f$ with no nontrivial automorphisms.

(2) If a prime divisor is not contained in the nonstable locus its projection will have codimension one. If it is contained in the nonstable locus, by (1) it will have to be one of the $(\hat{\mathscr{D}}_j^f)^{ss}$ or $(\hat{\mathscr{D}}_j^t)^{ss}$. We have already seen that the respective images of these in $\mathscr{P}$ are the $\mathscr{D}_j$. □

Consider the local universal family $\tilde{\mathscr{R}}'$ of Appendix B. The open subscheme $\mathscr{H}$ of $\tilde{\mathscr{R}}'$ is defined in §4a (Notation 4.3a).

**Lemma 6.10.** *There is a morphism* Det: $\mathscr{H} \to J_{\tilde{X}}^d$ *which extends the determinant morphism on the open set* $\tilde{\mathscr{R}}'_F$.

*Proof.* The determinant of $\mathscr{E}_{\tilde{\mathscr{R}}'}$ can be defined as the inverse of $\det \mathscr{K}$, where the vector bundle $\mathscr{K}$ is defined in Remark 6.7(a). This gives a morphism from $\tilde{\mathscr{R}}'$ to $J_{\tilde{X}}^d$. □

Restricted to $\tilde{\mathscr{R}}'^{ss}$ the map Det clearly factors through the quotient by the $SL(\tilde{n})$ action and yields a morphism $\mathscr{P} \to J_{\tilde{X}}^d$, which we again denote by Det.

**Lemma 6.11.** *The determinant morphism on the open set of stable torsion-free GPSs extends to a flat morphism* Det: $\mathscr{P} \to J_{\tilde{X}}^d$.

*Proof.* Note that $J_{\tilde{X}}^0$ does not act on $\mathscr{P}$. However, $J_X$ does. Given a GPS $(E, Q)$ and a line bundle $M$ on $X$, the action is defined by

$$(E, Q) \mapsto M * (E, Q) \equiv (E \otimes \pi^* M, Q \otimes M_{x_0}).$$

We have $\operatorname{Det} M * (E, Q) = \operatorname{Det}(E, Q) \otimes (\pi^* M)^2$. Now the pull-back map $J_X^0 \to J_{\tilde{X}}^0$ and the squaring map $J_{\tilde{X}}^0 \to J_{\tilde{X}}^0$ are surjective and $J_{\tilde{X}}^0$ acts transitively on $J_{\tilde{X}}^d$. By generic flatness it follows that the map Det: $\mathscr{P} \to J_{\tilde{X}}^d$ is flat. □

Let $\mathscr{H}^L$ denote the (reduced) fibre over $L \in J_{\tilde{X}}^d$. Similarly let $\mathscr{P}^L$ be the (reduced) fibre of Det above $L$. Clearly $\mathscr{P}^L$ is the GIT quotient of $\mathscr{H}^L$. All the properties of $\mathscr{H}$ and $\mathscr{P}$ continue to be valid for $\mathscr{H}^L$ and $\mathscr{P}^L$; the proofs require only minor modifications. We have

**Proposition 6.12.** *The dualising sheaf of* $\mathscr{P}^L$ *is the restriction of* $\Omega_{\mathscr{P}}$ *to* $\mathscr{P}^L$.

*Proof.* We first note that $\mathscr{P}^L$ is the scheme-theoretic fibre above $L$. For, by Bertini, the scheme-theoretic fibre is reduced for generic $L$, and then we can use the argument of the proof of the previous lemma to extend this to all $L$.

Next we use the following fact: Suppose $f: V \to U$ is a flat map of varieties, with $U$ smooth, and $V$ Gorenstein. Let $V_p$ be the scheme-theoretic fibre above $p \in U$. Then the dualising sheaf of $V_p$ is the restriction of the dualising sheaf of $V$. This in turn is proved by repeated use of Bertini (on $U$) and the adjunction formula. □

**Proposition 6.13.** (1) *We have a (canonical) isomorphism:*

$$H^0(\mathscr{P}^L, \theta_{\mathscr{P}}) \sim \bigoplus_{\tilde{\mu}} H^0((\mathscr{U}_{\tilde{X}}^{\tilde{\mu}})^L, \theta_{\tilde{\mu}}) .$$

*where $\tilde{\mu}$ runs through the integers $(\alpha, \beta)$, $0 \leqq \alpha \leqq \beta \leqq k$.*

(2) *Assume $\tilde{g} \geqq 3$. Then $H^1(\mathscr{P}^L, \theta_{\mathscr{P}}) = 0$.*

(We have used the obvious notation $(\mathscr{U}_{\tilde{X}}^{\tilde{\mu}})^L$ for the fibre above $L$ of the determinant morphism from $\mathscr{U}_{\tilde{X}}^{\tilde{\mu}}$ to $J_{\tilde{X}}^d$. The morphism itself will be denoted $\mathrm{Det}_{\tilde{\mu}}$ below.)

*Proof.* The first claim is proved exactly as Theorem 4. The proof of the second statement is along the lines of that of Theorem 6. Restricted to $\mathscr{H}^L$ we have the following equality (the analogue of (6.3)):

$$\hat{\theta}' = \hat{\theta}'_{\tilde{\omega}} \otimes \Omega_{\mathscr{H}} ,$$

for a suitable $\hat{\theta}'_{\tilde{\omega}}$, where we have to use Remark 6.7(b) to define this latter line bundle. The rest of the proof proceeds as before except that an analogue of Lemma 6.4 is not needed. Note that $\mathscr{H}$ has rational singularities, and is in particular Cohen–Macaulay, so that Hartogs-type extension theorems for cohomology are applicable. □

*Proof of Theorem 8.* Consider the map Det: $\mathscr{P} \to J_{\tilde{X}}^d$. Proposition 6.13 shows that $R^1(\mathrm{Det})_*(\theta_{\mathscr{P}}) = 0$. On the other hand the decomposition theorem for $\mathscr{P}^L$ shows that $R^0(\mathrm{Det})_*(\theta_{\mathscr{P}}) = \bigoplus_{\tilde{\mu}} R^0(\mathrm{Det}_{\tilde{\mu}})_*(\theta_{\tilde{\mu}})$. By Theorem 6 we have $H^1(R^0(\mathrm{Det})_*(\theta_{\mathscr{P}})) = 0$. □

### *6d. Codimension computations*

We have to do a number of codimension computations. We do the first in some detail.

**Lemma 6.14.** (1) *The complement in $\tilde{\mathscr{R}}^{ss}$ of the set $\tilde{\mathscr{R}}^{s}$ of stable points has codimension $\geqq \tilde{g}$ if $|I| > 0$, and codimension $\geqq \tilde{g} - 1$ if $|I| = 0$*

(2) *The complement in $\tilde{\mathscr{R}}_F$ of the set $\tilde{\mathscr{R}}^{ss}$ of semistable points has codimension $\geqq \tilde{g}$.*

*Proof.* The dimension of $\tilde{\mathscr{R}}_F$ is easily computed to be $4\tilde{g} - 3 + |I| + \dim PLS(\tilde{n})$. (At a point $0 \to K \to \mathcal{O}^{\tilde{n}} \to E \to 0$ of $\tilde{\mathbf{Q}}_F$ the tangent space is $H^0(\tilde{X}, K^* \otimes E)$. Using the exact sequence

$$0 \to H^0(E^* \otimes E) \to \mathbf{C}^{\tilde{n}} \otimes \mathbf{C}^{\tilde{n}} \to H^0(K^* \otimes E) \to H^1(E^* \otimes E) \to 0$$

and Riemann–Roch we get $\dim H^0(K^* \otimes E) = 4\tilde{g} - 3 + (\tilde{n}^2 - 1)$.)

We first prove (1). Consider a semistable, unstable bundle $E$. It is an extension $0 \to L_1 \to E \to L_2 \to 0$, with par degree $L_1 = 1/2$ (par degree $E$). (Equivalently, $2\,\mathrm{degree}\, L_1 - d = \sum_{R^c} (b_i - a_i) - \sum_R (b_i - a_i)$.) We will now describe a (countable) number of quasi-projective varieties parametrising such bundles. (For the present we do not require a variety to be irreducible.)

For $q = 1, 2$, let $d_q$ be integers such that $d_1 + d_2 = d$, and let $I = R_1 \cup R_2$ be a decomposition of $I$ such that $2\, d_1 - d = \sum_{R_2}(b_i - a_i) - \sum_{R_1}(b_i - a_i)$. Let $h > 0$ be an integer, and let $v = (d_1, R_1, h)$. Choose Poincaré bundles $\mathscr{L}_q$ on $\tilde{X} \times J^{d_q}_{\tilde{X}}$. Let $\mathscr{J} \equiv J^{d_1}_{\tilde{X}} \times J^{d_2}_{\tilde{X}}$ and let $\mathscr{L}$ denote the line bundle $\mathscr{L}_2^* \mathscr{L}_1$ on $\tilde{X} \times \mathscr{J}$. Let $\pi$ denote the projection $\tilde{X} \times \mathscr{J} \to \mathscr{J}$.

We define a variety $V(v) \equiv V(d_1, R_1, h)$ as follows.

(a) We first define varieties $V_2(v)$:

(1) If $h = 0$, set $V_2(v) = \mathscr{J}$. Define the bundle $\mathscr{E}_v$ on $\tilde{X} \times V_2$ to be $\mathscr{L}_1 \oplus \mathscr{L}_2$.

(2) Write $\operatorname{Supp} R^+\pi_* \mathscr{L} = \bigcup_{h>0} V_1(v)$ with $V_1(v)$ denoting the locally closed subscheme of $\mathscr{J}$ where $R^1\pi_*\mathscr{L}$ is locally free of rank $h$. Let $V_2(v)$ be the projective bundle $\mathbf{P}(\{R^1\pi_*\mathscr{L}\}^*)$ on $V_1(v)_{\text{red}}$. On $\tilde{X} \times V_2(v)$ there is an universal extension $0 \to \mathscr{L}_1(-1) \to \mathscr{E}_v \to \mathscr{L}_2 \to 0$.

(b) Let $V_3(v)$ be the fibre product

$$\times_{i \in R_2} V_2(v\mathbf{P}((\mathscr{E}_v)_{y_i})\,.$$

The sub-bundle $\mathscr{L}_1(-1) \subsetneq \mathscr{E}$ yields, for each $i \in I$, a divisor in $V_3$.

(c) Let $V(v) = V(d_1, R_1, h)$ be the complement of the union of these divisors for $i \in R_2$.

Each $V(v)$ parametrises a family of parabolic bundles $E$, which occur as extensions $0 \to L_1 \to L_2 \to 0$ (the extension being split if $h = 0$), with parabolic structures at the $\{y_i\}_{R_1}$ given by the sub-bundle $L_1$. The dimensions of the $V(v)$ are easily bounded. These are:

(1) $\dim V(v) = 2\tilde{g} + |R_2|$ if $h = 0$,

(2) $\dim V_v \leqq 2\tilde{g} + h - 1 + |R_2|$ otherwise.

Let $V(v)^{\text{ss}}$ be the open set of semistable parabolic bundles, and let $F(v)$ be the frame-bundle of the direct image of $\mathscr{E}$ on $V(v)^{\text{ss}}$.

There is a map from each $F(v)$ to $\mathscr{R}^{\text{ss}} \backslash \mathscr{R}^{\text{s}}$, and the union of the images covers the latter set. We now estimate the dimension of the (closure of the) image of $F(v)$. We have ([H, Exercise 3.22]) $\dim \operatorname{Im} F(v) = \dim F(v) - e$ where $e$ is the infimum of the dimensions of the irreducible components of the fibres. Since the $E$ are generated by sections, any automorphism of $E$ acts nontrivially on the frames of $H^0(E)$, and we compute

(1) $e \geqq 2 + \dim h_0$ if $h = 0$ and

(2) $e \geqq 1 + \dim h_0$ if $h > 0$,

where $h_0 = H^0(L_2^* L_1)$. In any case the codimension of the image is bounded below by $4\tilde{g} - 3 + |I| + \dim PSL(n) - \{2\tilde{g} + |R_2| + h - h_0 - 2 + \dim GL(n)\} = 2\tilde{g} - 2 + |R_1| + h_0 - h$. By Riemann–Roch this is equal to $\tilde{g} - 1 + |R_1| + 2d_1 - d = \tilde{g} - 1 + |R_1| + \sum_{R_2}(b_i - a_i) - \sum_{R_1}(b_i - a_i)$. This gives the required bound on the codimension.

We turn now to the second assertion of the lemma. The analysis is exactly as above, except that we replace the equality $2d_1 - d = \sum_{R_2}(b_i - a_i) - \sum_{R_1}(b_i - a_i)$ by $2d_1 - d > \sum_{R_2}(b_i - a_i) - \sum_{R_1}(b_i - a_i)$. □

**Lemma 6.15.** (1) *The complement in $\tilde{\mathscr{R}}'^{\text{ss}} \backslash \hat{\mathscr{D}}_1 \cup \hat{\mathscr{D}}_2$ of the set $\tilde{\mathscr{R}}'^{\text{s}}$ of stable points has codimension $\geqq \tilde{g} + 1$ if $|I| > 0$, and codimension $\geqq \tilde{g}$ if $|I| = 0$.*

(2) *The complement in $\mathscr{H}$ of the set $\tilde{\mathscr{R}}'^{\text{ss}}$ of semistable points has codimension $\geqq \tilde{g} + 1$.*

*Proof.* The dimension of $\mathscr{H}$ is easily computed to be $4\tilde{g} - 3 + |I| + 4 + \dim PSL(\tilde{n})$.

We first prove (2). Consider a point in $\mathscr{H}\backslash\tilde{\mathscr{R}}'^{ss}$. To such a point there corresponds a GPS $E$ with a rank subsheaf $L$ contradicting semistability. We can assume $L$ is rank 1, and that $E/L$ is torsion-free outside $\{x_1, x_2\}$. We have

$$d-2+\sum_{R^c} b_i - a_i - \sum_{R} b_i - a_i < 2\,\mathrm{degree}\,L - 2\dim Q^L. \tag{6.7}$$

In fact $E/L$ can be assumed torsion-free. Suppose it is not, and let $L' \supset L$ be the inverse image in $E$ of the torsion-subsheaf of $E/L$. Clearly the sets $R$ and $R^c$ are the same for $L$ and $L'$. Now if (degree $L'$ – degree $L$) – ($\dim Q^{L'} - \dim Q^L$) $< 0$ we have degree $L'$ – degree $L = 1$ and $\dim Q^{L'} = 2$, $\dim Q^L = 0$, which is not possible. This shows that $L'$ satisfies (6.7). Thus $E$ is an extension $0 \to L_1 \to E \to L_2 \to 0$ with $L_2$ torsion-free (i.e. a line bundle) and $L_1$ satisfying (6.7).

Fix an integer $r$, with $0 \leqq r \leqq 2$. Fix two nonnegative integers $s_1, s_2$ with $s_1 + s_2 \equiv s \leqq r$. For $q = 1, 2$, let $d_q$ be integers such that $d_1 + d_2 + s = d$, and let $I = R_1 \cup R_2$ be a decomposition of $I$ such that $2\,(d_1 + s) - d - 2r > -2 + \sum_{R_2}(b_i - a_i) - \sum_{R_1}(b_i - a_i)$. Let $r' = r'(r, s)$ be defined by $r' = 0$ if $r = 2$, $r' = 1 + s$ if $r = 1$ and $r' = 4 + 2s$ if $r = 0$.

Let $h > 0$ be an integer, and let $v = (r, s_1, s_2, d_1, R_1, h)$. Choose Poincaré bundles $\mathscr{L}'_q$ on $\tilde{X} \times J^{d_q}_{\tilde{X}}$. Let $\mathscr{J} \equiv J^{d_1}_{\tilde{X}} \times J^{d_2}_{\tilde{X}}$ and let $\mathscr{L}'$ denote the line bundle $(\mathscr{L}'_2)^*\mathscr{L}'_1$ on $\tilde{X} \times \mathscr{J}$. Let $\pi$ denote the projection $\tilde{X} \times \mathscr{J} \to \mathscr{J}$.

We define a variety $V(v) \equiv (r, s_1, s_2, d_1, R_1, h)$ as follows.

(a) We first define varieties $V_2(v)$:

(1) If $h = 0$, set $V_2(v) = \mathscr{J}$. Define the bundle $\mathscr{E}'_v$ on $\tilde{X} \times V_2$ to be $\mathscr{L}'_1 \oplus \mathscr{L}'_2$.

(2) Write $\mathrm{Supp}\, R^1\pi_*\mathscr{L}^1 = \bigcup_{h>0} V_1(v)$ with $V_1(v)$ denoting the locally closed subscheme of $\mathscr{J}$ where $R^1\pi_*\mathscr{L}^1$ is locally free of rank $h$. Let $V_2(v)$ be the projective bundle $\mathbf{P}(\{R^1\pi_*\mathscr{L}'\}^*)$ on $V_1(v)_{\mathrm{red}}$. On $\tilde{X} \times V_2(v)$ there is an universal extension $0 \to \mathscr{L}'_1(-1) \to \mathscr{E}'_v \to \mathscr{L}'_2 \to 0$.

In both cases let $\mathscr{E}_v = \mathscr{E}'_v \oplus {}_{x_1}\mathbf{C}^{s_1} \oplus {}_{x_2}\mathbf{C}^{s_2}$.

(b) Consider the bundle of two dimensional quotients $\mathscr{Q}$ of $\mathscr{E}_{x_1} \oplus \mathscr{E}_{x_2}$ such that the map ${}_{x_1}\mathbf{C}^{s_1} \oplus {}_{x_2}\mathbf{C}^{s_2} \to \mathscr{Q}$ is an injection and the map $\mathscr{L}'_{x_1} \oplus \mathscr{L}'_{x_2} \oplus {}_{x_1}\mathbf{C}^{s_1} \oplus {}_{x_2}\mathbf{C}^{s_2} \to \mathscr{Q}$ has rank $r$. Let $V_3(v)$ be the fibre product

$$\mathscr{Q} \times_{V_2(v)} \Big\{ \times_{V_2(v)}{}_{i \in R_2} \mathbf{P}((\mathscr{E}_v)_{y_i}) \Big\}.$$

The sub-bundle $\mathscr{L}_1(-1) \subsetneq \mathscr{E}$ yields, for each $i \in I$, a divisor in $V_3$.

(c) Let $V(v) = V(r, s_1, s_2, d_1, R_1, h)$ be the complement of the union of these divisors for $i \in R_2$.

Each $V(v)$ parametrises a family of generalised parabolic sheaves $E = E' \oplus {}_{x_1}\mathbf{C}^{s_1} \oplus {}_{x_2}\mathbf{C}^{s_2}$, where $E'$ occurs as an extension $0 \to L'_1 \to E' \to L'_2 \to 0$ (the extension being split if $h = 0$), with parabolic structures at the $\{y_i\}_{R_1}$ given by the sub-bundle $L'_1$. The dimensions of the $V(v)$ are easily bounded. These are:

(1) $\dim V(v) = 2\tilde{g} + |R_2| + 2s + 4 - r'(r, s)$ if $h = 0$,

(2) $\dim V(v) \leqq 2\tilde{g} + h - 1 + |R_2| + 2s + 4 - r'(r, s)$ otherwise.

Let $V(v)^{ss}$ be the open set of semistable parabolic bundles, and let $F(v)$ be the frame-bundle of the direct image of $\mathscr{E}$ on $V(v)^{ss}$.

As in the proof of the previous Lemma we take into account automorphisms, and find that the codimension is $\geqq \tilde{g} - 1 + |R_1| + 2d_1 - d + r' + 2s$, and hence strictly greater than $\tilde{g} - 1 + |R_1| + \sum_{R_2}(b_i - a_i) - \sum_{R_1}(b_i - a_i) + 2r - 2 + r'$. This proves (1). (Note that the sheaf $E' \oplus {}_{x_1}\mathbf{C}^{s_1} \oplus {}_{x_2}\mathbf{C}^{s_2}$ has an automorphism group of dimension $\geqq \dim \mathrm{aut}(E') + 2s + s$.)

The assertion (1) is proved similarly, the only change being that the inequality in (6.7) is replaced by an equality. This however does not affect the final bound. □

*Remark 6.16.* It is *not* true that $\tilde{\mathscr{R}}'^{ss}\backslash\tilde{\mathscr{R}}'^{s}$ has codimension $\geqq \tilde{g}$. Points on the $\mathscr{D}_j$ are never stable. The above codimension bound breaks down because one cannot assume that the sub-sheaf contradicting stability is rank 1.

We need next to consider two sets of parabolic weights $\omega$ and $\omega'$. Write $\omega \subset \omega'$ if the indexing set $I$ of the first set of weights is a subset of the indexing set $I'$ of the second set $\{y_i\}_I \subset \{y_i\}_{I'}$ compatibly, and the two sets of weights agree at the points $\{y_i\}_I$. We have

**Lemma 6.17.** *Suppose $g > 0$ and $\omega \subset \omega'$. Then*

(1) *if $\mathscr{U}^s_X(d, \omega)$ is nonempty so is $\mathscr{U}^s_X(d, \omega')$.*

(2) *if $X$ is irreducible with a node and there exist $\omega$-stable non-locally-free sheaves then there also exist $\omega'$-stable non-locally-free sheaves.*

*Proof.* We prove (1). The other statement has a similar proof.

Clearly it is enough to consider the case $I' = I \cup \{0\}$. For simplicity we assume that a Poincaré family $\mathscr{F}$ exists on $X \times \mathscr{U}^s_X(d, \omega)$. (By working with an étale open set in $\mathscr{U}_X$ one can avoid this assumption.) Consider the projective bundle $\mathbf{P} \equiv \mathbf{P}(\mathscr{F}_{y_0})$. This parametrises a $(4g - 3 + |I'|$-dimensional family of parabolic bundles with weights $\omega'$. We will show that there exist $\omega'$-stable bundles in this family.

Let $(F, Q_i, Q_0)$ be a bundle in the family which is *not* $\omega'$-stable. Then it has a line sub-bundle $L$ such that $L_{y_0} = \ker(E_{y_0} \to Q_0)$ and

$$\sum_{R^c}(b_i - a_i) - \sum_{R'}(b_i - a_i) - (b_0 - a_0) \leqq 2\,\text{degree}\, L - d < \sum_{R^c}(b_i - a_i) - \sum_{R}(b_i - a_i)$$

where $R \equiv R(L) \subset I$ is the subset where $L_{y_i} \subset \ker(F_{y_i} \to Q_i)$ and $R^c \equiv R^c(L)$ its complement. As in the proof of Lemma 6.14 we find that such bundles $(F, Q_i, Q_0)$ are parametrised by a (finite) number of subvarieties of $\mathbf{P}$ (labelled by $(R_1, d_1, h)$), of dimension $\leqq 2g + |I| - |R_1| + h - 1 - h_0$. The codimension is therefore greater than

$$2g - 1 + |R_1| + h_0 - h_1 = g + |R_1| + 2d_1 - d \geqq g + |R_1| + \sum_{R_2}(b_i - a_i) - \sum_{R_1}(b_i - a_i) - (b_0 - a_0).$$

Grouping the terms on the right as $\{|R_1| - \sum_{R_1}(b_i - a_i)\} + \sum_{R_2}(b_i - a_i) + \{g - (b_0 - a_0)\}$ we get a positive lower bound on the codimension. □

**Corollary 6.18.** *Suppose $X$ irreducible with one node. Then there exist stable (non-locally-free) sheaves on $X$ except when $g = 1$, $d$ even, $|I| = 0$.*

*Proof.* It is well-known that $\mathscr{U}_X(d, \omega)$ is nonempty when $X$ is smooth, $|I| = 0$, $g \geqq 2$. Now suppose $X$ irreducible with one node. Using Lemma 3.3 we get stable non-locally-free sheaves when

(1) $|I| = 0$, $g \geqq 3$.

If $g = 1$, $|I| = 0$ and $d$ is odd, we get stable sheaves by taking a nontrivial extension $0 \to \mathcal{O} \to E \to L \to 0$ where $L$ is a rank one torsion-free sheaf of degree 1. This covers the case

(2) $|I| = 0$, $g = 1$, $d$ odd.

Further, if $g = 1$ and $d$ is even, one constructs a stable parabolic bundle with parabolic structure at one point $y_1$ as follows. Take two different rank one torsion-free sheaves $L_1$ and $L_2$ (one is then necessarily locally free), let $E = L_1 \oplus L_2$, and take a quasi-parabolic structure $E_{y_1} \to Q_1 \to 0$ such that $(L_i)_{y_1} \neq \ker(E_{y_1} \to Q_1)$, $i = 1, 2$, and arbitrary weights $a_1 < b_1$. This yields a stable parabolic sheaf with

(3) $|I| = 1$, $g = 1$, $d$ even.

The above constructions of course work for nosingular $X$ as well, and again using Lemma 3.3, we can add the cases

(4) $|I| = 0$, $g = 2$, $d$ odd.

(5) $|I| = 1$, $g = 2$, $d$ even.

where again we get non-locally-free sheaves.

The case

(6) $|I| = 0$, $g = 2$, $d$ even,

can be covered by taking a suitable extension $0 \to L_1 \to E \to L_2 \to 0$, with degree $L_1 = -1$, degree $L_2 = +1$. We omit the details.

We now use Lemma 6.17 to finish the proof. □

*Remark 6.19.* Note that since stability is an open condition, if stable non-locally-free sheaves exist, stable locally free sheaves also must exist. Thus Corollary 6.18 implies that if $X$ is a nodal curve,

$$\emptyset \neq \mathscr{W} \neq \mathscr{U}_X, \tag{6.8}$$

except possibly when $g = 1$, $d$ even, $|I| = 0$. In fact in this case it is easy to see (normalising $d = -2$) that $\mathscr{U}_X = (X \times X)/\sim$ where $\sim$ is the involution exchanging the two factors, and that (6.8) holds in this case as well.

## Appendix A. The moduli space of parabolic sheaves

There exist two constructions of parabolic moduli spaces on curves – that of [M-S] and that of [B2]. Neither works in the case of a singular curve. We present in this Appendix a construction of the moduli space, which generalises the work of C. Simpson, and is applicable when the underlying curve has a nodal singularity (and presumably more generally). This approach to the construction of parabolic moduli spaces arose out of conversations with A. Ramanathan.

For ease of reference we have tried to make this Appendix self-contained, at the risk of some repetition.

Unless otherwise mentioned, $X$ will denote an irreducible projective curve of genus $g$ over $\mathbf{C}$, smooth but for one node $x_0$. Let $\mathcal{O}_X(1)$ be an ample line bundle on $X$ of degree 1, $\{y_i\}_I$ a finite set of smooth points on $X$. Let $d$ denote an integer, the degree (to be chosen below). Fix another integer $k > 0$, and also, for each $i \in I$ integers $0 < \alpha_i < \beta_i \leqq k$. We set $n = d + 2(1 - g)$ and let $l$ denote the number determined by

$$nk = 2k|I| + 2l - \sum_i (\alpha_i + \beta_i). \tag{A.1}$$

We assume that the data are such that $l$ is integer, i.e. that $dk + \sum_i(\alpha_i + \beta_i)$ is even Let $a_i = \alpha_i/k$, $b_i = \beta_i/k$. Set $\omega \equiv \{(a_i, b_i)\}_I$. Note that $0 < a_i < b_i \leqq 1$. The usual range assumed is $0 \leqq a_i < b_i < 1$. (This is not a significant difference since the definition of stability only involves the difference $b_i - a_i$. However, the construction below certainly requires $a_i > 0$.)

We wish to construct the moduli space $\mathscr{U}_X$ of $s$-equivalence classes of semistable rank 2 torsion-free sheaves on $X$ with parabolic structures at the $\{y_i\}_I$ (with weights $\omega$). It will be clear from the construction that it works for an irreducible curve with an arbitrary of nodes. In particular $X$ could be smooth.

**Definition A.1a.** Let $F$ be rank 2 torsion-free sheaf on $X$. By a *quasi-parabolic structure* on $F$ at a smooth point $x \in X$ we mean a choice of a one-dimensional quotient $F_x \to Q \to 0$ of the fibre of $F$ at the point $x$. If in addition real numbers ("weights") $a < b$ are given, this is a *parabolic structure.*

We shall refer to a torsion-free sheaf with parabolic structures at the $\{y_i\}_I$ (with weights $\omega$) as a "parabolic sheaf".

**Definition A.1b.** A parabolic sheaf $F$ is said to be *stable* (respectively, *semistable*) with if for every rank one subsheaf $L$ of $F$ such that $F/L$ is torsion-free we have

$$\text{par degree } L \underset{(\text{resp.} \leqq 2)}{<} \tfrac{1}{2}(\text{par degree } F').$$

The parabolic degree of $F$ is by definition par degree $F = d + \sum_i(a_i + b_i)$; given a rank one subsheaf $L \subset F$ such that $F/L$ is torsion-free, its parabolic degree is by definition par degree $L$ = degree $L + \sum_{R^c} a_i + \sum_R b_i$ where $R \subset I$ is the subset of $i \in I$ such that $L_{y_i} \subset \ker(F_{y_i} \to Q_i)$ and $R^c$ its complement.

*Remark A.2.* The condition for (semistability can be written

$$2\,\text{degree } L \underset{(\text{resp.} \leqq)}{<} d + \sum_{R^c}(b_i - a_i) - \sum_R (b_i - a_i). \tag{A.2}$$

In particular this implies

$$2\,\text{degree } L < d + |I|. \tag{A.3}$$

**Theorem X1.** *There exists a (coarse) moduli space $\mathscr{U}^s(X, d, \omega)$ of stable parabolic sheaves $F$. We have an open immersion $\mathscr{U}^s(X, d, \omega) \subsetneq \mathscr{U}(X, d, \omega)$ where $\mathscr{U}(X, d, \omega)$ denotes the space of s-equivalence classes of semistable parabolic sheaves. The latter is a projective variety. If $X$ is smooth, then $\mathscr{U}$ is normal, with rational singularities.*

(The notion of $s$-equivalence of parabolic sheaves is defined as in the case of vector bundles, using [S2, Troisième Partie, Theorem 12]. In the notation of that theorem we say that two parabolic sheaves $F_1$ and $F_2$ are $s$-equivalent if $Gr(F_1) = Gr(F_2)$.)

The rest of this Appendix will be devoted to a proof of Theorem X1. By Remark 2.2 we are free to choose $d$ as large as we wish.

**Lemma A.3.** *There exists an integer $N_1 > 0$ such that for any semistable parabolic sheaf $F$ of rank 2 and euler characteristic $> N_1$*

(1) *$F$ is generated by its sections, and*

(2) *$H^1(F) = 0$.*

*Proof.* One imitates the proof of [N, Lemma (5.2)′] and uses equation (A.3). Note that the constant $\delta'$ in the statement of the quoted lemma can be majorised by $q$ [N, page 165]. □

*Remark A.4.* The method of proof shows the following: Suppose $F$ is a rank 2 parabolic sheaf (not necessarily semistable) such that for every torsion-free quotient $F \to L \to 0$ we have $h^0(L) \geqq N_1/2 - |I|$. Then $H^1(F) = 0$.

Choose $d$ large enough that for any parabolic semistable $F$ of degree $d$, $H^0(F)$ generates $F$, $H^1(F) = 0$. (One can do this without loss of generality because of Remark 2.2). Let $\mathbf{Q}$ denote the Quot scheme [G] of coherent sheaves over $X$ which are quotients of $\mathcal{O}^n$, where $n = d + 2(1 - g)$, with Hilbert Polynomial $P$ equal to that of any such $F$, i.e. $P(m) = 2m + n$. Thus there is on $X \times \mathbf{Q}$ a sheaf $\mathscr{F}_{\mathbf{Q}}$, flat over $\mathbf{Q}$, and a surjection $\mathcal{O}^n \to \mathscr{F}_{\mathbf{Q}} \to 0$. The Quot scheme is a projective scheme [G]: there exists an integer $M_1(n)$ such that for $m \geqq M_1(n)$ we have (denoting the vector space $H^0(\mathcal{O}_X(m))$ by $W$):

(1) for every point $\mathcal{O}^n \to F \to 0$ in the Quot scheme, if we let $K$ be the kernel, we have $H^1(K(m)) = 0$, so that the map $\mathbf{C}^n \otimes W \to H^0(F(m))$ is onto, and

(2) the map $\mathbf{Q} \to Grass_{P(m)}(\mathbf{C}^n \otimes W)$ given by (1) is an closed embedding.

We define another (complete) scheme $\mathscr{R}$ as follows. For $i \in I$, consider the sheaf $\mathscr{F}_{y_i}$ on $\mathbf{Q}$ given by restricting $\mathscr{F}_{\mathbf{Q}}$ to $\{y_i\} \times \mathbf{Q}$, and let $\mathrm{Flag}_{(1,2)}(\mathscr{F}_{y_i})$ be the relative Flag variety of locally-free quotients of $\mathscr{F}_{y_i}$ of rank (1, 2) [EGA-I, 9.9.2]. The scheme $\mathscr{R}$ is then a fibre product over $\mathbf{Q}$:

$$\mathscr{R} = \underset{i \in I}{\times}{}_{\mathbf{Q}} \mathrm{Flag}_{(1,2)}(\mathscr{F}_{y_i}).$$

**Notation A.5.** A (closed) point of $\mathscr{R}$ will be given by a point $\mathcal{O}^n \xrightarrow{p} F \to 0$ in the Quot scheme, together with quotients $F \xrightarrow{Pr,i} Q_{r,i} \to 0$, where $Q_{r,i}$ is a skyscraper sheaf supported at the (reduced) point $y_i$, with $h^0(Q_{r,i}) = r$, $r =$ 1, 2, the $p_{r,i}$ satisfying $\ker p_{2,i} \subset \ker p_{1,i}$. We let $p_m$ denote the map $\mathcal{O}^n(m) \to F(m)$.

We have a $SL(n)$-equivariant embedding $\mathscr{R} \subsetneq \mathbf{G}$ where

$$\mathbf{G} \equiv Grass_{P(m)}(\mathbf{C}^n \otimes W) \times \underset{i}{\times} \{Grass_2(\mathbf{C}^n) \times Grass_1(\mathbf{C}^n)\}.$$

Each factor on the right has a canonical ample generator of the Picard group. We give $\mathbf{G}$ the polarisation (using the obvious notation):

$$\frac{l}{m} \times \underset{i}{\times} \{(k - \beta_i), (\beta_i - \alpha_i)\}. \tag{A.4}$$

This gives a linearisation of the $SL(\mathrm{n})$ action.

Let $\mathscr{R}^{ss}$ denote the subset of closed points of $\mathscr{R}$ such that the corresponding parabolic sheaves are semistable (in particular torsion-free), and the map $H^0(p)\colon \mathbf{C}^n \to H^0(F)$ is an isomorphism. We will prove below that for large enough choices of $n$ and $m$ these are precisely the semistable points for the action of $SL(n)$ on $\mathscr{R}$ (in the sense of Geometric Invariant Theory) w.r.t. this polarisation. This will yield the existence of $\mathscr{U}$ and also show, incidentally, that semistability is an open condition for parabolic sheaves and that $\mathscr{R}^{ss}$ is (the set of closed points of) an open subscheme.

At a point $(P, \{(P_{2,i}, P_{1,i})\} \in \mathbf{G}$ we shall denote by $(U, \{(U_{2,i}, U_{1,i})\}_I)$ the respective quotients. Note that if the point $(P, \{(P_{2,i}, P_{1,i})\}_I)$ is the image of $(p, \{(p_{2,i}, p_{1,i})\}_I) \in \mathscr{R}$ then $P = H^0(p_m)$, $P_{r,i} = H^0(p_{r,i})$ and $H^0(Q_{r,i}) = U_{r,i}$.

We have a straightforward generalisation of [N-T, Proposition 5.1.1] (see also [Si, Proposition 4.3]) whose proof we omit:

**Proposition A.6.** *A point $(P, \{(P_{2,i}, P_{1,i})\}_I) \in \mathbf{G}$ is stable (respectively, semistable) for the action of $SL(n)$, with respect to the polarisation (A.4) (we refer to this from now on as GIT-stability), iff for all nontrivial subspaces $H \subset \mathbf{C}^n$ we have (with $h \equiv \dim H$)*

$$\frac{l}{m}(hP(m) - n\dim P(H \otimes W)) + \sum_i (k - \beta_i)(2h - n\dim P_{2,i}(H))$$

$$+ \sum_i (\beta_i - \alpha_i)(h - n\dim P_{1,i}(H)) \underset{(\text{resp. } \leqq)}{<} 0. \qquad \text{(A.5)}$$

**Notation A.7.** Given a point $(p, \{(p_{2,i}, p_{1,i})\}_I) \in \mathscr{R}$ (as in A.5), and a subsheaf $F'$ of $F$ we set $Q^{F'}_{r,i} \equiv p_{r,i}(F')$. Similarly, given a quotient $F \xrightarrow{T} G \to 0$, set $G/T(\ker p_{r,i}) = Q/p_{r,i}(\ker T) = Q^G_{r,i}$.

**Lemma A.8.** *Suppose $(p, \{(p_{2,i}, p_{1,i})\}_I) \in \mathscr{R}$ is a point such that $F$ is torsionfree and let $m$ be a positive integer. Then $F$ is stable (respectively, semistable) iff for every subshear $0 \neq F' \neq F$ we have:*

$$\frac{l}{m}(\chi(F')P(m) - n\chi(F'(m))) + \sum_i (k - \beta_i)(2\chi(F') - n h^0(Q^{F'}_{2,i}))$$

$$+ \sum_i (\beta_i - \alpha_i)(\chi(F') - n h^0(Q^{F'}_{1,i})) \underset{(\text{resp. } \leqq)}{<} 0. \qquad \text{(A.6)}$$

*Proof.* For any subsheaf $F'$ of $F$ let $\mathrm{LHS}(F')$ denote the left-hand side of (A.6). Assume first that the inequality holds for every proper subsheaf. Let $F'$ be a proper nonzero subsheaf such that $F/F'$ is torsion-free. For any such $F'$ (which is necessarily of rank 1) we have by Riemann–Roch,

$$\begin{aligned}
\mathrm{LHS}(F') &= \frac{l}{m}(\chi(F')(2m+n) - n(m + \chi(F'))) + \chi(F')\Big(2\sum_i (k-\beta_i) + \sum_i(\beta_i - \alpha_i)\Big) \\
&\quad - n\Big(\sum_i (k-\beta_i) + \tfrac{1}{2}\sum_i(\beta_i - \alpha_i)\Big) - \frac{n}{2}\Big(\sum_{R^c}(\beta_i - \alpha_i) - \sum_R(\beta_i - \alpha_i)\Big) \\
&= l(2\chi(F') - n) + (2\chi(F') - n)\Big(\sum_i (k-\beta_i) + \tfrac{1}{2}\sum_i(\beta_i - \alpha_i)\Big) \\
&\quad - \frac{n}{2}\Big(\sum_{R^c}(\beta_i - \alpha_i) - \sum_R(\beta_i - \alpha_i)\Big) \\
&= \frac{1}{2}(2\,\mathrm{degree}\,F' - d)\Big(2l + 2k|I| - \sum_i(\beta_i + \alpha_i)\Big) \\
&\quad - \frac{n}{2}\Big(\sum_{R^c}(\beta_i - \alpha_i) - \sum_R(\beta_i - \alpha_i)\Big) \\
&= \frac{nk}{2}\Big\{(2\,\mathrm{degree}\,F' - d) + \sum_R(b_i - a_i) - \sum_{R^c}(b_i - a_i)\Big\},
\end{aligned}$$

where in the last step we use (A.1). Comparison with equation (A.2) shows that $F$ is (semi)stable if (A.6) is satisfied.

Suppose now that $F$ is (semi)stable and $F'$ is a proper subsheaf. The above computations yield the desired inequality when $F/F'$ is torsion-free. Suppose now that $F/F'$ is not torsion-free. We will show that $\mathrm{LHS}(F') < 0$. Write $0 \to F' \to \tilde{F}' \to \mathcal{T} \to 0$ where $\mathcal{T}$ is torsion, $\tilde{F}' \subsetneq F$ and $F/\tilde{F}'$ is torsion-free. Let $\mathcal{T} = \tilde{\mathcal{T}} + \sum_i \mathcal{T}_i$ where $\mathcal{T}_i$ is the subsheaf of $\mathcal{T}$ determined by the requirement that its stalk at $y_i$ is the same as that of $\mathcal{T}$. Clearly $\mathrm{LHS}(\tilde{F}') \leqq 0$. We will now show that $\mathrm{LHS}(F') - \mathrm{LHS}(\tilde{F}') < 0$:

$$\begin{aligned}\mathrm{LHS}(F') - \mathrm{LHS}(\tilde{F}') = &-\frac{l}{m}(h^0(\mathcal{T})(2m+n) - nh^0(\mathcal{T}))\\ &- h^0(\mathcal{T})\Big(2\sum_i(k-\beta_i) + \sum_i(\beta_i - \alpha_i)\Big)\\ &- n\Big\{\sum_i(k-\beta_i)(h^0(Q^{F'}_{2,i}) - h^0(Q^{\tilde{F}'}_{2,i}))\\ &+ \sum_i(\beta_i-\alpha_i))(h^0(Q^{F'}_{1,i}) - h^0(Q^{\tilde{F}'}_{1,i}))\Big\}\\ \leqq &- nkh^0(\tilde{\mathcal{T}}) - n\sum_i h^0(\mathcal{T}_i)\{k - (k-\beta_i) - (\beta_i - \alpha_i)\}\\ = &- nkh^0(\tilde{\mathcal{T}}) - n\sum_i h^0(\mathcal{T}_i)\alpha_i,\end{aligned}$$

where we have used $h^0(Q^{\tilde{F}'}_{r,i}) - h^0(Q^{F'}_{r,i}) \leqq h^0(\mathcal{T}_i)$. Since by assumption $\alpha_i > 0$ we have the required inequality. □

The next two lemmas are generalisations of [Si, Lemmas 2.8 and 2.9] respectively.

**Lemma A.9.** *There exists $M_2(n) \geqq M_1(n)$ such that for $M \geqq M_2(n)$ the following holds. Suppose $(p, \{(p_{2,i}, p_{1,i})\}_I) \in \mathcal{R}$ is a point such that $H^0(p)$: $\mathbf{C}^n \to H^0(F)$ is an isomorphism and, for every subsheaf $F'$ of $F$ generated by sections we have*

$$\begin{aligned}&\frac{l}{m}(h^0(F')P(m) - n\chi(F'(m))) + \sum_i(k-\beta_i)(2h^0(F') - nh^0(Q^{F'}_{2,i}))\\ &+ \sum_i(\beta_i - \alpha_i)(h^0(F') - nh^0(Q^{F'}_{1,i})) \underset{(\text{resp. } \leqq)}{<} 0.\end{aligned} \tag{A.7}$$

*Then the point is GIT-(semi)stable.*

*Proof.* For $H \subset \mathbf{C}^n$ let $F'_H$ denote the subsheaf of $F$ generated by $H$, define $K_H$ by the exact sequence: $0 \to K_H \to H \otimes \mathcal{O}_X \to F'_H \to 0$. Now, for all points of $\mathbf{Q}$ and all subspaces $H$ the sheaves $F'_H$ run over a bounded family, as do the sheaves $K_H$. Therefore we can find $M_2(n)$ such that for $m \geqq M_2(n)$ we have $h^1(F'_H(m)) = 0$ and $h'(K_H(m)) = 0$ for all such $F'_H$ and $K_H$.

Note now that

(1) $\dim H \leqq h^0(F'_H)$,

(2) $P_{r,i}(H)) = H^0(Q^{F'_H}_{2,i})$ and

(3) $\dim P(H \otimes W) = \chi(F'_H(m))$ for $m \geqq M_2(n)$ (by the previous paragraph).
The Lemma now follows from Proposition (A.6) □

**Lemma A.10.** *There exists $M_3(n) \geqq M_2(n)$ such that for $m \geqq M_3(n)$ the following holds. Suppose $(p, \{(p_{2,i}, p_{1,i})\}_I) \in \mathscr{R}$ is a point which is GIT-semistable then $\mathbf{C}^n \to H^0(F)$ is an isomorphism, and for all quotients $F \xrightarrow{T} G \to 0$ we have*

$$l(-2h^0(G) + nr(G)) + \sum_i (k - \beta_i)(-2h^0(G) + n h^0(Q^G_{2,i}))$$

$$+ \sum_i (\beta_i - \alpha_i)(-h^0(G) + n h^0(Q^G_{1,i})) \leqq 0. \tag{A.8}$$

*Proof.* Denote by $H_1$ the kernel of the map $\mathbf{C}^n \to H^0(F)$. Note that $P_{r,i}(H_1) = 0$, and $P(H_1 \otimes W) = 0$. But this implies, by (A.5), that $H_1 = 0$. This proves that $\mathbf{C}^n \to H^0(F)$ is an injection.

Suppose now that $G$ is a quotient contradicting (A.8), i.e.

$$l(2h^0(G) + nr(G)) + \sum_i (k - \beta_i)(2h^0(G) + n h^0(Q^G_{2,i}))$$

$$+ \sum_i (\beta_i - \alpha_i)(-h^0(G) - n h^0(Q^G_{1,i})) < 0. \tag{A.9}$$

Let $H$ be the kernel of the map $\mathbf{C}^n \to H^0(G)$ and let $F'$ be the subsheaf of $F$ generated by $H$. From (A.9) we conclude that $h^0(G) < n$, and from this and the definition of $H$ and $F'$ that

(1) $\dim H \geqq n - h^0(G) > 0$,
(2) $r(F') + r(G) \leqq 2$, and
(3) $h^0(Q^G_{r,i}) + h^0(Q^{F'}_{2,i}) \leqq r$,
(4) $h^0(Q^{F'}_{2,i}) = \dim P_{r,i}(H)$.

Combining these inequalities with (A.9) we get (with $h = \dim H$ as before)

$$-l(2h - nr(F')) - \sum_i (k - \beta_i)(2h - n \dim P_{2,i}(H))$$

$$-\sum_i (\beta_i - \alpha_i)(h - n \dim P_{1,i}(H)) < 0.$$

For large $m \geqq M(F')$ we can replace the first term by $l/m(h P(m) - n\chi(F'(m)))$, which equals $l/m(h P(m) - n \dim P(H \otimes W))$ provided $m \geqq M_2$. Since the $F'$'s range over a bounded family we can fine $M_3(n) \geqq M_2(n)$ so that $M_3(n) \geqq M(F')$ for all $F'$'s. Now, if $m \geqq M_3(n)$ we have

$$-\frac{l}{m}(hP(m) - n \dim P(H \otimes W)) - \sum_i (k - \beta_i)(2h - n \dim P_{2,i}(H))$$

$$-\sum_i (\beta_i - \alpha_i)(h - n \dim P_{1,i}(H)) < 0.$$

But this contradicts (A.5) which holds by GIT semistability. Thus (A.8) is now established.

Let now $L$ be a rank 1 torsion-free quotient of $F$. Then we have, by (A.8), $h^0(L) \geqq n/2 - |I|$. This implies, since $n > N_1$, that $H^1(F) = 0$ and therefore $h^0(F) = n$. (See Remark A.4). This proves that the map $\mathbf{C}^n \to H^0(F)$ is an isomorphism. □

We now state the analogue of [Si, Theorem 2]:

**Proposition A.11.** *There exists an integer $N > 0$, and given $n \geqq N$ and, an integer $M(n) > 0$ such that for $m \geqq M(n)$ the following is true. A point $(p, \{(p_{2,i}, p_{1,i})\}_I) \in \mathscr{R}$ is GIT-stable (respectively, GIT-semistable) iff the quotient $F$ is torsion-free and a stable (respectively, semistable) sheaf, and the map $\mathbf{C}^n \to H^0(F)$ is an isomorphism.*

We will need the following

**Lemma A.12.** *There exists $N_2 \geqq N_1$ such that the following holds. If $F$ is a semistable parabolic sheaf with Euler characterstic $n \geqq N_2$:*

(1) *$\forall F' \subset F$ we have*

$$l(2h^0(F') - r(F')n) + \sum_i (k - \beta_i)(2h^0(F') - n\,h^0(Q^{F'}_{2,i}))$$
$$+ \sum_i (\beta_i - \alpha_i)(h^0(F') - n\,h^0(Q^{F'}_{2,i})) \leqq 0. \tag{A.10}$$

(2) *If, for some $F' \subset F$, equality holds in* (A.10) *then, for any $m \geqq 1$,*

$$\frac{l}{m}(\chi(F')P(m) - n\chi(F'(m))) + \sum_i (k - \beta_i)(2\chi(F') - n\,h^0(Q^{F'}_{2,i}))$$
$$+ \sum_i (\beta_i - \alpha_i)(\chi(F') - n\,h^0(Q^{F'}_{1,i})) = 0. \tag{A.11}$$

*Proof.* Let $F'_i$ denote the terms in the canonical filtration [H-N] of $F'$ (the filtration being defined ignoring parabolic structures), let $Q_i = F'_i/F'_{i-1}$. Let $\mu(F)$ denote the slope (degree $F$)/(rank $F$). Then $h^0(F') \leqq \sum_i h^0(Q_i)$, $\mu(Q_i) \leqq \mu(F) + c|I|$ for some constant $c$. Also, by [Si, Corollary 2.5] we have, when $h^0(Q_i) > 0$ the inequality $h^0(Q_i) \leqq r(Q_i)(\mu(Q_i) + B_1)$ for some constant $B_1$. Let $v = \inf\{\mu(Q_i) \,\|\, h^0(Q_i) > 0\}$. Then $h^0(F') \leqq (r(F') - 1)(\mu(F) + c|I| + B_1) + (v + B_1)$. If $v \leqq \mu(F) - C$ ($C$ to be fixed below) this yields $h^0(F') \leqq r(F')2n + B_2 - C$ for some constant $B_2$; thus for such $F'$ the left hand side of (A.10) is less than or equal to

$$h^0(F')\left(2l + 2k|I| - \sum_i (\beta_i + \alpha_i)\right) - nlr(F')$$
$$\leqq \left(\frac{r(F')}{2}n + B_2 - C\right)\left(2l + 2k|I| - \sum_i (\beta_i + \alpha_i)\right) - nlr(F')$$
$$\leqq 2l(B_2 - C) + \left(2k|I| - \sum_i (\beta_i + \alpha_i)\right)(n + B_2 - C)$$
$$= nk(B_2 - C) + n\left(2k|I| - \sum_i (\beta_i + \alpha_i)\right) \quad \text{using (A.1)}$$
$$= nk(B_3 - C)$$

where the last equality defines $B_3$. Choosing $C > B_3$ we get the desired inequality (which is in fact strict – this will be relevant for the proof of part (2) of the lemma) for subsheaves $F'$ satisfying $\nu \leqq \mu(F) - C$. On the other hand we can arrange (by taking $n \geqq N_2$, $N_2$ large enough) that all stable bundles $Q$ with rank $\leqq 2$ and $\mu(Q) \geqq \mu(F) - C$ have $H'(Q) = 0$, yielding, for $F'$ contradicting $\nu \leqq \mu(F) - C$, the equality $\chi(F'(m)) = h^0(F'(m))$ for $m \geqq 0$. Then (A.6) implies (A.10) for such $F'$. Part (2) of the lemma now follows easily. □

*Proof of Proposition A.11.* Choose $N = N_2$ where $N_2$ is determined by the above lemma. The proof of the "if" part is now similar to the proof of [Si, Theorem 2], where the first step of the proof has been isolated in Lemma A.12.

We sketch the proof of the "only if" part. Suppose $(p, \{(p_{2,i}, p_{1,i})\}_I) \in \mathscr{R}$ is a point which is GIT-(semi)stable. Note that by Lemma A.10, $\mathbf{C}^n \to H^0(F)$ is an isomorphism. Let $\tau = \operatorname{Tor} F$, $G = F/\tau$ and apply Lemma A.10, noting that $h^0(G) = n - h^0(\tau)$, $h^0(Q^G_{2,i}) = 2 - h^0(Q^\tau_{2,i})$ and $h^0(Q^G_{1,i}) = 1 - h^0(Q^\tau_{1,i})$. We get

$$kh^0(\tau) \leqq \sum_i (k - \beta_i) h^0(Q^\tau_{2,i} + (\beta_i - \alpha_i) h^0(Q^\tau_{1,i}),$$

from which one can easily conclude (since $\alpha_i > 0$) that $\tau = 0$.

*Proof of Theorem X1.* The proof of the first part of the theorem (existence of $\mathscr{U}$) is now similar to the proof of [Si, Theorem 2]. That $\mathscr{U}$ is projective follows from the GIT construction. The other properties of $\mathscr{U}$ follow from the corresponding facts about $\mathscr{R}^{ss}$, again by GIT. Consider for example the case when $X$ is smooth. Define $\mathbf{Q}_F$ to be the open subscheme of $\mathbf{Q}$ consisting of locally-free quotients $\mathcal{O}^n \to F \to 0)$ such that

(1) $\mathbf{C}^n \to H^0(F)$ is an isomorphism, and
(2) $H^1(F) = 0$.

Let $\mathscr{R}_F$ be the inverse image of $\mathbf{Q}_F$ by the projection $\mathscr{R} \to \mathbf{Q}$. This is a bundle over $\mathbf{Q}_F$,

$$\mathscr{R}_F = \underset{i \in I}{\times_{\mathbf{Q}_F}} \operatorname{Flag}_{(1,2)}(\mathscr{F}_{y_i}).$$

The projection $\mathscr{R}_F \to \mathbf{Q}_F$ is smooth, and $\mathbf{Q}_F$ itself can be proved to be smooth (in particular irreducible) as in [N, Remark 5.5]. Thus $\mathscr{R}_F$ is smooth, and hence so is its open subscheme $\mathscr{R}^{ss}$. This yields irreducibility and normality of $\mathscr{U}$; it also follows that $\mathscr{U}$ has rational singularities [Bo].

For $X$ a nodal curve, $\mathscr{R}^{ss}$ can be similarly proved to be reduced and irreducible defining $\mathbf{Q}_F$ as above, but replacing "locally-free quotients" by "torsion-free quotients". In this case $\mathbf{Q}_F$ is not smooth. That it is irreducible can be seen as before. That it is reduced is the main result of [S2, Huitième Partie, III], where in fact it is proved that given $q \in \mathbf{Q}_F$ the completion $\hat{\mathcal{O}}_q$ is reduced. □

## Appendix B. Generalised parabolic sheaves

### *Ba. The moduli space of generalised parabolic sheaves*

The notation of the previous Appendix holds. In addition let $\tilde{X}$ be the normalisation of $X$, $\tilde{g} = g - 1$ the genus of $\tilde{X}$, and $\pi: \tilde{X} \to X$ the canonical map. Let $\{x_1, x_2\}$

be the inverse image of $x_0$ in $\tilde{X}$. Set $\tilde{n} \equiv d + 2(1 - \tilde{g})$, and define $\tilde{l}$ by

$$\tilde{n}k = 2k|I| - 2\tilde{l} - \sum_i (\alpha_i + \beta_i).$$

(Note $\tilde{n} = n + 2$, and $\tilde{l} = l + k$.)

We wish to construct the moduli space $\mathscr{P}$ of $s$-equivalence classes of semistable rank 2 sheaves on $\tilde{X}$ with parabolic structures at the $\{y_i\}_I$ (with weights $\omega$) and a generalised parabolic structure over $\{x_1, x_2\}$.

**Definition B.1.** Let $E$ be a rank 2 sheaf, torsion-free outside $\{x_1, x_2\}$, with parabolic structures over $\{y_i\}_I$. A *generalised parabolic structure* on $E$ over the divisor $\{x_1, x_2\}$ is a choice of a two-dimensional quotient $Q$ of $E_{x_1} \oplus E_{x_2}$. We do not define a generalised quasiparabolic structure since a certain choice of "generalised weights" is assumed. A parabolic sheaf with, in addition, a generalised parabolic structure over $\{x_1, x_2\}$, is a *generalised parabolic sheaf* (GPS). A GPS $E$ is said to be *stable* (respectively, *semistable*) with respect to the weights $\omega$ if for every proper subsheaf $E'$ such that $E/E'$ is torsion-free outside $\{x_1, x_2\}$, we have

$$\text{par degree } E' \underset{(\text{resp. } \leq)}{<} \frac{\text{rank } E'}{2} (\text{par degree } E) - (\text{rank } E' - \dim Q^{E'}) \qquad \text{(B.1)}$$

where, for any subsheaf $E'$ we denote by $Q^{E'}$ the image of $E'_{x_1} \oplus E'_{x_2}$ in $Q$.

**Theorem X2.** *There exists a (coarse) moduli space $\mathscr{P}^s(\tilde{X}, d, \omega)$ of stable GPSs on $X$. We have an open immersion $\mathscr{P}^s(\tilde{X}, d, \omega) \hookrightarrow \mathscr{P}(\tilde{X}, d, \omega)$ where $\mathscr{P}(\tilde{X}, d, \omega)$ denotes the space of s-equivalence classes of semistable GPS's. The former is a smooth variety; the latter a normal projective variety with rational singularities.*

### *B.2. Outline of Proof of Theorem X2*

(1) Lemma A.3 is replaced by the following result: *There exists an integer $N'_1 > 0$ such that for any semistable generalised parabolic sheaf $E$ of rank 2 and euler characteristic $> N'_1$ we have $H'(E(-x_1 - x_2 - x)) = 0$, $x \in \tilde{X}$.* This ensures that $H^1(E) = 0$, $E$ is generated by sections, $H^0(E) \to E_{x_1} \oplus E_{x_2}$ is onto, and $E(-x_1 - x_2)$ is generated by sections.

(2) Let $\tilde{P}(m) = 2m + \tilde{n}$. Define

$$\mathscr{R}' = \text{Grass}_2(\mathscr{E}_{x_1} \oplus \mathscr{E}_{x_2}) \times_{\tilde{Q}} \{ \times_{\tilde{Q}, i \in I} \text{Flag}_{(1,2)}(\mathscr{E}_{y_i}) \}$$

(3) Define

$$\mathbf{G}' \equiv \text{Grass}_{P(m)}(\mathbf{C}^{\tilde{n}} \otimes W) \times \text{Grass}_2(\mathbf{C}^{\tilde{n}} \otimes \mathbf{C}^2) \times \underset{i}{\times} \{ \text{Grass}_2(\mathbf{C}^{\tilde{n}}) \times \text{Grass}_1(\mathbf{C}^{\tilde{n}}) \}.$$

(4) Define the polaristion on $\mathbf{G}'$:

$$\frac{(\tilde{l} - k)}{m} \times k \times \underset{i}{\times} \{ (k - \beta_i), (\beta_i - \alpha_i) \}. \qquad \text{(B.2)}$$

(5) Replace (A.5) by

$$\frac{(\tilde{l}-k)}{m}(hP(m) - \tilde{n}\dim P(H\otimes W)) + k(2h - \tilde{n}\dim P_G(H\otimes \mathbf{C}^2))$$

$$+\sum_i (k-\beta_i)(2h - \tilde{n}\dim P_{2,i}(H)) + \sum_i (\beta_i - \alpha_i)(h - \tilde{n}\dim P_{1,i}(H)) \underset{(\text{resp. } \leqq)}{<} 0,$$

where $P_G$ is the projection in the second factor of $\mathbf{G}'$.

(6) Replace (A.6) by

$$\frac{(\tilde{l}-k)}{m}(\chi(E')P(m) - \tilde{n}\chi(E'(m))) + k(2\chi(E') - \tilde{n}h^0(Q^{E'}))$$

$$+\sum_i (k-\beta_i)(2\chi(E') - \tilde{n}h^0(Q^{E'}_{2,i})) + \sum_i (\beta_i - \alpha_i)(\chi(E') - \tilde{n}h^0(Q^{E'}_{1,i})) \underset{(\text{resp. } \leqq)}{<} 0.$$

The rest of the proof of Theorem X1 goes through with obvious modifications *except that we cannot assume that the sheaves involved are torsion-free at* $x_1$ *and* $x_2$. The fact that $\tilde{\mathscr{R}}'^{ss}$ is reduced, irreducible and normal is proved in Appendix C (Lemma C.2 and Proposition C.3).

For example, the analogue of Proposition A.11 is the following result. (We denote a point of $\mathrm{Grass}_2(\mathbf{C}^n\otimes\mathbf{C}^2)$ by $p_2$.)

**Proposition B.3.** *There exists* $N'$ *and* $M'$ *such that for* $\tilde{n}\geqq N'$ *and* $m\geqq M'$ *the following is true. A point* $(p, p_2, \{(p_{2,i}, p_{1,i})\}_I)\in\tilde{\mathscr{R}}'$ *is GIT-stable (respectively, GIT-semistable) iff the quotient* $E$ *is torsion-free outside* $\{x_1, x_2\}$ *and a stable (respectively, semistable) generalised parabolic sheaf, and the map* $\mathbf{C}^{\tilde{n}}\to H^0(E)$ *is an isomorphism.*

*Remark B.4.* Note that if $(E, Q)$ is a semistable GPS, $\mathrm{Tor}\,E$ is supported on the reduced subscheme $\{x_1, x_2\}$ and

$$(\mathrm{Tor}\,E)_{x_1}\oplus(\mathrm{Tor}\,E)_{x_2}\subsetneq Q. \tag{B.3}$$

This follows from (B.1).

*Remark B.5.* The above construction of the moduli space also shows that $\tilde{\mathscr{R}}^{ss}$ is open in $\tilde{\mathscr{R}}'$ and hence, by a standard argument, semistability is an open property for GPS's.

## *Bb. S-equivalence of generalised parabolic sheaves*

We enlarge the category of GPS's by adopting the following more general definition. For simplicity we assume that no "ordinary" parabolic points are present. It should be noted that the detailed description of $s$-equivalence given below (Proposition B.15) is not really needed. Corollary B.17 and Proposition C.7(4) are the 1only places where it is used; and one can give direct proofs of these without using Proposition B.15.

**Definition B.6.** A *generalised m-parabolic structure* on a sheaf $E$ over the divisor $\{x_1, x_2\}$ is a choice of an $m$-dimensional quotient $Q$ of $E_{x_1}\oplus E_{x_2}$. A sheaf with a generalised $m$-parabolic structure will be called a $m$-GPS, or GPS for short.

A GPS $E$ is said to be *stable* (respectively, *semistable*) if $E$ is torsion-free outside $\{x_1, x_2\}$, and

(1) if rank $E > 0$ then for every proper subsheaf $E'$ such that $E/E'$ is torsion-free outside $\{x_1, x_2\}$ we have

$$\operatorname{rank} E(\operatorname{degree} E' - \dim Q^{E'}) \underset{(\text{resp. } \leq)}{<} \operatorname{rank} E'(\operatorname{degree} E - m). \tag{B.4}$$

(2) If rank $E = 0$, then we have $E_{x_1} \oplus E_{x_2} = Q$ and $\dim Q = 1$ (respectively $E_{x_1} \oplus E_{x_2} = Q$).

(For any subsheaf $E'$, we denote by $Q^{E'}$ the image of $E'_{x_1} \oplus E'_{x_2}$ in $Q$).

**Definition B.7.** If $(E, Q)$ is a GPS, and rank $E > 0$ set

$$\mu_G[(E, Q)] = \frac{(\operatorname{degree} E - \dim Q)}{\operatorname{rank} E}$$

**Examples B.8.** (1) Any torsion-sheaf $\tau$ supported on $\{x_1, x_2\}$ is in a canonical way a semistable GPS: one takes $Q = \tau_{x_1} \oplus \tau_{x_2}$. Such a GPS is stable iff degree $\tau = 1$.

(2) A line bundle $L$ with a one-dimensional quotient $Q$ of $L_{x_1} \oplus L_{x_2}$ is a semistable GPS. It is stable iff each map $L_{x_j} \to Q$ is nonzero.

(3) A line bundle $L$ with a two-dimensional quotient $Q$ of $L_{x_1} \oplus L_{x_2}$ is a semistable 2-GPS. It is never stable.

It is useful to think of a $m$-GPS as a sheaf $E$ on $\tilde{X}$ together with a map $\pi_* E \to {}_{x_0}Q \to 0$, with $Q$ being thought of as a sheaf on $X$ supported on the reduced point $x_0$, with $h^0(Q) = m$. In this subsection we will omit the (pre-)subscript $x_0$. Let $K_E$ denote the kernel of the sheaf map $\pi_* E \to Q$.

**Definition B.9.** *A morphism of GPS's* $(E, Q) \to (E'', Q'')$ is a sheaf map $E \to E''$ which maps $K_E$ to $K_{E''}$ (and therefore induces a map $Q \to Q''$).

**Definition B.10.** Given an exact sequence $0 \to E' \to E \to E'' \to 0$ of sheaves on $\tilde{X}$, and $\pi_* E \to Q \to 0$ a GP structure on $E$ we define GP structures on $E'$ and $E''$ via the diagram:

$$\begin{array}{ccccccccc} 0 & \to & \pi_* E' & \to & \pi_* E & \to & \pi_* E'' & \to & 0 \\ & & \downarrow & & \downarrow & & \downarrow & & \\ 0 & \to & Q' & \to & Q & \to & Q'' & \to & 0 \end{array}$$

(The first horizontal sequence is exact because $\pi$ is finite, $Q'$ is defined as the image in $Q$ of $\pi_* E'$ so that the first vertical arrow is onto, $Q''$ is defined by demanding that the second horizontal sequence is exact, and finally the third vertical arrow is onto by the snake lemma.) We will sometimes write $0 \to (E', Q') \to (E, Q) \to (E'', Q'') \to 0$; the meaning of such a sequence is clear.

A morphism $(E, Q) \to (E'', Q'')$ of GPS's factors:

$$\begin{array}{ccccc} (E, Q) & \to & (E'', Q'') & & \\ \downarrow & & \uparrow & & \\ (W, Q_1) & \to & (W, Q_1'') & \to & 0 \\ \downarrow & & \uparrow & & \\ 0 & & 0 & & \end{array}$$

We have the following Lemmas, whose proofs we omit.

**Lemma B.11.** *Let $(E, Q)$ be a GPS with* rank $E > 0$, *and suppose E is torsion-free outside $\{x_1, x_2\}$. Then the following are equivalent:*

(1) *$(E, Q)$ is (semi)stable.*

(2) *For every proper sub-GPS $(E', Q')$ we have*

$$\operatorname{rank} E(\operatorname{degree} E' - \dim Q') \underset{(\text{resp. } \leqq)}{<} \operatorname{rank} E'(\operatorname{degree} E - \dim Q).$$

(3) *For every proper quotient GPS$(E'', Q'')$ we have*

$$\operatorname{rank} E(\operatorname{degree} E'' - \dim Q'') \underset{(\text{resp. } \geqq)}{<} \operatorname{rank} E''(\operatorname{degree} E - \dim Q).$$

**Lemma B.12.** *Let $(E, Q) \to (E'', Q'')$ be a morphism of semistable GPS's. Assume that if* rank $E \neq 0$ *and* rank $E'' \neq 0$, *then $\mu_G[(E, Q)] = \mu_G[(E'', Q'')]$. Then the kernel and cokernel are semistable GPS's. If both $(E, Q)$* and *$(E'', Q'')$ are stable GPS's the morphism must be an isomorphism or zero.*

**Proposition B.13.** *Fix $\mu$ a rational number. Then the category of semistable GPS's $(E, Q)$ such that* rank $E = 0$ *or,* rank $E > 0$, *with $\mu_G[(E, Q)] = \mu$, is an abelian, artinian, noetherian category whose simple objects are the stable GPS's in the category.*

One can conclude as usual that given a semi-stable GPS it has a Jordan-Holder filtration.

**Definition B.14.** Two semistable GPS's are said to be *s-equivalent* if they have the same "associated graded" GPS.

**Proposition B.15.** *The s-equivalence classes of* rank 2 *2-GPS's are the following:*

(1) *If $(E, Q)$ is a stable GPS then E is necessarily a vector bundle, and both maps $E_{x_j} \to Q$ are isomorphisms. Two such GPS's are s-equivalent iff they are isomorphic.*

(2) *If d is even, consider GPS's $(E, Q)$ such that E is an extension $0 \to L_1 \to E \to L_2 \to 0$ with* degree $L_p = d/2$, $p = 1, 2$, *and such that the induced parabolic structure on $L_1$ is stable (i.e. the maps $(L_1)_{x_j} \to Q$ have the same one-dimensional image $Q_1$ – denote by $Q_2$ the quotient $Q/Q_1$.) All such GPS's with $(L_1, Q_1)$ and $(L_2, Q_2)$ fixed form an s-equivalence class.*

(3) *Consider extensions $0 \to \tilde{E} \to E \to \tau \to 0$, or extensions $0 \to \tau \to E \to \tilde{E} \to 0$, with $\tau$ a torsion-sheaf of degree* 1 *supported at $x_1$, with the induced structure on $\tilde{E}$ that of a stable 1-GPS – denote by $(\tilde{E}, \tilde{Q})$ this structure. All such GPS's with $(\tilde{E}, \tilde{Q})$ fixed form an s-equivalence class. (Included in this equivalence class is the case when E is locally-free, the map $E_{x_2} \to Q$ has one-dimensional image $\tilde{Q}$, the map $E_{x_1} \to Q$ is an isomorphism, and $\tilde{E}$ is the kernel of the sheaf map $E \to Q/\tilde{Q} \to 0$, $Q/\tilde{Q}$ being thought of as a sheaf supported at $x_1$.)*

(4) *If d is even, consider extensions as in the previous case, with $\tilde{E}$ an extension $0 \to L_1 \to \tilde{E} \to L_2 \to 0$ or $0 \to L_2 \to E \to L_1 \to 0$* degree $L_1 = d/2$, *degree $L_1 = d/2 - 1$, the induced generalised parabolic structure on $L_1$ is stable, and that on $L_2$ trivial. Such GPS's with fixed $(L_1, \tilde{Q})$ and $L_2$ form an s-equivalence class.*

(5) *Same as* (3) *with $x_2$ in place of $x_1$.*

(6) *Same as* (4) *with $x_2$ in place of $x_1$.*

(7) (i) *Extensions $0 \to \tilde{E} \to E \to \tau_1 \oplus \tau_2 \to 0$, or extensions $0 \to \tau_1 \oplus \tau_2 \to E \to \tilde{E} \to 0$, with $\tau_j$ a torsion-sheaf of* degree 1 *supported at $x_j$, with the induced generalised parabolic structure on $\tilde{E}$ trivial, $\tilde{E}$ a stable bundle.* (ii) *Extensions*

$0 \to \tilde{E}_1 \to E \to \tau \to 0$, *or extensions* $0 \to \tau_1 \to E \to \tilde{E}_1 \to 0$, *with the induced structure on* $\tilde{E}_1$ *that of a unstable* 1-*GPS, with* $\tilde{E}_1$ *in turn an extension of* $\tau_2$ *by* $\tilde{E}$, *with the induced parabolic structure on* $\tilde{E}$ *trivial.* (iii) *The same as* (ii) *with the roles of* $x_1$ *and* $x_2$ *reversed. All such GPS's with a fixed* $\tilde{E}$ *form an s-equivalence class. (Included in this equivalence class are the cases when E is locally-free, the maps* $E_{x_j} \to Q$ *have one-dimensional images* $Q_j$, *and* $\tilde{E}$ *is the kernel of the sheaf map* $E \to Q_1 \oplus Q_2$, *the* $Q_j$ *being thought of as sheaves supported on the* $\{x_j\}$.)

(8) *If d is even, the same as above, with* $\tilde{E}$ *an extension* $0 \to L_1 \to E \to L_2 \to 0$, degree $L_p = d/2 - 1$.

(9) *Extensions* $0 \to \tilde{E} \to E \to \tau \to 0$, *or extensions* $0 \to \tau \to E \to \tilde{E} \to 0$, *with* $\tau$ *a torsion-sheaf of* degree 2 *supported at* $x_1$, *with the induced generalised parabolic structure on* $\tilde{E}$ *trivial,* $\tilde{E}$ *a stable bundle. All such extensions, with* $\tilde{E}$ *fixed, form an s-equivalence class. (Included in this equivalence class is the case when E is locally-free, the map* $E_{x_2} \to Q$ *is zero.)*

(10) *The same as above, with* $\tilde{E}$ *an extension* $0 \to L_1 \to E \to L_2 \to 0$, degree $L_p = d/2 - 1$, $p = 1, 2$.

(11) *Same as* (9) *with* $x_2$ *in place of* $x_1$.

(12) *Same as* (10) *with* $x_2$ *in place of* $x_1$.

*Remark B.16.* In case (3) above the Jordan–Holder filtration has two terms, with one of the factors a torsion sheaf of length one and the other a stable 1-GPS. In case (7) and (9) the filtration has three terms, with one term a stable rank two bundle and the other two torsion sheaves of length one each.

**Corollary B.17.** *Every semistable GPS*$(E', Q')$ *is sequivalent to a semistable GPS* $(E, Q)$ *with E locally free.*

## Appendix C. The singularities of moduli space of generalised parabolic sheaves

The notation of the previous Appendix holds. For simplicity we assume $|I| = 0$. Including ordinary parabolic points makes no difference to the following considerations.

**Notation C.1.** Define $\mathscr{H}$ to be the set of (closed) points $(\mathcal{O}^{\bar{n}} \to E \to 0, Q)$ in $\tilde{\mathscr{R}}'$, where $\mathbf{C}^{\bar{n}} \to H^0(E)$ is an isomorphism, $H^1(E(-x_1 - x_2 - x)) = 0$ for $x \in \tilde{X}$, and

(T) Tor $E$ is supported on the reduced subscheme $\{x_1, x_2\}$ and $(\text{Tor } E)_{x_1} \oplus (\text{Tor } E)_{x_2} \subsetneq Q$.

Requiring that $H^1(E(-x_1 - x_2 - x)) = 0$ ensures that $H^1(E) = 0$, $E$ is generated by sections, $H^0(E) \to E_{x_1} \oplus E_{x_2}$ is onto, and $E(-x_1 - x_2)$ is generated by sections.

We will see below that $\mathscr{H}$ is the set of closed points of an open subscheme of $\tilde{\mathscr{R}}'$. We will continue to denote this subscheme by $\mathscr{H}$. Clearly then

$$\tilde{\mathscr{R}}'^{ss} \underset{\text{open}}{\subsetneq} \mathscr{H} \underset{\text{open}}{\subsetneq} \tilde{\mathscr{R}}'.$$

**Lemma C.2.** *The set of points where the conditions of Notation C.1 hold is open.* $\mathscr{H}$ *is irreducible, as in* $\tilde{\mathscr{R}}'^{ss}$.

*Proof.* We first check that $\mathscr{H}$ is open. Consider the flat family of sheaves $F$ on $X$, parametrised by $\tilde{R}'$, constructed as in §4b via the sequence:

$$0 \to \mathscr{F} \to (\pi_\times I_{\tilde{\mathscr{R}}'})_* \mathscr{E} \to {}_{x_0}\mathscr{Q} \to 0.$$

Consideration (T) precisely determines the points $(E, Q)$ where $F$ is torsion-free on $X$ (Lemma 4.6(1)). This can be seen to be an open condition using [EGA-IV, (12.2.1)]. The other conditions in the definition of $\mathscr{H}$ are clearly open.

Next we prove the irreducibility of $\mathscr{H}$ (which, clearly, implies that of $\tilde{\mathscr{R}}'^{ss}$.) Let $\tilde{\mathscr{R}}'_F$ be the open subscheme of $\mathscr{H}$ consisting of locally-free sheaves. This is a grassmannian bundle over $\tilde{\mathbf{Q}}_F$ (4.5b). That $\tilde{\mathbf{Q}}_F$ is irreducible is easy to see by a standard argument [N, Remark 5.5]; hence so is $\tilde{\mathscr{R}}'_F$. We will show, in the course of the proof of the next proposition, that $\tilde{\mathscr{R}}'_F$ is dense in $\mathscr{H}$. □

**Proposition C.3.** *$\mathscr{H}$ is reduced, normal, Gorenstein and has rational singularities. Hence the same holds for $\tilde{\mathscr{R}}'^{ss}$.*

*Proof.* The claim is obvious at a point $(E, Q)$ corresponding to a torsion-free sheaf, where in fact the space is smooth.

We divide the rest of the proof into steps. Let $(\mathcal{O}^{\tilde{n}} \to E \to 0, Q)$ be a point of $\mathscr{H}$, with $P$ denoting the projection $E_{x_1} \oplus E_{x_2} \to Q$, and assume $E$ is not locally free. We shall use Lemma 4.18 and Proposition 4.19 without comment.

*Step 1.* The simplest nontrivial case is when $\tau_{x_2} = 0$ and the map $\mathscr{E}_{x_1} \to \mathscr{Q}$ is surjective at $(E, Q)$ and hence in an open neighbourhood $\mathbf{U} \subset \tilde{\mathscr{R}}'$. Define the sheaf $\tilde{\mathscr{E}}$ in this neighbourhood by the exact sequence $0 \to \tilde{\mathscr{E}} \to \mathscr{E} \to {}_{x1}\mathscr{Q} \to 0$ (where ${}_{x1}\mathscr{Q}$ is the sheaf on $\tilde{X} \times \tilde{\mathscr{R}}'$ got by pulling back $\mathscr{Q}$ from $\tilde{\mathscr{R}}'$ and then restricting to $\{x_1\} \times \tilde{\mathscr{R}}'$); at the point $(E, Q)$ we have, with obvious notation $0 \to \tilde{E} \to E \to {}_{x_1}Q \to 0$. It follows from the definition of $\mathscr{H}$ that $\tilde{E}$ is locally free, $\dim H^0(\tilde{E}) = \tilde{n} - 2$, $\tilde{E}$ is generated by global sections, and $H^1(\tilde{E}) = 0$ – all this will continue to be true in a possibly smaller neighbourhood, say $\mathbf{U}'$. (To see why $\tilde{E}$ is generated by sections use the following fact: there are exact sequences $0 \to \tilde{E}' \to \tilde{E} \to \tau_1 \to 0$ and $0 \to \tau_2 \to E(-x_1 - x_2) \to \tilde{E}' \to 0$ where $\tilde{E}'$ is the image of the map $E(-x_1 - x_2) \to E$, and $\tau_1$ and $\tau_2$ are torsion sheaves.)

Consider the fibre-product $\mathbf{B}$ of the frame bundle of the zeroth direct image of $\tilde{\mathscr{E}}$ onto $\mathbf{U}'$ with the frame bundle of $\mathscr{Q}$. One has a smooth morphism $\mathbf{B} \to \mathbf{U}'$.

Let now $\tilde{\mathbf{Q}}^1$ be the Quot scheme of rank 2 degree $d - 2$ quotients $\mathcal{O}^{\tilde{n}-2} \to \tilde{E} \to 0$, and $\tilde{\mathbf{Q}}^1_F$ the open subset of locally-free quotients with vanishing first cohomology such that the map $\mathbf{C}^{\tilde{n}-2} \to H^0(\tilde{E})$ is an isomorphism. The space $\tilde{\mathbf{Q}}^1_F$ is smooth. Consider on $\tilde{\mathbf{Q}}^1_F$ the bundle $\mathbf{E} \equiv \mathrm{Ext}^1({}_{x_1}\mathbf{C}^2, \tilde{\mathscr{E}})$ of extensions [La] where ${}_{x_1}\mathbf{C}^2$ is the skyscraper sheaf on the (reduced) point $x_1$ with $\mathbf{C}^2$ as fibre. On $\tilde{X} \times \mathbf{E}$ there is an exact sequence of sheaves flat over $\mathbf{E}$:$0 \to \tilde{\mathscr{E}} \to \mathscr{E} \to {}_{x_1}\mathbf{C}^2 \to 0$. Let $\mathbf{W}$ denote the total space of the vector bunle $\mathrm{Hom}(\tilde{\mathscr{E}}_{x_2}, \mathcal{O}^2)$ on $\mathbf{E}$.

There is a smooth morphism $\mathbf{B} \to \mathbf{W}$. On the other hand $\mathbf{W}$ is smooth which shows the same for the original point $(E, Q)$.

*Step 2.* We next turn to the case when $\tau_{x_2} = 0$, $\tau_{x_1} \neq 0$, and the map $(\mathscr{E})_{x_1} \to \mathscr{Q}$ is not surjective. Let $\mathbf{F}$ denote the frame-bundle of $\mathscr{Q}$, and consider a point $(Fr{:}Q \to \mathbf{C}^2) \in \mathbf{F}$ above $(E, Q)$ where $p_1 \circ Fr \circ P$: $\tau_{x_1} \to \mathbf{C}$ is an isomorphism ($p_1$ denoting the projection to the first co-ordinate $\mathbf{C}^2 \to \mathbf{C}$.). The map $P_1 \equiv p_1 \circ Fr \circ P$: $\mathscr{E}_{x_1} \to \mathbf{C}$ is nonzero in some neighbourhood, say $\mathbf{F}_1$. On $\tilde{X} \times \partial\mathbf{F}_1$ define $\tilde{\mathscr{E}}$ by the sequence $0 \to \tilde{\mathscr{E}} \to \mathscr{E} \xrightarrow{P_1} {}_{x1}\mathbf{C} \to 0$. As in Step 1 one sees that in a possibly smaller neighbourhood $\mathbf{F}_2$, $\tilde{\mathscr{E}}$ is locally-free, $H^0(\tilde{E})$ generates $\tilde{E}$, $h^0(\tilde{E}) = \tilde{n} - 1$ and $H^1(\tilde{E}) = 0$. Let $\mathbf{B} \to \mathbf{F}_2$ now be the bundle of frames of the direct image of $\tilde{\mathscr{E}}$ with respect to $\pi_{\mathbf{E}_2}$.

On the other hand let $\tilde{\mathbf{Q}}^1$ be the Quot scheme of rank 2 degree $d-1$ quotients $\mathcal{O}^{\tilde{n}-1}\tilde{\mathscr{E}} \to 0$, and let $\tilde{\mathbf{Q}}^1_F$ denote the open subset of locally-free quotients with vanishing first cohomology such that $\mathbf{C}^{\tilde{n}-1} \to H^0(\tilde{E})$ is an isomorphism. Let $\mathbf{E} \equiv \mathrm{Ext}^1({}_{x1}\mathbf{C}, \tilde{\mathscr{E}})$ be the bundle of extensions [La] where ${}_{x_1}\mathbf{C}$ is the skyscraper sheaf on the (reduced) point $x_1$ with $\mathbf{C}$ as fibre. On $\tilde{X} \times \mathbf{E}$ there is an exact sequence of sheaves flat over $\mathbf{E}$: $0 \to \tilde{\mathscr{E}} \to \mathscr{E} \to {}_{x1}\mathbf{C} \to 0$. Let $\mathbf{W}$ denote the total space of the vector bundle $\mathrm{Hom}(\tilde{\mathscr{E}}_{x_2}, \mathcal{O}^2)$ on $\mathbf{E}$. Finally let $\mathbf{V} \equiv \mathbf{V}(\mathscr{E}_{x_1}) = \mathrm{Spec}(\mathbf{S}(\mathscr{E}_{x_1}))$ be defined as in [EGA-I, (9.4.8)].

There is a smooth morphism $\mathbf{B} \to \mathbf{V} \times_{\mathbf{E}} \mathbf{W}$. We need, therefore, to analyse the singularities of $\mathbf{V}$. The map $\mathbf{V} \times_{\mathbf{E}} \mathbf{W} \to \tilde{\mathbf{Q}}^1_F$ is locally trivial, so clearly we can hold $\tilde{E}$ fixed for this purpose. Lemma C.4 concludes the proof in this case.

*Step 3.* We next consider the case when both $\tau_{x_1}$ and $\tau_{x_2}$ are one-dimensional. The nontrivial case is when $(\mathscr{E})_{x_j} \to \mathscr{Q}$ is not surjective at either point. (The other cases can be reduced to at most a combination of the two earlier ones.) We now imitate Step 2 and reduce the proof to Lemma C.6 below. □

**Lemma C.4.** *Let $\tilde{E}$ be a* rank 2 *locally-free sheaf on $\tilde{X}$, let $x \in \tilde{X}$ be a smooth point. Let $\mathbf{E} \equiv \mathrm{Ext}^1({}_x\mathbf{C}, \tilde{E})$ and consider the universal extension $0 \to \tilde{E} \to E \to {}_x\mathbf{C} \to 0$ on $\tilde{X} \times \mathbf{E}$. Then the space $\mathbf{V}(E_x)$* (cf., [EGA-I, (9.4.8)]) *is reduced, normal, Gorenstein, with rational singularities.*

*Proof.* Clearly we can replace $\tilde{X}$ by an affine neighbourhood of $x$ where $\tilde{E}$ is trivial, and then by using Noether normalisation, by the affine line $\mathbf{A}^1$. We let $w$ denote the affine co-ordinate, and identify $\tilde{E} \sim \mathcal{O}^2$. We have then natural co-ordinates $(u_1, u_2)$ on $\mathbf{E}$.

Let $E$ be the sheaf on $\mathbf{A}^1 \times \mathbf{E}$ defined by the exact sequence: $0 \to \mathcal{O} \to \mathcal{O}^3 \to E \to 0$, the map $\mathcal{O} \to \mathcal{O}^3$ being $\tilde{h} \mapsto (u_1\tilde{h}, u_2\tilde{h}, -\omega\tilde{h})$. The inclusion $\mathcal{O}^2 \to \mathcal{O}^3$ given by $(f, g) \mapsto (f, g, 0)$ induces an inclusion $\mathcal{O}^2 \to E$. We have thus the diagram on $U \times \mathbf{E}$, the middle horizontal sequence eing split:

$$\begin{array}{ccccccccc}
 & & & & 0 & & 0 & & \\
 & & & & \downarrow & & \downarrow & & \\
 & & & & \mathcal{O} & = & \mathcal{O} & & \\
 & & & & \downarrow & & \downarrow & & \\
0 & \to & \mathcal{O}^2 & \to & \mathcal{O}^3 & \to & \mathcal{O} & \to & 0 \\
 & & =\downarrow & & \downarrow & & \downarrow & & \\
0 & \to & \mathcal{O}^2 & \to & E & \to & {}_x\mathbf{C} & \to & 0 \\
 & & & & \downarrow & & \downarrow & & \\
 & & & & 0 & & 0 & &
\end{array}$$

It is clear that $E$ is the universal extension that we seek. (The map $\mathcal{O} \to \mathcal{O}$ is given by $\tilde{h} \mapsto wh$.)

Restricitng to $\{x\} \times \mathbf{E}$ we get

$$0 \to \mathcal{O} \xrightarrow{\mathrm{a}} \mathcal{O}^3 \to E_x \to 0. \qquad \text{(C.3)}$$

The map a is given by $\tilde{h} \mapsto \tilde{h}(u_1, u_2)$ (and is therefore an injection.) This shows that $\mathbf{V}(E_x)$ is the subscheme of $\mathbf{V}(\mathcal{O}^3)$ defined as follows. The scheme $\mathbf{V}(\mathcal{O}^3)$ is the total space of the dual bundle of $\mathcal{O}^3$; with respect to the natural co-ordinates $(u_1, u_2, v_1, v_2, s)$ on $\mathbf{V}(\mathcal{O}^3)$ the subscheme defined by the ideal $(u_1v_1 + u_2v_2)$ is $\mathbf{V}(E_x)$. This is the product of the affine line with the cone over the nonsingular quadric surface in $P^3$, and is easily seen to be reduced, normal and Gorenstein; also, it has a rational singularity at the vertex. □

*Remark C.5.* (a) $\mathscr{H}$ *is not* locally factorial. It is well-known that the cone over the nonsingular quadric in $\mathbf{P}^3$ is not factorial at the vertex, with class group equal to $\mathbf{Z}$.

(b) The canonical map $c$: $E_x \to \mathcal{O}$ on $\mathbf{V}(E_x)$ is induced by the map $\mathcal{O}^3 \to \mathcal{O}$, $(f, g, h) \mapsto fv_1 + gv_2 + hs$.

(c) The locus of non-locally-free extensions is given by the (non-Cartier) divisor defined by the ideal $(u_1, u_2)$.

(d) Let $c$ be the map $E_x \to \mathcal{O}$ on $\mathbf{V}(E_x)$ defined in (b) above, and let $b$ be the map obtained by restricting the map $E \to {}_x\mathbf{C}$. Consider the map $E_x \to \mathcal{O}^2 \sim Q$, given by $t \mapsto (c(t), b(t))$. In the complement of the non-free locus this map is of rank one precisely when $\ker c = \ker b = \mathcal{O}^2$. This yields the equation $v_1 = 0$, $v_2 = 0$.

(e) $\hat{\mathscr{D}}_1^f \backslash \hat{\mathscr{D}}_{1,F}$ has codimension $\geqq 3$ in $\hat{\mathscr{D}}_1^f$. This follows from (c–d).

**Lemma C.6.** *Let* $\tilde{E}$ *be a* rank 2 *locally-free sheaf on* $\tilde{X}$, *let* (*for* $j = 1, 2$) $x_j \in \tilde{X}$ *be smooth points. Let* $\mathbf{E}' \equiv \mathrm{Ext}^1({}_{x_1}\mathbf{C} \oplus {}_{x_2}\mathbf{C}, \tilde{E})$ *and consider the universal extension* $0 \to \tilde{E} \to E \to {}_{x_1}\mathbf{C} \oplus {}_{x_2}\mathbf{C} \to 0$ *on* $\tilde{X} \times \mathbf{E}'$. *Then the space* $\mathbf{V}(E_{x_1} \oplus E_{x_2})$ *is normal, Gorenstein, with rational singularities.*

*Proof.* Clear extension of the proof of Lemma C.4. □

We are now in a position to state

**Theorem X3.** $\mathscr{P}$ *is reduced, irreducible and normal, with rational singularities.*

*Proof.* We use Lemma 4.18 and Proposition 4.19. These are all then immediate consequences of well-known properties of GIT quotients. The relevant result about rational singularities is that of [Bo]. □

The codimension one subschemes $\hat{\mathscr{D}}_j^f$ and $\hat{\mathscr{D}}_j^t$ in $\mathscr{H}$ are defined in §4a, and also the subscheme $\hat{\mathscr{V}}_j^f$ of each $\hat{\mathscr{D}}_j^f$. The following description of the varieties $\mathscr{D}_j$ should be kept in mind; it follows easily from Proposition B.14: $\mathscr{D}_1$ consists of $s$-equivalence classes of GPS's such that the "associated graded" GPS has torsion at $x_2$.

**Proposition C.7.** (1) *The* $\hat{\mathscr{D}}_j^f$ *are reduced, irreducible, and normal.*

(2) *The* $\hat{\mathscr{D}}_j^t$ *are reduced, irreducible, and normal.*

(3) *The* $\hat{\mathscr{V}}_j^f$ *are smooth. We have* $\hat{\mathscr{V}}_j^f \cap \{\hat{\mathscr{D}}_1^f \cap \hat{\mathscr{D}}_2^f\} = \emptyset$.

(4) *The closed orbits in* $\hat{\mathscr{D}}_j^f$ *and* $\hat{\mathscr{D}}_j^t$ *are contained in* $\hat{\mathscr{D}}_j^f \cap \hat{\mathscr{D}}_j^t$.

*Proof.* We will prove these claims for $j = 1$. The proofs depend heavily on the local description of $\mathscr{H}$ obtained during the proof of Proposition C.3.

(1) We will give the proof of (1) in some detail. By definition $\hat{\mathscr{D}}_1^f$ is reduced. The divisor $\hat{\mathscr{D}}_{1,F}$ is irreducible, hence so is its closure. Normality of $\hat{\mathscr{D}}_{1,F}$ is also clear (because, for example, it is a complete intersection and the singular set, $\hat{\mathscr{V}}_{1,F}$ has codimension 2). It remains to prove normality of $\hat{\mathscr{D}}_1^f$ at points of $\hat{\mathscr{D}}_1^f \backslash \hat{\mathscr{D}}_{1,F}$. By semicontinuity, at such a point $(E, Q)$, the map $E_{x_1} \to Q$ must be either zero or have one-dimensional image.

(i) Suppose first that $E$ is locally free at $x_1$. Then it is not so at $x_2$ and the local model of $\mathscr{H}$ at such a point is either as in Step 1 (if $E_{x_2} \to Q$ is surjective) or as in Step 2 (if $E_{x_2} \to Q$ has one-dimensional image.) of the proof of C.3. Note, however, that the roles of $x_1$ and $x_2$ are reversed vis-à-vis that proof. In either case the inverse image of $\hat{\mathscr{D}}_1^f$ by the smooth map $\mathbf{B} \to \mathbf{U}'$ (respectively, $\mathbf{B} \to \mathbf{F}_2$) is the pull-back, in turn, of $\tilde{\mathscr{D}}$ via the map $\mathbf{B} \to \mathbf{W}$ (respectively, $\mathbf{B} \to \mathbf{V} \times_E \mathbf{W}$) where $\mathbf{W}$ is the total space of the vector bundle $\mathrm{Hom}(\tilde{\mathscr{E}}_{x_1}, \mathcal{O}^2)$ and $\tilde{D} \subset \mathbf{W}$ is defined by the determinantal ideal. In either case we have normality.

(ii) If $E$ is not locally free at $x_1$, $E_{x_1} \to Q$ must have one-dimensional image, and there are again two cases to consider: (1) If $E$ is locally free at $x_2$ the local model is the divisor given by the ideal $(x, y)$ in $\mathbf{C}[u, v, x, y]/(ux + vy)$ (C.5(d)). (2) If $E$ is not locally free at $x_2$ the local model is the product of the above with another normal variety.

(2) We prove irreducibility. Consider the open subset of $\hat{\mathscr{D}}_1^t$ where the torsion subsheaf has degree 1; this set is easily seen to be dense. Such a sheaf $E$ is necessarily of the form $\tilde{E} \oplus \mathbf{C}_{x_1}$, with $\tilde{E}$ generated by global sections. It is now straightforward to imitate the proof of [N, Remark 5.5]. The other facts are proved as in (ii) above. The relevant result is C.5 (4).

(3) It is easily seen that $\hat{\mathscr{V}}_1^f$ is the set of $(E, Q)$ such that the map $E_{x_1} \to \mathscr{Q}$ is zero. $E$ is therefore locally free at $x_1$, and the map $E_{x_2} \to \mathscr{Q}$ surjective. The local model is as in Step 1 if $E$ is not locally free at $x_2$. In any case it is clear that $\hat{\mathscr{V}}_1^f$ is smooth. The other statements have similar lproofs.

(4) This follows from Proposition B.15. □

*Acknowledgements.* It is a pleasure to thank U. Bhosle, R.C. Cowsik, J.M. Drezet, R.V. Gurjar, M. Maruyama, V.B. Mehta, N. Mohan Kumar, S. Ramanan, A. Ramanathan, C.S. Seshadri and R.R. Simha for very useful discussions. We also thank the referee for a careful reading of the manuscript. The first author is grateful to the I.C.T.P., Trieste, where part of this work was done, for hospitality. The second author similarly wishes to thank the I.C.T.P. and the M.S.R.I., Berkely.

## References

[A] Artin, M.: On the solutions of analytic equations. Invent. Math. **5**, 277–291 (1968)

[A-M] Atiyah, M.F., Macdonald, I.G.: Introduction to commutative algebra. Reading: Addison–Wesley 1969

[B] Bourbaki, N.: Commutative algebra. Paris: Hermann 1972

[Bo] Boutot, J.F.: Singularités rationnelles et quotients par les groupes réductifs, Invent. Math. **88**, 65–68 (1987)

[B2] Bhosle, U.: Parabolic vector bundles on curves, Arkiv för matematik **27**, 15–22 (1989)

[B1] Bhosle, U.: Generalised parabolic bundles and applications to torsion-free sheaves on nodal curves. TIFR preprint 1990, to appear in Arkiv för matematik

[D] Donaldson, S.K.: Polynomial invariants for smooth four-manifolds. Topology **29**, 257–315 (1990)

[D-N] Drezet, J.M., Narasimhan, M.S.: Groupe de Picard des variétés de fibrés semistables sur les courbes algébriques. Invent. Math. **97**, 53–94 (1989)

[G-R] Grauert, H., Riemenschneider, O.: Verschwindungssätze fur analytische Kohomologiegruppen auf komplexen Räumen. Invent. Math. **11**, 263–292 (1970)

[G] Grothendieck, A.: Techniques de construction et théorèmes d'existence en géométrie algébrique IV: les schemas de Hilbert., Exposé 221, Vol. 1960–61, Séminaire Bourbaki

[SGA-II] Grothendieck, A.: Cohomologie locale des faisceaux cohérents et théorèmes de lefschetz locaux et globaux (SGA-II). Amsterdam: North-Holland 1968

[SGA-I] Grothendieck, A.: Revêtements Étales et Groupe Fondamental (SGA-I). Springer Lecture Notes 224. Berlin-Heidelberg-New York: Springer 1971

[EGA-IV] Grothendieck, A., Dieudonné, J.: Étude locale des schémas et des morphismes de schémas (iii), Publ. Math. IHES **28** (1966)

[EGA-I] Grothendieck, A., Dieudonné, J.: Éléments de Géométrie Algébrique I. (new edn.) Grundlehren 166. Berlin-Heidelberg-New York: Springer 1971

[H-N] Harder, G., Narasimhan, M.S.: On the cohomology groups of the moduli space of vector bundles of on curves. Math. Ann. **212**, 215–248 (1975)

[H] Hartshorne, R.: Algebraic geometry. Berlin-Heidelberg-New York.: Springer 1977

[K] Kempf, G.: Cohomology and covexity, in Kempf G et al. (eds.). Toroidal Imbeddings, Springer Lecture Notes, Vol. 339, Berlin-Heidelberg-New York.: Springer, 1973

[Kn] Knop, F.: Der kanonische Moduleines Invariantenrings. J. Algebra **127**, 40–54 (1989)

[K-M] Knudsen, F., Mumford, D.: The projectivity of the moduli space of stable curves I: Preliminaries on "det" and "Div". Math. Scand. **39**, 19–55 (1976)

K-P] Kraft, H., Procesi, C.: On the geometry of conjugacy classes in classical groups. Comment. Math. Helv. **57**, 536–602 (1982)

[L] Lang, S., Introduction to Arakelov theory. Berlin-Heidelberg-New York: Springer, 1988

[La] Lange, H.: Universal families of extensions. J. Algebra **83**, 101–112 (1983)

[Lu] Luna, D.: Adehérences d'orbit et invariants. Invent. Math. **29**, 231–238 (1975)

[Ma] Matsumura, H.: Commutative ring theory. Cambridge University Press, Cambridge: Cambridge University Press, 1986

[M-S] Mehta, V.B., Seshadri, C.S.: Moduli of vector bundles on curves with parabolic structures. Math. Ann. **248**, 205–239 (1980)

[Mu] Mumfor, D.: Picard groups of moduli problems. in: Schilling, O.F.G. (ed.) Arithmetical algebraic geometry. New York: Harper and Row

[M-F] Mumford, D., Fogarty, J.: Geometric invariant theory. 2nd edn. Berlin-Heidelberg-New York.: Springer, 1982

[N-T] Narasimhan, M.S., Trautmann, G.: Compactification of $M_{\mathbf{P}_3}(0, 2)$ and Poncelet pairs of conics. Pacific J. Math. **145**, 255–365 (1990)

[N] Newstead, P.E.: Introduction to moduli problems and orbit spaces. TIFR lecture notes. New Delhi: Narosa 1978

[R] Ramanan, S.: The moduli spaces of vector bundles over an algebraic curve. Math. Ann. **200**, 69–84 (1973)

[S-S] Schiffman, B., Sommense, A.J.: Vanishing theorems on complex manifolds. Boston-Basel-Stuttgart: Birkhaüser 1985.

[GAGA] Serre, J.P.: Géométrie algébrique et géométrie analytique. Ann. Inst. Fourier **6**, 1–42 (1956)

[S1] Seshadri, C.S.: Quotient spaces modulo reductive algebraic groups. Ann. Math. **95**, 511–556 (1972)

[S2] Seshadri, C.S.: Fibrés vectoriels sur les courbes algébriques (Lectures at the E.N.S., notes by J.M. Drezet) Astérisque **96** (1982)

[Si] Simpson, C.: Moduli of representations of the fundamental group of a smooth projective variety. Princeton University. Preprint 1990

[Sw] Swan, R.G.: On seminormality. J. Algebra **67**, 210–229 (1980)

[T] Traverso, C.: Seminormality and Picard group. Ann. Scuola Norm. Sup. Pisa **24**, 585–595 (1970)

[T-U-Y] Tsuchiya, A., Ueno, K., Yamada, Y.: Conformal field theory on universal family of stable curves with gauge symmetries. Adv. Studies Pure Math. **19**, 459–566 (1989)

[Z] Zariski, O.: Sur la normalité analytique des variétés normales. Ann. Inst. Fourier (Grenoble) **2**, 161–164 (1950)

[Z-S] Zariski, O., Samuel, P.: Commutative algebra. II. Princeton: Van Nostrand 1960

Proc. Indian Acad. Sci. (Math. Sci.), Vol. 104, No. 1, February 1994, pp. 191–200.


# Vector bundles as direct images of line bundles

A HIRSCHOWITZ and M S NARASIMHAN*

Université de Nice Sophia-Antipolis, Parc Valrose, 06108 Nice Cedex 2, France
*International Centre for Theoretical Physics, P.O. Box 586, 34100 Trieste, Italy

Dedicated to the memory of Professor K G Ramanathan

**Abstract.** Let $X$ be a smooth irreducible projective variety over an algebraically closed field $K$ and $E$ a vector bundle on $X$. We prove that, if $\dim X \geqslant 1$, there exist a smooth irreducible projective variety $Z$ over $K$, a surjective separable morphism $f: Z \to X$ which is finite outside an algebraic subset of codimension $\geqslant 3$ in $X$ and a line bundle $L$ on $X$ such that the direct image of $L$ by $f$ is isomorphic to $E$. When $X$ is a curve, we show that $Z$, $f$, $L$ can be so chosen that $f$ is finite and the canonical map

$$H^1(Z, \mathcal{O}) \to H^1(X, \operatorname{End} E)$$

is surjective.

**Keywords.** Projective variety; algebraic vector bundle; line bundle; direct image; finite morphism.

## 1. Introduction

Let $X$ be a smooth irreducible projective variety over an algebraically closed field and $E$ a vector bundle on $X$. We prove in this paper first that, if $\dim X \geqslant 1$, $E$ is the direct image of a line bundle $L$ on a smooth irreducible projective variety $Z$ by a morphism $f: Z \to X$ which is finite outside an algebraic subset of codimension $\geqslant 3$ in $X$. Moreover one can choose the morphism $f$ to be separable and to have the property that all higher direct images of $L$ by $f$ are zero [Theorem 4.2].

In particular if $\dim X \leqslant 2$ the morphism $f$ may be chosen to be finite. In the case of surfaces this result has been proved by R.L.E. Schwarzenberger for rank two vector bundles [5, Theorem 3]. We also give an example of a vector bundle on $\mathbf{P}_3$ which cannot be obtained as the direct image of a line bundle on a smooth variety by a finite morphism.

In the second part of the paper we consider the case when $X$ is a curve. We prove in this case that $Z$, $L$ and $f$ can be so chosen that the canonical homomorphism (see 5.1)

$$H^1(Z, \mathcal{O}_Z) \to H^1(X, \operatorname{End} E)$$

is surjective (Theorem 6.4). This result was proved for a "very stable" vector bundle $E$ by Beauville-Narasimhan-Ramanan in the case of a curve over $\mathbf{C}$ by using the Hitchin map [3]. (For the significance of this result see Remark 6.5).

Let $\pi:\mathbf{P}(E)\to X$ be the projective bundle associated to $E$. The variety $Z$ is constructed as the subscheme (of $\mathbf{P}(E)$) of zeros of a generic section of the tangent bundle along the fibres of $\pi$ twisted by a suitable ample line bundle on $X$ pulled up to $\mathbf{P}(E)$; the line bundle $L$ is simply taken to be the restriction of $\mathcal{O}_{\mathbf{P}(E)}(1)$ to $Z$.

The scheme $Z$ is essentially the scheme of 'eigenstates' of a generic twisted endomorphism of $E$. However, in general, $Z$ is not the spectral variety of the twisted endomorphism; the canonical map from the spectral variety into $X$ is always a finite morphism.

## 2. Sections of the tangent bundle of a projective space

Let $V$ be a finite dimensional vector space of dimension $\geqslant 2$ over an algebraically closed field $K$ and $\mathbf{P}=\mathbf{P}(V)$ the projective space of hyperplanes in $V$. We have the exact sequences of vector bundles on $\mathbf{P}(E)$:

$$0\to\Omega^1(1)\to V_{\mathbf{P}}\to\mathcal{O}(1)\to 0$$

and

$$0\to\mathcal{O}_{\mathbf{P}}\to V^*_{\mathbf{P}}\otimes\mathcal{O}(1)\to\Theta\to 0,$$

where $\Theta$ denotes the tangent bundle of $\mathbf{P}$ and $V^*$ the dual of $V$. We obtain an exact sequence of vector spaces

$$0\to K\to V^*\otimes V\to H^0(\mathbf{P},\Theta)\to 0.$$

If $\mathrm{End}^0(V):=\mathrm{End}(V)/(\text{Scalar endomorphisms})$ we have $\mathrm{End}^0(V)=H^0(\mathbf{P},\Theta)$.

If an endomorphism $T$ of $V$ leaves a hyperplane $\xi$ invariant, $T$ induces an endomorphism of the one dimensional space $V/\xi$. The subspace $(V/\xi)^*$ of $V^*$ is an eigenspace of the transpose of $T$. If $s_T$ is the section of $\Theta$ defined by $T$, we have $s_T(\xi)=0$ if and only if $T(\xi)\subset\xi$. Thus we can view the subscheme $Z=Z_T$ of zeros of $s_T$ as the scheme of "eigenstates" of $T$. Moreover the "eigenvalue" of $T$ is a section of $\mathcal{O}_Z$; in fact it is the section of $\mathcal{O}_Z$ corresponding to the morphism $\mathcal{O}_Z(1)\to\mathcal{O}_Z(1)$ induced by $T$ from:

$$\begin{array}{ccccccc} 0\to\Omega^1(1)\to & V_{\mathbf{P}} & \to\mathcal{O}(1)\to 0 \\ & T\downarrow & \\ & V_{\mathbf{P}} & \to\mathcal{O}(1)\to 0. \end{array}$$

Observe that the scheme $Z_T$ has dimension $\geqslant 1$ if and only if the transpose of $T$ has an eigenspace of dimension $\geqslant 2$ corresponding to some eigenvalue.

PROPOSITION 2.1

*Consider the exact sequence of vector bundles*

$$0\to F\to\mathrm{End}^0(V)_{\mathbf{P}}\to\Theta\to 0$$

*($F$ being defined as the kernel of homomorphism $\mathrm{End}^0(V)_{\mathbf{P}}\to\Theta$). Let $p:F\to\mathrm{End}^0(V)$ be the restriction to $F$ of the projection $\mathrm{End}^0(V)\times\mathbf{P}\to\mathrm{End}^0(V)$. Then there exists an open subset $\Omega$ in $\mathrm{End}^0(V)$, whose complement is of codimension $\geqslant 3$, such that the morphism $p:p^{-1}(\Omega)\to\Omega$ is finite.*

*Proof*. Consider the commutative diagram

$$\begin{array}{ccccccccc} & & 0 & & 0 & & & & \\ & & \downarrow & & \downarrow & & & & \\ & & K_{\mathbf{P}} & & K_{\mathbf{P}} & & & & \\ & & \downarrow & & \downarrow & & & & \\ 0 & \to & F^1 & \to & \mathrm{End}(V)_{\mathbf{P}} & \to & \Theta & \to & 0 \\ & & \downarrow & & \downarrow & & \| & & \\ 0 & \to & F & \to & \mathrm{End}^0(V)_{\mathbf{P}} & \to & \Theta & \to & 0 \\ & & \downarrow & & \downarrow & & & & \\ & & 0 & & 0 & & & & \end{array}$$

Let $q:F^1 \to \mathrm{End}(V)$ be the projection. We shall show that there exists an open set $U$ of $\mathrm{End}(V)$ which is saturated for the map $\mathrm{End}(V) \to \mathrm{End}^0(V)$ and whose complement is of codimension $\geqslant 3$ such that the morphism $q:q^{-1}(U) \to U$ is finite. This will prove the proposition.

For each subspace $W$ of dimension $k \geqslant 2$ of $V^*$, consider the subspace of $\mathrm{End}(V^*)$ consisting of those endomorphisms whose restriction to $W$ is a scalar endomorphism of $W$. The dimension of this space is $1+(r-k)^2+k(r-k)$. Varying $W$ over the Grassmannian $G(r,k)$ we get a vector bundle $W(r,k)$ over $G(r,k)$ and the dimension of the total space of this bundle is

$$1+(r-k)^2+k(r-k)+k(r-k)=1+r^2-k^2.$$

We have a natural morphism $\pi_k$: $W(r,k) \to \mathrm{End}(V)$ which maps an endomorphism to its transpose. If $S_k := \overline{\pi_k(W(r,k))}$, we have $\dim S_k \leqslant (1+r^2-k^2)$ and $\operatorname{codim} S_k \geqslant k^2-1 \geqslant 3$. Let $S = \bigcup_{2\leqslant k\leqslant r} S_k$ and $U = \mathrm{End}(V) - S$. We have $\operatorname{codim} S \geqslant 3$ and $S$ is saturated for the map $\mathrm{End}(V) \to \mathrm{End}^0(V)$. The fibres of $q:q^{-1}(U) \to U$ are finite and $q$ is proper. Hence $q$ is a finite morphism.

## 3. Sections of the (twisted) relative tangent bundle of a projective bundle

Let $E$ be a vector bundle of rank $r \geqslant 2$ on a smooth irreducible projective variety $X$ of dimension $\geqslant 1$ over $K$. Let $\pi:\mathbf{P}(E) \to X$ be the associated projective bundle. We have the exact sequences on $\mathbf{P}(E)$:

$$0 \to \Omega^1_\pi(1) \to \pi^*(E) \to \mathcal{O}_{P(E)}(1) \to 0$$

and

$$0 \to \mathcal{O}_{\mathbf{P}(E)} \to \pi^*(E^*) \otimes \mathcal{O}(1) \to \Theta_\pi \to 0$$

where $\Theta_\pi$ (resp. $\Omega^1_\pi$) denotes the relative tangent (resp. cotangent) bundle along the fibres of $\pi$. Let $\mathrm{End}^0(E)$ denote the vector bundle $\mathrm{End}(E)/\mathcal{O}_X$. We have an exact sequence of vector bundles on $\mathbf{P}(E)$:

$$0 \to F \to \pi^*(\mathrm{End}^0(E)) \to \Theta_\pi \to 0.$$

Let $M$ be a line bundle on $X$. We obtain the exact sequence:

$$0 \to F \otimes \pi^*(M) \to \pi^*(\mathrm{End}^0(E) \otimes M) \to \Theta_\pi \otimes \pi^*(M) \to 0$$

PROPOSITION 3.1

*Let $p: F \otimes \pi^*(M) \to End^0(E) \otimes M$ be the canonical morphism (of total spaces of geometric vector bundles over $\mathbf{P}(E)$ and $X$ respectively). Then there exists an open subset $\Omega$ of $End^0(E) \otimes M$ whose complement is of codimension $\geqslant 3$ such that the morphism $p: p^{-1}(\Omega) \to \Omega$ is finite.*

*Proof.* This follows from Proposition 2.1.

PROPOSITION 3.2

*There exists an ample line bundle $M$ on $X$ such that a generic section $s$ of $\Theta_\pi \otimes \pi^*(M)$ (i.e. for $s$ in a non-empty open subset of $H^0(\mathbf{P}(E), \Theta_\pi \otimes \pi^*(M))$) satisfies the following conditions:*

a) *The scheme $Z$ of zeros of $s$ is smooth and irreducible.*
(b) *The morphism $\pi | Z: Z \to X$ is surjective and separable.*
(c) *There exists a closed subset $S$ of $X$ of codimension $\geqslant 3$ such that the morphism*

$$\pi: Z \backslash \pi^{-1}(S) \to X \backslash S$$

*is finite.*

*Proof.* Let $\xi$ be an ample line bundle on $X$. Then the line bundle $\pi^*(\xi^k) \otimes \mathcal{O}(1)$ is very ample on $\mathbf{P}(E)$ for $k \geqslant k_0$. [4, II, Prop. 7.10, p. 161]. We may also assume that for $k \geqslant k_0$, the bundle $\xi^k \otimes E^*$ is generated by its sections. Let $M = \xi^{2k}$. Since the sections of the very ample line bundle $\pi^*(\xi)^k \otimes \mathcal{O}(1)$ generate its first order jet bundle and $\pi^*(\xi^k \otimes E^*)$ is generated by its sections, we see that the sections of $\pi^*(M \otimes E^*) \otimes \mathcal{O}(1)$ generate its first order jet bundle [7, Lemma 5]. Since $\Theta_\pi \otimes \pi^*(M)$ is a quotient bundle of $\pi^*(M \otimes E^*) \otimes \mathcal{O}(1)$, the sections of $\Theta_\pi \otimes \pi^*(M)$ generate its first order jet bundle. Now by [7, Theorem 1] the zero scheme $Z$ of a generic section $s$ of $\Theta_\pi \otimes \pi^*(M)$ is smooth. We will prove in the next proposition (Proposition 4.1) that $Z$ is irreducible.

Let $x_0 \in X$ and $S := \pi^{-1}(x_0)$ the fibre over $x_0$. Let $W$ be the image of the homomorphism

$$H^0(\mathbf{P}(E), \Theta_\pi \otimes \pi^*(M)) \to H^0(S, \Theta_\pi \otimes \pi^*(M) | S).$$

(We may even assume that $W = H^0(S, \Theta_\pi \otimes \pi^*(M) | S)$, by choosing $k$ large enough). Then the first order jets of elements of $W$ generate the first order jet bundle of $\Theta_\pi \otimes \pi^*(M) | S$; hence, again by [7, Theorem 1], for a generic element $\sigma$ of $W$, the zero subscheme (of $S$) defined by $\sigma$ is smooth. Thus we see that for a generic section of $\Theta_\pi \otimes \pi^*(M)$ the zero scheme $Z$ is smooth and irreducible and $Z$ intersects $S$ transversally. It follows that there exists a point $z_0 \in Z \cap S$ such that the differential of $\pi | Z$ at $z_0$ is an isomorphism. This proves that $\pi | Z: Z \to X$ is surjective and (assuming $Z$ to be irreducible) separable. (Observe that $Z$ intersects every fibre $\pi^{-1}(x), x \in X$, for otherwise the tangent bundle of the projective space $\pi^{-1}(x)$ would contain a trivial line bundle.)

$$\begin{aligned} \text{Let } \Sigma &:= H^0(\mathbf{P}(E), \pi^*(M)\otimes\Theta_\pi) \\ &= H^0(X, M\otimes\pi_*(\Theta_\pi)) \\ &= H^0(X, M\otimes \mathrm{End}^0(E)). \end{aligned}$$

Consider the morphisms

$$\begin{array}{l} \varphi:\Sigma\times X \overset{\varphi}{\to} M\otimes\mathrm{End}^0(E) \\ \quad\ \downarrow p_X \\ \quad\ X. \end{array}$$

where the evaluation map $\varphi$ is a smooth morphism, being a surjection of vector bundles. Let $\Omega$ be the open subset of $M\otimes\mathrm{End}^0(E)$ defined in Proposition 3.1 and $N$ its complement. Then

$$\dim\varphi^{-1}(N)\leqslant\dim\Sigma+\dim X-3.$$

By considering $p_X:\varphi^{-1}(N)\to X$ we see that for a generic section $s$ of $M\otimes\mathrm{End}^0(E)$ we have

$$\dim(\varphi^{-1}(N)\cap\{s\times X\})\leqslant(\dim X)-3.$$

Let $S\subset X$ be defined to be $p_X(\varphi^{-1}(N)\cap\{s\times X\})$. Then $\dim S\leqslant(\dim X)-3$ and

$$\pi|Z:Z\backslash\pi^{-1}(S)\to X\backslash S$$

is finite.

Thus for a generic section $s$ of $\Theta_\pi\otimes\pi^*(M)$ all the conditions a), b) and c) are satisfied.

## 4. Koszul resolution of the zero scheme Z

*Proof of Theorem* 4.2

PROPOSITION 4.1

*Let $s$ be a section of $\Theta_\pi\otimes\pi^*(M)$ over $\mathbf{P}(E)$ such that the zero scheme $Z$ of $s$ is smooth. We then have*

a) $\pi_*(\mathcal{O}_Z\otimes\mathcal{O}_{\mathbf{P}(E)}(1))\simeq E$ *and* $R^i\pi_*(\mathcal{O}_Z(1))=0$ *for* $i\geqslant 1$.

b) $\pi_*(\mathcal{O}_Z)$ *has a filtration*
$0=F_0\subset F_1\subset F_2\subset\ldots\subset F_i\subset\ldots\subset F_r=\pi_*(\mathcal{O}_Z)$
*such that* $F_i/F_{i-1}\simeq M^{-(i-1)}(:=(M^*)^{\otimes(i-1)})$ *for* $1\leqslant i\leqslant r$ *(In particular* $F_1\simeq\mathcal{O}_Z$*).*

c) *$Z$ is irreducible (if $\dim X\geqslant 1$ and $M$ is ample).*

*Proof.* Using our assumption on $Z$, we have a Koszul resolution for $\mathcal{O}_z$ on $\mathbf{P}(E)$: [1, Ch I, Lemma 4.2 and Ch. III, Propositions 4.10 and 4.11]

$$(A)\ 0\to\overset{r-1}{\Lambda}(\Omega^1_\pi\otimes\pi^*(M^*))\to\cdots\to\overset{2}{\Lambda}(\Omega^1_\pi\otimes\pi^*(M^*))\to\Omega^1_\pi\otimes\pi^*(M^*)\to\mathcal{O}_{\mathbf{P}(E)}\to\mathcal{O}_Z\to 0$$

and also a resolution of $\mathcal{O}_Z(1)$:

$$(B)\ 0\to\Omega_\pi^{r-1}\otimes(\pi^*(M^*))^{r-1}\otimes\mathcal{O}(1)\to\cdots\to\Omega_\pi^1\otimes\pi^*(M^*)\otimes\mathcal{O}(1)\to\mathcal{O}(1)\to\mathcal{O}(1)|_Z\to 0$$

Now we have, for the projective space **P**, the Bott vanishing theorem:

$H^i(\mathbf{P},\Omega^p(1))=0$ for $p\geqslant 1$ and all $i$ [6, Theórème 1].

Hence we have

$$R^i\pi_*(\Omega_\pi^p\otimes\pi^*((M^*)^{\otimes p})\otimes\mathcal{O}(1))=0$$

for $p\geqslant 1$ and all $i$. Splitting $B$ into short exact sequences we deduce that

$$\pi_*(\mathcal{O}_Z(1))=\pi_*(\mathcal{O}_{P(E)}(1))=E$$

and $R^i\pi_*(\mathcal{O}_Z(1))=0$ for $i>0$.

For proving (b) we observe that $R^q\pi_*(\Omega_\pi^p)=0$ for $p\neq q$ and

$$R^p\pi_*(\Omega_\pi^p)=\mathcal{O}_X \text{ for } 0\leqslant p\leqslant(r-1)\ [6].$$

Splitting (A) into short exact sequences we obtain b).

To prove c), since $\dim X\geqslant 1$ and $M$ is ample we have $H^0(X,M^{-k})=0$ for $k>0$. Using the filtration of $\pi_*(\mathcal{O}_Z)$ given in b), we see that

$$H^0(Z,\mathcal{O}_Z)=H^0(X,\pi_*(\mathcal{O}_Z))=H^0(X,\mathcal{O}_X)=K.$$

Since $Z$ is smooth it follows that $Z$ is irreducible.

**Theorem 4.2** *Let $X$ be a smooth irreducible projective variety of dimension $\geqslant 1$ over an algebraically closed field $K$. Let $E$ be a vector bundle on $X$. Then there exist a smooth irreducible projective variety $Z$ over $K$, a line bundle $L$ on $Z$ and a surjective separable morphism $f:Z\to X$ having in addition the following properties:*

1) *there exists a closed subset $S$ in $X$ of codimension $\geqslant 3$ such that the morphism*

$$f:Z\backslash f^{-1}(S)\to X\backslash S$$

*is finite.*

2) *we have* $f_*(L)\simeq E$ and $R^if_*(L)=0$ *for* $i>0$.

*Proof.* We may assume that $E$ is of rank $\geqslant 2$. Choose an ample line bundle $M$ on $X$ satisfying the conditions of Proposition 3.2. Let $L$ be the restriction of $\mathcal{O}_{P(E)}(1)$ to $Z$ and $f$ be the restriction of $\pi:\mathbf{P}(E)\to X$ to $Z$. Then by Proposition 4.1(a) we have

$$f_*(L)\simeq E \text{ and } R^if_*(L)=0 \text{ for } i>0.$$

## 5. The map $D$

Let $f:Z\to X$ be a morphism and $L$ a line bundle on $Z$ such that $f_*(L)=E$ is a vector bundle of rank $r$ on $X$. The morphism $f^*(f_*(L))\to L$ gives rise to a morphism

$$f_*(\mathcal{O}_Z)\otimes E\to f_*(L)=E$$

which may be viewed as a morphism

$$D: f_*(\mathcal{O}_Z) \to E^* \otimes E.$$

($D$ gives the canonical $f_*(\mathcal{O}_Z)$-module structure on $f_*(L)$).

Suppose that $f: Z \to X$ is a finite surjective morphism of smooth varieties. Then $f$ is flat [1, Ch. V, Cor. 3.6]. We have

$$H^1(Z, \mathcal{O}_Z) = H^1(X, f_*(\mathcal{O}_Z)).$$

The homomorphism

$$D_*: H^1(Z, \mathcal{O}_Z) \to H^1(X, \operatorname{End} E) \tag{5.1}$$

induced by $D$ is the infinitesimal deformation map (at $L$) for the variation of the direct image bundles as $L$ deforms as a line bundle on $X$ [2, Lemma 1.3.1].

Since $f$ is finite, the map $f^*(E) \to L$ is surjective and we have an exact sequence

$$0 \to N \to f^*(E) \to L \to 0$$

of vector bundles on $Z$. From this we get an exact sequence of vector bundles.

$$0 \to \mathcal{O}_Z \to f^*(E^*) \otimes L \to N^* \otimes L \to 0.$$

Since $f$ is flat and finite, $f_*(\mathcal{O}_Z)$ is a vector bundle on $X$ of rank $r$ and $f_*(N^* \otimes L)$ is a vector bundle. So we have an exact sequence of vector bundles on $X$:

$$0 \to f_*(\mathcal{O}_Z) \to E^* \otimes E \to f_*(N^* \otimes L) \to 0.$$

Observe that $f_*(\mathcal{O}_Z)/\mathcal{O}_X$ is a vector subbundle of rank $(r-1)$ of the vector bundle $\operatorname{End}(E)/\mathcal{O}_X = \operatorname{End}^0(E)$. Thus we have

*Lemma* 5.2 *Let $f: Z \to X$ be a finite morphism of smooth varieties and $L$ a line bundle on $Z$. If $E := f_*(L)$ is a vector bundle of rank $r$, then the vector bundle $\operatorname{End}^0(E)$ contains a vector subbundle of rank $(r-1)$.*

Let us get back to the situation in § 3 and § 4.

PROPOSITION 5.3

*Let $s$ be a section of $\Theta_\pi \otimes \pi^*(M)$ and $Z$ the zero subscheme of $s$ with the property that the canonical map $E = \pi_*(\mathcal{O}_{\mathbf{P}(E)}(1)) \to f_*(\mathcal{O}_Z(1))$ is an isomorphism, where $f = \pi | Z$. Suppose that $T$ is a section of $\operatorname{End}(E) \otimes M$ such that the image of $T$ in $H^0(\mathbf{P}(E), \Theta_\pi \otimes \pi^*(M))$ is $s$. Then there is a homomorphism $\mu: M^{-1} \to f_*(\mathcal{O}_Z)$ and a commutative diagram*

$$\begin{array}{ccc} f_*(\mathcal{O}_Z) & \xrightarrow{D} & \operatorname{End} E \\ {\scriptstyle\mu}\nwarrow & & \nearrow{\scriptstyle T} \\ & M^{-1} & \end{array}$$

*where $D$ is defined at the beginning of this section* (§ 5).

*Proof.* Consider the diagram

$$\begin{array}{ccccccccc} 0 & \to & \Omega^1_\pi \otimes \pi^*(M)^{-1} & \xrightarrow{i} & \pi^*(E\otimes M^{-1}) & \to & \mathcal{O}(1)\otimes \pi^*(M^{-1}) & \to & 0 \\ & & & & \downarrow \tilde{T} & & & & \\ 0 & \longrightarrow & \Omega^1_\pi & \longrightarrow & \pi^*(E) & \xrightarrow{g} & \mathcal{O}(1) & \longrightarrow & 0 \end{array}$$

where $\tilde{T}$ is induced by $T$. The homomorphism $g\circ \tilde{T}\circ i:\Omega^1_\pi \otimes M^{-1} \to \mathcal{O}(1)$ gives the section $s$ of $\Theta_\pi \otimes \pi^*(M)$. So $\tilde{T}$ induces on $Z$ a homomorphism $\lambda:\mathcal{O}_Z(1)\otimes f^*(M^{-1})\to \mathcal{O}_Z(1)$ and we have a commutative diagram

$$\begin{array}{ccc} \pi^*(E)\otimes \pi^*(M^{-1}) & \xrightarrow{q\otimes 1} & \mathcal{O}_Z(1)\otimes \pi^*(M^{-1}) \\ \downarrow \tilde{T} & & \downarrow \lambda \\ \pi^*(E) & \xrightarrow{q} & \mathcal{O}_Z(1) \end{array}$$

Considering $\lambda$ as a section of $\mathcal{O}_Z \otimes \pi^*(M)$ we obtain the section $\pi_*(\lambda)$ of $M\otimes \pi_*(\mathcal{O}_Z)$ which we view as a homomorphism $\mu: M^{-1}\to f_*(\mathcal{O}_Z)$. Since by assumption $E \to \pi_*(\mathcal{O}_Z(1))$ is an isomorphism, it follows, from the above diagram, that $T$ corresponds to $\pi_*(\lambda)$: $M^{-1}\otimes E \to E$. But by the definitions of $D$ and $\mu$ we see that $\pi_*(\lambda)$ corresponds to $D\circ\mu$. This means that $T = D\circ \mu$.

## 6. Vector bundles on curves

*Lemma* 6.1. *Let $X$ be a smooth, projective irreducible curve over $K$ and $F$ a vector bundle of rank $\geqslant 2$ on $X$. Let $\xi$ be an ample line bundle on $X$. Then there exists an integer $l_0$ such that for $l \geqslant l_0$, $\xi^{-l}$ is a subbundle of $F$ and the induced homomorphism $H^1(X,\xi^{-l}) \to H^1(X,F)$ is surjective.*

*Proof.* Let first $F$ be of rank 2. Since for all large $l$, $\xi^l \otimes F^*$ contains a trivial line subbundle, we get an exact sequence

$$0 \to \xi^{-l} \to F \to \xi^l \otimes \det F \to 0.$$

Choose $l$ large enough so that $H^1(X,\xi^l\otimes \det F) = 0$.

Now suppose that $F$ is a vector bundle of rank $\geqslant 3$. Then we can find a filtration of $F$ by subbundles

$$0 \subset F_1 \subset F_2 \ldots \subset F_i \subset \ldots F_{r-1} = F$$

such that rank $(F_i) = i+1$ (in particular rank $F_1 = 2$) and such that $H^1(X, F_i/F_{i-1}) = 0$, for $i \geqslant 2$. Now choose a line subbundle $\xi^{-l}$ of $F_1$ with $H^1(F_1/\xi^{-l}) = 0$. We see easily, by induction on $i$, that $H^1(X, F/\xi^{-l}) = 0$.

*Remark* 6.2 Note that $H^1(X,\xi^{-l}) \to H^1(X,F)$ is surjective if and only if $H^1(F/\xi^{-l}) = 0$, as $H^2(X,\xi^{-l}) = 0$.

COROLLARY 6.3

*Let $F$ be a vector bundle on $X$. Then there exists an integer $l_0$ such that, for $l \geqslant l_0$, for a generic section $\sigma$ of $\xi^l \otimes F$ the map $H^1(X, \xi^{-l}) \to H^1(X, F)$ induced by $\sigma$ is surjective.*

**Theorem 6.4** *Let $X$ be a smooth projective irreducible curve over an algebraically closed field $K$ and let $E$ be a vector bundle on $X$. Then there exist a smooth projective irreducible curve $Z$ over $K$, a line bundle $L$ on $Z$ and a finite surjective separable morphism $f: Z \to X$ such that*

1) $f_*(L) \simeq E$
2) *and the homomorphism (defined in 5.1)*

$$H^1(Z, \mathcal{O}_Z) \to H^1(X, \operatorname{End} E)$$

*is surjective.*

*Proof*. Choose an ample line bundle $M$ as in the proof of Theorem 4.2. We may also choose $M$ to have the further properties:

a) $H^1(X, M) = 0$

and

b) a generic section $\sigma$ of $H^0(X, \operatorname{End} E \otimes M)$ verifies the condition that the homomorphism

$$H^1(X, M^{-1}) \to H^1(X, \operatorname{End} E)$$

is surjective (use Corollary 6.3).

By condition a) the map

$$H^0(X, \operatorname{End} E \otimes M) \to H^0(X, \operatorname{End}^0 E \otimes M)$$

is surjective.

Now a generic section $s$ of $H^0(\mathbf{P}(E), \Theta_\pi \otimes \pi^*(M)) = H^0(X, \operatorname{End}^0(E) \otimes M)$ is the image of a section $T$ of $\operatorname{End}(E) \otimes M$ with the property that the homomorphism

$$H^1(X, M^{-1}) \to H^1(X, \operatorname{End} E)$$

induced by $T$ is surjective and satisfies conditions a), b) and c) of Proposition 3.2. Choose $Z, L, f$ as in the proof of Theorem 4.2. Then $f_*(L) = E$.

To prove 2), observe that the factorisation, given in Proposition 5.3,

$$\begin{array}{ccc} f_*(\mathcal{O}_Z) & \xrightarrow{D} & \operatorname{End} E \\ {\scriptstyle \mu}\nwarrow & & \nearrow{\scriptstyle T} \\ & M^{-1} & \end{array}$$

induces a commutative diagram:

$$\begin{array}{ccc} H^1(Z, \mathcal{O}_Z) = H^1(X, f_*(\mathcal{O}_Z)) & \xrightarrow{D_*} & H^1(X, \operatorname{End} E) \\ {\scriptstyle \mu_*}\nwarrow & & \nearrow{\scriptstyle T_*} \\ & H^1(X, M^{-1}) & \end{array}$$

Since, by choice, $T_*$ is surjective, the homomorphism $D_*$ is forced to be surjective.

*Remark* 6.5 If $E$ is a stable bundle on $X$ and if $Z, f$ and $L$ are chosen as in Theorem 6.4, we see that $f_*$ gives a *dominant* separable rational morphism from an appropriate component of $Pic(Z)$ into the moduli space of vector bundles on $X$ of rank $r = rkE$ and degree $d = degree\, E$ (compare [3]). We thus obtain 'most' stable bundles of a given rank and degree as direct images of line bundles on a *fixed* covering $Z$ of $X$.

## 7. The example

We now give an example of a rank 2 vector bundle on the projective space $\mathbf{P}_3(\mathbf{C})$ which cannot be obtained as the direct image of a line bundle by a *finite* morphism $f: Z \to \mathbf{P}_3(\mathbf{C})$, with $Z$ smooth.

Let $E$ be a stable vector bundle of rank 2 on $\mathbf{P}_3(\mathbf{C})$ with $c_1(E) = 0$ and $c_2(E) > 0$. If $E$ were the direct image of a line bundle by $f: Z \to X$, with $Z$ smooth and $f$ finite, the bundle $\mathrm{End}^0(E)$ would contain a line *subbundle* $L$ by Lemma 5.2. If $\xi = L^{-1}$, we would have

$$c_3(\xi \otimes \mathrm{End}^0 E) = 0. \text{ We have}$$

$$c_3(\xi \otimes \mathrm{End}^0 E) = 4c_1(\xi)c_2(E) + c_1(\xi)^3.$$

But the bundle $L$ and hence $\xi$, is non-trivial, since $h^0(\mathbf{P}_3, \mathrm{End}^0 E) = 0$, $E$ being stable. So $c_1(\xi) \neq 0$ and we would have

$$c_1(\xi)(4c_2(E) + c_1(\xi)^2) = 0,$$

a contradiction.

## Acknowledgements

The first author (AH) carried out this work in the framework of the Vector Bundle group of Europroj. The second author (MSN) would like to thank CNRS and Université de Nice-Sophia Antipolis for hospitality when part of this work was done.

## References

[1] Altman A and Kleiman S, *Introduction to Grothendieck duality theory* LNM 146 (1970) Springer
[2] Beauville A, Fibrés de rang 2 sur une courbe, fibré determinant et fonction theta, *Bull. Soc. Math. France*, **116** (1988) 431–418
[3] Beauville A, Narasimhan M S and Ramanan S, Spectral curves and the generalised theta divisor, *J. Reine Angew. Math.* **398** (1989) 169–179
[4] Hartshorne R, *Algebraic Geometry* (Springer–Verlag) (1977)
[5] Schwarzenberger R L E, Vector bundles on the projective plane, *Proc. London Math. Soc.* (3) **11** (1961) 623–640
[6] Verdier J L, Le théorème de Le Potier, Séminaire de géometrie analytique, *Astérisque* **17** (1974)
[7] Walter C H, Transversality theorems in general characteristic and arithmetically Buchsbaum schemes *Int. J. Math.* (to appear)

Math. Ann. 300, 41–75 (1994)

Mathematische Annalen


# Infinite Grassmannians and moduli spaces of $G$-bundles

**Shrawan Kumar**[1], **M.S. Narasimhan**[2], **A. Ramanathan**[3,⋆]

[1] Department of Mathematics, University of North Carolina, Chapel Hill, NC 27599-3250, USA
[2] ICTP, P.O. Box 586, Strada Costiera 11, Miramare, I-34100 Trieste, Italy
[3] School of Mathematics, TIFR, Colaba, Bombay 400005, India

Received: 15 May 1993

*Mathematics Subject Classification (1991):* 14D20, 22E67

## Introduction

Let $C$ be a smooth projective irreducible algebraic curve over $\mathbb{C}$ of any genus and $G$ a connected simply-connected simple affine algebraic group over $\mathbb{C}$. In this paper we elucidate the relationship between

(1) the space of vacua ("conformal blocks") defined in conformal field theory, using an integrable highest weight representation of the affine Kac-Moody algebra associated to $G$ and

(2) the space of regular sections ("generalised theta functions") of a line bundle on the moduli space $\mathfrak{M}$ of semistable principal $G$-bundles on $C$.

Fix a point $p$ in $C$ and let $\hat{\mathcal{O}}_p$ (resp. $\hat{k}_p$) be the completion of the local ring $\mathcal{O}_p$ of $C$ at $p$ (resp. the quotient field of $\hat{\mathcal{O}}_p$). Let $\mathcal{G} := G(\hat{k}_p)$ (the $\hat{k}_p$-rational points of the algebraic group $G$) be the loop group of $G$ and let $\mathcal{P} := G(\hat{\mathcal{O}}_p)$ be the standard maximal parahoric subgroup of $\mathcal{G}$. Then the generalised flag variety $X := \mathcal{G}/\mathcal{P}$ is an inductive limit of projective varieties, in fact of generalised Schubert varieties. One has a natural $\hat{\mathcal{G}}$-equivariant line bundle $\mathfrak{L}(\chi_o)$ on $X$ (cf. Sect. 2.2), and the Picard group Pic($X$) is isomorphic to $\mathbb{Z}$ which is generated by $\mathfrak{L}(\chi_o)$ (Proposition 2.3), where $\hat{\mathcal{G}}$ is the universal central extension of $\mathcal{G}$ by the multiplicative group $\mathbb{C}^*$ (cf. Sect. 2.2). By an analogue of the Borel-Weil theorem proved in the Kac-Moody setting by Kumar (and also by Mathieu), the space $H^0(\mathcal{G}/\mathcal{P}, \mathfrak{L}(d\chi_o))$ of the regular sections of the line bundle $\mathfrak{L}(d\chi_o) := \mathfrak{L}(\chi_o)^{\otimes d}$ (for any $d \geqq 0$) is canonically isomorphic with the full vector space dual $V(d\chi_o)^*$ of the (integrable) highest weight (irreducible) module $V(d\chi_o)$ of the affine Kac-Moody group $\hat{\mathcal{G}}$ (or of the associated affine Kac-Moody algebra, which is a certain one dimensional central extension of the loop algebra (Lie $G) \otimes \hat{k}_p$) with highest weight $d\chi_o$ (cf. Sect. 6.1).

⋆ A. Ramanathan passed away on March 12, 1993

Using the fact that any principal $G$-bundle on $C\backslash p$ is trivial (Proposition 1.3), one sees easily that the set of isomorphism classes of principal $G$-bundles on $C$ is in bijective correspondence with the double coset space $\Gamma\backslash\mathscr{G}/\mathscr{P}$, where $\Gamma := \text{Mor}(C\backslash p, G)$ is the subgroup of $\mathscr{G}$ consisting of all the algebraic morphisms of $C\backslash p \to G$. Moreover $X$ parametrizes an algebraic family $\mathscr{U}$ of principal $G$-bundles on $C$ (cf. Proposition 2.8). As an interesting byproduct of this parametrization, we obtain that the moduli space $\mathfrak{M}$ of semistable principal $G$-bundles on $C$ is a unirational variety (cf. Corollary 6.3). Now, given a finite dimensional representation $V$ of $G$, let $\mathscr{U}(V)$ be the family of associated vector bundles on $C$ parametrized by $X$. We have then the determinant line bundle $\text{Det}(\mathscr{U}(V))$ on $X$, defined as the dual of the determinant of the cohomology of the family $\mathscr{U}(V)$ of vector bundles on $C$ (cf. Sect. 3.8). As we mentioned above, $\text{Pic}(X)$ is freely generated by the homogeneous line bundle $\mathfrak{L}(\chi_o)$ on $X$, in particular, there exists a unique integer $m_V$ (depending on the choice of the representation $V$) such that $\text{Det}(\mathscr{U}(V)) \simeq \mathfrak{L}(m_V\chi_o)$. We determine this number explicitly in Theorem 5.4, the proof of which makes use of Riemann-Roch theorem. It may be mentioned that the number $m_V$ is given explicitly in terms of the decomposition of $V$ under $sl(2)$ "passing through the highest root space" (cf. Sect. 5.1), and coincides with the Dynkin index of the representation $V$. For example, if we take $V$ to be the adjoint representation of $G$, then $m_V = 2\times$ dual Coxeter number of $G$ (cf. Lemma 5.2 and Remark 5.3). The number $m_V$ is also expressed in terms of the induced map at the third homotopy group level $\pi_3(G) \to \pi_3(\text{SL}(V))$ (cf. Corollary 5.6).

The subgroup $\Gamma \subset \mathscr{G}$ can canonically be thought of as a subgroup of $\hat{\mathscr{G}}$ (cf. Lemma 2.7). Suggested by conformal field theory, we consider the space $H^0(\mathscr{G}/\mathscr{P}, \mathfrak{L}(d\chi_o))^\Gamma$ of $\Gamma$-invariant regular sections of the $\hat{\mathscr{G}}$-equivariant (in particular $\Gamma$-equivariant) line bundle $\mathfrak{L}(d\chi_o)$ (for any $d \geqq 0$). This space of invariants is called the space of vacua. More precisely, in conformal field theory, the space of vacua is defined to be the space of invariants of the Lie algebra $\mathfrak{g} \otimes R$ in $V(d\chi_o)^*$, where $R$ is the ring of regular functions on the affine curve $C\backslash p$ and $\mathfrak{g}$ is the Lie algebra of the group $G$. We have (by Proposition 6.7) $[V(d\chi_o)^*]^\Gamma = [V(d\chi_o)^*]^{\mathfrak{g}\otimes R}$ and, as already mentioned above, $H^0(\mathscr{G}/\mathscr{P}, \mathfrak{L}(d\chi_o)) \simeq V(d\chi_o)^*$. Thus, by Theorem 6.6, we see that (for any $d \geqq 0$) the space $H^0(\mathfrak{M}, \Theta(V)^{\otimes d})$ of the regular sections of the $d$-th power of the $\Theta$-bundle $\Theta(V)$ (cf. Sect. 3.8) on the moduli space $\mathfrak{M}$ is isomorphic with the space of vacua $[V(dm_V\chi_o)^*]^{g\otimes R}$. This is the connection, alluded to in the beginning of the introduction, between the space of vacua and the space of generalised theta functions. (In the case $G = \text{SL}(n, \mathbb{C})$, this result has also independently been obtained recently by A. Beauville and Y. Laszlo by different methods.)

The proof of our Theorem 6.6 uses geometric invariant theory; in particular, we make crucial use of the following extension lemma (cf. Proposition 7.2):

Let $H$ be a reductive group and $Q$ be a projective scheme with a $H$-linearised ample line bundle $\mathfrak{L}$ on $Q$, and let $Q^s$ denote the (open) subset of semistable points of $Q$. Then, for any irreducible normal open $H$-invariant subscheme

$U \supseteqq Q^s$ of $Q$, the canonical restriction map $H^0(U, \mathfrak{L}^N)^H \to H^0(Q^s, \mathfrak{L}^N)^H$ is an isomorphism, for any $N \geqq 1$.

We also make crucial use of a "descent" lemma (cf. Proposition 4.1), in the proof of Theorem 6.6.

Our Theorem 6.6 can be generalised to the situation where the curve $C$ has $n$ marked points $\{p_1, \ldots, p_n\}$ together with finite dimensional $G$-modules $\{V_1, \ldots, V_n\}$ attached to them respectively, by bringing in moduli space of parabolic $G$-bundles on $C$.

It should be mentioned that Tsuchiya–Ueno–Yamada [TUY] have obtained a factorization theorem for the space of vacua, from which one gets the validity of the Verlinde's formula for the dimension of the space of vacua. In view of our identification of the space of generalised theta functions with the space of vacua, one gets the same formula for the dimension of the space of generalised theta functions (for general $G$). Recently G. Faltings has also announced a proof of the Verlinde's formula. A purely algebro-geometric study (which does not use loop groups) of generalised theta functions on the moduli space of (parabolic) rank two torsion-free sheaves on a nodal curve is made by Narasimhan and Ramadas [NRa]. A factorization theorem and a vanishing theorem for the theta line bundle are proved there. In addition, several other mathematicians (A. Bertram, S. Bradlow, S. Chang, G. Daskalopoulos, B. van Geemen, E. Previato, A. Szenes, M. Thaddeus, R. Wentworth, D. Zagier, ...) and physicists have studied the space of generalized theta functions (from different view points) in the case when $G = \mathrm{SL}(2)$, in the last few years.

The organization of the paper is as follows:

Apart from introducing some notation in Sect. 1, we realize the affine flag variety $X$ as a parameter set for $G$-bundles. Section 2 is devoted to recalling some basic facts (we need) about the affine Kac-Moody groups and their flag varieties. In this section we prove that the affine flag variety is the parameter space for an algebraic family of $G$-bundles on the curve $C$ (cf. Proposition 2.8). Section 3 is devoted to recalling some basic definitions and results on the moduli space of semistable $G$-bundles, including the definitions of the determinant line bundle and the $\Theta$-bundle on the moduli space. We prove a result (cf. Proposition 4.1) on algebraic descent in Sect. 4. Section 5 is devoted to identifying the determinant line bundle on the affine flag variety with a suitable power of the basic homogeneous line bundle. Section 6 contains the statement and the proof of the main result (Theorem 6.6). Finally in Sect. 7 we prove the basic extension result (Proposition 6.5), using Geometric Invariant Theory.

## 1 Affine flag variety as parameter set for $G$-bundles

(1.1) **Notation.** Throughout the paper $k$ denotes an algebraically closed field of char. 0. By a scheme we will mean a scheme over $k$. Let us fix a projective curve $C$ over $k$, and a smooth point $p \in C$. Let $C^*$ denote the open set $C \backslash p$. We also fix an affine algebraic connected reductive group $G$ over $k$.

For any $k$-algebra $A$, by $G(A)$ we mean the $A$-rational points of the algebraic group $G$. We fix the following notation to be used throughout the paper:

$$\mathscr{G} = \mathscr{G}_{\mathfrak{T}} = G(\hat{k}_p)\,,$$
$$\mathscr{P} = \mathscr{P}_{\mathfrak{T}} = G(\hat{\mathscr{O}}_p)\,, \quad \text{and}$$
$$\Gamma = \Gamma_{\mathfrak{T}} = G(k[C^*])\,,$$

where $\hat{\mathscr{O}}_p$ is the completion of the local ring $\mathscr{O}_p$ of $C$ at $p$, $\hat{k}_p$ is the quotient field of $\hat{\mathscr{O}}_p$, $k[C^*]$ is the ring of regular functions on the affine curve $C^*$ (which can canonically be viewed as a subring of $\hat{k}_p$), and $\mathfrak{T}$ is the triple $(G, C, p)$.

We recall the following

(1.2) **Definition.** *Let $G$ be any (not necessarily reductive) affine algebraic group over $k$. By a* **principal $G$-bundle** *(for short* **$G$-bundle***) on an algebraic variety $X$, we mean an algebraic variety $E$ on which $G$ acts algebraically from the right and a $G$-equivariant morphism $\pi: E \to X$ (where $G$ acts trivially on $X$), such that $\pi$ is locally isotrivial (i.e. locally trivial in the étale topology).*

Let $G$ act algebraically on a quasi-projective variety $F$ from the left. We can then form the **associated bundle with fiber** $F$, denoted by $E(F)$. Recall that $E(F)$ is the quotient of $E \times F$ under the $G$-action given by $g(e, f) = (eg^{-1}, gf)$, for $g \in G$, $e \in E$ and $f \in F$.

**Reduction of structure group of $E$ to a closed algebraic subgroup $H \subset G$** is, by definition, an $H$-bundle $E_H$ such that $E_H(G) \approx E$, where $H$ acts on $G$ by left multiplication. Reduction of structure group to $H$ can canonically be thought of as a section of the associated bundle $E(G/H) \to X$.

Let $\mathscr{X} = \mathscr{X}(G, C)$ denote the set of isomorphism classes of $G$-bundles on the base $C$, and $\mathscr{X}_0 = \mathscr{X}_0(\mathfrak{T}) \subset \mathscr{X}$ denote the subset consisting of those $G$-bundles on $C$ which are algebraically trivial restricted to $C^*$.

(1.3) **Proposition.** *Let $G$ be a connected reductive algebraic group over $k$. Then the structure group of a $G$-bundle on a smooth affine curve $Y$ can be reduced to the connected component $Z^0(G)$ of the centre $Z(G)$ of $G$.*

*In particular, if $G$ as above is semi-simple, then any $G$-bundle on $Y$ is trivial.*

*Proof.* This proposition is essentially proved in Harder's paper [H1, Satz 3.3 and the remarks following it], (as pointed out by the referee). We also need to use the following facts:

(a) An affine group scheme over $Y$ is rationally quasi-trivial (a result due to Springer and Steinberg, cf. [Se2, Chap. III, Sect. 2.3]).

(b) $\operatorname{Pic} Y$ is a divisible group. □

The following map is of basic importance for us in this paper. This provides a bridge between the moduli space of $G$-bundles and the affine (Kac-Moody) flag variety.

(1.4) **Definition** (of the map $\varphi: \mathscr{G} \to \mathscr{X}_0$). *Let $G$ be a connected reductive algebraic group over $k$. Consider the canonical morphisms $i_1$:* $\operatorname{Spec}(\hat{\mathscr{O}}_p) \to C$

*and* $i_2 : C^* \hookrightarrow C$. *The morphisms* $i_1$ *and* $i_2$ *together provide a flat cover of* $C$. *Let us take the trivial* $G$*-bundles on both the schemes* $\mathrm{Spec}(\hat{\mathscr{O}}_p)$ *and* $C^*$. *The fiber product*

$$F := \mathrm{Spec}(\hat{\mathscr{O}}_p) \times_C C^*$$

*of* $i_1$ *and* $i_2$ *can canonically be identified with* $\mathrm{Spec}(\hat{k}_p)$. *This identification* $\mathrm{Spec}(\hat{k}_p) \simeq F$ *is induced from the natural morphisms*

$$\begin{array}{ccccc} & & \mathrm{Spec}(\hat{k}_p) & & \\ & \swarrow & & \searrow & \\ \mathrm{Spec}(\hat{\mathscr{O}}_p) & & \downarrow \wr & & C^* . \\ & \nwarrow & & \nearrow & \\ & & F & & \end{array}$$

By a "glueing" lemma of Grothendieck [Mi, Part I, Theorem 2.23, p. 19], to give a $G$-bundle on $C$, it suffices to give an automorphism of the trivial $G$-bundle on $\mathrm{Spec}(\hat{k}_p)$, i.e., to give an element of $\mathscr{G} := G(\hat{k}_p)$. (Observe that since we have a flat cover of $C$ by only two schemes, the cocycle condition is vacuously satisfied.) This is, by definition, the map $\varphi : \mathscr{G} \to \mathscr{X}_0$.

(1.5) **Proposition.** *The map* $\varphi$ *(defined above) factors through the double coset space to give a bijective map (denoted by)*

$$\bar{\varphi} : \Gamma \backslash \mathscr{G} / \mathscr{P} \to \mathscr{X}_0 .$$

(Observe that, by Proposition 1.3, $\mathscr{X}_0 = \mathscr{X}$ if $G$ is assumed to be connected and semi-simple and $C$ is smooth and irreducible.)

*Proof.* From the above construction, it is clear that for $g, g' \in \mathscr{G}$, $\varphi(g)$ is isomorphic with $\varphi(g')$ (written $\varphi(g) \approx \varphi(g')$) if and only if there exist two $G$-bundle isomorphisms:

$$\begin{array}{ccc} \mathrm{Spec}(\hat{\mathscr{O}}_p) \times G & \xrightarrow[\sim]{\theta_1} & \mathrm{Spec}(\hat{\mathscr{O}}_p) \times G \\ & \searrow \quad \swarrow & \\ & \mathrm{Spec}(\hat{\mathscr{O}}_p) & \end{array}$$

and

$$\begin{array}{ccc} C^* \times G & \xrightarrow[\sim]{\theta_2} & C^* \times G \\ & \searrow \quad \swarrow & \\ & C^* & \end{array}$$

such that the following diagram is commutative:

$$(*)\qquad \begin{array}{ccc} \operatorname{Spec}(\hat{k}_p) \times G & \xrightarrow{\theta_1|_{\operatorname{Spec}(\hat{k}_p)}} & \operatorname{Spec}(\hat{k}_p) \times G \\ \downarrow g' & & \downarrow g \\ \operatorname{Spec}(\hat{k}_p) \times G & \xrightarrow{\theta_2|_{\operatorname{Spec}(\hat{k}_p)}} & \operatorname{Spec}(\hat{k}_p) \times G\,. \end{array}$$

Any $G$-bundle isomorphism $\theta_1$ (resp. $\theta_2$) as above is given by an element $h \in \mathscr{P}$ (resp. $\gamma \in \Gamma$). In particular, from the commutativity of the above diagram $(*)$, $\varphi(g) \approx \varphi(g')$ if and only if there exists $h \in \mathscr{P}$ and $\gamma \in \Gamma$ such that $gh = \gamma g'$, i.e., $\gamma^{-1}gh = g'$. This shows that the map $\varphi$ factors through $\Gamma\backslash\mathscr{G}/\mathscr{P}$ to give an injective map $\bar{\varphi}$. The surjectivity of $\bar{\varphi}$ follows immediately from the definition of $\mathscr{X}_0$, and the fact that any $G$-bundle on $\operatorname{Spec}(\hat{\mathscr{O}}_p)$ is trivial. □

(1.6) *Remarks.* (a) We will show (cf. Proposition 2.8) that $\mathscr{G}/\mathscr{P}$ in fact is a parameter space for an algebraic family of $G$-bundles.

(b) The correspondence given in the above proposition is parallel to the correspondence from the Adele group to bundles on a curve (cf. [H1, H2], also see [PS, Sect. 8.11]). Some other analogous constructions are given by Beilinson-Schechtman, Mulase [Mu], ....

(c) $\mathscr{G}/\mathscr{P}$ should be thought of as a the parameter space for $G$-bundles $E$ on $C$ together with a trivialization of $E_{|C^*}$ (cf. Proposition 2.8).

(1.7) *An alternative description of the map $\varphi$ for vector bundles.* We give an alternative description of the map $\varphi$ in the case when $G = \mathrm{GL}_n$. In this case $\mathscr{X}_0$ can also be thought of as the set of isomorphism classes of locally free $\mathscr{O}_C$-modules of rank $n$ (where $\mathscr{O}_C$ is the structure sheaf of $C$) which are free as $\mathscr{O}_{C^*}$-modules.

Let us denote by $E = E_n$ the $n$-dimensional standard representation of $\mathrm{GL}_n$. Then the group $\mathscr{G}$ has a canonical representation in $E(\hat{k}_p)$ and $\mathscr{P}$ is precisely the stabilizer of $E(\hat{\mathscr{O}}_p)$. Let $\mathfrak{E} := C \times E(k) \to C$ be the trivial rank-$n$ vector bundle over $C$. Fix any $g \in \mathscr{G}$, and define the presheaf $\tilde{\varphi}(g)$ of $\mathscr{O}_C$-modules on $C$ as follows: For any Zariski open $U \subset C$, set

$$\tilde{\varphi}(g)(U) = H^0(U, \mathfrak{E}), \quad \text{if} \quad p \notin U \quad \text{and}$$
$$\tilde{\varphi}(g)(U) = \{\sigma \in H^0(U\backslash p, \mathfrak{E}) : (\sigma)_p \in g(E(\hat{\mathscr{O}}_p))\}, \quad \text{if} \quad p \in U,$$

where $(\sigma)_p$ denotes the germ of the rational section $\sigma$ at $p$ viewed canonically as an element of $E(\hat{k}_p)$.

Now let $\varphi(g)$ be the associated sheaf of $\mathscr{O}_C$-modules on $C$. Since the representation of $\mathscr{G}$ in $E(\hat{k}_p)$ is $\hat{k}_p$-linear (in particular $\hat{\mathscr{O}}_p$-linear), it is easy to see that the sheaf $\varphi(g)$ is a locally free sheaf of $\mathscr{O}_C$-modules of rank $n$ and of course (by construction) $\varphi(g)_{|C^*}$ is trivial. It can be easily seen that the map $\varphi : \mathscr{G} \to \mathscr{X}_0$ thus obtained is the same as the map $\varphi$ defined in Sect. 1.4.

## 2 Affine Kac-Moody groups and their flag varieties

Let $\mathfrak{T} = (G, C, p)$ be as in Sect. 1.1. *In this section we will assume that the base field $k$ is $\mathbb{C}$ and further assume that $G$ is a connected simply-connected simple affine algebraic group over* $\mathbb{C}$. We fix a Borel subgroup $B \subset G$ and a maximal torus $T \subset B$, and define the **standard Borel subgroup** $\mathscr{B}$ of $\mathscr{G}$ as $ev_p^{-1}(B)$, where $ev_p : \mathscr{P} = G(\hat{\mathscr{O}}_p) \to G$ is the group homomorphism induced from the $\mathbb{C}$-algebra homomorphism $\hat{\mathscr{O}}_p \to \mathbb{C}$, which takes $f \mapsto f(p)$.

(2.1) *Generalized Schubert varieties.* The generalised flag variety $X := \mathscr{G}/\mathscr{P}$ (where $\mathscr{G}, \mathscr{P}$ are as in Sect. 1.1) has the following *Bruhat decomposition*:

$$(1) \qquad X = \bigcup_{\mathfrak{w} \in \tilde{W}/W} \mathscr{B}\mathfrak{w}\mathscr{P}/\mathscr{P},$$

where $W := N_G(T)/T$ is the (finite) Weyl group of $G$, $N_G(T)$ is the normalizer of $T$ in $G$, and $\tilde{W}$ is the affine Weyl group of $G$ (cf. [K, Sect. 6.6]). Moreover the union in (1) is disjoint.

The affine Weyl group $\tilde{W}$ is a Coxeter group and hence has a Bruhat partial order $\leqq$. This induces a partial order (again denoted by) $\leqq$ in $\tilde{W}/W$ defined by

$$\mathfrak{u} := u \bmod W \leqq \mathfrak{v} \quad (\text{for } u, v \in \tilde{W})$$

if and only if there exists a $w \in W$ such that

$$u \leqq vw .$$

We define the **generalized Schubert variety** $X_{\mathfrak{w}}$ (for any $\mathfrak{w} \in \tilde{W}/W$) by

$$(2) \qquad X_{\mathfrak{w}} = \bigcup_{\mathfrak{v} \leqq \mathfrak{w}} \mathscr{B}\mathfrak{v}\mathscr{P}/\mathscr{P}.$$

Then clearly $X_{\mathfrak{v}} \subseteqq X_{\mathfrak{w}}$ if and only if $\mathfrak{v} \leqq \mathfrak{w}$. The set $X_{\mathfrak{w}}$ has the structure of a (not necessarily smooth) finite dimensional projective variety over $\mathbb{C}$. Moreover, the inclusion $X_{\mathfrak{v}} \subseteqq X_{\mathfrak{w}}$ (for $\mathfrak{v} \leqq \mathfrak{w}$) is a closed immersion.

We put the inductive limit Hausdorff (resp. Zariski) topology on $\mathscr{G}/\mathscr{P}$, i.e., a set $U \subset \mathscr{G}/\mathscr{P}$ is open if and only if $U \cap X_{\mathfrak{w}}$ is open in $X_{\mathfrak{w}}$ in the Hausdorff (resp. Zariski) topology for all $\mathfrak{w} \in \tilde{W}/W$. The decomposition (1) provides a cellular decomposition of $\mathscr{G}/\mathscr{P}$, where $\mathscr{B}\mathfrak{w}\mathscr{P}/\mathscr{P}$ is biregular isomorphic with $\mathbb{C}^{\ell(\mathfrak{w})}$ and $\ell(\mathfrak{w})$ is the length of the smallest element in the coset $\mathfrak{w} := wW$.

(2.2) *Line bundles on* $\mathscr{G}/\mathscr{P}$. We define

$$(1) \qquad \operatorname{Pic}(\mathscr{G}/\mathscr{P}) = \varprojlim_{\mathfrak{w} \in \tilde{W}/W} \operatorname{Pic}(X_{\mathfrak{w}}),$$

where $\operatorname{Pic}(X_{\mathfrak{w}})$ is of course the set of isomorphism classes of (algebraic) line bundles on $X_{\mathfrak{w}}$. Clearly an element $\mathscr{L} \in \operatorname{Pic}(\mathscr{G}/\mathscr{P})$ is given by a collection of algebraic line bundles $\mathscr{L}_{\mathfrak{w}}$ on $X_{\mathfrak{w}}$ (for every $\mathfrak{w} \in \tilde{W}/W$) together with morphisms $i_{\mathfrak{w},\mathfrak{v}}$ (for all $\mathfrak{v} \leqq \mathfrak{w}$)

$$\begin{array}{ccc} \mathscr{L}_{\mathfrak{v}} & \overset{i_{\mathfrak{w},\mathfrak{v}}}{\hookrightarrow} & \mathscr{L}_{\mathfrak{w}} \\ \downarrow & & \downarrow \\ X_{\mathfrak{v}} & \hookrightarrow & X_{\mathfrak{w}}\,, \end{array}$$

satisfying $i_{\mathfrak{w},\mathfrak{v}} \circ i_{\mathfrak{v},\mathfrak{u}} = i_{\mathfrak{w},\mathfrak{u}}$, for all $\mathfrak{u} \leqq \mathfrak{v} \leqq \mathfrak{w}$.

One can similarly define the notion of vector bundles or principal bundles on $\mathscr{G}/\mathscr{P}$.

Let us recall that the group $\mathscr{G}$ admits a "canonical" one-dimensional central extension:

(2) $$1 \to \mathbb{C}^* \underset{i}{\to} \tilde{\mathscr{G}} \underset{\beta}{\to} \mathscr{G} \to 1\,.$$

The "Lie algebra" Lie($\tilde{\mathscr{G}}$) of $\tilde{\mathscr{G}}$ is described explicitly in [K, Chap. 7, Identity 7.2.1] and is denoted by $\tilde{L}(\mathfrak{g})$.

The composite map $\mathbb{C}^* \underset{i}{\to} \tilde{\mathscr{P}} \underset{q}{\to} \tilde{\mathscr{P}}/[\tilde{\mathscr{P}},\tilde{\mathscr{P}}]$ is an isomorphism, where $\tilde{\mathscr{P}} := \beta^{-1}(\mathscr{P})$ and $q$ is the canonical projection. In particular, identifying $\tilde{\mathscr{P}}/[\tilde{\mathscr{P}},\tilde{\mathscr{P}}]$ with $\mathbb{C}^*$ (under $q \circ i$), we get the character denoted $e^{\chi_o}: \tilde{\mathscr{P}} \to \mathbb{C}^*$. Alternatively, this is the unique character which is identically 1 restricted to the commutator $[\tilde{\mathscr{P}},\tilde{\mathscr{P}}]$, and restricted to the *standard maximal torus* $\tilde{T} := \beta^{-1}(T)$ it is got by exponentiating the "integral" weight $\chi_o$: Lie$(\tilde{T}) \to \mathbb{C}$, where $\chi_o$ is defined by

(3) $$\chi_o(\alpha_0^\vee) = 1\,, \quad \text{and} \qquad \chi_o(\alpha_i^\vee) = 0\,, \quad \text{for all} \quad 1 \leqq i \leqq \ell\,,$$

where $\{\alpha_0^\vee, \alpha_1^\vee, \ldots, \alpha_\ell^\vee\}$ (resp. $\{\alpha_1^\vee, \ldots, \alpha_\ell^\vee\}$) are the simple coroots for $\tilde{L}(\mathfrak{g})$ (resp. $\mathfrak{g} := \operatorname{Lie} G$) (cf. [K, Sect. 7.4]).

For any $d \in \mathbb{Z}$, let $\mathscr{L}(d\chi_o)$ be the homogeneous line bundle on the base $\tilde{\mathscr{G}}/\tilde{\mathscr{P}} \approx \mathscr{G}/\mathscr{P}$, which is associated to the principal $\tilde{\mathscr{P}}$-bundle $\tilde{\mathscr{G}} \to \tilde{\mathscr{G}}/\tilde{\mathscr{P}}$ by the character $(e^{\chi_o})^{-d}$. We denote its restriction to $X_{\mathfrak{w}}$ by $\mathscr{L}_{\mathfrak{w}}(d\chi_o)$. Then $\mathscr{L}_{\mathfrak{w}}(d\chi_o)$ has a canonical structure of an algebraic line bundle, which is compatible with respect to the inclusions, i.e., $\mathscr{L}_{\mathfrak{w}}(d\chi_o)_{|X_{\mathfrak{v}}} = \mathscr{L}_{\mathfrak{v}}(d\chi_o)$ for any $\mathfrak{v} \leqq \mathfrak{w}$ (cf. [S1, Sect. 2.7]). In particular, we get an element (again denoted by) $\mathscr{L}(d\chi_o) \in$ Pic$(\mathscr{G}/\mathscr{P})$.

We have the following proposition determining Pic$(\mathscr{G}/\mathscr{P})$.

(2.3) **Proposition.** *The map* $\mathbb{Z} \to$ Pic$(\mathscr{G}/\mathscr{P})$ *given by*

$$d \mapsto \mathscr{L}(d\chi_o)$$

*is an isomorphism.*

*Proof.* Since $X_{\mathfrak{w}}$ is a projective variety, by GAGA, the natural map

(1) $$\text{Pic}(X_{\mathfrak{w}}) \overset{\sim}{\to} \text{Pic}_{\text{an}}(X_{\mathfrak{w}})$$

is an isomorphism, where $\mathrm{Pic}_{\mathrm{an}}(X_{\mathfrak{w}})$ is the set of isomorphism classes of analytic line bundles on $X_{\mathfrak{w}}$.

We have the sheaf exact sequence:

(2) $$0 \to \mathbb{Z} \to \mathcal{O}_{\mathrm{an}} \to \mathcal{O}^*_{\mathrm{an}} \to 0,$$

where $\mathcal{O}_{\mathrm{an}}$ (resp. $\mathcal{O}^*_{\mathrm{an}}$) denotes the sheaf of analytic functions (resp. the sheaf of invertible analytic functions) on $X_{\mathfrak{w}}$. Taking the associated long exact cohomology sequence, we get

(3) $$\dots \to H^1(X_{\mathfrak{w}}, \mathcal{O}_{\mathrm{an}}) \to H^1(X_{\mathfrak{w}}, \mathcal{O}^*_{\mathrm{an}}) \xrightarrow{c_1} H^2(X_{\mathfrak{w}}, \mathbb{Z}) \to H^2(X_{\mathfrak{w}}, \mathcal{O}_{\mathrm{an}}) \to \dots,$$

where the map $c_1$ associates to any line bundle its first Chern class. Now

(4) $$H^i(X_{\mathfrak{w}}, \mathcal{O}) = 0, \quad \text{for all} \quad i > 0,$$

by [Ku, Theorem 2.16(3)] (also proved in [M]), and by GAGA

(5) $$H^i(X_{\mathfrak{w}}, \mathcal{O}) \approx H^i(X_{\mathfrak{w}}, \mathcal{O}_{\mathrm{an}}),$$

and hence the map $c_1$ is an isomorphism. But

(6) $$\mathrm{Pic}_{\mathrm{an}}(X_{\mathfrak{w}}) \approx H^1(X_{\mathfrak{w}}, \mathcal{O}^*_{\mathrm{an}}).$$

Hence, by combining (1) and (3)–(6), we get the isomorphism (again denoted by)

(7) $$c_1 : \mathrm{Pic}(X_{\mathfrak{w}}) \xrightarrow{\approx} H^2(X_{\mathfrak{w}}, \mathbb{Z}).$$

Further the following diagram is commutative (whenever $X_{\mathfrak{v}} \subseteq X_{\mathfrak{w}}$):

($\mathcal{D}$) $$\begin{array}{ccc} \mathrm{Pic}(X_{\mathfrak{w}}) & \xrightarrow[\sim]{c_1} & H^2(X_{\mathfrak{w}}, \mathbb{Z}) \\ \downarrow & & \downarrow \\ \mathrm{Pic}(X_{\mathfrak{v}}) & \xrightarrow[\sim]{c_1} & H^2(X_{\mathfrak{v}}, \mathbb{Z}), \end{array}$$

where the vertical maps are the canonical restriction maps. But from the Bruhat decomposition (1) of Sect. 2.1, for any $\mathfrak{w} \geqq \mathfrak{s}_o$, the restriction map

(8) $$H^2(X_{\mathfrak{w}}, \mathbb{Z}) \to H^2(X_{\mathfrak{s}_o}, \mathbb{Z})$$

is an isomorphism, where $\mathfrak{s}_o$ is the (simple) reflection corresponding to the simple coroot $\alpha_0^\vee$, and $\mathfrak{s}_o := s_o \bmod W$. Moreover, $X_{\mathfrak{s}_o}$ being isomorphic with the complex projective line $\mathbb{P}^1$, $H^2(X_{\mathfrak{s}_o}, \mathbb{Z})$ is a free $\mathbb{Z}$-module of rank 1, which is generated by the first Chern class $-1$ of the line bundle $\mathcal{L}_{\mathfrak{s}_o}(\chi_o)$.

Since any element $\mathfrak{w} \neq \mathfrak{e} \in \tilde{W}/W$ satisfies $\mathfrak{w} \geqq \mathfrak{s}_o$ (in particular the elements $\mathfrak{w} \geqq \mathfrak{s}_o$ are cofinal in $\tilde{W}/W$), taking the inverse limit of diagram ($\mathcal{D}$), we get the proposition. □

(2.4) *Topology on $\Gamma$.* We fix an embedding $G \hookrightarrow \mathrm{GL}_m(\mathbb{C})$ (for some large $m$), and define a filtration of $\Gamma$ as follows:

$$G = \Gamma_0 \subset \Gamma_1 \subset \dots ,$$

where $\Gamma_i := \{f: C^* \to G \subset \mathrm{GL}_m(\mathbb{C})$ such that all the matrix coefficients of $f$ have poles of order $\leqq i$ at $p\}$.

It is easy to see that $\Gamma_i$'s admit canonically a compatible structure of finite dimensional affine varieties. In particular, we have Hausdorff as well as Zariski topology on $\Gamma_i$'s. Now we define the corresponding (Hausdorff or Zariski) topology on $\Gamma$ as the inductive limit topology from $\Gamma_i$'s. It is easy to see that neither topology on $\Gamma$ depends upon the particular embedding of $G \hookrightarrow \mathrm{GL}_m(\mathbb{C})$.

We prove the following lemma.

(2.5) **Lemma.** *Let $X$ be a connected variety over $\mathbb{C}$. Then any regular map $X \to \mathbb{C}^*$, which is null-homotopic in the topological category, is a constant.*

(Observe that if the singular cohomology $H^1(X, \mathbb{Z}) = 0$, then any continuous map $X \to \mathbb{C}^*$ is null-homotopic.)

*Proof.* Assume, if possible, that there exists a null-homotopic non-constant regular map $\lambda: X \to \mathbb{C}^*$. Since $\lambda$ is algebraic, there exists a number $N > 0$ such that the number of irreducible components of $\lambda^{-1}(z) \leqq N$, for all $z \in \mathbb{C}^*$. Now we consider the $N'$-sheeted covering $\pi_{N'}: \mathbb{C}^* \to \mathbb{C}^*(z \mapsto z^{N'})$, for any $N' > N$. Since $\lambda$ is null-homotopic, there exists a (regular) lift $\tilde{\lambda}: X \to \mathbb{C}^*$, making the following diagram commutative:

$$\begin{array}{ccc} & & \mathbb{C}^* \\ & {}^{\tilde{\lambda}}\nearrow & \downarrow \pi_{N'} \\ X & \underset{\lambda}{\longrightarrow} & \mathbb{C}^* . \end{array}$$

Since $\tilde{\lambda}$ is regular and non-constant, by Chevalley's theorem, $\operatorname{Im} \tilde{\lambda}$ (being a constructible set) misses only finitely many points of $\mathbb{C}^*$. In particular, there exists a $z_o \in \mathbb{C}^*$ (in fact a Zariski open set of points) such that $\pi_{N'}^{-1}(z_o) \subset \operatorname{Im} \tilde{\lambda}$. But then the number of irreducible components of $\lambda^{-1}(z_o) = \tilde{\lambda}^{-1}(\pi_{N'}^{-1} z_o) \geqq N' > N$, a contradiction to the choice of $N$. This proves the lemma. □

(2.6) **Corollary.** *There does not exist any non-constant regular map $\lambda: \Gamma \to \mathbb{C}^*$.*

(A regular map $\lambda: \Gamma \to \mathbb{C}^*$ is, by definition, a map such that $\lambda|_{\Gamma_n}$ is regular for each $n$, cf. Sect. 2.4.)

*Proof.* By Segal [S] (see also [PS, Proposition 8.11.6(i), p. 157]), $\Gamma$ is connected and simply-connected, in particular, $H^1(\Gamma, \mathbb{Z}) = 0$ (where $H^*(\Gamma, \mathbb{Z})$ denotes the singular cohomology of the topological space $\Gamma$). This gives that the map $\lambda$ is null-homotopic. By using the above Lemma 2.5, $\lambda$ is constant on each connected component of $\Gamma_n$ (for any $n \geqq 0$) and hence $\lambda$ itself is constant. □

Restrict the central extension (2) of Sect. 2.2 to get a central extension

(1) $$1 \to \mathbb{C}^* \underset{i}{\to} \tilde{\Gamma} \underset{\beta}{\to} \Gamma \to 1 ,$$

where $\tilde{\Gamma}$ is by definition $\beta^{-1}(\Gamma)$. The group $\tilde{\Gamma}$ admits a canonical structure of an inductive limit of affine algebraic varieties.

(2.7) **Lemma.** *There exists a unique regular group homomorphism $\Gamma \to \tilde{\Gamma}$, which splits the above central extension.*

*In particular, we can canonically view $\Gamma$ as a subgroup of $\tilde{\mathscr{G}}$.*

*Proof.* The existence of a regular splitting on $\Gamma$ is well known (cf., e.g., [W, Sect. 4]). The uniqueness follows immediately from the above corollary. □

We have the following proposition.

(2.8) **Proposition.** (a) *There is an algebraic G-bundle $\mathscr{U} \to C \times \mathscr{G}/\mathscr{P}$ (i.e. $\mathscr{U}_{|C\times X_{\mathfrak{w}}}$ is algebraic for any $\mathfrak{w} \in \tilde{W}/W$) such that, for any $x \in \mathscr{G}/\mathscr{P}$ the G-bundle $\mathscr{U}_x := \mathscr{U}_{|C\times x}$ is isomorphic with $\varphi(x)$ (where $\varphi$ is the map of Sect. 1.4). Moreover the bundle $\mathscr{U}_{|C^*\times\mathscr{G}/\mathscr{P}}$ comes equipped with a trivialization $\alpha:\varepsilon \tilde{\to} \mathscr{U}_{|C^*\times\mathscr{G}/\mathscr{P}}$, where $\varepsilon$ is the trivial G-bundle on $C^* \times \mathscr{G}/\mathscr{P}$.*

(b) *Let $\mathscr{E} \to C \times T$ be an algebraic family of G-bundles (parametrized by an algebraic variety T), such that $\mathscr{E}$ is trivial over $C^* \times T$ and also over* (Spec $\hat{\mathscr{O}}_p$) $\times T$. *Then, if we choose a trivialization $\beta:\varepsilon' \tilde{\to} \mathscr{E}_{|C^*\times T}$, we get a Schubert variety $X_{\mathfrak{w}}$ and a unique morphism $f:T \to X_{\mathfrak{w}}$ together with a G-bundle morphism $\hat{f}:\mathscr{E} \to \mathscr{U}_{|C\times X_{\mathfrak{w}}}$ inducing the map* Id $\times f$ *at the base such that $\hat{f} \circ \beta = \alpha \circ \theta$, where $\varepsilon'$ is the trivial bundle on $C^* \times T$ and $\theta$ is the canonical G-bundle morphism $\varepsilon' \to \varepsilon$ inducing the map* Id $\times f$ *at the base.*

*Proof.* Let $R$ be a $\mathbb{C}$-algebra and let $T :=$ Spec $R$ be the corresponding scheme. Suppose $E \to C\times T$ is a $G$-bundle with trivializations $\alpha$ of $E$ over $C^* \times T$ and $\beta$ of $E$ over (Spec $\hat{\mathscr{O}}_p$)$\times T$. Note that the fiber product $(C^* \times T)\times_{C\times T}$(Spec $\hat{\mathscr{O}}_p \times T$) is canonically isomorphic with (Spec $\hat{k}_p$) $\times T$ (cf. Sect. 1.4). Therefore the trivializations $\alpha$ and $\beta$ give rise to an element $\alpha\beta^{-1} \in G(\hat{k}_p \otimes_{\mathbb{C}} R)$. Conversely, given an element $g \in G(\hat{k}_p \otimes R)$, we can construct the family $E \to C \times$ Spec $R$ by taking the trivial bundles on $C^* \times T$ and (Spec $\hat{\mathscr{O}}_p$) $\times T$ and glueing them via the element $g$. Moreover, if $g_1$ and $g_2$ are two elements of $G(\hat{k}_p \otimes R)$ such that $g_2 = g_1 h$ with $h \in G(\hat{\mathscr{O}}_p \otimes R)$, then $h$ induces a canonical isomorphism of the bundles corresponding to $g_1$ and $g_2$. All these assertions are easily verified.

To construct the family parametrized by $\mathscr{G}/\mathscr{P}$, we note that it is enough to construct the families $\mathscr{U}_{\mathfrak{w}} \to C \times X_{\mathfrak{w}}$ parametrized by the Schubert varieties $X_{\mathfrak{w}}$ together with certain isomorphisms $\phi_{\mathfrak{w},\mathfrak{v}}$ of $\mathscr{U}_{\mathfrak{w}}|_{C\times X_{\mathfrak{v}}}$ with $\mathscr{U}_{\mathfrak{v}}$, for any $X_{\mathfrak{v}} \subset X_{\mathfrak{w}}$, such that the isomorphisms $\phi_{\mathfrak{w},\mathfrak{v}}$ satisfy the cocycle condition $\phi_{\mathfrak{w},\mathfrak{v}}\phi_{\mathfrak{v},\mathfrak{u}} = \phi_{\mathfrak{w},\mathfrak{u}}$, for all $\mathfrak{w} \geqq \mathfrak{v} \geqq \mathfrak{u}$.

Choose a local parameter $t$ for $C$ at $p$ and set $\mathscr{N}^- = G(\mathbb{C}[t^{-1}])$. Then $\mathscr{N}^-$ can canonically be thought of as a subgroup of $\mathscr{G}$. Further the $\mathscr{N}^-$-orbit $U^-$ through the base point $\mathsf{e} \in X := \mathscr{G}/\mathscr{P}$ is open in the Zariski topology on $X$. In particular, by the Bruhat decomposition $\{\mathfrak{w}U^-\}_{\mathfrak{w}}$ provides an open cover of $X$. The map $U^- \to \mathscr{G}$, defined by $x.\mathsf{e} \mapsto x$ for $x \in \mathscr{N}^-$, provides a section $\sigma$ of the principal $\mathscr{P}$-bundle $\mathscr{G} \to \mathscr{G}/\mathscr{P}$ over the open set $U^-$, and by translating this section we also get sections $\sigma_{\mathfrak{w}}$ over any $\mathfrak{w}U^-$ ($\sigma_{\mathfrak{w}}$ does depend upon the coset representative $w$ of $\mathfrak{w} = wW$, but for each, $\mathfrak{w} \in \tilde{W}/W$ we fix one coset representative $w$ and then $\sigma_{\mathfrak{w}}$ really means $\sigma_w$). Now fix any Schubert variety $X_{\mathfrak{v}}$ and cover this by the affine open sets $\{(\mathfrak{w}U^-) \cap X_{\mathfrak{v}}\}$ and take the sections $\sigma_{\mathfrak{w}}$ over them. In view of the discussion above, this canonically gives rise to $G$-bundles $\mathscr{U}_{\mathfrak{w}} = \mathscr{U}_{\mathfrak{w}}^{\mathfrak{v}}$ on $C \times (\mathfrak{w}U^- \cap X_{\mathfrak{v}})$. Further, for any $x$ in the intersection $U_{\mathfrak{w}_1} \cap U_{\mathfrak{w}_2}$, where $U_{\mathfrak{w}_1} := (\mathfrak{w}_1 U^-) \cap X_{\mathfrak{v}}$, we have $\sigma_{\mathfrak{w}_2}(x) = \sigma_{\mathfrak{w}_1}(x) h_{\mathfrak{w}_1,\mathfrak{w}_2}(x)$ with $h_{\mathfrak{w}_1,\mathfrak{w}_2}(x) \in G(\hat{\mathscr{O}}_p)$. These $h_{\mathfrak{w}_1,\mathfrak{w}_2}$ canonically give rise to the isomorphisms $\mathscr{U}_{\mathfrak{w}_1} \to \mathscr{U}_{\mathfrak{w}_2}$ over the intersection $C \times (\mathscr{U}_{\mathfrak{w}_1} \cap \mathscr{U}_{\mathfrak{w}_2})$, which obviously satisfy the cocycle condition. Thus the bundles $\{\mathscr{U}_{\mathfrak{w}}^{\mathfrak{v}}\}_{\mathfrak{w}}$ patch-up to give the $G$-bundle $\mathscr{U} = \mathscr{U}^{\mathfrak{v}}$ on $C \times X_{\mathfrak{v}}$. Since the sections $\sigma_{\mathfrak{w}}$ are defined on the whole of $\mathfrak{w}U^-$, it is easy to see that $\mathscr{U}^{\mathfrak{v}_1}$ canonically restricts to $\mathscr{U}^{\mathfrak{v}_2}$ whenever $\mathfrak{v}_1 \geqq \mathfrak{v}_2$. This completes the (a)-part, i.e., the construction of the family $\mathscr{U}$ parametrized by $\mathscr{G}/\mathscr{P}$.

To prove the (b) part, let us choose a trivialization $\tau$ of the bundle $\mathscr{E}$ restricted to $(\operatorname{Spec} \hat{\mathscr{O}}_p) \times T$. As above, this (together with the trivialization $\beta$) gives rise to a map $f_\tau : T \to \mathscr{G}$ and hence a map $f : T \to \mathscr{G}/\mathscr{P}$. (It is easy to see that the map $f$ does not depend upon the choice of the trivialization $\tau$.) We claim that there exists a large enough $X_{\mathfrak{w}}$ such that $\operatorname{Im} f \subset X_{\mathfrak{w}}$ and moreover $f : T \to X_{\mathfrak{w}}$ is a morphism:

For both of these assertions, we can assume that $T$ is an affine variety $T = \operatorname{Spec} R$, for some $\mathbb{C}$-algebra $R$. Then the map $f_\tau$ can be thought of as an element (again denoted by) $f_\tau \in G(\hat{k}_p \otimes R)$. Choose an embedding $G \hookrightarrow \mathrm{GL}(N)$, and also choose a local parameter $t$ around $p \in C$. Then we can write $f_\tau = (f_\tau^{i,j})_{1 \leqq i,j \leqq N}$, with $f_\tau^{i,j} \in \hat{k}_p \otimes R$. In particular, there exists a large enough $l \geqq 0$ such that (for any $1 \leqq i, j \leqq N$) $f_\tau^{i,j} \in t^{-l}\mathbb{C}[[t]] \otimes R$. From this one can see that $\operatorname{Im} f$ is contained in a Schubert variety $X_{\mathfrak{w}}$. Now the assertion that $f : T \to X_{\mathfrak{w}}$ is a morphism follows from the description of the map $f_\tau$ as an element of $G(\hat{k}_p \otimes R)$ together with the explicit description of the variety structure on $\mathscr{G}/\mathscr{P}$, as given, e.g., in [KL, Sect. 5.2]. The remaining assertions of (b) are easy to verify, thereby completing the proof of (b). □

Let $X_0 \subset X_1 \subset X_2 \subset \dots$ be a sequence of algebraic varieties such that $X_i \subset X_{i+1}$ is a closed immersion, for all $i$. Let $X := \cup X_i$ be the corresponding ind-variety. For any $x \in X$, we define the **Zariski tangent space** $T_x(X) := \lim T_x(X_i)$, where $T_x(X_i)$ is the Zariski tangent space of $X_i$ at $x$. If $X$ as above is an algebraic ind-group, then $T_e(X)$ has a canonical structure of a Lie algebra (see [Sa, Sect. 1]). Endowed with this Lie algebra structure, $T_e(X)$ is denoted by $\operatorname{Lie} X$.

We define the map $\operatorname{Lie} \Gamma \to (\operatorname{Lie} G) \otimes_k k[C^*]$, by considering the differential of the evaluation map at each point of $C^*$, where (as in Sect. 1.1) $k[C^*]$ is the ring of regular functions on the affine curve $C^* := C \backslash p$. The following lemma determines the Lie algebra of the algebraic ind-group $\Gamma$.

(2.9) **Lemma.** *Under the above map,* $\operatorname{Lie} \Gamma$ *is isomorphic with* $\mathfrak{g} \otimes_k k[C^*]$ *as Lie algebras, where* $\mathfrak{g} := \operatorname{Lie} G$ *and the bracket in* $\mathfrak{g} \otimes k[C^*]$ *is defined as* $[X \otimes p, Y \otimes q] = [X, Y] \otimes pq$, *for* $X, Y \in \mathfrak{g}$ *and* $p, q \in k[C^*]$.

*Proof.* Embed $G$ as a (closed) algebraic subgroup $i: G \hookrightarrow \mathrm{SL}_N(k) (\subset M_N(k))$, for some $N > 0$. This gives rise to a closed immersion $\tilde{i}: G(R) = \Gamma \hookrightarrow M_N(R)$ (where $M_N(R)$ is the space of $N \times N$ matrices over the ring $R := k[C^*]$). In particular, it induces an injective map $d\tilde{i}: T_e(\Gamma) = \mathrm{Lie}\,\Gamma \hookrightarrow T_I(M_N(R)) = M_N(R)$, at the Zariski tangent space level (where $I$ is the identity matrix). We claim that $d\tilde{i}$ is a Lie algebra homomorphism, if we endow $M_N(R)$ with the standard Lie algebra structure. To prove this, consider the following commutative diagram (for any fixed $x \in C^*$):

$$\begin{array}{ccc} T_e(\Gamma) & \overset{d\tilde{i}}{\hookrightarrow} & M_N(R) \\ \downarrow & & \downarrow \\ T_e(G) & \overset{di}{\hookrightarrow} & M_N(k), \end{array}$$

where the vertical maps are induced by the evaluation map $e_x: R \to k$ given by $p \mapsto p(x)$. Since $di$ is a Lie algebra homomorphism, and so are the vertical maps, we obtain that $d\tilde{i}$ itself is a Lie algebra homomorphism. It is further clear, from the above commutative diagram, that the image of $d\tilde{i}$ is contained in $\mathfrak{g} \otimes R$, where $\mathfrak{g}$ is identified with its image in $M_N(k)$ via the Lie algebra homomorphism $di$.

Next, we prove that the image of $d\tilde{i}$ contains at least the set $\mathfrak{g} \otimes R$: Fix any ad-nilpotent vector $X \in \mathfrak{g}$ and $p \in R$, and define a morphism $\mathbb{A}^1 \to \Gamma$ by $z \mapsto \exp(zX \otimes p)$. (Since $X$ is ad-nilpotent, the image is indeed contained in $\Gamma$.) It is easy to see that the image of the induced map (at the tangent space level at 0) is precisely the space $k(X \otimes p)$. But since the image of $d\tilde{i}$ is a Lie subalgebra, and ad-nilpotent vectors $X \in \mathfrak{g}$ generate $\mathfrak{g}$ (as a Lie algebra), the assertion follows. This completes the proof of the lemma. □

## 3 Preliminaries on moduli space of $G$-bundles and the determinant bundle

*Throughout this section, $G$ denotes a connected reductive group over $\mathbb{C}$ and $C$ a smooth projective irreducible curve over $\mathbb{C}$.*

We recall some basic concepts and results on semistable $G$-bundles on $C$. The references are [NS, R1, R2, RR]. Recall the definition of $G$-bundles and reduction of structure group from Sect. 1.2.

(3.1) **Definition.** *Let $E \to C$ be a $G$-bundle. Then $E$ is said to be* **semistable** *(resp.* **stable***), if for any reduction $E_P$ of structure group of $E$ to any parabolic subgroup $P \subset G$ and any non-trivial character $\chi: P \to G_m$ which is dominant with respect to some Borel subgroup contained in $P$, the degree of the associated line bundle $E_P(\chi)$ is $\leqq 0$ (resp. $< 0$). (Note that, by definition, a dominant character is taken to be trivial on the connected component of the centre of $G$.)*

(3.2) *Remark.* When $G = \mathrm{GL}_n$, this definition coincides with the usual definition of semistability (resp. stability) due to Mumford (cf. [NS]) viz. a vector bundle $V \to C$ is semistable (resp. stable) if for every subbundle $W \subsetneqq V$, we have $\mu(W) \leqq \mu(V)$ (resp. $\mu(W) < \mu(V)$), where $\mu(V) := \deg V / \operatorname{rank} V$.

Let $V \to C$ be a semistable vector bundle. Then there exists a filtration by subbundles

$$V_0 = 0 \subsetneq V_1 \subsetneq V_2 \subsetneq \dots \subsetneq V,$$

such that $\mu(V_i) = \mu(V)$ and $V_i/V_{i-1}$ are stable. Though such a filtration in general is not unique, the associated graded

$$\mathrm{gr}(V) := \bigoplus_{i \geq 1} V_i/V_{i-1}$$

is uniquely determined by $V$ (upto an isomorphism).

We will now describe the corresponding notion of $\mathrm{gr}(E)$ for a semistable $G$-bundle $E$.

(3.3) **Definition.** *A reduction of structure group of a $G$-bundle $E \to C$ to a parabolic subgroup $P$ is called* **admissible** *if for any character of $P$, which is trivial on the connected component of the centre of $G$, the associated line bundle of the reduced $P$-bundle has degree 0.*

It is easy to see that if $E_P$ is an admissible reduction of structure group of $E$ to a parabolic subgroup $P$, then $E$ is semistable if and only if the $P/U$-bundle $E_P(P/U)$ is semistable, where $U$ is the unipotent radical of $P$. Moreover, a semistable $G$-bundle $E$ admits an admissible reduction to some parabolic subgroup $P$ such that $E_P(P/U)$ is, in fact, a stable $P/U$-bundle. Let $M$ be a Levi component of $P$. Then $M \approx P/U$ (as algebraic groups) and thus we get a stable $M$-bundle $E_P(M)$. Extend the structure group of this $M$-bundle to $G$ to get a semistable $G$-bundle denoted by $\mathrm{gr}(E)$. Then $\mathrm{gr}(E)$ is uniquely determined by $E$ (up to an isomorphism) (see [R1]).

Two semistable $G$-bundles $E_1$ and $E_2$ are said to be ***S*-equivalent** if $\mathrm{gr}(E_1) \approx \mathrm{gr}(E_2)$. We call a semistable $G$-bundle $E$ **quasistable** if $E \approx \mathrm{gr}(E)$. (It can be seen that a semistable vector bundle is quasistable if and only if it is a direct sum of stable vector bundles with the same $\mu$.)

Two $G$-bundles $E_1$ and $E_2$ on $C$ are said to be of the **same topological type** if they are isomorphic as $G$-bundles in the topological category. The topological types of all the algebraic $G$-bundles on $C$ are bijectively parametrized by the first fundamental group $\pi_1(G)$ (cf. [R2, Sect. 5]).

(3.4) **Theorem.** *The set $\mathfrak{M}$ of $S$-equivalence classes of all the semistable $G$-bundles on $C$ of a fixed topological type admits the structure of a normal, irreducible, projective variety over $k$, making it into a coarse moduli.*

*In particular, for any algebraic family $\mathcal{E} \to C \times T$ of semistable $G$-bundles of the same topological type (parametrized by a variety $T$), the set map $\beta : T \to \mathfrak{M}$, which takes $t \in T$ to the $S$-equivalence class of $\mathcal{E}_t$ in $\mathfrak{M}$ is a morphism.*

The details can be found in [NS, R1, R2, Ses, . . .].

(3.5) *Remarks.* (a) In general $\mathfrak{M}$ is not a *fine moduli*, i.e., there may not exist any family $\mathcal{F} \to C \times \mathfrak{M}$ (parametrized by $\mathfrak{M}$) such that $\mathcal{F}_m$ belongs to the $S$-equivalence class $m \in \mathfrak{M}$.

(b) For $G = \mathrm{GL}_n$, i.e., for the case of rank-$n$ vector bundles, the topological type is nothing but its degree. When the degree is coprime to the rank, the coarse moduli is in fact a fine moduli. (When the degree is not coprime to the rank, the coarse moduli is *not* a fine moduli.)

We prove a result on $\mathrm{gr}(E)$ which we will need in Sect. 6. We first prove the following:

(3.6) **Lemma.** *Let $H$ be a connected affine algebraic group and $C$ a smooth projective curve over $k$. Then any principal $H$-bundle on $C$ is locally trivial in the Zariski topology.*

*Proof.* Let $E$ be a principal $H$-bundle on $C$ and $U$ the unipotent radical of $H$. Since the group $M = H/U$ is connected and reductive, the $M$-bundle $E(M)$, obtained from $E$ by extension of structure group to $M$, is locally trivial in the Zariski topology [R3, Proposition 4.3].

Let $W$ be a non-empty affine open subset of $C$ such that the restriction of $E(M)$ to $W$ is trivial. We shall show that $E_{|W}$ is trivial (which will prove the lemma): Observe that a trivialization of $E(M)$ on $W$ gives a reduction of the structure group $H$ of $E_{|W}$ to the subgroup $U$. So, it suffices to show that any (principal) $U$-bundle on $W$ is trivial:

We may assume $U \neq e$. Then there exists a (finite) filtration of $U$ by closed normal subgroups such that the successive quotients are isomorphic to the additive group $G_a$. Now the assertion follows since any principal $G_a$-bundle on $W$ is trivial, $W$ being affine (see [Se1, Sect. 5.1]). □

Let $P$ be a parabolic subgroup of $G$ and $P = MU$ a Levi decomposition, where $U$ is the unipotent radical of $P$ and $M$ a Levi component. The next proposition will be used in Sect. 6 in the case of an admissible reduction of a semistable bundle $E$.

(3.7) **Proposition.** *Let $E$ be a $G$-bundle on $C$ and $E_P$ a reduction of the structure group of $E$ to $P$. Denote by $\mathrm{gr}(E_P)$ the $G$-bundle on $C$ obtained from the $P$-bundle $E_P$ by extension of the structure group via the homomorphism*

$$P \to P/U \approx M \hookrightarrow G.$$

Assume that $G$ is semisimple and connected.

*Then there exists a $G$-bundle $\mathscr{E}$ on $C \times \mathbb{A}^1$, where $\mathbb{A}^1$ is the affine line, such that we have*

(a) $\mathscr{E}_{|C\times(\mathbb{A}^1\backslash 0)} \approx p_C^*(E)$, $\mathscr{E}_{|C\times\{0\}} \approx \mathrm{gr}(E_P)$ and

(b) $\mathscr{E}_{|C^*\times\mathbb{A}^1}$ *is trivial and also the pull-back of* $\mathscr{E}$ *to* $(\mathrm{Spec}\,\hat{\mathscr{O}}_p) \times \mathbb{A}^1$ *is trivial, where $p_C$ is the projection on the $C$-factor.*

*Proof.* By [R1, Lemma 2.5.12], there exists a one-parameter group $\lambda : G_m (:= \mathbb{A}^1\backslash 0) \to M$, such that the regular map

$$G_m \times P \to P, \quad \text{given by} \quad (t,p) \mapsto \lambda(t)p\lambda(t)^{-1} \quad \text{for} \quad t \in G_m,\ p \in P,$$

extends to a regular map $\phi : \mathbb{A}^1 \times P \to P$ satisfying $\phi(0, mu) = m$, for $m \in M$, $u \in U$. By Lemma (3.6), the $P$-bundle $E_P$ is locally trivial in the Zariski topology. Let $\{U_i\}$ be an affine open covering of $C$ in which the bundle $E_P$

is given by the transition functions $p_{ij}:U_i \cap U_j \to P$. Let $\mathscr{F}$ be the (Zariski locally trivial) $P$-bundle on $C \times \mathbb{A}^1$ defined by the covering $\{U_i \times \mathbb{A}^1\}$ and the transition functions

$$h_{ij}:(U_i \cap U_j) \times \mathbb{A}^1 \to P,$$

where $h_{ij}(z,t) = \phi(t, p_{ij}(z))$, for $t \in \mathbb{A}^1$, $z \in U_i \cap U_j$. Now let $\mathscr{E}$ be the $G$-bundle obtained from the $P$-bundle $\mathscr{F}$ by extension of the structure group to $G$. Then clearly $\mathscr{E}$ satisfies condition (a).

We next show that for any non-empty affine open subset $W$ of $C$, the restriction of $\mathscr{E}$ to $W \times \mathbb{A}^1$ is trivial (this will, in particular, imply that condition (b) is satisfied): Note that, by our construction, there exists a (finite) open covering $W_i$ of $W$ such that $\mathscr{E}_{|W_i \times \mathbb{A}^1}$ is trivial, for every $i$. Now by an analogue of a result of Quillen (cf. [Ra, Theorem 2]) $\mathscr{E}_{|W \times \mathbb{A}^1}$ is the pull-back of a $G$-bundle on $W$. But, by Proposition (1.3), any $G$-bundle on $W$ is trivial. □

(3.8) *Determinant bundle and $\Theta$-bundle.* We now briefly recall a few definitions and facts on the determinant bundles and $\Theta$-bundles associated to families of bundles on $C$. We follow [DN, NRa].

In the case of the moduli $J_d$ of line bundles of fixed degree $d$ on $C$, i.e., the Jacobian, there is a natural divisor (on the Jacobian) called the $\Theta$-divisor. It is defined only up to an algebraic equivalence in general, but on the Jacobian $J_{g-1}$ it is canonically defined (where $g$ is the genus of $C$). Since we have chosen a base point $p$ on $C$, the $\Theta$-divisor on any $J_d$ is canonically defined.

To generalise this notion to the moduli of higher rank vector bundles, one makes use of the determinant bundle associated to any family of vector bundles.

Let $\mathscr{V} \to C \times T$ be a vector bundle. Then there exists a complex of vector bundles $\mathscr{V}_i$ on $T$ (with $\mathscr{V}_i = 0$, for all $i \geqq 2$):

$$\mathscr{V}_0 \to \mathscr{V}_1 \to 0 \to 0 \to \dots,$$

such that for any base change $f:Z \to T$, the $i^{\text{th}}$ direct image on $Z$ (under the projection $C \times Z \to Z$) of the pull back $(\text{id} \times f)^* \mathscr{V}$ is given by the $i^{\text{th}}$ cohomology of the pull back of the above complex to $Z$. We define the **determinant line bundle** $\operatorname{Det} \mathscr{V}$ on $T$ to be the product $\overset{\text{top}}{\bigwedge}(\mathscr{V}_1) \otimes \left(\overset{\text{top}}{\bigwedge}(\mathscr{V}_0)^*\right)$. (Notice that our $\operatorname{Det} \mathscr{V}$ is dual to the determinant line bundle as defined, e.g., in [L, Chap. 6, Sect. 1].)

The above base change property gives rise to the base change property for $\operatorname{Det} \mathscr{V}$, i.e., if $f:Z \to T$ is a morphism then $\operatorname{Det}((\text{id} \times f)^* \mathscr{V}) \cong f^*(\operatorname{Det} \mathscr{V})$.

Let $\mathfrak{L}$ be a line bundle on $T$, and let $p_2:C \times T \to T$ be the projection on the second factor. Then for the family $\mathscr{V} \otimes p_2^* \mathfrak{L} \to C \times T$, we have $\operatorname{Det}(\mathscr{V} \otimes p_2^* \mathfrak{L}) = (\operatorname{Det} \mathscr{V}) \otimes \mathfrak{L}^{-\mathscr{X}(\mathscr{V})}$, where $\mathscr{X}(\mathscr{V}) := h^0(\mathscr{V}_t) - h^1(\mathscr{V}_t)$ is the Euler characteristic and $\mathscr{V}_t := \mathscr{V}|_{C \times t}$. (Observe that $h^0(\mathscr{V}_t) - h^1(\mathscr{V}_t)$ remains constant on any connected component of $T$.)

We now define the **$\Theta$-bundle** $\Theta(\mathscr{V})$ of a family of rank $r$ and degree 0 bundles $\mathscr{V} \to C \times T$ to be the modified determinant bundle given by $(\operatorname{Det} \mathscr{V}) \otimes \det(\mathscr{V}_p)^{\mathscr{X}(\mathscr{V})/r}$, where $\mathscr{V}_p$ is the bundle $\mathscr{V}|_{p \times T}$ on $T$, and $\det \mathscr{V}_p$ is its usual determinant line bundle. It follows then that $\Theta(\mathscr{V}) = \Theta(\mathscr{V} \otimes p_2^* \mathfrak{L})$, for any line bundle $\mathfrak{L}$ on $T$. Moreover $\Theta(\mathscr{V})$ also has the functorial property $\Theta((\mathrm{id}\times f)^*\mathscr{V}) \cong f^*(\Theta(\mathscr{V}))$.

If $\mathscr{E} \to C\times T$ is a family of $G$-bundles and $V$ is a $G$-module, then $\mathrm{Det}(\mathscr{E}(V))$ and $\Theta(\mathscr{E}(V))$ are defined to be the corresponding line bundles of the associated family of vector bundles, via the representation $V$ of $G$, for $G$ semisimple.

For the family $\mathscr{U} \to C \times \mathscr{G}/\mathscr{P}$ (cf. Proposition 2.8), the line bundles $\Theta(\mathscr{U}(V))$ and $\mathrm{Det}(\mathscr{U}(V))$ coincide, since $\mathscr{U}_{|p\times \mathscr{G}/\mathscr{P}}$ is trivial.

It is known ([DN, NRa]; see also Remark 7.6) that there exists a line bundle $\Theta$ on the moduli space $\mathfrak{M}_o$ of rank $r$ and degree 0 (semistable) bundles, such that for any family $\mathscr{V}$ of rank $r$ and degree 0 semistable bundles parametrized by $T$ we have $f^*(\Theta) \cong \Theta(\mathscr{V})$, where $f:T \to \mathfrak{M}_o$ is the morphism given by the coarse moduli property of $\mathfrak{M}_o$ (cf. Theorem 3.4).

Assume that $G$ is semisimple (and connected) and let $V$ be a finite dimensional representation of $G$ of dimension $r$. Then for any semistable $G$-bundle on $C$, the associated vector bundle (via the representation $V$) is semistable (cf. [RR, Theorem 3.18]). Thus, given a family of semistable $G$-bundles on $C$ parametrized by $T$, we have a canonical morphism (induced from the representation $V$) $T \to \mathfrak{M}_o$ (where $\mathfrak{M}_o$ as above is the moduli space of semistable bundles of rank $r$ and degree 0). Let $\mathfrak{M}$ be the moduli space of semistable $G$-bundles on $C$. By the coarse moduli property of $\mathfrak{M}$, we see that we have a canonical morphism $\phi_V:\mathfrak{M} \to \mathfrak{M}_o$. We define the **theta bundle $\Theta(V)$ on $\mathfrak{M}$ associated to** $V$ to be the pull-back of the line bundle $\Theta$ on $\mathfrak{M}_o$ via the morphism $\phi_V$. It can be easily seen that for any family $\mathscr{V} \to C\times T$ of semistable $G$-bundles, $f^*(\Theta(V)) \simeq \Theta(\mathscr{V}(V))$, where $f:T \to \mathfrak{M}$ is the morphism (induced from the family $\mathscr{V}$) given by the coarse moduli property of $\mathfrak{M}$.

## 4 A result on algebraic descent

We prove the following technical result, which will crucially be used in the paper. Even though we believe that it should be known, we did not find a precise reference.

(4.1) **Proposition.** *Let $f:X \to Y$ be a surjective morphism between irreducible algebraic varieties $X$ and $Y$ over an algebraically closed field $k$ of* char 0. *Assume that $Y$ is normal and let $\mathscr{E} \to Y$ be an algebraic vector bundle on $Y$.*

*Then any set theoretic section $\sigma$ of the vector bundle $\mathscr{E}$ is regular if and only if the induced section $f^*(\sigma)$ of the induced bundle $f^*(\mathscr{E})$ is regular.*

*Proof.* The "only if" part is of course trivially true. So we come to the "if" part.

Since the question is local (in $Y$), we can assume that $Y$ is affine and moreover the vector bundle $\mathscr{E}$ is trivial, i.e., it suffices to show that any (set theoretic) map $\sigma:Y \to k$ is regular, provided $\bar{\sigma} := \sigma\circ f:X \to k$ is regular (under the assumption that $Y = \mathrm{Spec}\, R$ is irreducible normal and affine):

Since the map $f$ is surjective (in particular dominant), the ring $R$ is canonically embedded in $\Gamma(X) := H^0(X,\mathscr{O}_X)$. Let $R[\bar{\sigma}]$ denote the subring of $\Gamma(X)$ generated by $R$ and $\bar{\sigma} \in \Gamma(X)$. Then $R[\bar{\sigma}]$ is a (finitely generated) domain (as $X$

is irreducible by assumption), and we get a dominant morphism $\hat{f}: Z \to \operatorname{Spec} R$, where $Z := \operatorname{Spec}(R[\bar{\sigma}])$. Consider the commutative diagram:

$$\begin{array}{ccc} & X & \\ {}^{\theta}\swarrow & & \searrow^{f} \\ Z & \xrightarrow[\hat{f}]{} & Y \end{array}$$

where $\theta$ is the dominant morphism induced from the inclusion $R[\bar{\sigma}] \hookrightarrow \Gamma(X)$. In particular, $\operatorname{Im}\theta$ contains a non-empty Zariski open subset $U$ of $Z$. Let $x_1, x_2 \in X$ be closed points such that $f(x_1) = f(x_2)$. Then $r(x_1) = r(x_2)$, for all $r \in R$ and also $\bar{\sigma}(x_1) = \bar{\sigma}(x_2)$. This forces $\theta(x_1) = \theta(x_2)$, in particular, $\hat{f}_{|U}$ is injective on the closed points of $U$.

Since $\hat{f}$ is dominant, by cutting down $U$ if necessary, we can assume that $\hat{f}_{|U}: U \to V$ is a bijection, for some open subset $V \subset Y$. Now since $Y$ is (by assumption) normal and $Z$ is irreducible, by Zariski's main theorem (cf. [Mum, p. 288, I. Original form]), $\hat{f}_{|U}: U \to V$ is an isomorphism, and hence $\sigma$ is regular on $V$. Now we give two different proofs for the remaining part:

*First proof.* Assume, if possible, that $\sigma_{|V}$ does not extend to a regular function on the whole of $Y$. Then, by [B, Lemma 18.3, AG], there exists a point $y_0 \in Y$ and a regular function $h$ on a Zariski neighborhood $W$ of $y_0$ such that $h(y_0) = 0$ and $h\sigma = 1$ on $W \cap V$. But then $\bar{h}\bar{\sigma} = 1$ on $f^{-1}(W \cap V)$ (where $\bar{h} := h \circ f$) and hence, $\bar{\sigma}$ being regular on the whole of $X$, $\bar{h}\bar{\sigma} = 1$ on $f^{-1}(W)$. Taking $\bar{y}_0 \in f^{-1}(y_0)$ ($f$ is, by assumption, surjective), we get $\bar{h}(\bar{y}_0)\bar{\sigma}(\bar{y}_0) = 0$. This contradiction shows that $\sigma_{|V}$ does extend to some regular function (say $\sigma'$) on the whole of $Y$. Hence $\bar{\sigma} = \bar{\sigma}'$, in particular, by the surjectivity of $f$, $\sigma = \sigma'$. This proves the proposition.

*Second proof.* Let us define a subset $U_0 \subset Z$ by

$$U_0 = \{x \in Z : \dim e(x) = 0\},$$

where $e(x)$ is the union of all the irreducible components of $\hat{f}^{-1}(\hat{f}(x))$ containing $x$. Then, by Chevalley's theorem, $U_0$ is open (possibly empty) in $Z$ and the map $\hat{f}_{|U_0}: U_0 \to Y$ has all its fibers finite. But since $\hat{f}$ is birational, $U_0$ is non-empty. Further, by Zariski's main theorem, $V_0 := \hat{f}(U_0)$ is open in $Y$ and the map $\hat{f}_{|U_0}: U_0 \to V_0$ is an isomorphism. This gives that $\sigma_{|V_0}$ is a regular function.

Consider the surjective map

$$\hat{f}: Z \backslash \hat{f}^{-1}(V_0) \to Y \backslash V_0 .$$

Then, by the definition of $V_0$, every fiber of the above map has at least one irreducible component of dim $\geqq 1$ (actually of dim exactly 1). Hence

$$\dim(Y \backslash V_0) \leqq \dim(Z \backslash \hat{f}^{-1}(V_0)) - 1 \leqq \dim Y - 2$$

(since $\hat{f}: Z \to Y$ is birational and $Z$ is irreducible), i.e., $\operatorname{codim}_Y(Y \backslash V_0) \geqq 2$. But since $Y$ is assumed to be normal, the regular function $\sigma_{|V_0}$ admits a regular extension $\sigma'$ to the whole of $Y$. Now by the same argument as in the first proof, we get that $\sigma = \sigma'$ on the whole of $Y$. This completes the second proof as well. □

(4.2) *Remark.* Even though we do not need it, the same result as above is true in the analytic category if the underlying field $k$ is taken to be $\mathbb{C}$.

## 5 Identification of the determinant bundle

*In this section we consider the triple $\mathfrak{T} = (G, C, p)$, where $G$ is a connected, simply-connected, simple algebraic group over $\mathbb{C}$, $C$ is a smooth projective irreducible curve* over $\mathbb{C}$, *and $p$ is any point of $C$. We follow the notation as in Sect. 1.1.*

(5.1) Recall from Proposition 2.8 that $\mathcal{G}/\mathcal{P}$ is a parameter space for an algebraic family $\mathcal{U}$ of $G$-bundles on $C$. Let us fix a (finite dimensional) representation $V$ of $G$. In particular, we can talk of the determinant line bundle Det($\mathcal{U}(V)$) (cf. Sect. 3.8). Also recall the definition of the (fundamental) homogeneous line bundle $\mathfrak{L}(\chi_o)$ on $\mathcal{G}/\mathcal{P}$ from Sect. 2.2. Our aim in this section is to determine the line bundle Det($\mathcal{U}(V)$) in terms of $\mathfrak{L}(\chi_o)$. We begin with the following preparation.

Let $\theta$ be the highest root of $\mathfrak{g}$. Define the following Lie subalgebra $sl_2(\theta)$ of the Lie algebra $\mathfrak{g}$ of $G$:

(1) $$sl_2(\theta) := \mathfrak{g}_{-\theta} \oplus \mathbb{C}\theta^\vee \oplus \mathfrak{g}_\theta\,,$$

where $\mathfrak{g}_\theta$ is the $\theta$-th root space, and $\theta^\vee$ is the corresponding coroot. Clearly $sl_2(\theta) \approx sl_2$ as Lie algebras. Decompose

(2) $$V = \oplus_i V_i\,,$$

as a direct sum of irreducible $sl_2(\theta)$-submodules $V_i$ of dim $m_i$. Now we define

(3) $$m_V = \sum_i \binom{m_i + 1}{3}, \quad \text{where we set} \quad \binom{2}{3} = 0\,.$$

The number $m_V$ coincides with the *Dynkin index* of the representation $V$ (cf. [D, Sect. 2]). We give an expression for $m_V$ in the following lemma.

Write the formal character

(4) $$\operatorname{ch} V = \sum n_\lambda e^\lambda\,.$$

(5.2) **Lemma.**

(1) $$m_V = \frac{1}{2}\sum_\lambda n_\lambda \langle \lambda, \theta^\vee\rangle^2\,.$$

*In particular, for the adjoint representation* ad *of* $\mathfrak{g}$ *we have*

(2) $$m_{\text{ad}} = 2(1 + \langle \varrho, \theta^\vee\rangle)\,,$$

*where $\varrho$ as usual is the half sum of the positive roots of* $\mathfrak{g}$.

*Similarly, for the standard $n$-dim. representation $E_n$ of $sl_n$, $m_{E_n} = 1$.*

*Proof.* It suffices to show that, for the irreducible representation $W(m)$ (of $\dim m+1$) of $sl_2$

$$\frac{1}{2}\sum_{n=0}^{m}\langle m\varrho_1 - n\alpha, H\rangle^2 = \binom{m+2}{3}, \tag{3}$$

where $\alpha$ is the (unique) positive root of $sl_2$, $H$ the corresponding coroot and $\varrho_1 := \frac{1}{2}\alpha$.

Now the left side of (3) is equal to

$$\begin{aligned} 2\sum_{n=0}^{m}\left(\frac{m}{2}-n\right)^2 &= 4\sum_{k=1}^{k_o}k^2 = \frac{m(m+1)(m+2)}{6}, \quad \text{if } m=2k_o \text{ is even, and} \\ &= 2\sum_{n=0}^{m}\left(k_o-\frac{1}{2}-n\right)^2, \quad \text{if } m=2k_o-1 \text{ is odd} \\ &= 4\sum_{k=1}^{k_o}\left(k-\frac{1}{2}\right)^2 = \left(4\sum_{k=1}^{k_o}k^2\right)+k_o-4\sum_{k=1}^{k_o}k \\ &= \frac{m(m+1)(m+2)}{6}. \end{aligned}$$

So in either case the left side of (3) $= \dfrac{m(m+1)(m+2)}{6} = \dbinom{m+2}{3}$. This proves the first part of the lemma.

For the assertion regarding the adjoint representation, we have

$$\operatorname{ch}(\mathrm{ad}) = (\dim\mathfrak{h}).e^0 + \sum_{\beta\in\Delta_+}(e^{\beta}+e^{-\beta}).$$

So

$$\begin{aligned} m_{\mathrm{ad}} &= \sum_{\beta\in\Delta_+}\langle\beta,\theta^\vee\rangle^2 \\ &= 4 + \sum_{\beta\in\Delta_+\backslash\theta}\langle\beta,\theta^\vee\rangle, \quad \text{since } \langle\beta,\theta^\vee\rangle = 0 \text{ or } 1, \text{ for any } \beta\in\Delta_+\backslash\theta \\ &= 4+\langle 2\varrho-\theta,\theta^\vee\rangle \\ &= 2(1+\langle\varrho,\theta^\vee\rangle). \end{aligned}$$

The assertion about $m_{E_n}$ is easy to verify. □

(5.3) *Remark.* The number $(1+\langle\varrho,\theta^\vee\rangle)$ is called the **dual Coxeter number**(cf. [K, Sect. 6.1 and Exercise 6.2]) of $\mathfrak{g}$ (rather of the corresponding affine Kac-Moody algebra). Its value is given as below.

| Type of $\mathfrak{g}$ | Dual Coxeter number |
|---|---|
| $A_l$ | $l+1$ |
| $B_l$ | $2l-1$ |
| $C_l$ | $l+1$ |
| $D_l$ | $2l-2$ |
| $E_6$ | 12 |
| $E_7$ | 18 |
| $E_8$ | 30 |
| $G_2$ | 4 |
| $F_4$ | 9 |

Now we can state the main theorem of this section.

(5.4) **Theorem.** *With the notation as in Sect. 5.1 (as elements of* Pic $\mathscr{G}/\mathscr{P}$*)*

$$\mathrm{Det}(\mathscr{U}(V)) = \mathfrak{L}(m_V \chi_o),$$

*for any finite dimensional representation $V$ of $G$, where the number $m_V$ is defined by* (3) *of Sect. 5.1.*

*Proof.* By Proposition 2.3, there exists an integer $m$ such that

$$\mathrm{Det}(\mathscr{U}(V)) = \mathfrak{L}(m\chi_o) \in \mathrm{Pic}(\mathscr{G}/\mathscr{P}).$$

We want to prove that $m = m_V$:

Set $\mathscr{U}_o := \mathscr{U}(V)_{|C\times X_o}$ as the family restricted to the Schubert variety $X_o := X_{\mathfrak{s}_o}$ (cf. proof of Proposition 2.3). Denote by $\alpha$ (resp. $\beta$) the canonical generator of $H^2(X_o, \mathbb{Z})$ (resp. $H^2(C, \mathbb{Z})$). Then it suffices to show that Det $\mathscr{U}_o \simeq \mathfrak{L}_{\mathfrak{s}_o}(m_V \chi_o)$, which is equivalent to showing that the first Chern class

(1) $$c_1(\mathrm{Det}\,\mathscr{U}_o) = m_V \alpha :$$

From the definition of the determinant bundle we have

(2) $$c_1(\mathrm{Det}\,\mathscr{U}_o) = -c_1(\pi_{2*}\mathscr{U}_o),$$

where $\pi_2$ is the projection $C \times X_o \to X_o$, and the notation $\pi_{2*}$ is as in [F, Chap. 9].

Since $\mathscr{U}_{o|C^*\times X_o}$ as well as $\mathscr{U}_{o|C\times \mathrm{e}}$ is trivial (where $\mathrm{e}$ is the base point of $\mathscr{G}/\mathscr{P}$), we get

(3) $$c_1(\mathscr{U}_o) = 0.$$

Let $\tilde{\alpha}$ (resp. $\tilde{\beta}$) be the pull back of $\alpha$ (resp. $\beta$) under $\pi_2$ (resp. $\pi_1$). Now write

(4) $$c_2(\mathscr{U}_o) = l\tilde{\alpha}\tilde{\beta}, \quad \text{for some (unique) } l \in \mathbb{Z}.$$

Let $T_{\pi_2}$ be the relative tangent bundle along the fibers of $\pi_2$. Let us denote by $c_1$ (resp. $c_2$) the first (resp. second) Chern class of $\mathcal{U}_o$. By the Grothendieck's Riemann-Roch theorem [F, Sect. 9.1] applied to the (proper) map $\pi_2$, we get

$$\begin{aligned}\mathrm{ch}(\pi_{2*}\mathcal{U}_o) &= \pi_{2*}(\mathrm{ch}(\mathcal{U}_o).\,\mathrm{td}(T_{\pi_2}))\\ &= \pi_{2*}\left[\left(\mathrm{rk}\,\mathcal{U}_o + c_1 + \tfrac{1}{2}(c_1^2 - 2c_2)\right)\left(1 + \tfrac{1}{2}c_1(T_{\pi_2})\right)\right]\\ &= \pi_{2*}\left[(\mathrm{rk}\,\mathcal{U}_o - c_2)\left(1 + \tfrac{1}{2}c_1(T_{\pi_2})\right)\right], \quad \text{by (3)},\end{aligned}$$

where ch denotes the Chern character and td denotes the Todd class. Hence

$$\begin{aligned}(5)\qquad c_1(\pi_{2*}\mathcal{U}_o) &= \pi_{2*}(-c_2(\mathcal{U}_o))\\ &= \pi_{2*}(-l\tilde{\alpha}\tilde{\beta}), \quad \text{by (4)}\\ &= -l\alpha, \quad \text{since } \pi_{2*}(\tilde{\alpha}\tilde{\beta}) = \alpha.\end{aligned}$$

So to prove the theorem, by (1), (2) and (5), we need to show that $l = m_V$, where $l$ is given by (4):

It is easy to see (from its definition) that topologically the bundle $\mathcal{U}_o$ is pull back of the bundle $\mathcal{U}'_o$ on $\mathbb{P}^1 \times X_o$ (where $\mathcal{U}'_o$ is the same as $\mathcal{U}_o$ for $C = \mathbb{P}^1$) via the map

$$C \times X_o \overset{\delta\times \mathrm{Id}}{\to} \mathbb{P}^1 \times X_o,$$

where $\delta : C \to \mathbb{P}^1$ pinches all the points outside a small open disc around $p$ to a point. Of course the map $\delta$ is of degree 1, so the cohomology generator $\alpha$ pulls back to the generator $\beta$ (observe that $X_o \approx \mathbb{P}'$ as shown below). Hence it suffices to compute the second Chern class of the bundle $\mathcal{U}'_o$ on $\mathbb{P}^1 \times X_o$:

Choose $X_\theta \in \mathfrak{g}_\theta$ (where $\theta$ is the highest root of $\mathfrak{g}$) such that $\langle X_\theta, -\omega X_\theta\rangle = 1$, where $\omega$ is the Cartan involution of $\mathfrak{g}$ and $\langle\,,\rangle$ is the Killing form on $\mathfrak{g}$, normalized so that $\langle\theta,\theta\rangle = 2$. Set $Y_\theta = -\omega(X_\theta) \in \mathfrak{g}_{-\theta}$. Define a Lie algebra homomorphism $sl_2 \to \mathbb{C}[t, t^{-1}] \otimes \mathfrak{g}$, by

$$\begin{aligned} X &\mapsto t \otimes Y_\theta\\ Y &\mapsto t^{-1} \otimes X_\theta\\ H &\mapsto -1 \otimes \theta^\vee,\end{aligned}$$

where $\{X, Y, H\}$ is the standard basis of $sl_2$. The corresponding group homomorphism (choosing a local parameter $t$ around $p$) $\eta : \mathrm{SL}_2(\mathbb{C}) \to \mathcal{G}$ induces a biregular isomorphism $\bar{\eta} : \mathbb{P}^1 \approx \mathrm{SL}_2(\mathbb{C})/B_1 \overset{\sim}{\to} X_o$, where $B_1$ is the standard Borel subgroup of $\mathrm{SL}_2(\mathbb{C})$ consisting of upper triangular matrices. In what follows we will identify $X_o$ with $\mathbb{P}^1$ under $\bar{\eta}$. The representation $V$ of $G$ on restriction, under the decomposition (2) of Sect. 5.1, gives rise to a continuous group homomorphism

$$\psi : \mathrm{SU}_2(\theta) \to \prod_i (\mathrm{Aut}\, V_i),$$

where $\mathrm{SU}_2(\theta)$ is the (standard) compact form (induced from the involution $\omega$) of the group $\mathrm{SL}_2(\theta)$ (with Lie algebra $sl_2(\theta)$).

There is a principal $SU_2$-bundle $\mathscr{W}$ on $S^4$ (in the topological category) got by the clutching construction from the identity map $S^3 \approx SU_2 \to SU_2$. In particular, we obtain the vector bundle $\mathscr{W}(\psi) \to S^4$ associated to the principal bundle $\mathscr{W}$ via the representation $\psi$, which breaks up as a direct sum of subbundles $\mathscr{W}_i(\psi)$ (got from the representations $V_i$).

We further choose a degree 1 continuous map $\nu: \mathbb{P}^1 \times \mathbb{P}^1 \to S^4$. We claim that the vector bundle $\mathscr{U}'_o$ on $\mathbb{P}^1 \times \mathbb{P}^1$ is isomorphic (in the topological category) with the pull back $\nu^*(\mathscr{W}(\psi))$:

Define a map $\Phi: (SU_2/D) \times S^1 \to SU_2$ by

$$\left(\begin{pmatrix} a & b \\ c & d \end{pmatrix} \bmod D, t\right) \mapsto \begin{pmatrix} d & ct^{-1} \\ bt & a \end{pmatrix} \begin{pmatrix} d & c \\ b & a \end{pmatrix}^{-1},$$

for $\begin{pmatrix} a & b \\ c & d \end{pmatrix} \in SU_2$ and $t \in S^1$, where $D$ is the diagonal subgroup of $SU_2$. It is easy to see that the principal $SU_2$-bundle $\nu^*(\mathscr{W})$ on $\mathbb{P}^1 \times \mathbb{P}^1$ is isomorphic with the principal $SU_2$-bundle obtained by the clutching construction from the map $\Phi$ (by covering $\mathbb{P}^1 \times \mathbb{P}^1 = S^2 \times S^2 = S^2 \times H^+ \cup S^2 \times H^-$, where $H^+$ and $H^-$ are resp. the upper and lower closed hemispheres). By composing $\Phi$ with the isomorphism $SU_2 \to SU_2(\theta)$ (induced from the Lie algebra isomorphism $sl_2 \to sl_2(\theta)$ taking $X \mapsto X_\theta$, $Y \mapsto Y_\theta$, and $H \mapsto \theta^\vee$), and using the isomorphism $\bar{\eta}$ together with the definition of the vector bundle $\mathscr{U}_o$ we get the assertion that $\mathscr{U}'_o \approx \nu^*(\mathscr{W}(\psi))$.

So

$$\begin{aligned} c_2(\mathscr{U}'_o) &= \nu^*(c_2(\mathscr{W}(\psi))) = \nu^* \sum_i c_2(\mathscr{W}_i(\psi)) \\ &= \sum_i \binom{m_i + 1}{3} \tilde{\alpha}\tilde{\beta}, \text{ by the following lemma} \end{aligned}$$

(since $\nu$ is a map of degree 1).

Hence $l = \sum_i \binom{m_i + 1}{3} = m_V$, proving the theorem modulo the next lemma. □

(5.5) **Lemma.** *Let $W(m)$ be the $(m+1)$-dimensional irreducible representation of* $SU_2$ *and let $\mathscr{W}(m)$ be the vector bundle on $S^4$ associated to the principal* $SU_2$*-bundle $\mathscr{W}$ on $S^4$ (defined in the proof of Theorem 5.4) by the representation $W(m)$ of* $SU_2$. *Then*

(1)
$$c_2(\mathscr{W}(m)) = \binom{m+2}{3} \Omega,$$

*where $\Omega$ is the fundamental cohomology generator of $H^4(S^4, \mathbb{Z})$.*

*Proof.* By the Clebsch-Gordan theorem (cf. [Hu, p. 126]), we have the following decomposition as $SU_2$-modules:

$$W(m) \otimes W(1) = W(m+1) \oplus W(m-1), \quad \text{for any } m \geqq 1.$$

In particular, the Chern character

$$\text{(2)}\qquad \operatorname{ch}\mathscr{W}(m).\operatorname{ch}\mathscr{W}(1) = \operatorname{ch}\mathscr{W}(m+1) + \operatorname{ch}\mathscr{W}(m-1)\,.$$

Assume, by induction, that (1) is true for all $l \leqq m$. (The validity of (1) for $m = 1$ is trivial to see.) Then by (2) we get

$$\begin{aligned}\operatorname{ch}\mathscr{W}(m+1) &= \operatorname{ch}\mathscr{W}(m).\operatorname{ch}\mathscr{W}(1) - \operatorname{ch}\mathscr{W}(m-1)\\ &= ((m+1).1 - c_2\mathscr{W}(m))\,(2.1 - c_2\mathscr{W}(1))\\ &\quad - (m.1 - c_2\mathscr{W}(m-1))\,,\\ &\quad \text{since } c_1\mathscr{W}(l) = 0 \text{ as it is a } \mathrm{SU}_2\text{-bundle}\,.\end{aligned}$$

Hence (by induction)

$$\text{(3)}\qquad \begin{aligned}\operatorname{ch}\mathscr{W}(m+1) &= \left((m+1).1 - \binom{m+2}{3}\Omega\right)(2.1 - \Omega)\\ &\quad - \left(m.1 - \binom{m+1}{3}\Omega\right).\end{aligned}$$

Writing $\operatorname{ch}\mathscr{W}(m+1) = (m+2).1 - c_2\mathscr{W}(m+1)$, and equating the coefficients of (3), we get

$$\begin{aligned}c_2\mathscr{W}(m+1) &= \left(2\binom{m+2}{3} + m + 1 - \binom{m+1}{3}\right)\Omega\\ &= \binom{m+3}{3}\Omega\,.\end{aligned}$$

This completes the induction and hence proves the lemma. □

Recall that for any connected complex simple group $G$, the third homotopy group $\pi_3(G)$ is canonically isomorphic with $\mathbb{Z}$.

(5.6) **Corollary.** *For any representation $\varrho$ of $G$ in a finite dimensional (complex) vector space $V$, the induced map $\pi_3(G) \to \pi_3(\mathrm{SL}(V))$ is multiplication by the number $m_V$.*

*Proof.* We can clearly assume that $G$ is simply connected. The representation $\varrho: G \to \mathrm{SL}(V)$ gives rise to a morphism $\tilde{\varrho}: \mathscr{G}/\mathscr{P} \to \mathscr{G}_o/\mathscr{P}_o$, where $\mathscr{G}_o := \mathrm{SL}(V)(\hat{k}_p)$ and $\mathscr{P}_o := \mathrm{SL}(V)(\hat{\mathscr{O}}_p)$. Moreover, the family $\mathscr{U}_o(V)$ parametrized by $\mathscr{G}_o/\mathscr{P}_o$ (got from the standard representation of $\mathrm{SL}(V)$ in $V$) pulls back to the family $\mathscr{U}(V)$ (parametrized by $\mathscr{G}/\mathscr{P}$). In particular, from the functoriality of the determinant bundle (cf. Sect. 3.8), Theorem 5.4, and Lemma 5.2, we see that the induced map $\tilde{\varrho}^*: H^2(\mathscr{G}_o/\mathscr{P}_o, \mathbb{Z}) \to H^2(\mathscr{G}/\mathscr{P}, \mathbb{Z})$ is multiplication by the number $m_V$ (under the canonical identifications $H^2(\mathscr{G}_o/\mathscr{P}_o, \mathbb{Z}) \simeq \mathbb{Z} \simeq H^2(\mathscr{G}/\mathscr{P}, \mathbb{Z})$). But the flag variety $\mathscr{G}/\mathscr{P}$ is homotopic to the based loop group $\Omega_e(K)$ (where $K$ is a compact form of $G$), and similarly $\mathscr{G}_o/\mathscr{P}_o$ is homotopic to $\Omega_e(\mathrm{SU}(V))$. In particular, by the Hurewicz's theorem and the long exact homotopy sequence corresponding to the fibration $\Omega_e(K) \to P(K) \to K$ (where $P(K)$ is the path space of $K$ consisting of the paths starting at the base point $e$), the corollary follows. □

(5.7) *Remark.* J.-L. Brylinski has observed a direct proof of the above corollary using Lemma 5.2.

## 6 Statement of the main theorem and its proof

*Let the triple $\mathfrak{T} = (G, C, p)$ be as in the beginning of Sect. 5.*

(6.1) **Definition.** *Recall the definition of the homogeneous line bundle $\mathfrak{L}(m\chi_o)$ on $X := \mathscr{G}/\mathscr{P} \approx \tilde{\mathscr{G}}/\tilde{\mathscr{P}}$ (for any $m \in \mathbb{Z}$) from Sect. 2.2. Define, for any $p \in \mathbb{Z}$, (cf.* [Ku, *Sect. 3.8*]*)*

$$(1) \qquad H^p(X, \mathfrak{L}(m\chi_o)) = \varprojlim_{\mathfrak{w} \in \tilde{W}/W} H^p(X_{\mathfrak{w}}, \mathfrak{L}_{\mathfrak{w}}(m\chi_o)) .$$

Since $\mathfrak{L}(m\chi_o)$ is a $\tilde{\mathscr{G}}$-equivariant line bundle, $H^p(X, \mathfrak{L}(m\chi_o))$ is canonically a $\tilde{\mathscr{G}}$-module. This module is determined in [Ku, Corollary 3.11] (and also in [M]). We summarize the results:

$$(2) \qquad H^p(X, \mathfrak{L}(m\chi_o)) = 0 , \quad \text{if} \quad p > 0 \quad \text{and} \quad m \geqq 0 ,$$

$$(3) \qquad H^0(X, \mathfrak{L}(m\chi_o)) = 0 , \quad \text{if} \quad m < 0 , \quad \text{and}$$

$$(4) \qquad H^0(X, \mathfrak{L}(m\chi_o)) \simeq V(m\chi_o)^* \quad \text{for} \quad m \geqq 0 ,$$

where $V(m\chi_0)$ is the irreducible highest weight $\tilde{\mathscr{G}}$-module with highest weight $m\chi_o$, and $V(m\chi_o)^*$ denotes its full vector space dual. Recall from Lemma 2.7 that $\Gamma$ is canonically embedded in $\tilde{\mathscr{G}}$. By $H^p(X, \mathfrak{L}(m\chi_o))^\Gamma$ we mean the $\Gamma$-invariants in $H^p(X, \mathfrak{L}(m\chi_o))$.

Recall the definition of the map $\varphi: \mathscr{G} \to \mathscr{X}_o$ from Sect. 1.4, and the family $\mathscr{U}$ parametrized by $X$ from Proposition 2.8. Now define

$$\begin{aligned} X^s &= \{g\mathscr{P} \in \mathscr{G}/\mathscr{P} : \ \varphi(g) \text{ is semistable}\} \\ &= \{x \in X : \mathscr{U}_x := \ \mathscr{U}_{|C \times x} \text{ is semistable}\} , \end{aligned}$$

and set (for any $\mathfrak{w} \in \tilde{W}/W$)

$$X^s_{\mathfrak{w}} = X^s \cap X_{\mathfrak{w}} .$$

Then by the same proof as [R2, Proposition 4.1], $X^s_{\mathfrak{w}}$ is a Zariski open (and non-empty, since $\mathsf{e} \in X^s_{\mathfrak{w}}$) subset of $X_{\mathfrak{w}}$, in particular, $X^s$ is a Zariski open subset of $X$. Now define

$$(5) \qquad H^p(X^s, \mathfrak{L}(m\chi_o)) = \varprojlim_{\mathfrak{w} \in \tilde{W}/W} H^p(X^s_{\mathfrak{w}}, \mathfrak{L}_{\mathfrak{w}}(m\chi_o)) .$$

Clearly $\Gamma$ keeps $X^s$ stable, in particular, $\Gamma$ acts on the cohomology

$$H^p(X^s, \mathfrak{L}(m\chi_o)) ,$$

and we can talk of the $\Gamma$-invariants $H^p(X^s, \mathfrak{L}(m\chi_o))^\Gamma$.

The family $\mathscr{U}_{|X^s}$ yields a morphism $\psi: X^s \to \mathfrak{M}$, which maps any $x \in X^s$ to the $S$-equivalence class of the semistable bundle $\mathscr{U}_x$, where $\mathfrak{M}$ is the moduli space of semistable $G$-bundles on $C$ (cf. Theorem 3.4). (By a morphism $X^s \to \mathfrak{M}$ we mean a map which is a morphism restricted to any $X^s_{\mathfrak{w}}$.)

**(6.2) Lemma.** *There exists a* $\mathfrak{v}_o \in \tilde{W}/W$ *such that*

$$\psi(X^s_{\mathfrak{v}_o}) = \mathfrak{M}\,.$$

*Proof.* Since $\bigcup_{\mathfrak{w}} X^s_{\mathfrak{w}} = \mathcal{G}^s/\mathcal{P}$ and $\psi(\mathcal{G}^s/\mathcal{P}) = \mathfrak{M}$, we get $\mathfrak{M} = \bigcup_{\mathfrak{w}} \psi(X^s_{\mathfrak{w}})$. But by a result of Chevalley (cf. [B, Chap. AG, Corollary 10.2]), $\psi(X^s_{\mathfrak{w}})$ is a finite union of locally closed subvarieties $\{\mathfrak{M}^i_{\mathfrak{w}}\}$ of $\mathfrak{M}$, hence $\mathfrak{M}$ is a countable union $\bigcup \mathfrak{M}^i_{\mathfrak{w}}$ of locally closed subvarieties. But then, by a Baire category argument, $\mathfrak{M}$ is a certain finite union of (locally closed) subvarieties $\{\mathfrak{M}^{i_1}_{\mathfrak{w}_1}, \ldots, \mathfrak{M}^{i_n}_{\mathfrak{w}_n}\}$. Now choosing a $\mathfrak{v}_o \in \tilde{W}/W$ such that $\mathfrak{v}_o \geqq \mathfrak{w}_i$, for all $1 \leqq i \leqq n$, we get that $\mathfrak{M} = \psi(X^s_{\mathfrak{v}_o})$. This proves the lemma. □

**(6.3) Corollary.** *The moduli space* $\mathfrak{M}$ *is a unirational variety.*

*Proof.* Since $X^s_{\mathfrak{w}}$ is an open subset of $X_{\mathfrak{w}}$ and $X_{\mathfrak{w}}$ is a rational variety (by the Bruhat decomposition, cf. Sect. 2.1), the corollary follows from the above Lemma 6.2. □

**(6.4) Proposition.** *For any* $d \geqq 0$ *and any finite dimensional representation* $V$ *of* $G$*, the canonical map*

$$\psi^* : H^0(\mathfrak{M}, \Theta(V)^{\otimes d}) \to H^0(X^s, \psi^*(\Theta(V))^{\otimes d})^{\Gamma}$$

*is an isomorphism, where* $\Theta(V)$ *is the theta bundle on the moduli space* $\mathfrak{M}$ *associated to the representation* $V$ *(cf. Sect. 3.8), and the vector space on the right denotes the space of* $\Gamma$*-invariants under its natural action on the line bundle* $\psi^*(\Theta(V))$. *(Since the map* $\psi : X^s \to \mathfrak{M}$ *is* $\Gamma$*-equivariant, with trivial action of* $\Gamma$ *on* $\mathfrak{M}$*, the pull back bundle* $\psi^*(\Theta(V))$ *has a natural* $\Gamma$*-action.)*

*Proof.* Using Lemma 6.2, we see that the map $\psi^*$ is injective. Now part b) of Proposition 2.8, and Proposition 3.7 show that if $x$ and $y$ are two points in $X^s$ with $\mathcal{U}_y \simeq \mathrm{gr}(\mathcal{U}_x)$, then $y$ belongs to the Zariski closure of the $\Gamma$-orbit of $x$. In particular, two points in $X^s$ are in the same fiber of $\psi$ if and only if the closures of their $\Gamma$-orbits intersect. This, in turn, shows that if $\sigma$ is a $\Gamma$-invariant regular section of $\psi^*(\Theta(V))^{\otimes d}$ on $X^s$, it is induced from a set theoretic section $\underline{\sigma}$ of $\Theta(V)^{\otimes d}$ on $\mathfrak{M}$. That $\underline{\sigma}$ is regular, is seen by taking all those Schubert varieties $X_{\mathfrak{w}}$ such that $\psi(X^s_{\mathfrak{w}}) = \mathfrak{M}$ (cf. Lemma 6.2) and applying Proposition 4.1 to the morphism $\psi_{|X^s_{\mathfrak{w}}} : X^s_{\mathfrak{w}} \to \mathfrak{M}$. □

By the functorial property of the theta bundle, $\Theta(\mathcal{U}(V))_{|X^s}$ is canonically isomorphic to $\psi^*(\Theta(V))$, since $\psi$ is defined using the restriction of the family $\mathcal{U}(V)$ to $X^s$ (cf. Sect. 3.8). Moreover, as observed in Sect. 3.8, the line bundles $\Theta(\mathcal{U}(V))$ and $\mathrm{Det}(\mathcal{U}(V))$ coincide on the whole of $X$.

**(6.5) Proposition.** *Any* $\Gamma$*-invariant regular section of* $\psi^*(\Theta(V))^{\otimes d}$ *on* $X^s$ *extends uniquely to a regular section of* $(\mathrm{Det}\,\mathcal{U}(V))^{\otimes d}$ *on* $X$.

This proposition will be proved in the next section.

We now state and prove our main theorem, assuming the validity of Proposition 6.5.

(6.6) **Theorem.** *Let the triple* $\mathfrak{T} = (G, C, p)$ *be as in the beginning of Sect. 5 and let* $V$ *be a finite dimensional representation of* $G$. *Then, for any* $d \geqq 0$,

$$H^0(\mathfrak{M}, \Theta(V)^{\otimes d}) \simeq H^0(\mathcal{G}/\mathcal{P}, \mathfrak{L}(dm_V \chi_o))^{\Gamma},$$

*where the latter space of* $\Gamma$*-invariants is defined in Sect. 6.1, the integer* $m_V$ *is the same as in Theorem 5.4, and the moduli space* $\mathfrak{M}$ *and the theta bundle* $\Theta(V)$ *are as in Proposition 6.4.*

*In particular,* $H^0(\mathcal{G}/\mathcal{P}, \mathfrak{L}(dm_V\chi_o))^{\Gamma}$ *is finite dimensional.*

*(Observe that, by (4) of Sect. 6.1,* $H^0(\mathcal{G}/\mathcal{P}, \mathfrak{L}(dm_V\chi_o))^{\Gamma}$ *is isomorphic with the space of* $\Gamma$*-invariants in the dual space* $V(dm_V\chi_o)^*$*.)*

*Proof.* We first begin with some simple observations:

(a) *For any line bundle* $\mathfrak{L}$ *on* $X$, *the canonical restriction map* $H^0(X, \mathfrak{L}) \to H^0(X^s, \mathfrak{L}_{|X^s})$ *is injective:* This is seen by restricting any section to each Schubert variety $X_{\mathfrak{w}}$, and observing that (cf. Sect. 6.1) $X_{\mathfrak{w}}^s$ is non-empty (since the base point e corresponds to the trivial bundle, which is semistable), and open (and hence dense) in the irreducible variety $X_{\mathfrak{w}}$.

(b) *If* $\mathfrak{L}$ *is a* $\Gamma$*-equivariant line bundle on* $X$ *(with respect to the standard action of* $\Gamma$ *on* $X$*) and* $\sigma$ *is a regular section of* $\mathfrak{L}$ *such that its restriction to* $X^s$ *is* $\Gamma$*-invariant, then* $\sigma$ *itself is* $\Gamma$*-invariant:* By assumption, for $\gamma \in \Gamma$, the section $\gamma^*(\sigma) - \sigma$ vanishes on $X^s$ and hence on the whole of $X$.

(c) *Suppose that* $\mathfrak{L}'$ *and* $\mathfrak{L}''$ *are two* $\Gamma$*-equivariant line bundles on* $X^s$. *Then any algebraic isomorphism of line bundles* $\xi: \mathfrak{L}' \to \mathfrak{L}''$ *(inducing the identity on the base) in fact is* $\Gamma$*-equivariant. In particular,* $\xi$ *induces an isomorphism of the corresponding spaces of* $\Gamma$*-invariant sections:*

Define a map $\varepsilon: \Gamma \times X^s \to \mathbb{C}^*$ by

$$\varepsilon(\gamma, x) = L_{\gamma^{-1}} \xi_{\gamma x} L_\gamma (\xi_x)^{-1} \in \operatorname{Aut}(\mathfrak{L}''_x) = \mathbb{C}^*,$$

for $\gamma \in \Gamma$ and $x \in X^s$, where $L_\gamma$ is the action of $\gamma$ on the appropriate line bundles, and $\xi_x$ denotes the restriction of $\xi$ to the fiber over $x \in X^s$. It is easy to see that $\varepsilon$ is a regular map and of course $\varepsilon(1, x) = 1$, for all $x \in X^s$. In particular, by Corollary 2.6, $\varepsilon(\gamma, x) = 1$, for all $\gamma \in \Gamma$. This proves the assertion (c).

We now consider $(\operatorname{Det} \mathcal{U}(V))^{\otimes d}_{|X^s}$ as a $\Gamma$-equivariant line bundle by transporting the natural $\Gamma$-action on $\psi^*(\Theta(V))^{\otimes d}$ (cf. Proposition 6.4), via the canonical identification

$$\operatorname{Det} \mathcal{U}(V)_{|X^s} \simeq \psi^*(\Theta(V)). \tag{1}$$

Choose an isomorphism of line bundles on $X$

$$\xi: (\operatorname{Det} \mathcal{U}(V))^{\otimes d} \to \mathfrak{L}(\chi_o)^{\otimes dm_V},$$

which exists by Theorem 5.4. Recall from Sect. 2.2 that $\mathfrak{L}(\chi_o)^{\otimes dm_V}$ is a $\tilde{\mathcal{G}}$-equivariant line bundle, in particular, by Lemma 2.7, it is a $\Gamma$-equivariant line bundle on $X$. Hence by (c) above, the map $\xi_o := \xi_{|X^s}$ is automatically

$\Gamma$-equivariant. We have the following commutative diagram:

$$\begin{array}{ccc} H^0(X, (\operatorname{Det} \mathscr{U}(V))^{\otimes d}) & \xrightarrow[\sim]{\bar{\xi}} & H^0(X, \mathfrak{L}(\chi_o)^{\otimes d m_V}) \\ \downarrow & & \downarrow \\ H^0(X^s, (\operatorname{Det} \mathscr{U}(V))^{\otimes d}) & \xrightarrow[\sim]{\bar{\xi}_o} & H^0(X^s, \mathfrak{L}(\chi_o)^{\otimes d m_V}) \end{array}$$

where $\bar{\xi}$ (resp. $\bar{\xi}_o$) is induced from $\xi$ (resp. $\xi_o$), and the vertical maps are the canonical restriction maps. Observe that $\bar{\xi}_o$ is $\Gamma$-equivariant (since $\xi_o$ is so).

Further, we have

$$\begin{aligned} H^0(\mathfrak{M}, \Theta(V)^{\otimes d}) &\simeq H^0(X^s, (\operatorname{Det} \mathscr{U}(V))^{\otimes d})^{\Gamma} \quad \text{(by (1) and Proposition 6.4)} \\ &\simeq H^0(X^s, \mathfrak{L}(\chi_o)^{\otimes d m_V})^{\Gamma} \quad (\text{under } \bar{\xi}_o)\,. \end{aligned}$$

We complete the proof of the theorem by showing that the restriction map

$$H^0(X, \mathfrak{L}(\chi_o)^{\otimes d m_V})^{\Gamma} \to H^0(X^s, \mathfrak{L}(\chi_o)^{\otimes d m_V})^{\Gamma}$$

is an isomorphism:

It suffices to show that any $\Gamma$-invariant section $\sigma$ of $\mathfrak{L}(\chi_o)^{\otimes d m_V}$ over $X^s$ extends to a section over $X$, for then the extension will automatically be $\Gamma$-invariant by (b) and unique by (a). By the above commutative diagram, this is equivalent to showing that any $\Gamma$-invariant section $\sigma_o$ of $(\operatorname{Det} \mathscr{U}(V))^{\otimes d}$ over $X^s$ extends to the whole of $X$. But this is the content of Proposition 6.5, thereby completing the proof of the theorem. □

Recall the definition of the $\tilde{\mathscr{G}}$-module $V(m\chi_o)$ from Sect. 6.1, and the definition of Lie $\Gamma$ from Sect. 2.9.

(6.7) **Proposition.** *For any $m \geqq 0$, we have*

$$[V(m\chi_o)^*]^{\Gamma} = [V(m\chi_o)^*]^{\operatorname{Lie}\Gamma} = [V(m\chi_o)^*]^{\mathfrak{g}\otimes\mathbb{C}[C^*]},$$

*where $\mathfrak{g}$ is the Lie algebra of the group $G$ and (as in Sect. 1.1) $\mathbb{C}[C^*]$ is the ring of regular functions on the affine curve $C^*$.*

*Proof.* Abbreviate $V(m\chi_o)$ by $V$. Fix $v \in V$ and consider the morphism $\pi_v : \Gamma \to V$ given by $\pi_v(\gamma) = \gamma.v$ for $\gamma \in \Gamma$ (where $\Gamma$ is considered as a subgroup of $\tilde{\mathscr{G}}$, by Lemma 2.7). Recall that, by definition, the action of the Lie algebra Lie $\Gamma$ on $v \in V$ is given by the induced map $(d\pi_v)_e : T_e(\Gamma) = \operatorname{Lie}\Gamma \to T_v(V) = V$.

Fix $\theta \in V^*$. For any $v \in V$, define the map $\theta_v : \Gamma \to \mathbb{A}^1$ by $\theta_v(\gamma) = \theta(\gamma.v)$. The induced map $(d\theta_v)_e : T_e(\Gamma) = \operatorname{Lie}\Gamma \to T_{\theta(v)}(\mathbb{A}^1) = \mathbb{A}^1$ is given by

$$(1) \qquad (d\theta_v)_e(a) = \theta(a.v)\,, \quad \text{for} \quad a \in \operatorname{Lie}\Gamma\,.$$

For any $\gamma_o \in \Gamma$, we now determine the map $(d\theta_v)_{\gamma_o}$: Consider the right translation map $R_{\gamma_o} : \Gamma \to \Gamma$, given by $R_{\gamma_o}(\gamma) = \gamma\gamma_o$. Then we have

$$(2) \qquad (d\theta_v)_{\gamma_o} \circ (dR_{\gamma_o})_e = (d\theta_{\gamma_o.v})_e\,.$$

If $\theta \in [V^*]^{\Gamma}$, then $\theta_v$ (for any fixed $v \in V$) is the constant map $\gamma \mapsto \theta(v)$. In particular, $(d\theta_v)_e \equiv 0$, proving (by 1) that $\theta \in [V^*]^{\mathrm{Lie}\,\Gamma}$. Conversely, take $\theta \in [V^*]^{\mathrm{Lie}\,\Gamma}$. Then by (1) and (2), for any fixed $v \in V$, $(d\theta_v)_{\gamma_o} \equiv 0$ for any $\gamma_o \in \Gamma$. In particular, for any fixed $v \in V$ and $i \geqq 0$, the map $\theta_{v|\Gamma_i} : \Gamma_i \to \mathbb{A}^1$ ($\Gamma_i$ is as in Sect. 2.4) is constant on the irreducible components of $\Gamma_i$ (as the base field is of char. 0). But since $\Gamma$ is connected, $\theta_v$ itself is forced to be a constant. Thus, we have $(\gamma\theta - \theta)v = 0$, for every $v \in V$ and $\gamma \in \Gamma$; proving that $\theta \in [V^*]^{\Gamma}$. Moreover, by Lemma 2.9, we have $\mathrm{Lie}\,\Gamma = \mathfrak{g} \otimes \mathbb{C}[C^*]$. This proves the proposition. □

(6.8) *Remarks.* (a) From the proof it is clear that the above proposition is true with $V(m\chi_o)$ replaced by any algebraic representation of the algebraic group $\Gamma$.

(b) In conformal field theory, the space of vacua is defined to be the space of invariants $[V(m\chi_o)^*]^{\mathfrak{g}\otimes\mathbb{C}[C^*]}$ of the Lie algebra $\mathfrak{g} \otimes \mathbb{C}[C^*]$ in the affine Kac-Moody algebra module $V(m\chi_o)^*$ (cf. [TUY, Definition 2.2.2]). We see, by Theorem 6.6 and Proposition 6.7, that the space of vacua is isomorphic to the space of generalised theta functions.

As an immediate consequence of the above remark (b), we obtain the following.

(6.9) **Corollary.** *Let the notation and assumptions be as in Theorem 6.6. Then the space of coinvariants* $V(dm_V\chi_o)/((\mathfrak{g} \otimes_{\mathbb{C}} \mathbb{C}[C^*]).V(dm_V\chi_o))$ *is finite dimensional. (Cf.* [K, *Exercise 11.10, p. 209*] *for a purely algebraic proof of this Corollary.).*

## 7 Geometric invariant theory – Proof of Proposition 6.5

In this section $C$ is a smooth projective irreducible curve over $\mathbb{C}$ with a fixed base point $p$.

(7.1) **Lemma.** *Let $X$ be an irreducible normal variety, $U \subset X$ a non-empty open subset and $\mathfrak{L}$ a line bundle on $X$. Then any element of $\bigoplus_{n\in\mathbb{Z}} H^0(U, \mathfrak{L}^n)$ which is integral over $\bigoplus_n H^0(X, \mathfrak{L}^n)$ belongs to $\bigoplus_n H^0(X, \mathfrak{L}^n)$.*

*Proof.* Since the rings in question are graded, it suffices to prove the lemma only for homogeneous elements. Let $b \in H^0(U, \mathfrak{L}^{n_o})$ be integral over $\oplus H^0(X, \mathfrak{L}^n)$, i.e., $b$ satisfies a relation $b^m + a_1 b^{m-1} + \ldots + a_m = 0$ with $a_i \in \oplus H^0(X, \mathfrak{L}^n)$. Let $D$ be a prime divisor in $X \backslash U$ and let $b$ have a pole of order $\ell \geqq 0$ along $D$. Then the order of the pole of $b^m$ along $D$ is of course $\ell m$ and that of $a_i b^{m-i}$ is $\leqq \ell(m-1)$ for every $i \geqq 1$. But since $b^m + a_1 b^{m-1} + \ldots + a_{m-1} b$ is by assumption regular along $D$, we are forced to have $\ell = 0$, i.e., $b$ is regular along $D$. Hence $b \in H^0(X, \mathfrak{L}^{n_o})$. □

We state now a general result on the extendability of invariant sections for actions of reductive groups.

Let a reductive group $H$ operate on a projective scheme $Q$ along with a linearization with respect to an ample line bundle $\mathfrak{L}$ on $Q$. Let $Q^s$ denote the open subset of $Q$ of semistable points (with respect to the $H$-equivariant ample line bundle $\mathfrak{L}$) for the action of $H$. Recall that $Q^s := \{x \in Q : \exists \sigma \in H^0(Q, \mathfrak{L}^N)^H$ for some $N \geqq 1$ such that $\sigma(x) \neq 0\}$. We then have the following proposition (cf. [NRa, Se1]).

(7.2) **Proposition.** *Let $U \supset Q^s$ be a $H$-invariant open subset of $Q$, which (i.e. $U$) is normal and irreducible. Then, for $N \geqq 1$, any $H$-invariant section of $\mathfrak{L}^N$ on $Q^s$ can be extended to a $H$-invariant section of $\mathfrak{L}^N$ on $U$.*

*Proof.* We indicate the proof when $Q$ is normal and $U = Q$. (The general case can be reduced to this case by the arguments as in [NRa, Lemma 4.15].) Let $\sigma \in H^0(Q^s, \mathfrak{L}^N)^H$. If $D$ is an irreducible divisor in $Q \backslash Q^s$ along which $\sigma$ has a pole, then we can find a $\tau \in H^0(Q, \mathfrak{L}^{N'})^H$ (for some $N' \gg 0$) such that $\sigma^{N_1}\tau^{N_2}$ will not vanish identically on $D$ for suitable $N_1, N_2 > 0$. This is a contradiction since $D \subset Q \backslash Q^s$, in particular, any invariant section vanishes on $D$. □

(7.3) G.I.T. *and moduli of vector bundles.* We recall the construction of the moduli spaces of vector bundles on $C$ using G.I.T.. Let $r \geqq 1$ and $\delta$ be integers. For the fixed point $p \in C$ and for a coherent sheaf $\mathbf{F}$ on $C$, put $\mathbf{F}(m) = \mathbf{F} \otimes_{\mathcal{O}} \mathcal{O}(mp)$, for any $m \in \mathbb{Z}$, where $\mathcal{O} = \mathcal{O}_C$ is the structure sheaf of $C$. We can choose an integer $m_o = m_o(r, \delta)$ such that for any $m \geqq m_o$ and any semistable vector bundle $\mathbf{E}$ of rank $r$ and degree $\delta$ on $C$, we have $H^1(\mathbf{E}(m)) = 0$ and $\mathbf{E}(m)$ is generated by its global sections. Let $q = \dim H^0(\mathbf{E}(m)) = \delta + r(m+1-g)$ and consider the *Grothendieck quot scheme* $Q$ consisting of coherent sheaves on $C$ which are quotients of $\mathbb{C}^q \otimes \mathcal{O}$ with Hilbert polynomial (in the indeterminate $v$) $rv + q$ (where $g$ is the genus of $C$). The group $\mathrm{GL}(q, \mathbb{C})$ operates canonically on $Q$ and the action on $C \times Q$ (with the trivial action on $C$) lifts to an action of the tautological sheaf $\mathscr{E}$ on $C \times Q$.

We denote by $R_o$ the GL($q$)-invariant open subset of $Q$ consisting of those $x \in Q$ such that $\mathscr{E}_x = \mathscr{E}_{|C \times x}$ is locally free and such that the following canonical map is an isomorphism:

$$\mathbb{C}^q = H^0(\mathbb{C}^q \otimes \mathcal{O}) \xrightarrow{\sim} H^0(\mathscr{E}_x).$$

Then $R_o$ is smooth and irreducible. We still denote by $\mathscr{E}$ the restriction of the family to $R_o$.

We obtain a GL($q$)-linearized ample line bundle $\mathfrak{L}$ on $Q$ by embedding $Q$ in a suitable Grassmannian as follows: We choose an integer $k_o = k_o(m)$ such that for $k \geqq k_o$ the composite map

$$\mathbb{C}^q \otimes H^0(\mathcal{O}(k)) \to H^0(\mathscr{E}_x) \otimes H^0(\mathcal{O}(k)) \to H^0(\mathscr{E}_x(k))$$

is surjective for all $x \in Q$, and such that the morphism $Q \to$ Grass (taking $x \mapsto H^0(\mathscr{E}_x(k))$) is a closed embedding, where $\mathcal{O}(k) := \mathcal{O}(kp)$ and Grass denotes the Grassmannian of $\delta+1-g+r(m+k)$ dimensional quotient spaces of $\mathbb{C}^q \otimes H^0(\mathcal{O}(k))$. We define the ample line bundle $\mathfrak{L}$ on $Q$ to be the pull back of the natural ample line bundle on Grass, namely the determinant of the universal quotient bundle on Grass. The action of GL($q$) clearly lifts to $\mathfrak{L}$.

There exists a positive integer $m'_o$ with $m'_o \geqq m_o$ such that for any integer $m \geqq m'_o$ there is a positive integer $k'_o = k'_o(m) \geqq k_o(m)$ with the property that the following conditions are equivalent (for any $k \geqq k'_o$):

(1) A point $x \in Q$ is semistable in the sense of G.I.T. for the SL($q$)-linearized bundle $\mathfrak{L}$.

(2) $x \in R_o$ (in particular, the sheaf $\mathscr{E}_x$ is locally free) and the bundle $\mathscr{E}_x$ is a semistable vector bundle on $C$.

We denote by $R_o^s$, by abuse of notation, the set of semistable points (in the sense of G.I.T.) in $Q$. By the above equivalent conditions, we have $R_o^s \subset R_o$. Now the G.I.T. quotient $R_o^s//\,\mathrm{GL}(q)$ yields the *moduli space* $\mathfrak{M}_o$ *of vector bundles* of rank $r$ and degree $\delta$.

(For all this, see [NRa, Appendix A] or [Le].)

(7.4) We note that we can arrange the above construction in such a way that any fixed bounded family of vector bundles of rank $r$ and degree $\delta$ occurs in $R_o$. (This observation will be crucial for us.) More precisely, let $\mathscr{V}_o \to C \times T_o$ be a family of vector bundles of rank $r$ and degree $\delta$ (parametrized by a variety $T_o$). We can find an integer $m_{T_o}$ such that for $m \geqq m_{T_o}$, we have:

(1) $R^1 p_{T_o *}(\mathscr{V}_o(m)) = 0$,

(2) $p_{T_o *}(\mathscr{V}_o(m))$ is a vector bundle on $T_o$ (say of rank $q$).

(3) The canonical map $p^*_{T_o} p_{T_o *}(\mathscr{V}_o(m)) \to \mathscr{V}_o(m)$ is surjective, where $p_{T_o} : C \times T_o \to T_o$ is the projection on the second factor,

$$\mathscr{V}_o(m) := \mathscr{V}_o \otimes_{\mathscr{O}_{C\times T_o}} p^*_C \mathscr{O}(m)\,,$$

and $p_C : C \times T_o \to C$ is the projection on the first factor.

Choose $m > \max(m_{T_o}, m'_o)$, where $m'_o$ is as in Sect. 7.3. Let $\mathbf{P}_o$ be the frame bundle of $p_{T_o *}(\mathscr{V}_o(m))$ with the projection $\pi_o : \mathbf{P}_o \to T_o$. Then there exists a canonical GL($q$)-equivariant morphism $\varphi_o : \mathbf{P}_o \to R_o$ such that the families $\pi^*_o(\mathscr{V}_o)$ and $\varphi^*_o(\mathscr{E}(-m))$ are isomorphic, where the family $\mathscr{E}$ on $R_o$ is as in Sect. 7.3, $\mathscr{E}(-m) := \mathscr{E} \otimes_{\mathscr{O}_{C\times R_o}} \bar{p}^*_C(\mathscr{O}(-m))$ and $\bar{p}_C : C \times R_o \to C$ is the projection on the first factor.

(7.5) **Lemma.** *Suppose that* $\delta = 0$. *Let* $\Theta(\mathscr{F})$ *be the theta bundle (on* $R_o$*) of the family* $\mathscr{F} := \mathscr{E}(-m)$ *(cf. Sect. 3.8). Then there exist positive integers* $e$ *and* $f$ *such that, as* GL($q$) *equivariant line bundles,*

$$\Theta(\mathscr{F})^{\otimes e} \cong (\mathfrak{L}_{|R_o})^{\otimes f}\,,$$

*where* $\mathfrak{L}$ *is the ample line bundle on* $Q$ *defined in Sect. 7.3.*

*Proof.* For any integer $\ell \geqq 1$, we have

$$\mathrm{Det}(\mathscr{F}(\ell)) = (\mathrm{Det}\,\mathscr{F}) \otimes (\det(\mathscr{F}_{|p\times R_o}))^{-\ell}\,,$$

as is seen from the exact sequence

$$0 \to \mathscr{F} \to \mathscr{F}(\ell) \to \mathscr{F} \otimes_{\mathscr{O}_{C\times R_o}} \bar{p}^*_C(\mathscr{O}/\mathfrak{m}^\ell_p) \to 0\,,$$

where $\mathfrak{m}_p \subset \mathscr{O}$ is the sheaf of functions on $C$ vanishing at $p$. Observing that $\mathfrak{L}^{-1}_{|R_o} \simeq \mathrm{Det}(\mathscr{F}(k+m))$ (where $m$ and $k$ are as in Sect. 7.3) and $\mathrm{Det}\,\mathscr{F}(m)$ is trivial, we see that $\mathfrak{L}_{|R_o} \simeq (\det(\mathcal{F}_{|p \times R_o}))^k$ and $\Theta(\mathcal{F}) \simeq (\det(\mathcal{F}_{|p \times R_o}))^{m+1-g}$. (By choosing $m$ large enough in Sect. 7.3, we may assume that $m+1-g > 0$.) This proves the lemma. (Compare [NRa, Proof of Theorem 1(B)].) □

(7.6) *Remark.* One knows that $\Theta(\mathcal{F})_{|R^s_o}$ descends to a line bundle $\Theta$ on $\mathfrak{M}_o$ [DN], [NRa, Proof of Theorem 1(A)]. By G.I.T., some power of $\mathfrak{L}_{|R^s_o}$ descends as an ample line bundle on $\mathfrak{M}_o$. Using Lemma 7.5, we see that $\Theta$ is an ample line bundle on $\mathfrak{M}_o$.

(7.7) **Proposition.** *Let $f_o : R^s_o \to \mathfrak{M}_o = R^s_o // \mathrm{GL}(q)$ be the canonical map. Let $\sigma$ be a section of $\Theta^{\otimes \ell}$ over $\mathfrak{M}_o$ (for any $\ell \geqq 1$). Then the section $f^*_o(\sigma)$ over $R^s_o$ of the line bundle $f^*_o(\Theta^{\otimes \ell}) \simeq (\Theta.\mathcal{F})^{\otimes \ell}$ extends uniquely as a* GL($q$)*-invariant section of $(\Theta.\mathcal{F})^{\otimes \ell}$ over $R_o$, where, as in Lemma 7.5, $\mathcal{F} = \mathcal{E}(-m)$.*

*Proof.* By Proposition 7.2, any GL($q$)-invariant section of any positive power of $\mathfrak{L}$ over $R^s_o$ extends to $R_o$, as $R_o$ is smooth. Thus, by Lemma 7.5, some power of $f^*_o(\sigma)$ extends to $R_o$. Hence, by Lemma 7.1, $f^*_o(\sigma)$ itself extends. Observe that $R^s_o \neq \emptyset$, as the trivial bundle is semistable. Since $R_o$ is irreducible, the extension is unique and invariant. □

(7.8) *Moduli of principal G-bundles.* Assume that $G$ is a connected semisimple algebraic group. Let $T$ be a variety parametrizing a family $\mathcal{V}$ of $G$-bundles on $C$. Then there exists a smooth quasi-projective irreducible variety $R$ with an action of GL($N$) (for some $N$), a family $\mathcal{W}$ of $G$-bundles on $C$ parametrized by $R$ and a lift of the GL($N$)-action to $\mathcal{W}$ (as bundle automorphisms), such that the following holds:

(I) Let $R^s := \{x \in R : \mathcal{W}_x := \mathcal{W}_{|C \times x}$ is a semistable $G$-bundle$\}$ be the GL($N$)-invariant open subset of $R$. Then the canonical map $R^s \to \mathfrak{M}$ is surjective, where $\mathfrak{M}$ is the moduli space of semistable $G$-bundles.

(II) Moreover, there exists a principal GL($N$)-bundle $\pi : \mathbf{P} \to T$ and a GL($N$)-equivariant morphism $\varphi : \mathbf{P} \to R$ such that the families $\varphi^*(\mathcal{W})$ and $\pi^*(\mathcal{V})$ are isomorphic. (See [R1].)

Now if $V$ is a finite dimensional representation of $G$, we denote by $\Theta(\mathcal{W}(V))$ the theta bundle on $R$ of the family $\mathcal{W}(V)$, of vector bundles of rank $r$ ($r = \dim V$) and degree 0 parametrized by $R$, obtained from the family $\mathcal{W}$ of (principal) $G$-bundles via the representation $V$. Note that GL($N$) operates on $\Theta(\mathcal{W}(V))$. Let $\Theta(V)$ be the theta bundle on the moduli space $\mathfrak{M}$ associated to the representation $V$ of $G$ (cf. Sect. 3.8). If $f : R^s \to \mathfrak{M}$ is the canonical map, we have

$$f^*(\Theta(V)) \simeq \Theta(\mathcal{W}(V))_{|R^s} .$$

(7.9) **Proposition.** *Any section of $\Theta(V)^{\otimes \ell}$ over $\mathfrak{M}$ (for $\ell \geqq 1$), considered as a* GL($N$)*-invariant section of $(\Theta(\mathcal{W}(V)))^{\otimes \ell}$ over $R^s$, extends uniquely as an invariant section of $(\Theta(\mathcal{W}(V)))^{\otimes \ell}$ over $R$.*

*Proof.* We will prove the proposition by showing that such a section of $\Theta(\mathcal{W}(V))^{\otimes \ell}$ over $R^s$ is integral over $\bigoplus_n H^0(R, \Theta(\mathcal{W}(V))^n)$, and then applying Lemma 7.1:

We will apply the results of Sects. 7.3 and 7.4. With the notation of Sect. 7.4, choose for $T_o$ the variety $R$ and for $\mathscr{V}_o$ the vector bundle $\mathscr{W}(V)$ on $C \times R$ defined above in Sect. 7.8. Let $h = h_V : \mathfrak{M} \to \mathfrak{M}_o$ be the morphism defined by $V$, where (as in Sect. 7.3) $\mathfrak{M}_o$ is the moduli space of rank $r$ and degree 0 vector bundles on $C$. We have $h^*(\Theta) \simeq \Theta(V)$, where $\Theta$ is the theta bundle on $\mathfrak{M}_o$ (see Remark 7.6 and Sect. 3.8).

Since $\Theta$ is ample and $h$ is a projective morphism, we see that

$$\bigoplus_n H^0(\mathfrak{M}, \Theta(V)^{\otimes n})$$

is a module of finite type over $\bigoplus_n H^0(\mathfrak{M}_o, \Theta^{\otimes n})$. In particular, the former ring is integral over the latter. Let $\sigma$ be a section of $\Theta(V)^{\otimes \ell}$ over $\mathfrak{M}$. Then $\sigma$ satisfies an equation

$$\sigma^d + a_{d-1}\sigma^{d-1} + \ldots + a_1\sigma + a_0 = 0,$$

where $a_j \in \bigoplus_n H^0(\mathfrak{M}_o, \Theta^{\otimes n})$. Let $f_o : R_o^s \to R_o^s // \mathrm{GL}(q) = \mathfrak{M}_o$ be the canonical map (as in Proposition 7.7). If $\{a_{ij}\}_i$ are the homogeneous components of $a_j$, using Proposition 7.7, we can extend $f_o^*(a_{ij})$ to an invariant section (say) $\sigma_{ij}$ of some appropriate power of $\Theta(\mathscr{F})$ over $R_o$, where $\mathscr{F} = \mathscr{E}(-m)$ (as in Lemma 7.5). Pulling back $\sigma_{ij}$ via $\varphi_o : \mathbf{P}_o \to R_o$ (cf. Sect. 7.4) and descending them via the projection $\pi_o : \mathbf{P}_o \to R$ (cf. Sect. 7.4) to sections of some appropriate powers of $\Theta(\mathscr{W}(V))$ over $R$), we see that $f^*(\sigma)$ is integral over $\bigoplus_n H^0(R, \Theta(\mathscr{W}(V))^{\otimes n})$, where $f : R^s \to \mathfrak{M}$ is the canonical map as in Sect. 7.8. (Observe that $\varphi_o$ maps $\pi_o^{-1}(R^s)$ into $R_o^s$.) □

Finally we prove Proposition 6.5 and thus complete the proof of Theorem 6.6.

(7.10) *Proof of Proposition 6.5.* Let $\tilde{\sigma}$ be a $\Gamma$-invariant section of $\psi^*(\Theta(V))^{\otimes d}$ on $X^s$. By Proposition 6.4, there is a section $\sigma$ of $\Theta(V)^{\otimes d}$ over $\mathfrak{M}$ such that $\psi^*(\sigma) = \tilde{\sigma}$. Let $X_\mathfrak{w}$ be a Schubert variety. With the notation of Sect. 7.8 we construct $R$, where we take for $T$ the variety $X_\mathfrak{w}$ and for $\mathscr{V}$ the restriction of the family $\mathscr{U}$ (Proposition 2.8) to $X_\mathfrak{w}$. Now $\sigma$ can be viewed as an invariant section of $\Theta(\mathscr{W}(V))^{\otimes d}$ over $R^s$ and hence (by Proposition 7.9) extends to an invariant section $\sigma'$ of $\Theta(\mathscr{W}(V))^{\otimes d}$ over $R$. Pulling back $\sigma'$ via $\varphi : \mathbf{P} \to R$ (cf. Sect. 7.8) and descending via $\pi : \mathbf{P} \to T = X_\mathfrak{w}$, we obtain a section of $(\Theta(\mathscr{U}(V)_{|X_\mathfrak{w}}))^{\otimes d}$ which extends the section $\tilde{\sigma}_{|X_\mathfrak{w}^s}$. Moreover, this extension is unique as $X_\mathfrak{w}^s \neq \emptyset$ (cf. Sect. 6.1). Varying $X_\mathfrak{w}$ we see that $\tilde{\sigma}$ extends to a section of $\Theta(\mathscr{U}(V))^{\otimes d}$ over $X$. This completes the proof of the proposition. □

*Acknowledgement.* We thank R. R. Simha, J. Wahl for some helpful conversations, and J.-L. Brylinski, M. S. Raghunathan for some comments. The first author was partially supported by the NSF grant no. DMS-9203660.

## References

[B] Borel, A.: Linear algebraic groups. New York: Benjamin 1969
[BR] Bhosle, U., Ramanathan, A.: Moduli of parabolic $G$-bundles on curves. Math. Z. **202**, 161–180 (1989)
[D] Dynkin, E.B.: Semisimple subalgebras of semisimple Lie algebras. Am. Math. Soc. Transl., Ser. II, vol. **6**, 111–244 (1957)
[DN] Drezet, J.-M., Narasimhan, M.S.: Groupe de Picard des variétés de modules de fibrés semi-stables sur les courbes algébriques. Invent. Math. **97**, 53–94 (1989)
[F] Fulton, W.: Introduction to intersection theory in algebraic geometry. (Reg. Conf. Ser. Math., vol. 54) Providence, RI: Am. Math. Soc. 1984
[H1] Harder, G.: Halbeinfache Gruppenschemata über Dedekindringen. Invent. Math. **4**, 165–191 (1967)
[H2] Harder, G.: Halbeinfache Gruppenschemata über vollständigen Kurven. Invent. Math. **6**, 107–149 (1968)
[Ha] Hartshorne, R.: Algebraic geometry. Berlin Heidelberg New York: Springer 1977
[Hu] Humphreys, J.E.: Introduction to Lie algebras and representation theory. (Grad. Texts Math., vol. 9) Berlin Heidelberg New York: Springer 1972
[IM] Iwahori, N., Matsumoto, H.: On some Bruhat decomposition and the structure of the Hecke rings of p-adic Chevalley groups. Publ. Math., Inst. Hautes Étud. Sci. **25**, 237–280 (1965)
[K] Kac, V.G.: Infinite dimensional Lie algebras. (Third edition) Cambridge: Cambridge University Press 1990
[KL] Kazhdan, D., Lusztig, G.: Schubert varieties and Poincaré duality. Proc. Symp. Pure Math., vol. 36, pp. 185–203, 1980
[Ku] Kumar, S.: Demazure character formula in arbitrary Kac-Moody setting. Invent. Math. **89**, 395–423 (1987)
[L] Lang, S.: Introduction to Arakelov theory. Berlin Heidelberg New York: Springer 1988
[Le] Le Potier, J.: Fibrés vectoriels sur les courbes algébriques. Cours de DEA, Université Paris 7 (1991)
[M] Mathieu, O.: Formules de caractères pour les algèbres de Kac-Moody générales. Astérisque **159–160**, 1–267 (1988)
[Mi] Milne, J.S.: Étale cohomology. Princeton: Princeton University Press 1980
[Mu] Mulase, M.: A correspondence between an infinite grassmannian and arbitrary vector bundles on algebraic curves. Proc. Symp. Pure Math., vol. 49, pp. 39–50, 1989
[Mum] Mumford, D.: The red book of varieties and schemes. (Lect. Notes Math., vol. 1358) Berlin Heidelberg New York: Springer 1988
[NR] Narasimhan, M.S., Ramanan, S.: Moduli of vector bundles on a compact Riemann surface. Ann. Math. **89**, 14–51 (1969)
[NRa] Narasimhan, M.S., Ramadas, T.R.: Factorisation of generalised theta functions. I. Invent. Math. **114**, 565–623 (1993)
[NS] Narasimhan, M.S., Seshadri, C.S.: Stable and unitary vector bundles on a compact Riemann surface. Ann. Math. **82**, 540–567 (1965)
[PS] Pressley, A., Segal, G.: Loop groups. Oxford: Oxford Science Publications, Clarendon Press 1986
[Q] Quillen, D.: Determinants of Cauchy-Riemann operators over a Riemann surface. Funct. Anal. Appl. **19**, 31–34 (1985)
[R1] Ramanathan, A.: Stable principal bundles on a compact Riemann surface-construction of moduli space. Thesis, University of Bombay (1976)
[R2] Ramanathan, A.: Stable principal bundles on a compact Riemann surface. Math. Ann. **213**, 129–152 (1975)
[R3] Ramanathan, A.: Deformations of principal bundles on the projective line. Invent. Math. **71**, 165–191 (1983)
[Ra] Raghunathan, M.S.: Principal bundles on affine space. In: C.P. Ramanujam. A tribute (T.I.F.R. studies in Math., No. 8), pp. 187–206, Oxford University Press, 1978
[RR] Ramanan, S., Ramanathan, A.: Some remarks on the instability flag. Tohoku Math. J. **36**, 269–291 (1984)
[Sa] Šafarevič, I.R.: On some infinite-dimensional groups. II. Math. USSR-Izv. **18**, 185–194 (1982)

[S] Segal, G.: The topology of spaces of rational functions. Acta Math. **143**, 39–72 (1979)

[Se1] Serre, J.P.: Espaces fibrés algébriques. In: Anneaux de Chow et applications. Séminaire C. Chevalley, 1958

[Se2] Serre, J.P.: Cohomologie Galoisienne (Lect. Notes Math., vol. 5). Berlin Heidelberg New York: Springer 1986

[Ses] Seshadri, C.S.: Quotient spaces modulo reductive algebraic groups. Ann. Math. **95**, 511–556 (1972)

[Sl] Slodowy, P.: On the geometry of Schubert varieties attached to Kac-Moody Lie algebras. In: Can. Math. Soc. Conf. Proc. on "Algebraic Geometry", vol. 6, pp. 405–442, Vancouver 1984

[TUY] Tsuchiya, A., Ueno, K., Yamada, Y.: Conformal field theory on universal family of stable curves with gauge symmetries. Adv. Stud. Pure Math. **19**, 459–566 (1989)

[W] Witten, E.: Quantum field theory, Grassmannians, and algebraic curves. Commun. Math. Phys. **113**, 529–600 (1988)

Math. Ann. 308, 155–173 (1997)

Mathematische Annalen


# Picard group of the moduli spaces of $G$-bundles

**Shrawan Kumar**[1], **M.S. Narasimhan**[2]

[1] Department of Mathematics, University of North Carolina at Chapel Hill, Chapel Hill, NC 27599-3250, USA (e-mail: kumar@math.unc.edu)

[2] ICTP, P.O. Box 586, Strada Costiera 11, Miramare, I-34100 Trieste, Italy (e-mail: narasim@ictp.trieste.it)

Received: 29 November 1995/Revised version: 13 January 1996

*Mathematics Subject Classification (1991):* 14C22

## Introduction

Let $G$ be a simple simply-connected connected complex affine algebraic group and let $C$ be a smooth irreducible projective curve of genus $\geqq 2$ over the field of complex numbers $\mathbb{C}$. Let $\mathfrak{M}$ be the moduli space of semistable principal $G$-bundles on $C$ and let Pic $\mathfrak{M}$ be its Picard group, i.e., the group of isomorphism classes of algebraic line bundles on $\mathfrak{M}$. Following is our main result (which generalizes a result of Drezet-Narasimhan for $G = \mathrm{SL}(N)$ [DN] to any $G$).

**(A) Theorem.** *With the notation as above,* $\mathrm{Pic}\,(\mathfrak{M}) \approx \mathbb{Z}$.

A more precise result is obtained in Theorem (2.4) together with Theorem (4.9).

We use the above result and a result of Grauert-Riemenschneider to prove the following second main result of this paper.

**(B) Theorem.** *The dualizing sheaf $\omega$ of the moduli space $\mathfrak{M}$ is locally free. In particular, $\mathfrak{M}$ is a Gorenstein variety.*

*Further, for any finite dimensional representation $V$ of $G$, $H^i(\mathfrak{M}, \Theta(V)) = 0$, for all $i > 0$, where $\Theta(V)$ is the theta bundle on the moduli space $\mathfrak{M}$. In particular,*

$$\mathscr{X}(\mathfrak{M}, \Theta(V)) = \dim H^0(\mathfrak{M}, \Theta(V))\,,$$

*where $\mathscr{X}$ is the Euler-Poincaré characteristic.*

In fact, we have a sharper result than the above (cf. Theorem 2.8).

We make essential use of the generalized flag variety $X$ associated to the affine Kac-Moody group corresponding to $G$, which (i.e. $X$) parametrizes an algebraic family of $G$-bundles on $C$, and the fact that Pic $X \simeq \mathbb{Z}$. We also need to make use of the explicit construction of the moduli space $\mathfrak{M}$ via GIT.

## 1. Notation

Let $G$ be a simple simply-connected connected complex affine algebraic group and let $C$ be a smooth irreducible projective curve of genus $\geqq 2$ over the field of complex numbers $\mathbb{C}$. As in [KNR, Theorem 3.4], let $\mathfrak{M}$ be the moduli space of semistable principal $G$-bundles on $C$. Also, fix a point $p \in C$ and recall the definition of the generalized flag variety $X = \mathcal{G}/\mathcal{P}$ (associated to the affine Kac-Moody group $\mathcal{G}$ corresponding to the group $G$) from [KNR, Sect. 2.1], its open subset $X^s$ and the morphism $\psi : X^s \to \mathfrak{M}$ from [loc. cit., Definition 6.1]. Also, recall the notation $\Gamma$ from [loc. cit., Sect. 1.1] and the notation $\widetilde{W}, W, X_{\mathfrak{w}}$ from [loc. cit., Sect. 2.1].

For any ind-variety $Y$, by an *algebraic vector bundle of rank* $r$ over $Y$, we mean an ind-variety $E$ together with a morphism $\theta : E \to Y$ such that (for any $n$) $E_n \to Y_n$ is an algebraic vector bundle of rank $r$ over the (finite dimensional) variety $Y_n$, where $\{Y_n\}$ is the filtration of $Y$ giving the ind-variety structure and $E_n := \theta^{-1}(Y_n)$. If $r = 1$, we call $E$ an *algebraic line bundle* over $Y$. For an introduction to ind-varieties, see [Ku2, Appendix B].

Let $E$ and $F$ be two algebraic vector bundles over $Y$. Then a morphism (of ind-varieties) $\varphi : E \to F$ is called a *bundle morphism* if the following diagram is commutative:

$$\begin{array}{ccc} E & \xrightarrow{\varphi} & F \\ & \searrow \quad \swarrow & \\ & Y & \end{array}$$

and moreover $\varphi_{|E_n} : E_n \to F_n$ is a bundle morphism for all $n$. In particular, we have the notion of isomorphism of vector bundles over $Y$.

We define Pic $Y$ as the set of isomorphism classes of algebraic line bundles on $Y$. It is clearly an abelian group under the tensor product of line bundles.

For any set $Y$, $I_Y$ denotes the identity map of $Y$.

## 2. Statement of the main theorems

We follow the notation from Sect. 1.

**(2.1) Lemma.** *The morphism $\psi : X^s \to \mathfrak{M}$ induces an injective map*

$$\psi^* : \mathrm{Pic}\,(\mathfrak{M}) \longrightarrow \mathrm{Pic}\,(X^s)\,.$$

*Proof.* Let $\mathfrak{L} \in \mathrm{Pic}\,(\mathfrak{M})$ be in the kernel of $\psi^*$, i.e., $\psi^*(\mathfrak{L})$ admits a nowhere-vanishing regular section $\sigma$ on the whole of $X^s$. Fix $m \in \mathfrak{M}$ and a trivialization for $\mathfrak{L}_{|m}$. This canonically induces a trivialization for the bundle $\psi^*(\mathfrak{L})_{|\psi^{-1}(m)}$. In particular, the section $\sigma_{|\psi^{-1}(m)}$ can be viewed as a (regular) map $\sigma_m : \psi^{-1}(m) \to \mathbb{C}^*$. But $\psi^{-1}(m)$ is a certain union of $\Gamma$-orbits say $\psi^{-1}(m) = \bigcup_{i \in I} \Gamma x_i$, for $x_i \in X$ and moreover $\overline{\Gamma x_i} \cap \overline{\Gamma x_j} \neq \emptyset$, for any $i, j \in I$, where $\overline{\Gamma x_i}$ is the closure of $\Gamma x_i$ in $X^s$ (cf. [KNR, Proof of Proposition 6.4]). Fixing $i \in I$, we get a regular map $\sigma_{m,i} : \Gamma \to \mathbb{C}^*$, defined as $\sigma_{m,i}(\gamma) = \sigma_m(\gamma x_i)$, for

$\gamma \in \Gamma$. Now by [Ku2, Proposition 2.4], $\sigma_{m,i}$ is a constant map for any $i \in I$, and hence $\sigma_m : \psi^{-1}(m) \to \mathbb{C}^*$ itself is a constant map. Thus the section $\sigma$ descends to a set theoretic section $\hat{\sigma}$ of the line bundle $\mathfrak{L}$, which is regular by [KNR, Proposition 4.1 and Lemma 6.2]. Of course, the section $\hat{\sigma}$ does not vanish anywhere on $\mathfrak{M}$ (since $\sigma$ was chosen to be nowhere-vanishing on $X^s$). This proves that $\mathfrak{L}$ is a trivial line bundle on $\mathfrak{M}$, thereby proving the lemma. □

It is clear that for any ind-variety $Y$, we have a natural map $\alpha : \text{Pic}\, Y \to \varprojlim_n \text{Pic}\,(Y_n)$.

**(2.2) Lemma.** Pic $X \approx \varprojlim_{\mathfrak{w} \in \tilde{W}/W} \text{Pic}(X_{\mathfrak{w}}) \approx \mathbb{Z}$.

*Proof.* We will freely follow the notation from [KNR, Sect. 2.3]. Since the line bundles $\mathfrak{L}(d\chi_0)$ (for $d \in \mathbb{Z}$) (denoted in loc. cit. by $\mathscr{L}(d\chi_o)$) are, by construction, algebraic line bundles on $X$ and moreover, for any $\mathfrak{w} \geqq \mathfrak{s}_o$, $\mathfrak{L}(\chi_0)_{|X_{\mathfrak{w}}}$ freely generates $\text{Pic}(X_{\mathfrak{w}})$, the surjectivity of the map $\alpha$ follows. Now we come to the injectivity of $\alpha$:

Let $\mathfrak{L} \in \text{Ker}\, \alpha$. Fix a non-zero vector $v_o$ in the fiber of $\mathfrak{L}$ over the base point $e \in X$. Then $\mathfrak{L}_{|X_{\mathfrak{w}}}$ being a trivial line bundle on each $X_{\mathfrak{w}}$, we can choose a nowhere-vanishing section $s_{\mathfrak{w}}$ of $\mathfrak{L}_{|X_{\mathfrak{w}}}$ such that $s_{\mathfrak{w}}(e) = v_o$. We next show that for any $\mathfrak{v} \geqq \mathfrak{w}$, $s_{\mathfrak{v}|X_{\mathfrak{w}}} = s_{\mathfrak{w}}$: Clearly $s_{\mathfrak{v}|X_{\mathfrak{w}}} = f s_{\mathfrak{w}}$, for some algebraic function $f : X_{\mathfrak{w}} \to \mathbb{C}^*$. But $X_{\mathfrak{w}}$ being projective and irreducible, $f$ is constant and in fact $f \equiv 1$ since $s_{\mathfrak{v}}(e) = s_{\mathfrak{w}}(e)$. This gives rise to a nowhere-vanishing regular section $s$ of $\mathfrak{L}$ on the whole of $X$ such that $s_{|X_{\mathfrak{w}}} = s_{\mathfrak{w}}$. From this it is easy to see that $\mathfrak{L}$ is isomorphic with the trivial line bundle on $X$. This proves that $\alpha$ is an isomorphism. Now the second isomorphism is proved in [KNR, Proposition 2.3]. □

We state the following very crucial 'lifting' result, the proof of which will be given in the next section.

**(2.3) Proposition.** *There exists a map* $\overline{\psi^*} : \text{Pic}\,(\mathfrak{M}) \to \text{Pic}\,(X)$, *making the following diagram commutative:*

$$\begin{array}{ccc} & \text{Pic}\,(\mathfrak{M}) & \\ \overline{\psi^*} \swarrow & & \searrow \psi^* \\ \text{Pic}\,(X) & \xrightarrow[i^*]{} & \text{Pic}\,(X^s), \end{array}$$

*where* $i^*$ *is the canonical restriction map.*

As an easy consequence of the above proposition, Lemmas (2.1) and (2.2), we get the following main result of this paper.

**(2.4) Theorem.** *For any smooth projective irreducible curve C of genus* $\geqq 2$ *and simple simply-connected connected affine algebraic group G, the map* $\overline{\psi^*}$ *(as in the above proposition) is an injective group homomorphism.*

*In particular,* $\text{Pic}\,(\mathfrak{M}) \approx \mathbb{Z}$.

*Proof.* Injectivity of $\overline{\psi^*}$ follows from the injectivity of $\psi^*$ (cf. Lemma 2.1) and the commutativity of the diagram in Proposition (2.3). By Proposition (2.3), Image $\psi^* \subset$ Image $i^*$. But since Pic $X \approx \mathbb{Z}$ (by Lemma 2.2), Image $i^*$ is either finite or else Image $i^* \approx \mathbb{Z}$. Now since $\mathfrak{M}$ is a projective variety of dim $> 0$ (cf. [R1, Theorem 4.9]) and $\psi^*$ is injective, Image $i^*$ can not be finite, in particular, $i^*$ is injective. Since $\psi^*$ and $i^*$ are group homomorphisms and $i^*$ is injective, we get that $\overline{\psi^*}$ is a group homomorphism. This proves the theorem. □

**(2.5) Definition.** Let $n_{C,G} > 0$ be the least (positive) integer such that $\mathfrak{L}(n_{C,G}\chi_0) \in \text{Image}\,\overline{\psi^*}$. Then of course

$$\textit{Image}\,\overline{\psi^*} = \{\mathfrak{L}(dn_{C,G}\chi_0)\}_{d\in\mathbb{Z}}$$

We will be concerned with determining the number $n_{C,G}$ in Sect. 4.

*(2.6) Remark.* In the case when $G = SL(n, \mathbb{C})$, it is a result of Drezet–Narasimhan [DN] that Pic$(\mathfrak{M}) \approx \mathbb{Z}$.

We recall the following well known result. (We include a proof since we did not find it in the literature in this form.)

**(2.7) Lemma.** *Let $Y$ be a Cohen–Macaulay projective variety and let $U \subset Y$ be an open subset such that* $\text{codim}_Y(Y\backslash U) \geqq 2$. *Now let $\mathcal{S}_1$ and $\mathcal{S}_2$ be two reflexive sheaves on $Y$ such that $\mathcal{S}_{1|_U} \approx \mathcal{S}_{2|_U}$. Then the sheaf $\mathcal{S}_1$ is isomorphic with $\mathcal{S}_2$ on the whole of $Y$.*

*Proof*[1]. We recall the following two facts from Commutative Algebra.

*Fact 1*: If $M, N$ are modules over a noetherian local ring with depth $M, N > 1$, and $0 \to M \to N \to K \to 0$ is an exact sequence, then depth $K > 0$.

*Fact 2*: If $M$ is reflexive, then for any localisation $M_\mathfrak{p}$ of $M$ at a prime ideal $\mathfrak{p}$, depth $M_\mathfrak{p} > 1$, unless the dimension of the local ring itself is less than 2 (i.e. $M$ satisfies the 'Serre condition' $S_2$).

Let $i : U \hookrightarrow Y$ be the inclusion. Then from the above facts (and the assumptions of the lemma), one can check that $i_* i^* \mathcal{S}_j = \mathcal{S}_j$ (for $j = 1, 2$). Thus any homomorphism $i^*\mathcal{S}_1 \to i^*\mathcal{S}_2$ on U gives rise to a homomorphism $\mathcal{S}_1 \to \mathcal{S}_2$, i.e., Hom $(\mathcal{S}_1, \mathcal{S}_2) \to$ Hom $(i^*\mathcal{S}_1, i^*\mathcal{S}_2)$ is surjective. Injectivity is clear using reflexivity. This proves the lemma. □

We come to the following second main result of this paper.

**(2.8) Theorem.** *The dualizing sheaf $\omega$ of the moduli space $\mathfrak{M}$ is locally free. Moreover, $\overline{\psi^*}(\omega) = \mathfrak{L}(-2g\chi_0)$, where $g$ is the dual Coxeter number of the Lie algebra $\mathfrak{g}$ of $G$ (cf. [KNR, Remark 5.3]).*

*In particular, $\mathfrak{M}$ is a Gorenstein variety. Further, for any line bundle $\mathfrak{L}$ on $\mathfrak{M}$ such that $\overline{\psi^*}(\mathfrak{L}) = \mathfrak{L}(d\chi_0)$ for some $d > -2g$, $H^i(\mathfrak{M}, \mathfrak{L}) = 0$, for all $i > 0$ So, for any finite dimensional representation $V$ of $G$, $H^i(\mathfrak{M}, \Theta(V)) = 0$, for all $i > 0$, where $\Theta(V)$ is the theta bundle on the moduli space $\mathfrak{M}$.*

[1] This proof is due to N. Mohan Kumar.

*Proof.* Let $\mathfrak{M}^o := \{E \in \mathfrak{M}; E$ is a stable $G$-bundle and $\operatorname{Aut} E =$ centre of $G\}$. Then $\mathfrak{M}^o$ is an open subset of the smooth locus of $\mathfrak{M}$ and, for any $E \in \mathfrak{M}^o$, the tangent space $T_E(\mathfrak{M}^o)$ can be identified with $H^1(C, \operatorname{Ad} E)$, where $\operatorname{Ad} E$ is the vector bundle on $C$ associated to the principal $G$-bundle $E$ via the adjoint representation Ad of $G$ in its Lie algebra $\mathfrak{g}$. Also, on the set of stable bundles in the moduli space there are no identifications, i.e., if $E_1$ and $E_2$ are two stable $G$-bundles on $C$ such that $E_1$ is $S$-equivalent to $E_2$, then $E_1$ is isomorphic with $E_2$ (as follows from the definition of $S$-equivalence, cf. [KNR, Sect. 3.3]). Moreover, for any $E \in \mathfrak{M}^o$, $H^0(C, \operatorname{Ad} E) = 0$. In particular, the fiber of the canonical bundle of $\mathfrak{M}^o$ at $E$ can be identified with $\wedge^{\text{top}}(H^1(C, \operatorname{Ad} E)^*)$, where $\wedge^{\text{top}}$ is the top exterior power. This gives, from the definitions of the determinant bundle and the $\Theta$-bundle (cf. [KNR, Sect. 3.8]), that

$$\operatorname{Det}(\operatorname{Ad})^*_{|\mathfrak{M}^o} = \Theta(\operatorname{Ad})^*_{|\mathfrak{M}^o} = \omega_{|\mathfrak{M}^o} .$$

But $\Theta(\operatorname{Ad})^*$ is a line bundle on the whole of $\mathfrak{M}$ (cf. [loc. cit., Sect. 3.8]). Since any line bundle is a reflexive sheaf (cf. [H, Exercise 5.1, p. 123]), $\Theta(\operatorname{Ad})^*$ is a reflexive sheaf on $\mathfrak{M}$. Since the dualizing sheaf $\omega$ of a normal variety is always reflexive; the moduli space $\mathfrak{M}$ is Cohen–Macaulay and normal (cf. [R1, Theorem 4.9]); and $\operatorname{codim}_{\mathfrak{M}}(\mathfrak{M}\backslash\mathfrak{M}^o) \geqq 2$ (unless the curve $C$ is of genus 2 and $G = SL(2)$) (cf. [F, Theorem II.6]); we obtain from Lemma (2.7):

$$\omega \approx \Theta(\operatorname{Ad})^*, \text{ on the whole of } \mathfrak{M} . \tag{1}$$

(In the case of $G = SL(2)$ the validity of (1) is well known.) This of course gives that $\mathfrak{M}$ is a Gorenstein variety (by definition). Now the assertion that $\overline{\psi}^*(\omega) = \mathfrak{L}(-2g\chi_0)$ follows from [KNR, Theorem 5.4 and Lemma 5.2].

Finally we come to the proof of cohomology vanishing: By Serre duality [H, Corollary 7.7, Chap. III] (denoting $\dim \mathfrak{M} = n$),

$$\begin{aligned} H^i(\mathfrak{M}, \mathfrak{L})^* &\approx H^{n-i}(\mathfrak{M}, \mathfrak{L}^* \otimes \omega) \\ &= H^{n-i}(\mathfrak{M}, \mathfrak{L}^* \otimes \Theta(\operatorname{Ad})^*), \text{ by (1)} . \end{aligned} \tag{2}$$

But $\overline{\psi}^*(\mathfrak{L}^* \otimes \Theta(\operatorname{Ad})^*) = \mathfrak{L}((-d-2g)\chi_0)$. Now since $\operatorname{Pic}(\mathfrak{M}) \approx \mathbb{Z}$ (by Theorem 2.4), we get that the line bundle $\mathfrak{L} \otimes \Theta(\operatorname{Ad})$ is ample on $\mathfrak{M}$ (by assumption $d > -2g$).

The moduli space $\mathfrak{M}$ has rational singularities, as follows from [R1, Proof of Theorem 4.9] and a result of Boutot [Bo]. Now the vanishing of $H^i(\mathfrak{M}, \mathfrak{L})$ (for $i > 0$) follows from (2) and a result of Grauert-Riemenschneider [GR]. So the proof of the theorem is complete in view of [KNR, Theorem 5.4]. □

**(2.9) Corollary.** *For any finite dimensional representation $V$ of $G$,*

$$\mathcal{X}(\mathfrak{M}, \Theta(V)) = \dim H^0(\mathfrak{M}, \Theta(V)) ,$$

*where $\mathcal{X}$ is the Euler-Poincaré characteristic:*

$$\mathcal{X}(\mathfrak{M}, \Theta(V)) = \sum_i (-1)^i \dim H^i(\mathfrak{M}, \Theta(V)) .$$

## 3. Extension of line bundles. Proof of Proposition (2.3)

(3.1). Recall the definition of the map $\varphi : \mathcal{G} \to \mathcal{X}_0$ from [KNR, Sect. 1] (where $\mathcal{X}_0$ denotes the set of isomorphism classes of principal $G$-bundles on $C$ which are algebraically trivial restricted to $C^* := C\backslash p$). Fix an embedding $G \hookrightarrow SL(n)$, for some $n$. In particular, any principal $G$-bundle $E$ on $C$ gives rise to a vector bundle $\overline{E}$ of rank $n$ on $C$ (associated to the standard representation of $SL(n)$). For any integer $d \geqq 1$, define

$$X_d = \{g\mathcal{P} \in X : H^1(C, \overline{\varphi(g)} \otimes \mathcal{O}(-x+dp)) = 0 \text{ for all } x \in C\},$$

where $p \in C$ is the fixed base point. Then

$$X_1 \subset X_2 \subset \cdots .$$

**(3.2) Lemma.** *Each $X_d$ is open in $X$. Moreover $X^s \subset X_{2h}$, where $X^s := \{g\mathcal{P} \in X : \varphi(g)$ is a semistable $G$-bundle$\}$, and $h$ is the genus of the curve $C$.*

*Proof.* It suffices to prove that $X_d \cap X_{\mathfrak{w}}$ is open in $X_{\mathfrak{w}}$, for each $\mathfrak{w} \in \widetilde{W}/W$:

Recall the definition of the family of $G$-bundles $\mathcal{U} \to C \times X$ from [KNR, Proposition 2.8]. Consider the restriction $\mathcal{U}_{\mathfrak{w}}$ of the $G$-bundle $\mathcal{U} \to C \times X$ to $C \times X_{\mathfrak{w}}$ and let $\overline{\mathcal{U}_{\mathfrak{w}}}$ be the associated rank-$n$ vector bundle (corresponding to the embedding $G \hookrightarrow \mathrm{SL}(n)$). Define a vector bundle $\widetilde{\mathcal{U}_{\mathfrak{w}}}$ on $C \times C \times X_{\mathfrak{w}}$ such that $\widetilde{\mathcal{U}}_{\mathfrak{w}|x \times C \times X_{\mathfrak{w}}} = \mathcal{O}(-x+dp) \otimes \overline{\mathcal{U}_{\mathfrak{w}}}$ for each $x \in C$; and let $\pi: C \times C \times X_{\mathfrak{w}} \to C \times X_{\mathfrak{w}}$ be the projection on the two extreme factors. Applying the upper semi-continuity theorem [H, Chapter III, Sect. 12] to the morphism $\pi$ and the locally free sheaf $\widetilde{\mathcal{U}_{\mathfrak{w}}}$ on $C \times C \times X_{\mathfrak{w}}$, we get that the set

$$S := \{(x, g\mathcal{P}) : H^1(C, \overline{\varphi(g)} \otimes \mathcal{O}(-x+dp)) \neq 0\}$$

is a closed subset of $C \times X_{\mathfrak{w}}$. In particular, $\pi_2(S)$ is a closed subset of $X_{\mathfrak{w}}$, where $\pi_2 : C \times X_{\mathfrak{w}} \to X_{\mathfrak{w}}$ is the projection on the second factor. It is easy to see that $X_d \cap X_{\mathfrak{w}} = X_{\mathfrak{w}} \setminus \pi_2(S)$. This proves that $X_d$ is open in $X$.

For $g\mathcal{P} \in X^s$, $\overline{\varphi(g)}$ is a semistable vector bundle (cf. [RR, Theorem 3.18]), and hence the dual vector bundle $\overline{\varphi(g)}^*$ is also semistable. Now, by the Serre duality,

$$H^1(C, \overline{\varphi(g)} \otimes \mathcal{O}(-x+dp)) \approx H^0(C, \overline{\varphi(g)}^* \otimes \mathcal{O}(x-dp) \otimes K)^* .$$

Since $\overline{\varphi(g)}^*$ is semistable, $H^0(C, \overline{\varphi(g)}^* \otimes \mathcal{O}(x-dp) \otimes K) \neq 0$ implies that $d-1-\deg K \leqq 0$. In particular, if $d \geqq 2 + \deg K$, then $g\mathcal{P} \in X_d$. This proves the lemma since $\deg K = 2h-2$. □

We have

$$\bigcup_{d \geqq 1} X_d = X ,$$

since each Schubert variety $X_{\mathfrak{w}}$ is contained in some large enough $X_d$ ($d$ of course depending upon $\mathfrak{w}$). This follows by the upper semi-continuity theorem (using an argument similar to the one used in the proof of the above lemma).

(3.3). Fix any $d \geqq 2h$. For all $m \geqq d$ and $g\mathscr{P} \in X_d$, we have

(1) $H^1(C, \overline{\varphi(g)} \otimes \mathcal{O}(mp)) = 0$, and
(2) $H^0(C, \overline{\varphi(g)} \otimes \mathcal{O}(mp))$ generates the vector bundle $\overline{\varphi(g)} \otimes \mathcal{O}(mp)$ at every point of $C$.

Let $q_d := \dim H^0(C, \overline{\varphi(g)} \otimes \mathcal{O}(dp))$. Then by Riemann-Roch theorem, $q_d = n(d+1-h)$. Denote by $\pi_d : \mathscr{F}_d \to X_d$ the $\mathrm{GL}(q_d)$-bundle such that for $g\mathscr{P} \in X_d$, $\pi_d^{-1}(g\mathscr{P})$ is the set of all the frames of the vector space $H^0(C, \overline{\varphi(g)} \otimes \mathcal{O}(dp))$. We call $\mathscr{F}_d$ the *frame bundle associated to the family* $\mathscr{U}_{|X_d}$ (parametrized by $X_d$). Similarly, define the frame bundle $\pi_{d+1} : \mathscr{F}_{d+1} \to X_{d+1}$. Consider the parabolic subgroup $P = \{\theta \in \mathrm{GL}(q_{d+1}) : \theta\mathbb{C}^{q_d} = \mathbb{C}^{q_d}\}$ of $\mathrm{GL}(q_{d+1})$, where (for definiteness) $\mathbb{C}^{q_d} \hookrightarrow \mathbb{C}^{q_{d+1}}$ is sitting in the first $q_d$ coordinates. We define the principal $P$-subbundle $Q_d$ of $\mathscr{F}_{d+1|X_d}$ by

$$Q_d = \bigcup_{g\mathscr{P} \in X_d} \{s = (s_1, \ldots, s_{q_{d+1}}) \text{ a frame of } H^0(C, \overline{\varphi(g)} \otimes \mathcal{O}((d+1)p))$$
$$\text{such that } (s_1, \ldots, s_{q_d}) \text{ is a frame of } H^0(C, \overline{\varphi(g)} \otimes \mathcal{O}(dp))\} .$$

(Observe that $H^0(C, \overline{\varphi(g)} \otimes \mathcal{O}(dp))$ sits canonically inside $H^0(C, \overline{\varphi(g)} \otimes \mathcal{O}((d+1)p))$ induced from the embedding $\overline{\varphi(g)} \otimes \mathcal{O}(dp) \hookrightarrow \overline{\varphi(g)} \otimes \mathcal{O}((d+1)p)$.) Then we have the following commutative diagram:

$$\begin{array}{ccccc} \mathscr{F}_d & \overset{\beta_d}{\twoheadleftarrow} & Q_d & \hookrightarrow & \mathscr{F}_{d+1} \\ \pi_d \downarrow & & & & \downarrow \pi_{d+1} \\ X_d & & \hookrightarrow & & X_{d+1}, \end{array}$$

where $\beta_d$ takes any $s = (s_1, \ldots, s_{q_{d+1}}) \in Q_d$ to the frame $(s_1, \ldots, s_{q_d})$ of $H^0(C, \overline{\varphi(g)} \otimes \mathcal{O}(dp))$. It is clear that $\beta_d$ is a principal $U$-bundle, where $U := \{\theta \in \mathrm{GL}(q_{d+1}) : \theta_{|\mathbb{C}^{q_d}} = I\} \subset P$. Clearly $U$ is a normal subgroup of $P$.

As in [KNR, Sect. 7.8], we have an irreducible smooth quasi-projective variety $R_d$ with an action of $\mathrm{GL}(q_d)$, a family $\mathscr{W}_d$ of $G$-bundles on $C$ parametrized by $R_d$ and a lift of the $\mathrm{GL}(q_d)$-action to $\mathscr{W}_d$ (as bundle automorphisms), such that there exists a $\mathrm{GL}(q_d)$-equivariant morphism $\varphi_d : \mathscr{F}_d \to R_d$ with the property that the families $\pi_d^*(\mathscr{U}_{|X_d})$ and $\varphi_d^*(\mathscr{W}_d)$ are isomorphic. Moreover, let $R_d^s = \{x \in R_d : \mathscr{W}_d(x) := \mathscr{W}_{d|C\times x}$ is a semistable $G$-bundle$\}$ be the $\mathrm{GL}(q_d)$-invariant open subset of $R_d$. Then the canonical map $\theta_d : R_d^s \to \mathfrak{M}$ is surjective. Moreover, $\theta_d$ is $\mathrm{GL}(q_d)$-equivariant with respect to the trivial action of $\mathrm{GL}(q_d)$ on the moduli space $\mathfrak{M}$ (of semistable $G$-bundles on $C$). We recall the construction of $R_d$ for its use in the sequel [R1, Sects. 3.8, 3.13.3]:

Let $R_d^o$ be the set of locally free quotients $E$ of $\mathbb{C}^{q_d} \otimes_{\mathbb{C}} \mathcal{O}_C$ of rank $n$ and degree $nd$ such that the canonical map $\mathbb{C}^{q_d} \approx H^0(\mathbb{C}^{q_d} \otimes_{\mathbb{C}} \mathcal{O}_C) \to H^0(E)$ is an isomorphism. Then $R_d^o$ supports the tautological family $\widehat{\mathscr{W}_d}^o$ of rank-$n$ vector bundles on $C$. Set $\mathscr{W}_d^o = \widehat{\mathscr{W}_d}^o \otimes_{\mathscr{O}_{C\times R_d^o}} \mathscr{O}_C(-dp)$. Now let

$$R_d = \{(x,\sigma) : x \in R_d^o \text{ and } \sigma \text{ is a reduction of the structure group of } \mathscr{W}^o_{d|C\times x} \text{ to } G\} .$$

Then clearly $R_d$ supports a canonical family $\mathscr{W}_d$ of $G$-bundles on $C$ and moreover $\mathrm{GL}(q_d)$ acts on $\mathscr{W}_d$ via its action on $\mathbb{C}^{q_d}$.

Using $H^1(C,E)=0$, one proves that $R_d$ is smooth and that the infinitesimal deformation map $T_t(R_d) \to H^1(C, \mathrm{Ad}\,(\mathscr{W}_{d|C\times t}))$ is surjective, where $T_t(R_d)$ is the tangent space at $t$ to $R_d$.

**(3.4) Proposition.** *For any $d \geqq 2h$, the codimension of $R_d\backslash R_d^s$ in $R_d$ is at least 2, where $R_d$ is explicitly constructed as above.*

To prove the above proposition, we need the notion of the canonical reduction (or filtration) of a principal $G$-bundle on $C$. We choose a Borel subgroup $B$ of $G$ and a maximal torus $T \subset B$. By a *standard parabolic subgroup* we mean a parabolic subgroup $P$ containing $B$. The following result is due to Ramanathan [R2, Proposition 1] (see also [Be]).

**(3.5) Theorem.** *Let $E$ be a principal $G$-bundle on $C$. Then there exists a unique standard parabolic subgroup $P$ of $G$ and a unique reduction $E_P$ of $E$ to the subgroup $P$ such that the following conditions hold:*

(1) *If $U$ is the unipotent radical of $P$, then the $P/U$-bundle $E_{P/U}$, obtained from $E_P$ by extension of the structure group via $P \to P/U$, is semistable. (Observe that $P/U$ is reductive.)*
(2) *For any non-trivial character $\chi$ of $P$ which is a non-negative linear combination of simple roots of $B$, the line bundle on $C$ associated to $E_P$ by $\chi$ has strictly positive degree.*

The unique reduction $E_P$ of $E$ as above is called the *canonical reduction.*

**(3.6) Lemma.** *Let $E_P$ be the canonical reduction of a principal $G$-bundle $E$ on $C$. Let $\mathfrak{g}$ and $\mathfrak{p}$ be the Lie algebras of $G$ and $P$ respectively. Denote by $E_{\mathfrak{s}}$ the vector bundle associated to $E_P$ by the natural representation of $P$ on the vector space $\mathfrak{s} := \mathfrak{g}/\mathfrak{p}$. Then we have*

$$H^0(C, E_{\mathfrak{s}}) = 0 .$$

*Proof.* We may assume that $P \neq G$. Let $0 = V_0 \subset V_1 \subset \ldots \subset V_k = \mathfrak{s}$ be a filtration of $\mathfrak{s}$ by $P$-submodules $V_i$ such that, for any $1 \leqq i \leqq k$, the $P$-module $W_i := V_i/V_{i-1}$ is irreducible. In particular, $U$ acts trivially on $W_i$ (cf. [Ku, Lemma 1]). If $\mathscr{V}_i$ is the vector bundle on $C$ associated to $E_P$ by the representation of $P$ on $V_i$, then $E_{\mathfrak{s}}$ is filtered by the subbundles $\mathscr{V}_i$. We now show that $H^0(C, \mathscr{W}_i) = 0$ for all $1 \leqq i \leqq k$, where $\mathscr{W}_i := \mathscr{V}_i/\mathscr{V}_{i-1}$. This will of course prove the lemma.

Since the action of $U$ on $W_i$ is trivial, we obtain an (irreducible) representation of the reductive group $P/U$ on $W_i$. Since $E_{P/U}$ is semistable, the vector bundles $\mathscr{W}_i$ are semistable (cf. [RR, Theorem 3.18]), and hence it is sufficient to show that $\deg(\mathscr{W}_i) < 0$: Now the weights of $T$ on $\mathfrak{s}$ are of the form $\sum c_\alpha \alpha$ with $c_\alpha \leqq 0$ and $c_\alpha < 0$ for at least one $\alpha \notin I$, where $I$ is the subset of the set of simple roots $\Pi = \{\alpha\}$ defining the parabolic subgroup $P$ (i.e. $I$ is the set of simple roots for $P/U$). It follows from this that the character of $P$ defined by the determinant of the representation of $P$ on $W_i$ is non-trivial and is a non-positive linear combination of $\{\alpha\}_{\alpha\in\pi}$. By Condition (2) of Theorem (3.5), we see that $\deg(\mathscr{W}_i) < 0$. This completes the proof of the lemma. □

Let $P$ be a standard parabolic subgroup of $G$ and $E_P$ be a reduction of the $G$-bundle $E$ to $P$. For any character $\chi$ of $P$, denote by $E_{P,\chi}$ the line bundle on $C$ associated to $E_P$ by $\chi$. Let $X(P)$ (resp. $X(T)$) denote the character group of $P$ (resp. $T$). Then $X(T) = \oplus_{\alpha\in\Pi}\mathbb{Z}\omega_\alpha$, where $\omega_\alpha$ is the fundamental weight defined by $\omega_\alpha(\beta^\vee) = \delta_{\alpha,\beta}$, for any simple coroot $\beta^\vee$. Moreover (since $G$ is simply-connected) $X(P) = \oplus_{\alpha\notin I}\mathbb{Z}\omega_\alpha$. The map $\chi \mapsto \deg(E_{P,\chi})$ defines an element of $\mathrm{Hom}_{\mathbb{Z}}(X(P),\mathbb{Z})$, which in turn can be lifted to the element $\mu$ of $\mathrm{Hom}_{\mathbb{Z}}(X(T),\mathbb{Z})$ defined by $\mu(\omega_\alpha) = \deg(E_{P,\omega_\alpha})$ if $\alpha \notin I$ and $\mu(\omega_\alpha) = 0$ if $\alpha \in I$. We call $\mu$ the *type* of the reduction $E_P$.

Using the above lemma, one can prove the following proposition; the proof being similar to that of [PV, Theorem 4, p. 90].

**(3.7) Proposition.** *Let $\mathscr{W}$ be a family of $G$-bundles on $C$ parametrized by a smooth variety $S$. Assume that at each point $t \in S$ the infinitesimal deformation map*

$$T_t(S) \to H^1(C, \mathrm{Ad}(\mathscr{W}_t))$$

*is surjective, where $\mathscr{W}_t = \mathscr{W}_{|C\times t}$ and $T_t(S)$ is the tangent space at $t$ to $S$. For $\mu \in \mathrm{Hom}(X(T),\mathbb{Z})$, let $S_\mu$ be the subset of $S$ consisting of those points $t \in S$ such that the canonical reduction of $\mathscr{W}_t$ is of type $\mu$. Then $S_\mu$ is non-empty only for finitely many $\mu$. Moreover, $S_\mu$ is locally closed and smooth, and the normal space at $t \in S_\mu$ is given by $H^1(C, \mathscr{W}_{t,\mathfrak{s}})$, where $\mathscr{W}_{t,\mathfrak{s}}$ is the vector bundle associated to the canonical reduction $\mathscr{W}_{t,P}$ by the representation of $P$ on $\mathfrak{s} := \mathfrak{g}/\mathfrak{p}$.*

*(3.8) Proof of Proposition (3.4).* The family $\mathscr{W} = \mathscr{W}_d$ parametrized by $R_d$ satisfies the hypothesis of the above proposition (3.7). So it suffices to prove that for $t \in R_d \backslash R_d^s$, we have $\dim H^1(C, \mathscr{W}_{t,\mathfrak{s}}) \geqq 2$:

By Lemma (3.6), $H^0(C, \mathscr{W}_{t,\mathfrak{s}}) = 0$ and hence by Riemann-Roch theorem,

$$(1) \qquad \dim H^1(C, \mathscr{W}_{t,\mathfrak{s}}) = -\deg \mathscr{W}_{t,\mathfrak{s}} + \dim(\mathfrak{s})(h-1)\,,$$

where recall that $h$ is the genus of $C$. Further, since $t \in R_d \backslash R_d^s$, we have $\mathfrak{g} \neq \mathfrak{p}$. By the same argument, used in the proof of Lemma (3.6), $\deg \mathscr{W}_{t,\mathfrak{s}} < 0$: This gives (using 1 and the assumption that $h \geqslant 2$) that $\dim H^1(C, \mathscr{W}_{t,\mathfrak{s}}) \geqq 2$, proving Proposition (3.4). □

**(3.9) Lemma.** *Let $H$ be an affine algebraic group acting algebraically on a smooth variety $Y$ and let $U$ be a $H$-stable open subset such that* $\operatorname{codim}_Y(Y\backslash U) \geqq 2$. *Then the canonical restriction map* $\operatorname{Pic}^H(Y) \to \operatorname{Pic}^H(U)$ *is an isomorphism, where* $\operatorname{Pic}^H(Y)$ *denotes the set of isomorphism classes of $H$-equivariant line bundles on $Y$.*

*Proof.* Let $\mathscr{L}$ be an $H$-equivariant line bundle on $U$. Since $Y$ is smooth and $\operatorname{codim}_Y(Y\backslash U) \geqq 2$, $\mathscr{L}$ extends uniquely to a line bundle $\tilde{\mathscr{L}}$ on $Y$. We show that $\tilde{\mathscr{L}}$ is $H$-equivariant:

Fix $h \in H$ and an open subset $V \subset Y$ such that $\tilde{\mathscr{L}}_{|V}$ is a trivial line bundle. In particular, the line bundle $\tilde{\mathscr{L}}_{|hV}$ also is trivial (since by the $H$-equivariance of $\mathscr{L}$, $\tilde{\mathscr{L}}_{|h(U\cap V)}$ is trivial and moreover $\operatorname{codim}_V(V\backslash U) \geqq 2$). Take a nowhere-vanishing section $s_1$ of $\tilde{\mathscr{L}}_{|V}$ and $s_2$ of $\tilde{\mathscr{L}}_{|hV}$. Now for any $x \in U \cap V$, $f_h(x)s_2(hx) = h(s_1(x))$, for some (unique) $f_h(x) \in \mathbb{C}^*$. Clearly the map $U \cap V \to \mathbb{C}^*$, taking $x \mapsto f_h(x)$ is a regular map, which extends to a regular map $\tilde{f}_h : V \to \mathbb{C}^*$ (since $\operatorname{codim}_V(V\backslash U) \geqq 2$). Define an action of $h$ on $\tilde{\mathscr{L}}_{|V}$ by

$$h(s_1(x)) = \tilde{f}_h(x)s_2(hx), \qquad \text{for all } x \in V\,.$$

By the uniqueness of extension, this action of $h$ on $\tilde{\mathscr{L}}_{|V}$ patches-up to give an action of $h$ on the whole of $\tilde{\mathscr{L}}$. Further, as can be easily seen, this is a regular action of $H$ on $\tilde{\mathscr{L}}$.

The injectivity of $\operatorname{Pic}^H(Y) \to \operatorname{Pic}^H(U)$ is easy to see: An $H$-equivariant section, which does not vanish anywhere on $U$, extends to a nowhere-vanishing section on $Y$ (and by uniqueness of extension it is $H$-equivariant). □

*(3.10) Lifting of line bundles from $\mathfrak{M}$ to $X_d$.* Take any $d \geqq 2h$. Let $\mathfrak{L}$ be a line bundle on $\mathfrak{M}$. Pull back the line bundle $\mathfrak{L}$ via the $\mathrm{GL}(q_d)$-equivariant morphism $\theta_d : R_d^s \to \mathfrak{M}$ to get a $\mathrm{GL}(q_d)$-equivariant line bundle $\theta_d^*(\mathfrak{L})$ on $R_d^s$ (cf. Sect. 3.3). By the above Lemma (3.9) and Proposition (3.4), $\theta_d^*(\mathfrak{L})$ extends to a $\mathrm{GL}(q_d)$-equivariant line bundle $\widehat{\theta_d^*(\mathfrak{L})}$ on $R_d$. Consider the diagram, where all the maps are $\mathrm{GL}(q_d)$-equivariant morphisms (the map $i_d$ is the inclusion, $\varphi_d$ and $\pi_d$ are as in Sect. 3.3, and $\mathrm{GL}(q_d)$ acts trivially on $X_d$):

$$\begin{array}{ccccc} \mathscr{F}_d & \xrightarrow{\varphi_d} & R_d & \xhookleftarrow{i_d} & R_d^s \\ {\scriptstyle \pi_d}\downarrow & & & & \downarrow{\scriptstyle \theta_d} \\ X_d & & & & \mathfrak{M} \end{array}$$

Now $\varphi_d^*(\widehat{\theta_d^*(\mathfrak{L})})$ being a $\mathrm{GL}(q_d)$-equivariant line bundle (and $\pi_d$ being a principal $\mathrm{GL}(q_d)$-bundle) descends to give a line bundle (denoted) $\mathfrak{L}_d$ on $X_d$ (cf. [Kr, Proposition 6.4]).

**(3.11) Lemma.** *For any line bundle $\mathfrak{L}$ on $\mathfrak{M}$ and $d \geqq 2h$*

$$\mathfrak{L}_{d+1|X_d} \approx \mathfrak{L}_d\,, \quad \textit{and} \quad \mathfrak{L}_{d|X^s} \approx \psi^*(\mathfrak{L})\,,$$

*where $\psi : X^s \to \mathfrak{M}$ is the morphism as in Sect. 1 (cf. Lemma 3.2).*

*Proof.* We will freely use the notation from Sect. 3.3. Let $X_{\mathfrak{w}}$ be a fixed Schubert variety, and denote the (reduced) variety $X_{\mathfrak{w}} \cap X_d$ by $Y = Y_{d,\mathfrak{w}}$. Then $Y^s := Y \cap X^s$ is an open non-empty (irreducible) subvariety of $X_{\mathfrak{w}}$. We denote by $\mathscr{F}_{d,Y}$, $\mathscr{F}_{d+1,Y}$ and $Q_{d,Y}$ the restrictions of $\mathscr{F}_d$, $\mathscr{F}_{d+1}$ and $Q_d$ to $Y$, where $Q_d$ is the $P$-subbundle of $\mathscr{F}_{d+1|X_d}$ as in Sect. 3.3. We show that $\mathfrak{L}_{d|Y} \approx \mathfrak{L}_{d+1|Y}$ and $\mathfrak{L}_{d|Y^s} \approx \psi^*(\mathfrak{L})_{|Y^s}$. This will of course prove the lemma.

We first show that

$$(1) \qquad \mathfrak{L}_{d|Y^s} \approx \psi^*(\mathfrak{L})_{|Y^s} :$$

From the commutativity of the diagram (where $\mathscr{F}^s_{d,Y} := \pi_d^{-1}(Y^s)$, and $\pi_d, \varphi_d$, and $\psi$ are the corresponding maps got by restriction, which we denote by the same symbols)

$$(D_1) \qquad \begin{array}{ccc} & \mathscr{F}^s_{d,Y} & \\ {\scriptstyle \pi_d} \swarrow & & \searrow {\scriptstyle \varphi_d} \\ Y^s & & R^s_d \\ {\scriptstyle \psi} \searrow & & \swarrow {\scriptstyle \theta_d} \\ & \mathfrak{M} & \end{array}$$

we see that the $\mathrm{GL}(q_d)$-linearizations on $\pi_d^*(\psi^*\mathfrak{L})$ and $\varphi_d^*(\theta_d^*\mathfrak{L})$ are the same. This shows that $\mathfrak{L}_{d|Y^s} \approx \psi^*(\mathfrak{L})_{|Y^s}$ (since $\pi_d$ is a principal $\mathrm{GL}(q_d)$-bundle).

If $H$ is an affine algebraic group and $\mathscr{H}$ an $H$-linearized line bundle on a principal $H$-bundle, we denote by $\mathscr{H}^H$ the line bundle on the base space (of the $H$-bundle) obtained by descending $\mathscr{H}$.

Let $\widetilde{\mathscr{W}}^o_d$ be the vector bundle on $C \times R_d$ which is the pull-back of $\widehat{\mathscr{W}}^o_d$ by the map $I_C \times \beta : C \times R_d \to C \times R^o_d$, where $\beta : R_d \to R^o_d$ is the canonical map. Let $\pi''_d : \mathscr{F}''_d \to R_d$ (resp. $\pi'_d : \mathscr{F}'_d \to R_d$) be the frame bundle of the vector bundle $(p_{R_d})_*(\widetilde{\mathscr{W}}^o_d \otimes \mathcal{O}(p))$ (resp. $(p_{R_d})_*(\widetilde{\mathscr{W}}^o_d)$), where $p_{R_d} : C \times R_d \to R_d$ is the projection on the second factor. Just as in Sect. 3.3, the inclusion

$$(p_{R_d})_*(\widetilde{\mathscr{W}}^o_d) \hookrightarrow (p_{R_d})_*(\widetilde{\mathscr{W}}^o_d \otimes \mathcal{O}(p))$$

defines a $P$-subbundle $Q'_d \subset \mathscr{F}''_d$ on $R_d$ and a morphism $\beta'_d : Q'_d \to \mathscr{F}'_d$. Further, analogous to the map $\varphi_d : \mathscr{F}_d \to R_d$ there is a $\mathrm{GL}(q_{d+1})$-equivariant morphism $\varphi'_d : \mathscr{F}''_d \to R_{d+1}$. Thus we have the diagram:

$$(D_2) \qquad \begin{array}{ccc} & Q'_d & \\ {\scriptstyle \beta'_d} \swarrow & & \searrow \\ \mathscr{F}'_d & & \mathscr{F}''_d \\ {\scriptstyle \pi'_d} \downarrow & & \downarrow {\scriptstyle \varphi'_d} \\ R_d & & R_{d+1} \, . \end{array}$$

(Observe that $\beta'_d$ is a principal $U$-bundle, $\pi'_d$ is a principal $\mathrm{GL}(q_d)$-bundle and $\pi''_d$ is a principal $\mathrm{GL}(q_{d+1})$-bundle.) Considering the commutative diagram (where $\mathscr{F}''^{s}_{d} := \pi''^{-1}_{d}(R^{s}_{d})$)

$$(D_3)\qquad \begin{array}{ccccc} & & \mathscr{F}''^{s}_{d} & & \\ & \pi''_d \swarrow & & \searrow \varphi'_d & \\ R^s_d & & & & R^s_{d+1} \\ & \theta_d \searrow & & \swarrow \theta_{d+1} & \\ & & \mathfrak{M} & & \end{array}$$

we see, as above, that

$$(\varphi'^{*}_{d}\theta^{*}_{d+1}\mathfrak{L})^{\mathrm{GL}(q_{d+1})} \approx \theta^{*}_{d}(\mathfrak{L})\,.$$

Since $\mathrm{codim}_{R_d}(R_d \backslash R^s_d) \geqq 2$ and $R_d$ is smooth, we have

$$\widehat{\theta^{*}_{d}\mathfrak{L}} \approx (\varphi'^{*}_{d}(\widehat{\theta^{*}_{d+1}\mathfrak{L}}))^{\mathrm{GL}(q_{d+1})}\,.$$

Now

$$\begin{aligned} &(\varphi'^{*}_{d}(\widehat{\theta^{*}_{d+1}\mathfrak{L}}))^{\mathrm{GL}(q_{d+1})} \\ &\quad\approx (\gamma^{*}_{d}(\widehat{\theta^{*}_{d+1}\mathfrak{L}}))^{P} \\ &\quad\approx ((\gamma^{*}_{d}(\widehat{\theta^{*}_{d+1}\mathfrak{L}}))^{U})^{\mathrm{GL}(q_d)} \\ &\quad\approx \sigma^{*}((\gamma^{*}_{d}(\widehat{\theta^{*}_{d+1}\mathfrak{L}}))^{U})\,, \end{aligned}$$

where $\gamma_d : Q'_d \to R_{d+1}$ is the restriction of $\varphi'_d$ to $Q'_d$ and $\sigma : R_d \to \mathscr{F}'_d$ is the canonical section, given by the isomorphism

$$\mathbb{C}^{q_d} = H^0(C, \mathbb{C}^{q_d} \otimes \mathcal{O}_C) \xrightarrow{\sim} H^0(C, \widetilde{\mathscr{W}}^{o}_{d}|_{C\times t})$$

for $t \in R_d$. Thus

$$(2)\qquad \widehat{\theta^{*}_{d}\mathfrak{L}} \approx \sigma^{*}((\gamma^{*}_{d}(\widehat{\theta^{*}_{d+1}\mathfrak{L}}))^{U})\,.$$

Consider the following commutative diagram

$$(D_4)\qquad \begin{array}{ccccccc} Q_{d,Y} & \xrightarrow{\alpha} & Q'_d & \hookrightarrow & \mathscr{F}''_d \\ \beta_d \downarrow & & \downarrow \beta'_d & \searrow \gamma_d & \downarrow \varphi'_d \\ \mathscr{F}_{d,Y} & \xrightarrow{\delta} & \mathscr{F}'_d & & R_{d+1} \\ \pi_d \downarrow & \varphi_d \searrow & \downarrow \pi'_d & & \\ Y & & R_d & & \end{array}$$

where $\delta := \sigma \circ \varphi_d$, and the map $\alpha$ is defined as follows: Let $g\mathscr{P} \in Y$ and let $s = (s_1, \ldots, s_{q_d}, \ldots, s_{q_{d+1}})$ be a frame of $H^0(C, \overline{\varphi(g)} \otimes \mathcal{O}((d+1)p))$ such that $\bar{s} := (s_1, \ldots, s_{q_d})$ is a frame of $H^0(C, \overline{\varphi(g)} \otimes \mathcal{O}(dp))$. We have a commutative diagram:

$$\begin{array}{ccccc} 0 \longrightarrow & H^0(C, \overline{\varphi(g)} \otimes \mathcal{O}(dp)) & \longrightarrow & H^0(C, \overline{\varphi(g)} \otimes \mathcal{O}((d+1)p)) \\ & \big\downarrow & & \big\downarrow \\ 0 \longrightarrow & H^0(C, \widetilde{\mathscr{W}}^{o}_{d|C\times\varphi_d(\bar{s})}) & \longrightarrow & H^0(C, \widetilde{\mathscr{W}}^{o}_{d|C\times\varphi_d(\bar{s})} \otimes \mathcal{O}(p))\,, \end{array}$$

where the vertical maps are isomorphisms. Observe that, under the first vertical isomorphism, the frame $\bar{s}$ is mapped to the frame $\delta(\bar{s})$. Now define $\alpha(s)$ to be the frame in $H^0(C, \mathscr{W}^o_{d|C\times\varphi_d(\bar{s})} \otimes \mathcal{O}(p))$ which is the image of the frame $s$ under the second vertical isomorphism. Then $\alpha$ is a $U$-equivariant morphism.

We claim that (as line bundles on $\mathscr{F}_{d,Y}$)

$$(3) \qquad \varphi_d^*(\widehat{\theta_d^*\mathfrak{L}}) \approx (\alpha^*\gamma_d^*(\widehat{\theta_{d+1}^*\mathfrak{L}}))^U \ :$$

This follows since

$$\begin{aligned} &(\alpha^*\gamma_d^*(\widehat{\theta_{d+1}^*\mathfrak{L}}))^U \\ &\approx \delta^*((\gamma_d^*(\widehat{\theta_{d+1}^*\mathfrak{L}}))^U) \\ &\approx \varphi_d^*\sigma^*((\gamma_d^*(\widehat{\theta_{d+1}^*\mathfrak{L}}))^U) \\ &\approx \varphi_d^*(\widehat{\theta_d^*\mathfrak{L}}), \text{ using } (2). \end{aligned}$$

Now the bundle $\varphi_d^*(\widehat{\theta_d^*\mathfrak{L}})$ has a $\mathrm{GL}(q_d)$-linearization coming from the action of $GL(q_d)$ on $\widehat{\theta_d^*\mathfrak{L}}$ and (by definition of $\mathfrak{L}_d$) the bundle on $Y$ obtained by descent is $\mathfrak{L}_{d|Y}$. On the other hand, the bundle $(\alpha^*\gamma_d^*(\widehat{\theta_{d+1}^*\mathfrak{L}}))^U$ has a $\mathrm{GL}(q_d)$-action given by the action of $P/U \approx \mathrm{GL}(q_d)$ arising from the action of $P$ on $\alpha^*\gamma_d^*(\widehat{\theta_{d+1}^*\mathfrak{L}})$ (which in turn comes from the action of $\mathrm{GL}(q_{d+1})$, in particular, $P$ on $\widehat{\theta_{d+1}^*\mathfrak{L}}$; observe that even though $\alpha$ is only $U$-equivariant, $\gamma_{d^{\circ\alpha}}$ is $P$-equivariant) and the bundle on $Y$ obtained by descent via $\pi_d$ is $\mathfrak{L}_{d+1|Y}$. Now over $\mathscr{F}^s_{d,Y} := \pi_d^{-1}(Y^s)$, these two actions of $\mathrm{GL}(q_d)$ coincide (i.e. the isomorphism $\eta$ of line bundles on $\mathscr{F}_{d,Y}$ as guaranteed by (3) is $\mathrm{GL}(q_d)$-equivariant on $\mathscr{F}^s_{d,Y}$), as is seen from the following commutative diagram (got from the diagrams $D_1$ and $D_4$) (where $Q^s_{d,Y} := \beta_d^{-1}(\mathscr{F}^s_{d,Y})$):

$$\begin{array}{ccccc} & & Q^s_{d,Y} & & \\ & \overset{\beta_d}{\swarrow} & & \searrow^{\gamma_d\circ\alpha} & \\ \mathscr{F}^s_{d,Y} & \xrightarrow{\varphi_d} & R^s_d & & R^s_{d+1} \\ {\scriptstyle\pi_d}\downarrow & & {\scriptstyle\theta_d}\downarrow & \swarrow{\scriptstyle\theta_{d+1}} & \\ Y^s & \xrightarrow{\psi} & \mathfrak{M}. & & \end{array}$$

Since $Y^s$ is dense in $Y$, we have that $\mathscr{F}^s_{d,Y}$ is dense in $\mathscr{F}_{d,Y}$; in particular, the isomorphism $\eta$ is $\mathrm{GL}(q_d)$-equivariant on the whole of $\mathscr{F}_{d,Y}$. Hence $\mathfrak{L}_{d|Y} \approx \mathfrak{L}_{d+1|Y}$. Denote this isomorphism by $\mu$. Then the restriction of $\mu$ to $Y^s$ is the identity map under the identification (1). From this it is easy to see that $\mathfrak{L}_d \approx \mathfrak{L}_{d+1|X_d}$. This completes the proof of the lemma. □

Finally we come to the

*(3.12) Proof of Proposition (2.3).* For any Schubert variety $X_{\mathfrak{w}}$, there exists a large enough $d(\mathfrak{w})$ such that $X_{\mathfrak{w}} \subset X_{d(\mathfrak{w})}$. Fix a line bundle $\mathfrak{L}$ on $\mathfrak{M}$. Let $\widehat{\mathfrak{L}_{\mathfrak{w}}}$ be the line bundle on $X_{\mathfrak{w}}$ defined by $\widehat{\mathfrak{L}_{\mathfrak{w}}} = \mathfrak{L}_{d(\mathfrak{w})|X_{\mathfrak{w}}}$. By Lemma (3.11), $\widehat{\mathfrak{L}_{\mathfrak{w}}}$ is

well defined and $\widehat{\mathfrak{L}_{\mathfrak{w}}}_{|X^s_{\mathfrak{w}}} \approx \psi^*(\mathfrak{L})_{|X^s_{\mathfrak{w}}}$, where $\psi : X^s \to \mathfrak{M}$ is the morphism as in Sect. 1. Moreover, for $\mathfrak{v} \leqq \mathfrak{w}$, $\widehat{\mathfrak{L}_{\mathfrak{w}}}_{|X_{\mathfrak{v}}} \approx \widehat{\mathfrak{L}_{\mathfrak{v}}}$. In particular, by Lemma (2.2), we get a line bundle $\widehat{\mathfrak{L}}$ on $X$ with $\widehat{\mathfrak{L}}_{|X^s} \approx \psi^*(\mathfrak{L})$. This proves the proposition. □

## 4. Determination of Pic($\mathfrak{M}$)

**(4.1) Definition [D, Sect. 2].** Let $\mathfrak{g}_1$ and $\mathfrak{g}_2$ be two (finite dimensional) complex simple Lie algebras and $\varphi : \mathfrak{g}_1 \to \mathfrak{g}_2$ be a Lie algebra homomorphism. There exists a unique number $m_\varphi \in \mathbb{C}$, called the ***Dynkin index*** of the homomorphism $\varphi$, satisfying

$$\langle \varphi(x), \varphi(y) \rangle = m_\varphi \langle x, y \rangle\ , \quad \text{for all } x, y \in \mathfrak{g}_1\ ,$$

where $\langle , \rangle$ is the Killing form on $\mathfrak{g}_1$ (and $\mathfrak{g}_2$) normalized so that $\langle \theta, \theta \rangle = 2$ for the highest root $\theta$.

It is easy to see from [KNR, Lemma 5.2] that for a finite dimensional representation $V$ of $\mathfrak{g}_1$ given by a Lie algebra homomorphism $\varphi : \mathfrak{g}1 \to sl(V)$, we have $m_\varphi = m_V$, where $m_V$ is as in [KNR, Sect. 5.1] and $sl(V)$ is the Lie algebra of trace 0 endomorphisms of $V$.

By taking a representation $V$ of $G_2$ such that $m_V \neq 0$, and using [KNR, Corollary 5.6], the following proposition follows easily.

**(4.2) Proposition.** *Let $G_1, G_2$ be two connected complex simple algebraic groups. Then for any algebraic group homomorphism $\phi : G_1 \to G_2$, the induced map at the third homotopy group level*

$$\phi_* : \pi_3(G_1) \approx \mathbb{Z} \longrightarrow \pi_3(G_2) \approx \mathbb{Z}$$

*is given by the multiplication via the Dynkin index $m_{d\phi}$ of the induced Lie algebra homomorphism $d\phi : \mathfrak{g}_1 \to \mathfrak{g}_2$, where $\mathfrak{g}_1$ (resp. $\mathfrak{g}_2$) is the Lie algebra of $G_1$ (resp. $G_2$).*

*In particular, $m_{d\phi}$ is an integer.*

*(4.3) Remark.* The integrality of $m_\phi$ is proved by Dynkin [D, Theorem 2.2], and so is the following lemma [D, Theorem 2.5], by a quite different (and long) argument.

**(4.4) Lemma.** *Let $\mathfrak{g}$ be a complex simple Lie algebra and let $V(\lambda)$ be an irreducible representation of $\mathfrak{g}$ with highest weight $\lambda$. Then the Dynkin index $m_{V(\lambda)}$ of the representation $V(\lambda)$ is given by*

$$m_{V(\lambda)} = (\|\lambda + \rho\|^2 - \|\rho\|^2) \frac{\dim_{\mathbb{C}} V(\lambda)}{\dim_{\mathbb{C}} \mathfrak{g}}\ ,$$

*where $\rho$ is the half sum of positive roots and the Killing form on $\mathfrak{g}$ is normalized (as earlier) so that $\|\theta\|^2 = 2$ for the highest root $\theta$.*

*Proof.* The representation $V = V(\lambda)$ of course gives rise to a Lie algebra homomorphism $\varphi = \varphi_V : \mathfrak{g} \to sl(V)$. Since $m_V = m_\varphi$ (cf. Sect. 4.1), for any $x, y \in \mathfrak{g}$

$$(1) \qquad m_V\langle x, y\rangle = \text{trace}\,(\varphi(x) \circ \varphi(y)) .$$

Choose a basis $\{e_i\}$ of $\mathfrak{g}$ and let $\{e^i\}$ be the dual basis of $\mathfrak{g}$ with respect to the Killing form $\langle , \rangle$. Consider the Casimir element $\Omega := \sum_i e_i e^i \in U(\mathfrak{g})$. Then $\Omega$ acts on $V$ via

$$(2) \qquad \Omega_V := \sum_i \varphi(e_i) \circ \varphi(e^i) .$$

But $V$ being irreducible of highest weight $\lambda$,

$$(3) \qquad \Omega_V = (\|\lambda + \rho\|^2 - \|\rho\|^2) I_V ,$$

where $I_V$ is the identity operator of $V$. In particular,

$$\begin{aligned} m_V &= \frac{1}{\dim \mathfrak{g}} \sum_i \text{trace}\,(\varphi(e_i) \circ \varphi(e^i)) , && \text{by (1)} \\ &= \frac{1}{\dim \mathfrak{g}} \text{trace}\, \Omega_V , && \text{by (2)} \\ &= \frac{1}{\dim \mathfrak{g}} (\|\lambda + \rho\|^2 - \|\rho\|^2) \dim V , && \text{by (3)} . \end{aligned}$$

This proves the lemma. □

We also need the following

**(4.5) Lemma.** *Let* $\mathfrak{g}$ *be a complex simple Lie algebra and let* $V$ *and* $W$ *be two finite dimensional representations of* $\mathfrak{g}$. *Then*

$$m_{V \otimes W} = m_V \dim W + m_W \dim V .$$

*Proof.* Write the characters

$$\begin{aligned} \text{ch}\, V &= \sum_\lambda n_\lambda e^\lambda , && \text{and} \\ \text{ch}\, W &= \sum_\mu m_\mu e^\mu , && \text{for some } n_\lambda,\ m_\mu \in \mathbb{Z}_+ . \end{aligned}$$

Then

$$\text{ch}\,(V \otimes W) = \sum_{\lambda, \mu} n_\lambda m_\mu e^{\lambda + \mu} .$$

Hence by [KNR, Lemma 5.2],

$$\begin{aligned} 2m_{V\otimes W} &= \sum_{\lambda,\mu} n_\lambda m_\mu \langle \lambda+\mu, \theta^\vee\rangle^2 \\ &= \sum n_\lambda m_\mu \langle \lambda, \theta^\vee\rangle^2 + \sum n_\lambda m_\mu \langle \mu, \theta^\vee\rangle^2 \\ &\quad +2\sum n_\lambda m_\mu \langle \lambda, \theta^\vee\rangle\langle \mu, \theta^\vee\rangle \\ &= 2\left(\sum_\mu m_\mu\right) m_V + 2\left(\sum_\lambda n_\lambda\right) m_W \\ &\quad +2\left(\sum_\lambda n_\lambda \langle \lambda, \theta^\vee\rangle\right)\left(\sum_\mu m_\mu \langle \mu, \theta^\vee\rangle\right) \\ &= 2(\dim W)m_V + 2(\dim V)m_W \qquad (1) \\ &\quad +2\left(\sum_\lambda n_\lambda \langle \lambda, \theta^\vee\rangle\right)\left(\sum_\mu m_\mu \langle \mu, \theta^\vee\rangle\right). \end{aligned}$$

For any $h \in \mathfrak{h}$, define $\beta_V(h) = \sum_\lambda n_\lambda \langle \lambda, h\rangle$. Then the map $\beta_V : \mathfrak{h} \to \mathbb{C}$, $h \mapsto \beta_V(h)$ is $W$-equivariant (with the trivial action of $W$ on $\mathbb{C}$). Hence, $\mathfrak{h}$ being an irreducible $W$-module,

(2) $$\beta_V \equiv 0 .$$

Combining (1) and (2), the lemma follows. □

**(4.6) Definition.** Let $\mathfrak{g}$ be a complex simple Lie algebra and let $\theta$ be the highest root (with respect to some choice of the set of positive roots). Express the associated coroot $\theta^\vee$ in terms of the simple coroots:

$$\theta^\vee = \sum_{i=1}^{\ell} m_i \alpha_i^\vee .$$

Now define $d = d(\mathfrak{g})$ to be the least common multiple of $\{m_i\}_{i=1,\dots,\ell}$. Then the number $d$ is given as follows:

| **Type of** $\mathfrak{g}$ | $d(\mathfrak{g})$ |
|---|---|
| $A_\ell$ ($\ell \geqq 1$), $C_\ell$ ($\ell \geqq 2$) | 1 |
| $B_\ell$ ($\ell \geqq 3$) | 2 |
| $D_\ell$ ($\ell \geqq 4$) | 2 |
| $G_2$ | 2 |
| $F_4$ | 6 |
| $E_6$ | 6 |
| $E_7$ | 12 |
| $E_8$ | 60 |

**(4.7) Proposition.** *For any finite dimensional representation $V$ of $\mathfrak{g}$, the number $d(\mathfrak{g})$ divides $m_V$. Moreover, there exists an irreducible representation $V_o$ of $\mathfrak{g}$ such that $d(\mathfrak{g}) = m_{V_o}$.*

*Proof.* Unfortunately, our proof is case by case. We follow the indexing convention as in [B, Planche I-IX]. We denote the $i$-th fundamental weight $(1 \leqq i \leqq \ell)$ by $\omega_i$.

*Case 1.* $A_\ell(\ell \geqq 1)$, $C_\ell(\ell \geqq 2)$: As in [KNR, Lemma 5.2], $m_{V_o} = 1$, for the standard $(\ell+1)$-dimensional representation $V_o$ of $A_\ell$. Similarly for the standard $2\ell$-dimensional representation $V_o$ of $C_\ell$ (with highest weight $\omega_1$), $m_{V_o} = 1$ (as can be seen from Lemma 4.4).

For a simply-connected group $G$, since the fundamental representations $\{V(\omega_i)\}_{1\leqq i\leqq \ell}$ generate the representation ring $R(G)$ as an algebra (cf. [A, Theorem 6.41]), to prove that $d(\mathfrak{g})$ divides $m_V$ for any $\mathfrak{g}$-module $V$, it suffices to show that $d(\mathfrak{g})$ divides $m_i := m_{V(\omega_i)}$ for all $1 \leqq i \leqq \ell$ (cf. Lemma 4.5). In the following calculations, we make use of Lemma (4.4) and [B, Planche I-IX] freely.

*Case 2.* $B_\ell$ $(\ell \geqq 3)$: For $1 \leqq i \leqq \ell - 1$, $m_i = 2\binom{2\ell-1}{i-1}$, since $\dim V(\omega_i) = \binom{2\ell+1}{i}$; and $m_\ell = 2^{\ell-2}$.

In particular, $m_1 = 2$, so take $V_o = V(\omega_1)$.

*Case 3.* $D_\ell$ $(\ell \geqq 4)$: For $1 \leqq i \leqq \ell - 2$, $m_i = 2\binom{2\ell-2}{i-1}$, since $\dim V(\omega_i) = \binom{2\ell}{i}$; and $m_{\ell-1} = m_\ell = 2^{\ell-3}$.

In particular, $m_1 = 2$.

In the following calculations, $\dim V(\omega_i)$ is taken from [BMP].

*Case 4.* $G_2$: $m_1 = 2$, $m_2 = 8$.

(Observe that $V(\omega_2)$ is the adjoint representation of $G_2$ and hence $m_2$ can be calculated from [KNR, Lemma 5.2 and Remark 5.3].)

*Case 5.* $F_4$: $m_1, \ldots, m_4$ are respectively 18, 9×98, 126, and 6.

*Case 6.* $E_6$: $m_1, \ldots, m_6$ are respectively 6, 24, 150, 1800, 150, and 6.

*Case 7.* $E_7$: $m_1, \ldots, m_7$ are respectively 36, 360, 65×72, 2750×108, 104×165, 8×81, and 12.

*Case 8.* $E_8$: $m_1, \ldots, m_8$ are respectively 12×125, 4750×18, 49×108000, 75×111275472, 30×4720170, 45×39520, 15×980, and 60. □

*(4.8) Remark.* The values of $m_i$ given above are also contained in [D], but some of his values are incorrect.

Combining Proposition (4.7) and Theorem (2.4) with the chart in Definition (4.6), we get the following strengthening of Theorem (2.4).

**(4.9) Theorem.** *With the notation and assumptions as in Theorem* (2.4), *consider the injective map* $\overline{\psi}^* : \text{Pic}(\mathfrak{M}) \hookrightarrow \text{Pic}(X) \approx \mathbb{Z}$. *Then*

*(1)* $\overline{\psi}^*$ *is surjective in the case where $G$ is of type* $A_\ell$ $(\ell \geqq 1)$, *and* $C_\ell(\ell \geqq 2)$.

*(2) The order* $\gamma = \gamma_G$ *of the cokernel of* $\overline{\psi^*}$ *is bounded as follows:*

(a) $G = B_\ell$ $(\ell \geqq 3)$, $\gamma \leqq 2$
(b) $G = D_\ell$ $(\ell \geqq 4)$, $\gamma \leqq 2$
(c) $G = G_2$, $\gamma \leqq 2$
(d) $G = F_4$, $\gamma \leqq 6$
(e) $G = E_6$, $\gamma \leqq 6$
(f) $G = E_7$, $\gamma \leqq 12$
(g) $G = E_8$, $\gamma \leqq 60$. □

*(4.10) Remarks (added at the time of revision).* (a) We have received a preprint by Y. Laszlo and C. Sorger " The line bundles on the stack of parabolic $G$-bundles over curves and their sections", which has some overlap with our paper. In particular, they calculate the Picard group of the moduli *stack* of parabolic $G$-bundles for the classical groups and $G_2$.
(b) We have shown that the order $\gamma = \gamma_G$ of the cokernel of $\overline{\psi^*}$ is precisely equal to 2 in the cases of $G = B_\ell$ $(\ell \geqq 3)$ and $G = D_\ell$ $(\ell \geqq 4)$, i.e., the theta bundle $\Theta(V_n)$ on the moduli space $\mathfrak{M} = \mathfrak{M}(n)$ of semistable Spin$(n)$-bundles $(n \geqq 7)$ on the curve $C$ does not admit a square root as a line bundle on the whole of $\mathfrak{M}$, where the standard $SO(n)$-representation $V_n$ is to be considered as a representation of Spin$(n)$ via $SO(n)$. We are not including the proof here as we have been informed that this has also been obtained recently by Beauville-Laszlo-Sorger (and prior to us). It is very likely that the bounds for $\gamma$ given in Theorem (4.9) for all the other groups ($G$ of type $G_2, F_4, E_6, E_7$, and $E_8$) are sharp as well.

*Acknowledgements.* We thank J. Wahl for some helpful conversations, G. Faltings for a helpful communication, and the referee for some comments. This work was partially done while the first author was visiting the Max Planck Institut für Mathematik (Bonn) in June, 1993 and the Tata Institute of Fundamental Research (Bombay) in Dec., 1993; hospitality of which is gratefully acknowledged. The first author lectured on the contents of this paper at the University of North Carolina (Chapel Hill) in the fall of 1994, at the Rutgers University (New Brunswick) in Jan., 1995, and at the University of Aarhus in May, 1995. The first author was supported by the NSF grant no. DMS-9203660.

## References

[A] Adams, J.F.: *Lectures on Lie Groups*, W.A. Benjamin, Inc., 1969.
[B] Bourbaki, N.: *Groupes et Algèbres de Lie*, Chap. IV–VI, Hermann, Paris, 1968.
[Be] Behrend, K.A.: *Semi-stability of reductive group schemes over curves*, Math. Ann. **301** (1995), 281–305.
[BMP] Bremner, M.R., Moody, R.V., Patera, J.: *Tables of Dominant Weight Multiplicities for Representations of Simple Lie Algebras*, Marcel Dekker, Inc., New York-Basel, 1985.
[Bo] Boutot, J.F.: *Singularités rationnelles et quotients par les groupes réductifs*, Invent. Math. **88** (1987), 65–68.
[D] Dynkin, E.B.: *Semisimple subalgebras of semisimple Lie algebras*, Amer. Math. Soc. Transl., Ser. II, **6** (1957), 111–244.
[DN] Drezet, J.-M., Narasimhan, M.S.: *Groupe de Picard des variétés de modules de fibrés semi-stables sur les courbes algébriques*, Invent. Math. **97** (1989), 53–94.

[F] Faltings, G.: *Stable G-bundles and projective connections*, J. Alg. Geom. **2** (1993), 507–568.

[GR] Grauert, H., Riemenschneider, O.: *Verschwindungssätze für analytische Kohomologiegruppen auf komplexen Räumen*, In: Several Complex Variables I, Springer Lecture Notes no. 155, Maryland, 1970, pp. 97–109.

[H] Hartshorne, R., *Algebraic Geometry*, Springer Verlag, Berlin-Heidelberg-New York, 1977.

[K] Kac, V.G.: *Infinite Dimensional Lie Algebras*, Third Edition, Cambridge University Press, 1990.

[KNR] Kumar, S., Narasimhan, M.S., Ramanathan, A.: *Infinite Grassmannians and moduli spaces of G-bundles*, Math. Ann. **300** (1994), 41–75.

[Kr] Kraft, H.: *Algebraic automorphisms of affine space*, In: Proceedings of the Hyderabad Conference on Algebraic Groups, ed. S. Ramanan, Manoj Prakashan, 1991, 251–274.

[Ku] Kumar, S.: *Symmetric and exterior powers of homogeneous vector bundles*, Math. Annalen **299** (1994), 293–298.

[Ku2] Kumar, S.: Infinite Grassmannians and moduli spaces of $G$-bundles, Preprint 1995, (to appear as CIME Lecture Notes).

[NR] Narasimhan, M.S., Ramadas, T.R.: *Factorisation of generalised theta functions. I*, Invent. Math. **114** (1993), 565–623.

[PV] Verdier, J.L., Potier, J. Le: *Module des fibrés stables sur les courbes algébriques, exposé 4*, Progress in Mathematics **54** (1985), Birkhäuser.

[R1] Ramanathan, A.: *Stable principal bundles on a compact Riemann surface- Construction of moduli space*, Thesis (University of Bombay, Bombay, India), (1976).

[R2] Ramanathan, A.: *Moduli for principal bundles*, L. N. M. **732** (1979), (Algebraic Geometry, Proceedings Copenhagen), 527–533.

[RR] Ramanan, S., Ramanathan, A.: *Some remarks on the instability flag*, Tohoku Math. J. **36** (1984), 269–291.

J. ALGEBRAIC GEOMETRY
6 (1997) 697 715



# HODGE CLASSES OF MODULI SPACES OF PARABOLIC BUNDLES OVER THE GENERAL CURVE

INDRANIL BISWAS AND M. S. NARASIMHAN

## 1. Introduction

It is known that the algebraic cohomology classes (i.e., rational cohomology classes represented by algebraic cycles) of the Jacobian of a general smooth curve are spanned by the cohomology classes of the theta divisor and its powers; moreover, they span the rational Hodge classes of this variety, so that, in particular, the Hodge conjecture is true for these abelian varieties. We generalise these results in this paper to the case of moduli spaces of vector bundles of arbitrary rank on curves, when the rank and degree are co-prime. In fact, we prove these results in the case of certain moduli spaces of parabolic bundles [Theorem 5.1].

Similar results are also proved for moduli spaces of parabolic bundles with fixed determinant [Theorem 5.4].

The result concerning algebraic cohomology classes was proved in the case of usual vector bundles of rank two with fixed determinant by V. Balaji, A. King, and P. Newstead, using a different method [BKN].

Let $X$ be a smooth projective curve of genus $g \geq 2$ over $\mathbb{C}$ and let $\mathcal{M}$ be the moduli space of stable bundles over $X$ of rank $n$ and degree $d$ with $(n, d) = 1$, or more generally a parabolic moduli space satisfying the conditions stated in Section 2. We define in Section 2 explicitly a subalgebra $I$ of the cohomology algebra $H^*(\mathcal{M}, \mathbb{Q})$ and show that the elements in $I$ are represented by algebraic cycles [Proposition 2.4]. (In the case of the Jacobian, $I$ is the subalgebra generated by the cohomology class of a theta divisor.) Our main theorem is that when $X$ is a general curve, any Hodge class (i.e., a rational cohomology class of type $(p, p)$) is contained in $I$.

Our proof is, as in the case of the Jacobian, by a monodromy argument. Consider the monodromy action on $H^*(\mathcal{M}, \mathbb{Q})$ given by the variation of $X$ (and hence of $\mathcal{M}$). It turns out that this action is an action of the Siegel

Received November 30, 1995.

modular group $Sp(g,\mathbb{Z})$ on $H^*(\mathcal{M},\mathbb{Q})$. The crucial results we prove, which make the monodromy argument applicable, are:

(1) $I$ is the algebra of invariants in $H^*(\mathcal{M},\mathbb{Q})$, for the monodromy action [Proposition 4.1].

(2) Any linear subspace of $H^*(\mathcal{M},\mathbb{C})$ invariant under the action of $Sp(g,\mathbb{Z})$ and consisting entirely of classes of type $(p,p)$ must be contained in the space of $Sp(g,\mathbb{Z})$ invariants in $H^*(\mathcal{M},\mathbb{C})$ [Theorem 4.2].

Taking into account known generators of the cohomology algebra, [BR], these results are proved by studying the action of the complex symplectic group $Sp(g,\mathbb{C})$ on the exterior algebra $\wedge(V\oplus\cdots\oplus V)$ ($n$ copies of $V$), where $V$ is the symplectic vector space $H_1(X,\mathbb{C})$. Some key representation theoretic inputs are provided by a paper of R. Howe [Ho].

In Section 6 we indicate how the above results can be generalised to the $m$-fold product of $\mathcal{M}$ with itself. As a consequence it is shown that the cohomology class in $\mathcal{M}\times\mathcal{M}$ defined by the operator $\Lambda : H^*(\mathcal{M}) \longrightarrow H^*(\mathcal{M})$ associated to a polarization is algebraic, where $\mathcal{M}$ is the moduli space of parabolic bundles on any smooth curve (not necessarily general). In particular, homological and numerical equivalence of algebraic cycles on $\mathcal{M}$ coincide.

We are grateful to Richard Hain for sending us a proof for the case of the Jacobian. His argument is used in Section 5 of this paper.

## 2. Cohomology of moduli spaces and the basic algebraic classes

Let $X$ be a smooth projective irreducible algebraic curve over $\mathbb{C}$, of genus $g\geq 2$. Let

$$S := \{p_1,p_2,\ldots,p_h\}\subset X\,,\qquad h\geq 1$$

be a finite (ordered) subset of $X$. A parabolic data consists of a positive integer $n$, an integer $d$ and for each $p_j$, a sequence of positive integers $\{m_{j1},\ldots,m_{jl_j}\}$, with $\sum_{i=1}^{l_j} m_{ji} = n$ and a sequence of real numbers $\alpha_{j,i}$ satisfying the condition

$$0\leq \alpha_{j,1} < \alpha_{j,2} < \cdots < \alpha_{j,l_j} < 1\,.$$

Let $\mathcal{M}$ be the moduli space of semi-stable parabolic bundles of rank $n$, degree $d$ and with parabolic structure of the given type [MS]. We will suppose in this paper that the following conditions are satisfied.

(a) Any semi-stable parabolic bundle (of the given type) is stable.

(b) The greatest common divisor of the set of integers $(n,d,\{m_{ji}\})$ ($1\leq j\leq h$, $1\leq i\leq l_j$) is 1, where $n$ is the rank and $d$ is the degree of a vector bundle $E\in\mathcal{M}$.

The moduli space of stable bundles of rank $n$ and degree $d$ with $(n,d)=1$ is a special case with one parabolic point and trivial flag.

Under conditions (a) and (b) the space $\mathcal{M}$ is a smooth (projective) variety and there exists a universal bundle $(\mathcal{E},\{\mathcal{F}_{ji}\})$, $1\le j\le h$ and $1\le i\le l_j$, consisting of a vector bundle $\mathcal{E}$ on $X\times\mathcal{M}$ and, for each $p_j\in S$, a flag of vector sub-bundles

$$\mathcal{E}|_{p_j\times\mathcal{M}}=\mathcal{F}_{j,1}\supset\cdots\supset\mathcal{F}_{j,l_j}\supset\mathcal{F}_{j,l_j+1}=0$$

of $\mathcal{E}|_{p_j\times\mathcal{M}}$. (We may view $\mathcal{F}_{ji}$ as vector bundles on $\mathcal{M}$.)

Let $\mathcal{L}'$ be a Poincaré bundle on $X\times J^d$, with $\mathcal{L}'|_{p_1\times J^d}$ trivial, where $J^d$ is the space of equivalence classes of line bundles of degree $d$ ($d$ is the degree of a vector bundle in $E\in\mathcal{M}$). Let

$$det\,:\,X\times\mathcal{M}\longrightarrow X\times J^d$$

be the morphism defined by $(x,E)\longmapsto(x,\det E)$ and let

$$\mathcal{L}:=(det)^*\mathcal{L}'\,.$$

We denote by

$$a_1\in H^2(X\times\mathcal{M},\mathbb{Q})$$

the first Chern class of $\mathcal{L}$ and for $m=2,\ldots,n$, by

$$a_m\in H^{2m}(X\times\mathcal{M},\mathbb{Q})$$

the $m$-th characteristic class of the projective bundle $P(\mathcal{E})$ [BR]. Let $c_{k,j,i}$ denote the $k$-th Chern class of the bundle $\mathrm{Hom}(\mathcal{F}_{j,i},\mathcal{F}_{j,i-1})$ on $\mathcal{M}$, $1\le j\le h$, $2\le i\le l_j+1$, $1\le k\le\max\{\mathrm{rank}(\mathrm{Hom}(\mathcal{F}_{j,i},\mathcal{F}_{j,i-1}))\}$.

We observe that the bundles $P(\mathcal{E})$ , $\mathrm{Hom}(\mathcal{F}_{j,i},\mathcal{F}_{j,i-1})$ are uniquely determined, although $\mathcal{E}$ and $\mathcal{F}_{j,i}$ are not.

Let $N$ be the number of triplets $(i,j,k)$ satisfying the above condition. The classes $c_{k,j,i}$ define a $\mathbb{Q}$-linear map

$$f_1\,:\,\mathbb{Q}^N\longrightarrow H^*(\mathcal{M},\mathbb{Q})\,.$$

Let $H_k$, $k=0,1,2$ denote the $k$-th homology group of $X$ with coefficients in $\mathbb{Q}$. If $\delta\in H^p(X\times\mathcal{M},\mathbb{Q})$, Künneth decomposition of $H^*(X\times\mathcal{M},\mathbb{Q})$ and the duality between homology and cohomology of $X$ define a homomorphism

$$\gamma\,:\,H_k(X,\mathbb{Q})\longrightarrow H^{p-k}(X\times\mathcal{M},\mathbb{Q}),\qquad k=0,1,2.$$

( The "slant product" with $\delta$) If $\alpha\in H_k(X,\mathbb{Q})$ we denote $\gamma(\alpha)$ by $\langle\delta,\alpha\rangle$.

For $1\le l\le n$, the characteristic classes $\{a_l\}$, defined above, define homomorphisms

$$f_{2,l}\;:\;H_0(X,\mathbb{Q})\oplus H_2(X,\mathbb{Q})\longrightarrow H^*(\mathcal{M},\mathbb{Q})$$

and

$$f_l \ : \ H_1(X,\mathbb{Q}) \ \longrightarrow \ H^*(\mathcal{M},\mathbb{Q}),$$

by "slanting" with $a_l$. We combine $f_1$ and $\{f_{2,l}\}$ into a homomorphism

$$f' \ : \ W' \ := \ \mathbb{Q}^N \oplus \Big(H_0(X,\mathbb{Q}) \oplus H_2(X,\mathbb{Q}) \otimes \mathbb{Q}^n\Big) \ \longrightarrow \ H^{\mathrm{ev}}(\mathcal{M},\mathbb{Q})$$

and the $\{f_l\}$ into a homomorphism

$$f \ : \ W \ := \ H_1(X,\mathbb{Q}) \bigotimes_{\mathbb{Q}} \mathbb{Q}^n \ \longrightarrow \ H^{\mathrm{odd}}(\mathcal{M},\mathbb{Q}) \,.$$

The homomorphisms $f$ and $f'$ together induce homomorphisms of algebras

$$F \ : \ S(W') \otimes \bigwedge(W) \ \longrightarrow \ H^*(\mathcal{M},\mathbb{Q}) \tag{2.1}$$

and

$$F_{\mathbb{C}} \ : \ S(W'_{\mathbb{C}}) \bigotimes_{\mathbb{C}} \bigwedge(W_{\mathbb{C}}) \ \longrightarrow \ H^*(\mathcal{M},\mathbb{C}) \tag{2.2}$$

where $S$ (resp. $\wedge$) denotes the symmetric (resp. exterior) algebra, and $W_{\mathbb{C}} = W \otimes_{\mathbb{Q}} \mathbb{C}$.

By Theorem 1.4 of [BR], the maps $F$ and $F_{\mathbb{C}}$ are surjective.

We now define a subalgebra $I_0$ (resp. $I_{0,\mathbb{C}}$) of $\wedge W$ (resp. $\wedge W_{\mathbb{C}}$). Let $V := H_1(X,\mathbb{Q})$; we have

$$W \ = \ V \otimes \mathbb{Q}^n \ = \ \bigoplus_{i=1}^{n} V_i \,,$$

where $V_i = V \otimes \xi_i$, where $\xi_i$ is the subspace of $\mathbb{Q}^n$ spanned by $e_i$, $\{e_1, \ldots, e_n\}$ being the canonical basis of $\mathbb{Q}^n$. The cup-product in $H^1(X,\mathbb{Q})$ $(=H_1(X,\mathbb{Q})^*)$ defines an element in $\overset{2}{\wedge} V$. Write

$$\overset{2}{\wedge} W \ = \ \bigoplus_{k=1}^{n} \overset{2}{\wedge} V_i \bigoplus_{i<j} (V_i \otimes V_j).$$

For $(i,j)$ with $1 \leq i \leq j \leq n$, the cup-product determines an element $\tau_{ij}$, with $\tau_{ii} \in \overset{2}{\wedge} V_i$ and $\tau_{ij} \in V_i \otimes V_j$, for $i < j$. We consider $\tau_{ij}$ as elements of $\overset{2}{\wedge} W$, and $I_0$ (resp. $I_{0,\mathbb{C}}$) will be the $\mathbb{Q}$ subalgebra (resp. $\mathbb{C}$-subalgebra) of $\wedge W$ (resp. $\wedge W_{\mathbb{C}}$) generated by $\{\tau_{ij}\}$. Observe that

$$I_{0,\mathbb{C}} \ = \ I \otimes_{\mathbb{Q}} \mathbb{C} \,.$$

**Remark 2.3.** Choose a symplectic basis

$$(\alpha_1, \ldots, \alpha_g, \alpha_{g+1}, \ldots, \alpha_{2g})$$

of $H^1(X,\mathbb{Q})$ satisfying

$$\alpha_l \cup \alpha_{g+l} \ = \ -\alpha_{g+l} \cup \alpha_l \ = \ 1_X \quad (1 \leq l \leq g)$$

(where $1_X$ denotes the canonical generator of $H^2(X,\mathbb{Q})$) and $\alpha_k \cup \alpha_l = 0$ if $|k-l| \neq g$. Let

$$(\beta_1, \ldots, \beta_g, \beta_{g+1}, \ldots, \beta_{2g})$$

be the dual basis in $V = H^1(X,\mathbb{Q})^*$. The element in $\overset{2}{\wedge} V$ defined by the cup-product is given by

$$\Omega := -\sum_{l=1}^{g} \beta_l \wedge \beta_{l+g}$$

and we can identify $\Omega$ with the anti-symmetric tensor

$$\Omega = -\frac{1}{2}\sum_{l=1}^{g}(\beta_l \otimes \beta_{l+g} - \beta_{l+g} \otimes \beta_l)$$

in $V \otimes V$. Let denote by $\{\beta_k^i\}$, $i = 1, \ldots, n$ the basis of $V_i$ given by $\beta_k^i = \beta_l \otimes e_i$, $l = 1, \ldots, 2g$. Then the tensor $\tau_{ij}$ in $V_i \otimes V_j$ is given by

$$\tau_{ij} = -\frac{1}{2}\sum_{l=1}^{g}(\beta_l^i \otimes \beta_{l+g}^j - \beta_{l+g}^i \otimes \beta_l^j) .$$

Recall that $I_0$ is defined to be the subalgebra of $\wedge W$ generated by $\{\tau_{ij}\}$; and recall the homomorphism $F$ obtained in (2.1).

**Proposition 2.4.** *The image by $F$ of the subalgebra $S(W') \otimes_{\mathbb{Q}} I_0$ is contained in the space of algebraic classes in $H^{\mathrm{ev}}(\mathcal{M},\mathbb{Q})$.*

*Proof.* Let 1 and $[X]$ be the canonical generators of $H_0(X,\mathbb{Q})$ and $H_2(X,\mathbb{Q})$ respectively. Let $\gamma \in H^{2m}(X \times \mathcal{M},\mathbb{Q})$ be an algebraic class and let

$$\gamma = \langle \gamma, 1\rangle \otimes 1 + \gamma^{1,1} + \langle \gamma, [X]\rangle \otimes [X]$$

be the Künneth decomposition of $\gamma$, where $\gamma^{1,1} \in H^1(X,\mathbb{Q}) \otimes H^{2m-1}(\mathcal{M},\mathbb{Q})$. The classes $\langle \gamma, 1\rangle$ and $\langle \gamma, [X]\rangle$ are algebraic : $\langle \gamma, 1\rangle$ is the restriction of $\gamma$ to $x \times \mathcal{M}$, $x \in X$, and $\langle \gamma, [X]\rangle$ is the image of $\gamma$ by the Gysin homomorphism for the projection $X \times \mathcal{M} \longrightarrow \mathcal{M}$. (The Gysin homomorphism corresponds to the push-forward of cycles.) It follows that $\gamma^{1,1}$ is also algebraic.

Since algebraic classes form an algebra, it suffices to prove that the image (by $F$) of any element in $W'$ and the image of any $\tau_{ij}$ are algebraic classes. Since the classes $a_m$, $m = 1, \ldots, n$ and $c_{k,j,i}$ are algebraic, it is clear that for $w \in W'$, the class $F(w)$ is algebraic by the remark in the last paragraph.

We will now prove that

$$F(\tau_{ij}) = -\langle\, a_i^{1,1} \cup a_j^{1,1}, [X]\rangle, \tag{2.5}$$

where $a_i^{1,1}$ (resp. $a_j^{1,1}$) is the Künneth component in $H^1(X) \otimes H^{2i-1}(\mathcal{M})$ (resp. in $H^1(X) \otimes H^{2j-1}(\mathcal{M})$). Since $a_i^{1,1} \cup a_j^{1,1}$ is algebraic, this would imply that $F(\tau_{ij})$ is algebraic.

Write

$$a_i^{1,1} = \sum_{m=1}^{2g} b_{i,m} \otimes \alpha_m \quad \text{and} \quad a_j^{1,1} = \sum_{m=1}^{2g} c_{j,m} \otimes \alpha_m$$

where $(\alpha_1, \dots, \alpha_g, \alpha_{g+1}, \dots, \alpha_{2g})$ is the symplectic basis in Remark 2.3. Then

(2.6)

$$a_i^{1,1} \cup a_j^{1,1} = \frac{1}{2}\Big( \sum_{m=1}^{g} b_{i,m} \cup c_{j,m+g} - b_{i,m+g} \cup c_{j,m} \Big) \otimes 1_{[X]}.$$

For $\xi \in H_1(X, \mathbb{Q})$ we have $\langle a_i, \xi \rangle = \sum_{m=1}^{2g} \alpha_m(\xi) b_{i,m}$ and similar equality for $\langle a_j, g \rangle$. Now

(2.7)

$$\begin{aligned} F(\tau_{ij}) &= -\frac{1}{2} F\Big( \sum_{m=1}^{g} (\beta^i_m \otimes \beta^j_{m+g} - \beta^i_{m+g} \otimes \beta^j_m) \Big) \qquad \text{(Remark 2.3)} \\ &= -\frac{1}{2}\Big( \sum_{m=1}^{g} \langle a_i^{1,1}, \beta^i_m \rangle \langle a_j^{1,1}, \beta^j_{m+g} \rangle - \sum_{m+1}^{g} \langle a_i^{1,1}, \beta^i_{m+g} \rangle \langle a_j^{1,1}, \beta^j_m \rangle \Big) \\ &= -\frac{1}{2}\Big( \sum_{m=1}^{g} b_{i,m} c_{j,m+g} - \sum_{m=1}^{g} b_{i,m+g} c_{j,m} \Big) \end{aligned}$$

since $\{\beta_k\}$ is the dual basis of $\{\alpha_k\}$. The assertion (2.5) follows from (2.6) and (2.7). Thus proposition 2.4 is proved. □

Let $H^1(X, \mathbb{C}) = H^{1,0} \oplus H^{0,1}$ be the Hodge decomposition of $H^1$. Using the isomorphism $V_{\mathbb{C}} := H_1(X, \mathbb{C}) = H^1(X, \mathbb{C})$ given by the cup-product, we obtain a decomposition

$$V_{\mathbb{C}} = V^{1,0} \oplus V^{0,1}.$$

Write $W_{\mathbb{C}} := V_{\mathbb{C}} \otimes \mathbb{C}^n = W^{1,0} \oplus W^{0,1}$, where $W^{1,0} = V^{1,0} \otimes \mathbb{C}^n$ and $W^{0,1} = V^{0,1} \otimes \mathbb{C}^n$. Then

$$\bigwedge W_{\mathbb{C}} = \bigoplus_{p,q} W^{p,q}$$

where $W^{p,q} = \overset{p}{\wedge} W^{1,0} \otimes \overset{q}{\wedge} W^{0,1}$.

**Lemma 2.8.** *The image of $S(W'_{\mathbb{C}}) \otimes_{\mathbb{C}} W^{p,q}$ by the map $F_{\mathbb{C}}$ (see 2.2) is contained in*

$$\bigoplus_{k \in \mathbb{Z}} H^{p+k,q+k}(\mathcal{M}, \mathbb{C}).$$

*Proof.* Since $F(S(W'))$ are algebraic classes, it suffices to prove the statement for $F_{\mathbb{C}}(W^{p,q})$. Now $a_i \in H^{2i}(X \times \mathcal{M}, \mathbb{Q})$ are of type $(i,i)$. For $\xi \in V^{1,0}$ (resp. $V^{0,1}$), $\langle a_i, \xi \rangle \in H^{i,i-1}(\mathcal{M})$ (resp. $H^{i-1,i}(\mathcal{M})$). The lemma follows. □

## 3. The representation of the symplectic group on $\wedge W_{\mathbb{C}}$

We keep the notation of the previous section. In particular, $V = H_1(X, \mathbb{Q})$, $W = V \otimes \mathbb{Q}^n$. Let $\mathcal{A} = \wedge W_{\mathbb{C}}$, the exterior algebra of $W_{\mathbb{C}}$. The integral symplectic form on $H_1(X, \mathbb{Z})$ gives a symplectic form on $V_{\mathbb{C}}$. Let $G$ be the corresponding complex symplectic group and let $G_{\mathbb{Z}}$ be the subgroup of $G$ leaving $H_1(X, Z)$ invariant. We note that $G_{\mathbb{Z}}$ is isomorphic to the Siegel modular group $Sp(g, \mathbb{Z})$. Consider the natural action of $G$ on $W_{\mathbb{C}}$, the action on $\mathbb{C}^n$ being trivial (the diagonal action on $V_{\mathbb{C}} \oplus \cdots \oplus V_{\mathbb{C}}$) and the induced action on $\mathcal{A}$. Recall the definition of $I_0$ and $I_{0,\mathbb{C}}$ in Section 2.

**Proposition 3.1.** *$I_0$ (resp. $I_{0,\mathbb{C}}$) coincides with the algebra of invariants in $\wedge W$ (resp. $\mathcal{A}$) for the action of $G_{\mathbb{Z}}$ (resp. $G$).*

*Proof.* The statement in the case of $I_{0,\mathbb{C}}$ is a special case of a theorem of Howe [Theorem 2 of [Ho] and a remark preceding the theorem]. Since $G_{\mathbb{Z}}$ is Zariski dense in $G$ [Bo], the result follows since $\tau_{ij} \in \wedge W$. □

Recall that we have a decomposition $\mathcal{A} = \bigoplus W^{p,q} := \overset{p}{\wedge} W^{1,0} \otimes \overset{q}{\wedge} W^{0,1}$. We shall say, by abuse of language, that an element of $W^{p,q}$ is a form of type $(p,q)$. The following theorem is crucial for the proof of the main theorem.

**Theorem 3.2.** *Any $G$ invariant subspace of $\mathcal{A}$ must either be contained in the space $I_{0,\mathbb{C}}$ of $G$ invariants or must contain a nonzero element of the direct sum $\bigoplus_k W^{k+p,k}$ with $p \geq 1$.*

We need some preliminaries for the proof of this theorem. Write $(V_{\mathbb{C}} \otimes \mathbb{C}^n)^* = \bigoplus_{i=1}^n V_i^*$ and let, for $1 \leq i \leq g$, $\bar{\tau}_{ij} \in V_i^* \otimes V_j^*$, the element given by the skew-symmetric form on $H_1(X, \mathbb{C})$. We may regard $\bar{\tau}_{ij}$ as elements of $\overset{2}{\wedge} W_{\mathbb{C}}^*$. Let $T_{ij} \in \mathrm{End}(\mathcal{A})$ be the operator of interior multiplication by $\bar{\tau}_{ij}$. Following [Ho, page 555] define

$$\mathcal{H} := \bigcap_{i,j} \mathrm{kernel}(T_{i,j}) \subset \mathcal{A}.$$

**Lemma 3.3.** (1) *The space of $(p,0)$ forms in $\mathcal{A}$ is contained in $\mathcal{H}$.*

(2) *Let $W_{\mathbb{C}} = \bigoplus_{i=1}^n V_{\mathbb{C},i}$ be the decomposition of $W_{\mathbb{C}} = V_{\mathbb{C}} \otimes \mathbb{C}^n$. Consider the decomposition*

$$\overset{r}{\wedge} W_{\mathbb{C}} = \bigoplus_{\substack{0 \leq a_i \leq 2g}}^{\sum a_i = r} \left( \overset{a_1}{\wedge} V_{\mathbb{C},1} \otimes \overset{a_2}{\wedge} V_{\mathbb{C},2} \otimes \overset{a_3}{\wedge} \ldots \otimes \overset{a_n}{\wedge} V_{\mathbb{C},n} =: W_{a_1,\ldots,a_n} \right).$$

*Then if some $a_i > g$, we have $W_{a_1,\ldots,a_n} \cap \mathcal{H} = 0$. (That is, if some $a_i > g$ then there are no nonzero "primitive" forms in $W_{a_1,\ldots,a_n}$.)*

*Proof.* We may choose a symplectic basis of $H^1(X, \mathbb{C})$,

$$\alpha_1, \alpha_2, \ldots, \alpha_g, \alpha_{g+1}, \ldots, \alpha_{2g}$$

such that $\alpha_1, \dots, \alpha_g \in H^{1,0}(X, \mathbb{C})$ and $\alpha_{g+1}, \dots, \alpha_{2g} \in H^{0,1}(X, \mathbb{C})$ and satisfy the relations $\alpha_l \cup \alpha_{g+l} = -\alpha_{g+l} \cup \alpha_l = 1_X$, $1 \le l \le g$, and $\alpha_k \cup \alpha_l = 0$ if $|k-l| \neq g$. Let $(\alpha^i_m)_{1 \le m \le 2g}$ be the corresponding basis in $V_{\mathbb{C},i}$, $1 \le i \le n$; then $(\alpha^i_m)^{1 \le i \le n}_{1 \le m \le 2g}$ form a basis of $W^*_{\mathbb{C}} = H^1(X, \mathbb{C}) \otimes \mathbb{C}^n$. We have

$$\bar{\tau}_{ij} = \frac{1}{2} \sum_{l=1}^{g} \left( \alpha^i_l \wedge \alpha^j_{l+g} - \alpha^i_{l+g} \wedge \alpha^j_l \right) \in \overset{2}{\wedge} W^*_{\mathbb{C}}.$$

Let $(\beta^i_m)$ be the dual basis of $W_{\mathbb{C}}$. Then $(\beta^i_m)$, $1 \le i \le n$, $1 \le m \le g$ (resp. $(\beta^i_m)$, $1 \le i \le n$, $g < m \le 2g$) form a basis of $W^{1,0}$ (resp. $W^{0,1}$). Now the first part of the lemma follows from the explicit formula for the interior product [Bou1, Ch. III, §11, no. 9, Formula 68, also §11, no. 10, Formula 76].

The operator $T_{ii}$ leaves $\wedge V_{\mathbb{C},i}$ invariant.

The explicit formula for the interior product shows that, for $\xi_j \in \overset{a_j}{\wedge} V_{\mathbb{C},j}$, we have

$$T_{ii}(\xi_1 \otimes \xi_2 \otimes \dots \otimes \xi_n) = \xi_1 \otimes \dots \otimes T_{ii}(\xi_i) \otimes \dots \otimes \xi_n \in W_{a_1, \dots, a_n}.$$

But if $a_i > g$, it is well known that $T_{ii}$ is injective on $\overset{a_i}{\wedge} V_{\mathbb{C},i}$. (This can be seen by the primitive decomposition [Bou2, Ch. VIII, §13, no. 3, IV, pp. 202–203]: with the notation there, if $r > l$, the weight of $H$ on $\overset{r}{\wedge} V$, namely $l - r$, is strictly negative, so that $X_+$ is injective on $\overset{r}{\wedge} V$.) The second part of the lemma follows. □

For notational simplicity we write $D = \mathbb{C}^n$ so that $W_{\mathbb{C}} = V_{\mathbb{C}} \otimes D$. We now consider the "twisted" action $\rho$ of $GL(D^*)$ on $\mathcal{A} = \wedge W_{\mathbb{C}}$ defined as follows: For $T \in GL(D^*)$, let $\tilde{T}$ be the element of $GL(\mathcal{A})$ obtained by extending $Id(V_{\mathbb{C}}) \otimes (T^*)^{-1}$ as an algebra automorphism of $\mathcal{A}$; then

$$\rho(T) = (\det T)^g \cdot \tilde{T}.$$

[The corresponding action of the Lie algebra $\mathrm{End}(D^*)$ is given as follows : Let $S \in \mathrm{End}(D^*)$ and let $\tilde{S} \in \mathrm{End}(\mathcal{A})$ be the extension of $Id(V_{\mathbb{C}}) \otimes (-S^*)$ as a derivation of the algebra $\mathcal{A}$; then the action of $S$ is given by $(g \cdot \mathrm{trace}(S) + \tilde{S})$.]

The actions of $G$ and $GL(D^*)$ on $\mathcal{A}$ commute.

We now state a special case of a theorem of Howe [Ho, Theorem 9]. In our special case (where we have only the exterior algebra and no symmetric algebra), the space $\Gamma'^{(0,2)}$ defined on page 555 of [Ho], is the vector space in $\mathrm{End}(\mathcal{A})$ generated by the operators $T_{ij}$; moreover the space $\Gamma'^{(1,1)}$ is the algebra $\mathrm{End}(D^*)$ of endomorphisms of $D^*$. (There are some misprints on page 554 of [Ho], in the definition of $\mathfrak{gr}^{0,2}$ (resp. $\mathfrak{gr}^{2,0}$); $\overset{2}{\wedge}(U^*)$ (resp. $\overset{2}{\wedge}(U)$) should be replaced by $\overset{2}{\wedge}(W^*)$ (resp. $\overset{2}{\wedge}(W)$).)

**Theorem 3.4** ([Ho, Theorem 9]). *The action of $G$ leaves $\mathcal{H}$ invariant. Let $\mathcal{A} = \bigoplus_j I_j$ be the isotypical decomposition of the $G$-module $\mathcal{A}$, $I_{0,\mathbb{C}}$ being the space of invariants. Then*

(1) *The map $I_{0,\mathbb{C}} \otimes \mathcal{H} \longrightarrow \mathcal{A}$ given by the multiplication is surjective.*
(2) *If $\mathcal{H}_j := \mathcal{H} \cap I_j$, then $\mathcal{H} = \bigoplus_j H_j$, $\mathcal{H}_j \neq 0$, and $I_j = I_{0,\mathbb{C}}\mathcal{H}_j$.*
(3) *Each $\mathcal{H}_j$ is invariant under $\operatorname{End}(D^*)$ and forms an irreducible module for the action of the group $G \times GL(D^*)$, so that $\mathcal{H}_j = \sigma_j \otimes \tau_j$, where $\sigma_j$ (resp. $\tau_j$) is an irreducible representation of $G$ (resp. $GL(D^*)$). The resulting correspondence between $G$-modules and $GL(D^*)$-modules is bijective.*

**Lemma 3.5.** *With the notation of Theorem 3.4, let $S_j$ (resp. $T_j$) denote the underlying subspaces of the representation $\sigma_j$ (resp. $\tau_j$) (with $S_j \otimes T_j = \mathcal{H}_j$). Then the algebra $\operatorname{End}(T_j)$ maps the space $\mathcal{H}_j^{p,0} := \mathcal{H}_j \cap W^{p,0}$ into itself, where $W^{p,0}$ is the space of $(p,0)$ forms.*

*Proof.* Let $\tau_j : GL(D^*) \longrightarrow GL(T_j)$ be the homomorphism giving the irreducible representation on $T_j$. Since $\tau_j$ is irreducible, any $\phi \in \operatorname{End}(T_j)$ is of the form $\phi = \sum_i \lambda_i \tau_j(g_i)$, where $g_i \in GL(D^*)$, $\lambda_i \in \mathbb{C}$, by Burnside's theorem [Bou3, §4, no. 3, Cor. 1]. Now $GL(D^*)$ maps $W^{1,0} = V_{\mathbb{C}}^{1,0} \otimes D$ into itself and hence $W^{p,0} = \overset{p}{\wedge} W^{1,0}$ into itself and the result follows. □

**Lemma 3.6.** *If an irreducible $G$-component $F$ of $\mathcal{H}_j$ contains a nonzero form of type $(p,0)$, so does every component of $\mathcal{H}_j$.*

*Proof.* Let $F'$ be another irreducible component of $\mathcal{H}_j$. There exists $\psi \in GL(T_j)$ (with the notation of the above lemma) which maps $F$ into $F'$ isomorphically [Bou3, §1, no. 5, also §4 no. 4]. But by Lemma 3.5, $\psi$ maps $(p,0)$ forms in $\mathcal{H}_j$ into $(p,0)$ forms. □

**Lemma 3.7.** *Each irreducible component of $\mathcal{H}_j$ with $j$ nontrivial (i.e. $\mathcal{H}_j \neq \mathcal{H}_0$) contains a nonzero form of type $(p,0)$ with $p \geq 1$.*

*Proof.* We have the decomposition

$$\mathcal{A} = \bigoplus_r \Big( \bigoplus_{0 \leq a_i \leq 2g}^{\sum a_i = r} W_{a_1,\dots,a_n} \Big) \qquad \text{(for the notation see Lemma 3.3).}$$

Note that this decomposition is the weight space decomposition of $\mathcal{A}$ with respect to the Cartan subgroup of $GL(D^*)$ ($= GL(\mathbb{C}^n)$) consisting of the diagonal matrices, the weights being $(g - a_1, g - a_2, \dots, g - a_n)$. By Lemma 3.3, we see that $\mathcal{H}$ is contained in

$$\bigoplus_{0 \leq a_i \leq g} W_{a_1,\dots,a_n}.$$

In particular, the highest weight $(g-a_1,\dots,g-a_n)$ (with respect to upper triangular matrices as the Borel subgroup) of any irreducible $GL(D^*)$-submodule of $\mathcal{H}$ satisfies the condition that $g \geq a_n \geq a_{n-1} \geq \cdots \geq a_2 \geq a_1 \geq 0$.

Given integers $\underline{a} := (a_1,\dots,a_n)$ with

$$g \geq a_n \geq a_{n-1} \geq \cdots \geq a_2 \geq a_1 \geq 0$$

we shall now construct an irreducible representation $\rho_{\underline{a}} = \sigma \otimes \tau$ of $G \times GL(D^*)$ such that :

(a) The underlying subspace $R_{\underline{a}}$ of $\rho_{\underline{a}}$ is contained in $\mathcal{H}$.
(b) $R_{\underline{a}}$ contains a nonzero form of type $(p,0)$ where $p = \sum_{i=1}^{n} a_i$.
(c) The highest weight of $\tau$ is $(g-a_1,\dots,g-a_n)$ (with respect to the Borel subgroup of $GL(D^*)$ consisting of the upper triangular matrices).

Choose a basis $(\omega_1,\omega_2,\dots,\omega_g)$ of $V_{\mathbb{C}}^{1,0}$. Let $B_1$ be the subgroup of $G$ leaving invariant the flag

$$(\omega_1) \subset (\omega_1,\omega_2) \subset (\omega_1,\omega_2,\omega_3) \subset \cdots \subset (\omega_1,\omega_2,\dots,\omega_g).$$

Since $V_{\mathbb{C}}^{1,0}$ is a maximal isotropic space for the symplectic form, we see that $B_1$ is a Borel subgroup of $G$. Then

$$B := B_1 \times B_2 ,$$

where $B_2$ is the group of upper triangular matrices in $GL(D^*)$, is a Borel subgroup of $G \times GL(D^*)$. Let

$$\phi_k = \omega_1 \wedge \cdots \wedge \omega_k \in \overset{k}{\wedge} V^{1,0} \subset \overset{k}{\wedge} V_{\mathbb{C}} ,\ 1 \leq k \leq g, \text{ and } \phi_0 = 1.$$

Let $\phi_k^i$ be the corresponding element in $\overset{k}{\wedge} V_{\mathbb{C},i}$. For $\underline{a} := (a_1,\dots,a_n)$, $g \geq a_1 \geq a_2 \geq \cdots \geq a_n \geq 0$ consider the vector

$$\phi_{\underline{a}} := \phi_{a_1}^1 \otimes \phi_{a_2}^2 \otimes \cdots \otimes \phi_{a_n}^n \in W_{a_1,\dots,a_n} \subset \mathcal{A}.$$

It may be verified that the one-dimensional subspace $(\phi_{\underline{a}})$ of $\mathcal{A}$ spanned by $\phi_{\underline{a}}$ is left invariant by the Borel subgroup $B$ of $G \times GL(D^*)$. In fact, it is clear that $(\phi_{\underline{a}})$ is invariant under $B_1$. As for $B_2$, first observe that the diagonal Cartan subgroup of $GL(D^*)$ leaves $(\phi_{\underline{a}})$ invariant (and the weight of this group on $(\phi_{\underline{a}})$ is $(g-a_1,\dots,g-a_n)$). So in order to show that $B_2$ leaves $(\phi_{\underline{a}})$ invariant, it is sufficient to verify that the element $\phi_{\underline{a}}$ is annihilated (under the action of the Lie algebra of $GL(D^*)$ on $\mathcal{A}$) by the elementary upper triangular matrices $E_{ij}$ whose only nonzero entry is 1 at the $(i,j)$-th place with $j > i$. Observing that the action of $E_{ij}$ on $W_{\mathbb{C}}$ is given by : $E_{ij}(V_{\mathbb{C},k}) = 0$ if $k \neq i$ and the action of $E_{ij}$ on $V_{\mathbb{C},i}$ is $(-1)$ times the natural identification of $V_{\mathbb{C},i}$ with $V_{\mathbb{C},j}$, and using the condition $a_n \geq a_{n-1} \geq \cdots \geq a_1$ it is easily checked that $E_{ij}$ annihilates $\phi_{\underline{a}}$. Since $\phi_{a_i}$ is of type $(a_i,0)$, $\phi_{\underline{a}}$ is of type $(\sum a_i, 0)$. By Lemma

3.3(1), $\phi_{\underline{a}} \in \mathcal{H}$. Since the subspace $(\phi_{\underline{a}})$ is left invariant by the Borel subgroup $B$ of $G \times GL(D^*)$, we see that we obtain an irreducible representation $\rho_{\underline{a}}$ of $G \times GL(D^*)$, on the subspace $R_{\underline{a}}$ of $\mathcal{A}$ spanned by the transforms of $\phi_{\underline{a}}$ by elements of $G \times GL(D^*)$. Since $\phi_{\underline{a}} \in \mathcal{H}$, we have $R_{\underline{a}} \subset \mathcal{H}$. Thus conditions (a), (b), and (c) are fulfilled.

In view of Lemma 3.6, every $G$-irreducible component of $\rho_{\underline{a}}$ will contain a nonzero form of type $(p, 0)$, $p = \sum a_i$, since the space spanned by $G$-transforms of $\phi_{\underline{a}}$ is $G$-irreducible, and $\phi_{\underline{a}}$ is of type $(p, 0)$.

Thus, in order to complete the proof of the lemma, it suffices to show that any $\mathcal{H}_j$ is of the form $R_{\underline{a}}$, for some $\underline{a}$ with $g \geq a_n \geq \cdots \geq a_1 \geq 0$, and that if $j \neq 0$, we have $a_n \geq 1$.

Let then $\mathcal{H}_j = \sigma_j \otimes \tau_j$. Let $\underline{a} := (a_1, \ldots, a_n)$ be the highest weight of $\tau_j$. As already observed $\underline{a}$ satisfies the condition $g \geq a_n \geq \cdots \geq a_1 \geq 0$ [Lemma 3.3]. Consider the irreducible representation $\rho_{\underline{a}}$ on $R_{\underline{a}}$ constructed above; it is of the form $\sigma \otimes \tau_j$, for some irreducible representation $\sigma$ of $G$. Let the class of $\sigma$ (in the isotypical decomposition) be denoted by $k$; so $R_{\underline{a}} \subset I_k$. Hence $R_{\underline{a}} \subset I_k \cap \mathcal{H} = \mathcal{H}_k$. Since both $R_{\underline{a}}$ and $\mathcal{H}_k$ are irreducible representations of $G \times GL(D^*)$, we must have $R_{\underline{a}} = \mathcal{H}_k$. Thus $\sigma \otimes \tau_j = \sigma_k \otimes \tau_k$. Hence $\sigma = \sigma_k = \sigma_j$ by the bijective correspondence [Theorem 3.4(3)] between $\sigma_j$ and $\tau_j$. Thus $R_{\underline{a}} = I_k = I_j$.

Finally, the (one-dimensional) subspace of $\mathcal{A}$ spanned by 1 is pointwise fixed by $G$, and $GL(D^*)$ acts on this space with weight $(g, g, \ldots, g)$. Using the bijective correspondence, we see that if $j$ is not equal to zero then not all $a_i$ could be zero. This completes the proof of the lemma. □

**Remark 3.8.** The proof of the lemma gives the highest weights of irreducible representations of $G$ occurring in $\mathcal{H}$ and their multiplicity.

*Proof of Theorem 3.2.* It is sufficient to show that any $G$-invariant subspace $R \subset \mathcal{A}$ on which the representation is nontrivial must contain a nonzero element of $\bigoplus_k W^{k+p,k}$ with $p \geq 1$. Let $R \subset I_j$, $j \neq 0$. Write as before, $\mathcal{H}_j = S_j \otimes T_j$, with irreducible representations $\sigma_j$ (resp. $\tau_j$) of $G$ (resp. $GL(D^*)$) on $S_j$ (resp. $T_j$). Consider $S_j \otimes T_j \otimes I_{0,\mathbb{C}}$ as a $G$-module (with trivial action on $I_{0,\mathbb{C}}$). The homomorphism of isotypical $G$-modules

$$\mu : S_j \otimes T_j \otimes I_{0,\mathbb{C}} \longrightarrow I_j$$

is surjective, since by Theorem 3.4 (2) we have $I_j = I_{0,\mathbb{C}}\mathcal{H}_j$ and $I_{0,\mathbb{C}}$ is in the center of $\mathcal{A}$. There exists a one-dimensional subspace $\xi$ of $T_j \otimes I_{0,\mathbb{C}}$ such that $\mu$ maps $S_j \otimes \xi$ isomorphically onto $R$ [Bou3, §1, no. 5, also §4 no. 4]. Let $0 \neq v \in \xi$. Let

$$v = \sum_\alpha c_\alpha \otimes d_\alpha$$

with $c_\alpha \in T_j$ and $d_\alpha \in I_{0,\mathbb{C}}$.

Let $\mathcal{H}_j^{p,0} := \mathcal{H}_j \cap W^{p,0}$. Since $j \neq 0$, we have $\mathcal{H}_j^{p,0} \neq 0$ for some $p \geq 1$ [Lemma 3.7]; moreover $GL(D^*)$ operates on $\mathcal{H}_j^{p,0}$. Since $\mathcal{H}_j$ is an isotypical module of type $T_j$ for the action of $GL(D^*)$, there exists a one-dimensional subspace $\eta$ of $S_j$ with $\eta \otimes T_j \subset \mathcal{H}_j^{p,0}$. Choose $0 \neq w \in \eta$. Thus for any $t \in T_j$, $\mu(w \otimes t \otimes 1) \in W^{p,0}$; in particular any $\mu(w \otimes c_\alpha \otimes 1) \in W^{p,0}$. Since $I_{0,\mathbb{C}} \subset \oplus_k W^{k,k}$ we have

$$\mu(w \otimes v) \subset \bigoplus_k W^{k+p,k} \text{ with } p \geq 1.$$

This completes the proof of the theorem. □

**Remark 3.9.** Let $R \subset \mathcal{A}$ be a nontrivial and irreducible sub-representation of $G$. Then $R$ cannot consist entirely of forms of type $(p,p)$. In fact, let $P'$ be a $G$-invariant subspace of $\mathcal{A}$ consisting entirely of $(p,p)$-forms. If $P'$ were not contained in the space of invariants $I_{0,\mathbb{C}}$, then $P'$ would contain an irreducible and nontrivial $G$-representation $P$, and $P$ would consist entirely of $(p,p)$-forms. But this would contradict Theorem 3.2.

## 4. Monodromy action on the cohomology of the moduli space

Let $\mathcal{U}_g^h$ be the moduli space of smooth curves of genus $g \geq 2$ with $h$ ordered marked points ($h \geq 1$). Let $U \subset \mathcal{U}_g^h$ be the Zariski open subset consisting of pointed curves $(Y, \{y_1, \ldots, y_h\})$ such that any automorphism of $Y$ that fixes $\{y_1, \ldots, y_h\}$ point-wise is the identity map. Since for a generic curve of genus 2 the only nontrivial automorphism is the hyperelliptic involution and a generic curve of $g \geq 3$ has no nontrivial automorphism, it follows that $U$ is nonempty.

Let $u_0 := (X, \{p_1, \ldots, p_h\}) \in U$. Let

$$\pi : \mathcal{X}_U \longrightarrow U$$

be the universal curve, along with sections $\sigma_i : U \to \mathcal{X}_U$, $i = 1, \ldots, h$. The local systems on $U$ corresponding to the zero-th and second homology groups of fibers of $\pi$ are canonically trivial and the local system corresponding to the first homology group is given by a homomorphism

$$\rho' : \pi_1(U, u_0) \longrightarrow \mathrm{Aut}(H_1(X, \mathbb{Z})).$$

The group $\mathrm{Aut}(H_1(X,\mathbb{Z}))$, which will also be denoted by $G_{\mathbb{Z}}$, is the group of all automorphisms of $H_1(X,\mathbb{Z})$ preserving the cap product. After choosing a symplectic basis of $H_1(X,\mathbb{Z})$, the group $G_{\mathbb{Z}}$ can be identified with the symplectic group $Sp(g,\mathbb{Z})$. The above homomorphism $\rho'$ is a surjection onto the Siegel modular group $G_{\mathbb{Z}}$. To see this, consider for $l \geq 1$ the moduli space $\mathcal{U}_g^h[l]$ of pointed curves with level structure of level $l$ [i.e., the quotient of the pointed Teichmüller space $\mathcal{T}_g^h$ by the subgroup of the mapping class group $\Gamma_l^h$ which acts as identity on the homology groups with coefficients in $\mathbb{Z}/l$]. For $l \geq 3$ the space $\mathcal{U}_g^h[l]$ is a smooth variety; assume that $l \geq 3$. Let $q : \mathcal{U}_g^h[l] \to \mathcal{U}_g^h$ be the canonical projection. The map $q^{-1}(U) \xrightarrow{q} U$ is an étale Galois cover with Galois group $G_{\mathbb{Z}}/G[l]$ where $G[l] \subset G_{\mathbb{Z}}$ is the normal subgroup which acts as identity on homology groups with coefficients in $\mathbb{Z}/l$. Since $\mathcal{U}_g^h[l]$ is smooth, the fundamental group $\pi_1(q_l^{-1}(U))$ surjects onto $\pi_1(\mathcal{U}_g^h[l])$. On the other hand, the fact that $\Gamma_l^h$ surjects onto $G_{\mathbb{Z}}$ implies that $\pi_1(\mathcal{U}_g^h[l])$ surjects onto the kernel of $p : G_{\mathbb{Z}} \to \mathrm{Aut}(H_1(X, \mathbb{Z}/l))$. So the image of $\rho'$ contains the kernel of the homomorphism $p$ and also surjects onto $G_{\mathbb{Z}}/G[l]$. This implies that $\rho'$ surjects onto $G_{\mathbb{Z}}$.

Let $\psi : \mathcal{M}_U \longrightarrow U$ be the family of parabolic bundles (of the given type) along the fibers of $\pi : \mathcal{X}_U \to U$. Let $\psi^{-1}(u_0) = \mathcal{M}_0$. Consider the monodromy of the local system $\bigoplus_i R^i\psi_*\mathbb{Q}$ given by a homomorphism

$$\rho_0 \,:\, \pi_1(U, u_0) \longrightarrow \mathrm{Aut}(H^*(\mathcal{M}_0, \mathbb{Q})).$$

By Theorem 3.4 of [B], $\rho_0$ is the identity on the kernel of the (surjective) homomorphism $\rho' : \pi_1(U, u_0) \to G_{\mathbb{Z}}$, so that $\rho_0$ factors through a homomorphism

$$\rho \,:\, G_{\mathbb{Z}} \longrightarrow \mathrm{Aut}(H^*(\mathcal{M}_0, \mathbb{Q})).$$

Moreover, consider the fiber product

$$\mathcal{X}_U \times_U \mathcal{M}_U \longrightarrow U$$

and the corresponding monodromy homomorphism

$$\pi_1(U, u_0) \longrightarrow \mathrm{Aut}(H^*(X \times \mathcal{M}_0, \mathbb{Q})).$$

This homomorphism factors through a homomorphism

$$\rho_1 \,:\, G_{\mathbb{Z}} \longrightarrow \mathrm{Aut}(H^*(X \times \mathcal{M}_0, \mathbb{Q})).$$

Let us recall that we defined in Section 2, characteristic classes $a_i \in H^{2i}(X \times \mathcal{M}_0, \mathbb{Q})$ and $c_{k,j,i} \in H^{2k}(\mathcal{M}_0, \mathbb{Q})$. The classes $a_i$ (resp. $c_{k,j,i}$) are invariant under the monodromy action $\rho_1$ (resp. $\rho$) by [B, Lemma 2.7]. Hence the homomorphisms

$$f' \,:\, W' := \mathbb{Q}^N \oplus (H_0(X, \mathbb{Q}) \oplus H_2(X, \mathbb{Q}) \otimes \mathbb{Q}^n) \longrightarrow H^*(\mathcal{M}_0, \mathbb{Q}),$$

$$f \,:\, W := H_1(X, \mathbb{Q}) \otimes_{\mathbb{Q}} \mathbb{Q}^n \longrightarrow H^*(\mathcal{M}_0, \mathbb{Q})$$

commute with the action of the modular group (the action on $H_*(X, \mathbb{Q})$ and $H^*(\mathcal{M}_0, \mathbb{Q})$ being the monodromy action and the actions on $\mathbb{Q}^N$ and $\mathbb{Q}^n$

being trivial) and the surjective homomorphism $F$ in (2.1) commutes with the actions of the monodromy group (the action on $S(W')$ being trivial).

Consider the surjective homomorphism $F_{\mathbb{C}}$ in (2.2). The action of $G_{\mathbb{Z}}$ on $S(W'_{\mathbb{C}})\otimes\wedge W_{\mathbb{C}}$ is the restriction of an action of the complex symplectic group. We claim that the action of $G_{\mathbb{Z}}$ on $H^*(\mathcal{M}_0,\mathbb{C})$ extends uniquely to a rational action of $G$ on $H^*(\mathcal{M}_0,\mathbb{C})$ and $F_{\mathbb{C}}$ commutes with these $G$-actions. This follows from the following two facts :

(1) If $\rho$ is a finite-dimensional rational representation of $G$ on a finite-dimensional complex vector space $R$, a subspace of $R$ is invariant under $G_{\mathbb{Z}}$ if and only if it is invariant under $G$, since $G_{\mathbb{Z}}$ is Zariski dense in $G$.

(2) $S(W'_{\mathbb{C}})\otimes\wedge W_{\mathbb{C}}$ is a union of finite-dimensional $G$-invariant subspaces on which the action of $G$ is rational. (Note that the action of $G$ on $S(W'_{\mathbb{C}})$ is trivial.)

**Proposition 4.1.** *The space of invariants in $H^*(\mathcal{M}_0,\mathbb{C})$ (resp. $H^*(\mathcal{M}_0,\mathbb{Q})$) under the action of $G_{\mathbb{Z}}$ coincides with $F_{\mathbb{C}}(S(W'_{\mathbb{C}})\otimes I_{0,\mathbb{C}})$ (resp. $F(S(W')\otimes I_0)$).*

($I_{0,\mathbb{C}}$ and $I_0$ are defined in Section 3.)

*Proof.* By Proposition 3.1, $S(W'_{\mathbb{C}})\otimes I_{0,\mathbb{C}}$ is the space of invariants of $G$ in $S(W'_{\mathbb{C}})\otimes\wedge W_{\mathbb{C}}$. We have $H^*(\mathcal{M}_0,\mathbb{C})^G = H^*(\mathcal{M}_0,\mathbb{C})^{G_{\mathbb{Z}}}$. Since finite-dimensional representations of $G$ are completely reducible, if $T : V_1 \longrightarrow V_2$ is a surjective linear $G$-morphism of $G$-modules we should have $T(V_1^G) = V_2^G$. So the assertion for $H^*(\mathcal{M}_0,\mathbb{C})^{G_{\mathbb{Z}}}$ follows.

As for $H^*(\mathcal{M}_0,\mathbb{Q})^{G_{\mathbb{Z}}}$, observe that $H^*(\mathcal{M}_0,\mathbb{C})^{G_{\mathbb{Z}}} = H^*(\mathcal{M}_0,\mathbb{Q})^{G_{\mathbb{Z}}}\otimes_{\mathbb{Q}}\mathbb{C}$ and

$$\big(S(W')\otimes_{\mathbb{Q}}I_0\big)\otimes_{\mathbb{Q}}\mathbb{C} \;=\; S(W'_{\mathbb{C}})\otimes I_{0,\mathbb{C}}\,.$$

Since $F_{\mathbb{C}}$ maps, by above, $\big(S(W')\otimes_{\mathbb{Q}}I_0\big)\otimes_{\mathbb{Q}}\mathbb{C}$ onto $H^*(\mathcal{M}_0,\mathbb{Q})^{G_{\mathbb{Z}}}\otimes_{\mathbb{Q}}\mathbb{C}$, we see that $F$ should map $S(W')\otimes I_0$ onto $H^*(\mathcal{M}_0,\mathbb{Q})^{G_{\mathbb{Z}}}$. □

**Theorem 4.2.** *Let $\Sigma$ be a complex linear subspace of $H^*(\mathcal{M}_0,\mathbb{C})$ consisting entirely of classes of type $(p,p)$ such that $\Sigma$ is invariant under the action of the modular group $G_{\mathbb{Z}}$. Then $\Sigma$ is contained in the space of invariants in $H^*(\mathcal{M}_0,\mathbb{C})$ under the action of $G_{\mathbb{Z}}$.*

*Proof.* Consider the surjective homomorphism

$$F_{\mathbb{C}} \;:\; \bigwedge(W_{\mathbb{C}})\bigotimes_{\mathbb{C}}S(W'_{\mathbb{C}}) \;\longrightarrow\; H^*(\mathcal{M}_0,\mathbb{C})\,,$$

defined by the maps $f$ and $f'$ of Section 2. Note that the image of $S(W'_{\mathbb{C}})$ is contained in the centre of $H^*(\mathcal{M}_0,\mathbb{C})$. Since $G_{\mathbb{Z}}$ is Zariski dense in $G$, we may assume that $\Sigma$ is a $G$-submodule of $H^*(\mathcal{M}_0,\mathbb{C})$ (for the action of $G$ on $H^*(\mathcal{M}_0,\mathbb{C})$ defined earlier). Let $\Sigma'$ be an irreducible $G$-submodule of $\Sigma$. We can find a finite-dimensional $G$-submodule $M$ of $\wedge W_{\mathbb{C}}\otimes S(W'_{\mathbb{C}})$ such that $F_{\mathbb{C}} : M \longrightarrow \Sigma'$ is an isomorphism. If the representation $\Sigma'$ were nontrivial, we shall find a nonzero element $\theta$ in $M$ such that $F_{\mathbb{C}}(\theta)$ belongs to $\bigoplus_r H^{r+l,r}(\mathcal{M}, \mathbb{C})$ for some $l \geq 1$. Since $\Sigma$ consists entirely of classes of type $(p,p)$ it would lead to a contradiction.

We shall find the element $\theta$ by the method used in the proof of Theorem 3.2. We use the notation of Section 3. Now $M$ is contained in $I_j \otimes S(W'_{\mathbb{C}})$ with $j \neq 0$ (since we assume that $\Sigma'$ is nontrivial). As in the proof of Theorem 3.2, consider $\mathcal{H}_j = S_j \otimes T_j$ and choose $0 \neq \omega \in S_j$ with

$$\omega \otimes T_j \subset \mathcal{H}_j^{l,0} := \mathcal{H}_j \cap W^{l,0}$$

for some $l \geq 1$. (By Lemma 3.7, $\mathcal{H}_j^{l,0} \neq 0$ with $l \geq 1$, since $j \neq 0$.)

Consider the surjective homomorphism of $G$-modules

$$\nu := \mu \otimes 1 : S_j \otimes T_j \otimes I_{0,\mathbb{C}} \otimes S(W'_{\mathbb{C}}) \longrightarrow I_j \otimes S(W'_{\mathbb{C}})$$

where the map $\mu$ was defined in the proof of Theorem 3.2. We can find a $G$-submodule $M'$ of $S_j \otimes T_j \otimes I_{0,\mathbb{C}} \otimes S(W'_{\mathbb{C}})$ which is mapped isomorphically onto $M$. Since $M$ is an irreducible $G$-module, $M'$ is also an irreducible $G$-module, and hence $M'$ is of the form $S_j \otimes \xi$, where $\xi$ is a one-dimensional subspace of $T_j \otimes I_{0,\mathbb{C}} \otimes S(W'_{\mathbb{C}})$. Let $0 \neq v \in \xi$; then $\omega \otimes v \in M'$. Let

$$\theta := \nu(\omega \otimes v) \in M ;$$

clearly $\theta \neq 0$. Write $v = \sum_\alpha c_\alpha \otimes d_\alpha$ with $c_\alpha \in T_j$ and $d_\alpha \in I_{0,\mathbb{C}} \otimes S(W'_{\mathbb{C}})$. Now $F_{\mathbb{C}}(d_\alpha)$ is a sum of elements of pure type, since, by Proposition 2.4, the image under $F_{\mathbb{C}}$ of the rational classes in $I_{0,\mathbb{C}} \otimes S(W'_{\mathbb{C}})$ are algebraic, and hence are sum of elements of pure type. The element $\mu(\omega \otimes c_\alpha)$ is of type $(l,0)$ for each $\alpha$, by choice of $\omega$. Hence, it follows from Lemma 2.8 that $F_{\mathbb{C}}(\theta) \in \bigoplus_r H^{r+l,r}(\mathcal{M}, \mathbb{C})$ with $l \geq 1$. This completes the proof. □

## 5. The main theorem

We continue with the notation of the previous section. For $u \in U$, let $\mathcal{M}_u$ be the moduli space of parabolic bundles on $\pi^{-1}(u)$ satisfying the conditions (1) and (2) in Section 1.

**Theorem 5.1.** *There is a no-where dense subset (in Euclidean topology) $B$ of $U$ such that for $u \in U \backslash B$ we have*

1. *The Hodge classes in $H^*(\mathcal{M}_u, \mathbb{Q})$ (i.e., rational classes represented by forms of type $(p,p)$) are contained in $F(S(W') \otimes I_0)$.*
2. *The ring of algebraic cohomology classes in $H^*(\mathcal{M}_u, \mathbb{Q})$ coincides with the ring $F(S(W') \otimes I_0)$.*

*In particular, the Hodge conjecture is true for the varieties $\mathcal{M}_u$, $u \in U\backslash B$.*

**Remark 5.2.** The proof shows that $B$ may be taken to be locally a countable union of analytic sets. In fact, using a theorem of Cattani-Deligne-Kaplan [CDK, page 484, Corollary 1.3] we may even take $B$ to be a countable union of proper algebraic sub-varieties.

*Proof of Theorem 5.1.* Given the results of the previous sections, the proof is essentially the same as the proof, communicated to us by Richard Hain, for the case of the Jacobian. For the sake of completeness we give the proof.

Let $\alpha$ be a Hodge class in $H^{2p}(\mathcal{M}_0, \mathbb{Q})$ of type $(p,p)$. Let $p : \tilde{U} \longrightarrow U$ be the universal cover of $U$. Let

$$\tilde{\psi} : \mathcal{M}_{\tilde{U}} \longrightarrow \tilde{U}$$

be the pull-back to $\tilde{U}$ of the universal moduli space $\psi:\mathcal{M}_U \longrightarrow U$. Let $\tilde{\alpha}$ be the section of $R^{2p}\tilde{\psi}_*\mathbb{Q}$ defined by $\alpha$. The set of points $\tilde{u} \in \tilde{U}$, where $\tilde{\alpha}(\tilde{u})$ is of type $(p,p)$ is an analytic set $A_\alpha$ (in fact, since $\alpha$ is a real cohomology class—this condition is the same as saying $\tilde{\alpha}(\tilde{u})$ belongs to an appropriate part of the flag given by the Hodge filtration of $R^{2p}\tilde{\psi}_*\mathbb{C}$—which is a holomorphic condition.)

If $\alpha$ does not belong to $F(S(W')\otimes I_0)$ we shall show that the set $A_\alpha$ is proper (i.e., $A_\alpha \neq \tilde{U}$). Suppose that $\tilde{\alpha}(\tilde{u})$ were of type $(p,p)$ for all $\tilde{u} \in \tilde{U}$. The $\mathbb{C}$-subspace $\Sigma$ generated by $G_{\mathbb{Z}}\alpha$ would be contained in $H^{p,p}(\mathcal{M}_0)$. But then by Theorem 4.2, $\Sigma$ would be contained in $H^*(\mathcal{M}_0, \mathbb{C})^{G_{\mathbb{Z}}}$. But by Proposition 4.1, $F(S(W')\otimes I_0) = H^*(\mathcal{M}_0, \mathbb{Q})^{G_{\mathbb{Z}}}$. So $\alpha \in F(S(W')\otimes I_0)$, a contradiction.

So if $\alpha \notin F(S(W')\otimes I_0)$, $A_\alpha$ is a proper analytic set and $p(A_\alpha)$ (where $p : \tilde{U} \to U$ is the projection) is locally a countable union of analytic sets. (Corollary 1.3 of [CDK] would say that $p(A_\alpha)$ is an algebraic variety.) Define

$$B := \bigcup_{\alpha \in H^*(\mathcal{M}_0,\mathbb{Q})\backslash F(S(W')\otimes I_0)} p(A_\alpha) . \tag{5.3}$$

Since $H^{2p}(\mathcal{M}_0, \mathbb{Q})$ is a countable set we see that $B$ is a nowhere dense set, being locally a countable union of analytic sets. So for $u \in U\backslash B$, rational Hodge classes are contained in $F(S(W')\otimes I_0)$. This proves (1), and (2) follows from Proposition 2.4 which says that $F(S(W')\otimes I_0)$ is contained in the set of algebraic classes. □

In the rest of this section we will show how Theorem 5.1 extends to the case of moduli of parabolic bundles with fixed determinant.

Let $\mathcal{N} \subset \mathcal{M}$ be the sub-variety such that for any $E \in \mathcal{N}$ the top exterior product $\overset{n}{\wedge} E = \mathcal{O}_X(d \cdot p_1)$. Let $J$ be the Jacobian of degree zero line bundles on $X$. The morphism

$$P : \mathcal{N} \times J \longrightarrow \mathcal{M}$$

defined by $(E, L) \longmapsto E \otimes L$ is an étale Galois cover.

The homomorphism $P^* : H^*(\mathcal{M}, \mathbb{Q}) \longrightarrow H^*(\mathcal{N} \times J, \mathbb{Q})$ is an isomorphism [N, Remark 3.11; also [BR], Section 4]. Using the Künneth decomposition of $H^*(\mathcal{N} \times J, \mathbb{Q})$ we have

$$f_1 : H^*(\mathcal{N}, \mathbb{Q}) \longrightarrow H^*(\mathcal{N} \times J, \mathbb{Q})$$

given by $\theta \longmapsto \theta \otimes 1$, where 1 is the canonical generator of $H^0(J, \mathbb{Z})$. Define

$$f := (P^*)^{-1} \circ f_1 : H^*(\mathcal{N}, \mathbb{Q}) \longrightarrow H^*(\mathcal{M}, \mathbb{Q}).$$

Let $i : \mathcal{N} \longrightarrow \mathcal{M}$ be the embedding of $\mathcal{N}$ in $\mathcal{M}$. It can be checked that

$$i^* \circ f : H^*(\mathcal{N}, \mathbb{Q}) \longrightarrow H^*(\mathcal{N}, \mathbb{Q})$$

is $\deg(P) \cdot Id$. Define

$$R := i^*(F(I_0)) \, .$$

[In [BR] it is proved that the cohomology ring $H^*(\mathcal{N}, \mathbb{Q})$ is generated by the image of the slant product of the characteristic classes of the projective universal bundle $\{a_i(P(\mathcal{E}))\}$. In view of this $R$ can be directly defined in a fashion similar to the way we defined $I_0$.] Now the homomorphism $f$ preserves the Hodge decompositions of $H^*(\mathcal{M}, \mathbb{C})$ and $H^*(\mathcal{N}, \mathbb{C})$. For $u \in U$, let $\mathcal{N}_u$ be the moduli space of parabolic bundles with fixed determinant on $\psi^{-1}(u)$. So Theorem 5.1 implies

**Theorem 5.4.** *For any $u \in U \backslash B$ ($B$ is defined in* 5.3*), the Hodge classes in $H^*(\mathcal{N}_u, \mathbb{Q})$ are contained in the algebra $R$ defined above. The ring of algebraic cohomology classes in $H^*(\mathcal{N}_u, \mathbb{Q})$ coincides with $R$. In particular the Hodge conjecture is true for the varieties $\mathcal{N}_u$, $u \in U \backslash B$.*

## 6. Products of moduli spaces

We shall briefly indicate in this section how the above methods can be used to generalise Propositions 2.4 and 3.1 to the case of the $m$-fold product of the moduli space for any smooth curve and Theorem 5.1 to the case when the curve is general. As a consequence we will see that for a suitable polarization (hence for all polarizations [K, Corollary 2.10]) on the moduli space of parabolic bundles on any smooth curve (not necessarily general), the cohomology class in $H^{2N-2}(\mathcal{M} \times \mathcal{M}, \mathbb{Q})$, $N = \dim_{\mathbb{C}}\mathcal{M}$, given by the associated operator $\Lambda : H^*(\mathcal{M}) \longrightarrow H^*(\mathcal{M})$ is algebraic. (See [K, 1.4.2.1] for the definition of $\Lambda$.) For the Jacobian of a smooth curve the algebraicity of $\Lambda$ follows from a theorem of Lieberman, [Li] (also [K, Theorem 2A11]), where he proves that the operator $\Lambda$ is algebraic for any abelian variety.

For simplicity let us consider the case $m = 2$, that is, the space $\mathcal{M} \times \mathcal{M}$. We now define certain algebraic cohomology classes in $\mathcal{M} \times \mathcal{M}$. Let $F$ be the homomorphism obtained in (2.1). The homomorphism

$$\begin{aligned} F \otimes F : S(W') \otimes S(W') \textstyle\bigwedge(W \oplus W) = S(W') \otimes \textstyle\bigwedge(W) \otimes S(W') \otimes \textstyle\bigwedge(W) \\ \longrightarrow H^*(\mathcal{M}, \mathbb{Q}) \otimes H^*(\mathcal{M}, \mathbb{Q}) = H^*(\mathcal{M} \times \mathcal{M}, \mathbb{Q}) \end{aligned}$$

is surjective (the last tensor product being the tensor product of graded anti-commutative algebras). Let $I^0$ be the subalgebra of $\bigwedge(W \oplus W)$ generated by $\tau_{ij} \in V_i \otimes V_j$, $1 \leq i \leq j \leq 2n$ (see Section 2). Define $I \subset H^*(\mathcal{M} \times \mathcal{M}, \mathbb{Q})$ to be the image of $S(W) \otimes S(W) \otimes I^0$ by $F \otimes F$. Then $I$ is contained in the space of algebraic classes in $H^*(\mathcal{M} \times \mathcal{M}, \mathbb{Q})$. To see this, consider the two projections $p_1$ and $p_2$ from $X \times \mathcal{M} \times \mathcal{M}$ to $X \times \mathcal{M}$; then one verifies, as in the proof of Proposition 2.4, that for any pair of indices $(i, j)$ with $i \leq n$ and $j \geq n + 1$,

$$\begin{aligned} (F \otimes F)(\tau_{ij}) &= \langle p_1^* a_i \cup p_2^* a_j, [X] \rangle \\ &- \langle p_1^* a_i, [X] \rangle \cup \langle p_2^* a_j, 1 \rangle - \langle p_1^* a_i, 1 \rangle \cup \langle p_2^* a_j, [X] \rangle; \end{aligned}$$

and for $j \leq n$ (resp. $i \geq n+1$) $(F \otimes F)(\tau_{ij})$ is the pull-back of $F(\tau_{ij})$ (resp. $F(\tau_{i-n,j-n})$), where $F(\tau_{kl})$ are the cohomology classes defined in Section 2, using the projection of $\mathcal{M} \times \mathcal{M}$ onto the first factor (resp. the second factor). Then, as in the proof of Theorem 5.1, one sees that, for a general curve, the Hodge classes of $\mathcal{M} \times \mathcal{M}$ are contained in $I$.

Now for each smooth curve $X$ not necessarily general, the corresponding moduli space $\mathcal{M}$ has a natural polarization $L$. (For the case of ordinary vector bundles $H^*(\mathcal{M}, \mathbb{Q}) = H^*(J, \mathbb{Q}) \otimes H^*(\mathcal{N}, \mathbb{Q})$; since the second Betti number of $\mathcal{N}$ is one there is a natural polarization on $\mathcal{N}$ and, on $J$, we take the polarization given by the theta divisor. For the parabolic case we take the parabolic determinant bundle [NR, Theorem 1; BR1, Definition 4.8] as the polarization.) This polarization $L$, being canonical, is invariant under the monodromy action; see also Proposition 4.14 of [BR1]. Hence the operator $\Lambda : H^*(\mathcal{M}, \mathbb{C}) \longrightarrow H^{*-2}(\mathcal{M}, \mathbb{C})$ defined by $L$ [K, 1.4.2.1] commutes with the natural action of the Siegel modular group. (The monodromy homomorphism $\rho'$ in Section 4 is surjective.) Thus the class in $H^{2N-2}(\mathcal{M} \times \mathcal{M}, \mathbb{Q})$ defined by $\Lambda$ is invariant under the natural action of the Siegel modular group. By the analogue of Proposition 3.1, $\Lambda$ is contained in $I$ and hence $\Lambda$ defines an algebraic class in $\mathcal{M} \times \mathcal{M}$.

**Remark 6.1.** (i) It is known that if $\Lambda$ is algebraic for a smooth projective variety over $\mathbb{C}$, then homological equivalence of algebraic cycles on the variety is equal to numerical equivalence [K]. So homological and numerical equivalence of algebraic cycles on the moduli space $\mathcal{M}$ coincide.

(ii) Using Proposition 3.1, Theorem 3.2, and the monodromy argument in 5.1, one can determine the Hodge classes of the $m$-fold product of the moduli of parabolic bundles with fixed determinant of the general curve (resp. $m$-fold product of the general curve) and verify the Hodge conjecture for these varieties.

## References

[BKN] Balaji, V., King, A., and Newstead, P., *Algebraic cohomology of the moduli space of rank 2 vector bundles on a curve*, Preprint, alg-geom/9502019.

[B] Biswas, I., *On the mapping class group action on the cohomology of the representation space of a surface*, Proc. Amer. Math. Soc. **124** (1996), 1959–1965.

[BR] Biswas, I. and Raghavendra, N., *Canonical generators of the cohomology of moduli of parabolic bundles on curves*, Math. Ann. **306** (1996) 1–14.

[BR1] Biswas, I. and Raghavendra, N., *Determinants of parabolic bundles on Riemann surfaces*, Proc. Indian Acad. Sci. Math. Sci. **103** (1993) 41–71.

[Bo] Borel, A., *Density properties for certain subgroups of semi-simple groups without compact components*, Ann. Math. **72** (1960) 179-188.

[Bou1] Bourbaki, N., *Eléments de Mathématiques Algèbre*, Ch. 1 à 3, Hermann, 1970.

[Bou2] Bourbaki, N., *Eléments de Mathématiques, Groupes et algèbres de Lie*, Ch. 7 et 9, Hermann, 1975.

[Bou3] Bourbaki, N., *Eléments de Mathématiques, Algèbre, Ch. 8, Modules et anneaux semi-simples*, Hermann, 1958.

[CDK] Cattani, E., Deligne, P., and Kaplan, A., *On the locus of Hodge classes*, J. Amer. Math. Soc. **8** (1995) 483-506.

[G] Grothendieck, A., *Standard conjectures on algebraic cycles*, Algebraic Geometry – Bombay Colloquium, 1968, Oxford, 193-199.

[H] Hain, R. M., *Torelli groups and geometry of moduli spaces of curves*, In *Current topics in Complex Algebraic Geometry*, ed. C. H. Clemens and J. Kollar, M.S.R.I. Publications no. 28, Cambridge University Press, 1995.

[Ho] Howe, R., *Remarks on classical invariant theory*, Trans. Amer. Math. Soc. **313** (1989) 539-570.

[K] Kleiman, S, *Algebraic cycles and the Weil conjectures. Dix Exposeś sur la cohomologie des Schémas*, North-Holland, Amsterdam, 1968, pp. 359-386.

[Li] Lieberman, D., *Higher Picard varieties*, Amer. J. Math. **90** (1968) 1165-1199.

[MS] Mehta, V. and Seshadri, C.S., *Moduli of vector bundles on curves with parabolic structure*, Math. Ann. **248** (1980) 205-239.

[NR] Narasimhan, M.S. and Ramadas, T.R., *Factorisation of generalised theta functions I*, Invent. Math. **114** (1993) 565-623.

[N] Nitsure, N., *Cohomology of the moduli of parabolic vector bundles*, Proc. Indian Acad. Sci. Math. Sci. **95** (1986) 61-77.

School of Mathematics, Tata Institute of Fundamental Research, Homi Bhabha Road, Bombay 400005, INDIA
*E-mail address*: indranil@math.tifr.res.in

Mathematics Division, International Centre for Theoretical Physics, P.O. Box 586, I-34100, Miramare, Trieste, ITALY
*E-mail address*: narasim@ictp.trieste.it

Acta Mathematica Sinica, English Series
1999, Jan., Vol.15, No.1, p. 93–114

**Acta Mathematica Sinica, English Series**


# Hermitian-Einstein Metrics on Parabolic Stable Bundles

**Jiayu Li**
*Institute of Mathematics, Academia Sinica, Beijing* 100080, *P. R. China*
*E-mail: lijia@math03.math.ac.cn*

**M. S. Narasimhan**
*Mathematics Section, International Centre for Theoretic Physics, P.O.Box* 586, 34100 *Trieste, Italy*
*E-mail: narasim@ictp.trieste.it*

**Abstract** Let $\overline{M}$ be a compact complex manifold of complex dimension two with a smooth Kähler metric and $D$ a smooth divisor on $\overline{M}$. If $E$ is a rank 2 holomorphic vector bundle on $\overline{M}$ with a stable parabolic structure along $D$, we prove the existence of a metric on $E' = E|_{\overline{M}\setminus D}$ (compatible with the parabolic structure) which is Hermitian-Einstein with respect to the restriction of the Kähler metric to $\overline{M} \setminus D$. A converse is also proved.

**Keywords** Hermitian-Einstein metric, Parabolic stable bundle, Kähler manifold
**1991MR Subject Classification** 58E15, 32L07, 53C07

## 1 Introduction

Let $\overline{M}$ be a compact complex manifold of complex dimension two with a smooth Kähler metric and let $D$ be a smooth divisor on $\overline{M}$. If $E$ is a holomorphic vector bundle of rank 2 on $\overline{M}$ we have the notion of a stable parabolic structure on $E$ (along $D$) and the notion of a Hermitian-Einstein metric on $E' := E|_{\overline{M}\setminus D}$ with respect to the restriction of the Kähler metric to $\overline{M} \setminus D$. We prove in this paper the essential equivalence of these concepts (For a precise statement see Theorem 7.3).

The corresponding result for ordinary vector bundles is well known [1–5]. The case of parabolic bundles over Riemann surfaces is treated in [6–12].

We now give a brief summary of the contents of the paper. The notion of parabolic stability is introduced in Section 2 and it is observed that this notion is the same as that of Maruyama

Received September 12, 1998, Accepted October 12, 1998

and Yokogawa [13]. If $V$ is a rank 1 coherent subsheaf of $E$ with quotient torsion free and $K$ is a metric on $E'$, we define the analytic degree $d(V, K)$ (Lemma 3.6). In Lemma 5.2 we construct a metric $K_0$ on $E'$ with $|\Lambda F_{K_0}|_{K_0} \in L^p(M)$ for some $p > 2$ where $\Lambda F_{K_0}$ is the contraction of the curvature $F_{K_0}$ with respect to the restriction of the Kähler form to $\overline{M} \setminus D$. Moreover we prove that the parabolic degree of $E$ = the analytic degree $d(E, K_0)$ (defined in 3.2). We show in Lemma 5.3 that the parabolic degree of $V$ is equal to $d(V, K_0)$. We then consider the class of metrics on $E'$ compatible with the parabolic structure (Definition 7.1): these are the metrics belonging to $\mathcal{A}_{K_0}$ in the sense of Definition 3.3. We prove in Lemma 5.4 that for any metric $K$ compatible with the parabolic structure we have $d(V, K) = d(V, K_0)$. These two lemmas are proved by working on an appropriate blow up of the surface $\overline{M}$ which is provided by the following result of Schwarzenberger [14]: there exists a blow up $q : \widetilde{\overline{M}} \to \overline{M}$ such that $q^*(V)$ becomes a line *subbundle* of $q^*(E)$ twisted by a line bundle defined by the exceptional divisor. In Section 6 we prove, using a result of Siu, an extension lemma for coherent subsheaves of $E'$; this says that if $V'$ is a coherent subsheaf of $E'$ (with quotient torsion free) and the curvature form of $V'$ is in $L^1$, then $V'$ extends as a coherent subsheaf of $E$. In Proposition 7.2, we prove that the heat flow $K(t)$ for Hermitian-Einstein metrics with initial condition $K_0$ exists for all time and that $|\Lambda F_{K(t)}|_{K(t)}$ is bounded for $t > 0$. Given the above results, a general scheme of Simpson (proof of Theorem 1 in [11]) allows us to prove the existence of a Hermitian-Einstein metric compatible with the parabolic structure on the parabolic stable bundle (Theorem 7.3).

We finished writing this paper and submitted it to J. Diff. Geom. in May 1995. The editor recommended it to Comm. Anal. Geom. in Jan. 1997. Then we received the referee's report in March 1998. In the report, the referee said that a similar result was proved by Biquard. We found that Biquard's paper appeared in Proc. London Math. Soc. in 1996, so our work is independent of Biquard's paper.

Even comparing with Biquard's article, we think the following are new. We use blowing up and Schwarzenberger's theorem to prove that the analytic degree equals the geometric one. This technique is new and is applicable and useful in higher dimensions too . We prove the existence of a H-E metric under the assumption that the curvature of the initial metric is in $L^p$ ($p > 2$), while in Simpson's theorem he assumes that it belongs to $L^\infty$.

## 2 Parabolic Stability

Let $\overline{M}$ be a compact Kähler manifold of complex dimension 2, let $\omega$ be a Kähler metric on $\overline{M}$. Let $D$ be a smooth irreducible divisor in $\overline{M}$, and let $M = \overline{M} \setminus D$. The restriction of $\omega$ to $M$ gives a Kähler metric on $M$, which we fix once for all.

For simplicity, in this paper we assume that $E$ is a rank 2 holomorphic vector bundle over $\overline{M}$ and let $E' = E|_M$.

**Definition 2.1** *A parabolic structure on $E$ with respect to $D$ consists of*

(a) *a flag of $E|_D$ $0 \longrightarrow F_2 \longrightarrow E|_D \longrightarrow Q \longrightarrow 0$ where $F_2$ and $Q$ are line bundles over $D$.*

(b) *weights* $\alpha_1$, $\alpha_2$ *attached to* $E|_D$ *and* $F_2$ *satisfying* $0 \leq \alpha_1 < \alpha_2 < 1$.

A holomorphic vector bundle $E$ with a parabolic structure is called parabolic bundle, we write it as $(E, D, \alpha_1, \alpha_2)$, or simply $E_*$.

**Definition 2.2** *We define the parabolic degree of* $(E, D, \alpha_1, \alpha_2)$ *by*

$$par \deg E = \deg E + (\alpha_1 + \alpha_2) \deg[D]$$

*where* $[D]$ *is the line bundle defined by the divisor* $D$, $\deg E$ *(resp.* $\deg[D]$*) is the degree of* $E$ *(resp. the degree of* $[D]$*) in the usual sense using the Kähler form* $\omega$.

**Definition 2.3** *Suppose that* $V$ *is a rank* 1 *coherent subsheaf of* $E$ *with quotient torsion free, we define the parabolic degree of* $V$ *by*

$$par \deg V = \begin{cases} \deg V + \alpha_2 \deg[D], & \text{if } V|_D \text{ is equal to } F_2 \text{ outside} \\ & \text{an analytic set of} \\ & \text{codimension} \geq 1 \text{ in } D, \\ \deg V + \alpha_1 \deg[D], & \text{otherwise.} \end{cases}$$

**Definition 2.4** *We say that* $(E, D, \alpha_1, \alpha_2)$ *is parabolic stable if for every rank* 1 *coherent subsheaf of* $E$ *with quotient torsion free we have*

$$par \deg V < \frac{1}{2} par \deg E.$$

**Remark 2.5** The notion of parabolic stability in our paper is the same as parabolic $\mu$-stability in [13].

Now we prove the remark. If $F_2(E)$ is the inverse image of $F_2$ by the map $E \to E|_D$ we have a filtration of $E$:

$$E = F_1(E) \supset F_2(E) \supset E(-D).$$

We recall the definition of the parabolic degree in [13]. Let

$$E_\alpha = \begin{cases} E, & 0 \leq \alpha \leq \alpha_1, \\ F_2 E, & \alpha_1 < \alpha \leq \alpha_2, \\ E(-D), & \alpha_2 < \alpha \leq 1. \end{cases}$$

The parabolic degree defined in [13] is

$$par \deg E = \int_0^1 \deg E_\alpha d\alpha + 2 \deg D.$$

Suppose that $V$ is a rank 1 coherent subsheaf of $E$ with quotient torsion free, we set $V_\alpha = V \cap E_\alpha$, then

$$V_\alpha = \begin{cases} V, & 0 \leq \alpha \leq \alpha_1, \\ V \cap F_2 E, & \alpha_1 < \alpha \leq \alpha_2, \\ V(-D), & \alpha_2 < \alpha \leq 1 \end{cases}$$

and $par \deg V = \int_0^1 \deg V_\alpha d\alpha + \deg D$.

A simple computation gives

$$par \deg E = \alpha_1 \deg E + (\alpha_2 - \alpha_1) \deg F_2 E + (1 - \alpha_2) \deg E(-D) + 2 \deg D.$$

On the other hand, for instance by applying the Riemann-Roch Theorem for the embedding of $D$ in $\overline{M}$ (see [15]), we see that $\deg F_2 E = \deg E - \mathrm{rank}(E/F_2 E|_D) \deg D = \deg E - \deg D$, and $\deg E(-D) = \deg E - 2 \deg D$, we have

$$par \deg E = \deg E + (\alpha_1 + \alpha_2) \deg D.$$

Similarly

$$par \deg V = \alpha_1 \deg V + (\alpha_2 - \alpha_1) \deg V \cap F_2 E + (1 - \alpha_2) \deg V(-D) + \deg D,$$

$$\deg V(-D) = \deg V - \deg D,$$

and

$$\deg V \cap F_2 E = \deg V - \mathrm{rank}(V/(V \cap F_2 E)|_D) \deg D.$$

So,

$$par \deg V = \begin{cases} \deg V + \alpha_2 \deg[D] & \text{if } \mathrm{rank}(V|_D \cap F_2) = 1, \\ \deg V + \alpha_1 \deg[D] & \text{if } \mathrm{rank}(V|_D \cap F_2) = 0, \end{cases}$$

thus the remark follows.

## 3 Analytic Stability

Let $E' = E|_M$. Suppose that $K$ is a Hermitian metric on $E'$, then we have an operator $\partial_K$ such that $d_K = \partial_K + \overline{\partial}$ is the Hermitian connection of $K$. Let $F_K$ be the curvature of $d_K$.

**Definition 3.1** *$\mathcal{M} = \{K$ is a Hermitian metric on $E'$ such that $|\Lambda F_K|_K \in L^1(M)$ $\}$, where $\Lambda$ is the contraction with respect to the Kähler form $\omega$.*

**Definition 3.2** *If $K \in \mathcal{M}$, we define the analytic degree of $(E, K)$ by*

$$d(E,K) = \frac{\sqrt{-1}}{2\pi} \int_M \mathrm{tr}(\Lambda F_K) dV = \frac{\sqrt{-1}}{2\pi} \int_M \mathrm{tr} F_K \wedge *\omega = \frac{\sqrt{-1}}{2\pi} \int_M \mathrm{tr} F_K \wedge \omega.$$

In general, $d(E,K)$ will depend on $K$. If $H$ is another Hermitian metric on $E'$, we write $H = Kh$, where $h$ is a global section of the endomorphism bundle $\mathrm{End}(E)$ of $E$ over $M$, which is Hermitian with respect to $K$ and $H$. Then we have $d_H - d_K = h^{-1}\partial_K h$, $F_H = F_K + \overline{\partial}(h^{-1}\partial_K h)$, and $\mathrm{tr} F_H = \mathrm{tr} F_K + \overline{\partial}\partial \log \det h$.

Let $S_K = S_K(E')$ denote the vector bundle of self-adjoint endomorphisms of $E'$ with respect to $K$. For $1 \leq p < \infty$, let $L^p(S_K)$ (resp. $L_1^p(S_K)$) denote the space of sections $s$ of the vector bundle $S_K$ satisfying $|s|_K \in L^p(M)$ (resp. $s \in L^p(S_K)$ and $\overline{\partial} s \in L^p(\Omega^1(S_K))$ where $\Omega^1(S_K)$ is the space of 1-forms with values in $S_K$).

**Definition 3.3** *Suppose that* $K, H \in \mathcal{M}$, $H = Kh$. *If*

(a) $h \in L_1^2(S_K)$,

(b) $H$, $K$ *are mutually bounded (that is, for any* $x \in M$, $s_x \in E'_x$, $c|s_x|_K \leq |s_x|_H \leq C|s_x|_K$ *where* $c$ *and* $C$ *are two positive constants), then we say that* $H$ *and* $K$ *are compatible.*

*Let* $\mathcal{A}_K$ *denote the set of Hermitian metrics on* $E'$ *compatible with* $K$.

**Lemma 3.4** *Suppose that* $K, H \in \mathcal{M}$, *and assume that* $H \in \mathcal{A}_K$, *then* $d(E, H) = d(E, K)$.

*Proof* Since $H \in \mathcal{M}$, we can define $d(E, H)$. The formula $\mathrm{tr}\Lambda F_H - \mathrm{tr}\Lambda F_K = \sqrt{-1}\,\triangle \log\det h$ implies $\triangle \log\det h \in L^1(M)$. By $\overline{\partial}\log\det h = \mathrm{tr}(\overline{\partial}hh^{-1})$, we know that $\log\det h \in L_1^2(M)$. By a result of Gaffney ([16] p. 145, also see Lemma 5.2 in [11]) we have $\int_M \triangle \log\det h dV = 0$. So the lemma follows.

Suppose that $V$ is a rank 1 coherent subsheaf of $E$ with quotient torsion-free, then it is a line subbundle of $E$ in an open set which is the complement of a finite set of points. Choose a Hermitian metric $K \in \mathcal{M}$, and let $\pi$ denote the orthogonal projection with respect to the metric $K$ onto $V$ in this open set. The metric $K$ restricts to a metric on $V$ in the open set.

**Definition 3.5** *For any Hermitian metric* $K$ *on* $E'$, *let* $\mathcal{V}_K = \{V$ *is a rank* 1 *coherent subsheaf of* $E$ *with quotient torsion-free and* $\pi \in L_1^2(S_K)\}$, *where* $\pi$ *is the orthogonal projection onto* $V$ *with respect to* $K$.

For any $V \in \mathcal{V}_K$, we can define the analytic degree $d(V, K)$ of $V$ by integrating outside the finite set of points where $V$ is not a subbundle of $E$, and we have the following formula.

**Lemma 3.6** *If* $K \in \mathcal{M}$ *and* $V \in \mathcal{V}_K$,

$$d(V,K) = \frac{\sqrt{-1}}{2\pi}\int_M \mathrm{tr}\Lambda \mathrm{F}_{\mathrm{K_V}}\mathrm{dV} = \frac{\sqrt{-1}}{2\pi}\int_{\mathrm{M}} \mathrm{tr}\pi\Lambda \mathrm{F}_{\mathrm{K}}\mathrm{dV} - \frac{1}{2\pi}\int_{\mathrm{M}} |\overline{\partial}\pi|_{\mathrm{K}}^2\mathrm{dV}.$$

**Definition 3.7** *Suppose that* $K \in \mathcal{M}$. *We say that* $(E, K)$ *is analytic stable if for every* $V \in \mathcal{V}_K$, $d(V, K) < \frac{1}{2}d(E, K)$.

**Remark 3.8** This notion of analytic stability is not the same as that of Simpson [11, p. 878], since we demand that $V$ is a subsheaf of $E$ (not just $E'$, see Proposition 6.5 below).

## 4 Hermitian-Einstein Metrics

We first recall the following definition.

**Definition 4.1** *A Hermitian metric* $H$ *on* $E' = E|_M$ *is called Hermitian-Einstein, if* $\Lambda F_H = \lambda I$ *where* $I$ *is the identity endomorphism of* $E'$, $\lambda$ *is a constant.*

**Remark 4.2** If $H$ is a Hermitian-Einstein metric on $E'$, then $|\Lambda F_H| \in L^\infty(M)$ and $\lambda = \frac{\pi d(E,H)}{\sqrt{-1}\mathrm{Vol}(M)}$.

Let us consider the trace free part of $F_H$, defined by $F_H^\perp = F_H - \frac{1}{2}\mathrm{tr}F_H I$.

**Proposition 4.3** *Suppose that $H$ satisfies $\Lambda F_H^{\perp} = 0$, and $|\Lambda \mathrm{tr} F_H| \in L^p(M)$ $(p > 2)$. Then there is $H' \in \mathcal{A}_H$ such that $H'$ is a Hermitian-Einstein metric.*

*Proof* We set $H' = e^{-\phi}H$, where $\phi$ is a real valued function on $M$, we have

$$\sqrt{-1}\Lambda F_{H'} = \sqrt{-1}\Lambda F_H + \Delta\phi I = (\Delta\phi + \frac{\sqrt{-1}}{2}\mathrm{tr}\Lambda F_H)I.$$

Clearly $f = \frac{\sqrt{-1}}{2\mathrm{Vol}(M)}\int_M \mathrm{tr}\Lambda F_H dV - \frac{\sqrt{-1}}{2}\mathrm{tr}\Lambda F_H \in L^p(M)$ with $p > 2$. Since $\int_{\overline{M}} f dV = 0$, we can get $\phi \in L_1^2(\overline{M})$ such that $\Delta\phi = f$ in the sense of distributions (on $\overline{M}$). Moser's estimate [17] (also see [18] Ch. 8) implies that $\phi \in L^\infty(M)$ (recalling that $\dim_C \overline{M} = 2$), and regularity results imply that $\phi$ is smooth in $M$. Hence $H' \in \mathcal{A}_H$ and $\Lambda F_{H'} = \lambda I$.

**Remark 4.4** If $H$ is a Hermitian-Einstein metric, then $\Lambda F_H^{\perp} = 0$.

## 5 Construction of Initial Metrics and Equivalence Between Parabolic Stability and Analytic Stability

We shall first construct a class of metrics.

Let $\exp : T\overline{M} \to \overline{M}$ be the exponential map defined by the real part of the Kähler metric. Let $N$ be the orthogonal complement of $TD$ in $T\overline{M}|_D$. Then $\exp|_N$ maps diffeomorphically a neighborhood $W$ of the image of the zero section of $N$ onto a tubular neighborhood $U$ of $D$. We define the projection $p : U \to D$ by $p := \eta(\exp|_N)^{-1}$, where $\eta : N \to D$ is the canonical projection. We can see that the differential $dp_x : T_x\overline{M} \to T_x D$ is the orthogonal projection if $x \in D$.

Since the real part $g$ of the Kähler metric is invariant under $J$, where $J$ is the complex structure on $T\overline{M}$, we see that, for $x \in D$, $N_x(D) = (T_x D)^{\perp}$ is invariant under $J$.

Since $p : U \to D$ is a deformation retract, the smooth bundles $E|_U$ and $p^*(E|_D)$ are isomorphic. However, for our purposes, we choose a specific isomorphism $\psi : p^*(E|_D) \to E|_U$ as follows. Choose a Hermitian metric $h$ on $E$, and let $d_h = \partial_h + \overline{\partial}$ be the Hermitian connection. Let $\gamma$ be the geodesic connecting $p(x)$ and $x$. For any $v \in (E|_D)_{p(x)}$, we define $\psi(v) \in E|_x$ to be the parallel transport of $v$ with respect to $d_h$ along $\gamma$. We claim that, for any smooth section $s$ of $E|_D$, we have

$$(\overline{\partial}\psi(p^*s))|_D = \overline{\partial}s, \tag{1}$$

where $\overline{\partial}\psi(p^*s)$ is formed using the holomorphic structure on $E$, $(\overline{\partial}\psi(p^*s))|_D$ and $\overline{\partial}s$ are considered as sections of the restriction of $E \otimes T^*\overline{M} \otimes C$ to $D$.

In fact, we choose a local coordinate system of $x \in D$ denoted by $(z_1, z_2) \in \{|z_1| < 1\} \times \{|z_2| < 1\}$ such that $D \cap \{|z_1| < 1\} \times \{|z_2| < 1\} = \{z_1 = 0\} \times \{|z_2| < 1\}$, $x$ corresponds to $(0, 0)$, and $g_{ij} = \delta_{ij}$ at $(0, 0)$. To prove (1), it is clear that it suffices to show that $\frac{\partial}{\partial \overline{z}_1}\psi(p^*s)|_D = 0$.

For any $v \in N_x$, we consider the geodesic $\gamma(t) = \exp_x tv$; since

$$\langle d_h\psi(p^*s), \gamma'(t)\rangle = 0,$$

it holds that $\langle d_h\psi(p^*s), v\rangle|_x = 0$. So, for any $v \in N_x \otimes C$, we also have $\langle d_h\psi(p^*s), v\rangle|_x = 0$.

We write $z_1 = x_1 + \sqrt{-1}y_1$, because $\frac{\partial}{\partial x_1}, \frac{\partial}{\partial y_1} \in N_x$ and $N_x$ is $J$-invariant, we can see that $\frac{\partial}{\partial \bar{z}_1} \in N_x \otimes C$. Therefore $\langle d_h\psi(p^*s), \frac{\partial}{\partial \bar{z}_1}\rangle|_x = 0$, and the claim follows.

We choose a metric on the line bundle $[D]$ defined by the divisor $D$. Let $\sigma$ be the canonical section of $D$ which vanishes on $D$. We may assume that its length $|\sigma| < 1$.

**Definition 5.1** *Suppose that $K$ is a Hermitian metric on $E' = E|_M$. We say that $K$ is of polynomial growth with respect to $E$(or with respect to $K_1$), if $|K|_{K_1} = O(|\sigma|^{a_1})$ and $|K_1|_K = O(|\sigma|^{a_2})$ for some $a_1,\ a_2 \in R$, where $K_1$ is a Hermitian metric on $E$ over $\overline{M}$.*

Clearly, if $K_2$ is another Hermitian metric on $E$ over $\overline{M}$, $K$ is of polynomial growth with respect to $K_1$, then it is also of polynomial growth with respect to $K_2$.

**Lemma 5.2** *Suppose that $(E, D, \alpha_1, \alpha_2)$ is a parabolic bundle. Then there exists a Hermitian metric $K_0$ on $E' = E|_M$ such that*

(a) *$K_0$ is of polynomial growth with respect to a metric $K_1$ on $E$,*

(b) *If $d_{K_0}$(resp. $d_{K_1}$) is the Hermitian connection of $K_0$(resp. $K_1$) and $F_{K_0}$ is its curvature form, then we have $|d_{K_0} - d_{K_1}|_{K_0} \in L^p(M)$ for any $1 \le p < 2$, and $|F_{K_0}|_{K_0} \in L^p(M)$ for any $1 \le p < p_0 = \min\{\frac{2}{1-(\alpha_2-\alpha_1)}, \frac{2}{\alpha_2-\alpha_1}\}$.*

(c) *$|\mathrm{tr}F_{K_0}| \in L^\infty(M)$ and par $\deg E_*$ = the analytic degree $d(E, K_0)$.*

*Proof* Suppose that $U'$ is a tubular neighborhood of $D$, that $p : U' \longrightarrow D$ is the projection, $\psi : p^*(E|_D) \to E|_{U'}$ is the smooth isomorphism defined at the beginning of this section. Choose a $C^\infty$ splitting $E|_D = F_2 \oplus Q$, let $h_1$ and $h_2$ be smooth metrics on $Q$ and $F_2$ respectively. Then we define a metric on $E|_{U'}$ by

$$K_1' = \phi^*p^*h_1 \oplus \phi^*p^*h_2$$

and define a metric on $E|_{U'\setminus D}$ by

$$K_0' = |\sigma|^{2\alpha_1}\phi^*p^*h_1 \oplus |\sigma|^{2\alpha_2}\phi^*p^*h_2,$$

where $\phi = \psi^{-1}$.

We can construct metrics $K_1$ and $K_0$ on $E$ and $E'$ such that $K_1$ and $K_0$ coincide with $K_1'$ and $K_0'$ respectively on a smaller tubular neighborhood $U$.

Clearly, $K_0$ satisfies (a).

To prove that $K_0$ satisfies (b), for any $x \in D$, we choose a holomorphic frame $e_2$ for $F_2$ and a smooth frame $e_1$ for $Q$ in a neighborhood of $x$ in $D$, then $\psi(p^*e_1)$ and $\psi(p^*e_2)$ is a frame for $E|_U$ in a neighborhood of $x$ in $\overline{M}$. Suppose that $f_1,\ f_2$ is a holomorphic frame of $E|_U$ in the neighborhood such that $f_2|_D = e_2$. Assume that $(f_1, f_2)^T = g_1(\psi(p^*e_1), \psi(p^*e_2))^T$, where

$$g_1 = \begin{pmatrix} \alpha & \beta \\ \gamma & \delta \end{pmatrix}.$$

Then

$$f_2 = \gamma\psi(p^*e_1) + \delta\psi(p^*e_2)$$

and

$$\bar{\partial}f_2 = (\bar{\partial}\gamma)\psi(p^*e_1) + \gamma\bar{\partial}\psi(p^*e_1) + \bar{\partial}\delta\psi(p^*e_2) + \delta\bar{\partial}\psi(p^*e_2).$$

Note that $(\overline{\partial}\psi(p^*e_2))|_D = \overline{\partial}e_2 = 0$ by (1), so that $\gamma|_D = 0$, $\delta|_D = 1$ and $\overline{\partial}\gamma|_D = \overline{\partial}\delta|_D = 0$ considered as sections of $T^*\overline{M} \otimes C|_D$.

With respect to the holomorphic frame $f_1$, $f_2$, the matrix of $K_1$ is $g_1 h_0 g_1^*$ and the matrix of $K_0$ is $g_1 h_0 h g_1^*$ where

$$h_0 = \begin{pmatrix} \langle e_1, e_1\rangle_{h_1} & 0 \\ 0 & \langle e_2, e_2\rangle_{h_2} \end{pmatrix},$$

$$h = \begin{pmatrix} |\sigma|^{2\alpha_1} & 0 \\ 0 & |\sigma|^{2\alpha_2} \end{pmatrix}.$$

We set $g = g_1\sqrt{h_0}$, then $g_1 h_0 g_1^* = gg^*$ and $g_1 h_0 h g_1^* = ghg^*$.

If we write

$$g = \begin{pmatrix} \alpha_1 & \beta_1 \\ \gamma_1 & \delta_1 \end{pmatrix},$$

we also have $\gamma_1|_D = 0$ and $\overline{\partial}\gamma_1|_D = 0$ in the above sense.

If we write $g^{-1}\overline{\partial}g$ as

$$\begin{pmatrix} a & b \\ c & d \end{pmatrix},$$

then we have

$$\partial g^*(g^*)^{-1} = \begin{pmatrix} \overline{a} & \overline{c} \\ \overline{b} & \overline{d} \end{pmatrix}.$$

Since $c = \frac{1}{\det(g)}(-\gamma_1\overline{\partial}\alpha_1 + \alpha_1\overline{\partial}\gamma_1)$, we have

$$c|_D = 0. \tag{2}$$

Let $A_{K_1}$ (resp. $A_{K_0}$) be the connection form of $K_1$ (resp. $K_0$) with respect to the frame $f_1$, $f_2$. Then

$$\begin{aligned} A_{K_0} &= \partial(ghg^*)(ghg^*)^{-1} \\ &= \partial gg^{-1} + g\partial hh^{-1}g^{-1} + gh\partial g^*(g^*)^{-1}h^{-1}g^{-1} \\ A_{K_1} &= \partial(gg^*)(gg^*)^{-1} = \partial gg^{-1} + g\partial g^*(g^*)^{-1}g^{-1}. \end{aligned}$$

Hence,

$$A_{K_0} - A_{K_1} = g\partial hh^{-1}g^{-1} + gh\partial g^*(g^*)^{-1}h^{-1}g^{-1} - g\partial g^*(g^*)^{-1}g^{-1}.$$

$$\begin{aligned} F_{K_0} = \; & g\overline{\partial}(\partial hh^{-1})g^{-1} + \overline{\partial}g \wedge \partial hh^{-1}g^{-1} - g\partial hh^{-1} \wedge \overline{\partial}g^{-1} \\ & + \overline{\partial}gh \wedge \partial g^*(g^*)^{-1}h^{-1}g^{-1} + g\overline{\partial}h \wedge \partial g^*(g^*)^{-1}h^{-1}g^{-1} \\ & + gh\overline{\partial}(\partial g^*(g^*)^{-1})h^{-1}g^{-1} - gh\partial g^*(g^*)^{-1} \wedge \overline{\partial}h^{-1}g^{-1} \\ & - gh\partial g^*(g^*)^{-1}h^{-1} \wedge \overline{\partial}g^{-1} + \overline{\partial}(\partial gg^{-1}). \end{aligned}$$

With respect to the frame $(s_1, s_2)^T = h^{-\frac{1}{2}}g^{-1}(f_1, f_2)^T$, we have

$$A_{K_0} - A_{K_1} = \partial hh^{-1} + h^{\frac{1}{2}}\partial g^*(g^*)^{-1}h^{-\frac{1}{2}} - h^{-\frac{1}{2}}\partial g^*(g^*)^{-1}h^{\frac{1}{2}}.$$

$$\begin{aligned}F_{K_0} &= h^{-\frac{1}{2}}\bar\partial(\partial h h^{-1})h^{\frac{1}{2}} + h^{\frac{1}{2}}\bar\partial(\partial g^*(g^*)^{-1})h^{-\frac{1}{2}} + h^{-\frac{1}{2}}g^{-1}\bar\partial(\partial g g^{-1})gh^{\frac{1}{2}} \\ &\quad + h^{-\frac{1}{2}}g^{-1}\bar\partial g\wedge\partial hh^{-1}h^{\frac{1}{2}} + h^{-\frac{1}{2}}\partial hh^{-1}\wedge g^{-1}\bar\partial gh^{\frac{1}{2}} \\ &\quad + h^{\frac{1}{2}}h^{-1}\bar\partial h\wedge\partial g^*(g^*)^{-1}h^{-\frac{1}{2}} + h^{\frac{1}{2}}\partial g^*(g^*)^{-1}\wedge\bar\partial hh^{-1}h^{-\frac{1}{2}} \\ &\quad + h^{-\frac{1}{2}}g^{-1}\bar\partial gh^{\frac{1}{2}}\wedge h^{\frac{1}{2}}\partial g^*(g^*)^{-1}h^{-\frac{1}{2}} \\ &\quad + h^{\frac{1}{2}}\partial g^*(g^*)^{-1}h^{-\frac{1}{2}}\wedge h^{-\frac{1}{2}}g^{-1}\bar\partial gh^{\frac{1}{2}}. \end{aligned} \tag{3}$$

We will show that

$$|\sigma||A_{K_0} - A_{K_1}| \le C \tag{4}$$

and

$$|\sigma|^{\beta}|F_{K_0}|_{K_0} \le C, \tag{5}$$

where $\beta = \max\{1-(\alpha_2-\alpha_1), (\alpha_2-\alpha_1)\}$.

Suppose that $(z_1, z_2)$ is a coordinate system in the neighborhood $U_x$ and $D$ is defined by $z_1 = 0$. If $|\sigma|^2 = |z_1|^2 e$ in $U_x$, where $e$ is a positive smooth function, then

$$\bar\partial hh^{-1} = \begin{pmatrix} \alpha_1\frac{d\bar z_1}{\bar z_1} + \partial\log e & 0 \\ 0 & \alpha_2\frac{d\bar z_1}{\bar z_1} + \partial\log e\end{pmatrix}.$$

Without loss of generality, we may assume that

$$\bar\partial hh^{-1} = \begin{pmatrix} \alpha_1\frac{d\bar z_1}{\bar z_1} & 0 \\ 0 & \alpha_2\frac{d\bar z_1}{\bar z_1}\end{pmatrix},$$

$$\partial hh^{-1} = \begin{pmatrix} \alpha_1\frac{dz_1}{z_1} & 0 \\ 0 & \alpha_2\frac{dz_1}{z_1}\end{pmatrix}.$$

It is obvious that we have (4). And it is also clear that the norm of the first three terms on the right hand side of (3) is bounded above by $C|\sigma|^{(\alpha_1-\alpha_2)}$. To deal with the last six terms, we need some computations which we will give in the following, in which we essentially use Equation (2), which is due to our particular choice of isomorphism $\psi$ from $p^*(E|_D)$ to $E|_U$.

We have

$$h^{-\frac{1}{2}}g^{-1}\bar\partial g\wedge\partial hh^{-1}h^{\frac{1}{2}} + h^{-\frac{1}{2}}\partial hh^{-1}\wedge g^{-1}\bar\partial gh^{\frac{1}{2}}$$

$$= \begin{pmatrix} 0 & (\alpha_2-\alpha_1)|\sigma|^{\alpha_2-\alpha_1}b\wedge\frac{dz_1}{z_1} \\ (\alpha_2-\alpha_1)|\sigma|^{\alpha_1-\alpha_2}\frac{dz_1}{z_1}\wedge c & 0\end{pmatrix},$$

$$h^{\frac{1}{2}}\partial g^*(g^*)^{-1}\wedge\bar\partial hh^{-1}h^{-\frac{1}{2}} + h^{\frac{1}{2}}h^{-1}\bar\partial h\wedge\partial g^*(g^*)^{-1}h^{-\frac{1}{2}}$$

$$= \begin{pmatrix} 0 & (\alpha_2-\alpha_1)|\sigma|^{\alpha_1-\alpha_2}\frac{d\bar z_1}{\bar z_1}\wedge\bar c \\ (\alpha_2-\alpha_1)|\sigma|^{\alpha_2-\alpha_1}\bar b\wedge\frac{d\bar z_1}{\bar z_1} & 0\end{pmatrix}$$

and

$$h^{-\frac{1}{2}}g^{-1}\bar\partial gh^{\frac{1}{2}}\wedge h^{\frac{1}{2}}\partial g^*(g^*)^{-1}h^{-\frac{1}{2}}$$

$$= \begin{pmatrix} a\wedge\bar a + b\wedge\bar b|\sigma|^{2(\alpha_2-\alpha_1)} & a\wedge\bar c|\sigma|^{\alpha_1-\alpha_2} + b\wedge\bar d|\sigma|^{\alpha_2-\alpha_1} \\ c\wedge\bar a|\sigma|^{\alpha_1-\alpha_2} + d\wedge\bar b|\sigma|^{\alpha_2-\alpha_1} & c\wedge\bar c|\sigma|^{2(\alpha_1-\alpha_2)} + d\wedge\bar d\end{pmatrix},$$

$$h^{\frac{1}{2}}\partial g^*(g^*)^{-1}h^{-\frac{1}{2}} \wedge h^{-\frac{1}{2}}g^{-1}\overline{\partial}gh^{\frac{1}{2}}$$

$$= \begin{pmatrix} \overline{a}\wedge a + \overline{c}\wedge c|\sigma|^{2(\alpha_1-\alpha_2)} & \overline{a}\wedge b|\sigma|^{\alpha_2-\alpha_1} + \overline{c}\wedge d|\sigma|^{\alpha_1-\alpha_2} \\ \overline{b}\wedge a|\sigma|^{\alpha_2-\alpha_1} + \overline{d}\wedge c|\sigma|^{\alpha_1-\alpha_2} & \overline{b}\wedge b|\sigma|^{2(\alpha_2-\alpha_1)} + \overline{d}\wedge d \end{pmatrix}.$$

Then (5) follows from (2). By (4) and (5) we see that $|d_{K_0} - d_{K_1}|_{K_0} \in L^p(M)$ for $1 \le p < 2$ and $|F_{K_0}|_{K_0} \in L^p(M)$ for $1 \le p < p_0$. This proves (b).

We have

$$\mathrm{tr}F_{K_0} = \overline{\partial}\partial \log \det K_0 = \overline{\partial}\partial \log \det K_1 + \overline{\partial}\partial \log \det h = \mathrm{tr}F_{K_1} + (\alpha_1+\alpha_2)\overline{\partial}\partial \log|\sigma|^2$$

in $U \setminus D$.

By the Poincaré-Lelong formula ([19], Ch.II Section 1, Theorem 2), one has

$$\frac{1}{2\pi\sqrt{-1}}\partial\overline{\partial}\log|\sigma|^2 + \delta_D = C_1([D]),$$

where $C_1([D])$ is the first Chern form of $[D]$ and $\delta_D$ is the current defined by $D$. Hence $\overline{\partial}\partial\log|\sigma|^2$ is in $L^\infty(M)$, being the restriction to $M$ of the smooth form $\frac{2\pi}{\sqrt{-1}}C_1([D])$ on $\overline{M}$. This proves the first part of (c).

Since

$$\int_M \overline{\partial}\partial \log \det K_1^{-1}K_0 \wedge *\omega$$
$$= \int_M \overline{\partial}\partial \log(|\sigma|^{-2(\alpha_1+\alpha_2)} \det K_1^{-1}K_0) \wedge *\omega + \int_M \overline{\partial}\partial \log|\sigma|^{2(\alpha_1+\alpha_2)} \wedge *\omega$$

and the first term on the right hand side of the last identity clearly vanishes, we have

$$\int_M \overline{\partial}\partial \log \det K_1^{-1}K_0 \wedge *\omega = \int_M \overline{\partial}\partial \log|\sigma|^{2(\alpha_1+\alpha_2)} \wedge *\omega.$$

Hence

$$\begin{aligned} d(E,K_0) &= \frac{\sqrt{-1}}{2\pi}\int_M \mathrm{tr}F_{K_0}\wedge *\omega \\ &= \frac{\sqrt{-1}}{2\pi}\int_M \mathrm{tr}F_{K_1}\wedge *\omega + (\alpha_1+\alpha_2)\frac{\sqrt{-1}}{2\pi}\int_M \overline{\partial}\partial\log|\sigma|^2\wedge *\omega \\ &= \frac{\sqrt{-1}}{2\pi}\int_M \mathrm{tr}F_{K_1}\wedge *\omega + (\alpha_1+\alpha_2)\frac{1}{2\pi\sqrt{-1}}\int_M \partial\overline{\partial}\log|\sigma|^2\wedge *\omega. \end{aligned}$$

Using the Poincaré-Lelong formula again, we have

$$\frac{1}{2\pi\sqrt{-1}}\int_M \partial\overline{\partial}\log|\sigma|^2 \wedge *\omega = \deg[D].$$

This completes the proof of the lemma.

Now we consider the relationship between parabolic stability and analytic stability.

**Lemma 5.3** *Let $V$ be a rank 1 coherent subsheaf of $E$ with quotient torsion-free.*

(1) *If* $K_1$ *is a metric on* $E$

$$\deg(V) = \frac{\sqrt{-1}}{2\pi}\int_M (F_{K_1|_V}) \wedge \omega.$$

(2) *If* $K_0$ *is the metric defined in Lemma* 5.2, *we have* $V \in \mathcal{V}_{K_0}$ ($\mathcal{V}_{K_0}$ *is defined in* 3.5).

(3) *And par* $\deg V = d(V, K_0)$.

*Proof* Observe that $V$ is a line bundle and outside a finite set of points, $V$ is a line subbundle of $E$.

For simplicity we assume that $x_0 \in \overline{M}$ is the only singular point of $V$, that is, $V$ is a line subbundle of $E$ outside $x_0$. By a theorem of Schwarzenberger ([14], Proposition 2) we can find a smooth surface $\widetilde{M}$, which is obtained by successively blowing up points over $x_0$, such that the following holds. Let $q : \widetilde{M} \to M$ be the canonical map and let $\{P_i\}$ $(i = 1, \ldots, l)$ be the components of the exceptional divisor $S = q^{-1}(x_0)$; then there exist positive integers $\{m_i\}$ $(i = 1, \ldots, l)$ such that the canonical map from $q^*(V)$ to $q^*(E)$ maps $q^*(V)$ isomorphically onto a line subbundle of $q^*(E) \otimes \mathcal{O}(-m_i P_i)$.

We first prove (1), i.e.

$$\deg(V) = \frac{\sqrt{-1}}{2\pi}\int_M (F_{K_1|_V}) \wedge \omega.$$

Note that for any choice of a metric $\gamma$ on $q^*(V)$ we have

$$\deg(V) = \frac{\sqrt{-1}}{2\pi}\int_{\widetilde{M}} F_\gamma \wedge q^*(\omega),$$

as is seen by using a metric on $q^*(V)$ which is a pullback of a metric on $V$. Now consider the metric on $q^*(V)$ obtained from the metric on $q^*(E) \otimes \mathcal{O}(-m_i P_i)$, where we use a pullback of $K_1$ on $q^*(E)$ and on each $\mathcal{O}(P_i)$ some metric. Let $|\sigma_i|$ be the norm of the canonical section of $\mathcal{O}(P_i)$ with respect to the metric. We have

$$\begin{aligned}
\deg V &= \frac{\sqrt{-1}}{2\pi}\int_{\widetilde{M}} F_\gamma \wedge q^*(\omega) \\
&= \frac{\sqrt{-1}}{2\pi}\int_{\widetilde{M}\setminus S} (q^*(F_{K_1|_V})) \wedge q^*\omega \\
&\quad + \frac{\sqrt{-1}}{2\pi}\int_{\widetilde{M}\setminus S} \Big(-\frac{m_i}{2}\Big)\overline{\partial}\partial \log|\sigma_i|^2 \wedge q^*\omega \\
&= \frac{\sqrt{-1}}{2\pi}\int_M (F_{K_1|_V}) \wedge \omega \\
&\quad + \frac{\sqrt{-1}}{2\pi}\int_{\widetilde{M}\setminus S} \Big(-\frac{m_i}{2}\Big)\overline{\partial}\partial \log|\sigma_i|^2 \wedge q^*\omega \\
&= \frac{\sqrt{-1}}{2\pi}\int_M (F_{K_1|_V}) \wedge \omega - \frac{m_i}{2}\int_{\widetilde{M}} C_1(\mathcal{O}(P_i)) \wedge q^*(\omega) \\
&= \frac{\sqrt{-1}}{2\pi}\int_M (F_{K_1|_V}) \wedge \omega - \frac{m_i}{2}\int_{P_i} q^*(\omega) \\
&= \frac{\sqrt{-1}}{2\pi}\int_M (F_{K_1|_V}) \wedge \omega,
\end{aligned}$$

since $\int_{P_i} q^*(\omega) = 0$, where $C_1(\mathcal{O}(P_i))$ is the first Chern form of $\mathcal{O}(P_i)$.

From now on, we suppose that $K_1$ is the metric of $E$ defined in Lemma 5.2. Since $F_{K_0|_V} = F_{K_1|_V} + \bar{\partial}\partial \log((K_1|_V)^{-1}(K_0|_V))$, using (1) we have

$$d(V, K_0) = \deg V + \frac{\sqrt{-1}}{2\pi} \int_M \bar{\partial}\partial \log((K_1|_V)^{-1}(K_0|_V)) \wedge \omega.$$

In the following we calculate the second term on the right hand side of the last identity.

We consider the case where $x_0 \in D$. The other case where $x_0 \in M$ is easier. In fact, we expand the exprssion above and compute

$$\frac{\sqrt{-1}}{2\pi} \int_{\widetilde{M}\setminus q^{-1}(D)} \bar{\partial}\partial \log((q^*(K_1)|_{q^*(V)})^{-1}(q^*(K_0)|_{q^*(V)})) \wedge q^*\omega.$$

We choose a local coordinate system $(z_1, z_2) \in U = \{|z_1| < 1\} \times \{|z_2| < 1\}$ such that $x_0$ corresponds to $(0,0)$. Suppose that $e_1$ is a frame for $Q$ and $e_2$ is a holomorphic frame for $F_2$, then $\psi(p^*e_1)$, $\psi(p^*e_2)$ is a frame for $E$ in the neighborhood. Using this frame, the matrices of $K_1$ and $K_0$ are respectively $h_0$ and $h_0 h$. Here we use the same notation as that in the proof of Lemma 5.2.

Suppose that $f_1$ $f_2$ is a holomorphic frame of $E$ in the neighborhood and $f_2|_D = e_2$. Assume that $m$ is the metric on $\mathcal{O}(-m_i P_i)$ and that $e_3$ is its holomorphic frame, then $q^*(f_1) \otimes e_3$ and $q^*(f_2) \otimes e_3$ is a holomorphic frame for $q^*(E) \otimes \mathcal{O}(-m_i P_i)$.

We set $(q^*(f_1), q^*(f_2))^T = q^*(g)(q^*(\psi(p^*e_1)), q^*(\psi(p^*e_2)))^T$ where

$$q^*(g) = \begin{pmatrix} g_{11} & g_{12} \\ g_{21} & g_{22} \end{pmatrix}.$$

Then

$$q^*(f_1) = g_{11}q^*(\psi(p^*e_1)) + g_{12}q^*(\psi(p^*e_2)),$$

$$q^*(f_2) = g_{21}q^*(\psi(p^*e_1)) + g_{22}q^*(\psi(p^*e_2)).$$

Since $f_2|_D = e_2$, we have $g_{21}|_{q^{-1}(D)} = 0$ and $g_{22}|_{q^{-1}(D)} = 1$.

Suppose that $s$ is a holomorphic frame of $V$ in the neighborhood $U$, then $q^*(s)$ is a frame of $q^*(V)$. Since $q^*(V)$ is a subbundle of $q^*(E) \otimes \mathcal{O}(-m_i P_i)$, we have $q^*(s) = aq^*(f_1) \otimes e_3 + bq^*(f_2) \otimes e_3$, where $a$ and $b$ do not vanish simultaneously at any point in $q^{-1}(U)$.

We denote by $D^*$ the proper transform of $D$. We may assume that $\{D^*, P_i\}$ forms a divisor with normal crossings.

Now we consider the first case, that is $V|_D = F_2$ outside $x_0$, in this case we have $a|_{D^*} = 0$, $b$ has no zero on $D^*$ and only finitely many zeroes on the exceptional divisor; for simplicity, we assume that there is only one zero $x_1$.

Set $h_1 = \langle q^*(e_1), q^*(e_1)\rangle_{q^*(K_1)}$, $h_2 = \langle q^*(e_2), q^*(e_2)\rangle_{q^*(K_1)}$, $h_3 = \langle e_3, e_3\rangle_m$, and

$$A = h_1 h_3(|a|^2|g_{11}|^2 + 2\mathrm{Re}(a\bar{b}g_{11}\overline{g_{21}}) + |b|^2|g_{21}|^2),$$

$$B = h_2 h_3(|a|^2|g_{12}|^2 + 2\mathrm{Re}(a\bar{b}g_{12}\overline{g_{22}}) + |b|^2|g_{22}|^2),$$

then

$$|q^*s|^2_{q^*(K_0)} = A|q^*\sigma|^{2\alpha_1} + B|q^*\sigma|^{2\alpha_2} = (A|q^*\sigma|^{2(\alpha_1-\alpha_2)} + B)|q^*\sigma|^{2\alpha_2},$$

$$|q^*s|^2_{q^*(K_1)} = A + B.$$

So, in $q^{-1}(U)$ we have

$$\log(|q^*\sigma|^{-2\alpha_2}(q^*(K_1)|_{q^*(V)})^{-1}(q^*(K_0)|_{q^*(V)})) = \log\left(\frac{A|q^*\sigma|^{2(\alpha_1-\alpha_2)} + B}{A+B}\right).$$

In the following calculation, we need the local expression of $q^*(\omega)$. Suppose that $(u, v)$ is a local coordinate system around a point in the exceptional divisor $S$, and $S$ is defined by $v = 0$. Then

$$q^*(\omega) = |v|^2 a_{1\bar{1}} du \wedge d\bar{u} + v a_{1\bar{2}} du \wedge d\bar{v} + \bar{v} a_{2\bar{1}} dv \wedge d\bar{u} + a_{2\bar{2}} dv \wedge d\bar{v}. \tag{6}$$

Suppose that $\sigma^*$ is the canonical section of $D^*$, then $|q^*\sigma|^2 = |\sigma^*|^2\Pi_i|\sigma_i|^2$. Note that $a|_{D^*} = g_{21}|_{D^*} = 0$ and $b|_{q^{-1}(D)}$ does not vanish except at $x_1$, by a simple computation in local coordinates, using (6) we can see that

$$\bar{\partial}\partial\log(|q^*\sigma|^{-2\alpha_2}(q^*(K_1)|_{q^*(V)})^{-1}(q^*(K_0)|_{q^*(V)})) \wedge q^*(\omega)$$
$$\in L^1(\widetilde{M} \setminus q^{-1}(D) \setminus B_\delta(x_1)),$$

where $B_\delta(x_1)$ is a small neighborhood of $x_1$. However in $B_\delta(x_1)$, we have

$$\log\left(\frac{A|q^*\sigma|^{2(\alpha_1-\alpha_2)} + B}{A+B}\right) = \sum_i \log|\sigma_i|^{2(\alpha_1-\alpha_2)}$$
$$+ \log\left(\frac{A|\sigma^*|^{2(\alpha_1-\alpha_2)} + B\Pi_i|\sigma_i|^{2(\alpha_2-\alpha_1)}}{A+B}\right).$$

Note that $A$ is bounded from below by a positive constant in $B_\delta(x_1)$ when $\delta$ is sufficiently small. By a simple computation and the Poincaré-Lelong formula we can get

$$\bar{\partial}\partial\log\left(\frac{A|q^*\sigma|^{2(\alpha_1-\alpha_2)} + B}{A+B}\right) \wedge q^*(\omega) \in L^1(B_\delta(x_1) \setminus q^{-1}(D)).$$

Therefore

$$\bar{\partial}\partial\log(|q^*\sigma|^{-2\alpha_2}(q^*(K_1)|_{q^*(V)})^{-1}(q^*(K_0)|_{q^*(V)})) \wedge q^*(\omega) \in L^1(\widetilde{M} \setminus q^{-1}(D)).$$

Now we compute

$$\int_{\widetilde{M}\setminus q^{-1}(D)} \bar{\partial}\partial\log(|q^*\sigma|^{-2\alpha_2}(q^*(K_1)|_{q^*(V)})^{-1}(q^*(K_0)|_{q^*(V)})) \wedge q^*\omega$$

and show that it is zero.

Let $\widetilde{M}_\beta = \{x \in \widetilde{M} | |\sigma^*| > \beta \text{ and } |\sigma_i| > \beta \text{ for every } i\}$. Choose a local coordinate system $(u_1, u_2) \in \{|u_1| < 1\} \times \{|u_2| < 1\}$ such that $x_1$ corresponds to $(0, 0)$ and the exceptional divisor is defined by $u_2 = 0$. Then

$$
\begin{aligned}
&\int_{\widetilde{M}\setminus q^{-1}(D)} \bar{\partial}\partial \log(|q^*\sigma|^{-2\alpha_2}(q^*(K_1)|_{q^*(V)})^{-1}(q^*(K_0)|_{q^*(V)})) \wedge q^*\omega \\
&= \lim_{\beta\to 0}\int_{\widetilde{M}_\beta} \bar{\partial}\partial \log(|q^*\sigma|^{-2\alpha_2}(q^*(K_1)|_{q^*(V)})^{-1}(q^*(K_0)|_{q^*(V)})) \wedge q^*\omega \\
&= \lim_{\beta\to 0}\int_{\partial\widetilde{M}_\beta} \partial \log(|q^*\sigma|^{-2\alpha_2}(q^*(K_1)|_{q^*(V)})^{-1}(q^*(K_0)|_{q^*(V)})) \wedge q^*\omega \qquad (7)\\
&= \lim_{\beta\to 0}\int_{\partial\widetilde{M}_\beta\setminus\{|u_1|>\delta\}} \partial \log(|q^*\sigma|^{-2\alpha_2}(q^*(K_1)|_{q^*(V)})^{-1}(q^*(K_0)|_{q^*(V)})) \wedge q^*\omega \\
&\quad + \lim_{\beta\to 0}\int_{\partial\widetilde{M}_\beta\cap\{|u_1|\le\delta\}} \partial \log(|q^*\sigma|^{-2\alpha_2}(q^*(K_1)|_{q^*(V)})^{-1}(q^*(K_0)|_{q^*(V)})) \wedge q^*\omega
\end{aligned}
$$

for any $\delta > 0$.

Since

$$
\begin{aligned}
&\log(|q^*\sigma|^{-2\alpha_2}(q^*(K_1)|_{q^*(V)})^{-1}(q^*(K_0)|_{q^*(V)})) \\
&= \sum_i \log|\sigma_i|^{2(\alpha_1-\alpha_2)} + \log\left(\frac{A|\sigma^*|^{2(\alpha_1-\alpha_2)} + B\Pi_i|\sigma_i|^{2(\alpha_2-\alpha_1)}}{A+B}\right)
\end{aligned}
$$

and $A$ is bounded from below by a positive constant in the neighborhood $\{|u_1| < \delta\}\times\{|u_2| < \beta\}$ when $\delta$ is sufficiently small, by (6) we have

$$
\lim_{\beta\to 0}\int_{\partial\widetilde{M}_\beta\cap\{|u_1|\le\delta\}} \partial \log(|q^*\sigma|^{-2\alpha_2}(q^*(K_1)|_{q^*(V)})^{-1}(q^*(K_0)|_{q^*(V)})) \wedge q^*\omega = 0
$$

for sufficiently small fixed $\delta > 0$. Then similarly we can see that the first term on the right hand side of the last identity in (7) is also 0.

Thus,

$$
\int_{\widetilde{M}\setminus q^{-1}(D)} \bar{\partial}\partial \log(|q^*\sigma|^{-2\alpha_2}(q^*(K_1)|_{q^*(V)})^{-1}(q^*(K_0)|_{q^*(V)})) \wedge q^*\omega = 0.
$$

It is obvious that

$$
\frac{\sqrt{-1}}{2\pi}\int_{\widetilde{M}\setminus q^{-1}(D)} \bar{\partial}\partial \log(|q^*\sigma|^{2\alpha_2}) \wedge q^*\omega = \frac{\sqrt{-1}}{2\pi}\int_M \bar{\partial}\partial\log(|\sigma|^{2\alpha_2})\wedge\omega = \alpha_2\deg[D].
$$

Therefore

$$
\begin{aligned}
d(V, K_0) &= \deg V + \frac{\sqrt{-1}}{2\pi}\int_{\widetilde{M}\setminus q^{-1}(D)} \bar{\partial}\partial \log(|q^*\sigma|^{-2\alpha_2}(q^*(K_1)|_{q^*(V)})^{-1} \\
&\quad \cdot (q^*(K_0)|_{q^*(V)})) \wedge q^*\omega + \frac{\sqrt{-1}}{2\pi}\int_{\widetilde{M}\setminus q^{-1}(D)} \bar{\partial}\partial\log(|q^*\sigma|^{2\alpha_2})\wedge q^*\omega \\
&= \deg V + \alpha_2\deg[D].
\end{aligned}
$$

In the second case, we may assume that, in the formula for $q^*(s)$, $b|_{D^*} = 0$ and $a$ does not vanish on $D^*$. In this case we write

$$
|q^*s|^2_{q^*K_0} = (A + B|q^*\sigma|^{2(\alpha_2-\alpha_1)})|q^*\sigma|^{2\alpha_1}.
$$

By an argument similar to the one used above we have

$$d(V, K_0) = \deg V + \alpha_1 \deg[D].$$

As a consequence of Lemma 3.6, we see from above, that $V \in \mathcal{V}_{K_0}$. This completes the proof of the lemma.

**Lemma 5.4** *Suppose that $K_0$ is the metric constructed in Lemma 5.2 and $K \in \mathcal{A}_{K_0}$. Then $\mathcal{V}_K = \mathcal{V}_{K_0}$, and for any $V \in \mathcal{V}_{K_0}$, $d(V, K) = d(V, K_0)$.*

*Proof* Let $h = KK_0^{-1}$; It suffices to show that $|\bar\partial(h|_V)| \in L^2(M)$ because $\pi_K = (h|_V)^{-1}\pi_{K_0}h$. We adopt the notation in the proof of Lemma 5.3. In particular $q : \widetilde{M} \to \overline{M}$ is the expansion. Note that $|h|_{K_0} \in L^\infty(M)$ and $|\bar\partial h|_{K_0} \in L^2(M)$. We shall prove that

$$\left|\bar\partial\Big(\frac{\langle q^*hq^*s, q^*s\rangle_{q^*(K_0)}}{\langle q^*s, q^*s\rangle_{q^*(K_0)}}\Big)\right| \in L^2(\widetilde{M} \setminus q^{-1}(D)). \tag{8}$$

which will imply $|\bar\partial(h|_V)| \in L^2(M)$.

Assume that $q^*hq^*(f_i) = \sum_j h_{ij}q^*(f_j)$, then

$$\begin{aligned}
&\langle q^*hq^*s, q^*s\rangle_{q^*(K_0)}\\
&= (|a|^2h_{11} + \bar a b h_{21})(|g_{11}|^2h_1|q^*\sigma|^{2\alpha_1} + |g_{21}|^2h_2|q^*\sigma|^{2\alpha_2})h_3\\
&\quad +(a\bar b h_{11} + |b|^2h_{21})(g_{11}\overline{g_{21}}h_1|q^*\sigma|^{2\alpha_1} + g_{12}\overline{g_{22}}h_2|q^*\sigma|^{2\alpha_2})h_3\\
&\quad +(|a|^2h_{12} + \bar a b h_{22})(\overline{g_{11}}g_{21}h_1|q^*\sigma|^{2\alpha_1} + g_{22}\overline{g_{12}}h_2|q^*\sigma|^{2\alpha_2})h_3\\
&\quad +(a\bar b h_{12} + |b|^2h_{22})(|g_{21}|^2h_1|q^*\sigma|^{2\alpha_1} + |g_{22}|^2h_2|q^*\sigma|^{2\alpha_2})h_3.
\end{aligned}$$

We consider, as in the proof of Lemma 5.3, the case where $V|_D = F_2$ outside $x_0$.

In the neighborhood of a point where $b$ does not vanish, we have

$$\begin{aligned}
&\left|\bar\partial\left(\frac{\langle q^*hq^*s, q^*s\rangle_{q^*(K_0)}}{\langle q^*s, q^*s\rangle_{q^*(K_0)}}\right)\right|\\
&\le \left|\frac{\bar\partial(|q^*\sigma|^{-2\alpha_2}\langle q^*hq^*s, q^*s\rangle_{q^*(K_0)})}{|q^*\sigma|^{-2\alpha_2}\langle q^*s, q^*s\rangle_{q^*(K_0)}}\right| + C\left|\frac{\bar\partial(|q^*\sigma|^{-2\alpha_2}\langle q^*s, q^*s\rangle_{q^*(K_0)})}{|q^*\sigma|^{-2\alpha_2}\langle q^*s, q^*s\rangle_{q^*(K_0)}}\right|.
\end{aligned}$$

We make the computations in a neighborhood of $p \in D^* \cap S$. Choose a local coordinate system $(u, v)$ such that $p$ corresponds to $(0,0)$, $D^* = \{u = 0\}$ and $S = \{v = 0\}$. Then $|a| = O(|u|)$ and $|g_{21}| = O(|uv|)$.

We have

$$\begin{aligned}
&\left|\frac{(\bar\partial|a|^2)h_{11}|g_{11}|^2h_1h_3|q^*\sigma|^{2(\alpha_1-\alpha_2)}}{A|q^*\sigma|^{2(\alpha_1-\alpha_2)} + B}\right|^2_{q^*(\omega)}\\
&\le \frac{|\partial a|^2|a|^2|h_{11}|^2|g_{11}|^4|h_1h_3|^2|q^*\sigma|^{4(\alpha_1-\alpha_2)}}{(A|q^*\sigma|^{2(\alpha_1-\alpha_2)} + B)^2|v|^2}\\
&\le C\frac{h_1h_3|a|^2|g_{11}|^2|q^*\sigma|^{4(\alpha_1-\alpha_2)}}{(A|q^*\sigma|^{2(\alpha_1-\alpha_2)} + B)^2|v|^2}.
\end{aligned}$$

Note that

$$\begin{aligned}
&\frac{h_1h_3|a|^2|g_{11}|^2|q^*\sigma|^{4(\alpha_1-\alpha_2)}}{(A|q^*\sigma|^{2(\alpha_1-\alpha_2)}+B)^2|v|^2}\\
&\quad\le \frac{A|q^*\sigma|^{4(\alpha_1-\alpha_2)}}{(A|q^*\sigma|^{2(\alpha_1-\alpha_2)}+B)^2|v|^2}\\
&\quad\quad+\frac{2h_1h_3|abg_{11}g_{21}||q^*\sigma|^{4(\alpha_1-\alpha_2)}}{(A|q^*\sigma|^{2(\alpha_1-\alpha_2)}+B)^2|v|^2}\\
&\quad\le C\frac{|q^*\sigma|^{2(\alpha_1-\alpha_2)}}{|v|^2}+\frac{\epsilon h_1h_3|a|^2|g_{11}|^2|q^*\sigma|^{4(\alpha_1-\alpha_2)}}{(A|q^*\sigma|^{2(\alpha_1-\alpha_2)}+B)^2|v|^2}\\
&\quad\quad+\frac{4}{\epsilon}\frac{h_1h_3|b|^2|g_{21}|^2|q^*\sigma|^{4(\alpha_1-\alpha_2)}}{(A|q^*\sigma|^{2(\alpha_1-\alpha_2)}+B)^2|v|^2}
\end{aligned}$$

for any $\epsilon > 0$. Hence we have

$$\frac{h_1h_3|a|^2|g_{11}|^2|q^*\sigma|^{4(\alpha_1-\alpha_2)}}{(A|q^*\sigma|^{2(\alpha_1-\alpha_2)}+B)^2|v|^2}\le C\frac{|q^*\sigma|^{2(\alpha_1-\alpha_2)}}{|v|^2}.$$

A similar proof shows that

$$\left|\frac{(\overline{\partial}|a|^2)h_{12}\overline{g_{11}}g_{21}h_1h_3|q^*\sigma|^{2(\alpha_1-\alpha_2)}}{A|q^*\sigma|^{2(\alpha_1-\alpha_2)}+B}\right|^2_{q^*(\omega)}\le C\frac{|q^*\sigma|^{2(\alpha_1-\alpha_2)}}{|v|^2}.$$

It is easier to estimate the other terms, for example we have

$$\begin{aligned}
&\left|\frac{\overline{a}b(\overline{\partial}h_{21})|g_{11}|^2h_1h_3|q^*\sigma|^{2(\alpha_1-\alpha_2)}}{A|q^*\sigma|^{2(\alpha_1-\alpha_2)}+B}\right|_{q^*(\omega)}\\
&\quad\le C|a||q^*\sigma|^{2(\alpha_1-\alpha_2)}|\overline{\partial}h_{21}|_{q^*(\omega)}\\
&\quad\le C\frac{1}{|v|^{\alpha_2-\alpha_1}}|q^*\sigma|^{(\alpha_1-\alpha_2)}|\overline{\partial}h_{21}|_{q^*(\omega)}.
\end{aligned}$$

The above estimates show that in a neighborhood of a point where $b$ does not vanish, (8) holds.

In the neighborhood of a point where $a$ does not vanish, we write

$$\frac{\langle q^*hq^*s,q^*s\rangle_{q^*(K_0)}}{\langle q^*s,q^*s\rangle_{q^*(K_0)}}=\frac{|q^*\sigma|^{-2\alpha_1}\langle q^*hq^*s,q^*s\rangle_{q^*(K_0)}}{|q^*\sigma|^{-2\alpha_1}\langle q^*s,q^*s\rangle_{q^*(K_0)}}$$

and prove (8) as above.

The following proposition can be easily derived from Lemma 3.4, Lemma 5.2, Lemma 5.3, and Lemma 5.4.

**Proposition 5.5** *Suppose that $(E, D, \alpha_1, \alpha_2)$ is a parabolic bundle, and assume that $K \in \mathcal{A}_{K_0}$ where $K_0$ is defined in Lemma 5.2. Then $E_*$ is parabolic stable if and only if $(E, K)$ is analytic stable.*

## 6 An Extension Lemma

In this section, we prove a converse of Lemma 5.3(2).

We shall use the following theorem which was proved in [20] (Theorem 2.2).

**Theorem 6.1** *If $\mathcal{F} \subset \mathcal{G}$ are coherent analytic sheaves on a complex space $X$ and $A$ is a subvariety, then the gap sheaf $\mathcal{F}[A]$ is coherent, where the subsheaf $\mathcal{F}[A]$ of $\mathcal{G}$ is defined by the following presheaf: $U \longmapsto \{\ s \in \Gamma(U, \mathcal{G}) \mid s|_{U \setminus A} \in \Gamma(U \setminus A, \mathcal{F})\ \}$.*

Using this theorem we can prove

**Lemma 6.2** *Suppose that $F$ is a coherent subsheaf of $E' = E|_M$. Assume that $F$ extends a coherent subsheaf of $E$ locally along $D$, i.e. for every point $x \in D$ we have a neighborhood $N$ of $x$ and an extension as a coherent subsheaf of $E|_N$. Then $F$ extends as a coherent subsheaf of $E$ globally.*

*Proof* Define a subsheaf $F'$ of $E$ on $\overline{M}$. For any open set $U \subset \overline{M}$, $F'(U) = \{$ section $s$ of $E$ on $U$, $s(x) \in F$, if $x \in U \cap M$ $\}$. Clearly $F'|_M = F$. By the theorem of Siu-Trautmann, for every point $x \in D$, there is a neighborhood $N$ of $x$, such that $F'|_N$ is coherent. So $F'$ is coherent.

**Proposition 6.3** *Suppose that $K$ is a metric on $E' = E|_M$ with polynomial growth with respect to $E$, assume that the curvature form $F_K$ belongs to $L^1$. If $V$ is a rank 1 coherent subsheaf of $E'$ with quotient torsion-free and $\pi \in L_1^2(S_K)$, then it extends to a coherent subsheaf of $E$.*

*Proof* Lemma 6.2 implies that the extension is a local problem, so we may assume that $\overline{M} = U_1 \times U_2$ is a polycylinder where $U_1 = \{z_1 \mid |z_1| < r_1\}$, $U_2 = \{z_2 \mid |z_2| < r_2\}$ and that $M = \overline{M} \setminus D = U_1^* \times U_2$ where $U_1^* = \{z_1 \mid 0 < |z_1| < r_1\}$. Furthermore we may assume that the Kähler metric on $\overline{M}$ is Euclidean, and $E \cong \mathcal{O}_{\overline{M}}^2$. Let $H$ denote the constant metric on $E$ obtained from this identification.

$V$ is a line bundle and outside some points $x_i = (z_1^i, z_2^i)$, it is a line subbundle of $E'$. This line bundle is trivial, as the divisor $D$ is smooth. Hence we can choose a non-vanishing section $v$ such that with respect to the induced metric, the curvature of $V$ is $f = -\partial\overline{\partial}\log|v|_K$ outside the singular set of points.

The Chern-Weil formula yields $f = \mathrm{tr}\pi F_K - |\overline{\partial}\pi|^2\omega$; we have $\int_{\overline{M}}|f|dV < \infty$ by our hypotheses, that is $\int_{|z_2|<r_2}\int_{|z_1|<r_1}|f(z_1,z_2)|dV_1dV_2 < \infty$.

We set $A = \{z_2 \in U_2 \mid \int_{U_1}|f(z_1,z_2)|dV_1 < \infty\ \} \setminus \bigcup_i\{z_2^i\}$, then the measure of $U_2 \setminus A$ is zero, so that $A$ is a thick set in $U_2$. [A subset of an open set $G$ in $C^n$ is called a thin set if it is contained in $\bigcup_{i=1}^{\infty} A_i$ and $A_i$ is a subvariety of codimension $\geq 1$ in some open subset of $G$. And a subset of $G$ i called thick if it is not thin.] Now we shall use a theorem of Siu ([21], Theorem 4.5); the following lemma is a corollary of the theorem.

**Lemma 6.4** *Suppose that $A$ is a thick set in $U_2$, assume that $\mathcal{G}$ is a coherent analytic sheaf on $U_1 \times U_2$ and that $\mathcal{F}$ is a coherent analytic subsheaf of $\mathcal{G}$ with quotient torsion free on $U_1^* \times U_2$. If for every $z_2 \in A$, $\mathcal{F}|_{U_1^* \times \{z_2\}}$ can be extended to $U_1 \times \{z_2\}$ as a coherent analytic subsheaf of*

*$\mathcal{G}|_{U_1\times\{z_2\}}$, then $\mathcal{F}$ can be extended uniquely to a coherent analytic subsheaf of $\mathcal{G}$ on $U_1 \times U_2$.*

The above lemma implies that it suffices to prove that for every $z_2 \in A$, $V|_{U_1^*\times\{z_2\}}$ can be extended as a coherent subsheaf of $E|_{U_1\times\{z_2\}}$; but this is proved by Simpson in [11] (Lemma 10.6). This completes the proof of the proposition.

Therefore we have

**Proposition 6.5** *Suppose that $K$ is a metric on $E' = E|_M$ with polynomial growth with respect to $E$, and assume that the curvature form $F_K$ belongs to $L^1$. The notion of analytic stability for $(E, K)$ defined in this paper is the same as that of* [11, *p.* 878].

Let $K_0$ be the metric constructed in Lemma 5.2. Since $K_0$ is of polynomial growth with respect to $E$ and $|F_{K_0}|_{K_0} \in L^1(M)$, we have the following proposition.

**Proposition 6.6** *Suppose that $(E, D, \alpha_1, \alpha_2)$ is a parabolic bundle, and assume that $K_0$ is the metric defined in Lemma* 5.2. *Then $E_*$ is parabolic stable if and only if $(E, K_0)$ is analytic stable in the sense of Simpson* [11].

## 7 Equivalence Between the Parabolic Stability and the Existence of a H-E Metric

In this section we shall prove our main theorem.

We recall that a holomorphic vector bundle $E$ over $\overline{M}$ is said to be undecomposable if there do not exist two proper subbundles $V$, $W$ such that $E = V \oplus W$.

**Definition 7.1** *Suppose that $(E, D, \alpha_1, \alpha_2)$ is a parabolic bundle, $H$ is a Hermitian metric on $E' = E|_M$, we say that $H$ is compatible with the parabolic structure if $H \in \mathcal{A}_{K_0}$ where $K_0$ is defined in Lemma* 5.2

In the following proposition we show the long time existence of the heat equation for Hermitian-Einstein metrics with the initial data $K_0$.

**Proposition 7.2** *Let $(E, D, \alpha_1, \alpha_2)$ be a parabolic bundle. Then there exists a Hermitian metric $K \in \mathcal{A}_{K_0}$ on $E' = E|_M$ satisfying the heat equation*

$$\begin{cases} K^{-1}\frac{dK}{dt} = -\sqrt{-1}\Lambda F_K^{\perp}, \\ K|_{t=0} = K_0 \quad \text{and} \quad \det K = \det K_0 \end{cases}$$

*on $M$ and $|\Lambda F_K|_K \in L^\infty(M)$ for any $t > 0$, where $K_0$ is the metric constructed in Lemma* 5.2 *($\mathcal{A}_{K_0}$ is defined in* 3.3).

*Proof* Let $M_\beta = \{x \in M \mid |\sigma(x)| > \beta\}$, where $0 < \beta < 1$, and consider the above heat equation on $M_\beta$ with Dirichlet boundary condition. More precisely, we consider, writing $h = K_0^{-1}K$, the equation

$$(\triangle_{K_0} - \frac{\partial}{\partial t})h = \sqrt{-1}h\Lambda F_{K_0}^{\perp} - \sqrt{-1}\Lambda\bar{\partial}hh^{-1}\partial_{K_0}h$$

on $M_\beta$ with $h|_{t=0} = I$, $\det h = 1$, and $h|_{\partial M_\beta} = I$, where $\triangle_{K_0} = -\sqrt{-1}\Lambda\bar{\partial}\partial_{K_0}$.

It was proved in [11] that this heat equation with Dirichlet boundary condition has a solution for all time.

Let $G_\beta(x,y,t)$ be the heat kernel of $M_\beta$ with Dirichlet boundary condition, we then have $|G_\beta(x,y,t)| \le C_{\delta,R}$ for all $(x,y,t) \in M_\beta \times M_\beta \times [\delta, R]$ where $0 < \delta < R < \infty$.

For each $\beta$, let $K_\beta$ be the solution on $M_\beta$ with Dirichlet boundary condition. By Lemma 6.1 in [11] we have $(\triangle - \frac{\partial}{\partial t})|\Lambda F^{\perp}_{K_\beta}|^2_{K_\beta} = 2|\bar\partial \Lambda F^{\perp}_{K_\beta}|^2_{K_\beta} \ge 0$.

Define $L(x,t) = \int_{M_\beta} G_\beta(x,y,t)|\Lambda F^{\perp}_{K_0}(y)|^2_{K_0} dy$, then $(\triangle - \frac{\partial}{\partial t})(|\Lambda F^{\perp}_{K_\beta}(x,t)|^2_{K_\beta} - L(x,t)) \ge 0$ and $|\Lambda F^{\perp}_{K_\beta}(x,t)|^2_{K_\beta} - L(x,t)$ satisfies the Dirichlet boundary condition. The maximum principle implies

$$|\Lambda F^{\perp}_{K_\beta}|^2_{K_\beta} \le \int_{M_\beta} |\Lambda F^{\perp}_{K_0}(y)|^2_{K_0} G_\beta(x,y,t)dy \le C_{\delta,R}||\ |\Lambda F^{\perp}_{K_0}|_{K_0}||^2_{L^2(M)} \le C_{\delta,R} \tag{9}$$

for all $(x,t) \in M_\beta \times [\delta, R]$.

Using an argument similar to the one used in obtaining Lemma 3.1 (d) in [11] we have

$$(\triangle - \frac{\partial}{\partial t}) \log \frac{1}{2}\mathrm{tr} h_\beta \ge -|\Lambda F^{\perp}_{K_0}|_{K_0},$$

where $h_\beta = K_0^{-1}K_\beta$. On the other hand, by Lemma 5.2 we have $|\Lambda F^{\perp}_{K_0}|_{K_0} \in L^p(M)$ for any $2 < p < p_0$. Multiplying the above inequality by $\log^{p-1}\frac{1}{2}\mathrm{tr}h_\beta$ and integrating, we obtain

$$\begin{aligned} \frac{1}{p}\frac{\partial}{\partial t}\int_{M_\beta} \log^p \frac{1}{2}\mathrm{tr}h_\beta dV + C_p \int_{M_\beta} | \nabla \log^{\frac{p}{2}} \frac{1}{2}\mathrm{tr}h_\beta|^2 dV \\ \le \int_{M_\beta} (\log^{p-1}\frac{1}{2}\mathrm{tr}h_\beta)|\Lambda F^{\perp}_{K_0}|_{K_0} dV \\ \le \left(\int_{M_\beta} \log^p \frac{1}{2}\mathrm{tr}h_\beta dV\right)^{\frac{p-1}{p}} ||\ |\Lambda F^{\perp}_{K_0}|_{K_0}||_p. \end{aligned}$$

So,

$$\int_{M_\beta} \log^p \frac{1}{2}\mathrm{tr}h_\beta dV \le C_p t^p. \tag{10}$$

In $M_\beta$, we have ([11], Lemma 3.1 (c))

$$\triangle \log \frac{1}{2}\mathrm{tr}h_\beta \ge -(|\Lambda F^{\perp}_{K_\beta}|_{K_\beta} + |\Lambda F^{\perp}_{K_0}|_{K_0}), \quad \log\frac{1}{2}\mathrm{tr}h_\beta|_{\partial M_\beta} = 0,$$

by Lemma 5.2 and (9), $||\ |\Lambda F^{\perp}_{K_\beta}|_{K_\beta} + |\Lambda F^{\perp}_{K_0}|_{K_0}||_{L^p(M_\beta)} \le C_{\delta,p,R}$ for some $p > 2$ and any $t \in [\delta, R]$. Hence by Moser's iterative argument [17](also see [18] Ch.8), we have

$$\sup_{M_\beta} \log \frac{1}{2}\mathrm{tr}h_\beta \le C_{\delta,p,R}(1 + \left(\int_{M_\beta} \log^p \frac{1}{2}\mathrm{tr}h_\beta dV)^{\frac{1}{p}}\right).$$

recalling that $\dim_C M = 2$.

Therefore we can conclude, using (10), that

$$||\log \mathrm{tr}h_\beta||_{L^\infty(M_\beta)} \le C_{\delta,p,R} \tag{11}$$

for all $t \in [\delta, R]$.

Thus $\mathrm{tr} h_\beta$ and $|K_\beta|_{K_0}$ are bounded on $M_\beta \times [\delta, R]$. The bounds are independent of $\beta$.

Note that $\det h_\beta = 1$, we have $\mathrm{tr} h_\beta \geq 2 = \mathrm{tr} h_\beta|_{\partial M_\beta}$, so $\frac{\partial}{\partial n}\mathrm{tr} h_\beta|_{\partial M_\beta} \leq 0$, where $\frac{\partial}{\partial n}$ denotes the differentiation in the direction perpendicular to the boundary using the Hermitian connection $d_{K_0}$. Since [11] (Lemma 3.1 (b) and (c))

$$\triangle \mathrm{tr} h_\beta = -\sqrt{-1}\mathrm{tr}(h_\beta(\Lambda F_{K_\beta} - \Lambda F_{K_0})) + |\overline{\partial} h_\beta h_\beta^{-\frac{1}{2}}|^2,$$

we can obtain

$$\int_{M_\beta} |\overline{\partial} h_\beta|^2 dV \leq C_{\delta,p,R} \tag{12}$$

for $t \in [\delta, R]$ by integrating on both sides of the last identity. Therefore there exists a sequence $\beta_i \longrightarrow 0$ such that $K_{\beta_i} \longrightarrow K$ in $L^2(M_\tau \times [\delta, R])$ for $0 < \tau < 1$.

Suppose $0 < \beta_i,\ \ \beta_j < \tau$, then in $M_\tau$,

$$(\triangle - \frac{\partial}{\partial t})\sigma(K_{\beta_i}, K_{\beta_j}) \geq 0,$$

where $\sigma(K_{\beta_i}, K_{\beta_j}) = \mathrm{tr} K_{\beta_i}^{-1} K_{\beta_j} + \mathrm{tr} K_{\beta_j}^{-1} K_{\beta_i} - 4$.

By Moser's[22] iterative argument for parabolic equations we have

$$||\sigma(K_{\beta_i}, K_{\beta_j})||_{L^\infty(M_{\frac{\tau}{2}} \times [2\delta, R])} \leq C_{\tau,\delta,R} ||\sigma(K_{\beta_i}, K_{\beta_j})||_{L^2(M_\tau \times [\delta, R])}$$

and hence we get $C^0$ convergence of $K_{\beta_i}$ on compact sets.

Now we can repeat the argument used in the last paragraph in the proof of Proposition 6.6 in [11] to get the global existence of the heat equation on $M$. Indeed, Lemma 6.4 and the subsequent remark in [11] imply that on a fixed relatively compact subset $Z \subset M$, $K_{\beta_i}$ is bounded in $L_2^p(Z)$ as $\beta_i \longrightarrow 0$ for any $1 < p < \infty$. The bound is uniform in $t \in [2\delta, R]$. Let $L^p_{2/1}(Z \times [2\delta, R])$ denote the space of metrics with two $L^p$ derivatives in the space direction and one in the time direction. By the heat equation, the time derivative of $K_{\beta_i}$ is bounded in $L^p$, so $K_{\beta_i}$ is bounded in $L^p_{2/1}$. Subsequently we may assume that for each relatively compact open subset $Z$, $K_{\beta_i} \rightharpoonup K$ in $L^p_{2/1}(Z \times [2\delta, R])$ for any $1 < p < \infty$. By the Sobolev embedding $K_{\beta_i} \longrightarrow K$ in $C^{1/0}(Z \times [2\delta, R])$. Therefore the limit $K$ satisfies the heat equation on $M \times (0, \infty)$ and thus belongs to $C^\infty(M \times (0, \infty))$.

By (11) and (12) we know that, for any $t > 0$, $K \in \mathcal{A}_{K_0}$. By (9), we have $|\Lambda F_K^\perp|_K \in L^\infty(M)$ for any $t > 0$. This completes the proof of the proposition.

**Theorem 7.3** *Let $\overline{M}$ be a compact Kähler manifold of complex dimension 2 and $D$ a smooth irreducible divisor of $\overline{M}$. Let $E$ be a rank 2 holomorphic vector bundle on $\overline{M}$ with a parabolic structure $E_* = (E, D, \alpha_1, \alpha_2)$. If $E_*$ is parabolic stable there exists a Hermitian-Einstein metric on $E'$ compatible with the parabolic structure. Conversely, if $E$ is indecomposable and $E'$ admits a Hermitian-Einstein metric compatible with the parabolic structure, then $E_*$ is parabolic stable.*

*Proof* Suppose that $E_*$ is parabolic stable, by Proposition 6.6 we know that $(E, K_0)$ is analytic stable in the sense of Simpson ([11, p. 878]), where $K_0$ is defined in Lemma 5.2.

Since we use a Kähler metric $\omega$ defined on $\overline{M}$, the three assumptions in Section 2 of [11] are verified as shown by Simpson ([11, Proposition 2.2]).

We would like to apply Theorem 1 in [11]; however it does not quite apply to our situation. We show below how the proof can be modified in our case. [This amounts to the observation that the condition $|\Lambda F_K|_K \in L^\infty(X)$ in Theorem 1 of [11] can be weakened to the condition $|\Lambda F_K|_K \in L^p(X)$ for some $p > n = \dim_C X$]

Let $K_\beta(t)$ $(\beta > 0)$ be the metric constructed in the proof of Proposition 7.2, we have (see the first inequality in (9))

$$|\Lambda F^{\perp}_{K_\beta}|^2_{K_\beta} \le \int_{M_\beta} |\Lambda F^{\perp}_{K_0}(y)|^2_{K_0} G_\beta(x, y, t) dy.$$

By Young's inequality and the fact that $\int_{M_\beta} G_\beta dy = 1$, we have $|||\Lambda F^{\perp}_{K_\beta}|_{K_\beta}||_p \le |||\Lambda F^{\perp}_{K_0}|_{K_0}||_p$ for all $t > 0$, where $2 < p < p_0$ ($p_0$ is defined in Lemma 5.2).

Therefore $K(t)$, the limit of $K_\beta(t)$ , also satisfies $|||\Lambda F^{\perp}_K|_K||_p \le |||\Lambda F^{\perp}_{K_0}|_{K_0}||_p$ for all $t > 0$.

Then we can see that the main estimate Proposition 5.3 in [11] holds (recalling $\dim_C M = 2$). In fact, using Moser's iterative estimate, we have (see p. 885 and the proof of Proposition 2.1 in [11]) $\sup_M |s| \le C_1 + C_2 ||s||_{L^1}$ where $s$ is defined by $e^s = K_0^{-1} K$.

By Proposition 7.2, we know that $\sup_M |\Lambda F_K|_K < \infty$, for any $t > 0$, so that Lemma 7.1 in [11] holds. Hence (see the proof of [11, Theorem 1, p. 895]) $K(t)$ convergences to a Hermitian metric $H'$ on $E'$ as $t \to \infty$ and $H' \in \mathcal{A}_{K_0}$, $\Lambda F^{\perp}_{H'} = 0$ and $\det K_0^{-1} H' = 1$. Since $\mathrm{tr} F_{H'} - \mathrm{tr} F_{K_0} = \overline{\partial}\partial \log \det K_0^{-1} H' = 0$, we have $\mathrm{tr} F_{H'} = \mathrm{tr} F_{K_0} \in L^\infty(M)$. By Proposition 4.3 we have a Hermitian-Einstein metric $H \in \mathcal{A}_{K_0}$. This proves the first part.

Conversely, suppose that $H$ is a Hermitian-Einstein metric compatible with the parabolic structure.

By Lemma 3.4 and Lemma 5.2 we have $par \deg E = d(E, H)$.

Suppose that $V$ is a rank 1 coherent subsheaf of $E$ with quotient torsion-free, then by Lemma 5.3 and Lemma 5.4, $par \deg V = d(V, H)$.

Since $H$ is a Hermitian-Einstein metric, Lemma 3.6 yields that $par \deg V \le \frac{1}{2} par \deg E$ and if the equality holds then $\overline{\partial}\pi = 0$ on $M$ outside a finite set of points. If $E$ were not parabolic stable, then there would exist $V$ with $\overline{\partial}\pi_V = 0$ outside a finite set of points in $M$. Since $|\pi_V|_H$ is bounded in a neighborhood of each of these points, $\pi_V$ extends to $M$ holomorphically.

Now we show that $\pi_V$ has a holomorphic extension $\widetilde{\pi_V} : E \to E$ over $\overline{M}$. To see this, we choose, for any $x_0 \in D$, a neighborhood $U_{x_0} \subset U$ of $x_0$ and a frame $\psi(p^* e_1)$, $\psi(p^* e_2)$ of $E$ with the notation in the proof of Lemma 5.2.

We represent $\pi_V$ in $U_{x_0} \setminus D$ by

$$\pi_V \frac{\psi(p^* e_i)}{|\sigma|^{\alpha_i}} = \sum_{j=1}^{2} a^i_j \frac{\psi(p^* e_j)}{|\sigma|^{\alpha_j}},$$

where $a^i_j$ is bounded in $U_{x_0} \setminus D$ because $\pi_V$ is an orthogonal projection. Since $\alpha_1 - \alpha_2 > -1$ and

$$\pi_V(\psi(p^* e_i)) = \sum_j |\sigma|^{\alpha_i - \alpha_j} a^i_j \psi(p^* e_j),$$

we conclude that $\pi_V$ has a holomorphic extension $\widetilde{\pi_V}$ to $\overline{M}$.

Since $\widetilde{\pi_V}^2 = \widetilde{\pi_V}$, we see that $E = V \oplus W$ on $\overline{M}$ with holomorphic subbundles $V$ and $W$. This contradicts our assumption that $E$ is undecomposable.

## References

[1] M S Narasimhan, C S Seshadri. Stable and unitary vector bundles on a compact Riemann surface. Ann of Math, 1965, 82: 540–567

[2] S K Donaldson. A new proof of a theorem of Narasimhan and Seshadri. J Diff Geom, 1983, 18: 269–277

[3] S K Donaldson. Anti-self-dual Yang-Mills connections over complex algebraic surfaces and stable bundles. Proc London Math Soc, 1985, 50: 1–26

[4] S K Donaldson. Infinite determinants, stable bundles and curvature. Duke Math J, 1987, 54: 231–247

[5] K K Uhlenbeck, S T Yau. On the existence of Hermitian Yang-Mills connections in stable bundles. Comm Pure Appl Math, 1986, 39(S): 257–293

[6] V Mehta, C S Seshadri. Moduli of vector bundles on curves with parabolic structures. Math Ann, 1980, 248: 205–239

[7] O Biquard. Fibrés parabolique stables et connexions singulières plates. France: Bull Soc Math, 1991, 119: 231–157

[8] H Konno. Construction of the moduli space of stable parabolic Higgs bundles on a Riemann surface. Japan: J Math Soc, 1993, 45: 253–276

[9] E B Nasatyr, B Steer. The Narasimhan-Seshadri theorem for parabolic bundles with irrational weights. preprint

[10] J A Poritz. Parabolic vector bundles and Hermitian-Yang-Mills connections over a Riemann surface. Inter J Math, 1993, 4: 467–501

[11] C T Simpson. Constructing variation of Hodge structure using Yang-Mills theory and applications to uniformization. J Amer Math Soc, 1988, 1: 867–918

[12] C T Simpson. Harmonic bundles on noncompact curves. J Amer Math Soc, 1990, 3: 713–770

[13] M Maruyama, K Yokogawa. Moduli of parabolic stable sheaves. Math Ann, 1992, 293: 77–99

[14] R L E Schwarzenberger. Vector bundles on algebraic surfaces. Proc London Math Soc, 1961, 11: 601–622

[15] A Borel, Serre J P. Le théorème de Riemann-Roch. France: Bull Soc Math, 1958, 86: 97–136

[16] M P Gaffney. A special stokes theorem for complete Riemannian manifolds. Ann of Math, 1945, 60: 140–145

[17] J Moser. On Harnack's theorem for elliptic differential equations. Comm Pure Appl Math, 1961, 14: 577–591

[18] D Gilbarg, N S Trudinger. Elliptic Partial Differential Equations of Second Order. Springer-Verlag, 1977

[19] C Soulé, D Abramovich, J-F Burnol , J Kramer. Lectures on Arakelov Geometry. Cambridge studies in Advanced Math 33. Cambridge University Press, 1992

[20] Y T Siu, G Trautmann. Gap-Sheaves and Extension of Coherent Analytic Subsheaves. Lecture Notes in Mathematics 172, Springer-Verlag, 1971

[21] Y T Siu. Techniques of Extension of Analytic Objects. Lect. Notes in Pure and Appl Math, Vol. 8, Marcel Dekker, Inc, New York, 1974

[22] J Moser. A Harnack inequality for parabolic differential equations. Comm Pure Appl Math, 1964, 17: 101–134

Acta Mathematica Sinica, English Series
Jan., 2001, Vol.17, No.1, pp. 77–80

**Acta Mathematica Sinica, English Series**


# A Note on Hermitian-Einstein Metrics on Parabolic Stable Bundles

**Jia Yu LI**
*Institute of Mathematics, Academy of Mathematics and System Sciences, Academia Sinica, Beijing* 100080, *P. R. China*
*Institute of Mathematics, Fudan University, Shanghai* 200433, *P. R. China*
*E-mail: lijia@math03.math.ac.cn*

**M. S. NARASIMHAN**
*Mathematics Section, International Centre for Theoretic Physics, P.O.Box* 586, 34100 *Trieste, Italy*
*E-mail: narasim@ictp.trieste.it*

**Abstract** Let $\overline{M}$ be a compact complex manifold of complex dimension two with a smooth Kähler metric and $D$ a smooth divisor on $\overline{M}$. If $E$ is a rank 2 holomorphic vector bundle on $\overline{M}$ with a stable parabolic structure along $D$, we prove that there exists a Hermitian-Einstein metric on $E' = E|_{\overline{M}\setminus D}$ compatible with the parabolic structure, whose curvature is square integrable.

**Keywords** Hermitian-Einstein metric, Parabolic stable bundle, Kähler manifold
**1991MR Subject Classification** 58E15, 32L07, 53C07

## 1 Introduction

Let $\overline{M}$ be a compact Kähler manifold of complex dimension 2, let $\omega$ be a Kähler metric on $\overline{M}$. Let $D$ be a smooth irreducible divisor in $\overline{M}$, and let $M = \overline{M} \setminus D$. The restriction of $\omega$ to $M$ gives a Kähler metric on $M$. For simplicity, we assume in this paper that $E$ is a rank 2 holomorphic vector bundle over $\overline{M}$ and let $E' = E|_M$ be the restriction of the bundle $E$ to $M$.

We defined [1] the notion of a stable parabolic structure on $E$ (along $D$) and the notion of a Hermitian-Einstein metric on $E'$ with respect to the restriction of the Kähler metric $\omega$ to $M$. We proved in [1] that there exists a Hermitian-Einstein metric on $E'$ compatible with the parabolic structure. We prove in this paper that in fact there exists a Hermitian-Einstein (H-E) metric on $E'$ (compatible with the parabolic structure) with the property that the curvature of the metric is square integrable (Theorem 2.2). In the case of a projective surface, the square

Received February 18, 2000, Accepted September 6, 2000
Supported by the NSF of China

integrability was proved by Biquard [2, (4.2)] using a result of Simpson, while our proof is valid with the Kähler case as well.

Once we know the curvature of the H-E metric is in $L^2$, it is in fact in $L^p$ for $p > 2$ (Remark 2.4), and hence the metric defines a parabolic bundle on $\overline{M}$ as in [2, Theorem 1.1]. Since the metric is also compatible with the given parabolic structure, both parabolic structures are the same. Therefore proving the result that the curvature form of the H-E metric is in $L^2$ completes our earlier paper and this is the motivation for this article.

## 2 The Existence of an H-E Metric

In this section we shall prove our main theorem. See [1] for the definitions, such as Hermitian-Einstein metrics, parabolic bundles, etc.

We need the following result proved in [1], regarding the initial metric $K_0$ on $E'$.

**Lemma 2.1** ([1] Lemma 5.2 and Proposition 6.6) *Let $(E, D, \alpha_1, \alpha_2)$ be a parabolic bundle. Then there exists a Hermitian metric $K_0$ on $E' = E|_M$ such that*

a) *The curvature form of $K_0$, $F_{K_0}$ satisfies that $|F_{K_0}|_{K_0} \in L^p(M)$ for any $1 \leq p < p_0$ where $p_0 = \min\{\frac{2}{1-(\alpha_2-\alpha_1)}, \frac{2}{\alpha_2-\alpha_1}\}$ and $|\mathrm{tr} F_{K_0}| \in L^\infty(M)$.*

b) *$\mathrm{Par}\deg E_* =$ the analytic degree $d(E, K_0)$ and $(E, D, \alpha_1, \alpha_2)$ is parabolic stable if and only if $(E, K_0)$ is analytic stable.*

**Theorem 2.2** *Let $\overline{M}$ be a compact Kähler manifold of complex dimension 2 and $D$ a smooth irreducible divisor of $\overline{M}$. Let $E$ be a rank 2 holomorphic vector bundle on $\overline{M}$ with a parabolic structure $E_* = (E, D, \alpha_1, \alpha_2)$. If $E_*$ is parabolic stable there exists a Hermitian-Einstein metric $H$ on $E'$ compatible with the parabolic structure whose curvature form is square integrable over $M$.*

We shall modify Proposition 7.2 in [1], and its proof, to prove the theorem. The main additional point is the derivation of an $L^2$ estimate for the curvature of the metrics arising in the heat flow.

**Proposition 2.3** *Let $(E, D, \alpha_1, \alpha_2)$ be a parabolic bundle. Then there exists a Hermitian metric $K \in \mathcal{A}_{K_0}$ on $E' = E|_M$ satisfying the heat equation*

$$\begin{cases} K^{-1}\dfrac{dK}{dt} = -\sqrt{-1}\Lambda F_K^{\perp} \\ K|_{t=0} = K_0 \quad \text{and} \quad \det K = \det K_0 \end{cases}$$

*on $M$, with $\||F_K|_K\|_{L^2(M)} \leq C$, $\||\Lambda F_K^{\perp}|_K\|_{L^p(M)} \leq \||\Lambda F_{K_0}^{\perp}|_{K_0}\|_{L^p(M)}$, and $|\Lambda F_K|_K \in L^\infty(M)$ for any $t > 0$, $2 < p < p_0$, where $K_0$ is the metric constructed in Lemma 2.1, $p_0$ is the constant in Lemma 2.1, $C > 0$ is a constant depending only on $K_0$.*

*Proof* Let $M_\beta = \{x \in M \mid |\sigma(x)| > \beta\}$, where $0 < \beta < 1$, and consider the above heat equation on $M_\beta$ with the Neumann boundary condition. More precisely, we consider, writing

$h = K_0^{-1}K$, the equation

$$\left(\triangle_{K_0} - \frac{\partial}{\partial t}\right)h = \sqrt{-1}h\Lambda F_{K_0}^{\perp} - \sqrt{-1}\Lambda\bar{\partial}hh^{-1}\partial_{K_0}h$$

on $M_\beta$ with $h|_{t=0} = I$, $\det h = 1$, and $\frac{\partial}{\partial n}h|_{\partial M_\beta} = 0$, where $\triangle_{K_0} = -\sqrt{-1}\Lambda\bar{\partial}\partial_{K_0}$, $\frac{\partial}{\partial n}$ denotes the differentiation in the direction perpendicular to the boundary using the operator $\partial_{K_0}$.

In [1] we used the Dirichlet boundary condition. We use the Neumann boundary condition here so that we have the fact that $\frac{\partial}{\partial n}\Lambda F_K^{\perp}|_{\partial M_\beta} = 0$, obtained by applying $\frac{\partial}{\partial n}$ to both sides of the heat equation; this fact will enable us to apply Stokes theorem for deriving relation (2) below.

It was proved in [3] that this heat equation with the Neumann boundary condition has a solution at all times. We denote the solution by $K_\beta$ for each $\beta$. Let $h_\beta = K_0^{-1}K_\beta$.

By an argument similar to the one used in the proof of Proposition 7.2 in [1], we can show that, there exist a sequence $\beta_i \to 0$ and a Hermitian metric $K \in \mathcal{A}_{K_0}$ such that for any relatively compact open subset $Z$, any $\delta > 0$, and any $R > 0$, $K_{\beta_i} \rightharpoonup K$ in $L^p_{2/1}(Z\times[\delta, R])$ for any $1 < p < \infty$. From the Sobolev embedding, we have $K_{\beta_i} \longrightarrow K$ in $C^{1/0}(Z \times [\delta, R])$. Therefore the limit $K$ satisfies the heat equation on $M \times (0, \infty)$ and thus belongs to $C^\infty(M \times (0, \infty))$. We can also show that $\||\Lambda F_K^{\perp}|_K\|_{L^p(M)} \leq \||\Lambda F_{K_0}^{\perp}|_{K_0}\|_{L^p(M)}$ and $|\Lambda F_K|_K \in L^\infty(M)$ for any $t > 0$, $2 < p < p_0$, as we did in [1].

Now we derive the $L^2$ bound of the curvature.

From the formula $\partial_{K_\beta} = \partial_{K_0} + h_\beta^{-1}\partial_{K_0}h_\beta$ and the fact that $\frac{\partial}{\partial n}\Lambda F_K^{\perp}|_{\partial M_\beta} = 0$ we can see that

$$\frac{\partial}{\partial n_\beta}\Lambda F_K^{\perp}|_{\partial M_\beta} = 0, \tag{1}$$

where $\frac{\partial}{\partial n_\beta}$ denotes the differentiation in the direction perpendicular to the boundary using the operator $\partial_{K_\beta}$.

Because $\det h_\beta = 1$, we have $\mathrm{tr}F_{K_\beta} = \mathrm{tr}F_{K_0}$ for all $t$, and

$$\frac{d}{dt}F_{K_\beta}^{\perp} = \frac{d}{dt}F_{K_\beta} = \sqrt{-1}\bar{\partial}\partial_{K_\beta}K_\beta^{-1}\frac{d}{dt}K_\beta = \bar{\partial}\partial_{K_\beta}\Lambda F_{K_\beta}^{\perp}.$$

Using the above identity we get

$$\begin{aligned}
\frac{d}{dt}\int_{M_\beta}|F_{K_\beta}^{\perp}|^2_{K_\beta}dV &= 2\mathrm{Re}\int_{M_\beta}\left(\frac{d}{dt}F_{K_\beta}^{\perp}, F_{K_\beta}^{\perp}\right)_{K_\beta}dV \\
&= 2\mathrm{Re}\int_{M_\beta}(\bar{\partial}\partial_{K_\beta}\Lambda F_{K_\beta}^{\perp}, F_{K_\beta}^{\perp})_{K_\beta}dV \\
&= 2\mathrm{Re}\int_{M_\beta}\nabla_{\bar{k}}\nabla_l\,(F_{K_\beta}^{\perp})_\gamma{}^\delta{}_{i\bar{i}} \cdot (F_{K_\beta}^{\perp})_\delta{}^\gamma{}_{k\bar{l}}dV \\
&= -2\mathrm{Re}\int_{M_\beta}\nabla_l(F_{K_\beta}^{\perp})_\gamma{}^\delta{}_{i\bar{i}} \cdot \nabla_{\bar{k}}(F_{K_\beta}^{\perp})_\delta{}^\gamma{}_{k\bar{l}}dV \text{ (from (1) and Stokes theorem)} \\
&= -2\mathrm{Re}\int_{M_\beta}\nabla_i(F_{K_\beta}^{\perp})_\gamma{}^\delta{}_{l\bar{i}} \cdot \nabla_{\bar{k}}(F_{K_\beta}^{\perp})_\delta{}^\gamma{}_{k\bar{l}}dV \text{ (from the Bianchi identity)} \\
&\leq 0.
\end{aligned} \tag{2}$$

Letting $\beta \to \infty$, using Fatou's lemma, we get

$$\int_M |F_K^\perp|_K^2 dV \leq \int_M |F_{K_0}^\perp|_{K_0}^2 dV.$$

Since $|\mathrm{tr}F_{K_0}| \in L^\infty(M)$ and $\mathrm{tr}F_K = \mathrm{tr}F_{K_0}$, we have

$$\int_M |F_K|_K^2 dV \leq C.$$

This completes the proof of the proposition.

*Proof of the Main Theorem* As in the proof of Theorem 7.3 in [1], the metric $K(t)$ converges to a Hermitian-Einstein metric $H$ (as $t \to \infty$) compatible with the parabolic structure. On the other hand, from Proposition 2.3 we have $\||F_K|_K\|_{L^2(M)} \leq C$. It follows from Fatou's lemma that $|F_H|_H \in L^2(M)$.

**Remark 2.4** Once we know that the curvature form of the H-E metric is in $L^2$, then it belongs in fact to $L^p$, for $p \geq 2$, as implied by the result of Sibner-Sibner [4, Theorem 5.1 and Theorem 5.2] (see [2, (4.2)]).

**Remark 2.5** Conversely, if $E'$ is a holomorphic vector bundle over $M$ and admits a Hermitian-Einstein metric $H$ with $\||F_H|_H\|_{L^p(M)} < \infty$, for some $p > 2$, one can show (cf. [2], Theorem 1.1) that $E'$ can be extended to a holomorphic vector bundle $E$ over $\overline{M}$ with a parabolic structure along $D$ such that $H$ is compatible with the parabolic structure. Moreover $E$ is parabolic polystable (cf. [2] and [1]).

**Remark 2.6** We can use our existence theorem to derive a Bogomolov Chern number inequality for parabolic bundles (cf. [5]). For the case of projective varieties see Biswas [6].

## References


[1] J. Li, M. S. Narasimhan, Hermitian-Einstein metrics on parabolic stable bundles, *Acta Math. Sinica, English Series*, 1999, **15**: 93–114.

[2] O. Biquard, Sur les fibrés paraboliques sur une surface complexe, *J. London Math. Soc.*, 1996, **53**: 302–316.

[3] C. T. Simpson, Constructing variation of Hodge structure using Yang-Mills theory and applications to uniformization, *J. Amer. Math. Soc.*, 1998, **1**: 867–918.

[4] L. Sibner, R. Sibner, Classification of singular Sobolev connections by their holonomy, *Comm. Math. Phys.*, 1992, **144**: 337–350.

[5] J. Li, Hermitian-Einstein metrics and Chern number inequality on parabolic stable bundles over Kähler manifolds, *Comm. Anal. Geom.*, in press.

[6] I. Biswas, Parabolic bundles as orbifold bundles, *Duke Math. J.*, 1997, **88**: 305–325.

Compositio Mathematica **127**: 321–332, 2001.



# A Generalisation of Nagata's Theorem on Ruled Surfaces

YOGISH I. HOLLA[1] and M. S. NARASIMHAN[2]
[1] *School of Mathematics, Tata Institute of Fundamental Research, Homi Bhabha Road, Mumbai 400 005, India. e-mail: yogi@math.tifr.res.in*
[2] *ICTP, PO Box 586, Strada Costiera 11, Miramare, I-34100, Italy*

(Received: 18 January 2000; accepted: 6 June 2000)

**Abstract.** We prove a generalisation of a theorem of Nagata on ruled surface to the case of the fiber bundle $E/P \to X$, associated to a principal $G$-bundle $E$. Using this we prove boundedness for the isomorphism classes of semi-stable $G$-bundles in all characteristics.

**Mathematics Subject Classifications (2001).** 14D20, (14H60).

**Key words.** Principal bundles, Algebraic curves, Reductive algebraic groups

## 1. Introduction

Let $X$ be a smooth projective irreducible curve of genus $g$ over an algebraically closed field $k$, $G$ a connected reductive algebraic group over $k$ and $P$ a parabolic subgroup. For a principal $G$-bundle $E$ over $X$, consider the associated $G/P$-bundle $\pi\colon E/P \longrightarrow X$. If $\sigma$ is a section of $\pi$ we denote by $N_\sigma$ the normal bundle of $\sigma(X)$ in $E/P$. The first result proved in this paper is the following.

THEOREM 1.1. *There exist a section $\sigma$ of $\pi\colon E/P \longrightarrow X$ such that*

$$\deg(N_\sigma) \leqslant g \cdot \dim(G/P),$$

*where $g$ is the genus of $X$ and $\deg(N_\sigma)$ denotes the degree of the normal bundle $N_\sigma$ considered as a vector bundle on $X$.*

The above result was classically known in the case of $G = GL(2)$ and $P$ a maximal parabolic, in the form of the theorem of M. Nagata [8] and C. Segre, which asserts that a ruled surface on $X$ admits a section whose self intersection number is $\leqslant g$. It has also been proved for $G = GL(n)$ and $P$ a maximal parabolic subgroup by Mukai and Sakai [12], and for $G$ a classical group and $P$ a maximal parabolic subgroup by Nitsure [9]. For a general survey of the topic in the case of vector bundles one may refer to Lange [7].

The main idea of our proof of the Theorem 1.1 is a 'no-ghosts theorem' for the Hilbert scheme of $E/P$, which asserts that every point of the Hilbert scheme which lies in an irreducible component containing the Hilbert point of a minimal section (i.e. for which $\deg(N_\sigma)$ is minimum), is itself the Hilbert point of a section (Proposition 2.3). We then adapt an argument of Mukai–Sakai to complete the proof of the theorem.

In the second part of the paper, we prove the following theorem:

THEOREM 1.2. *Let G be a connected reductive algebraic group and X a smooth projective irreducible curve over an algebraically closed field k of arbitrary characteristic. Then the set of isomorphism classes of semi-stable G-bundles on the curve X with a given degree is bounded. In particular, if G is semi-simple then semi-stable G-bundles form a bounded family.*

For a precise definition of degree see Section 3. In characteristic 0, the above theorem is due to Ramanathan [3]. For the classical groups, the result follows in all characteristics (except in characteristic 2 for $G = \mathrm{SO}(n)$) from the observation of Ramanan (see [13], Proposition 4.2) that a $G$-bundle is semi-stable if and only if the underlying vector bundle is so.

## 2. Minimal Sections

Let $X$ be a smooth projective irreducible curve over an algebraically closed field $k$. Let $G$ be a connected reductive algebraic group over $k$ and $P$ a parabolic subgroup of $G$.

Given a principal $G$ bundle $E$ over $X$, denote by $\pi\colon M \longrightarrow X$ the associated bundle $E/P$ with $G/P$ as fibres. If $\sigma$ is a section of $\pi\colon M \longrightarrow X$, we denote by $N_\sigma$, the vector bundle on X obtained by pulling back by $\sigma$ the normal bundle of $\sigma(X)$ in $M$. Observe that $N_\sigma$ is the pullback $\sigma^*(T_\pi)$ where $T_\pi$ is the tangent bundle along the fibres of $\pi\colon M \longrightarrow X$.

In the following lemma we prove that the degree $\deg(N_\sigma)$ of the vector bundle $N_\sigma$ on $X$ is bounded below.

LEMMA 2.1. *Given a principal G-bundle $E \longrightarrow X$, there exists a constant C such that $\deg(N_\sigma) \geqslant C$ for all sections of the associated bundle $\pi\colon M \longrightarrow X$.*

*Proof.* Let $T_\pi$ be the tangent bundle along the fibres of $\pi$. As already observed, $N_\sigma \cong \sigma^*(T_\pi)$. If $\mathfrak{g}$ (resp. $\mathfrak{p}$) are the Lie algebras of $G$ (resp. $P$) we have an exact sequence of $P$-modules

$$0 \longrightarrow \mathfrak{p} \longrightarrow \mathfrak{g} \longrightarrow \mathfrak{g}/\mathfrak{p} \longrightarrow 0.$$

Note that $\mathfrak{g}/\mathfrak{p}$ is the tangent space of $G/P$ at $e$. On $M$, we have the principal $P$-bundle $E \longrightarrow M$, and the above short exact sequence of $P$-modules gives a short exact sequence of vector bundles on $M$. Pulling it back under $\sigma$ gives us a short exact sequence of vector bundles on $X$, whose middle term is the adjoint bundle of $E$ and the last term is $\sigma^*(T_\pi)$. This implies that $\sigma^*(T_\pi)$ is a quotient of a fixed vector bundle (independent of $\sigma$). □

It is known that $\pi: M \longrightarrow X$ admits sections. This follows from a theorem of Springer (see, for example, Ramanathan [3], 2.11, p. 306).

Suppose $\sigma$ is a section of $\pi: M \longrightarrow X$. We say $\sigma$ is a *minimal section* if $\deg(N_\sigma)$ is minimum. As sections exist, and as their degrees are bounded below by Lemma 2.1, there exists a minimal section.

We will now prove a lemma which is a crucial step in the proof of Theorem 1.1. Let $Y$ be a one-dimensional projective scheme over $k$. If $L$ is a line bundle (locally free sheaf of rank one) on $Y$, we define the degree of $L$ by

$$\deg(Y, L) = \chi(Y, L) - \chi(Y, \mathcal{O}_Y).$$

Note that this is consistent with the usual definition of the degree of a line bundle on a non-singular projective curve.

It is well known that if $L$ is ample on $Y$, then $\deg(Y, L) > 0$ (see for example Iitaka [10], 8.4). Observe that $\deg(Y, L)$ is the sum of $\deg(Y_i, L)$ where $Y_i$ are the connected components of $Y$ (regarded as open subschemes), where the zero dimensional components of $Y$ contribute 0 to the degree.

LEMMA 2.2. *Let $X$ be a smooth projective irreducible curve over $k$ and $Y$ a projective one dimensional scheme over $k$.*

*Let $f : Y \longrightarrow X$ be a morphism. Assume that*

(a) $\chi(X, \mathcal{O}_X) = \chi(Y, \mathcal{O}_Y)$.
(b) *For some line bundle $L$ of degree 1 on $X$, we have $\chi(X, L) = \chi(Y, f^*(L))$.*

*Then we have the following:*

(i) *There exists a unique irreducible component $D$ of $Y$ which dominates $X$. Let $D_{red}$ be the reduced subscheme structure on $D$ induced from $Y$. Then $f|_{D_{red}} : D_{red} \longrightarrow X$ is an isomorphism.*
(ii) *Suppose that the component $D$ given by (i) above is the only irreducible component of $Y$ of dimension one. Then $f: Y \longrightarrow X$ is an isomorphism (in particular, $Y$ has no zero-dimensional components).*
(iii) *Let $\xi$ be a line bundle on $Y$. Suppose that $Y$ has more than one irreducible component of dimension one. Let $D_1, D_2, \ldots, D_k$ be the one-dimensional irreducible components other than $D$ and let $D_{i,red}$ be the corresponding reduced subscheme of $Y$. Suppose $\xi|_{D_{i,red}}$ is ample for all $i$. Then we have $\deg(D_{red}, \xi) < \deg(Y, \xi)$.*

*Proof.* We prove the proposition in several steps:
*Step (1).* $R^1 f_*(\mathcal{O}_Y)$ is a torsion sheaf, in particular, $H^1(X, R^1 f_*(\mathcal{O}_Y)) = 0$.

*Proof of Step (1).* Let $S \subset X$ be the set of points of $X$ over which the fibres of $Y \longrightarrow X$ are positive-dimensional. As $Y$ is one-dimensional, it follows from the semi-continuity of the dimension of fibres that $S$ is a finite set of points of $X$. If $U = X - S$, then we observe that $f|_{f^{-1}(U)}$ is quasi-finite and proper, hence it is a finite map. Therefore $R^1\psi_*(\mathcal{O}_{f^{-1}(U)}) = 0$. Now it is clear that $R^1 f_*(\mathcal{O}_Y)$ is supported over $S$, hence it is a torsion sheaf.

*Step (2).* $\deg(Y, f^*(L)) = \chi(X, f_*(\mathcal{O}_Y) \otimes L) - \chi(X, f_*(\mathcal{O}_Y))$

*Proof of Step (2).* We have $H^0(Y, \mathcal{O}_Y) = H^0(X, f_*(\mathcal{O}_Y))$, and $H^0(Y, f^*(L)) = H^0(X, f_* f^*(L)) = H^0(X, f_*(\mathcal{O}_Y) \otimes L)$ by the projection formula. Since $\dim(X) = 1$, the Leray spectral sequence gives us the following exact sequences

$$0 \longrightarrow H^1(X, f_*(\mathcal{O}_Y)) \longrightarrow H^1(Y, \mathcal{O}_Y) \longrightarrow H^0(X, R^1 f_*(\mathcal{O}_Y)) \longrightarrow 0$$

and

$$0 \longrightarrow H^1(X, f_*(\mathcal{O}_Y) \otimes L) \longrightarrow H^1(Y, f^*(L)) \longrightarrow H^0(X, R^1 f_*(\mathcal{O}_Y) \otimes L) \longrightarrow 0.$$

Hence

$$\chi(Y, \mathcal{O}_Y) = \chi(X, f_*(\mathcal{O}_Y)) - h^0(X, R^1 f_*(\mathcal{O}_Y))$$

and

$$\chi(Y, f^*(L)) = \chi(X, f_*(\mathcal{O}_Y) \otimes L) - h^0(X, R^1 f_*(\mathcal{O}_Y) \otimes L).$$

Note that as $R^1 f_*(\mathcal{O}_Y)$ is torsion by step (1), we have

$$h^0(X, R^1 f_*(\mathcal{O}_Y)) = h^0(X, R^1 f_*(\mathcal{O}_Y) \otimes L).$$

Hence

$$\begin{aligned}\deg(Y, f^*(L)) &= \chi(Y, f^*(L)) - \chi(Y, \mathcal{O}_Y) \\ &= \chi(X, f_*(\mathcal{O}_Y) \otimes L) - \chi(X, f_*(\mathcal{O}_Y)).\end{aligned}$$

*Step (3).* Rank $f_*(\mathcal{O}_Y) = 1$, in particular, $f$ is dominant.

*Proof of Step (3).* If $T$ is the torsion submodule of $f_*(\mathcal{O}_Y)$, we have the short exact sequence

$$0 \longrightarrow T \longrightarrow f_*(\mathcal{O}_Y) \longrightarrow Q \longrightarrow 0,$$

$Q$ being locally free. Since we have

$$\begin{aligned}\deg(Y, f^*(L)) &= \chi(Y, f^*(L)) - \chi(Y, \mathcal{O}_Y) \\ &= \chi(X, L) - \chi(X, \mathcal{O}_X) \text{(by (a) and (b) of the lemma)} \\ &= 1,\end{aligned}$$

it follows from step (2) that

$$1 = \chi(X, f_*(\mathcal{O}_Y) \otimes L) - \chi(X, f_*(\mathcal{O}_Y)).$$

From the short exact sequence $0 \longrightarrow T \otimes L \longrightarrow f_*(\mathcal{O}_Y) \otimes L \longrightarrow Q \otimes L \longrightarrow 0$ we see that

$$\chi(X, f_*(\mathcal{O}_Y) \otimes L) - \chi(X, f_*(\mathcal{O}_Y)) = \chi(X, Q \otimes L) - \chi(X, Q),$$

as $\chi(X, T \otimes L) = \chi(X, T)$ since $T$ is a torsion sheaf.

Thus $\chi(X, Q \otimes L) - \chi(X, Q) = 1$, in particular we have $Q \neq 0$. Note that this implies that $f$ is dominant. Let $r$ be the rank of $Q$. Since $\deg(L) = 1$, Riemann–Roch on $X$ gives

$$\begin{aligned}\chi(X, Q \otimes L) - \chi(X, Q) \\ &= (r + \deg(Q) + r(1-g)) - (\deg(Q) + r(1-g)) \\ &= r.\end{aligned}$$

Thus $r = 1$.

*Step (4).* Proof of (i)

Since $(f, f^\sharp) : (Y, \mathcal{O}_Y) \longrightarrow (X, \mathcal{O}_X)$ is dominant (by step (3)) and $X$ is reduced, the corresponding homomorphism $f^\sharp : \mathcal{O}_X \longrightarrow f_*(\mathcal{O}_Y)$ is injective (see EGA [2], Proposition 5.4.3, p. 284). Since $\operatorname{rank}(f_*(\mathcal{O}_Y)) = 1$ (by step (3)), we have a short exact sequence

$$0 \longrightarrow \mathcal{O}_X \longrightarrow f_*(\mathcal{O}_Y) \longrightarrow T' \longrightarrow 0$$

where $T'$ is torsion. Let $V = X - \operatorname{Supp}(T')$ and $U$ a non-empty open subscheme of $X$ such that $f^{-1}(U) \longrightarrow U$ is finite (see step (1)). Then $f' : f^{-1}(U \cap V) \longrightarrow U \cap V$, $(f' = f|_{f^{-1}(U \cap V)})$ is finite (and, hence, affine) and $f'_*(f^{-1}(U \cap V), \mathcal{O}_{f^{-1}(U \cap V)}) = \mathcal{O}_{U \cap V}$. Hence $f'$ is an isomorphism. Let $Y_0$ be the schematic image (see EGA [2], 6.10, pp. 324–325) of the open inclusion $f^{-1}(U \cap V) \hookrightarrow Y$. Since $f^{-1}(U \cap V)$ is reduced, $Y_0$ is the reduced structure on $\overline{f^{-1}(U \cap V)}$ induced by $Y$. Then $Y_0$ is also irreducible and, hence, by Zariski's main theorem, $f|_{Y_0} : Y_0 \longrightarrow X$ is an isomorphism. Since $f' : f^{-1}(U \cap V) \longrightarrow U \cap V$ is an isomorphism, we see that $Y_0$ is the only component of $Y$ which dominates $X$. In the notation of the statement (i) of the lemma, we have $D_{red} = Y_0$.

*Step (5).* Proof of (ii)

Suppose now that $Y$ has only one irreducible component $D$ of dimension 1. Let $D_{red}$ be the reduced subscheme of $Y$ with support $D$. Then we have a short exact sequence

$$0 \longrightarrow I \longrightarrow \mathcal{O}_Y \longrightarrow \mathcal{O}_{D_{red}} \longrightarrow 0.$$

Since $f^{-1}(U \cap V)$ is reduced and the other components, if any, are zero-dimensional, we see that $I$ is supported at finitely many points. Now by hypothesis, $\chi(Y, \mathcal{O}_Y) = \chi(X, \mathcal{O}_X) = \chi(D_{red}, \mathcal{O}_{D_{red}})$ as $D_{red} \longrightarrow X$ is an isomorphism. Since

$$\chi(Y, \mathcal{O}_Y) = \chi(Y, \mathcal{O}_{D_{red}}) + h^0(Y, I) = \chi(X, \mathcal{O}_X) + h^0(Y, I),$$

we see that $h^0(Y, I) = 0$, and since $I$ is torsion, $I = 0$. Thus in this case $f : Y \longrightarrow X$ is an isomorphism.

*Step (6).* Proof of (iii)

Suppose that $Y$ has other one-dimensional components apart from $D$.

Let $D_1, \ldots, D_k$ ($k \geqslant 1$) be the other one dimensional components by $P_1, \ldots, P_l$ the 0-dimensional components. Let $Y' = Y - \{P_1, \ldots, P_l\}$, considered as an open subscheme of $Y$. Let $W = Y' - \{$points of intersection of two distinct components$\}$, considered as an open subscheme of $Y$. Let $W^s$ be the schematic closure of $W$ in $Y'$. Similarly define $D_i^s$ for any component to be the schematic closure in $Y'$ of $D_i - \{$points of intersection of $D_i$ with the other components$\}$. Observe that $D^s = D_{red}$ (see step (4)). Note that $D^s$ and $D_i^s$ are closed subschemes of $W^s$. We have a short exact sequence

$$0 \longrightarrow T_1 \longrightarrow \mathcal{O}_{Y'} \longrightarrow \mathcal{O}_{W^s} \longrightarrow 0$$

and

$$0 \longrightarrow \mathcal{O}_{W^s} \longrightarrow \mathcal{O}_{D^s_{red}} \oplus \mathcal{O}_{D^s_1} \oplus \ldots \oplus \mathcal{O}_{D^s_k} \longrightarrow T_2 \longrightarrow 0.$$

where $T_1$ and $T_2$ are supported at finite number of points.

For the line bundle $\xi$ on $Y$, we have

$$\begin{aligned} \deg(Y, \xi) &= \deg(Y', \xi) = \deg(W^s, \xi) \\ &= \deg(D^s_{red}, \xi) + \sum_{i=1}^{k} \deg(D_i^s, \xi). \end{aligned}$$

Now $(D_i^s)_{red}$ is the same as the reduced scheme structure $D_{i,red}$ induced on $D_i$ by $Y'$. Since by hypothesis, $\xi|_{D_{i,red}}$ is ample, $\xi|_{D_i^s}$ is ample too as can be seen. Hence $\deg(D_i^s, \xi) > 0$ for each $i$. Thus, as $D_{red} = D^s$, we get

$$\deg(D_{red}, \xi) = \deg(D^s, \xi) < \deg(Y, \xi).$$

This completes the proof of the Proposition 2.2. □

We now go back to proving the Theorem 1.1. The above Lemma is used in the proof of the following proposition.

PROPOSITION 2.3. *Let $\sigma$ be a minimal section of $\pi$: $M \longrightarrow X$ as defined earlier. Let $\mathcal{H}$ be the Hilbert scheme of closed subschemes of $M$ (we may restrict ourselves to $\mathrm{Hilb}^P(M)$ where $P$ is the Hilbert polynomial of $\sigma$, with respect to an ample line bundle). Let $Y$ be the closed subscheme of $M$, represented by a point of $\mathcal{H}$ which lies in an irreducible component containing the Hilbert point of $\sigma_0(X)$. Then $\pi|_Y : Y \longrightarrow X$ is an isomorphism.*

*Proof.* Let $L$ be a line bundle of degree 1 on $X$. Let $\eta$ be the line bundle $\det(T_\pi)$ on $M$, where $T_\pi$ is the tangent bundle along the fibres of $\pi$. Consider the diagram

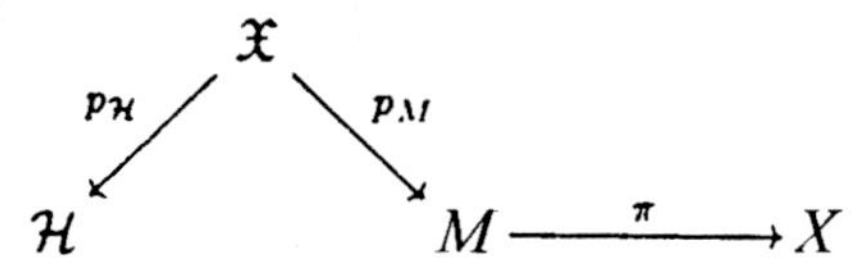


where $p_{\mathcal{H}} : \mathfrak{X} \longrightarrow \mathcal{H}$ is the universal family which is a flat morphism. By considering the line bundles $\mathcal{O}_{\mathfrak{X}}$, $p_M^*\pi^*(L)$ and $p_M^*(\eta)$ and using the fact that Euler characteristics are locally constant in a flat family of coherent sheaves, we see that

$$\chi(Y, \mathcal{O}_Y) = \chi(X, \mathcal{O}_X) \quad \text{and} \quad \chi(Y, f^*(L)) = \chi(X, L) = 1,$$

where $f = \pi|_Y$ and $\chi(Y, \xi) = \chi(X, \sigma_0^*(\eta))$, where $\xi = \eta|_Y$.

Now apply Lemma 2.2 to the morphism $f$. Using the notation of that proposition, if $D$ is the only irreducible component of dimension one which dominates $X$, then by (ii) of the proposition, $f : Y \longrightarrow X$ is an isomorphism. Suppose there were other one-dimensional components $D_1, D_2, \ldots, D_k$. Now $D_{i,red}$ is contained as a closed subschemes of a fibre of $f$. Since the restriction of $\eta$ to any fibre of the map $M$ is ample we conclude that $\xi|_{D_{i,red}}$ is ample. Let $\tau$ be the section of $M \longrightarrow X$ defined by the inverse of the isomorphism $f|_{D_{red}} : D_{red} \longrightarrow X$. We would then have

$$\deg(X, \tau^*(T_\pi)) = \deg(D_{red}, \xi) < \deg(Y, \xi) = \deg(X, \sigma^*(T_\pi)),$$

by (iii) of the Lemma 2.2 if there are other components. But this would contradict that $\sigma$ is a minimal section. Hence $D$ is the only component of $Y$. This completes the proof of the proposition. □

*Remark.* The Hilbert scheme $\mathcal{H}$ has an open subscheme $\Pi(M/X)$, which consists of Hilbert points of all sections of $\pi : M \longrightarrow X$ (see FGA [1], TDTE, IV, SS 4c, pp. 19–20).

LEMMA 2.4. *Suppose that $S$ is an irreducible component of $\Pi(M/X)$ which is proper over $k$. Then* $\dim(S) \leqslant \dim(G/P)$.

*Proof.* The restriction $s : S \times X \longrightarrow M$ of the universal morphism $\Pi(M/X) \times X \longrightarrow M$ makes the following diagram commute.

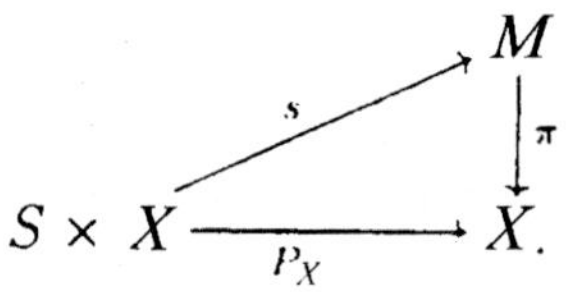


We claim that $s$ is a finite morphism. As by assumption $S$ is proper over $k$, the morphism $s$ is proper, it is enough to check that the fibres of $s$ are zero-dimensional. Suppose $y_0 \in M$, such that $\dim(s^{-1}(y_0)) \geqslant 1$. Then $s^{-1}(y_0)$ is of the form $B_0 \times \{x_0\}$ where $x_0 = \pi(y_0)$, $B_0 \subset S$. We can find a closed sub-variety $B$ of $B_0$ with $\dim(B) \geqslant 1$, with $s(B \times \{x_0\}) = y_0$. Consider the morphism $s|_{B \times X} : B \times X \longrightarrow M$.

Now $B$ is complete, since $S$ is so. Since $s(B \times \{x_0\}) = y_0$, by rigidity Lemma (Mumford [5], II.4, p. 43) $s$ factors through $X$, that is, there exists a morphism $\phi : X \longrightarrow M$ such that $s = \phi \circ p_X$. This is a contradiction as $\dim(B) \geqslant 1$ (compare Mukai and Sakai [12], pp. 254–255). □

LEMMA 2.5. *Let $S$ be an irreducible component of the Hilbert scheme $\mathcal{H}$ which contains the Hilbert point of a minimal section $\sigma$. Then $S$ lies in $\Pi(M/X)$, and* $\dim(S) \leqslant \dim(G/P)$.

*Proof.* As $S$ is closed in $\mathcal{H}$, and $\mathcal{H}$ is proper over $k$, it follows that $S$ is proper over $k$. By Proposition 2.3, every point of $S$ is the Hilbert point of some section of $\pi : M \longrightarrow X$, hence $S$ is contained in the open subscheme $\Pi(M/X)$ of $\mathcal{H}$. Therefore, $S$ is an irreducible component of $\Pi(M/X)$. Hence, by Proposition 2.4, we have the desired conclusion.

*Proof of the Theorem 1.1.* Let $\sigma$ be a minimal section, and $N_\sigma$ be the normal bundle of $\sigma(X)$ in $M$. By deformation theory, it is known that the dimension of the Hilbert scheme $\mathcal{H}$ at a point $\sigma(X)$ satisfies the inequality (see Mori [11], Proposition 3)

$$\dim_{[\sigma(X)]}(\mathcal{H}) \geqslant h^0(X, N) - h^1(X, N).$$

By Lemma 2.5 we have $\dim_{[\sigma(X)]}(\mathcal{H}) \leqslant \dim(G/P)$. On the other hand by Riemann–Roch we have

$$h^0(X, N_\sigma) - h^1(X, N_\sigma) = \deg(N_\sigma) + \dim(G/P)(1 - g).$$

Hence it follows that $\deg(N_\sigma) \leqslant g \cdot \dim(G/P)$. This completes the proof of the Theorem 1.1. □

*Remark 2.6.* In the case of a Borel subgroup it is easier to prove the existence of a section $\sigma$ such that $\deg(N_\sigma) \leqslant C$, where $C$ is a constant which depends on the genus of the curve and the group $G$, but not on the particular $G$-bundle. In fact by a result of Harder ([6], satz 2.2.6) there exists a reduction $\sigma$ to $B$, a Borel subgroup, such that if $L_{\alpha_i}$ is the line bundle associated to a simple root $\alpha_i$ we have $\deg(L_{\alpha_i}) \geqslant 2g$. Now $\det(N_\sigma)$ is the line bundle associated to the character of $B$ defined by $(-\sum_{\alpha>0} \alpha)$, sum over all positive roots. Now

$$\left(-\sum_{\alpha>0} \alpha\right) = \left(-\sum m_i \alpha_i\right),$$

where $\alpha_i$'s are simple roots taken with respect to a fixed maximal torus contained in $B$ and $m_i > 0$ depending only on the group $G$. Hence

$$\begin{aligned} \deg(N_\sigma) &= -\sum m_i \deg(L_{\alpha_i}) \\ &\leqslant \left(\sum m_i\right) \cdot 2g. \end{aligned}$$

This remark is sufficient for the applications we have in mind.

## 3. Boundedness for Semi-Stable $G$-Bundles

In this section we use the results of the previous section to prove the boundedness of semi-stable $G$-bundles of a fixed degree on a smooth projective curve $X$ over an algebraically closed field $k$ of arbitrary characteristic, where $G$ is a connected reductive algebraic group over $k$.

For any algebraic group $G$, a set $\mathcal{S}$ of principal $G$-bundles on $X$ is called *bounded* if there exists a scheme $S$ of finite type over $k$, and a family of principal $G$-bundles parametrised by $S$, such that each element of $\mathcal{S}$ is isomorphic on $X$ to the $G$-bundle on $X$ obtained by restriction of the given family to some point of $S$.

PROPOSITION 3.1. *Let $B$ be a Borel subgroup of the reductive group $G$ and $T = B/B_u$, where $B_u$ the unipotent radical of $B$. Let $\mathcal{B}_T$ be a bounded set of $T$-bundles on $X$, and let $\mathcal{B}$ be a set of $G$-bundles on $X$ such that each member of $\mathcal{B}$ admits a reduction of structure group to $B$ such that the associated $T$-bundle is isomorphic to a member of $\mathcal{B}_T$. Then $\mathcal{B}$ is a bounded set of $G$-bundles.*

*Proof.* We first prove it in the case of $G = GL(n)$ and $B =$ upper triangular matrices. We may identify principal $GL(n)$-bundles with their associated vector bundles. By hypothesis, each vector bundle $E$ in $\mathcal{B}$ admits a full flag $0 \subset E_1 \subset E_2 \subset \ldots \subset E_n = E$ such that the degrees of the line bundles $E_i/E_{i-1} (i = 1, \ldots, n)$ are bounded. Since line bundles of a given degree form a bounded family and extensions of vector bundles in bounded families form a bounded family (see FGA [1], 4, Proposition 1.2, p. 221), the proposition is proved in this case.

In the general case, we embed $G$ as a closed subgroup of $GL(n)$ for some $n$. Let $B_1$ (resp. $B$) be a Borel subgroup of $GL(n)$ (resp. $G$) with $B \subset B_1$. Since $B_u$ is contained in $B_{1,u}$, we have an induced homomorphism of $T$ into $T_1$, where $T = B/B_u$ and $T_1 = B_1/B_{1,u}$.

Let $\mathcal{B}'$ be the set of $GL(n)$ bundles obtained from $\mathcal{B}$ by extension of structure group via $G \hookrightarrow GL(n)$. From the commutative diagram

$$\begin{array}{ccc} G & \longrightarrow & GL(n) \\ \uparrow & & \uparrow \\ B & \longrightarrow & B_1 \\ \downarrow & & \downarrow \\ T & \longrightarrow & T_1 \end{array}$$

we see that each bundle in $\mathcal{B}'$ has a reduction to $B_1$, such that the corresponding $T_1$ bundle is obtained by extension of structure group from an element of the set $\mathcal{B}_T$. Since by hypothesis $\mathcal{B}_T$ is a bounded set, by what has been proved above for $GL(n)$-bundles, $\mathcal{B}_T$ is a bounded set.

Let $\mathcal{P} \longrightarrow X \times W$ be a family of principal $GL(n)$-bundles on $X$ parametrised by a scheme $W$ of finite type over $k$, such that up to isomorphism all the bundles in

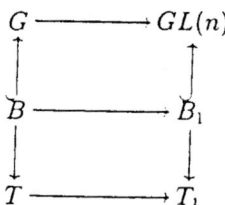

$\mathcal{B}'$ occur in this family. Using $\mathcal{P}$ we shall now construct a family of $G$-bundles on $X$ parametrised by a scheme $S$ of finite type over $k$, such that every bundle in $\mathcal{B}$ occurs in this family.

By the results of Grothendieck (see FGA, 221, 4.c), there exists a $W$-scheme $S = \Pi_{W\times X/W}((\mathcal{P}/G)/W \times X)$ which has the following universal property: for any $W$-scheme $U \longrightarrow W$, the set of sections of $(\mathcal{P}/G)_U \longrightarrow X \times_W U$ is in bijective correspondence with the set of sections of $S_U$ over $U$. In particular, for $w \in W$, the fibres of $S \longrightarrow W$ consists of the sections of the fibre bundle $\mathcal{P}_w/G \longrightarrow X$, where $\mathcal{P}_w = \mathcal{P}|w \times X$, and these are exactly the reductions of the $GL(n)$ bundle $\mathcal{P}_w$ to $G$.

Therefore, the universal section of $(\mathcal{P}/G)_S \longrightarrow X \times_W S$ gives a family of $G$-bundles parametrised by $S$, in which each bundle from the set $\mathcal{B}$ occurs. Finally, as $G$ and $GL(n)$ are reductive groups, $GL(n)/G$ is affine, and there is a representation of $GL(n)$ on a vector space $V$ which gives a $GL(n)$-equivariant closed embedding of $GL(n)/G \hookrightarrow V$. Now it is clear that the scheme $S$ is a closed subscheme of the scheme $S' = \Pi_{W\times X/W}(\tilde{V}/W \times X)$, where $\tilde{V}$ is the vector bundle associated to $\mathcal{P}$ by the representation of $GL(n)$ on $V$. Hence, $S$ is of finite type over $k$ (see Ramanathan [4], Remark 4.8.2, p. 425). This completes the proof of the Proposition 3.1. □

Let $G$ be a connected reductive group. Let $\mathcal{X}^*(G) = \mathrm{Hom}(G, k^*)$. Let $Z$ be the center of $G$ and $Z^0$ its connected component of identity. Then $G = Z^0 \cdot [G, G]$ and $Z^0 \cap [G, G]$ is finite. Thus $\mathcal{X}^*(G)$ is a subgroup of $\mathcal{X}^*(Z^0)$ of maximal rank.

If $E$ is a $G$-bundle on $X$, we have a homomorphism $d_E : \mathcal{X}^*(G) \longrightarrow \mathbb{Z}$ given by $\chi \mapsto \deg(E_\chi)$, where $E_\chi$ is the line bundle associated to $E$ by $\chi$.

DEFINITION 3.2. *We shall call the element $d_E \in \mathrm{Hom}(\mathcal{X}^*(G), \mathbb{Z})$ the degree of $E$.* When $G = GL(n)$, the above definition reduces to the usual definition of the degree of the associated rank $n$ vector bundle, as $\mathcal{X}^*(GL(n)) = \mathbb{Z}$. Also note that if $G$ is semi-simple then $d_E$ is zero as $\mathrm{Hom}(\mathcal{X}^*(G), \mathbb{Z}) = 0$. We have the following:

LEMMA 3.3. *Let $T = GL(1)^l$ be a torus and $L \subset \mathcal{X}^*(T)$ be a subgroup of maximal rank. For a $T$-bundle $F$ on $X$, let $d_F : \mathcal{X}^*(T) \longrightarrow \mathbb{Z}$ be the homomorphism as above, and $d'_F : L \longrightarrow \mathbb{Z}$ be the restriction of $d_F$ to $L$. If $\mathcal{S}$ is a set of $T$-bundles on $X$ such that the set $\{d'_F | F \in \mathcal{S}\}$ is a finite set, then $\mathcal{S}$ is a bounded set of $T$-bundles.*

*Proof.* We reduce the proof to the case where $L = \mathcal{X}^*(T)$ as follows. If $L \subset \mathcal{X}^*(T)$ is an arbitrary subgroup of maximal rank, then there exists a basis $\{\chi_1, \ldots, \chi_l\}$ of $\mathcal{X}^*(T)$ such that $\{\lambda_1\chi_1, \ldots, \lambda_l\chi_l\}$ forms a basis for $L$, with $\lambda_i \in \mathbb{Z}$, $\lambda_i \neq 0$ for each $i$. Since $d_F(\chi_i) = \lambda_i^{-1} d_F(\lambda_i\chi_i)$, the result is true for $L$ if it is true for $\mathcal{X}^*(T)$.

Hence we can assume that $L = \mathcal{X}^*(T)$. Let $\{\chi_1, \ldots, \chi_l\}$ be a basis of $\mathcal{X}^*(T)$. Then by our hypothesis the set $N_0 = \{d'_F(\chi_i) | F \in \mathcal{S}, 1 \leqslant i \leqslant l\}$ is a finite set of integers. Thus the set $\mathcal{S}$ can be considered as a subset of the set of all $l$-tuples ($l = \dim(T)$)

$$\{(L_1, \ldots, L_l) | L_i \in \mathrm{Pic}(X) \quad \text{with} \quad \deg(L) \in N_0\}.$$

Hence $\mathcal{S}$ is a bounded set. □

PROPOSITION 3.4. *Let $G$ be a connected semi-simple group. Then the family of semi-stable $G$-bundles on $X$ is bounded.*

*Proof.* Let $\mathcal{S}$ be the set of $G$-bundles such that every element is semi-stable. We shall show that each member $E$ of $\mathcal{S}$ admits a reduction of structure group to $B$ such that the associated $T$-bundles $E_T$ (as $E$ varies in $\mathcal{S}$) form a bounded family. We then apply the Proposition 3.1 to complete the proof. For any principal $G$-bundle $E$, by Remark 2.6, we can choose a reduction $\sigma$ of the structure group to $B$ such that $\deg(N_\sigma) \leqslant C$, where $C$ is a constant independent of $E$. To show that the set of associated $T$-bundles $\{E_T\}$ is bounded, we will show that there is a subgroup $L$ of $\mathcal{X}^*(T)$ of maximal rank with the property that $\{(d_{E_T} | L) \mid E \in \mathcal{S}\}$ is a finite set and then use Lemma 3.3.

Let $\Lambda_1, \ldots, \Lambda_l$ be the set of fundamental weights with respect to a maximal torus contained in $B$ and the positive roots being contained in the Lie algebra of $B$. Let $m$ be a positive integer with the property that $m\Lambda_i$ is a character of $T$ for every $i$. Let $L$ be the subgroup of $\mathcal{X}^*(T)$ generated by $\{m\Lambda_i \mid 1 \leqslant i \leqslant l\}$. Then we observe that $L$ is of maximal rank. Now the line bundle $\det(N_\sigma)^{\otimes m}$ is associated to the character

$$-2\sum_{i=1}^{l}(m\Lambda_i) = -\sum_{\alpha>0\ \text{root}} m\alpha.$$

Hence for each $E_T$ as above we have the condition

$$\sum_{i=1}^{l} d_{E_T}(m\Lambda_i) = -\deg(\det(N_\sigma)^{\otimes m})/2 \geqslant -mC/2,$$

where $d_{E_T}(m\Lambda_i)$ is the degree of the line bundle associated to $E_T$ by the character $m\Lambda_i$. On the other hand, if $E$ is semi-stable then for any reduction of structure group to $B$ the degree of the line bundle associated to a dominant character of $B$ is $\leqslant 0$ (see Ramanathan [3]). Thus we have $d_{E_T}(m\Lambda_i) \leqslant 0$. This together with the above inequality implies that $-mC/2 \leqslant d_{E_T}(m\Lambda_i) \leqslant 0$ for each $i$. Hence $\{(d_{E_T} | L) \mid E \in \mathcal{S}\}$ is a finite set. This completes the proof. □

*Proof of the Theorem 1.2.* Let $\mathcal{S}'$ be the set of semi-stable $G$-bundles with a fixed degree. For each element $E$ of $\mathcal{S}'$ we choose a reduction $\sigma$ of structure group to $B$ with $\deg(N_\sigma) \leqslant C$, $C$ independent of $E$. We shall show that the associated $T$-bundles form a bounded family and apply Proposition 3.1. This will be shown by proving that there is a subgroup $L$ of maximal rank in $\mathcal{X}^*(T)$ such that $\{(d_{E_T} | L) \mid E \in \mathcal{S}'\}$ is a finite set and then using the Lemma 3.3.

Note that $T' = T/Z^0$ is a maximal torus of $G' = G/Z^0$, contained in its Borel subgroup $B' = B/Z^0$. As we have the isomorphism $G/B \cong G'/B'$, it follows that the $G'$-bundle $E'$ obtained from $E$ by extension of structure group is semi-stable, and $\sigma$ gives rise to a reduction $\sigma'$ of structure group of $E$ to $B'$. We also observe that any dominant character vanishes on $Z^0$.

Consider the following diagram:

$$\begin{array}{ccccccccc} 0 & \longrightarrow & \mathcal{X}^*(T/Z^0) & \longrightarrow & \mathcal{X}^*(T) & \longrightarrow & \mathcal{X}^*(Z^0) & \longrightarrow & 0 \\ & & & & \uparrow & \nearrow & & & \\ & & & & \mathcal{X}^*(G) & & & & \end{array}$$

where the row is exact. As already remarked, $\mathcal{X}^*(G)$ is a subgroup of maximal rank in $\mathcal{X}^*(Z^0)$. Hence the subgroup $L$ generated by $\mathcal{X}^*(G)$ and $\mathcal{X}^*(G/Z^0)$ is of maximal rank in $\mathcal{X}^*(T)$. The set $\{d_{E_T}|_L \mid E \in \mathcal{S}'\}$ is finite because $d_{E_T}|_{\mathrm{im}(\mathcal{X}^*(G))}$ is fixed while $\{d_{E_T}|_{\mathrm{im}(T/Z^0)} \mid E \in \mathcal{S}'\}$ is a finite set as shown in Lemma 3.4, since $G'$ is semi-simple and $E'$ is semi-stable. Now the theorem follows from the Lemma 3.3 and Proposition 3.1. □


## Acknowledgement

The first-named author would like to thank the Abdus Salam International Centre for Theoretical Physics, Trieste, where this work was done.


## References


1. Grothendieck, A.: *Fondements de la géométrie algébrique*, Séminaire Bourbaki, Secretariat Mathematique, Paris, 1962.
2. Grothendieck, A. and Dieudonné, J.: *Elements de géométrie algébriques I*, 1960.
3. Ramanathan, A.: Moduli for principal bundles over algebraic curves I. *Proc. Indian Acad. Sci. (Math. Sci.)*, **106**(3) (1997), 301–328.
4. Ramanathan, A.: Moduli for principal bundles over algebraic curves II. *Proc. Indian Acad. Sci. (Math. Sci.)*, **106**(4) (1997), 421–449.
5. Mumford, D.: *Abelian Varieties*, Oxford Univ. Press, Oxford, 1970.
6. Harder, G.: Halbeinfache Gruppenschemate über völlstandigen Kurven, *Invent. Math.* **6** (1968), 107–149.
7. Lange, H.: Some geometrical aspects of vectorbundles on curves, *Aportaciones Mat.* **5** (1992) 53–72.
8. Nagata, M.: On self-intersection number of a section on a ruled surface, *Nagoya Math. J.* **37** (1970) 191–196.
9. Nitsure, N.: Boundedness for the moduli of G-bundles for classical groups, Algebraic Geometry Preprint (9802090), 1998.
10. Itaka, S.: *Algebraic Geometry*, Springer-Verlag, Berlin, 1982.
11. Mori, S.: Projective manifolds with ample tangent bundles, *Ann. of Math.* **110** (1979), 593–606.
12. Mukai, S. and Sakai, F.: Maximal sub-bundles of a vector bundles on a curve, *Manuscripta Math. 52* (1985), 251–256.
13. Ramanan, S.: Orthogonal and spin bundles over hyperelliptic curves, *Proc. Indian Acad. Sci.* **90(2)** (1981), 151–166.

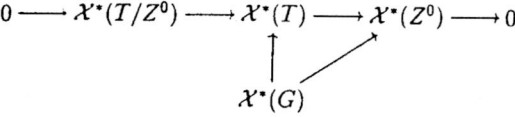

# Acknowledgements

The Tata Institute of Fundamental Research gratefully acknowledges the kindness of the following institutions and individuals in granting permissions to reproduce the following.

*The problem of limits on a Riemannian manifold*
Reprinted with permission from Journal of the Indian Mathematical Society (N.S.) **20** (1956), 291–297.
Copyright © 1956 Indian Mathematical Society.

*The identity of the weak and strong extensions of a linear elliptic differential operator*
Reprinted with permission from Proceedings of the Natural Academy of Sciences U.S.A. **43** (1957), 513–514.
Copyright © M.S. Narasimhan.

*The identity of the weak and strong extensions of a linear elliptic differential operator. II*
Reprinted with permission from Proceedings of the Natural Academy of Sciences U.S.A. **43** (1957), 620.
Copyright © M.S. Narasimhan.

*The type and the Green's kernel of an open Riemann surface*
Reprinted with permission from Annales de L'Institut Fourier **10** (1960), 285–296.
Copyright © Annales de L'Institut Fourier.

*Variations of complex structures on an open Riemann surface*
Reprinted with permission from Annales de L'Institut Fourier **11** (1961), 493–514, XVI–XVII.
Copyright © Annales de L'Institut Fourier.

*Existence of universal connections* (with S. Ramanan)
American Journal of Mathematics **83** (1961), 563–572.
Reprinted with permission from Johns Hopkins University Press.
Copyright © Johns Hopkins University Press.

*Existence of universal connections, II*, (with S. Ramanan)
American Journal of Mathematics **85:2** (1963), 223–231.
Reprinted with permission from the Johns Hopkins University Press.
Copyright © Johns Hopkins University Press.

*Regularity theorems for fractional powers of a linear elliptic operator* (with Takeshi Kotake)
Reprinted with permission from Bulletin de la Societe Mathematique de France **90** (1962), 449–471.
Copyright © French Mathematical Society.

*Stable bundles and unitary bundles on a compact Riemann surface* (with C.S. Seshadri)
Reprinted with permission from Proceedings of Natural Academy of Sciences U.S.A. **52** (1964), 207–211.
Copyright © M.S. Narasimhan and C.S. Seshadri.

*Holomorphic vector bundles on a compact Riemann surface* (with C.S. Seshadri)
Reprinted with permission from Mathematicshe Annalen **155** (1964), 69–80.
Copyright © 1964 Springer-Verlag.

*Stable and unitary vector bundles on a compact Riemann surface* (with C.S. Seshadri)
Reprinted with permission from Annals of Mathematics (2) **82** (1965), 540–567.
Copyright © Annals of Mathematics.

*Manifolds with ample canonical class* (with R.R. Simha)
Reprinted with permission from Inventiones Mathematicae **5** (1968), 120–128.
Copyright © 1968 Springer-Verlag.

*Vector bundles on curves* (with S. Ramanan)
Algebraic Geometry (Internat. Colloq., Tata Inst. Fund. Res., Bombay, 1968) pp. 335–346 Oxford Univ. Press, London, 1968.
Copyright © Tata Institute of Fundamental Research.

*Moduli of vector bundles on a compact Riemann surface*
Reprinted with permission from Annals of Mathematics (2) **89** (1969), 14–51.
Copyright © Annals of Mathematics.

*An analogue of the Borel-Weil-Bott theorem for hermitian symmetric pairs of non-compact type*
Reprinted with permission from Annals of Mathematics (2) **91** (1970), 486–511.
Copyright © Annals of Mathematics.

*Geometry of moduli spaces of vector bundles*
Reprinted from Actes du Congres International des Mathematiciens (Nice, 1970), Tome 2, M.S. Narasimhan, pp. 199–201, Copyright © (1971) Gauthier-Villars, Paris, with permission from Elsevier.

*On the cohomology groups of moduli spaces of vector bundles on curves* (with G. Harder)
Reprinted with permission from Mathematicshe Annalen **212** (1975), 215–248.
Copyright © 1975 Springer-Verlag.

*Deformations of the moduli space of vector bundles over an algebraic curve*
Reprinted with permission from Annals of Mathematics (2) **101** (1975), 391–417.


*Generalised Prym varieties as fixed points*
Reprinted with permission from Journal of the Indian Mathematical Society (N.S.) **39** (1975), 1–19.


*Geometry of Hecke cycles. I*
C. P. Ramanujam—A Tribute, pp. 291–345, Tata Inst. Fund. Res. Studies in Math., **8** Springer, Berlin-New York, 1978.


*Geometry of* SU(2) *gauge fields* (with T.R. Ramadas)
Reprinted with permission from Communications in Mathematical Physics **67** (1979), no. 2, 121–136.


*Polarisations on an abelian variety*
Reprinted with permission from Proc. Indian Acad. Sci. Math. Sci. **90** (1981), no. 2, 125–128.


*Fibres de 't Hooft speciaux et applications* (with A. Hirschowitz)
Enumerative geometry and classical algebraic geometry (Nice, 1981), pp. 143–164, Progr. Math., **24** Birkhauser, Boston, Mass., 1982.
Reprinted with permission from Birkhauser Publishing Ltd., Basel.


*Projective bundles on a complex torus*
Reprinted with permission from Journal für die reine und angewandte Mathematik **340** (1983), 1–5.


*Maximal subbundles of rank two vector bundles on curves* (with H. Lange)
Reprinted with permission from Mathematicshe Annalen **266** (1983), no. 1, 55–72.


*Survey of vector bundles on curves*
Singularities, representation of algebras, and vector bundles (Lambrecht, 1985), 1–8, Lecture Notes in Mathematics, **1273** Springer, Berlin, 1987.

*$2\theta$-linear systems on abelian varieties*
Vector bundles on algebraic varieties (Bombay, 1984), 415–427, Tata Inst. Fund. Res. Stud. Math., **11** Tata Inst. Fund. Res., Bombay, 1987.
Copyright © Tata Institute of Fundamental Research.

*Squares of ample line bundles on abelian varieties*
Reprinted from Expositiones Mathematicae, **7** no. 3, H. Lange and M.S. Narasimhan, pp. 275–287, Copyright © (1989), with permission from Elsevier.

*Spectral curves and the generalised theta divisor*
Reprinted with permission from Journal für die reine und angewandte Mathematik **398** (1989), 169–179.
Copyright © 1989 Walter de Gruyter GmbH & Co. KG

*Groupe de Picard des varieties de modules de fibres semi-stables sur les courbes algebriques* (with J.-M. Drezet)
Reprinted with permission from Inventiones Mathematicae **97** (1989), no. 1, 53–94.
Copyright © 1989 Springer-Verlag.

*Compactification of $M_{P_3}(0,2)$ and Poncelet pairs of conics*
Reprinted with permission from Pacific Journal of Mathematics **145** (1990), no. 2, 255–365.
Copyright ©1990 Pacific Journal of Mathematics

*Rank 2 vector bundles on $P_4$ with $c_1$ odd and contact curves* (with W. Decker and F.-O. Schreyer)
Reprinted with permission from Mathematische Zeitschrift **205** (1990), no. 1, 123–136.
Copyright © 1990 Springer-Verlag.

*The Picard group of the compactification of $M_{P_3}(0,2)$*
Reprinted with permission from Journal für die reine und angewandte Mathematik **422** (1991), 21–44.
Copyright © 1991 Walter de Gruyter GmbH & Co. KG

*Factorisation of generalised theta functions. I* (with T.R. Ramadas)
Reprinted with permission from Inventiones Mathematicae **114** (1993), no. 3, 565–623.
Copyright © 1993 Springer-Verlag.

*Vector bundles as direct images of line bundles*
Reprinted with permission from Indian Academy of Sciences
Copyright © 1994 Proc. Indian Acad. Sci. Math. Sci.
(K. G. Ramanathan memorial issue) **104** (1994), no. 1, 191–200.

*Infinite Grassmannians and moduli spaces of G-bundles* (with Shrawan Kumar and A. Ramanathan)
Reprinted with permission from Mathematicshe Annalen **300** (1994), no. 1, 41–75.


*Picard group of the moduli spaces of G-bundles* (with Shrawan Kumar)
Reprinted with permission from Mathematicshe Annalen **308** (1997), no. 1, 155–173.


*Hodge classes of moduli spaces of parabolic bundles over the general curve*
Reprinted with permission from Journal of Algebraic Geometry **6** (1997), no. 4, 697–715.


*Hermitian-Einstein metrics on parabolic stable bundles* (with (with Jia Yu Li)
Reprinted with permission from Acta Mathematica Sinica (English Series) **15** (1999), no. 1, 93–114.


*A note on Hermitian-Einstein metrics on parabolic stable bundles* (with Jia Yu Li)
Reprinted with permission from Acta Mathematica Sinica (English Series) **17** (2001), no. 1, 77–80.


*A generalisation of Nagata's theorem on ruled surfaces*
Reprinted with permission from Foundation Compositio Mathematica


We thank the Abdus Salam International Centre for Theoretical Physics, Trieste, and Red Temática de Geometría y Física and Residencia de Estudiantes, Madrid for kindly providing some of the photographs reproduced in this collection.